Index for Computer and Calculator Instructions

Index of Applications

Index of Interactivities

DUXBURY

ELEMENTARY STATISTICS

NINTH EDITION

Robert Johnson
Patricia Kuby

Monroe Community College

Australia • Canada • Mexico • Singapore • Spain
United Kingdom • United States

Statistics Editor: Carolyn Crockett
Development Editor: Cheryll Linthicum
Assistant Editor: Ann Day
Editorial Assistant: Julie Bliss
Technology Project Manager: Burke Taft
Marketing Manager: Joseph Rogove
Marketing Assistant: Jessica Perry
Advertising Project Manager: Tami Strang
Project Manager, Editorial Production: Hal Humphrey
Print/Media Buyer: Jessica Reed
Permissions Editor: Joohee Lee
Production Service: Susan L. Reiland
Text Designer: Stephen Rapley and John Edeen
Photo Researcher: Sue C. Howard
Copy Editor: Carol Reitz
New Illustrations: Lori Heckelman
Cover Designer: William Stanton
Compositor: Graphic World
Text and Cover Printer: R. R. Donnelley/Willard
Table of Contents Photo Credits: Ch. 1, © USA Today; Ch. 2, © Jules Frazier/Getty Images; Ch. 3, © Brian Bahr/Getty Images; Ch. 4, © Rachel Epstein/The Image Works; Ch. 5, © David Hanover/Getty Images; Ch. 6, © Bob Daemmrich/The Image Works; Ch. 7, © Michael Dwyer/Stock Boston; Ch. 8, © Joe Sohm/Stock Boston; Ch. 9, © Journal Courier/Steve Warmowski/The Image Works; Ch. 10, © Wayne Scarberry/The Image Works; Ch. 11, © Dion Ogust/The Image Works; Ch. 12, © Kent Meireis/The Image Works; Ch. 13, © Francis de Richemond/The Image Works; Ch. 14, © Peter Hvizdak/The Image Works.

Printed in the United States of America
1 2 3 4 5 6 7 07 06 05 04 03

For more information about our products, contact us at:
Thomson Learning Academic Resource Center
1-800-423-0563

For permission to use material from this text, contact us by:
Phone: 1-800-730-2214
Fax: 1-800-730-2215
Web: http://www.thomsonrights.com

Library of Congress Control Number: 2003104394

Student Edition with InfoTrac College Edition:
ISBN 0-534-39915-0
Student Edition without InfoTrac College Edition:
ISBN 0-534-39922-3
Annotated Instructor's Edition: ISBN 0-534-39924-X

Brooks/Cole—Thomson Learning
10 Davis Drive
Belmont, CA 94002
USA

Asia
Thomson Learning
5 Shenton Way #01-01
UIC Building
Singapore 068808

Australia/New Zealand
Thomson Learning
102 Dodds Street
Southbank, Victoria 3006
Australia

Canada
Nelson
1120 Birchmount Road
Toronto, Ontario M1K 5G4
Canada

Europe/Middle East/Africa
Thomson Learning
High Holborn House
50/51 Bedford Row
London WC1R 4LR
United Kingdom

Latin America
Thomson Learning
Seneca, 53
Colonia Polanco
11560 Mexico D.F.
Mexico

Spain/Portugal
Paraninfo
Calle/Magallanes, 25
28015 Madrid, Spain

Preface

Our Objectives

When *Elementary Statistics* was first published, the objective was to create a truly readable introductory textbook that would promote learning, understanding, and motivation by presenting statistics in a real-world context for students. Over the course of nine editions, *Elementary Statistics* has responded to the gradual recognition by almost every discipline that statistics is a most valuable tool for them. As a result, the applications, examples, case studies, and exercises contain data from a wide variety of areas of interest, including the physical and social sciences, public opinion and political science, business, economics, and medicine.

Now, more than 30 years after *Elementary Statistics* was first published, at least one statistics course is recommended for students in all disciplines, because statistics today is seen as reaching into multiple areas of daily life. Despite this change in perception, our objectives have not changed. We continue to strive for readability, and a common-sense tone that will appeal to students who are increasingly more interested in application than in theory.

One major change in the revisions of the text has been the gradual integration of various technologies. A strong computer component is built upon exercises that rely on technology, clear instruction in the use of those technologies, numerous examples of graphical output, and a focus on interpreting that output. We recognize that the amount of technology integrated into each course depends upon instructor preference and the availability of particular types of hardware and software. In the interests of flexibility, then, we provide integration of graphing calculator, spreadsheet, and statistical technologies, and position them in such a way that they can be included or omitted as necessary.

New to This Edition:

- Java applets—called **Interactivities**—appear throughout the text and are included on the Student's Suite CD-ROM. These Interactivities help students visualize statistical concepts, and redirect their attention from simple memorization of formulas to interpretation of data.
- The role of **case studies** has been enhanced—they are an integral part of each chapter. Each chapter case study applies the techniques of that chapter to a real-life situation. The chapter's first **Answer Now** exercise asks the student to apply prior knowledge to the situation, thereby forming a foundation and motivation for the material about to be studied. At each chapter's end, a **Return to Chapter Case Study** help students assess what they've learned. Here, students are asked to bring their learning process full-circle by answering three exercises: the **First Thoughts** exercise reviews where the chapter started, the **Putting Chapter to Work** exercise applies the methods just learned, and **Your Own Mini-Study** offers them an individualized learning experience.

- A vastly improved **Student's Suite CD-ROM** is included with every new text, containing lab manuals for MINITAB, Excel, and theTI-83 graphing calculator, as well as animated tutorials in Microsoft® *PowerPoint*® format that will guide students through essential concepts presented in the text.
- **Answer Now** exercises, strategically placed to encourage students to practice with a concept as soon as it is introduced.
- Streamlined discussions of topics and other **student-friendly features,** including a new text design, enhance the learning process.
- **Three hundred new and revised exercises** provide instructors with additional homework sets for years to come. In addition, exercises from the Eighth Edition are available on the Book Companion Web Site at www.duxbury.com.
- **Additional critical thinking opportunities for interpreting solutions** are woven throughout the text. Instructors can implement these new elements into their focus as much (or as little) as they wish. Questions that ask for more than "the answer" have been added for those who want more emphasis on interpretation.
- **Basic Principles of Counting** and **Bayes' Rule** have been moved to the Book Companion Web Site.

Ongoing Key Features

- Instructions for MINITAB, Excel, and TI-83 are incorporated throughout the text, not relegated to an end-of-chapter appendix.
- The *p*-value and classical approaches to hypothesis testing are introduced separately, but are thereafter presented side-by-side to emphasize their comparability.
- In response to classroom feedback, descriptive regression and correlation are introduced early, in Chapter 3.
- Real applications of statistics used in examples, exercises, and chapter case studies enhance the relevance of the material for students.
- "Working with Your Own Data" sections, included at the end of each of the four major parts of the book, are designed to encourage further exploration, independent student learning, and critical thinking.

Supplements

Student's Suite CD-ROM

0-534-39928-2

Free with every new copy of the text, the Student's Suite CD-ROM contains:

- Lab manuals for MINITAB, Excel, and the TI-83 graphing calculator
- Animated tutorials in PowerPoint format
- Video tutorials
- Microsoft PowerPoint Presentation
- CyberStats interactive Java applets
- Data Analysis Plus Microsoft Excel Add-in
- Data sets formatted for Excel, MINITAB, JMP, SPSS, and ASCII
- Appendix A: Basic Principles of Counting
- Student Web Resources and Tutorial Quiz

Statistical Tutor

0-534-39916-9

To package with the text at no additional charge, use ISBN 0-534-08091-X.

The *Statistical Tutor* contains complete solutions for all of the Answer Now and odd-numbered exercises as well as helpful hints and other information for students. Sections covering introductory concepts and review lessons on various algebraic or statistical concepts appear at the end of the manual.

Internet Companion for Statistics

Michael Larsen, University of Chicago
0-534-42356-6

This brief, practical book helps students use the Internet to increase their understanding of statistics. Organized by key topics covered in the introductory course, the text offers a brief review of a topic, listings of appropriate Web sites, and study questions designed to build students' analytical skills.

CyberStats

Developed by CyberGnostics, Inc. **www.cyberk.com**

CyberStats: An Introduction to Statistics is an entirely Web-delivered resource that allows students to visualize the behavior of statistical concepts through the use of conceptual software. *CyberStats* provides more than 400 active simulations and hundreds of immediate-feedback practice items to create a learning opportunity unlike any delivered in print—equally effective for both on campus and distance learning courses.

Instructor's Suite CD-ROM

0-534-39919-3

Available to qualified adopters, the *Instructor's Suite CD-ROM* includes everything featured on the *Student's Suite CD-ROM, plus*:

- Instructor's Resource Manual
- The Test Bank in Microsoft® *Word* format

BCA Testing

0-534-39921-5

With a balance of efficiency and high-performance, simplicity, and versatility, BCA gives you the power to transform the learning and teaching experience. BCA is a fully integrated testing and course management software accessible by instructors and students anytime, anywhere. BCA uses correct statistical notation, and is delivered in a browser-based format that does not require any proprietary software or plug-ins. Results flow automatically to a grade book for tracking so that you will be better able to assess student understanding of the material—even prior to class or an actual test.

Printed Test Bank

0-534-39920-7

The printed *Test Bank*, available to qualified adopters, includes an array of true/false, multiple choice, short answer, and applied and computational questions.

Annotated Instructor's Edition—featuring the *Resource Integration Guide!*

0-534-39924-X

The Annotated Instructor's Edition of the Ninth Edition includes an essential tool for instructors—the ***Resource Integration Guide***. This guide links the outline of every chapter—topic by topic—to instructional ideas and corresponding supplement resources. See at a glance which specific videos, Microsoft® PowerPoint® slides, test questions, and lecture suggestions are appropriate for each key chapter topic.

InfoTrac® College Edition

Included with every new text, *InfoTrac College Edition* gives students online access to thousands of journals and periodicals, including *American Statistician*, *Journal of the American Statistical Association*, and *Technometrics*.

WebTutor™ on WebCT and Blackboard

On WebCT 0-534-39926-6
On Blackboard 0-534-39927-4

WebTutor extends the vision, concepts, and pedagogy of your text into a highly visual, hands-on learning environment available any time you or your students need it. The ***WebTutor*** flexible platform allows you to assign pre-formatted, text-specific content that is available as soon as you log on, or to customize its environment in any way you choose—from uploading images and other resources, to adding Web links, to creating your own practice materials. WebTutor makes it all possible—whether you simply want to post your syllabus, or you want to connect with your students through virtual office hours, asynchronous discussion, real-time chat sessions, or the integrated e-mail system. Backed by ***WebTutor's*** T.O.T.A.L. support, training and service, this rich resource is available to you for the entire length of your ***WebTutor*** adoption. New to ***WebTutor:*** seamless access to both *InfoTrac® College Edition*, an online database of millions of full-text articles, and *NewsEdge*, an authoritative online news service that delivers customized news feeds daily.

The Duxbury Resource Center
www.duxbury.com

The Duxbury site makes teaching and learning an interactive experience, providing you and your students with a rich array of resources, including links to relevant Web sites, Web-based teaching and learning resources, and discussion forums.

Book Companion Web Site
At the Duxbury Resource Center

Click on Online Book Companions at the Duxbury Resource Center and you'll find a comprehensive, text-specific Web site that includes learning objectives, tutorial quizzes, data sets, a glossary, Internet exercises, *InfoTrac College Edition* exercises, and other information relevant to both students and instructors.

ACKNOWLEDGMENTS

We owe a debt to many other books. Many of the ideas, principles, examples, and developments that appear in this text stem from thoughts provoked by these sources.

It is a pleasure to acknowledge the aid and encouragement we have received throughout the development of this text from students and colleagues at Monroe Community College. In addition, special thanks to all the reviewers who read and offered suggestions about this and previous editions:

Nancy Adcox
Mt. San Antonio College

Paul Alper
College of St. Thomas

William D. Bandes
San Diego Mesa College

Tim Biehler
Fingerlakes Community College

Barbara Jean Blass
Oakland Community College

Austin Bonis
Rochester Institute of Technology

Nancy Bowers
Pennsylvania College of Technology

Robert Buck
Slippery Rock University

Louis F. Bush
San Diego City College

Ronnie Catipon
Franklin University

Rodney E. Chase
Oakland Community College

Pinyuen Chen
Syracuse University

Wayne Clark
Parkland College

David M. Crystal
Rochester Institute of Technology

Joyce Curry and Frank C. Denny
Chabot College

Larry Dorn
Fresno Community College

Shirley Dowdy
West Virginia University

Kenneth Fairbanks
Murray State University

Joan Garfield
University of Minnesota General College

David Gurney
Southeastern Louisiana University

Carol Hall
New Mexico State University

Silas Halperin
Syracuse University

Noal Harbertson
California State University, Fresno

Hank Harmeling
North Shore Community College

Bryan A. Haworth
California State College at Bakersfield

Harold Hayford
Pennsylvania State University, Altoona

Marty Hodges
Colorado Technical University

John C. Holahan
Xerox Corporation

James E. Holstein
University of Missouri

Soon B. Hong
Grand Valley State University

Robert Hoyt
Southwestern Montana State

Peter Intarapanach
Southern Connecticut State University

T. Henry Jablonski, Jr.
East Tennessee State University

Brian Jean
Bakersfield University

Jann Huei Jinn
Grand Valley State University

Sherry Johnson

Meyer M. Kaplan
The William Patterson College of New Jersey

Michael Karelius
American River College

Anand S. Katiyar
McNeese State University

Jane Keller
Metropolitan Community College

Gayle S. Kent
Florida Southern College

Andrew Kim
Westfield State College

Amy Kimchuk
University of the Sciences in Philadelphia

Raymond Knodel
Bemidji State University

Larry Lesser
University of Northern Colorado

Robert O. Maier
El Camino College

Mark Anthony McComb
Mississippi College

Carolyn Meitler
Concordia University Wisconsin

John Meyer
Muhlenberg College

Jeffrey Mock
Diablo Valley College

David Naccarato
University of New Haven

Harold Nemer
Riverside Community College

Dennis O'Brien
University of Wisconsin, LaCrosse

Chandler Pike
University of Georgia

Daniel Powers
University of Texas, Austin

Janet M. Rich
Miami-Dade Junior College

Larry J. Ringer
Texas A & M University

John T. Ritschdorff
Marist College

John Rogers
California State Polytechnic Institute
at San Luis Obispo

Neil Rogness
Grand Valley State University

Thomas Rotolo
University of Arizona

Barbara F. Ryan and Thomas A. Ryan
Pennsylvania State University

Robert J. Salhany
Rhode Island College

Sherman Sowby
California State University, Fresno

Roger Spalding
Monroe County Community College

Timothy Stebbins
Kalamazoo Valley Community College

Howard Stratton
State University of New York at Albany

Larry Stephens
University of Nebraska-Omaha

Paul Stephenson
Grand Valley State University

Richard Stockbridge
University of Wisconsin-Milwaukee

Thomas Sturm
College of St. Thomas

Edward A. Sylvestre
Eastman Kodak Co.

William K. Tomhave
Concordia College
Moorhead, MN

Bruce Trumbo
California State University, Hayward

Richard Uschold
Canisius College

John C. Van Druff
Fort Steilacoom Community College

Philip A. Van Veidhuizen
University of Alaska

John Vincenzi
Saddleback College

Kenneth D. Wantling
Montgomery College

Mary Wheeler
Monroe Community College

Sharon Whitton
Hofstra University

Don Williams
Austin College

Pablo Zafra
Kean University

We would like to extend our thanks to the many authors and publishers who so generously granted reproduction permissions for the news articles and tables used in the text. These acknowledgments are specified individually throughout the text. In addition, a special acknowledgment goes to Gwen Terwilliger and Edwin Hackleman, whose assistance in locating data sets and writing new, current, and interesting exercises was greatly valued and appreciated.

And finally, the most significant acknowledgment of all—thank you to our spouses, Barbara and Joe, for your assistance and just "being there."

Robert Johnson
Patricia Kuby

CONTENTS

PART THREE INFERENTIAL STATISTICS 340

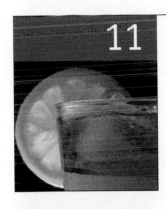

The Statistical Process

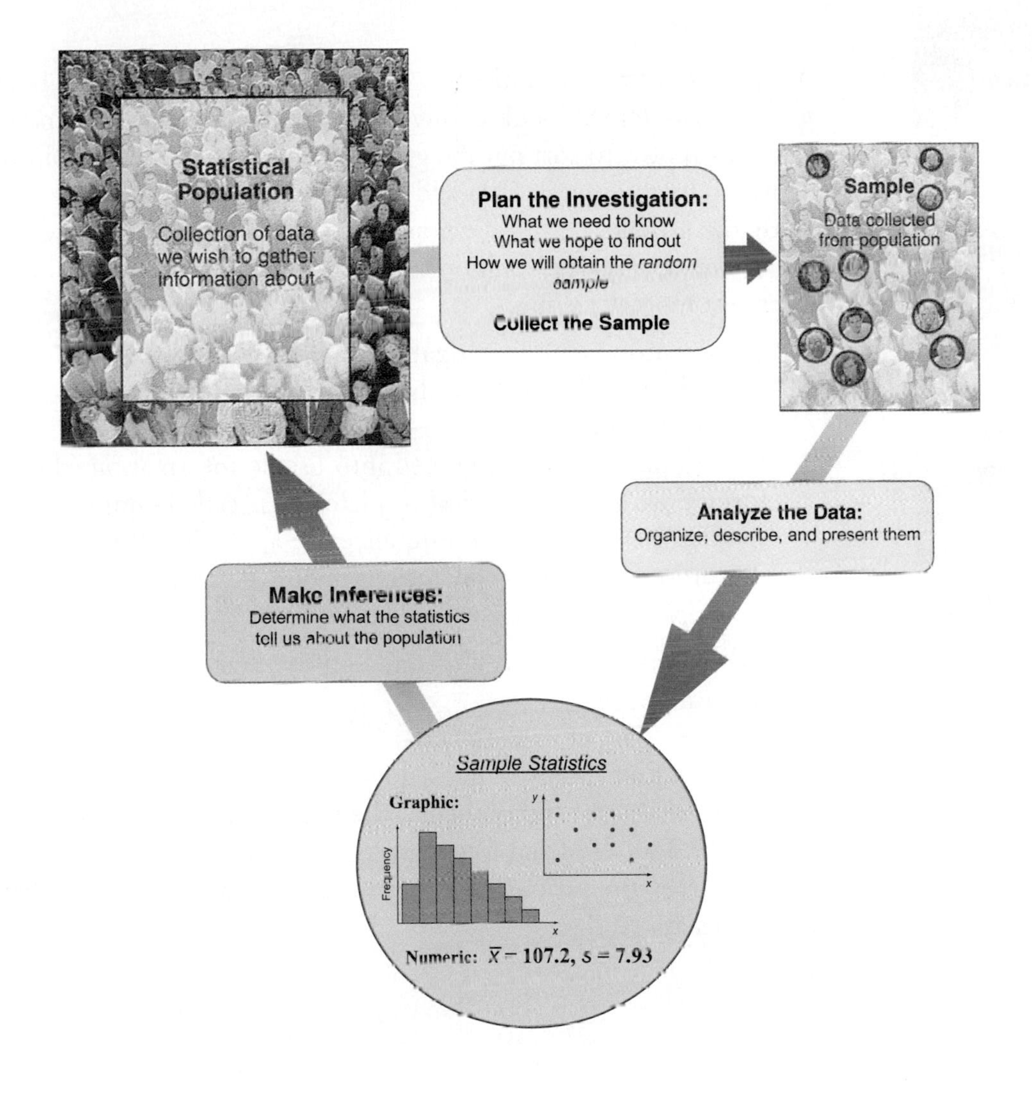

PART 1

Descriptive Statistics

© StockTrek/Getty Images

A typical objective in statistics is to describe "the population" based on information obtained by observing relatively few individual elements. We must learn how to sort out the generalizations contained within the clues provided by the sample data and "paint" a picture of the population. We study the sample, but it's the population that is of primary interest to us.

When a statistical solution to a problem is sought, a certain sequence of events must develop:

1. the situation being investigated is carefully and fully defined,
2. a sample is collected from the population following an established and appropriate procedure,
3. the sample data are converted into usable information (this usable information, either numerical or pictorial, is called sample statistics), and
4. the theories of statistical inference are applied to the sample information in order to draw conclusions about the sampled population (these conclusions or answers are called inferences).

This sequence of events is illustrated on the Statistical Process diagram on the opposite page.

The first part of this textbook, Chapters 1–3, concentrates on the first three of the four events in the sequence above. The second part, Chapters 4–7, deals with probability theory, the theory on which statistical inferences rely. The third and fourth parts, Chapters 8–10 and Chapters 11–14, survey the various types of inferences that can be made from sample information.

Sir Francis Galton

Sir Francis Galton, English anthropologist and a pioneer of human intelligence studies, was born on February 16, 1822, in a village near Birmingham, England, to Samuel Tertiles and Anne (Violetta) Galton.

Galton's family included men and women of exceptional ability, one of whom was his cousin, Charles Darwin. Although his family life was happy and he was grateful to his parents for all they had done for him, he felt little use for the conventional religious and classical education he was given.

As a teen, Galton toured a number of medical institutions in Europe and began his medical training in hospitals in Birmingham and London. He continued his medical studies until the death of his father. Having been left a sizable fortune, he decided to discontinue medical training and pursue his love of traveling. Several years of these travels led him to the exploration of primitive parts of southwestern Africa, where he gained valuable information that earned him recognition and a fellowship in the Royal Geographical Society. In 1853, at the age of only 31, Galton was awarded a gold medal from the Society in recognition of his hard work and many achievements. It was also in 1853 that Galton married Louisa Butler and they settled in London; their marriage remained childless.

Although Galton made important contributions to many fields of knowledge, he was best known for his work in eugenics; he spent most of the latter part of his life researching and promoting his belief that inheritance played a major role in the intelligence of man. Galton, a pioneer in the development of some of the refined statistical techniques that we use today, used the laws of probability to support his theory. His application of research techniques (curves of the normal distribution, correlation coefficients, and percentile grading) to a large population revealed important facts about the intellectual and physical characteristics that are passed from one generation to the next and the ways in which children differ from their parents. Galton also discovered, with the use of the graph, that characteristics of two different generations could be plotted against each other to reveal important information. In his original paper on correlation ("Co-relations and Their Measurement"), Galton used r for the "index of co-relation," a statistic we now call the correlation coefficient. He calculated it from the regression of y on x or of x on y after both variables are standardized. More information about this paper can be found on the Internet site http://www.galton.org.

Sir Francis Galton died in England on January 17, 1911, leaving behind a collection of 9 books, approximately 200 articles and lectures, and a valuable legacy of statistical techniques and knowledge.

Here are some links to Internet sites with more information about Galton: http://www.galton.org, http://www.mugu.com/galton/, http://www.maps.jcu.edu.au/hist/stats/galton/, and http://www.psych.usyd.edu.au/difference5/scholars/galton.html.

CHAPTER

1 Statistics

Chapter Outline and Objectives

CHAPTER CASE STUDY
Statistics are used to **describe** every aspect of our daily life.

1.1 WHAT IS STATISTICS?
It's **more than just numbers!** An image of the field of statistics will be created—an image that will grow and develop.

1.2 INTRODUCTION TO BASIC TERMS
Population, sample, variable, data, experiment, parameter, statistic, qualitative data, and **quantitative data** are some of the basic vocabulary used in studying statistics.

1.3 MEASURABILITY AND VARIABILITY
Statistics is a **study of the variability** that takes place in observed values of a variable.

1.4 DATA COLLECTION
Initial ideas and concerns are presented about the processes used to obtain sample data. Specifically, the problem of selecting a **representative sample** from a defined population can be solved using the **random method.**

1.5 COMPARISON OF PROBABILITY AND STATISTICS
Probability is related to statistics as the phrase "**likelihood** of rain today" is related to the phrase "**actual amount** of rain that falls."

1.6 STATISTICS AND TECHNOLOGY
Today's state of the art.

RETURN TO CHAPTER CASE STUDY

Chapter Case Study

Americans, Here's Looking at You

The U.S. Census Bureau annually publishes the *Statistical Abstract of the United States,* and the 1000+-page book provides us with a statistical insight into many of the most obscure and unusual facets of our lives. That is only one of thousands of sources for all kinds of things you have always wanted to know about and never thought to ask. Are you interested in how many hours we work and play? How much we spend on snack foods? How much the price of Red Delicious apples has gone up? All this and more, much more can be found in the *Statistical Abstract* (www.census.gov/statab/www).

The collection of statistical tidbits shown here all happened to be reported in *USA Today* but come from a variety of sources and represent only a tiny sampling of what can be learned about Americans statistically.

USA SNAPSHOTS®

Eliminate the penny?

3% Don't know

32% Eliminate it

65% Keep it

Source: Coinstar USA Today 08/23/01

Source: Response Insurance USA Today 08/28/01

USA SNAPSHOTS®

Prime time conditions of road rage

- Time: 4-6 p.m. Friday
- Season: Sunny, Summer
- Location: Congested, urban freeway

Source: The AAA Foundation for Traffic Safety USA Today 05/09/00

Source: Opinion Research Corp. for Steelcase USA Today 01/08/01

Source: Gallup Poll of 800 adults March 16-18 USA Today 06/14/01

The above and thousands of other measures are used to describe life in the United States.

ANSWER NOW 1.1

a. What population is being studied in all of these Snapshots?

b. Identify one specific Snapshot. Describe the information that was collected and used to determine the statistics reported in your Snapshot.

c. Identify one specific statistic that is reported in one of the Snapshots and describe what that statistic tells you.

d. Consider "Eliminate the penny?" If you had been asked, "Should the penny be eliminated from daily business transactions?" how would you have answered? Do you believe your answer is represented accurately in the diagram? What does the percentage associated with your answer really mean? Explain.

After completing Chapter 1, further investigate the Chapter Case Study in the Return to Chapter Case Study section with Exercises 1.74, 1.75, and 1.76 (p. 32).

FYI

It is important that you complete the Answer Now exercises as you encounter them. They are an important part of learning.

1.1 What Is Statistics?

Statistics is the universal language of the sciences. As potential users of statistics, we need to master both the "science" and the "art" of using statistical methodology correctly. Careful use of statistical methods will enable us to obtain accurate information from data. Statistical methods include (1) carefully defining the situation, (2) gathering data, (3) accurately summarizing the data, and (4) deriving and communicating meaningful conclusions.

Statistics involves information, numbers and visual graphics to summarize this information, and their interpretation. The word **statistics** has different meanings to people of varied backgrounds and interests. To some people it is a field of "hocus-pocus" in which a person attempts to overwhelm others with incorrect information and conclusions. To others it is a way of collecting and dis-

Application 1.1 Telling Us Where We Snack

Most of us, young and old, regardless of our weight, enjoy snacking. You've most likely seen summaries describing what we snack on or reasons we give for snacking, but this survey reports where we like to be when we do our snacking. Is there no privacy!

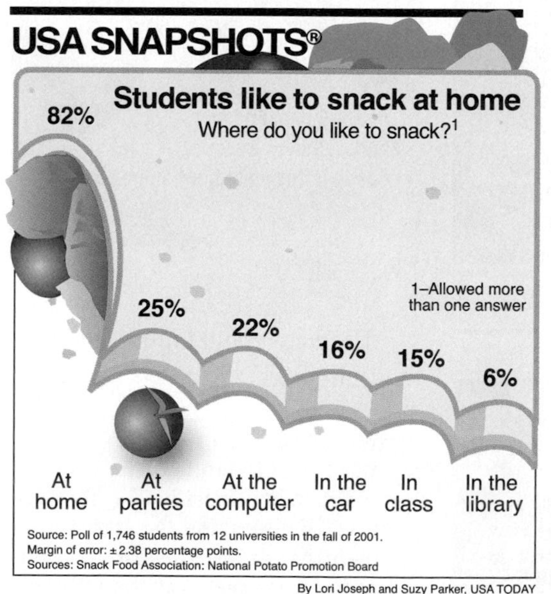

ANSWER NOW 1.2
Refer to "Students like to snack at home."

a. **Who was surveyed?**

b. **How many were surveyed?**

c. **Explain the meaning of "At home: 82%."**

d. **Why do the percentages add up to more than 100%?**

playing information. And to still another group it is a way of "making decisions in the face of uncertainty." In the proper perspective, each of these points of view is correct.

The field of statistics can be roughly subdivided into two areas: descriptive statistics and inferential statistics. **Descriptive statistics** is what most people think of when they hear the word *statistics*. It includes the collection, presentation, and description of sample data. The term **inferential statistics** refers to the technique of interpreting the values that result from the descriptive techniques and making decisions and drawing conclusions about the population.

Statistics is more than just numbers: It is data, what is done to data, what is learned from the data, and the resulting conclusions. Let's use the following definition: *(continued on p. 8)*

Application 1.2 Explaining One of Life's Mysteries

How much do people earn at their jobs? The U.S. government and its many branches continuously collect information and publish reports about many aspects of our daily lives. They can explain how we spend our time, how we earn our money, how we spend our money, how they spend our tax money, and so on; the list is endless.

Sharper Picture of Pay

GANNETT NEWS SERVICE, 1-6-98

The government issues its most detailed report yet of wages for each occupation.

1996 Median Wage Samples

The Labor Department's survey of 1996 pay in 764 occupations showed the majority of 19 managerial and 211 professional occupations paid workers in the upper wage range, while service and agricultural jobs paid the lowest.

Occupation	Employment	Hourly median wages
Athletes, coaches, umpires, and related workers	19,710	$8.33
Automotive mechanics	647,560	$12.35
Cashiers	3,262,120	$5.75
Child-care workers	377,980	$6.12
Dentists	85,250	$47.66

Complete tables are available on the Internet at http://stats.bls.gov/oes/

Source: U.S. Bureau of Labor Statistics

FYI
Log on the Internet site and look at the complete tables. How does your wage rate compare to others of the same occupation?

ANSWER NOW 1.3
Refer to "1996 Median Wage Samples."

a. Who was surveyed?

b. Who did the surveying?

c. Explain the meaning of "Cashiers; 3,262,120; and $5.75."

STATISTICS
The science of collecting, describing, and interpreting data.

ANSWER NOW 1.4
To see another perspective on what statistics is all about, view the video "The Statistical Process" on your Student Suite CD (found inside the back cover of this textbook). Using your own words, write a sentence describing each of the three main statistics activities used to solve a statistical ("puzzling") problem referred to in this video.

Before we begin our detailed study, let's look at a few illustrations of how and when statistics can be applied.

Application 1.3 Describing Our Romance with the Road

The U.S. Department of Transportation takes a snapshot of the country's travelers and finds that America's love affair with the car is still in bloom. Americans are logging 800 billion miles a year on long-distance trips and even on trips up to 2,000 miles. The typical traveler would rather drive than fly.

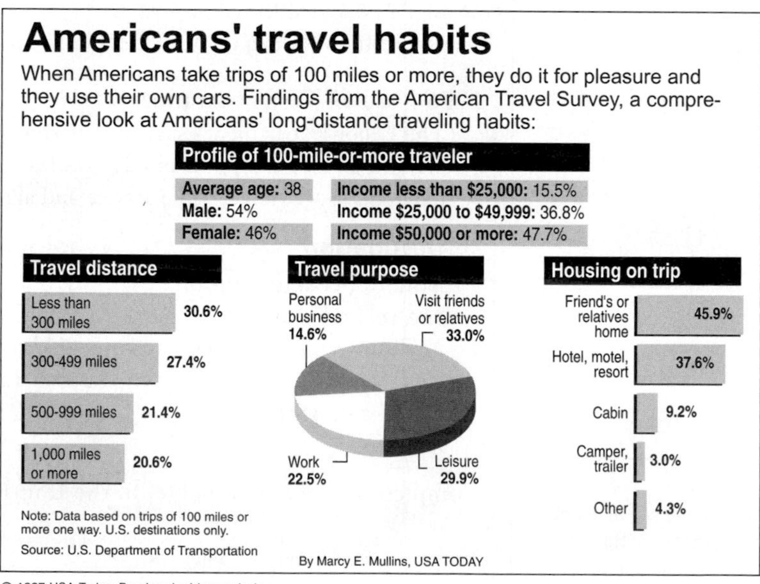

ANSWER NOW 1.5
Refer to "Americans' travel habits."

a. **Who was surveyed?**

b. **Based on this information, how would you describe the "typical" long-distance traveler?**

c. **How do you compare to this "typical" long-distance traveler?**

The uses of statistics are unlimited. It is much harder to name a field in which statistics is not used than to name one in which statistics plays an integral part. Here are a few examples of how and where statistics are used:

- In education descriptive statistics are frequently used to describe test results.
- In science the data resulting from experiments must be collected and analyzed.
- In government many kinds of statistical data are collected all the time. In fact, the U.S. government is probably the world's greatest collector of statistical data.

A very important part of the statistical process is studying the statistical results and formulating appropriate conclusions. These conclusions must then be communicated accurately; nothing is gained from research unless the findings are shared with others. Statistics are being reported everywhere: newspapers, magazines, radio, and television. We read and hear about all kinds of new research results, especially in the health-related fields.

Application 1.4 Telling Us "What We Think"

Newspapers frequently publish articles containing statistics that tell us how we, as a population, think collectively. Did you ever wonder how much of what we think is directly influenced by the information we read in these articles?

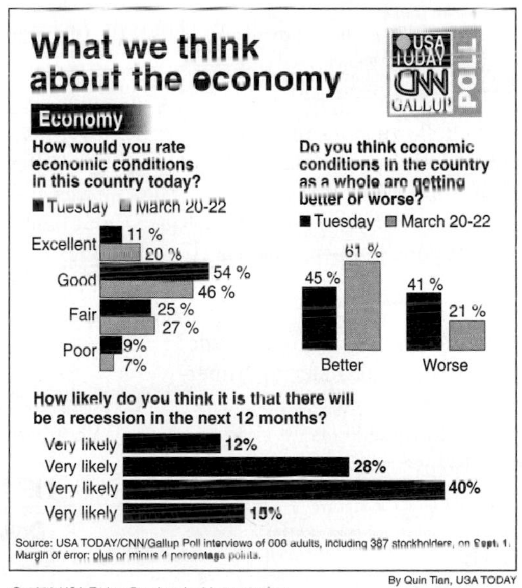

ANSWER NOW 1.6
Refer to "What we think about the economy."

a. **Who was surveyed?**

b. **How accurate are the reported percentages believed to be?**

c. **What do you think the 54% combined with the margin of error means?**

Application 1.5 Statistics Is Tricky Business

One ounce of statistics technique requires one pound of common sense for proper application.

Harvard Health Letter, Special Supplement, October 1994

Because we are mortal and . . . live in an imperfect world, risk will always be with us. . . . As we go about our lives, we weigh the relative risks and benefits of our actions all the time. Most often we act on imperfect and incomplete information.

Common Sense

So, . . . how can responsible individuals with no special expertise make intelligent decisions about all the information and misinformation that bombards us? The same humble horse sense that keeps us from sticking our hand into the fire is an invaluable tool for sorting out what we read and hear. It's important to remember that news, by its very definition, is something new and unusual. After all, the hundredth study showing a relationship between cholesterol and heart disease is hardly news, but the one study that fails to make such a connection is likely to become a headline. Clearly it would be silly for people to drastically change their lives on the basis of one newspaper article or a lone study.

That doesn't mean we should throw out everything we read or hear.

ANSWER NOW 1.7

Find a recent newspaper article that illustrates an "apples are bad" type of report.

Exercises

1.8 Determine which of the following statements is descriptive in nature and which is inferential. Refer to "Students like to snack at home" in Application 1.1 (p. 6).

a. 82% of all U.S. students prefer to eat snacks at home.

b. 82% of the 1746 U.S. students polled prefer to eat snacks at home.

1.9 Determine which of the following statements is descriptive in nature and which is inferential. Refer to "Americans' travel habits" in Application 1.3 (p. 8).

a. The average age of the surveyed travelers is 38 years.

b. 54% of all American long-distance travelers are men.

USA SNAPSHOTS®

A look at statistics that shape your finances

Cars, trucks older than ever

The median age of cars on U.S. roads – half are older, half younger – is a record high. How our vehicles have aged:

	Median age in years Cars	Trucks[1]
1997	8.1	7.8
1995	7.7	7.6
1990	6.5	6.5
1985	6.9	7.6
1980	6.0	6.3
1975	5.4	5.8
1970	4.9	5.9

1 – Pickups, sport utilities and minivans

Source: Polk Co. By Anne R. Carey and Jerry Mosemak, USA TODAY

1.10 Refer to "Cars, trucks older than ever."

a. Construct a graph of the information: Use "years" as the x-axis and "age" as the y-axis; use red to plot the car ages and blue to plot the truck ages.

b. Does your graph support the claim made by the title: "Cars, trucks older than ever"? Explain.

c. State an inference about the quality of cars and trucks that is suggested by your graph.

1.11 Refer to "Drivers using cellphones have problems."

a. What group of people was polled?

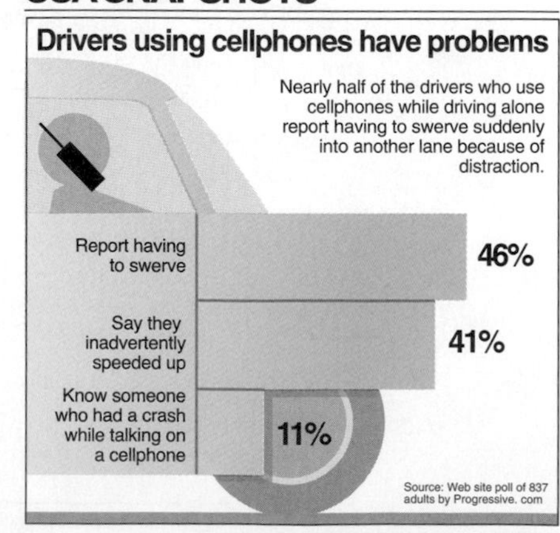
USA SNAPSHOTS®
Drivers using cellphones have problems
Nearly half of the drivers who use cellphones while driving alone report having to swerve suddenly into another lane because of distraction.
Report having to swerve
46%
Say they inadvertently speeded up
41%
Know someone who had a crash while talking on a cellphone
11%
Source: Web site poll of 837 adults by Progressive. com
By Lori Joseph and Sam Ward, USA TODAY

b. How many people were polled?
c. What information was obtained from each person?
d. Explain the meaning of "41% say they inadvertently speeded up."
e. How many people said they inadvertently speeded up?

1.12 EX01-12 Business investors find it valuable to compare the performances of competitive firms before deciding which ones to support by buying stock. This table was extracted from *Fortune* magazine, which reported information gathered from four rental-car companies:

Company	Stock Price	Price/ Earnings Ratio	Earnings Growth Rate
Hertz Corp.	43 ½	21.6	17%
Avis Rent A Car	26 ¼	23.6	17%
Budget Group	33 ¾	28.4	25%
Dollar Thrifty	19 ¼	13.0	17%

Note: Stock prices as of April 29, 1998. P/Es are on estimated 1998 earnings.

Source: *Fortune*, May 25, 1998. Reprinted with permission.

a. Define the nature of the businesses being studied.
b. What information was collected from each company? Is the *information* descriptive or inferential?
c. How many representative companies were included in the study?
d. If you had $5000 to invest in any one of these four companies and no other information available, which stock would you buy? Why?

1.13 During a radio broadcast on August 16, 1998, David Essel reported the following three statistics: (1) the U.S. divorce rate is 55%; and when married adults were asked whether they would remarry their spouse, (2) 75% of the women said yes and (3) 65% of the men said yes.
a. What is the "stay married" rate?
b. There seems to be a contradiction in this information. How is it possible for all three of these statements to be correct? Explain.

1.14 A working knowledge of statistics is very helpful when you want to understand the statistics reported in the news. The news media and our government often make statements like "Crime rate jumps 50% in your city."
a. Does an increase in the crime rate from 4% to 6% represent an increase of 50%? Explain.
b. Why would anybody report an increase from 4% to 6% as a "50% rate jump"?

1.2 Introduction to Basic Terms

In order to begin our study of statistics we first need to define a few basic terms.

POPULATION
A collection, or set, of individuals or objects or events whose properties are to be analyzed.

The population is the complete collection of individuals or objects that are of interest to the sample collector. The concept of a population is the most fundamental idea in statistics. The population of concern must be carefully defined and is considered fully defined only when its membership list of elements is specified. The set of "all students who have ever attended a U.S. college" is an example of a well-defined population.

Typically, we think of a population as a collection of people. However, in statistics the population could be a collection of animals, or manufactured objects, or whatever. For example, the set of all redwood trees in California could be a population.

There are two kinds of populations: finite and infinite. When the membership of a population can be (or could be) physically listed, the population is said to be **finite**. When the membership is unlimited, the population is **infinite**. The books in your college library form a finite population; the OPAC (Online Public Access Catalog, the computerized card catalog) lists the exact membership. All the registered voters in the United States form a very large finite population; if necessary, a composite of all voter lists from all voting precincts across the United States could be compiled. On the other hand, the population of all people who might use aspirin and the population of all 40-watt light bulbs to be produced by Sylvania are infinite. Large populations are difficult to study; therefore, it is customary to select a *sample* and study the data in the sample.

SAMPLE
A subset of a population.

A sample consists of the individuals, objects, or measurements selected by the sample collector from the population.

VARIABLE (OR RESPONSE VARIABLE)
A characteristic of interest about each individual element of a population or sample.

A student's age at entrance into college, the color of the student's hair, the student's height, and the student's weight are four variables.

DATA (SINGULAR)
The value of the variable associated with one element of a population or sample. This value may be a number, a word, or a symbol.

For example, Bill Jones entered college at age "23," his hair is "brown," he is "71 inches" tall, and he weighs "183 pounds." These four pieces of data are the values for the four variables as applied to Bill Jones.

DATA (PLURAL)
The set of values collected for the variable from each of the elements that belong to the sample.

The set of 25 heights collected from 25 students is an example of a set of data.

EXPERIMENT
A planned activity whose results yield a set of data.

An experiment includes the activities for both selecting the elements and obtaining the data values.

ANSWER NOW 1.15
The Chesapeake Bay is the subject of the "Data Collection" video on your Student Suite CD (found inside the back cover of this textbook). View the video clip and answer these questions:

a. **How did the fishermen go about getting their data?**

b. **How were the results used?**

PARAMETER
A numerical value summarizing all the data of an entire population.

FYI
Parameters describe the **p**opulation; notice that both words begin with the letter **p**.

The "average" age at time of admission for all students who have ever attended our college and the "proportion" of students who were over 21 years of age when they entered college are examples of two population parameters. A parameter is a value that describes the entire population. Often a Greek letter is used to symbolize the name of a parameter. These symbols will be assigned as we study specific parameters.

For every parameter there is a *corresponding sample statistic.* The statistic describes the sample the same way the parameter describes the population.

STATISTIC
A numerical value summarizing the sample data.

FYI
A statistic describes the **s**ample; notice that both words begin with the letter **s**.

The "average" height, found by using the set of 25 heights, is an example of a sample statistic. A statistic is a value that describes a sample. Most sample statistics are found with the aid of formulas and are typically assigned symbolic names that are letters of the English alphabet (for example, $\bar{x}$, s, and r).

Illustration 1.1 Applying the Basic Terms

A statistics student is interested in finding out something about the average dollar value of cars owned by the faculty members of our college. Each of the eight terms just described can be identified in this situation.

1. The *population* is the collection of all cars owned by all faculty members at our college.
2. A *sample* is any subset of that population. For example, the cars owned by members of the mathematics department is a sample.
3. The *variable* is the "dollar value" of each individual car.
4. One *data* is the dollar value of a particular car. Mr. Jones's car, for example, is valued at $9400.
5. The *data* are the set of values that correspond to the sample obtained (9400; 8700; 15,950; . . .).
6. The *experiment* consists of the methods used to select the cars that form the sample and to determine the value of each car in the sample. It could be carried out by questioning each member of the mathematics department, or in other ways.
7. The *parameter* about which we are seeking information is the "average" value of all cars in the population.
8. The *statistic* that will be found is the "average" value of the cars in the sample. ■

FYI
Parameters are fixed in value, while statistics vary in value.

NOTE: If a second sample were to be taken, it would result in a different set of people being selected—say, the English department—and therefore a different value would be anticipated for the statistic "average value." The average value for "all faculty-owned cars" would not change, however.

ANSWER NOW 1.16
Thirty-six percent of the U.S. adult population has an allergy. A sample of 1200 randomly selected adults resulted in 33.2% having an allergy. Describe each of the eight terms in this context.

ANSWER NOW 1.17
In your own words, explain why the parameter is fixed and the statistic varies.

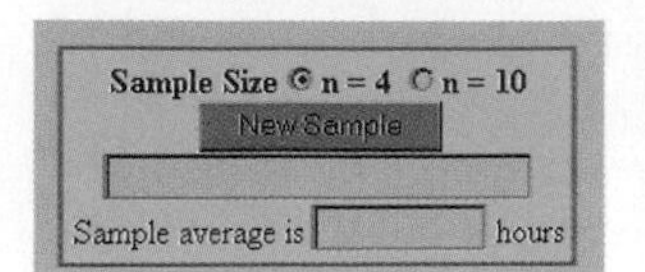

ANSWER NOW 1.18 INTERACTIVITY
Interactivity 1-A simulates sampling from a population of college students.

a. **Take several samples of size 4 and estimate the average number of hours per week that students study.**

b. **Take several samples of size 10 and estimate the average number of hours per week that students study.**

c. **Explain and comment on the effect the two sample sizes have on the sample averages.**

There are basically two kinds of variables: (1) variables that result in *qualitative* information, and (2) variables that result in *quantitative* information.

QUALITATIVE, OR ATTRIBUTE, OR CATEGORICAL, VARIABLE
A variable that describes or categorizes an element of a population.

A sample of four hair-salon customers was surveyed for their "hair color," "hometown," and "level of satisfaction" with the results of their salon treatment. All three variables are examples of qualitative (attribute) variables, because they describe some characteristic of the person and all people with the same attribute belong to the same category. The data collected were {blonde, brown, black, brown}, {Brighton, Columbus, Albany, Jacksonville}, and {very satisfied, satisfied, very satisfied, somewhat satisfied}.

NOTE: Arithmetic operations, such as addition and averaging, are not meaningful for data that result from a qualitative variable.

ANSWER NOW 1.19
Name two attribute variables about its customers that a newly opened department store might find informative to study.

QUANTITATIVE, OR NUMERICAL, VARIABLE
A variable that quantifies an element of a population.

The "total cost" of textbooks purchased by each student for this semester's classes is an example of a quantitative (numerical) variable. A sample resulted in the following data: $238.87, $94.57, $139.24. [To find the "average cost," simply add the three numbers and divide by 3: (238.87 + 94.57 + 139.24)/3 = $157.56.]

NOTE: Arithmetic operations, such as addition and averaging, are meaningful for data that result from a quantitative variable.

ANSWER NOW 1.20
Name two numerical variables about its customers that a newly opened department store might find informative to study.

Each of these types of variables (qualitative and quantitative) can be further subdivided.

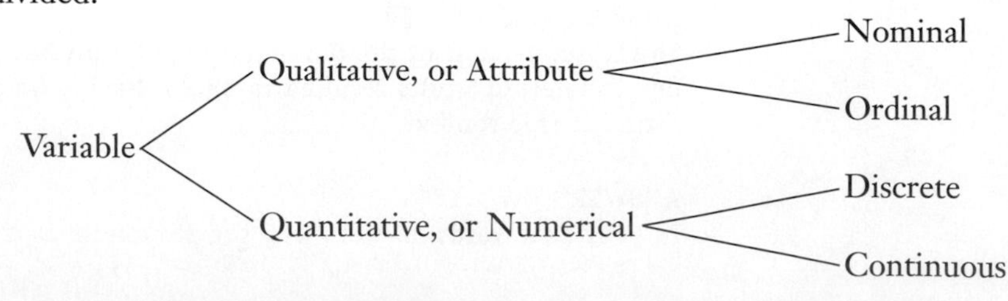

NOMINAL VARIABLE
A qualitative variable that categorizes (or describes, or names) an element of a population. Not only are arithmetic operations not meaningful for data that result from a nominal variable, but also an order cannot be assigned to the categories.

In the survey of four hair-salon customers, two of the variables, "hair color" and "hometown," are examples of nominal variables because both name some characteristic of the person and it would be meaningless to find the sample average by adding and dividing by 4. For example, (blonde + brown + black + brown)/4 is undefined. Further, color of hair and hometown do not have an order to their categories.

ANSWER NOW 1.21
Name two nominal variables about its customers that a newly opened department store might find informative to study.

ORDINAL VARIABLE
A qualitative variable that incorporates an ordered position, or ranking.

In the survey of four hair-salon customers, the variable "level of satisfaction" is an example of an ordinal variable because it does incorporate an ordered ranking: "Very satisfied" ranks ahead of "satisfied," which ranks ahead of "somewhat satisfied." Another illustration of an ordinal variable is the ranking of five landscape pictures according to someone's preference; first choice, second choice, and so on.

ANSWER NOW 1.22
Name two ordinal variables about its customers that a newly opened department store might find informative to study.

Quantitative or numerical variables can also be subdivided into two classifications: *discrete* variables and *continuous* variables.

DISCRETE VARIABLE
A quantitative variable that can assume a countable number of values. Intuitively, the discrete variable can assume the values corresponding to isolated points along a line interval. That is, there is a gap between any two values.

CONTINUOUS VARIABLE
A quantitative variable that can assume an uncountable number of values. Intuitively, the continuous variable can assume any value along a line interval, including every possible value between any two values.

In many cases, the two types of variables can be distinguished by deciding whether the variables are related to a count or a measurement. The variable "number of courses for which you are currently registered" is an example of a discrete variable; the values of the variable may be found by counting the courses. (When we count, fractional values cannot occur; thus, there are gaps, or fractional numbers, between the values that can occur.) The variable "weight of books and supplies you are carrying as you attend class today" is an example of a continuous random variable; the values of the variable may be found by measuring the weight. (When we measure, any fractional value can occur; thus, every value along the number line is possible.)

When trying to determine whether a variable is discrete or continuous, remember to look at the variable and think about the values that might occur. Do not look at only data values that have been recorded; they can be very misleading.

Consider the variable "judge's score" at a figure-skating competition. If we look at some scores that have previously occurred, 9.9, 9.5, 8.8, 10.0, and we see the presence of decimals, we might think that all fractions are possible and conclude that the variable is continuous. This is not true, however. A score of 9.134 is not possible; thus, there are gaps between the possible values and the variable is discrete.

ANSWER NOW 1.23

a. **Explain why the variable "score" for the home team at a basketball game is discrete.**

b. **Explain why the variable "number of minutes to commute to work" is continuous.**

NOTE: Don't let the appearance of the data fool you in regard to their type. Qualitative variables are not always easy to recognize; sometimes they appear as numbers. The sample of hair colors could be coded: 1 = black, 2 = blonde, 3 = brown. The sample data would then appear as {2, 3, 1, 3}, but they are still nominal data. Calculating the "average hair color" [(2 + 3 + 1 + 3)/4 = 9/4 = 2.25] is still meaningless. The hometowns could be identified using ZIP codes. The average of the ZIP codes doesn't make sense either; therefore, ZIP code numbers are nominal, too.

Let's look at another example. Suppose that after surveying a parking lot, I summarized the sample data by reporting 5 red, 8 blue, 6 green, and 2 yellow cars. You must look at each individual source to determine the kind of information being collected. One specific car was red; "red" is the data from that one car, and red is an attribute. Thus, this collection (5 red, 8 blue, and so on) is a summary of nominal data.

Another example of information that is deceiving is an identification number. Flight #249 and Room #168 both appear to be numerical data. The numeral 249 does not describe any property of the flight—late or on time, quality of snack served, number of passengers, or anything else about the flight. The flight number only identifies a specific flight. Driver's license numbers, Social Security numbers, and bank account numbers are all identification numbers used in the nominal sense, not in the quantitative sense.

Remember to inspect the individual variable and one individual data, and you should have little trouble distinguishing among the various types of variables.

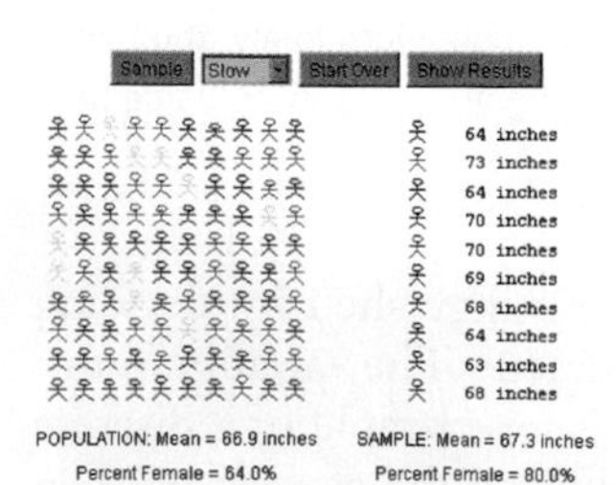

ANSWER NOW 1.24 INTERACTIVITY

Interactivity 1-B simulates taking a sample of size 10 from a population of 100 college students. Take one sample and note the outcome.

a. **Name the attribute variable involved in this experiment. Is it nominal or ordinal?**

b. **Name the numerical variable involved in this experiment. Is it discrete or continuous?**

Application 1.6 Census Data

Census information often makes news whether it is the local census or the national census. The results of the census have a variety of uses—from helping to determine legislative seats and appropriation of taxes to visitor information (as shown here). We are all part of the population census and have all seen reports similar to this one.

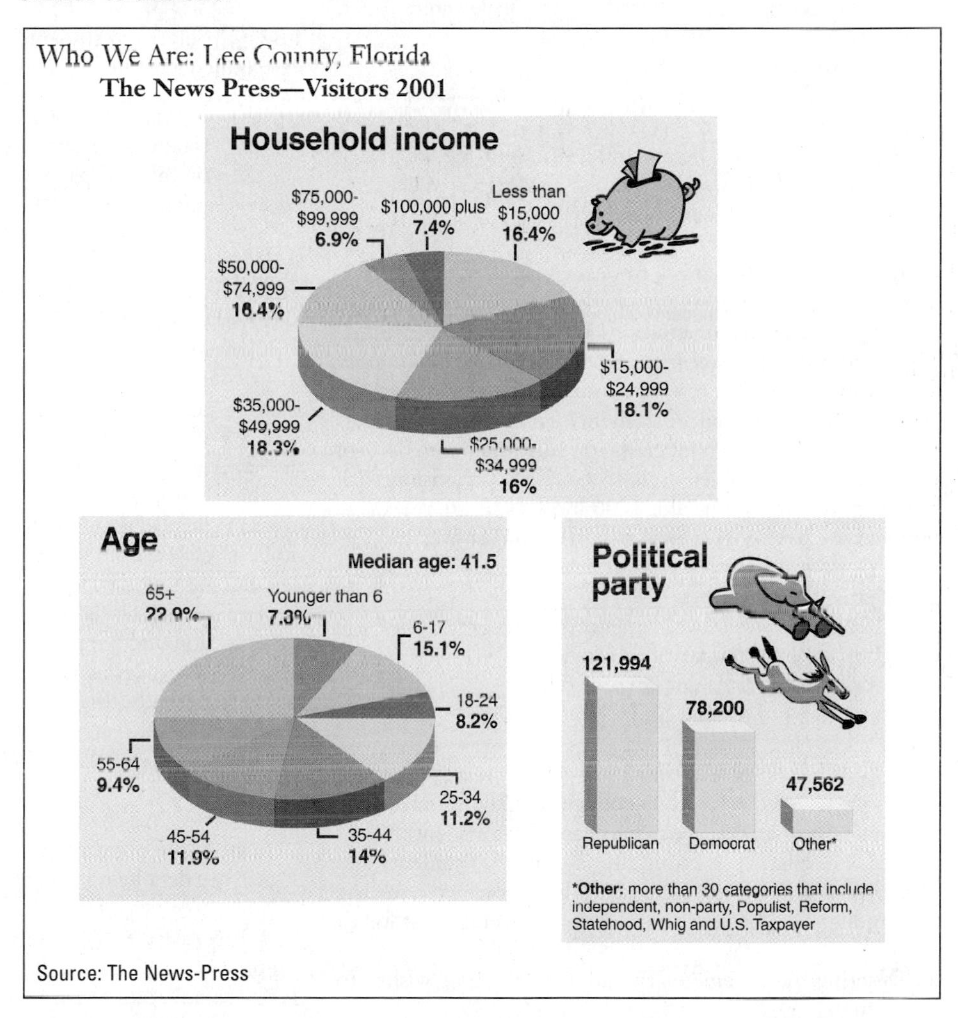

ANSWER NOW 1.25

a. Describe the general population of interest in the census information.

b. Describe the statistical population being used in each of the summaries, being sure to identify the individual elements that make up that specific statistical population.

c. List the variable used to collect that data for each of the summaries.

d. Identify each variable as qualitative (nominal or ordinal) or quantitative (discrete or continuous).

e. Describe how the data collected were used to create each of the three summaries.

Exercises

1.26 Is football jersey number a quantitative or a categorical variable? Support your answer with a detailed explanation.

1.27 The severity of side effects experienced by patients while being treated with a particular medicine is under study. The severity is measured on the scale: none, mild, moderate, severe, very severe.
a. Name the variable of interest.
b. Identify the type of variable.

1.28 Students are being surveyed about the weight of books and supplies they are carrying as they attend class.
a. Identify the variable of interest.
b. Identify the type of variable.
c. List a few values that might occur in a sample.

1.29 A drug manufacturer is interested in the proportion of persons who have hypertension (elevated blood pressure) whose condition can be controlled by a new drug the company has developed. A study involving 5000 individuals with hypertension is conducted, and it is found that 80% of the individuals are able to control their hypertension with the drug. Assuming that the 5000 individuals are representative of the group who have hypertension, answer the following questions:
a. What is the population?
b. What is the sample?
c. Identify the parameter of interest.
d. Identify the statistic and give its value.
e. Do we know the value of the parameter?

1.30 The admissions office wants to estimate the cost of textbooks for students at our college. Let the variable x be the total cost of all textbooks purchased by a student this semester. The plan is to randomly identify 100 students and obtain their total textbook costs. The average cost for the 100 students will be used to estimate the average cost for all students.
a. Describe the parameter the admissions office wishes to estimate.
b. Describe the population.
c. Describe the variable involved.
d. Describe the sample.
e. Describe the statistic and how you would use the 100 data collected to calculate the statistic.

1.31 A quality-control technician selects assembled parts from an assembly line and records the following information concerning each part:

A: defective or nondefective
B: the employee number of the individual who assembled the part
C: the weight of the part

a. What is the population?
b. Is the population finite or infinite?
c. What is the sample?
d. Classify the three variables as either attribute or numerical.

1.32 Select ten students currently enrolled at your college and collect data for these three variables:

X: number of courses enrolled in
Y: total cost of textbooks and supplies for courses
Z: method of payment used for textbooks and supplies

a. What is the population?
b. Is the population finite or infinite?
c. What is the sample?
d. Classify the three variables as nominal, ordinal, discrete, or continuous.

1.33 A study was conducted by Zeneca, Inc., to measure the adverse side effects of Zomig™, a new drug being used for the treatment of migraine headaches. A sample of 2633 migraine headache sufferers was given various dosages of the drug in tablet form for one year. The patients reported whether or not during the period they experienced relief from migraines along with any atypical sensations such as hypertension and paresthesia, pain and pressure sensations, digestive disorders, neurological disorders such as dizziness and vertigo, and other adverse side effects (sweating, palpitations, etc.). (*Life*, June 1998, p. 92)
a. What is the population being studied?
b. What is the sample?
c. What are the characteristics of interest about each element in the population?
d. Are the data being collected qualitative or quantitative?

1.34 Interactivity Interactivity 1-B simulates taking a sample of size 10 from a population of 100 college students. Take a sample of size 10.

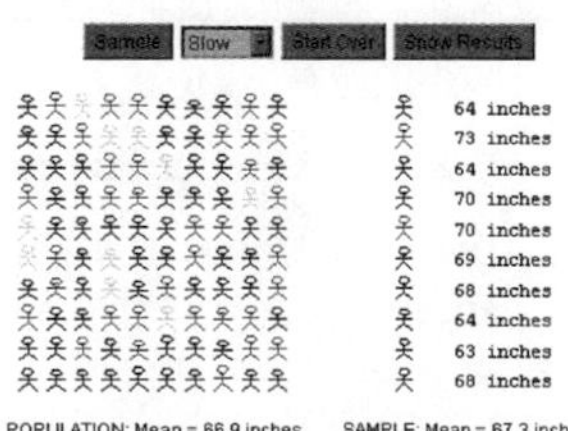

a. What is the population?
b. Is the population finite or infinite?
c. Name two parameters and give their values.
d. What is the sample?
e. Name the two corresponding statistics and give their values.
f. Take another sample of size 10. Which of the items above remain fixed and which change?

1.35 Identify each of the following as an example of (1) attribute (qualitative) or (2) numerical (quantitative) variables:
a. the breaking strength of a given type of string
b. the hair color of children auditioning for the musical *Annie*
c. the number of stop signs in towns of less than 500 people
d. whether or not a faucet is defective
e. the number of questions answered correctly on a standardized test
f. the length of time required to answer a telephone call at a certain real estate office

1.36 Identify each of the following as an example of (1) nominal, (2) ordinal, (3) discrete, or (4) continuous variables:
a. a poll of registered voters about which candidate they support
b. the length of time required for a wound to heal when a new medicine is used
c. the number of telephone calls arriving at a switchboard per 10-minute period
d. the distance first-year college women can kick a football
e. the number of pages per job coming off a computer printer
f. the kind of tree used as a Christmas tree

1.37 Suppose a 12-year-old asked you to explain the difference between a sample and a population.
a. What information should your answer include?
b. What reasons would you give for why one would take a sample instead of surveying every member of the population?

1.38 Suppose a 12-year-old asked you to explain the difference between a statistic and a parameter.
a. What information should your answer include?
b. What reasons would you give for why one would report the value of a statistic instead of a parameter?

1.3 Measurability and Variability

Within a set of data, we always expect variation. If little or no variation is found, we would guess that the measuring device is not calibrated with a small enough unit. For example, we take a carton of a favorite candy bar and weigh each bar individually. We observe that each of the 24 candy bars weighs $\frac{7}{8}$ ounce, to the nearest $\frac{1}{8}$ ounce. Does this mean that the bars are all identical in weight? Not really! Suppose we were to weigh them on an analytical balance that weighs to the nearest ten-thousandth of an ounce. Now the 24 weights will most likely show **variability**.

It does not matter what the response variable is; there will most likely be variability in the data if the tool of measurement is precise enough. One of the primary objectives of statistical analysis is measuring variability. For example, in the study of quality control, measuring variability is absolutely essential. Controlling (or reducing) the variability in a manufacturing process is a field all its own—namely, statistical process control.

Exercises

1.39 Suppose we measure the weights (in pounds) of the individuals in each of the following groups:

Group 1: cheerleaders for National Football League teams
Group 2: players for National Football League teams

For which group would you expect the data to have more variability? Explain why.

1.40 Suppose you were trying to decide which of two machines to purchase. Furthermore, suppose the length to which the machines cut a particular product part was important. If both machines produced parts that had the same length on the average, what other consideration regarding the lengths would be important? Why?

1.41 Consumer activist groups for years have encouraged retailers to use unit pricing of products. They argue that food prices, for example, should always be labeled in \$/ounce, \$/pound, \$/gram, \$/liter, and so on, in addition to \$/package, \$/can, \$/box, \$/bottle. Explain why.

1.42 A coin-operated coffee vending machine dispenses, on the average, 6 oz of coffee per cup. Can this statement be true of a vending machine that occasionally dispenses only enough to fill the cup half full (say, 4 oz)? Explain.

1.43 Interactivity Interactivity 1-A simulates sampling from a population of college students.

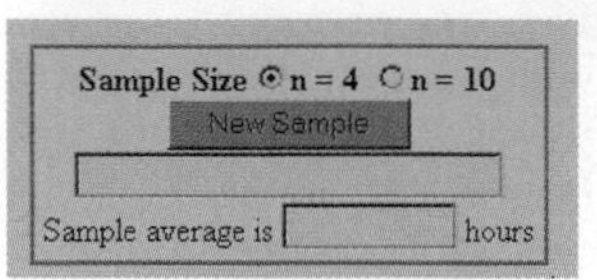

a. Take ten samples of size 4, keeping track of the sample averages of hours per week that students study. Find the range of these averages by subtracting the lowest average from the highest average.
b. Take ten samples of size 10, keeping track of the sample averages of hours per week that students study. Find the range of these averages by subtracting the lowest average from the highest average.
c. Which sample size demonstrated more variability?
d. If the population average is about 15 hours per week, which sample size demonstrated this most accurately? Why?

1.4 Data Collection

One of the first problems a statistician faces is obtaining data. Data don't just happen; data must be collected. It is important to obtain "*good data*," since the inferences ultimately made will be based on the statistics obtained from the data. These inferences can be only as good as the data.

While it is relatively easy to define "good data" as data that accurately represent the population from which they were taken, it is not easy to guarantee that a particular sampling method will produce "good data." We want to use sampling (data collection) methods that are *unbiased*.

BIASED SAMPLING METHOD
A sampling method that produces values that systematically differ from the population being sampled. An **unbiased** sampling method is one that is not biased.

Two commonly used sampling methods that often result in biased samples are *convenience* and *volunteer samples*.

A **convenience sample** occurs when a sample is selected from elements of a population that are easily accessible, whereas a **volunteer sample** consists of results collected from those elements of the population that chose to contribute the needed information on their own initiative.

Did you ever buy a basket of fruit at the market based on the "good appearance" of the fruit on top, only to discover later that the fruit on the bottom was not so nice? It was not convenient to inspect the bottom fruit, so you trusted a convenience sample.

Have you ever taken part in a volunteer survey? Under what conditions would you take the time to complete such a questionnaire? Most people's immediate attitude is to ignore the survey. Those with strong feelings will make the effort to respond; therefore, representative samples should not be expected when volunteer samples are collected.

ANSWER NOW 1.44
The way to ensure meaningful and reliable results is not always to just get a "larger" sample, as demonstrated in the video "Representative Sampling" on your Student Suite CD. View the video clip and then write a short paragraph describing what type of sampling was used and what went wrong.

ANSWER NOW 1.45
***USA Today* regularly asks readers, "Have a complaint about airline baggage, refunds, advertising, customer service? Write: . . . ." What kind of sampling method is this? Are the results likely to be biased? Explain.**

The collection of data for statistical analysis is an involved process and includes the following steps:

1. Define the objectives of the survey or experiment.
   Examples: compare the effectiveness of a new drug to the effectiveness of the standard drug; estimate the average household income in our county
2. Define the variable and the population of interest.
   Examples: length of recovery time for patients suffering from a particular disease; total income for households in our county
3. Define the data-collection and data-measuring schemes.
   Includes sampling procedures, sample size, and the data-measuring device (questionnaire, telephone, interview, and so on)
4. Determine the appropriate descriptive or inferential data-analysis techniques.

Often an analyst is stuck with data already collected, possibly even data collected for other purposes, which makes it impossible to determine whether or not the data are "good." Using approved techniques to collect your own data is much preferred. Although this text will be concerned chiefly with various data-analysis techniques, you should be aware of the concerns of data collection.

The following illustration describes the population and the variable of interest for a specific investigation.

Illustration 1.2 Population and Variable of Interest

The admissions dean at our college wishes to estimate the current "average" cost of textbooks per semester, per student. The population of interest is the "currently enrolled student body," and the variable is the "total amount spent for textbooks" by each student this semester. ■

The two methods used to collect data for a statistical analysis are an *experiment* and an *observational study*. In an **experiment**, the investigator controls or modifies the environment and observes the effect on the variable under study. We often read about laboratory results obtained when white rats are used to test different doses of a new medication and their effect on blood pressure. The experimental treatments were designed specifically to obtain the data needed to study the effect on the variable. In an **observational study**, the investigator does not modify the environment and does not control the process being observed. The data are obtained by sampling some of the population of interest. **Surveys** are often observational studies of people.

If every element in the population can be listed, or enumerated, and observed, then a census is compiled. A **census** is a 100% survey. A census for the population in Illustration 1.2 (p. 21) could be obtained by contacting each student on the registrar's computer printout of all registered students. Censuses are seldom used, however, because they are often difficult and time-consuming to compile and therefore expensive. Imagine the task of compiling a census of every woman who has the potential of using HRT as discussed in Application 1.8. Instead of a census, a sample survey is usually conducted.

When a sample is selected for a survey, it is necessary to construct a *sampling frame*.

SAMPLING FRAME
A list of the elements that belong to the population from which the sample will be drawn.

Ideally, the sampling frame should be identical to the population with every element of the population listed once and only once. In Illustration 1.2, the registrar's computer list will serve as a sampling frame for the admissions office. In this case, the list of all registered students becomes the sampling frame. In other situations, a 100% list may not be so easy to obtain. Lists of registered voters and the telephone directory are sometimes used as sampling frames of the general public. Depending on the nature of the information sought, the list of registered voters or the telephone directory may or may not be an unbiased sampling frame. Since only the elements in the frame have a chance to be selected as part of the sample, it is important that the sampling frame be **representative** of the population.

Once a representative sampling frame has been established, we proceed to select the sample elements from the sampling frame. This selection process is called the **sample design**. There are many different types of sample designs; however, they all fit into two categories: *judgment samples* and *probability samples*.

JUDGMENT SAMPLES
Samples that are selected on the basis of being "typical."

When a judgment sample is drawn, the person selecting the sample chooses items that he or she thinks are representative of the population. The validity of the results from a judgment sample reflects the soundness of the collector's judgment.

Application 1.7 Rain Again This Weekend!

It does, in fact, rain more on weekends

The Associated Press, 8-6-98

It's maddening but true: More rain does fall on weekends, a study says. Saturdays receive an average of 22% more precipitation than Mondays, climatologists at Arizona State University report in today's issue of *Nature*. "We were quite surprised to see weekends are substantially wetter than weekdays," said Randall Cerveny, one of the study's principal authors. . . .

ANSWER NOW 1.46
Was this study an experiment or an observational study? Explain.

PROBABILITY SAMPLES
Samples in which the elements to be selected are drawn on the basis of probability. Each element in a population has a certain known probability of being selected as part of the sample.

One of the most common probability sampling methods used to collect data is the *simple random sample.*

SIMPLE RANDOM SAMPLE
A sample selected in such a way that every element in the population has an equal probability of being chosen. Equivalently, all samples of size *n* have an equal chance of being selected. Simple random samples are obtained either by sampling with replacement from a finite population or by sampling without replacement from an infinite population.

Inherent in the concept of randomness is the idea that the next result (or occurrence) is not predictable. When a simple random sample is drawn, every effort must be made to ensure that each element has an equal probability of being selected and that the next result does not become predictable. The proper procedure for selecting a simple random sample is to use a random-number generator or a table of random numbers. Mistakes are frequently made because the term *random* (equal chance) is confused with **haphazard** (without pattern).

Application 1.8 The Importance of Proper Sampling

HRT and your heart—a new concern

For years women have been told that HRT (hormone replacement therapy) will help prevent cardiovascular (heart and blood vessel) disease. Doctors had good reason to recommend HRT for this purpose. Many studies observed that women taking HRT seem to have a much lower risk of heart disease than women not taking HRT. The problem with relying on results from these studies, however, is that they're observational trials. That means researchers looked at groups of people who decided for themselves whether or not to take HRT. And women who chose to take HRT are healthier to begin with and may already be at lower risk of heart disease.

A more precise way to determine the actual benefits of HRT is to look at a group of women with similar risks and then evenly and randomly divide them into two groups—called a prospective study. One group receives HRT, the other doesn't. One prospective study, called the Heart and Estrogen-Progestin Replacement Study (HERS), was done to assess HRT and heart disease risk. All of the women in the study were older (the average age was 71) and—more importantly—already had heart disease. They were expected to derive a huge benefit from HRT. That didn't happen. In fact, researchers found no difference in the number of heart attacks in hormone users and nonusers. . . .

Source: Mayo Clinic Women's HealthSource, Special Report November 2002

ANSWER NOW 1.47

a. **Did the observational study described in the first paragraph of Application 1.8 result in biased or unbiased results? Explain.**
b. **Was the HERS study an experiment or an observational study? Explain.**

ANSWER NOW 1.48
The Ann Landers' survey of voters is a classic example of a biased sample. View the "Sampling Bias" video clip on your Student Suite CD (found inside the back cover). Why do you think Landers' survey sample percent was so different from the random sample's percent?

To select a simple random sample, first assign a number to each element in the sampling frame. Numbers are usually assigned sequentially using the same number of digits for each element. Then go to a table of random numbers and select as many numbers with that number of digits as are needed for the sample size desired. Each numbered element in the sampling frame that corresponds to a selected random number is chosen for the sample.

Illustration 1.3 Using the Random Number Table

Let's return to Illustration 1.2 (p. 21). Mr. Clar, who works in the admissions office, has obtained a computer list of this semester's full-time enrollment. There are 4265 student names on the list. He numbered the students 0001, 0002, 0003, and so on, up to 4265; then, using four-digit random numbers, he identified a sample: 1288, 2177, 1952, 2463, 1644, 1004, and so on were selected. (See the *Statistical Tutor* for a discussion of the use of the random-number table.) ■

NOTE: In this text, all the statistical methods assume that random sampling has been used to collect the data.

There are many ways to approximate random sampling; four of these methods are briefly discussed below. They are presented here to illustrate some of the methodology involved in data collection. The topic of survey sampling is a complete textbook in itself.

One of the easiest methods for approximating a simple random sample is the *systematic sampling method.*

SYSTEMATIC SAMPLE
A sample in which every *k*th item of the sampling frame is selected, starting from a randomly selected first element.

To select an x percent (%) systematic sample, first we randomly select one element from the first $\frac{100}{x}$ elements and then we proceed to select every $\frac{100}{x}$th item thereafter until we have the desired number of data for our sample.

For example, if we desire a 3% systematic sample, we would locate the first item by randomly selecting an integer between 1 and 33 ($\frac{100}{x} = \frac{100}{3} = 33.33$, which when rounded becomes 33). Suppose 23 was randomly selected. This means that our first data is the value obtained from the subject in the 23rd position in the sampling frame. The second data will come from the subject in the 56th ($23 + 33 = 56$) position. The third from the 89th ($56 + 33$) position and so on until our sample is complete.

The systematic technique is easy to describe and execute; however, it has some inherent dangers when the sampling frame is repetitive or cyclical in nature. In these situations the results may not approximate a simple random sample.

When sampling is from very large populations, sometimes it is possible (and helpful) to divide the population into subpopulations on the basis of some characteristic. These subpopulations are called **strata**. These smaller, easier-to-work-with strata are sampled separately. One of the sample designs that starts by stratifying the sampling frame is the *stratified random sampling method.*

STRATIFIED RANDOM SAMPLE

A sample obtained by stratifying the sampling frame and then selecting a fixed number of items from each of the strata by means of a simple random sampling technique.

When a stratified random sample is drawn, the sampling frame is subdivided into various strata, usually some subdivisions that already occur naturally, and then a subsample is drawn from each of these strata. These subsamples may be drawn from the various strata by using random or systematic methods. The subsamples are first summarized separately and then combined to draw conclusions about the whole population.

ANSWER NOW 1.49

What body of the federal government is an example of a stratified sampling of the people (a random selection process is not used)?

An alternative to selecting the same number of items from each strata is to select from each strata proportionally to the size of the strata. This method is called a *proportional sampling*.

PROPORTIONAL SAMPLE

A sample obtained by stratifying the sampling frame and then selecting a number of items in proportion to the size of the strata from each strata by means of a simple random sampling technique.

When a proportional random sample is drawn, the sampling frame is subdivided into various strata and then a subsample is drawn from each strata. A convenient way to express the idea of proportional sampling is to establish a proportion. For example, the proportion "one for every 150" directs you to select one (1) element out of every 150 elements in the strata. That way, the size of the strata determines the size of the subsample. The subsamples are first summarized separately and then combined to draw conclusions about the whole population.

ANSWER NOW 1.50

What body of the federal government is an example of a proportional sampling of the people (a random selection process is not used)?

Another sampling method that starts by stratifying the sampling frame is a *cluster sample*.

CLUSTER SAMPLE

A sample obtained by sampling some of, but not all of, the possible subdivisions within a population. These subdivisions, called clusters, often occur naturally within the population.

A cluster sample can be obtained by using either random or systematic methods to identify the clusters to be sampled. Either all elements of each cluster or a simple random sample of some of the elements can then be selected. The subsamples are summarized separately and the information is then combined. A cluster sample becomes a judgment sample when the clusters are selected on the basis of being typical or reliable.

Sample design is not a simple matter. Many colleges and universities offer separate courses in sample surveying and experimental design. It is intended that the information presented here will provide you with an overview of sampling and put its role in perspective.

ANSWER NOW 1.51
Explain why the polls that are so frequently quoted during early returns on election day TV coverage are an example of cluster sampling.

Application 1.9 Family Dinner Together

National surveys in the 1990s indicated that a large majority of parents consider eating dinner with their children very important. Compared with other children-related activities, more than 80% of parents rank eating dinner together as being very important. Despite the importance placed on family dinner by parents, however, the proportion of children who eat dinner with their families is not high.

Family Dinner and Diet Quality Among Older Children and Adolescents
Matthew W. Gillman, MD; and others
(*Arch Fam Med.* 2000;9:235-240)

Context—The proportion of children eating dinner with their families declines with age.

Objective—To examine the associations between frequency of eating dinner with family and measures of diet quality.

Setting—A national convenience sample.

Participants—There were 8677 girls and 7525 boys in the study, aged 9 to 14 years, who were children of the participants in the ongoing Nurses' Health Study II.

Main Outcome Measures—We collected data from a self-administered mailed survey.

Results—Approximately 17% of participants ate dinner with members of their family never or some days, 40% on most days, and 43% every day. More than half of the 9-year-olds ate family dinner every day, whereas only about one third of 14-year-olds did so.

Age	Frequency of Family Dinner		
	Never or Some Days	Most Days	Every Day
9	12.1%	37.3%	50.7%
10	13.1%	38.2%	48.8%
11	15.5%	40.1%	44.4%
12	17.8%	40.8%	41.4%
13	20.5%	40.6%	38.9%
14	23.7%	40.9%	35.4%
All participants	17%	40%	43%

Activity	Mean Number		
	Never or Some Days	Most Days	Every Day
Make own dinner (x/wk)	2.3	1.9	1.6
Ready-made dinner (x/wk)	1.9	1.5	1.4
Physical activity (hr/wk)	15.9	15.8	15.2
Team sports (seasons/yr)	2.6	2.7	2.6
TV watching (hr/day)	2.5	2.3	2.1

ANSWER NOW 1.52
Refer to "Family Dinner and Diet Quality Among Older Children and Adolescents."

a. **Describe the population of interest.**

b. **Describe the sample used for this report.**

c. **Identify all seven variables used to collect the information being summarized in the table.**

d. **How were the age data used in summarizing the information shown on the table?**

e. **Was the sampling done randomly? That is, did all members of the population have an equal chance of being selected as part of the sample? Explain.**

f. **What kind of sample was used?**

ANSWER NOW 1.53
Discuss each of the following statements with regard to the information summarized in "Family Dinner and Diet Quality Among Older Children and Adolescents." Be specific about whether you believe the statement is justified or not, and why.

a. **The information collected in this sample could be biased.**

b. **A decrease in the frequency of dinner together is expected to result in an increase in the number of "make your own dinners."**

c. **An increase in TV watching causes a decrease in the frequency of dinner together.**

d. **The number of team sports played by a participant was unchanged by the frequency of dinner together.**

Exercises

1.54 *USA Today* conducted a survey asking readers, "What is the most hilarious thing that has ever happened to you en route to or during a business trip?"
a. What kind of sampling method is this?
b. Are the results likely to be biased? Explain.

1.55 Consider this question taken from CNN Quick Vote on the Internet on July 24, 1998: "In this last season of *Seinfeld* should Jerry Seinfeld have been nominated for an Emmy?" The response was 34% yes and 66% no.
a. What kind of survey was used?
b. Do you think these results could be biased? Why?

1.56 Consider a simple population consisting of only the numbers 1, 2, and 3 (an unlimited number of each). Nine different samples of size 2 could be drawn from this population: (1, 1), (1, 2), (1, 3), (2, 1), (2, 2), (2, 3), (3, 1), (3, 2), (3, 3).
a. If the population consists of the numbers 1, 2, 3, and 4, list all the samples of size 2 that could possibly be selected.
b. If the population consists of the numbers 1, 2, and 3, list all the samples of size 3 that could possibly be selected.

1.57
a. What is a sampling frame?
b. What did Mr. Clar use for a sampling frame in Illustration 1.3 (p. 24)?
c. Where did the number 1288 come from, and how was it used?

1.58 A random sample may be very difficult to obtain. Why?

1.59 A wholesale food distributor in a large metropolitan area would like to test the demand for a new food product. He distributes food through five large supermarket chains. The food distributor selects a sample of stores located in areas where he believes the shoppers are receptive to trying new products. What type of sampling does this represent?

1.60 Why is the random sample so important in statistics?

1.61 An article titled "Surface Sampling in Gravel Streams" (*Journal of Hydraulic Engineering*, April 1993) discusses grid sampling and areal sampling. Grid sampling involves the removal by hand of stones found at specific points. These points are established on the gravel surface by using either a wire mesh or predetermined distances on a survey tape. The material collected by grid sampling is usually analyzed as a frequency distribution. An areal sample is collected by removing all the particles found in a predetermined area of a channel bed. The material recovered is most often analyzed as a frequency distribution by weight. Would you categorize these sample designs as judgment samples or probability samples?

1.62 Sheila Jones works for an established marketing research company in Cincinnati, Ohio. Her supervisor just handed her a list of 500 four-digit random numbers extracted from a statistical table of random digits. He told Sheila to conduct a survey by calling 500 Cincinnati residents on the telephone, provided the last four digits of their phone numbers matched one of the numbers on the list. If Sheila follows her supervisor's instructions, is he assured of obtaining a random sample of respondents? Explain.

1.63 One question that people often ask is whether or not the risk of heart disease and other cardiovascular disorders is increased by high levels of LDL ("bad") cholesterol. LDL is far more prevalent in some foods than in others. A recent study by the University of Texas Health Sciences Center in San Antonio examined the arteries of 1400 males and females throughout the United States between the ages of 15 and 34 who had died of accidents, homicides, and suicides. The more LDL cholesterol and the less HDL ("good") cholesterol that were found in their blood, the more lesions were found in their arteries, irrespective of race, gender, or age. What type of sampling does this represent? (*Nutrition Action HealthLetter*, "Kids Can't Wait," April 1997, p. 2)

1.64 Describe in detail how you would select a 4% systematic sample of the adults in a nearby large city in order to complete a survey about a political issue.

1.65 Suppose you have been hired by a group of all-sports radio stations to determine the age distribution of their listeners. Describe in detail how you would select a random sample of 2500 from the 35 listening areas involved.

1.66 The telephone book might not be a representative sampling frame. Explain why.

1.67 The election board's voter registration list is not a census of the adult population. Explain why.

1.5 Comparison of Probability And Statistics

Probability and **statistics** are two separate but related fields of mathematics. It has been said that "probability is the vehicle of statistics." That is, if it were not for the laws of probability, the theory of statistics would not be possible.

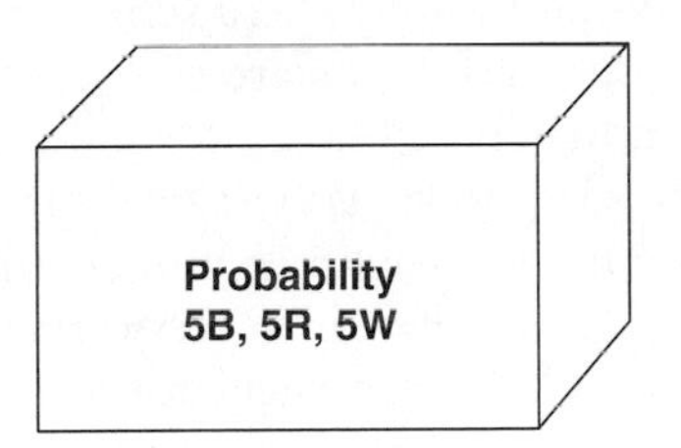

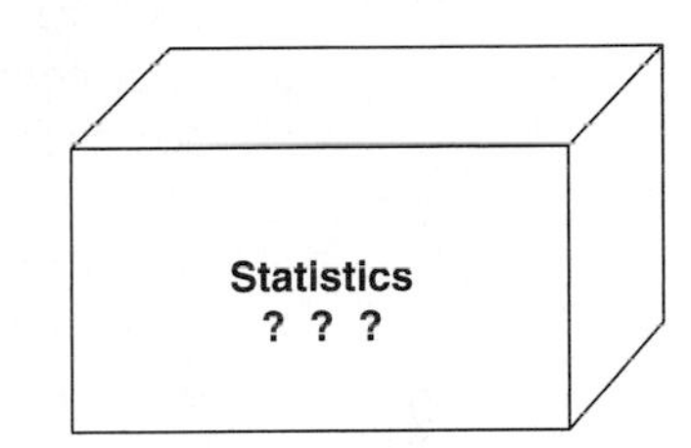

Let's illustrate the relationship and the difference between these two branches of mathematics by looking at two boxes. On one hand, we know the probability box contains five blue, five red, and five white poker chips. Probability tries to answer questions such as "If one chip is randomly drawn from this box, what is the chance that it will be blue?" On the other hand, in the statistics box we don't know what the combination of chips is. We draw a sample and, based on the findings in the sample, make conjectures about what we believe to be in the box. Note the difference: Probability asks you about the chance that something specific, like drawing a blue chip, will happen when you know the possibilities (that is, you know the population). Statistics, in contrast, asks you to draw a sample, describe the sample (descriptive statistics), and then make inferences about the population based on the information found in the sample (inferential statistics).

ANSWER NOW 1.68
Which of the following illustrates probability? Statistics?

a. **How likely is heads to occur when a coin is tossed?**
b. **The weights of 35 babies are studied to estimate weight gain in the first month after birth.**

Exercises

1.69 Classify each of the following as a probability or a statistics problem:
a. determining whether a new drug shortens the recovery time from a certain illness
b. determining the chance that heads will result when a coin is tossed
c. determining the amount of waiting time required to check out at a certain grocery store
d. determining the chance that you will be dealt a "blackjack"

1.70 Classify each of the following as a probability or a statistics problem:
a. determining how long it takes to handle a typical telephone inquiry at a real estate office
b. determining the length of life for the 100-watt light bulbs a company produces
c. determining the chance that a blue ball will be drawn from a bowl that contains 15 balls, of which 5 are blue
d. determining the shearing strength of the rivets that your company just purchased for building airplanes
e. determining the chance of getting "doubles" when you roll a pair of dice

1.6 Statistics and Technology

In recent years, electronic technology has had a tremendous impact on almost every aspect of life. The field of statistics is no exception. As you will see, the field of statistics uses many techniques that are repetitive in nature: calculations of numerical statistics, procedures for constructing graphic displays of data, and procedures that are followed to formulate statistical inferences. Computers and calculators are very good at performing these sometimes long and tedious operations. If your computer has one of the standard statistical packages on line or if you have a statistical calculator, then it will make the analysis easy to perform.

Throughout this textbook, as statistical procedures are studied, you will find the information you need to have a computer complete the same procedures using MINITAB (Release 13) and Excel XP software. Calculator procedures will also be demonstrated for the TI-83 Plus calculator.

An explanation of the most common typographical conventions that will be used in this textbook is given below. As additional explanations or selections are needed, they will be given.

Technology Instructions: Basic Conventions

MINITAB (Release 13)

Choose: tells you to make a menu selection by a mouse "point and click" entry.
For example: **Choose: Stat > SPC > Pareto Chart** instructs you to, in sequence, "point and click on" **Stat** on the menu bar, "followed by" **SPC** on the pulldown, and then "followed by" **Pareto Chart** on the pulldown.
Select: indicates that you should click on the small box or circle to the left of a specified item.
Enter: instructs you to type or select information needed for a specific item.

FYI
For information about obtaining MINITAB, check the Internet at www.minitab.com.

Excel XP

Choose: tells you to make a menu or tab selection by a mouse "point and click" entry.
For example: **Choose: Chart Wizard > XY(Scatter) > 1st graph picture > Next** instructs you to, in sequence, "point and click on" the **Chart Wizard** icon, followed by **XY(Scatter)** under Chart type, followed by **1st graph picture** on the Chart subtype, and then followed by **Next** on the dialog window.
Select: indicates that you should click on the small box or circle to the left of a specified item. It is often followed by a "point and click on" **Next** or **Finish** on the dialog window.
Enter: instructs you to type or select information needed for a specific item.

FYI
Excel is part of Microsoft Office and can be found on many personal computers.

TI-83 Plus

Choose: tells you which keys to press or menu selections to make.
For example: **Choose: Zoom > 9:ZoomStat > Trace > > >** instructs you to press the **Zoom** key, followed by selecting **9:ZoomStat** from the menu, followed by pressing the **Trace** key; > > > indicates to press arrow keys repeatedly to move along a graph to obtain important points.
Enter: instructs you to type or select information needed for a specific item.
Screen Capture: gives pictures of what your calculator screen should look like with chosen specifications are highlighted.

FYI
For information about TI-83 Plus, check the Internet at www.ti.com/calc.

Additional details about the use of MINITAB and Excel are available by using the Help system in the MINITAB and Excel software. Additional details for the TI-83 are contained in its corresponding *TI-83 Plus Graphing Calculator Guidebook*. Specific details on the use of computers and calculators available to you need to be obtained from your instructor or from your local computer lab person.

Your local computer center can provide you with a list of what is available to you. Some of the more readily available packaged programs are: MINITAB, JMP-IN, and SPSS (Statistical Package for the Social Sciences).

NOTE: *There is a great temptation to use the computer or calculator to analyze any and all sets of data and then treat the results as though the statistics are correct. Remember the old adage: "Garbage-in, garbage-out!" Responsible use of statistical methodology is very important. The burden is on the user to ensure that the appropriate methods are correctly applied and that accurate conclusions are drawn and communicated to others.*

Exercises

1.71 How have computers increased the usefulness of statistics to professionals such as researchers, government workers who analyze data, statistical consultants, and others?

1.72 How might computers help you in statistics?

1.73 What is meant by the saying "Garbage-in, garbage-out!" and how have computers increased the probability that studies may be victimized by the adage?

CHAPTER SUMMARY

RETURN TO CHAPTER CASE STUDY

Americans, Here's Looking at You

Let's return to the Chapter Case Study (p. 5) as a way to assess what we have learned in this chapter. Study the statistical information presented by the graphs and charts, and ask yourself how the terms (population, sample, variable, statistic, type of variable) studied in this chapter apply to each and how you compare to the statistical story being told. If you haven't viewed them yet, watch the videos on your CD about Ann Landers and the Gallup Poll.

FIRST THOUGHTS 1.74

a. The name "USA Today Snapshots" seems to suggest that information presented in these graphics will be about what population? Is that the case?
b. Describe the information that was collected and used to determine the statistics reported in "Movie mania ebbing?"
c. "53%—Pen / pencil / paper" was one specific statistic reported in "Tools workers use at meetings." Describe what that statistic tells you.
d. Consider "Excuses to get out of tickets." If you had just been given a ticket, how would you have responded? Do you believe your answer is represented accurately in the diagram? What does the percentage associated with your answer really mean? Explain.

PUTTING CHAPTER 1 TO WORK 1.75

a. What statistical population is of concern for the "USA Today Snapshots"?
b. Identify one specific Snapshot. What variable(s) was used to collect the information needed to determine the statistics reported?
c. Name one statistic that is reported in your Snapshot.
d. To obtain the data for your Snapshot, what method(s) do you think was used: convenience sample, volunteer sample, random sample, survey, observational study, experiment, or judgment sample?
e. Considering the method, how much faith do you have in the printed statistics? Describe possible biases.

YOUR OWN MINI-STUDY 1.76

Select one of the Snapshots. Then, using the students at your school or college as the population of concern, collect sample data from 30 students and produce your own version of the Snapshot. Write a paragraph describing how your results compare with those reported in the selected Snapshot in the Chapter Case Study.

In Retrospect

You should now have a general feeling for what statistics is about, an image that will grow and change as you work your way through this book. You know what a sample and a population are and the distinction between qualitative (attribute) and quantitative (numerical) variables. You even know the difference between statistics and probability (although we will not study probability in detail until Chapter 4). You should also have an appreciation for and a partial understanding of how important random samples are in statistics.

Throughout the chapter you have seen numerous articles that represent various aspects of statistics. The USA Snapshots® picture a variety of information about ourselves as we describe ourselves and other aspects of the world around us. Statistics can even be entertaining. The examples are endless. Look around and find some examples of statistics in your daily life (see Exercises 1.87 and 1.88, p. 34).

Chapter Exercises

1.77 We want to describe the so-called typical student at your college. Describe a variable that measures some characteristic of a student and results in:

a. attribute data
b. numerical data

1.78 A candidate for a political office claims that he will win the election. A poll is conducted, and 35 of 150 voters indicate that they will vote for the candidate, 100 voters indicate that they will vote for his opponent, and 15 voters are undecided.

a. What is the population parameter of interest?
b. What is the value of the sample statistic that might be used to estimate the population parameter?
c. Would you tend to believe the candidate based on the results of the poll?

1.79 A researcher studying consumer buying habits asks every 20th person entering Publix Supermarket how many times per week he or she goes grocery shopping. She then records the answer as T.

a. Is $T = 3$ an example of a sample, a variable, a statistic, a parameter, or a piece of data?

Suppose the researcher questions 427 shoppers during the survey.

b. Give an example of a question that can be answered using the tools of descriptive statistics.
c. Give an example of a question that can be answered using the tools of inferential statistics.

1.80 A researcher studying the attitudes of parents of preschool children interviews a random sample of 50 mothers, each having one preschool child. He asks each mother, "How many times did you compliment your child yesterday?" He records the answer as C.

a. Is $C = 4$ an example of a piece of data, a statistic, a parameter, a variable, or a sample?
b. Give an example of a question that can be answered using the tools of descriptive statistics.
c. Give an example of a question that can be answered using the tools of inferential statistics.

1.81 *Ladies' Home Journal*, in a June 1998 article titled "States of Health," presented the results of a study that analyzed data collected by the U.S. Census Bureau in 1997. Results reveal that for both men and women in the United States, heart disease remains the number one killer, victimizing 500,000 people annually. Age, obesity, and inactivity all contribute to heart disease, and all three of these factors vary considerably from one location to the next. The highest mortality rates (deaths per 100,000 people) were in New York, Florida, Oklahoma, and Arkansas, whereas the lowest were reported in Alaska, Utah, Colorado, and New Mexico. (*Ladies' Home Journal*, "States of Health," June 1998, p. 152)

a. What is the population?
b. What are the characteristics of interest?
c. What is the parameter?
d. Classify all the variables in the study as either attribute or numerical.

1.82 A USA Snapshot® from *USA Today* (June 4, 2002) described how executives feel about looking for a new job while still employed? According to the Snapshot, a survey of 150 executives from the nation's 1000 largest companies resulted in the following responses: 36% felt very comfortable, 33% felt somewhat comfortable, 26% felt somewhat uncomfortable, and 5% felt very uncomfortable. Would you classify the data collected and used to determine these percentages as qualitative (nominal or ordinal) or quantitative (discrete or continuous)?

1.83 An article titled "Want a Job in Food?" in *Parade* magazine (November 13, 1994) referenced a study at the University of California involving 2000 young men. The study found that in 2000 young men who did not go to college, of those who took restaurant jobs (typically as fast-food counter workers), one in two reached a higher-level blue-collar job and one in four reached a managerial position within four years.

a. What is the population?
b. What is the sample?
c. Is this a judgment sample or a probability sample?

1.84 Teachers use examinations to measure a student's knowledge about the subject. Explain how a lack of variability in the students' scores might indicate that the exam was not a very effective measuring device.

1.85 Describe, in your own words, and give an example of each of the following terms. Your examples should not be ones given in class or in the textbook.

a. variable b. data c. sample
d. population e. statistic f. parameter

1.86 Describe, in your own words, and give an example of the following terms. Your examples should not be ones given in class or in the textbook.

a. random sample b. probability sample
c. judgment sample

1.87 Find an article or an advertisement in a newspaper or magazine that exemplifies the use of statistics.

a. Identify and describe one statistic reported in the article.
b. Identify and describe the variable related to the statistic in part a.
c. Identify and describe the sample related to the statistic in part a.
d. Identify and describe the population from which the sample in part c was taken.

1.88

a. Find an article in a newspaper or magazine that exemplifies the use of statistics in a way that might be considered "entertainment" or "recreational." Describe why you think this article fits one of these categories.
b. Find an article in a newspaper or magazine that exemplifies the use of statistics and is presenting an unusual finding as the result of a study. Describe why these results are (or are not) "newsworthy."

Vocabulary and Key Concepts

Be able to define each term. Pay special attention to the key terms, which are printed in **red**. In addition, describe each term in your own words and give an example of each. Your examples should not be ones given in class or in the textbook. Page numbers indicate the first appearance of the term.

attribute variable (p. 14)
biased sampling method (p. 20)
categorical variable (p. 14)
census (p. 22)
cluster sample (p. 25)
continuous variable (p. 15)
convenience sample (p. 20)
data (p. 12)
descriptive statistics (pp. 2, 6)
discrete variable (p. 15)
experiment (pp. 12, 21)
finite population (p. 11)
haphazard (p. 23)
inferential statistics (p. 6)
infinite population (p. 11)
judgment sample (p. 23)
nominal variable (p. 14)
numerical data (p. 14)
observational study (p. 21)
ordinal variable (p. 15)
parameter (p. 12)
population (p. 11)
probability (p. 28)
probability sample (p. 23)
proportional sample (p. 25)
qualitative variable (p. 14)
quantitative variable (p. 14)
random sample (p. 23)
representative sampling frame (p. 23)
sample (p. 12)
sample design (p. 23)
sampling frame (p. 23)
simple random sample (p. 23)
statistic (p. 13)
statistics (pp. 6, 28)
strata (p. 25)
stratified random sample (p. 25)
survey (p. 21)
systematic sample (p. 24)
unbiased sampling method (p. 20)
variability (p. 19)
variable (p. 12)
volunteer sample (p. 20)

Chapter Practice Test

PART I: KNOWING THE DEFINITIONS

Answer "True" if the statement is always true. If the statement is not always true, replace the words printed in bold with words that make the statement always true.

1.1 **Inferential** statistics is the study and description of data that result from an experiment.

1.2 **Descriptive statistics** is the study of a sample that enables us to make projections or estimates about the population from which the sample is drawn.

1.3 A **population** is typically a very large collection of individuals or objects about which we desire information.

1.4 A statistic is the calculated measure of some characteristic of a **population**.

1.5 A parameter is the measure of some characteristic of a **sample**.

1.6 As a result of surveying 50 freshmen, it was found that 16 had participated in interscholastic sports, 23 had served as officers of classes and clubs, and 18 had been in school plays during their high school years. This is an example of **numerical data**.

1.7 The "number of rotten apples per shipping crate" is an example of a **qualitative** variable.

1.8 The "thickness of a sheet of sheet metal" used in a manufacturing process is an example of a **quantitative** variable.

1.9 A **representative** sample is a sample obtained in such a way that all individuals had an equal chance to be selected.

1.10 The basic objectives of **statistics** are obtaining a sample, inspecting this sample, and then making inferences about the unknown characteristics of the population from which the sample was drawn.

PART II: APPLYING THE CONCEPTS

The owners of Corner Convenience Store are concerned about the quality of service their customers receive. In order to study the service, they collected samples for each of several variables.

1.11 Classify each of the following variables as nominal, ordinal, discrete, or continuous:
a. method of payment for purchases (cash, credit card, check)
b. ZIP code for the customer's home mailing address
c. amount of sales tax on purchase
d. number of items purchased
e. customer's driver's license number

1.12 The mean checkout time for all customers at Corner Convenience Store is to be estimated by using the mean checkout time for 75 randomly selected customers. Match the items in column 2 with the statistical terms in column 1.

1	2
____ data (one)	(a) the 75 customers
____ data (set)	(b) the mean time for all customers
____ experiment	(c) 2 minutes, one customer's checkout time
____ parameter	(d) the mean time for the 75 customers
____ population	(e) all customers at Corner Convenience Store
____ sample	(f) the checkout time for one customer
____ statistic	(g) the 75 times
____ variable	(h) the process used to select 75 customers and measure their times

PART III: UNDERSTANDING THE CONCEPTS

Write a brief paragraph in response to each question.

1.13 The population and the sample are both sets of objects. Describe the relationship between them and give an example.

1.14 The variable and the data for a specific situation are closely related. Explain this relationship and give an example.

1.15 The data, the statistic, and the parameter are all values used to describe a statistical situation. How does one distinguish among these three terms? Give an example.

1.16 What conditions are required in order for a sample to be a random sample? Explain and include an example of a sample that is random and one that is not random.

CHAPTER

2 Descriptive Analysis and Presentation of Single-Variable Data

Chapter Outline and Objectives

CHAPTER CASE STUDY
Graphic displays of statistical information make interesting and easy reading.

Graphic Presentation of Data

2.1 GRAPHS, PARETO DIAGRAMS, AND STEM-AND-LEAF DISPLAYS
A **picture** is often worth a thousand words.

2.2 FREQUENCY DISTRIBUTIONS AND HISTOGRAMS
An **increase** in the amount of data requires us to modify our techniques and leads to graphic methods that express the data's distribution.

Numerical Descriptive Statistics

2.3 MEASURES OF CENTRAL TENDENCY
The four measures of central tendency—mean, median, mode, and midrange—are **average** values.

2.4 MEASURES OF DISPERSION
Measures of dispersion—range, variance, and standard deviation—assign numerical values to the **amount of spread** in a set of data.

2.5 MEAN AND STANDARD DEVIATION OF FREQUENCY DISTRIBUTION
The **frequency distribution** is an aid in calculating mean and standard deviation.

2.6 MEASURES OF POSITION
Measures of position allow us to **compare** one data to the set of data.

2.7 INTERPRETING AND UNDERSTANDING STANDARD DEVIATION
A standard deviation is the length of a **standardized yardstick**.

2.8 THE ART OF STATISTICAL DECEPTION
How the unwitting or the unscrupulous can use **"tricky" graphs** and **insufficient information** to mislead the unwary.

RETURN TO CHAPTER CASE STUDY

Chapter Case Study

What Do People Do When They Are on the Internet?

©Jules Frazier/Getty Images

Stanford Institute for the Quantitative Study of Society (SIQSS) supported a study that looked at how people utilize the Internet. Four thousand respondents were asked to select which of 17 common Internet activities they did or did not do. E-mail was the most common use of the Internet for about 90% of the respondents. Other common uses were information gathering, entertainment, chat rooms, and business transactions.

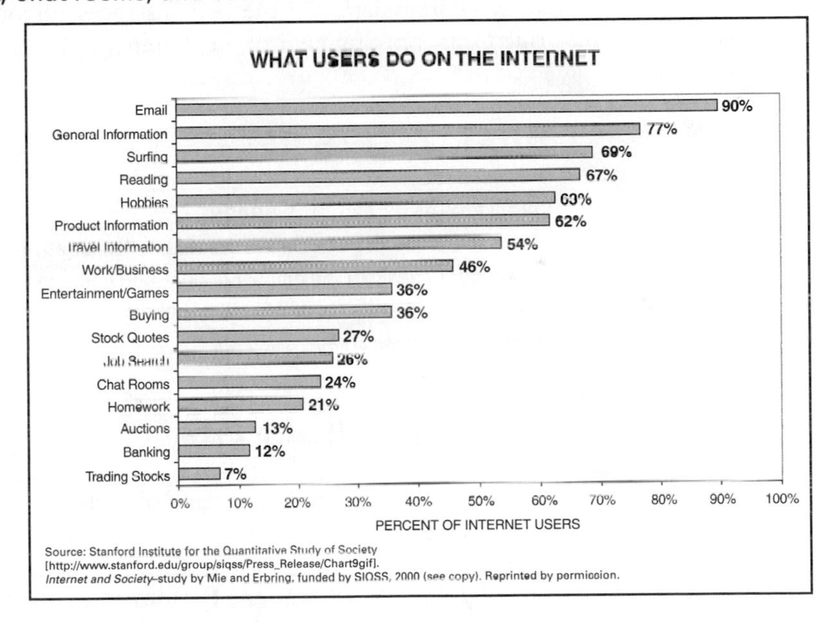

FYI

It is important that you complete the Answer Now exercises as you encounter them. They are an important part of learning.

ANSWER NOW 2.1 CS02

Students in a statistics course offered on the Internet were asked how many different Internet activities they engaged in during a typical week. The following data are the numbers of activities.

6	7	3	6	9	10	8	9	9	6
4	9	4	9	4	2	3	5	13	12
4	6	4	9	5	6	9	11	5	6
5	3	7	9	6	5	12	2	6	9

a. **If you were asked to present these data, how would you organize and summarize them?**

b. **This chapter will discuss a variety of methods for displaying and describing data. What type of information or conclusions would you like to know about these data if one of the pieces of data were from you?**

After completing Chapter 2, further investigate the Chapter Case Study in the Return to Chapter Case Study section with Exercises 2.160, 2.161, and 2.162. (pp. 115–116).

Graphic Presentation of Data

2.1 Graphs, Pareto Diagrams, and Stem-and-Leaf Displays

Once the sample data have been collected, we must "get acquainted" with them. One of the most helpful ways to become acquainted with the data is to use an initial exploratory data-analysis technique that will result in a pictorial representation of the data. The display will visually reveal patterns of behavior of the variable being studied. There are several graphic (pictorial) ways to describe data. The type of data and the idea to be presented determine which method is used.

NOTE: There is no single correct answer when constructing a graphic display. The analyst's judgment and the circumstances surrounding the problem play a major role in the development of the graphic.

QUALITATIVE DATA

CIRCLE GRAPHS AND BAR GRAPHS
Graphs that are used to summarize qualitative, or attribute, or categorical, data. Circle graphs (pie diagrams) show the amount of data that belong to each category as a proportional part of a circle. Bar graphs show the amount of data that belong to each category as a proportionally sized rectangular area.

Illustration 2.1 **Graphing Qualitative Data**

Table 2.1 lists the number of cases of each type of operation performed at General Hospital last year.

TABLE 2.1 **Operations Performed at General Hospital Last Year** Ta02-01

Type of Operation	Number of Cases
Thoracic	20
Bones and joints	45
Eye, ear, nose, and throat	58
General	98
Abdominal	115
Urologic	74
Proctologic	65
Neurosurgery	23
Total	498

The data in Table 2.1 are displayed on a circle graph in Figure 2.1, with each type of operation represented by a relative proportion of the circle, found by dividing the number of cases by the total sample size—namely, 498. The proportions

FIGURE 2.1
Circle Graph

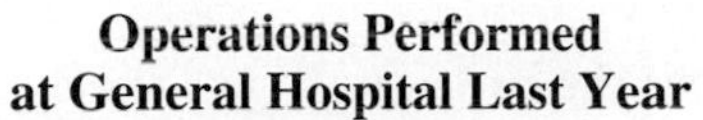

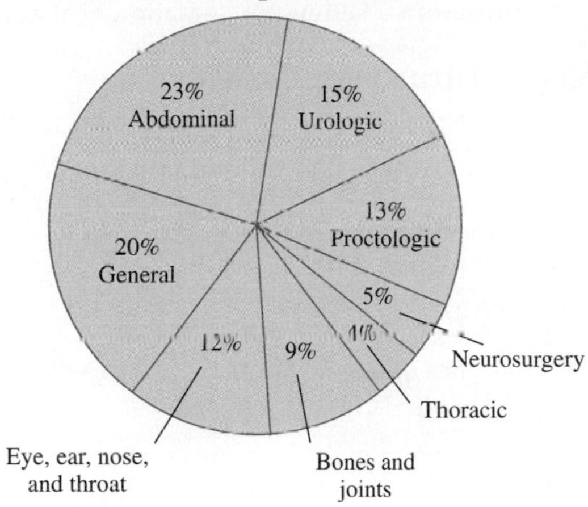

ANSWER NOW 2.2
Construct a circle graph showing what U.S. adults would choose if stranded on an island, as reported in *USA Today* (March 6, 2000): Computer connected to the Internet—66%, Telephone—23%, Television—8%, Don't know—3%.

are then reported as percentages (for example, 25% is ¼ of the circle). Figure 2.2 displays the same "type of operation" data but in the form of a bar graph. Bar graphs of attribute data should be drawn with a space between bars of equal width.

FIGURE 2.2
Bar Graph

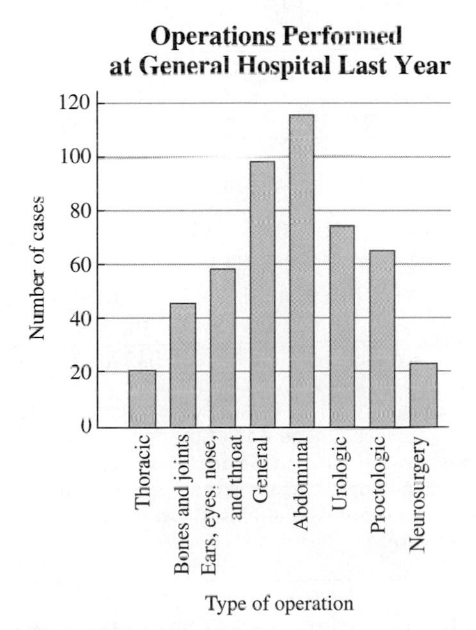

■

ANSWER NOW 2.3
Construct a bar graph showing what U.S. adults would choose if stranded on an island, as reported in *USA Today* (March 6, 2000): Computer connected to the Internet—66%, Telephone—23%, Television—8%, Don't know—3%.

ANSWER NOW 2.4
In your opinion, does the circle graph (in Answer Now 2.2) or the bar graph (in Answer Now 2.3) result in a better representation of the information? Explain.

Technology Instructions: Circle Graph

MINITAB (Release 13)

Input the categories into C1 and the corresponding frequencies into C2; then continue with:

```
Choose:   Graph > Pie Chart...
Select:   Chart table
Enter:    Categories in:  C1
          Frequencies in: C2
          Your title
```

Excel XP

Input the categories into column A and the corresponding frequencies into column B; then continue with:

```
Choose:   Chart Wizard > Pie > 1st picture (usually) > Next
Enter:    Data range: (A1:B5 or select cells)
Check:    Series in: columns > Next
Choose:   Titles
Enter:    Chart title: Your title
Choose:   Data Labels
Select:   Category name and Percentage > Next > Finish
```

To edit the pie chart:

```
Click On: Anywhere clear on the chart
          —use handles to size
          Any cell in the category or frequency column
          and type in different name or amount > ENTER
```

TI-83 Plus

Input the frequencies for the various categories into L1; then continue with:

```
Choose:   PRGM > EXEC > CIRCLE*
Enter:    LIST: L1 > ENTER
          DATA DISPLAYED?: 1:PERCENTAGES
                                 OR
                               2:DATA
```

*The TI-83 Plus program 'CIRCLE' and others can be downloaded from the Web site www.duxbury.com. Select 'Online Book Companions' followed by 'Student Resources' for 'Elementary Statistics, 9th edition'. Select 'TI-83 Programs' under 'Course Resources'. The TI-83 Plus programs and data files are jkprogs.zip and jklists.zip. Copy the files from the site to your computer. Uncompress the files. Download the programs to your calculator using TI-Graph Link Software.

When the bar graph is presented in the form of a *Pareto diagram*, it presents additional and very helpful information.

PARETO DIAGRAM
A bar graph with the bars arranged from the most numerous category to the least numerous category. It includes a line graph displaying the cumulative percentages and counts for the bars.

Illustration 2.2 Pareto Diagram of Hate Crimes

The FBI reported the number of hate crimes by category for 1993 (*USA Today*, June 29, 1994). The Pareto diagram in Figure 2.3 shows the 6746 categorized hate crimes, their percentages, and cumulative percentages.

FIGURE 2.3
Pareto Diagram

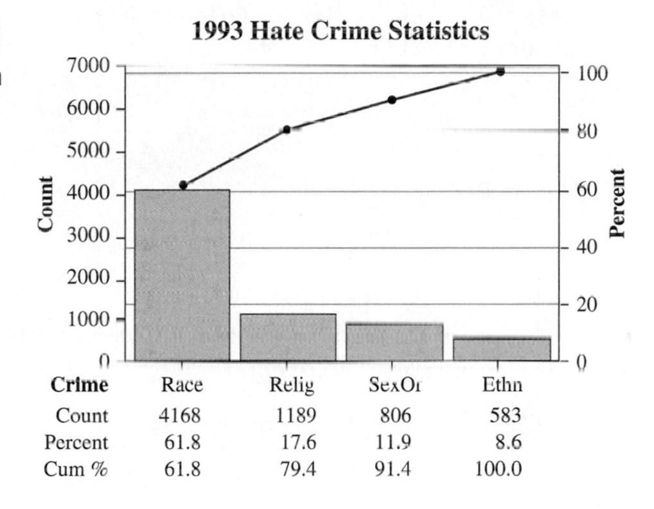

Crime	Race	Relig	SexOr	Ethn
Count	4168	1189	806	583
Percent	61.8	17.6	11.9	8.6
Cum %	61.8	79.4	91.4	100.0

■

ANSWER NOW 2.5
A shirt inspector at a clothing factory categorized the last 500 defects as: 67—missing button, 153—bad seam, 258—improperly sized, 22—fabric flaw. Construct a Pareto diagram for this information.

The Pareto diagram is popular in quality-control applications. A Pareto diagram of types of defects will show the ones that have the greatest effect on the defective rate in order of effect. It is then easy to see which defects should be targeted in order to most effectively lower the defective rate.

Technology Instructions: Pareto Diagram

MINITAB (Release 13)

Input the categories into C1 and the corresponding frequencies into C2; then continue with:

```
Choose:   Stat > Quality Tools > Pareto Chart
Select:   Chart defects table
Enter:    Labels in:          C1
          Frequencies in:     C2
          Your Title
```

Excel XP

Input the categories into column A and the corresponding frequencies into column B (column headings are optional); then continue with:

First, sorting the table:

```
Activate both columns of the distribution
Choose:   Data > Sort > Sort by: Column B (freq or rel freq
          col.)
Select:   Descending
          My list has: Header row or No Header row > OK
```

Technology Instructions: Pareto Diagram (continued)

Excel XP (continued)

```
Choose:   Chart Wizard > Column > 1st picture (usually) >
          Next
Choose:   Data Range
Enter:    Data Range:   (A1:B5 or select cells)
Select:   Series in:    Columns > Next
Choose:   Titles
Enter:    Chart title: your title
          Category (x) axis: title for x-axis
          Value (y) axis: title for y-axis > Next > Finish
```

To edit the Pareto diagram:

```
Click on: Anywhere clear on the chart
          —use handles to size
          Any title name to change
          Any cell in the category column and type in a
          name > Enter
```

Excel does not include the line graph.

TI-83 Plus

Input the numbered categories into L1 and the corresponding frequencies into L2; then continue with:

```
Choose:   PRGM > EXEC > PARETO*
Enter:    LIST: L2 > ENTER
  Ymax:   at least the sum of the frequencies > ENTER
  Yscl:   increment for y-axis > ENTER
```

*Program 'PARETO' is one of many programs that are available for downloading from a Web site. See page 40 for specific instructions.

QUANTITATIVE DATA

One major reason for constructing a graph of quantitative data is to display its *distribution.*

DISTRIBUTION

The pattern of variability displayed by the data of a variable. The distribution displays the frequency of each value of the variable.

One of the simplest graphs used to display a distribution is the *dotplot.*

DOTPLOT DISPLAY

Displays the data of a sample by representing each data with a dot positioned along a scale. This scale can be either horizontal or vertical. The frequency of the values is represented along the other scale.

Illustration 2.3

Dotplot of Exam Grades II02-03

A sample of 19 exam grades was randomly selected from a large class:

76	74	82	96	66	76	78	72	52	68
86	84	62	76	78	92	82	74	88	

Figure 2.4 is a dotplot of the 19 exam scores.

FIGURE 2.4
Dotplot

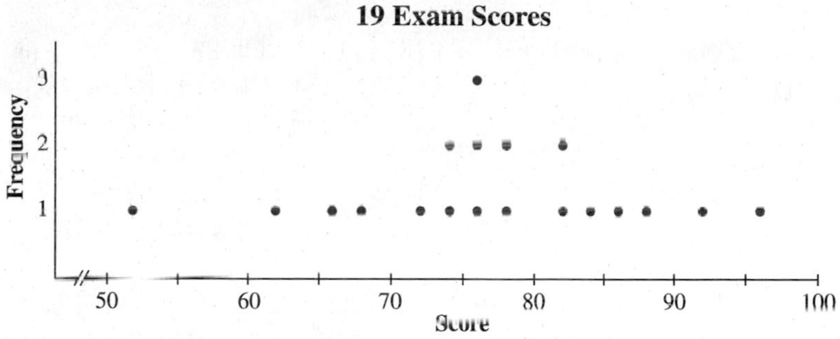

19 Exam Scores

Frequency
Score

Notice how the data in Figure 2.4 are "bunched" near the center and more "spread out" near the extremes. ■

The dotplot display is a convenient technique to use as you first begin to analyze the data. It results in a picture of the data as well as sorts the data into numerical order. (To *sort* data is to list the data in rank order according to numerical value.)

ANSWER NOW 2.6 Ex02-006
Here are the numbers of points scored during each of 16 games by a high school basketball team last season:

56	54	61	71	46	61	55	68
60	66	54	61	52	36	64	51

Construct a dotplot of these data.

Technology Instructions: Dotplot

MINITAB (Release 13)

Input the data into C1; then continue with:

```
Choose:   Graph > Character Graph > Dotplot...
Enter:    Variables: C1
```

Excel XP

The dotplot display is not available, but the initial step of ranking the data can be done. Input the data into column A and activate the column of data; then continue with:

```
Choose:   Data > Sort
Enter:    Sort by: Column A
Select:   Ascending > My list has: Header row or No Header
          row
```

Use the sorted data to finish constructing the dotplot display.

TI-83 Plus

Input the data into L1; then continue with:

```
Choose:   PRGM > EXEC > DOTPLOT*
```

(continued)

*Program 'DOTPLOT' is one of many programs that are available for downloading from a Web site. See page 40 for specific instructions.

Technology Instructions: Dotplot (continued)

TI-83 Plus (continued)

```
Enter:    LIST: L1 > ENTER
          Xmin: at most the lowest x value
          Xmax: at least the highest x value
          Xscl: 0 or increment
          Ymax: at least the highest frequency
```

In recent years a technique known as the *stem-and-leaf display* has become very popular for summarizing numerical data. It is a combination of a graphic technique and a sorting technique. These displays are simple to create and use, and they are well suited to computer applications.

STEM-AND-LEAF DISPLAY

Displays the data of a sample using the actual digits that make up the data values. Each numerical value is divided into two parts: The leading digit(s) becomes the stem, and the trailing digit(s) becomes the leaf. The stems are located along the main axis, and a leaf for each data value is located so as to display the distribution of the data.

Illustration 2.4

Constructing a Stem-and-Leaf Display

Let's construct a stem-and-leaf display for the 19 exam scores given in Illustration 2.3. II02-03

76	74	82	96	66	76	78	72	52	68
86	84	62	76	78	92	82	74	88	

At a quick glance we see that there are scores in the 50s, 60s, 70s, 80s, and 90s. Let's use the first digit of each score as the stem and the second digit as the leaf. Typically, the display is constructed vertically. We draw a vertical line and place the stems, in order, to the left of the line.

```
5 |
6 |
7 |
8 |
9 |
```

Next we place each leaf on its stem. This is done by placing the trailing digit on the right side of the vertical line opposite its corresponding leading digit. Our first data value is 76; 7 is the stem and 6 is the leaf. Thus, we place a 6 opposite the stem 7:

```
7 | 6
```

The next data value is 74, so a leaf of 4 is placed on the 7 stem next to the 6.

```
7 | 6 4
```

The next data value is 82, so a leaf of 2 is placed on the 8 stem.

```
7 | 6 4
8 | 2
```

ANSWER NOW 2.7
What stem values and leaf values will be used to represent the next pieces of data, 96 and 66, in Illustration 2.4?

We continue until each of the other 16 leaves is placed on the display. Figure 2.5A shows the resulting stem-and-leaf display; Figure 2.5B shows the completed stem-and-leaf display after the leaves have been ordered.

FIGURE 2.5A
Unfinished Stem-and-Leaf

19 Exam Scores

5	2
6	6 8 2
7	6 4 6 8 2 6 8 4
8	2 6 4 2 8
9	6 2

FIGURE 2.5B
Final Stem-and-Leaf

19 Exam Scores

5	2
6	2 6 8
7	2 4 4 6 6 6 8 8
8	2 2 4 6 8
9	2 6

FIGURE 2.6
Stem-and-Leaf

19 Exam Scores

(50–54)	5	2
(55–59)	5	
(60–64)	6	2
(65–69)	6	6 8
(70–74)	7	2 4 4
(75–79)	7	6 6 6 8 8
(80–84)	8	2 2 4
(85–89)	8	6 8
(90–94)	9	2
(95–99)	9	6

From Figure 2.5B, we see that the grades are centered around the 70s. In this case, all scores with the same tens digit were placed on the same branch, but this may not always be desired. Suppose we reconstruct the display; this time instead of grouping ten possible values on each stem, let's group the values so that only five possible values could fall on each stem. Do you notice a difference in the appearance of Figure 2.6? The general shape is approximately symmetrical about the high 70s. Our information is a little more refined, but basically we see the same distribution. ■

ANSWER NOW 2.8 Ex02-006
Construct a stem-and-leaf display of the numbers of points scored during each of 16 basketball games last season:

56	54	61	71	46	61	55	68
60	66	54	61	52	36	64	51

Technology Instructions: Stem-and-Leaf Diagram

MINITAB (Release 13)

Input the data into C1; then continue with:

```
Choose:   Graph > Char.Graph > Stem-and-Leaf...
Enter:    Variables: C1
          Increment value for stem (optional)
```

Excel XP

Input the data into column A; then continue with:

```
Choose:   Tools > Data Analysis Plus* > Stem and Leaf
          Display > OK
Enter:    Input Range: (A2:A6 or select cells)
          Increment: Stem Increment
```

*Data Analysis Plus is a collection of statistical macros for Excel. They can be downloaded onto your computer from your Student Suite CD.

Technology Instructions: Stem-and-Leaf Diagram (continued)

TI-83 Plus

Input the data into L1; then continue with:

```
Choose:   STAT > EDIT > 2:SortA(
Enter:    L1
```

Use sorted data to finish constructing the stem-and-leaf diagram by hand.

It is fairly typical of many variables to display a distribution that is concentrated (mounded) about a central value and then in some manner dispersed in one or both directions. Often a graphic display reveals something that the analyst may or may not have anticipated. Illustration 2.5 demonstrates what generally occurs when two populations are sampled together.

Illustration 2.5 Overlapping Distributions

A random sample of 50 college students was selected. Their weights were obtained from their medical records. The resulting data are listed in Table 2.2.

Notice that the weights range from 98 to 215 pounds. Let's group the weights on stems of ten units using the hundreds and the tens digits as stems and the units digit as the leaf. See Figure 2.7. The leaves have been arranged in numerical order.

ANSWER NOW 2.9
What do you think "Leaf Unit = 1.0" means in Figure 2.7?

Close inspection of Figure 2.7 suggests that two overlapping distributions may be involved. That is exactly what we have: a distribution of female weights and a

TABLE 2.2 Weights of 50 College Students Ta02-02

Student	1	2	3	4	5	6	7	8	9	10
Male/Female	F	M	F	M	M	F	F	M	M	F
Weight	98	150	108	158	162	112	118	167	170	120
Student	11	12	13	14	15	16	17	18	19	20
Male/Female	M	M	M	F	F	M	F	M	M	F
Weight	177	186	191	128	135	195	137	205	190	120
Student	21	22	23	24	25	26	27	28	29	30
Male/Female	M	M	F	M	F	F	M	M	M	M
Weight	188	176	118	168	115	115	162	157	154	148
Student	31	32	33	34	35	36	37	38	39	40
Male/Female	F	M	M	F	M	F	M	F	M	M
Weight	101	143	145	108	155	110	154	116	161	165
Student	41	42	43	44	45	46	47	48	49	50
Male/Female	F	M	F	M	M	F	F	M	M	M
Weight	142	184	120	170	195	132	129	215	176	183

FIGURE 2.7
Stem-and-Leaf

Weights of 50 College Students (lb)
Stem-and-Leaf of WEIGHT
$N = 50$ Leaf Unit = 1.0

9	8
10	1 8 8
11	0 2 5 5 6 8 8
12	0 0 0 8 9
13	2 5 7
14	2 3 5 8
15	0 4 4 5 7 8
16	1 2 2 5 7 8
17	0 0 6 6 7
18	3 4 6 8
19	0 1 5 5
20	5
21	5

FIGURE 2.8
"Back-to-Back" Stem-and-Leaf

Weights of 50 College Students (lb)

Female		Male
8	09	
1 8 8	10	
0 2 5 5 6 8 8	11	
0 0 0 8 9	12	
2 5 7	13	
2	14	3 5 8
	15	0 4 4 5 7 8
	16	1 2 2 5 7 8
	17	0 0 6 6 7
	18	3 4 6 8
	19	0 1 5 5
	20	5
	21	5

FIGURE 2.9
Dotplots with Common Scale

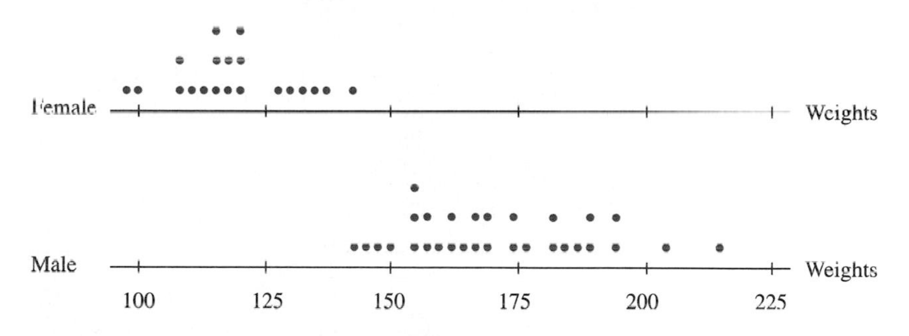

distribution of male weights. Figure 2.8 shows a "back-to-back" stem-and-leaf display of this set of data and makes it obvious that two distinct distributions are involved.

Figure 2.9, a "side-by-side" dotplot (same scale) of the same 50 weight data, shows the same distinction between the two subsets.

Based on the information shown in Figures 2.8 and 2.9, and on what we know about people's weight, it seems reasonable to conclude that female college students weigh less than male college students. Situations involving more than one set of data are discussed further in Chapter 3. ■

Technology Instructions: Multiple Dotplots

MINITAB (Release 13)

Input the data into C1 and the corresponding numerical categories into C2; then continue with:

```
Choose:   Graph > Character Graph > Dotplot...
Enter:    Variables: C1
```

(continued)

Technology Instructions: Multiple Dotplots (continued)

MINITAB (Release 13) (continued)

```
Select:   By variable:
Enter:    C2
Select:   Same scale for all variables
```

If the various categories are in separate columns, enter all of the columns under Variables and deselect 'By variable'.

Excel XP

Multiple dotplots are not available, but the initial step of ranking the data can be done. Use the commands as shown with the dotplot display on page 43, then finish constructing the dotplots by hand.

TI-83 Plus

Input the data for the first dotplot into L1 and the data for the second dotplot into L3; then continue with:

```
Choose:   STAT > EDIT > 2:SortA(
Enter:    L1 > ENTER
          In L2, enter counting numbers for each category.
          Ex.    L1     L2
                 15     1
                 16     1
                 16     2
                 17     1
Choose:   STAT > EDIT > 2:SortA(
Enter:    L3 > ENTER
          In L4, enter counting numbers (a higher set*) for
          each category;
          *for example: use 10,10,11,10,10,11,12, ... (off-
          sets the two dotplots).
Choose:   2nd > FORMAT > AxesOff (Optional—must return to
          AxesOn)
Choose: 2nd > STAT PLOT > 1:PLOT1
```

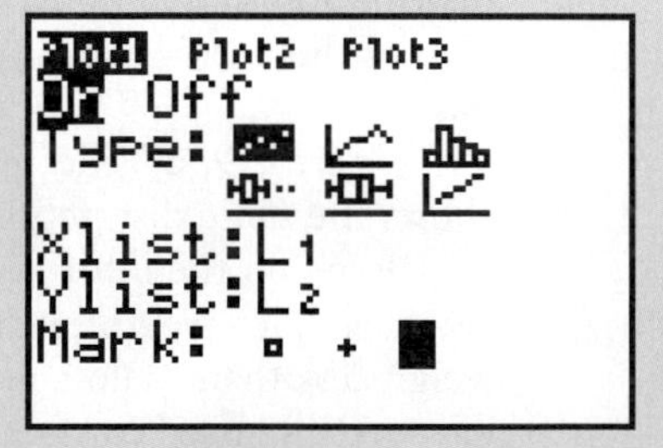


```
Choose:   2nd > STAT PLOT > 2:PLOT2
```

```
Choose:   Window
Enter:    at most lowest value for both, at least highest
          value for both, 0 or increment, -2, at least
          highest counting number,1,1
Choose:   Graph > Trace > > > (gives data values)
```

Exercises

2.10 Ex02-010 The number of points scored by the winning teams on October 29, 2002, opening night of the 2002–2003 NBA season, are listed below:

Team	Orlando	San Antonio	Sacramento
Score	95	87	94

http://sports.espn.go.com/nba/scoreboard

a. Draw a bar graph of these scores using a vertical scale ranging from 80 to 100.

b. Draw a bar graph of the scores using a vertical scale ranging from 50 to 100.

c. In which bar graph does it appear that the NBA scores vary more? Why?

d. How could you create an accurate representation of the relative size and variation between these scores?

2.11 Ex02-011 An article in *Fortune* magazine entitled "What Really Goes on in Your Doctor's Office?" showed a breakdown of how patients' fees are used to support the operations of various clinics and health care facilities. For each $100 in fees collected by the doctors in the study, the following eight categories of expenses were isolated to show how the $100 was deployed:

Expense Category	**Amount**
1. Doctor's personal income	$55.60
2. Nonphysician personnel	15.70
3. Office expenses	10.90
4. Medical supplies	4.00
5. Malpractice insurance premiums	3.50
6. Employee physicians	2.30
7. Medical equipment	1.50
8. All other	6.50

Source: *Fortune*, August 17, 1998, p. 168

a. Construct a circle graph of this breakdown.

b. Construct a bar graph of this breakdown.

c. Compare the two graphs you constructed in parts a and b. Which one seems to be more informative? Explain why.

2.12 Ex02-012 A sample of student-owned General Motors automobiles was identified and the make of each noted. The resulting sample follows (Ch = Chevrolet, P = Pontiac, O = Oldsmobile, B = Buick, Ca = Cadillac):

Ch	B	Ch	P	Ch	O	B	Ch	Ca	Ch
B	Ca	P	O	P	P	Ch	P	O	O
Ch	B	Ch	B	Ch	P	O	Ca	P	Ch
O	Ch	Ch	B	P	Ch	Ca	O	Ch	B
B	O	Ch	Ch	O	Ch	Ch	B	Ch	B

a. Find the number of cars of each make in the sample.

b. What percentage of these cars were Chevrolets? Pontiacs? Oldsmobiles? Buicks? Cadillacs?

c. Draw a bar graph showing these percentages.

2.13 "Monster cookies" shows adults' choices of favorite cookie.

a. Draw a bar graph picturing the percentages of adults for each kind of cookie.

b. Draw a Pareto diagram picturing adults' favorite cookies. How can the "Other" category be handled so that the resulting graph does not misrepresent the situation?

c. If a store is to stock only four varieties of cookies, which varieties should it carry if it wishes to please the greatest

number of customers? How does the Pareto diagram show this?

d. If 300 adults are to be surveyed, what frequencies would you expect to occur for each cookie based on the "Monster cookies" graph?

FYI

Try the PARETO commands on your computer or calculator.

2.14 Ex02-014 The 2000–2001 annual report given to members of the State Teachers Retirement System of Ohio (STRS) outlined the following breakdown of benefits (in thousands) paid to retired members:

Disability retirement	$160,775
Survivor benefits	$65,591
Supplemental benefit	$50,386
Health care	$369,354
Service retirement	$2,203,280

Source: State Teachers Retirement System of Ohio, "Report to Members 2001," 275 East Broad St., Columbus, OH 43215-3771

a. Construct a Pareto diagram of this information.

b. Write a paragraph describing what the Pareto diagram dramatically shows to its reader.

2.15 Ex02-015 A study by Bruskin-Goldring for Whirlpool Corp. lists the major chores mothers say they would like to have the family help with. The most popular response was cleaning (53%), followed by laundry (18%), cooking (9%), dishes (8%), and other (12%). http://pqasb.pqarchiver.com/ USAToday/

a. Construct a Pareto diagram displaying this information.

b. Because of the size of the "other" category, the Pareto diagram may not be the best graph to use. Explain why. What additional information is needed to make the Pareto diagram more appropriate?

2.16 The final-inspection defect report for assembly line A12 is reported on a Pareto diagram.

a. What is the total defect count in the report?

b. Verify the 30.0% listed for "Scratch."

Product Defects

Defect	Blem	Scratch	Chip	Bend	Dent	Others
Count	56	45	23	12	8	6
Percent	37.3	30.0	15.3	8.0	5.3	4.0
Cum %	37.3	67.3	82.7	90.7	96.0	100.0

c. Explain how the "Cum % for bend" value of 90.7% was obtained and what it means.

d. Management has given the production line the goal of reducing defects by 50%. What two defects would you suggest they give special attention to in working toward this goal? Explain.

2.17 Ex02-017 *Sports Illustrated* published a mock draft list of 29 future players for the NBA one week before the 1998 draft actually took place. Shown here are the heights (in inches) of the basketball players who the editorial staff expected to be first-round picks by the professional teams:

74	85	79	83	81	77	79	80	83	83
81	82	77	77	73	82	84	78	80	78
83	81	81	75	78	72	82	81	81	

a. Construct a dotplot of the heights of these players.

b. Use the dotplot to uncover the shortest and the tallest players.

c. What is the most common height and how many players share that height?

d. What feature of the dotplot illustrates the most common height?

2.18 Ex02-018 As baseball players, Hank Aaron, Babe Ruth, and Roger Maris were well known for their ability to hit home runs. Mark McGwire, Sammy Sosa, and Ken Griffey, Jr., became well known for their ability to hit home runs during the "great home run chase" of 1998. Barry Bonds gained his fame in 2001. Listed here are the numbers of home runs each player hit in each major league season in which he played in 75 or more games:

Aaron	13	27	26	44	30	39
	40	34	45	44	24	32
	44	39	29	44	38	47
	34	40	20	12	10	
Ruth	11	29	54	59	35	41
	46	25	47	60	54	46
	49	46	41	34	22	
Maris	14	28	16	39	61	33
	23	26	13	9	5	
McGwire	49	32	33	39	22	42
	39	52	58	70	65	32
	29					
Bonds	16	25	24	19	33	25
	34	46	37	33	42	40
	37	34	49	73	46	
Griffey	16	22	22	27	45	40
	49	56	56	48	40	22
	8					
Sosa	15	10	33	25	36	40
	36	66	63	50	64	49

a. Construct a dotplot of the data for Aaron, Ruth, Maris, and McGwire, using the same axis.
b. Using the dotplots from part a, make a case for each of the following statements with regard to past players: "Aaron is the home run king!" "Ruth is the home run king!" "McGwire is the home run king!" "Maris is not the home run king!"
c. Construct a dotplot of the data for Bonds, Sosa, and Griffey, using the same axis.
d. Using the dotplots from part c, make a case for the statement "Griffey is not currently the homerun king!" with regard to present players. In what way do the dotplots support the statement?

FYI
If you use your computer or calculator, use the commands on page 47.

2.19 A computer was used to construct the dotplot at the bottom of the page.
a. How many data are shown?
b. List the values of the five smallest data.
c. What is the value of the largest data?
d. What value occurred the greatest number of times? How many times did it occur?

2.20 Ex02-020 Delco Products, a division of General Motors, produces commutators designed to be 18.810 mm in overall length. (A commutator is a device used in the electrical system of an automobile.) The following sample of 35 commutator lengths was taken while monitoring the manufacturing process:

18.802 18.810 18.780 18.757 18.824 18.827 18.825
18.809 18.794 18.787 18.844 18.824 18.829 18.817
18.785 18.747 18.802 18.826 18.810 18.802 18.780
18.830 18.874 18.836 18.758 18.813 18.844 18.861
18.824 18.835 18.794 18.853 18.823 18.863 18.808

Source: With permission of Delco Products Division, GMC

Use a computer to construct a dotplot of these data.

2.21 Ex02-021 *Forbes* magazine (January 7, 2002) reported "The Best Big Companies in America." One classification was for the banking industry. The following is a list of the banks in the top 400 companies with their corresponding last 12 months profitability percentages:

Bank	Profitability for last 12 months (%)
AmSouth Bancorp	14.8
Bank of New York	15.1
BB & T	15.7
Charter One Financial	17.9
Comerica	12.8
Compass Bancshares	14.3
Dime Bancorp	14.8
First Tennessee National	18.4
Firth Third Bancorp	13.4
Golden State Bancorp	11.3
Golden West Financial	18.2
Green Point Financial	12.5
Hibernia	12.2
M&T Bank	10.4
Marshall & Ilsley	12.5
Northern Trust	14.6
PNC Financial Services	13.8
Popular	11.7
SouthTrust	12.9
SunTrust Banks	13.9
Synovus Financial	18.7
Washington Mutual	20.2
Zions Bancorp	13.6

a. Construct a stem-and-leaf display of the data.
b. Based on the stem-and-leaf display, describe the distribution of the percentages of profitability.

2.22 Ex02-022 These amounts are the fees charged by Quik Delivery for the 40 small packages it delivered last Thursday afternoon:

4.03 3.56 3.10 6.04 5.62 3.16 2.93 3.82 4.30 3.86
4.57 3.59 4.57 6.16 2.88 5.03 5.46 3.87 6.81 4.91
3.62 3.62 3.80 3.70 4.15 2.07 3.77 5.77 7.86 4.63
4.81 2.86 5.02 5.24 4.02 5.44 4.65 3.89 4.00 2.99

Construct a stem-and-leaf display.

2.23 Given the following stem-and-leaf display:

```
Stem-and-Leaf of C1 N = 16
Leaf Unit = 0.010
1 59 7
  4    60   148
 (5)   61   02669
  7    62   0247
  3    63   58
  1    64   3
```

(continued)

Dotplot for Exercise 2.19

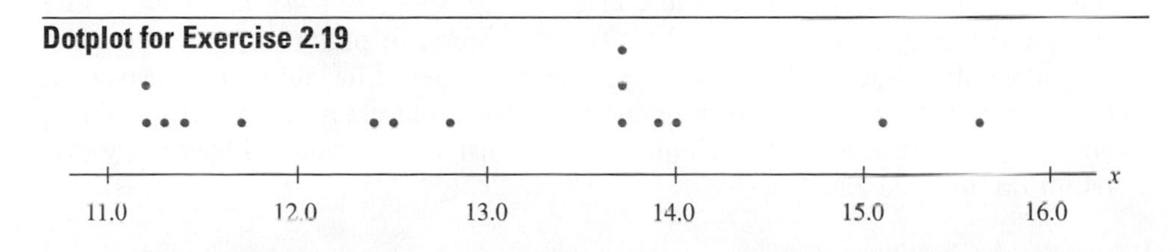

a. What is the meaning of "Leaf Unit = 0.010"?
b. How many data are shown on this stem-and-leaf?
c. List the first four data values.
d. What is the column of numbers down the left-hand side of the figure?

2.24 Ex02-024 A term often used in solar energy research is *heating-degree-days*. This concept is related to the difference between an indoor temperature of 65°F and the average outside temperature for a given day. An average outside temperature of 5°F gives 60 heating-degree-days. The annual heating-degree-day normals for several Nebraska locations are shown on the accompanying stem-and-leaf display constructed using MINITAB (data are in column C1).
a. What is the meaning of "Leaf Unit = 10"?
b. List the first four data values.
c. List all the data values that occurred more than once.

```
Stem-and-leaf of C1 N = 25
Leaf Unit = 10
  2     60 78
  7     61 03699
  9     62 69
 11     63 26
 (3)    64 233
 11     65 48
  9     66 8
  8     67 249
  5     68 18
  3     69 145
```

2.2 Frequency Distributions and Histograms

Lists of large sets of data do not present much of a picture. Sometimes we want to condense the data into a more manageable form. This can be accomplished with the aid of a *frequency distribution*.

FREQUENCY DISTRIBUTION
A listing, often expressed in chart form, that pairs each value of a variable with its frequency.

To demonstrate the concept of a frequency distribution, let's use this set of data:

3	2	2	3	2	4	4	1	2	2
4	3	2	0	2	2	1	3	3	1

If we let x represent the variable, then we can use a frequency distribution to represent this set of data by listing the x values with their frequencies. For example, the value 1 occurs in the sample three times; therefore, the **frequency** for $x = 1$ is 3. The complete set of data is shown in the frequency distribution in Table 2.3.

TABLE 2.3

Ungrouped Frequency Distribution

x	f
0	1
1	3
2	8
3	5
4	3

ANSWER NOW 2.25
Form an ungrouped frequency distribution of these data:
1, 2, 1, 0, 4, 2, 1, 1, 0, 1, 2, 4.

The frequency f is the number of times the value x occurs in the sample. Table 2.3 is an *ungrouped frequency distribution*—"ungrouped" because each value of x in the distribution stands alone. When a large set of data has many different x values instead of a few repeated values, as in the previous example, we can group the values into a set of classes and construct a *grouped frequency distribution*. The stem-and-leaf display in Figure 2.5B (p. 45) shows, in picture form, a grouped frequency distribution. Each stem represents a class. The number of leaves on each stem is the same as the frequency for that same **class** (sometimes called a *bin*). The data represented in Figure 2.5B are listed as a grouped frequency distribution in Table 2.4.

TABLE 2.4 Grouped Frequency Distribution

		Class	Frequency
50 or more to less than 60	⟶	$50 \le x < 60$	1
60 or more to less than 70	⟶	$60 \le x < 70$	3
70 or more to less than 80	⟶	$70 \le x < 80$	8
80 or more to less than 90	⟶	$80 \le x < 90$	5
90 or more to less than 100	⟶	$90 \le x < 100$	2
			19

ANSWER NOW 2.26
Refer to Table 2.4.

a. **Explain what $f = 8$ represents.**

b. **What is the sum of the frequency column?**

c. **What does this sum represent?**

The stem-and-leaf process can be used to construct a frequency distribution; however, the stem representation is not compatible with all **class widths**. For example, class widths of 3, 4, and 7 are awkward to use. Thus, sometimes it is advantageous to have a separate procedure for constructing a grouped frequency distribution.

Illustration 2.6

Grouping Data to Form a Frequency Distribution

To illustrate this grouping (or classifying) procedure, let's use a sample of 50 final exam scores taken from last semester's elementary statistics class. Table 2.5 lists the 50 scores.

These are the basic guidelines to follow in constructing a grouped frequency distribution:

1. Each class should be of the same width.
2. Classes (sometimes called bins) should be set up so that they do not overlap and so that each data belongs to exactly one class.
3. For the exercises given in this textbook, 5 to 12 classes are most desirable, since all samples contain fewer than 125 data. (The square root of n is a reasonable guideline for the number of classes with samples of fewer than 125 data.)
4. Use a system that takes advantage of a number pattern to guarantee accuracy. (This is demonstrated below.)
5. When it is convenient, an even class width is often advantageous.

Procedure

1. Identify the high score ($H = 98$) and the low score ($L = 39$), and find the range:

$$\text{range} = H - L = 98 - 39 = 59$$

TABLE 2.5 Statistics Exam Scores Ta02-05

60	47	82	95	88	72	67	66	68	98
90	77	86	58	64	95	74	72	88	74
77	39	90	63	68	97	70	64	70	70
58	78	89	44	55	85	82	83	72	77
72	86	50	94	92	80	91	75	76	78

2. Select a number of classes ($m = 7$) and a class width ($c = 10$) so that the product ($mc = 70$) is a bit larger than the range (range = 59).
3. Pick a starting point. This starting point should be a little smaller than the lowest score L. Suppose we start at 35; counting from there by tens (the class width), we get 35, 45, 55, 65, . . . , 95, 105. These are called the **class boundaries.** The classes for the data in Table 2.5 are:

35 or more to less than 45	⟶	$35 \leq x < 45$
45 or more to less than 55	⟶	$45 \leq x < 55$
55 or more to less than 65	⟶	$55 \leq x < 65$
65 or more to less than 75	⟶	$65 \leq x < 75$
	⋮	$75 \leq x < 85$
		$85 \leq x < 95$
95 or more to and including 105	⟶	$95 \leq x \leq 105$

NOTES:

1. At a glance you can check the number pattern to determine whether the arithmetic used to form the classes was correct (35, 45, 55, . . . , 105).
2. For the interval $35 \leq x < 45$, the 35 is the lower class boundary and 45 is the upper class boundary. Observations that fall on the lower class boundary stay in that interval; observations that fall on the upper class boundary go into the next higher interval, except for the last class.
3. The class width is the difference between the upper and lower class boundaries.
4. Many combinations of class widths, numbers of classes, and starting points are possible when classifying data. There is no one best choice. Try a few different combinations, and use good judgment to decide on the one to use.

Once the classes are set up, we need to sort the data into those classes. The method used to sort will depend on the current format of the data: If the data are ranked, the frequencies can be counted; if the data are not ranked, we will **tally** the data to find the frequency numbers. When classifying data, it helps to use a standard chart (see Table 2.6).

TABLE 2.6 Standard Chart for Frequency Distribution

Class Number	Class Tallies	Boundaries	Frequency
1	II	$35 \leq x < 45$	2
2	II	$45 \leq x < 55$	2
3	IIIII II	$55 \leq x < 65$	7
4	IIIII IIIII III	$65 \leq x < 75$	13
5	IIIII IIIII I	$75 \leq x < 85$	11
6	IIIII IIIII I	$85 \leq x < 95$	11
7	IIII	$95 \leq x \leq 105$	4
			50

NOTES:

1. If the data have been ranked (list form, dotplot, or stem-and-leaf), tallying is unnecessary; just count the data that belong to each class.
2. If the data are not ranked, be careful as you tally.

3. The frequency f for each class is the number of pieces of data that belong in that class.
4. The sum of the frequencies should equal the number of pieces of data n ($n = \sum f$). This summation serves as a good check.

NOTE: See the *Statistical Tutor* for information about $\sum$ **notation** (read **"summation notation"**).

Each class needs a single numerical value to represent all the data values that fall into that class. The **class midpoint** (sometimes called the *class mark*) is the numerical value that is exactly in the middle of each class. It is found by adding the class boundaries and dividing by 2. Table 2.7 shows an additional column for the class midpoint, x. As a check of your arithmetic, successive class midpoints should be a class width apart, which is 10 in this illustration (40, 50, 60, . . ., 100 is a recognizable pattern).

Application 2.1 Cleaning House

"Hours spent cleaning" presents a circle graph version of a relative frequency distribution. Each sector of the circle represents an amount of time spent cleaning weekly by each person, and the "width" of the sector represents the percentage or relative frequency.

ANSWER NOW 2.27

a. How is the variable "time spent cleaning" represented on the graph?

b. How is the relative frequency represented on the graph?

c. Express the category "2–4 hours" shown on the graph using the interval notation $a \le x < b$.

d. Express the information on the circle graph on a distribution chart.

TABLE 2.7 Frequency Distribution with Class Midpoints

Class Number	Class Boundaries	Frequency f	Class Midpoints x
1	$35 \le x < 45$	2	40
2	$45 \le x < 55$	2	50
3	$55 \le x < 65$	7	60
4	$65 \le x < 75$	13	70
5	$75 \le x < 85$	11	80
6	$85 \le x < 95$	11	90
7	$95 \le x \le 105$	4	100
		50	

ANSWER NOW 2.28

a. **65 belongs to which class?**

b. **Explain the meaning of "$65 \le x < 75$."**

c. **Explain what class width is, and describe four ways it can be identified in Table 2.7.**

NOTE: Now you can see why it is helpful to have an even class width. An odd class width would have resulted in a class midpoint with an extra digit. (For example, the class 45–54 is 9 wide and the class midpoint is 49.5.) ■

When we classify data into classes, we lose some information. Only when we have all the raw data do we know the exact values that were actually observed for each class. For example, we put a 47 and a 50 into class 2, with class boundaries of 45 and 55. Once they are placed in the class, their values are lost to us and we use the class midpoint, 50, as their representative value.

HISTOGRAM

A bar graph that represents a frequency distribution of a quantitative variable. A histogram is made up of the following components:

1. A title, which identifies the population or sample of concern.
2. A vertical scale, which identifies the frequencies in the various classes.
3. A horizontal scale, which identifies the variable x. Values for the class boundaries or class midpoints may be labeled along the x-axis. Use whichever method of labeling the axis best presents the variable.

The frequency distribution from Table 2.7 appears in histogram form in Figure 2.10.

ANSWER NOW 2.29

Draw a frequency histogram of the annual salaries for resort-club managers. Label the class boundaries.

Annual Salary ($1000)	15–25	25–35	35–45	45–55	55–65
Number of Managers	12	37	26	19	6

Sometimes the **relative frequency** of a value is important. The relative frequency is a proportional measure of the frequency for an occurrence. It is found by dividing the class frequency by the total number of observations. Relative frequency can be expressed as a common fraction, in decimal form, or as a percentage. For example, in Illustration 2.6 the frequency associated with the third class

FYI
Be sure to identify both scales so that the histogram tells the complete story.

FIGURE 2.10
Frequency Histogram

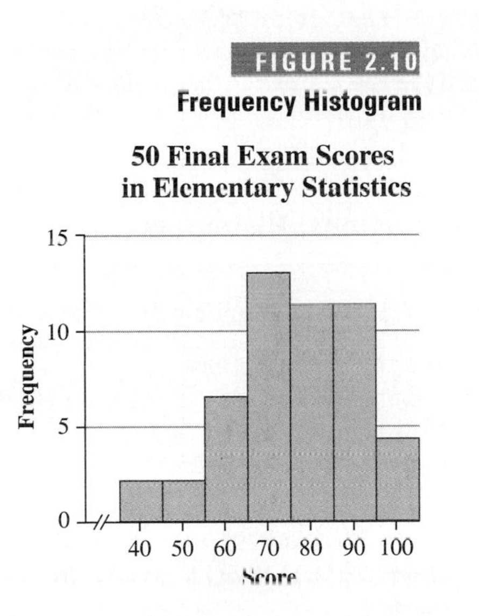

FIGURE 2.11
Relative Frequency Histogram

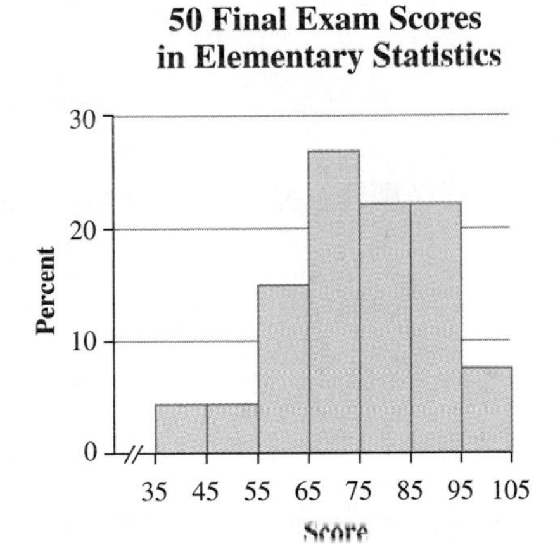

(55–65) is 7. The relative frequency for the third class is $\frac{7}{50}$, or 0.14, or 14%. Relative frequencies are often useful in a presentation because nearly everybody understands fractional parts when expressed as percents. Relative frequencies are particularly useful when comparing the frequency distributions of two different size sets of data. Figure 2.11 is a **relative frequency histogram** of the sample of the 50 final exam scores from Table 2.7.

ANSWER NOW 2.30
Explain the similarities and the differences between Figures 2.10 and 2.11.

A stem-and-leaf display contains all the information needed to create a histogram. Figure 2.5B (p. 45) shows the stem-and-leaf display constructed in Illustration 2.4. In Figure 2.12A the stem-and-leaf has been rotated 90° and labels have been added to show its relationship to a histogram. Figure 2.12B shows the same set of data as a completed histogram.

FIGURE 2.12A
Modified Stem-and-Leaf

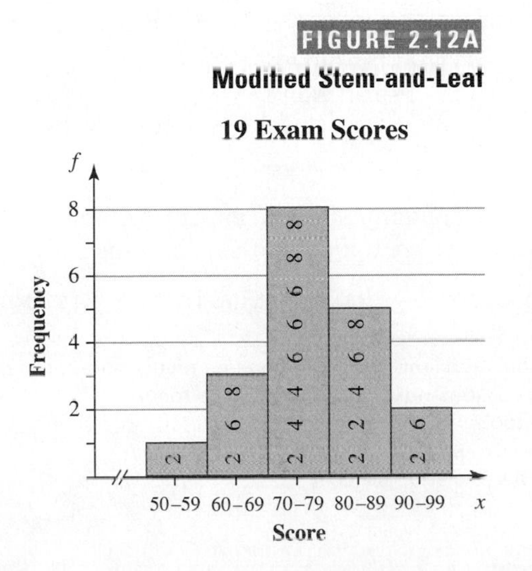

FIGURE 2.12B
Histogram of 19 Exam Scores

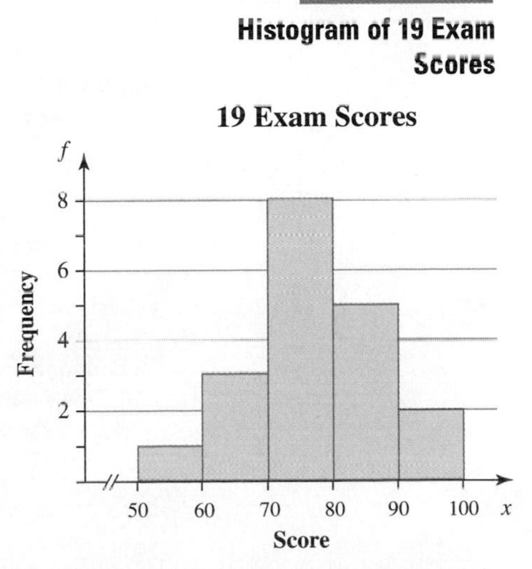

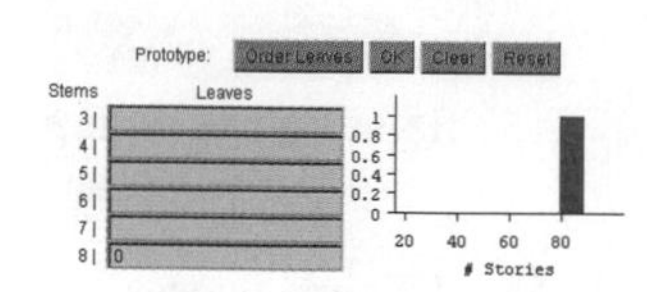

ANSWER NOW 2.31 INTERACTIVITY

Interactivity 2-A demonstrates the procedure of transforming a stem-and-leaf display into a histogram. Type the leaves for the number of stories into the stem-and-leaf. Click OK to see the corresponding histogram. Comment on the similarities and differences.

Technology Instructions: Histogram

MINITAB (Release 13)

Input the data into C1; then continue with:

```
Choose:   Graph > Histogram
Enter:    Graph variables: Graph 1: C1
Choose:   Annotation > Title
Enter:    Your title > OK
Choose:   Options
Select:   Type of histogram: Frequency or Percent
          Type of interval: Midpoint or Cutpoint
          Definition of intervals:
                   Automatic
               or, Number of intervals; Enter: N
               or, Midpt/cutpt positions; Enter: A:B/C
```

FYI
Choose one of the three.

NOTES:

1. Midpoints are the class midpoints, and cutpoints are the class boundaries.
2. Percent is relative frequency.
3. Automatic means MINITAB will make all the choices; N = number of intervals—that is, the number of classes you want used.
4. A = smallest class midpoint or boundary, B = largest class midpoint or boundary, C = class width you want to specify.

The following commands will draw the histogram of a frequency distribution. The end classes can be made full width by adding an extra class with frequency zero to each end of the frequency distribution. Input the class midpoints into C1 and the corresponding frequencies into C2.

```
Choose:   Graph > Plot
Enter:    Graph variables: Graph 1: Y: C2
                                    X: C1
          Data display: Item 1: Display: AREA
Choose:   Edit Attributes
Enter:    Fill Type: None
Select:   Connection Function: Step > OK
```

Excel XP

Input the data into column A and the upper class limits* into column B (optional) and (column headings are optional); then continue with:

```
Choose:   Tools > Data Analysis** > Histogram > OK
```

* If boundary = 50, then limit = 49.9 (depending on the number of decimal places in the data).
** If Data Analysis does not show on the Tools menu:
Choose: Tools > Add-Ins
Select: Analysis ToolPak
Analysis ToolPak-VBA

```
Enter:    Input Range: Data (A1:A6 or select cells)
          Bin Range: upper class limits (B1:B6 or select
          cells)
          [leave blank if Excel determines the intervals]
Select:   Labels (if column headings are used)
          Output Range
Enter:    area for freq. distr. & graph (C1 or select cell)
Select:   Chart Output
```

To remove gaps between bars

```
Click on: Any bar on graph
Click on: Right mouse button
Choose:   Format Data Series > Options
Enter:    Gap Width: 0
```

To edit histogram:

```
Click on: Anywhere clear on the chart
            -use handles to size
          Any title or axis name to change
          Any upper class limit* or frequency in the fre-
          quency distribution to change value > Enter
```

*Note that the upper class limits appear in the center of the bars. Replace with class midpoints. The "More" cell in the frequency distribution may also be deleted.

For tabled data, input the classes into column A (ex. 30–40) and the frequencies into column B; then continue with:

```
Choose:   Chart Wizard > Column > 1st picture (usually) >
          Next
Enter:    Data Range: (A1:B4 or select cells)
Select:   Series in: Columns > Next
Choose:   Titles
Enter:    Chart title: your title
          Category (x) axis: title for x-axis
          Value (y) axis: title for y-axis > Next > Finish
```

Do as above to remove gaps and adjust.

TI-83 Plus Input the data into L1; then continue with:

```
Choose:   2nd > STAT PLOT > 1:Plot1
```

(continued)

Technology Instructions: Histogram (continued)

TI-83 Plus (continued)

Calculator selects classes:

```
Choose: Zoom > 9:ZoomStat > Trace > > >
```

Individual selects classes:

```
Choose:   Window
Enter:    at most lowest value, at least highest value,
          class width, -1, at least highest frequency, 1
          (depends on frequency numbers), 1
Choose:   Graph > Trace (use values to construct frequency
          distribution)
```

For tabled data, input the class midpoints into L1 and the frequencies into L2; then continue with:

```
Choose:   2nd > STAT PLOT > 1:Plot1
```

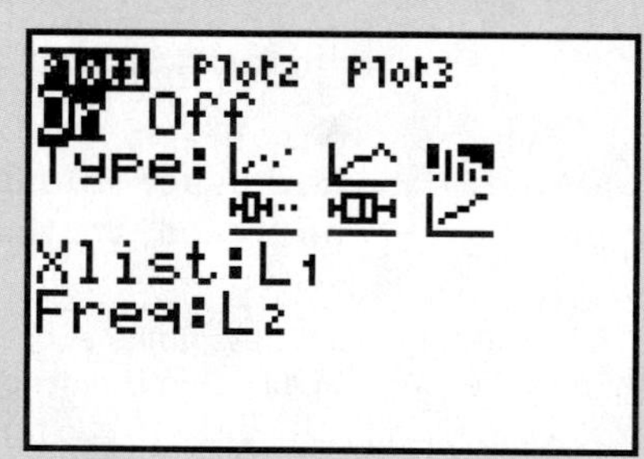

```
Choose:   Window
Enter:    smallest lower class boundary, largest upper class
          boundary, class width, -ymax/4, highest frequency,
          0 (for no tick marks), 1
Choose:   Graph > Trace > > >
```

To obtain a relative frequency histogram of tabled data instead:

```
Choose:      STAT > EDIT > 1:EDIT...
Highlight:   L3
Enter:       L3 = L2/SUM(L2) [SUM - 2nd LIST > MATH >
             5:sum]
Choose:      2nd > STAT PLOT > 1:Plot1
```

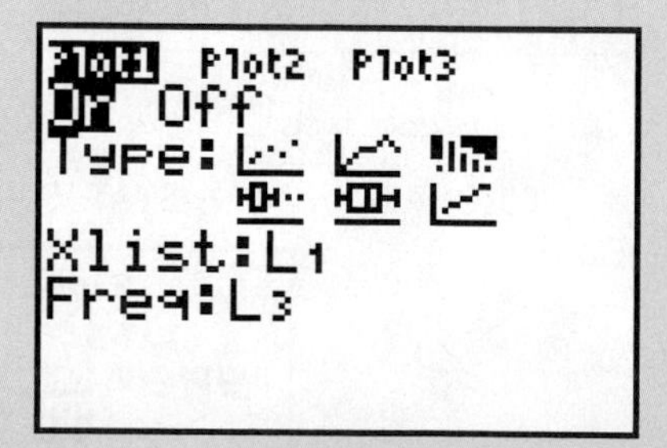

```
Choose:   Window
Enter:    smallest lower class boundary, largest upper class
          boundary, class width, -ymax/4, highest rel. fre-
          quency, 0 (for no tick marks), 1
Choose:   Graph > Trace > > >
```

Histograms are valuable tools. For example, the histogram of a sample should have a distribution shape very similar to that of the population from which the sample was drawn. If the reader of a histogram is at all familiar with the variable involved, he or she will usually be able to interpret several important facts. Figure 2.13 presents histograms with descriptive labels resulting from their geometric shape.

FIGURE 2.13
Shapes of Histograms

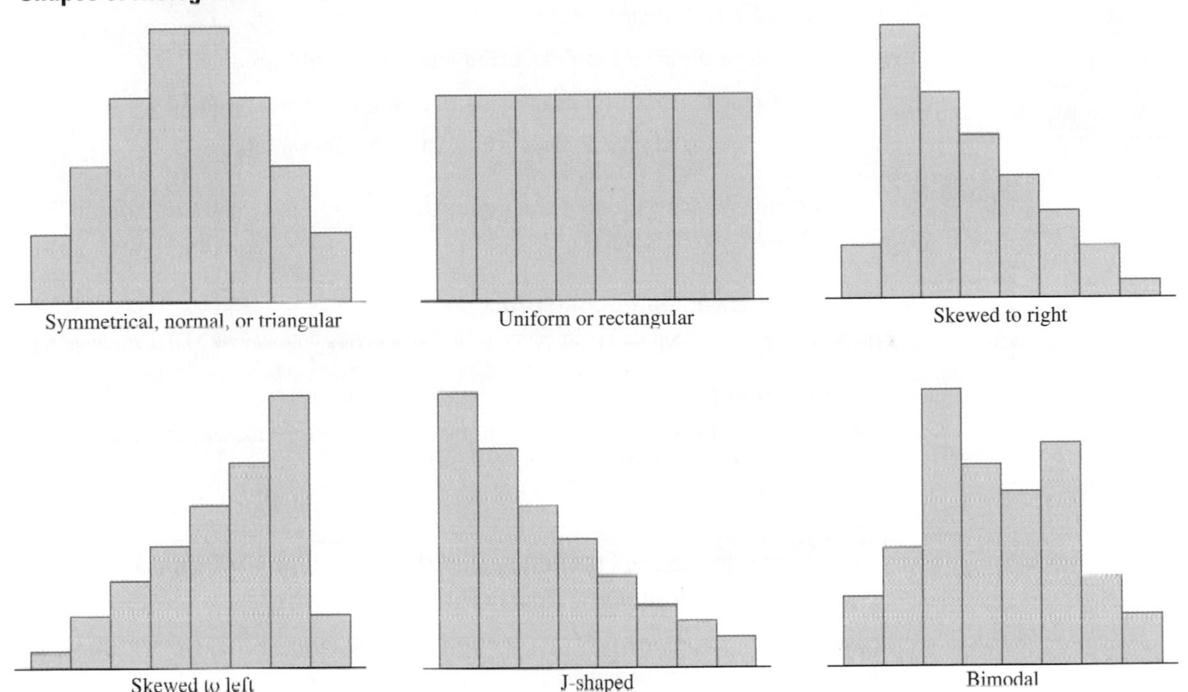

Briefly, the terms used to describe histograms are as follows:

Symmetrical: Both sides of this distribution are identical (halves are mirror images).
Uniform (rectangular): Every value appears with equal frequency.
Skewed: One tail is stretched out longer than the other. The direction of skewness is on the side of the longer tail.
J-shaped: There is no tail on the side of the class with the highest frequency.
Bimodal: The two most populous classes are separated by one or more classes. This situation often implies that two populations are being sampled. (See Figure 2.7, p. 47.)
Normal: A symmetrical distribution is mounded up about the mean and becomes sparse at the extremes. (Additional properties are discussed later.)

ANSWER NOW 2.32
Can you think of variables whose distribution might yield histograms like each of the histograms in Figure 2.13?

NOTES:

1. The **mode** is the value of the data that occurs with the greatest frequency. (Mode will be discussed in Section 2.3, p. 72.)
2. The **modal class** is the class with the highest frequency.

3. A **bimodal distribution** has two high-frequency classes separated by classes with lower frequencies. It is not necessary for the two high frequencies to be the same.

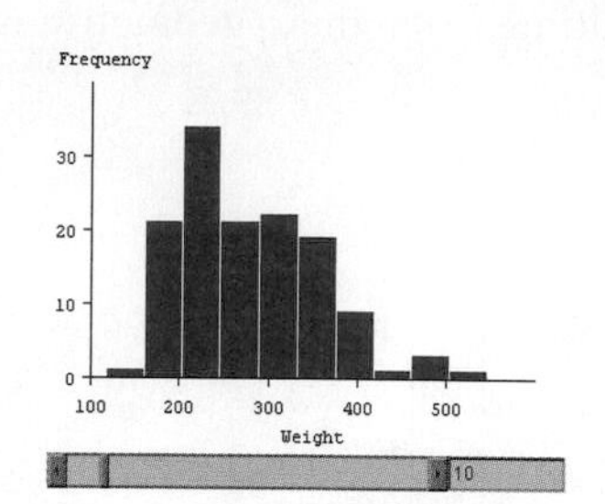

ANSWER NOW 2.33 INTERACTIVITY

Interactivity 2-B demonstrates the effect that the number of classes or bins has on the shape of a histogram.

a. **What shape distribution does using one class or bin produce?**

b. **What shape distribution does using two classes or bins produce?**

c. **What shape distributions does using 10 or 20 bins produce?**

Another way to express a frequency distribution is to use a *cumulative frequency distribution.*

CUMULATIVE FREQUENCY DISTRIBUTION

A frequency distribution that pairs cumulative frequencies with values of the variable.

The **cumulative frequency** for any given class is the sum of the frequency for that class and the frequencies of all classes of smaller values. Table 2.8 shows the cumulative frequency distribution from Table 2.7.

TABLE 2.8 Using Frequency Distribution to Form a Cumulative Frequency Distribution

Class Number	Class Boundaries	Frequency f	Cumulative Frequency
1	$35 \le x < 45$	2	2 (2)
2	$45 \le x < 55$	2	4 (2 + 2)
3	$55 \le x < 65$	7	11 (7 + 4)
4	$65 \le x < 75$	13	24 (13 + 11)
5	$75 \le x < 85$	11	35 (11 + 24)
6	$85 \le x < 95$	11	46 (11 + 35)
7	$95 \le x \le 105$	4	50 (4 + 46)
		50	

ANSWER NOW 2.34

Prepare a cumulative frequency distribution for this frequency distribution:

Annual Salary ($1000)	15–25	25–35	35–45	45–55	55–65
Number of Managers	12	37	26	19	6

The same information can be presented by using a *cumulative relative frequency distribution* (see Table 2.9). This combines the cumulative frequency and the relative frequency ideas.

ANSWER NOW 2.35

Prepare a cumulative relative frequency distribution for this frequency distribution:

Annual Salary ($1000)	15–25	25–35	35–45	45–55	55–65
Number of Managers	12	37	26	19	6

TABLE 2.9 Cumulative Relative Frequency Distribution

Class Number	Class Boundaries	Cumulative Relative Frequency	*Cumulative frequencies are for the interval 35 up to the upper boundary of that class.*
1	$35 \le x < 45$	2/50, or 0.04	← *from 35 up to less than 45*
2	$45 \le x < 55$	4/50, or 0.08	← *from 35 up to less than 55*
3	$55 \le x < 65$	11/50, or 0.22	← *from 35 up to less than 65*
4	$65 \le x < 75$	24/50, or 0.48	.
5	$75 \le x < 85$	35/50, or 0.70	:
6	$85 \le x < 95$	46/50, or 0.92	
7	$95 \le x \le 105$	50/50, or 1.00	← *from 35 up to and including 105*

OGIVE (PRONOUNCED Ō'JĪV)

A line graph of a cumulative frequency or cumulative relative frequency distribution. An ogive has the following components:

1. A title, which identifies the population or sample.
2. A vertical scale, which identifies either the cumulative frequencies or the cumulative relative frequencies. (Figure 2.14 shows an ogive with cumulative relative frequencies.)
3. A horizontal scale, which identifies the upper class boundaries. Until the upper boundary of a class has been reached, you cannot be sure you have accumulated all the data in that class. Therefore, the horizontal scale for an ogive is always based on the upper class boundaries.

FIGURE 2.14
Ogive

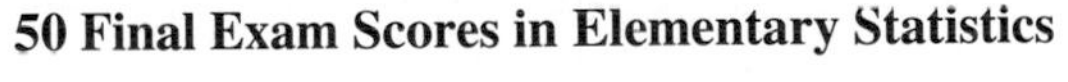

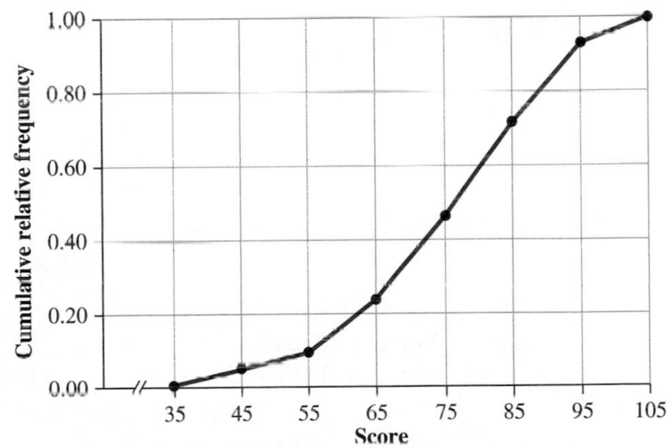

NOTE: Every ogive starts on the left with a relative frequency of zero at the lower class boundary of the first class and ends on the right with a cumulative relative frequency of 100% at the upper class boundary of the last class.

NOTE All graphic representations of sets of data need to be completely self-explanatory. That includes a descriptive, meaningful title and proper identification of the vertical and horizontal scales.

ANSWER NOW 2.36
Construct an ogive for the cumulative relative frequency distribution in the answer to Answer Now 2.35.

Technology Instructions: Ogive

MINITAB (Release 13)

Input the class boundaries into C1 and the cumulative percentages into C2 [enter 0 (zero) for the percentage paired with the lower boundary of the first class and pair each cumulative percentage with the class upper boundary]. Use percentages; that is, use 25% in place of 0.25.

```
Choose:   Graph > Plot
Enter:    Graph variables: Graph 1: Y, C2 and X, C1
          Data display: Item 1: Display: Connect
Choose:   Annotation > Title
Enter:    your title > OK
```

Excel XP

Input the data into column A and the upper class limits* into column B (include an additional class at the beginning).

```
Choose:   Tools > Data Analysis > Histogram > OK
Enter:    Input Range: data (A1:A6 or select cells)
          Bin Range: upper class limits (B1:B6 or select
          cells)
Select:   Labels (if column headings were used)
          Output Range
          Enter: area for freq. distr. & graph: (C1 or se-
          lect cell)
          Cumulative Percentage
          Chart Output
```

To close gaps and edit, see the histogram commands on page 59.

For tabled data, input the upper class boundaries into column A and the cumulative relative frequencies into column B [include an additional class boundary at the beginning with a cumulative relative frequency equal to 0(zero)]; then continue with:

```
Choose:   Chart Wizard > Line > 4th picture (usually) > Next
Choose:   Series (have more control on input) > Remove (re-
          move all columns except column B)
Enter:    Name: (B1 or select name cell - cum. rel. freq.)
          Values: (B2:B6 or select cells)
          Category (x) axis labels: (A2:A8 or select cells)
          > Next
Choose:   Titles
Enter:    Chart title: your title
          Category (x) axis: title for x-axis
          Value (y) axis: title for y-axis > Next > Finish
```

For editing, see the histogram commands on page 59.

* If the boundary = 50, then the limit = 49.9 (depending on the number of decimal places in the data).

TI-83 Plus Input the class boundaries into L1 and the frequencies into L2 (include an extra class boundary at the beginning with a frequency of zero); then continue with:

```
Choose:      STAT > EDIT > 1:EDIT...
Highlight:   L3
Enter:       L3 = 2nd > LIST > OPS > 6:cum sum(L2)
Highlight:   L4
Enter:       L4 -L3 / 2nd > LIST > Math > 5:sum (L2)
Choose:      2nd > STAT PLOT > 1:Plot
```

```
Plot1 Plot2 Plot3
On Off
Type:
Xlist:L1
Ylist:L4
Mark: ▫ + ·
```

```
Choose:      Zoom > 9:ZoomStat > Trace >>>
```

Adjust window if needed for better readability.

ANSWER NOW 2.37

The levels of various compounds is the subject of the "Frequency Distributions & Histograms" video on your Student Suite CD (found inside the back cover of this textbook). View the video clip and answer these questions:

a. **For which histogram (upper right or lower left) would you anticipate the numerical measure of spread to be the largest? The smallest?**

b. **Which of the two histograms would you anticipate have about the same difference between their smallest values and their largest values?**

Exercises

2.38 Ex02-038 The players on the Rochester Raging Rhinos professional soccer team scored 38 goals during the 2002 season.

Player	1	2	3	4	5	6	7	8	9	10	11	12	13
Goals	2	8	1	2	2	6	2	1	5	2	3	2	2

http://www.rhinossoccer.com/team.asp

a. If you want to show the number of goals scored by each player, would it be more appropriate to display this information on a bar graph or a histogram? Explain.
b. Construct the appropriate graph for part a.
c. If you wanted to show (emphasize) the distribution of scoring by the team, would it be more appropriate to display this information on a bar graph or a histogram? Explain.
d. Construct the appropriate graph for part c.

2.39 Ex02-039 The California Department of Education gives an annual report on the Advanced Placement (AP) test results for each year. In the 2000–2001 school year, Mariposa County students had the following scores:

2	3	2	1	2	3	3	2	2	3
3	2	2	2	4	5	1	2	2	4
2	2	5	4	1	2	2	2	5	2
3	2	3	2	3	3	5			

Source: http://data1.cde.ca.gov/dataquest/

a. Construct an ungrouped frequency distribution for the test scores.
b. Construct a frequency histogram of this distribution.
c. Prepare a relative frequency distribution for these same data.
d. AP scores of at least 3 are often required for college transferability. What percent of Mariposa AP scores will receive college credit?

2.40 The success of a batter against a pitcher in baseball varies with the count (balls and strikes) and, in particular, the number of strikes already called against a batter. *Sports Illustrated*, in the article "The Book on Maddux" (July 6, 1998, p. 46), tabulated the existing strike count against the number of hits batters obtained on the next pitch and the number of times the at bat ended on the next pitch. This information is for batters who faced Greg Maddux during the first half of the 1998 season. The results are summarized as follows:

Strikes	Hits	At Bats
0	38	137
1	43	138
2	25	223

The writer concluded, ". . . as the numbers show, let Maddux get two strikes on you, and you don't stand a chance."

a. Construct a frequency histogram for the strike count when the player made a hit.
b. Construct a frequency histogram for the strike count when the player's turn at bat ended.
c. Do you agree or disagree with the writer's conclusion? Explain why or why not.

2.41 Ex02-041 Results from the 1999 Central Oregon Household Telecommunications Survey found this information on the number of people per household in central Oregon:

Number in Household	1	2	3	4	5	6+
Percent	18.2	38.2	13.2	17.2	8.1	5.1

a. Draw a relative frequency histogram for the number of people per household.
b. What shape distribution does the histogram suggest?
c. Based on the graph, what do you know about the households in central Oregon?

2.42 Ex02-042 Results from the 1999 Central Oregon Household Telecommunications Survey found this information on the number of children per household in central Oregon:

Number of Children	0	1	2	3	4	5
Percent	62.1	12.2	15.2	7.2	2.2	1.1

a. Draw a relative frequency histogram for the number of children per household.
b. What shape distribution does the histogram suggest?
c. Based on the graph, what do you know about the number of children per household in central Oregon?

2.43 EX02-043 Here are the ages of 50 dancers who responded to a call to audition for a musical comedy:

21 19 22 19 18 20 23 19 19 20
19 20 21 22 21 20 22 20 21 20
21 19 21 21 19 19 20 19 19 19
20 20 19 21 21 22 19 19 21 19
18 21 19 18 22 21 24 20 24 17

a. Prepare an ungrouped frequency distribution of these ages.
b. Prepare an ungrouped relative frequency distribution of the same data.
c. Prepare a relative frequency histogram of these data.
d. Prepare a cumulative relative frequency distribution of the same data.
e. Prepare an ogive of these data.

2.44 Ex02-044 The opening-round scores for the Ladies' Professional Golf Association tournament at Locust Hill Country Club were posted as follows:

69 73 72 74 77 80 75 74 72 83 68 73
75 78 76 74 73 68 71 72 75 79 74 75
74 74 68 79 75 76 75 77 74 74 75 75
72 73 73 72 72 71 71 70 82 77 76 73
72 72 72 75 75 74 74 74 76 76 74 73
74 73 72 72 74 71 72 73 72 72 74 74
67 69 71 70 72 74 76 75 75 74 73 74
74 78 77 81 73 73 74 68 71 74 78 70
68 71 72 72 75 74 76 77 74 74 73 73
70 68 69 71 77 78 68 72 73 78 77 79
79 77 75 75 74 73 73 72 71 68 70 71
78 78 76 74 75 72 72 72 75 74 76 77
78 78

a. Form an ungrouped frequency distribution of these scores.
b. Draw a histogram of the first-round golf scores. Use the frequency distribution from part a.

2.45 The KSW computer science aptitude test was given to 50 students. The following frequency distribution resulted from their scores:

KSW Test Score	0–4	4–8	8–12	12–16	16–20	20–24	24–28
Frequency	4	8	8	20	6	3	1

a. What are the class boundaries for the class with the largest frequency?
b. Give all the class midpoints associated with this frequency distribution.
c. What is the class width?
d. Give the relative frequencies for the classes.
e. Draw a relative frequency histogram of the test scores.

2.46 Ex02-046 The USA Snapshot® "Nuns an aging order" reports that the median age of the 94,022 Roman Catholic nuns in the United States is 65 years and the percentages of U.S. nuns by age group are:

Under 50	51–70	Over 70	Refused to give age
16%	42%	37%	5%

This information is based on a survey of 1049 Roman Catholic nuns. Suppose the survey had resulted in the following frequency distribution (52 ages unknown):

Age	20–30	30–40	40–50	50–60	60–70	70–80	80–90
Frequency	34	58	76	187	254	241	147

a. Draw and completely label a frequency histogram.
b. Draw and completely label a relative frequency histogram of the same distribution.
c. Carefully examine the two histograms in parts a and b, and explain why one of them might be easier to understand. (Retain these solutions to use in Exercise 2.115, p. 93.)

FYI

Use the computer or calculator commands on pages 58–60 to construct a histogram of a frequency distribution.

2.47 Ex02-047 The speeds of 55 cars were measured by a radar device on a city street:

27	23	22	38	43	24	35	26	28	18	20
25	23	22	52	31	30	41	45	29	27	43
29	28	27	25	29	28	24	37	28	29	18
26	33	25	27	25	34	32	36	22	32	33
21	23	24	18	48	23	16	38	26	21	23

a. Classify these data into a grouped frequency distribution by using class boundaries 12–18, 18–24, . . ., 48–54.
b. Find the class width.
c. For the class 24–30, find the class midpoint, the lower class boundary, and the upper class boundary.
d. Construct a frequency histogram of these data.

FYI

Use the computer or calculator commands on pages 58–60 to construct a histogram for a given set of data.

2.48 Ex02-048 The hemoglobin A_{1c} test, a blood test given to diabetics during their periodic checkups, indicates the level of control of blood sugar during the past two to three months. The following data were obtained for 40 different diabetics at a university clinic:

6.5	5.0	5.6	7.6	4.8	8.0	7.5	7.9	8.0	9.2
6.4	6.0	5.6	6.0	5.7	9.2	8.1	8.0	6.5	6.6
5.0	8.0	6.5	6.1	6.4	6.6	7.2	5.9	4.0	5.7
7.9	6.0	5.6	6.0	6.2	7.7	6.7	7.7	8.2	9.0

a. Classify these A_{1c} values into a grouped frequency distribution using the classes 3.7–4.7, 4.7–5.7, and so on.
b. What are the class midpoints for these classes?
c. Construct a frequency histogram of these data.

2.49 Ex02-049 People have marveled for years about the continuing eruptions of the geyser Old Faithful in Yellowstone National Park. The times of duration, in minutes, for a sample of 50 eruptions of Old Faithful are listed below.

4.00	3.75	2.25	1.67	4.25
3.92	4.53	1.85	4.63	2.00
1.80	4.00	4.33	3.77	3.67
3.68	1.88	1.97	4.00	4.50
4.43	3.87	3.43	4.13	4.13
2.33	4.08	4.35	2.03	4.57
4.62	4.25	1.82	4.65	4.50
4.10	4.28	4.25	1.68	3.43
4.63	2.50	4.58	4.00	4.60
4.05	4.70	3.20	4.60	4.73

Source: http://www.stat.sc.edu/~west/javahtml/Histogram.html

a. Draw a dotplot displaying the eruption-length data.
b. Draw a histogram of the eruption-length data using class boundaries 1.6–2.0–2.4– · · · –4.8.
c. Draw another histogram of the data using different class boundaries and widths.
d. Repeat part c.
e. Repeat parts a and b using the larger set of 107 eruptions found on the CD as file DS-Old Faithful1 .
f. Which graph, in your opinion, does the best job of displaying the distribution? Why?
g. Write a short paragraph describing the distribution.

2.50 Ex02-050 The Office of Coal, Nuclear, Electric and Alternate Fuels reported the following data as the costs (in cents) of the average revenue per kilowatt-hour for sectors in Arkansas:

6.61	7.61	6.99	7.48	5.10	7.56
6.65	5.93	7.92	5.52	7.47	6.79
8.27	7.50	7.44	6.36	5.20	5.48
7.69	8.74	5.75	6.94	7.70	6.67
4.59	5.96	7.26	5.38	8.88	7.49
6.89	7.25	6.89	6.41	5.86	8.04

a. Prepare a grouped frequency distribution for the average revenue per kilowatt-hour using class boundaries 4, 5, 6, 7, 8, 9.

b. Find the class width.
c. List the class midpoints.
d. Construct a relative frequency histogram of these data.

FYI
Use the computer or calculator commands on pages 64–65 to construct an ogive for a given set of data

2.51
a. Prepare a cumulative relative frequency distribution for the variable "AP score" in Exercise 2.39.
b. Construct an ogive of the distribution.

2.52
a. Prepare a cumulative relative frequency distribution for the variable "KSW test score" in Exercise 2.45.
b. Construct an ogive of the distribution.

Numerical Descriptive Statistics

2.3 Measures of Central Tendency

Measures of central tendency are numerical values that locate, in some sense, the center of a set of data. The term *average* is often associated with all measures of central tendency.

MEAN (ARITHMETIC MEAN)
The average with which you are probably most familiar. The sample mean is represented by $\bar{x}$ (read "*x*-bar" or "sample mean"). The mean is found by adding all the values of the variable *x* (this sum of x values is symbolized Σx) and dividing the sum by the number of these values, *n* (the "sample size"). We express this in formula form as

$$\text{sample mean:} \quad \text{x-bar} = \frac{\text{sum of all x}}{\text{number of x}}$$

$$\bar{x} = \frac{\sum x}{n} \tag{2.1}$$

NOTES:

1. See the *Statistical Tutor* for information about $\sum$ notation ("summation notation").
2. The population mean, μ (lowercase mu, Greek alphabet), is the mean of all x values for the entire population.

Illustration 2.7 Finding the Mean

A set of data consists of the five values 6, 3, 8, 6, and 4. Find the mean.

SOLUTION

Using formula (2.1), we find

$$\bar{x} = \frac{\sum x}{n} = \frac{6 + 3 + 8 + 6 + 4}{5} = \frac{27}{5} = 5.4$$

Therefore, the mean of this sample is **5.4**. ■

FYI
See Animated Tutorial Mean on CD for assistance with this calculation.

ANSWER NOW 2.53
The numbers of children, x, who belong to each of eight families registering for swimming are 1, 2, 1, 3, 2, 1, 5, and 3. Find the mean, $\bar{x}$.

FYI
The mean is the middle point by weight.

A physical representation of the mean can be constructed by thinking of a number line balanced on a fulcrum. A weight is placed on a number on the line corresponding to each number in the sample of Illustration 2.7. In Figure 2.15 there is one weight each on the 3, 8, and 4 and two weights on the 6, since there are two 6s in the sample. The mean is the value that balances the weights on the number line—in this case, 5.4.

FIGURE 2.15
Physical Representation of the Mean

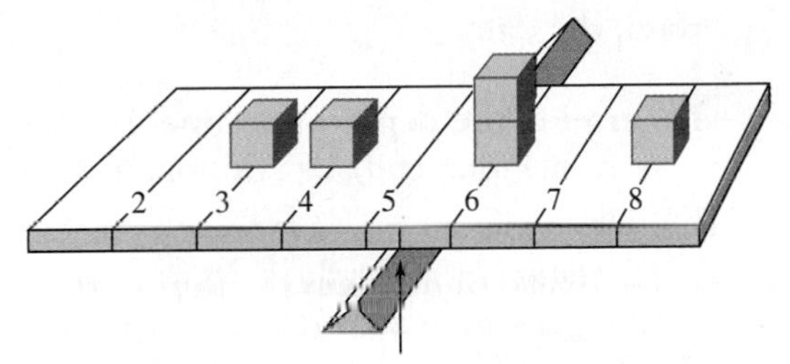

$\bar{x} = 5.4$ (the center of gravity, or balance point)

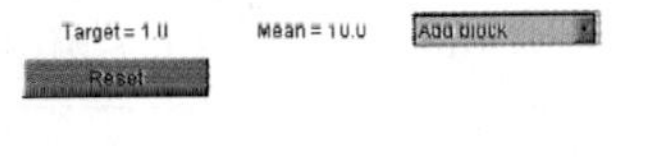

ANSWER NOW 2.54 INTERACTIVITY
Interactivity 2-C demonstrates the balancing effect of the mean. A plot is given with one data point at 10. Add more blocks by pointing and clicking on the desired locations on the plot until you achieve a mean of 1.

a. **How many blocks were required to balance for a mean of 1?**

b. **At what value are these blocks located?**

Technology Instructions: Mean

MINITAB (Release 13)

Input the data into C1; then continue with:

```
Choose:  Calc > Column Statistics
Select:  Mean
Enter:   Input variable: C1
```

Excel XP

Input the data into column A and activate a cell for the answer; then continue with:

```
Choose:  Insert Function, fx > Statistical > AVERAGE > OK
Enter:   Number 1: (A2:A6 or select cells)
         [Start at A1 if no header row (column title) is
         used.]
```

TI-83 Plus

Input the data into L1; then continue with:

```
Choose:  2nd > LIST > Math > 3:mean(
Enter:   L1
```

MEDIAN
The value of the data that occupies the middle position when the data are ranked in order according to size. The sample median is represented by $\tilde{x}$ (read "x-tilde" or "sample median").

NOTE: The population median, M (uppercase mu in the Greek alphabet), is the data value in the middle position of the entire ranked population.

Procedure for Finding the Median

Step 1: **Rank the data.**

Step 2: **Determine the depth of the median.** The **depth** (number of positions from either end), or position, of the median is determined by the formula

depth of median: $$\textit{depth of median} = \frac{\textit{number} + 1}{2}$$

$$d(\tilde{x}) = \frac{n+1}{2} \qquad (2.2)$$

The median's depth (or position) is found by adding the position numbers of the smallest data (1) and the largest data (n) and dividing the sum by 2 (n is the number of pieces of data).

Step 3: **Determine the value of the median.** Count the ranked data, locating the data in the $d(\tilde{x})$th position. The median will be the same regardless of which end of the ranked data (high or low) you count from. In fact, counting from both ends will serve as an excellent check.

The following two illustrations demonstrate this procedure as it applies to both odd-numbered and even-numbered sets of data.

Illustration 2.8 Median for Odd n

Find the median for the set of data {6, 3, 8, 5, 3}.

SOLUTION

Step 1 The data, ranked in order of size, are 3, 3, 5, 6, and 8.

Step 2 Depth of the median: $d(\tilde{x}) = \frac{n+1}{2} = \frac{5+1}{2} = 3$ (the "3rd" position).

Step 3 The median is the third number from either end in the ranked data, or $\tilde{x} = \mathbf{5}$. ■

FYI See Animated Tutorial Median on CD for assistance with this calculation.

FYI The value of $d(\tilde{x})$ is the depth of the median, NOT the value of the median, $\tilde{x}$.

Notice that the median essentially separates the ranked set of data into two subsets of equal size (see Figure 2.16).

FIGURE 2.16
Median of {3, 3, 5, 6, 8}

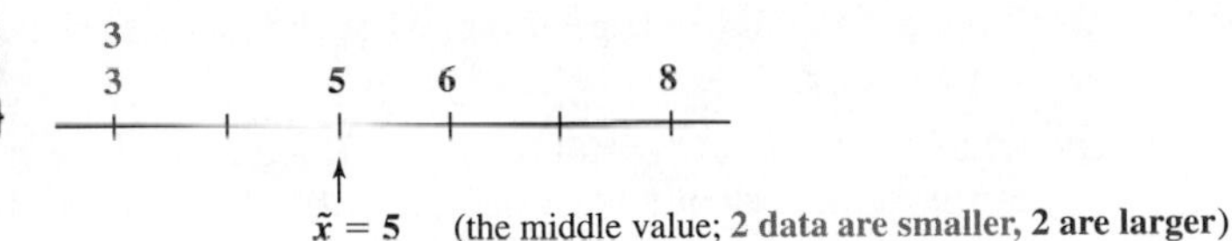

ANSWER NOW 2.55
Find the median height of a team of basketball players: 73, 76, 72, 70, and 74 inches.

As in Illustration 2.8, when n is odd, the depth of the median, $d(\tilde{x})$, will always be an integer. When n is even, however, the depth of the median, $d(\tilde{x})$, will always be a half-number, as shown in Illustration 2.9.

Illustration 2.9 Median of Even n

Find the median of the sample 9, 6, 7, 9, 10, 8.

SOLUTION

Step 1 The data, ranked in order of size, are 6, 7, 8, 9, 9, and 10.

Step 2 Depth of the median: $d(\tilde{x}) = \frac{n+1}{2} = \frac{6+1}{2} = 3.5$ (the "3.5th" position).

Step 3 The median is halfway between the third and fourth data. To find the number halfway between any two values, add the two values together and divide the sum by 2. In this case, add the third value (8) and the fourth value (9) and then divide the sum (17) by 2. The median is $\tilde{x} = \frac{8+9}{2} = \mathbf{8.5}$, a number halfway between the "middle" two numbers (see Figure 2.17). Notice that the median again separates the ranked set of data into two subsets of equal size.

FYI
The median is the middle point by count.

FYI
See Animated Tutorial Median on CD for assistance with this calculation.

FIGURE 2.17
Median of {6, 7, 8, 9, 9, 10}

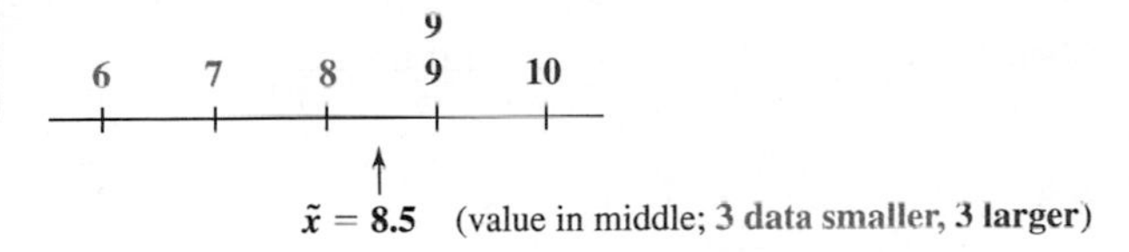

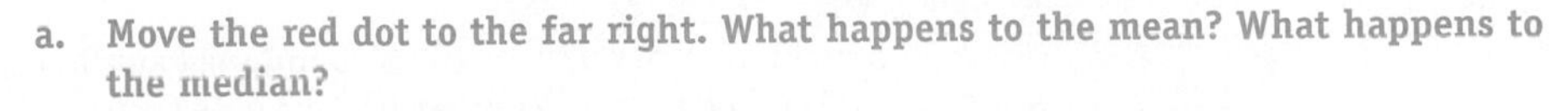

■

ANSWER NOW 2.56
Find the median rate paid at Jim's Burgers if the workers' hourly rates (in dollars) are 4.25, 4.15, 4.90, 4.25, 4.60, 4.50, 4.60, and 4.75.

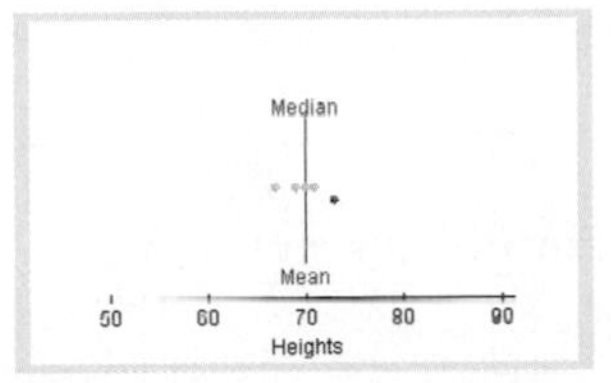

ANSWER NOW 2.57 INTERACTIVITY
Interactivity 2-D demonstrates the effect one data can have on the mean and on the median.

a. Move the red dot to the far right. What happens to the mean? What happens to the median?

b. **Move the red dot to the far left. What happens to the mean? What happens to the median?**

c. **Which measure of central tendency, the mean or the median, gives a better sense of the center when a maverick (or outlier) is present in the data?**

Technology Instructions: Median

MINITAB (Release 13)

Input the data into C1; then continue with:

```
Choose:   Calc > Column Statistics
Select:   Median
Enter:    Input variable: C1
```

Excel XP

Input the data into column A and activate a cell for the answer; then continue with:

```
Choose:   Insert Function, fx > Statistical > MEDIAN > OK
Enter:    Number 1: (A2:A6 or select cells)
```

TI-83 Plus

Input the data into L1; then continue with:

```
Choose:   2nd > LIST > Math > 4:median(
Enter:    L1
```

MODE

The mode is the value of x that occurs most frequently.

In the set of data from Illustration 2.8, {3, 3, 5, 6, 8}, the mode is 3 (see Figure 2.18). In the sample 6, 7, 8, 9, 9, 10, the mode is 9. In this sample, only the 9 occurs more than once; in the data from Illustration 2.8, only the 3 occurs more than once. If two or more values in a sample are tied for the highest frequency (number of occurrences), we say there is no mode.

FIGURE 2.18
Mode of {3, 3, 5, 6, 8}

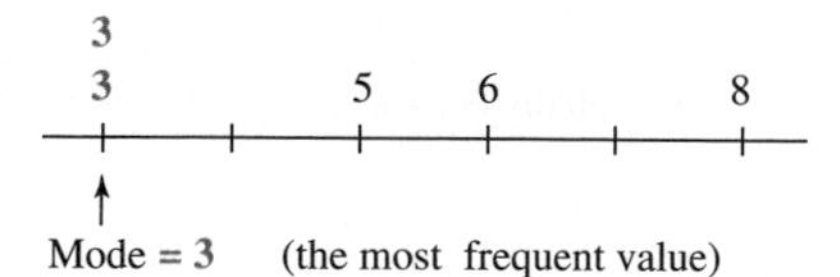

For example, in the sample 3, 3, 4, 5, 5, 7, the 3 and the 5 appear an equal number of times. There is no one value that appears most often; thus, this sample has no mode.

ANSWER NOW 2.58

The numbers of cars per apartment owned by a sample of tenants in a large complex are 1, 2, 1, 2, 2, 2, 1, 2, 3, 2. What is the mode?

MIDRANGE
The number exactly midway between a lowest valued data L and a highest valued data H. It is found by averaging the low and the high values:

$$\text{midrange} = \frac{\text{low value} + \text{high value}}{2}$$

$$\text{midrange} = \frac{L + H}{2} \tag{2.3}$$

For the set of data from Illustration 2.8, {3, 3, 5, 6, 8}, $L = 3$ and $H = 8$ (see Figure 2.19).

FIGURE 2.19
Midrange of {3, 3, 5, 6, 8}

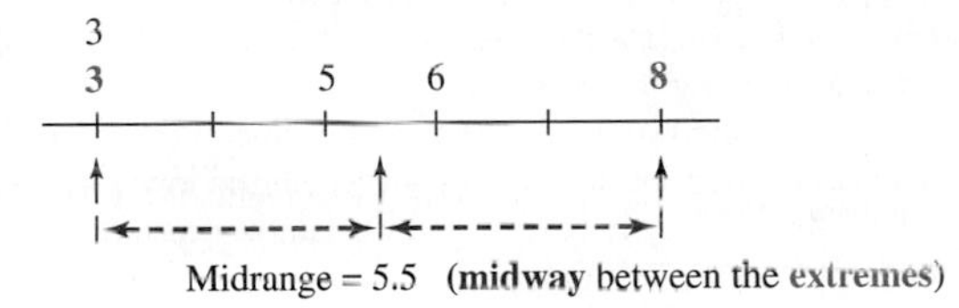

Therefore, the midrange is

$$\text{midrange} = \frac{L + H}{2}$$

$$= \frac{3 + 8}{2} = 5.5$$

ANSWER NOW 2.59
USA Today (July 1998) reported on "What airlines spent on food" per passenger 1st quarter 1998: least—\$0.20, most—\$10.76. Find the midrange.

The four measures of central tendency represent four different methods of describing the middle. These four values may be the same, but more likely they will be different.

For the sample data from Illustration 2.9, the mean $\bar{x}$ is 8.2, the median $\tilde{x}$ is 8.5, the mode is 9, and the midrange is 8. Their relationship to one another and to the data is shown in Figure 2.20.

FIGURE 2.20
Measures of Central Tendency for {6, 7, 8, 9, 9, 10}

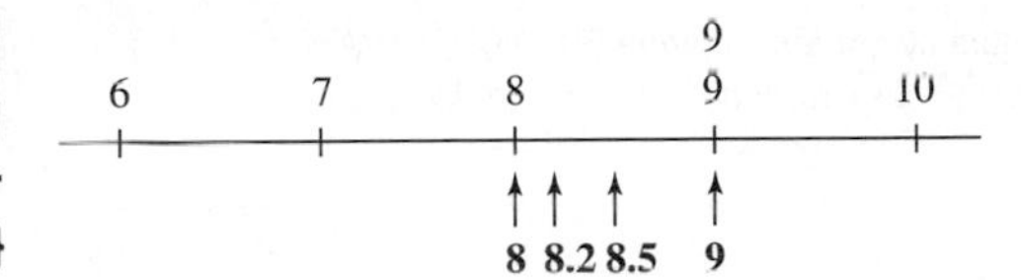

ANSWER NOW 2.60
Find the mean, median, mode, and midrange for the sample data 9, 6, 7, 9, 10, 8.

Application 2.2 "Average" Means Different Things

When it comes to convenience, few things can match that wonderful mathematical device called *averaging*. With an **average** you can take a fistful of figures on any subject and compute one figure that will represent the whole fistful.

But there is one thing to remember. There are several kinds of measures ordinarily known as averages. And each gives a different picture of the figures it is called on to represent.

Take an example. Here are the annual incomes of ten families:

Ap02-2

$54,000	$39,000	$37,500	$36,750	$35,250
$31,500	$31,500	$31,500	$31,500	$25,500

What would this group's "typical" income be? Averaging would provide the answer, so let's compute the typical income by the simpler and most frequently used kinds of averaging.

The arithmetic mean. It is the most common form of average, obtained by adding items in the series and then dividing by the number of items: $35,400. The mean is representative of the series in the sense that the sum of the amounts by which the higher figures exceed the mean is exactly the same as the sum of the amounts by which the lower figures fall short of the mean.

The median. As you may have observed, six families earn less than the mean, four earn more. You might very well wish to represent this varied group by the income of the family that is right smack dab in the middle of the whole bunch. The median works out to $33,375.

The midrange. Another number that might be used to represent the group is the midrange, computed by calculating the figure that lies halfway between the highest and lowest incomes: $39,750.

The mode. So, three kinds of averages, and not one family actually has an income matching any of them. Say you want to represent the group by stating the income that occurs most frequently. That is called a mode. $31,500 would be the modal income.

Four different averages, each valid, correct, and informative in its way. But how they differ!

arithmetic mean	median	midrange	mode
$35,400	$33,375	$39,750	$31,500

And they would differ still more if just one family in the group were a millionaire—or one were jobless!

So there are three lessons: First, when you see or hear an average, find out which average it is. Then you'll know what kind of picture you are being given. Second, think about the figures being averaged so you can judge whether the average used is appropriate. And third, don't assume that a literal mathematical quantification is intended every time somebody says "average." It isn't. All of us often say "the average person" with no thought of implying a mean, median, or mode. All we intend to convey is the idea of other people who are in many ways a great deal like the rest of us.

ANSWER NOW 2.61

Application 2.2 uses a sample of ten annual incomes to discuss the four averages.

a. **Calculate the mean, median, mode, and midrange for the ten incomes. Compare your results with those found in the article.**

b. **What is there about the distribution of these ten data that causes the values of these four averages to be so different?**

c. **Comment on "the sum of the amounts by which the higher figures exceed the mean is exactly the same as the sum of the amounts by which the lower figures fall short of the mean."**

ROUND-OFF RULE

When rounding off an answer, let's agree to keep one more decimal place in our answer than was present in the original information. To avoid round-off buildup, round off only the final answer, not the intermediate steps. That is, avoid using a rounded value to do further calculations. In our previous examples, the data were composed of whole numbers; therefore, those answers that have decimal values should be rounded to the nearest tenth. See the *Statistical Tutor* for specific instructions on how to perform the rounding off.

ANSWER NOW 2.62

Lightning strikes are the subject of the "Descriptive Statistics" video on your Student Suite CD (found inside the back cover of this textbook). View the video clip and answer these questions:

a. **Data were collected for what variable?**

b. **What does each bar (interval) represent?**

c. **What conclusion was reached?**

d. **What characteristic(s) of the graph support the conclusion?**

Exercises

2.63 Explain why it is possible to find the mean for the data of a quantitative variable, but not for a qualitative variable.

2.64 Consider the sample 2, 4, 7, 8, 9. Find the following:
a. mean $\bar{x}$ b. median $\tilde{x}$ c. mode d. midrange

2.65 Consider the sample 6, 8, 7, 5, 3, 7. Find the following:
a. mean $\bar{x}$ b. median $\tilde{x}$ c. mode d. midrange

2.66 Fifteen randomly selected college students were asked to state the number of hours they slept last night. The resulting data are 5, 6, 6, 8, 7, 7, 9, 5, 4, 8, 11, 6, 7, 8, 7. Find the following:
a. mean $\bar{x}$ b. median $\tilde{x}$ c. mode d. midrange

2.67 Ex02-067 The article "More hours in class logged by college profs" in *The Blade* (January 23, 2000) reported on a study done at the four-year colleges in Ohio about the total hours spent by college professors in instructing and advising students. The times reported for the 13 schools were:

28.4	44.4	36.4	33.5	34.9	30.5	32.7
24.6	26.9	34.7	33.4	29.3	33.3	

a. Find the mean time Ohio professors spend teaching and advising students.
b. Find the median time Ohio professors spend teaching and advising students.
c. Find the midrange of time Ohio professors spend teaching and advising students.
d. Find the mode, if one exists, per professor.

2.68 Ex02-068 "America West leads monthly on-time arrivals" was an article in *USA Today* (February 12, 2002) about overall flight on-time rates. For December 2001 the on-time arrival rates of domestic flights at the 25 largest U.S. airports were as follows:

Atlanta	79.7
Baltimore	82.5
Boston	86.5
Charlotte	82.3
Chicago (O'Hare)	78.4
Dallas/Fort Worth	84.9
Denver	81.5
Detroit	81.4
Houston	82.8
Las Vegas	76.9
Los Angeles	81.0
Miami	76.0
Minneapolis/St. Paul	82.5
New York (LaGuardia)	89.7
New York (Newark)	82.7

(continued)

Orlando	82.2
Philadelphia	77.8
Phoenix	82.3
Pittsburgh	84.0
Salt Lake City	74.2
San Diego	78.4
San Francisco	64.6
Seattle/Tacoma	69.8
St. Louis	80.0
Tampa	79.9

a. Find the mean rate for December 2001.
b. Find the median rate for December 2001.
c. Construct a stem-and-leaf display of the data.
d. Describe the relationship between the mean and the median and what properties of the data cause the mean to be lower than the median.

2.69 *USA Today* (May 29, 2002) reported these statistics about the average annual taxes paid per person by state:

Highest: Connecticut—$3092
Lowest: South Dakota—$1292

a. Based on this information, find the "average" of the 50 state average taxes.
b. Explain why your answer in part a is the only average value you can determine from the given information.
c. If you were told that the mean value of the 50 state averages is $1842, what could you tell about their distribution?

2.70 The "average" is a commonly reported statistic. This single bit of information can be very informative or very misleading, with the mean and median being the two most commonly reported.

a. The mean is a useful measure, but it can be misleading. Describe a circumstance when the mean is very useful as the average and a circumstance when the mean is very misleading as the average.
b. The median is a useful measure, but it can be misleading. Describe a circumstance when the median is very useful as the average and a circumstance when the median is very misleading as the average.

2.71 Ex02-071 All the third-graders at Roth Elementary School were given a physical fitness strength test. These data resulted:

12	22	6	9	2	9	5	9	3	5	16	1	22
18	6	12	21	23	9	10	24	21	17	11	18	19
17	5	14	16	19	19	18	3	4	21	16	20	15
14	17	4	5	22	12	15	18	20	8	10	13	20
6	9	2	17	15	9	4	15	14	19	3	24	

a. Construct a dotplot.
b. Find the mode.
c. Prepare a grouped frequency distribution using classes 1–4, 4–7, and so on, and draw a histogram of the distribution.
d. Describe the distribution; specifically, is the distribution bimodal (about what values)?
e. Compare your answers in parts a and c, and comment on the relationship between the mode and the modal values in these data.
f. Could the discrepancy found in the comparison in part e occur when using an ungrouped frequency distribution? Explain.
g. Explain why, in general, the mode of a set of data does not necessarily give us the same information as the modal values do.

2.72 Ex02-072 Consumers are frequently cautioned against eating too much food that is high in calories, fat, and sodium for numerous health and fitness reasons. *Nutrition Action HealthLetter* published a list of popular low-fat brands of hot dogs commonly labeled "fat-free," "reduced fat," "low-fat," "light," and so on, together with their calories, fat content, and sodium. All quantities are for one dog:

Hot Dog Brand	Calories	Fat (g)	Sodium (mg)
Ball Park Fat Free Beef Franks	50	0	460
Butterball Fat Free Franks	40	0	490
Oscar Meyer Free Hot Dogs	40	0	490
Jennie-O Fat Free Turkey Franks	40	0	520
Ball Park Fat Free Smoked White Turkey Franks	40	0	530
Eckrich Fat Free Beef Franks	50	0	570
Hormel Fat Free Beef Hot Dogs	50	0	590
Healthy Choice	60	2	430
Empire Kosher Turkey Franks	50	2	320
Hebrew National 97% Fat Free	50	2	440
Hillshire Farm Lean & Hearty Roast Turkey Hot Dogs	80	5	650
Eckrich Lite Bunsize Franks	110	7	480
Ball Park Smart Creations Lite Beef Franks	100	7	630
Louis Rich Bun-Length Cheese Franks	90	7	540
Gwaltney Great Dogs	140	10	730
Hillshire Farm Lean & Hearty Hot Dogs	80	6	610
Oscar Meyer Light Beef Franks	110	8	610
Butterball Bun Size Franks	130	10	600
Shelton's Smoked Uncured Turkey Franks	200	16	800

(continued)

Mr Turkey Cheese Franks	110	9	540
Hebrew National Reduced Fat	120	10	350
Mr Turkey Bun Size Franks	130	11	670
Jennie-O Jumbo Turkey Franks	130	11	600
Shelton's Uncured Franks	80	7	350
Jennie-O Turkey Franks	80	7	360
Wampler Foods Turkey Franks	90	8	430
Frankfurter, beef, typical	180	16	580
Wampler Foods Chicken Franks	120	11	480

Source: *Nutrition Action HealthLetter*, "On the Links," July/August 1998, pp. 12–13

a. Find the mean, median, mode, and midrange of the calories, fat, and sodium contents of all the frankfurters listed. Use a table to summarize your results.
b. Construct a dotplot of the fat contents. Locate the mean, median, mode, and midrange on the plot.
c. In the summer of 1997, the winner of Nathan's Famous Fourth of July Hot Dog Eating Contest consumed 24.5 hot dogs in 12 minutes. If he had been served the median hot dog, how many calories, grams of fat, and milligrams of sodium did he consume in the single sitting? If the recommended daily allowance for sodium intake is 2400 mg, did he likely exceed it? Explain.

2.73 Ex02-073 The number of runs that baseball teams score is likely influenced by the ballparks in which they play. In an attempt to measure differences between stadiums, *Sports Illustrated* collected data for nearly every major league team over a three-year span. The numbers of runs scored per game by teams and their opponents while playing on their home field were compared with the numbers of runs scored per game while playing away (at the opponents' fields). The table summarizes the data.

Stadium (Team)	Home	Away
Coors Field (Rockies)	13.65	8.76
The Ballpark (Rangers)	10.90	9.78
Three Rivers Stadium (Pirates)	9.65	8.95
County Stadium (Brewers)	10.17	9.63
Wrigley Field (Cubs)	9.59	9.15
Metrodome (Twins)	10.65	10.20
Fenway Park (Red Sox)	10.82	10.50
Veterans Stadium (Phillies)	9.26	9.03
Kauffman Stadium (Royals)	9.60	9.43
Olympic Stadium (Expos)	8.78	8.67
Tiger Stadium (Tigers)	10.55	10.44
Kingdome (Mariners)	10.97	10.93
Jacobs Field (Indians)	10.33	10.33
Cinergy Field (Reds)	9.18	9.24
Skydome (Blue Jays)	9.29	9.35
Edison International Field (Angels)	10.20	10.29
Busch Stadium (Cardinals)	8.83	9.09
Camden Yards (Orioles)	9.82	10.18
Yankee Stadium (Yankees)	9.77	10.17
Oakland Coliseum (A's)	10.36	10.83
3Com Park (Giants)	9.43	10.03
Pro Player Stadium (Marlins)	8.70	9.32
Shea Stadium (Mets)	8.47	9.44
Comiskey Park (White Sox)	9.85	11.06
Astrodome (Astros)	8.63	10.19
Qualcomm Stadium (Padres)	8.66	10.23
Dodger Stadium (Dodgers)	7.64	9.38

(continued)

Source: *Sports Illustrated*, "The Coors Curse," July 20, 1998, p. 40

a. Find the mean, median, maximum, minimum, and midrange of the combined runs scored by the teams while playing at home and away.
b. Subtract the combined runs scored away from the combined runs scored at home to obtain the difference for each stadium. Then find the mean, median, maximum, minimum, and midrange of the differences between the combined runs scored.
c. Compare each of the measures you found in parts a and b with the data collected for just Coors Field, home of the Rockies. What can you conclude?

2.74 You are responsible for planning the parking needed for a new 256-unit apartment complex, and you're told to base the needs on the statistic "average number of vehicles per household is 1.9."
a. Which average (mean, median, mode, midrange) will be helpful to you? Explain.
b. Explain why 1.9 cannot be the median, the mode, or the midrange for the variable "number of vehicles."
c. If the owner wants parking that will accommodate 90% of all the tenants who own vehicles, how many spaces must you plan for?

2.75 Starting with the data values 70 and 100, add three data values to the sample so that the sample has: (Justify your answer in each case.)
a. a mean of 100
b. a median of 70
c. a mode of 87
d. a midrange of 70
e. a mean of 100 and a median of 70
f. a mean of 100 and a mode of 87
g. a mean of 100 and a midrange of 70
h. a mean of 100, a median of 70, and a mode of 87

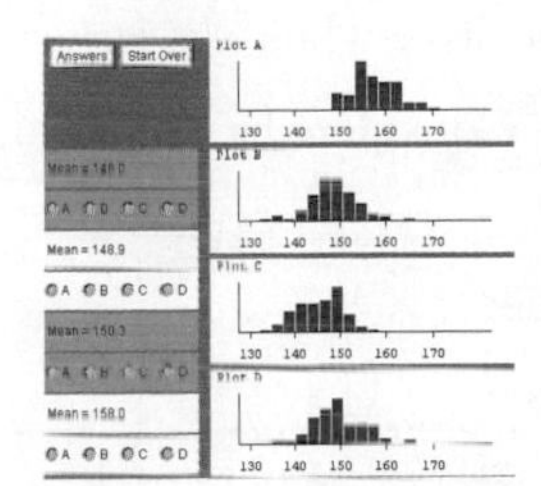

2.76 Interactivity Interactivity 2-E matches means with corresponding histograms. After several practice rounds using 'New Plots,' explain your method of matching.

2.4 Measures of Dispersion

Having located the "middle" with the measures of central tendency, our search for information from data sets now turns to the measures of dispersion (spread). The **measures of dispersion** include the *range*, *variance*, and *standard deviation*. These numerical values describe the amount of spread, or variability, that is found among the data: Closely grouped data have relatively small values, and more widely spread-out data have larger values. The closest possible grouping occurs when the data have no dispersion (all data are the same value); in this situation, the measure of dispersion will be zero. There is no limit to how widely spread out the data can be; therefore, measures of dispersion can be very large.

RANGE

The difference in value between the highest-valued (H) and the lowest-valued (L) data:

$$\textit{range} = \textit{high value} - \textit{low value}$$

$$\text{range} = H - L \tag{2.4}$$

The sample 3, 3, 5, 6, 8 has a range of $H - L = 8 - 3 = 5$. The range of 5 tells us that these data all fall within a 5-unit interval (see Figure 2.21).

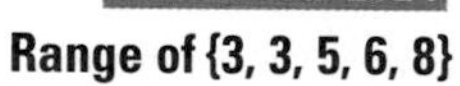

FIGURE 2.21
Range of {3, 3, 5, 6, 8}

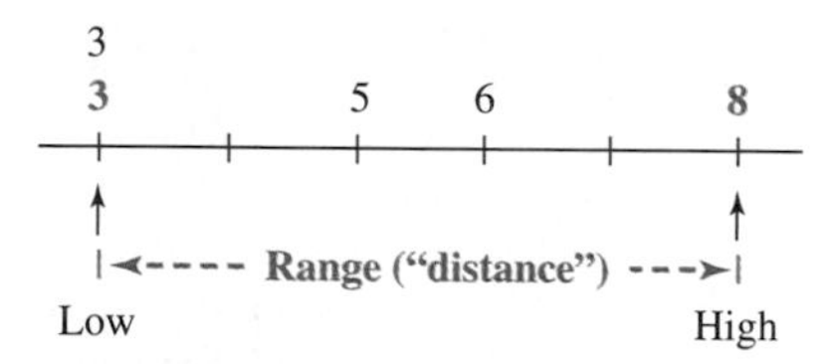

ANSWER NOW 2.77

USA Today **(May 29 and 30, 2002) reported on annual taxes. South Dakota: lowest amount per person—$1292; Connecticut: highest amount per person—$3092. Find the range.**

The other measures of dispersion to be studied in this chapter are measures of dispersion about the mean. To develop a measure of dispersion about the mean, let's first answer the question: How far is each x from the mean?

DEVIATION FROM THE MEAN

A deviation from the mean, $x - \bar{x}$, is the difference between the value of x and the mean $\bar{x}$.

Each individual value x deviates from the mean by an amount equal to $(x - \bar{x})$. This deviation $(x - \bar{x})$ is zero when x is equal to the mean $\bar{x}$. The deviation $(x - \bar{x})$ is positive when x is larger than $\bar{x}$ and negative when x is smaller than $\bar{x}$.

Consider the sample 6, 3, 8, 5, 3. Using formula (2.1), $\bar{x} = \frac{\Sigma x}{n}$, we find that the mean is 5. Each deviation, $(x - \bar{x})$, is then found by subtracting 5 from each x value:

Data	x	6	3	8	5	3
Deviation	$x - \bar{x}$	1	−2	3	0	−2

FIGURE 2.22
Deviations from the Mean

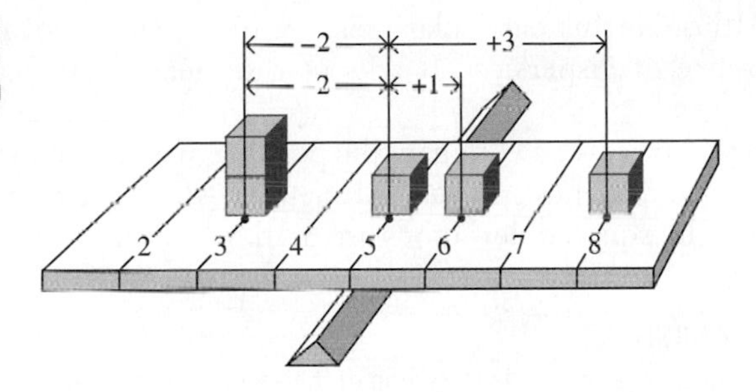

Figure 2.22 shows the four nonzero deviations from the mean.

ANSWER NOW 2.78

a. **The data value x = 45 has a deviation value of 12. Explain the meaning of this.**

b. **The data value x = 84 has a deviation value of −20. Explain the meaning of this.**

To describe the "average" value of these deviations, we might use the mean deviation, the sum of the deviations divided by n, $\frac{\Sigma(x - \bar{x})}{n}$. However, since the sum of the deviations, $\Sigma(x - \bar{x})$, is exactly zero, the mean deviation will also be zero. As a matter of fact, it will always be zero. Since the mean deviation is always zero, it is not a useful statistic.

ANSWER NOW 2.79
The summation $\Sigma(x - \bar{x})$ is always zero. Why? Think back to the definition of the mean (p. 68) and see if you can justify this statement.

The sum of the deviations, $\Sigma(x - \bar{x})$, is always zero because the deviations of x values smaller than the mean (which are negative) cancel out those x values larger than the mean (which are positive). We can remove this neutralizing effect if we do something to make all the deviations positive. This can be accomplished in two ways. First, by using the absolute value of the deviation, $|x - \bar{x}|$, we can treat each deviation as its "size" or distance only. For our illustration we obtain the following *absolute deviations.*

Data	x	6	3	8	5	3
Absolute Value of Deviation	$\lvert x - \bar{x}\rvert$	1	2	3	0	2

MEAN ABSOLUTE DEVIATION
The mean of the absolute values of the deviations from the mean:

$$\textit{mean absolute deviation} = \frac{\textit{sum of (absolute values of deviations)}}{\textit{number}}$$

$$\text{mean absolute deviation} = \frac{\Sigma|x - \bar{x}|}{n} \quad (2.5)$$

For our example, the sum of the absolute deviations is 8 (1 + 2 + 3 + 0 + 2) and

$$\text{mean absolute deviation} = \frac{\Sigma|x - \bar{x}|}{n} = \frac{8}{5} = 1.6$$

Although this particular measure of spread is not used very frequently, it is a measure of dispersion. It tells us the mean "distance" the data are from the mean.

A second way to eliminate the positive–negative neutralizing effect is to square each of the deviations; squared deviations will all be nonnegative (positive or zero) values. The squared deviations are used to find the *variance*.

SAMPLE VARIANCE

The sample variance, s^2, is the mean of the squared deviations, calculated using $n - 1$ as the divisor:

$$\text{sample variance:} \quad \textit{s-squared} = \frac{\textit{sum of (deviations squared)}}{\textit{number} - 1}$$

$$s^2 = \frac{\Sigma(x - \bar{x})^2}{n - 1} \qquad (2.6)$$

where n is the sample size—that is, the number of data in the sample.

The variance of the sample 6, 3, 8, 5, 3 is calculated in Table 2.10 using formula (2.6).

TABLE 2.10 Calculating Variance Using Formula (2.6)

Step 1. Find Σx	Step 2. Find $\bar{x}$	Step 3. Find each $x - \bar{x}$	Step 4. Find $\Sigma(x - \bar{x})^2$	Step 5. Find s^2
6	$\bar{x} = \frac{\Sigma x}{n}$	$6 - 5 = 1$	$(1)^2 = 1$	$s^2 = \frac{\Sigma(x - \bar{x})^2}{n - 1}$
3		$3 - 5 = -2$	$(-2)^2 = 4$	
8	$\bar{x} = \frac{25}{5}$	$8 - 5 = 3$	$(3)^2 = 9$	$s^2 = \frac{18}{4}$
5		$5 - 5 = 0$	$(0)^2 = 0$	
3		$3 - 5 = -2$	$(-2)^2 = 4$	
$\Sigma x = 25$	$\bar{x} = 5$	$\Sigma(x - \bar{x}) = 0$ (ck)	$\Sigma(x - \bar{x})^2 = 18$	$s^2 = \mathbf{4.5}$

NOTES:

1. The sum of all the x values is used to find $\bar{x}$.
2. The sum of the deviations, $\Sigma(x - \bar{x})$, is always zero, provided the exact value of $\bar{x}$ is used. Use this fact as a check in your calculations, as was done in Table 2.10 (denoted by (ck)).
3. If a rounded value of $\bar{x}$ is used, then $\Sigma(x - \bar{x})$ will not always be exactly zero. It will, however, be reasonably close to zero.
4. The sum of the squared deviations is found by squaring each deviation and then adding the squared values.

FYI See Animated Tutorial Variance on CD for assistance with this calculation.

ANSWER NOW 2.80
Use formula (2.6) to find the variance for the sample {1, 3, 5, 6, 10}.

The set of data in Answer Now 2.80 is more dispersed than the data in Table 2.10, and therefore its variance is larger. A comparison of these two samples is shown in Figure 2.23.

FIGURE 2.23
Comparison of Data

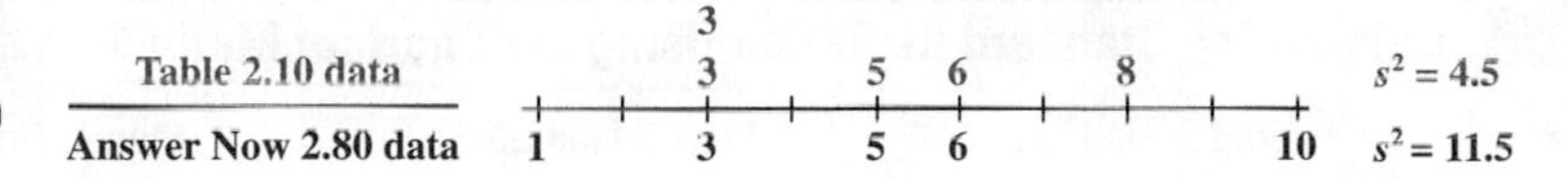

SAMPLE STANDARD DEVIATION
The standard deviation of a sample, s, is the positive square root of the variance:

sample standard deviation: s = *square root of sample variance*

$$s = \sqrt{s^2} \tag{2.7}$$

For the samples shown in Figure 2.23, the standard deviations are $\sqrt{4.5}$ or **2.1**, and $\sqrt{11.5}$ or **3.4**.

NOTE: The numerator for the sample variance, $\Sigma(x - \bar{x})^2$, is often called the *sum of squares for x* and symbolized by SS(x). Thus, formula (2.6) can be expressed as

$$\text{sample variance:} \quad s^2 = \frac{SS(x)}{n - 1}, \quad \text{where } SS(x) = \Sigma(x - \bar{x})^2 \tag{2.8}$$

The formulas for variance can be modified into other forms for easier use in various situations. For example, suppose we have the sample 6, 3, 8, 5, 2. The variance for this sample is computed in Table 2.11.

TABLE 2.11 Calculating Variance Using Formula (2.6)

Step 1. Find Σx	Step 2. Find $\bar{x}$	Step 3. Find each $x - \bar{x}$	Step 4. Find $\Sigma(x - \bar{x})^2$	Step 5. Sample variance
6	$\bar{x} = \frac{\Sigma x}{n}$	$6 - 4.8 = 1.2$	$(1.2)^2 = 1.44$	$s^2 = \frac{\Sigma(x - \bar{x})^2}{n - 1}$
3		$3 - 4.8 = -1.8$	$(-1.8)^2 = 3.24$	
8	$\bar{x} = \frac{24}{5}$	$8 - 4.8 = 3.2$	$(3.2)^2 = 10.24$	$s^2 = \frac{22.80}{4}$
5		$5 - 4.8 = 0.2$	$(0.2)^2 = 0.04$	
2		$2 - 4.8 = -2.8$	$(-2.8)^2 = 7.84$	$s^2 = \mathbf{5.7}$
$\Sigma x = 24$	$\bar{x} = 4.8$	$\Sigma(x - \bar{x}) = 0$ (ck)	$\Sigma(x - \bar{x})^2 = 22.80$	

The arithmetic for this example has become more complicated because the mean contains nonzero digits to the right of the decimal point. However, the "sum of squares for x," the numerator of formula (2.6), can be rewritten so that $\bar{x}$ is not included:

$$\text{sum of squares:} \quad SS(x) = \Sigma x^2 - \frac{(\Sigma x)^2}{n} \tag{2.9}$$

Combining formulas (2.8) and (2.9) yields the "shortcut formula" for sample variance:

$$\textit{s-squared} = \frac{(\textit{sum of } x^2) - \left[\frac{(\textit{sum of } x)^2}{\textit{number}}\right]}{}$$

FYI
See page 84 for an explanation of icons.

$$\text{sample variance:} \quad s^2 = \frac{\Sigma x^2 - \frac{(\Sigma x)^2}{n}}{n - 1} \tag{2.10}$$

Formulas (2.9) and (2.10) are called "shortcut" because they bypass the calculation of $\bar{x}$.

TABLE 2.12 Calculating Standard Deviation Using the Shortcut Method

Step 1. Find Σx	Step 2. Find Σx^2	Step 3. Find SS(x)	Step 4. Find s^2	Step 5. Find s
6	$6^2 = 36$	$SS(x) = \Sigma x^2 - \frac{(\Sigma x)^2}{n}$	$s^2 = \frac{\Sigma x^2 - \frac{(\Sigma x)^2}{n}}{n-1}$	$s = \sqrt{s^2}$
3	$3^2 = 9$			$s = \sqrt{5.7}$
8	$8^2 = 64$	$SS(x) = 138 - \frac{(24)^2}{5}$		$s = \mathbf{2.4}$
5	$5^2 = 25$		$s^2 = \frac{22.80}{4}$	
2	$2^2 = 4$	$SS(x) = 138 - 115.2$		
$\Sigma x = 24$	$\Sigma x^2 = 138$	$SS(x) = 22.8$	$s^2 = 5.7$	

FYI
See Animated Tutorial Variance Shortcut on CD for assistance with this calculation.

The computations for SS(x), s^2, and s using formulas (2.9), (2.10), and (2.7) are performed as shown in Table 2.12.

ANSWER NOW 2.81
Use formula (2.10) to find the variance of the sample {1, 3, 5, 6, 10}. Compare the results to Answer Now 2.80.

The unit of measure for the standard deviation is the same as the unit of measure for the data. For example, if our data are in pounds, then the standard deviation s will also be in pounds. The unit of measure for variance might then be thought of as *units squared*. In our example of pounds, this would be *pounds squared*. As you can see, the unit has very little meaning.

Technology Instructions: Standard Deviation

MINITAB (Release 13)

Input the data into C1; then continue with:

```
Choose:   Calc > Column Stats
Select:   Standard deviation
Enter:    Input variable: C1
```

Excel XP

Input the data into column A and activate a cell for the answer; then continue with:

```
Choose:   Insert Function, fx > Statistical > STDEV > OK
Enter:    Number 1: (A2:A6 or select cells)
```

TI-83 Plus

Input the data into L1; then continue with:

```
Choose:   2nd > LIST > Math > 7:StdDev(
Enter:    L1
```

Technology Instructions: Additional Statistics

MINITAB (Release 13)

Input the data into C1, then continue with:

```
Choose:   Calc > Column Statistics
Then one at a time select the desired statistic
```

```
Select:   N total     Number of data in column
          Sum         Sum of the data in column
          Minimum     Smallest value in column
          Maximum     Largest value in column
          Range       Range of values in column
          SSQ         Sum of squared x-values, Σx²
Enter:    Input variable: C1
```

Excel XP Input the data into column A and activate a cell for the answer; then continue with:

```
Choose:   Insert Function, fx > Statistical > COUNT
                                            > MIN
                                            > MAX
     OR                         > All > SUM
                                      > SUMSQ
Enter:    Number 1: (A2:A6 or select cells)
For range, write a formula: Max( ) - Min( )
```

TI-83 Plus Input the data into L1; then continue with:

```
Choose:   2nd > LIST > Math > 5:sum(
                            > 1:min(
                            > 2:max(
Enter:    L1
```

STANDARD DEVIATION ON YOUR CALCULATOR

Most calculators have two formulas for finding standard deviation and mindlessly calculate both, fully expecting the user to decide which one is correct for the given data. How do you decide?

The sample standard deviation is denoted by s and uses the "divide by $n - 1$" formula.
The population standard deviation is denoted by σ and uses the "divide by n" formula.

When you have sample data, always use the s or "divide by $n - 1$" formula. Having the population data is a situation that will probably never occur, other than in a textbook exercise. If you don't know whether you have sample data or population data, it is a "safe bet" that they are sample data—use the s or "divide by $n - 1$" formula!

MULTIPLE FORMULAS

Statisticians have multiple formulas for convenience—that is, convenience relative to the situation. The following statements will help you decide which formula to use:

1. When you are working on a computer and using statistical software, you will generally store all the data values first. The computer handles repeated operations easily and can "revisit" the stored data as often as necessary to complete a procedure. The computations for sample variance will be done using formula (2.6), following the process shown in Table 2.10.

(continued)

MULTIPLE FORMULAS (CONTINUED)

2. When you are working on a calculator with built-in statistical functions, the calculator must perform all necessary operations on each data as the values are entered (most handheld nongraphing calculators do not have the ability to store data). Then after all data have been entered, the computations will be completed using the appropriate summations. The computations for sample variance will be done using formula (2.10), following the procedure shown in Table 2.12.
3. If you are doing the computations either by hand or with the aid of a calculator, but not using statistical functions, the most convenient formula to use will depend on how many data there are and how convenient the numerical values are to work with.

When a formula has multiple forms, look for one of these icons:

is used to identify the formula most likely to be used by a computer.
is used to identify the formula most likely to be used by a calculator.
is used to identify the formula most likely to be convenient for hand calculations.
D is used to identify the "definition" formula.

Exercises

2.82 All measures of variation are nonnegative in value for all sets of data.

a. What does it mean for a value to be "nonnegative"?
b. Describe the conditions necessary for a measure of variation to have the value zero.
c. Describe the conditions necessary for a measure of variation to have a positive value.

2.83 Consider the sample 2, 4, 7, 8, 9. Find the following:

a. range
b. variance s^2, using formula (2.6)
c. standard deviation, s

2.84 Consider the sample 6, 8, 7, 5, 3, 7. Find the following:

a. range
b. variance s^2, using formula (2.6)
c. standard deviation, s

2.85 Given the sample 7, 6, 10, 7, 5, 9, 3, 7, 5, 13, find the following:

a. variance s^2 using formula (2.6)
b. variance s^2, using formula (2.10)
c. standard deviation, s

2.86 Fifteen randomly selected college students were asked to state the number of hours they slept last night. The resulting data are 5, 6, 6, 8, 7, 7, 9, 5, 4, 8, 11, 6, 7, 8, 7. Find the following:

a. variance s^2, using formula (2.6)
b. variance s^2, using formula (2.10)
c. standard deviation, s

2.87 Ex02-067 The article "More hours in class logged by college profs" in *The Blade* (January 23, 2000) reported on a study done at the four-year colleges in Ohio about the total hours spent by college professors in instructing and advising students. The times reported for the 13 schools were:

28.4	44.4	36.4	33.5	34.9	30.5	32.7
24.6	26.9	34.7	33.4	29.3	33.3	

a. Find the range.
b. Find the variance.
c. Find the standard deviation.

2.88 Adding (or subtracting) the same number from each value in a set of data does not affect the measures of variability for that set of data.

a. Find the variance of this set of annual heating-degree-day data: 6017, 6173, 6275, 6350, 6001, 6300.
b. Find the variance of this set of data (obtained by subtracting 6000 from each value in part a): 17, 173, 275, 350, 1, 300.

2.89 Ex02-089 Recruits for a police academy were required to undergo a test that measures their exercise capacity. The exercise capacity (in minutes) was obtained for each of 20 recruits:

25	27	30	33	30	32	30	34	30	27
26	25	29	31	31	32	34	32	33	30

a. Draw a dotplot of the data.
b. Find the mean.
c. Find the range.
d. Find the variance.

e. Find the standard deviation.
f. Using the dotplot from part a, draw a line representing the range. Then draw a line starting at the mean with a length that represents the value of the standard deviation.
g. Describe how the distribution of data, the range, and the standard deviation are related.

2.90 Ex02-090 Since 1981, *Fortune* magazine has been tracking what it judges to be the "best 100 companies to work for." The companies must be at least ten years old and employ no fewer than 500 people. The list is the top 25 in 1998, together with each company's percentage of women employees, percentage of job growth over a two-year span, and number of hours of professional training required each year by the employer.

Company	Women (%)	Job Growth (%)	Training (hr/yr)
Southwest Airlines	55	26	15
Kingston Technology	48	54	100
SAS Institute	53	34	32
FEL-Pro	36	10	60
TDIndustries	10	31	40
MBNA	58	48	48
W. L. Gore	43	26	27
Microsoft	29	22	8
Merck	52	24	40
Hewlett-Packard	37	10	—
Synovus Financial	65	23	13
Goldman Sachs	40	13	20
MOOG	19	17	25
DeLoitte & Touche	45	23	70
Corning	38	9	80
Wegmans Food Products	54	3	30
Harley-Davidson	22	15	50
Federal Express	32	11	40
Procter & Gamble	40	1	25
Peoplesoft	44	122	—
First Tennessee Bank	70	1	60
J. M. Smucker	48	1	24
Granite Rock	17	29	43
Patagonia	52	−5	62
Cisco Systems	25	189	80

Source: *Fortune*, "The 100 Best Companies to Work For in America," January 12, 1998, pp. 84–86

a. Find the mean, range, variance, and standard deviation for each of the three variables shown in the list. Present your results in a table.
b. Using your results from part a, compare the distributions for job growth percentage and percentage of women employed. What can you conclude?

2.91 Ex02-068 "America West leads monthly on-time arrivals" was an article in *USA Today* (February 12, 2002) about overall flight on-time rates. For December 2001 the on-time arrival rates of domestic flights at the 25 largest U.S. airports were as follows:

Atlanta	79.7
Baltimore	82.5
Boston	86.5
Charlotte	82.3
Chicago (O'Hare)	78.4
Dallas/Fort Worth	84.9
Denver	81.5
Detroit	81.4
Houston	82.8
Las Vegas	76.9
Los Angeles	81.0
Miami	76.0
Minneapolis/St. Paul	82.5
New York (LaGuardia)	89.7
New York (Newark)	82.7
Orlando	82.2
Philadelphia	77.8
Phoenix	82.3
Pittsburgh	84.0
Salt Lake City	74.2
San Diego	78.4
San Francisco	64.6
Seattle/Tacoma	69.8
St. Louis	80.0
Tampa	79.9

a. Find the range and the standard deviation for the on-time arrival rates.
b. Draw lines on the stem-and-leaf diagram you drew in Exercise 2.68 (p. 75) that represent the range and standard deviation. Remember that the standard deviation is a measure of the spread about the mean.
c. Describe the relationship among the distribution of the data, the range, and the standard deviation.

2.92 Ex02-092 The *Financial Times* publishes the midday temperatures at well-known cities and nations throughout the world. The table at the top of page 86 lists the midday temperatures (in °F) on August 4, 1998, in 100 world locations. Find the mean, range, variance, and standard deviation for the temperatures of the cities on the list. (Retain the solutions to use in Exercise 2.128, p. 105.)

Location	°F	Location	°F	Location	°F	Location	°F
Abu Dhabi	115	Cologne	73	Lisbon	99	Paris	75
Accra	82	Dakar	88	London	73	Perth	63
Algiers	90	Dallas	98	Los Angeles	81	Prague	75
Amsterdam	68	Delhi	86	Luxembourg	72	Rangoon	88
Athens	90	Dubai	111	Lyon	75	Reykjavik	55
Atlanta	90	Dublin	70	Madeira	81	Rio de Janeiro	79
Bangkok	91	Dubrovnik	95	Madrid	93	Rome	90
Barcelona	86	Edinburgh	78	Majorca	86	San Francisco	80
Beijing	86	Faro	91	Malta	93	Seoul	90
Belfast	70	Frankfurt	77	Manchester	68	Singapore	88
Belgrade	99	Geneva	77	Manila	91	Stockholm	68
Berlin	77	Gibraltar	81	Melbourne	61	Strasbourg	77
Bermuda	88	Glasgow	63	Mexico City	86	Sidney	66
Birmingham	70	Hamburg	68	Miami	92	Tangiers	86
Bogota	68	Helsinki	68	Milan	88	Tel Aviv	91
Bombay	90	Hong Kong	91	Montreal	82	Tokyo	86
Brussels	70	Honolulu	90	Moscow	79	Toronto	83
Budapest	97	Istanbul	95	Munich	75	Vancouver	76
Buenos Aires	59	Jacarta	90	Nairobi	73	Venice	91
Copenhagen	68	Jersey	66	Naples	93	Vienna	91
Cairo	99	Johannesburg	59	Nassau	99	Warsaw	77
Caracas	90	Karachi	99	New York	84	Washington, DC	86
Cardiff	68	Kuwait	120	Nice	82	Wellington	54
Casablanca	84	Las Palmas	82	Nicosia	93	Winnipeg	79
Chicago	79	Lima	66	Oslo	70	Zurich	73

Source: *Financial Times*, August 4, 1998, p. 16

2.93 Consider these two sets of data:

Set 1:	46	55	50	47	52
Set 2:	30	55	65	47	53

Both sets have the same mean, 50. Compare these measures for both sets: $\Sigma(x - \bar{x})$, $\Sigma|x - \bar{x}|$, SS(x), and range. Comment on the meaning of these comparisons.

2.94 Comment on the statement: "The mean loss for customers at First State Bank (which was not insured) was \$150. The standard deviation of the losses was −\$125."

2.95 Start with $x = 100$ and add four x values to make a sample of five data such that:
a. $s = 0$ b. $0 < s < 1$ c. $5 < s < 10$ d. $20 < s < 30$

2.96 Each of two samples has a standard deviation of 5. If the two sets of data are made into one set of ten data, will the new sample have a standard deviation that is less than, about the same as, or greater than the original standard deviation of 5? Make up two sets of five data, each with a standard deviation of 5, to justify your answer. Include the calculations.

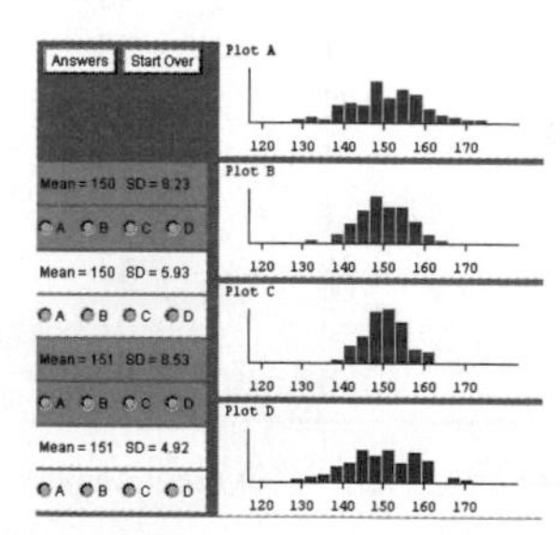

2.97 Interactivity Interactivity 2-F matches means and standard deviations with corresponding histograms. After several practice rounds using 'Start Over,' explain your method of matching.

2.5 Mean and Standard Deviation of Frequency Distribution

When the sample data are in the form of a frequency distribution, we need to make a slight adaptation to formulas (2.1) and (2.10) in order to find the mean, the variance, and the standard deviation.

Illustration 2.10 Calculations Using a Frequency Distribution

TABLE 2.13

Ungrouped Frequency Distribution

x	f
1	5
2	9
3	8
4	6
	$\Sigma f = 28$

Find the mean, the variance, and the standard deviation for the sample data represented by the frequency distribution in Table 2.13.

NOTE This frequency distribution represents a sample of 28 values: five 1's, nine 2's, eight 3's, and six 4's.

In order to calculate the sample mean $\bar{x}$ and the sample variance s^2 using formulas (2.1) and (2.10), we need the sum of the 28 x values, Σx, and the sum of the 28 x-squared values, Σx^2.

The summations, Σx and Σx^2, could be found as follows:

$$\Sigma x = \underbrace{1 + 1 + \cdots + 1}_{\text{5 of them}} + \underbrace{2 + 2 + \cdots + 2}_{\text{9 of them}} + \underbrace{3 + 3 + \cdots + 3}_{\text{8 of them}} + \underbrace{4 + 4 + \cdots + 4}_{\text{6 of them}}$$

$$= (5)(1) + (9)(2) + (8)(3) + (6)(4)$$

$$= 5 + 18 + 24 + 24 = \mathbf{71}$$

$$\Sigma x^2 = \underbrace{1^2 + \cdots + 1^2}_{\text{5 of them}} + \underbrace{2^2 + \cdots + 2^2}_{\text{9 of them}} + \underbrace{3^2 + \cdots + 3^2}_{\text{8 of them}} + \underbrace{4^2 + \cdots + 4^2}_{\text{6 of them}}$$

$$= (5)(1) + (9)(4) + (8)(9) + (6)(16)$$

$$= 5 + 36 + 72 + 96 = \mathbf{209}$$

However, we will use the frequency distribution to determine these summations by expanding it to become an *extensions table*. The extensions xf and x^2f are formed by multiplying across the columns row by row and then adding to find three column totals. The objective of the extensions table is to obtain these three column totals. (See Table 2.14.)

TABLE 2.14 Ungrouped Frequency Distribution: Extensions xf and x^2f

x	f	xf	x^2f
1	5	5	5
2	9	18	36
3	8	24	72
4	6	24	96
	$\Sigma f = 28$	$\Sigma xf = 71$	$\Sigma x^2f = 209$

number of data (→ Σf); *sum of x, using frequencies* (→ Σxf); *sum of x^2, using frequencies* (→ Σx^2f)

NOTES:

1. The extensions in the xf column are the subtotals of the like x values.
2. The extensions in the x^2f column are the subtotals of the like x-squared values.
3. The three column totals, Σf, Σxf, and Σx^2f, are the values previously known as n, Σx, and Σx^2, respectively. That is, $\Sigma f = n$, the number of pieces of data; $\Sigma xf = \Sigma x$, the sum of the data; and $\Sigma x^2f = \Sigma x^2$, the sum of the squared data.

4. Think of the f in the summation expressions Σxf and $\Sigma x^2 f$ as an indication that the sums were obtained with the use of a frequency distribution.
5. The sum of the x column is NOT a meaningful number. The x column lists each possible value of x once, which does not account for the repeated values.

ANSWER NOW 2.98
A survey asked for the "number of telephones" per household, x. The results are shown as a frequency distribution.

x	f
0	1
1	3
2	8
3	5
4	3

a. **Complete the extensions table.**
b. **Find the three summations, Σf, Σxf, and $\Sigma x^2 f$, for the frequency distribution.**
c. **Describe what each of the following represents: $x = 4$, $f = 8$, Σf, Σxf.**

ANSWER NOW 2.99
Explain each statement.

a. **The "sum of the x column" has no relationship to the "sum of the data."**
b. **Σxf represents the "sum of the data" represented by the frequency distribution in Answer Now 2.98.**

To find the **mean** of a frequency distribution, we modify formula (2.1) on page 68 to indicate the use of the frequency distribution:

mean of frequency distribution: $$x\ bar = \frac{\text{sum of all x, using frequencies}}{\text{number using frequencies}}$$

$$\bar{x} = \frac{\Sigma xf}{\Sigma f} \tag{2.11}$$

The mean value of x for the frequency distribution in Table 2.14 is found by using formula (2.11):

mean: $$\bar{x} = \frac{\Sigma xf}{\Sigma f} = \frac{71}{28} = 2.536 = \mathbf{2.5}$$

FYI See Animated Tutorial Frequency Distribution on CD for assistance with these calculations.

ANSWER NOW 2.100
Find the mean of the data shown in the frequency distribution in Answer Now 2.98.

To find the **variance** of the frequency distribution, we modify formula (2.10) on page 81 to indicate the use of the frequency distribution:

$$\text{s-squared} = \frac{(\text{sum of } x^2\text{, using frequencies}) - \left[\frac{(\text{sum of x, using frequencies})^2}{\text{number, using frequencies}}\right]}{\text{number, using frequencies} - 1}$$

$$s^2 = \frac{\Sigma x^2 f - \frac{(\Sigma xf)^2}{\Sigma f}}{\Sigma f - 1} \tag{2.12}$$

The variance of x for the frequency distribution in Table 2.14 is found by using formula (2.12):

$$\text{variance:} \quad s^2 = \frac{\Sigma x^2 f - \frac{(\Sigma xf)^2}{\Sigma f}}{\Sigma f - 1} = \frac{209 - \frac{(71)^2}{28}}{28 - 1} = \frac{28.964}{27} = 1.073 = \mathbf{1.1}$$

ANSWER NOW 2.101
Find the variance for the data shown in the frequency distribution in Answer Now 2.98.

The **standard deviation** of x for the frequency distribution in Table 2.14 is found by using formula (2.7), the positive square root of variance.

$$\text{standard deviation:} \quad s = \sqrt{s^2} = \sqrt{1.073} = 1.036 = \mathbf{1.0}$$ ■

ANSWER NOW 2.102
Find the standard deviation for the data shown in the frequency distribution in Answer Now 2.98.

Illustration 2.11 Calculations Using Grouped Frequencies

Find the mean, variance, and standard deviation of the sample of 50 exam scores using the grouped frequency distribution in Table 2.7 (p. 56).

SOLUTION
We will use an extensions table to find the three summations in the same manner we did in Illustration 2.10. The class midpoints will be used as the representative values for the classes.

The mean value of x for the frequency distribution in Table 2.15 is found by using formula (2.11):

$$\text{mean:} \quad \bar{x} = \frac{\Sigma xf}{\Sigma f} = \frac{3780}{50} = \mathbf{75.6}$$

The variance of x for the frequency distribution in Table 2.15 is found by using formula (2.12):

$$\text{variance:} \quad s^2 = \frac{\Sigma x^2 f - \frac{(\Sigma xf)^2}{\Sigma f}}{\Sigma f - 1} = \frac{296{,}600 - \frac{3780^2}{50}}{50 - 1} = \frac{10{,}832}{49}$$
$$= 221.0612 = \mathbf{221.1}$$

FYI
See Animated Tutorial Frequency Distribution on CD for assistance with these calculations.

TABLE 2.15 Frequency Distribution of 50 Exam Scores

Class Number	Class Midpoints, x	f	xf	x^2f
1	40	2	80	3,200
2	50	2	100	5,000
3	60	7	420	25,200
4	70	13	910	63,700
5	80	11	880	70,400
6	90	11	990	89,100
7	100	4	400	40,000
		$\Sigma f = 50$	$\Sigma xf = 3780$	$\Sigma x^2f = 296{,}600$

The standard deviation of x for the frequency distribution in Table 2.15 is found by using formula (2.7):

$$\text{standard deviation:} \quad s = \sqrt{s^2} = \sqrt{221.0612} = 14.868 = \mathbf{14.9} \quad \blacksquare$$

ANSWER NOW 2.103
Find the mean, variance, and standard deviation of the data in this frequency distribution.

Class	f
2–6	2
6–10	10
10–14	12
14–18	9
18–22	7

Technology Instructions: Frequency Distribution Statistics

MINITAB (Release 13)

Input the class midpoints or data values into C1 and the corresponding frequencies into C2; then continue with the following commands to obtain the extensions table:

```
Choose:   Calc > Calculator...
Enter:    Store result in variable: C3
          Expression: C1*C2 > OK
Repeat the above commands, replacing the variable with C4
and the expression with C1*C3.
Choose:   Calc > Column Statistics
Select:   Sum
Enter:    Input variable: C2
          Store result in: K1 > OK
Repeat above 'sum' commands, replacing variable with C3 and
result with K2.
Repeat above 'sum' commands, replacing variable with C4 and
result with K3.
Choose:   Manip > Display data
Enter:    Columns to display: C1-C4 K1-K3
```

To find the mean, variance, and standard deviation, respectively, continue with:

```
Choose:   Calc > Calculator
Enter:    Store result in variable: K4
          Expression: K2/K1
Repeat above 'mean' commands, replacing variable with K5
and expression with (K3-(K2**2/K1))/(K1-1).
Repeat above 'mean' commands, replacing variable with K6
and expression with SQRT(K5)  (select square root from
functions).
Choose:   Manip > Display data
Enter:    Columns to display: K4-K6
```

Excel XP

Input the class midpoints or data values into column A and the corresponding frequencies into column B; activate C1 or C2 (depending on whether column headings are used or not); then continue with the following commands to obtain the extensions table:

```
Enter:     = A2*B2 (if column headings are used)
Drag:      Bottom right corner of C2 down to give other
           products
Activate D2 and repeat above commands, replacing the for-
mula with = A2*C2.
Activate the data in columns B, C, and D.
Choose:    AutoSum  (sums will appear at the bottom of the
           columns)
```

To find the mean, activate E2; then continue with:

```
Enter:     = (column C total/column B total) (ex.  = C9/B9)
```

To find the variance, activate E3 and repeat above 'mean' commands, replacing the formula with = (D9-(C9^2/B9))/(B9-1).
To find the standard deviation, activate E4 and repeat above 'mean' commands, replacing the formula with = SQRT(E3).

TI-83 Plus

Input the class midpoints or data values into L1 and the frequencies into L2; then continue with:

```
Highlight:  L3
Enter:      L3 = L1*L2
Highlight:  L4
Enter:      L4 = L1*L3
Highlight:  L5(1)  (first position in L5 column)
Enter:      L5(1) = sum(L2)    [Σf]
            [sum = 2nd LIST > MATH > 5:sum(]
            L5(2) = sum(L3)    [Σxf]
            L5(3) = sum(L4)    [Σx²f]
            L5(4) = L5(2)/L5(1) [to find mean]
            L5(5) = (L5(3)-((L5(2))²/L5(1)))/(L5(1)-1)
            [to find variance]
            L5(6) = 2nd √ (L5(5))
            [to find standard deviation]
```

If the extensions table is not needed, just use:

```
Choose:    STAT  > CALC  > 1:1-VAR STATS
Enter:     L1, L2
```

Exercises

2.104 Pediatric dentists say a child's first dental exam should occur between ages 6 months and 1 year. The ages at first dental exam for a sample of children are shown in the distribution:

Age at First Dental Exam, x	1	2	3	4	5
Number of Children, f	9	11	23	16	21

a. Find the mean age of first dental exam for these children.
b. Find the median age.
c. Find the standard deviation.

2.105 A survey of medical doctors asked the number of children each had fathered. The results are summarized by this ungrouped frequency distribution:

Number of Children	0	1	2	3	4	6
Number of Doctors	15	12	26	14	4	2

Calculate the sample mean, variance, and standard deviation for the number of children the doctors had fathered.

2.106 The weight gains (in grams) for chicks fed on a high-protein diet were as follows:

Weight Gain	12.5	12.7	13.0	13.1	13.2	13.8
Frequency	2	6	22	29	12	4

a. Find the mean.
b. Find the variance.
c. Find the standard deviation.

2.107 Find the mean, variance, and standard deviation for this grouped frequency distribution:

Class Boundaries	f
3–6	2
6–9	10
9–12	12
12–15	9
15–18	7

2.108 Find the mean and the variance for this grouped frequency distribution:

Class Boundaries	f
2–6	7
6–10	15
10–14	22
14–18	14
18–22	2

2.109 The following distribution of commuting distances was obtained for a sample of Mutual of Nebraska employees:

Distance (miles)	Frequency
1.0–3.0	2
3.0–5.0	6
5.0–7.0	12
7.0–9.0	50
9.0–11.0	35
11.0–13.0	15
13.0–15.0	5

Find the mean and the standard deviation for the commuting distances.

2.110 A quality-control technician selected 25 1-pound boxes from a production process and found the following distribution of weights (in ounces):

Weight	Frequency
15.95–15.98	2
15.98–16.01	4
16.01–16.04	15
16.04–16.07	3
16.07–16.10	1

Find the mean and the standard deviation for this weight distribution.

2.111 It has been found that 35.2 million Americans 16 years and older fish our waters. A sample of freshwater fishermen produced the following age distribution:

Age of Fishermen, x	15–25	25–35	35–45	45–55	55–65	65–75
Number of Fishermen, f	13	20	28	20	10	9

Find the mean and the standard deviation for this distribution.

2.112 Private industry reported that more than 31,000 workers were absent from work in 1995 because of carpal tunnel syndrome (a nerve disorder causing arm, wrist, and hand pain). The length of time (in days) workers are absent due to this problem varies greatly.

Days of Absence, x	0–10	10–20	20–30	30–40	40–50
Number of Workers, f	37	24	38	32	27

Find the mean and the standard deviation for this distribution.

2.113 Ex02-113 The California Department of Education issues a yearly report on SAT scores for students in various school districts. The frequency table shows SAT verbal test results for 2000–2001 for Fresno County school districts.

District	Number Tested	Verbal Average
Caruthers Unified	23	441
Central Unified	172	437
Clovis Unified	976	514
Coalinga/Huron Joint Unified	31	501
Firebaugh–Las Deltas Joint Unified	31	420
Fowler Unified	35	439
Fresno Unified	1169	454
Golden Plains Unified	37	367
Kerman Unified	68	420
Kings Canyon Joint Unified	144	459
Kingsburg Joint Union High	76	491
Mendota Unified	34	369
Parlier Unified	18	367
Riverdale Joint Unified	45	427
Sanger Unified	107	482
Selma Unified	76	458
Sierra Unified	106	520
Washington Union High	51	429

a. What do the entries 23 and 441 for Caruthers Unified mean?
b. What is the total of all student scores at Caruthers Unified?
c. How many student test results are shown on this table?
d. What is the total of all student scores shown on the table?
e. Find the mean SAT verbal test results.

2.114 Ex02-092
a. Using the worldwide midday temperature readings listed in Exercise 2.92 (page 86), construct a grouped frequency distribution using 10°F class boundaries of 51, 61, 71, ..., 121. Show the class midpoints and the associated frequency counts in your table.
b. Find the mean, variance, and standard deviation of the grouped frequency distribution.
c. Compare these values with the corresponding ungrouped statistics using the percent error in each case, and present all your results in a table.

2.115 Ex02-046 The USA Snapshot® "Nuns an aging order" reports that the median age of 94,022 American Roman Catholic nuns is 65 years and the percentage of U.S. nuns by age group is:

Under 50	51–70	Over 70	Refused to give age
16%	42%	37%	5%

This information is based on a survey of 1049 Roman Catholic nuns. Suppose the survey had resulted in the following frequency distribution (52 ages unknown):

Age	20–30	30–40	40–50	50–60	60–70	70–80	80–90
Frequency	34	58	76	187	254	241	147

(See the histogram drawn in Exercise 2.46, p. 67.)
a. Find the mean, median, mode, and midrange for this distribution of ages.
b. Find the variance and standard deviation.

2.116 The amount of money adults say they will spend on gifts this holiday was described in "What 'Santa' Will Spend" (*USA Today*, November 23, 1994):

Amount	Nothing	\$1–\$300	\$301–\$600	\$601–\$1000	Over \$1000	Didn't know
Percent Who Said	1%	24%	30%	20%	14%	11%

Average: \$734

The following distribution represents that part of the sample who did know:

Amount, x	0	150	450	800	1500
Frequency, f	1	24	30	20	14

a. Find the mean of the frequency distribution.
b. Do you believe the average reported could have been the mean? Explain.
c. Find the median of the frequency distribution.
d. Do you believe the average reported could have been the median? Explain.
e. Find the mode of the frequency distribution.
f. Do you believe the average reported could have been the mode? Explain.
g. Could the average reported have been the midrange? If so, what was the largest amount of money reported?

2.117 Ex02-117 A Senior PGA professional golfer is not expected to play in all the available tournaments during a season. The numbers of tournaments played by each of the 2001 tour's top 50 money leaders are listed here.

a. Construct a grouped frequency distribution of the number of tournaments played using group intervals 9–11, 11–13, . . . 21–23, the class midpoints, and the associated frequency counts.

b. Find the mean, variance, and standard deviation of the number of tournaments played both with and without using the grouped distribution.

c. Compare the two sets of answers you obtained in part b. What percent is the error in each case?

Senior PGA Player	Tournaments	Senior PGA Player	Tournaments
Allen Doyle	34	John Bland	30
Bruce Fleisher	31	John Schroeder	28
Hale Irwin	26	Stewart Ginn	28
Larry Nelson	28	Terry Mauney	31
Gil Morgan	24	Hugh Baiocchi	34
Jim Thorpe	35	Isao Aoki	20
Doug Tewell	28	Graham Marsh	31
Bob Gilder	30	Bobby Wadkins	10
Dana Quigley	37	Ray Floyd	14
Tom Kite	23	Steven Veriato	31
Walter Hall	35	Dave Stockton	24
Mike McCullough	35	Jay Sigel	18
Ed Dougherty	36	J.C. Snead	26
Jose Maria Canizares	30	Bobby Walzel	25
Tom Jenkins	36	Bob Eastwood	29
Bruce Lietzke	10	Dave Eichelberger	31
Tom Watson	13	Joe Inman	34
Sammy Rachels	27	John Mahaffey	25
Jim Colbert	29	Howard Twitty	29
Bruce Summerhays	34	Mike Smith	31
Leonard Thompson	31	Walter Morgan	34
Vincente Fernandez	29	Ted Goin	30
Gary McCord	20	Andy North	24
Jim Ahern	31	Jim Holtgrieve	32
John Jacobs	36	Mike Hill	16

Source: *bestcourses.com*

2.6 Measures of Position

Measures of position are used to describe the position a specific data value possesses in relation to the rest of the data. *Quartiles* and *percentiles* are two of the most popular measures of position.

QUARTILES

Values of the variable that divide the ranked data into quarters; each set of data has three quartiles. The *first quartile*, Q_1, is a number such that at most 25% of the data are smaller in value than Q_1 and at most 75% are larger. The *second quartile* is the median. The *third quartile*, Q_3, is a number such that at most 75% of the data are smaller in value than Q_3 and at most 25% are larger. (See Figure 2.24.)

FIGURE 2.24
Quartiles

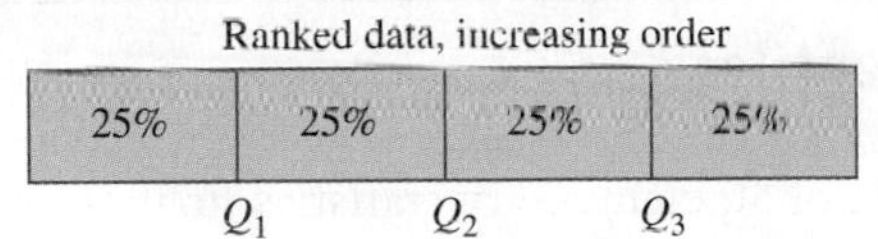

The procedure for determining the values of the quartiles is the same as that for percentiles and is shown in the following description of *percentiles*.

PERCENTILES
Values of the variable that divide a set of ranked data into 100 equal subsets; each set of data has 99 percentiles (see Figure 2.25). The *k*th percentile, P_k, is a value such that at most *k*% of the data are smaller in value than P_k and at most $(100 - k)$% of the data are larger (see Figure 2.26).

FIGURE 2.25
Percentiles

FIGURE 2.26
***k*th Percentile**

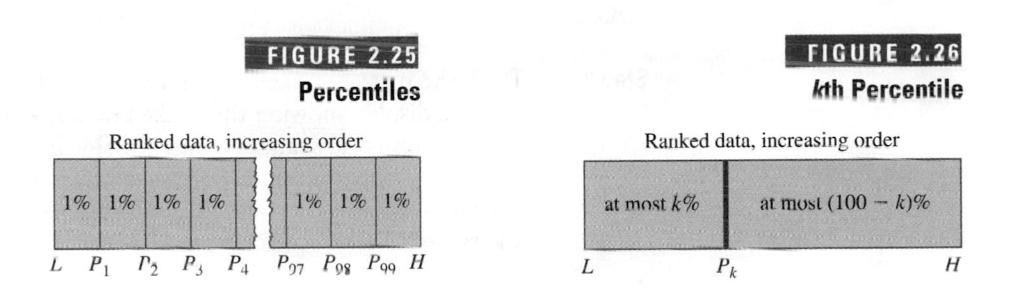

NOTES

1. The first quartile and the 25th percentile are the same; that is, $Q_1 = P_{25}$. Also, $Q_3 = P_{75}$.
2. The median, the second quartile, and the 50th percentile are all the same: $\tilde{x} = Q_2 = P_{50}$. Therefore, when asked to find P_{50} or Q_2, use the procedure for finding the median.

The procedure for determining the value of any *k*th percentile (or quartile) involves four basic steps as outlined on the diagram in Figure 2.27. Illustration 2.12 demonstrates the procedure.

FIGURE 2.27
Finding P_k Procedure

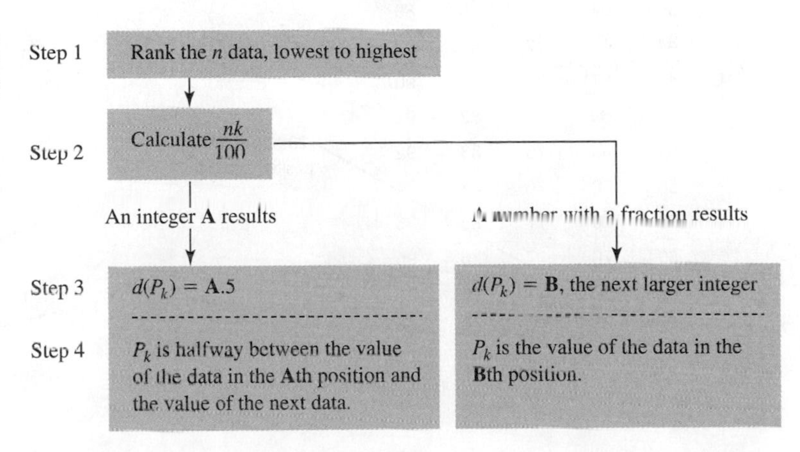

Illustration 2.12 Finding Percentiles

Using the sample of 50 elementary statistics final exam scores listed in Table 2.16, find the first quartile Q_1, the 58th percentile P_{58}, and the third quartile Q_3.

TABLE 2.16 Raw Scores for Elementary Statistics Exam

60	47	82	95	88	72	67	66	68	98
90	77	86	58	64	95	74	72	88	74
77	39	90	63	68	97	70	64	70	70
58	78	89	44	55	85	82	83	72	77
72	86	50	94	92	80	91	75	76	78

SOLUTION

Step 1 Rank the data: A ranked list may be formulated (see Table 2.17), or a graphic display showing the ranked data may be used. The dotplot and the stem-and-leaf are handy for this purpose. The stem-and-leaf is especially helpful, since it gives depth numbers counted from both extremes when it is computer generated (see Figure 2.28). Step 1 is the same for all three statistics.

ANSWER NOW 2.118
Using the concept of depth, describe the position of 91 in the set of 50 exam scores in two different ways.

Find Q_1:

Step 2 Find $\frac{nk}{100}$: $\frac{nk}{100} = \frac{(50)(25)}{100} = \mathbf{12.5}$

($n = 50$ and $k = 25$, since $Q_1 = P_{25}$.)

TABLE 2.17 Ranked Data: Exam Scores

39	64	72	78	89
44	66	72	80	90
47	67	74	82	90
50	68	74	82	91
55	68	75	83	92
58	70	76	85	94
58	70	77	86	95
60	70	77	86	95
63	72	77	88	97
64	72	78	88	98

13th position from *L*

29th and 30th positions from *L*

13th position from *H*

FIGURE 2.28
Final Exam Scores

Stem-and-leaf of score *N* 50
Leaf Unit = 1.0

1	3	9
2	4	4
3	4	7
4	5	0
7	5	588
11	6	0344
15	6	6788
24	7	000222244
(7)	7	5677788
19	8	0223
15	8	566889
9	9	00124
4	9	5578

Step 3 Find the depth of Q_1: $d(Q_1) = \mathbf{13}$ (Since 12.5 contains a fraction, **B** is the next larger integer, 13.)

Step 4 Find Q_1: Q_1 is the 13th value, counting from L (see Table 2.17 or Figure 2.28), $Q_1 = \mathbf{67}$

Find P_{58}:

Step 2 Find $\frac{nk}{100}$: $\frac{nk}{100} = \frac{(50)(58)}{100} = \mathbf{29}$ ($n = 50$ and $k = 58$ for P_{58}.)

Step 3 Find the depth of P_{58}: $d(P_{58}) = \mathbf{29.5}$ (Since **A** = 29, an integer, add 0.5 and use 29.5.)

Step 4 Find P_{58}: P_{58} is the value halfway between the values of the 29th and the 30th pieces of data, counting from L (see Table 2.17 or Figure 2.28), so

$$P_{58} = \frac{77 + 78}{2} = \mathbf{77.5}$$

FYI See Animated Tutorial Percentile on CD for assistance with this calculation.

Therefore, it can be stated that "at most 58% of the exam grades are smaller in value than 77.5." This is also equivalent to stating that "at most 42% of the exam grades are larger in value than 77.5."

ANSWER NOW 2.119
Find P_{20} and P_{35} for the exam scores in Table 2.17.

Optional technique: When k is greater than 50, subtract k from 100 and use $(100 - k)$ in place of k in Step 2. The depth is then counted from the largest-valued data H.

Find Q_3 using the optional technique:

Step 2 Find $\frac{nk}{100}$: $\frac{nk}{100} = \frac{(50)(25)}{100} = \mathbf{12.5}$ ($n = 50$ and $k = 75$, since $Q_3 = P_{75}$, and $k > 50$; use $100 - k = 100 - 75 = 25$.)

Step 3 Find the depth of Q_3 from H: $d(Q_3) = \mathbf{13}$

Step 4 Find Q_3: Q_3 is the 13th value, counting from H (see Table 2.17 or Figure 2.28), $Q_3 = \mathbf{86}$

Therefore, it can be stated that "at most 75% of the exam grades are smaller in value than 86." This is also equivalent to stating that "at most 25% of the exam grades are larger in value than 86." ■

ANSWER NOW 2.120
Find P_{80} and P_{95} for the exam scores in Table 2.17.

An additional measure of central tendency, the *midquartile*, can now be defined.

MIDQUARTILE
The numerical value midway between the first quartile and the third quartile.

$$\text{midquartile} = \frac{Q_1 + Q_3}{2} \tag{2.13}$$

Illustration 2.13 Finding the Midquartile

Find the midquartile for the set of 50 exam scores given in Illustration 2.12.

SOLUTION

$Q_1 = 67$ and $Q_3 = 86$, as found in Illustration 2.12. Thus,

$$\text{midquartile} = \frac{Q_1 + Q_3}{2} = \frac{67 + 86}{2} = \mathbf{76.5}$$

■

The median, the midrange, and the midquartile are not necessarily the same value. Each is the middle value, but by different definitions of "middle." Figure 2.29 summarizes the relationship of these three statistics as applied to the 50 exam scores from Illustration 2.12.

FIGURE 2.29
Final Exam Scores

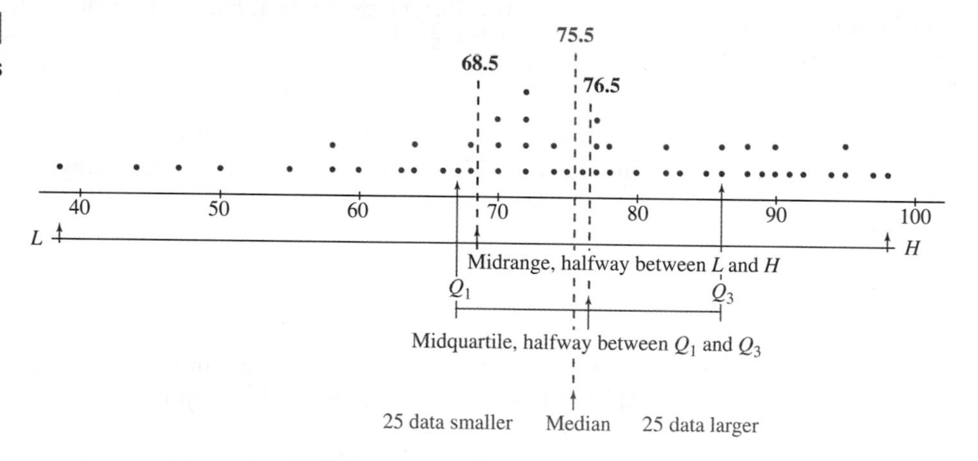

ANSWER NOW 2.121
What property does the distribution need for the median, the midrange, and the midquartile to all be the same value?

A *5-number summary* is very effective in describing a set of data. It is easy information to obtain and is very informative to the reader.

5-NUMBER SUMMARY
The 5-number summary is composed of:

1. L, the smallest value in the data set,
2. Q_1, the first quartile (also called P_{25}, the 25th percentile),
3. $\tilde{x}$, the median,
4. Q_3, the third quartile (also called P_{75}, the 75th percentile), and
5. H, the largest value in the data set.

The 5-number summary for the set of 50 exam scores in Illustration 2.12 is

39	67	75.5	86	98
L	Q_1	$\tilde{x}$	Q_3	H

Notice that these five numerical values divide the set of data into four subsets, with one-quarter of the data in each subset. From the 5-number summary we can observe how much the data are spread out in each of the quarters. We can now define an additional measure of dispersion.

INTERQUARTILE RANGE
The difference between the first and third quartiles. It is the range of the middle 50% of the data.

The 5-number summary is even more informative when it is displayed on a diagram drawn to scale. A computer-generated graphic display that accomplishes this is known as the *box-and-whiskers display*.

BOX-AND-WHISKERS DISPLAY
A graphic representation of the 5-number summary. The five numerical values (smallest, first quartile, median, third quartile, and largest) are located on a scale, either vertical or horizontal. The box is used to depict the middle half of the data that lies between the two quartiles. The whiskers are line segments used to depict the other half of the data: One line segment represents the quarter of the data that is smaller in value than the first quartile, and a second line segment represents the quarter of the data that is larger in value than the third quartile.

Figure 2.30 is a box-and-whiskers display of the 50 exam scores.

FIGURE 2.30
Box-and-Whiskers

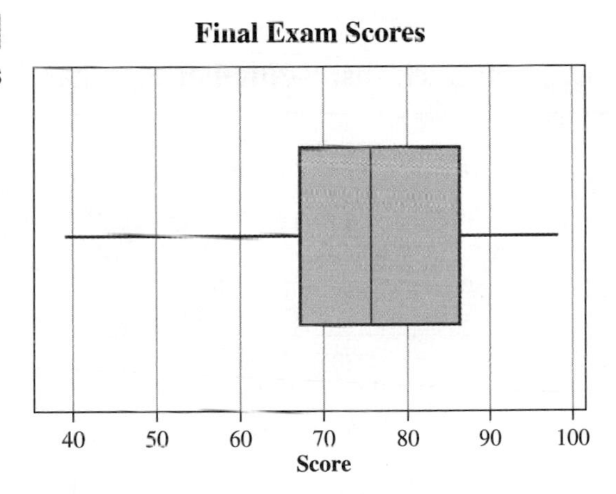

ANSWER NOW 2.122
Draw a box-and-whiskers display for the set of data with the 5-number summary 42–62–72–82–97.

Technology Instructions: Percentiles

MINITAB (Release 13)

Input the data into C1; then continue with:

```
Choose:    Manip > Sort...
Enter:     Sort column(s): C1
           Store sorted column(s) in: C2
           Sort by column: C1
```

A ranked list of data will be obtained in C2. Determine the depth position and locate the desired percentile.

(continued)

Technology Instructions: Percentiles (continued)

Excel XP

Input the data into column A and activate a cell for the answer; then continue with:

Choose: Insert Function, f_x > Statistical > PERCENTILE > OK
Enter: Array: (A2:A6 or select cells)
k: K (desired percentile; ex. .95, .47)

TI-83 Plus

Input the data into L1; then continue with:

Choose: STAT > EDIT > 2:SortA(
Enter: L1
Enter: percentile x sample size (ex. .25 × 100)
Based on product, determine the depth position; then continue with:
Enter: L1(depth position) > Enter

Technology Instructions: 5-Number Summary

MINITAB (Release 13)

Input the data into C1; then continue with:

Choose: Stat > Basic Statistics > Display Descriptive Statistics ...
Enter: Variables: C1

Excel XP

Input the data into column A; then continue with:

Choose: Tools > Data Analysis* > Descriptive Statistics > OK
Enter: Input Range: (A2:A6 or select cells)
Select: Labels in First Row (if necessary)
Output Range
Enter: (B1 or select cell)
Select: Summary Statistics > OK
To make output readable:
Choose: Format > Column > Autofit Selection

*If Data Analysis does not show on the Tools menu; see page 58.

TI-83 Plus

Input the data into L1; then continue with:

Choose: STAT > CALC > 1:1-VAR STATS
Enter: L1

Technology Instructions: Box-and-Whiskers Diagram

MINITAB (Release 13)

Input the data into C1; then continue with:

```
Choose:    Graph > Boxplot
Enter:     Graph 1, Y: C1
Optional:
Choose:    Annotations > Title
Enter:     your title    > OK
Choose:    Options
Select:    Transpose > OK
```

For multiple boxplots, enter additional set of data into C2; then do as above plus:

```
Enter:     Graph 2, Y: C2
Choose:    Frame > Multiple Graphs
Select:    Overlay graphs on the same page > OK
```

Excel XP

Input the data into column A; then continue with:

```
Choose:    Tools > Data Analysis Plus* > BoxPlot > OK
Enter:     (A2:A6 or select cells)
```

To edit the boxplot, review options shown with editing histograms on page 59.

*Data Analysis Plus is a collection of statistical macros for EXCEL. They can be downloaded onto your computer from your Student Suite CD.

TI-83 Plus

Input the data into L1; then continue with:

```
Choose:    2nd > STAT PLOT > 1:Plot1...
```

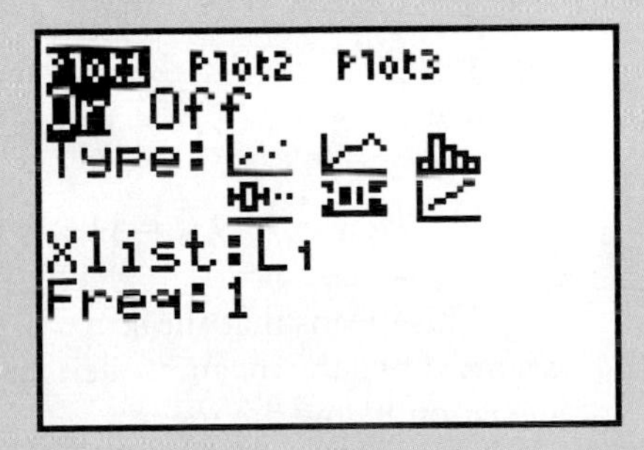


```
Choose:    ZOOM > 9:ZoomStat > TRACE >>>
```

If class midpoints are in L1 and frequencies are in L2, do as above except for:

```
Enter:     Freq: L2
```

Technology Instructions: Box-and-Whiskers Diagram (continued)

TI–83 Plus (continued)

For multiple boxplots, enter additional set of data into L2 or L3; do as above plus:

```
Choose: 2nd > STAT PLOT > 2:Plot2...
```

```
Plot1 Plot2 Plot3
On Off
Type:
Xlist:L3
Freq:1
```

The position of a specific value can be measured in terms of the mean and standard deviation using the *standard score*, commonly called the *z-score*.

STANDARD SCORE OR *z*-SCORE

The position a particular value of *x* has relative to the mean, measured in standard deviations. The *z*-score is found by the formula

$$z = \frac{\text{value} - \text{mean}}{\text{st. dev.}} = \frac{x - \bar{x}}{s} \qquad (2.14)$$

Illustration 2.14

Finding *z*-scores

Find the standard scores for (a) 92 and (b) 72 with respect to a sample of exam grades that have a mean score of 75.9 and a standard deviation of 11.1.

SOLUTION

a. $x = 92, \bar{x} = 75.9, s = 11.1$. Thus, $z = \frac{x - \bar{x}}{s} = \frac{92 - 75.9}{11.1} = \frac{16.1}{11.1} = \mathbf{1.45}$.

b. $x = 72, \bar{x} = 75.9; s = 11.1$. Thus, $z = \frac{x - \bar{x}}{s} = \frac{72 - 75.9}{11.1} = \frac{-3.9}{11.1} = \mathbf{-0.35}$.

This means that the score 92 is approximately one and one-half standard deviations above the mean, while the score 72 is approximately one-third of a standard deviation below the mean. ■

NOTES

1. Typically, the calculated value of z is rounded to the nearest hundredth.
2. z-scores typically range in value from approximately -3.00 to $+3.00$.

ANSWER NOW 2.123

Find the *z*-score for test scores of 92 and 63 on a test that has a mean of 72 and a standard deviation of 12.

Since the z-score is a measure of relative position with respect to the mean, it can be used to help us compare two raw scores that come from separate populations. For example, suppose you want to compare a grade you received on a test with a friend's grade on a comparable exam in her course. You received a raw score of 45 points; she got 72 points. Is her grade better? We need more information before we can draw a conclusion. Suppose the mean on the exam you took was 38 and the mean on her exam was 65. Your grades are both 7 points above the mean, but we still can't draw a definite conclusion. The standard deviation on the exam you took was 7 points, and it was 14 points on your friend's exam. This means that your score is one (1) standard deviation above the mean ($z = 1.0$), whereas your friend's grade is only one-half of a standard deviation above the mean ($z = 0.5$). Since your score has the "better" relative position, you conclude that your score is slightly better than your friend's score. (Again, this is speaking from a relative point of view.)

Technology Instructions: Additional Commands

MINITAB (Release 13)

Input the data into C1; then:
To sort the data into ascending order and store them in C2; continue with

```
Choose:   Manip > Sort
Enter:    Sort column(s): C1
          Store sorted column(s) in: C2
          Sort by column: C1
```

To form an ungrouped frequency distribution of integer data; continue with

```
Choose:   Stat > Tables > Tally
Enter:    Variables: C1
Select:   Counts
```

To print data on the session window; continue with

```
Choose:   Manip > Display Data
Enter:    Columns to display: C1 or C1 C2 or C1-C4
```

Excel XP

Input the data into column A; then continue with the following to sort the data:

```
Choose:   Data > Sort
Enter:    Sort by:  (A2:A6 or select cells)
Select:   Ascending  or  Descending
          Header row  or  No header row
```

TI-83 Plus

Input the data into L1; then continue with the following to sort the data:

```
Choose:   2nd > STAT > OPS > 1:SortA(
Enter:    L1
```

To form a frequency distribution of the data in L1; continue with:

```
Choose:   PRGM > EXEC > FREQDIST*
Enter:    L1 > ENTER
          LW BOUND =first lower class boundary
          UP BOUND =last upper class boundary
          WIDTH =class width (use 1 for ungrouped distribu-
          tion)
```

*Program 'FREQDIST' is one of many programs that are available for downloading from a Web site. See page 40 for specific instructions.

Technology Instructions: Generate Random Samples

MINITAB (Release 13)

The data will be put into C1:

```
Choose:     Calc > Random Data > {Normal, Uniform, etc.}
Enter:      Generate: K rows of data
            Store in column(s): C1
            Population parameters needed: (μ, σ, L, H, A,
            or B)
            (Required parameters will vary depending on the
            distribution)
```

Excel XP

```
Choose:     Tools > Data Analysis > Random Number Generation
            > OK
Enter:      Number of Variables: 1
            Number of Random Numbers: (desired quantity)
Select:     Distribution: Normal or others
Enter:      Parameters: (μ, σ, L, H, A, or B)
            (Required parameters will vary depending on the
            distribution.)
Select:     Output Range
Enter:      (A1 or select cell)
```

TI-83 Plus

```
Choose:     STAT > 1:EDIT
Highlight:  L1
Choose:     MATH > PRB > 6:randNorm(  or  5:randInt(
Enter:      μ, σ, # of trials  or  L, H, # of trials
```

Technology Instructions: Select Random Samples

MINITAB (Release 13)

The existing data to be selected from should be in C1; then continue with:

```
Choose: Calc > Random Data > Sample from Columns
Enter: Sample: K  rows from column(s): C1
       Store samples in: C2
Select: Sample with replacement (optional)
```

Excel XP

The existing data to be selected from should be in column A; then continue with:

```
Choose:   Tools > Data Analysis > Sampling > OK
Enter:    Input range: (A2:A10 or select cells)
Select:   Labels (optional)
          Random
            Enter: Number of Samples: K
          Output range:
            Enter: (B1 or select cell)
```

Exercises

2.124 Ex02-124 Fifteen countries were randomly selected from the *World Factbook 2001* list of world countries and the number of televisions per 1000 inhabitants for each country was recorded.

374	209	19	10	65
25	32	52	429	1
610	281	13	66	525

Source: http://www.outfo.org/almanac/world_factbook_01/

a. Find the first and the third quartiles for the number of televisions per 1000.
b. Find the midquartile.

2.125 Ex02-125 The following data are the yields (in pounds) of hops:

3.9	3.4	5.1	2.7	4.4	7.0	5.6	2.6	4.8	5.6
7.0	4.8	5.0	6.8	4.8	3.7	5.8	3.6	4.0	5.6

a. Find the first and the third quartiles of the yields.
b. Find the midquartile.
c. Find and explain the percentiles P_{15}, P_{33}, and P_{90}.

2.126 Ex02-126 Henry Cavendish, an English chemist and physicist (1731–1810), approached many of his experiments using quantitative measurements. He was the first to accurately measure the density of the Earth. Below are 29 measurements (ranked for your convenience) of the density of the earth done by Cavendish in 1798 using a torsion balance. Density is presented as a multiple of the density of water. (Measurements are in g/cm^3.)

4.88	5.07	5.10	5.26	5.27	5.29	5.29	5.30	5.34	5.34
5.36	5.39	5.42	5.44	5.46	5.47	5.50	5.53	5.55	5.57
5.58	5.61	5.62	5.63	5.65	5.68	5.75	5.79	5.85	

Source: The data and descriptive information are based on material from "Do robust estimators work with real data?" by Stephen M. Stigler, *Annals of Statistics* 5 (1977), 1055–1098.

a. Describe the data set by calculating the mean, median, and standard deviation.
b. Construct a histogram and explain how it demonstrates the values of the descriptive statistics in part a.
c. Find the 5-number summary.
d. Construct a box-and-whiskers display and explain how it demonstrates the values of the descriptive statistics in part c.
e. Based on the two graphs, what "shape" is this distribution of measurements?
f. Assuming that earth density measurements have an approximately normal distribution, approximately 95% of the data should fall within two standard deviations of the mean. Is this true?

2.127 Ex02-127 The U.S. Geological Survey collected atmospheric deposition data in the Rocky Mountains. Part of the sampling process was to determine the concentration of ammonium ions (in percentages). Here are the results from the 52 samples:

2.9	4.1	2.7	3.5	1.4	5.6	13.3	3.9	4.0
2.9	7.0	4.2	4.9	4.6	3.5	3.7	3.3	5.7
3.2	4.2	4.4	6.5	3.1	5.2	2.6	2.4	5.2
4.8	4.8	3.9	3.7	2.8	4.8	2.7	4.2	2.9
2.8	3.4	4.0	4.6	3.0	2.3	4.4	3.1	5.5
4.1	4.5	4.6	4.7	3.6	2.6	4.0		

Find:
a. Q_1
b. Q_2
c. Q_3
d. the midquartile
e. P_{30}
f. the 5-number summary
g. Draw the box-and-whiskers display.

2.128 Consider the worldwide temperature data shown in Exercise 2.92 pp. 85–86.
a. Omit names and build a 10 × 10 table of ranked temperatures in ascending order, reading vertically in each column.
b. Construct a 5-number summary table.
c. Find the midquartile temperature reading and the interquartile range.
d. What are the z-scores for Kuwait, Berlin, Washington, London, and Bangkok?

2.129 A sample has a mean of 50 and a standard deviation of 4.0. Find the z-score for each value of x:
a. $x = 54$ b. $x = 50$
c. $x = 59$ d. $x = 45$

2.130 An exam produced grades with a mean score of 74.2 and a standard deviation of 11.5. Find the z-score for each test score x:
a. $x = 54$ b. $x = 68$
c. $x = 79$ d. $x = 93$

2.131 A nationally administered test has a mean of 500 and a standard deviation of 100. If your standard score on this test was 1.8, what was your test score?

2.132 A sample has a mean of 120 and a standard deviation of 20.0. Find the value of x that corresponds to each of these standard scores:
a. $z = 0.0$ b. $z = 1.2$ c. $z = -1.4$ d. $z = 2.05$

2.133

a. What does it mean to say that $x = 152$ has a standard score of $+1.5$?

b. What does it mean to say that a particular value of x has a z-score of -2.1?

c. In general, the standard score is a measure of what?

2.134 The ACT Assessment® is designed to assess high school students' general educational development and their ability to complete college-level work. For 2001, the ACT English test had a mean score of 20.5 with a standard deviation of 5.6, and the Mathematics test had a mean score of 20.7 with a standard deviation of 5.0. Convert the following ACT test scores to z-scores for both English and Math. Compare placement between the two tests.

a. $x = 30$ b. $x = 23$ c. $x = 12$

d. Explain why the relative positions in English and Math changed for the ACT scores of 30 and 12.

2.135 Which x value has the higher position relative to the set of data from which it comes?

A: $x = 85$, where mean = 72 and standard deviation = 8
B: $x = 93$, where mean = 87 and standard deviation = 5

2.136 Which x value has the lower position relative to the set of data from which it comes?

A: $x = 28.1$, where $\bar{x} = 25.7$ and $s = 1.8$
B: $x = 39.2$, where $\bar{x} = 34.1$ and $s = 4.3$

2.7 Interpreting and Understanding Standard Deviation

Standard deviation is a measure of variation (dispersion) in the data. It has been defined as a value calculated with the use of formulas. Even so, you may be wondering what it really is and how it relates to the data. It is a kind of yardstick by which we can compare the variability of one set of data with another. This particular "measure" can be understood further by examining two statements that tell us how the standard deviation relates to the data: the *empirical rule* and *Chebyshev's theorem*.

THE EMPIRICAL RULE AND TESTING FOR NORMALITY

EMPIRICAL RULE

If a variable is normally distributed, then: within one standard deviation of the mean there will be approximately 68% of the data; within two standard deviations of the mean there will be approximately 95% of the data; and within three standard deviations of the mean there will be approximately 99.7% of the data. [This rule applies specifically to a normal (bell-shaped) distribution, but it is frequently applied as an interpretive guide to any mounded distribution.]

ANSWER NOW 2.137

Instructions for an essay assignment include the statement "The length is to be within 25 words of 200." What values of x, number of words, satisfy these instructions?

Figure 2.31 shows the intervals of one, two, and three standard deviations about the mean of an approximately normal distribution. Usually these proportions

FIGURE 2.31
Empirical Rule

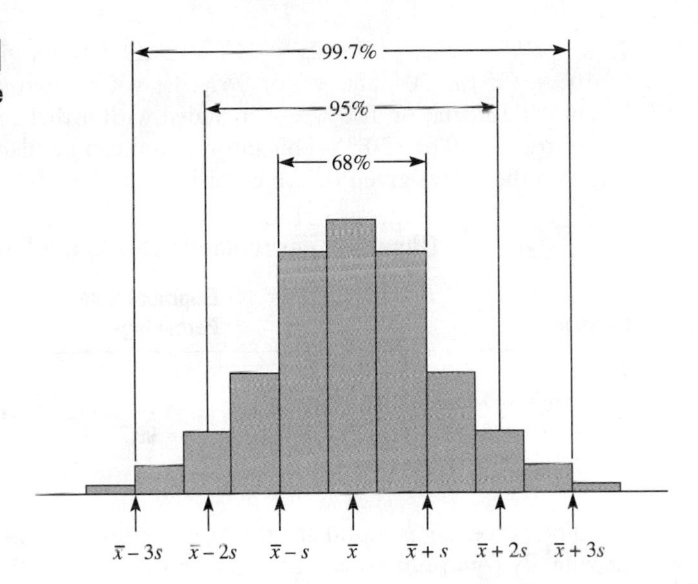

do not occur exactly in a sample, but your observed values will be close when a large sample is drawn from a normally distributed population.

If a distribution is approximately normal, it will be nearly symmetrical and the mean will divide the distribution in half (the mean and the median are the same in a symmetrical distribution). This allows us to refine the empirical rule, as shown in Figure 2.32.

FIGURE 2.32
Refinement of Empirical Rule

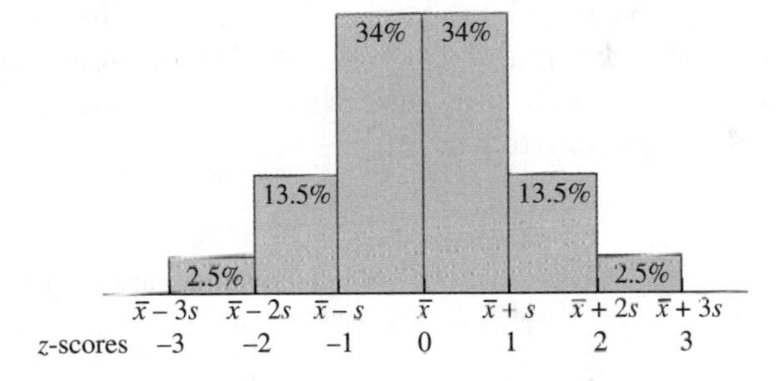

ANSWER NOW 2.138

a. What proportion of a normal distribution is greater than the mean?

b. What proportion is within one standard deviation of the mean?

c. What proportion is greater than a value that is one standard deviation below the mean?

The empirical rule can be used to determine whether or not a set of data is approximately normally distributed. Let's demonstrate this application by working with the distribution of final exam scores that we have been using throughout this chapter. The mean, $\bar{x}$, was found to be 75.6, and the standard deviation, s, was 14.9. The interval from one standard deviation below the mean, $\bar{x} - s$, to one standard deviation above the mean, $\bar{x} + s$, is $75.6 - 14.9 = 60.7$ to $75.6 + 14.9 = 90.5$. This interval (60.7 to 90.5) includes 61, 62, 63, . . . , 89, 90. Upon inspection of the ranked data (Table 2.17, p. 96), we see that 35 of the 50 data, or 70%, lie within one standard deviation of the mean. Furthermore, $\bar{x} - 2s = 75.6 - (2)(14.9) =$

$75.6 - 29.8 = 45.8$ to $\bar{x} + 2s = 75.6 + 29.8 = 105.4$ gives the interval from 45.8 to 105.4. Of the 50 data, 48, or 96%, lie within two standard deviations of the mean. All 50 data, or 100%, are included within three standard deviations of the mean (from 30.9 to 120.3). This information can be placed in a table for comparison with the values given by the empirical rule (see Table 2.18).

TABLE 2.18 Observed Percentages versus the Empirical Rule

Interval	Empirical Rule Percentage	Percentage Found
$\bar{x} - s$ to $\bar{x} + s$	≈ 68	70
$\bar{x} - 2s$ to $\bar{x} + 2s$	≈ 95	96
$\bar{x} - 3s$ to $\bar{x} + 3s$	≈ 99.7	100

The percentages found are reasonably close to those predicted by the empirical rule. By combining this evidence with the shape of the histogram, we can safely say that the final exam data are approximately normally distributed.

There is another way to test for normality—by drawing a probability plot (an ogive drawn on probability paper*) using a computer or graphing calculator. For our illustration, a probability plot of the statistics final exam scores is shown on Figure 2.33. The test for normality, at this point in our study of statistics, is simply to compare the graph of the data (the ogive) with the straight line drawn from the lower left corner to the upper right corner of the graph. If the ogive lies close to this straight line, the distribution is said to be approximately normal. The vertical scale used to construct the probability plot is adjusted so that the ogive for an exactly normal distribution will trace the straight line. The ogive of the exam scores follows the straight line quite closely, suggesting that the distribution of exam scores is approximately normal.

*On probability paper the vertical scale is not uniform; it has been adjusted to account for the mounded shape of a normal distribution and its cumulative percentages.

FIGURE 2.33 Probability Plot of Statistics Exam Scores

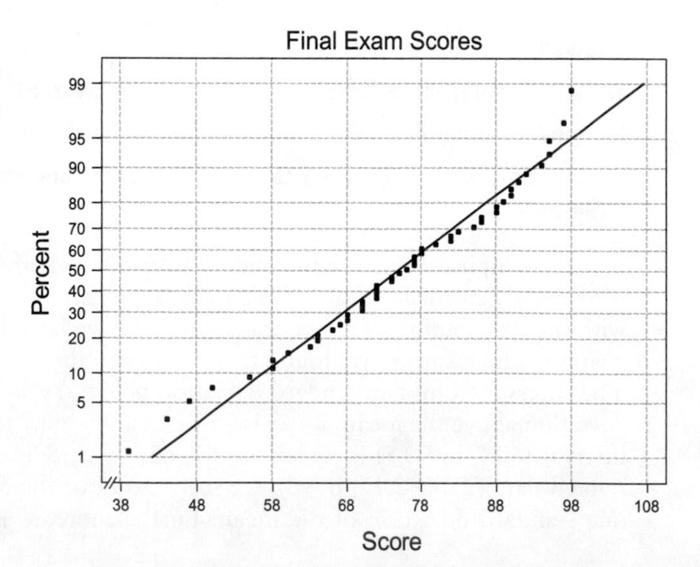

Technology Instructions: Testing for Normality

MINITAB (Release 13)

Input the data into C1; then continue with:

```
Choose:   Stat  >  Basic Statistics  >  Normality Test
Enter:    Variable: C1
          Title: your title  > OK
```

Excel XP

Excel uses a test for normality, not the probability plot.

Input the data into column A; then continue with:

```
Choose:   Tools  >  Data Analysis Plus >  Chi-Squared Test
          of Normality > OK
Enter:    Input Range: data (A1:A6 or select cells)
Select:   Labels (if column headings were used) > OK
```

Expected values for a normal distribution are given versus the given distribution. If the p-value is greater than .05, then the given distribution is approximately normal.

TI-83 Plus

Input the data into L1; then continue with:

```
Choose:   Window
Enter:    at most the smallest
          data value, at least the
          largest data value, x
          scale, -5, 5, 1,1
Choose:   2nd > STAT PLOT  >
          1:Plot
```

CHEBYSHEV'S THEOREM

In the event that the data do not display an approximately normal distribution, Chebyshev's theorem gives us information about how much of the data will fall within intervals centered at the mean for all distributions.

CHEBYSHEV'S THEOREM

The proportion of any distribution that lies within k standard deviations of the mean is at least $1 - \frac{1}{k^2}$, where k is any positive number greater than 1. This theorem applies to all distributions of data.

This theorem says that within two standard deviations of the mean ($k = 2$), you will always find at least 75% (that is, 75% or more) of the data:

$$1 - \frac{1}{k^2} = 1 - \frac{1}{2^2} = 1 - \frac{1}{4} = \frac{3}{4} = 0.75, \textbf{ at least 75\%}$$

Figure 2.34 (p. 110) shows a mounded distribution that illustrates at least 75%.

If we consider the interval enclosed by three standard deviations on either side of the mean ($k = 3$), the theorem says that we will always find at least 89% (that is, 89% or more) of the data:

$$1 - \frac{1}{k^2} = 1 - \frac{1}{3^2} = 1 - \frac{1}{9} = \frac{8}{9} = 0.89, \textbf{ at least 89\%}$$

Figure 2.35 (p. 110) shows a mounded distribution that illustrates at least 89%.

FIGURE 2.34
Chebyshev's Theorem with $k = 2$

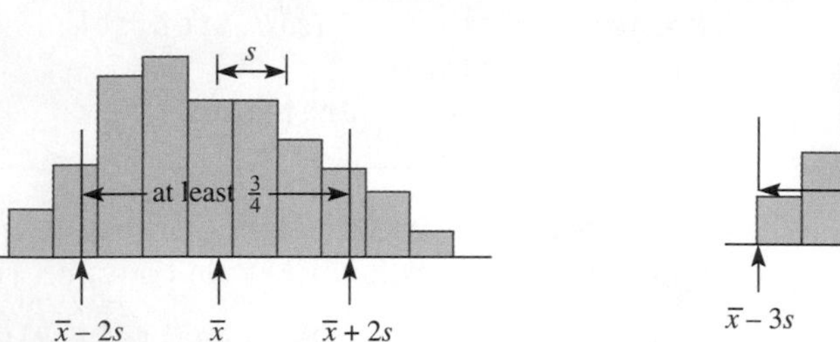

FIGURE 2.35
Chebyshev's Theorem with $k = 3$

s
at least $\frac{8}{9}$
$\bar{x} - 3s$
$\bar{x}$
$\bar{x} + 3s$

ANSWER NOW 2.139
According to Chebyshev's theorem, what proportion of a distribution will be within $k = 4$ standard deviations of the mean?

Exercises

2.140 The empirical rule indicates that we can expect to find what proportion of the sample included between the following:

a. $\bar{x} - s$ and $\bar{x} + s$
b. $\bar{x} - 2s$ and $\bar{x} + 2s$
c. $\bar{x} - 3s$ and $\bar{x} + 3s$

2.141 Why is it that the z-score for a value that belongs to a normal distribution usually lies between -3 and $+3$?

2.142 The mean lifetime of a certain tire is 30,000 miles and the standard deviation is 2500 miles.

a. If we assume the mileages are normally distributed, approximately what percentage of all such tires will last between 22,500 and 37,500 miles?
b. If we assume nothing about the shape of the distribution, approximately what percentage of all such tires will last between 22,500 and 37,500 miles?

2.143 The average clean-up time for a crew of a medium-size firm is 84.0 hours and the standard deviation is 6.8 hours. Assume the empirical rule is appropriate.

a. What proportion of the time will it take the clean-up crew 97.6 or more hours to clean the plant?
b. Within what interval will the total clean-up time fall 95% of the time?

2.144 Using the empirical rule, determine the approximate percentage of a normal distribution that is expected to fall within the interval described.

a. greater than the mean
b. greater than one standard deviation above the mean
c. less than one standard deviation above the mean
d. between one standard deviation below the mean and two standard deviations above the mean

2.145 According to the empirical rule, almost all the data should lie between $(\bar{x} - 3s)$ and $(\bar{x} + 3s)$. The range accounts for all the data.

a. What relationship should hold (approximately) between the standard deviation and the range?
b. How can you use the results of part a to estimate the standard deviation in situations when the range is known?

2.146 Chebyshev's theorem guarantees that what proportion of a distribution will be included between the following:

a. $\bar{x} - 2s$ and $\bar{x} + 2s$
b. $\bar{x} - 3s$ and $\bar{x} + 3s$

2.147 Chebyshev's theorem can be stated in an equivalent form to that given on page 109. For example, to say "at least 75% of the data fall within two standard deviations of the mean" is equivalent to stating "at most, 25% will be more than two standard deviations away from the mean."

a. At most, what percentage of a distribution will be three or more standard deviations from the mean?
b. At most, what percentage of a distribution will be four or more standard deviations from the mean?

2.148 In an article titled "Development of the Breast-Feeding Attrition Prediction Tool" (*Nursing Research*, March/April 1994, Vol. 43, No. 2), the results of administering the Breast-Feeding Attrition Prediction Tool (BAPT) to 72 women with prior successful experience breast-feeding are reported. The mean score on the Positive Breast-feeding Sentiment (PBS) score for this sample is reported to equal 356.3 with a standard deviation of 65.9.

a. According to Chebyshev's theorem, at least what percent of the PBS scores are between 224.5 and 488.1?
b. If it is known that the PBS scores are normally distributed, what percent of the PBS scores are between 224.5 and 488.1?

2.149 Ex02-149 The top 50 Nike Tour money leaders and their total earnings through the Utah Classic on September 8, 2002, are listed here:

Player	Money	Player	Money
Arron Oberholser	$262,227	Ken Green	$102,747
Doug Barron	232,207	Joel Kribel	101,624
Cliff Kresge	218,814	Tom Carter	99,745
Patrick Moore	216,428	Gary Hallberg	97,332
Todd Fischer	205,682	Mark Hensby	93,968
Darron Stiles	200,839	Mike Heinen	91,839
Marco Dawson	191,384	Eric Meeks	91,066
Patrick Sheehan	167,323	Tommy Biershenk	90,890
Jason Buha	151,877	Aaron Baddeley	87,756
Gavin Coles	146,995	Hunter Haas	84,835
Steven Alker	145,499	John Maginnes	84,407
Tyler Williamson	129,316	Rich Barcelo	84,368
Tag Ridings	128,855	Jeff Klauk	82,669
Jay Delsing	128,751	Andrew McLardy	82,547
Charles Warren	127,693	Rob McKelvey	82,163
Jace Bugg	124,067	Andy Sanders	80,937
Todd Barranger	123,481	Jeff Hart	80,879
Jason Caron	123,231	Brad Ott	80,683
Peter O'Malley	121,125	Brian Wilson	79,905
Omar Uresti	120,791	Roger Tambellini	73,616
Anthony Painter	115,693	Steve Ford	73,106
Zoran Zorkic	110,389	Jeff Freeman	72,151
Keoke Cotner	107,242	Bryce Molder	72,064
Brian Claar	104,833	Chris Tidland	68,801
Scott Sterling	103,780	Steve Haskins	68,569

Source: *PGATour.com* and *GolfWeb.com*, September 14, 2002

a. Calculate the mean and standard deviation of the earnings of the Nike Tour golf players.
b. Find the values of $\bar{x} - s$ and $\bar{x} + s$.
c. How many of the 50 pieces of data have values between $\bar{x} - s$ and $\bar{x} + s$? What percentage of the sample is this?
d. Find the values of $\bar{x} - 2s$ and $\bar{x} + 2s$.
e. How many of the 50 pieces of data have values between $\bar{x} - 2s$ and $\bar{x} + 2s$? What percentage of the sample is this?
f. Find the values of $\bar{x} - 3s$ and $\bar{x} + 3s$.
g. What percentage of the sample has values between $\bar{x} - 3s$ and $\bar{x} + 3s$?
h. Compare the answers found in parts e and g to the results predicted by Chebyshev's theorem.
i. Compare the answers found in parts c, e, and g to the results predicted by the empirical rule. Do the results suggest an approximately normal distribution?
j. Verify your answer to part i using one of the sets of technology instructions.

2.150 Ex02-150 On the first day of class last semester, 50 students were asked for the one-way distance from home to college (to the nearest mile). The resulting data follow:

6	5	3	24	15	15	6	2	1	3
5	10	9	21	8	10	9	14	16	16
10	21	20	15	9	4	12	27	10	10
3	9	17	6	11	10	12	5	7	11
5	8	22	20	13	1	8	13	4	18

a. Construct a grouped frequency distribution of the data by using 1–4 as the first class.
b. Calculate the mean and the standard deviation.
c. Determine the values of $\bar{x} \pm 2s$, and determine the percentage of data within two standard deviations of the mean.

2.151 The empirical rule states that the one, two, and three standard deviation intervals about the mean will contain 68%, 95%, and 99.7% respectively.

a. Use the computer or calculator commands on page 104 to randomly generate a sample of 100 data from a normal distribution with mean 50 and standard deviation 10. Construct a histogram using class boundaries that are multiples of the standard deviation 10; that is, use boundaries from 10 to 90 in intervals of 10 (see the commands on p. 58). Calculate the mean and the standard deviation using the commands found on pages 69 and 82; then inspect the histogram to determine the percentage of the data that fell within each of the one, two, and three standard deviation intervals. How closely do the three percentages compare to the percentages claimed in the empirical rule?
b. Repeat part a. Did you get results similar to those in part a? Explain.
c. Consider repeating part a several more times. Are the results similar each time? If so, in what way?
d. What do you conclude about the truth of the empirical rule?

2.152 Chebyshev's theorem states that "at least $1 - \frac{1}{k^2}$" of the data of a distribution will lie within k standard deviations of the mean.

a. Use the computer commands on page 104 to randomly generate a sample of 100 data from a uniform (nonnormal) distribution that has a low value of 1 and a high value of 10. Construct a histogram using class boundaries of 0 to 11 in increments of 1 (see the commands on p.58). Calculate the mean and the standard deviation using the commands found on pages 69 and 82; then inspect the histogram to determine the percentage of the data that fell within each of the one, two, three, and four standard deviation intervals. How closely do these percentages compare to the percentages claimed in Chebyshev's theorem and in the empirical rule?

b. Repeat part a. Did you get results similar to those in part a? Explain.

c. Consider repeating part a several more times. Are the results similar each time? If so, in what way are they similar?

d. What do you conclude about the truth of Chebyshev's theorem and the empirical rule?

2.8 The Art of Statistical Deception

"There are three kinds of lies—lies, damned lies, and statistics." These remarkable words spoken by Benjamin Disraeli (19th-century British prime minister) represent the cynical view of statistics held by many people. Most people are on the consumer end of statistics and therefore have to "swallow" them.

GOOD ARITHMETIC, BAD STATISTICS

Let's explore an outright statistical lie. Suppose a small business employs eight people who earn between \$300 and \$350 per week. The owner of the business pays himself \$1250 per week. He reports to the general public that the average wage paid to the employees of his firm is \$430 per week. That may be an example of good arithmetic, but it is bad statistics. It is a misrepresentation of the situation because only one employee, the owner, receives more than the mean salary. The public will think that most of the employees earn about \$430 per week.

ANSWER NOW 2.153

Is it possible for eight employees to earn between \$300 and \$350, and one earn \$1250 per week, and for the mean to be \$430? Verify your answer.

INSUFFICIENT INFORMATION

Application 2.3 The Information That Was Almost Presented

FYI
The bite taken out of the graph looks more like a human bite mark than a dog's. Just what is this graph depicting?

This is an attractive, colorful graph about the age of dog-bite victims, I guess. But what does the height of the bars on the graph represent?

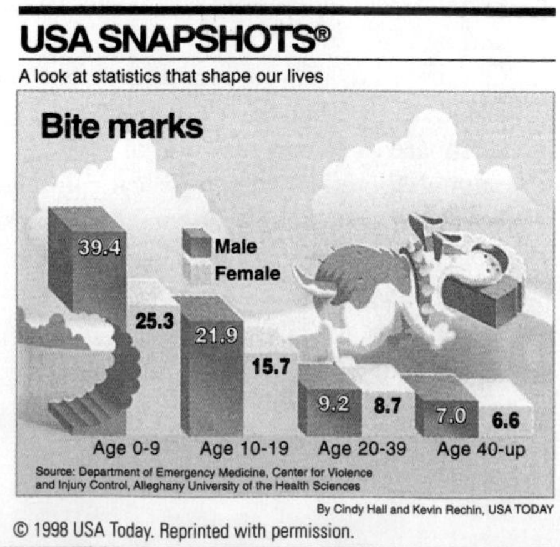

USA SNAPSHOTS®
A look at statistics that shape our lives
Bite marks
Male
Female
39.4
25.3
21.9
15.7
9.2
8.7
7.0
6.6
Age 0-9
Age 10-19
Age 20-39
Age 40-up
Source: Department of Emergency Medicine, Center for Violence and Injury Control, Alleghany University of the Health Sciences
By Cindy Hall and Kevin Rechin, USA TODAY

Graphic representations can be tricky and misleading. The frequency scale (which is usually the vertical axis) should start at zero in order to present a total picture. Usually, graphs that do not start at zero are used to save space. Nevertheless, this can be deceptive. Graphs in which the frequency scale starts at zero tend to emphasize the size of the numbers involved, whereas graphs that are chopped off may tend to emphasize the variation in the numbers without regard to the actual size of the numbers. The labeling of the horizontal scale can be misleading also. You need to inspect graphic presentations very carefully before you draw any conclusions from the "story being told." The following four case studies will demonstrate some of these misrepresentations.

SUPERIMPOSED MISREPRESENTATION

Application 2.4 Claiming What The Reader Expects/ Claiming Anticipated Bad News

Source: http://www.math.yorku.ca/SCS/Gallery/context.html
Courtesy of the Ithaca Times

This "clever" graphic overlay, from the *Ithaca Times* (December 7, 2000), has to be the worst graph ever to make a front page. The cover story, "Why does college have to cost so much?" pictures two graphs superimposed on a Cornell University campus scene. The two broken lines represent "Cornell's Tuition" and "Cornell's Ranking," with the tuition steadily increasing and the ranking staggering and falling. A very clear image is created: Students get less, pay more!

Now view the two graphs separately. Notice: (1) The graphs cover two different time periods. (2) The vertical scales differ. (3) The "best" misrepresentation comes from the impression that a "drop in rank" represents a lower quality of education. Wouldn't a rank of 6 be better than a rank of 15?

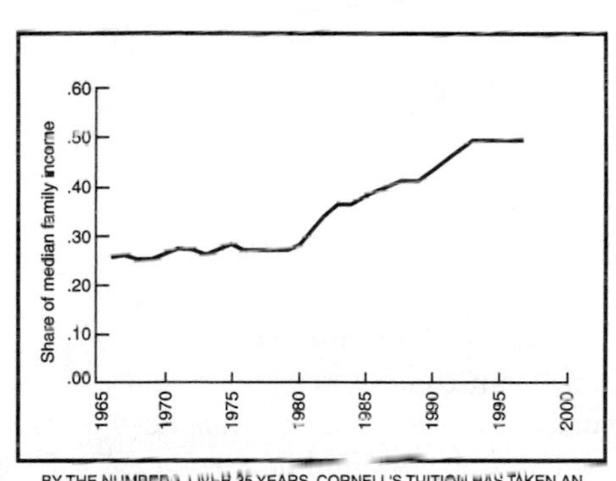

BY THE NUMBERS: OVER 35 YEARS, CORNELL'S TUITION HAS TAKEN AN INCREASINGLY LARGER SHARE OF ITS MEDIAN STUDENT FAMILY INCOME

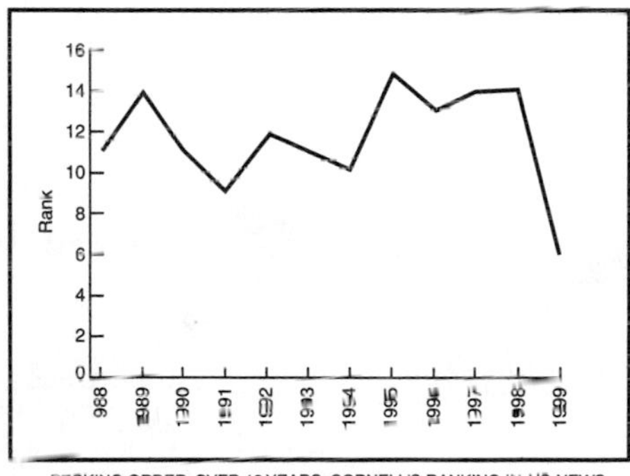

PECKING ORDER: OVER 12 YEARS, CORNELL'S RANKING IN *US NEWS & WORLD REPORT* HAS RISEN AND FALLEN ERRATICALLY

Source: http://www.math.yorku.ca/SCS/Gallery/context.html

ANSWER NOW 2.154
What's wrong with this picture?

a. **Find and describe at least four features about the front-page graph that are incorrectly used.**

b. **Find and describe at least two features about the "Pecking Order" graph that are misrepresenting.**

What it all comes down to is that statistics, like all languages, can be and is abused. In the hands of the careless, the unknowledgeable, or the unscrupulous, statistical information can be as false as "damned lies."

TRUNCATED SCALE

Application 2.5 Simple Is Not Always Best

This graphic is neat and very readable, but does it represent the information being shown? Truncating scales on graphs often leads to misleading visual impressions. For example, in "Service complaints," it appears that "Take too long" is twice as likely to be the complaint as "Messy." Look for other visual misrepresentations.

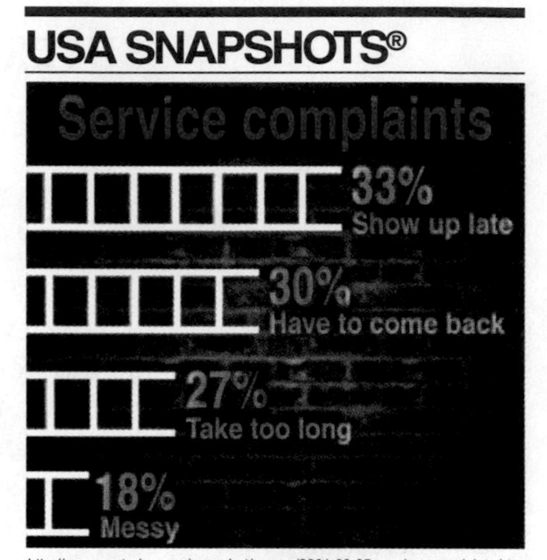

http://www.usatoday.com/snapshot/money/2001-09-05-service-complaints.htm

ANSWER NOW 2.155

a. **Find and describe at least four incorrect impressions created by the truncated horizontal axis on "Service complaints."**

b. **Redraw the bar graph in Application 2.5 starting the horizontal scale at zero.**

c. **Comment on the effect your graph has on the impression presented.**

VARIABLE SCALES

Application 2.6 S-t-r-e-t-c-h-ing the Graph

Sometimes deception is unintentional and sometimes it is difficult to tell intent from mistake. The "Overweight Adults" diagram is deceiving. Study it carefully and find its errors; then do Answer Now 2.156 to be sure you understand the deception that has occurred and how it was accomplished.

Overweight Adults

Percentage of adults who are overweight: 40, 35, 30, 25, 20

Female

Male

1962 1974 1980 1994

ANSWER NOW 2.156

a. **Plot the points (62, 23), (74, 23.5), (80, 24), and (94, 33) on a coordinate axis system using graph paper. Connect the points with line segments.**

b. **Compare the graph in part a to the graph of overweight adult males in Application 2.6. Comment on the difference in appearance.**

FYI

Darrell Huff's *How to Lie with Statistics* is an easy, fun to read, clever book that every student of statistics should read.

Exercises

2.157 Exercise 2.116 (p. 93) displays a relative frequency distribution and a frequency distribution. A histogram of either distribution would qualify as a "tricky graph." How does the graph violate the guidelines for drawing histograms?

2.158 *PC World* magazine periodically publishes performance test results and overall ratings of the top 20 power desktop computers. No machine in the top 20 received an overall rating below 75, and the maximum rating any machine could have received was 100. The top four manufacturers, together with the performance ratings of their machines, are listed in the accompanying table. Prepare two bar graphs to depict the performance rating data. Scale the vertical axis on the first graph from 80 to 89. Scale the axis on the second graph from 0 to 100.

Manufacturer	Overall Rating
1. Gateway	88
2. Dell	87
3. Quantex	84
4. NEC	82

What is your conclusion concerning how the performance ratings of the four computers stack up based on the two bar graphs, and what would you recommend, if anything, to improve the presentations?

2.159 Find an article or an advertisement containing a graph that in some way misrepresents the information of statistics. Describe how this graph misrepresents the facts.

CHAPTER SUMMARY

RETURN TO CHAPTER CASE STUDY

©Jules Frazier/Getty Images

What Do People Do When They Are on the Internet?

Let's return to the Chapter Case Study (page 37) as a way to assess what we have learned in this chapter. Recall that students were asked, "How many different Internet activities do you engage in during a typical week?" Inspect the data again. CS02

6	7	3	6	9	10	8	9	9	6
4	9	4	9	4	2	3	5	13	12
4	6	4	9	5	6	9	11	5	6
5	3	7	9	6	5	12	2	6	9

FIRST THOUGHTS 2.160

a. How many different Internet activities did you engage in last week?
b. How do you think you compare to the 40 Internet users in the sample above?
c. How do you think you compare to all Internet users?

PUTTING CHAPTER 2 TO WORK 2.161

a. List all the types of charts and graphs shown in Chapter 2 that would be appropriate for use with the set of 40 data listed above.
b. What types of graphs would not be appropriate? Explain why.
c. Display the data using each of the charts and graphs listed in part a.
d. Which graph do you think best represents the data? Explain why.
e. Find the five measures of central tendency for these data (mean, median, mode, midrange, and midquartile).
f. Find the three measures of dispersion for the data (range, variance, and standard deviation).

g. Find the value of these measures of position: P_5, P_{10}, Q_1, Q_3, P_{90}, and P_{98}.
h. How many different Internet activities do you engage in during a typical week? Use the mean and standard deviation calculated in parts e and f to determine your z-score. What does it tell you about yourself with respect to the students' Internet usage?
i. Use one graph from part c plus at least one measure of central tendency and one measure of dispersion to write a description of the students' number of Internet activities per week.
j. According to the empirical rule, if the distribution is normal, approximately 68% of the number of different Internet activities engaged in by students will fall between what two values? Is this true? Why or why not?
k. According to Chebyshev's theorem, at least 75% of the number of different Internet activities engaged in by students will fall within what two values? Is this true? Why or why not?
l. The sample information pictured in the Chapter Case Study graph on page 37 is different from, but related to, the sample information you have been working with in parts a–k. Describe the data collected for the Chapter Case Study graph and explain how they are different from the data listed on page 115.

YOUR OWN MINI-STUDY 2.162

a. Design your own study of Internet usage. Define a specific population that you will sample, describe your sampling plan, collect your data, and complete parts c–l in Putting Chapter 2 to Work 2.161.
b. Discuss the differences and similarities between the Internet usage described by the sample of 40 statistics students (given above) and your own sample.

In Retrospect

You have been introduced to some of the more common techniques of descriptive statistics. There are far too many specific types of statistics used in nearly every specialized field of study for us to review here. We have outlined the uses of only the most universal statistics. Specifically, you have seen several basic graphic techniques (circle and bar graphs, Pareto diagrams, dotplots, stem-and-leaf displays, histograms, and box-and-whiskers) that are used to present sample data in picture form. You have also been introduced to some of the more common measures of central tendency (mean, median, mode, midrange, and midquartile), measures of dispersion (range, variance, and standard deviation), and measures of position (quartiles, percentiles, and *z*-scores).

You should now be aware that an average can be any one of five different statistics, and you should understand the distinctions among the different types of averages. The article "'Average' Means Different Things" in Application 2.2 (p. 73) discusses four of the averages studied in this chapter. You might reread it now and find that it has more meaning and is of more interest. It will be time well spent!

You should also have a feeling for, and an understanding of, the concept of a standard deviation. You were introduced to the empirical rule and Chebyshev's theorem for this purpose.

The exercises in this chapter (as in others) are extremely important; they will reinforce the concepts studied before you go on to learn how to use these ideas in later chapters. A good understanding of the descriptive techniques presented in this chapter is fundamental to your success in the later chapters.

Chapter Exercises

2.163 Near the end of the 20th century the corporate world predicted that thousands of desktop computer programs would fail to handle the new year 2000 because of installed instruction code that depended upon a dating scheme designed for the previous century. For example, it was estimated by Tangram Enterprise Solutions, Inc., that 4593 desktops running CalcuPro software prior to version 3.0 under MS-Windows were at risk to fail in the year 2000. The breakdown within corporate divisions was as follows:

Corporate Area	At-Risk Frequency
1. Sales	700
2. Research & Development	1550
3. Finance	843
4. Administration	1100
5. Other	400

Source: *Fortune*, "Teaching Chipmunks to Dance," May 25, 1998, p. S8. Extracted from *Teaching Chipmunks to Dance*, Kendall/Hunt Publishing, 1-800-228-0810.

a. Construct a relative frequency distribution of these data.
b. Construct a bar graph from these data.
c. Explain why the graph drawn in part b is not a histogram.

2.164 Identify each of the following as an example of attribute (qualitative) variables or numerical (quantitative) variables:

a. scores registered by people taking their written state automobile driver's license examination
b. whether or not a motorcycle operator possesses a valid motorcycle operator's license
c. the number of television sets installed in a house
d. the brand of bar soap used in a bathroom
e. the value of a cents-off coupon used with the purchase of a box of cereal
f. the amount of weight lost in the past month by a person following a strict diet
g. batting averages of major league baseball players
h. decisions by the jury in felony trials
i. sun screen usage before going in the sun (always, often, sometimes, seldom, never)
j. reason manager failed to act against an employee's poor performance

2.165 Consider samples A and B. Notice that the two samples are the same except that the 8 in A has been replaced by a 9 in B.

A:	2	4	5	5	7	8
B:	2	4	5	5	7	9

What effect does changing the 8 to a 9 have on each of the following statistics?

a. mean b. median c. mode d. midrange
e. range f. variance g. std. dev.

2.166 Consider samples C and D. Notice that the two samples are the same except for two values.

C:	20	60	60	70	90
D:	20	30	70	90	90

What effect does changing the two 60's to 30 and 90 have on each of the following statistics?

a. mean b. median c. mode d. midrange
e. range f. variance g. std. dev.

2.167 Ex02-167 The addition of a new accelerator is claimed to decrease the drying time of latex paint by more than 4%. Several test samples were conducted with the following percentage decreases in drying time:

5.2	6.4	3.8	6.3	4.1	2.8	3.2	4.7

a. Find the sample mean.
b. Find the sample standard deviation.
c. Do you think these percentages average 4 or more? Explain.
(Retain these solutions to use in Exercise 9.34, p. 428.)

2.168 Ex02-168 Gasoline pumped from a supplier's pipeline is supposed to have an octane rating of 87.5. On 13 consecutive days, a sample of octane ratings was taken and analyzed, with the following results:

88.6	86.4	87.2	88.4	87.2	87.6	86.8
86.1	87.4	87.3	86.4	86.6	87.1	

a. Find the sample mean.
b. Find the sample standard deviation.
c. Do you think these readings average 87.5? Explain.
(Retain these solutions to use in Exercise 9.46, p. 429.)

2.169 Ex02-169 These data are the ages of 118 known offenders who committed an auto theft last year in Garden City, Michigan:

11	14	15	15	16	16	17	18	19	21	25	36
12	14	15	15	16	16	17	18	19	21	25	39
13	14	15	15	16	17	17	18	20	22	26	43
13	14	15	15	16	17	17	18	20	22	26	46
13	14	15	16	16	17	17	18	20	22	27	50
13	14	15	16	16	17	17	19	20	23	27	54
13	14	15	16	16	17	18	19	20	23	29	59
13	15	15	16	16	17	18	19	20	23	30	67
14	15	15	16	16	17	18	19	21	24	31	
14	15	15	16	16	17	18	19	21	24	34	

a. Find the mean.
b. Find the median.
c. Find the mode.
d. Find Q_1 and Q_3.
e. Find P_{10} and P_{95}.

2.170 Ex02-170 A survey of 32 workers at building 815 of Eastman Kodak Company was taken last May. Each worker was asked: "How many hours of television did you watch yesterday?" The results were as follows:

0	0	$\frac{1}{2}$	1	2	0	3	$2\frac{1}{2}$
0	0	1	$1\frac{1}{2}$	5	$2\frac{1}{2}$	0	2
$2\frac{1}{2}$	1	0	2	0	$2\frac{1}{2}$	4	0
6	$2\frac{1}{2}$	0	$\frac{1}{2}$	1	$1\frac{1}{2}$	0	2

a. Construct a stem-and-leaf display.
b. Find the mean.
c. Find the median.
d. Find the mode.
e. Find the midrange.
f. Which measure of central tendency would best represent the average viewer if you were trying to portray the typical television viewer? Explain.
g. Which measure of central tendency would best describe the amount of television watched? Explain.
h. Find the range.
i. Find the variance.
j. Find the standard deviation.

2.171 Ex02-171 The stopping distance on a wet surface was determined for 25 cars, each traveling at 30 miles per hour. The data (in feet) are shown on the following stem-and-leaf display:

6	3 7 6 3 9
7	4 2 0 1 1 2 0 5
8	5 4 5 5 6
9	4 1 0 0 5
10	5 4

Find the mean and the standard deviation of these stopping distances.

2.172 Ex02-172 *Forbes* magazine published information about "The Best Big Companies in America" in the January 7, 2002, issue. Part of the article gave the net incomes (in $ millions) for the last 12 months for 23 U.S. banks:

522	1384	921	475	705	259	348	951
312	391	740	290	180	348	314	511
1232	304	532	1369	301	2694	284	

a. Find the mean net income for the banks.
b. Find the median net income for the banks.
c. Find the midrange of the net income for the banks.
d. Write a discussion comparing the results from parts a, b, and c.
e. Find the standard deviation of these net incomes.
f. Find the percentage of data that are within one standard deviation of the mean.
g. Find the percentage of data that are within two standard deviations of the mean.
h. Based on the above results, discuss whether you think the data are normally distributed, and why.

2.173 The number of lost luggage reports filed per 1000 airline passengers during July 1998 was reported by *USA Today*. On September 23, 1998, they reported the airlines with the fewest: Continental—3.50, US Airways—3.95, Delta—4.07, American—4.22. On September 24, 1998, they reported the airlines with the most: Alaska—8.64, United—7.63, Northwest—6.76, TWA—5.12. The industry average was 5.09.
a. Define the terms *population* and *variable* with regard to this information.
b. Are the numbers reported (3.50, 3.95, ..., 8.64) data or statistics? Explain.
c. Is the average, 5.09, a data, a statistic, or a parameter? Explain why.
d. Is the average value the midrange?

2.174 Ex02-174 One of the first scientists to study the density of nitrogen was Lord Raleigh. He noticed that the density of nitrogen produced from the air seemed to be greater than the density of nitrogen produced from chemical compounds. Do his conclusions seem to be justified even though he has so little data?

Lord Raleigh's measurements, which first appeared in *Proceedings, Royal Society* (London, 55, 1894, pp. 340–344) are listed below. The units are the mass of nitrogen filling a certain flask under specified pressure and temperature.

Atmospheric	Chemical
2.31017	2.30143
2.30986	2.29890
2.31010	2.29816
2.31001	2.30182
2.31024	2.29869
2.31010	2.29940
2.31028	2.29849
2.31163	2.29889
2.30956	2.30074
	2.30054

http://exploringdata.cqu.edu.au/datasets/nitrogen.xls

a. Construct side-by-side dotplots of the two sets of data, using a common scale.
b. Calculate mean, median, standard deviation, and first and third quartiles for each set of data.
c. Construct side-by-side boxplots of the two sets of data, using a common scale.
d. Discuss how these two sets of data compare. Do these two very small sets of data show convincing evidence of a difference?

FYI
The differences between these sets of data helped lead to the discovery of argon.

2.175 Ex02-175 The distribution of credit hours, per student, taken this semester at a certain college was as follows:

Credit Hours	Frequency
3	75
6	150
8	30
9	50
12	70
14	300
15	400
16	1050
17	750
18	515
19	120
20	60

a. Draw a histogram of the data.
b. Find the five measures of central tendency.
c. Find Q_1 and Q_3. d. Find P_{15} and P_{12}.
e. Find the three measures of dispersion (range, s^2, and s).

2.176 Ex02-176 An article in *Therapeutic Recreation Journal* reports a distribution for the variable "number of persistent disagreements." Sixty-six patients and their therapeutic recreation specialist each answered a checklist of problems with yes or no. Disagreement occurs when the specialist and the patient did not respond identically to an item on the checklist. It becomes a persistent disagreement if the item remains in disagreement after a second interview.

Number of Disagreements	Frequency
0	2
1	2
2	4
3	10
4	7
5	9
6	8
7	11
8	7
9	3
10	1
11	2

Source: Data reprinted with permission of the National Recreation and Park Association, Alexandria, VA, from Pauline Petryshen and Diane Essex-Sorlie, "Persistent Disagreement Between Therapeutic Recreation Specialists and Patients in Psychiatric Hospitals," *Therapeutic Recreation Journal*, Vol. XXIV, Third Quarter, 1990

a. Draw a dotplot of these sample data.
b. Find the median number of persistent disagreements.
c. Find the mean number of persistent disagreements.
d. Find the standard deviation of the number of persistent disagreements.
e. Draw a vertical line on the dotplot at the mean.
f. Draw a horizontal line segment on the dotplot whose length represents the standard deviation (start at the mean).

2.177 Ex02-177 *USA Today* (October 25, 1994) reported in the USA Snapshot® "Mystery of the remote" that 44% of the families surveyed never misplaced the family television remote control, 38% misplaced it one to five times weekly, and 17% misplaced it more than five times weekly. One percent of the families surveyed didn't know. Suppose you took a survey that resulted in the following data. Let x be the number of times per week that the family's television remote control gets misplaced.

x	0	1	2	3	4	5	6	7	8	9
f	220	92	38	21	24	30	34	20	16	5

a. Construct a histogram.
b. Find the mean, median, mode, and midrange.
c. Find the variance and standard deviation.
d. Find Q_1, Q_3, and P_{90}.
e. Find the midquartile.
f. Find the 5-number summary and draw a box-and-whiskers display.

2.178 Ex02-178 The table shows the age distribution of heads of families:

Age of Head of Family	Number
20–25	23
25–30	38
30–35	51
35–40	55
40–45	53
45–50	50
50–55	48
55–60	39
60–65	31
65–70	26
70–75	20
75–80	16
	450

a. Find the mean age of the heads of families.
b. Find the standard deviation.

2.179 Ex02-179 The lifetimes of 220 incandescent 60-watt lamps were obtained and yielded the frequency distribution shown in this table:

Class Limits	Frequency
500–600	3
600–700	7
700–800	14
800–900	28
900–1000	64
1000–1100	57
1100–1200	23
1200–1300	13
1300–1400	7
1400–1500	4

a. Construct a histogram of these data using a vertical scale for the relative frequencies.
b. Find the mean lifetime.
c. Find the standard deviation of the lifetimes.

2.180 "Grandparents are grand" appeared as a USA Snapshot® in *USA Today* (December 9, 1994) and reported "how much grandparents say they spend annually on gifts and entertainment for each of their grandchildren."

Dollar Interval	\$0–\$100	\$101–\$200	\$201–\$500	\$501 or more
Percent	54.4	16.5	15.7	4.7

Suppose that this information was obtained from a sample of 1000 grandparents and that the 8.7% who did not answer spent nothing. Use values of \$0, \$50, \$150, \$350, and \$750 as class marks, and estimate the sample mean and the standard deviation for the variable x, amount spent.

2.181 Ex02-181 The earnings per share for 40 firms in the radio and transmitting equipment industry follow:

4.62	0.10	1.29	7.25	6.04	3.20	9.56	4.90	4.22	3.71
0.25	1.34	2.11	5.39	0.84	−0.19	3.72	2.27	2.08	1.12
1.07	2.50	2.14	3.46	1.91	7.05	5.10	1.80	0.91	0.50
5.56	1.62	1.36	1.93	2.05	2.75	3.58	0.44	3.15	1.93

a. Prepare a frequency distribution and a frequency histogram for these data.
b. Which class of your frequency distribution contains the median?

2.182 Ex02-182 The lengths (in millimeters) of 100 brown trout in pond 2-B at Happy Acres Fish Hatchery on June 15 of last year were as follows:

15.0	15.3	14.4	10.4	10.2	11.5	15.4	11.7	15.0	10.9
13.6	10.5	13.8	15.0	13.8	14.5	13.7	13.9	12.5	15.2
10.7	13.1	10.6	12.1	14.9	14.1	12.7	14.0	10.1	14.1
10.3	15.2	15.0	12.9	10.7	10.3	10.8	15.3	14.9	14.8
14.9	11.8	10.4	11.0	11.4	14.3	15.1	11.5	10.2	10.1
14.7	15.1	12.8	14.8	15.0	10.4	13.5	14.5	14.9	13.9
10.1	14.8	13.7	10.9	10.6	12.4	14.5	10.5	15.1	15.8
12.0	15.5	10.8	14.4	15.4	14.8	11.4	15.1	10.3	15.4
15.0	14.0	15.0	15.1	13.7	14.7	10.7	14.5	13.9	11.7
15.1	10.9	11.3	10.5	15.3	14.0	14.6	12.6	15.3	10.4

a. Find the mean.
b. Find the median.
c. Find the mode.
d. Find the midrange.
e. Find the range.
f. Find Q_1 and Q_3.
g. Find the midquartile.
h. Find P_{35} and P_{64}.
i. Construct a grouped frequency distribution that uses 10.0–10.5 as the first class.
j. Construct a histogram of the frequency distribution.
k. Construct a cumulative relative frequency distribution.
l. Construct an ogive of the cumulative relative frequency distribution.
m. Find the mean of the frequency distribution.
n. Find the standard deviation of the frequency distribution.

2.183 Ex02-183 The dollar amounts listed here are the average hourly earnings of production or nonsupervisory workers on private nonfarm payrolls by major industry. Five years of information is listed below and eleven years of information is listed on the CD inside the back cover. Investigate this information, looking for any pattern that might exist. Find both numerical and graphic statistics by months and by years. Describe all patterns found:

Year	Jan.	Feb.	Mar.	Apr.	May	Jun.	Jul.	Aug.	Sep.	Oct.	Nov.	Dec
1998	12.54	12.60	12.64	12.69	12.73	12.77	12.79	12.85	12.88	12.92	12.95	12.99
1999	13.04	13.06	13.10	13.15	13.19	13.24	13.28	13.30	13.35	13.38	13.40	13.44
2000	13.51	13.55	13.59	13.64	13.67	13.73	13.76	13.81	13.86	13.93	13.97	14.03
2001	14.05	14.12	14.17	14.21	14.24	14.29	14.33	14.38	14.43	14.46	14.52	14.56
2002	14.59	14.62	14.65	14.68	14.70	14.75	14.78	14.82	14.87			

http://www.bls.gov/

a. Use the five years listed here.
b. Use the eleven years listed on the CD. DS-HourlyEarnings

2.184 Ex02-184 Who ate the M&M's®? The table below gives the color counts and net weight (in grams) for a sample of 30 bags of M&M's®. The advertised net weight is 47.9 grams per bag.

Case	Red	Gr.	Blue	Or.	Yel.	Br.	Weight
1	15	9	3	3	9	19	49.79
2	9	17	19	3	3	8	48.98
3	14	8	6	8	19	4	50.40
4	15	7	3	8	16	8	49.16
5	10	3	7	9	22	4	47.61
6	12	7	6	5	17	11	49.80
7	6	7	3	6	26	10	50.23
8	14	11	4	1	14	17	51.68
9	4	2	10	6	18	18	48.45
10	9	9	3	9	8	15	46.22
11	9	11	13	0	7	18	50.43
12	8	8	6	5	11	20	49.80
13	12	9	13	2	6	13	46.94
14	9	7	7	2	18	7	47.98
15	6	6	6	4	21	13	48.49
16	4	6	9	4	12	20	48.33
17	3	5	11	12	11	16	48.72
18	14	5	6	6	21	6	49.69
19	5	5	16	12	7	12	48.95
20	8	9	13	4	15	11	51.71
21	8	7	7	13	7	18	51.53
22	9	8	3	8	23	8	50.97
23	20	2	7	5	13	9	50.01
24	12	6	1	12	6	19	48.28
25	8	9	4	6	21	7	48.74
26	4	6	7	6	14	19	46.72
27	10	12	11	6	11	7	47.67
28	5	4	2	9	18	16	47.70
29	15	11	4	13	7	8	49.40
30	11	6	7	12	12	13	52.06

Source: http://www.math.uah.edu/stat/
Christine Nickel and Jason York, ST 687 project, Fall 1998

There is something about one case in this data set that is suspiciously inconsistent with the rest of the data. Find the inconsistency.

(continued)

a. Construct two different graphs for the weights.
b. Calculate several numerical statistics for the weight data.
c. Did you find any potential inconsistencies in parts a and b? Explain.
d. Find the number of M&M's® in each bag.
e. Construct two different graphs for the number of M&M's® per bag.
f. Calculate several numerical statistics for the number of M&M's® per bag.
g. What inconsistency did you find in parts e and f? Explain.
h. Give a possible explanation as to why the inconsistency does not show up in the weight data but does show up in the number data.

2.185 For a normal (or bell-shaped) distribution, find the percentile rank that corresponds to:

a. $z = 2$ b. $z = -1$

c. Sketch the normal curve, showing the relationship between the z-score and the percentiles for parts a and b.

2.186 For a normal (or bell-shaped) distribution, find the z score that corresponds to the kth percentile:

a. $k = 20$ b. $k = 95$

c. Sketch the normal curve, showing the relationship between the z-score and the percentiles for parts a and b.

2.187 Bill and Rob are good friends, although they attend different high schools in their city. The city school system uses a battery of fitness tests to test all high school students. After completing the fitness tests, Bill and Rob are comparing their scores to see who did better in each event. They need help.

	Sit-ups	Pull-ups	Shuttle Run	50-Yard Dash	Softball Throw
Bill	$z = -1$	$z = -1.3$	$z = 0.0$	$z = 1.0$	$z = 0.5$
Rob	61	17	9.6	6.0	179 ft
Mean	70	8	9.8	6.6	173 ft
Std. Dev.	12	6	0.6	0.3	16 ft

Bill received his test results in z-scores, whereas Rob was given raw scores. Since both boys understand raw scores, convert Bill's z-scores to raw scores in order to make an accurate comparison.

2.188 Twins Jean and Joan Wong are in fifth grade (different sections), and the class has been given a series of ability tests. If the scores for these ability tests are approximately normally distributed, which girl has the higher relative score on each of the skills listed? Explain your answers.

Skill	Jean: z-Score	Joan: Percentile
Fitness	2.0	99
Posture	1.0	69
Agility	1.0	88
Flexibility	−1.0	35
Strength	0.0	50

2.189 Ex02-189 Manufacturing specifications are often based on the results of samples taken from satisfactory pilot runs. The following data resulted from just such a situation, in which eight pilot batches were completed and sampled. The resulting particle sizes are in angstroms (where 1 Å = 10^{-8} cm):

3923 3807 3786 3710 4010 4230 4226 4133

a. Find the sample mean.
b. Find the sample standard deviation.
c. Assuming that particle size has an approximately normal distribution, determine the manufacturing specification that bounds 95% of the particle sizes (that is, find the 95% interval, $\bar{x} \pm 2s$).

2.190 Ex02-190 Delco Products, a division of General Motors, produces a bracket that is used as part of a power doorlock assembly. The length of this bracket is constantly being monitored. A sample of 30 power door brackets had the following lengths (in millimeters):

11.86 11.88 11.88 11.91 11.88 11.88 11.88 11.88 11.88 11.86
11.88 11.88 11.88 11.88 11.86 11.83 11.86 11.86 11.88 11.88
11.88 11.83 11.86 11.86 11.86 11.88 11.88 11.86 11.88 11.83

Source: With permission of Delco Products Division, GMC

a. Without doing any calculations, what would you estimate for the sample mean?
b. Construct an ungrouped frequency distribution.
c. Draw a histogram of this frequency distribution.
d. Use the frequency distribution and calculate the sample mean and standard deviation.
e. Determine the limits of the $\bar{x} \pm 3s$ interval and mark this interval on the histogram.
f. The product specification limits are 11.7–12.3. Does the sample indicate that production is within these requirements? Justify your answer.

Soup Brand	Calories	Sodium (mg)	Soup Brand	Calories	Sodium (mg)
Health Valley Organic	90	150	Healthy Choice Clam Chowder	120	480
Taste Adventure Minestrone	140	210	Campbell's HR Chicken with Rice	60	480
Health Valley Fat-Free	90	230	Campbell's HR Clam Chowder	120	480
Pritkin Lentil or Split Pea	160	290	Campbell's HR Cream of Chicken	70	480
Health Valley Healthy Pasta	110	290	Taste Adventure Golden Pea	210	490
Pritkin Hearty Vegetable	90	290	Westbrae Natural	90	580
Arrowhead Mills Red Lentil	100	230	Lipton Recipe Secrets Onion	20	610
Taste Adventure Navy Bean	160	390	Progresso Chicken Noodle	90	620
Healthy Choice Minestrone	110	400	Campbell's FF Ramen, Chicken	140	720
Healthy Choice Chicken Noodle	120	410	Lipton Noodle with Chicken Broth	60	720
Baxters Mediterranean Tomato	70	420	Campbell's Tomato	100	730
Shari's Organic	130	420	Knorr Vegetable	30	730
Healthy Choice Vegetable Beef	130	420	Progresso Lentil	140	750
Healthy Choice Bean and Ham	160	430	Maruchan Ramen, Chicken	190	780
Baxters Italian Bean & Pasta	80	430	Campbell's Cream of Mushroom	110	870
Mayacamas Black Bean	60	450	Campbell's Cream of Chicken	130	890
Healthy Choice Cream of Mushroom	80	450	Nissin Top Ramen, Chicken	190	910
Healthy Choice Beef & Potato	120	450	Campbell's Vegetable	90	920
Progresso Beef Barley or Lentil	140	460	Progresso Minestrone	120	960
Campbell's HR Bean with Ham & Bacon	150	480	Campbell's 98% FF Clam Chowder	150	970
Campbell's HR Southwestern Vegetable	160	480	Campbell's Chicken Noodle	70	980
Hain Healthy Naturals	120	480	Campbell's French Onion	70	980
Campbell's HR Hearty Minestrone	120	480	Knorr French Onion	45	980
Hain Home Style Chunky Tomato	120	480	Progresso Clam Chowder	200	1,050
Campbell's HR Chicken Noodle	120	480	Campbell's Cream of Mushroom (jar)	260	1,130

Source: *Nutrition in Action*, "Soups: The Middle Ground," December 1997, pp. 14–15

2.191 Ex02-191 Americans love soups, which are one of the most popular foods for lunch and as an appetizer before dinner. The manufacturers provided the calorie and sodium contents of 50 popular brands of soups that were published in the December 1997 issue of *Nutrition in Action*. The data for multiserving (8 oz) cans and mixes, most of which were low-fat varieties, appear in the table above.

a. Compute the mean and the standard deviation of both calories and sodium contents of the soups listed in the table.
b. Use your answers in part a to test Chebyshev's theorem that at least 75% of the soups' calories and sodium contents will fall within ±2 standard deviations from the mean. Is this the case?
c. Find the limits for ±1 standard deviation from the mean for the soups' sodium contents. Do the sodium contents of soups appear to be normally distributed? Explain.

2.192 Ex02-192 The manager of Jerry's Barber Shop recently asked his last 50 customers to punch a time card when they first arrived at the shop and to punch out right after they paid for their hair cut. He then used the data on the cards to measure how long it took Jerry and his barbers to cut hair in order to schedule their appointment intervals. The following times (in minutes) were tabulated:

50	21	36	35	35	27	38	51	28	35
32	32	27	25	24	38	43	46	29	45
40	27	36	38	35	31	28	38	33	46
35	31	38	48	23	35	43	31	32	38
43	32	18	43	52	52	49	53	46	19

a. Construct a stem-and-leaf plot of these data.
b. Compute the mean, median, mode, range, midrange, variance, and standard deviation of the haircut service times.
c. Construct a 5-number summary table.
d. According to Chebyshev's theorem, at least 75% of the haircut service times will fall between what two values? Is this true? Explain why or why not.
e. How far apart would you recommend that Jerry schedule his appointments to keep his shop operating at a comfortable pace?

2.193 Ex02-193 Each year stock car drivers compete for the NASCAR Winston Cup. Points are earned on the basis of finishes in sanctioned races scheduled on the circuit. Partway through the 2002 season, the standings posted at *SportingNews.com* of the top 40 drivers are shown in the table:

Driver	Points	Driver	Points
Sterling Marlin	3439	Kyle Petty	2611
Mark Martin	3430	Robby Gordon	2604
Jimmie Johnson	3367	Terry Labonte	2603
Jeff Gordon	3357	Ward Burton	2511
Tony Stewart	3321	Elliott Sadler	2440
Rusty Wallace	3293	Jimmy Spencer	2388
Bill Elliott	3255	Jeremy Mayfield	2377
Matt Kenseth	3191	Bobby Hamilton	2345
Ricky Rudd	3191	John Andretti	2230
Ryan Newman	3168	Ken Schrader	2191
Dale Jarrett	3094	Mike Skinner	2064
Kurt Busch	3078	Casey Atwood	2045
Jeff Burton	2948	Brett Bodine	1895
Dale Earnhardt	2915	Johnny Benson	1889
Michael Waltrip	2856	Jerry Nadeau	1837
Ricky Craven	2827	Hut Stricklin	1781
Jeff Green	2773	Joe Nemechek	1734
Bobby Labonte	2731	Steve Park	1656
Dave Blaney	2637	Todd Bodine	1448
Kevin Harvick	2632	Stacy Compton	1393

Source: *SportingNews.com*, September 2002

a. Calculate the mean and the standard deviation of the points accumulated by the Winston Cup stock car drivers.
b. Construct a 5-number summary table.
c. According to Chebyshev's theorem, at least 75% of the points will fall between what two values? Is this the case?
d. According to the empirical rule, approximately 68% of the points will fall between what two values? Is this the case?
e. Compare your answers in parts c and d with the results predicted by the empirical rule. Does your comparison suggest that the distribution of Winston Cup points approximates the normal distribution? Explain.

2.194 The dotplot below shows the number of attempted passes thrown by the quarterbacks for 22 of the NFL teams that played on one particular Sunday afternoon.

a. Describe the distribution, including how points *A* and *B* relate to the others.
b. If you remove point *A*, and maybe point *B*, would you say the remaining data have an approximately normal distribution? Explain.
c. Based on the information about distributions that Chebyshev's theorem and the empirical rule give us, how typical an event do you think point *A* represents? Explain.

2.195 Starting with the data values of 70 and 85, add three data values to your sample so that the sample has: (Justify your answer in each case.)

a. a standard deviation of 5
b. a standard deviation of 10
c. a standard deviation of 15
d. Compare your three samples and the variety of values needed to obtain each of the required standard deviations.

2.196 Make up a set of 18 data (think of them as exam scores) so that the sample meets each of these sets of criteria:

a. Mean is 75, standard deviation is 10.
b. Mean is 75, maximum is 98, minimum is 40, standard deviation is 10.
c. Mean is 75, maximum is 98, minimum is 40, standard deviation is 15.
d. How are the data in the sample for part b different from those in part c?

2.197 Construct two different graphs of the points (62, 2), (74, 14), (80, 20), and (94, 34).

a. On the first graph, along the horizontal axis, lay off equal intervals and label them 62, 74, 80, and 94; lay off equal intervals along the vertical axis and label them 0, 10, 20, 30, and 40. Plot the points and connect them with line segments.
b. On the second graph, along the horizontal axis, lay off equally spaced intervals and label them 60, 65, 70, 75, 80, 85, 90, and 95; mark off the vertical axis in equal intervals and label them 0, 10, 20, 30, and 40. Plot the points and connect them with line segments.
c. Compare the effect that scale has on the appearance of the graphs in parts a and b. Explain the impression presented by each graph.
d. Explain how your graphs demonstrate the inappropriateness of the graph in Application 2.6 (p. 114).

Dotplot for Exercise 2.194

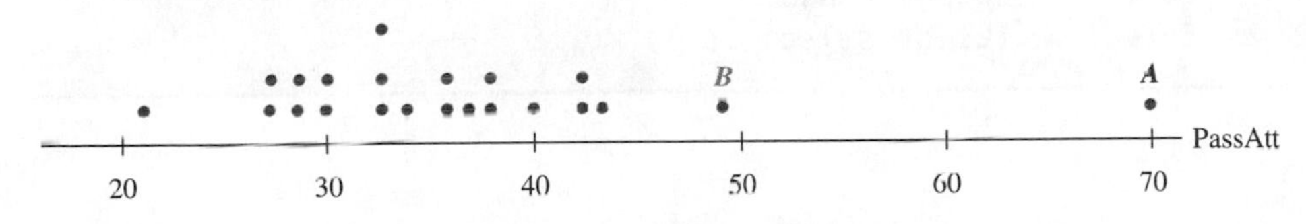

2.198 Ex02-198 When the Internet study (Chapter Case Study, p. 37) was performed, it appeared that the variable x, the number of Internet activities in a week, had an approximately normal distribution. That distribution is approximated by this relative frequency distribution:

Number of Internet Activities in Week, x	Relative Frequency
1	0.01
2	0.03
3	0.05
4	0.09
5	0.10
6	0.14
7	0.13
8	0.14
9	0.11
10	0.08
11	0.05
12	0.04
13	0.03

a. Select a random sample of size 40 from this relative frequency representation of the population of all Internet users.
b. Construct a histogram of the sample obtained in part a. Do not group the data. (See instructions below.)
c. Find the mean, median, and standard deviation of the sample obtained in part a.
d. Repeat parts a–c three more times, being sure to keep the answers for each set of data together.
e. Describe the similarities and differences between the distributions shown on the four histograms.
f. Make a chart displaying the numerical statistics for each of the four samples and describe the variability from sample to sample of each statistic.
g. The four samples were all drawn randomly from the same distribution. Write a statement describing the overall variability between these four random samples.

2.199 Use a computer to generate a random sample of 500 values of a normally distributed variable x with a mean of 100 and a standard deviation of 20. Construct a histogram of the 500 values.

a. Use the computer commands on page 104 to randomly generate a sample of 500 data from a normal distribution with mean 100 and standard deviation 20.

MINITAB (Release 13)

Input the x values into C1 and the corresponding relative frequencies into C2; then continue with:

```
Choose:   Calc > Random Data > Discrete
Enter:    Generate: 40
          Store in: C3
          Values (of x) in: C1
          Probabilities in: C2
```

Excel XP

Input the x values into column A and the corresponding relative frequencies into column B; then continue with:

```
Choose:   Tools > Data Analysis > Random Number Generation
          > OK
Enter:    Number of Variables: 1
          Number of Random Numbers: 40
          Distribution: Discrete
          Value & Prob. Input Range:  (A2:B5 select data
          cells not labels)
Select:   Output Range
Enter:    (C1 or select cell)
```

Construct a histogram using class boundaries that are multiples of the standard deviation 20; that is, use boundaries from 20 to 180 in intervals of 20 (see commands on p. 58).

Let's consider the 500 x values found in part a as a population.

b. Use the computer commands on page 104 to randomly select a sample of 30 values from the population found in part a. Construct a histogram of the sample with the same class intervals used in part a.
c. Repeat part b three times.
d. Calculate several values (mean, median, maximum, minimum, standard deviation, etc.) that describe the population and each of the four samples. (See p. 100 for commands.)
e. Do you think a sample of 30 data adequately represents a population? (Compare each of the four samples found in parts b and c to the population.)

2.200 Repeat Exercise 2.199 using a different sample size. You might try a few different sample sizes: $n = 10$, $n = 15$, $n = 20$, $n = 40$, $n = 50$, $n = 75$. What effect does increasing the sample size have on the effectiveness of the sample in depicting the population? Explain.

2.201 Repeat Exercise 2.199 using populations with different shaped distributions.

a. Use a uniform or rectangular distribution. (Replace the subcommands used in Exercise 2.199; in place of NORMAL use: UNIFORM with a low of 50 and a high of 150, and use class boundaries of 50 to 150 in increments of 10.)
b. Use a skewed distribution. (Replace the subcommands used in Exercise 2.199; in place of NORMAL use: POISSON 50 and use class boundaries of 20 to 90 in increments of 5.)
c. Use a J-shaped distribution. (Replace the subcommands used in Exercise 2.199; in place of NORMAL use: EXPONENTIAL 50 and use class boundaries of 0 to 250 in increments of 10.)
d. Does the shape of the distribution of the population have an effect on how well a sample of size 30 represents the population? Explain.
e. What effect do you think changing the sample size has on the effectiveness of the sample to depict the population? Try a few different sample sizes. Do the results agree with your expectations? Explain.

Vocabulary and Key Concepts

Be able to define each term. Pay special attention to the key terms, which are printed in **red**. In addition, describe each term in your own words and give an example of each. Your examples should not be ones given in class or in the textbook. Page numbers indicate the first appearance of the term.

bar graph (p. 38)
bell-shaped distribution (p. 106)
bimodal frequency distribution (p. 61)
box-and-whiskers plot (p. 99)
Chebyshev's theorem (p. 109)
circle graph (p. 38)
class (p. 52)
class boundary (p. 54)
class midpoint (class mark) (p. 55)
class width (p. 53)
depth (p. 70)
deviation from the mean (p. 78)
distribution (p. 42)
dotplot (p. 42)
empirical rule (p. 106)
5-number summary (p. 98)
frequency (p. 52)
frequency distribution (p. 52)
frequency histogram (p. 56)
grouped frequency distribution (p. 52)
histogram (p. 56)
interquartile range (p. 99)
mean (pp. 68, 88)
measure of central tendency (p. 68)
measure of dispersion (p. 78)
measure of position (p. 94)
median (p. 70)
midquartile (p. 98)
midrange (p. 73)
modal class (p. 61)
mode (p. 61)
normal distribution (pp. 61, 106)
ogive (p. 63)
Pareto diagram (p. 40)
percentile (p. 95)
qualitative data (p. 38)
quantitative data (p. 42)
quartile (p. 94)
range (p. 78)
rectangular distribution (p. 61)
relative frequency (p. 56)
relative frequency distribution (p. 57)
relative frequency histogram (p. 57)
single-variable data (p. 36)
skewed distribution (p. 61)
standard deviation (pp. 81, 89)
standard score (p. 102)
stem-and-leaf display (p. 44)

(continued)

summation (p. 55)
tally (p. 54)
ungrouped frequency distribution (p. 52)
variance (pp. 80, 88)
x-bar ($\bar{x}$) (p. 68)
z-score (p. 102)

Chapter Practice Test

PART I: KNOWING THE DEFINITIONS

Answer "True" if the statement is always true. If the statement is not always true, replace the words in bold with the words that make the statement always true.

2.1 The **mean** of a sample always divides the data into two halves (half larger and half smaller in value than itself).

2.2 A measure of **central tendency** is a quantitative value that describes how widely the data are dispersed about a central value.

2.3 The sum of the squares of the deviations from the mean, $\Sigma(x - \bar{x})^2$, will **sometimes** be negative.

2.4 For any distribution, the sum of the deviations from the mean equals **zero.**

2.5 The standard deviation for the set of values 2, 2, 2, 2, and 2 is **2.**

2.6 On a test John scored at the 50th percentile and Jorge scored at the 25th percentile; therefore, John's test score was **twice** Jorge's test score.

2.7 The frequency of a class is the number of pieces of data whose values fall within the **boundaries** of that class.

2.8 **Frequency distributions** are used in statistics to present large quantities of repeating values in a concise form.

2.9 The unit of measure for the standard score is always **standard deviations.**

2.10 For a bell-shaped distribution, the range will be approximately equal to **six standard deviations.**

PART II: APPLYING THE CONCEPTS

2.11 The results of a consumer study completed at Corner Convenience Store are reported in the accompanying histogram. Answer each question.

a. What is the class width?
b. What is the class mark for the class 31–61?

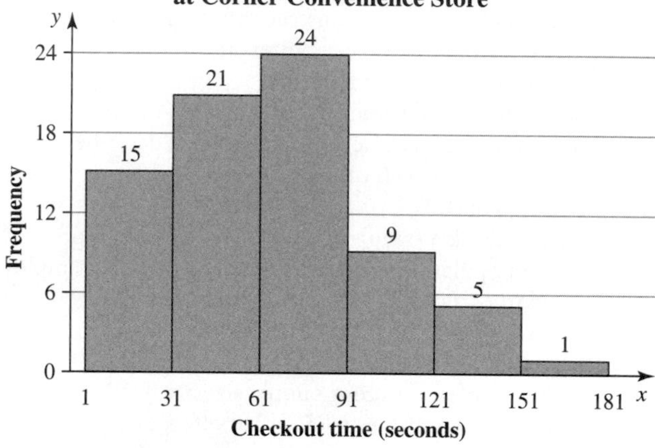

c. What is the upper boundary for the class 61–91?
d. What is the frequency of the class 1–31?
e. What is the frequency of the class that contains the largest observed value of x?
f. What is the lower boundary of the class with the largest frequency?
g. How many pieces of data are shown in this histogram?
h. What is the value of the mode?
i. What is the value of the midrange?
j. Estimate the value of the 90th percentile, P_{90}.

2.12 A sample of the purchases of several Corner Convenience Store customers resulted in the following sample data (x = number of items purchased per customer):

x	f
1	6
2	10
3	9
4	8
5	7

a. What does the 2 represent?
b. What does the 9 represent?
c. How many customers were used to form this sample?
d. How many items were purchased by the customers in this sample?
e. What is the largest number of items purchased by one customer?

Find each of the following (show formulas and work):

f. mode g. median h. midrange
i. mean j. variance k. standard deviation

2.13 Given the set of data 4, 8, 9, 8, 6, 5, 7, 5, 8, find each of the following sample statistics:

a. mean b. median c. mode
d. midrange e. first quartile f. P_{40}
g. variance h. standard deviation i. range

2.14

a. Find the standard score for the value $x = 452$ relative to its sample, where the sample mean is 500 and the standard deviation is 32.
b. Find the value of x that corresponds to the standard score of 1.2, where the mean is 135 and the standard deviation is 15.

PART III: UNDERSTANDING THE CONCEPTS

Answer all questions.

2.15 The Corner Convenience Store kept track of the number of paying customers it had during the noon hour each day for 100 days. The resulting statistics are rounded to the nearest integer:

mean = 95
median = 97
mode = 98
first quartile = 85
third quartile = 107
midrange = 93
range = 56
standard deviation = 12

a. The Corner Convenience Store served what number of paying customers during the noon hour more often than any other number? Explain how you determined your answer.
b. On how many days were there between 85 and 107 paying customers during the noon hour? Explain how you determined your answer.
c. What was the greatest number of paying customers during any one noon hour? Explain how you determined your answer.
d. For how many of the 100 days was the number of paying customers within three standard deviations of the mean ($\bar{x} \pm 3s$)? Explain how you determined your answer.

2.16 Mr. VanCott started his own machine shop several years ago. His business grew and has become very successful in recent years. Currently he employs 14 people, including himself, and pays the following annual salaries:

Owner, President	$80,000	Worker	$25,000
Business Manager	50,000	Worker	25,000
Production Manager	40,000	Worker	25,000
Shop Foreman	35,000	Worker	20,000
Worker	30,000	Worker	20,000
Worker	30,000	Worker	20,000
Worker	28,000	Worker	20,000

a. Calculate the four "averages": mean, median, mode, and midrange.
b. Draw a dotplot of the salaries and locate each of the four averages on it.
c. Suppose you were the feature writer assigned to write this week's feature story on Mr. VanCott's machine shop, one of a series on local small businesses that are prospering. You plan to interview Mr. VanCott, his business manager, the shop foreman, and one of his newer workers. Which statistical average do you think each will give when asked, "What is the average annual salary paid to the employees here at VanCott's?" Explain why each person interviewed has a different perspective and why this viewpoint may cause each to cite a different statistical average.
d. What is there about the distribution of these salaries that causes the four "average values" to be so different?

2.17 Create a set of data containing three or more values:

a. where the mean is 12 and the standard deviation is zero
b. where the mean is 20 and the range is 10
c. where the mean, median, and mode are all equal
d. where the mean, median, and mode are all different
e. where the mean, median, and mode are all different and the median is the largest and the mode is the smallest
f. where the mean, median, and mode are all different and the mean is the largest and the median is the smallest

2.18 A set of test papers was machine scored. Later it was discovered that two points should be added to each score. Student A said, "The mean score should also be increased by two points." Student B added, "The standard deviation should also be increased by 2 points." Who is right? Justify your answer.

2.19 Student A stated, "Both the standard deviation and the variance preserve the same unit of measurement as the data." Student B disagreed, arguing, "The unit of measurement for variance is a meaningless unit of measurement." Who is right? Justify your answer.

CHAPTER

3 Descriptive Analysis and Presentation of Bivariate Data

Chapter Outline and Objectives

CHAPTER CASE STUDY

A **statistical relationship** in the NBA is investigated.

3.1 BIVARIATE DATA

Two variables are paired together for analysis. Techniques of tabling and graphing bivariate data are used to represent the data.

3.2 LINEAR CORRELATION

Are these two variables related? Does an increase in the value of one variable **indicate a change** in the value of the other?

3.3 LINEAR REGRESSION

The **line of best fit** is a mathematical expression for the relationship between two variables. When two variables have a mathematical relationship, it does not imply the existence of a cause-and-effect relationship.

RETURN TO CHAPTER CASE STUDY

We will restrict our discussion in this chapter to descriptive techniques for the most basic form of correlation and regression analysis—the bivariate linear case.

Chapter Case Study

Duncan Wins First MVP

© Brian Bahr/Getty Images

Forward Tim Duncan, among the NBA's leaders in scoring, rebounding, and blocked shots, won the league's Most Valuable Player award, edging out Nets guard Jason Kidd in one of the closest votes ever. Tim Duncan led the NBA with 67 double-doubles this season and finished in the top five in the league in rebounds, blocks, and points per game. He also joined Billy Cunningham, Elgin Baylor, and Bob Pettit as the only forwards in league history with at least 2000 points and 1000 rebounds in a single season.

Duncan was the San Antonio Spurs team statistical leader. He led his team in several of the season statistics: games played, games started, minutes played, points scored, offensive rebounds, defensive rebounds, blocked shots, turnovers, personal fouls, technical fouls, field goal attempts, field goals made, foul shot attempts, foul shots made, two-point shot attempts, and two-point shots made.

San Antonio Spurs

2001–2002 Regular Season CS03

Player	Personal Fouls per Game	Points per Game
Duncan	2.6	25.5
Robinson	2.5	12.2
Smith	2.1	11.6
Rose	2.5	9.4
Daniels	1.3	9.2
Parker	2.2	9.2
Smith	1.9	7.4
Bowen	2.0	7.0
Porter	1.2	5.5
Ferry	1.4	4.6
Jackson	1.3	3.9
Hart	1.6	2.0
McCaskill	0.9	1.9
Bryant	1.3	1.9
Parks	0.8	1.5

Source: http://sports.espn.go.com/nba/teamstats

FYI
It is important that you complete the Answer Now exercises as you encounter them. They are an important part of learning.

ANSWER NOW 3.1

a. Is there a relationship (pattern) between the two variables: points scored per game and personal fouls committed per game? Explain why or why not.

b. Do you think it is reasonable (or possible) to predict the number of points scored based on the number of personal fouls committed per game for a San Antonio Spurs player? Explain why or why not.

After completing Chapter 3, further investigate the Chapter Case Study in the Return to Chapter Case Study section with Exercises 3.67, 3.68, and 3.69 (pp. 167–168).

3.1 Bivariate Data

BIVARIATE DATA
The values of two different variables that are obtained from the same population element.

Each of the two variables may be either *qualitative* or *quantitative.* As a result, three combinations of variable types can form bivariate data:

1. Both variables are qualitative (attribute).
2. One variable is qualitative (attribute) and the other is quantitative (numerical).
3. Both variables are quantitative (both numerical).

In this section we present tabular and graphic methods for displaying each of these combinations of bivariate data.

TWO QUALITATIVE VARIABLES

When bivariate data result from two qualitative (attribute or categorical) variables, the data are often arranged on a **cross-tabulation** or **contingency table.** Let's look at an illustration.

Illustration 3.1 Constructing Cross-Tabulation Tables

Thirty students from our college were randomly identified and classified according to two variables: gender (M/F) and major (liberal arts, business administration, technology), as shown in Table 3.1. These 30 bivariate data can be summarized on a 2×3 cross-tabulation table, where the two rows represent the two genders male and female, and the three columns represent the three major categories of liberal arts (LA), business administration (BA), and technology (T). The entry in each cell is found by determining how many students fit into each category. Adams is male (M) and liberal arts (LA) and is classified in the cell in the first row, first column.

TABLE 3.1 Genders and Majors of 30 College Students Ta03-01

Name	Gender	Major	Name	Gender	Major
Adams	M	LA	Kee	M	BA
Argento	F	BA	Kleeberg	M	LA
Baker	M	LA	Light	M	BA
Bennett	F	LA	Linton	F	LA
Brock	M	BA	Lopez	M	T
Brand	M	T	McGowan	M	BA
Chun	F	LA	Mowers	F	BA
Crain	M	T	Ornt	M	T
Cross	F	BA	Palmer	F	LA
Ellis	F	BA	Pullen	M	T
Feeney	M	T	Rattan	M	BA
Flanigan	M	LA	Sherman	F	LA
Hodge	F	LA	Small	F	T
Holmes	M	T	Tate	M	BA
Jopson	F	T	Yamamoto	M	LA

See the red tally mark in Table 3.2. The other 29 students are classified (tallied, shown in black) in a similar fashion.

TABLE 3.2 Cross-Tabulation of Gender and Major (tallied)

		Major		
		Liberal Arts	Business Administration	Technology
Gender	Male	||||| (5)	|||||| (6)	||||||| (7)
	Female	|||||| (6)	|||| (4)	|| (2)

The resulting 2 × 3 cross-tabulation (contingency table), Table 3.3, shows the frequency for each cross category of the two variables along with the row and column totals, called *marginal totals* (or *marginals*). The total of the marginal totals is the *grand total* and is equal to n, the *sample size.*

TABLE 3.3 Cross-Tabulation of Gender and Major (frequencies)

		Major			
		Liberal Arts	Business Administration	Technology	Row Total
Gender	Male	5	6	7	18
	Female	6	4	2	12
	Column Total	11	10	9	30

Contingency tables often show percentages (relative frequencies). These percentages can be based on the entire sample or on the subsample (row or column) classifications.

Percentages Based on the Grand Total (Entire Sample)

The frequencies in the contingency table shown in Table 3.3 can easily be converted to percentages of the grand total by dividing each frequency by the grand total and multiplying the result by 100. For example, 6 becomes 20% $\left[\left(\frac{6}{30}\right) \times 100 = 20\right]$. See Table 3.4.

TABLE 3.4 Cross-Tabulation of Gender and Major (relative frequencies; % of grand total)

		Major			
		Liberal Arts	Business Administration	Technology	Row Total
Gender	Male	17%	20%	23%	60%
	Female	20%	13%	7%	40%
	Column Total	37%	33%	30%	100%

From the table of percentages of the grand total, we can easily see that 60% of the sample were male, 40% were female, 30% were technology majors, and so on. These same statistics (numerical values describing sample results) can be shown in a bar graph (see Figure 3.1).

FIGURE 3.1
Bar Graph

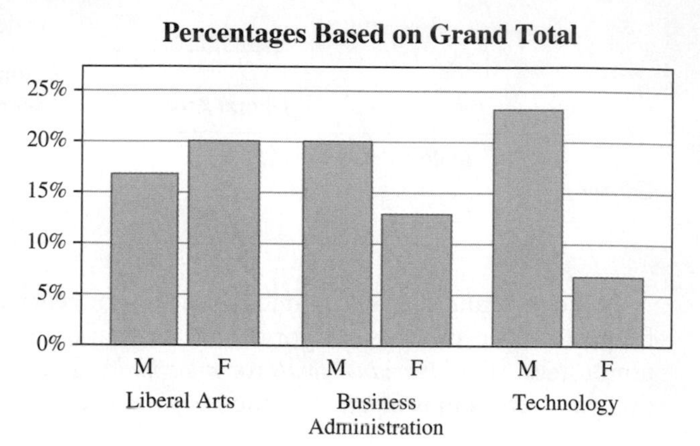

Table 3.4 and Figure 3.1 show the distribution of male liberal arts students, female liberal arts students, male business administration students, and so on relative to the entire sample.

ANSWER NOW 3.2
In a national survey of 500 business and 500 leisure travelers, each was asked where they would most like "more space."

	Airplane	Hotel Room	All Other
Business	355	95	50
Leisure	250	165	85

Express the frequencies in the table as percentages of the total.

Percentages Based on Row Totals

The frequencies in the same contingency table, Table 3.3, can be expressed as percentages of the row totals (or gender) by dividing each row entry by that row's total and multiplying the results by 100. Table 3.5 is based on row totals.

From Table 3.5 we see that 28% of the male students were majoring in liberal arts, while 50% of the female students were majoring in liberal arts. These same statistics are shown in the bar graph in Figure 3.2.

TABLE 3.5 Cross-Tabulation of Gender and Major (% of row totals)

		Major			
		Liberal Arts	Business Administration	Technology	Row Total
Gender	Male	28%	33%	39%	100%
	Female	50%	33%	17%	100%
	Column Total	37%	33%	30%	100%

FIGURE 3.2
Bar Graph

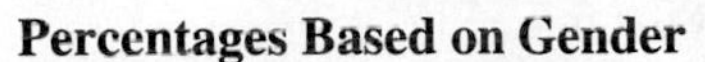

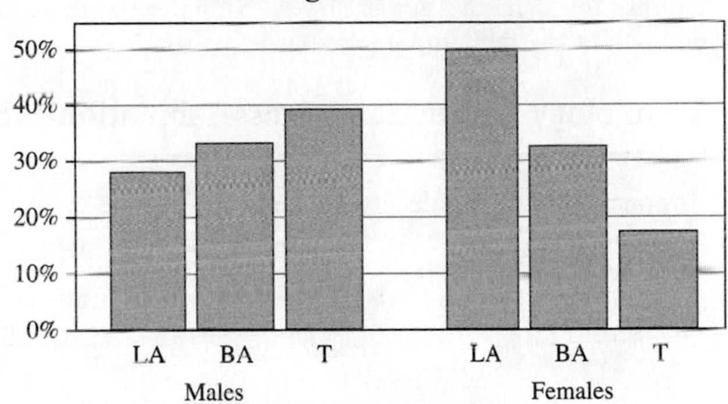

Table 3.5 and Figure 3.2 show the distribution of the three majors for male and female students separately.

ANSWER NOW 3.3
Express the table in Answer Now 3.2 in percentages of the row totals. Why might one prefer the table to be expressed this way?

Percentages Based on Column Totals

The frequencies in the contingency table, Table 3.3, can be expressed as percentages of the column totals (or major) by dividing each column entry by that column's total and multiplying the result by 100. Table 3.6 is based on column totals.

From Table 3.6 we see that 45% of the liberal arts students were male, while 55% of the liberal arts students were female. These same statistics are shown in the bar graph in Figure 3.3.

Table 3.6 and Figure 3.3 show the distribution of male and female students for each major separately.

TABLE 3.6 **Cross-Tabulation of Gender and Major (% of column totals)**

		Major			
		Liberal Arts	Business Administration	Technology	Row Total
Gender	Male	45%	60%	78%	60%
	Female	55%	40%	22%	40%
	Column Total	100%	100%	100%	100%

FIGURE 3.3
Bar Graph

Percentages Based on Major

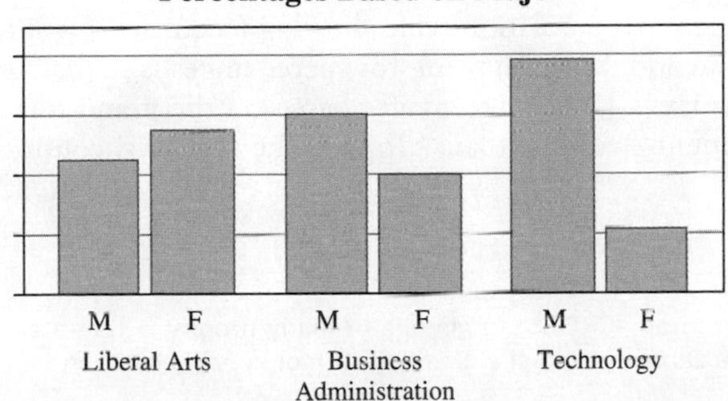

ANSWER NOW 3.4
Express the table in Answer Now 3.2 in percentages of the column totals. Why might one prefer the table to be expressed this way? ■

Technology Instructions: Cross-Tabulation Tables

MINITAB (Release 13)

Input the row-variable categorical values into C1 and the corresponding column-variable categorical values into C2; then continue with:

```
Choose:   Stat > Tables > Cross Tabulation
Enter:    Classification variables: C1 C2
Select:   Counts
          Row Percents
          Column Percents
          Total Percents
```

Suggestion: The four subcommands that are available for 'Display' can be used together; however, the resulting table will be much easier to read if one subcommand at a time is used.

Excel XP

Using column headings or titles, input the row-variable categorical values into column A and the corresponding column-variable categorical values into column B; then continue with:

```
Choose:   Data > Pivot Table and PivotChart Report ...
Select:   Microsoft Excel list or database > Next
Enter:    Range: (A1:B5 or select cells) > Next
Select:   Existing Worksheet
Enter:    (C1 or select cell) > Finish
Drag:     Headings to row or column (depends on preference)
          One heading into data area*
```

* For other summations, double click "Count of" in data area box; then continue with:

```
Choose:   Summarize by: Count > Options
          Show data as: % of row or % of column or % of total > OK
```

TI-83 Plus

The categorical data must be numerically coded first; use 1, 2, 3, . . . for the various row variables and 1, 2, 3, . . . for the various column variables. Input the numeric row-variable values into L1 and the corresponding numeric column-variable values into L2; then continue with:

```
Choose:   PRGM > EXEC > CROSSTAB*
Enter:    ROWS: L1 > ENTER
          COLS: L2 > ENTER
```

The cross-tabulation table showing frequencies is stored in matrix [A], the cross-tabulation table showing row percentages is in matrix [B], column percentages in matrix [C], and percentages based on the grand total in matrix [D]. All matrices contain marginal totals. To view the matrices, continue with:

```
Choose:   MATRX > NAMES
Enter:    1:[A] or 2:[B] or 3:[C] or 4:[D] > ENTER
```

*Program 'CROSSTAB' is one of many programs that are available for downloading from the Duxbury Web site. See page 40 for specific instructions.

ONE QUALITATIVE AND ONE QUANTITATIVE VARIABLE

When bivariate data result from one qualitative and one quantitative variable, the quantitative values are viewed as separate samples, each set identified by levels of the qualitative variable. Each sample is described using the techniques from Chapter 2, and the results are displayed side by side for easy comparison.

Illustration 3.2 Constructing Side-by-Side Comparisons

The distance required to stop a 3000-pound automobile on wet pavement was measured to compare the stopping capabilities of three tire tread designs (see Table 3.7). Tires of each design were tested repeatedly on the same automobile on a controlled wet pavement.

TABLE 3.7 Stopping Distances (in feet) for Three Tread Designs Ta03 07

Design A ($n = 6$)		Design B ($n = 6$)		Design C ($n = 6$)	
37	36	33	35	40	39
34	40	34	42	41	41
38	32	38	34	40	43

The design of the tread is a qualitative variable with three levels of response, and the stopping distance is a quantitative variable. The distribution of the stopping distances for tread design A is to be compared with the distribution of stopping distances for each of the other tread designs. This comparison may be made with both numerical and graphic techniques. Some of the available options are shown in Figure 3.4, Table 3.8, and Table 3.9.

FIGURE 3.4 Dotplot and Box-and-Whiskers Using a Common Scale

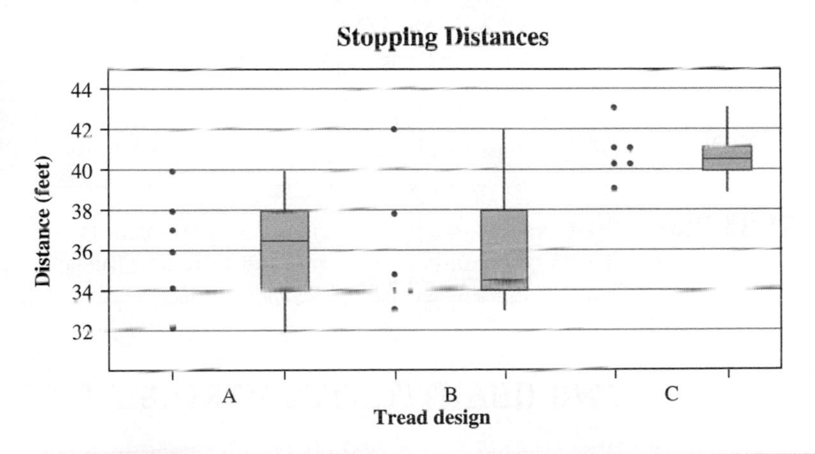

TABLE 3.8 5-Number Summary for Each Design

	Design A	Design B	Design C
High	40	42	43
Q_3	38	38	41
Median	36.5	34.5	40.5
Q_1	34	34	40
Low	32	33	39

TABLE 3.9 Mean and Standard Deviation for Each Design

	Design A	Design B	Design C
Mean	36.2	36.0	40.7
Standard deviation	2.9	3.4	1.4

■

Much of the information presented here can be demonstrated using many other statistical techniques, such as stem-and-leaf displays or histograms.

ANSWER NOW 3.5

Here are the January unemployment rates for selected U.S. cities:

Eastern: 5.5, 6.6, 5.3, 5.4, 5.8 Western: 5.4, 6.0, 6.4, 5.6, 6.9

Display these rates as two dotplots using the same scale. Compare the means and medians.

Technology Instructions: Side-by-Side Boxplots and Dotplots

MINITAB (Release 13)

Input the numerical values into C1 and the corresponding categories into C2; then continue with:

```
Choose:   Graph > Boxplot
Enter:    Y: C1   X: C2
```

MINITAB commands to construct side-by-side dotplots for data in this form are located on pages 47–48.

If the data for the various categories are in separate columns, use the MINITAB commands for multiple boxplots on page 101. If sideby-side dotplots are needed for data in this form, continue with:

```
Choose:   Graph > Character Graphs > Dotplot
Enter:    Variables: C1 C2 C3...
Select:   Same scale for all variables
```

Excel XP

Excel commands to construct a single boxplot are on page 101.

TI-83 Plus

TI-83 commands to construct multiple boxplots are on page 102.
TI-83 commands to construct multiple dotplots are on page 48.

TWO QUANTITATIVE VARIABLES

When the bivariate data are the result of two quantitative variables, it is customary to express the data mathematically as **ordered pairs** (x, y), where x is the **input variable** (sometimes called the **independent variable**) and y is the **output variable** (sometimes called the **dependent variable**). The data are said to be *ordered* because one value, x, is always written first. They are called *paired* because for each x value, there is a corresponding y value from the same source. For example, if x is height and y is weight, then a height and a corresponding weight are recorded for each person. The input variable x is measured or controlled in order to predict the output variable y.

Suppose some research doctors are testing a new drug by prescribing different dosages and observing the lengths of the recovery times of their patients. Since the researcher can control the amount of drug prescribed, the amount of drug is referred to as x. In the case of height and weight, either variable could be treated as input and the other as output, depending on the question being asked. However, different results will be obtained from the regression analysis, depending on the choice made.

ANSWER NOW 3.6
Which variable, height or weight, would you use as the input variable? Explain why.

In problems that deal with two quantitative variables, we present the sample data pictorially on a *scatter diagram.*

SCATTER DIAGRAM
A plot of all the ordered pairs of bivariate data on a coordinate axis system. The input variable x is plotted on the horizontal axis, and the output variable y is plotted on the vertical axis.

NOTE When you construct a scatter diagram, it is convenient to construct scales so that the range of the y values along the vertical axis is equal to or slightly shorter than the range of the x values along the horizontal axis. This creates a "window of data" that is approximately square.

ANSWER NOW 3.7
Draw a coordinate axis and plot the points (0, 6), (3, 5), (3, 2), and (5, 0).

Illustration 3.3 Constructing a Scatter Diagram

In Mr. Chamberlain's physical fitness course, several fitness scores were taken. The following sample is the numbers of push-ups and sit-ups done by ten randomly selected students:

(27, 30) (22, 26) (15, 25) (35, 42) (30, 38)
(52, 40) (35, 32) (55, 54) (40, 50) (40, 43)

Table 3.10 shows these sample data, and Figure 3.5 (p. 138) shows a scatter diagram of the data.

TABLE 3.10 Data for Push-ups and Sit-ups Ta03-10

Student	1	2	3	4	5	6	7	8	9	10
Push-ups, x	27	22	15	35	30	52	35	55	40	40
Sit-ups, y	30	26	25	42	38	40	32	54	50	43

ANSWER NOW 3.8
Does studying for an exam pay off?

a. **Using the data given, draw a scatter diagram of the number of hours studied, x, compared to the exam grade, y, received.**

x	2	5	1	4	2
y	80	80	70	90	60

b. **What can you conclude based on the pattern of data shown on the scatter diagram drawn in part a?**

FIGURE 3.5
Scatter Diagram

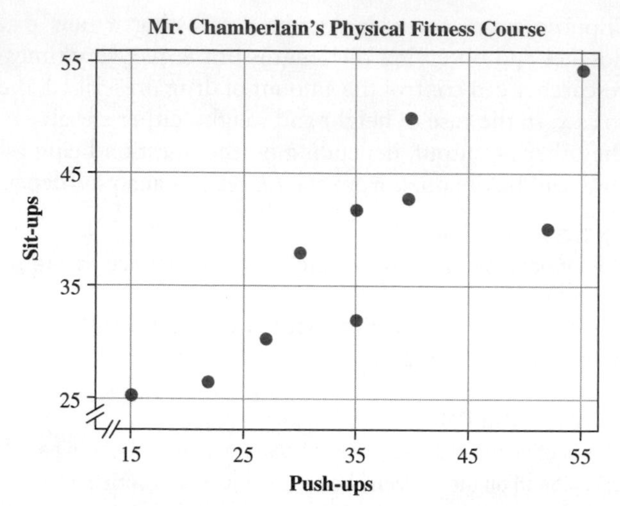

■

Application 3.1 Northwest Ohio Schools and How They Rate

Poverty predicts scores

Each of 2,025 elementary schools in Ohio analyzed by The Blade is represented on this chart as a single dot. The dots were located on the chart based on each school's poverty level compared with each school's overall passage rate on the state's fourth-grade reading proficiency test.

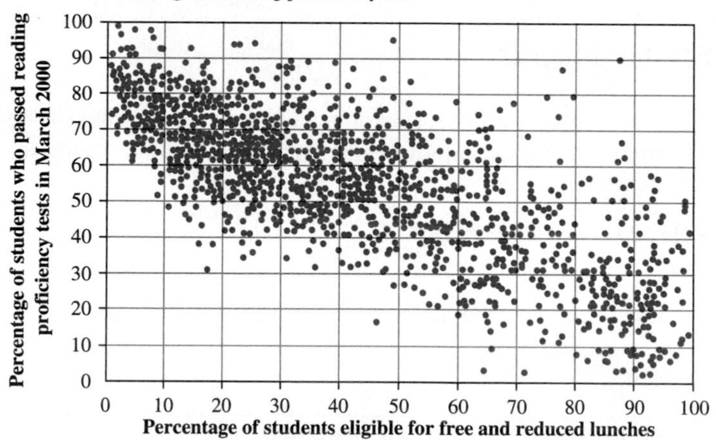

Reprinted with permission of the (Toledo) Blade, August 5, 2001.

It has long been known that a student's ability to pass the state's fourth-grade proficiency tests is closely related to the income level in the student's home.

This chart shows how individual elementary schools performed in the March 2000 proficiency tests in fourth-grade math and reading—and whether the schools performed better or worse than could be predicted based on the poverty level of the students attending the school. . . .

The percentage of children receiving free or reduced-price lunch was used as the measure of poverty.

Each of 2025 elementary schools in Ohio analyzed by *The Blade* is represented on this chart as a single dot. The dots were located on the chart based on each school's poverty level compared with each school's overall passage rate on the state's fourth-grade reading proficiency test.

ANSWER NOW 3.9
Refer to "Northwest Ohio Schools and How They Rate."

a. **Name the two variables examined.**

b. **Does the scatter diagram suggest a relationship between the two variables? Explain.**

c. **What conclusion, if any, can you draw from the appearance of the scatter diagram?**

Technology Instructions: Scatter Diagram

MINITAB (Release 13)

Input the x-variable values into C1 and the corresponding y-variable values into C2; then continue with:

```
Choose:    Graph > Plot
Enter:     Graph 1: Y: C2 X: C1
Choose:    Annotation > Title
Enter:     your title > OK
```

Excel XP

Input the x-variable values into column A and the corresponding y-variable values into column B; then continue with:

```
Choose:    Chart Wizard > XY(Scatter) > 1st picture (usu-
           ally) > Next
Enter:     Data Range: (A1:B12 or select cells(if necessary))
           > Next
Choose:    Titles
Enter:     Chart title: your title; Value(x) axis: title for
           x axis; Value(y) axis: title for y axis* >
           Finish
```

*To remove gridlines:

```
Choose:    Gridlines
Unselect:  Value(Y) axis: Major Gridlines > Finish
```

To edit the scatter diagram, follow the basic editing commands shown for a histogram on page 59.
To change the scale, double click on the axis; then continue with:

```
Choose:    Scale
Unselect:  any Auto boxes
Enter:     new values > OK
```

TI-83 Plus

Input the x-variable values into L1 and the corresponding y-variable values into L2; then continue with:

```
Choose:    2nd > STATPLOT > 1:Plot1
```

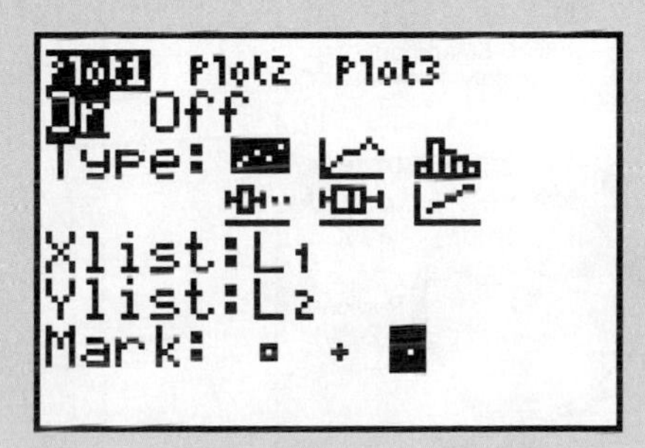

(continued)

Technology Instructions: Scatter Diagram (Continued)

TI-83 Plus (continued)

```
Choose:   ZOOM > 9:ZoomStat > TRACE  >>>
          or
          WINDOW
          Enter: at most lowest x value, at least highest x
          value, x-scale, - y-scale, at least highest y
          value, y-scale,1
          TRACE >>>
```

ANSWER NOW 3.10

Sarah's growth is the subject of the "Bivariate Data" video on your Student Suite CD (found inside the back cover of this textbook). View the video clip and answer these questions:

a. **What are the two variables shown on the graph?**

b. **What information does the ordered pair (3, 87) represent?**

c. **What conclusion did the doctor reach?**

d. **How does the graph indicate that Sarah's growth rate was below normal?**

Exercises

3.11 The USA Snapshot ® "Can't get enough of school" shows the results from a 2 × 3 contingency table of two qualitative variables.

a. Identify the population and name the two variables.

b. Construct a contingency table using entries of percentages based on row totals.

USA SNAPSHOTS®

A look at statistics that shape your finances

Can't get enough of school

Among employed college graduates age 30-55 and out of college 10 or more years, 57% have taken college-level courses since graduation. Reasons:

Tech workers: Professional 28%, Personal 31%, Both 41%

Other workers: Professional 47%, Personal 20%, Both 33%

Source: Market Research Institute for George Mason University, Potomac KnowledgeWay

By Anne R. Carey and Jerry Mosemak, USA TODAY

3.12 The USA Snapshot ® "I don't want to grow up" shows the results from a 9 × 2 contingency table for one qualitative and one quantitative variable.

a. Identify the population; name the qualitative and the quantitative variables.

b. Construct a bar graph showing the two distributions side by side.

c. Does there seem to be a big difference between the genders on this subject?

USA SNAPSHOTS®
A look at statistics that shape our lives
I don't want to grow up
The age adults say they'd like to remain for the rest of their lives if they could:
Source: IRC Research for Walt Disney
By Cindy Hall and Genevieve Lynn, USA TODAY

Age	Men	Women
1-4	0%	2%
5-10	8%	8%
11-14	4%	6%
15-20	34%	20%
21-25	29%	28%
26-30	8%	10%
31-35	7%	10%
36-40	3%	7%
41-up	7%	9%

3.13 Ex03-13 Under the National Highway System Designation Act passed in 1995, states were allowed to set their own highway speed limits. Most of the states raised their limits. As of early 1998, the maximum speed limits (in miles per hour) on interstate highways for cars and trucks are given in the table.

State	Cars	Trucks	State	Cars	Trucks
Alabama	70	70	Montana	65	65
Alaska	65	65	Nebraska	75	75
Arizona	75	75	Nevada	75	75
Arkansas	70	65	New Hampshire	65	65
California	70	55	New Jersey	65	65
Colorado	75	75	New Mexico	75	75
Connecticut	55	55	New York	65	65
Delaware	65	65	North Carolina	70	70
Florida	70	70	North Dakota	70	70
Georgia	70	70	Ohio	65	55
Hawaii	55	55	Oklahoma	75	75
Idaho	75	75	Oregon	65	55
Illinois	65	55	Pennsylvania	65	65
Indiana	65	60	Rhode Island	65	65
Iowa	65	65	South Carolina	65	65
Kansas	70	70	South Dakota	75	65
Kentucky	65	65	Tennessee	65	65
Louisiana	70	70	Texas	70	60
Maine	65	65	Utah	75	75
Massachusetts	65	65	Vermont	65	65
Maryland	65	65	Virginia	65	65
Michigan	70	55	Washington	70	60
Minnesota	70	70	West Virginia	70	70
Mississippi	70	70	Wisconsin	65	65
Missouri	70	70	Wyoming	75	75

Source: *The World Almanac and Book of Facts 1998*, p. 214. Data provided by National Motorists Association.

a. Build a cross-tabulation of the two variables, vehicle type and maximum speed limit on interstate highway. Express the results in frequencies, showing marginal totals.
b. Express the contingency table you derived in part a in percentages based on the grand total.
c. Draw a bar graph showing the results from part b.

FYI
If you are using a computer or a calculator, try the cross-tabulation table commands on page 134.

3.14
a. Express the contingency table you derived in Exercise 3.13, part a, in percentages based on the marginal total for speed limit.
b. Draw a bar graph showing the results from part a.

3.15 Ex03-15 A statewide survey was conducted to investigate the relationship between viewers' preferences for ABC, CBS, NBC, or PBS for news information and their political party affiliation. The results are shown in tabular form.

	ABC	CBS	NBC	PBS
Democrat	200	200	250	150
Republican	450	350	500	200
Other	150	400	100	50

a. How many viewers were surveyed?
b. Why are these bivariate data? What type of variable is each one?
c. How many preferred to watch CBS?
d. What percentage of the survey was Republicans?
e. What percentage of the Democrats preferred ABC?

3.16 Ex03-16 Consider the accompanying contingency table, which presents the results of an advertising survey about the use of credit by Martan Oil Company customers.
a. How many customers were surveyed?
b. Why are these bivariate data? What type of variable is each one?
c. How many customers preferred to use an oil-company credit card?
d. How many customers made 20 or more purchases last year?
e. How many customers preferred to use an oil-company credit card and made between five and nine purchases last year?
f. What does the 80 in the fourth cell in the second row mean?

Table for Exercise 3.16

	Number of Purchases at Gasoline Station Last Year					
Preferred Method of Payment	**0–4**	**5–9**	**10–14**	**15–19**	**20 and Over**	**Sum**
Cash	150	100	25	0	0	275
Oil-Company Card	50	35	115	80	70	350
National or Bank Credit Card	50	60	65	45	5	225
Sum	250	195	205	125	75	850

3.17 Ex03-17 What effect does the minimum amount have on the interest rate being offered on three-month CDs? The following are advertised rates of return, y, for a minimum deposit of \$500, \$1000, or \$2500, x. Note that x is in \$100 and y is annual percent of return.

x	5	5	25	10	25	10	5	5	5	10
y	1.79	1.60	1.99	0.98	1.99	1.29	1.74	2.16	1.34	2.15

x	10	25	5	5	25	5	5	5	25	10
y	1.73	1.00	1.24	1.19	2.00	1.24	2.16	2.16	1.45	2.15

x	25	25	10	25	25	25	10	10	10	10
y	1.99	1.29	2.13	1.35	1.39	1.99	1.41	2.13	1.73	1.83

x	10	10	10	10	25	10	5	10	10	5
y	1.49	1.83	1.09	2.23	1.49	1.19	1.49	1.49	1.59	1.49

Source: http://www.bankrate.com, September 2002

a. Prepare a dotplot of the three sets of data using a common scale.
b. Prepare a 5-number summary and a boxplot of the three sets of data. Use the same scale for the boxplots.
c. Describe any differences you see between the three sets of data.

FYI
If you are using a computer or calculator for Exercise 3.17, try the commands on pages 100 and 136.

3.18 Ex03-18 Can a woman's height be predicted from her mother's height? The heights of some mother–daughter pairs are listed; x is the mother's height and y is the daughter's height.

x	63	63	67	65	61	63	61	64	62	63
y	63	65	65	65	64	64	63	62	63	64

x	64	63	64	64	63	67	61	65	64	65	66
y	64	64	65	65	62	66	62	63	66	66	65

a. Draw two dotplots using the same scale and showing the two sets of data side by side.
b. What can you conclude from seeing the two sets of heights as separate sets in part a? Explain.
c. Draw a scatter diagram of these data as ordered pairs.
d. What can you conclude from seeing the data presented as ordered pairs? Explain.

3.19 Ex03-19 The table lists the heights (in meters), weights (in kilograms), and the ages of the players on the two teams that played in the 2002 World Cup finals: Brazil and Germany.

	Brazil			Germany		
Player	Height	Weight	Age	Height	Weight	Age
1	1.93	86	28	1.88	88	33
2	1.95	85	29	1.90	87	33
3	1.88	85	29	1.91	91	28
4	1.76	74	32	1.83	90	33
5	1.88	81	24	1.87	85	30
6	1.86	73	26	1.87	79	27
7	1.85	73	26	1.86	82	30
8	1.68	70	29	1.86	80	22
9	1.74	69	26	1.93	84	22
10	1.82	73	23	1.85	80	28
11	1.70	63	29	1.89	76	29
12	1.73	67	26	1.78	72	26
13	1.85	78	26	1.89	80	26
14	1.86	75	30	1.77	76	28
15	1.80	76	22	1.78	75	28
16	1.75	64	23	1.76	74	29
17	1.82	78	28	1.82	80	26
18	1.65	58	29	1.71	64	29
19	1.83	77	26	1.93	93	28
20	1.77	65	25	1.82	74	24
21	1.85	73	32	1.80	85	24
22	1.76	77	27	1.89	85	33
23	1.83	73	20	1.91	87	34

Source: http://worldcup.espnsoccernet.com/index

a. Compare each of the three variables—height, weight, and age—using either a dotplot or a histogram (use the same scale).
b. Based on what you see in the graphs in part a, can you detect a substantial difference between the two teams in regard to these three variables? Explain.
c. Explain why the data, as used in part a, are not bivariate data.

3.20

a. Draw a scatter diagram showing height, x, and weight, y, for the Brazilian World Cup soccer team using the data in Exercise 3.19.
b. Draw a scatter diagram showing height, x, and weight, y, for the German World Cup soccer team using the data in Exercise 3.19.
c. Explain why the data, as used in parts a and b, are bivariate data.

FYI
If you are using a computer or calculator, try the commands on page 139.

3.21 Ex03-21 The accompanying data show the number of hours, x, studied for an exam and the grade, y, received in the exam (y is measured in tens; that is, $y = 8$ means that the grade, rounded to the nearest ten points, is 80). Draw the scatter diagram. (Retain this solution to use in Exercise 3.39, p. 152.)

x	2	3	3	4	4	5	5	6	6	6	7	7	7	8	8
y	5	5	7	5	7	7	8	6	9	8	7	9	10	8	9

3.22 Ex03-22 An experimental psychologist asserts that the older a child is, the fewer irrelevant answers he or she will give during a controlled experiment. To investigate this claim, the following data were collected. Draw a scatter diagram. (Retain this solution to use in Exercise 3.40, p 152.)

Age, x	2	4	5	6	6	7	9	9	10	12
Number of Irrelevant Answers, y	12	13	9	7	12	8	6	9	7	5

3.23 Ex03-23 The table lists the percents of students who receive free or reduced lunches compared with the percents who passed the reading portion of a state exam. The results are for Sandusky County, Ohio, and were reported in *The Blade*, a Toledo newspaper, on August 5, 2001. Sandusky County has a combination of 13 rural and urban schools.

School	% Free/Reduced Lunches	% Passing Reading
1	29	66
2	29	59
3	23	62
4	60	53
5	57	53
6	50	57
7	49	54
8	47	58
9	29	88
10	17	68
11	22	60
12	38	47
13	15	62

Construct a scatter diagram of these data. (Retain to use in Exercise 3.37, p. 152.)

3.24 Ex03-24 A sample of 15 upper-class students who commute to classes was selected at registration. The students were asked to estimate the distance, x, and the time, y, required to commute each day to class (see the table). Construct a scatter diagram depicting these data.

Distance, x (nearest mile)	Time, y (nearest 5 min)	Distance, x (nearest mile)	Time, y (nearest 5 min)
18	20	2	5
8	15	15	25
20	25	16	30
5	20	9	20
5	15	21	30
11	25	5	10
9	20	15	20
10	25		

3.25 Ex03-25 Are people stronger today than they used to be? Can they run faster? Let's compare the performances of the Olympic gold medal winners for the last century as a way to decide. The distances (in inches) for the gold medal performances in the long jump, high jump, and discus throw are given in the table. The event year is coded, with 1900 = 0.

Year	Long Jump	High Jump	Discus Throw
−4	249.75	71.25	1147.5
0	282.875	74.8	1418.9
4	289	71	1546.5
8	294.5	75	1610
12	299.25	76	1780
20	281.5	76.25	1759.25
24	293.125	78	1817.125
28	304.75	76.375	1863
32	300.75	77.625	1948.875
36	317.3125	79.9375	1987.375
48	308	78	2078
52	298	80.32	2166.85
56	308.25	83.25	2218.5
60	319.75	85	2330
64	317.75	85.75	2401.5
68	350.5	88.25	2550.5
72	324.5	87.75	2535
76	328.5	88.5	2657.4
80	336.25	92.75	2624
84	336.25	92.5	2622
88	343.25	93.5	2709.25
92	342.5	92	2563.75
96	334.65	94.09	2732.3
100	336.62	92.52	2728.34

Source: http://www.ex.ac.uk/cimt/data/olympics/olymindx.htm

a. Plot the data for each event on a separate scatter diagram using year, x.

(continued)

b. Describe the shape of the distribution. Does the relationship between year and performance appear to be linear?
c. How do the three scatter diagrams answer the question: Are people stronger today? Explain.
d. On each of the three scatter diagrams, draw the straight line that seems to best trace the pattern of points from 1896 to 2000. Use this line as an aid to predict the Olympic gold-medal-winning performance for each event at the Athens 2004 games.
e. Investigate the relationship between high and long jumps with the aid of a scatter diagram. Describe your findings.

3.26 Ex03-26 Ronald Fisher, an English statistician (1890–1962), collected measurements for a sample of 150 irises. Of concern were these variables: species, petal width (PW), petal length (PL), sepal width (SW), and sepal length (SL) (all in mm). Sepals are the outermost leaves that encase the flower before it has opened. The goal of Fisher's experiment was to produce a simple function that could be used to classify flowers correctly. A random sample of his complete data set is given in the table.

Type	PW	PL	SW	SL	Type	PW	PL	SW	SL
0	2	15	35	52	1	24	51	28	58
2	18	48	32	59	1	19	50	25	63
1	19	51	27	58	0	1	15	31	49
0	3	13	35	50	1	23	59	32	68
0	3	15	38	51	2	13	44	23	63
2	12	44	26	55	2	15	42	30	59
1	20	64	38	79	1	25	57	33	67
2	15	49	31	69	1	21	57	33	67
2	15	45	29	60	0	2	15	37	54
2	12	39	27	58	1	18	49	27	63
1	22	56	28	64	1	17	45	25	49
1	13	52	30	67	1	24	56	34	63
0	2	14	29	44	0	2	14	36	50
2	16	51	27	60	2	10	50	22	60
0	5	17	33	51	0	2	12	32	50

a. Construct a scatter diagram of petal length, x, and petal width, y. Use different symbols to represent the three species.*
b. Construct a scatter diagram of sepal length, x, and sepal width, y. Use different symbols to represent the three species.
c. Explain what the scatter diagrams in parts a and b portray.

How well does a random sample represent the data from which it was selected?

d. Do parts a and b using all 150 of Fisher's data on your CD. DS-FisherslrisesDS-FishersIrises
e. Aside from the fact that the scatter diagrams in parts a and b have less data, comment on the similarities and differences between the distributions shown for 150 data and for the 30 randomly selected data.

*In addition to using the commands on page 139, use:

for MINITAB:	`Data display:`	`For each: Select:` **Group**
	`Group variable:`	`Select:` **Type**
For TI-83:	Enter different groups into separate x, y columns. Use a separate Stat Plot and "Mark" for each group.	

3.2 Linear Correlation

The primary purpose of **linear correlation analysis** is to measure the strength of a linear relationship between two variables. Let's examine some scatter diagrams that demonstrate different relationships between input, or independent variables, x, and output, or dependent variables, y. If as x increases there is no definite shift in the values of y, we say there is **no correlation,** or no relationship between x and y. If as x increases there is a shift in the values of y, then there is a correlation. The correlation is **positive** when y tends to increase and **negative** when y tends to decrease. If the ordered pairs (x, y) tend to follow a straight-line path, there is a linear correlation. The preciseness of the shift in y as x increases determines the strength of the **linear correlation.** The scatter diagrams in Figure 3.6 demonstrate these ideas.

Perfect linear correlation occurs when all the points fall exactly along a straight line, as shown in Figure 3.7. The correlation can be either positive or negative, de-

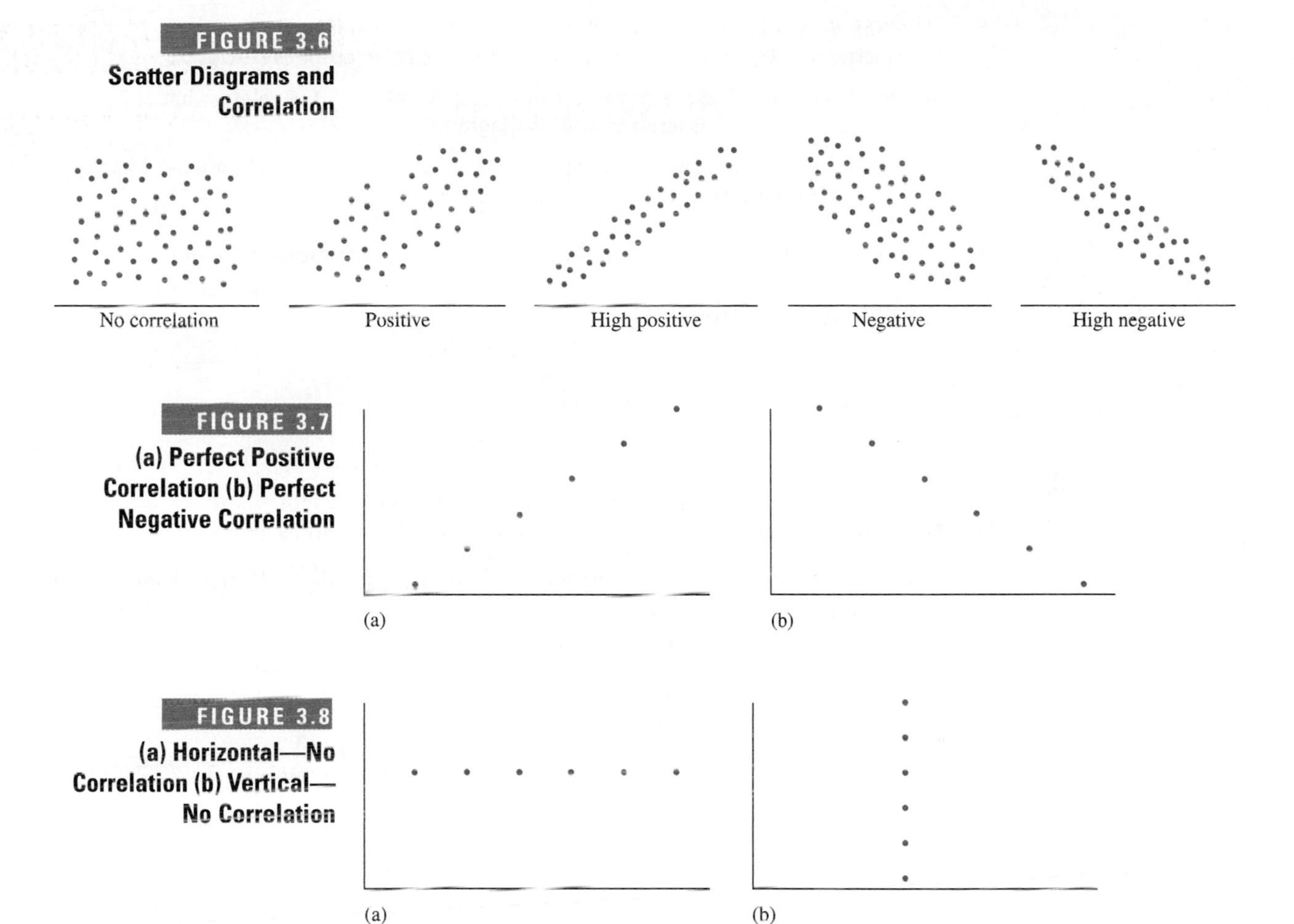

FIGURE 3.6
Scatter Diagrams and Correlation

FIGURE 3.7
(a) Perfect Positive Correlation (b) Perfect Negative Correlation

FIGURE 3.8
(a) Horizontal—No Correlation (b) Vertical—No Correlation

pending on whether y increases or decreases as x increases. If the data form a straight horizontal or vertical line, there is no correlation, since one variable has no effect on the other, as shown in Figure 3.8.

FIGURE 3.9
No Linear Correlation

Scatter diagrams do not always appear in one of the forms shown in Figures 3.6, 3.7, and 3.8. Sometimes they suggest relationships other than linear, as in Figure 3.9. There appears to be a definite pattern; however, the two variables are not related linearly, and therefore there is no linear correlation.

The **coefficient of linear correlation, r,** is the numerical measure of the strength of the linear relationship between two variables. The coefficient reflects the consistency of the effect that a change in one variable has on the other. The value of the linear correlation coefficient helps us to answer the question: Is there a linear correlation between the two variables under consideration? The linear correlation coefficient, r, always has a value between -1 and $+1$. A value of $+1$ signifies a perfect positive correlation, and a value of -1 shows a perfect negative correlation. If as x increases there is a general increase in the value of y, then r will be positive in value. For example, a positive value of r would be expected for the age and height of children because as children grow older, they grow taller. Also, consider the age, x, and resale value, y, of an automobile. As the car ages, its resale value decreases. Since as x increases, y decreases, the relationship results in a negative value for r.

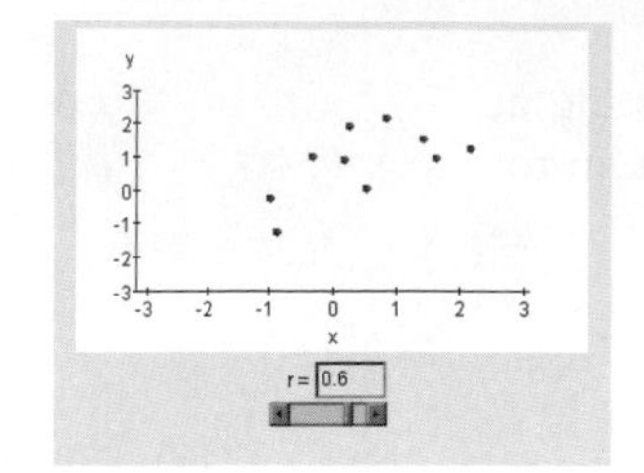

ANSWER NOW 3.27 INTERACTIVITY

Interactivity 3-A provides scatter plots for various correlation coefficients.

a. **Starting at $r = 0$, move the slider to the right until $r = 1$. Explain what is happening to the corresponding scatter diagrams.**

b. **Starting at $r = 0$, move the slider to the left until $r = -1$. Explain what is happening to the corresponding scatter diagrams.**

The value of r is defined by **Pearson's product moment formula:**

DEFINITION FORMULA

$$r = \frac{\sum(x - \bar{x})(y - \bar{y})}{(n - 1)s_x s_y} \tag{3.1}$$

NOTES

1. s_x and s_y are the standard deviations of the x and y variables.
2. The development of this formula is discussed in Chapter 13.

To calculate r, we will use an alternative formula, formula (3.2), that is equivalent to formula (3.1). As preliminary calculations, we will separately calculate three sums of squares and then substitute them into formula (3.2) to obtain r.

COMPUTATIONAL FORMULA

linear correlation coefficient = $\frac{\textit{sum of squares for xy}}{\sqrt{(\textit{sum of squares for x})(\textit{sum of squares for y})}}$

$$r = \frac{SS(xy)}{\sqrt{SS(x)SS(y)}} \tag{3.2}$$

FYI
SS(x) is the numerator of the variance.

Recall the SS(x) calculation from formula (2.9) for sample variance (p. 81):

sum of squares for x = *sum of* x^2 − $\frac{(\textit{sum of x})^2}{n}$

$$SS(x) = \sum x^2 - \frac{(\sum x)^2}{n} \tag{2.9}$$

We can also calculate:

sum of squares for y = *sum of* y^2 − $\frac{(\textit{sum of y})^2}{n}$

$$SS(y) = \sum y^2 - \frac{(\sum y)^2}{n} \tag{3.3}$$

sum of squares for xy = *sum of xy* − $\frac{(\textit{sum of x})(\textit{sum of y})}{n}$

$$SS(xy) = \sum xy - \frac{\sum x \sum y}{n} \tag{3.4}$$

Illustration 3.4 Calculating the Linear Correlation Coefficient, r

Find the linear correlation coefficient for the push-up/sit-up data in Illustration 3.3 (p. 137).

SOLUTION
First, we construct an extensions table (Table 3.11) listing all the pairs of values (x, y) to aid us in finding x^2, xy, and y^2 and the five column totals.

FYI See Animated Tutorial Preliminary Calculations on CD for assistance with these calculations.

TABLE 3.11 Extensions Table for Finding Five Summations Ta03-10

Student	Push-ups, x	x^2	Sit-ups, y	y^2	xy
1	27	729	30	900	810
2	22	484	26	676	572
3	15	225	25	625	375
4	35	1,225	42	1,764	1,470
5	30	900	38	1,444	1,140
6	52	2,704	40	1,600	2,080
7	35	1,225	32	1,024	1,120
8	55	3,025	54	2,916	2,970
9	40	1,600	50	2,500	2,000
10	40	1,600	43	1,849	1,720
	$\sum x = 351$ sum of x	$\sum x^2 = 13{,}717$ sum of x^2	$\sum y = 380$ sum of y	$\sum y^2 = 15{,}298$ sum of y^2	$\sum xy = 14{,}257$ sum of xy

Second, to complete the preliminary calculations, we substitute the five summations (the five column totals) from the extensions table into formulas (2.9), (3.3), and (3.4), and calculate the three sums of squares:

$$SS(x) = \sum x^2 - \frac{(\sum x)^2}{n} = 13{,}717 - \frac{(351)^2}{10} = 1396.9$$

$$SS(y) = \sum y^2 - \frac{(\sum y)^2}{n} = 15{,}298 - \frac{(380)^2}{10} = 858.0$$

$$SS(xy) = \sum xy - \frac{\sum x \sum y}{n} = 14{,}257 - \frac{(351)(380)}{10} = 919.0$$

FYI The $\sum$ and SS values will be needed for regression in Section 3.3. Be sure to save them!

Third, we substitute the three sums of squares into formula (3.2) to find the value of the correlation coefficient:

FYI See Animated Tutorial Correlation on CD for assistance with this calculation.

$$r = \frac{SS(xy)}{\sqrt{SS(x)SS(y)}} = \frac{919.0}{\sqrt{(1396.9)(858.0)}} = 0.8394 = \mathbf{0.84}$$ ■

NOTE Typically, r is rounded to the nearest hundredth.

ANSWER NOW 3.28
Does studying for an exam pay off? The table gives the numbers of hours studied, x, compared with the exam grades, y.

x	2	5	1	4	2
y	80	80	70	90	60

a. **Complete the preliminary calculations: extensions, five sums, and SS(x), SS(y), and SS(xy).**

b. **Find r.**

The value of the linear correlation coefficient helps us answer the question: Is there a linear correlation between the two variables under consideration? When the calculated value of r is close to zero, we conclude that there is little or no linear correlation. As the calculated value of r changes from 0.0 toward either +1.0 or

-1.0, it indicates an increasingly stronger linear correlation between the two variables. From a graphic viewpoint, when we calculate r, we are measuring how well a straight line describes the scatter diagram of ordered pairs. As the value of r changes from 0.0 toward $+1.0$ or -1.0, the data points creating a pattern move closer to a straight line.

ANSWER NOW 3.29 INTERACTIVITY

Interactivity 3-B matches correlation coefficients with their scatter diagrams. After several practice rounds using 'New Plots,' explain your method of matching.

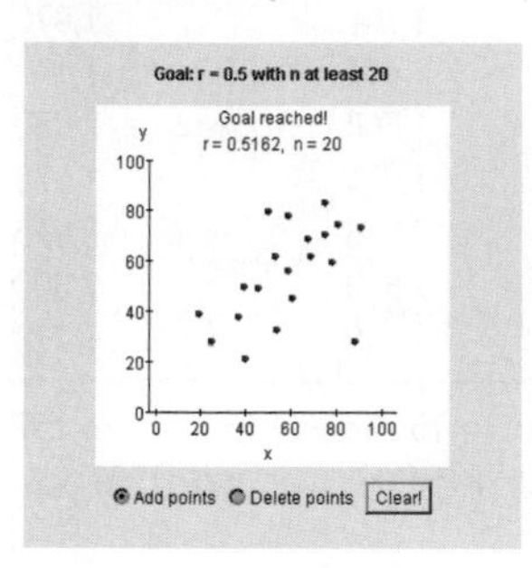

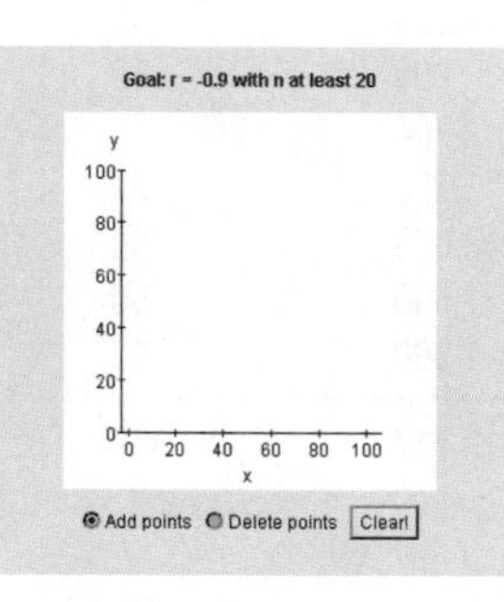

ANSWER NOW 3.30 INTERACTIVITY

Interactivities 3-C and 3-D provide practice in constructing scatter diagrams to match given correlation coefficients.

a. After you have placed just two points, what is the calculated r value for each applet? Why?

b. Which scatter diagram did you find easier to construct? Why?

Technology Instructions: Correlation Coefficient

MINITAB (Release 13)

Input the x-variable data into C1 and the corresponding y-variable data into C2; then continue with:

```
Choose:  Stat > Basic Statistics > Correlation...
Enter:   Variables: C1 C2
```

Excel XP

Input the x-variable data into column A and the corresponding y-variable data into column B, activate a cell for the answer; then continue with:

```
Choose:  Insert function, fx > Statistical > CORREL >
         OK
Enter:   Array 1: x data range
         Array 2: y data range > OK
```

TI-83 Plus

Input the x-variable data into L1 and the corresponding y-variable data into L2; then continue with:

```
Choose:  2nd > CATALOG > DiagnosticOn* > ENTER >
         ENTER
Choose:  STAT > CALC > 8:LinReg(a+bx)
Enter:   L1, L2
```

*DiagnosticOn must be selected for r and r^2 to show. Once set, omit this step.

UNDERSTANDING THE LINEAR CORRELATION COEFFICIENT

The following method will create: (1) a visual meaning for correlation, (2) a visual meaning for what the linear coefficient is measuring, and (3) an estimate for r. The method is quick and generally yields a reasonable estimate when the "window of data" is approximately square.

NOTE: This estimation technique does not replace the calculation of r. It is very sensitive to the "spread" of the diagram. However, if the "window of data" is approximately square, this approximation will be useful as a mental estimate or check.

Procedure

1. Construct a scatter diagram of your data, being sure to scale the axes so that the resulting graph has an approximately square "window of data," as demonstrated in Figure 3.10 by the red frame. The window may not be the same region as determined by the bounds of the two scales, shown as a blue rectangle on Figure 3.10.
2. Lay two pencils on your scatter diagram. Keeping them parallel, move them to a position so that they are as close together as possible yet have all the points on the scatter diagram between them. (See Figure 3.11.)
3. Visualize a rectangular region that is bounded by the two pencils and that ends just beyond the points on the scatter diagram. (See the shaded portion of Figure 3.11.)
4. Estimate the number of times longer the rectangle is than it is wide. An easy way to do this is to mentally mark off squares in the rectangle. (See Figure 3.12.) Call this number of multiples k.

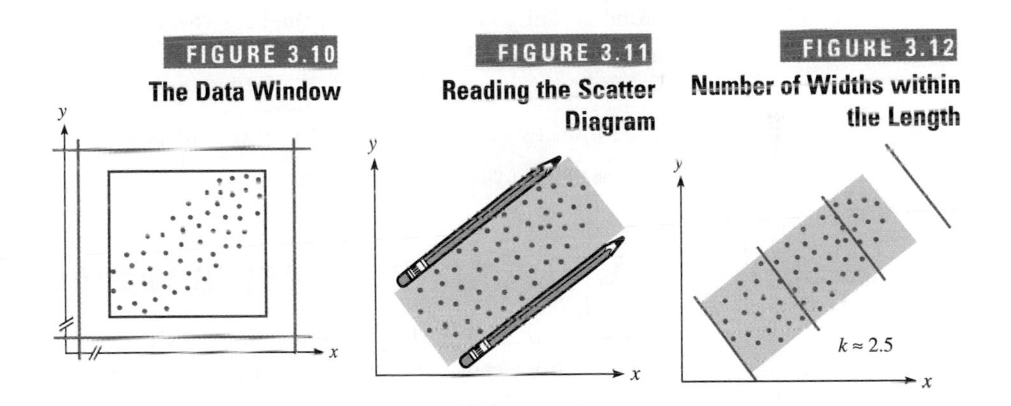

FIGURE 3.10 The Data Window

FIGURE 3.11 Reading the Scatter Diagram

FIGURE 3.12 Number of Widths within the Length

5. The value of r may be estimated as $\pm\left(1 - \frac{1}{k}\right)$.
6. The sign assigned to r is determined by the general position of the length of the rectangular region. If it lies in an increasing position, r will be positive; if it lies in a decreasing position, r will be negative (see Figure 3.13). If the rectangle is in either a horizontal or a vertical position, then r will be zero, regardless of the length–width ratio.

FIGURE 3.13

(a) Increasing Position ⇒ Positive r
(b) Decreasing Position ⇒ Negative r

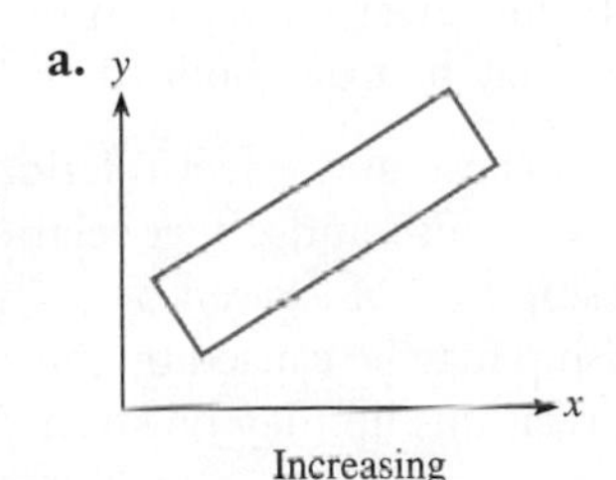

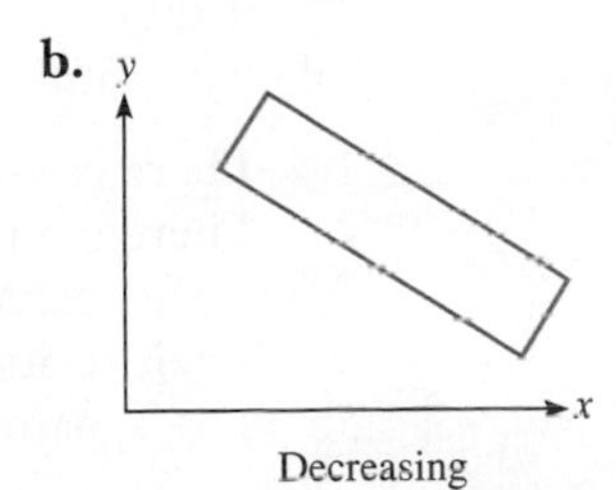

FIGURE 3.14
Push-ups Versus Sit-ups for Ten Students

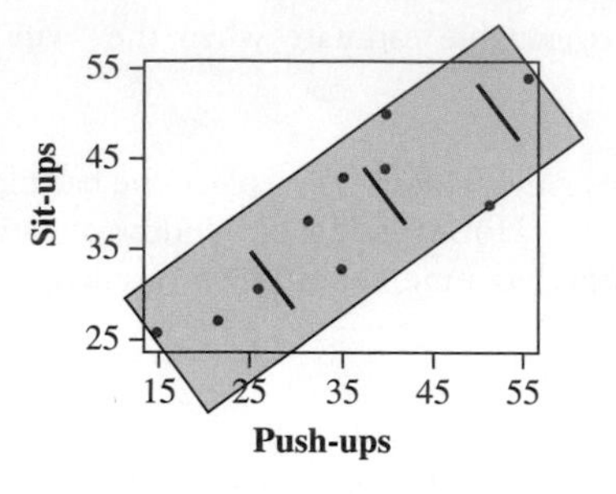

Let's use this method to estimate the value of the linear correlation coefficient for the relationship between the numbers of push-ups and sit-ups. As shown in Figure 3.14, we find that the rectangle is approximately 3.5 times longer than it is wide—that is, $k \approx 3.5$—and the rectangle lies in an increasing position. Therefore, our estimate for r is

$$r \approx +\left(1 - \frac{1}{3.5}\right) \approx \mathbf{+0.7}$$

ANSWER NOW 3.31
Estimate the correlation coefficient for each graph.

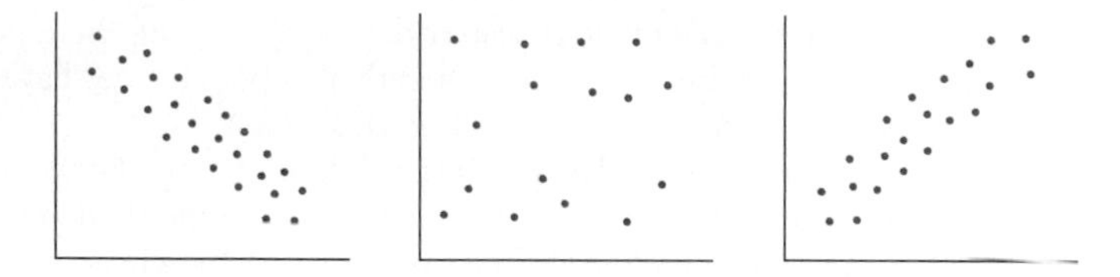

ANSWER NOW 3.32
Manatees and powerboats are the subjects of the "Linear Correlation" video on your Student Suite CD (found inside the back cover of this textbook). View the video clip and answer these questions:

a. **What two groups of subjects are being compared?**

b. **What two variables are used to make the comparison?**

c. **What did the graph tell them?**

CAUSATION AND LURKING VARIABLES

As we try to explain the past, understand the present, and estimate the future, judgments about cause and effect are necessary because of our desire to impose order on our environment.

The *cause-and-effect relationship* is fairly straightforward. You may focus on a situation, the *effect* (e.g., a disease or social problem) and try to determine its *cause(s)*, or you may begin with a *cause* (unsanitary conditions or poverty) and discuss its *effect(s)*. To determine the cause of something, ask yourself **why** it happened. To determine the effect, ask yourself **what** happened.

LURKING VARIABLE
A variable that has an important effect on the relationship between the variables of a study but is not included in the study.

If there is a strong linear correlation between two variables, then one of the following situations may be true about the relationship between the two variables:

1. There is a direct cause-and-effect relationship between the two variables.
2. There is a reverse cause-and-effect relationship between the two variables.
3. Their relationship may be caused by a third variable.
4. Their relationship may be caused by the interactions of several other variables.
5. The apparent relationship may be strictly a coincidence.

Remember that a strong correlation does not necessarily imply causation. Here are some pitfalls to avoid:

1. In a direct cause-and-effect relationship, an increase (or decrease) in one variable causes an increase (or decrease) in another. Suppose there is a strong positive correlation between weight and height. Does an increase in weight *cause* an increase in height? Not necessarily. Or to put it another way, does a decrease in weight *cause* a decrease in height? Many other possible variables are involved, such as gender, age, and body type. These other variables are called *lurking variables.*
2. In Application 3.1, a negative correlation existed between the percent of students who received free or reduced price lunches and the percent of students who passed the mathematics proficiency test. Shall we hold back on the free lunches so that more students pass the mathematics test? A third variable is the motivation for this relationship—namely, poverty level.
3. Don't reason from *correlation* to *cause:* Just because all people who move to the city get old doesn't mean that the city *causes* aging. The city may be a factor, but you can't base your argument on the correlation.

Application 3.2 Life Insurance Rates Ap03-2

Does a high linear correlation coefficient, r, imply that the data are linear in nature? The age of the insured and the non-tobacco monthly rate for life insurance are highly correlated.

In less than a minute, you can receive **INSTANT** life insurance quotes from top-rated insurance companies!

Non-Tobacco Monthly Rates for Life Insurance

Issue Age	$100,000 Male	$100,000 Female	$250,000 Male	$250,000 Female	$500,000 Male	$500,000 Female
30	$8.27	$6.79	$14.36	$11.14	$25.67	$19.14
35	8.27	7.48	14.36	12.62	25.67	22.19
40	10.35	9.05	18.97	16.10	34.80	27.41
45	15.57	12.09	30.28	22.62	56.55	40.89
50	20.71	15.66	45.07	31.15	84.83	56.99
55	31.76	22.71	66.82	45.24	130.50	85.26
60	49.59	33.50	104.84	67.60	204.45	130.07

All of the rates listed above are for each carrier's best non-tobacco classifications.

ANSWER NOW 3.33

a. **Calculate the correlation coefficient, r, for the variables: issue age, x, and monthly rate, y, for $250,000 for males in Application 3.2.**

b. **Draw a scatter diagram of the insurance data for males at the $250,000 level based on age, x.**

c. **Explain how a nonlinear data pattern can have a high linear correlation coefficient.**

d. **Explain why you should have anticipated this nonlinear pattern.**

ANSWER NOW 3.34

Summertime activities are the subject of the "Linear Regression" video on your Student Suite CD (found inside the back cover of this textbook). View the video clip and answer these questions:

a. What two variables seem to have a relationship?

b. As one of the variables increases, the other also increases. Does it mean that one variable causes the other? Is there a better explanation?

Exercises

3.35

a. How would you interpret the findings of a correlation study that reported a linear correlation coefficient of −1.34?

b. How would you interpret the findings of a correlation study that reported a linear correlation coefficient of +0.3?

3.36 Explain why it makes sense for a set of data to have a correlation coefficient of zero when the data show a very definite pattern, as in Figure 3.9 (p. 145).

3.37 Ex03-23 The table lists the percents of students who receive free or reduced lunches compared with the percents who passed the reading portion of the state exam. The results are for Sandusky County, Ohio, and were reported in *The Blade*, a Toledo newspaper, on August 5, 2001. Sandusky County is a combination of 13 rural and urban schools. (Same data as in Exercise 3.23, p. 143.)

School	% Free/Reduced Lunches	% Passing Reading
1	29	66
2	29	59
3	23	62
4	60	53
5	57	53
6	50	57
7	49	54
8	47	58
9	29	88
10	17	68
11	22	60
12	38	47
13	15	62

Find: a. SS(x) b. SS(y) c. SS(xy) d. r

3.38 Ex03-38 Australia is becoming well known for Shiraz wines. The United States is catching on to these bold red wines. Listed in the table are five varieties with their *Wine Spectator* score and price per bottle. *Wine Spectator* rates wines on a 100-point scale and all wines are blind-tested.

Name	Score	Price
Rosemont Shiraz Mudgee Hill of Gold 2000	90	$19.00
D'Arenberg Shiraz-Grenache McLaren Vale d'Arry's Orginal 2000	88	$18.00
De Bortoli Shiraz South Eastern Australia Vat 8 2000	87	$10.00
Grant Burge Shiraz Barossa Barossa Vines 2000	86	$11.00
Rosemont Shiraz-Cabernet South Eastern Australia 2001	86	$9.00

a. Construct a scatter diagram with price as the dependent variable, y, and score as the independent variable, x.

b. Find SS(x).

c. Find SS(y).

d. Find SS(xy).

e. Find Pearson's product moment, r.

3.39

a. Use the scatter diagram you drew in Exercise 3.21 (p. 143) to estimate r for the sample data on the number of hours studied and the exam grade.

b. Calculate r.

3.40

a. Use the scatter diagram you drew in Exercise 3.22 (p. 143) to estimate r for the sample data on the number of irrelevant answers and the child's age.

b. Calculate r.

FYI

Have you tried to use the correlation commands on your computer or calculator?

3.41 Ex03-41 A marketing firm wished to determine whether or not the numbers of television commercials broadcast were linearly correlated with the sales of its product. The data, obtained from each of several cities, are shown in the table.

City	A	B	C	D	E	F	G	H	I	J
Number of Commercials, x	12	6	9	15	11	15	8	16	12	6
Sales Units, y	7	5	10	14	12	9	6	11	11	8

a. Draw a scatter diagram. b. Estimate r. c. Calculate r.

3.42 Ex03-42 "Olé? No Way," in the March 2000 *Consumer Reports*, compared several brands of supermarket enchiladas in cost and sodium content.

Brand and Type	Cost (per serving)	Sodium Content (mg)
Amy's Black Bean	$3.03	780
Patio Cheese	$1.07	1570
Banquet Cheese	$1.28	1500
El Charrito Beef	$1.53	1370
Patio Beef	$1.05	1700
Banquet Beef	$1.27	1330
Healthy Choice Chicken	$2.34	440
Lean Cuisine Chicken	$2.47	520
Weight Watchers Chicken	$2.09	660

a. Draw a scatter diagram of these data.
b. Does there appear to be a linear relationship?
c. Calculate the linear correlation coefficient, r.
d. What does this correlation coefficient tell us? Explain.

3.43 Ex03-43 The National Adoption Information Clearinghouse at naic@calib.com tracks and posts information about child adoptions in the United States. Sixteen U.S. states were randomly identified and the numbers of adoptions for 1993 and 1997 were recorded. Is there a linear relationship between the 1993 and the 1997 data? Use graphic and numerical statistics to support your answer.

State	1993	1997
Washington	2364	2699
Delaware	176	199
New York	7587	9436
Rhode Island	471	598
Colorado	1737	2423
Kentucky	2139	1582
Michigan	5679	6118
Utah	1217	1483
Wyoming	444	404
Ohio	4895	4972
Puerto Rico	501	471
DC	353	492
Kansas	1785	1792
Vermont	466	357
Massachusetts	2773	3029
South Dakota	432	358

Source: http://www.calib.com/naic/pubs/s_flang3.htm

3.44 Ex03-44 Americans spend more than $10,000 a day on bottled water. Many studies have been done to determine whether the water is actually better than tap water. Most studies are concerned with contaminants. A study comparing the cost per glass of bottled water with its flavor was done by *Consumer Reports* in August 2000. Cost per glass was recorded in cents; flavor was rated on a scale from 0 through 100, where 0 represented poor flavor and 100 represented excellent flavor.

Cost/Glass, x	Flavor, y	Cost/Glass, x	Flavor, y
29	94	14	72
12	67	12	63
12	63	12	54
12	57	14	45
28	54	12	54
23	54	12	54
15	45	11	50
13	50	19	50
19	49	14	48
11	48	14	48
13	45	12	36

a. Draw a scatter diagram. Be sure to include a title and label the axes.
b. Estimate the value of the correlation coefficient by examining the scatter plot.
c. Calculate the correlation coefficient, r.
d. Interpret the correlation coefficient with respect to the data. Comment on the direction and the strength. Is this a reasonable relationship?

3.45 In many communities there is a strong positive correlation between the amount of ice cream sold in a given month and the number of drownings that occur in that month. Does this mean that ice cream causes drowning? If not, can you think of an alternative explanation for the strong association? Write a few sentences addressing these questions.

3.46 Explain why one would expect to find a positive correlation between the number of fire engines that respond to a fire and the amount of damage done in the fire. Does this mean that the damage would be less extensive if fewer fire engines were dispatched? Explain.

3.3 Linear Regression

Although the correlation coefficient measures the strength of a linear relationship, it does not tell us about the mathematical relationship between the two variables. In Section 3.2, the correlation coefficient for the push-up/sit-up data was found to be 0.84 (see p. 147). This implies that there is a linear relationship between the number of push-ups and the number of sit-ups a student does. However, the correlation coefficient does not help us predict the number of sit-ups a person can do based on knowing he or she can do 28 push-ups. **Regression analysis** finds the equation of the line that best describes the relationship between the two variables. One use of this equation is to make predictions. We make use of these predictions regularly—for example, predicting the success a student will have in college based on high school results and predicting the distance required to stop a car based on its speed. Generally, the exact value of y is not predictable and we are usually satisfied if the predictions are reasonably close.

The relationship between these two variables will be an algebraic expression describing the mathematical relationship between x and y. Here are some examples of various possible relationships, called *models* or **prediction equations:**

Linear (straight-line):	$\hat{y} = b_0 + b_1x$
Quadratic:	$\hat{y} = a + bx + cx^2$
Exponential:	$\hat{y} = a(b^x)$
Logarithmic:	$\hat{y} = a \log_b x$

Figures 3.15, 3.16, and 3.17 show patterns of bivariate data that appear to have a relationship, whereas in Figure 3.18 the variables do not seem to be related.

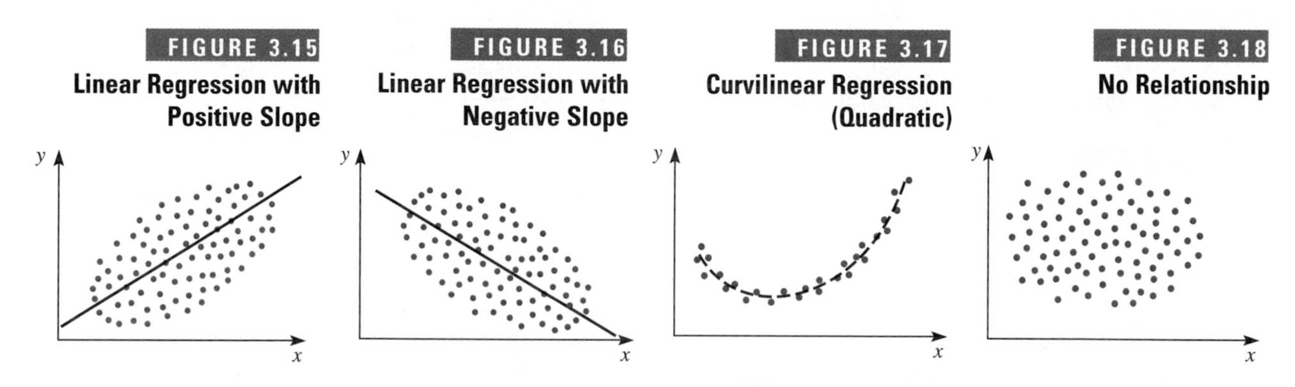

FIGURE 3.15 Linear Regression with Positive Slope

FIGURE 3.16 Linear Regression with Negative Slope

FIGURE 3.17 Curvilinear Regression (Quadratic)

FIGURE 3.18 No Relationship

If a straight-line model seems appropriate, the best-fitting straight line is found by using the **method of least squares.** Suppose that $\hat{y} = b_0 + b_1x$ is the equation of a straight line, where $\hat{y}$ (read "y-hat") represents the **predicted value of y** that corresponds to a particular value of x. The **least squares criterion** requires that we find the constants b_0 and b_1 such that the sum $\sum(y - \hat{y})^2$ is as small as possible.

Figure 3.19 shows the distance of an observed value of y from a **predicted value of $\hat{y}$.** The length of this distance represents the value $(y - \hat{y})$ (shown as the red line segment in Figure 3.19). Note that $(y - \hat{y})$ is positive when the point (x, y) is above the line and negative when (x, y) is below the line.

Figure 3.20 shows a scatter diagram with what appears to be the **line of best fit,** along with ten individual $(y - \hat{y})$ values. (Positive values are shown in red; negative in green.) The sum of the squares of these differences is minimized (made as small as possible) if the line is indeed the line of best fit.

FIGURE 3.19
Observed and Predicted Values of y

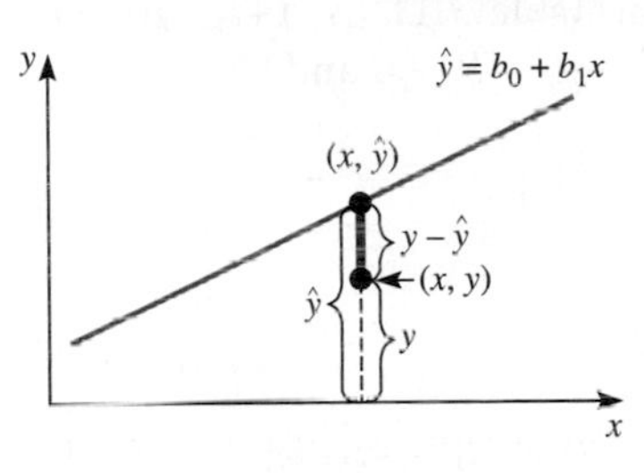

FIGURE 3.20
The Line of Best Fit

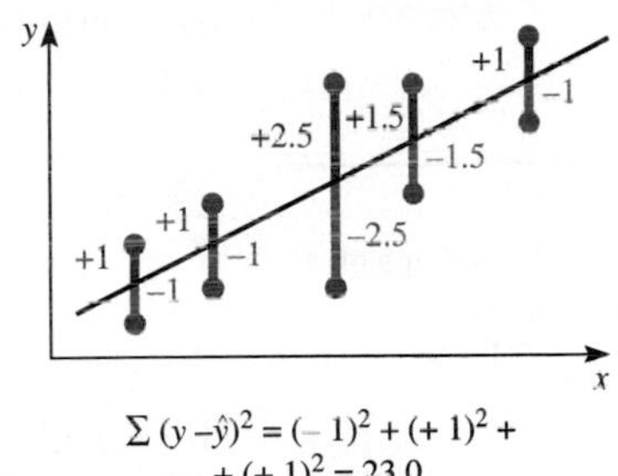

$\Sigma (y - \hat{y})^2 = (-1)^2 + (+1)^2 + \ldots + (+1)^2 = 23.0$

FIGURE 3.21
Not the Line of Best Fit

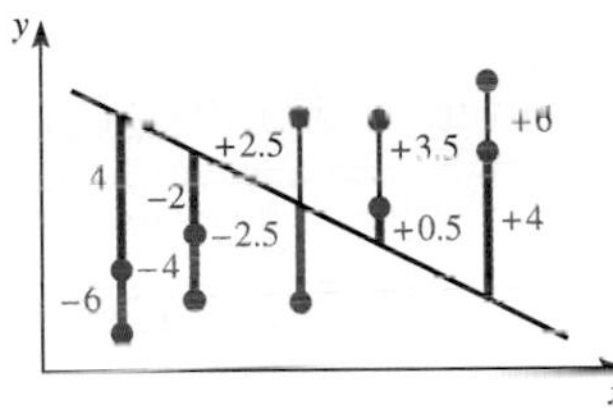

$\Sigma (y - \hat{y})^2 = (-6)^2 + (-4)^2 + \ldots + (+6)^2 = 149.0$

Figure 3.21 shows the same data points as Figure 3.20. The ten individual values of $(y - \hat{y})$ are plotted with a line that is definitely not the line of best fit. [The value of $\Sigma(y - \hat{y})^2$ is 149, much larger than 23 from Figure 3.20.] Every different line drawn through this set of ten points will result in a different value for $\Sigma(y - \hat{y})^2$. Our job is to find the one line that will make $\Sigma(y - \hat{y})^2$ the smallest possible value.

The equation of the line of best fit is determined by its **slope (b_1)** and its ***y*-intercept (b_0)**. (See the *Statistical Tutor* for a review of the concepts of slope and intercept of a straight line.) The values of the constants—slope and *y*-intercept—that satisfy the least squares criterion are found by using the formulas presented next:

DEFINITION FORMULA

$$\text{slope:} \quad b_1 = \frac{\Sigma(x - \bar{x})(y - \bar{y})}{\Sigma(x - \bar{x})^2} \tag{3.5}$$

We will use a mathematical equivalent of formula (3.5) for the slope b_1 that uses the sums of squares found in the preliminary calculations for correlation:

COMPUTATIONAL FORMULA

$$\text{slope:} \quad b_1 = \frac{SS(xy)}{SS(x)} \tag{3.6}$$

Notice that the numerator of formula (3.6) is the SS(*xy*) formula (3.4) and the denominator is formula (2.9) from the correlation coefficient calculations. Thus, if you have previously calculated the linear correlation coefficient using the procedure outlined on pages 146–147, you can easily find the slope of the line of best fit. If you did not previously calculate *r*, set up a table similar to Table 3.11 (p. 147) and complete the necessary preliminary calculations.

For the *y*-intercept, we have:

COMPUTATIONAL FORMULA

$$y\text{-intercept} = \frac{(\text{sum of } y) - [(\text{slope})(\text{sum of } x)]}{\text{number of data}}$$

$$b_0 = \frac{\Sigma y - (b_1 \cdot \Sigma x)}{n} \tag{3.7}$$

ALTERNATIVE COMPUTATIONAL FORMULA

$$y\text{-intercept} = y\text{-bar} - (\text{slope} \cdot x\text{-bar})$$

$$b_0 = \bar{y} - (b_1 \cdot \bar{x}) \tag{3.7a}$$

ANSWER NOW 3.47
Show that formula (3.7a) is equivalent to formula (3.7).

Now let's consider the data in Illustration 3.3 (p. 137) and the question of predicting a student's number of sit-ups based on the number of push-ups. We want to find the line of best fit, $\hat{y} = b_0 + b_1x$. The preliminary calculations have already been completed in Table 3.11 (p. 147). To calculate the slope, b_1, using formula (3.6), recall that $SS(xy) = 919.0$ and $SS(x) = 1396.9$. Therefore,

FYI See Animated Tutorial Line of Best Fit on CD for assistance with these calculations.

$$\text{slope:} \quad b_1 = \frac{SS(xy)}{SS(x)}$$

$$= \frac{919.0}{1396.9} = 0.6579 = \mathbf{0.66}$$

To calculate the y-intercept, b_0, using formula (3.7), recall that $\sum x = 351$ and $\sum y = 380$ from the extensions table. We have

$$y\text{-intercept:} \quad b_0 = \frac{\sum y - (b_1 \cdot \sum x)}{n}$$

$$= \frac{380 - (0.6579)(351)}{10}$$

$$= \frac{380 - 230.9229}{10} = 14.9077 = \mathbf{14.9}$$

By placing the two values just found into the model $\hat{y} = b_0 + b_1x$, we get the equation of the line of best fit:

$$\hat{y} = \mathbf{14.9 + 0.66}x$$

NOTES

1. Remember to keep at least three extra decimal places while doing the calculations to ensure an accurate answer.
2. When rounding off the calculated values of b_0 and b_1, always keep at least two significant digits in the final answer.

ANSWER NOW 3.48

The formulas for finding the slope and the y-intercept of the line of best fit use both summations, $\sum$'s, and sums of squares, SS( )'s. It is important to know the difference. Refer to Illustration 3.4 (p. 146).

a. **Find three pairs of values: $\sum x^2$, $SS(x)$; $\sum y^2$, $SS(y)$; and $\sum xy$, $SS(xy)$.**

b. **Explain the difference between the numbers for each pair of numbers.**

Now that we know the equation for the line of best fit, let's draw the line on the scatter diagram so that we can see the relationship between the line and the data. We need two points in order to draw the line on the diagram. Select two convenient x values, one near each extreme of the domain ($x = 10$ and $x = 60$ are good choices for this illustration), and find their corresponding y values.

For $x = 10$: $\hat{y} = 14.9 + 0.66x = 14.9 + 0.66(10) = 21.5$; **(10, 21.5)**

For $x = 60$: $\hat{y} = 14.9 + 0.66x = 14.9 + 0.66(60) = 54.5$; **(60, 54.5)**

These two points, (10, 21.5) and (60, 54.5), are then located on the scatter diagram (we use a blue + to distinguish them from data points) and the line of best fit is drawn (shown in red in Figure 3.22.)

FIGURE 3.22

Line of Best Fit for Push-ups Versus Sit-ups

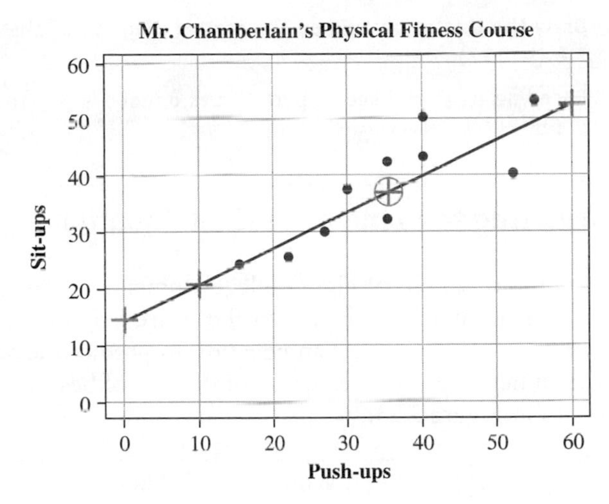

ANSWER NOW 3.49

The values of x used to find points for graphing the line $\hat{y} = 14.9 + 0.66x$ are arbitrary. Suppose you use $x = 20$ and $x = 50$.

a. **What are the corresponding $\hat{y}$ values?**

b. **Locate these two points on Figure 3.22. Are these points on the line of best fit? Explain why or why not.**

There are some additional facts about the least squares method that we need to discuss.

1. The slope, b_1, represents the predicted change in y per unit increase in x. In our example, where $b_1 = 0.66$, if a student can do an additional ten push-ups (x), we predict that he or she would be able to do approximately an additional seven (0.66×10) sit-ups (y).
2. The y-intercept is the value of y where the line of best fit intersects the y-axis. (The y-intercept is easily seen on the scatter diagram, shown as a green + in Figure 3.22, when the vertical scale is located above $x = 0$.) First, however, in interpreting b_0, you must consider whether $x = 0$ is a realistic x value before you conclude that you would predict $\hat{y} = b_0$ if $x = 0$. To predict that if a student did no push-ups, he or she would still do approximately 15 sit-ups ($b_0 = 14.9$) is probably incorrect. Second, the x value of zero may be outside the domain of the data on which the regression line is based. In predicting y based on an x value, check to be sure that the x value is within the domain of the x values observed.
3. The line of best fit will always pass through the *centroid*, the point $(\bar{x}, \bar{y})$. When drawing the line of best fit on your scatter diagram, use this point as a check. For our illustration,

$$\bar{x} = \frac{\sum x}{n} = \frac{351}{10} = 35.1, \qquad \bar{y} = \frac{\sum y}{n} = \frac{380}{10} = 38.0$$

Therefore, $(\bar{x}, \bar{y}) = (35.1, 38.0)$, as shown in green $\oplus$ in Figure 3.22.

ANSWER NOW 3.50

Does it pay to study for an exam? The table gives the numbers of hours studied, x, compared with the exam grades, y.

x	2	5	1	4	2
y	80	80	70	90	60

a. **Find the equation for the line of best fit.**

(continued)

b. **Draw the line of best fit on the scatter diagram of the data drawn in Answer Now 3.8 (p. 137).**

c. **Based on what you see in parts a and b, does it pay to study for an exam? Explain.**

Illustration 3.5 Calculating the Line of Best Fit Equation

In a random sample of eight college women, each was asked her height (to the nearest inch) and her weight (to the nearest 5 pounds). The data obtained are shown in Table 3.12. Find an equation to predict the weight of a college woman based on her height (the equation of the line of best fit), and draw it on the scatter diagram in Figure 3.23.

TABLE 3.12 College Women's Heights and Weights Ta03-12

	1	2	3	4	5	6	7	8
Height, x	65	65	62	67	69	65	61	67
Weight, y	105	125	110	120	140	135	95	130

FIGURE 3.23
Scatter Diagram

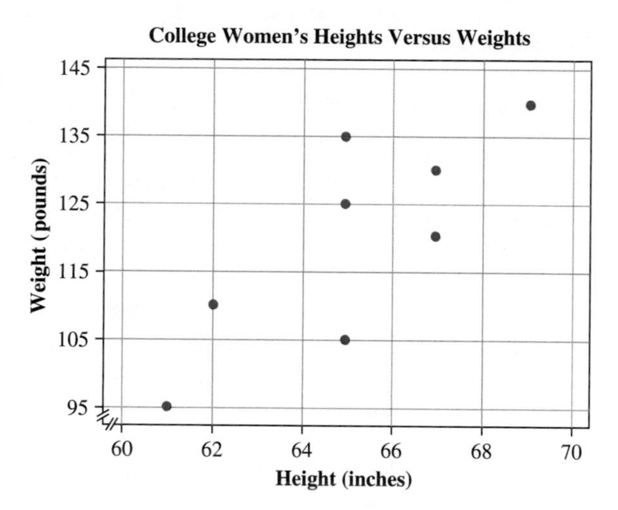

SOLUTION

Before we start to find the equation for the line of best fit, it is often helpful to draw the scatter diagram, which provides visual insight into the relationship between the two variables. The scatter diagram for the data on the heights and weights of college women, shown in Figure 3.23, indicates that the linear model is appropriate.

To find the equation for the line of best fit, we first need to complete the preliminary calculations, as shown in Table 3.13. The other preliminary calculations include finding SS(x) from formula (2.9) and SS(xy) from formula (3.4):

$$\text{SS}(x) = \sum x^2 - \frac{(\sum x)^2}{n} = 33{,}979 - \frac{(521)^2}{8} = 48.875$$

$$\text{SS}(xy) = \sum xy - \frac{\sum x \sum y}{n} = 62{,}750 - \frac{(521)(960)}{8} = 230.0$$

TABLE 3.13 Preliminary Calculations Needed to Find b_1 and b_0

Student	Height, x	x^2	Weight, y	xy
1	65	4,225	105	6,825
2	65	4,225	125	8,125
3	62	3,844	110	6,820
4	67	4,489	120	8,040
5	69	4,761	140	9,660
6	65	4,225	135	8,775
7	61	3,721	95	5,795
8	67	4,489	130	8,710
	$\sum x = 521$	$\sum x^2 = 33{,}979$	$\sum y = 960$	$\sum xy = 62{,}750$

Second, we need to find the slope and the y-intercept using formulas (3.6) and (3.7):

slope: $$b_1 = \frac{SS(xy)}{SS(x)} = \frac{230.0}{48.875} = 4.706 = \mathbf{4.71}$$

y-intercept: $$b_0 = \frac{\sum y - (b_1 \cdot \sum x)}{n} = \frac{960 - (4.706)(521)}{8} = -186.478 = \mathbf{-186.5}$$

Thus, the equation of the line of best fit is $\hat{y} = \mathbf{-186.5 + 4.71x}$.

To draw the line of best fit on the scatter diagram, we need to locate two points. Substitute two values for x—for example, 60 and 70—into the equation for the line of best fit and obtain two corresponding values for $\hat{y}$:

$$\hat{y} = -186.5 + 4.71x = -186.5 + (4.71)(60) = -186.5 + 282.6 = 96.1 \approx 96$$

$$\hat{y} = -186.5 + 4.71x = -186.5 + (4.71)(70) = -186.5 + 329.7 = 143.2 \approx 143$$

The values (60, 96) and (70, 143) represent two points (designated by a red + in Figure 3.24) that enable us to draw the line of best fit.

FIGURE 3.24
Scatter Diagram with Line of Best Fit

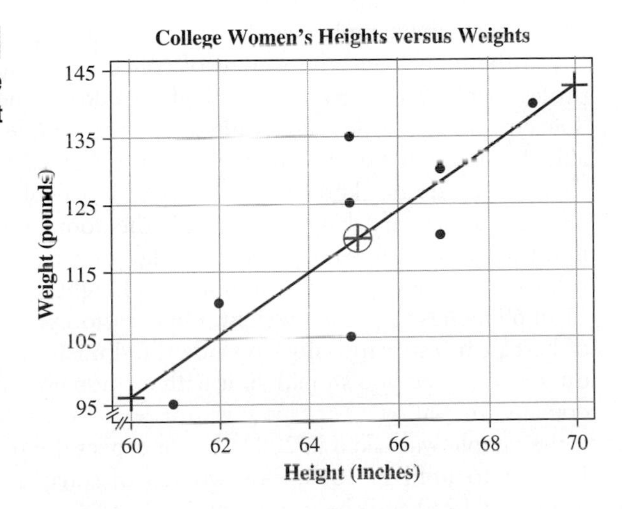

(continued)

NOTE: In Figure 3.24, $(\bar{x}, \bar{y}) = (65.1, 120)$ is also on the line of best fit. It is the green cross in the circle. Use $(\bar{x}, \bar{y})$ as a check on your work. ■

ANSWER NOW 3.51
Refer to the scatter diagram in Figure 3.24.

a. **Explain how the slope of 4.71 can be seen.**

b. **Explain why the y-intercept of -186.5 cannot be seen.**

MAKING PREDICTIONS

One of the main reasons for finding a regression equation is to make predictions. Once a linear relationship has been established and the value of the input variable x is known, we can predict a value of y, $\hat{y}$. For example, in the physical fitness illustration, the equation was found to be $\hat{y} = 14.9 + 0.66x$. If a particular student can do 25 push-ups, how many sit-ups do you therefore predict that she will be able to do? The predicted value is

$$\hat{y} = 14.9 + 0.66x = 14.9 + (0.66)(25) = 14.9 + 16.5 = 31.4 \approx \mathbf{31}$$

You should not expect this predicted value to occur exactly; rather, it is the average number of sit-ups that you would expect from all students who can do 25 push-ups.

ANSWER NOW 3.52
All students who can do 40 push-ups are asked to do as many sit-ups as possible.

a. **How many sit-ups do you expect each can do?**

b **Will they all be able to do the same number of sit-ups?**

c. **Explain the meaning of the answer to part a.**

When you make predictions based on the line of best fit, observe the following restrictions:

1. The equation should be used to make predictions only about the population from which the sample was drawn. For example, using the relationship between the height and the weight of college women to predict the weight of professional athletes given their height would be questionable.
2. The equation should be used only within the sample domain of the input variable. We know the data demonstrate a linear trend within the domain of the x data, but we do not know what the trend is outside this interval. Hence, predictions can be very dangerous outside the domain of the x data. For example, in Illustration 3.5 it is nonsense to predict that a college woman of height zero will weigh -186.5 pounds. Do not use a height outside the sample domain of 61 to 69 inches to predict weight. On occasion you might wish to use the line of best fit to estimate values outside the domain interval of the sample. This can be done, but you should do it with caution and only for values close to the domain interval.
3. If the sample was taken in 2002, do not expect the results to have been valid in 1929 or to hold in 2010. The women of today may be different from the women of 1929 and the women of 2010.

Technology Instructions: Line of Best Fit

MINITAB (Release 13)

Input the x values into C1 and the corresponding y values into C2; then continue with:

Method 1—

```
Choose:   Stat > Regression > Regression...
Enter:    Response (y): C2
          Predictors (x): C1
Choose:   Storage
Select:   FITS (calculates predicted y values and stores
          them in C3 or the first
          available column)
```

To draw the scatter diagram with the line of best fit superimposed on the data points, FITS must have been selected above; then continue with:

```
Choose:   Graph > Plot
Enter:    Graph 1: Y: C2 X2: C1
Choose:   Annotation > Title
Enter:    your title
Choose:   Annotation > Line
Enter:    Points: C1 C3
          Type: Solid
```

OR

Method 2—

```
Choose:   Stat > Regression > Fitted Line Plot
Enter:    Response (Y): C2
          Response (X): C1
Select:   Linear > OK
```

Excel XP

Input the x-variable data into column A and the corresponding y-variable data into column B; then continue with:

```
Choose:   Tools > Data Analysis > Regression > OK
Enter:    Input Y Range: (B1:B10 or select cells)
          Input X Range: (A1:A10 or select cells)
Select:   Labels (if necessary)
          Output Range
          Enter: (C1 or select cell)
          Line Fits Plots > OK
```

To make the output readable; continue with:

```
Choose: Format > Column > Autofit Selection
```

To form the regression equation, the y-intercept is located at the intersection of intercept and coefficients columns, whereas the slope is located at the intersection of the x variable and the coefficients columns.

(continued)

Technology Instructions: Line of Best Fit (Continued)

Excel XP (continued)

To draw the line of best fit on the scatter diagram, activate the chart; then continue with:

```
Choose:   Chart > Add Trendline > Linear > OK
```

(This command also works with the scatter diagram Excel commands on p. 139.)

TI-83 Plus

Input the x-variable data into L1 and the corresponding y-variable data into L2; then continue with:

If just the equation is desired:

```
Choose:   STAT > CALC > 8:LinReg(a+bx)
Enter:    L1, L2*
```

*If the equation and graph on the scatter diagram are desired, use:

```
Enter:    L1, L2, Y1**
```

then continue with the same commands for a scatter diagram as shown on page 139.

**To enter Y1; use:

```
Choose:   VARS > Y-VARS > 1:Function > 1:Y1 > ENTER
```

UNDERSTANDING THE LINE OF BEST FIT

The following method will create:

(1) a visual meaning for the line of best fit,
(2) a visual meaning for what the line of best fit is describing, and
(3) an estimate for the slope and y-intercept of the line of best fit.

As with the approximation of r, estimations of the slope and y-intercept of the line of best fit should be used only as a mental estimate or check.

NOTE: This estimation technique *does not* replace the calculations for b_1 and b_0.

Procedure

1. On the scatter diagram of the data, draw the straight line that appears to be the line of best fit. (*Hint:* If you draw a line parallel to and halfway between the two pencils described in Section 3.2 on page 149 (Figure 3.11), you will have a reasonable estimate for the line of best fit.) The two pencils border the "path" demonstrated by the ordered pairs, and the line down the center of this path approximates the line of best fit. Figure 3.25 shows the pencils and the resulting estimated line for Illustration 3.5.

ANSWER NOW 3.53

The graph's y-intercept is -250, not approximately 80, as might be read from Figure 3.25. Why?

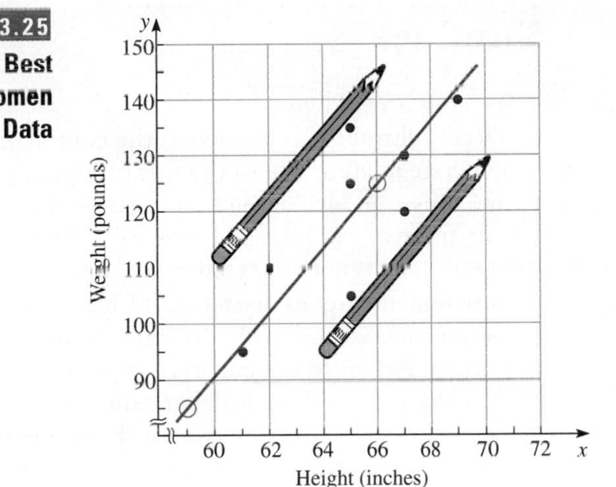

FIGURE 3.25
Estimate the Line of Best Fit for the College Women Data

2. This line can now be used to approximate the equation. First, locate any two points (x_1, y_1) and (x_2, y_2) along the line and determine their coordinates. Two such points, circled in Figure 3.25, have the coordinates (59, 85) and (66, 125). These two pairs of coordinates can now be used in the following formula to estimate the slope b_1:

estimate of the slope, b_1:

$$b_1 \approx \frac{y_2 - y_1}{x_2 - x_1}$$

$$= \frac{125 - 85}{66 - 59} = \frac{40}{7} = 5.7$$

3. Using this result, the coordinates of one of the points, and the following formula, we can determine an estimate for the y-intercept, b_0:

estimate of the y-intercept, b_0:

$$b_0 \approx y - b_1 \cdot x = 85 - (5.7)(59)$$

$$= 85 - 336.3 = -251.3$$

Thus b_0 is approximately -250.

4. We now can write the estimated equation for the line of best fit:

$$\hat{y} = -250 + 5.7x$$

This should serve as a crude estimate.

ANSWER NOW 3.54
The choice of the two points, (x_1, y_1) and (x_2, y_2), is somewhat arbitrary. When different points are selected, slightly different values for b_0 and b_1 will result, but the values should be approximately the same. Use the points (61, 95) and (67, 130) to find the slope and the y-intercept.

Application 3.3 Concrete Shrinkage

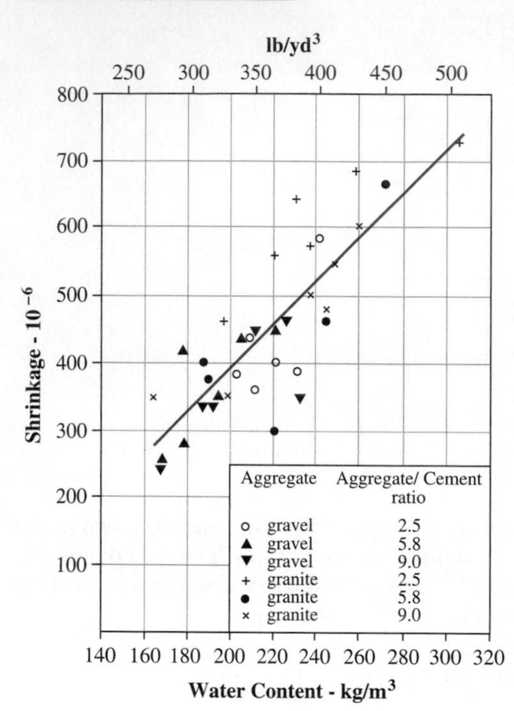

Drying Shrinkage

Drying shrinkage is defined as the contracting of a hardened concrete mixture due to the loss of capillary water. This shrinkage causes an increase in tensile stress, which may lead to cracking, internal warping, and external deflection, before the concrete is subjected to any kind of loading. All portland cement concrete undergoes drying shrinkage, or hydral volume change, as the concrete ages. The hydral volume change in concrete is very important to the engineer in the design of a structure. . . .

Drying shrinkage is dependent upon several factors. These factors include the properties of the components, proportions of the components, mixing manner, amount of moisture while curing, dry environment, and member size. . . . Drying shrinkage happens mostly because of the reduction of capillary water by evaporation and the water in the cement paste. The higher amount of water in the fresh concrete, the greater the drying shrinkage affects. . . .

The concrete properties' influence on drying shrinkage depends on the ratio of water to cementitious materials content, aggregate content, and total water content. The total water content is the most important of these. The relationship between the amount of water content of fresh concrete and the drying shrinkage is linear. Increase of the water content by one percent will approximately increase the drying shrinkage by three percent.

ANSWER NOW 3.55

These data were obtained from the Web site where the article in Application 3.3 is posted. Ap03-3

Water Content (kg/m³)	202	210	220	231	242	167	178	178	193	204	220	166	187	191	210	225	232
Shrinkage (10^{-6})	380	360	400	390	580	255	280	420	350	440	450	240	340	340	450	460	350

a. **Sketch a scatter diagram showing shrinkage, *y*, and water content, *x*, for the concrete data.**

b. **Do the two variables appear to have a linear relationship, as claimed in the article? Explain.**

c. **Find the line of best fit for these data.**

Source: http://www.engr.psu.edu/ce/concrete_clinic/expansionscontractions/dryshrinkage/links/linearrelationship.htm

Exercises

3.56 Ex03-56 Twenty-four countries were randomly selected from the *World Factbook 2001* list of world countries. Data on the number of televisions owned per 1000 people and the life expectancy were collected for those countries.

Country	TVs/1000	Life Expectancy
Samoa	61.43	69.50
Iran	69.71	69.95
Uzbekistan	254.42	63.81
Fiji	24.87	68.25
Greenland	532.37	68.37
Poland	337.79	73.42
Angola	14.47	39.00
Nigeria	54.49	51.07
Saint Pierre, Miquelon	577.37	77.77
Saint Helena	275.25	77.01
Netherlands Antilles	325.13	74.94
Austria	521.42	78.00
Nauru	41.36	61.20
Sao Tome, Principe	139.37	65.59
Anguilla	82.43	76.00
Colombia	113.76	70.57
Moldova	284.32	64.60
Albania	115.37	72.00
French Guiana	168.96	76.30
Venezuela	171.43	73.31
Guadeloupe	273.67	77.16
Bulgaria	429.45	71.20
Taiwan	393.38	76.54
American Samoa	208.69	75.00

Source: http://www.outfo.org/almanac/world_factbook_01/

a. Construct a scatter diagram of the number of TVs per 1000 people, x, and the life expectancy, y.
b. Does it appear that these two variables are correlated? Explain.
c. Are we justified in using the techniques of linear regression on these data? Explain.

3.57 Ajay used linear regression to help understand his monthly telephone bill. The line of best fit was found to be $\hat{y} = 23.65 + 1.28x$; x is the number of long-distance calls made during a month and y is the total telephone cost for a month.

a. Explain the meaning of the y-intercept, 23.65.
b. Explain the meaning of the slope, 1.28.

3.58 A study was conducted to investigate the relationship between the cost, y (in tens of thousands of dollars), per unit of equipment manufactured and the number of units produced per run, x. The resulting equation for the line of best fit was $\hat{y} = 7.31 - 0.01x$, with x being observed for values between 10 and 200. If a production run was scheduled to produce 50 units, what would you predict the cost per unit to be?

3.59 A study was conducted to investigate the relationship between the resale price, y (in hundreds of dollars), and the age, x (in years), of midsize luxury American automobiles. The equation of the line of best fit was determined to be $\hat{y} = 185.7 - 21.52x$.

a. Find the resale value of such a car when it is three years old.
b. Find the resale value of such a car when it is six years old.
c. What is the average annual decrease in the resale price of these cars?

3.60 A study of the tipping habits of restaurant-goers was completed. The data for two of the variables—x, the amount of the restaurant check, and y, the amount left as a tip for the servers—were used to construct a scatter diagram. What do you expect the scatter diagram to reveal?

a. Do you expect the two variables will show a linear relationship? Explain.
b. What will the scatter diagram suggest about linear correlation? Explain.
c. What value do you expect for the slope of the line of best fit? Explain.
d. What value do you expect for the y-intercept of the line of best fit? Explain.

The data are used to determine the equation for the line of best fit: $\hat{y} = 0.02 + 0.177x$.

e. What does the slope of this line represent as applied to the actual situation? Does the value 0.177 make sense? Explain.
f. What does the y-intercept of this line represent as applied to the actual situation? Does the value 0.02 make sense? Explain.
g. If the next restaurant check was for \$30, what would the line of best fit predict for the tip?
h. Using the line of best fit, predict the tip for a check of \$31. What is the difference between this amount and the amount in part g for a \$30 check? Does this difference make sense? Where do you see it in the equation for the line of best fit?

3.61 Ex03-61 Stride rate (number of steps per second) is important to the serious runner. Stride rate is closely related to speed, and a runner's goal is to achieve the optimum stride rate. As part of a study, researchers measured the

PUTTING CHAPTER 3 TO WORK 3.68

a. Construct a scatter diagram, using points scored per game, y, and personal fouls committed per game, x. Explain why you believe there is, or is not, a relationship.
b. Are the two variables—points scored per game and personal fouls committed per game—correlated? Use the correlation coefficient to justify your answer.
c. Express the relationship between the two variables—points scored, y, and personal fouls committed, x—as a linear equation.
d. From the results in part c, if a San Antonio Spur committed two fouls in a game, how many points would you expect him to score?
e. If the player in part d committed a third personal foul, how many extra points would you expect him to score?
f. How does the slope of the line of best fit relate to the number of additional points expected when the player commits one extra personal foul?
g. Do your results show a cause-and-effect relationship between points scored and personal fouls committed? Explain.
h. Should the team coach instruct a player to commit an extra personal foul so that he will score more points? Explain.
i. What is the lurking variable here? Demonstrate it with convincing evidence in the team statistics.

For parts (j) and (k), use the data in the accompanying table. CS03Re

Player	Minutes per Game	Personal Fouls per Game	Points per Game
Duncan	40.6	2.6	25.5
Robinson	29.5	2.5	12.2
Smith	28.7	2.1	11.6
Rose	21.0	2.5	9.4
Daniels	26.5	1.3	9.2
Parker	29.4	2.2	9.2
Smith	19.0	1.9	7.4
Bowen	28.8	2.0	7.0
Porter	18.0	1.2	5.5
Ferry	16.0	1.4	4.6
Jackson	9.9	1.3	3.9
Hart	9.2	1.6	2.6
McCaskill	5.7	0.9	1.9
Bryant	6.9	1.3	1.9
Parks	5.6	0.8	1.5

Source: http://sports.espn.go.com/nba/teamstats

j. What effect do you think "minutes played" has on the variable "points scored per game"? Use statistics to demonstrate the effect.
k. What effect do you think "minutes played" has on the variable "personal fouls committed per game"? Use statistics to demonstrate the effect.

YOUR OWN MINI-STUDY 3.69

a. Is our analysis unique to the San Antonio Spurs during the 2001–2002 regular season? Use the Internet to obtain the season team statistics for your favorite professional or intercollegiate basketball team, or see the coach of a local high school or college team. [Search for "team name."]
b. Do problem 3.68 for your selected team.
c. Discuss the differences and similarities between the San Antonio Spurs and your own team. Consider other lurking variables.

In Retrospect

To sum up what we have just learned: There is a distinct difference between the purpose of regression analysis and the purpose of correlation. In regression analysis, we seek a relationship between the variables. The equation that represents this relationship may be the answer that is desired, or it may be the means to the prediction that is desired. In correlation analysis, we measure the strength of the linear relationship between the two variables.

The applications show a variety of uses for the techniques of correlation and regression. These articles are worth reading again. When bivariate data appear to fall along a straight line on the scatter diagram, they suggest a linear relationship. But this is not proof of cause and effect. Clearly, if a basketball player commits too many fouls, he (or she) will not be scoring more points. Players in foul trouble are "riding the pine" with no chance to score. It also seems reasonable that the more game time they have, the more points they will score and the more fouls they will commit. Thus, a positive correlation and a positive regression relationship will exist between these two variables. Time is a lurking variable here.

The bivariate linear methods we have studied thus far have been presented as a first, descriptive look. More details must, by necessity, wait until additional developmental work has been done. After completing this chapter, you should have a basic understanding of bivariate data, how they are different from just two sets of data, how to present them, what correlation and regression analysis are, and how each is used.

Chapter Exercises

3.70 EX03-70 Fear of the dentist (or the dentist's chair) is an emotion felt by many people of all ages. A survey of 100 individuals in five age groups was conducted about this fear, and these were the results:

	Elementary	Jr. High	Sr. High	College	Adult
Fear	37	28	25	27	21
Do Not Fear	63	72	75	73	79

a. Find the marginal totals.
b. Express the frequencies as percentages of the grand total.
c. Express the frequencies as percentages of each age group's marginal totals.
d. Express the frequencies as percentages of those who fear and those who do not fear.
e. Draw a bar graph based on age groups.

3.71 The accompanying USA Snapshot ® "Rainy day savings" lists in percentages the distributions for the amount both genders have saved for emergencies.

a. Identify the population, the variables, and the type of variables.
b. Construct a bar graph showing the two distributions side by side.
c. Do the distributions seem to differ for the genders? Explain.

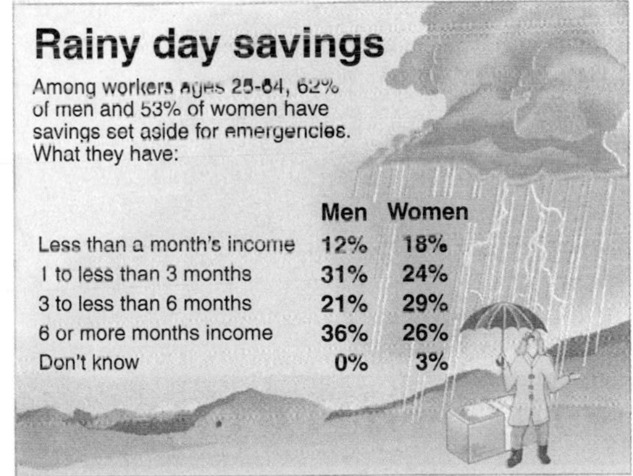

USA SNAPSHOTS®
A look at statistics that shape the nation
Rainy day savings
Among workers ages 25-64, 62% of men and 53% of women have savings set aside for emergencies. What they have:
Men Women
Less than a month's income 12% 18%
1 to less than 3 months 31% 24%
3 to less than 6 months 21% 29%
6 or more months income 36% 26%
Don't know 0% 3%
Source: Merrill Lynch
By Anne R. Carey and Grant Jerding, USA TODAY

3.72 Ex03-72 Six breeds of dogs have been popular in the United States over the past few years. The table lists each breed along with the numbers of registrations filed with the American Kennel Club in 1995 and 1996.

Breed	1995	1996
Labrador Retriever	132,051	149,505
Rottweiler	93,656	89,867
German Shepherd	78,088	79,076
Golden Retriever	64,107	68,993
Beagle	57,063	58,946
Poodle	54,784	56,803

Source: American Kennel Club, New York, NY

a. Build a cross-tabulation of the two variables: year (rows) and dog breed (columns). Express the results in frequencies, showing marginal totals.
b. Express the frequencies you found in part a in percentages based on the grand total.
c. Draw a bar graph showing the results from part b.
d. Express the frequencies you found in part a in percentages based on the marginal total for the year.
e. Draw a bar graph showing the results from part d.

3.73 Ex03-73 When was the last time you saw your doctor? That question was asked for the survey summarized in the table.

		Time Since Last Consultation with Your Physician		
		6 Months to Less Than 6 Months	Less Than 1 Year	1 Year or More
Age	Under 28 years	413	192	295
	28–40	574	208	218
	Over 40	653	288	259

a. Find the marginal totals.
b. Express the frequencies as percentages of the grand total.
c. Express the frequencies as percentages of each age group's marginal totals.
d. Express the frequencies as percentages of each time period.
e. Draw a bar graph based on the grand total.

3.74 Ex03-74 Part of quality control is keeping track of what is occurring. The contingency table shows the numbers of rejected castings last month categorized by their cause and the work shift during which they occurred.

	1st Shift	2nd Shift	3rd Shift
Sand	87	110	72
Shift	16	17	4
Drop	12	17	16
Corebreak	18	16	33
Broken	17	12	20
Other	8	18	22

a. Find the marginal totals.
b. Express the numbers as percentages of the grand total.
c. Express the numbers as percentages of each shift's marginal total.
d. Express the numbers as percentages of each type of rejection.
e. Draw a bar graph based on the shifts.

3.75 Determine whether each of the following questions requires correlation analysis or regression analysis to obtain an answer.
a. Is there a correlation between the grades a student attained in high school and the grades he or she attained in college?
b. What is the relationship between the weight of a package and the cost of mailing it first class?
c. Is there a linear relationship between a person's height and shoe size?
d. What is the relationship between the number of worker-hours and the number of units of production completed?
e. Is the score obtained on a certain aptitude test linearly related to a person's ability to perform a certain job?

3.76 An automobile owner records the number of gallons of gasoline, x, required to fill the gasoline tank and the number of miles traveled, y, between fill-ups.
a. If she does a correlation analysis on the data, what would be her purpose and what would be the nature of her results?
b. If she does a regression analysis on the data, what would be her purpose and what would be the nature of her results?

3.77 These data were generated using the equation $y = 2x + 1$.

x	0	1	2	3	4
y	1	3	5	7	9

A scatter diagram of the data results in five points that fall perfectly on a straight line. Find the correlation coefficient and the equation of the line of best fit.

3.78 Consider this set of bivariate data:

x	1	1	3	3
y	1	3	1	3

a. Draw a scatter diagram.
b. Calculate the correlation coefficient.
c. Calculate the line of best fit.

3.79 Start with the point (5, 5) and add at least four ordered pairs, (x, y), to make a set of ordered pairs that display the following properties. Show that your sample satisfies the requirements.
a. The correlation of x and y is 0.0.
b. The correlation of x and y is $+1.0$.
c. The correlation of x and y is -1.0.
d. The correlation of x and y is between -0.2 and 0.0.
e. The correlation of x and y is between $+0.5$ and $+0.7$.

3.80 A scatter diagram is drawn showing the data for x and y, two normally distributed variables. The data fall within the intervals $20 \leq x \leq 40$ and $60 \leq y \leq 100$. Where would you expect to find the data on the scatter diagram, if:

a. the correlation coefficient is 0.0
b. the correlation coefficient is 0.3
c. the correlation coefficient is 0.8
d. the correlation coefficient is -0.3
e. the correlation coefficient is -0.8

3.81 Start with the point (5, 5) and add at least four ordered pairs, (x, y), to make a set of ordered pairs that display the following properties. Show that your sample satisfies the requirements.

a. The correlation of x and y is between $+0.9$ and $+1.0$, and the slope of the line of best fit is 0.5.
b. The correlation of x and y is between $+0.5$ and $+0.7$, and the slope of the line of best fit is 0.5.
c. The correlation of x and y is between -0.7 and -0.9, and the slope of the line of best fit is -0.5.
d. The correlation of x and y is between $+0.5$ and $+0.7$, and the slope of the line of best fit is 1.0.

3.82 Ex03-82 Twenty-four countries were randomly selected from the *World Factbook 2001* list of world countries. The life expectancies for males and females were recorded for those countries.

Country	Male Life Expectancy	Female Life Expectancy
Samoa	66.77	72.37
Iran	68.61	71.37
Uzbekistan	60.24	67.56
Fiji	65.83	70.78
Greenland	64.82	72.01
Poland	69.26	77.82
Angola	37.36	39.87
Nigeria	51.07	51.07
Saint Pierre, Miquelon	75.51	80.13
Saint Helena	74.13	80.04
Netherlands Antilles	72.76	77.22
Austria	74.60	81.15
Nauru	57.70	64.88
Sao Tome, Principe	64.15	67.07
Anguilla	73.41	79.29
Colombia	66.71	74.55
Moldova	60.15	69.26
Albania	69.01	74.87
French Guiana	72.97	79.79
Venezuela	70.29	76.56
Guadeloupe	74.01	80.48
Bulgaria	67.72	74.89
Taiwan	73.81	79.51
American Samoa	70.89	80.02

Source: http://www.outfo.org/almanac/world_factbook_01/

a. Construct a scatter diagram of life expectancy for males, x, and life expectancy for females, y.
b. Does it appear that these two variables are correlated? Explain.
c. Find the line of best fit.
d. What does the numerical value of the slope represent?

3.83 Ex03-83 The sound of crickets chirping is a welcome summer night's sound. In fact, those chirping crickets may be telling you the temperature. In the book *The Song of Insects*, George W. Pierce, a Harvard physics professor, presented real data relating the number of chirps per second for striped ground crickets to the temperature in °F. The table gives real cricket and temperature data. It appears that the number of chirps represents an average, since it is given to the nearest tenth.

Chirps/sec	Temperature (°F)
20.0	88.6
16.0	71.6
19.8	93.3
18.4	84.3
17.1	80.6
15.5	75.2
14.7	69.7
17.1	82.0
15.4	69.4
16.2	83.3
15.0	79.6
17.2	82.6
16.0	80.6
17.0	83.5
14.4	76.3

Source: George W. Pierce, *The Song of Insects*, Harvard University Press, 1948

a. Draw a scatter diagram of the number of chirps per second, x, and the air temperature, y.
b. Describe the pattern displayed.
c. Find the equation for the line of best fit.
d. Using the equation from part c, find the temperatures that correspond to 14 and 20 chirps, the approximate bounds for the domain of the study.
e. Does the range of temperature values bounded by the temperature values found in part d seem reasonable for this study? Explain.
f. The next time you are out where crickets chirp on a summer night and you find yourself without a thermometer, just count the chirps and you will be able to tell the temperature. ☺ If the count is 16, what temperature would you suspect it is?

Data Table for Exercise 3.84 Ex03-84

Lake	Area (sq mi)	Max. Depth (ft)	Lake	Area (sq mi)	Max. Depth (ft)
Caspian Sea	143,244	3,363	Chad	6,300	24
Superior	31,700	1,330	Maracaibo	5,217	115
Victoria	26,828	270	Onega	3,710	328
Aral Sea	24,904	220	Titicaca	3,200	922
Huron	23,000	750	Nicaragua	3,100	230
Michigan	22,300	923	Athabasca	3,064	407
Tanganyika	12,700	4,823	Reindeer	2,568	720
Baykal	12,162	5,315	Turkana	2,473	240
Great Bear	12,096	1,463	Issyk Kul	2,355	2,303
Nyasa	11,150	2,280	Vanern	2,156	328
Great Slave	11,031	2,015	Winnipegosis	2,075	38
Erie	9,910	210	Albert	2,075	168
Winnipeg	9,417	60	Kariba	2,050	390
Ontario	7,340	802	Nipigon	1,872	540
Balkhash	7,115	85	Urmia	1,815	49
Ladoga	6,835	738	Manitoba	1,799	12

Source: Geological Survey, U.S. Department of the Interior

3.84 Ex03-84 Lakes are bodies of water surrounded by land and may include seas. The table above lists the areas and maximum depths of 32 lakes throughout the world.

a. Draw a scatter diagram showing area, x, and maximum depth, y, for the lakes.

b. Find the linear correlation coefficient between area and maximum depth. What does the value of this linear correlation imply?

3.85 Ex03-85 Wildlife populations are monitored with aerial photographs. Numbers of animals and their locations relative to areas inhabited by the human population are useful information. Sometimes it is possible to monitor the physical characteristics of the animals. The length of an alligator can be estimated quite accurately from aerial photographs, but its weight cannot. The data are the lengths (in inches) and weights (in pounds) of alligators captured in central Florida and can be used to predict the weight of an alligator based on its length.

Weight	Length	Weight	Length
130	94	83	86
51	74	70	8
640	147	61	72
28	58	54	74
80	86	44	61
110	94	106	90
33	63	84	89
90	86	39	68
36	69	42	76
38	72	197	114
366	12	102	90
84	85	57	78
80	82		

Source: http://exploringdata.cqu.edu.au/stories.htm#alligatr

a. Construct a scatter diagram for length, x, and weight y.

b. Does it appear that the weight of an alligator is predictable from its weight? Explain.

c. Is the relationship linear?

d. Explain why the line of best fit, as described in this chapter, is not adequate for estimating weight based on length.

e. Find the value of the linear correlation coefficient.

f. Explain why the value of r can be so high for a set of data that is so obviously not linear in nature.

3.86 Ex03-86 Sugar cane growers are interested in the relationship between the acres of crop harvested and the total sugar cane production of those acres. The data listed are for the 2001 crop from 14 randomly selected sugar-cane-producing counties in Louisiana.

Acres Harvested	Production (tons)
33,700	940,000
15,200	460,000
14,400	440,000
2,300	65,000
30,200	830,000
13,100	380,000
29,600	860,000
20,200	590,000
33,800	1,020,000
20,500	585,000
33,100	1,020,000
8,000	200,000
41,100	1,130,000
17,900	570,000

Source: http://www.usda.gov/nass/graphics/county01/data/

a. These data values have many zeros that will be in the way. Change acres harvested to 100s of acres and production to 1000s of tons of production before continuing.
b. Construct a scatter diagram of acres harvested, x, and tons of production, y.
c. Does the relationship between the variables appear to be linear? Explain.
d. Find the equation for the line of best fit.
e. What is the slope for the line of best fit? What does the slope represent? Explain what it means to the sugar cane grower.

3.87 Ex03-87 Cicadas are flying, plant-eating insects. One particular species, 13-year cicadas (*Magicicada*), spends five juvenile stages in underground burrows. During the 13 years underground, the cicadas grow from approximately the size of a small ant to nearly the size of an adult cicada. Every 13 years, the animals then emerge from their burrows as adults. The table presents three different species of these 13-year cicadas with their adult body weight (BW), in grams, and wing length (WL), in millimeters.

Species	BW	WL	Species	BW	WL
tredecula	0.15	28	tredecula	0.18	29
tredecim	0.29	32	tredecassini	0.21	27
tredecim	0.17	27	tredecula	0.15	30
tredecula	0.18	30	tredecula	0.17	27
tredecim	0.39	35	tredecassini	0.13	27
tredecim	0.26	31	tredecassini	0.17	29
tredecassini	0.17	29	tredecassini	0.23	30
tredecassini	0.16	28	tredecim	0.12	22
tredecassini	0.14	25	tredecula	0.26	30
tredecassini	0.14	28	tredecula	0.19	30
tredecassini	0.28	25	tredecassini	0.20	30
tredecim	0.12	28	tredecula	0.14	23

Source: http://insects/ummz.lsa.umich.edu

a. Construct a scatter diagram of the body weights, x, and the corresponding wing lengths, y. Use a different symbol to represent the ordered pairs for each species.
b. Describe what the scatter diagram displays with respect to relationship and species.
c. Calculate the correlation coefficient, r.
d. Find the equation for the line of best fit.
e. Suppose the body weight of a cicada is 0.20 gram. What wing length would you predict? Which species do you think this cicada might be?

3.88 Ex03-88 Yellowstone National Park's Old Faithful has been a major tourist attraction for a long time. Understanding the duration of eruptions and the time between eruptions is necessary to predict the timing of the next eruption. The Old Faithful data set variables are: Date: an index of the date the observation was taken (days 1, 2, and 3 are given here—all 16 days are on your CD); Duration: the duration of an eruption of the geyser, in minutes; and Intereruption: the time until the next eruption, in minutes.

Day 1		Day 2		Day 3	
Duration	Intereruption	Duration	Intereruption	Duration	Intereruption
4.4	78	4.3	80	4.5	76
3.9	74	1.7	56	3.9	82
4.0	68	3.9	80	4.3	84
4.0	76	3.7	69	2.3	53
3.5	80	3.1	57	3.8	86
4.1	84	4.0	90	1.9	51
2.3	50	1.8	42	4.6	85
4.7	93	4.1	91	1.8	45
1.7	55	1.8	51	4.7	88
4.9	76	3.2	79	1.8	51
1.7	58	1.9	53	4.6	80
4.6	74	4.6	82	1.9	49
3.4	75	2.0	51	3.5	82

Source: http://comp.uark.edu/~jtubbs/Biostat/Labs/Oldfaithful/oldfaithful.html

a. Construct a scatter diagram of the 39 durations, x, and intereruptions, y. Use a different symbol to represent the ordered pairs for each day.
b. Describe the pattern displayed by all 39 ordered pairs.
c. Do the data for the individual days show the same pattern as one another and as the total data set?
d. Based on information in the scatter diagram, if Old Faithful's last eruption lasted 4 minutes, how long would you predict we will need to wait until the next eruption starts?
e. Find the line of best fit for the data listed in the table.
f. Based on the line of best fit, if Old Faithful's last eruption lasted 4 minutes, how long would you predict we will need to wait until the next eruption starts?
g. What effect do you think the distinctive pattern shown on the scatter diagram has on the line of best fit? Explain.
h. Repeat parts a and g using the data set for 16 days of observations. The data are on your CD. DS-OldFaithful2.
i. Compare the results found in part h to the results in parts a–g. Discuss your conclusions.

3.89

a. Verify, algebraically, that formula (3.2) for calculating r is equivalent to the definition formula (3.1).
b. Verify, algebraically, that formula (3.6) is equivalent to formula (3.5).

3.90 This equation gives a relationship that exists between b_1 and r:

$$r = b_1 \sqrt{\frac{SS(x)}{SS(y)}}$$

a. Verify the equation for these data:

x	4	3	2	3	0
y	11	8	6	7	4

b. Verify this equation using formulas (3.2) and (3.6).

Vocabulary and Key Concepts

Be able to define each term. Pay special attention to key terms, which are printed in **red.** In addition, describe each term in your own words and give an example of each. Your examples should not be the ones given in class or in the textbook. Page numbers indicate the first appearance of the term.

bivariate data (p. 130)
cause-and-effect (p. 150)
coefficient of linear correlation (p. 145)
contingency table (p. 130)
correlation (p. 144)
correlation analysis (p. 144)
cross-tabulation (p. 130)
dependent variable (p. 136)
independent variable (p. 136)
input variable (p. 136)
least squares criterion (p. 154)
line of best fit (p. 154)
linear correlation (p. 144)
linear regression (p. 154)
lurking variable (p. 150)
method of least squares (p. 154)
negative correlation (p. 144)
ordered pair (p. 136)
output variable (p. 136)
Pearson's product moment, r (p. 146)
positive correlation (p. 144)
predicted value (p. 154)
prediction equation (p. 154)
regression (p. 154)
regression analysis (p. 154)
scatter diagram (p. 137)
slope, b_1 (p. 155)
y-intercept, b_0 (p. 155)

Chapter Practice Test

PART I: KNOWING THE DEFINITIONS

Answer "True" if the statement is always true. If the statement is not always true, replace the words shown in bold with words that make the statement always true.

3.1 **Correlation** analysis is a method of obtaining the equation that represents the relationship between two variables.

3.2 The linear correlation coefficient is used to determine the **equation that represents** the relationship between two variables.

3.3 A correlation coefficient of **zero** means that the two variables are perfectly correlated.

3.4 Whenever the slope of the regression line is zero, the **correlation coefficient** will also be zero.

3.5 When r is positive, b_1 will always be **negative.**

3.6 The **slope** of the regression line represents the amount of change expected to take place in y when x increases by one unit.

3.7 When the calculated value of r is positive, the calculated value of b_1 will be **negative.**

3.8 Correlation coefficients range between **0 and +1.**

3.9 The value being predicted is called the **input variable.**

3.10 The line of best fit is used to predict the **average value of y** that can be expected to occur at a given value of x.

PART II: APPLYING THE CONCEPTS

3.11 Refer to the scatter diagram at the top of page 175.

a. Match the descriptions in column 2 with the terms in column 1.

_____ population (a) the horsepower rating for an automobile

_____ sample (b) all 1995 American-made automobiles

_____ input variable (c) the EPA mileage rating for an automobile

_____ output variable (d) the 1995 automobiles with ratings shown on the scatter diagram

Scatter Diagram for 3.11

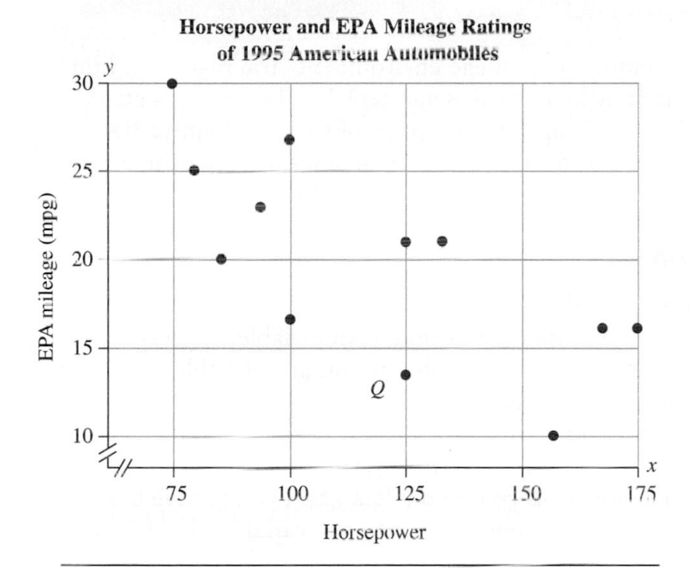

b. Find the sample size.
c. What is the smallest value reported for the output variable?
d. What is the largest value reported for the input variable?
e. Does the scatter diagram suggest a positive, negative, or zero linear correlation coefficient?
f. What are the coordinates of point Q?
g. Will the slope for the line of best fit be positive, negative, or zero?
h. Will they intercept for the line of best fit be positive, negative, or zero?

3.12 A research group reports a 2.3 correlation coefficient for two variables. What can you conclude from this information?

3.13 For the bivariate data, the extensions, and the totals shown on the table, find the following:

x	y	x^2	xy	y^2
2	6	4	12	36
3	5	9	15	25
3	7	9	21	49
4	7	16	28	49
5	7	25	35	49
5	9	25	45	81
6	8	36	48	64
28	49	124	204	353

a. SS(x)
b. SS(y)
c. SS(xy)
d. the linear correlation coefficient, r
e. the slope, b_1
f. the y-intercept, b_0
g. the equation of the line of best fit

PART III: UNDERSTANDING THE CONCEPTS

3.14 A test was administered to measure the mathematics ability of the people in a certain town. Some of the townspeople were surprised to find out that their test results and their shoe sizes correlated strongly. Explain why a strong positive correlation should not have been a surprise.

3.15 Student A collected a set of bivariate data and calculated r, the linear correlation coefficient. Its value was -1.78. Student A proclaimed that there was no correlation between the two variables because the value of r was not between -1.0 and $+1.0$. Student B argued that -1.78 was impossible and that only values of r near zero implied no correlation. Who is correct? Justify your answer.

3.16 The linear correlation coefficient, r, is a numerical value that ranges from -1.0 to $+1.0$. Write a sentence or two describing the meaning of r for each of these values:
a. -0.93
b. $+0.89$
c. -0.03
d. $+0.08$
e. -2.3

3.17 Make up a set of three or more ordered pairs such that:
a. $r = 0.0$
b. $r = +1.0$
c. $r = -1.0$
d. $b_1 = 0.0$

WORKING WITH YOUR OWN DATA

© StockTrek/Getty Images

Each semester, new students enter your college environment. You may have wondered, What will the student body be like this semester? As a beginning statistics student, you have just finished studying three chapters of basic descriptive statistical techniques. You can use some of these techniques to describe some characteristics of your college's student body.

A SINGLE-VARIABLE DATA

1. Define the population to be studied.
2. Choose a variable to define. (You may define your own variable, or you may use one of the variables in the accompanying table if you are not able to collect your own data. Ask your instructor for guidance.)
3. Collect 35 pieces of data for your variable.
4. Construct a stem-and-leaf display of your data. Be sure to label it.
5. Calculate the value of the measure of central tendency that you believe best answers the question: What is the average value of your variable? Explain why you chose this measure.
6. Calculate the sample mean for your data (unless you used the mean in question 5).
7. Calculate the sample standard deviation for your data.
8. Find the value of the 85th percentile, P_{85}.
9. Construct a graphic display (other than a stem-and-leaf) that you believe "best" displays your data. Explain why the graph best presents your data.
10. Write a summary paragraph describing your findings.

B BIVARIATE DATA

1. Define the population to be studied.
2. Choose and define two quantitative variables that will produce bivariate data. (You may define your own variables, or you may use two of the variables in the accompanying table if you are not able to collect your own data. Ask your instructor for guidance.)
3. Collect 15 ordered pairs of data.
4. Construct a scatter diagram of your data. (Be sure to label it completely.)
5. Using a table to assist with the organization, calculate the extensions x^2, xy, and y^2, and the summations of x, y, x^2, xy, and y^2.
6. Calculate the linear correlation coefficient, r.
7. Calculate the equation of the line of best fit.
8. Draw the line of best fit on your scatter diagram.
9. Write a summary paragraph describing your findings.

The following table of data was collected on the first day of class last semester. You may use it as a source for your data if you are not able to collect your own.

Variable *A:* student's gender (male/female)
Variable *B:* student's age at last birthday
Variable *C:* number of completed credit hours toward degree
Variable *D:* "Do you have a job (full/part time)?" (yes/no)
Variable *E:* number of hours worked last week, if D = yes
Variable *F:* wages (before taxes) earned last week, if D = yes

FYI
The computer will select your random sample (see p. 104).

DS-1

Student	A	B	C	D	E	F	Student	A	B	C	D	E	F
1	M	21	16	No			51	F	42	34	Yes	40	244
2	M	18	0	Yes	10	34	52	M	25	60	Yes	60	503
3	F	23	18	Yes	46	206	53	M	39	32	Yes	40	500
4	M	17	0	No			54	M	29	13	Yes	39	375
5	M	17	0	Yes	40	157	55	M	19	18	Yes	51	201
6	M	40	17	No			56	M	25	0	Yes	48	500
7	M	20	16	Yes	40	300	57	F	18	0	No		
8	M	18	0	No			58	M	32	68	Yes	44	473
9	F	18	0	Yes	20	70	59	F	21	0	No		
10	M	29	9	Yes	8	32	60	F	26	0	Yes	40	320
11	M	20	22	Yes	38	146	61	M	24	11	Yes	45	330
12	M	34	0	Yes	40	340	62	F	19	0	Yes	40	220
13	M	19	31	Yes	29	105	63	M	19	0	Yes	10	33
14	M	18	0	No			64	F	35	59	Yes	25	88
15	M	20	0	Yes	48	350	65	F	24	0	Yes	40	300
16	F	27	3	Yes	40	130	66	F	20	33	Yes	40	170
17	M	19	10	Yes	40	202	67	F	26	0	Yes	52	300
18	F	18	16	Yes	40	140	68	F	17	0	Yes	27	100
19	M	19	4	Yes	6	22	69	M	25	18	Yes	41	355
20	F	29	9	No			70	M	24	0	No		
21	F	21	0	Yes	20	80	71	M	21	0	Yes	30	150
22	F	39	6	No			72	M	30	12	Yes	48	555
23	M	23	34	Yes	42	415	73	F	19	0	Yes	38	169
24	F	31	0	Yes	48	325	74	M	32	45	Yes	40	385
25	F	22	7	Yes	40	195	75	M	26	90	Yes	40	340
26	F	27	75	Yes	20	130	76	M	20	64	Yes	10	45
27	F	19	0	No			77	M	24	0	Yes	30	150
28	M	22	20	Yes	40	470	78	M	20	14	No		
29	F	80	0	Yes	40	390	79	M	21	70	Yes	40	340
30	M	25	14	No			80	F	20	13	Yes	40	206
31	F	24	45	No			81	F	33	3	Yes	32	246
32	M	34	4	No			82	F	25	68	Yes	40	330
33	M	29	48	No			83	F	29	48	Yes	40	525
34	M	22	80	Yes	40	336	84	F	40	0	Yes	40	400
35	M	21	12	Yes	26	143	85	F	36	3	Yes	40	300
36	F	18	0	No			86	F	35	0	Yes	40	280
37	M	18	0	Yes	13	65	87	F	28	0	Yes	40	350
38	M	40	64	Yes	40	390	88	F	27	9	Yes	40	260
39	F	31	0	Yes	40	200	89	F	26	3	Yes	40	240
40	F	32	0	Yes	40	270	90	F	23	9	Yes	40	330
41	F	37	0	Yes	24	150	91	M	41	3	Yes	23	253
42	F	35	0	Yes	40	350	92	M	39	0	Yes	40	110
43	M	21	72	Yes	45	470	93	M	21	0	Yes	40	246
44	F	27	0	Yes	40	550	94	F	32	0	Yes	40	350
45	F	42	47	Yes	37	300	95	F	48	58	Yes	40	714
46	F	41	21	Yes	40	250	96	F	26	0	Yes	32	200
47	M	36	0	Yes	40	400	97	F	27	0	Yes	40	350
48	M	25	16	Yes	40	480	98	F	52	56	Yes	40	390
49	F	18	0	Yes	45	189	99	F	34	27	Yes	8	77
50	M	22	0	Yes	40	385	100	F	49	3	Yes	24	260

PART

2 Probability

© StockTrek/Getty Images

Before continuing our study of statistics, we must make a slight detour and study some basic probability. Probability is often called the "vehicle" of statistics; that is, the probability associated with chance occurrences is the underlying theory for statistics. Recall that in Chapter 1 we described *probability* as the science of making statements about what will occur when samples are drawn from known populations. *Statistics* was described as the science of selecting a sample and making inferences about the unknown population from which the sample is drawn. To make these inferences, we need to study sample results in situations in which the population is known so that we will be able to understand the behavior of chance occurrences.

In Part Two we study the basic theory of probability (Chapter 4), probability distributions of discrete variables (Chapter 5), and probability distributions for continuous random variables (Chapter 6). Following this brief study of probability, we will study the techniques of inferential statistics in Part Three.

Karl Pearson

Karl Pearson, known as one of the fathers of modern statistics, was born on March 27, 1857, in London, the second son of prominent attorney William Pearson and his wife, Fanny Smith. Karl was tutored at home until, at age nine, he entered University College School in London. In 1875, following a year of illness that required him to be tutored privately, he was awarded a scholarship to King's College, Cambridge. There, in May 1879, he earned his B.A. (with honors) in mathematics; he then went on to earn an M.A. in law in 1882.

After receiving his law degree, Pearson moved to Heidelberg, Germany, where he became proficient in literature, philosophy, physics, and metaphysics, as well as in German history and folklore.

Pearson returned to University College, where, in 1884, he was appointed Goldsmid Professor of Applied Mathematics and Mechanics. In addition, Pearson lectured in geometry at Gresham College, London, from 1891 to 1894. Later, in 1911, he relinquished the Goldsmid chair to become the first Galton Professor of Eugenics.

In 1896 Pearson was elected to the Royal Society and was awarded the society's Darwin Medal in 1898. In 1900 Pearson developed the chi-square (denoted by χ^2); it is the oldest inference procedure still used in its original form and is often used in today's economics and business applications. Around that time, Pearson also verified the concept of random phenomena (or probability) by tossing a coin 24,000 times to determine the frequency of its landing "heads up" as opposed to "tails up." Result: 12,012 heads, a relative frequency of 0.5005.

It was during Pearson's association with Sir Francis Galton that he developed the linear correlation coefficient (sometimes referred to as the Pearson product moment correlation coefficient, in his honor). Pearson was editor of, and a major contributor to, the statistical journal *Biometrika*, which he cofounded with fellow statisticians Galton and Weldon. Pearson returned to London in 1933, where he died on April 27, 1936.

Pearson's only son, Egon S. (second oldest of three children), born in 1895 to Karl and his wife, Maria Sharpe, also became a well-known statistician.

More information about Karl Pearson can be found at these Web sites:
http://www-gap.dcs.st-and.ac.uk/~history/Mathematicians/Pearson.html
http://www.economics.soton.ac.uk/staff/aldrich/main.htm
http://www.ucl.ac.uk/Stats/history/pearson.html
http://www.fordham.edu/halsall/mod/1900pearsonl.html

CHAPTER

4 Probability

Chapter Outline and Objectives

CHAPTER CASE STUDY

Probability is part of our everyday culture even when it comes to M&M's.

Concepts of Probability

4.1 THE NATURE OF PROBABILITY

Probability can be thought of as the **relative frequency** with which an event occurs, such as: "There is a 40% chance of rain this afternoon" or "I have a 50–50 chance of passing today's chemistry exam."

4.2 PROBABILITY OF EVENTS

The three methods for assigning probabilities to an event are **empirical, theoretical**, and **subjective.**

4.3 SIMPLE SAMPLE SPACES

All **possible outcomes** of an experiment are listed.

4.4 RULES OF PROBABILITY

The probabilities for the outcomes of a sample space **total exactly one.**

Calculating Probability of Compound Events

4.5 MUTUALLY EXCLUSIVE EVENTS AND THE ADDITION RULE

Events are mutually exclusive if they **cannot** both occur at the **same time.**

4.6 INDEPENDENCE, THE MULTIPLICATION RULE, AND CONDITIONAL PROBABILITY

Events are independent if the **occurrence** of one event does **not change** the **probability** of the other event.

4.7 COMBINING THE RULES OF PROBABILITY

The addition and multiplication rules are often used together to calculate the probability of **compound events.**

RETURN TO CHAPTER CASE STUDY

Chapter Case Study

Statistics Students' Favorite Candy

Where did all these colorful candies come from?

Did you know that the idea for M&M's Plain Chocolate Candies was born in the backdrop of the Spanish Civil War? Legend has it that on a trip to Spain, Forrest Mars Sr. encountered soldiers who were eating pellets of chocolate encased in a hard sugary coating to prevent them from melting. Inspired by this idea, Mr. Mars went back to his kitchen and invented the recipe for M&M's Plain Chocolate Candies.

A brief colorful history:*

- M&M's were first sold in 1941 in cardboard tubes.
- In 1954 the famous slogan "The milk chocolate melts in your mouth, not in your hand" debuted.
- In 1960 three new colors—red, green, and yellow—were added to the original brown color.
- In 1976 the color orange was added and red was removed from the traditional color blend.
- In 1987 red was added back into the traditional color blend as a result of overwhelming requests from consumers.
- In 1995 Americans voted for a new color to appear in the traditional mix. Blue won by a landslide over pink and purple, with 54% of the more than 10 million votes cast.
- And now—Here's Purple!

---

*Source: http://global.mms.com/us/index.jsp

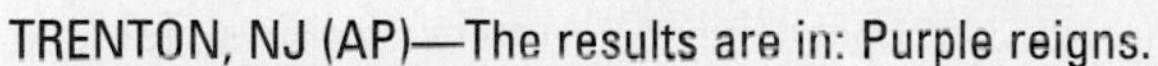
PURPLE REIGNS: FANS OF M&M'S CHOOSE NEW COLOR

TRENTON, NJ (AP)—The results are in: Purple reigns.

The maker of M&M's said Wednesday that purple will be the latest color for the little chocolate candies. The decision came after more than 10 million people in 200 countries voted on the Internet and over the phone on whether to add purple, pink, or aqua to the mix.

Purple received 41% of the vote. Aqua had 37% and pink had 19%. Purple will join the existing mix of red, blue, brown, green, orange, and yellow in August.

Source: http://www.usatoday.com/news/nation/2002/06/19/mandm.htm

Here are the percentages Mars, Incorporated, uses to mix the colors for M&M's Milk Chocolate Candies:

Before GCV (Global Color Vote):

10% blue, 30% brown, 10% green, 10% orange, 20% red, 20% yellow

After GCV:

10% blue, 10% brown, 10% green, 10% orange, 20% red, 20% yellow, 20% purple

Mars, Incorporated, at http://global.mms.com/us/index.jsp has a link to a world map where you can check the GCV results for the various countries.

ANSWER NOW 4.1

a. If you bought a bag of M&M's, what color would you expect to see the most? What color the least? Why?

b. If you bought a bag of M&M's, would you expect to find the percentages listed above? If not, why and what would you expect?

After completing Chapter 4, further investigate the Chapter Case Study in the Return to Chapter Case Study section with Exercises 4.108, 4.109, and 4.110 (pp. 223–224).

Concepts of Probability

4.1 The Nature of Probability

Let's consider an experiment in which we toss two coins simultaneously and record the number of heads that occur. The only possible outcomes are 0H (zero heads), 1H (one head), and 2H (two heads). Let's toss the two coins ten times and record our findings:

2H 1H 1H 2H 1H 0H 1H 1H 1H 2H

Here is the summary:

Outcome	Frequency
2H	3
1H	6
0H	1

Suppose we repeat this experiment 19 times. Table 4.1 shows the totals for 20 sets of ten tosses. (Trial 1 shows the totals from our first experiment.)

The 200 tosses of the pair of coins resulted in 2H on 53 occasions, 1H on 104 occasions, and 0H on 43 occasions. We can express these results in terms of relative frequencies and show the results on a histogram, as in Figure 4.1.

TABLE 4.1 Experimental Results of Tossing Two Coins Ta04-1

	Trial																				
Outcome	1	2	3	4	5	6	7	8	9	10	11	12	13	14	15	16	17	18	19	20	Total
2H	3	3	5	1	4	2	4	3	1	1	2	5	6	3	1	4	1	0	3	1	53
1H	6	5	5	5	5	7	5	5	5	5	8	4	3	7	5	1	5	4	5	9	104
0H	1	2	0	4	1	1	1	2	4	4	0	1	1	0	4	5	4	6	2	0	43

FIGURE 4.1
Relative Frequency Histogram for Coin-Tossing Experiment

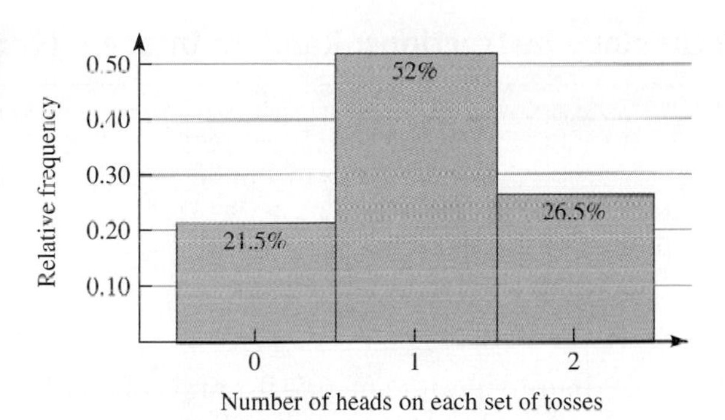

What conclusions can we reach? If we look at the individual sets of ten tosses, we notice a large variation in the number of times each of the events (2H, 1H, and 0H) occurred. In both the 0H and the 2H categories, there were as many as 6 occurrences and as few as 0 occurrences in a set of ten tosses. In the 1H category, there were as few as 1 occurrence and as many as 9 occurrences.

If we continued this experiment for several hundred more tosses, what would you expect to happen to the relative frequencies of these three events? It looks as if we have approximately a 1:2:1 ratio in the totals of Table 4.1. We might therefore expect to find the **relative frequency** for 0H to be approximately $\frac{1}{4}$, or 25%; the relative frequency for 1H to be approximately $\frac{1}{2}$, or 50%; and the relative frequency for 2H to be approximately $\frac{1}{4}$, or 25%. These relative frequencies accurately reflect the concept of probability.

Many probability experiments can be simulated by using a random-number table or by having a computer/calculator randomly generate number values that represent the various experimental outcomes. For example, the preceding tossing of two coins experiment could be simulated by letting odd integers represent heads (H) and even integers, tails (T). Since we are tossing two coins, we need a two-digit random integer—the first digit for the first coin and the second digit for the second coin—for each toss of the two coins. The key to either method is to maintain the probabilities. [$P(\text{H}) = \frac{1}{2}$ and $P(\text{odd}) = \frac{1}{2}$, $P(\text{T}) = \frac{1}{2}$ and $P(\text{even}) = \frac{1}{2}$; therefore, we have assigned random digits to the events so as to maintain the equal probabilities of heads and tails.] Thus, the random integer 45 would represent TH, or 1H. This two-digit method would work with both the random-number table and the computer. However, if the computer is used, then by letting 1 = H and 0 = T on each toss of a coin, and by randomly generating two columns, one for each coin, the computer will do all the tally work, too. By adding the two columns together, we will find the total to be the number of heads seen in each toss of the two coins. See the commands that follow.

Technology Instructions: Random Integers

MINITAB (Release 13)

```
Choose:   Calc > Random Data > Integer
Enter:    Generate: 200
          Store in: C1 C2
          Minimum value: 0
          Maximum value: 1 > OK
Choose:   Calc > Calculator
```

(continued)

Technology Instructions: Random Integers (Continued)

MINITAB (Release 13) (continued)

Enter: Store result in variable: C3
Expression: C1 + C2 > OK
Choose: Stat > Tables > Tally
Enter: Variable: C3
Check: Counts and Percents

Excel XP

Enter 0, 1 into column A and 0.5, 0.5 into column B. Label column C as Coin1, column D as Coin2, and column E as Sum.

Choose: Tools > Data Analysis > Random Number Generation
Enter: Number of Variables: 2
Number of Random Numbers: 200
Distribution: Discrete
Value and Probability Input Range: (A1:B2 or select cells)
Select: Output Range
Enter: (C2 or select cells) > OK

Activate the cell E2.

Enter: = C2 + D2 > Enter
Drag: Bottom right corner of E2 down to give other sums
Choose: Data > Pivot Table and PivotChart Report...
Select: Microsoft Excel list or database > Next
Enter: Range: (E1:E201 or select cells) > Next
Select: Existing Worksheet
Enter: (F1 or select cell)
Choose: Layout
Drag: "Sum" heading into both row & data areas

Double click "sum of sum" in data area box; then continue with:

Choose: Summarize by: Count > Options
Show data as: % of totals or Normal > OK > OK > Finish

TI-83 Plus

Choose: MATH > PRB > 5:randInt(
Enter: 0,1,200)
Choose: STO→ > 2nd L1

Repeat above commands but store data in L2.

Choose: STAT > EDIT > 1:Edit
Highlight: L3 (column heading)
Enter: L1 + L2
Choose: 2nd > STAT PLOT > 1:Plot1
Choose: WINDOW
Enter: -.5, 2.5, 1, -1, 120, 20,1
Choose: TRACE > > > (use to find counts; percents can be calculated)

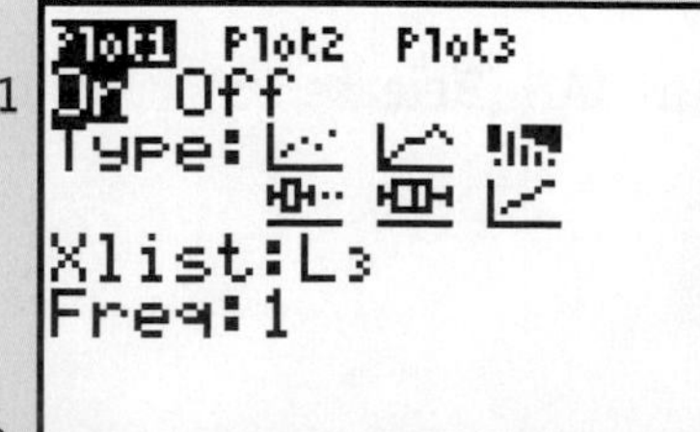

Exercises

4.2

a. Toss a single coin 10 times and record H (head) or T (tail) after each toss. Using your results, find the relative frequency of heads and of tails.
b. Roll a single die 20 times and record a 1, 2, 3, 4, 5, or 6 after each roll. Using your results, find the relative frequency of each number: 1, 2, 3, 4, 5, and 6.

4.3 Place three coins in a cup, shake and dump them out, and observe the number of heads showing. Record 0H, 1H, 2H, or 3H after each trial. Repeat the process 25 times. Using your results, find the relative frequency of each result: 0H, 1H, 2H, and 3H.

4.4 Place a pair of dice in a cup, shake and dump them out, and observe the sum of the dots. Record 2, 3, 4, . . . , 12. Repeat the process 25 times. Using your results, find the relative frequency for each of the values: 2, 3, 4, 5, . . . , 12.

4.5 Use either the random-number table (Appendix B) or a computer to simulate:

a. the rolling of a die 50 times; express your results as relative frequencies.
b. the tossing of a coin 100 times; express your results as relative frequencies.

4.6 Use either the random-number table (Appendix B) or a computer to simulate the random selection of 100 single-digit numbers, 0 through 9.

a. List the 100 digits.
b. Prepare a relative frequency distribution of the 100 digits.
c. Prepare a relative frequency histogram of the distribution in part b.

4.2 Probability of Events

We are now ready to define what is meant by *probability*. Specifically, we talk about "the probability that a certain event will occur."

PROBABILITY THAT AN EVENT WILL OCCUR
The relative frequency with which that event can be expected to occur.

The probability of an event may be obtained in three different ways: (1) *empirically*, (2) *theoretically*, and (3) *subjectively*. The first method was illustrated in the experiment in Section 4.1 and might be called **experimental,** or **empirical probability.** This probability is the *observed relative frequency with which an event occurs.* In our coin-tossing illustration, we observed exactly one head (1H) on 104 of the 200 tosses of the pair of coins. The observed empirical probability for the occurrence of 1H was 104/200, or 0.52.

When the value assigned to the probability of an event results from experimental data, we will identify the probability of the event with the symbol $P'(\ )$.

NOTE The *prime notation* is used to denote empirical probabilities.

The value assigned to the probability of event A as a result of experimentation can be found by means of the formula

EMPIRICAL (OBSERVED) PROBABILITY: $P'(A)$

$$\text{probability of } A = \frac{\text{number of times A occurred}}{\text{number of trials}}$$

$$P'(A) = \frac{n(A)}{n} \tag{4.1}$$

ANSWER NOW 4.7
If you roll a die 40 times and 9 of the rolls result in a "5," what empirical probability was observed for the event "5"?

ANSWER NOW 4.8
Explain why an empirical probability, an observed proportion, and a relative frequency are actually three different names for the same thing.

Consider rolling a die. Define event A as the occurrence of a 1. A single roll of a die has six possible outcomes. Assuming that the die is symmetrical, each number should have an equal likelihood of occurring. Intuitively, the probability of A, or the expected relative frequency of a 1, is $\frac{1}{6}$. (Later we will formalize this calculation.)

What does this mean? Does it mean that once in every six rolls a 1 will occur? No, it does not. Saying that the probability of a 1, $P(1)$, is $\frac{1}{6}$ means that in the long run the proportion of times that a 1 occurs is approximately $\frac{1}{6}$. How close to $\frac{1}{6}$ can we expect the observed relative frequency to be?

Table 4.2 shows the number of 1's observed in each set of six rolls of a die (column 1), the observed relative frequency for each set of six rolls (column 2), and the cumulative relative frequency (column 3). Each trial is a set of six rolls. Figure 4.2a shows the fluctuation of the observed probability for event A on each of the 20 trials (column 2, Table 4.2). Figure 4.2b shows the fluctuation of the cumulative relative frequency (column 3, Table 4.2). Notice that the observed relative frequency on each trial of six rolls of a die tends to fluctuate about $\frac{1}{6}$. Notice also that the observed values on the cumulative graph seem to become more stable; in fact, they become relatively close to the expected $\frac{1}{6}$, or $0.1666\overline{6} = 0.167$.

TABLE 4.2 Experimental Results of Rolling a Die Six Times in Each Trial

Trial	Column 1: Number of 1's Observed	Column 2: Relative Frequency	Column 3: Cumulative Relative Frequency
1	1	1/6	1/6 = 0.17
2	2	2/6	3/12 = 0.25
3	0	0/6	3/18 = 0.17
4	1	1/6	4/24 = 0.17
5	0	0/6	4/30 = 0.13
6	1	1/6	5/36 = 0.14
7	2	2/6	7/42 = 0.17
8	2	2/6	9/48 = 0.19
9	0	0/6	9/54 = 0.17
10	0	0/6	9/60 = 0.15
11	1	1/6	10/66 = 0.15
12	0	0/6	10/72 = 0.14
13	2	2/6	12/78 = 0.15
14	1	1/6	13/84 = 0.15
15	1	1/6	14/90 = 0.16
16	3	3/6	17/96 = 0.18
17	0	0/6	17/102 = 0.17
18	1	1/6	18/108 = 0.17
19	0	0/6	18/114 = 0.16
20	1	1/6	19/120 = 0.16

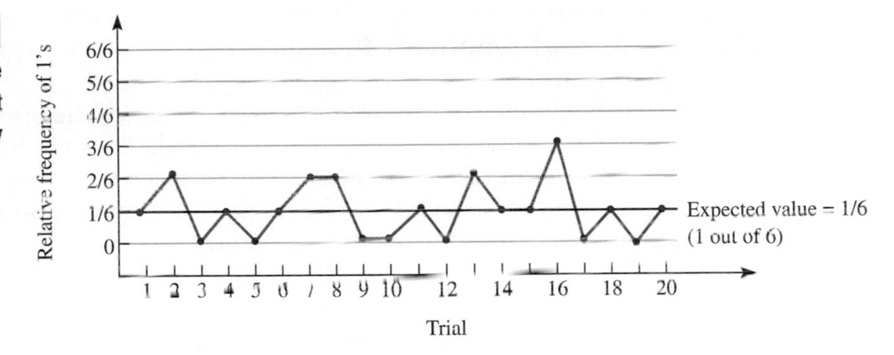

FIGURE 4.2a
Fluctuations Found in the Die-Tossing Experiment (a) Relative Frequency

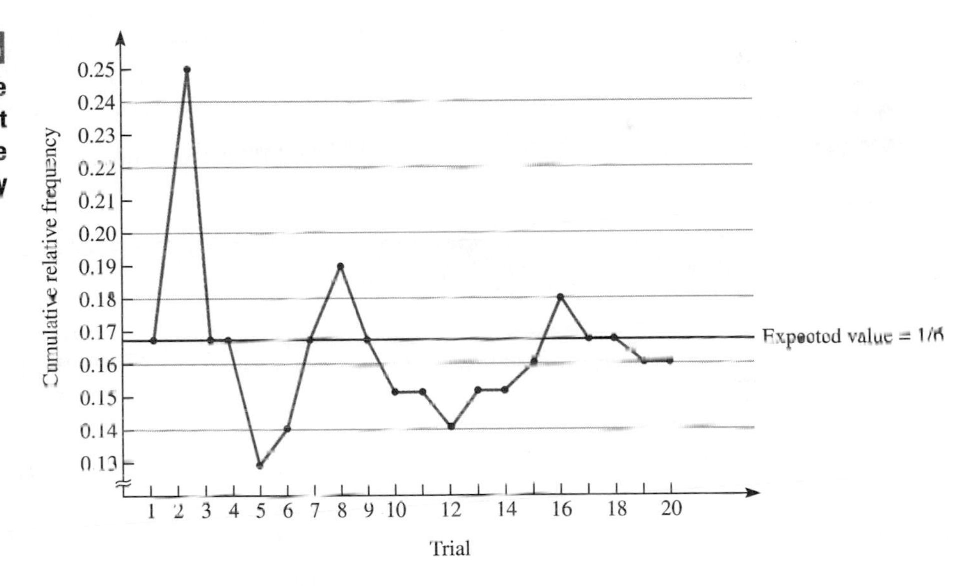

FIGURE 4.2b
Fluctuations Found in the Die-Tossing Experiment (b) Cumulative Relative Frequency

A cumulative graph such as Figure 4.2b demonstrates the idea of long-term average. When only a few rolls were observed (as on each trial, Figure 4.2a), the probability $P'(A)$ fluctuated between 0 and $\frac{1}{2}$. As the experiment was repeated, however, the cumulative graph suggests a stabilizing effect on the observed cumulative probability. This stabilizing effect, or **long-term average** value, is often referred to as the *law of large numbers.*

LAW OF LARGE NUMBERS

As the number of times an experiment is repeated increases, the ratio of the number of successful occurrences to the number of trials will tend to approach the theoretical probability of the outcome for an individual trial.

The law of large numbers is telling us that the larger the number of experimental trials n, the closer the empirical probability $P'(A)$ is expected to be to the true or theoretical probability $P(A)$. There are many applications of this concept. The die-tossing experiment is an example in which we can easily compare the results to what we expect to happen; this gives us a chance to verify the claim of the law of large numbers. Illustration 4.1 shows how we use the results obtained from large sets of data when the theoretical expectation is unknown.

Illustration 4.1 Uses of Empirical Probabilities

In establishing life insurance rates, actuaries use the probabilities that the insureds will live one, two, three years, and so forth, from the time they purchase policies. These probabilities are derived from actual life and death statistics and hence are empirical probabilities. They are published by the government and are extremely important to the life insurance industry. ■

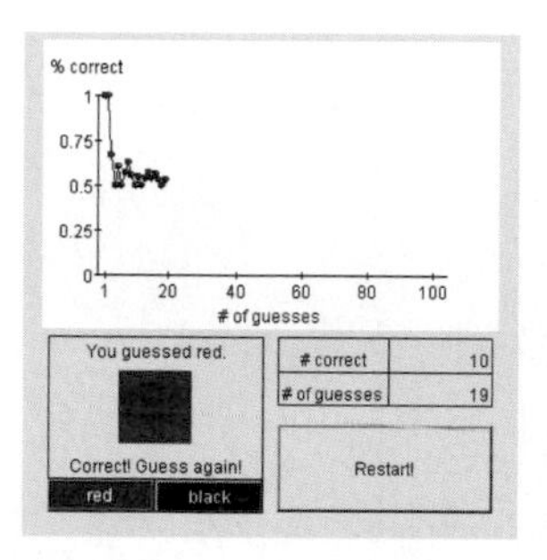

ANSWER NOW 4.9 INTERACTIVITY

Interactivity 4-A demonstrates the law of large numbers and also allows you to see whether you have psychic powers. Repeat the simulations at least 50 times, guessing between picking a red card or a black card from a deck of cards.

a. **What proportion of the time did you guess correctly?**

b. **As you made more and more guesses, did your proportions start to stabilize? If so, at what value? Does this value make sense for the experiment? Why?**

c. **How might you know if you have ESP?**

Exercises

4.10

a. Explain what is meant by the statement: "When a single die is rolled, the probability of a 1 is $\frac{1}{6}$."

b. Explain what is meant by the statement: "When one coin is tossed one time, there is a 50–50 chance of getting a tail."

4.11 According to the National Climatic Data Center, San Juan, Puerto Rico, experienced 197 days of measurable precipitation (rain) in 1995. This was more days of rain than any of the other weather stations reported that year, but it was not considered unusual for San Juan. By comparison, the station at Phoenix, Arizona, reported 31 days of precipitation during the same year (again, not unusual). (National Climatic Data Center, NESDIS, NOAA, U.S. Department of Commerce)

a. Suppose you were taking a one-day business trip to San Juan. What is the probability that it will rain while you are there?

b. Suppose you decide instead to take your trip to Phoenix. What is the probability that measurable precipitation will occur while you are there?

4.12 The September 1998 issue of *Visual Basic Programmer's Journal* reported the results of a survey of 500 subscribers who indicated the number of hours they worked per week at their jobs. Results are summarized in the table.

Hours Worked Per Week	Number of Respondents	Percentage of Respondents
29 or less	18	3.6
30–39	98	19.6
40–49	281	56.2
50–59	69	13.8
60 or more	34	6.8
Totals	500	100.0

Source: *Visual Basic Programmer's Journal,* "Fall 1998 Salary Survey Results," September 1998.

Suppose one of the respondents from the sample of returns is selected at random. Find the probability of the following events:

a. The respondent works less than 40 hours per week.

b. The respondent works 50 or more hours per week.

c. The respondent works between 30 and 59 hours per week.

4.13 Take two dice (one white and one colored) and roll them 50 times, recording the results as ordered pairs [(white, color); for example, (3, 5) represents 3 on the white die and 5 on the colored die]. (You could simulate these 50 rolls using a random-number table or a computer.) Then calculate each observed probability:

a. P'(white die is an odd number)

b. P'(sum is 6)

c. P'(both dice show odd number)

d. P'(number on color die is larger than number on white die)

4.14 Use a random-number table or a computer to simulate rolling a pair of dice 100 times.

a. List the results of each roll as an ordered pair and the sum.

b. Prepare an ungrouped frequency distribution and a histogram of the results.

c. Describe how these results compare with what you expect to occur when two dice are rolled.

Technology Instructions: Simulate Dice

MINITAB (Release 13)

```
Choose:    Calc > Random Data > Integer
Enter:     Generate: 100
           Store in: C1 C2
           Minimum value: 1
           Maximum value: 6 > OK
Choose:    Calc > Calculator
Enter:     Store result in variable: C3
           Expression: C1 + C2 > OK
Choose:    Stat > Tables > Tally
Enter:     Variable: C3
Select:    Counts
```

Use the MINITAB commands on page 58 to construct a frequency histogram of the data in C3 with the midpoints 2:12/1.

Excel XP

Enter 1, 2, 3, 4, 5, 6 into column A, label C1: **Die1**; D1: **Die2**; E1: **Dice**, and activate B1.

```
Choose:    Format > Cells > Number > Number
Enter:     Decimal places: 8 > OK
Enter:     1/6 in B1
Drag:      Bottom right corner of B1 down for 6 entries
Choose:    Tools > Data Analysis > Random Number Generation > OK
Enter:     Number of Variables: 2
           Number of Random Numbers: 100
           Distribution: Discrete
           Value and Probability Input Range: (A1:B6 or select cells)
Select:    Output Range
Enter:     (C2 or select cells) > OK
```

Activate the **E2** cell.

```
Enter:     = C2 + D2 > Enter
Drag:      Bottom right corner of E2 down for 100 entries
Choose:    Data > Pivot and PivotChart Table Report ...
Select:    Microsoft Excel list or database > Next
Enter:     Range: (E1:E101 or select cells) > Next
Select:    Existing Worksheet
Enter:     (F1 or select cell)
Choose:    Layout
Drag:      "Dice" heading into both row & data areas
```

Double click the "sum of dice" in data area box; then continue with:

```
Choose:    Summarize by: Count > OK > OK > Finish
```

Label column J "sums" and input the numbers 2, 3, 4, . . . , 12 into it. Use the Excel histogram commands on pages 58–59 with column E as the input range and column J as the bin range.

(continued)

Technology Instructions: Simulate Dice (Continued)

TI-83 Plus

```
Choose:      MATH > PRB > 5:randInt(
Enter:       1,6,100)
Choose:      STO→ > 2nd L1
Repeat above for L2.
Choose:      STAT > EDIT >1:Edit
Highlight:   L3
Enter:       L3 =L1 + L2
Choose:      2nd  > STAT PLOT > 1:Plot1
```

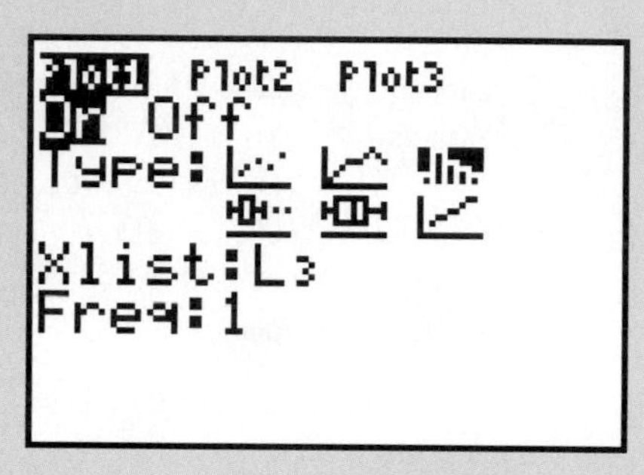


```
Choose:      WINDOW
Enter:       -.5, 12.5, 1, -10, 40, 10,1
Choose:      TRACE > > >
```

4.15 Using a coin, perform the experiment discussed on pages 186–187. Toss a coin ten times, observe the number of heads (or put ten coins in a cup, shake and dump them into a box, and use each toss for a block of ten), and record the results. Repeat until you have 200 tosses. Chart and graph the data as individual sets of ten and as cumulative relative frequencies. Do your data tend to support the claim that $P(\text{head}) = \frac{1}{2}$? Explain.

4.16 Let's estimate the probability that a thumbtack lands "point up" (as opposed to "point down") when tossed and lands on a hard surface. Using a thumbtack, perform the die experiment discussed on pages 186–187. Toss the thumbtack ten times, observe the number of "point up" (or put ten identical thumbtacks in a cup, shake and dump them into a box, and use each toss for a block of ten), and record the results. Repeat until you have 200 tosses. Chart and graph the data as individual sets of ten and as cumulative relative frequencies. What do you believe the $P'(\;)$ to be? Explain.

4.3 Simple Sample Spaces

Let's return to an earlier question: What values might we expect to be assigned to the three events (0H, 1H, 2H) associated with our coin-tossing experiment? As we inspect these three events, we see that they do not tend to occur with the same relative frequency. Why? Suppose the experiment of tossing two pennies and observing the number of heads had actually been carried out using a penny and a nickel—two distinct coins. Would this have changed our results? No, it would have had no effect on the experiment. However, it does show that there are more than three possible outcomes.

When a penny is tossed, it may land as heads or tails. When a nickel is tossed, it may also land as heads or tails. If we toss them simultaneously, we see that there are actually four different possible outcomes. These four outcomes match up with the previous events as follows:

1. heads on penny and heads on nickel—2H
2. heads on penny and tails on nickel—1H
3. tails on penny and heads on nickel—1H
4. tails on penny and tails on nickel—0H

In this experiment with the penny and the nickel, let's use **ordered pair** notation. The first listing will correspond to the penny, and the second will correspond to the nickel. Thus, (H, T) represents the event that a head occurs on the penny and a tail occurs on the nickel. Our listing of events for tossing a penny and a nickel looks like this:

(H, H) (H, T) (T, H) (T, T)

What we have here is a listing of what is known as the *sample space* for this experiment.

EXPERIMENT
Any process that yields a result or an observation.

OUTCOME
A particular result of an experiment.

SAMPLE SPACE
The set of all possible outcomes of an experiment. The sample space is typically denoted by S and may take any number of forms: a list, a tree diagram, a lattice grid system, and so on. The individual outcomes in a sample space are called **sample points**, and $n(S)$ is the number of sample points in sample space S.

EVENT
Any subset of the sample space. If A is an event, then $n(A)$ is the number of sample points that belong to event A.

Regardless of the form in which they are presented, the outcomes in a sample space can never overlap. Also, all possible outcomes must be represented. These characteristics are called **mutually exclusive** and **all inclusive,** respectively. A more detailed explanation of these characteristics will be presented later; for the moment, however, an intuitive grasp of their meaning is sufficient.

Now let's look at some probability experiments and their associated sample spaces.

Experiment 4.1

A single coin is tossed once and the outcome—a head (H) or a tail (T)—is recorded. The sample space is: $S = \{H, T\}$ and $n(S) = 2$.

Experiment 4.2

Two coins, one penny and one nickel, are tossed simultaneously, and the outcome for each coin is recorded using ordered pair notation: (penny, nickel). The sample space is shown here in two different ways:

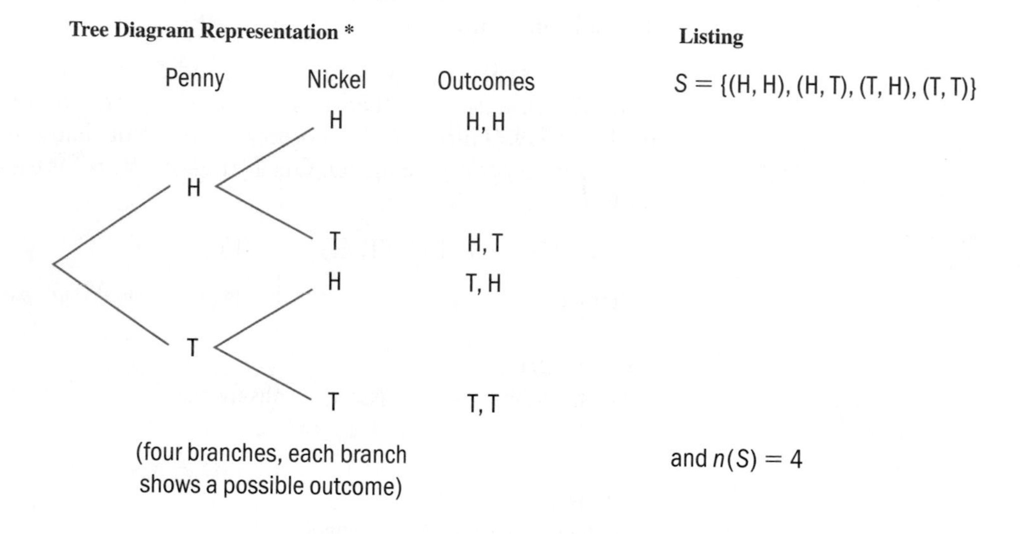

*See the *Statistical Tutor* for information about tree diagrams.

Notice that both representations show the same four possible outcomes. For example, the top branch on the tree diagram shows heads on both coins, as does the first ordered pair in the listing. ■

ANSWER NOW 4.17
Select one single-digit number randomly. List the sample space.

Experiment 4.3

A die is rolled one time and the number of spots on the top face observed. The sample space is $S = \{1, 2, 3, 4, 5, 6\}$ and $n(S) = 6$.

Experiment 4.4

A box contains three poker chips (one red, one blue, one white), and two are drawn *with replacement.* (This means that one chip is selected, its color is observed, and then the chip is put back into the box.) The chips are scrambled before a second chip is selected and its color observed. The sample space may be shown in two different ways:

Tree Diagram Representation

First drawing	Second drawing	Outcomes
R	R	R, R
	B	R, B
	W	R, W
B	R	B, R
	B	B, B
	W	B, W
W	R	W, R
	B	W, B
	W	W,W

Listing

$S = \{(R, R), (R, B), (R, W), (B, R), (B, B), (B, W), (W, R), (W, B), (W, W)\}$

and $n(S) = 9$

Experiment 4.5

A box contains one red, one blue, and one white poker chip. Two chips are drawn simultaneously or one at a time *without replacement* (meaning one chip is selected and then a second is selected without replacing the first). The sample space is shown in two ways:

Tree Diagram Representation

First drawing	Second drawing	Outcomes
R	B	R, B
	W	R, W
B	R	B, R
	W	B, W
W	R	W, R
	B	W, B

Listing

$S = \{(R, B), (R, W), (B, R), (B, W), (W, R), (W, B)\}$

and $n(S) = 6$

This experiment is the same as Experiment 4.4 except that the first chip is not replaced before the second selection is made. ■

ANSWER NOW 4.18
A penny is tossed and a die is rolled. List the sample space as a tree diagram.

ANSWER NOW 4.19
Draw a tree diagram representing the possible arrangements of boys and girls from oldest to youngest for a family of:

a. two children

b. three children

Experiment 4.6

Three coins are tossed, or one coin is tossed three times, with head (H) or tail (T) observed on each coin. The sample space is shown here:

Tree Diagram Representation

First toss	Second toss	Third toss	Outcomes
H	H	H	H, H, H
		T	H, H, T
	T	H	H, T, H
		T	H, T, T
T	H	H	T, H, H
		T	T, H, T
	T	H	T, T, H
		T	T, T, T

and $n(S) = 8$

Experiment 4.7

Two dice (one white and one black) are rolled one time, and the number of dots showing on each die is observed. The sample space is shown in a chart representation:

Chart Representation

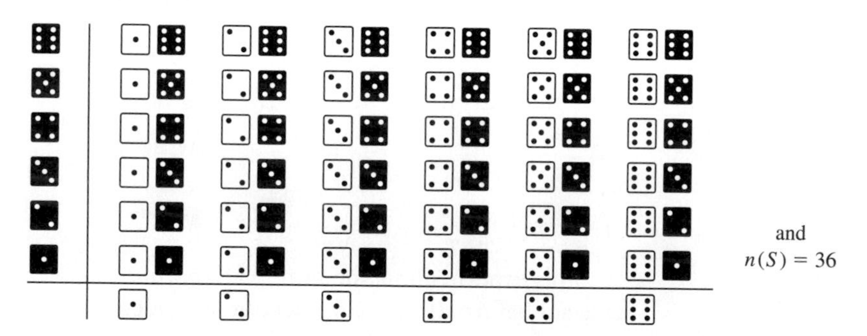

Experiment 4.8

Two dice are rolled and the sum of their dots is observed. The sample space is $S = \{2, 3, 4, 5, 6, 7, 8, 9, 10, 11, 12\}$ and $n(S) = 11$ (or the 36-point sample space shown in Experiment 4.7).

You will notice that two different sample spaces are suggested for Experiment 4.8. Both of these sets satisfy the definition of a sample space, and thus either may be used. We will learn later why the 36-point sample space is more useful than the other. ■

ANSWER NOW 4.20

A box contains one each of $1, $5, $10, and $20 bills.

a. **One bill is selected at random. List the sample space.**

b. **Two bills are drawn at random (without replacement). List the sample space as a tree diagram.**

c. **Two bills are drawn at random (without replacement). List the sample space as a chart.**

Experiment 4.9

A weather forecaster predicts that there will be a measurable amount of precipitation or no precipitation on a given day. The sample space is $S = \{\text{precipitation, no precipitation}\}$ and $n(S) = 2$.

Experiment 4.10

The 6024 students at a nearby college have been cross-tabulated according to their gender and full-time/part-time enrollment status. One student is to be picked at random from the student body.

	Full-Time	Part-Time
Female	2136	548
Male	2458	882

and $n(S) = 6024$

Experiment 4.11

A lucky customer will get to randomly select one key from a barrel that contains keys to every car on Used Car Charlie's lot. The **Venn diagram*** summarizes Charlie's inventory.

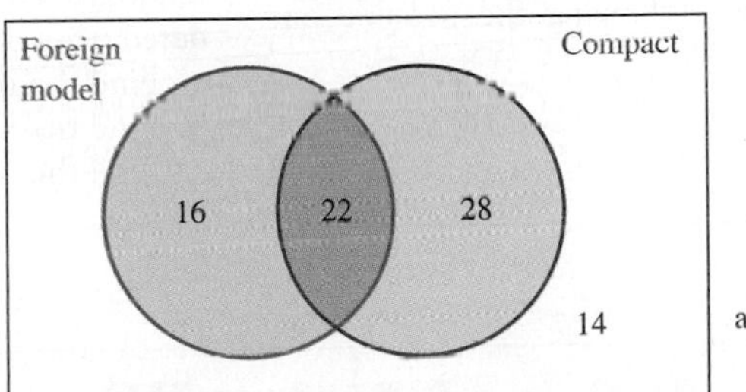

and $n(S) = 80$

Special attention should always be given to the sample space. Like the statistical population, the sample space must be well defined. Once the sample space is defined, you will find the remaining work much easier.

*See the *Statistical Tutor* for information about Venn diagrams.

Exercises

4.21 The face cards are removed from a regular deck and then one card is selected from this set of 12 face cards. List the sample space for this experiment.

4.22 An experiment consists of drawing one marble from a box that contains a mixture of red, yellow, and green marbles.

a. List the sample space.
b. Can we be sure that each outcome in the sample space is equally likely?
c. If two marbles are drawn from the box, list the sample space.

4.23 *Visual Basic Programmer's Journal* reported the results of a survey of 500 subscribers who were asked whether the training they had undergone at work affected their current salary level: 54.9% said yes, and 45.1% said no. Suppose three respondents are selected randomly. Give the sample space and the possible combinations of selected responses to the question. (*Visual Basic Programmer's Journal*, "Fall 1998 Salary Survey Results," September 1998, p. 78)

4.24 Marty Wilson cut out four equilateral triangles the same size from a sheet of cardboard. He labeled the triangles A, B, C, and D. Then he taped them together at their edges to form a perfect tetrahedron. If Marty throws the tetrahedron into the air, the outcome is that any one of the four sides will be face down when it lands.

a. Draw a tree diagram depicting two successive throws. Show all the outcomes and the size of the sample space.
b. What is the size of the sample space after six successive throws?

4.25

a. A balanced coin is tossed twice. List a sample space showing the possible outcomes.
b. A biased coin (it favors heads in a ratio of 3 to 1) is tossed twice. List a sample space showing the possible outcomes.

4.26 A computer generates (in random fashion) pairs of integers. The first integer is between 1 and 5, inclusive, and the second is between 1 and 4, inclusive. Represent the sample space on a coordinate axis system, where x is the first number and y is the second number.

(Retain your answer to use in Exercise 4.30.)

4.27 An experiment consists of two trials. The first is tossing a penny and observing heads or tails; the second is rolling a die and observing 1, 2, 3, 4, 5, or 6. Construct the sample space.

(Retain your answer to use in Exercise 4.29.)

4.28 A box stored in a warehouse contains 100 units of a specific part, of which 10 are defective and 90 are nondefective. Three parts are selected without replacement. Construct a tree diagram representing the sample space.

4.29 Use a computer (or a random-number table) to simulate 200 trials of the experiment described in Exercise 4.27: the tossing of a penny and the rolling of a die. Let 1 = H and 2 = T for the penny, and 1, 2, 3, 4, 5, 6 for the die. Report your results using a cross-tabulated table showing the frequency of each outcome.

a. Find the relative frequency for heads.
b. Find the relative frequency for 3.
c. Find the relative frequency for (H, 3).

FYI
Adjust the computer and calculator commands on pages 189 and 190.

4.30 Use a computer (or a random-number table) to simulate the experiment described in Exercise 4.26; x is an integer 1 to 5, and y is an integer 1 to 4. Generate a list of 100 random x values and 100 y values. List the resulting 100 ordered pairs of integers.

a. Find the relative frequency for $x = 2$.
b. Find the relative frequency for $y = 3$.
c. Find the relative frequency for the ordered pair (2, 3).

4.4 Rules of Probability

Let's return now to the concept of probability and relate it to the sample space. Recall that the probability of an event was defined as the relative frequency with which the event could be expected to occur.

In the sample space associated with Experiment 4.1, tossing a single coin once, we find two possible outcomes: heads (H) and tails (T). We have an "intuitive feeling" that these two events will occur with approximately the same frequency. The coin is a symmetrical object and therefore is not expected to favor either of the two outcomes. We expect heads to occur 1/2 of the time. Thus, the probability that a head will occur on a single toss of a coin is thought to be 1/2.

This description is the basis for the second technique for assigning the probability of an event. In a sample space containing *sample points that are* ***equally likely*** *to occur*, the probability $P(A)$ of an event A is the ratio of the number $n(A)$ of points that satisfy the definition of event A to the number $n(S)$ of sample points in the entire sample space:

THEORETICAL (EXPECTED) PROBABILITY: $P(A)$

$$\text{probability of A} = \frac{\text{number of times A occurs in sample space}}{\text{number of elements in sample space}}$$

$$P(A) = \frac{n(A)}{n(S)} \tag{4.2}$$

This formula gives a **theoretical probability** value of event A's occurrence.

If we apply formula (4.2) to the equally likely sample space for the tossing of two coins (Experiment 4.2), we find the theoretical probabilities $P(0H)$, $P(1H)$, and $P(2H)$ discussed earlier:

$$P(0H) = \frac{n(0H)}{n(S)} = \frac{1}{4}, \quad P(1H) = \frac{n(1H)}{n(S)} = \frac{2}{4} = \frac{1}{2}, \quad P(2H) = \frac{n(2H)}{n(S)} = \frac{1}{4}$$

The use of formula (4.2) requires a sample space in which each outcome is equally likely. Thus, when dealing with experiments that have more than one possible sample space, we need to construct a sample space in which the sample points are equally likely.

Consider Experiment 4.8, where two dice were rolled. If you list the sample space as the 11 sums, the sample points are not equally likely. If you use the 36-point sample space, all the sample points are equally likely, as in Experiment 4.7. For example, the sum of 2 represents {(1, 1)}; the sum of 3 represents {(2, 1), (1, 2)}; and the sum of 4 represents {(1, 3), (3, 1), (2, 2)}. Thus, we can use formula (4.2) and the 36-point sample space to obtain the probabilities for the 11 sums:

$$P(2) = \frac{n(2)}{n(S)} = \frac{1}{36}, \quad P(3) = \frac{n(3)}{n(S)} = \frac{2}{36}, \quad P(4) = \frac{n(4)}{n(S)} = \frac{3}{36},$$

and so forth.

ANSWER NOW 4.31

Find the probabilities $P(5)$, $P(6)$, $P(7)$, $P(8)$, $P(9)$, $P(10)$, $P(11)$, and $P(12)$ for the sum of two dice.

In many cases the assumption of equally likely events does not make sense. The sample points in Experiment 4.8 are not equally likely, and there is no reason to believe that the sample points in Experiment 4.9 are equally likely. What do we do when the sample space elements are not equally likely or not a combination of equally likely events? We could use empirical probabilities. But what do we do when no experiment has been done or can be performed?

Let's look again at Experiment 4.9. The weather forecaster often assigns a probability to the event "precipitation." For example, "there is a 20% chance of rain today" or "there is a 70% chance of snow tomorrow." In such cases the only basis for assigning probabilities is personal judgment. These probability assignments are called **subjective probabilities.** The accuracy of subjective probabilities depends on the individual's ability to correctly assess the situation.

Often, personal judgment of the probability of the possible outcomes of an experiment is expressed by comparing the likelihoods of the various outcomes. For example, the weather forecaster's personal assessment might be: "It is five times more likely to rain (R) tomorrow than not rain (NR)"; then $P(\text{R}) = 5 \cdot P(\text{NR})$. If this is the case, what values should be assigned to $P(\text{R})$ and $P(\text{NR})$? To answer this question, we need to review some of the ideas about probability that we've already discussed.

1. Probability represents a relative frequency.
2. $P(\text{A})$ is the ratio of the number of times an event can be expected to occur divided by the number of trials.
3. The numerator of the probability ratio must be a positive number or zero.
4. The denominator of the probability ratio must be a positive number (greater than zero).
5. The number of times an event can be expected to occur in n trials is always less than or equal to the total number of trials, n.

Thus, it is reasonable to conclude that a probability is always a numerical value between zero and one.

PROPERTY 1

$$0 \leq \text{each } P(\text{A}) \leq 1$$

NOTES

1. The probability is zero if the event cannot occur.
2. The probability is one if the event occurs every time.

PROPERTY 2

$$\sum_{\text{all outcomes}} P(\text{A}) = 1$$

Property 2 states that if we add the probabilities of each sample point in the sample space, the total probability must equal one. This makes sense because when we sum all the probabilities, we are asking, "What is the probability the experiment will yield an outcome?" and this will happen every time.

Now we are ready to assign probabilities to $P(\text{R})$ and $P(\text{NR})$. The events R and NR cover the sample space, and the weather forecaster's personal judgment was

$$P(\text{R}) = 5 \cdot P(\text{NR})$$

From Property 2, we know that

$$P(\text{R}) + P(\text{NR}) = 1$$

By substituting $5 \cdot P(\text{NR})$ for $P(\text{R})$, we get

$$5 \cdot P(\text{NR}) + P(\text{NR}) = 1$$
$$6 \cdot P(\text{NR}) = 1$$
$$P(\text{NR}) = \frac{1}{6}$$

$$P(\text{R}) = 5 \cdot P(\text{NR}) = 5\left(\frac{1}{6}\right) = \frac{5}{6}$$

ANSWER NOW 4.32
If four times as many students pass a statistics course as fail, and one statistics student is selected at random, what is the probability that the student will pass statistics?

ODDS

The statement "It is five times more likely to rain tomorrow (R) than not rain (NR)" is often expressed as "The odds are 5 to 1 in favor of rain tomorrow" (also written 5:1). Odds are simply another way of expressing probabilities. The relationships among odds for an event, odds against an event, and the probability of an event are expressed in the following rules:

If the odds in favor of an event A are ***a* to *b*,** then:

1. The odds against event A are ***b* to *a*** (or ***b*: *a***).
2. The probability of event A is $P(\mathrm{A}) = \dfrac{a}{a + b}$.
3. The probability that event A will not occur is $P(\mathrm{A\ does\ not\ occur}) = \dfrac{b}{a + b}$.

To illustrate these rules, consider the statement: "The odds favoring rain tomorrow are 5 to 1." Using the preceding notation, we have $a = 5$ and $b = 1$. Therefore, the probability of rain tomorrow is $\dfrac{5}{5+1}$, or $\dfrac{5}{6}$. The odds against rain tomorrow are 1 to 5 (or 1:5), and the probability that there is no rain tomorrow is $\dfrac{1}{5+1}$, or $\dfrac{1}{6}$.

ANSWER NOW 4.33
The odds of the Patriots winning next year's Super Bowl are 1 to 12.

a. What is the probability the Patriots will win next year's Super Bowl?

b. What are the odds against the Patriots winning next year's Super Bowl?

COMPLEMENT OF AN EVENT
The set of all sample points in the sample space that do not belong to event A. The complement of event A is denoted by $\overline{\mathrm{A}}$ (read "A *complement*").

For example, the complement of the event "success" is "failure"; the complement of "heads" is "tails" for the tossing of one coin; the complement of "at least one head" on ten tosses of a coin is "no heads."

By combining the information in the definition of complement with Property 2 (p. 198), we can say that

$$P(\mathrm{A}) + P(\overline{\mathrm{A}}) = 1.0 \quad \text{for any event A}$$

The next formula then follows:

THE COMPLEMENT

$$P(\overline{\mathrm{A}}) = 1 - P(\mathrm{A}) \tag{4.3}$$

NOTE Every event A has a complementary event $\overline{A}$. Complementary probabilities are useful when the question asks for the probability of *"at least one."* Generally this represents a combination of several events, but the complementary event "none" is a single outcome. It is easier to solve for the single complementary event and then get the answer by using formula (4.3).

Illustration 4.2

Using Complements to Find Probabilities

Two dice are rolled. What is the probability that the sum is at least 3 (that is, 3 or larger)?

SOLUTION

Rather than finding the probability for each of the sums 3 and larger, it is much simpler to find the probability that the sum is 2 (less than 3) and then use formula (4.3), letting "at least 3" be $\overline{A}$:

$$P(A) = \frac{1}{36} \qquad \text{(as found on p. 197)}$$

$$P(\overline{A}) = 1 - P(A) = 1 - \frac{1}{36} = \mathbf{\frac{35}{36}} \qquad \text{[using formula (4.3)]}$$

■

ANSWER NOW 4.34
Find the probability of drawing a non-face card from a well-shuffled deck of 52 playing cards.

Application 4.1

Beating the Odds

Many young men aspire to become professional athletes. Only a few make it to the big time, as indicated in the accompanying graph. For every 2400 college senior basketball players, only 64 make a professional team; that translates to a probability of only 0.027 (64/2400).

USA SNAPSHOTS®
A look at statistics that shape the sports world
Trying to beat the odds
64 make a pro team
2,400 play as college seniors
3,800 make college team
150,000 high school seniors
The odds against a high school senior basketball player making a professional team are 2,344 to 1 based on 1989 numbers:
Source: NCAA
By Julie Stacey, USA TODAY
© 1990 USA Today. Reprinted by permission.

ANSWER NOW 4.35
Refer to "Trying to beat the odds."

a. **Find the probability that a high school senior basketball player makes a pro team.**

b. **What are the odds that a player who makes a college basketball team plays as a senior?**

c. **What are the odds against a college senior basketball player making a pro team?**

Exercises

4.36 A box contains marbles of five different colors: red, green, blue, yellow, and purple. There is an equal number of each color. Assign probabilities to each color in the sample space.

4.37 Suppose a box of marbles contains equal numbers of red marbles and yellow marbles but twice as many green marbles as red marbles. Draw one marble from the box and observe its color. Assign probabilities to the elements in the sample space.

4.38 As of 1998 the two professional football coaches who won the most games during their careers were Don Shula and George Halas. Shula's teams (Colts and Dolphins) won 347 games and tied 6 of the 526 games that he coached, whereas Halas's team (Bears) won 324 games and tied 31 of the 506 games that he coached (*World Almanac and Book of Facts 1998*, p. 876). Suppose one filmstrip from every game each man coached is thrown into a bin and mixed. You select one filmstrip from the bin and load it into a projector. What is the probability that the film you select shows:

a. a tie game
b. a losing game
c. one of Shula's teams winning a game
d. Halas's team winning a game
e. one of Shula's teams losing a game
f. Halas's team losing a game
g. one of Shula's teams playing to a tie
h. Halas's team playing to a tie
i. a game coached by Halas
j. a game coached by Shula

4.39 A single die is rolled. Find the probability the number on top is:

a. a 3
b. an odd number
c. a number less than 5
d. a number no greater than 3

4.40 A transportation engineer in charge of a new traffic-control system expresses the subjective probability that the system functions correctly 99 times as often as it malfunctions.

a. Based on this belief, what is the probability that the system functions properly?
b. Based on this belief, what is the probability that the system malfunctions?

4.41 Events A, B, and C are defined on sample space *S*. Their corresponding sets of sample points do not intersect and their union is *S*. Furthermore, event B is twice as likely to occur as event A, and event C is twice as likely to occur as event B. Determine the probability of each of the three events.

4.42 Three coins are tossed, and the number of heads observed is recorded. Find the probability for each of the possible results: 0H, 1H, 2H, and 3H.

4.43 Let x be the success rating of a new television show. The table lists the subjective probabilities assigned to each x for a particular new show by three different media critics. Which of these sets of probabilities are inappropriate because they violate a basic rule of probability? Explain.

	Judge		
Success Rating, x	**A**	**B**	**C**
Highly Successful	0.5	0.6	0.3
Successful	0.4	0.5	0.3
Not Successful	0.3	0.1	0.3

4.44 Two dice are rolled. Find the probabilities in parts b–e. Use the sample space given in Experiment 4.7 (p. 194).

a. Why is the set {2, 3, 4, . . . , 12} not a useful sample space?
b. *P*(white die is an odd number)
c. *P*(sum is 6)
d. *P*(both dice show odd numbers)
e. *P*(number on black die is larger than number on white die)
f. Explain why these answers and the answers found in Exercise 4.13 (p. 188) are not exactly the same.

4.45 A group of files in a medical clinic classifies the patients by gender and type of diabetes (I or II). The table gives the number in each classification.

	Diabetes	
Gender	**I**	**II**
Male	25	20
Female	35	20

If one file is selected at random, find the probability that:

a. the selected individual is female
b. the selected individual has Type II diabetes

4.46 The odds against being dealt a contract bridge hand containing 13 cards of the same suit are 158,753,389,899 to 1. The odds against being dealt a royal flush while playing poker are 649,739 to 1.

a. What is the probability of being dealt a contact bridge hand containing 13 cards all of the same suit?
b. What is the probability of being dealt a royal flush poker hand?
c. Express the answers to parts a and b in scientific notation (powers of 10).

The probability of the event "A and B," $P(\text{A and B})$, is represented by the area contained in region II. The probability $P(\text{A})$ is represented by the area of circle A; that is, $P(\text{A}) = P(\text{region I}) + P(\text{region II})$. Furthermore, $P(\text{B}) = P(\text{region II}) + P(\text{region III})$. And $P(\text{A or B})$ is the sum of the probabilities associated with the three regions:

$$P(\text{A or B}) = P(\text{I}) + P(\text{II}) + P(\text{III})$$

However, if $P(\text{A})$ is added to $P(\text{B})$, we have

$$\begin{aligned} P(\text{A}) + P(\text{B}) &= [P(\text{I}) + P(\text{II})] + [P(\text{II}) + P(\text{III})] \\ &= P(\text{I}) + 2P(\text{II}) + P(\text{III}) \end{aligned}$$

This is the double count we mentioned. We need to subtract one measure of region II from this total so that we will be left with the correct value.

The addition formula (4.4b) is a special case of the more general rule stated in formula (4.4a). If A and B are mutually exclusive events, then $P(\text{A and B}) = 0$. (They cannot both happen at the same time.) Thus, the last term in formula (4.4a) is zero when events are mutually exclusive.

Illustration 4.7 Calculating Probabilities Using the Addition Rule

One white die and one black die are rolled. Find the probability that the white die shows a number smaller than 3 or the sum of the dice is greater than 9.

SOLUTION 1

The events are A—white die shows a 1 or a 2 and B—sum of both dice is 10, 11, or 12. We have

$$P(\text{A}) = \frac{12}{36} = \frac{1}{3} \quad \text{and} \quad P(\text{B}) = \frac{6}{36} = \frac{1}{6}$$

$$\begin{aligned} P(\text{A or B}) &= P(\text{A}) + P(\text{B}) - P(\text{A and B}) \\ &= \frac{1}{3} + \frac{1}{6} - 0 = \mathbf{\frac{1}{2}} \end{aligned}$$

$P(\text{A and B}) = 0$ because the events do not intersect.

SOLUTION 2

$$P(\text{A or B}) = \frac{n(\text{A or B})}{n(S)} = \frac{18}{36} = \mathbf{\frac{1}{2}}$$

Look at the sample space in Figure 4.3 (p. 204) and count. ■

Illustration 4.8 Calculating Probabilities and the Addition Rule

A pair of dice is rolled. Event T is defined as a "total of 10 or 11," and event D is "doubles." Find the probability $P(\text{T or D})$.

SOLUTION

Look at the sample space of 36 ordered pairs for the roll of two dice in Figure 4.3. Event T occurs if any one of five ordered pairs occurs: (4, 6), (5, 5), (6, 4), (5, 6), (6, 5). Therefore, $P(\text{T}) = \frac{5}{36}$. Event D occurs if any one of six ordered pairs occurs: (1, 1), (2, 2), (3, 3), (4, 4), (5, 5), (6, 6). Therefore, $P(\text{D}) = \frac{6}{36}$. Notice, however, that these two events are not mutually exclusive. The two events "share" the point

(5, 5). Thus, the probability $P(\text{T and D}) = \frac{1}{36}$. As a result, $P(\text{T or D})$ is found using formula (4.4a):

$$P(\text{T or D}) = P(\text{T}) + P(\text{D}) - P(\text{T and D})$$

$$= \frac{5}{36} + \frac{6}{36} - \frac{1}{36} = \frac{10}{36} = \frac{5}{18}$$

Look at the sample space in Figure 4.3 and verify that $P(\text{T or D}) = \frac{5}{18}$. ■

ANSWER NOW 4.52

A pair of dice are rolled. The events are: A—sum of 7, C—doubles, and E—sum of 8.

a. Which pairs of events, A and C, A and E, C and E, are mutually exclusive? Explain.

b. Find the probabilities *P*(A or C), *P*(A or E), and *P*(C or E).

Exercises

4.53 Determine whether or not each of the following pairs of events is mutually exclusive.

a. Five coins are tossed: "one head is observed," "at least one head is observed."
b. A salesperson calls on a client and makes a sale: "the sale exceeds \$100," "the sale exceeds \$1000."
c. One student is selected at random from a student body: the person selected is "male," the person selected is "over 21 years of age."
d. Two dice are rolled: the total showing is "less than 7," the total showing is "more than 9."

4.54 Determine whether each of the following sets of events is mutually exclusive.

a. Five coins are tossed: "no more than one head is observed," "two heads are observed," "three or more heads are observed."
b. A salesperson calls on a client and makes a sale: the amount of the sale is "less than \$100," is "between \$100 and \$1000," is "more than \$500."
c. One student is selected at random from the student body: the person selected is "female," is "male," is "over 21."
d. Two dice are rolled: the numbers of dots showing on the dice are "both odd," "both even," "total seven," "total 11."

4.55 Explain why $P(\text{A and B}) = 0$ when events A and B are mutually exclusive.

4.56 Explain why $P(\text{A occurring when B has occurred}) = 0$ when events A and B are mutually exclusive.

4.57 If $P(\text{A}) = 0.3$ and $P(\text{B}) = 0.4$, and if A and B are mutually exclusive events, find:

a. $P(\overline{\text{A}})$ b. $P(\overline{\text{B}})$ c. $P(\text{A or B})$ d. $P(\text{A and B})$

4.58 If $P(\text{A}) = 0.4$, $P(\text{B}) = 0.5$, and $P(\text{A and B}) = 0.1$, find $P(\text{A or B})$.

4.59 One student is selected at random from a student body. Suppose the probability that this student is female is 0.5 and the probability that this student works part time is 0.6. Are the two events "female" and "working" mutually exclusive? Explain.

4.60 SAT scores are divided between verbal and math, both of which average near 500 year after year. The table shows the frequency counts from the 50 states and the District of Columbia for verbal and math scores that averaged above and below 500 in 2000. Let A be the event that a state (or DC) selected at random "has an average score above 500" and let B be the event "math score."

	500 or less	Over 500	Total
Verbal	12	39	51
Math	6	45	51
Total	18	84	102

Source: College Board, 2002.

a. Find $P(\text{A})$.
b. Find $P(\text{B})$.
c. Find $P(\text{A and B})$.
d. Find $P(\text{A or B})$ and $P(\text{not (A or B)})$.
e. Draw the Venn diagram and show the probabilities in parts a–c and $P(\text{not (A or B)})$. Are events A and B mutually exclusive? Explain.

4.61 The U.S. population doubled between 1930 and 2000. The table (p. 208) lists the frequencies of the 50 states and the District of Columbia for 1930 and 2000 that were either above or below 2,000,000. Let A be the event that a state (or DC) selected at random "has a population under 2,000,000" and let B be the event "2000 census."

	Under 2,000,000	Over 2,000,000	Total
1930 Census	28	23	51
2000 Census	17	34	51
Total	45	57	102

Source: Bureau of Census, U.S. Department of Commerce

a. Find $P(\text{A})$.
b. Find $P(\text{B})$.
c. Find $P(\text{A and B})$.
d. Find $P(\text{A or B})$ and $P(\text{not (A or B)})$.
e. Draw the Venn diagram and show the probabilities in parts a–c and $P(\text{not (A or B)})$. Are events A and B mutually exclusive? Explain.

4.62 An aquarium at a pet store contains 40 orange swordfish (22 females and 18 males) and 28 green swordtails (12 females and 16 males). You randomly net one of the fish.
a. What is the probability that it is an orange swordfish?
b. What is the probability that it is a male fish?
c. What is the probability that it is an orange female swordfish?
d. What is the probability that it is a female or a green swordtail?
e. Are the events "male" and "female" mutually exclusive? Explain.
f. Are the events "male" and "swordfish" mutually exclusive? Explain.

4.63 A parts store sells both new and used parts. Sixty percent of the parts in stock are used. Sixty-one percent are used or defective. If 5% of the store's parts are defective, what percentage are both used and defective?

4.64 Union officials report that 60% of the workers at a large factory belong to the union, 90% earn over \$12 per hour, and 40% belong to the union and earn over \$12 per hour. Do you believe these percentages? Explain.

4.6 Independence, the Multiplication Rule, and Conditional Probability

Consider this example. The event that a 2 shows on a white die is A, and the event that a 2 shows on a black die is B. If both die are rolled once, what is the probability that two 2's occur?

$$P(\text{A}) = \frac{1}{6} \quad \text{and} \quad P(\text{B}) = \frac{1}{6}$$

$$P(\text{A and B}) = \frac{n(\text{A and B})}{n(S)} = \frac{1}{36}$$

Notice that by multiplying the probabilities of the simple events, we find the correct value for $P(\text{A and B})$. Multiplication does not always work, however. For example, $P(\text{sum of 7 and double})$ when two dice are rolled is zero (as seen in Figure 4.3). However, if $P(7)$ is multiplied by $P(\text{double})$, we obtain $(\frac{6}{36})(\frac{6}{36}) = (\frac{1}{6})(\frac{1}{6}) = \frac{1}{36}$.

Multiplication does not work for $P(\text{sum of 10 and double})$ either. By definition and by inspection of the sample space, we know that $P(\text{10 and double}) = \frac{1}{36}$; the point (5, 5) is the only element. However, if we multiply $P(10)$ by $P(\text{double})$, we obtain $(\frac{3}{36})(\frac{6}{36}) = \frac{1}{72}$. The probability of this event cannot be both values.

The property that is required for multiplying probabilities is **independence.** Multiplication worked in the one foregoing example because the events were independent. In the other two examples, the events were not independent and multiplication gave us incorrect answers.

NOTE Several situations result in the compound event "and." Some of the more common ones are: (1) A followed by B, (2) A and B occurred simultaneously, (3) the intersection of A and B, (4) both A and B, and (5) A but not B (equivalent to A and not B).

INDEPENDENCE AND CONDITIONAL PROBABILITIES

INDEPENDENT EVENTS

Two events A and B are independent events if and only if the occurrence (or nonoccurrence) of one does not affect the probability assigned to the occurrence of the other.

Sometimes independence is easy to determine—for example, if the two events being considered have to do with unrelated trials, such as the tossing of a penny and a nickel. The results on the penny in no way affect the probability of heads or tails on the nickel. Similarly, the results on the nickel have no effect on the probability of heads or tails on the penny. Therefore, the results on the penny and the results on the nickel are independent. However, if events are defined as combinations of outcomes from separate trials, the independence of the events may or may not be so easy to determine. The separate results of each trial (dice in the next example) may be independent, but the compound events defined using both trials (both dice) may or may not be independent.

Lack of independence, called **dependence,** is demonstrated by the following example: Reconsider the experiment of rolling two dice and observing the two events "sum of 10" and "double." As stated previously, $P(10) = \frac{3}{36} = \frac{1}{12}$ and $P(\text{double}) = \frac{6}{36} = \frac{1}{6}$. Does the occurrence of 10 affect the probability of a double? Think of it this way. A sum of 10 has occurred; it must be one of the following: {(4, 6), (5, 5), (6, 4)}. One of these three possibilities is a double. Therefore, we must conclude that *P*(double, knowing 10 has occurred), written $P(\text{double} \mid 10)$, is $\frac{1}{3}$. Since $\frac{1}{3}$ does not equal the original probability of a double, $\frac{1}{6}$, we conclude that the event "10" has an effect on the probability of a double. Therefore, "double" and "10" are dependent events.

Whether or not events are independent often becomes clear when we examine the events in question. Rolling one die does not affect the outcome of a second roll. However, in many cases, independence is not self-evident, and the question of independence itself may be of special interest. Consider the events "having a checking account at a bank" and "having a loan account at the same bank." Having a checking account at a bank may increase the probability that the same person has a loan account. This situation has practical implications. For example, it would make sense to advertise loan programs to checking-account clients if they are more likely to apply for loans than are people who are not customers of the bank.

One approach to the problem is to *assume* independence or dependence. The correctness of the probability analysis depends on the truth of the assumption. In practice, we often assume independence and then compare the calculated probabilities with actual frequencies of outcomes in order to infer whether the assumption of independence is warranted.

CONDITIONAL PROBABILITY

The symbol $P(A \mid B)$ represents the probability that A will occur given that B has occurred. This is called a conditional probability.

We can now write the previous definition of independent events in a more formal manner.

INDEPENDENT EVENTS
Two events A and B are independent events if and only if

probability of A, knowing B = probability of A or
probability of B, knowing A = probability of B

$$P(\text{A} \mid \text{B}) = P(\text{A}) \quad \text{or} \quad P(\text{B} \mid \text{A}) = P(\text{B}) \tag{4.5}$$

Let's consider conditional probability. Take, for example, the experiment in which a single die is rolled: $S = \{1, 2, 3, 4, 5, 6\}$. Two events that can be defined for this experiment are A, "a 4 occurs," and B, "an even number occurs." Then $P(\text{A}) = \frac{1}{6}$. Event A is satisfied by exactly one of the six equally likely sample points in S. The conditional probability of A given B, $P(\text{A} \mid \text{B})$, is found in a similar manner, but the list of possible events is no longer the sample space. Think of it this way: A die is rolled out of your sight, and you are told the number showing is even; that is, event B {2, 4, 6} has occurred. Knowing this condition, you are asked to assign a probability to the event that the even number is a 4. There are only three possibilities in the *current (or reduced) sample space*, {2, 4, 6}. Each of the three outcomes is equally likely; thus, $P(\text{A} \mid \text{B}) = \frac{1}{3}$.

We can write this case as

$$\textit{probability of A, knowing B} = \frac{\textit{number of elements in intersection of A and B}}{\textit{number of elements in B}}$$

$$P(\text{A} \mid \text{B}) = \frac{n(\text{A} \cap \text{B})}{n(\text{B})} \tag{4.6a}$$

or equivalently

$$P(\text{A} \mid \text{B}) = \frac{P(\text{A and B})}{P(\text{B})} \tag{4.6b}$$

Thus, for our example,

$$P(\text{A} \mid \text{B}) = \frac{\frac{1}{6}}{\frac{1}{2}} = \mathbf{\frac{1}{3}}$$

Illustration 4.9 Using Conditional Probabilities to Determine Independence

In a sample of 150 residents, each person was asked whether he or she favored the concept of having a single countywide police agency. The county is composed of one large city and many suburban townships. The residence (city or outside the city) and the responses of the residents are summarized in Table 4.3. If one of these residents was selected at random, what is the probability that the person will (a) fa-

TABLE 4.3 Sample Results

	Opinion		
Residence	**Favor (F)**	**Oppose ($\bar{\text{F}}$)**	**Total**
In City (C)	80	40	120
Outside of City ($\bar{\text{C}}$)	20	10	30
Total	100	50	150

vor the concept? (b) favor the concept if the person selected is a city resident? (c) favor the concept if the person selected is a resident from outside the city? (d) Are the events F (favor the concept) and C (reside in city) independent?

SOLUTION

(a) $P(\text{F})$ is the proportion of the total sample that favor the concept. Therefore,

$$P(\text{F}) = \frac{n(\text{F})}{n(S)} = \frac{100}{150} = \frac{2}{3}$$

(b) $P(\text{F} \mid \text{C})$ is the probability that the person selected favors the concept given that he or she lives in the city. The sample space is reduced to the 120 city residents in the sample. Of these, 80 favored the concept; therefore,

$$P(\text{F} \mid \text{C}) = \frac{n(\text{F and C})}{n(\text{C})} = \frac{80}{120} = \frac{2}{3}$$

(c) $P(\text{F} \mid \overline{\text{C}})$ is the probability that the person selected favors the concept given that the person lives outside the city. The sample space is reduced to the 30 noncity residents; therefore,

$$P(\text{F} \mid \overline{\text{C}}) = \frac{n(\text{F and } \overline{\text{C}})}{n(\overline{\text{C}})} = \frac{20}{30} = \frac{2}{3}$$

(d) All three probabilities have the same value, $\frac{2}{3}$. Therefore, we can say that the events F (favor) and C (reside in city) are independent. The location of residence did not affect $P(\text{F})$. ■

ANSWER NOW 4.65

Pollsters asked 300 viewers if they were satisfied with the TV coverage of a recent disaster. The table lists the results. One viewer is to be selected randomly from those surveyed.

	Female	Male
Satisfied	80	55
Not Satisfied	120	45

a. **Find $P(\text{satisfied})$.**

b. **Find $P(\text{satisfied} \mid \text{female})$.**

c. **Find $P(\text{satisfied} \mid \text{male})$.**

d. **Is the event "satisfied" independent of gender? Explain.**

MULTIPLICATION RULE

GENERAL MULTIPLICATION RULE

Let A and B be two events defined in sample space S. Then

probability of A and B = probability of A × probability of B, knowing A

$$P(\text{A and B}) = P(\text{A}) \cdot P(\text{B} \mid \text{A}) \tag{4.7a}$$

or

probability of A and B = probability of B × probability of A, knowing B

$$P(\text{A and B}) = P(\text{B}) \cdot P(\text{A} \mid \text{B}) \tag{4.7b}$$

ANSWER NOW 4.66
R and H are events with $P(R) = 0.6$ and $P(H \mid R) = 0.25$. Find $P(R \text{ and } H)$.

If events A and B are independent, then the general multiplication rule [formula (4.7)] reduces to the *special multiplication rule*, formula (4.8).

SPECIAL MULTIPLICATION RULE

Let A and B be two events defined in sample space *S*. A and B are independent events if and only if

$$P(\text{A and B}) = P(\text{A}) \cdot P(\text{B}) \qquad (4.8a)$$

This formula can be expanded. A, B, C, . . . , G are independent events if and only if

$$P(\text{A and B and C and} \ldots \text{and G}) = P(\text{A}) \cdot P(\text{B}) \cdot P(\text{C}) \cdot \cdots \cdot P(\text{G}) \qquad (4.8b)$$

ANSWER NOW 4.67
A and B are independent events, and $P(A) = 0.7$ and $P(B) = 0.4$. Find $P(A \text{ and } B)$.

Illustration 4.10 Determining Independence and Using the Multiplication Rule

One student is selected at random from a group of 200 known to consist of 140 full-time (80 female and 60 male) students and 60 part-time (40 female and 20 male) students. (See Illustration 4.3.) Event A is "the student selected is full-time," and event C is "the student selected is female."

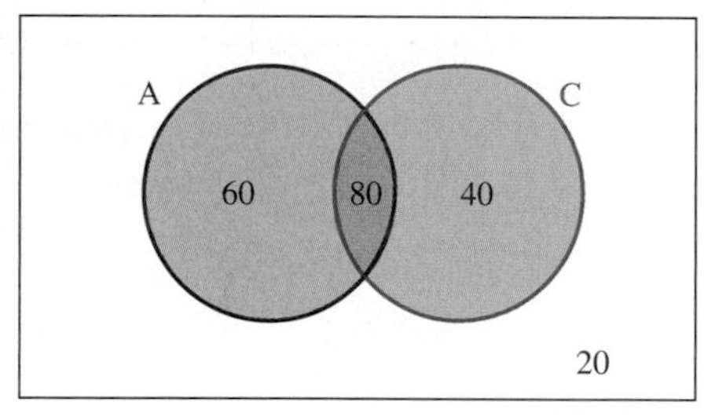

(a) Are events A and C independent?
(b) Find the probability $P(\text{A and C})$ using the multiplication rule.

FYI
This is a test for independence.

SOLUTION 1

(a) First find the probabilities $P(\text{A})$, $P(\text{C})$, and $P(\text{A} \mid \text{C})$:

$$P(\text{A}) = \frac{n(\text{A})}{n(S)} = \frac{140}{200} = 0.7$$

$$P(\text{C}) = \frac{n(\text{C})}{n(S)} = \frac{120}{200} = 0.6$$

$$P(\text{A} \mid \text{C}) = \frac{n(\text{A and C})}{n(C)} = \frac{80}{120} = 0.67$$

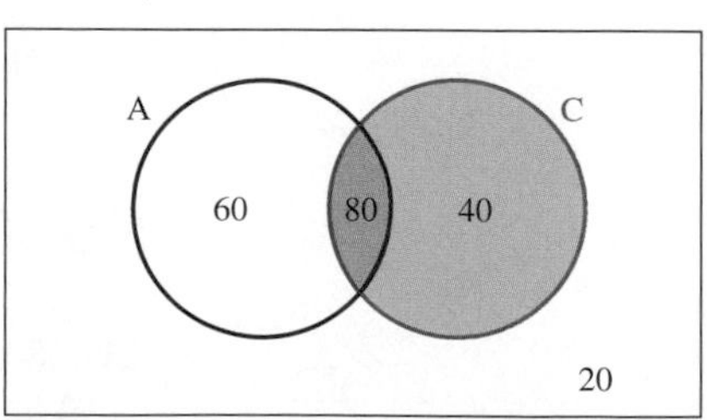

A and C are dependent events because $P(\text{A}) \neq P(\text{A} \mid \text{C})$.

(b) $P(\text{A and C}) = P(\text{C}) \cdot P(\text{A} \mid \text{C}) = \frac{120}{200} \cdot \frac{80}{120} = \frac{80}{200} = \mathbf{0.4}$

SOLUTION 2

(a) First find the probabilities $P(A)$, $P(C)$, and $P(C \mid A)$.

$$P(A) = \frac{n(A)}{n(S)} = \frac{140}{200} = 0.7$$

$$P(C) = \frac{n(C)}{n(S)} = \frac{120}{200} = 0.6$$

$$P(C \mid A) = \frac{n(C \text{ and } A)}{n(A)} = \frac{80}{140} = 0.57$$

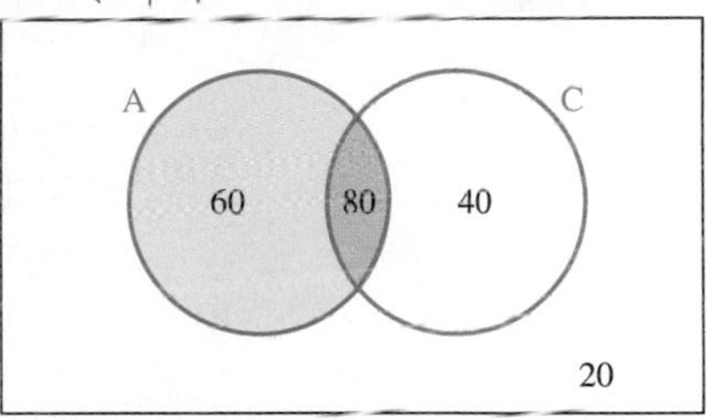

A and C are dependent events because $P(C) \neq P(C \mid A)$.

(b) $P(C \text{ and } A) = P(A) \cdot P(C \mid A) = \frac{140}{200} \cdot \frac{80}{140} = \frac{80}{200} = \mathbf{0.4}$

ANSWER NOW 4.68

a. **Find P(A and C) using the same sample space and formula (4.2).**

b. **Does your answer in part a agree with the solution in the illustration? Explain.**

■

Illustration 4.11 Using the Multiplication Rule with Dependent Events

One white and one black die are rolled. Find the probability that the sum of their numbers is 7 and that the number on the black die is larger than the number on the white die.

SOLUTION

We have A, "sum is 7," and B, "black number larger than white number." The "and" indicates that we use the multiplication rule. However, we do not yet know whether events A and B are independent.

(Refer to Figure 4.3 for the sample space of this experiment.) We have $P(A) = \frac{6}{36} = \frac{1}{6}$. We see that $P(A \mid B)$ is obtained from the reduced sample space, which includes 15 points above the gray diagonal line. Of the 15 equally likely points, 3 of them—(1, 6), (2, 5), and (3, 4)—satisfy event A. Therefore, $P(A \mid B) = \frac{n(A \text{ and } B)}{n(B)} = \frac{3}{15} = \frac{1}{5}$. Since this is a different value than $P(A)$, the events are dependent. So we must use formula (4.7b) to obtain P(A and B):

$$P(A \text{ and } B) = P(B) \cdot P(A \mid B) = \frac{15}{36} \cdot \frac{3}{15} = \frac{3}{36} = \mathbf{\frac{1}{12}}$$

■

ANSWER NOW 4.69

Space flight reliability is the subject of the "Probability" video on your Student Suite CD (found inside the back cover of this textbook). View the video clip and answer these questions:

a. **What does it mean to say that the six joints work independently?**

b. **If you multiply all six 0.9777's together, what is the product?**

c. **What was this product telling them?**

FYI
VERY INFORMATIVE!

NOTES

1. Independence and mutually exclusive are two very different concepts.
   a. Mutually exclusive means the two events cannot occur together; that is, they have no intersection.
   b. Independence means each event does not affect the other event's probability.
2. $P(\text{A and B}) = P(\text{A}) \cdot P(\text{B})$ when A and B are independent.
   a. Since $P(\text{A})$ and $P(\text{B})$ are not zero, $P(\text{A and B})$ is nonzero.
   b. Thus, independent events have an intersection.
3. Events cannot be both mutually exclusive and independent. Therefore:
   a. If two events are independent, then they are not mutually exclusive.
   b. If two events are mutually exclusive, then they are not independent.

Application 4.2 Probability

In the box of most standard decks of 52 cards you will find a card giving the probabilities of poker hands. The left-hand column lists the possible events, and the right-hand column lists the odds (probabilities) of various outcomes.

POKER

AVERAGES IN THE DRAW

Chances Against Making . . .without JOKER

HOLDING ONE PAIR, drawing three cards,	Two Pair	5 to 1
	Threes	8 to 1
	Full House	97 to 1
	Fours	359 to 1
HOLDING ONE PAIR, AND AN ACE KICKER; drawing two cards,	Aces up	$7\frac{1}{2}$ to 1
	Another pair	17 to 1
	Threes	12 to 1
	Full House	119 to 1
	Fours	1080 to 1
HOLDING TWO PAIR, drawing one card,	Full House	11 to 1
HOLDING THREES; drawing two cards,	Full House	$15\frac{1}{2}$ to 1
	Fours	$22\frac{1}{2}$ to 1
HOLDING THREES AND ONE OTHER CARD, drawing one card;	Full House	$14\frac{2}{3}$ to 1
	Fours	46 to 1
Drawing one card to a Four Straight, both ends open	Straight	5 to 1
one end or interior open	"	11 to 1
Four Flush	Flush	$4\frac{1}{4}$ to 1
Four Straight Flush, both ends open.	Straight Flush	$22\frac{1}{2}$ to 1
one end or interior open.	" "	46 to 1

Holding 1 ace and drawing 4 cards . . .odds are 3 to 1 against making a pair of aces and 14 to 1 against aces up (a pair of aces and another pair of anything).

ANSWER NOW 4.70

You have carefully shuffled the deck of cards and randomly selected four cards. Assume you have drawn two pairs (say, two kings and two 8's).

a. **What are the odds that with the next selection you will complete a full house (one pair and three of a kind)?**

b. **Verify the probability.**

c. **Would this probability be the same if you and a friend were both drawing from the same deck? Explain.**

Exercises

4.71

a. Describe in your own words what it means for two events to be mutually exclusive.
b. Describe in your own words what it means for two events to be independent.
c. Explain how mutually exclusive and independent are two very different properties.

4.72

a. Describe in your own words why two events cannot be independent if they are already known to be mutually exclusive.
b. Describe in your own words why two events cannot be mutually exclusive if they are already known to be independent.

4.73 Determine whether or not each of the following pairs of events is independent:

a. rolling a pair of dice and observing a "1" on the first die and a "1" on the second die
b. drawing a "spade" from a regular deck of playing cards and then drawing another "spade" from the same deck without replacing the first card
c. same as part b except the first card is returned to the deck before the second drawing
d. owning a red automobile and having blonde hair
e. owning a red automobile and having a flat tire today
f. studying for an exam and passing the exam

4.74 Determine whether or not each of the following pairs of events is independent:

a. rolling a pair of dice and observing a "2" on one of the dice and having a "total of 10"
b. drawing one card from a regular deck of playing cards and having a "red" card and having an "ace"
c. raining today and passing today's exam
d. raining today and playing golf today
e. completing today's homework assignment and being on time for class

4.75 If $P(A) = 0.3$ and $P(B) = 0.4$ and A and B are independent events, what is the probability of each of the following:

a. $P(A \text{ and } B)$ b. $P(B \mid A)$ c. $P(A \mid B)$

4.76 Suppose that $P(A) = 0.3$, $P(B) = 0.4$, and $P(A \text{ and } B) = 0.12$.

a. What is $P(A \mid B)$?
b. What is $P(B \mid A)$?
c. Are A and B independent?

4.77 Suppose that $P(A) = 0.3$, $P(B) = 0.4$, and $P(A \text{ and } B) = 0.20$.

a. What is $P(A \mid B)$?
b. What is $P(B \mid A)$?
c. Are A and B independent?

4.78 Suppose that A and B are events and the following probabilities are known: $P(A) = 0.3$, $P(B) = 0.4$, and $P(A \mid B) = 0.2$. Find $P(A \text{ or } B)$.

4.79 A single card is drawn from a standard deck. Let A be the event that "the card is a face card" (a jack, a queen, or a king), B is a "red card," and C is "the card is a heart." Determine whether the following pairs of events are independent or dependent:

a. A and B b. A and C c. B and C

4.80 A box contains four red and three blue poker chips. What is the probability that when three are selected randomly, all three will be red if we select each chip

a. with replacement b. without replacement

4.81 Excluding job benefit coverage, approximately 49% of adults have purchased life insurance. The likelihood that those aged 18–24 without life insurance will purchase life insurance in the next year is 15% and for those aged 25–34 it is 26%. (Opinion Research)

a. Find the probability that a randomly selected adult has not purchased life insurance.
b. What is the probability that an adult aged 18–24 will purchase life insurance within the next year?
c. Find the probability that a randomly selected adult will be 25 to 34 years old, does not currently have life insurance, and will purchase it within the next year?

4.82 You've seen the headline: "MEGA MILLIONS jackpot worth \$300 million!" Mega Millions tickets cost \$1.00 per play and players pick **six** numbers from two separate pools of numbers: five different numbers from 1 to 52, and one number from 1 to 52. You win the jackpot by matching all six winning numbers in a drawing. The advertised chances of winning are: 1 in 135,145,920.

a. If you had to guess the specific order in which the numbers from 1 to 52 are drawn, what would be the probability of your winning the jackpot per play?
b. You do not need to guess the specific order in which the numbers from 1 to 52 are drawn. What is the probability of your winning the jackpot per play?
c. Does your answer in part b agree with the advertised chances of winning?

4.83 Of households in the United States, 18 million or 17% have three or more vehicles, as stated in *USA Today* (June 12, 2002), quoting the Census Bureau as the source.

a. If two U.S. households are randomly selected, find the probability that both will have three or more vehicles.
b. If two U.S. households are randomly selected, find the probability that neither of the two has three or more vehicles.
c. If four households are selected, what is the probability that all four have three or more vehicles?

4.84 A *USA Today* article, "Survey: Records tainted—Fans want drug tests for baseball players" (June 12, 2002), quotes a USA Today/CNN Gallup Poll as finding 86% of baseball fans saying they favor testing ballplayers for steroids or other performance-enhancing drugs. If five baseball fans are randomly selected, what is the probability that all five will be in favor of drug testing?

4.85 The July 8, 2002, issue of *Democrat & Chronicle* gave the results from the 2000 census that 42% of grandparents are responsible for "most of the basic needs" of a grandchild in the home. If three American grandparents are contacted, what is the probability that all three are the primary caregiver for their grandchildren?

4.86 You have applied for two scholarships: a merit scholarship (M) and an athletic scholarship (A). Assume the probability that you receive the athletic scholarship is 0.25, the probability you receive both scholarships is 0.15, and the probability you get at least one of the scholarships is 0.37. Use a Venn diagram to answer the questions:

a. What is the probability you receive the merit scholarship?
b. What is the probability you do not receive either of the two scholarships?
c. What is the probability you receive the merit scholarship given that you have been awarded the athletic scholarship?
d. What is the probability you receive the athletic scholarship given that you have been awarded the merit scholarship?
e. Are the events "receiving an athletic scholarship" and "receiving a merit scholarship" independent events? Explain.

4.87 The owners of a two-person business make their decisions independently of each other and then compare their decisions. If they agree, the decision is made; if they do not agree, then further consideration is necessary before a decision is reached. If each has a history of making the right decision 60% of the time, what is the probability that together they:

a. make the right decision on the first try
b. make the wrong decision on the first try
c. delay the decision for further study

4.88 The odds against throwing a pair of dice and getting a total of 5 are 8 to 1. The odds against throwing a pair of dice and getting a total of 10 are 11 to 1. What is the probability of throwing the dice twice and getting a total of 5 on the first throw and 10 on the second throw?

4.89 Consider the set of integers 1, 2, 3, 4, and 5.

a. One integer is selected at random. What is the probability that it is odd?
b. Two integers are selected at random (one at a time with replacement so that each of the five is available for a second selection). Find the probability that neither is odd; exactly one of them is odd; both are odd.

4.90 A box contains 25 parts, of which 3 are defective and 22 are nondefective. If 2 parts are selected without replacement, find the probabilities:

a. *P*(both are defective) b. *P*(exactly one is defective)
c. *P*(neither is defective)

4.91 Graduation rates reached a record low in 2001. The percentage of students who graduate within five years was 41.9% for public and 55.1% for private colleges. One of the reasons for this might be that 42% of the students attend only part time. (ACT)

a. What additional information do you need to determine the probability that a student selected at random is part time and will graduate within five years?
b. Is it likely that these two events have the needed property? Explain.
c. If appropriate, find the probability that a student selected at random is part time and will graduate within five years.

4.92 From a survey of adults, 48% plan to buy candy this year at Easter. The types of candy they will buy are described in the table below.

a. What additional information do you need to determine the probability that a customer selected at random will buy candy and it will be chocolate?
b. Is it likely that these two events have the needed property? Explain.
c. If appropriate, find the probability that a customer selected at random will buy candy and it will be chocolate.

Table for Exercise 4.92

Chocolate	Nonchocolate	Jellybeans	Cream-filled	Marshmallow	Malted	Don't Know
30%	25%	13%	11%	8%	7%	6%

Source: International Mass Retail Association

4.7 Combining the Rules of Probability

Tree diagrams can be used to represent many probability problems. In these instances, the addition and multiplication rules can be applied readily. To illustrate the use of tree diagrams in solving probability problems, let's use Experiment 4.5 (p. 193). Two poker chips are drawn from a box that contains one red, one blue, and one white chip. The tree diagram representing this experiment (Figure 4.5) shows a first drawing and then a second drawing. One chip was drawn on each drawing and was not replaced.

After the tree has been drawn and labeled, we need to assign probabilities to each branch of the tree. If we assume it is equally likely that any chip will be drawn at each stage, we can assign a probability to each branch segment of the tree as shown in Figure 4.6. Notice that a set of branches that initiate from a single point has a total probability of 1. In this diagram there are four such sets of branches. The tree diagram shows six different outcomes. Reading down, we have branch (1) shows (R, B), branch (2) shows (R, W), and so on.

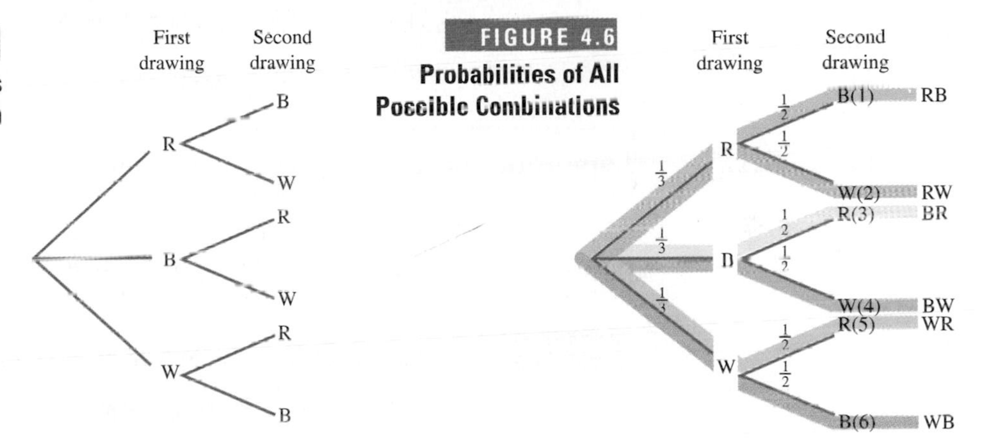

FIGURE 4.5
All Possible Combinations That Can Be Drawn

FIGURE 4.6
Probabilities of All Possible Combinations

NOTE: Each outcome for the experiment is represented by a branch that begins at the common starting point and ends at the terminal points at the right.

The probability associated with outcome (R, B), P(R on first drawing and B on second drawing), is found by multiplying P(R on first drawing) by P(B on second drawing | R on first drawing). These are the two probabilities $\frac{1}{3}$ and $\frac{1}{2}$ shown on the two branch segments of branch (1) in Figure 4.6. The $\frac{1}{2}$ is the conditional probability asked for by the multiplication rule. Thus, we multiply along the branches.

Some events are made up of more than one outcome from our experiment. For example, suppose we had asked for the probability that one red chip and one blue chip are drawn. Two outcomes satisfy this event: branch (1) and branch (3). With "or" we will use the addition rule, formula (4.4b). Since the branches of a tree diagram represent mutually exclusive events, we have

$$P(\text{one R and one B}) = P[(\text{R1 and B2}) \text{ or } (\text{B1 and R2})]$$

$$= \left(\frac{1}{3}\right)\left(\frac{1}{2}\right) + \left(\frac{1}{3}\right)\left(\frac{1}{2}\right) = \frac{1}{6} + \frac{1}{6} = \mathbf{\frac{1}{3}}$$

FIGURE 4.7
Multiply Along the Branches

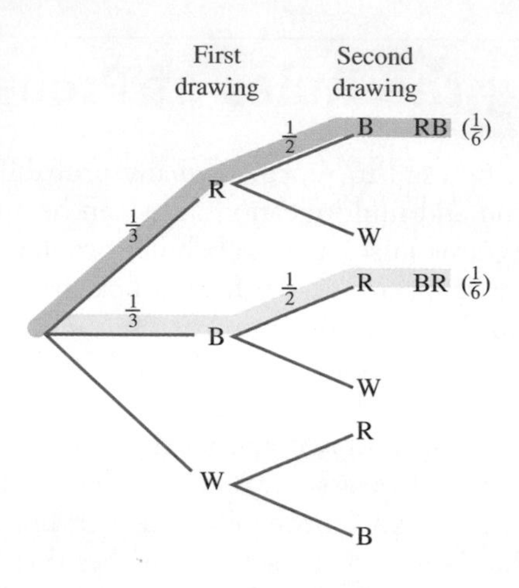

NOTES

1. Multiply along the branches (on Figure 4.7: $\frac{1}{3} \cdot \frac{1}{2}$).
2. Add across the branches (on Figure 4.7: $\frac{1}{6} + \frac{1}{6}$).

Now let's consider an example that places all the rules in perspective.

Illustration 4.12 Combining Probability Rules to Determine Independence

A firm plans to test a new product in one randomly selected market area. The market areas can be categorized on the basis of location and population density. The numbers of markets in each category are presented in Table 4.4.

TABLE 4.4 Number of Markets by Location and Population Density

	Population Density		
Location	**Urban (U)**	**Rural (R)**	**Total**
East (E)	25	50	75
West (W)	20	30	50
Total	45	80	125

What is the probability that the test market selected is in the East, $P(\text{E})$? In the West, $P(\text{W})$? What is the probability that the test market is in an urban area, $P(\text{U})$? In a rural area, $P(\text{R})$? What is the probability that the market is a western rural area, $P(\text{W and R})$? What is the probability it is an eastern or urban area, $P(\text{E or U})$? What is the probability that if it is in the East, it is an urban area, $P(\text{U} \mid \text{E})$? Are "location" and "population density" independent? (What do we mean by independence or dependence in this situation?)

SOLUTION

The first four probabilities—$P(\text{E})$, $P(\text{W})$, $P(\text{U})$, and $P(\text{R})$—represent "or" questions. For example, $P(\text{E})$ means that the area is an eastern urban area or an eastern rural area. Since in this and the other three cases, the two components are mutually exclusive (an area can't be both urban and rural), the desired probabilities can be found by simply adding. In each case the probabilities are added across all the rows or columns of the table. Thus, the totals are found in the total column or row.

$$P(\text{E}) = \frac{75}{125} \quad \text{(total for East divided by total number of markets)}$$

$$P(\text{W}) = \frac{50}{125} \quad \text{(total for West divided by total number of markets)}$$

$$P(\text{U}) = \frac{45}{125} \quad \text{(total for urban divided by total number of markets)}$$

$$P(\text{R}) = \frac{80}{125} \quad \text{(total for rural divided by total number of markets)}$$

Now we solve for $P(\text{W and R})$. There are 30 western rural markets and a total of 125 markets. Thus,

$$P(\text{W and R}) = \frac{n(\text{W and R})}{n(S)} = \frac{30}{125}$$

Note that $P(\text{W}) \cdot P(\text{R})$ does not give the right answer $\left[\left(\frac{50}{125}\right)\left(\frac{80}{125}\right) = \frac{32}{125}\right]$. Therefore, "location" and "population density" are dependent events.

$P(\text{E or U})$ can be found in several different ways. The most direct way is simply to examine the table and count the number of markets that satisfy the condition of being in the East or urban. We find 95 [25 + 50 + 20]. Thus,

$$P(\text{E or U}) = \frac{n(\text{E or U})}{n(S)} = \frac{95}{125}$$

Note that the first 25 markets were both in the East and urban; thus, E and U are not mutually exclusive events. Another way to solve for $P(\text{E or U})$ is to use the addition formula:

$$P(\text{E or U}) = P(\text{E}) + P(\text{U}) - P(\text{E and U})$$

which yields

$$\frac{75}{125} + \frac{45}{125} - \frac{25}{125} = \frac{95}{125}$$

A third way is to recognize that the complement of (E or U) is (W and R). Thus, $P(\text{E or U}) = 1 - P(\text{W and R})$. Using the previous calculation, we get $1 - \frac{30}{125} = \frac{95}{125}$.

Finally, we solve for $P(\text{U} \mid \text{E})$. From Table 4.4 we see that there are 75 markets in the East. Of the 75 eastern markets, 25 are urban. Thus,

$$P(\text{U} \mid \text{E}) = \frac{n(\text{U and E})}{n(\text{E})} = \frac{25}{75}$$

The conditional probability formula could also be used:

$$P(\text{U} \mid \text{E}) = \frac{P(\text{U and E})}{P(\text{E})} = \frac{\frac{25}{125}}{\frac{75}{125}} = \frac{25}{75}$$

"Location" and "population density" are not independent events. They are dependent. This means that the probability of these events is affected by the occurrence of each other. ■

Although each rule for computing compound probabilities has been discussed separately, you should not think they are used separately. In many cases they are combined to solve problems. Consider the next two illustrations.

Illustration 4.13 Using Several Probability Rules

A production process manufactures an item. On the average, 20% of all items produced are defective. Each item is inspected before it is shipped. The inspector misclassifies an item 10% of the time; that is,

$$P(\text{classified good} \mid \text{defective item}) = P(\text{classified defective} \mid \text{good item}) = 0.10$$

What proportion of the items will be "classified good"?

FYI
Misclassification can happen two ways!

SOLUTION
What do we mean by the event "classified good"? We define these events:

G: The item is good.
D: The item is defective.
CG: The item is classified good by the inspector.
CD: The item is classified defective by the inspector.

FIGURE 4.8
Using Probability Rules

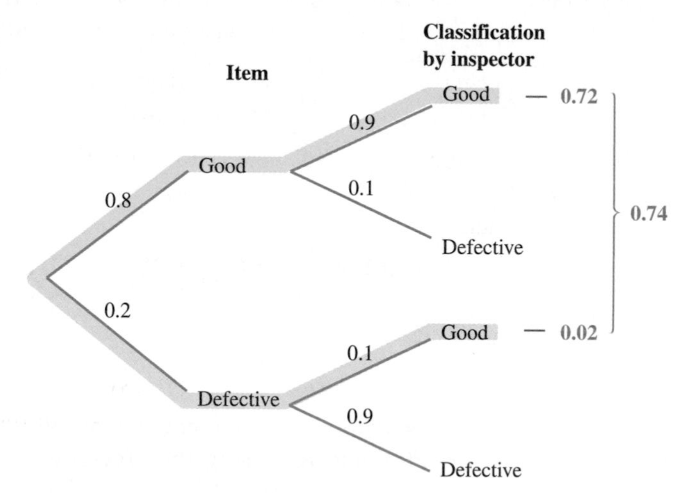

CG consists of two possibilities: "the item is good and is correctly classified good" and "the item is defective and is misclassified good." Thus,

$$P(CG) = P[(CG \text{ and } G) \text{ or } (CG \text{ and } D)]$$

Since the two possibilities are mutually exclusive, we can start by using the addition rule, formula (4.4b):

$$P(CG) = P(CG \text{ and } G) + P(CG \text{ and } D)$$

The condition of an item and its classification by the inspector are not independent. We must use the multiplication rule for dependent events, formula (4.7). Therefore,

$$P(CG) = [P(G) \cdot P(CG \mid G)] + [P(D) \cdot P(CG \mid D)]$$

Substituting the known probabilities in Figure 4.8, we get

$$\begin{aligned} P(\text{CG}) &= [(0.8)(0.9)] + [(0.2)(0.1)] \\ &= 0.72 + 0.02 \\ &= \mathbf{0.74} \end{aligned}$$

That is, 74% of the items are classified good. ■

Illustration 4.14 A Conditional Probability Within a Conditional Probability

Reconsider Illustration 4.13. Suppose only items that pass inspection are shipped, and items not classified good are scrapped. What is the quality of the shipped items? That is, what percentage of the items shipped are good, $P(\text{G} \mid \text{CG})$?

SOLUTION

We use the conditional probability equation, formula (4.6b):

$$\begin{aligned} P(\text{G} \mid \text{CG}) &= \frac{P(\text{G and CG})}{P(\text{CG})} \\ &= \frac{P(\text{G}) \cdot P(\text{CG} \mid \text{G})}{P(\text{CG})} \qquad \text{(see Illustration 4.13)} \\ &= \frac{(0.8)(0.9)}{0.74} = 0.97297 = \mathbf{0.973} \end{aligned}$$

In other words, 97.3% of all items shipped will be good. Inspection increases the quality of items shipped from 80% good to 97.3% good. ■

Application 4.3 Is What You Read, What Was Printed!

USA Today (July 23, 1998) published a profile of affluence that showed 17 million Americans live in households with annual incomes of at least $100,000. Of these, 75% owned a house, 70% were married, and 40% of the houses occupied were valued at over $200,000. Suppose one person from this group is selected at random. What is the probability that the person selected:

a. owns a house valued at over $200,000 b. is a married homeowner
c. is a married homeowner living in a house valued at over $200,000

The answer to all these questions cannot be determined from the data supplied because there is no way of ascertaining the conditional probabilities. The individual events—owning a house and being married—are not independent. Houses can be rented or owned, and married couples can rent or own a house, irrespective of value. This situation illustrates the potential pitfall of misinterpreting survey results.

ANSWER NOW 4.93

What conditional probabilities would you need to find the answers to the Application 4.3 questions?

Exercises

4.94 If $P(A) = 0.4$ and $P(B) = 0.5$, and if A and B are mutually exclusive events, find $P(A \text{ or } B)$.

4.95 If $P(A) = 0.4$ and $P(B) = 0.5$, and if A and B are independent events, find $P(A \text{ or } B)$.

4.96 $P(G) = 0.5$, $P(H) = 0.4$, and $P(G \text{ and } H) = 0.1$ (see the diagram).

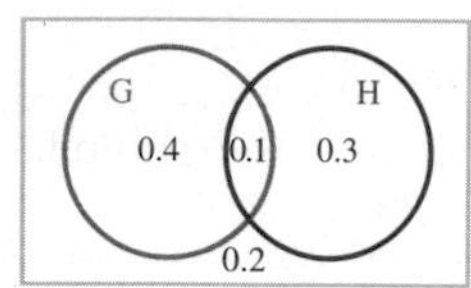

a. Find $P(G \mid H)$.
b. Find $P(H \mid G)$.
c. Find $P(\overline{H})$.
d. Find $P(G \text{ or } H)$.
e. Find $P(G \text{ or } \overline{H})$.
f. Are events G and H mutually exclusive? Explain.
g. Are events G and H independent? Explain.

4.97 $P(R) = 0.5$, $P(S) = 0.3$, and events R and S are independent.
a. Find $P(R \text{ and } S)$. b. Find $P(R \text{ or } S)$. c. Find $P(\overline{S})$.
d. Find $P(R \mid S)$. e. Find $P(\overline{S} \mid R)$.
f. Are events R and S mutually exclusive? Explain.

4.98 $P(M) = 0.3$, $P(N) = 0.4$, and events M and N are mutually exclusive.
a. Find $P(M \text{ and } N)$. b. Find $P(M \text{ or } N)$.
c. Find $P(M \text{ or } \overline{N})$. d. Find $P(M \mid N)$.
e. Find $P(M \mid \overline{N})$.
f. Are events M and N independent? Explain.

4.99 Two flower seeds are randomly selected from a package that contains five seeds for red flowers and three seeds for white flowers.
a. What is the probability that both seeds will result in red flowers?
b. What is the probability that one of each color is selected?
c. What is the probability that both seeds are for white flowers?

FYI
Draw a tree diagram.

4.100 The probability that a certain door is locked is 0.6. The key to the door is one of five unidentified keys hanging on a key rack. You randomly select two keys before approaching the door. What is the probability that you can open the door without returning for another key?

4.101 Alex, Bill, and Chen each in turn toss a balanced coin. The first one to throw a head wins.
a. What are their respective chances of winning if each tosses only one time?
b. What are their respective chances of winning if they continue, given a maximum of two tosses each?

FYI
Draw a tree diagram.

4.102 A coin is flipped three times.
a. Draw a tree diagram that represents all possible outcomes.
b. Identify all branches that represent the event "exactly one head occurred."
c. Find the probability of "exactly one head occurred."

4.103 Box 1 contains two red balls and three green balls, and Box 2 contains four red balls and one green ball. One ball is randomly selected from Box 1 and placed in Box 2. Then one ball is randomly selected from Box 2. What is the probability that the ball selected from Box 2 is green?

4.104 A company that manufactures shoes has three factories. Factory 1 produces 25% of the company's shoes, Factory 2 produces 60%, and Factory 3 produces 15%. One percent of the shoes produced by Factory 1 are mislabeled, 0.5% of those produced by Factory 2 are mislabeled, and 2% of those produced by Factory 3 are mislabeled. If you purchase one pair of shoes manufactured by this company, what is the probability that the shoes are mislabeled?

4.105 Your local art museum has planned next year's 52-week calendar by scheduling a mixture of one-week and two-week shows that feature the works of 22 painters and 20 sculptors. There is a showing scheduled for every week of the year, and only one artist is featured at a time. There are 42 different shows scheduled for next year. You have randomly selected one week to attend and have been told the probability of it being a two-week show of sculpture is 3/13.
a. What is the probability that the show you have selected is a painter's showing?
b. What is the probability that the show you have selected is a sculptor's showing?
c. What is the probability that the show you have selected is a one-week show?
d. What is the probability that the show you have selected is a two-week show?

4.106 The *Regional Economic Digest* listed the total deposits in commercial banks and thrifts in the states of Missouri and Nebraska during the fourth quarter of 1997. The deposits were further subdivided into two major categories: checkable deposits and time/savings deposits. Results are shown in the table in millions of dollars.

Table for Exercise 4.106

	Missouri		Nebraska		
	Commercial Banks	Thrifts	Commercial Banks	Thrifts	Total
Checkable	5,111	1,407	5,017	406	11,941
Time/Savings	12,428	128	15,447	5,312	33,315
Total	17,539	1,535	20,464	5,718	45,256

Source: *Regional Economic Digest*, First Quarter 1998, Vol. 9, No. 1, p. 39

Consider these events: A—"checkable deposit," B—"Nebraska deposit," and C—"deposit in a commercial bank." Suppose $1 is withdrawn randomly from the grand total of all deposits. Find each of the following probabilities and describe what they mean in your own words:

a. *P*(A), *P*(B), and *P*(C)
b. *P*(A and B), *P*(A and C), *P*(B and C), and *P*(A and B and C)
c. *P*(A or B), *P*(A or C), *P*(B or C), and *P*(not (A or B or C))
d. Sketch the Venn diagram and show all the related probabilities. Are A, B, and C all independent events? Explain.

4.107 S One thousand employees at the Russell Microprocessor Company were polled about worker satisfaction. One employee is selected at random.

	Male		Female		
	Skilled	Unskilled	Skilled	Unskilled	Total
Satisfied	350	150	25	100	625
Unsatisfied	150	100	75	50	375
Total	500	250	100	150	1000

a. Find the probability that an unskilled worker is satisfied with work.
b. Find the probability that a skilled woman employee is satisfied with work.
c. Is satisfaction for women employees independent of their being skilled or unskilled?

CHAPTER SUMMARY

RETURN TO CHAPTER CASE STUDY

© Rachel Epstein/The Image Works

Statistics Students' Favorite Candy

FIRST THOUGHTS 4.108

A theoretical look at the expected:

a. Construct a bar graph showing the expected (theoretical) proportion of M&M's for each color.
b. If a single bag of M&M's had 40 candies in it, how many of each color would you "expect" to find?
c. If you opened a bag of M&M's right now and counted the number of each color, would you be surprised to find the numbers differ from those given as answers in part b? Explain.
d. Would you be surprised to find the numbers are exact matches to those found in part b? Explain.

PUTTING CHAPTER 4 TO WORK 4.109

An empirical look at this situation:

a. Buy a pack of M&M's (at least a 1.69-oz size—costs approximately $0.50).
b. Record the number of each color in a frequency distribution with the headings "Color" and "Frequency."
c. Verify the total number of M&M's with the sum of the frequency column.
d. Now you may snack! ☺
e. Present the frequency distribution as a relative frequency distribution, using the heading "Probability."
f. Verify that the sum of the "Probability" column is equal to one. Explain the meaning of this sum.

(continued)

g. Construct a bar graph showing the relative frequency for each color. Use the same color order as in part a of problem 4.108.
h. What other statistical displays could you use to present the data from the pack of M&M's? Present them.
i. Compare your empirical (experimental) findings to the expectations (theoretical) expressed in part a of problem 4.108.

YOUR-OWN MINI STUDY 4.110

a. Use a computer (or random-number table) to generate a random sample of 56 M&M's, using the corresponding theoretical probabilities for each color.
b. Form a frequency distribution of the random data.
c. Construct a bar graph showing the relative frequency for each color. Use the same color order as in part a of problem 4.108.
d. Compare your experimental findings with the theoretical expectations.
e. Repeat parts a–d three more times.
f. Describe the variability you observe between the samples.
g. Consolidate your four frequency distributions into one frequency distribution that has a frequency total of 224 M&M's.
h. Construct a bar graph of the consolidation showing the relative frequencies for the colors. Use the same color order as in part a of problem 4.108.
i. Compare these experimental findings with the theoretical expectations.
j. Compare the consolidated findings with the four individual findings above.
k. How does the law of large numbers affect this mini-study?

MINITAB and Excel can generate only random numbers. Therefore, it is common practice to use numbers in place of the colors (words). If using "Before GVC," the numbers 1, 2, 3, 4, 5, 6 will correspond to blue, brown, . . . , yellow, respectively. If using "After GVC," the numbers 1, 2, 3, 4, 5, 6, 7 will correspond to blue, brown, . . . , purple, respectively.

MINITAB (Release 13)

a. Input the numbers 1–6 or 1–7 into C1 and the corresponding probabilities into C2; then continue with:

```
Choose:   Calc > Random Data > Discrete
Enter:    Generate: 56 (# of M&M's in a pack)
          Store in column(s): C3
          Values in: C1 (color numbers)
          Probabilities in: C2 > OK
```

b. To obtain the frequency distribution, continue with:

```
Choose:   Stat > Tables > Cross Tabulation
Enter:    Classification variables: C3
Select:   Display: Counts and Column percents
```

c. To construct a bar graph, enter the actual colors into C4 and the corresponding probabilities (percents) found in step b into C5:

```
Choose:   Graph > Chart
Enter:    Y: C5 X: C4
Select:   Frame > Min and Max...
Enter:    Minimum for Y: 0 > OK > OK
```

Excel XP

a. Input the numbers 1–6 or 1–7 into column A and the corresponding probabilities into column B; then continue with:

```
Choose: Tools > Data Analysis > Random Number Generation > OK
Enter:     Number of Variables: 1
           Number of Random Numbers: 56 (# of M&M's in a
           pack)
           Distribution: Discrete
           Value & Prob. Input Range: (A1:B7 select data
           cells)
Select:    Output range
Enter      (C1 or select cell) > OK
```

b. The frequency distribution is given with the histogram of the generated data. Use the histogram Excel commands on pages 58–59 using the data in column C and the bin range in column A.

c. Divide the frequencies by 56 to obtain the corresponding probabilities. Enter the actual colors in column D (ex. D13:D18) and the corresponding probabilities in column E (ex. E13:E18). To construct a bar graph, continue with:

```
Choose:    Chart Wizard > Column > 1st picture(usually) >
           Next
Enter:     Data range: (D13:E18 or select cells) > Next
Enter:     Chart and axes titles > Finish (Edit as needed)
```

In Retrospect

You have been studying the basic concepts of probability. These fundamentals need to be mastered before we continue with our study of statistics. Probability is the vehicle of statistics, and we have begun to see how probabilistic events occur. We have explored theoretical and experimental probabilities for the same event. Does the experimental probability turn out to have the same value as the theoretical? Not exactly, but we have seen that over the long run it does have approximately the same value.

Upon completion of this chapter, you should understand the properties of mutual exclusive and independence, and be able to apply the multiplication and addition rules to "and" and "or" compound events. You should also be able to calculate conditional probabilities.

In the next three chapters we will look at distributions associated with probabilistic events. This will prepare us for the statistics that follow. We must be able to predict the variability that the sample will show with respect to the population before we can be successful at "inferential statistics," in which we describe the population based on the sample statistics available.

Chapter Exercises

4.111 Ex04-111 The Federal Highway Administration periodically tracks licensed vehicle drivers by gender and age. The table shows the results of the administration's findings in 1995. Suppose you encountered a driver of a vehicle at random. Find the probabilities of the following events:

Age Group	Male	Female
19 or less	4,761,567	4,362,558
20–24	8,016,601	7,508,844
25–29	9,234,547	8,822,290
30–34	10,255,668	10,028,055
35–39	10,381,712	10,227,348
40–44	9,512,860	9,465,126
45–49	8,469,713	8,401,960
50–54	6,493,069	6,397,959
55–59	5,167,725	5,057,785
60–64	4,530,005	4,428,256
65–69	4,248,092	4,234,797
70–74	3,582,678	3,702,020
75–79	2,465,550	2,577,527
80 and over	2,094,581	2,149,589
Total	89,214,367	87,414,115

Source: Federal Highway Administration, U.S. Department of Transportation

a. The driver is a male, over the age of 59.
b. The driver is a female, under the age of 30.
c. The driver is under the age of 25.
d. The driver is a female.
e. The driver is a male between the ages of 35 and 49.
f. The driver is a female between the ages of 25 and 44.
g. The driver is over the age of 69.

4.112 Probabilities for events A, B, and C are distributed as shown in the figure. Find:
a. $P(A \text{ and } B)$ b. $P(A \text{ or } C)$ c. $P(A \mid C)$

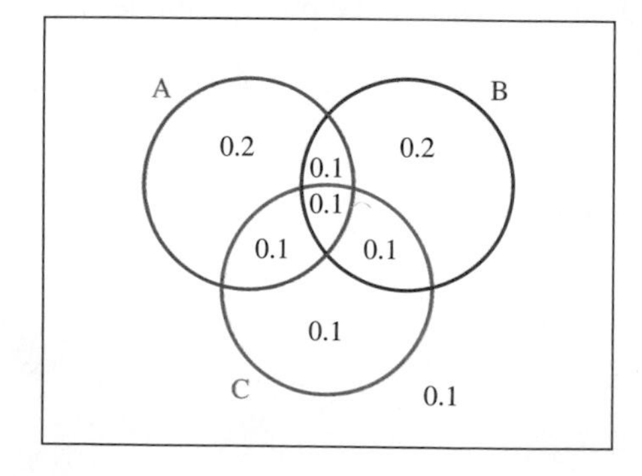

4.113 Suppose a certain ophthalmic trait is associated with eye color. Three hundred randomly selected individuals are studied, with results given in the table.

	Eye Color			
Trait	Blue	Brown	Other	Total
Yes	70	30	20	120
No	20	110	50	180
Total	90	140	70	300

a. What is the probability that a person selected at random has blue eyes?
b. What is the probability that a person selected at random has the trait?
c. Are events A (has blue eyes) and B (has the trait) independent? Justify your answer.
d. How are the two events A (has blue eyes) and C (has brown eyes) related—independent, mutually exclusive, complementary, or all inclusive? Explain why or why not each term applies.

4.114 A study was conducted in 1997 to measure the total fat content, calories, and sodium content of vegetable burgers available at supermarkets and commonly used as a meat substitute. Measurements were taken on 54 different brands of "veggie burgers," and the results were used to develop the contingency table.

	Under 130 Calories		130 Calories or More	
Fat Content	Under 320 g Sodium	320 g Sodium or More	Under 320 g Sodium	320 g Sodium or More
Under 3 g	6	12	1	1
3–4 g	7	4	3	5
Over 4 g	1	1	7	6

Source: *Nutrition Action Health Letter,* "Where's the Beef?", Vol. 24, No. 1, 1997

Consider these events: A—"under 130 calories," B—"320 grams of sodium or more," and C—"over 4 grams fat." A vegetable burger is selected randomly from the group.
a. Find $P(A)$, $P(B)$, and $P(C)$.
b. Find $P(A \text{ and } B)$, $P(A \text{ and } C)$, $P(B \text{ and } C)$, and $P(A \text{ and } B \text{ and } C)$.
c. Find $P(A \text{ or } B)$, $P(A \text{ or } C)$, and $P(B \text{ or } C)$.
d. Sketch the Venn diagram and show all the related probabilities. Are A, B, and C all independent events? Explain.

4.115 Events R and S are defined on the sample space. If $P(R) = 0.2$ and $P(S) = 0.5$, explain why each of the following statements is either true or false:

a. If R and S are mutually exclusive, then $P(R \text{ or } S) = 0.10$.
b. If R and S are independent, then $P(R \text{ or } S) = 0.6$.
c. If R and S are mutually exclusive, then $P(R \text{ and } S) = 0.7$.
d. If R and S are mutually exclusive, then $P(R \text{ or } S) = 0.6$.

4.116 Show that if event A is a subset of event B, then $P(A \text{ or } B) = P(B)$.

4.117 Let's assume there are three traffic lights between your house and a friend's house. As you arrive at each light, it may be red (R) or green (G).

a. List the sample space showing all possible sequences of red and green lights that could occur on a trip from your house to your friend's. (RGG represents red at the first light and green at the other two.)

Assume that each element of the sample space is equally likely to occur.

b. What is the probability that on your next trip to your friend's house you will have to stop for exactly one red light?
c. What is the probability that you will have to stop for at least one red light?

4.118 Assuming that a woman is equally likely to bear a boy as a girl, use a tree diagram to compute the probability that a four-child family consists of one boy and three girls.

4.119 Interactivity Interactivity 4-B simulates generating a family. The family will stop having children when there is one boy or three girls, whichever comes first. Assuming that a woman is equally likely to bear a boy as a girl, perform the simulation 24 times. What is the probability that the family will have a boy?

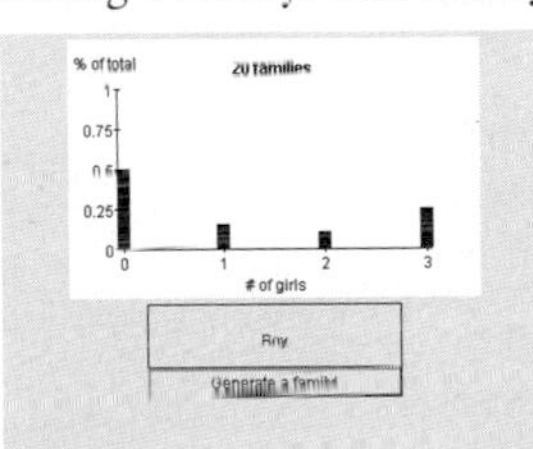

4.120 A traffic analysis at a busy traffic circle in Washington, DC, showed that 0.8 of the autos using the circle entered from Connecticut Avenue. Of those entering the traffic circle from Connecticut Avenue, 0.7 continued on Connecticut Avenue at the opposite side of the circle. What is the probability that a randomly selected auto observed in the traffic circle entered from Connecticut and will continue on Connecticut?

4.121 Suppose that when a job candidate comes to interview for a job at RJB Enterprises, the probability that he or she will want the job (A) after the interview is 0.68. Also, the probability that RJB wants the candidate (B) is 0.36. The probability $P(A \mid B)$ is 0.88.

a. Find $P(A \text{ and } B)$.
b. Find $P(B \mid A)$.
c. Are events A and B independent? Explain.
d. Are events A and B mutually exclusive? Explain.
e. What would it mean to say A and B are mutually exclusive events in this exercise?

4.122 The probability that thunderstorms are in the vicinity of a particular midwestern airport on an August day is 0.70. When thunderstorms are in the vicinity, the probability that an airplane lands on time is 0.80. Find the probability that thunderstorms are in the vicinity and the plane lands on time.

4.123 Tires salvaged from a train wreck are on sale at the Getrich Tire Company. Of the 15 tires offered in the sale, five tires have suffered internal damage and the remaining ten are damage free. You randomly selected and purchased two of these tires.

a. What is the probability that the tires you purchased are both damage free?
b. What is the probability that exactly one of the tires you purchased is damage free?
c. What is the probability that at least one of the tires you purchased is damage free?

4.124 According to automobile accident statistics, one out of every six accidents results in an insurance claim of \$100 or less in property damage. Three cars insured by an insurance company are involved in different accidents. Consider these two events:

A: The majority of claims exceed \$100.
B: Exactly two claims are \$100 or less.

a. List the sample points for this experiment.
b. Are the sample points equally likely?
c. Find $P(A)$ and $P(B)$.
d. Are events A and B independent? Justify your answer.

4.125 One thousand persons screened for a certain disease are given a clinical exam. As a result of the exam, the sample of 1000 persons is classified according to height and disease status.

	Disease Status				
Height	**None**	**Mild**	**Moderate**	**Severe**	**Total**
Tall	122	78	139	61	400
Medium	74	51	90	35	250
Short	104	71	121	54	350
Total	300	200	350	150	1000

Use the information in the table to estimate the probability of being medium or short and of having moderate or severe disease status.

4.126 The table shows the sentiments of 2500 wage-earning employees at the Spruce Company on a proposal to emphasize fringe benefits rather than wage increases during their impending contract discussions.

	Opinion			
Employee	**Favor**	**Neutral**	**Opposed**	**Total**
Male	800	200	500	1500
Female	400	100	500	1000
Total	1200	300	1000	2500

a. Calculate the probability that an employee selected at random from this group will be opposed.
b. Calculate the probability that an employee selected at random from this group will be female.
c. Calculate the probability that an employee selected at random from this group will be opposed, given that the person is male.
d. Are the events "opposed" and "female" independent? Explain.

4.127 A shipment of grapefruit arrived containing the following proportions of types: 10% pink seedless, 20% white seedless, 30% pink with seeds, and 40% white with seeds. A grapefruit is selected at random from the shipment. Find the probability of these events:
a. It is seedless.
b. It is white.
c. It is pink and seedless.
d. It is pink or seedless.
e. It is pink, given that it is seedless.
f. It is seedless, given that it is pink.

4.128 Salespersons Adams and Jones call on three and four customers, respectively, on a given day. Adams could make 0, 1, 2, or 3 sales, whereas Jones could make 0, 1, 2, 3, or 4 sales. The sample space listing the number of possible sales for each person on a given day is shown in the table. (3, 1 stands for 3 sales by Jones and 1 sale by Adams.)

	Jones				
Adams	**0**	**1**	**2**	**3**	**4**
0	0, 0	1, 0	2, 0	3, 0	4, 0
1	0, 1	1, 1	2, 1	3, 1	4, 1
2	0, 2	1, 2	2, 2	3, 2	4, 2
3	0, 3	1, 3	2, 3	3, 3	4, 3

Assume that each sample point is equally likely. Consider these events:

A: At least one of the salespersons made no sales.
B: Together they made exactly three sales.
C: Each made the same number of sales.
D: Adams made exactly one sale.

Find the probabilities by counting sample points:
a. $P(A)$ b. $P(B)$ c. $P(C)$
d. $P(D)$ e. $P(A \text{ and } B)$ f. $P(B \text{ and } C)$
g. $P(A \text{ or } B)$ h. $P(B \text{ or } C)$ i. $P(A \mid B)$
j. $P(B \mid D)$ k. $P(C \mid B)$ l. $P(B \mid \overline{A})$
m. $P(C \mid \overline{A})$ n. $P(A \text{ or } B \text{ or } C)$

Are the following pairs of events mutually exclusive? Explain.

o. A and B p. B and C q. B and D

Are the following pairs of events independent? Explain.

r. A and B s. B and C t. B and D

4.129 A testing organization wishes to rate a particular brand of television. Six TV's are selected at random from stock. If nothing is found wrong with any of the six, the brand is judged satisfactory.
a. What is the probability that the brand will be rated satisfactory if 10% of the TV's actually are defective?
b. What is the probability that the brand will be rated satisfactory if 20% of the TV's actually are defective?
c. What is the probability that the brand will be rated satisfactory if 40% of the TV's actually are defective?

4.130 Coin A is loaded in such a way that $P(\text{heads})$ is 0.6. Coin B is a balanced coin. Both coins are tossed. Find:
a. the sample space that represents this experiment; assign a probability measure to each outcome.
b. $P(\text{both show heads})$
c. $P(\text{exactly one head shows})$
d. $P(\text{neither coin shows a head})$
e. $P(\text{both show heads} \mid \text{coin A shows a head})$
f. $P(\text{both show heads} \mid \text{coin B shows a head})$
g. $P(\text{heads on coin A} \mid \text{exactly one head shows})$

4.131 Professor French forgets to set his alarm with a probability of 0.3. If he sets the alarm, it rings with a probability of 0.8. If the alarm rings, it will wake him on time to make his first class with a probability of 0.9. If the alarm does not ring, he wakes in time for his first class with a probability of 0.2. What is the probability that Professor French wakes in time to make his first class tomorrow?

4.132 A two-page typed report contains an error on one of the pages. Two proofreaders review the copy. Each has an 80% chance of catching the error. What is the probability that the error will be identified if:

a. each reads a different page
b. they each read both pages
c. the first proofreader randomly selects a page to read and then the second proofreader randomly selects a page unaware of which page the first selected

4.133 In sports, championships are often decided by two teams playing in a championship series. Often the fans of the losing team claim they were unlucky and their team is actually the better team. Suppose Team A is the better team, and the probability it will defeat Team B in any one game is 0.6. What is the probability that the better Team A will lose the series if it is:

a. a one-game series
b. a best out of three series
c. a best out of seven series
d. Suppose the probability that A would beat B in any given game were actually 0.7. Recompute parts a–c.
e. Suppose the probability that A would beat B in any given game were actually 0.9. Recompute parts a–c.
f. What is the relationship between the "best" team winning and the number of games played? The best team winning and the probabilities that each will win?

4.134 A woman and a man (unrelated) each have two children. At least one of the woman's children is a boy, and the man's older child is a boy. Is the probability that the woman has two boys greater than, equal to, or less than the probability that the man has two boys?

a. Demonstrate the truth of your answer using a simple sample to represent each family.
b. Demonstrate the truth of your answer by taking two samples, one from men with two-children families and one from women with two-children families.
c. Demonstrate the truth of your answer using computer simulation. Using the Bernoulli probability function with $p = 0.5$ (let $0 =$ girl and $1 =$ boy), generate 500 "families of two children" for the man and the woman. Determine which of the 500 satisfy the condition for each and determine the observed proportion with two boys.
d. Demonstrate the truth of your answer by repeating the computer simulation several times. Repeat the simulation in part c several times.
e. Do the above procedures seem to yield the same results? Explain.

Vocabulary List and Key Concepts

Be able to define each term. Pay special attention to the key terms, which are printed in red. In addition, describe each term in your own words and give an example of each. Your examples should not be ones given in class or in the textbook. The bracketed numbers indicate the chapter in which the term first appeared, but you should define the terms again to show increased understanding of their meaning. Page numbers indicate the first appearance of the term in Chapter 4.

addition rule (p. 204)
all-inclusive events (p. 191)
complementary event (p. 199)
compound event (p. 202)
conditional probability (p. 209)
dependent events (p. 209)
empirical probability (p. 185)
equally likely events (p. 197)
event (p. 191)
experiment [1] (p. 191)
experimental probability (p. 185)
general addition rule (p. 205)
general multiplication rule (p. 211)
independence (p. 208)
independent events (pp. 209, 210)
intersection (p. 203)
law of large numbers (p. 187)
long-term average (p. 187)
multiplication rule (p. 211)
mutually exclusive events (pp. 191, 202)
odds (p. 199)
ordered pair (p. 191)
outcome (p. 191)
probability of an event (p. 185)
relative frequency [2] (p. 183)
sample point (p. 191)
sample space (p. 191)
special addition rule (p. 205)
special multiplication rule (p. 212)
subjective probability (p. 198)
theoretical probability (p. 197)
tree diagram (p. 192)
Venn diagram (p. 195)

Chapter Practice Test

PART I: KNOWING THE DEFINITIONS

Answer "True" if the statement is always true. If the statement is not always true, replace the words shown in bold with the words that make the statement always true.

4.1 The probability of an event is a **whole number.**

4.2 The concepts of probability and relative frequency as related to an event are **very similar.**

4.3 The **sample space** is the theoretical population for probability problems.

4.4 The sample points of a sample space are **equally likely** events.

4.5 The value found for experimental probability will **always be** exactly equal to the theoretical probability assigned to the same event.

4.6 The probabilities of complementary events always **are equal.**

4.7 If two events are mutually exclusive, they are also **independent.**

4.8 If events A and B are **mutually exclusive,** the sum of their probabilities must be exactly one.

4.9 If the sets of sample points that belong to two different events do not intersect, the events are **independent.**

4.10 A compound event formed with the word "and" requires the use of the **addition rule.**

PART II: APPLYING THE CONCEPTS

4.11 A computer is programmed to generate the eight single-digit integers 1, 2, 3, 4, 5, 6, 7, and 8 with equal frequency. Consider the experiment "the next integer generated" and these events:

A: odd number, {1, 3, 5, 7}
B: number greater than 4, {5, 6, 7, 8}
C: 1 or 2, {1, 2}

Find:

a. $P(A)$ b. $P(B)$ c. $P(C)$ d. $P(\overline{C})$
e. $P(A \text{ and } B)$ f. $P(A \text{ or } B)$ g. $P(B \text{ and } C)$
h. $P(B \text{ or } C)$ i. $P(A \text{ and } C)$ j. $P(A \text{ or } C)$
k. $P(A \mid B)$ l. $P(B \mid C)$ m. $P(A \mid C)$
n. Are events A and B mutually exclusive? Explain.
o. Are events B and C mutually exclusive? Explain.
p. Are events A and C mutually exclusive? Explain.
q. Are events A and B independent? Explain.
r. Are events B and C independent? Explain.
s. Are events A and C independent? Explain.

4.12 Events A and B are mutually exclusive and $P(A) = 0.4$ and $P(B) = 0.3$.
a. Find $P(A \text{ and } B)$.
b. Find $P(A \text{ or } B)$.
c. Find $P(A \mid B)$.
d. Are events A and B independent? Explain.

4.13 Events C and D are independent and $P(C) = 0.2$ and $P(D) = 0.7$.
a. Find $P(C \text{ and } D)$.
b. Find $P(C \text{ or } D)$.
c. Find $P(C \mid D)$.
d. Are events C and D mutually exclusive? Explain.

4.14 Events E and F have probabilities $P(E) = 0.5$, $P(F) = 0.4$, and $P(E \text{ and } F) = 0.2$.
a. Find $P(E \text{ or } F)$.
b. Find $P(E \mid F)$.
c. Are E and F mutually exclusive? Explain.
d. Are E and F independent? Explain.

4.15 Events G and H have probabilities $P(G) = 0.3$, $P(H) = 0.2$, and $P(G \text{ and } H) = 0.1$.
a. Find $P(G \text{ or } H)$.
b. Find $P(G \mid H)$.
c. Are G and H mutually exclusive? Explain.
d. Are G and H independent? Explain.

4.16 Janice wants to become a police officer. She must pass a physical exam and then a written exam. Records show that the probability of passing the physical exam is 0.85 and that once the physical is passed, the probability of passing the written exam is 0.60. What is the probability that Janice passes both exams?

PART III: UNDERSTANDING THE CONCEPTS

4.17 Explain briefly how you decide which of the following two events is the more unusual: A—a 90-degree day in Vermont or B—a 100-degree day in Florida.

4.18 Student A says that independence and mutually exclusive are basically the same thing; namely, both mean neither event has anything to do with the other one. Student B argues that although Student A's statement has some truth in it, Student A has missed the point of these two properties. Student B is correct. Carefully explain why.

4.19 Using complete sentences, describe in your own words:

a. mutually exclusive events
b. independent events
c. the probability of an event
d. a conditional probability

4.20 The probability that there are no winners on any one Powerball Lottery game is approximately 0.15.

a. Interpret the meaning of that probability in terms of how often one might expect a drawing to have no winner.
b. Explain why you would or would not consider no winner on one Powerball game to be a rare occurrence.
c. What is the approximate likelihood of no winners for two consecutive Powerball games? Interpret the meaning of your answer.
d. Explain why you would or would not consider no winners on two consecutive Powerball games to be a rare occurrence.

CHAPTER

5 Probability Distributions (Discrete Variables)

Chapter Outline and Objectives

CHAPTER CASE STUDY

A **statistical distribution** is applied to an everyday situation.

5.1 RANDOM VARIABLES

To study probability distributions, a **numerical value** will be assigned to each outcome in the sample space.

5.2 PROBABILITY DISTRIBUTIONS OF A DISCRETE RANDOM VARIABLE

The probability of a value of the random variable is expressed by a **probability function.**

5.3 MEAN AND VARIANCE OF A DISCRETE PROBABILITY DISTRIBUTION

Population parameters are used to measure the probability distribution.

5.4 THE BINOMIAL PROBABILITY DISTRIBUTION

Binomial probability distributions occur in situations where each trial of an experiment has **two possible outcomes.** The binomial probability distribution is the most important discrete random variable encountered in many fields of application.

5.5 MEAN AND STANDARD DEVIATION OF THE BINOMIAL DISTRIBUTION

Two simple **formulas** are used to measure the binomial distribution.

RETURN TO CHAPTER CASE STUDY

Chapter Case Study

Family Values and Family Togetherness

© David Hanover/Getty Images

Family values and family togetherness are important aspects of our everyday lives and have often been the topics of newspaper headlines in recent years. The number of activities that a family participates in together is a measure of a family's strength and well-being. One daily activity that many families pay particular attention to is the evening meal, and for many families the food served at dinner is the focal point. "What's for dinner?" (*USA Today*, October 7, 1996) shows the frequency of home-cooked meals each week in many American homes.

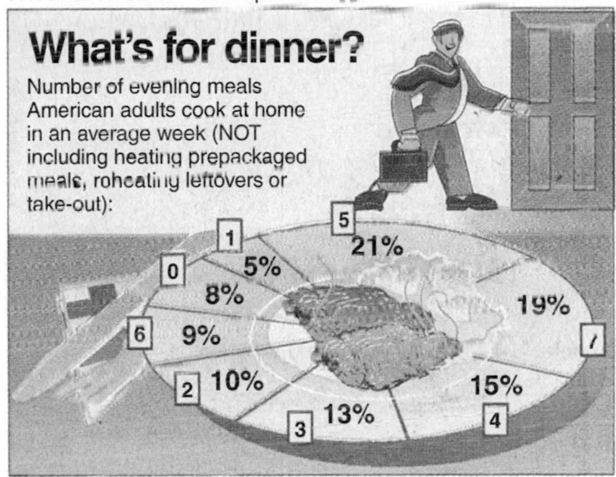
USA SNAPSHOTS®
A look at statistics that shape our lives
What's for dinner?
Number of evening meals American adults cook at home in an average week (NOT including heating prepackaged meals, reheating leftovers or take-out):
0 8%
1 5%
2 10%
3 13%
4 15%
5 21%
6 9%
7 19%

Source: Millward Brown for Whirlpool By Cindy Hall and Web Bryant, USA TODAY

CS05

Chapter 2 dealt with frequency distributions of data sets, and Chapter 4 dealt with the fundamentals of probability. Now we are ready to combine these ideas to form probability distributions, which are much like relative frequency distributions. The basic difference between probability and relative frequency distributions is that probability distributions are theoretical probabilities (populations), whereas relative frequency distributions are empirical probabilities (samples).

ANSWER NOW 5.1

a. What percentage of the families eat home cooked meals on all seven evenings? On no evenings?

b. What number of nights has the highest likelihood of occurrence?

c. What variable could be used to describe all eight of the events shown on the graph?

d. What other statistical graph could be used to picture or display this information? Draw it.

e. What other statistical methods can be used to describe the information shown on the pie chart?

After completing Chapter 5, further investigate the Chapter Case Study in the Return to Chapter Case Study section with Exercises 5.93, 5.94, and 5.95 (p. 265).

5.1 Random Variables

If each outcome of a probability **experiment** is assigned a numerical value, then as we observe the results of the experiment we are observing the values of a random variable. This numerical value is the *random variable value.*

RANDOM VARIABLE
A variable that assumes a unique numerical value for each of the outcomes in the sample space of a probability experiment.

In other words, a random variable is used to denote the outcomes of a probability experiment. The random variable can take on any numerical value that belongs to the set of all possible outcomes of the experiment. (It is called "random" because the value it assumes is the result of a chance, or random event.) Each event in a probability experiment must also be defined in such a way that only one value of the random variable is assigned to it **(mutually exclusive events),** and every event must have a value assigned to it (all-inclusive events).

The following illustrations demonstrate random variables.

Illustration 5.1

We toss five coins and observe the "number of heads" visible. The random variable x is the number of heads observed and may take on integer values from 0 to 5. ■

Illustration 5.2

Let the "number of phone calls received" per day by a company be the random variable. Integer values ranging from zero to some very large number are possible values. ■

Illustration 5.3

Let the "length of the cord" on an electrical appliance be a random variable. The random variable is a numerical value between 12 and 72 inches for most appliances. ■

Illustration 5.4

Let the "qualifying speed" for racecars trying to qualify for the Indianapolis 500 be a random variable. Depending on how fast the driver can go, the speeds are approximately 220 and faster and are measured in miles per hour (to the nearest thousandth of a mile). ■

ANSWER NOW 5.2
Survey your classmates about the number of siblings they have and the length of the last conversation they had with their mother. Identify the two random variables of interest and list their possible values.

Numerical random variables can be subdivided into two classifications: *discrete random variables* and *continuous random variables*.

FYI
Discrete and continuous variables were defined on page 15.

DISCRETE RANDOM VARIABLE
A quantitative random variable that can assume a countable number of values.

CONTINUOUS RANDOM VARIABLE
A quantitative random variable that can assume an uncountable number of values.

ANSWER NOW 5.3

a. **Explain why the variables in Illustrations 5.1 and 5.2 are discrete.**

b. **Explain why the variables in Illustrations 5.3 and 5.4 are continuous.**

Exercises

5.4
a. Are the variables in Answer Now 5.2 discrete or continuous? Why?
b. Explain why the variable "number of dinner guests for Thanksgiving" is discrete.
c. Explain why the variable "number of miles to your grandmother's house" is continuous.

5.5 A social worker is involved in a study about family structure. From census data she obtains information regarding the number of children per family for a certain community. Identify the random variable of interest, determine whether it is discrete or continuous, and list its possible values.

5.6 The staff at *Fortune* recently identified what they considered to be the 100 best companies in America to work for. Of these, the top four had the most new jobs in the past two years.

Company	New Jobs
Lowe's	10,000
Intel	11,196
FedEx	6,000
Marriott	5,936

Source: *Fortune*, "The 100 Best Companies to Work for in America," January 12, 1998

a. What is the random variable involved in this study?
b. Is the random variable discrete or continuous? Explain.

5.7 An archer shoots arrows at the bull's-eye of a target and measures the distance from the center of the target to the arrow. Identify the random variable of interest, determine whether it is discrete or continuous, and list its possible values.

5.8 A USA Snapshot®, "Are you getting a summer job?" (July 8, 2002), reported that 49% of high school students answered, "Getting? I already have one"; 26% said, "Maybe. Depends on my cash situation"; and 25% said, "No! Nothing interferes with my beach time."
a. What is the variable involved, and what are the possible values?
b. Why is this variable not a random variable?

5.9 The year 1998 may go down in baseball history as the year of the home run. On the way to the "great home-run chase," several "most homers in a month" records were broken or tied. The table lists the monthly records for home runs.

Month	Player (Team)	Home Runs	Year
April	Ken Griffey, Jr. (Seattle)	13	1997
	Luis Gonzalez (Arizona)	13	2001
May	Barry Bonds (San Francisco)	17	2001
June	Sammy Sosa (Chicago)	20	1998
July	Albert Belle (Chicago)	16	1998
	Mark McGwire (St. Louis)	16	1999
August	Rudy York (Detroit)	18	1937
September	Albert Belle (Cleveland)	17	1995
	Babe Ruth (New York)	17	1927

Source: http://www.baseball-almanac.com

a. What is the random variable involved in this study?
b. Is the random variable discrete or continuous? Explain.

5.10 Midway through 1998 three players were chasing the single-season home run records set by Babe Ruth in 1927 with 60 and Roger Maris in 1961 with 61. Did any one of the five sluggers have an unfair advantage by facing weaker pitchers? The percentage of wins and the earned run average (ERA) of opposing pitchers who gave up home runs to each player are shown in the table.

a. What are the two random variables involved in this study?

b. Are these random variables discrete or continuous? Explain.

	Opposing Pitcher Statistics	
Player (Year)	Won/Loss %	ERA
Babe Ruth (1927)	48.3	4.10
Roger Maris (1961)	48.2	4.06
Sammy Sosa (1998)	53.7	4.59
Ken Griffey Jr. (1998)	48.1	4.81
Mark McGwire (1998)	55.9	4.10

Source: *Sports Illustrated,* "Servin' Up Taters," July 6, 1998

5.2 Probability Distributions of a Discrete Random Variable

Recall the coin-tossing experiment we used at the beginning of Section 4.1. Two coins were tossed and no heads, one head, or two heads were observed. If we define the random variable x to be the number of heads observed when two coins are tossed, x can take on the value 0, 1, or 2. The probability of each of these three events is the same as we calculated in Chapter 4 (p. 197):

$$P(x = 0) = P(0\text{H}) = \frac{1}{4}$$

$$P(x = 1) = P(1\text{H}) = \frac{1}{2}$$

$$P(x = 2) = P(2\text{H}) = \frac{1}{4}$$

These probabilities can be listed in any number of ways. One of the most convenient is a table format known as a *probability distribution* (see Table 5.1).

TABLE 5.1

Probability Distribution: Tossing Two Coins

x	$P(x)$
0	0.25
1	0.50
2	0.25

FYI
Can you see why the name "probability distribution" is used?

PROBABILITY DISTRIBUTION
A distribution of the probabilities associated with each of the values of a random variable. The probability distribution is a theoretical distribution; it is used to represent populations.

ANSWER NOW 5.11
Express the tossing of one coin as a probability distribution of x, the number of heads that occur (that is, x = 0 for T and x = 1 for H).

In an experiment in which a single die is rolled and the number of dots on the top surface is observed, the random variable is the number observed. The probability distribution for this random variable is shown in Table 5.2.

FYI
The values of the random variable are mutually exclusive events.

TABLE 5.2 **Probability Distribution: Rolling a Die**

x	1	2	3	4	5	6
$P(x)$	$\frac{1}{6}$	$\frac{1}{6}$	$\frac{1}{6}$	$\frac{1}{6}$	$\frac{1}{6}$	$\frac{1}{6}$

Sometimes it is convenient to write a rule that algebraically expresses the probability of an event in terms of the value of the random variable. This expression is typically written in formula form and is called a *probability function.*

PROBABILITY FUNCTION

A rule that assigns probabilities to the values of the random variables.

A probability function can be as simple as a list that pairs the values of a random variable with their probabilities. Tables 5.1 and 5.2 show two such listings. However, a probability function is most often expressed in formula form.

Consider a die that has been modified so that it has one face with one dot, two faces with two dots, and three faces with three dots. Let x be the number of dots observed when this die is rolled. The probability distribution for this experiment is presented in Table 5.3.

TABLE 5.3

Probability Distribution: Rolling the Modified Die

x	$P(x)$
1	$\frac{1}{6}$
2	$\frac{2}{6}$
3	$\frac{3}{6}$

Each of the probabilities can be represented by the value of x divided by 6; that is, each $P(x)$ is equal to the value of x divided by 6, where $x = 1, 2,$ or 3. Thus,

$$P(x) = \frac{x}{6} \quad \text{for} \quad x = 1, 2, \text{ or } 3$$

is the formula for the probability function of this experiment.

The probability function for the experiment of rolling one ordinary die is

$$P(x) = \frac{1}{6} \quad \text{for} \quad x = 1, 2, 3, 4, 5, \text{ or } 6$$

This particular function is called a **constant function** because the value of $P(x)$ does not change as x changes.

ANSWER NOW 5.12

Express $P(x) = \frac{1}{6}$ for $x = 1, 2, 3, 4, 5,$ or 6 in distribution form.

ANSWER NOW 5.13

Explain how the various values of x in a probability distribution form a set of mutually exclusive events.

Every probability function must display the two basic properties of probability (see p. 198). These two properties are: (1) the probability assigned to each value of the random variable must be between zero and one, inclusive—that is,

FYI
These properties were presented in Chapter 4.

$$\textbf{Property 1} \qquad 0 \leq \text{Each } P(x) \leq 1$$

and (2) the sum of the probabilities assigned to all the values of the random variable must equal one—that is,

$$\textbf{Property 2} \qquad \sum_{\text{all } x} P(x) = 1$$

Illustration 5.5 Determining a Probability Function

Is $P(x) = \frac{x}{10}$ for $x = 1, 2, 3,$ or 4 a probability function?

SOLUTION

To answer this question we need only test the function in terms of the two basic properties. The probability distribution is shown in Table 5.4.

Property 1 is satisfied because 0.1, 0.2, 0.3, and 0.4 are all numerical values between zero and one. (See the ✓ showing each value was checked.) Property 2 is also satisfied because the sum of all four probabilities is exactly one. (See the ⓒⓚ showing the sum was checked.) Since both properties are satisfied, we can conclude that $P(x) = \frac{x}{10}$ for $x = 1, 2, 3,$ or 4 is a probability function. ■

TABLE 5.4

Probability Distribution for $P(x) = \frac{x}{10}$ for $x = 1, 2, 3,$ or 4

x	$P(x)$
1	$\frac{1}{10} = 0.1$ ✓
2	$\frac{2}{10} = 0.2$ ✓
3	$\frac{3}{10} = 0.3$ ✓
4	$\frac{4}{10} = 0.4$ ✓
	$\frac{10}{10} = 1.0$ ⓒⓚ

What about $P(x = 5)$ (or any value other than $x = 1, 2, 3,$ or 4) for the function $P(x) = \frac{x}{10}$ for $x = 1, 2, 3,$ or 4? $P(x = 5)$ is considered to be zero. That is, the probability function provides a probability of zero for all values of x other than the values specified as part of the domain.

ANSWER NOW 5.14
How does the property "all inclusive" relate to the probability distribution shown in Table 5.4?

FYI
The values of the random variable are all inclusive.

Probability distributions can be presented graphically. Regardless of the specific graphic representation used, the values of the random variable are plotted on the horizontal scale, and the probability associated with each value of the random variable is plotted on the vertical scale. The probability distribution of a discrete random variable could be presented by a set of line segments drawn at the values of x with lengths that represent the probability of each x. Figure 5.1 shows the probability distribution of $P(x) = \frac{x}{10}$ for $x = 1, 2, 3,$ or 4.

FIGURE 5.1
Line Representation: Probability Distribution for $P(x) = \frac{x}{10}$ for $x = 1, 2, 3,$ or 4

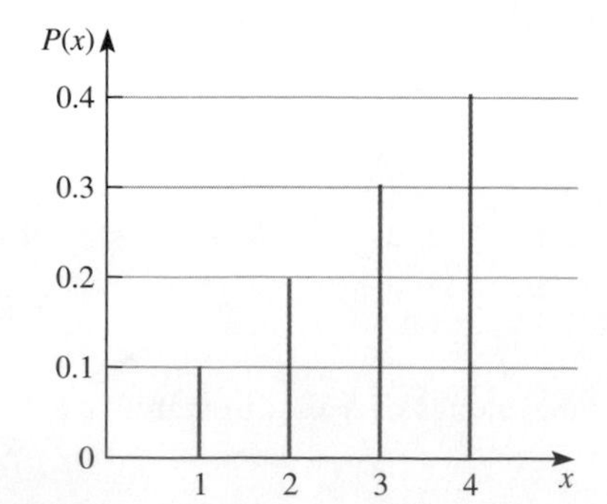

FYI
The graph in Figure 5.1 is sometimes called a "needle graph."

A regular histogram is used more frequently to present probability distributions. Figure 5.2 shows the probability distribution of Figure 5.1 as a **probability histogram.** The histogram of a probability distribution uses the physical area of each bar to represent its assigned probability. The bar for $x = 2$ is 1 unit wide (from 1.5 to 2.5) and 0.2 unit high. Therefore, its area (length $\times$ width) is $(1)(0.2) = 0.2$, the probability assigned to $x = 2$. The areas of the other bars can be determined in similar fashion. This area representation will be an important concept in Chapter 6 when we begin to work with continuous random variables.

FIGURE 5.2

Histogram: Probability Distribution for $P(x) = \frac{x}{10}$ for $x = 1, 2, 3,$ or 4

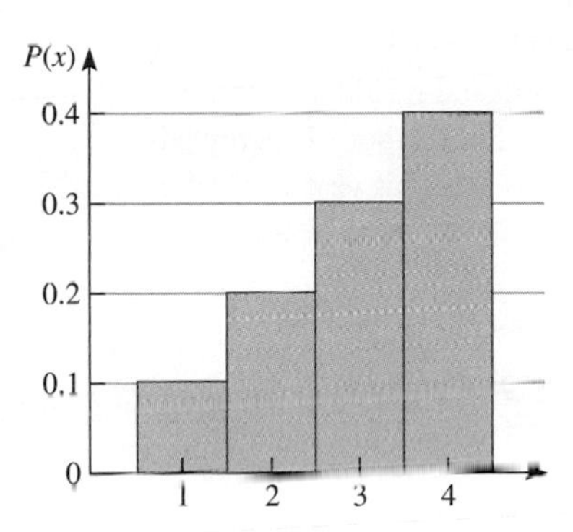

ANSWER NOW 5.15

a. **Construct a histogram of the probability distribution $P(x) = \frac{1}{6}$ for x = 1, 2, 3, 4, 5, or 6.**

b. **Describe the shape of the histogram in part a.**

Technology Instructions: Generate Random Data

MINITAB (Release 13)

Input the possible values of the random variable into C1 and the corresponding probabilities into C2; then continue with:

```
Choose:   Calc > Random Data > Discrete
Enter:    Generate: 25 (number wanted)
          Store in: C3
          Values (of x) in: C1
          Probabilities in: C2
```

Excel XP

Input the possible values of the random variable into column A and the corresponding probabilities into column B; then continue with:

```
Choose:   Tools > Data Analysis > Random Number Generation >
          OK
Enter:    Number of Variables: 1
          Number of Random Numbers: 25 (# wanted)
          Distribution: Discrete
          Value & Prob. Input Range: (A2:B5 select data
          cells, not labels)
Select:   Output Range
Enter     (C1 or select cell)
```

Application 5.1 Applying for Admission

Colleges Strive to Fill Dorms

By Mary Beth Marklein, *USA Today*

Colleges and universities will mail their last batch of admission offers in the next few days, but the process is far from over.

Now, students have until May 1 to decide where they'll go this fall. And with lingering concerns about the economy and residual fears about travel and security since September 11, many admissions officials are less able this year to predict how students will respond.

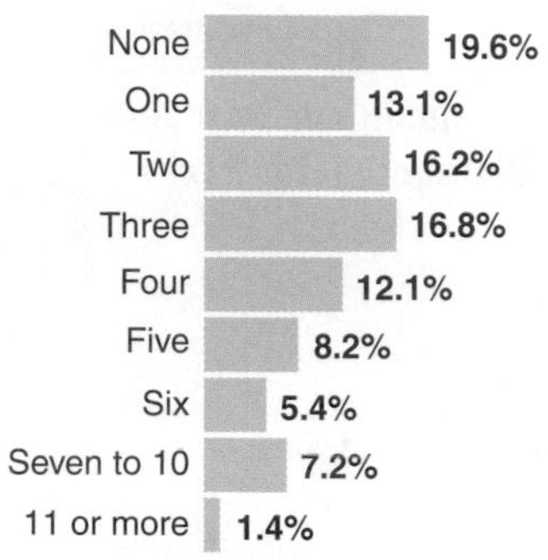
Students hedge their bets
Most students apply to more than one school, making it difficult for colleges to predict how many will actually enroll. Last fall's freshman class was asked:
To how many colleges, other than the one where you enrolled, did you apply for admission this year?
None 19.6%
One 13.1%
Two 16.2%
Three 16.8%
Four 12.1%
Five 8.2%
Six 5.4%
Seven to 10 7.2%
11 or more 1.4%
Source: The American Freshman: National Norms for Fall 2001, survey of 281,064 freshmen entering 421 four-year colleges and universities.
By Julie Snider, USA TODAY

ANSWER NOW 5.16

a. **Using the variable x, "number of additional applications for admission completed," express the information on the bar graph "Students hedge their bets" as a discrete probability distribution.**

b. **Explain how the distribution supports the article's opening statement: ". . . . but the process is far from over."**

Exercises

5.17 Census data are often used to create probability distributions for various random variables. Census data for families with a combined income of \$50,000 or more in a particular state show that 20% have no children, 30% have one child, 40% have two children, and 10% have three children. From this information, construct the probability distribution for x, where x represents the number of children per family for this income group.

5.18 Test the following function to determine whether it is a probability function. If it is not, try to make it into a probability function. $R(x) = 0.2$ for $x = 0, 1, 2, 3,$ or 4.

a. List the distribution of probabilities.
b. Sketch a histogram.

5.19 Test this function to determine whether it is a probability function:

$$P(x) = \frac{x^2 + 5}{50} \quad \text{for } x = 1, 2, 3, \text{ or } 4$$

a. List the probability distribution.
b. Sketch a histogram.

5.20 Test the function to determine whether it is a probability function. If it is not, try to make it into a probability function.

$$S(x) = \frac{6 - |x - 7|}{36} \quad \text{for } x = 2, 3, 4, 5, 6, 7, \ldots, 12$$

a. List the distribution of probabilities and sketch a histogram.
b. Do you recognize $S(x)$? If so, identify it.

5.21 Ex05-021 Commissions have often emphasized the importance of strong working relationships between audit committees and internal auditing in preventing financial reporting problems. A study was conducted to analyze the number of audit committee meetings per year that companies held with their chief internal auditor. A survey of 71 responding companies yielded the results shown here.

Meetings per Year	Percentage
0	8.5
1	11.3
2	21.1
3	5.6
4	35.2
5 or more	18.3

Source: *Accounting Horizons*, Vol. 12, No. 1, March 1998, p. 56

a. Is this a probability distribution? Explain.
b. Draw a relative frequency histogram to depict the results shown in the table.

5.22 Ex05-022 More than 44% of Americans take some form of prescription drugs daily. On January 31, 2001, the following figures were presented based on information from the American Society of Health-System Pharmacies.

Number of Daily Prescriptions	Percentage
1	16.1
2	9.8
3	6.2
4	3.5
5 or more	8.8

Source: *USA Today*, January 31, 2001

a. Is this a probability distribution? Explain.
b. What information could you add to make it a probability distribution?
c. Draw a relative frequency histogram to depict the results shown in the table plus part b.

5.23 Ex05-023 Accounting auditing results tend to vary from one occasion to the next because of inherent differences between companies, industries, and methods used when the audit is conducted. A 1998 study of 2221 audits provided a classification table of audit differences by industry.

Industry	Number of Audits	Percentage
Agriculture	104	4.7
High technology	205	9.2
Manufacturing	714	32.1
Merchandising	476	21.4
Real estate	83	3.7
Other	639	28.8

Source: *Auditing: A Journal of Practice and Theory*, Spring 1998, Vol. 17, No. 1, p. 19

a. Is this a probability distribution? Explain.
b. Draw a graph picturing the results shown in the table.

5.24 *USA Today* (July 23, 1998) presented a table depicting a profile of affluence in today's society. The statistics were derived from 17 million adults living in households with annual incomes of at least $100,000. Is this a probability distribution? Explain.

Characteristic	Percentage
Own house	75
Married	70
Age 35–54	58
Children under age 18	45
Value of home over $200,000	40

5.25

a. Use a computer* (or a random-number table) to generate a random sample of 25 observations drawn from this discrete probability distribution:

x	1	2	3	4	5
$P(x)$	0.2	0.3	0.3	0.1	0.1

Compare the resulting data to your expectations.

b. Form a relative frequency distribution of the random data.
c. Construct a probability histogram of the given distribution and a relative frequency histogram of the observed data using class midpoints of 1, 2, 3, 4, and 5.
d. Compare the observed data with the theoretical distribution. Describe your conclusions.
e. Repeat parts a–d several times with $n = 25$. Describe the variability you observe between samples.
f. Repeat parts a–d several times with $n = 250$. Describe the variability you see between samples of this much larger size.

***MINITAB (Release 13)**

a. Input the x values of the random variable into C1 and the corresponding probabilities $P(x)$ into C2; then continue with the generating random data MINITAB commands on page 239.
b. To obtain the frequency distribution, continue with:

```
Choose:    Stat > Tables > Cross Tabulation
Enter:     Classification variables: C3
Select:    Display: Total percents
```

c. To construct the histogram of the generated data in C3, continue with the histogram MINITAB commands on page 58, selecting the options: percent and midpoint with intervals 1:5. To construct a bar graph of the given distribution, continue with:

```
Choose:    Graph > Chart
Enter:     Y: C2 X: C1 > OK
```

Excel XP

a. Input the x values of the random variable into column A and the corresponding probabilities $P(x)$ into column B; then continue with the generating random data Excel commands on page 239 for an $n = 25$.

b, c. The frequency distribution is given with the histogram of the generated data. Use the histogram Excel commands on pages 58–59 using the data in column C and the bin range in column A.

To construct a histogram of the given distribution, continue with:

```
Choose:    Chart Wizard > Column > 1st pic-
           ture(usually) > Next
Enter:     Data range: (A1:B6 or select cells)
Choose:    Series > Remove (Series 1: x col-
           umn) > Next > Titles
Enter:     Chart and axes titles > Finish
           (Edit as needed)
```

5.26

a. Use a computer (or a random-number table) to generate a random sample of 100 observations drawn from the discrete probability function $P(x) = \frac{5-x}{10}$ for $x = 1, 2, 3$, or 4. List the resulting sample.

b. Form a relative frequency distribution of the random data.

c. Form a probability distribution of the expected probability distribution. Compare the resulting data to your expectations.

d. Construct a probability histogram of the given distribution and a relative frequency histogram of the observed data using class midpoints of 1, 2, 3, and 4.

e. Compare the observed data with the theoretical distribution. Describe your conclusions.

f. Repeat parts a–d several times with $n = 100$. Describe the variability you observe between samples.

FYI

Use the computer commands in Exercise 5.25; just change the arguments.

5.3 Mean and Variance of a Discrete Probability Distribution

Recall that in Chapter 2 we calculated several numerical sample statistics (mean, variance, standard deviation, and others) to describe empirical sets of data. Probability distributions may be used to represent theoretical populations, the counterpart to samples. We use **population parameters** (mean, variance, and standard deviation) to describe these probability distributions just as we use **sample statistics** to describe samples.

NOTES

1. $\bar{x}$ is the mean of the sample.
2. s^2 and s are the variance and standard deviation of the sample, respectively.
3. $\bar{x}$, s^2, and s are called *sample statistics.*
4. μ (lowercase Greek letter mu) is the mean of the population.
5. σ^2 (sigma squared) is the variance of the population.
6. σ (lowercase Greek letter sigma) is the standard deviation of the population.
7. μ, σ^2, and σ are called *population parameters.* (A parameter is a constant; μ, σ^2, and σ are typically unknown values in real statistics problems. About the only time they are known is in a textbook problem setting for the purpose of learning and understanding.)

The *mean of the probability distribution* of a discrete random variable, or the *mean of a discrete random variable,* is found in a manner somewhat similar to that used to find the mean of a frequency distribution. The mean of a discrete random variable is often referred to as its *expected value.*

MEAN OF A DISCRETE RANDOM VARIABLE (EXPECTED VALUE)
The mean, μ, of a discrete random variable x is found by multiplying each possible value of x by its own probability and then adding all the products together:

mean of x: *mu = sum of (each x multiplied by its own probability)*

$$\mu = \sum[xP(x)] \tag{5.1}$$

The variance of a discrete random variable is defined in much the same way as the variance of sample data, the mean of the squared deviations from the mean.

VARIANCE OF A DISCRETE RANDOM VARIABLE
The variance, σ^2, of a discrete random variable x is found by multiplying each possible value of the squared deviation from the mean, $(x - \mu)^2$, by its own probability and then adding all the products together:

variance: *sigma squared = sum of (squared deviation times probability)*

$$\sigma^2 = \sum[(x - \mu)^2 P(x)] \tag{5.2}$$

Formula (5.2) is often inconvenient to use; it can be reworked into the following form(s):

variance: *sigma squared = sum of (x^2 times probability) − [sum of (x times probability)]^2*

$$\sigma^2 = \sum[x^2 P(x)] - \{\sum[xP(x)]\}^2 \tag{5.3a}$$

or

$$\sigma^2 = \sum[x^2 P(x)] - \mu^2 \tag{5.3b}$$

ANSWER NOW 5.27
Verify that formulas (5.3a) and (5.3b) are equivalent to formula (5.2).

STANDARD DEVIATION OF A DISCRETE RANDOM VARIABLE
The positive square root of variance.

(5.4) standard deviation: $\sigma = \sqrt{\sigma^2}$ (5.4)

Illustration 5.6 Statistics for a Probability Function (Distribution)

Find the mean, variance, and standard deviation of the probability function

$$P(x) = \frac{x}{10} \quad \text{for } x = 1, 2, 3, \text{ or } 4$$

SOLUTION
We will find the mean using formula (5.1), the variance using formula (5.3a), and the standard deviation using formula (5.4). The most convenient way to organize the products and find the totals we need is to expand the probability distribution into an extensions table (see Table 5.5 on page 244).

TABLE 5.5 Extensions Table: Probability Distribution, $P(x) = \frac{x}{10}$ for $x = 1, 2, 3,$ or 4

x	$P(x)$	$xP(x)$	x^2	$x^2P(x)$
1	$\frac{1}{10} = 0.1$ ✓	0.1	1	0.1
2	$\frac{2}{10} = 0.2$ ✓	0.4	4	0.8
3	$\frac{3}{10} = 0.3$ ✓	0.9	9	2.7
4	$\frac{4}{10} = 0.4$ ✓	1.6	16	6.4
	$\frac{10}{10} = 1.0$ (ck)	$\sum[xP(x)] = 3.0$		$\sum[x^2P(x)] = 10.0$

Find the mean of x: The $xP(x)$ column contains each value of x multiplied by its corresponding probability, and the sum at the bottom is the value needed in formula (5.1):

$$\mu = \sum[xP(x)] = \mathbf{3.0}$$

Find the variance of x: The totals at the bottom of the $xP(x)$ and $x^2P(x)$ columns are substituted into formula (5.3a):

$$\sigma^2 = \sum[x^2P(x)] - \{\sum[xP(x)]\}^2$$
$$= 10.0 - \{3.0\}^2 = \mathbf{1.0}$$

Find the standard deviation of x: Use formula (5.4):

$$\sigma = \sqrt{\sigma^2} = \sqrt{1.0} = \mathbf{1.0}$$

■

NOTES

1. The purpose of the extensions table is to organize the process of finding the three column totals: $\sum[P(x)]$, $\sum[xP(x)]$, and $\sum[x^2P(x)]$.
2. The other columns, x and x^2, should not be totaled; they are not used.
3. $\sum[P(x)]$ will always be 1.0; use this only as a check.
4. $\sum[xP(x)]$ and $\sum[x^2P(x)]$ are used to find the mean and variance of x.

ANSWER NOW 5.28

a. **Form the probability distribution table for $P(x) = \frac{x}{6}$ for $x = 1, 2,$ or 3.**

b. **Find the extensions $xP(x)$ and $x^2P(x)$ for each x.**

c. **Find $\sum[xP(x)]$ and $\sum[x^2P(x)]$.**

d. **Find the mean for $P(x) = \frac{x}{6}$ for $x = 1, 2,$ or 3.**

e. **Find the variance for $P(x) = \frac{x}{6}$ for $x = 1, 2,$ or 3.**

f. **Find the standard deviation for $P(x) = \frac{x}{6}$ for $x = 1, 2,$ or 3.**

Illustration 5.7 Mean, Variance, and Standard Deviation of a Discrete Random Variable

A coin is tossed three times. Let the "number of heads" that occur in those three tosses be the random variable x. Find the mean, variance, and standard deviation of x.

SOLUTION

There are eight possible outcomes to this experiment: {HHH, HHT, HTH, HTT, THH, THT, TTH, TTT}. One outcome results in $x = 0$, three in $x = 1$, three in $x = 2$, and one in $x = 3$. Therefore, the probabilities for this random variable are $\frac{1}{8}$, $\frac{3}{8}$, $\frac{3}{8}$, and $\frac{1}{8}$. The probability distribution associated with this experiment is shown in Figure 5.3 and in Table 5.6. The necessary extensions and summations for the calculation of the mean, variance, and standard deviation are also shown in Table 5.6.

FIGURE 5.3
Probability Distribution: Three Coins

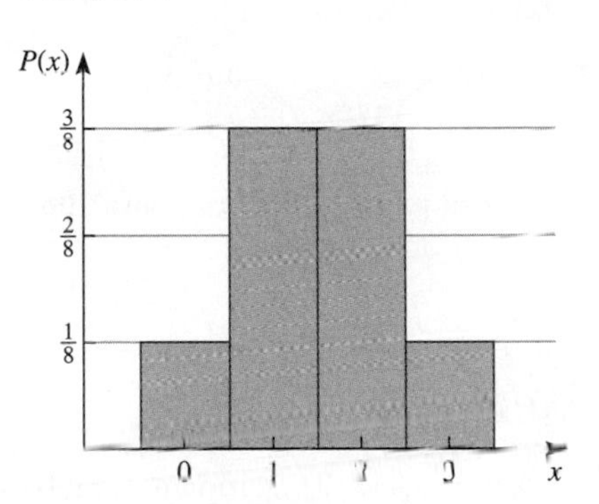

TABLE 5.6 **Extensions Table of Probability Distribution of Three Coins**

x	$P(x)$	$xP(x)$	x^2	$x^2P(x)$
0	$\frac{1}{8}$ ✓	$\frac{0}{8}$	0	$\frac{0}{8}$
1	$\frac{3}{8}$ ✓	$\frac{3}{8}$	1	$\frac{3}{8}$
2	$\frac{3}{8}$ ✓	$\frac{6}{8}$	4	$\frac{12}{8}$
3	$\frac{1}{8}$ ✓	$\frac{3}{8}$	9	$\frac{9}{8}$
	$\sum[P(x)] = \frac{8}{8} = 1.0$ (ck)	$\sum[xP(x)] = \frac{12}{8} = 1.5$		$\sum[x^2P(x)] = \frac{24}{8} = 3.0$

The mean is found using formula (5.1):

$$\mu = \sum[xP(x)] = \mathbf{1.5}$$

This result, 1.5, is the mean number of heads expected per experiment of three coins.

The variance is found using formula (5.3a):

$$\sigma^2 = \sum[x^2P(x)] - \{\sum[xP(x)]\}^2$$
$$= 3.0 - \{1.5\}^2 = 3.0 - 2.25 = \mathbf{0.75}$$

The standard deviation is found using formula (5.4):

$$\sigma = \sqrt{\sigma^2} = \sqrt{0.75} = 0.866 = \mathbf{0.87}$$

That is, 0.87 is the standard deviation expected among the number of heads observed per experiment of three coins. ■

ANSWER NOW 5.29

If you find the sum of the x and x² columns on the extensions table, exactly what have you found?

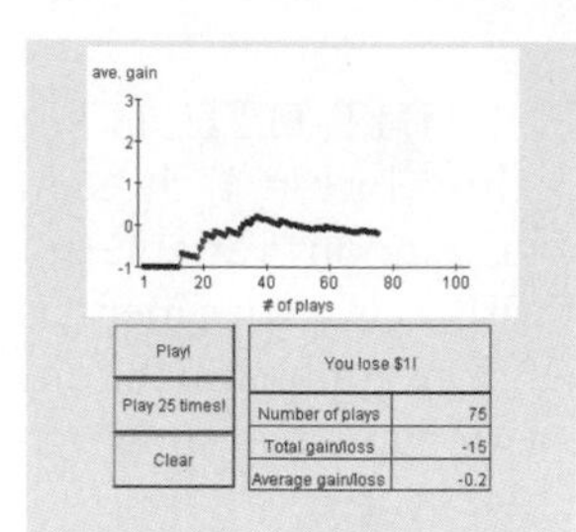

ANSWER NOW 5.30 INTERACTIVITY

Interactivity 5-A simulates playing a game where a player has a 0.2 probability of winning $3 and a 0.8 probability of losing $1. Repeat the simulations for several sets of 100 plays using the "Play 25 times" button.

a. What do you estimate for your expected value (average gain or loss) from the results?

b. Use this probability distribution to calculate the mean.

x	$P(x)$
$3	0.2
−$1	0.8

c. How do your answers to parts a and b compare? Do you consider this a fair game? Why?

Exercises

5.31 Given the probability function $P(x) = \frac{x-5}{10}$ for x = 1, 2, 3, or 4, find the mean and standard deviation.

5.32 Given the probability function $R(x) = 0.2$ for x = 0, 1, 2, 3, or 4, find the mean and standard deviation.

5.33

a. Draw a histogram of the probability distribution for the single-digit random numbers (0, 1, 2, . . . , 9).

b. Calculate the mean and standard deviation associated with the population of single-digit random numbers.

c. Represent the location of the mean on the histogram with a vertical line and the magnitude of the standard deviation with a line segment.

d. How much of this probability distribution is within two standard deviations of the mean?

5.34 The Air Transport Association of America tracks the number of fatal accidents experienced by all commercial airlines each year. The table shows the frequency distribution from 1981 to 1996.

Fatal Accidents	Frequency	Relative Frequency
1	2	0.1250
2	2	0.1250
3	2	0.1250
4	8	0.5000
6	1	0.0625
8	1	0.0625

Source: *The World Almanac and Book of Facts 1998*, p. 175

a. Build an extensions table of the probability distribution and use it to find the mean and standard deviation of the number of fatal accidents experienced annually by commercial airlines.

b. Draw the histogram of the relative frequencies.

5.35 The random variable A has the following probability distribution:

A	1	2	3	4	5
$P(A)$	0.6	0.1	0.1	0.1	0.1

a. Find the mean and standard deviation of A.

b. How much of the probability distribution is within two standard deviations of the mean?

c. What is the probability that A is between $\mu - 2\sigma$ and $\mu + 2\sigma$?

5.36 The random variable $\bar{x}$ has the following probability distribution:

$\bar{x}$	1	2	3	4	5
$P(\bar{x})$	0.6	0.1	0.1	0.1	0.1

a. Find the mean and standard deviation of $\bar{x}$.

b. What is the probability that $\bar{x}$ is between $\mu - \sigma$ and $\mu + \sigma$?

5.37 As reported in *USA Today* (June 12, 2002), the Census Bureau describes the number of vehicles per household in the United States:

Number	Percent	Number (millions)
0	10.3	10.9
1	34.2	36.1
2	38.4	40.5
3 or more	17.1	18.0

a. Replace the category "3 or more" with exactly "3," and find the mean and standard deviation of the number of vehicles per household.

b. Explain the effect that replacing the category "3 or more" with "3" had on the mean and standard deviation.

5.38 Every Tuesday, Jason's Video has "roll-the-dice" day. A customer may roll two fair dice and rent a second movie for an amount (in cents) determined by the numbers showing on the dice, the larger number first. For example, if the customer rolls a one and a five, a second movie may be rented for $0.51. Let x represent the amount paid for a second movie on roll-the-dice Tuesday.

a. Use the sample space for the rolling of a pair of dice and express the rental cost x as a probability distribution.
b. What is the expected mean rental cost (mean of x) of the second movie on roll-the-dice Tuesday?
c. What is the standard deviation of x?
d. Using a computer and the probability distribution from part a, generate a random sample of 30 values for x and determine the total cost of renting the second movies for the 30 days.
e. Using the computer, obtain an estimate for the probability that the total amount paid for the second movies will exceed $15.00 by repeating part d 500 times and using the 500 results.

5.4 The Binomial Probability Distribution

Consider the following probability experiment: Your instructor gives the class a surprise four-question multiple-choice quiz. You have not studied the material, and therefore you decide to answer the four questions by randomly guessing the answers without reading the questions or the answers.

ANSWER PAGE TO QUIZ

Directions: Circle the best answer to each question.

1. A B C
2. A B C
3. A B C
4. A B C

FYI
That's right, guess!

Circle your answers before continuing.

Before we look at the correct answers to the quiz and find out how you did, let's think about some of the things that might happen if you answer a quiz this way.

1. How many of the four questions are you likely to have answered correctly?
2. How likely are you to have more than half of the answers correct?
3. What is the probability that you selected the correct answers to all four questions?
4. What is the probability that you selected wrong answers for all four questions?
5. If an entire class answers the quiz by guessing, what do you think the class "average" number of correct answers will be?

To find the answers to these questions, let's start with a tree diagram of the sample space, showing all 16 possible ways to answer the four-question quiz. Each of the four questions is answered with the correct answer (C) or with a wrong answer (W). See Figure 5.4 on page 248.

We can convert the information on the tree diagram into a probability distribution. Let x be the "number of correct answers" on one person's quiz when the quiz was taken by randomly guessing. The random variable x may take on any one of the values 0, 1, 2, 3, or 4 for each quiz. Figure 5.4 shows 16 branches representing five different values of x. Notice that the event $x = 4$, "four correct answers," is represented by the top branch of the tree diagram, and the event $x = 0$, "zero correct answers," is shown on the bottom branch. The other events, "one correct answer," "two correct answers," and "three correct answers," are each represented

FIGURE 5.4

Tree Diagram: Possible Answers to a Four-Question Quiz

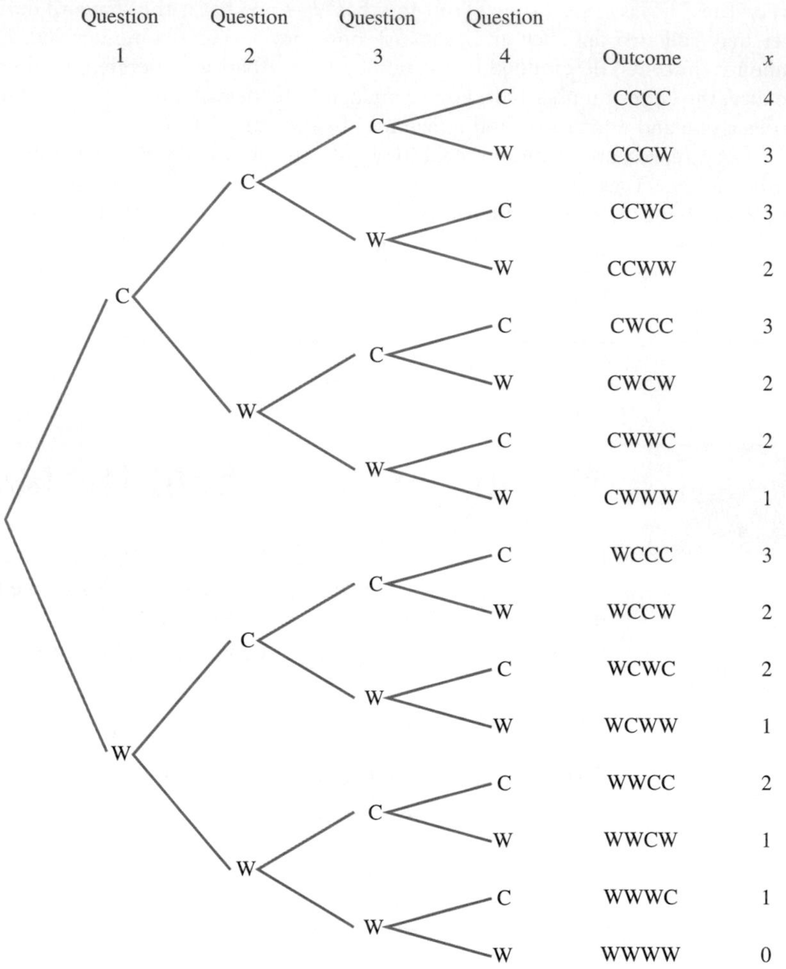

Question 1 | Question 2 | Question 3 | Question 4

Outcome	x
CCCC	4
CCCW	3
CCWC	3
CCWW	2
CWCC	3
CWCW	2
CWWC	2
CWWW	1
WCCC	3
WCCW	2
WCWC	2
WCWW	1
WWCC	2
WWCW	1
WWWC	1
WWWW	0

FYI
WWWW represents wrong on 1 and wrong on 2 and wrong on 3 and wrong on 4; therefore, its probability is found using the multiplication rule, formula (4.8b).

by several branches of the tree. We find that the event $x = 1$ occurs on four different branches, event $x = 2$ occurs on six branches, and event $x = 3$ occurs on four branches.

Since each individual question has only one correct answer among the three possible answers, the probability of selecting the correct answer to an individual question is $\frac{1}{3}$. The probability that a wrong answer is selected on each question is $\frac{2}{3}$. The probability of each value of x can be found by calculating the probabilities of all the branches and then combining the probabilities for branches that have the same x values. The calculations follow, and the resulting probability distribution appears in Table 5.7.

TABLE 5.7

Probability Distribution for the Four-Question Quiz

x	$P(x)$
0	0.198
1	0.395
2	0.296
3	0.099
4	0.012
	1.000 ck

$P(x = 0)$ is the probability that zero questions are answered correctly and four are answered wrong (there is only one branch on Figure 5.4 where all four are wrong—WWWW):

$$P(x = 0) = \frac{2}{3} \times \frac{2}{3} \times \frac{2}{3} \times \frac{2}{3} = \left(\frac{2}{3}\right)^4 = \frac{16}{81} = \mathbf{0.198}$$

NOTE Answering each individual question is a separate and independent event, thereby allowing us to use formula (4.8b) and multiply the probabilities.

ANSWER NOW 5.39
Explain why the four questions represent four independent trials.

$P(x = 1)$ is the probability that exactly one question is answered correctly and the other three are answered wrong (there are four branches on Figure 5.4 where this occurs—namely, CWWW, WCWW, WWCW, WWWC—and each has the same probability):

$$P(x = 1) = (4) \times \frac{1}{3} \times \frac{2}{3} \times \frac{2}{3} \times \frac{2}{3} = (4)\left(\frac{1}{3}\right)^1\left(\frac{2}{3}\right)^3 = \mathbf{0.395}$$

ANSWER NOW 5.40

Explain why the number 4 is a factor in $P(x = 1)$.

$P(x = 2)$ is the probability that exactly two questions are answered correctly and the other two are answered wrong (there are six branches on Figure 5.4 where this occurs—CCWW, CWCW, CWWC, WCCW, WCWC, WWCC—and each has the same probability):

$$P(x = 2) = (6) \times \frac{1}{3} \times \frac{1}{3} \times \frac{2}{3} \times \frac{2}{3} = (6)\left(\frac{1}{3}\right)^2\left(\frac{2}{3}\right)^2 = \mathbf{0.296}$$

$P(x = 3)$ is the probability that exactly three questions are answered correctly and the other one is answered wrong (there are four branches on Figure 5.4 where this occurs—CCCW, CCWC, CWCC, WCCC—and each has the same probability):

$$P(x = 3) = (4) \times \frac{1}{3} \times \frac{1}{3} \times \frac{1}{3} \times \frac{2}{3} = (4)\left(\frac{1}{3}\right)^3\left(\frac{2}{3}\right)^1 = \mathbf{0.099}$$

$P(x = 4)$ is the probability that all four questions are answered correctly (there is only one branch on Figure 5.4 where all four are correct—CCCC):

$$P(x = 4) = \frac{1}{3} \times \frac{1}{3} \times \frac{1}{3} \times \frac{1}{3} = \left(\frac{1}{3}\right)^4 = \frac{1}{81} = \mathbf{0.012}$$

Now we can answer the five questions that were asked about the four-question quiz (p. 247).

Answer 1: The most likely occurrence would be to get one answer correct; it has a probability of 0.395. Zero, one, or two correct answers are expected to result approximately 89% of the time (0.198 + 0.395 + 0.296 = 0.889).

Answer 2: To have more than half correct is represented by $x = 3$ or 4; their total probability is 0.099 + 0.012 = 0.111. (You will pass this quiz only 11% of the time by random guessing.)

Answer 3: $P(\text{all four correct}) = P(x = 4) = 0.012$ (All correct occurs only 1% of the time.)

Answer 4: $P(\text{all four wrong}) = P(x = 0) = 0.198$. (That's almost 20% of the time.)

Answer 5: The class average is expected to be $\frac{1}{3}$ of 4, or 1.33 correct answers.

The correct answers to the quiz are B, C, B, A. How many correct answers did you have? Which branch of the tree in Figure 5.4 represents your quiz results? You might ask several people to answer this same quiz by guessing the answers. Then construct an observed relative frequency distribution and compare it to the distribution shown in Table 5.7.

ANSWER NOW 5.41

In answer 5, where did $\frac{1}{3}$ and 4 come from? Why multiply them to find an expected average?

Many experiments are composed of repeated trials whose outcomes can be classified into one of two categories: **success** or **failure.** Examples of such experiments are coin tosses, right/wrong quiz answers, and other more practical experiments such as determining whether a product did or did not do its prescribed job, and whether a candidate gets elected or not. There are experiments in which the trials have many outcomes that, under the right conditions, may fit this general description of being classified in one of two categories. For example, when we roll a single die, we usually consider six possible outcomes. However, if we are interested only in knowing whether a "one" shows or not, there are really only two outcomes: the "one" shows or "something else" shows. The experiments just described are called *binomial probability experiments.*

BINOMIAL PROBABILITY EXPERIMENT

An experiment that is made up of repeated trials that possess the following properties:

1. There are n repeated identical independent trials.
2. Each trial has two possible outcomes (success, failure).
3. $P(\text{success}) = p$, $P(\text{failure}) = q$, and $p + q = 1$.
4. The binomial random variable x is the count of the number of successful trials that occur; x may take on any integer value from zero to n.

NOTES

1. Properties 1 and 2 are the two basic properties of any binomial experiment.
2. Property 3 gives the algebraic notation for each trial.
3. Property 4 concerns the algebraic notation for the complete experiment.
4. It is of utmost importance that both x and p be associated with "success."

The four-question quiz qualifies as a binomial experiment made up of four trials when all four of the answers are obtained by random guessing.

Property 1: A trial is the answering of one question, and it is repeated $n = 4$ times. The trials are independent because the probability of a correct answer on any one question is not affected by the answers on other questions.

Property 2: The two possible outcomes on each trial are success = C, correct answer, and failure = W, wrong answer.

Property 3: For each trial (each question): $p = P(\text{correct}) = \frac{1}{3}$ and $q = P(\text{wrong}) = \frac{2}{3}$. [$p + q = 1$ ⓒⓚ]

Property 4: For the total experiment (the quiz): x = number of correct answers and can be any integer value from zero to $n = 4$.

NOTE Independent trials means that the result of one trial does not affect the probability of success on any other trial in the experiment. In other words, the probability of success remains constant throughout the entire experiment.

Illustration 5.8 Demonstrating the Properties of a Binomial Probability Experiment

Consider the experiment of rolling a die 12 times and observing a "one" or "something else." At the end of all 12 rolls, the number of "ones" is reported. The random variable x is the number of times that a "one" is observed in the $n = 12$ trials. Since "one" is the outcome of concern, it is considered "success"; therefore, $p = P(\text{one}) = \frac{1}{6}$ and $q = P(\text{not one}) = \frac{5}{6}$. This experiment is binomial. ∎

Illustration 5.9

Demonstrating the Properties of a Binomial Probability Experiment

If you were an inspector on a production line in a plant where television sets are manufactured, you would be concerned with identifying the number of defective television sets. You probably would define "success" as the occurrence of a defective television. This is not what we normally think of as success, but if we count "defective" sets in a binomial experiment, we must define "success" as a "defective." The random variable x indicates the number of defective sets found per lot of n sets; $p = P(\text{television is defective})$ and $q = P(\text{television is good})$. ■

ANSWER NOW 5.42
Identify the properties that make flipping a coin 50 times and keeping track of heads a binomial experiment.

The key to working with any probability experiment is its probability distribution. All binomial probability experiments have the same properties, and therefore the same organization scheme can be used to represent all of them. The *binomial probability function* allows us to find the probability for each possible value of x.

BINOMIAL PROBABILITY FUNCTION

For a binomial experiment, let p represent the probability of a "success" and q represent the probability of a "failure" on a single trial. Then $P(x)$, the probability that there will be exactly x successes in n trials, is

$$P(x) = \binom{n}{x} (p^x) (q^{n-x}) \quad \text{for } x = 0, 1, 2, \ldots, n \tag{5.5}$$

When you look at the probability function, you notice that it is the product of three basic factors:

1. the number of ways that exactly x successes can occur in n trials, $\binom{n}{x}$,
2. the probability of exactly x successes, p^x, and
3. the probability that failure will occur on the remaining $(n - x)$ trials, q^{n-x}.

The number of ways that exactly x successes can occur in a set of n trials is represented by the symbol $\binom{n}{x}$, which must always be a positive integer. This term is called the **binomial coefficient** and is found by using the formula

$$\binom{n}{x} = \frac{n!}{x!(n-x)!} \tag{5.6}$$

NOTES

1. $n!$ ("*n factorial*") is an abbreviation for the product of the sequence of integers starting with n and ending with one. For example, $3! = 3 \cdot 2 \cdot 1 = 6$ and $5! = 5 \cdot 4 \cdot 3 \cdot 2 \cdot 1 = 120$. There is one special case, $0!$, that is defined to be 1. For more information about **factorial notation,** see the *Statistical Tutor.*
2. The values for $n!$ and $\binom{n}{x}$ can be found readily using most scientific calculators.
3. The binomial coefficient $\binom{n}{x}$ is equivalent to the number of combinations ${}_nC_x$, the symbol most likely on your calculator.
4. See the *Statistical Tutor* for general information on the binomial coefficient.

ANSWER NOW 5.43
Find the value of: a. 4! b. $\binom{4}{3}$

Let's reconsider Illustration 5.7 (pp. 244–245): A coin is tossed three times and we observe the number of heads that occur in the three tosses. This is a binomial experiment because it displays all the properties of a binomial experiment:

1. There are $\underline{n = 3}$ repeated independent trials (each coin toss is a separate trial, and the outcome of any one trial has no effect on the probability of another).
2. Each trial (each toss of the coin) has two outcomes: success = heads (what we are counting) and failure = tails.
3. The probability of success is $p = P(H) = \underline{0.5}$, and the probability of failure is $q = P(T) = \underline{0.5}$. $[p + q = 0.5 + 0.5 = 1$ ⓒⓚ$]$
4. The random variable x is the number of heads that occur in the three trials. x will assume exactly one of the values 0, 1, 2, or 3 when the experiment is complete.

The binomial probability function for the tossing of three coins is

$$P(x) = \binom{n}{x}(p^x)(q^{n-x}) = \binom{3}{x}(0.5)^x(0.5)^{n-x} \qquad \text{for } x = 0, 1, 2, \text{ or } 3$$

FYI
In Table 5.6 (p. 245), $P(1) = \frac{3}{8}$. Here, $P(1) = 0.375$ and $\frac{3}{8} = 0.375$.

Let's find the probability of $x = 1$ using the preceding binomial probability function:

$$P(x = 1) = \binom{3}{1}(0.5)^1(0.5)^2 = 3(0.5)(0.25) = \mathbf{0.375}$$

Compare this to the value found in Illustration 5.7 (p. 245).

ANSWER NOW 5.44
Use the probability function for three coin tosses and verify the probabilities for $x =$ 0, 2, and 3.

Illustration 5.10 Determining a Binomial Experiment and Its Probabilities

Consider an experiment that calls for drawing five cards, one at a time with replacement, from a well-shuffled deck of playing cards. The drawn card is identified as a spade or not a spade, it is returned to the deck, the deck is reshuffled, and so on. The random variable x is the number of spades observed in the set of five drawings. Is this a binomial experiment? Let's identify the four properties.

1. There are five repeated drawings; $\underline{n = 5}$. These individual trials are independent because the drawn card is returned to the deck and the deck is reshuffled before the next drawing.
2. Each drawing is a trial, and each drawing has two outcomes: "spade" or "not spade."
3. $p = P(\text{spade}) = \frac{13}{52}$ and $q = P(\text{not spade}) = \frac{39}{52}$. $[p + q = 1$ ⓒⓚ$]$
4. x is the number of spades recorded upon completion of the five trials; the possible values are 0, 1, 2, . . . , 5.

The binomial probability function is

$$P(x) = \binom{5}{x}\left(\frac{13}{52}\right)^x\left(\frac{39}{52}\right)^{5-x} = \binom{5}{x}\left(\frac{1}{4}\right)^x\left(\frac{3}{4}\right)^{5-x} = \binom{5}{x}(0.25)^x(0.75)^{5-x}$$

for $x = 0, 1, \ldots, 5$

$$P(0) = \binom{5}{0}(0.25)^0(0.75)^5 = (1)(1)(0.2373) = \mathbf{0.2373}$$

$$P(1) = \binom{5}{1}(0.25)^1(0.75)^4 = (5)(0.25)(0.3164) = \mathbf{0.3955}$$

$$P(2) = \binom{5}{2}(0.25)^2(0.75)^3 = (10)(0.0625)(0.421875) = \mathbf{0.2637}$$

$$P(3) = \binom{5}{3}(0.25)^3(0.75)^2 = (10)(0.015625)(0.5625) = \mathbf{0.0879}$$

The two remaining probabilities are left for you in Answer Now 5.45. ■

ANSWER NOW 5.45

a. Calculate $P(4)$ and $P(5)$.

b. Verify that the six probabilities $P(0), P(1), P(2), \ldots, P(5)$ form a probability distribution.

FYI
Answer: five

The preceding distribution of probabilities indicates that the single most likely value of x is one, the event of observing exactly one spade in a hand of five cards. What is the least likely number of spades that would be observed?

Illustration 5.11 Binomial Probability of "Bad Eggs"

The manager of Steve's Food Market guarantees that none of his cartons of a dozen eggs will contain more than one bad egg. If a carton contains more than one bad egg, he will replace the whole dozen and allow the customer to keep the original eggs. If the probability that an individual egg is bad is 0.05, what is the probability that the manager will have to replace a given carton of eggs?

SOLUTION

Assume this is a binomial experiment, let x be the number of bad eggs found in a carton of a dozen eggs, let $p = P(\text{bad}) = 0.05$, and let the inspection of each egg be a trial that results in finding a "bad" or "not bad" egg. There will be $n = 12$ trials. To find the probability that the manager will have to make good on his guarantee, we need the probability function associated with this experiment:

$$P(x) = \binom{12}{x}(0.05)^x(0.95)^{12-x} \quad \text{for } x = 0, 1, 2, \ldots, 12$$

The probability that the manager will replace a dozen eggs is the probability that $x = 2, 3, 4, \ldots, 12$. Recall that $\sum P(x) = 1$; that is,

$$\mathbf{P(0) + P(1)} + P(2) + \cdots + P(12) = 1$$

$$P(\text{replacement}) = P(2) + P(3) + \cdots + P(12) = 1 - \mathbf{[P(0) + P(1)]}$$

Finding $P(x = 0)$ and $P(x = 1)$ and subtracting their total from 1 is easier than finding all of the other probabilities. We have

$$P(x) = \binom{12}{x}(0.05)^x(0.95)^{12-x}$$

$$P(0) = \binom{12}{0}(0.05)^0(0.95)^{12} = \mathbf{0.540}$$

$$P(1) = \binom{12}{1}(0.05)^1(0.95)^{11} = \mathbf{0.341}$$

NOTE The value of many binomial probabilities for values of $n \le 15$ and common values of p are found in Table 2 of Appendix B. In this example, we have $n = 12$ and $p = 0.05$, and we want the probabilities for $x = 0$ and 1. We need to locate the section of Table 2 where $n = 12$, find the column headed $p = 0.05$, and read the numbers across from $x = 0$ and $x = 1$. We find .540 and .341, as shown in Table 5.8. (Look up these values in Table 2 in Appendix B.)

TABLE 5.8 Excerpt of Table 2 in Appendix B, Binomial Probabilities

		p													
n	*x*	0.01	0.05	0.10	0.20	0.30	0.40	0.50	0.60	0.70	0.80	0.90	0.95	0.99	*x*
	⋮														
12	0	.886	.540	.282	.069	.014	.002	0+	0+	0+	0+	0+	0+	0+	0
	1	.107	.341	.377	.206	.071	.017	.003	0+	0+	0+	0+	0+	0+	1
	2	.006	.099	.230	.283	.168	.064	.016	.002	0+	0+	0+	0+	0+	2
	3	0+	.017	.085	.236	.240	.142	.054	.012	.001	0+	0+	0+	0+	3
	4	0+	.002	.021	.133	.231	.213	.121	.042	.008	.001	0+	0+	0+	4
	⋮														

Now let's return to Steve's Food Market:

$$P(\text{replacement}) = 1 - (0.540 + 0.341) = \mathbf{0.119}$$

If $p = 0.05$ is correct, then the manager will be busy replacing cartons of eggs. If he replaces 11.9% of all the cartons of eggs he sells, he certainly will be giving away a substantial proportion of his eggs. This suggests that he should adjust his guarantee (or market better eggs). For example, if he were to replace a carton of eggs only when four or more were found to be bad, he would expect to replace only three out of 1000 cartons [1.0 − (0.540 + 0.341 + 0.099 + 0.017)], or 0.3% of the cartons sold. Notice that the manager will be able to control his "risk" (probability of replacement) if he adjusts the value of the random variable stated in his guarantee.

NOTE A convenient notation to identify the binomial probability distribution for a binomial experiment with $n = 12$ and $p = 0.30$ is $B(12, 0.30)$.

ANSWER NOW 5.46
What would the manager's "risk" be if he bought "better" eggs—say, with $P(\text{bad}) = 0.01$—and used the "more than one" guarantee? ■

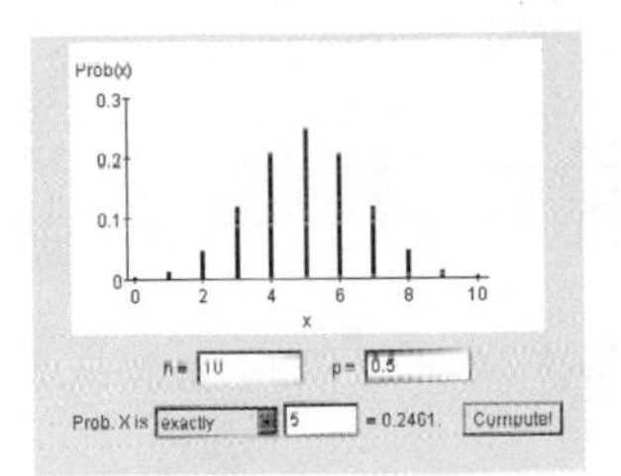

ANSWER NOW 5.47 INTERACTIVITY

Interactivity 5-B demonstrates calculating a binomial probability along with a visual interpretation. Suppose you buy 20 plants from a nursery and the nursery claims that 95% of its plants survive when planted. Inputting $n = 20$ and $p = 0.95$, compute the following:

a. **the probability that all 20 survive**

b. **the probability that at most 16 survive**

c. **the probability that at least 18 survive**

Technology Instructions: Binomial and Cumulative Binomial Probabilities

MINITAB (Release 13)

For binomial probabilities, input x values into C1; then continue with:

```
Choose:   Calc > Probability Distributions > Binomial
Select:   Probability *
Enter:    Number of trials: n
          Probability of success: p
Select:   Input column
Enter:    C1
          Optional Storage: C2 (not necessary)
Or
Select:   Input constant
Enter:    One single x value
```

*For cumulative binomial probabilities, repeat the commands above but replace the probability selection with:

```
Select:   Cumulative Probability
```

Excel XP

For binomial probabilities, input x values into column A and activate the column B cell across from the first x value; then continue with:

```
Choose:   Insert function, fx > Statistical > BINOMDIST > OK
Enter:    Number_s: (A1:A4 or select 'x value' cells)
          Trials: n
          Probability_s: p
          Cumulative: false* (gives individual probabili-
          ties) > OK
Drag:     Bottom right corner of probability value cell in
          column B down to give other probabilities
```

*For cumulative binomial probabilities, repeat the commands above but replace the false cumulative with:

```
Cumulative: true (gives cumulative probabilities) > OK
```

TI-83 Plus

To obtain a complete list of probabilities for a particular n and p, continue with:

```
Choose:   2nd > DISTR > 0:binompdf(
Enter:    n, p)
```

(continued)

Technology Instructions: Binomial and Cumulative Binomial Probabilities (Continued)

TI-83 Plus

Use the right arrow key to scroll through the probabilities.
To scroll through a vertical list in L1:

```
Choose:    STO→ > L1 > ENTER
           STAT > EDIT > 1:Edit
```

To obtain individual probabilities for a particular n, p, and x, continue with:

```
Choose:    2nd > DISTR > 0:binompdf(
Enter:     n, p, x)
```

To obtain cumulative probabilities for $x = 0$ to $x = n$ for a particular n and p, continue with:

```
Choose:    2nd > DISTR > A:binomcdf(
Enter:     n, p)* (see above for scrolling through probabili-
           ties)
```

*To obtain individual cumulative probabilities for a particular n, p, and x, repeat the commands above but replace the enter with:

```
Enter:     n, p, x)
```

ANSWER NOW 5.48

a. **Use a calculator or computer to find the probability that $x = 3$ in a binomial experiment where $n = 12$ and $p = 0.30$: $P(x = 3 \mid B(12, 0.30))$.**

b. **Use Table 5.8 to verify the answer in part a.**

Application 5.2 Living with the Law

What Is an Affirmative Action Program (AAP)?

As a condition of doing business with the federal government, federal contractors meeting certain contract and employee population levels agree to prepare, in accordance with federal regulations at 41 CFR 60-1, 60-2, etc., an Affirmative Action Program (AAP). A contractor's AAP is a combination of numerical reports, commitments of action, and description of policies. A quick overview of an AAP based on the federal regulations (41 CFR 60-2.10) is as follows:

AAPs must be developed for

- Minorities and women (41 CFR 60-1 and 60-2)
- Special disabled veterans, Vietnam-era veterans, and other covered veterans (41 CFR 60-250)
- Individuals with disabilities (41 CFR 60-741)

Source: http://eeosource.peopleclick.com/maintopic/default.asp?MainTopicID=1

The AAP regulations do not endorse the use of a specific test for determining whether the percentage of minorities or women is less than would be reasonably expected. However, several tests are commonly used. One of the tests is called the "exact binomial test" as defined next.

Exact Binomial Test

The variables used are:

T = the total number of employees in the job group
M = the number of females or minorities in the job group
A = the availability percentage of females or minorities for the job group

This test involves the calculation of a probability denoted as P and the comparison of that probability to 0.05. If P is less than or equal to 0.05, the percentage of minorities or women is considered to be "less than would be reasonably expected." The formula for calculating P is as follows:

1. Calculate the probability Q, the cumulative binomial probability for the binomial probability distribution with $n = T$, $x = M$, and $p = A/100$.
2. If Q is less than or equal to 0.5, then $P = 2Q$; otherwise, $P = Q$.

For example, $T = 50$ employees, $M = 2$ females, and $A = 6\%$ female availability. Using a computer, the value Q is found: $Q = 0.41625$. Since Q is less than 0.5, $P = 2Q = 0.8325$. And since P, 0.8325, is greater than 0.05, the percentage of women is found to be "not less than would be reasonably expected."

ANSWER NOW 5.49

a. **When using the Exact Binomial Test, what is the interpretation of the situation when the calculated value of P is less than or equal to 0.05?**

b. **When using the Exact Binomial Test, what is the interpretation of the situation when the calculated value of P is greater than 0.05?**

c. **An employer has 15 employees in a very specialized job group of which two are minorities. Based on 2000 census information, the proportion of minorities available for this type of work has 5% availability. With the binomial test, is the percentage of minorities what would be reasonably expected?**

d. **For this same employer and the same job group, there are three female employees. The female availability for this position is 50%. Does it appear that the percentage of females is what would be reasonably expected?**

Exercises

5.50 State a very practical reason why the defective item in an industrial situation is defined to be the "success" in a binomial experiment.

5.51 Evaluate each expression:

a. $4!$ b. $7!$ c. $0!$ d. $\frac{6!}{2!}$ e. $\frac{5!}{2!3!}$

f. $\frac{6!}{4!(6-4)!}$ g. $(0.3)^4$ h. $\binom{7}{3}$ i. $\binom{5}{2}$ j. $\binom{3}{0}$

k. $\binom{4}{1}(0.2)^1(0.8)^3$ l. $\binom{5}{0}(0.3)^0(0.7)^5$

5.52 Show that each of the following is true for any values of n and k. Use two specific sets of values for n and k to show that each is true.

a. $\binom{n}{0} = 1$ and $\binom{n}{n} = 1$ b. $\binom{n}{1} = n$ and $\binom{n}{n-1} = n$

c. $\binom{n}{k} = \binom{n}{n-k}$

5.53 S A carton that contains 100 T-shirts is inspected. Each T-shirt is rated "first quality" or "irregular." After all 100 T-shirts are inspected, the number of irregulars is reported as a random variable. Explain why x is a binomial random variable.

5.54 A die is rolled 20 times and the number of "fives" that occurred is reported as the random variable. Explain why x is a binomial random variable.

5.55 Four cards are selected, one at a time, from a standard deck of 52 cards. Let x represent the number of aces drawn in the set of four cards.

a. If this experiment is completed without replacement, explain why x is not a binomial random variable.
b. If this experiment is completed with replacement, explain why x is a binomial random variable.

5.56 The employees at a General Motors assembly plant are polled as they leave work. Each is asked, "What brand of automobile are you riding home in?" The random variable to be reported is the number of each brand mentioned. Is x a binomial random variable? Justify your answer.

5.57 Consider a binomial experiment made up of three trials with outcomes of success S and failure F, where $P(S) = p$ and $P(F) = q$.

a. Complete the accompanying tree diagram. Label all branches completely.

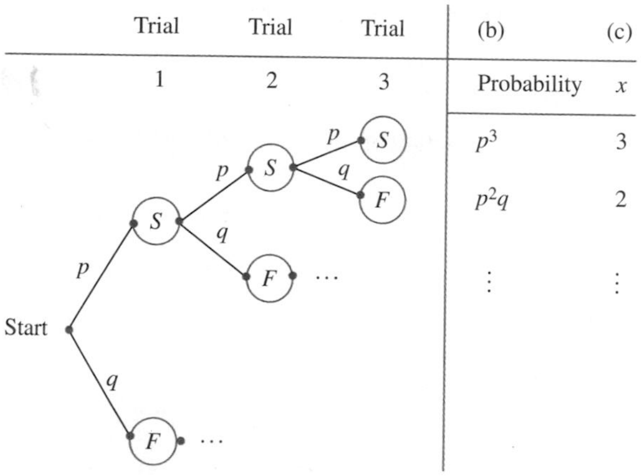

Trial 1	Trial 2	Trial 3	(b) Probability	(c) x
		p S	p^3	3
	p S	q F	p^2q	2
p S	q F …		⋮	⋮
q F …				

Start

b. In column (b) of the tree diagram, express the probability of each outcome represented by the branches as a product of powers of p and q.
c. Let x be the random variable, the number of successes observed. In column (c), identify the value of x for each branch of the tree diagram.
d. Notice that all the products in column (b) are made up of three factors and that the value of the random variable is the same as the exponent for the number p.
e. Write the equation for the binomial probability function for this situation.

5.58 Draw a tree diagram of a binomial experiment with four trials.

5.59 If x is a binomial random variable, calculate the probability of x for each case:

a. $n = 4, x = 1, p = 0.3$ b. $n = 3, x = 2, p = 0.8$
c. $n = 2, x = 0, p = \frac{1}{4}$ d. $n = 5, x = 2, p = \frac{1}{3}$
e. $n = 4, x = 2, p = 0.5$ f. $n = 3, x = 3, p = \frac{1}{6}$

5.60 If x is a binomial random variable, use Table 2 in Appendix B to determine the probability of x for each case:

a. $n = 10, x = 8, p = 0.3$ b. $n = 8, x = 7, p = 0.95$
c. $n = 15, x = 3, p = 0.05$ d. $n = 12, x = 12, p = 0.99$
e. $n = 9, x = 0, p = 0.5$ f. $n = 6, x = 1, p = 0.01$
g. Explain the meaning of the symbol 0+ that appears in Table 2.

5.61 Test the following function to determine whether or not it is a binomial probability function. List the distribution of the probabilities and sketch a histogram.

$$T(x) = \binom{5}{x}\left(\frac{1}{2}\right)^x\left(\frac{1}{2}\right)^{5-x} \quad \text{for } x = 0, 1, 2, 3, 4, \text{ or } 5$$

5.62 Let x be a random variable with this probability distribution:

x	0	1	2	3
$P(x)$	0.4	0.3	0.2	0.1

Does x have a binomial distribution? Justify your answer.

5.63 Ninety percent of the trees planted by a landscaping firm survive. What is the probability that eight or more of the ten trees just planted will survive? (Find the answer by using a table.)

5.64 Results from the 2000 census show that 42% of U.S. grandparents are the primary caregivers for their grandchildren (*Democrat & Chronicle*, "Grandparents as ma and pa," July 8, 2002). In a group of 20 grandparents, what is the probability that exactly half are primary caregivers for their grandchildren?

5.65 Nearly half of the nation's 10 million teens between the ages of 15 and 17 have jobs at some time during the year; 67% of adults say it's a good idea for teens to hold a part-time job while in school, even if they don't need the money. (Yankelovich Partners for Lutheran Brotherhood, U.S. Department of Labor, June 20, 2001)

a. What is the probability that exactly two of the next three randomly selected adults say it's a good idea for teens to hold a part-time job while in school, even if they don't need the money?
b. What is the probability that exactly 8 of the next 12 randomly selected adults say it's a good idea for teens to hold a part-time job while in school, even if they don't need the money?
c. What is the probability that exactly 20 of the next 30 randomly selected adults say it's a good idea for teens to hold a part-time job while in school, even if they don't need the money?

5.66 e-Marketer is a leading provider of Internet and e-business statistics. Based on research, they predict that in 2003 about 30% of the people who use the Internet will be 18 to 34 years old. A group of ten Internet users are randomly selected.
a. What is the probability that exactly three users will be in the 18–34 age group?
b. What is the probability that at least five will be in the 18–34 age group?

5.67 In the biathlon event of the Olympic games, a participant skis cross-country and on intermittent occasions stops at a rifle range and shoots a set of five shots. If the center of the target is hit, no penalty points are assessed. If a particular man has a history of hitting the center of the target with 90% of his shots, what is the probability that he will hit the center of the target with
a. all five of his next set of five shots
b. at least four of his next set of five shots (Assume independence.)

5.68 The survival rate during a risky operation for patients with no other hope of survival is 80%. What is the probability that exactly four of the next five patients survive this operation?

5.69 A machine produces parts of which 0.5% are defective. If a random sample of ten parts produced by this machine contains two or more defectives, the machine is shut down for repairs. Find the probability that the machine will be shut down for repairs based on this sampling plan.

5.70 A May 22, 2002, study by Progressive Insurance showed that nearly 20% of car renters always buy rental-car insurance (http://biz.yahoo.com/prnews/020522/nyw054_2.html). A group of ten individuals are renting cars.
a. What is the probability that none of the ten always buys rental-car insurance?
b. What is the probability that exactly three always buy rental-car insurance?
c. What is the probability that at least four always buy rental-car insurance?
d. What is the probability that no more than two always buy rental-car insurance?

5.71 If boys and girls are equally likely to be born, what is the probability that in a randomly selected family of six children, there will be boys? (Find the answer using a formula.)

5.72 One-fourth of a certain breed of rabbits are born with long hair. What is the probability that in a litter of six rabbits, exactly three will have long hair? (Find the answer by using a formula.)

5.73 San Francisco baseball player Barry Bonds' league-leading batting average (ratio of hits to at-bats) reached 0.373 after 378 times at bat during the 2001 season. Suppose Bonds has five official times at bat during his next game. Assuming no extenuating circumstances and that the binomial model will produce reasonable approximations, what is the probability that Bonds:
a. gets less than two hits
b. gets more than three hits
c. goes five-for-five (all hits)

5.74 As a quality-control inspector for toy trucks, you have observed that wooden wheels are bored off-center about 3% of the time. If six wooden wheels are used on each toy truck produced, what is the probability that a randomly selected set of wheels has no off-center wheels?

5.75 According to the USA Snapshot® "Knowing drug addicts," 45% of Americans know somebody who became addicted to a drug other than alcohol. If we assume this to be true, what are the probabilities of these events?
a. Exactly three of a random sample of five know someone who became addicted. Calculate the value.
b. Exactly 7 of a random sample of 15 know someone who became addicted. Estimate using Table 2 in Appendix B.
c. At least 7 of a random sample of 15 know someone who became addicted. Estimate using Table 2.
d. No more than 7 of a random sample of 15 know someone who became addicted. Estimate using Table 2.

5.76 It is found that 48% of all mortgage foreclosures in this country are caused by disability. When employees get injured or ill, they can't work and then they lose their job and thus their income. With no income, they can't make their mortgage payment and the bank forecloses (http://www.ricedelman.com, June 11, 2002). Twenty mortgage foreclosures are audited by a large lending institution.
a. Find the probability that five or fewer of the foreclosures are due to a disability.
b. Find the probability that at least three are due to a disability.

5.77 Use a computer to find the probabilities for all possible x values for a binomial experiment where $n = 30$ and $p = 0.35$.

MINITAB (Release 13)

Choose:	`Calc > Make Patterned Data > Simple Set of Numbers`

Enter:	`Store patterned data in: C1`
	`From first value: 0`
	`To last value: 30`
	`In steps of: 1 > OK`

Continue with the binomial probability MINITAB commands on page 255, using $n = 30$, $p = 0.35$, and C2 for optional storage.

(continued)

Excel XP

Enter: 0,1,2, ... ,30 into column A

Continue with the binomial probability Excel commands on page 255, using $n = 30$ and $p = 0.35$.

TI-83 Plus

Use the binomial probability TI-83 commands on pages 255–256, with $n = 30$ and $p = 0.35$.

5.78 Use a computer to find the cumulative probabilities for all possible x values for a binomial experiment where $n = 45$ and $p = 0.125$.

a. Explain why so many 1.000's are listed.
b. Explain what each number listed represents.

MINITAB (Release 13)

Choose: Calc > Make Patterned Data > Simple Set of Numbers
Enter: Store patterned data in: C1
From first value: 0
To last value: 45
In steps of: 1 > OK

Continue with the cumulative binomial probability MINITAB commands on pages 255, using $n = 45$, $p = 0.125$, and C2 as optional storage.

Excel XP

Enter: 0, 1, 2, ... , 45 into column A

Continue with the cumulative binomial probability Excel commands on page 255, using $n = 45$ and $p = 0.125$.

TI-83 Plus

Use the cumulative binomial probability TI-83 commands on page 256, with $n = 45$ and $p = 0.125$.

5.79 Results of a 1997 study conducted by Scarborough Research and published in *Fortune* ("Where Lotto Is King," January 12, 1998) showed that San Antonio leads all major cities in lottery participation. When adult respondents in San Antonio were asked whether they had purchased a lottery ticket within the last seven days, 50% said yes. In contrast, only 35% answered yes to the same question in Salt Lake City. (*Note:* The state of Utah has no lottery, so more travel is required.) Suppose you randomly select 25 adults in San Antonio and 25 adults in Salt Lake City.

a. What is the probability that the San Antonio sample contains ten or fewer lottery players?
b. What is the probability that the San Antonio sample and the Salt Lake City sample both contain 13 or fewer lottery players?

5.80 An article titled "Mom, I Want to Live with My Boyfriend" appeared in *Reader's Digest* (February 1994). The article quoted a Columbia University study that found that only 19% of the men who lived with their girlfriends eventually walked down the aisle with them. Suppose 25 men are interviewed who have lived with a girlfriend in the past. What is the probability that five or fewer of them married the girlfriend?

5.81 In Game 7 on the road in the 2002 NBA playoffs, the two-time defending champions Los Angeles Lakers did what they do best—thrived when the pressure was at its highest. Both of the Lakers' star players had their chance at the foul line late in overtime.

a. With 1:27 minutes left in overtime and the game tied at 106–106, Shaq was at the line for two free-throw attempts. During this game, prior to these shots, he had made 9 of his 13 attempts, and he has a history of making 0.555 of his free-throw attempts. Justify the statement, "The law of averages is working against him."
b. With 0:06 seconds left in overtime and the game score standing at 110–106, Kobe was at the line for two free throws. During this game, prior to these shots, he had made 6 of his 8 attempts, and he has a history of making 0.829 of his free throws. Justify the statement, "The law of averages is working for him."

Both players made both shots, and the series with the Sacramento Kings was over.

5.82 If the binomial $(q + p)$ is squared, the result is $(q + p)^2 = q^2 + 2qp + p^2$. For the binomial experiment with $n = 2$, the probability of no successes in two trials is q^2 (the first term in the expansion), the probability of one success in two trials is $2qp$ (the second term in the expansion), and the probability of two successes in two trials is p^2 (the third term). Find $(q + p)^3$ and compare its terms to the binomial probabilities for $n = 3$ trials.

5.5 Mean and Standard Deviation of the Binomial Distribution

The mean and standard deviation of a theoretical binomial probability distribution can be found by using these two formulas:

$$\mu = np \tag{5.7}$$

and

$$\sigma = \sqrt{npq} \tag{5.8}$$

The formula for the mean, μ, seems appropriate: the number of trials multiplied by the probability of "success." [Recall that the mean number of correct answers on the binomial quiz (Answer 5, p. 249) was expected to be $\frac{1}{3}$ of 4, $4(\frac{1}{3})$, or np.] The formula for the standard deviation, σ, is not as easily understood. Thus, at this point it is appropriate to look at an example, which demonstrates that formulas (5.7) and (5.8) yield the same results as formulas (5.1), (5.3a), and (5.4).

In Illustration 5.7 (pp. 244–245), x is the number of heads in three coin tosses, $n = 3$, and $p = \frac{1}{2} = 0.5$. Using formula (5.7), we find the mean of x to be

$$\mu = np = (3)(0.5) = \mathbf{1.5}$$

Using formula (5.8), we find the standard deviation of x to be

$$\sigma = \sqrt{npq} = \sqrt{(3)(0.5)(0.5)} = \sqrt{0.75} = 0.866 = \mathbf{0.87}$$

ANSWER NOW 5.83
Find the mean and standard deviation for the binomial random variable x with $n = 30$ and $p = 0.6$.

Now look back at the solution for Illustration 5.7 (p. 245). Note that the results are the same, regardless of the formula you use. However, formulas (5.7) and (5.8) are much easier to use when x is a binomial random variable.

Illustration 5.12 Calculating the Mean and Standard Deviation of a Binomial Distribution

Find the mean and standard deviation of the binomial distribution when $n = 20$ and $p = \frac{1}{5}$ (or 0.2, in decimal form). Recall that the "binomial distribution where $n = 20$ and $p = 0.2$" has the probability function

$$P(x) = \binom{20}{x}(0.2)^x(0.8)^{20-x} \quad \text{for } x = 0, 1, 2, \ldots, 20$$

and a corresponding distribution with 21 x values and 21 probabilities, as shown in the distribution chart, Table 5.9, and on the histogram in Figure 5.5 on page 262.

Let's find the mean and the standard deviation of this distribution of x using formulas (5.7) and (5.8):

$$\mu = np = (20)(0.2) = \mathbf{4.0}$$

$$\sigma = \sqrt{npq} = \sqrt{(20)(0.2)(0.8)} = \sqrt{3.2} = \mathbf{1.79}$$

TABLE 5.9

Binomial Distribution: $n = 20, p = 0.2$

x	$P(x)$
0	0.012
1	0.058
2	0.137
3	0.205
4	0.218
5	0.175
6	0.109
7	0.055
8	0.022
9	0.007
10	0.002
11	0+
12	0+
13	0+
⋮	⋮
20	0+

FIGURE 5.5
Histogram of Binomial Distribution $B(20, 0.2)$

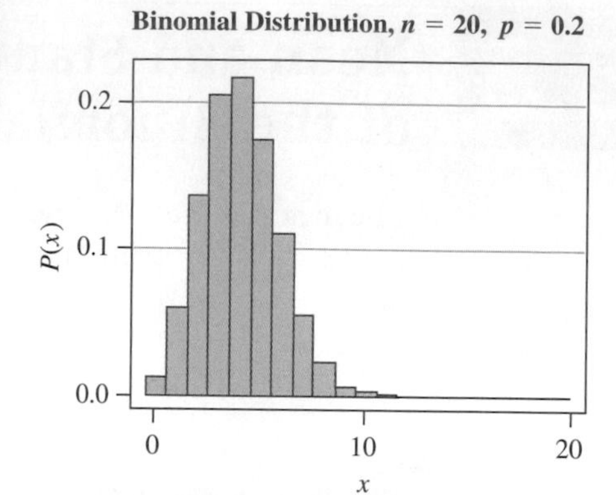

FIGURE 5.6
Histogram of Binomial Distribution $B(20, 0.2)$

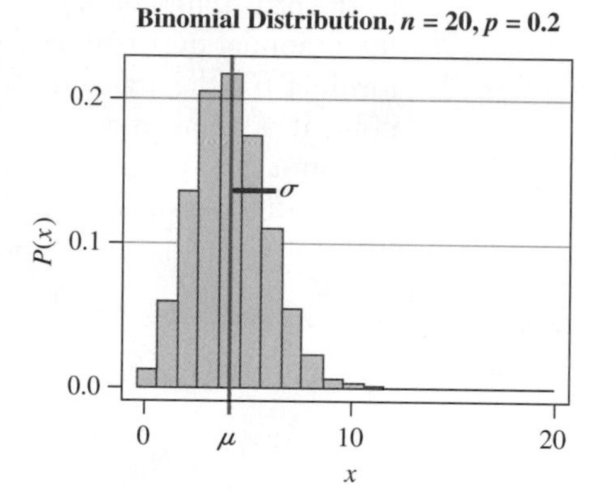

Figure 5.6 shows the location of the mean (vertical blue line) and the size of the standard deviation (horizontal red line segment) relative to the probability distribution of the variable x. ■

ANSWER NOW 5.84
Consider the binomial distribution where $n = 11$ and $p = 0.05$.

a. **Find the mean and standard deviation using formulas (5.7) and (5.8).**
b. **Using Table 2 in Appendix B, list the probability distribution and draw a histogram.**
c. **Locate μ and σ on the histogram.**

Exercises

5.85 Consider the binomial distribution where $n = 11$ and $p = 0.05$ (see Answer Now 5.84).

a. From the distribution (Answer Now 5.84b or Table 2) find the mean and standard deviation using formulas (5.1), (5.3a), and (5.4).
b. Compare the results of part a to the answers found in Answer Now 5.84a.

5.86 Consider this binomial probability function:

$$P(x) = \binom{5}{x}\left(\frac{1}{2}\right)^x\left(\frac{1}{2}\right)^{5-x} \quad \text{for } x = 0, 1, 2, 3, 4, 5$$

a. Calculate the mean and standard deviation of the random variable by using formulas (5.1), (5.3a), and (5.4).
b. Calculate the mean and standard deviation using formulas (5.7) and (5.8).
c. Compare the results of parts a and b.

5.87 Find the mean and standard deviation of x for each of the following binomial random variables:

a. the number of tails seen in 50 tosses of a quarter
b. the number of aces seen in 100 draws from a well-shuffled bridge deck (with replacement)
c. the number of cars found to have unsafe tires among the 400 cars stopped at a roadblock for inspection (Assume that 6% of all cars have one or more unsafe tires.)
d. the number of melon seeds that germinate when a package of 50 seeds is planted (The package states that the probability of germination is 0.88.)

5.88 Find the mean and standard deviation for each of the following binomial random variables:

a. the number of sixes seen in 50 rolls of a die
b. the number of defective television sets in a shipment of 125 (The manufacturer claimed that 98% of the sets were operative.)
c. the number of operative television sets in a shipment of 125 (The manufacturer claimed that 98% of the sets were operative.)
d. How are parts b and c related? Explain.

5.89 A binomial random variable has a mean of 200 and a standard deviation of 10. Find the values of n and p.

5.90 The probability of success on a single trial of a binomial experiment is known to be $\frac{1}{4}$. The random variable x, number of successes, has a mean value of 80. Find the number of trials involved in this experiment and the standard deviation of x.

5.91 A binomial random variable x is based on 15 trials with the probability of success equal to 0.3. Find the probability that this variable will take on a value more than two standard deviations from the mean.

5.92 Imprints Galore buys T-shirts from a manufacturer with the guarantee that the shirts have been inspected and that no more than 1% are imperfect in any way. The shirts arrive in boxes of 12. Let x be the number of imperfect shirts found in any one box.

a. List the probability distribution and draw the histogram of x.
b. What is the probability that any one box has no imperfect shirts?
c. What is the probability that any one box has no more than one imperfect shirt?
d. Find the mean and standard deviation of x.
e. What proportion of the distribution is between $\mu - \sigma$ and $\mu + \sigma$?
f. What proportion of the distribution is between $\mu - 2\sigma$ and $\mu + 2\sigma$?
g. How does this information relate to the empirical rule and Chebyshev's theorem? Explain.
h. Use a computer to simulate Imprints Galore buying 200 boxes of shirts and observing x, the number of imperfect shirts per box of 12. Describe how the information from the simulation compares to what was expected (answers a–g describe the expected results).
i. Repeat part h several times. Describe how these results compare to those of parts a–g and h.

MINITAB Release 13

```
a. Choose:   Calc > Make Patterned Data > Simple
             Set of Numbers
   Enter:    Store patterned data in: C1
             From first value: -1 (see note)
             To last value: 12
             In steps of: 1 > OK
```

Continue with the binomial probability MINITAB commands on page 255, using $n = 12$, $p = 0.01$, and C2 for optional storage.

```
   Choose:   Graph > Plot
   Enter:    Graph variable: Y: C2 X: C1
             Data display: Display: Area
   Choose:   Edit Attributes
   Enter:    Graph: Fill Type: None
   Select:   Connection function: Step > OK
```

c. Continue with the cumulative binomial probability MINITAB commands on page 255, using $n = 12$, $p = 0.01$, and C3 for optional storage.

```
h. Choose:   Calc > Random Data > Binomial
   Enter:    Generate: 200 rows of data
             Store in column C4
             Number of trials: 12
             Probability: .01 > OK
   Choose:   Stat > Tables > Cross Tabulation
   Enter:    Classification variables: C4
   Select:   Display: Total percents  > OK
   Choose:   Calc > Column Statistics
   Select:   Statistic: Mean
   Enter:    Input variable: C4          > OK
   Choose:   Calc > Column Statistics
   Select:   Statistic: Standard deviation
   Enter:    Input variable: C4 > OK
```

Continue with the histogram MINITAB commands on page 58, using the data in C4 and selecting the options: percent and midpoint with intervals 0:12/1.

NOTE The binomial variable x cannot take on the value -1. The use of -1 (the next would-be class midpoint to left of zero) allows MINITAB to draw the histogram of a probability distribution. Without -1, PLOT will draw only half of the bar representing $x = 0$.

Excel XP

```
a. Enter:    0, 1, 2, ... , 12 into column A
```

Continue with the binomial probability Excel commands on page 255, using $n = 12$ and $p = 0.01$.

(continued)

Excel XP (continued)

Activate columns A and B, then continue with:

Choose:	Chart Wizard > Column > 1st picture(usually) > Next > Series
Choose:	Series 1 > Remove
Enter:	Category (*x*)axis labels: (A1:A13 or select 'x value' cells)
Choose:	Next > Finish
Click on:	Anywhere clear on the chart —use handles to size so *x* values fall under corresponding bars

c. Continue with the cumulative binomial probability Excel commands on page 255, using $n = 12$, $p = 0.01$, and column C for the activated cell.

h.

Choose:	Tools > Data analysis > Random Number Generation > OK
Enter:	Number of Variables: 1
	Number of Random Numbers: 200
	Distribution: Binomial
	p Value = 0.01
	Number of Trials = 12
Select:	Output Options: Output Range
Enter	(D1 or select cell) > OK

Activate the E1 cell, then:

Choose:	Insert function, f_x > Statistical > AVERAGE > OK
Enter:	Number 1: D1:D200 > OK

Activate the E2 cell, then:

Choose:	Insert function, f_x > Statistical > STDEV > OK
Enter:	Number 1: D1:D200 > OK

Continue with the histogram Excel commands on pages 58–59, using the data in column D and the bin range in column A.

TI-83 Plus

a.

Choose:	STAT > EDIT > 1:Edit
Enter:	L1: 0,1,2,3,4,5,6,7,8,9,10,11,12
Choose:	2nd QUIT > 2nd DISTR > 0:binompdf(
Enter:	12, 0.01) > ENTER
Choose:	STO→ > L2 > ENTER
Choose:	2nd > STAT PLOT > 1:Plot1

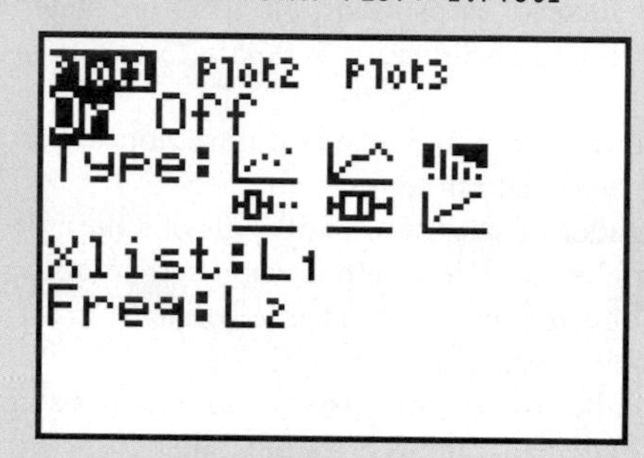

Choose:	WINDOW
Enter:	0, 13, 1, -.1, .9, .1, 1
Choose:	TRACE > > >

c.

Choose:	2nd > DISTR > A:binomcdf(
Enter:	12, 0.01)
Choose:	STO→ > L3 > ENTER
	STAT > EDIT > 1:Edit

h.

Choose:	MATH > PRB > 7:randBin(
Enter:	12, .01, 200) (takes a while to process)
Choose:	STO→ > L4 > ENTER
Choose:	2nd LIST > Math > 3:mean(
Enter:	L4
Choose:	2nd LIST > Math > 7:StdDev(
Enter:	L4

Continue with the histogram TI-83 commands on pages 59–60, using the data in column L4 and adjusting the window after the initial look using ZoomStat.

CHAPTER SUMMARY

RETURN TO CHAPTER CASE STUDY

© David Hanover/Getty Images

Family Values and Family Togetherness

Let's take a second look at the questions asked in the Chapter Case Study (p. 233) and test our knowledge of the material presented in this chapter.

FIRST THOUGHTS 5.93

a. What percentage of the families eat home-cooked meals on all seven evenings?
b. What is the percentage for no home-cooked meals?
c. What number of nights has the highest likelihood of occurrence?
d. What variable could be used to describe all eight of the events shown on the graph?
e. What characteristics of a circle graph make it appropriate for use with a probability distribution? Be specific.
f. What other statistical graph could be used to picture (display) the information shown on the circle (pie) chart? Construct it.
g. What other statistical methods can be used to describe the information shown on the pie (circle) chart?

PUTTING CHAPTER 5 TO WORK 5.94

a. Is the variable in First Thoughts 5.93 part d discrete or continuous? Why?
b. Did you draw a histogram in First Thoughts 5.93 part f? If not, draw a histogram. Describe the histogram. Is it a normal distribution? Explain.
c. Express the information on the circle graph as a probability distribution.
d. Assuming the information on the circle graph represents the population, find the mean and the standard deviation of the variable described in part c.
e. Locate the mean and standard deviation found in part d on the histogram drawn in part b.
f. Do the empirical rule and Chebyshev's theorem apply? Justify your answer.

YOUR OWN MINI-STUDY 5.95

Design your own study of family dinner routines.

a. Define a specific population that you will sample, describe your sampling plan, and collect your data.
b. Express your sample as a relative frequency distribution and draw a histogram.
c. Express your sample as a frequency distribution and find the sample mean and sample standard deviation.
d. Discuss the differences and similarities between your sample and the distribution shown in the USA Snapshot® "What's for dinner?"

In Retrospect

In this chapter we combined concepts of probability with some of the ideas presented in Chapter 2. We now are able to deal with distributions of probability values and to find means, standard deviations, and other statistics.

In Chapter 4 we explored the concepts of mutually exclusive events and independent events. We used the addition and multiplication rules on several occasions in this chapter, but very little was said about mutual exclusiveness or independence. Recall that every time we add probabilities, as we did in each of the probability distributions, we need to know that the associated events are mutually exclusive. If you look back over the chapter, you will notice that the random variable actually requires events to be mutually exclusive; therefore, no real emphasis was placed on this concept. The same basic comment can be made in reference to the multiplication of probabilities and the concept of independent events. Throughout this chapter, probabilities were multiplied and occasionally independence was mentioned. Independence, of course, is necessary to be able to multiply probabilities.

Now, after completing Chapter 5, if we were to take a close look at some of the sets of data in Chapter 2, we would see that several problems could be reorganized to form probability distributions. Here are some examples: (1) Let x be the number of credit hours for which a student is registered this semester, paired with the percentage of the entire student body reported for each value of x. (2) Let x be the number of correct passageways through which an experimental laboratory animal passes before taking a wrong one, paired with the probability of each x value. (3) Let x be the number of college applications made other than the one where you enrolled (Application 5.1), paired with the probability of each x value. The list of examples is endless.

We are now ready to extend these concepts to continuous random variables in Chapter 6.

Chapter Exercises

5.96 What are the two basic properties of every probability distribution?

5.97

a. Explain the difference and the relationship between a probability distribution and a probability function.
b. Explain the difference and the relationship between a probability distribution and a frequency distribution, and explain how they relate to a population and a sample.

5.98 Verify whether or not each of the following is a probability function. State your conclusion and explain.

a. $f(x) = \dfrac{\frac{3}{4}}{x!(3-x)!}$ for $x = 0, 1, 2, 3$

b. $f(x) = 0.25$ for $x = 9, 10, 11, 12$

c. $f(x) = \dfrac{3-x}{2}$ for $x = 1, 2, 3, 4$

d. $f(x) = \dfrac{x^2+x+1}{25}$ for $x = 0, 1, 2, 3$

5.99 The number of ships to arrive at a harbor on any given day is a random variable represented by x. The probability distribution for x is

x	10	11	12	13	14
P(x)	0.4	0.2	0.2	0.1	0.1

Find the probability that on a given day:

a. exactly 14 ships arrive
b. at least 12 ships arrive
c. at most 11 ships arrive

5.100 Did you ever wonder how many times buyers see an infomercial before they purchase its product or service? The USA Snapshot® "Television's hard sell" (October 21, 1994) answers that question. These data are from the National Infomercial Marketing Association:

Times Watched Before Buy	1	2	3	4	5 or more
Proportion of Buyers	0.27	0.31	0.18	0.09	0.15

a. What is the probability that a buyer watched only once before buying?
b. What is the probability that a viewer watching for the first time will buy?
c. What percentage of the buyers watched the infomercial three or more times before purchasing?
d. Is this a binomial probability experiment?
e. Let x be the number of times a buyer watched before making a purchase. Is this a probability distribution?
f. Assign $x = 5$ for "5 or more," and find the mean and standard deviation of x.

5.101 A doctor knows from experience that 10% of the patients to whom he gives a certain drug will have undesirable side effects. Find the probabilities that among the ten patients to whom he gives the drug:

a. at most two will have undesirable side effects
b. at least two will have undesirable side effects

5.102 In a recent survey of women, 90% admitted that they had never looked at a copy of *Vogue* magazine. Assuming that this is accurate information, what is the probability that in a random sample of three women, fewer than two will have looked at the magazine?

5.103 Seventy percent of those applying for a driver's license admitted that they would not report someone if he or she copied some answers during the written exam. You have just entered the room and see ten people waiting to take the written exam. What is the probability that, if the incident happened, five of the ten would not report what they saw?

5.104 The engines on an airliner operate independently. The probability that an individual engine operates for a given trip is 0.95. A plane will be able to complete a trip successfully if at least half of its engines operate for the entire trip. Determine whether a four-engine or a two-engine plane has the higher probability of a successful trip.

5.105 e-Marketer conducted a survey of Internet shoppers aged 35 to 54. Of this group, 10% said that they shop online more now than they did last year. A group of 15 Internet shoppers are selected randomly.

a. What is the probability that more than four will say they shop on the Internet more now than last year?
b. What is the probability that exactly two will say that they shop on the Internet more now than last year?
c. What is the probability that fewer than two will say that they shop on the Internet more now than last year?

5.106 EX05-106 Results of a study published in *Newsweek* during the summer of 1998 revealed what people do when they go away on vacations. The study also revealed that 60% of U.S. adults under 30 years of age think that going away on vacation is very important.

What People Do on Vacations	Percentage
Shopping	32
Outdoor activity	17
Historical / museum	14
Beach	11
Visit national / state parks	10
Cultural events / festivals	9
Theme / amusement parks	8
Night life / dancing	8
Gambling	7
Sporting events	6
Golfing / tennis / skiing	4

Source: *Newsweek*, "Why Can't We Get Away?," July 27, 1998

a. Is this a probability distribution? Explain why or why not.
b. If you sampled 30 adults under 30 who completed the questionnaire, what is the probability that fewer than 16 would agree with the survey's results that going away on vacation is very important?

5.107 The town council has nine members. A proposal must have at least two-thirds of the votes to be accepted. A proposal to establish a new industry in this town has been tabled. If we know that two members of the town council are opposed and that the others randomly vote "in favor" and "against," what is the probability that the proposal will be accepted?

5.108 The "Health Update" section of *Better Homes and Gardens* (July 1990) reported that patients who take long-half-life tranquilizers are 70% more likely to suffer falls resulting in hip fractures than those who take similar drugs with a short half-life. It was also reported that in a Massachusetts study, 30% of nursing-home patients who used tranquilizers used the long-half-life ones. Suppose that in a survey of 15 nursing-home patients in New York who used tranquilizers, it was found that 10 of the 15 used long-half-life tranquilizers.

a. If the 30% figure for Massachusetts also holds in New York, find the probability of finding 10 or more in a random sample of 15 who use long-half-life tranquilizers.
b. What might you infer from your answer in part a?

5.109 A box contains ten items of which three are defective and seven are nondefective. Two items are selected without replacement, and x is the number of defectives in the sample of two. Explain why x is not a binomial random variable.

5.110 A large shipment of radios is accepted upon delivery if an inspection of ten randomly selected radios yields no more than one defective radio.

a. Find the probability that this shipment is accepted if 5% of the total shipment is defective.
b. Find the probability that this shipment is not accepted if 20% of this shipment is defective.
c. The binomial probability distribution is often used in situations similar to this one—namely, large populations sampled without replacement. Explain why the binomial yields a good estimate.

5.111 A discrete random variable has a standard deviation of 10 and a mean of 50. Find $\sum x^2P(x)$.

5.112 A binomial random variable is based on $n = 20$ and $p = 0.4$. Find $\sum x^2P(x)$.

5.113 In a germination trial, 50 seeds were planted in each of 40 rows. The number of seeds germinating in each row was recorded in the table.

Seeds Germinating	Rows
39	1
40	2
41	3
42	4
43	6
44	7
45	8
46	4
47	3
48	1
49	1

a. Use the frequency distribution table to determine the observed rate of germination for these seeds.
b. The binomial probability experiment with its corresponding probability distribution can be used with the

variable "number of seeds germinating per row" when 50 seeds are planted in every row. Identify the specific binomial function and list its distribution using the germination rate found in part a. Justify your answer.

c. Suppose you are planning to repeat this experiment by planting 40 rows of the same seeds, with 50 seeds in each row. Use your probability model from part b to find the frequency distribution for x that you would "expect" to result from your planned experiment.

d. Compare your answer in part c to the results that were given above. Describe any similarities and differences.

5.114 In another germination experiment involving old seeds, 50 rows of seeds were planted and the number of seeds germinating in each row was recorded in the table (each row contained the same number of seeds).

Seeds Germinating	Rows
0	17
1	20
2	10
3	2
4	1
5 or more	0

a. What probability distribution (or function) would be helpful in modeling the variable "number of seeds germinating per row"? Justify your choice.

b. What information is missing in order to apply the probability distribution you chose in part a?

c. Based on the information you have, what is the highest or lowest rate of germination that you can estimate for these seeds? Explain.

5.115 A business firm is considering two investments, and it will choose the one that promises the greater payoff. Which of the investments should it accept? (Let the mean profit measure the payoff.)

Invest in Tool Shop		Invest in Book Store	
Profit	**Probability**	**Profit**	**Probability**
\$100,000	0.10	\$400,000	0.20
50,000	0.30	90,000	0.10
20,000	0.30	−20,000	0.40
−80,000	0.30	−250,000	0.30
Total	1.00	Total	1.00

5.116 Bill has completed a ten-question multiple-choice test on which he answered seven questions correctly. Each question had one correct answer to be chosen from five alternatives. Bill says that he answered the test by randomly guessing the answers without reading the questions or answers.

a. Define the random variable x to be the number of correct answers on this test, and construct the probability distribution if the answers were obtained by random guessing.

b. What is the probability that Bill guessed seven of the ten answers correctly?

c. What is the probability that anybody can guess six or more answers correctly?

d. Do you believe that Bill actually randomly guessed as he claims? Explain.

5.117 A random variable that can assume any one of the integer values 1, 2, . . . , n with equal probabilities of $\frac{1}{n}$ is said to have a *uniform* distribution. The probability function is written $P(x) = \frac{1}{n}$ for $x = 1, 2, 3, \ldots, n$. Show that $\mu = \frac{n+1}{2}$. (Hint: $1 + 2 + 3 + \cdots + n = [n(n + 1)]/2$.)

Vocabulary List and Key Concepts

Be able to define each term. Pay special attention to the key terms, which are printed in **red.** In addition, describe each term in your own words and give an example of each. Your examples should not be the ones given in class or in the textbook. The bracketed numbers indicate the chapter in which the term first appeared, but you should define the terms again to show increased understanding of their meaning. Page numbers indicate the first appearance of the term in Chapter 5.

binomial coefficient (p. 251)
binomial experiment (p. 250)
binomial probability function (p. 251)
binomial random variable (p. 250)
constant function (p. 237)
continuous random variable (p. 235)
discrete random variable (p. 235)
experiment [1, 4] (p. 234)
failure (p. 250)
independent trials (p. 250)
mean of discrete random variable (p. 243)
mutually exclusive events [4] (p. 234)
population parameter [1] (p. 242)
probability distribution (p. 236)
probability function (p. 237)
probability histogram (p. 239)
random variable (p. 234)
sample statistic [1] (p. 242)
standard deviation of discrete random variable (p. 243)
success (p. 250)
trial (p. 250)
variance of discrete random variable (p. 243)

Chapter Practice Test

PART I: KNOWING THE DEFINITIONS

Answer "True" if the statement is always true. If the statement is not always true, replace the words shown in bold with words that make the statement always true.

5.1 The number of hours you waited in line to register this semester is an example of a **discrete** random variable.

5.2 The number of automobile accidents you were involved in as a driver last year is an example of a **discrete** random variable.

5.3 The sum of all the probabilities in any probability distribution is always exactly **two.**

5.4 The various values of a random variable form a list of **mutually exclusive events.**

5.5 A binomial experiment always has **three or more** possible outcomes to each trial.

5.6 The formula $\mu = np$ may be used to compute the mean of a **discrete** population.

5.7 The binomial parameter p is the probability of **one success occurring in n trials** when a binomial experiment is performed.

5.8 A parameter is a statistical measure of some aspect of a **sample.**

5.9 **Sample statistics** are represented by letters from the Greek alphabet.

5.10 The probability of event A or B is equal to the sum of the probability of event A and the probability of event B when A and B are **mutually exclusive events.**

PART II: APPLYING THE CONCEPTS

5.11

a. Show that the following is a probability distribution:

x	$P(x)$
1	0.2
3	0.3
4	0.4
5	0.1

b. Find $P(x = 1)$.
c. Find $P(x = 2)$.
d. Find $P(x > 2)$.
e. Find the mean of x.
f. Find the standard deviation of x.

5.12 A T-shirt manufacturing company advertises that the probability of an individual T-shirt being irregular is 0.1. A box of 12 such T-shirts is randomly selected and inspected.

a. What is the probability that exactly 2 of these 12 T-shirts are irregular?
b. What is the probability that exactly 9 of these 12 T-shirts are not irregular?

Let x be the number of T-shirts that are irregular in all such boxes of 12 T-shirts.

c. Find the mean of x.
d. Find the standard deviation of x.

PART III: UNDERSTANDING THE CONCEPTS

5.13 What properties must an experiment possess in order for it to be a binomial probability experiment?

5.14 Student A uses a relative frequency distribution for a set of sample data and calculates the mean and standard deviation using formulas from Chapter 5. Student A justifies her choice of formulas by saying that since relative frequencies are empirical probabilities, her sample is represented by a probability distribution and therefore her choice of formulas was correct. Student B argues that since the distribution represented a sample, the mean and standard deviation involved are known as $\bar{x}$ and s and must be calculated using the corresponding frequency distribution and formulas from Chapter 2. Who is correct, A or B? Justify your choice.

5.15 Student A and Student B were discussing one entry in a probability distribution chart:

x	$P(x)$
−2	0.1

Student B thought this entry was okay because $P(x)$ was a value between 0.0 and 1.0. Student A argued that this entry was impossible for a probability distribution because x was −2 and negatives are not possible. Who is correct, A or B? Justify your choice.

CHAPTER

6

Normal Probability Distributions

Chapter Outline and Objectives

CHAPTER CASE STUDY

What is **normal** with respect to aptitude test scores?

6.1 NORMAL PROBABILITY DISTRIBUTIONS

The domain of **bell-shaped** distributions is the set of all real numbers.

6.2 THE STANDARD NORMAL DISTRIBUTION

To work with normal distributions we need the **standard score.**

6.3 APPLICATIONS OF NORMAL DISTRIBUTIONS

The normal distribution can help us determine **probabilities.**

6.4 NOTATION

The z **notation** is critical in the use of normal distributions.

6.5 NORMAL APPROXIMATION OF THE BINOMIAL

Binomial probabilities can be **estimated** by using a normal distribution.

RETURN TO CHAPTER CASE STUDY

Chapter Case Study

Aptitude Tests and Their Interpretation

© Bob Daemmrich/The Image Works

There are many kinds of aptitude tests. Some are for specific purposes, such as measurement of finger dexterity, something that might be important on a particular job. Others are of more general aptitudes. So-called intelligence tests are examples of general aptitude tests.

The Binet Intelligence Scale. Alfred Binet, who devised the first general aptitude test at the beginning of the 20th century, defined intelligence as *the ability to make adaptations.* The general purpose of the test was to determine which children in Paris could benefit from school. Binet's test, like its subsequent revisions, consists of a series of progressively more difficult tasks which children of different ages can successfully complete. A child who can solve problems typically solved by children at a particular age level is said to have that mental age. For example, if a child can successfully do the same tasks that an average eight-year-old can do, he or she is said to have a mental age of eight. The *intelligence quotient,* or IQ, is defined by the formula:

$$\text{Intelligence Quotient} = 100 \times \frac{\text{Mental age}}{\text{Chronological age}}$$

There has been a great deal of controversy in recent years over what intelligence tests measure. Many of the test items depend on either language or other specific cultural experiences for correct answers. Nevertheless, such tests can rather effectively predict school success. If school requires language and the tests measure language ability at a particular point of time in a child's life, then the test is a better-than-chance predictor of school performance.

Deviation IQ Scores. Present-day tests of intelligence or other abilities use *deviation scores.* These scores represent the deviation of a particular person from the average score for similar persons. Suppose you take a "general aptitude test" and get a score of 115. This does not mean that your mental age is greater than your chronological age; it means that you are "above average" in some degree. Because we have become accustomed to thinking of an IQ score of 100 as average, most general aptitude tests are scored in such a way that 100 is average. A person scoring 115 would generally have a score higher than the scores of about 85 percent of people who take the test; a score of 84 would be better than about 16 percent. The exact interpretation of a test score depends on the particular test, but Figure 2-2 shows how the scores on a number of commonly used aptitude tests are interpreted in terms of how an individual compares with a group.

Source: Robert C. Beck, Applying Psychology, Critical and Creative Thinking, 3rd ed. (Englewood Cliffs, NJ: Prentice-Hall, 1992)

FIGURE 2-2
Comparison of several deviation scores and the normal distribution. Standard scores have a mean of zero and a standard deviation of 1.0. Scholastic Aptitude Test scores have a mean of 500 and a standard deviation of 100. Binet Intelligence Scale scores have a mean of 100 and a standard deviation of 16. In each case there are 34 percent of the scores between the mean and one standard deviation, 14 percent between one and two standard deviations, and 2 percent beyond two standard deviations.

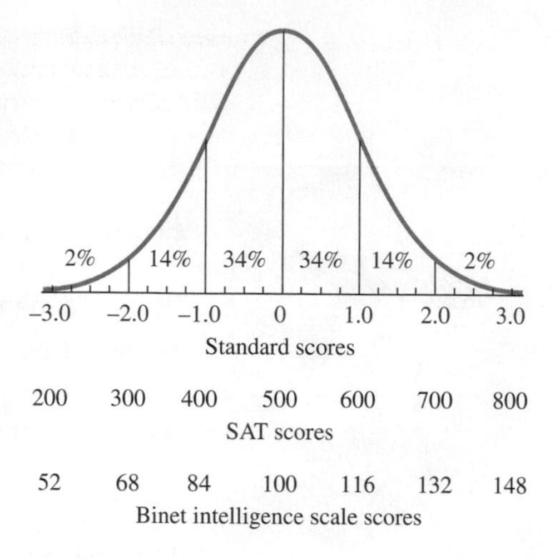

ANSWER NOW 6.1

a. Explain why IQ score is a continuous variable.

b. What are the mean and the standard deviation for the distribution of IQ scores? SAT scores? Standard scores?

c. Express, algebraically or as an equation, the relationship between standard scores and IQ scores. Standard scores and SAT scores.

d. What standard score is two standard deviations above the mean? What IQ score is two standard deviations above the mean? What SAT score is two standard deviations above the mean?

e. Compare the information about percentages of the distribution in Figure 2-2 above with the empirical rule studied in Chapter 2. Explain the similarities.

After completing Chapter 6, further investigate the Chapter Case Study in the Return to Chapter Case Study section with Exercises 6.86, 6.87, and 6.88 (p. 304).

6.1 Normal Probability Distributions

The **normal probability distribution** is considered the single most important probability distribution. An unlimited number of **continuous random variables** have either a normal or an approximately normal distribution. Other probability distributions of both discrete and continuous random variables are also approximately normal under certain conditions.

Recall that in Chapter 5 we learned how to use a probability function to calculate the probabilities associated with **discrete random variables**. The normal probability distribution has a continuous random variable and it uses two functions: one function to determine the ordinates (y values) of the graph picturing the distribution and a second to determine the probabilities. Formula (6.1) expresses the ordinate (y value) that corresponds to each abscissa (x value).

NORMAL PROBABILITY DISTRIBUTION FUNCTION

$$y = f(x) = \frac{e^{-\frac{1}{2}\left(\frac{x-\mu}{\sigma}\right)^2}}{\sigma\sqrt{2\pi}} \quad \text{for all real } x \qquad (6.1)$$

NOTE Each different pair of values for the mean, μ, and standard deviation, σ, will result in a different normal probability distribution function.

When a graph of all such points is drawn, the **normal (bell-shaped) curve** will appear as shown in Figure 6.1.

FIGURE 6.1
The Normal Probability Distribution

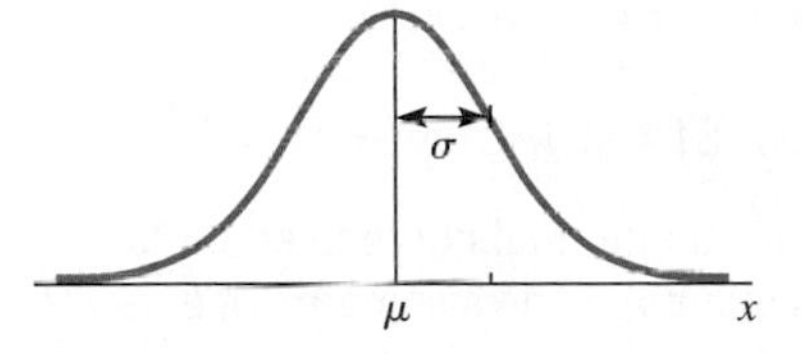

Formula (6.2) yields the probability associated with the interval from $x = a$ to $x = b$:

$$P(a \leq x \leq b) = \int_a^b f(x)\,dx \tag{6.2}$$

The probability that x is within the interval from $x = a$ to $x = b$ is shown as the shaded area in Figure 6.2.

FIGURE 6.2
Shaded area: $P(a \leq x \leq b)$

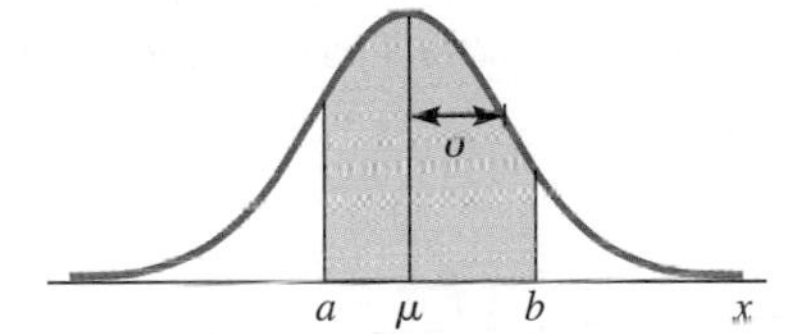

We will not be using the preceding formulas to calculate probabilities for normal distributions. The definite integral of formula (6.2) is a calculus topic and is mathematically beyond what is expected in elementary statistics. (These formulas often appear at the top of normal probability tables as identification.) Instead of using formulas (6.1) and (6.2), we will use a table to find probabilities for normal distributions. Before we learn to use the table, however, it must be pointed out that the table is expressed in "standardized" form. It is standardized so that this one table can be used to find probabilities for all combinations of mean, μ, and standard deviation, σ, values. That is, the normal probability distribution with mean 38 and standard deviation 7 is similar to the normal probability distribution with mean 123 and standard deviation 32. Recall the empirical rule and the percentages of the distribution that fall within certain intervals of the mean (p. 106). The same three percentages hold true for all normal distributions.

NOTE: **Percentage, proportion,** and **probability** are basically the same concepts. Percentage (25%) is usually used when talking about a proportion ($\frac{1}{4}$) of a population. Probability is usually used when talking about the chance that the next individual item will possess a certain property. Area is the graphic representation of all three when we draw a picture to illustrate the situation. The empirical rule is a fairly crude measuring device; with it we are able to find probabilities associated only with whole-number multiples of the standard deviation (within one, two, or three standard deviations of the mean). We will often be interested in the probabilities associated with fractional parts of the standard deviation. For example, we might want to know the probability that x is within 1.37 standard deviations of the mean. Therefore, we must refine the empirical rule so that we can deal with more precise measurements. This refinement is discussed in the next section.

6.2 The Standard Normal Distribution

There are an unlimited number of normal probability distributions, but fortunately they are all related to one distribution: the **standard normal distribution.** The standard normal distribution is the normal distribution of the standard variable z (called "**standard score**" or "**z-score**").

PROPERTIES OF THE STANDARD NORMAL DISTRIBUTION

1. The total area under the normal curve is equal to one.
2. The distribution is mounded and symmetric; it extends indefinitely in both directions, approaching but never touching the horizontal axis.
3. The distribution has a mean of 0 and a standard deviation of 1.
4. The mean divides the area in half—0.50 on each side.
5. Nearly all the area is between $z = -3.00$ and $z = 3.00$.

Table 3 in Appendix B lists the probabilities associated with the intervals from the mean (located at $z = 0.00$) to a specific value of z. Probabilities of other intervals may be found by using the table entries and the operations of addition and subtraction, in accordance with the preceding properties. Let's look at several illustrations demonstrating how to use Table 3 to find probabilities of the standard normal score, z.

Illustration 6.1 Finding Area to the Right of $z = 0$

Find the area under the standard normal curve between $z = 0$ and $z = 1.52$ (see Figure 6.3).

FIGURE 6.3
Area from $z = 0$ to $z = 1.52$

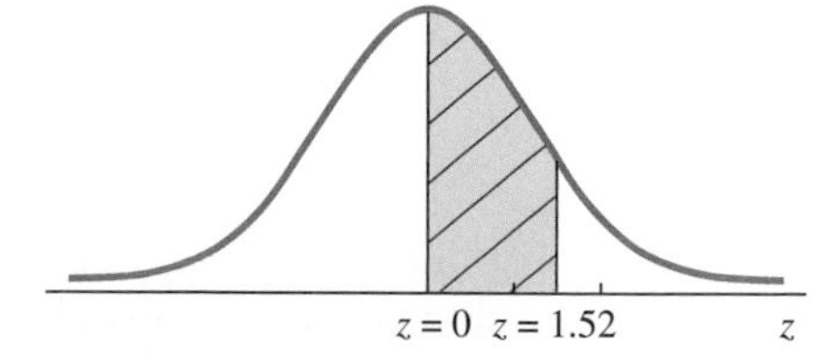

SOLUTION

Table 3 is designed to give the area between $z = 0$ and $z = 1.52$ directly. The z-score is located on the margins, with the units and tenths digit along the left side and the hundredths digit across the top. For $z = 1.52$, locate the row labeled 1.5 and the column labeled 0.02; at their intersection you will find 0.4357, the measure of the area or the probability for the interval $z = 0.00$ to $z = 1.52$ (see Table 6.1). Expressed as a probability: $P(0.00 < z < 1.52) = \mathbf{0.4357}$.

TABLE 6.1 **A Portion of Table 3**

z	0.00	0.01	0.02	...
⋮				
1.5			0.4357	...
⋮				

■

ANSWER NOW 6.2
Find the area under the standard normal curve between $z = 0$ and $z = 1.37$.

Recall that one of the basic properties of probability is that the sum of all probabilities is exactly 1.0. Since the area under the normal curve represents the measure of probability, the total area under the bell-shaped curve is exactly 1 unit. This distribution is also symmetric with respect to the vertical line drawn through $z = 0$, which cuts the area in half at the mean. Can you verify this fact by inspecting formula (6.1)? That is, the area under the curve to the right of the mean is exactly one-half unit, 0.5, and the area to the left is also one-half unit, 0.5. Areas (probabilities) not given directly in the table can be found by relying on these facts.

Now let's look at some illustrations.

Illustration 6.2 Finding Area in the Right Tail of a Normal Curve

Find the area under the normal curve to the right of $z = 1.52$: $P(z > 1.52)$.

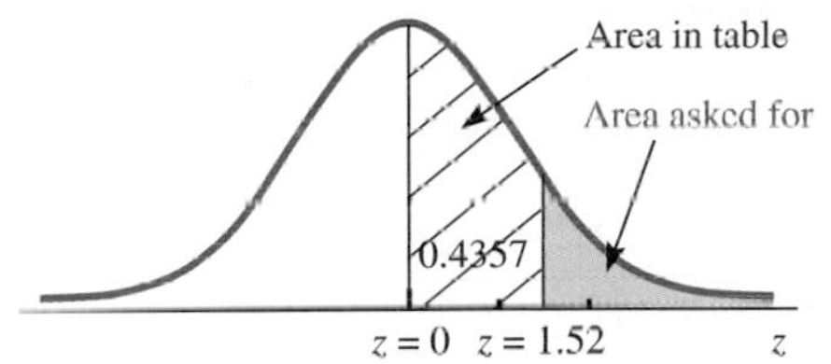

SOLUTION
The area to the right of the mean (all the shading in the figure) is exactly 0.5000. The problem asks for the shaded area that is not included in the 0.4357. Therefore, we subtract 0.4357 from 0.5000:

$$P(z > 1.52) = 0.5000 - 0.4357 = \mathbf{0.0643}$$ ■

NOTE As we have done here, always draw and label a sketch. It is most helpful.

Illustration 6.3 Finding Area to the Left of a Positive z Value

Find the area to the left of $z = 1.52$: $P(z < 1.52)$.

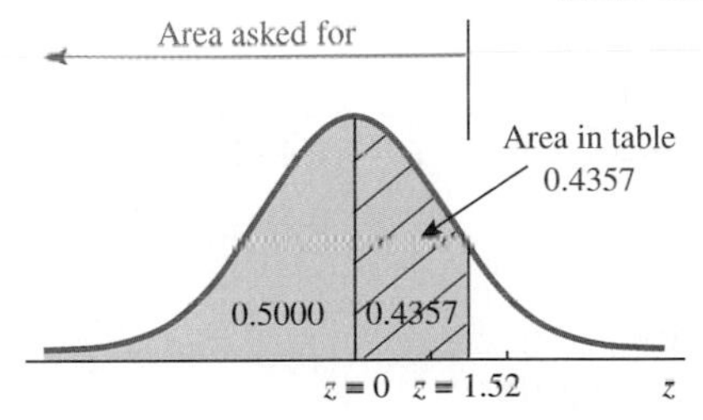

SOLUTION
The total shaded area is made up of 0.4357 found in the table and the 0.5000 that is to the left of the mean. Therefore, we add 0.4357 to 0.5000:

$$P(z \leq 1.52) = P(z < 0) + P(0 < z < 1.52)$$
$$= 0.4357 + 0.5000 = \mathbf{0.9357}$$ ■

ANSWER NOW 6.3
Find the area under the standard normal curve to the right of $z = 2.03$: $P(z > 2.03)$.

ANSWER NOW 6.4
Find the area under the standard normal curve to the left of $z = 1.73$: $P(z < 1.73)$.

NOTE The addition and subtraction done in Illustrations 6.2 and 6.3 are correct because the "areas" represent mutually exclusive events (discussed in Section 4.5).

The symmetry of the normal distribution is a key factor in determining probabilities associated with values below (to the left of) the mean. The area between the mean and $z = -1.52$ is exactly the same as the area between the mean and $z = +1.52$. This fact allows us to find values related to the left side of the distribution.

Illustration 6.4 Finding Area from a Negative z to $z = 0$

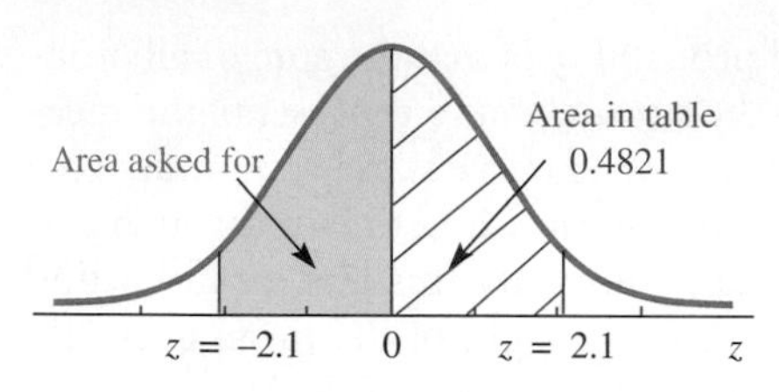

The area between the mean ($z = 0$) and $z = -2.1$ is the same as the area between $z = 0$ and $z = +2.1$; that is,

$$P(-2.1 < z < 0) = P(0 < z < 2.1)$$

Thus, we have

$$P(-2.1 < z < 0) = P(0 < z < 2.1) = \mathbf{0.4821}$$

■

Illustration 6.5 Finding Area in the Left Tail of a Normal Curve

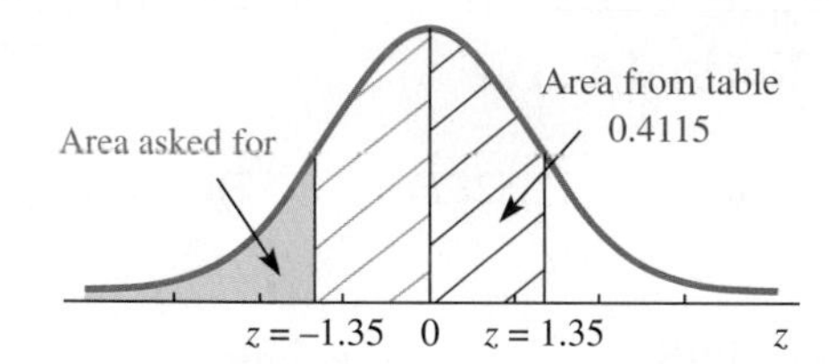

The area to the left of $z = -1.35$ is found by subtracting 0.4115 from 0.5000.

Therefore, we obtain

$$P(z < -1.35) = P(z < 0) - P(-1.35 < z < 0)$$
$$= 0.5000 - 0.4115 = \mathbf{0.0885}$$

■

ANSWER NOW 6.5
Find the area under the standard normal curve between −1.39 and the mean: $P(-1.39 < z < 0.00)$.

ANSWER NOW 6.6
Find the area under the standard normal curve to the left of $z = -1.53$: $P(z < -1.53)$.

Illustration 6.6 Finding Area from a Negative z to a Positive z

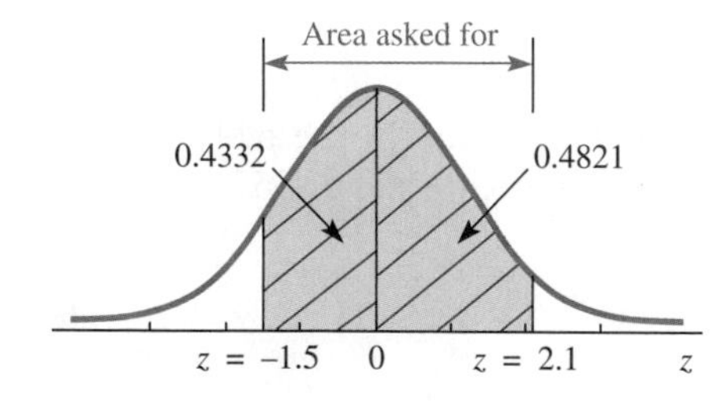

The area between $z = -1.5$ and $z = 2.1$, $P(-1.5 < z < 2.1)$, is found by adding the two areas together. Both probabilities are read directly from Table 3.

Therefore, we obtain

$$P(-1.5 < z < 2.1) = P(-1.5 < z < 0) + P(0 < z < 2.1)$$
$$= 0.4332 + 0.4821 = \mathbf{0.9153}$$

■

Illustration 6.7 Finding Area Between Two Nonzero z Values

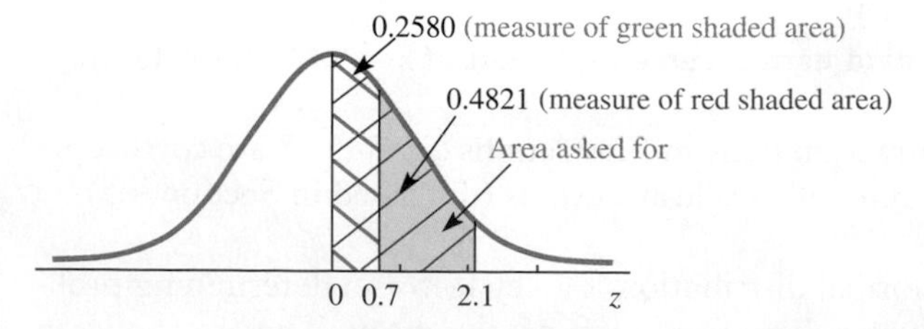

The area between $z = 0.7$ and $z = 2.1$, $P(0.7 < z < 2.1)$, is found by subtracting. The area between $z = 0$ and $z = 2.1$ includes all the area between $z = 0$ and $z = 0.7$. Therefore, we subtract the area between $z = 0$ and $z = 0.7$ from the area between $z = 0$ and $z = 2.1$.

Thus, we have

$$P(0.7 < z < 2.1) = P(0 < z < 2.1) - P(0 < z < 0.7)$$
$$= 0.4821 - 0.2580 = \mathbf{0.2241}$$

■

ANSWER NOW 6.7
Find the area under the standard normal curve between $z = -1.83$ and $z = 1.23$: $P(-1.83 < z < 1.23)$.

ANSWER NOW 6.8
Find the area under the standard normal curve between $z = 0.75$ and $z = 2.25$: $P(0.75 < z < 2.25)$.

The normal distribution table can also be used to find a z-score when we are given an area. The next illustration considers this idea.

Illustration 6.8 Finding z-scores Associated with a Percentile

What is the z-score associated with the 75th percentile? (Assume the distribution is normal.) See Figure 6.4.

SOLUTION

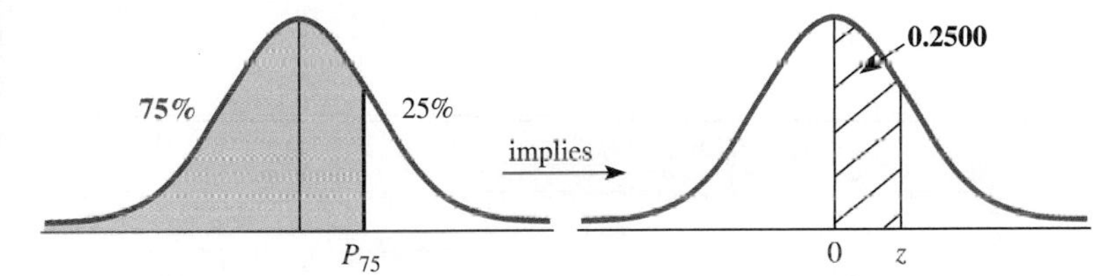

FIGURE 6.4
P_{75} and Its Associated z-Score

To find this z-score, look in Table 3 in Appendix B and find the "area" entry that is closest to 0.2500; this area entry is 0.2486. Now read the z-score that corresponds to this area.

z	...	0.07		0.08	...
⋮					
0.6		0.2486	0.2500	0.2517	...
⋮					

From the table, the z-score is found to be $z = 0.67$. This says that the 75th percentile in a normal distribution is 0.67 (approximately $\frac{2}{3}$) standard deviation above the mean. ■

Illustration 6.9 Finding z-scores That Bound an Area

What z-scores bound the middle 95% of a normal distribution? See Figure 6.5.

SOLUTION

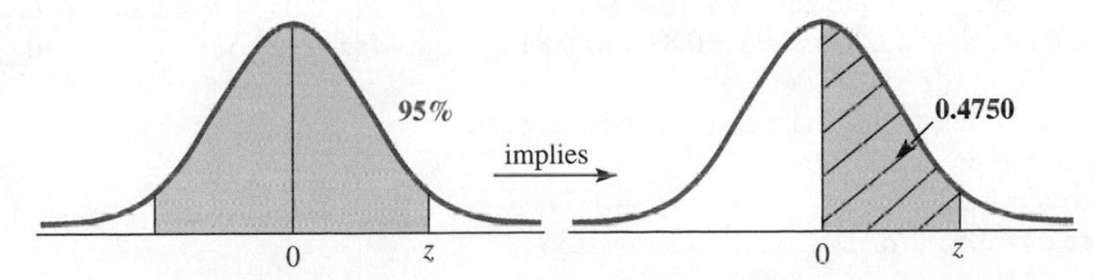

FIGURE 6.5
Middle 95% of Distribution and Its Associated z-score

(continued)

The 95% is split into two equal parts by the mean, so 0.4750 is the area (percentage) between $z = 0$, the mean, and the z-score at the right boundary. Since we have the area, we look for the entry in Table 3 closest to 0.4750 (it happens to be exactly 0.4750) and read off the z-score. We obtain $z = 1.96$.

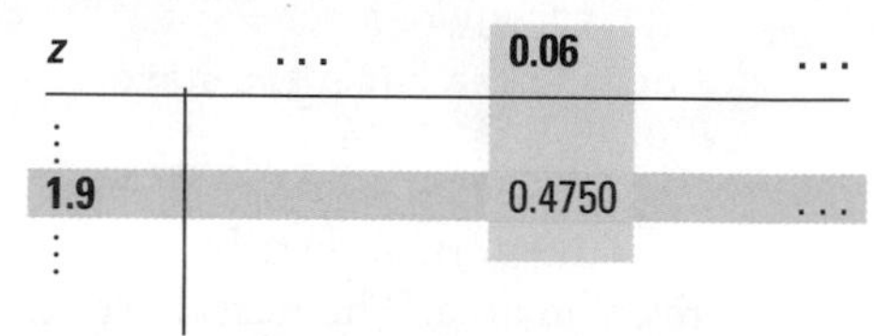

z	...	0.06	...
⋮			
1.9		0.4750	...
⋮			

Therefore, $\mathbf{z = -1.96}$ and $\mathbf{z = 1.96}$ bound the middle 95% of a normal distribution. ■

ANSWER NOW 6.9

a. **Find the z-score for the 80th percentile of the standard normal distribution.**

b. **Find the z-scores that bound the middle 75% of the standard normal distribution.**

Exercises

6.10
a. Describe the distribution of the standard normal score z.
b. Why is this distribution called *standard normal*?

6.11 Find the area under the normal curve that lies between the following pairs of z values:
a. $z = 0$ to $z = 1.30$ b. $z = 0$ to $z = 1.28$
c. $z = 0$ to $z = -3.20$ d. $z = 0$ to $z = -1.98$

6.12 Find the probability that a piece of data picked at random from a normal population will have a standard score, z, that lies between the following pairs of z values:
a. $z = 0$ to $z = 2.10$ b. $z = 0$ to $z = 2.57$
c. $z = 0$ to $z = -1.20$ d. $z = 0$ to $z = -1.57$

6.13 Find the area under the standard normal curve that corresponds to the following z values:
a. between 0 and 1.55 b. to the right of 1.55
c. to the left of 1.55 d. between -1.55 and 1.55

6.14 Find the probability that a piece of data picked at random from a normal population will have a standard score, z, that lies:
a. between 0 and 0.84 b. to the right of 0.84
c. to the left of 0.84 d. between -0.84 and 0.84

6.15 Find the area under the normal curve that lies between the following pairs of z values:
a. $z = -1.20$ to $z = 1.22$ b. $z = -1.75$ to $z = 1.54$
c. $z = -1.30$ to $z = 2.58$ d. $z = -3.5$ to $z = -0.35$

6.16 Find the probability that a piece of data picked at random from a normal population will have a standard score, z, that lies between the following pairs of z values:
a. $z = -2.75$ to $z = 1.38$ b. $z = 0.67$ to $z = 2.95$
c. $z = -2.95$ to $z = -1.18$

6.17 Find the following areas under the normal curve:
a. to the right of $z = 0.00$ b. to the right of $z = 1.05$
c. to the right of $z = -2.30$ d. to the left of $z = 1.60$
e. to the left of $z = -1.60$

6.18 Find the probability that a piece of data picked at random from a normally distributed population will have a standard score that is:
a. less than 3.00 b. greater than -1.55
c. less than -0.75 d. less than 1.25
e. greater than -1.25

6.19 Find the probabilities:
a. $P(0.00 < z < 2.35)$ b. $P(-2.10 < z < 2.34)$
c. $P(z > 0.13)$ d. $P(z < 1.48)$

6.20 Find the probabilities:
a. $P(-2.05 < z < 0.00)$ b. $P(-1.83 < z < 2.07)$
c. $P(z < -1.52)$ d. $P(z < -0.43)$

6.21 Find the z-score for the standard normal distribution shown on each diagram:

a.

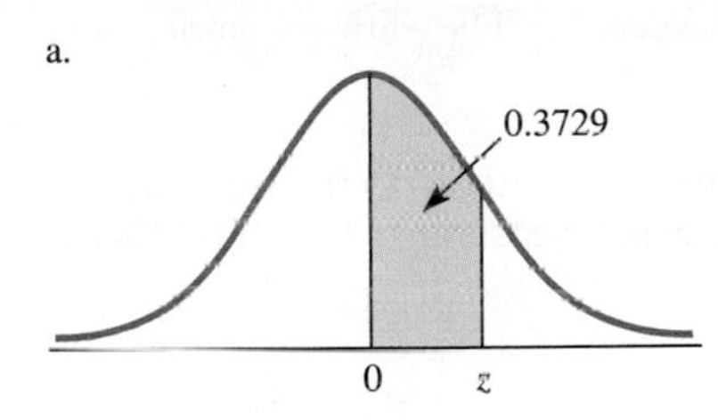

b.

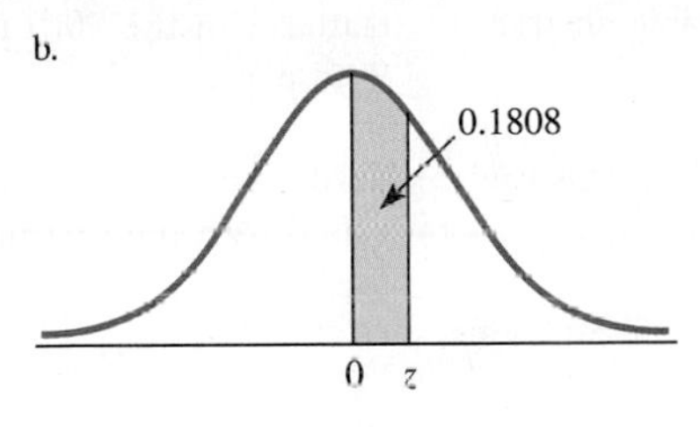

c.

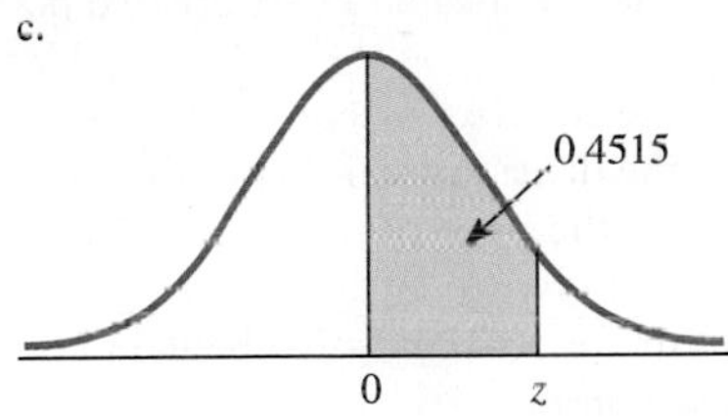

d.

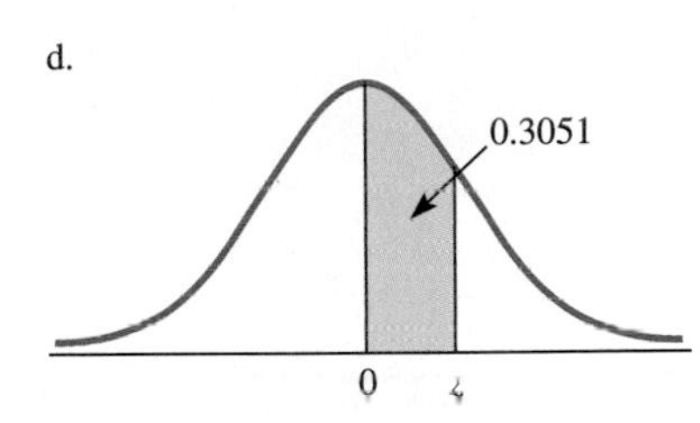

e.

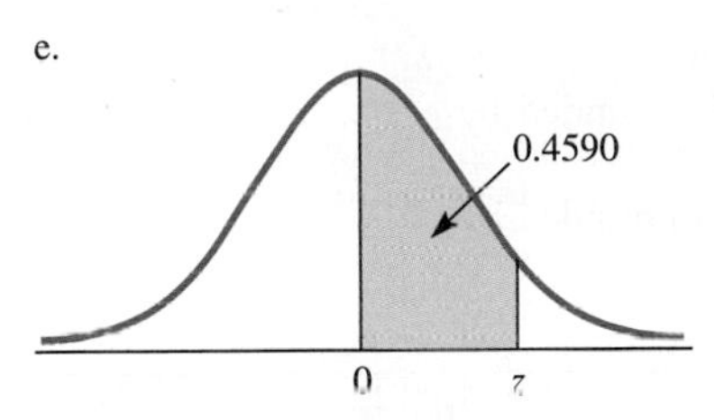

f.

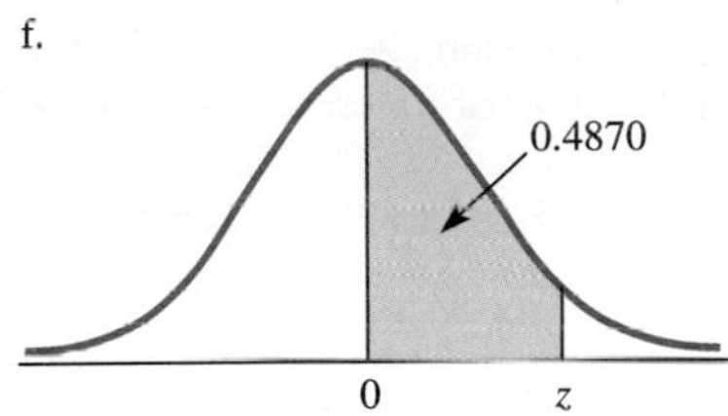

6.22 Find the z-score for the standard normal distribution shown in each diagram:

a.

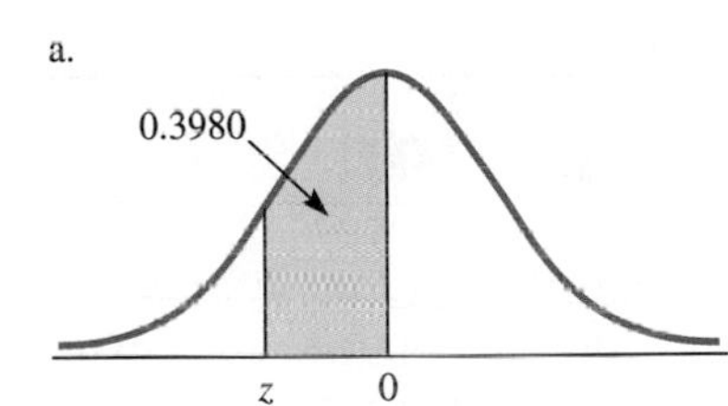

b.

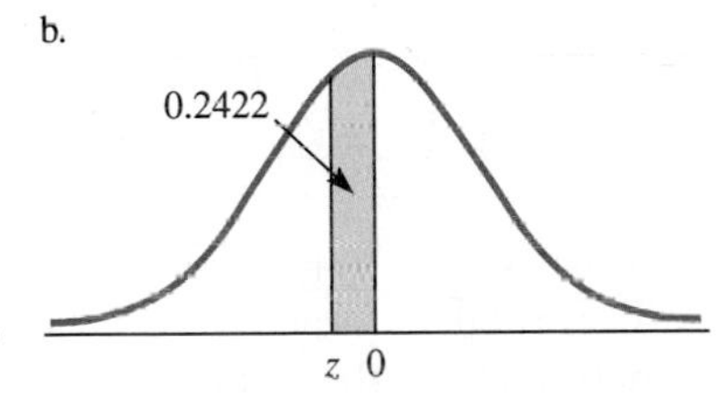

c.

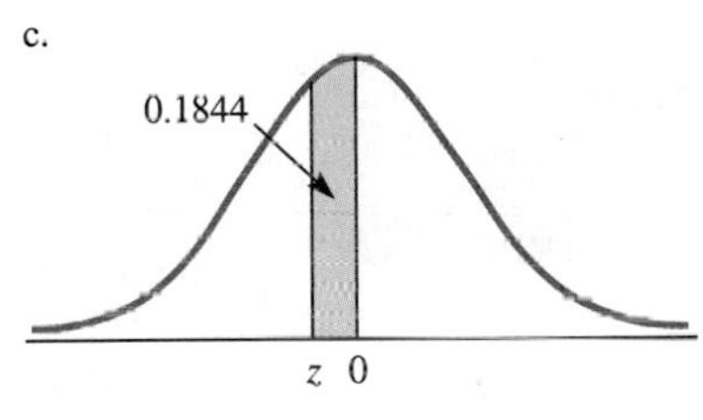

d.

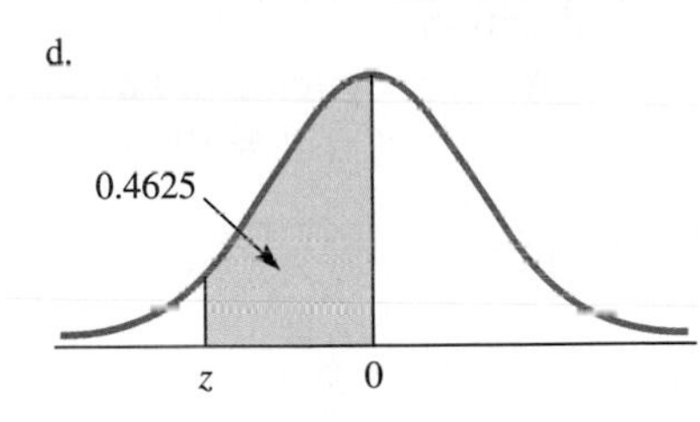

e.

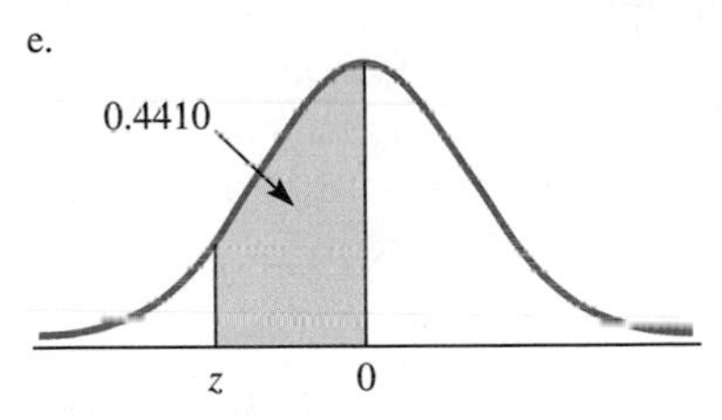

f.

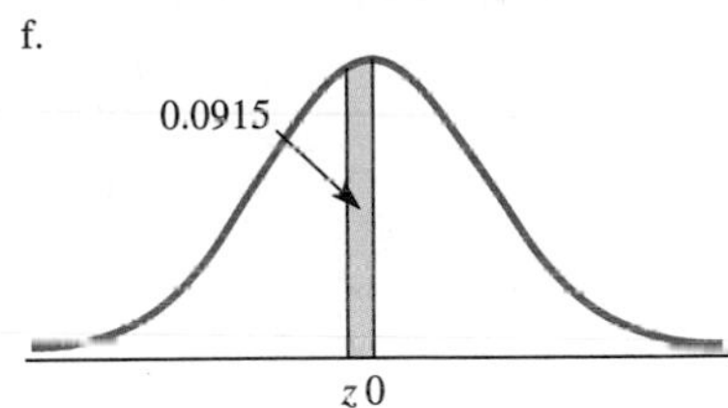

6.23 Find the standard score, z, shown on each diagram:

a.

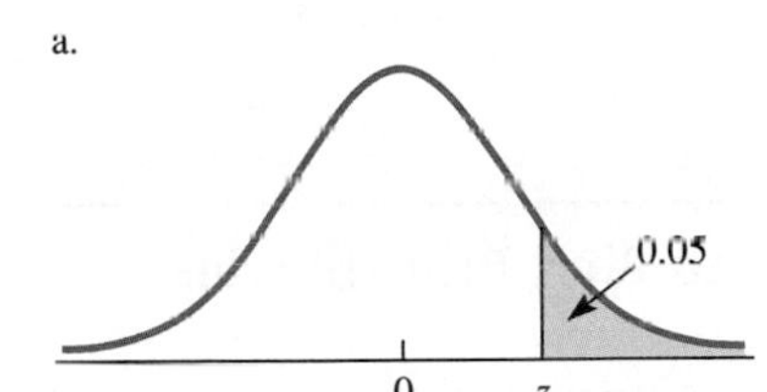

b.

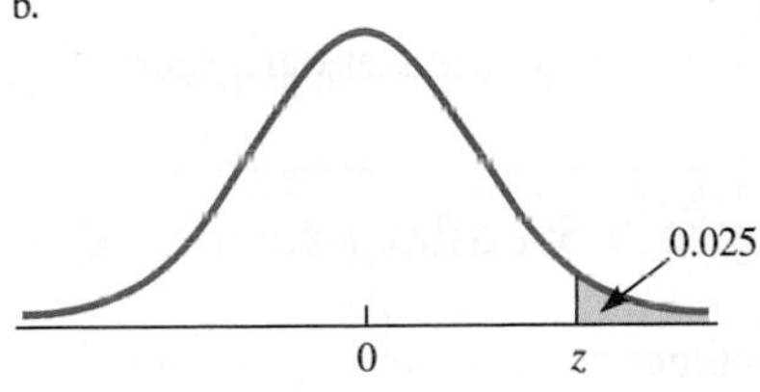

c.

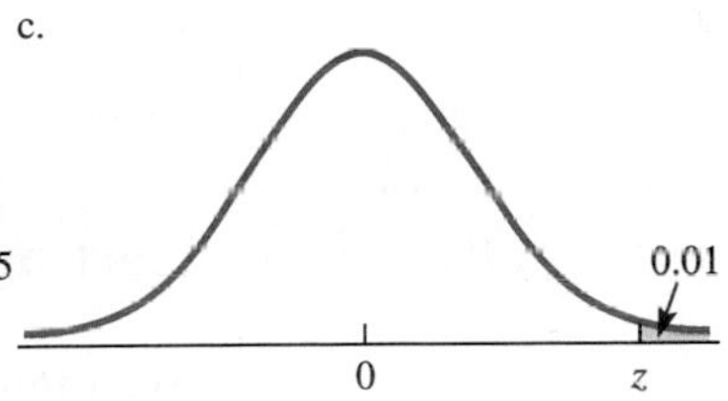

6.24 Find the standard score, z, shown on each diagram:

a.

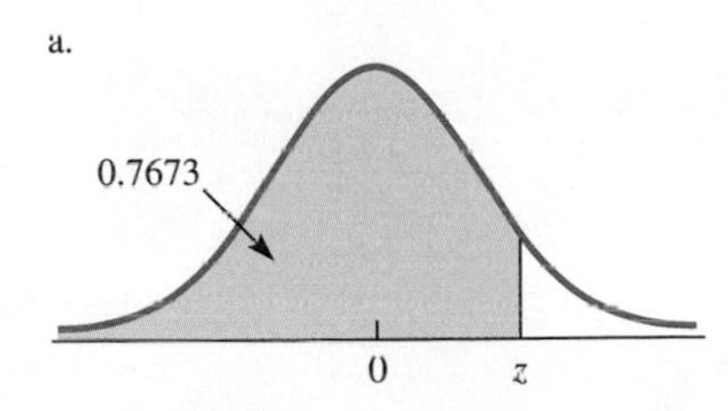

b.

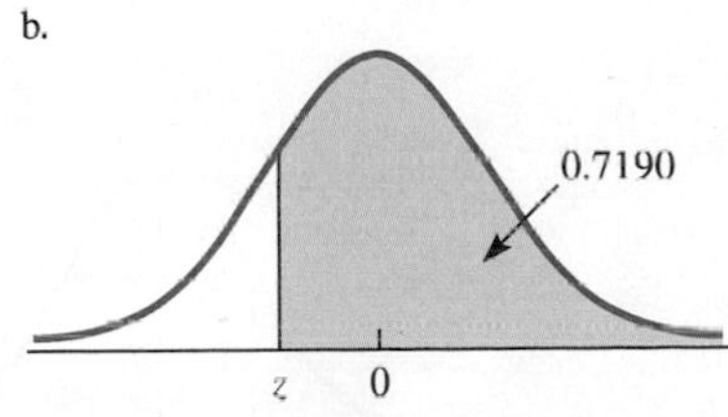

c.

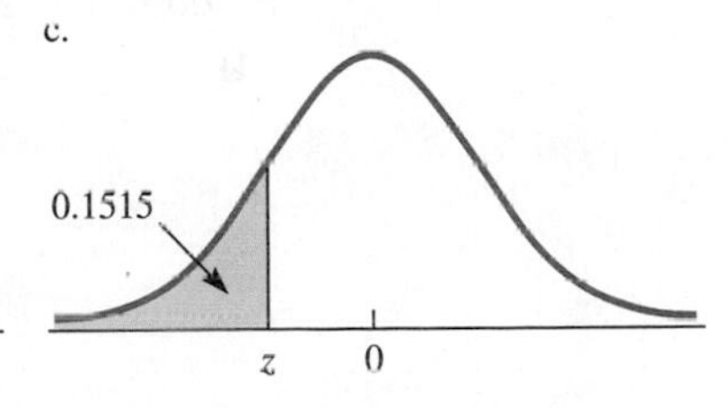

6.25 Find a value of z such that 40% of the distribution lies between it and the mean. (There are two possible answers.)

6.26 Find the standard z-score such that:
a. 80% of the distribution is below (to the left of) this value
b. the area to the right of this value is 0.15

6.27 Find the two z-scores that bound the middle 50% of a normal distribution.

6.28 Find the two standard scores, z, such that:
a. the middle 90% of a normal distribution is bounded by them
b. the middle 98% of a normal distribution is bounded by them

6.29 Assuming a normal distribution, what z-score is associated with the 90th percentile? The 95th percentile? The 99th percentile?

6.30 Assuming a normal distribution, what z-score is associated with the first quartile? Second quartile? Third quartile?

6.3 Applications of Normal Distributions

In Section 6.2 we learned how to use Table 3 in Appendix B to convert information about the standard normal variable z into probability or, the opposite, to convert probability information about the standard normal distribution into z-scores. Now we are ready to apply this methodology to all normal distributions. The key is the standard score, z. The information associated with a normal distribution will be in terms of x values or probabilities. We will use the z-score and Table 3 as the tools to "go between" the given information and the desired answer.

Recall that the standard score, z, was defined in Chapter 2:

standard score: $$z = \frac{x - (\text{mean of } x)}{\text{standard deviation of } x}$$

$$z = \frac{x - \mu}{\sigma} \tag{6.3}$$

(Note that when $x = \mu$, the standard score $z = 0$.)

Illustration 6.10 Converting to a Standard Normal Curve to Find Probabilities

Consider intelligence quotient (IQ) scores. IQ scores are normally distributed with a mean of 100 and a standard deviation of 16. If a person is picked at random, what is the probability that his or her IQ is between 100 and 115; that is, what is $P(100 < x < 115)$?

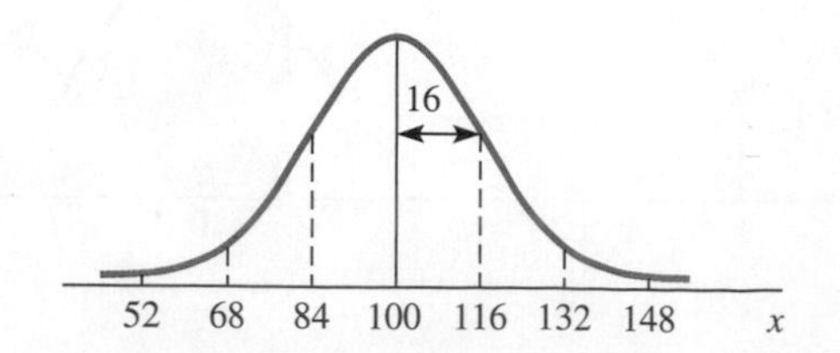

SOLUTION

$P(100 < x < 115)$ is represented by the shaded area in the figure.

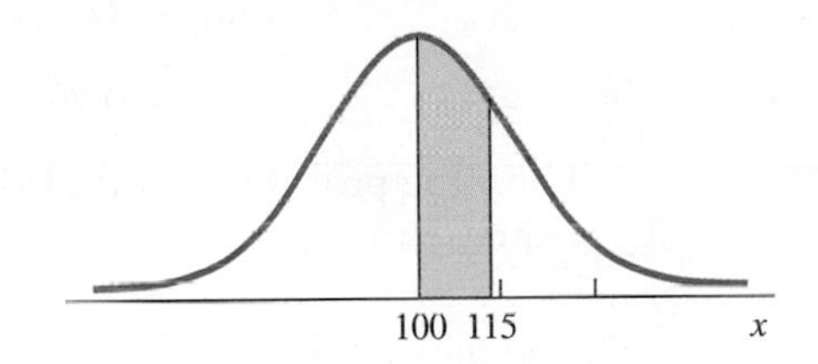

The variable x must be standardized using formula (6.3). The z values are shown on the next figure.

$$z = \frac{x - \mu}{\sigma}$$

$$\text{when } x = 100: \quad z = \frac{100 - 100}{16} = \mathbf{0.00}$$

$$\text{when } x = 115: \quad z = \frac{115 - 100}{16} = \mathbf{0.94}$$

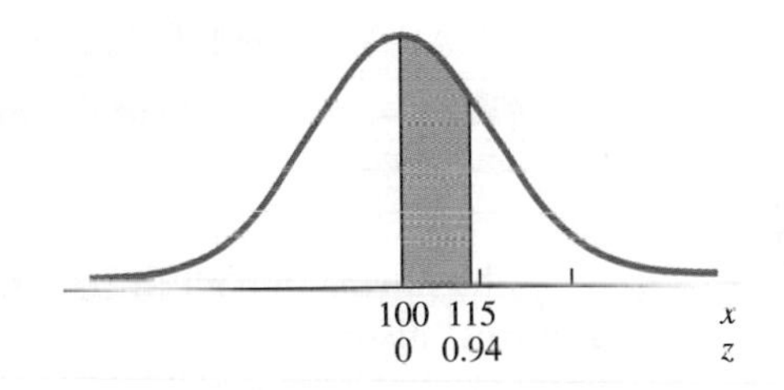

FYI
The value 0.3264 is found by using Table 3 in Appendix B.

Therefore,

$$P(100 < x < 115) = P(0.00 < z < 0.94) = \mathbf{0.3264}$$

Thus, the probability is 0.3264 that a person picked at random has an IQ between 100 and 115. ■

ANSWER NOW 6.31
Given $x = 58$, $\mu = 43$, and $\sigma = 5.2$, find z.

Illustration 6.11 Calculating Probability Under "Any" Normal Curve

Find the probability that a person selected at random will have an IQ greater than 90.

SOLUTION

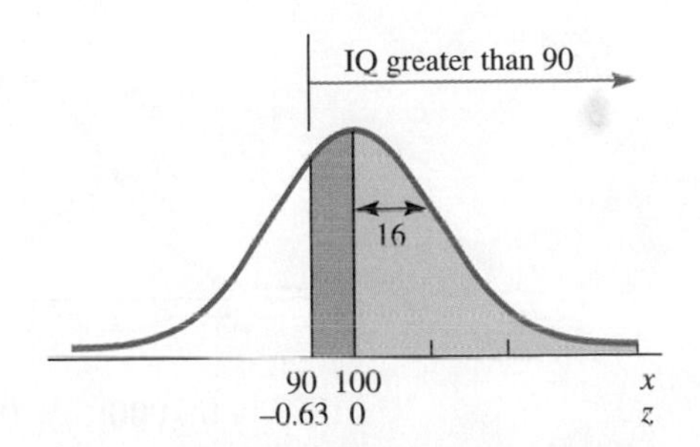

$$z = \frac{x - \mu}{\sigma} = \frac{90 - 100}{16} = \frac{-10}{16} = -0.625 = -0.63$$

$$P(x > 90) = P(z > -0.63)$$
$$= 0.2357 + 0.5000 = \mathbf{0.7357}$$

Thus, the probability is 0.7357 that a person selected at random will have an IQ greater than 90. ■

ANSWER NOW 6.32
Use the information given in Illustration 6.10.

a. **Find the probability that a randomly selected person will have an IQ score between 100 and 120.**

b. **Find the probability that a randomly selected person will have an IQ score greater than 80.**

ANSWER NOW 6.33
The nitrous oxide emitted by automobile engines is the subject of the "Normal Probability Distributions" video on your Student Suite CD (found inside the back cover of this textbook). View the video clip and answer these questions:

a. **Do all engines emit the same amount of nitrous oxide ("NOX")?**

b. **What information about the distribution did they need to estimate before they could find the percentage of cars that will exceed the limit?**

The normal table can be used to answer many kinds of questions that involve a normal distribution. Many times a problem will call for the location of a "cutoff point"—that is, a particular value of x such that exactly a certain percentage is in a specified area. The following illustrations concern some of these problems.

Illustration 6.12 Using the Normal Curve and z to Determine Data Values

In a large class, suppose your instructor tells you that you need to obtain a grade in the top 10% of your class to get an A on a particular exam. From past experience she is able to estimate that the mean and standard deviation on this exam will be 72 and 13, respectively. What will be the minimum grade needed to obtain an A? (Assume that the grades will be approximately normally distributed.)

SOLUTION
Start by converting the 10% to information that is compatible with Table 3 by subtracting:

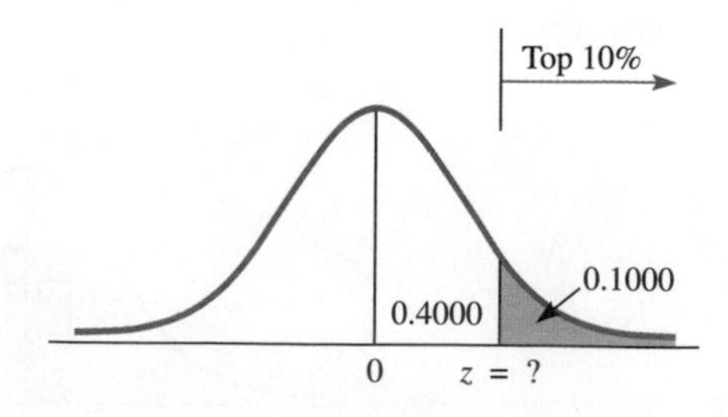

$$10\% = 0.1000; \quad 0.5000 - 0.1000 = 0.4000$$

FYI
Why is 0.5000 used?

Look in Table 3 to find the value of z associated with the area entry closest to 0.4000; it is $z = 1.28$. Thus,

$$P(z > 1.28) = 0.10$$

Now find the x value that corresponds to $z = 1.28$ by using formula (6.3):

$$z = \frac{x - \mu}{\sigma}: \quad 1.28 = \frac{x - 72}{13}$$

$$x - 72 = (13)(1.28)$$

$$x = 72 + (13)(1.28) = 72 + 16.64 = 88.64, \text{ or } \mathbf{89}$$

Thus, if you receive an 89 or higher, you can expect to be in the top 10% (which means an A). ■

Illustration 6.13

Using the Normal Curve and z to Determine Percentiles

Find the 33rd percentile for IQ scores ($\mu = 100$ and $\sigma = 16$ from Illustration 6.10, p. 280).

SOLUTION

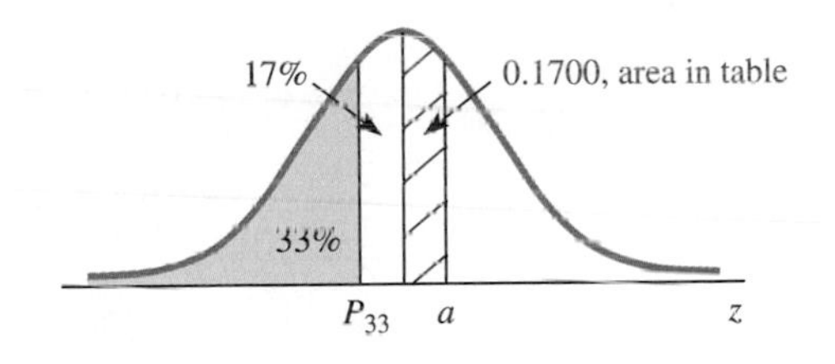

z	...	0.04	...
⋮			
0.4	...	0.1700	...

$$P(0 < z < a) = 0.17$$

$$a = 0.44 \quad \text{(cutoff value of } z \text{ from Table 3)}$$

$$\text{33rd percentile of } z = -0.44 \text{ (below mean)}$$

Now we convert the 33rd percentile of the z-scores, -0.44, to an x score using formula (6.3):

$$z = \frac{x - \mu}{\sigma}: \quad -0.44 = \frac{x - 100}{16}$$

$$x - 100 = 16(-0.44)$$

$$x = 100 - 7.04 = \mathbf{92.96}$$

Thus, 92.96 is the 33rd percentile for IQ scores. ■

Illustration 6.14 concerns a situation in which you are asked to find the mean μ when given related information.

Illustration 6.14 Using the Normal Curve and z to Determine Population Parameters

The incomes of junior executives in a large corporation are normally distributed with a standard deviation of \$1,200. A cutback is pending, at which time those who earn less than \$28,000 will be discharged. If such a cut represents 10% of the junior executives, what is the current mean salary of the group of junior executives?

SOLUTION

If 10% of the salaries are less than \$28,000, then 40% (or 0.4000) are between \$28,000 and the mean μ. Table 3 indicates that $z = -1.28$ is the standard score that occurs at $x = \$28,000$. Using formula (6.3), we can find the value of μ:

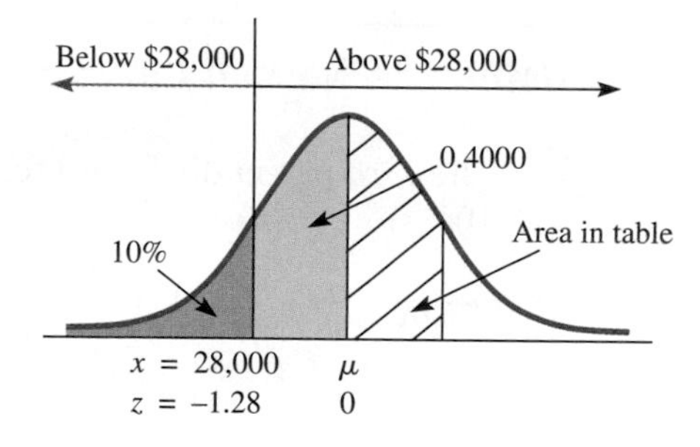

$$z = \frac{x - \mu}{\sigma}: \quad -1.28 = \frac{28{,}000 - \mu}{1{,}200}$$

$$-1{,}536 = 28{,}000 - \mu$$

$$\mu = 28{,}000 + 1{,}536 = \mathbf{\$29{,}536}$$

That is, the current mean salary of junior executives is \$29,536. ■

ANSWER NOW 6.34
Use the standard normal curve and z.

a. **Find the minimum score to receive an A if the instructor in Illustration 6.12 said the top 15% were to get A's.**

b. **Find the 25th percentile for IQ scores in Illustration 6.10.**

c. **If 20% of the salaries in Illustration 6.14 are less than \$28,000, find the current mean salary.**

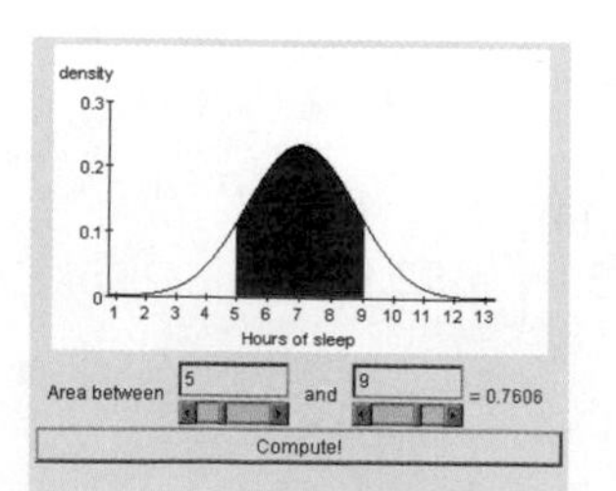

ANSWER NOW 6.35 INTERACTIVITY
Interactivity 6-A demonstrates that probability is equal to area under a curve. Given that college students sleep an average of 7 hours per night with a standard deviation equal to 1.7 hours, use the scroll bar in the applet to find:

a. **P(a student sleeps between 5 and 9 hours)**

b. **P(a student sleeps less than 4 hours)**

c. **P(a student sleeps between 8 and 11 hours)**

Referring again to IQ scores, what is the probability that a person picked at random has an IQ of 125: $P(x = 125)$? (IQ scores are normally distributed with a mean of 100 and a standard deviation of 16.) This situation has two interpretations: theoretical and practical. Let's look at the theoretical interpretation first. Recall that the probability associated with an interval for a continuous random variable is represented by the area under the curve; that is, $P(a \leq x \leq b)$ is equal to the area between a and b under the curve. $P(x = 125)$ (that is, x is exactly 125) is then $P(125 \leq x \leq 125)$, or the area of the vertical line segment at $x = 125$. This area is zero. However, this is not the practical meaning of $x = 125$. It generally means 125 to the nearest integer value. Thus, $P(x = 125)$ would most likely be interpreted as

$$P(124.5 < x < 125.5)$$

The interval from 124.5 to 125.5 under the curve has a measurable area and is then nonzero. In situations of this nature, you must be sure what meaning is being used.

NOTE: A standard notation used to abbreviate "normal distribution with mean μ and standard deviation σ" is $N(\mu, \sigma)$. That is, $N(58, 7)$ represents "normal distribution, mean = 58 and standard deviation = 7."

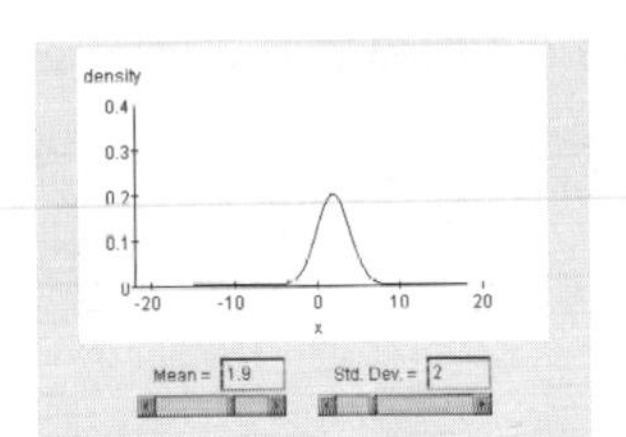

ANSWER NOW 6.36 INTERACTIVITY

Interactivity 6-B demonstrates the effects that the mean and standard deviation have on a normal curve.

a. **Leaving the standard deviation at 1, increase the mean to 3. What happens to the curve?**

b. **Returning the mean to 0, increase the standard deviation to 2. What happens to the curve?**

c. **If you could decrease the standard deviation to 0.5, what do you think would happen to the normal curve?**

Technology Instructions: Generate Random Data from a Normal Distribution

MINITAB (Release 13)

Choose: Calc > Random Data > Normal
Enter: Generate: n rows of data
Store in column(s): C1
Mean: μ
Stand. dev.: σ

If multiple samples (say, 12), all of the same size, are wanted, modify the above commands: Store in column(s): C1–C12.

NOTE: To find descriptive statistics for each of these samples, use the commands: Stat > Basic Statistics > Display Descriptive Statistics for C1–C12.

Excel XP

Choose: Tools > Data Analysis > Random Number Generation > OK
Enter: Number of Variables: 1
Number of Random Numbers: n
Distribution: Normal
Mean = : μ
Standard Deviation = : σ

(continued)

Technology Instructions: Generate Random Data from a Normal Distribution (Continued)

Excel XP (continued)

```
Select:   Output Options: Output Range
Enter:    (A1 or select cell) > OK
```

If multiple samples (say, 12), all of the same size, are wanted, modify the above commands: Number of variables: 12.

NOTE: To find descriptive statistics for each of these samples, use the commands: Tools > Data Analysis > Descriptive Statistics for columns A through L.

TI-83 Plus

```
Choose:   MATH > PRB > 6:randNorm(
Enter:    μ, σ, # of trials)
Choose:   STO→ > L1 > ENTER
```

If multiple samples (say, six), all of the same size, are wanted, repeat the above commands six times and store in L1 through L6.

NOTE: To find descriptive statistics for each of these samples, use the commands: STAT > CALC > 1:1-Var Stats for L1–L6.

ANSWER NOW 6.37
Generate a random sample of 100 data from a normal distribution with mean 50 and standard deviation 12.

ANSWER NOW 6.38
Generate ten random samples, each of size 25, from a normal distribution with mean 75 and standard deviation 14.

Technology Instructions: Calculating Ordinate Values (*y*'s) for a Normal Distribution Curve

MINITAB (Release 13)

Input the desired abscissas (*x*'s) into C1; then continue with:

```
Choose:   Calc > Probability Distributions > Normal
Select:   Probability Density
Enter:    Mean:          μ
          Stand. dev.: σ
          Input column: C1
          Optional Storage: C2
```

To draw the graph of a normal probability curve with the *x* values in C1 and the *y* values in C2, continue with:

```
Choose:   Graph > Plot
Enter:    Graph variables: Y: C2 X: C1
          Data display: Display: Connect
```

Application 6.1 A Predictive Failu

The standard normal prob
variables. This article show

The normal curve is commonly used for analysis of process capability and percentage out of specification. Additionally, some engineers use it to calculate probability of occurrence of an event. But there is another use for the normal curve that is seldom taken advantage of—process cost analysis. Using a normal curve to estimate scrap and rework is an accurate method of estimating those costs, and the calculations are simple.

A factory, for example, makes a process change that is expected to reduce scrap and save the company \$60,000 a year. Interestingly enough, neither the new or old process is capable of producing 100 percent of product to specifications, and the company has accepted sorting as a way of doing business.

The new process was started at the beginning of a week. Within three days, there was enough data for normal curve cost analysis to show that the process would not save \$60,000. In fact, it would cost an additional \$30,000 over the older process because, even though scrap had been reduced as predicted, the new process generated more rework at higher rework cost. But cost accounting didn't recognize the problem for a full month until it had enough data to identify the problem with the process' performance.

To use normal curve analysis for estimating cost, measurement error must be disregarded and the process must:

- Not be capable of meeting specifications.
- Be in statistical control.
- Produce a normal distribution.

A widget manufacturer, for example, shows how this statistical technique works.

Figure 1

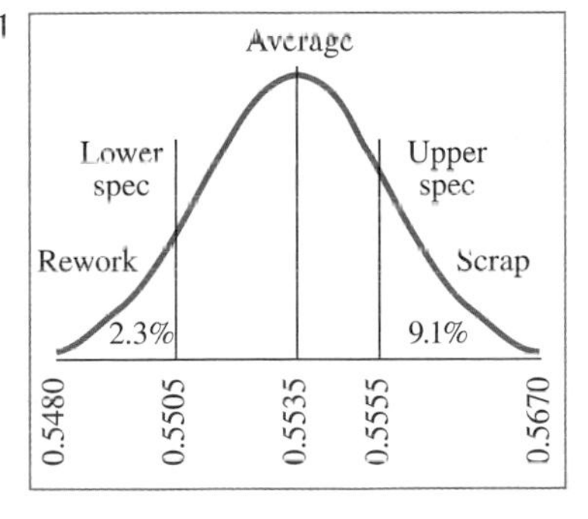

Figure 1 shows that the process is not capable of meeting specification. The process parameters indicate that all undersize parts are reworked (at a cost of \$5.25 each) and oversize parts are scrapped (at a cost of \$15.34 each).

To determine rework costs, calculate the area under the normal curve (z-score) that represents the percent of undersized parts.

$$z \text{ score} = \frac{\text{lower spec} - \text{average}}{\text{std. dev.}}$$

Excel XP Input the desired abscissas (x's) into column A and activate B1; then continue with:

Choose:	Insert function f_x > Statistical > NORMDIST > OK
Enter:	X: (A1:A100 or select 'x value' cells)
	Mean: μ
	Standard dev.: σ
	Cumulative: False > OK
Drag:	Bottom right corner of the ordinate value box down to give other ordinates

To draw the graph of a normal probability curve with the x values in column A and the y values in column B, continue with:

Choose:	Chart Wizard > XY(Scatter) > 1st picture > Next > Data Range
Enter:	Data range: (A1:B100 or select x & y cells)
Choose:	Next > Finish

TI-83 Plus The ordinate values can be calculated for individual abscissa values, x:

Choose:	2nd > DISTR > 1:normalpdf(
Enter:	x, μ, σ)

To draw the graph of the normal probability curve for a particular μ and σ, continue with:

Choose:	WINDOW
Enter:	$\mu - 3\sigma$, $\mu + 3\sigma$, σ, −.05, 1, .1, 0)
Choose:	Y= > 2nd > DISTR > 1:normalpdf(
Enter:	x, μ, σ)

After an initial graph, adjust with 0:ZoomFit from the ZOOM menu.

ANSWER NOW 6.39

Use the random sample of 100 data found in Answer Now 6.37 and find the 100 corresponding y values for the normal distribution curve with mean 50 and standard deviation 12.

ANSWER NOW 6.40

Use the 100 ordered pairs found in Answer Now 6.39 and draw the curve for the normal distribution with mean 50 and standard deviation 12.

Technology Instructions: Cumulative Probability for Normal Distributions

MINITAB (Release 13) Input the desired abscissas (x's) into C1; then continue with:

Choose:	Calc > Probability Distributions > Normal
Select:	Cumulative probability
Enter:	Mean: μ
	Stand. dev.: σ
	Input column: C1
	Optional Storage: C3

(continued)

Technology Instructions
for Normal Distribution

MINITAB (Release 13) (continued)

NOTES
1. To find the probability be
the above commands, and
2. To draw a graph of the
PLOT commands on pag

Excel XP

Input the desired abscissas
with:

```
Choose:   Insert fu
Enter:    X: (A1:A1
          Mean: μ
          Standard
          Cumulative
Drag:     Bottom ri
          box down t
```

NOTES
1. To find the probability bet
A, use the above command
2. To draw a graph of the
Chart Wizard commands
column C as the *y* values a

TI-83 Plus

The cumulative probabilities

```
Choose:   2nd > DISTR
Enter:    -1 EE 99,
```

NOTES
1. To find the probability bet
−1 EE 99 and the *x*.
2. To draw a graph of the cum
Scatter command under ST
probabilities in a pair of list

ANSWER NOW 6.41
Find the probability that a rand
mean 50 and standard deviation
using Table 3.

b. z-scores associated with the left-hand tail: Given the area B, find $z(B)$.

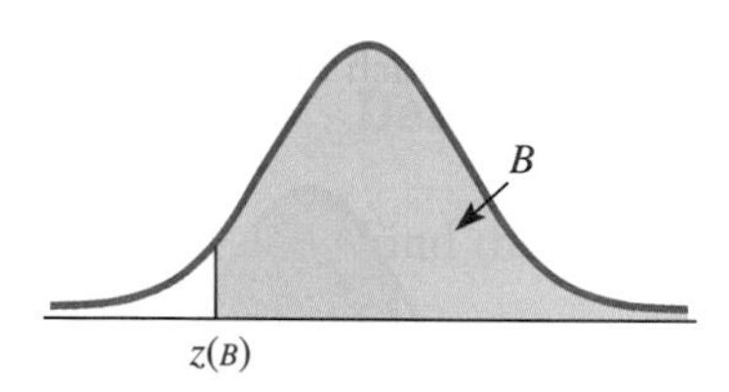

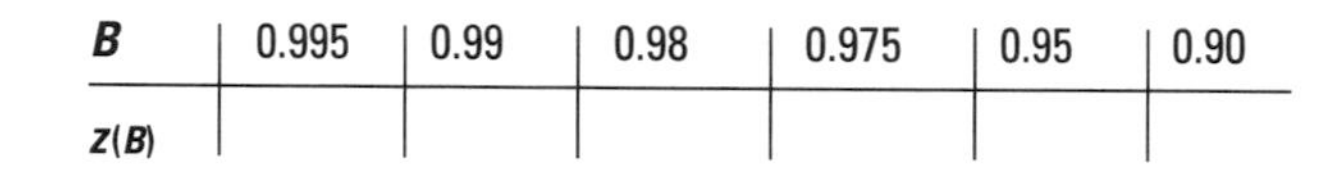

B	0.995	0.99	0.98	0.975	0.95	0.90
$z(B)$						

6.68
a. Find the area under the normal curve for z between $z(0.95)$ and $z(0.025)$.
b. Find $z(0.025) - z(0.95)$.

6.69 The z notation, $z(\alpha)$, combines two related concepts—the z-score and the area to the right—into a mathematical symbol. Identify the letter in each of the following as being a z-score or an area. Then, with the aid of a diagram, explain what both the given number and the letter represent on the standard curve.

a. $z(A) = 0.10$
b. $z(0.10) = B$
c. $z(C) = -0.05$
d. $-z(0.05) = D$

6.70 Understanding the z notation, $z(\alpha)$, requires us to know whether we have a z-score or an area. The following expressions use the z notation in a variety of ways, some typical and some not so typical. Find the value asked for in each of the following, and then with the aid of a diagram explain what your answer represents.

a. $z(0.08)$
b. the area between $z(0.98)$ and $z(0.02)$
c. $z(1.00 - 0.01)$
d. $z(0.025) - z(0.975)$

6.5 Normal Approximation of the Binomial

In Chapter 5 we introduced the **binomial distribution.** Recall that the binomial distribution is a probability distribution of the discrete random variable x, the number of successes observed in n repeated independent trials. We will now see how **binomial probabilities**—that is, probabilities associated with a binomial distribution—can be reasonably estimated by using the normal probability distribution.

Let's look first at a few specific binomial distributions. Figure 6.13 shows the probabilities of x for 0 to n for three situations: $n = 4$, $n = 8$, and $n = 24$. For each

FIGURE 6.13
Binomial Distributions

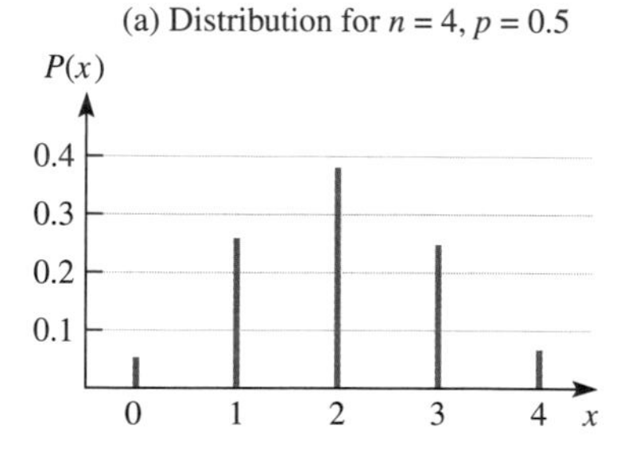

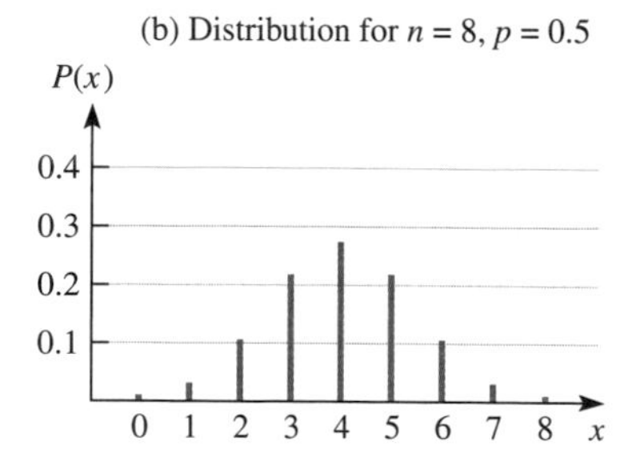

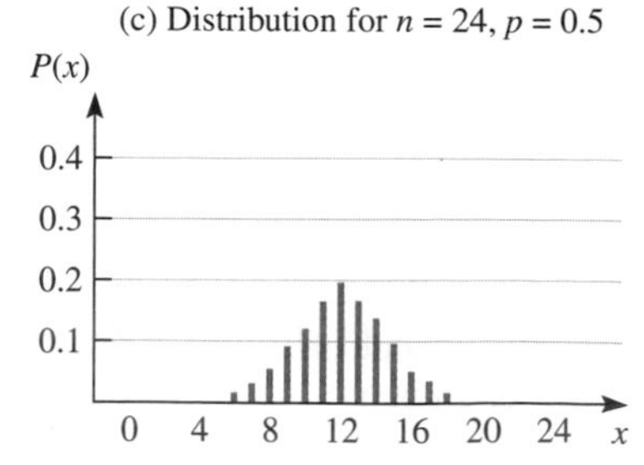

of these distributions, the probability of success for one trial is 0.5. Notice that as n becomes larger, the distribution appears more and more like the normal distribution.

To make the desired approximation, we need to take into account one major difference between the binomial and the normal probability distribution. The binomial random variable is **discrete,** whereas the normal random variable is **continuous.** Recall that Chapter 5 demonstrated that the probability assigned to a particular value of x should be shown on a diagram by means of a straight-line segment whose length represents the probability (as in Figure 6.13). Chapter 5 suggested, however, that we can also use a histogram in which the area of each bar is equal to the probability of x.

Let's look at the distribution of the binomial variable x, where $n = 14$ and $p = 0.5$. The probabilities for each x value can be obtained from Table 2 in Appendix B. This distribution of x is shown in Figure 6.14. We see the very same distribution in Figure 6.15 in histogram form.

FIGURE 6.14

The Distribution of *x* when $n = 14$, $p = 0.5$

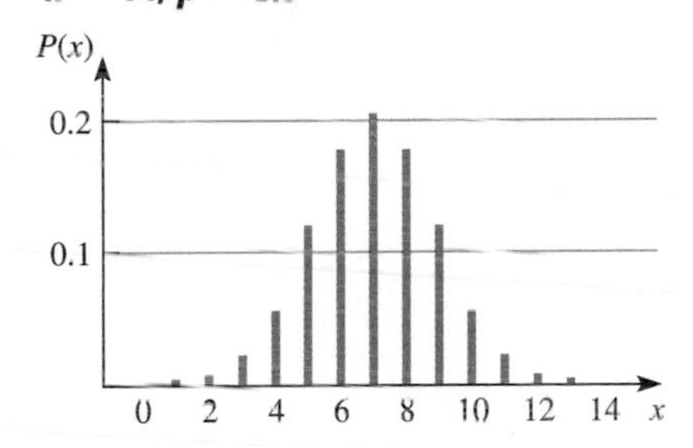

FIGURE 6.15

Histogram for the Distribution of *x* when $n = 14$, $p = 0.5$

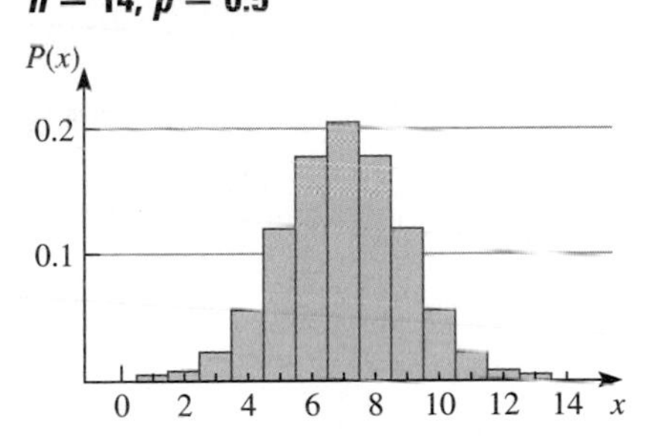

Let's examine $P(x = 4)$ for $n = 14$ and $p = 0.5$ to study the approximation technique. $P(x = 4)$ is equal to 0.061 (see Table 2 in Appendix B), the area of the bar above $x = 4$ in Figure 6.16. Area is the product of width and height. In this case the height is 0.061 and the width is 1.0, so the area is 0.061. Let's take a closer look at the width. For $x = 4$, the bar starts at 3.5 and ends at 4.5, so we are looking at an area bounded by $x = 3.5$ and $x = 4.5$. The addition and subtraction of 0.5 to the x value is commonly called the **continuity correction factor.** It is our method of converting a discrete variable into a continuous variable.

FIGURE 6.16

Area of Bar Above $x = 4$ Is 0.061, for $B(n = 14, p = 0.5)$

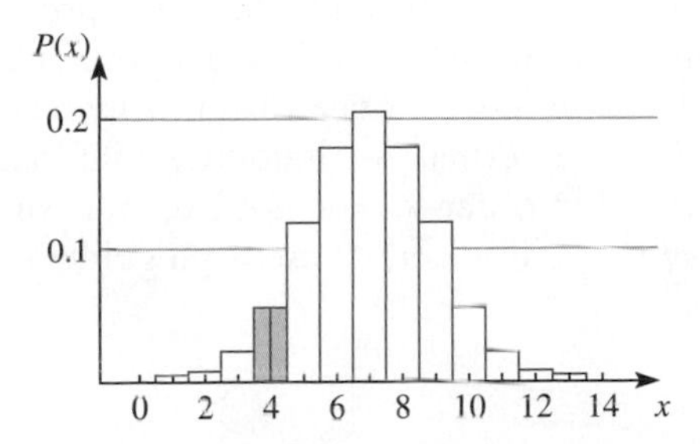

Now let's look at the normal distribution related to this situation. We will first need a normal distribution with a mean and a standard deviation equal to those of the binomial distribution we are discussing. Formulas (5.7) and (5.8) give us these values:

$$\mu = np = (14)(0.5) = \mathbf{7.0}$$

$$\sigma = \sqrt{npq} = \sqrt{(14)(0.5)(0.5)} = \sqrt{3.5} = \mathbf{1.87}$$

The probability that $x = 4$ is approximated by the area under the normal curve between $x = 3.5$ and $x = 4.5$, as shown in Figure 6.17. Figure 6.18 shows the entire distribution of the binomial variable x with a normal distribution of the same mean and standard deviation superimposed. Notice that the bars and the interval areas under the curve cover nearly the same area.

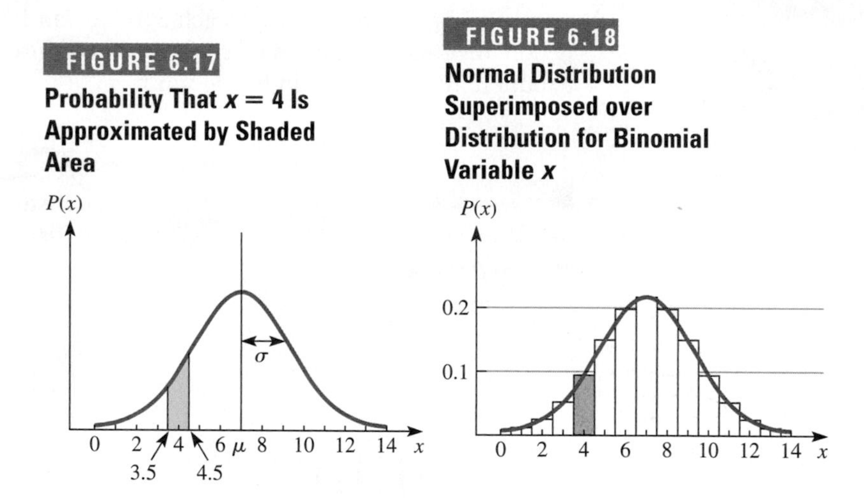

FIGURE 6.17 **Probability That $x = 4$ Is Approximated by Shaded Area**

FIGURE 6.18 **Normal Distribution Superimposed over Distribution for Binomial Variable x**

The probability that x is between 3.5 and 4.5 under this normal curve is found by using formula (6.3), Table 3, and the methods outlined in Section 6.3:

$$z = \frac{x - \mu}{\sigma}: \qquad P(3.5 < x < 4.5) = P\left(\frac{3.5 - 7.0}{1.87} < z < \frac{4.5 - 7.0}{1.87}\right)$$

$$= P(-1.87 < z < -1.34)$$

$$= 0.4693 - 0.4099 = \mathbf{0.0594}$$

Since the binomial probability of 0.061 and the normal probability of 0.0594 are reasonably close, the normal probability distribution seems to be a reasonable approximation of the binomial distribution.

The normal approximation of the binomial distribution is also useful for values of p that are not close to 0.5. The binomial probability distributions shown in Figures 6.19 and 6.20 suggest that binomial probabilities can be approximated using the normal distribution. Notice that as n increases, the binomial distribution begins to look like the normal distribution. As the value of p moves away from 0.5, a larger n is needed in order for the normal approximation to be reasonable. The following *rule of thumb* is generally used as a guideline:

RULE

The normal distribution provides a reasonable approximation to a binomial probability distribution whenever the values of np and $n(1 - p)$ both equal or exceed 5.

FIGURE 6.19
Binomial Distributions

(a) Distribution for $n = 4$, $p = 0.3$

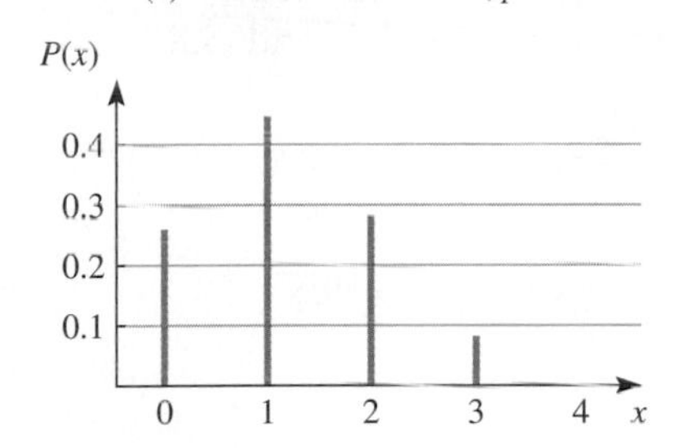

(b) Distribution for $n = 8$, $p = 0.3$

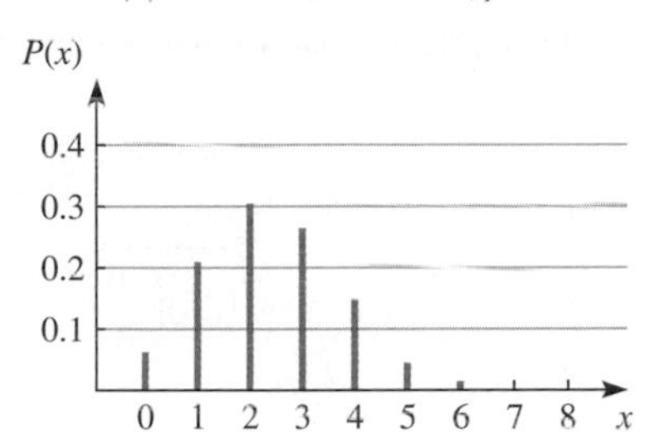

(c) Distribution for $n = 24$, $p = 0.3$

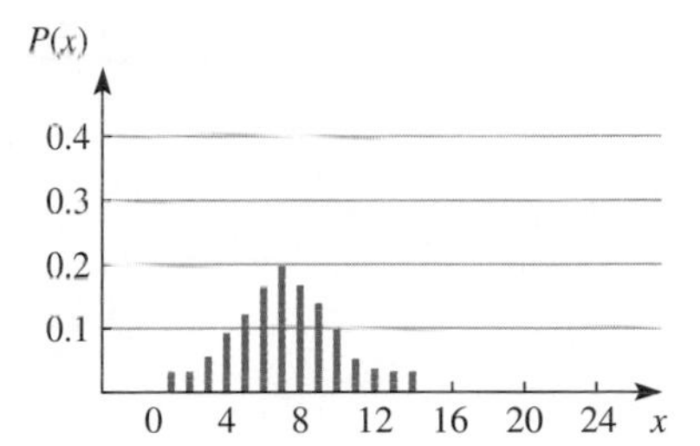

FIGURE 6.20
Binomial Distributions

(a) Distribution for $n = 4$, $p = 0.1$

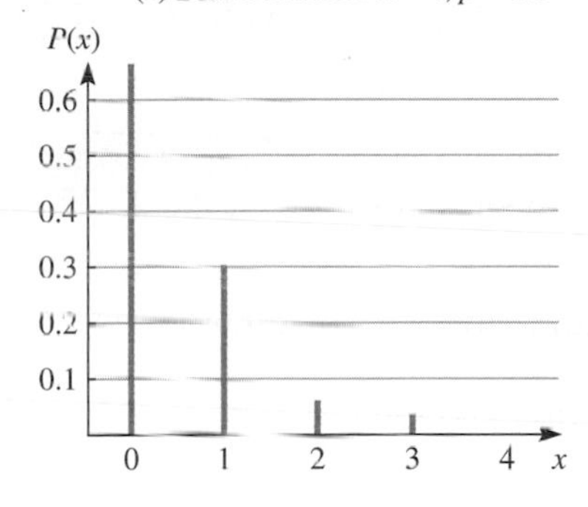

(b) Distribution for $n = 8$, $p = 0.1$

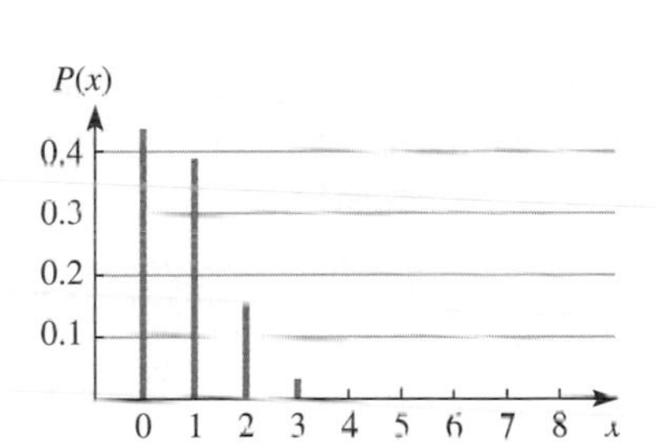

(c) Distribution for $n = 50$, $p = 0.1$

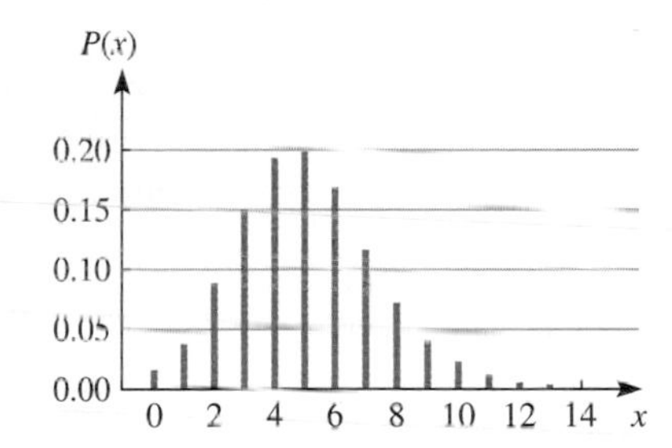

ANSWER NOW 6.71
Find the values np and nq for a binomial experiment with $n = 100$ and $p = 0.02$. Does this binomial distribution satisfy the rule of thumb for a normal approximation? Explain.

By now you may be thinking, "So what? I will just use the binomial table and find the probabilities directly and avoid all the extra work." But consider for a moment the situation presented in Illustration 6.21.

Illustration 6.21 Solving a Binomial Probability Problem with the Normal Distribution

An unnoticed mechanical failure has caused $\frac{1}{3}$ of a machine shop's production of 5000 rifle firing pins to be defective. What is the probability that an inspector will find no more than three defective firing pins in a random sample of 25?

SOLUTION

In this illustration of a binomial experiment, x is the number of defectives found in the sample, $n = 25$, and $p = P(\text{defective}) = \frac{1}{3}$. To answer the question using the binomial distribution, we will need to use the binomial probability function, formula (5.5):

$$P(x) = \binom{25}{x}\left(\frac{1}{3}\right)^x\left(\frac{2}{3}\right)^{25-x} \quad \text{for } x = 0, 1, 2, \ldots, 25$$

We must calculate the values for $P(0)$, $P(1)$, $P(2)$, and $P(3)$, since they do not appear in Table 2. This is a very tedious job because of the size of the exponent. In situations such as this, we can use the normal approximation method.

Now let's find $P(x \leq 3)$ by using the normal approximation method. We first need to find the mean and standard deviation of x, formulas (5.7) and (5.8):

$$\mu = np = (25)\left(\frac{1}{3}\right) = \mathbf{8.333}$$

$$\sigma = \sqrt{npq} = \sqrt{(25)\left(\frac{1}{3}\right)\left(\frac{2}{3}\right)} = \sqrt{5.55556} = \mathbf{2.357}$$

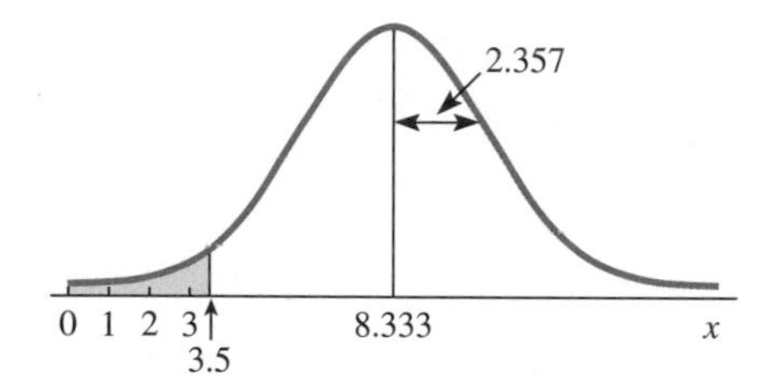

These values are shown in the figure. The measure of the shaded area ($x < 3.5$) represents the probability of $x = 0, 1, 2$, or 3. Remember that $x = 3$, the discrete binomial variable, covers the continuous interval from 2.5 to 3.5.

$$P(x \text{ is no more than } 3) = P(x \leq 3) \quad \text{(for a discrete variable } x\text{)}$$

$$= P(x < 3.5) \quad \text{(for a continuous variable } x\text{)}$$

$$z = \frac{x - \mu}{\sigma}: \qquad P(x < 3.5) = P\left(z < \frac{3.5 - 8.333}{2.357}\right) = P(z < -2.05)$$

$$= 0.5000 - 0.4798 = \mathbf{0.0202}$$

Thus, P(no more than three defectives) is approximately 0.02. ■

ANSWER NOW 6.72

a. **Calculate** $P\left(x \leq 3 \,\middle|\, B\left(25, \frac{1}{3}\right)\right)$.

b. **How good was the normal approximation? Explain. (*Hint:* If you use a computer or calculator, use the commands on p. 255.)**

Exercises

6.73 In which of the following binomial distributions does the normal distribution provide a reasonable approximation? Use computer commands to generate a graph of the distribution and compare the results to the rule of thumb. State your conclusions.

a. $n = 10, p = 0.3$
b. $n = 100, p = 0.005$
c. $n = 500, p = 0.1$
d. $n = 50, p = 0.2$

MINITAB (Release 13)

Insert the specific n and p as needed in the procedure below.

Use the Make Patterned Data commands in Exercise 6.58a, replacing the first value with 0, the last value with n, and the steps with 1.

Use the Binomial Probability Distribution commands on page 255, using C2 as optional storage.

Use the PLOT commands on page 286 for the data in C1 and C2, replacing the data display with Project, then choosing Frame > Grid, and entering Y for Direction.

Excel XP

Insert the specific n and p as needed in the procedure below.
Use the RANDOM NUMBER GENERATION Patterned Distribution commands in Exercise 6.58a, replacing the first value with 0, the last value with n, the steps with 1, and the output range with A1.
Activate cell B1; then use the Binomial Probability Distribution commands on page 255.
Use the Chart Wizard Column commands for the data in columns A and B. Choosing the Series subcommand, input column B for the y values and column A for the category (x) axis labels.

6.74 In order to see what happens when the normal approximation is used improperly, consider the binomial distribution with $n = 15$ and $p = 0.05$. Since $np = 0.75$, the rule of thumb ($np > 5$ and $nq > 5$) is not satisfied. Using the binomial tables, find the probability of one or fewer successes and compare this with the normal approximation.

6.75 Find the normal approximation for the binomial probability $P(x = 6)$, where $n = 12$ and $p = 0.6$. Compare this to the value of $P(x = 6)$ obtained from Table 2.

6.76 Find the normal approximation for the binomial probability $P(x = 4, 5)$, where $n = 14$ and $p = 0.5$. Compare this to the value of $P(x = 4, 5)$ obtained from Table 2.

6.77 Find the normal approximation for the binomial probability $P(x \leq 8)$, where $n = 14$ and $p = 0.4$. Compare this to the value of $P(x \leq 8)$ obtained from Table 2.

6.78 Find the normal approximation for the binomial probability $P(x \geq 9)$, where $n = 13$ and $p = 0.7$. Compare this to the value of $P(x \geq 9)$ obtained from Table 2.

6.79 Melanoma is the most serious form of skin cancer and its incidence is increasing at a rate higher than any other cancer in the United States. If it is caught in its early stage, the survival rate for patients is almost 90% in the United States. What is the probability that 200 or more of some group of 250 early-stage patients will survive melanoma?

6.80 If 30% of all students who enter a certain university drop out during or at the end of their first year, what is the probability that more than 600 of this year's entering class of 1800 will drop out during or at the end of their first year?

6.81 A 1999 EPA survey of Chesapeake Bay residents (www.epa.gov/npdes/menuofbmps/poll_3.htm) found that 59% of dog walkers clean up after their dogs most or all of the time.

a. What is the probability that no more than half (25) of the next 50 dog walkers you encounter clean up after their dog?
b. What is the probability that at least three-fourths (38) of the next 50 dog walkers you encounter clean up after their dog?

6.82 CyberAtlas reported results from the *Pew Internet and American Life Project*, which found that 48% of all Internet users were females. Assuming the percentage is correct, use the normal approximation to the binomial to find the probability that in a national survey of 2000 American Internet users:

a. at least 925 will be females
b. at least 980 will be females
c. at most 950 will be females
d. at most 975 will be females

6.83 Not all NBA coaches who enjoyed lengthy careers were consistently putting together winning seasons with the teams they coached. For example, Bill Fitch, who coached for 25 seasons of professional basketball after starting his coaching career at the University of Minnesota, won 944 games but lost 1106 while working with the Cavaliers, Celtics, Rockets, Nets, and Clippers. If you were to randomly select 60 box scores from the historical records of games in which Bill Fitch coached one of the teams, what is the probability that less than half of them show his team winning? To obtain your answer, use the normal approximation to the binomial distribution. (*Sports Illustrated*, "Who Is Tim Floyd?," August 3, 1998)

6.84 Researchers at the Annenberg Public Policy Center found that in a national study of young people (14 to 22 year olds), 57% thought their popular peers drank alcohol. The results were quoted in a July 9, 2002, article, "Young People Equate Drinking with Popularity," at www.jointogether.org. Use the normal approximation to the binomial distribution to find the probability that, in a poll of 1200 young people, between 600 and 900 inclusive will equate popularity with drinking.

a. Solve using the normal approximation and Table 3.
b. Solve using a computer or calculator and the normal approximation method.
c. Solve using a computer or calculator and the binomial probability function.

FYI

The commands on pages 255 and 287 may be helpful.

6.85 According to the June 13, 1994, issue of *Time* magazine, the proportion of all workers who are union members is 15.8%. Use the normal approximation to the binomial distribution to find the probability that, in a national survey of 2500 workers, at most 450 will be union members.

a. Solve using the normal approximation and Table 3.
b. Solve using a computer or calculator and the normal approximation method.

CHAPTER SUMMARY

RETURN TO CHAPTER CASE STUDY

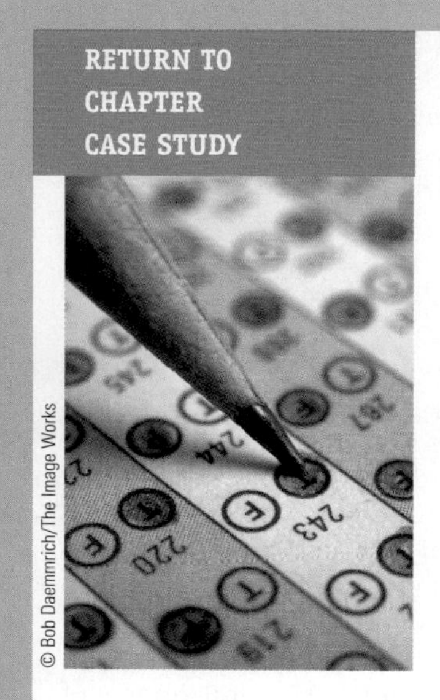
© Bob Daemmrich/The Image Works

Aptitude Tests and Their Interpretation

All normal probability distributions have the same shape and distribution relative to the mean and standard deviation. In this chapter we learned how to use the standard normal probability distribution to answer questions about all normal distributions. Let's return to the distribution of IQ scores discussed in the Chapter Case Study (p. 271) and try out some of our new knowledge.

FIRST THOUGHTS 6.86

Using your Answer Now 6.1 solutions (p. 272) as a basis, let's take a second look at the normally distributed intelligence scores in the Chapter Case Study.

a. How is an IQ score converted to a standard score?
b. What is the standard score for an IQ score of 90? 110? 120?
c. What is the standard score for an SAT score of 465? 575? 650?
d. Use Figure 2-2 (p. 272) with the empirical rule. What percentage of IQ scores is greater than 132? What percentage of SAT scores is less than 700?

PUTTING CHAPTER 6 TO WORK 6.87

Use Table 3 in Appendix B.

a. What is the probability that an IQ score is greater than 132?
b. What is the probability than an SAT score is less than 700?
c. Compare parts a and b with your answers in part d of First Thoughts that used the empirical rule and Figure 2-2 (p. 272). Explain any similarities.
d. What proportion of the IQ scores fall within the range of 80–120?
e. What proportion of the IQ scores exceed 125?
f. What percentage of the SAT scores are below 450?
g. What percentage of the SAT scores are above 575?
h. What SAT score is at the 95th percentile? Explain what this means.

YOUR OWN MINI-STUDY 6.88

Intelligence Tests

American psychologists quickly saw the value of Binet's test. In 1916, Lewis Terman and others at Stanford University revised it for use in this country. After several more revisions, the Stanford-Binet Intelligence Scale is still widely used. The Stanford-Binet assumes that intellectual ability in childhood improves as age increases. As a result, the Stanford-Binet is really a graded set of more difficult tests, one for each age group. The age-ranked questions allow a person's mental age to be measured.

The Wechsler Tests. Wechsler Adult Intelligence Scale—Revised (WAIS-R) and the Wechsler Intelligence Scale for Children (WISC-III) are widely used alternatives to the Stanford-Binet. The Wechsler tests rate performance (nonverbal) intelligence in addition to verbal intelligence and can be broken down to reveal strengths and weaknesses in various areas.

Based on scores from a large number of randomly selected people, IQ ranges have been classified as shown in Table 10-4. A look at the percentages reveals a definite pattern. The distribution of IQs approximates a normal curve, in which the majority of scores fall close to the average, with fewer at the extremes.

TABLE 10-4 Distribution of Adult IQ Scores on WAIS-R

IQ	Description	Percent
Above 130	Very superior	2.2
120–129	Superior	6.7
110–119	Bright normal	16.1
90–109	Average	50.0
80–89	Dull normal	16.1
70–79	Borderline	6.7
Below 70	Mentally retarded	2.2

Figure 10-15 shows this characteristic of measured intelligence.

FIGURE 10-15
Distribution of Stanford-Binet Intelligence Test Scores for 3184 Children

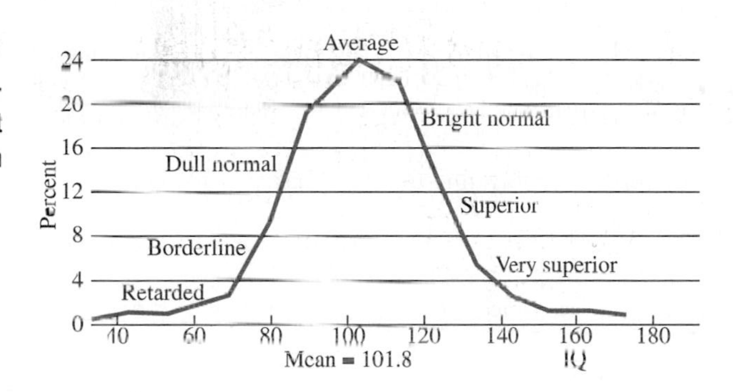

Source: Dennis Coon, *Essentials of Psychology, Exploration and Application,* 8th ed. (Belmont, CA. Wadsworth, 1999)

a. Use the information in Table 10-4 and estimate the standard deviation for adult WAIS-R scores. Use at least two different pieces of information to obtain two separate estimates. Determine your answer.
b. Does the IQ score discussed here seem to have a normal distribution? Give reasons to support your answer.
c. What percentage of the adult population has "superior" intelligence?
d. What is the probability of randomly selecting one adult from this population who is classified below "average"?
e. What IQ score is at the 95th percentile? Explain what this means.

In Retrospect

We have learned about the standard normal probability distribution, the most important family of continuous random variables. We have learned to apply it to all other normal probability distributions and how to use it to estimate probabilities of binomial distributions. We have seen a wide variety of problems (variables) that have this normal distribution or are reasonably well approximated by it.

In the next chapter we will examine sampling distributions and learn how to use the standard normal probability to solve additional applications.

Chapter Exercises

6.89 According to Chebyshev's theorem, at least how much area is under the standard normal distribution between $z = -2$ and $z = +2$? What is the actual area under the standard normal distribution between $z = -2$ and $z = +2$?

6.90 The middle 60% of a normally distributed population lies between what two standard scores?

6.91 Find the standard score z such that the area above the mean and below z under the normal curve is:
a. 0.3962 b. 0.4846 c. 0.3712

6.92 Find the standard score z such that the area below the mean and above z under the normal curve is:
a. 0.3212 b. 0.4788 c. 0.2700

6.93 Given that z is the standard normal variable, find the value of k such that:
a. $P(|z| > 1.68) = k$ b. $P(|z| < 2.15) = k$

6.94 Given that z is the standard normal variable, find the value of c such that:
a. $P(|z| > c) = 0.0384$ b. $P(|z| < c) = 0.8740$

6.95 Find the values of z:
a. $z(0.12)$ b. $z(0.28)$ c. $z(0.85)$ d. $z(0.99)$

6.96 Find the area under the normal curve that lies between each pair of z values:
a. $z = -3.00$ and $z = 3.00$ b. $z(0.975)$ and $z(0.025)$
c. $z(0.10)$ and $z(0.01)$

6.97 The length of life of a certain type of refrigerator is approximately normally distributed with a mean of 4.8 years and a standard deviation of 1.3 years.
a. If this machine is guaranteed for 2 years, what is the probability that the machine you purchased will require replacement under the guarantee?
b. What period of time should the manufacturer give as a guarantee if it is willing to replace only 0.5% of the machines?

6.98 Based on data from the ACT in 2001, the average science reasoning test score was 21.0 with a mean of 4.6. Assume that the scores are normally distributed.
a. Find the probability that a randomly selected student has a science reasoning ACT score of least 25.
b. Find the probability that a randomly selected student has a science reasoning ACT score between 20 and 26.
c. Find the probability that a randomly selected student has a science reasoning ACT score less than 16.

6.99 A machine is programmed to fill 10-oz containers with a cleanser. However, the variability inherent in any machine causes the actual amounts of fill to vary. The distribution is normal with a standard deviation of 0.02 oz. What must the mean amount, μ, be so that only 5% of the containers receive less than 10 oz?

6.100 In a large industrial complex, the maintenance department has been instructed to replace light bulbs before they burn out. It is known that the lifetimes of light bulbs are normally distributed with a mean of 900 hours and a standard deviation of 75 hours. When should the light bulbs be replaced so that no more than 10% of them will burn out while in use?

6.101 Suppose that x has a binomial distribution with $n = 25$ and $p = 0.3$.
a. Explain why the normal approximation is reasonable.
b. Find the mean and standard deviation of the normal distribution that is used in the approximation.

6.102 Let x be a binomial random variable for $n = 30$ and $p = 0.1$.
a. Explain why the normal approximation is not reasonable.
b. Find the function used to calculate the probability of any x from $x = 0$ to $x = 30$.
c. Use a computer or calculator to list the probability distribution.

6.103
a. Use a computer or calculator to list the binomial probabilities for the distribution where $n = 50$ and $p = 0.1$.
b. Use the results from part a and find $P(x \leq 6)$.
c. Find the normal approximation for $P(x \leq 6)$ and compare the results with those in part b.

6.104
a. Use a computer or calculator to list both the probability distribution and the cumulative probability distribution for the binomial probability experiment with $n = 40$ and $p = 0.4$.
b. Explain the relationship between the two distributions found in part a.
c. If you could use only one of these lists when solving problems, which one would you prefer and why?

6.105 Consider the binomial experiment with $n = 300$ and $p = 0.2$.
a. Set up, but do not evaluate, the probability expression for 75 or fewer successes in the 300 trials.
b. Use a computer or calculator to find $P(x \leq 75)$ using the binomial probability function.
c. Use a computer or calculator to find $P(x \leq 75)$ using the normal approximation.
d. Compare the answers in parts b and c.

FYI
Use the cumulative probability commands.

6.106 The grades on an examination whose mean is 525 and whose standard deviation is 80 are normally distributed.

a. Anyone who scores below 350 will be retested. What percentage does this represent?
b. The top 12% are to receive a special commendation. What score must be surpassed to receive this special commendation?
c. The interquartile range of a distribution is the difference between Q_1 and Q_3: $Q_3 - Q_1$. Find the interquartile range for the grades on this examination.
d. Find the grade such that only 1 out of 500 will score above it.

6.107 A soft-drink vending machine can be regulated so as to dispense an average of μ ounces of soft drink per glass.

a. If the ounces dispensed per glass are normally distributed with a standard deviation of 0.2 oz, find the setting for μ that will allow a 6-oz glass to hold (without overflowing) the amount dispensed 99% of the time.
b. Use a computer or calculator to simulate drawing a sample of 40 glasses of soft drink from the machine (set using your answer to part a).

(continued, next column)

MINITAB (Release 13)
Use the generate RANDOM DATA commands on page 285, replacing *n* with 40, store in with C1, mean with the value calculated in part a, and standard deviation with 0.2.
Use the HISTOGRAM commands on page 58 for the data in C1, entering cutpoints from 5 to 6.2 in increments of 0.05 (5:6.2/0.05).

Excel XP
Use the Normal RANDOM NUMBER GENERATION commands on page 285, replacing *n* with 40, the mean with the value calculated in part a, the standard deviation with 0.2, and the output range with A1.
Use the RANDOM NUMBER GENERATION Patterned Distribution on page 292, replacing the first value with 5, the last value with 6.2, the steps with 0.05, and the output range with B1.
Use the HISTOGRAM commands on page 58, with column A as the input range and column B as the bin range.

TI-83 Plus
Use the 6:randNorm commands on page 286 replacing the mean with the value calculated in part a, the standard deviation with 0.2, and the number of trials with 40. Store in L1.
Use the HISTOGRAM commands on page 59 for the data in L1, entering WINDOW VALUES: 5, 6.2, 0.05, −1, 10, 1, 1.

c. What percentage of your sample would have overflowed the cup?
d. Does your sample seem to indicate that the setting for μ is going to work? Explain.

FYI
Repeat part b a few times. Try a different value for *A* and repeat part b. Observe how many would overflow in each set of 40.

6.108 A company asserts that 80% of the customers who purchase its special lawn mower will have no repairs during the first two years of ownership. Your personal study has shown that only 70 of the 100 in your sample lasted the two years without repair expenses. What is the probability of your sample outcome or less if the actual expenses-free percentage is 80%?

6.109 A test-scoring machine is known to record an incorrect grade on 5% of the exams it grades. Find, by the appropriate method, the probability that the machine records

a. exactly 3 wrong grades in a set of 5 exams
b. no more than 3 wrong grades in a set of 5 exams
c. no more than 3 wrong grades in a set of 15 exams
d. no more than 3 wrong grades in a set of 150 exams

6.110 It is believed that 58% of married couples with children agree on methods of disciplining their children. Assuming this to be the case, what is the probability that in a random survey of 200 married couples, we would find

a. exactly 110 couples who agree
b. fewer than 110 couples who agree
c. more than 100 couples who agree

6.111 Infant mortality rates are often used to assess quality of life and adequacy of health care. The rate is the number of infant deaths that occur for every 1000 births. Prior to President Clinton's trips to China and Russia in the summer of 1998, *Newsweek* magazine published a table showing the infant mortality rates of eight nations throughout the world, including China and Russia, in 1996. Suppose the next 2000 births in each nation are tracked for the occurrence of infant deaths.

Nation	Infant Mortality
China	33
Germany	5
India	65
Japan	4
Mexico	32
Russia	17
South Africa	49
United States	7

Source: *Newsweek*, "China by the Numbers: Portrait of a Nation," June 29, 1998

a. Construct a table showing the mean and standard deviation of the associated binomial distributions.
b. In the final column of the table, find the probability that at least 70 infants from the samples in each nation will become casualties that contribute to the nation's mortality rate. Show all your work.

6.112 Princeton Survey Research Associates conducted telephone interviews between March 1 and December 22, 2000, of adults 18 years of age and older. The results of the survey found that approximately 10% of adult Internet users are between ages 55 and 64. Use the normal approximation to the binomial distribution to find the probability that in a random sample of 100 Internet users, no more than 12 will be between the ages of 55 and 64.

6.113 A survey conducted by the Association of Executive Search Consultants revealed that 75% of all chief executive officers believe that corporations should have fast-track training programs installed to help develop especially talented employees. At the same time, the study found that only 47% of the companies actually have such programs operating at their companies. Average annual sales of the companies in the sample were $2.3 billion (*Fortune*, "How to Tame the Fiercest Headhunter," July 20, 1998).

Suppose you randomly selected 50 of the questionnaires returned by the collection of CEOs. Use the normal approximation to the binomial distribution to find the probability that from within your collection:
a. more than 35 of the CEOs think that corporations should have a fast-track program installed
b. fewer than 25 of the companies have a fast-track program in operation
c. between 30 and 40 of the CEOs think that corporations should have a fast-track program installed
d. between 20 and 30 of the companies have a fast-track program in operation

6.114 The triangular distribution shown here provides an approximation to the normal distribution. Line segment l_1 has the equation $y = x/9 + 1/3$, and line segment l_2 has the equation $y = -x/9 + 1/3$.

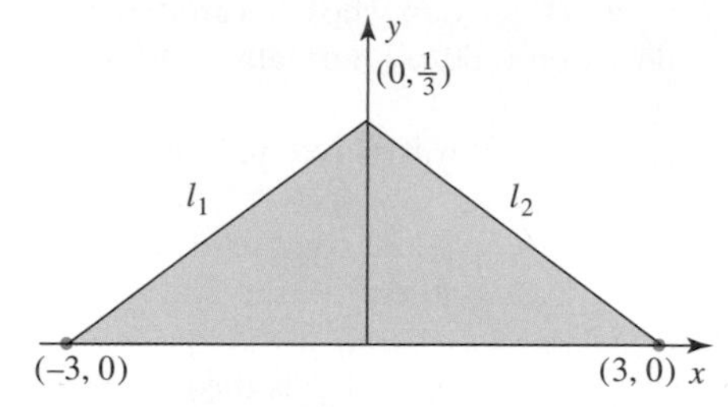

a. Find the area under the entire triangular distribution.
b. Find the area under the triangular distribution between 0 and 2.
c. Find the area under the standard normal distribution between 0 and 2.
d. Discuss the effectiveness of this "triangular" approximation.

Vocabulary and Key Concepts

Be able to define each term. Pay special attention to the key terms, which are printed in **red**. In addition, describe each term in your own words and give an example of each. Your examples should not be the ones given in class or in the textbook. The bracketed numbers indicate the chapter in which the term first appeared, but you should define the terms again to show increased understanding. Page numbers indicate the first appearance of the term in Chapter 6.

area representation for probability (p. 273)
bell-shaped curve (p. 273)
binomial distribution [5] (p. 298)
binomial probability (p. 298)
continuity correction factor (p. 299)
continuous random variable (pp. 272, 299)
discrete random variable [1, 5] (pp. 272, 299)
normal approximation of binomial (p. 298)
normal curve (p. 273)
normal distribution (p. 272)
percentage (p. 273)
probability [4] (p. 273)
proportion (p. 273)
random variable [5] (p. 272)
standard normal distribution (pp. 274, 280, 293)
standard score [2] (pp. 274, 280)
z-score [2] (pp. 274, 280)

Chapter Practice Test

PART I–KNOWING THE DEFINITIONS

Answer "True" if the statement is always true. If the statement is not always true, replace the words shown in bold with words that make the statement always true.

6.1 The normal probability distribution is symmetric about **zero.**

6.2 The total area under the curve of any normal distribution is **1.0.**

6.3 The theoretical probability that a particular value of a **continuous** random variable will occur is exactly zero.

6.4 The unit of measure for the standard score is the **same as the unit of measure of the data.**

6.5 All **normal** distributions have the same general probability function and distribution.

6.6 In the notation $z(0.05)$, the number in parentheses is the measure of the area to the **left** of the z-score.

6.7 Standard normal scores have a mean of **one** and a standard deviation of **zero.**

6.8 Probability distributions of **all** continuous random variables are normally distributed.

6.9 We are able to add and subtract the areas under the curve of a continuous distribution because these areas represent probabilities of **independent** events.

6.10 The most common distribution of a continuous random variable is the **binomial** probability.

PART II–APPLYING THE CONCEPTS

6.11 Find the following probabilities for z, the standard normal score:
a. $P(0 < z < 2.42)$ b. $P(z < 1.38)$
c. $P(z < -1.27)$ d. $P(-1.35 < z < 2.72)$

6.12 Find the value of each z-score:
a. $P(z > ?) = 0.2643$ b. $P(z < ?) = 0.17$
c. $z(0.04)$

6.13 Use the symbolic notation $z(\alpha)$ to give the symbolic name for each z-score shown in the figure at the bottom of the page.

6.14 The lifetimes of flashlight batteries are normally distributed about a mean of 35.6 hr with a standard deviation of 5.4 hr. Kevin selected one of these batteries at random and tested it. What is the probability that this one battery will last less than 40.0 hr?

6.15 The lengths of time, x, spent commuting daily, one-way, to college by students are believed to have a mean of 22 min with a standard deviation of 9 min. If the lengths of time spent commuting are approximately normally distributed, find the time, x, that separates the 25% who spend the most time commuting from the rest of the commuters.

6.16 Thousands of high school students take the SAT each year. The scores attained by the students in a certain city are approximately normally distributed with a mean of 490 and a standard deviation of 70. Find:
a. the percentage of students who score between 600 and 700
b. the percentage of students who score less than 650
c. the third quartile
d. the 15th percentile, P_{15}
e. the 95th percentile, P_{95}

PART III–UNDERSTANDING THE CONCEPTS

6.17 In 50 words, describe the standard normal distribution.

6.18 Describe the meaning of the symbol $z(\alpha)$.

6.19 Explain why the standard normal distribution, as computed in Table 3 in Appendix B, can be used to find probabilities for all normal distributions.

Figure for 6.13

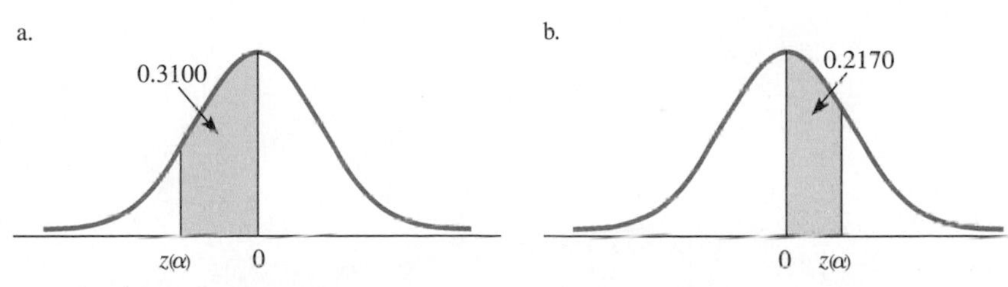

CHAPTER

7 Sample Variability

Chapter Outline and Objectives

CHAPTER CASE STUDY

Sampling from the 2000 U.S. census.

7.1 SAMPLING DISTRIBUTIONS

A distribution of values for a **sample statistic** is obtained by repeated sampling.

7.2 THE SAMPLING DISTRIBUTION OF SAMPLE MEANS

The theorem describes the sampling distribution of **sample means.**

7.3 APPLICATION OF THE SAMPLING DISTRIBUTION OF SAMPLE MEANS

The behavior of the sample means is **predictable.**

RETURN TO CHAPTER CASE STUDY

Recall our primary question: What can be deduced about the statistical population from which the sample is taken? The objective of this chapter is to study the measures and the patterns of variability for the distribution formed by repeatedly observed values of a sample mean.

Chapter Case Study

© Michael Dwyer/Stock Boston

The U.S. Census and Sampling It

The population of the United States, according to the 2000 census, consists of more than 275 million people. We read and hear about this population as the results of samples are reported nearly every day by the news media. One variable of interest to many is the age of Americans.

According to the 2000 census, the approximately 275 million Americans have a mean age of 37.09 years and a standard deviation of 22.56 years. The ages are distributed as shown in the accompanying histogram.

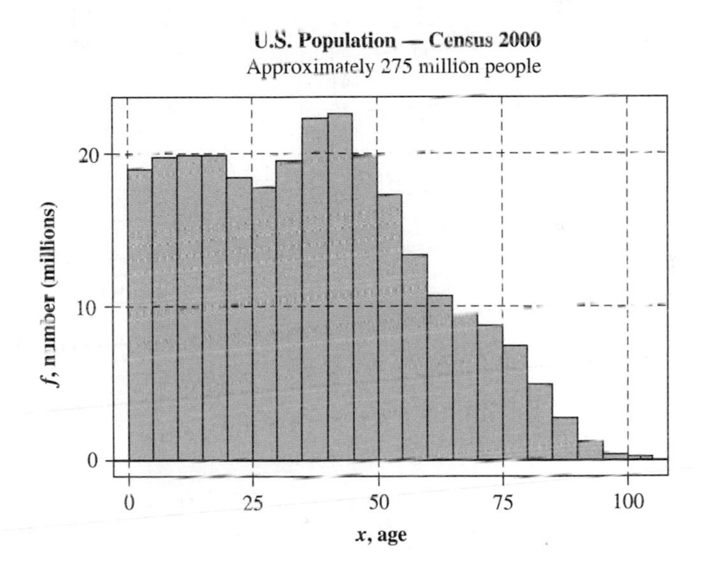

A census in the United States is done only every ten years. It is an enormous and overwhelming job, but the information that is obtained is vital to our country's organization and structure. Issues come up and times change; new information is needed and a census is impractical. This is where a representative sample comes in.

THE SAMPLING ISSUE

The fundamental goal of a survey is to come up with the same results that would have been obtained had every single member of a population been interviewed. For national Gallup polls, in other words, the objective is to present the opinions of a sample of people which are exactly the same opinions that would have been obtained had it been possible to interview all adult Americans in the country.

The key to reaching this goal is a fundamental principle called *equal probability of selection,* which states that if every member of a population has an equal probability of being selected in a sample, then that sample will be representative of the population. It's that straightforward. *(continued)*

Thus, it is Gallup's goal in selecting samples to allow every adult American an equal chance of falling into the sample. How that is done, of course, is the key to the success or failure of the process.

Source: http://www.gallup.com/help/FAQs/poll1.asp

This random sample of 100 ages was taken from the 2000 census distribution: CS07

45	78	55	15	47	85	93	46	13	41
87	78	7	7	94	48	11	41	81	32
59	8	15	20	49	66	11	61	16	19
39	74	34	6	46	8	46	21	44	41
52	84	27	53	33	48	80	6	62	21
47	11	17	3	31	43	46	23	52	20
35	24	30	37	54	90	26	55	89	2
58	44	30	45	15	25	47	13	28	10
80	41	30	57	63	79	75	7	26	4
2	10	21	19	5	62	32	59	40	16

ANSWER NOW 7.1

a. **How would you describe the sample data? (Consider both visual and numerical measures that are appropriate to organize and present the data.)**

b. **How well did the sample describe the population? Explain.**

c. **Using the graph that you constructed in part a, describe the shape of the distribution of sample data.**

d. **If another sample were to be collected, would you expect the same results? Explain.**

After completing Chapter 7, further investigate the Chapter Case Study in the Return to Chapter Case Study section with Exercises 7.41, 7.42, and 7.43 (p. 331).

7.1 Sampling Distributions

To make inferences about a population, we need to discuss sample results a little more. A sample mean $\bar{x}$ is obtained from a sample. Do you expect that this value, $\bar{x}$, is exactly equal to the value of the population mean μ? Your answer should be no. We do not expect the means to be identical, but we will be satisfied with our sample results if the sample mean is "close" to the value of the population mean. Let's consider a second question: If a second sample is taken, will the second sample have a mean equal to the population mean? Equal to the first sample mean? Again, no, we do not expect the sample mean to be equal to the population mean, nor do we expect the second sample mean to be a repeat of the first one. We do, however, again expect the values to be "close." (This argument should hold for any other sample statistic and its corresponding population value.)

The next questions should already have come to mind: What is "close"? How do we determine (and measure) this closeness? Just how will **repeated sample statistics** be distributed? To answer these questions we must look at a *sampling distribution*.

SAMPLING DISTRIBUTION OF A SAMPLE STATISTIC
The distribution of values for a sample statistic obtained from repeated samples, all of the same size and all drawn from the same population.

ANSWER NOW 7.2
The making of potato chips is the subject of the "Variability" video on your Student Suite CD (found inside the back cover of this textbook). View the video clip and answer these questions:

a. **Do independent random samples lead to the same results?**

b. **What characteristic of random samples must we learn about?**

Illustration 7.1 Forming a Sampling Distribution of Means and Ranges

FYI
Samples are drawn with replacement.

Let's consider a very small, finite population to illustrate the concept of a sampling distribution: the set of single-digit even integers, {0, 2, 4, 6, 8}, and all possible samples of size 2. We will look at two different sampling distributions that might be formed: the sampling distribution of sample means and the sampling distribution of sample ranges.

First we need to list all possible samples of size 2; there are 25 possible samples:

{0, 0}	{2, 0}	{4, 0}	{6, 0}	{8, 0}
{0, 2}	{2, 2}	{4, 2}	{6, 2}	{8, 2}
{0, 4}	{2, 4}	{4, 4}	{6, 4}	{8, 4}
{0, 6}	{2, 6}	{4, 6}	{6, 6}	{8, 6}
{0, 8}	{2, 8}	{4, 8}	{6, 8}	{8, 8}

Each of these samples has a mean $\bar{x}$. These means are, respectively:

0	1	2	3	4
1	2	3	4	5
2	3	4	5	6
3	4	5	6	7
4	5	6	7	8

TABLE 7.1

Probability Distribution: Sampling Distribution of Sample Means

$\bar{x}$	$P(\bar{x})$
0	0.04
1	0.08
2	0.12
3	0.16
4	0.20
5	0.16
6	0.12
7	0.08
8	0.04

Each of these samples is equally likely, and thus each of the 25 sample means can be assigned a probability of $\frac{1}{25} = 0.04$. (Why? See Answer Now 7.3.) The **sampling distribution of sample means** is shown in Table 7.1 as a **probability distribution** and shown in Figure 7.1 as a histogram.

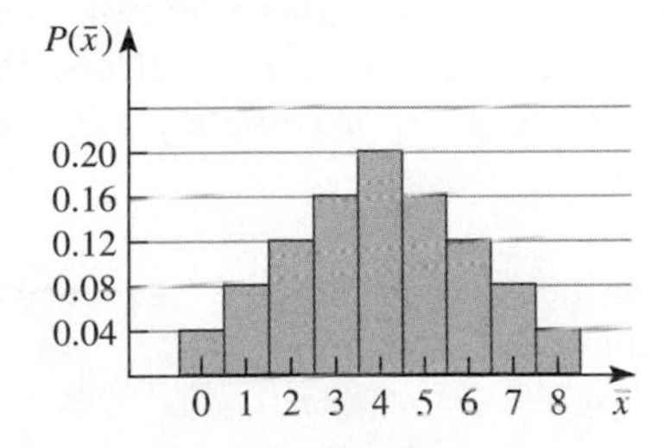

FIGURE 7.1
Histogram: Sampling Distribution of Sample Means

ANSWER NOW 7.3
Explain why the samples are equally likely; why $P(0) = 0.04$; and why $P(2) = 0.12$ (see Illustration 7.1).

For the same set of all possible samples of size 2, let's find the sampling distribution of sample ranges. Each sample has a range R. The ranges are:

0	2	4	6	8
2	0	2	4	6
4	2	0	2	4
6	4	2	0	2
8	6	4	2	0

Again, each of these 25 sample ranges has a probability of 0.04. Table 7.2 shows the sampling distribution of sample ranges as a probability distribution, and Figure 7.2 shows the sampling distribution as a histogram.

TABLE 7.2 Probability Distribution: Sampling Distribution of Sample Ranges

R	0	2	4	6	8
$P(R)$	0.20	0.32	0.24	0.16	0.08

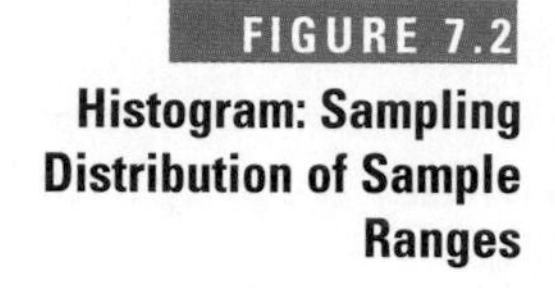

FIGURE 7.2 Histogram: Sampling Distribution of Sample Ranges

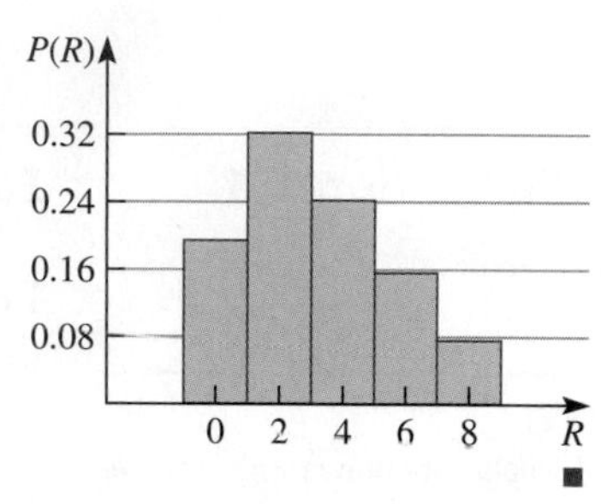

■

Illustration 7.1 is theoretical in nature and therefore expressed in probabilities. Since this population is small, it is easy to list all 25 possible samples of size 2 (a sample space) and assign probabilities. It is not always possible to do this.

Now, let's empirically (that is, by experimentation) investigate another sampling distribution.

Illustration 7.2 Creating a Sampling Distribution of Sample Means

Let's consider a population that consists of five equally likely integers: 1, 2, 3, 4, and 5. We can observe a portion of the sampling distribution of sample means when 30 samples of size 5 are randomly selected. Figure 7.3 shows a histogram representation of the population.

ANSWER NOW 7.4
Verify μ and σ for the population.

Table 7.3 shows 30 samples and their means. The resulting sampling distribution, a **frequency distribution**, of sample means is shown in Figure 7.4. Notice that this distribution of sample means does not look like the population. Rather, it seems to display the characteristics of a normal distribution; it is mounded and nearly symmetric about its mean (approximately 3.0).

ANSWER NOW 7.5
Table 7.3 lists 30 $\bar{x}$ values. Construct a grouped frequency distribution to verify the frequency distribution shown in Figure 7.4.

ANSWER NOW 7.6
Find the mean and standard deviation of the 30 $\bar{x}$ values in Table 7.3 to verify the values for $\bar{\bar{x}}$ and $s_{\bar{x}}$. Explain the meaning of the two symbols $\bar{\bar{x}}$ and $s_{\bar{x}}$.

FIGURE 7.3

The Population: Theoretical Probability Distribution

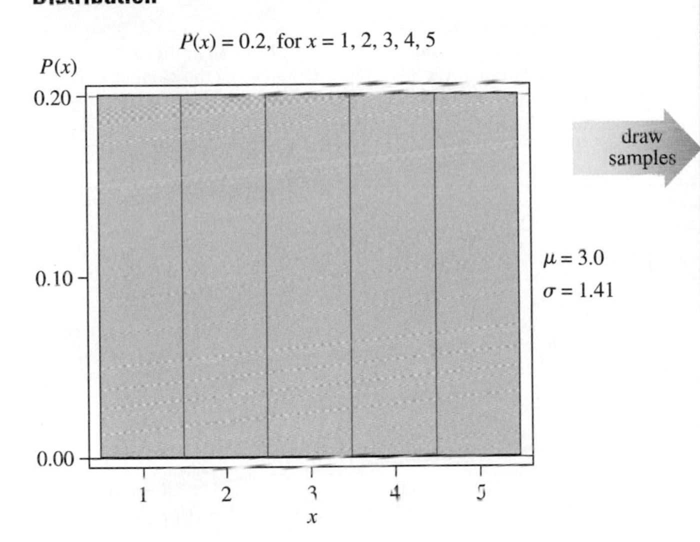

TABLE 7.3 **30 Samples of Size 5** Ta07-03

No.	Sample	$\bar{x}$	No.	Sample	$\bar{x}$
1	4,5,1,4,5	3.8	16	4,5,5,3,5	4.4
2	1,1,3,5,1	2.2	17	3,3,1,2,1	2.0
3	2,5,1,5,1	2.8	18	2,1,3,2,2	2.0
4	4,3,3,1,1	2.4	19	4,3,4,2,1	2.8
5	1,2,5,2,4	2.8	20	5,3,1,4,2	3.0
6	4,2,2,5,4	3.4	21	4,4,2,2,5	3.4
7	1,4,5,5,2	3.4	22	3,3,5,3,5	3.8
8	4,5,3,1,2	3.0	23	3,4,4,2,2	3.0
9	5,3,3,3,5	3.8	24	3,3,4,5,3	3.6
10	5,2,1,1,2	2.2	25	5,1,5,2,3	3.2
11	2,1,4,1,3	2.2	26	3,3,3,5,2	3.2
12	5,4,3,1,1	2.8	27	3,4,4,4,4	3.8
13	1,3,1,5,5	3.0	28	2,3,2,4,1	2.4
14	3,4,5,1,1	2.8	29	2,1,1,2,4	2.0
15	3,1,5,3,1	2.6	30	5,3,3,2,5	3.6

using the 30 means

FIGURE 7.4

Empirical Distribution of Sample Means

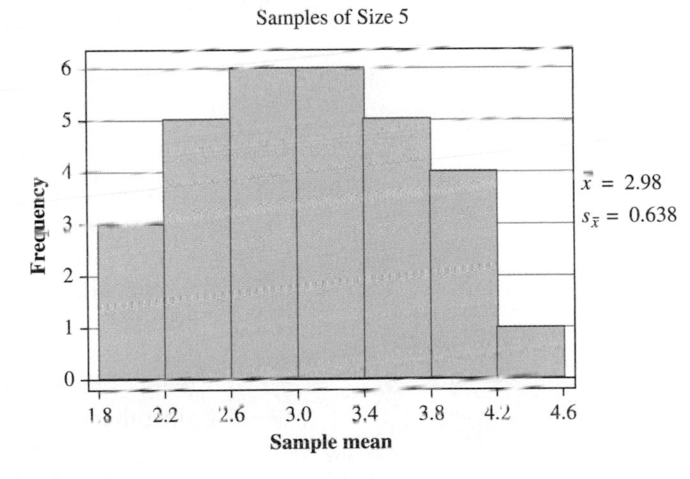

■

NOTE The variable for the sampling distribution is $\bar{x}$; therefore, the mean of the $\bar{x}$'s is $\bar{\bar{x}}$ and the standard deviation of $\bar{x}$ is $s_{\bar{x}}$.

The theory involved with sampling distributions that will be described in the remainder of this chapter requires *random sampling.*

RANDOM SAMPLE

A sample obtained in such a way that each possible sample of fixed size n has an equal probability of being selected (see p. 23).

Figure 7.5 shows how the sampling distribution of sample means is formed.

FIGURE 7.5 The Sampling Distribution of Sample Means

Statistical population being studied	Repeated sampling is needed to form the sampling distribution.	All possible samples of size n	One value of the sample statistic ($\bar{x}$ in this case) corresponding to the parameter of interest (μ in this case) is obtained from each sample	Then all of these values of the sample statistic, $\bar{x}$, are used to form the sampling distribution.

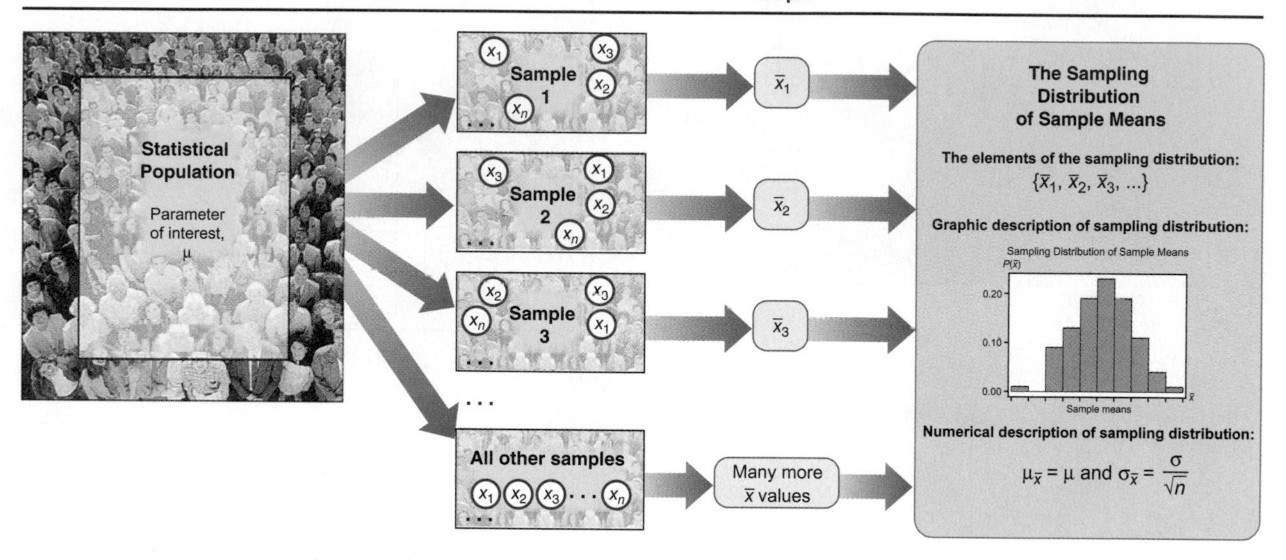

Application 7.1 Average Aircraft Age

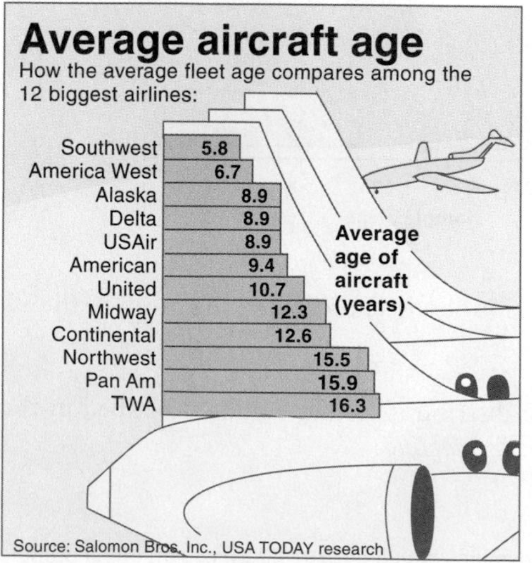

USA Today's "Average aircraft age" shows the average ages of the aircraft that make up the fleets of the 12 biggest airline companies. Each company reported its own average fleet age. The information is shown in the form of a bar graph so that each company's average fleet age can be easily visualized and compared to the others. With the variable "average fleet age," a frequency distribution or a histogram could have been used to present the data; however, that distribution would not be part of a sampling distribution.

ANSWER NOW 7.7

a. **Construct a frequency histogram of average fleet age using class boundaries 4.5, 6.5, ...**

b. **Explain why this distribution of "average fleet age" is not part of a sampling distribution.**

Exercises

7.8
a. What is the sampling distribution of sample means?
b. A sample of size 3 is taken from a population and the sample mean is found. Describe how this sample mean is related to the sampling distribution of sample means.

7.9 Consider the set of odd single-digit integers {1, 3, 5, 7, 9}.
a. Make a list of all samples of size 2 that can be drawn from this set of integers. (Sample with replacement; that is, the first number is drawn, observed, and then replaced before the next drawing.)
b. Construct the sampling distribution of sample means for samples of size 2 selected from this set.
c. Construct the sampling distribution of sample ranges for samples of size 2.

7.10 Consider the set of even single-digit integers {0, 2, 4, 6, 8}.
a. Make a list of all the possible samples of size 3 that can be drawn from this set of integers. (Sample with replacement; that is, the first number is drawn, observed, and then replaced before the next drawing.)
b. Construct the sampling distribution of the sample medians for samples of size 3.
c. Construct the sampling distribution of the sample means for samples of size 3.

7.11 Using the telephone numbers listed in your local directory as your population, obtain 20 random samples of size 3. From each telephone number identified as a source, take the fourth, fifth, and sixth digits. (For example, for 245-8268, you would take the 8, the 2, and the 6 as your sample of size 3.)
a. Calculate the mean of the 20 samples.
b. Draw a histogram showing the 20 sample means. (Use classes −0.5 to 0.5, 0.5 to 1.5, 1.5 to 2.5, and so on.)
c. Describe the distribution of $\bar{x}$'s that you see in part b (shape of distribution, center, amount of dispersion).
d. Draw 20 more samples and add the 20 new $\bar{x}$'s to the histogram in part b. Describe the distribution that seems to be developing.

7.12 Using a set of five dice, roll the dice and determine the mean number of dots showing on the five dice. Repeat the experiment until you have 25 sample means.
a. Draw a dotplot showing the distribution of the 25 sample means. (See Illustration 7.2 on p. 314.)
b. Describe the distribution of $\bar{x}$'s in part a.
c. Repeat the experiment to obtain 25 more sample means and add these 25 $\bar{x}$'s to your dotplot. Describe the distribution of 50 means.

7.13 From the table of random numbers in Appendix B, construct another table showing 20 sets of five randomly selected single digit integers. Find the mean of each set, the grand mean, and compare this value to the theoretical population mean μ using the absolute difference and the percent of error. Show all your work.

7.14
a. Simulate (using a computer or a random-number table) the drawing of 100 samples, each size 5, from the uniform probability distribution of single-digit integers, 0 to 9.
b. Find the mean of each sample.
c. Construct a histogram of the sample means. (Use integer values as class marks.)
d. Describe the sampling distribution shown in the histogram.

MINITAB (Release 13)
a. Use the Integer RANDOM DATA commands on page 183, replacing generate with 100, store in with C1–C5, minimum value with 0, and maximum value with 9.

b.
```
Choose:   Calc > Row Statistics
Select:   Mean
Enter:    Input variables: C1-C5
          Store result in: C6
```

c. Use the HISTOGRAM commands on page 58 for the data in C6, entering midpoints from 0 to 9 in increments of 1.

Excel XP
a. Input 0 through 9 into column A and the corresponding 0.1's into column B; then continue with:

```
Choose:   Tools > Data Analysis > Random
          Number Generation > OK
Enter:    Number of Variables: 5
          Number of Random Numbers: 100
          Distribution: Discrete
          Value and Probability Input Range:
          (A1:B10 or select cells)
Select:   Output Range:
Enter:    (C1 or select cell) > OK
```

b. Activate cell H1:

```
Choose:   Insert function, fx > Statistical
          > AVERAGE > OK
Enter:    Number1: (C1:G1 or select cells)
Drag:     Bottom right corner of average value
          box down to give other averages
```

c. Use the HISTOGRAM commands on page 59, with column H as the input range and column A as the bin range.

TI-83 Plus
a. Use the Integer RANDOM DATA and STO commands on page 184, replacing the Enter with 0,9,100).

(continued)

TI-83 Plus (Continued)

Repeat the above commands four more times, storing data in L2, L3, L4, L5, respectively.

```
b.  Choose:     STAT > EDIT > 1:Edit
    Highlight:  L6 (column heading)
    Enter:          (L1+L2+L3+L4+L5)/5
c.  Choose:     2nd > STAT PLOT > 1:Plot1
```

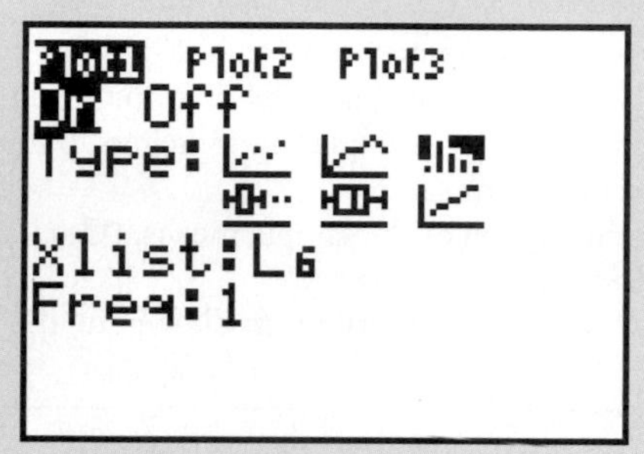


```
    Choose:     Window
    Enter:      0, 9, 1, 0, 30, 5, 1
    Choose:     Trace >>>
```

MINITAB (Release 13)

a. Use the Normal RANDOM DATA commands on page 285, replacing generate with 200, store in with C1–C10, mean with 100, and standard deviation with 20.

```
b.  Choose:     Calc > Row Statistics
    Select:     Mean
    Enter:      Input variables: C1-C10
                Store result in: C11
```

c. Use the HISTOGRAM commands on page 58 for the data in C11, entering cutpoints from 74.8 to 125.2 in increments of 6.3.

Excel XP

a. Use the Normal RANDOM NUMBER GENERATION commands on page 285, replacing number of variables with 10, number of random numbers with 200, mean with 100, and standard deviation with 20.

b. Activate cell K1:

```
    Choose:     Insert function, fx > Statistical
                > AVERAGE > OK
    Enter:      Number1: (A1:J1 or select cells)
    Drag:       Bottom right corner of average value
                box down to give other averages
```

c. Use the RANDOM NUMBER GENERATION Patterned Distribution commands in Exercise 6.58a on page 292, replacing the first value with 74.8, the last value with 125.2, the steps with 6.3, and the output range with L1. Use the HISTOGRAM commands on page 58, with column K as the input range and column L as the bin range.

7.15

a. Use a computer to draw 200 random samples, each of size 10, from the normal probability distribution with mean 100 and standard deviation 20.
b. Find the mean for each sample.
c. Construct a frequency histogram of the 200 sample means.
d. Describe the sampling distribution shown in the histogram.

7.2 The Sampling Distribution of Sample Means

On the preceding pages we discussed the sampling distributions of two statistics: sample means and sample ranges. Many others could be discussed; however, the only sampling distribution of concern to us at this time is the sampling distribution of sample means.

SAMPLING DISTRIBUTION OF SAMPLE MEANS (SDSM)

If all possible random samples, each of size n, are taken from any population with mean μ and standard deviation σ, then the sampling distribution of sample means will:

1. have a mean $\mu_{\bar{x}}$ equal to μ, and
2. have a standard deviation $\sigma_{\bar{x}}$ equal to $\frac{\sigma}{\sqrt{n}}$.

Furthermore, if the sampled population has a normal distribution, then the sampling distribution of $\bar{x}$ will also be normal for samples of all sizes.

NOTE This is very useful information!

This is a very interesting two-part statement. The first part tells us about the relationship between the population mean and standard deviation, and the sampling distribution mean and standard deviation for all sampling distributions of sample means. The second part indicates that this information is not always useful. Stated differently, it says that the mean value of only a few observations will be normally distributed when samples are drawn from a normally distributed population, but will not be normally distributed when the sampled population is uniform, skewed, or otherwise not normal. However, the central limit theorem gives us some additional and very important information about the sampling distribution of sample means.

FYI
Truly amazing: $\bar{x}$ is normally distributed when n is large enough, no matter what shape the population is!

CENTRAL LIMIT THEOREM (CLT)
The sampling distribution of sample means will more closely resemble the normal distribution as the sample size increases.

If the sampled distribution is normal, then the sampling distribution of sample means (SDSM) is normal, as stated above, and the central limit theorem (CLT) does not apply. But, if the sampled population is not normal, the sampling distribution will still be approximately normally distributed under the right conditions. If the sampled distribution is nearly normal, the $\bar{x}$ distribution is approximately normal for fairly small n (possibly as small as 15). When the sampled distribution lacks symmetry, n may have to be quite large (maybe 50 or more) before the normal distribution provides a satisfactory approximation.

STANDARD ERROR OF THE MEAN ($\sigma_{\bar{x}}$)
The standard deviation of the sampling distribution of sample means.

By combining the preceding information, we can describe the sampling distribution of $\bar{x}$ completely: (1) the location of the center (mean), (2) a measure of spread indicating how widely the distribution is dispersed (standard deviation), and (3) an indication of how it is distributed.

1. $\mu_{\bar{x}} = \mu$; the mean of the sampling distribution ($\mu_{\bar{x}}$) is equal to the mean of the population (μ).
2. $\sigma_{\bar{x}} = \dfrac{\sigma}{\sqrt{n}}$; the standard error of the mean ($\sigma_{\bar{x}}$) is equal to the standard deviation of the population (σ) divided by the square root of the sample size, n.
3. The distribution of sample means is normal when the parent population is normally distributed, and the CLT tells us that the distribution of sample means becomes approximately normal (regardless of the shape of the parent population) when the sample size is large enough.

NOTE The n referred to is the size of each sample in the sampling distribution. (The number of repeated samples used in an empirical situation has no effect on the standard error.)

We do not show the proof for the preceding three facts in this text; however, their validity will be demonstrated by examining two illustrations. For the first illustration, let's consider a population for which we can construct the theoretical sampling distribution of all possible samples.

Illustration 7.3 Constructing a Sampling Distribution of Sample Means

Let's consider all possible samples of size 2 that could be drawn from a population that contains the three numbers 2, 4, and 6. First let's look at the population itself: Construct a histogram to picture its distribution, Figure 7.6; calculate the mean μ and the standard deviation σ, Table 7.4. (Remember: We must use the techniques from Chapter 5 for discrete probability distributions.)

Table 7.5 lists all the possible samples of size 2 that can be drawn from this population. (One number is drawn, observed, and then returned to the population before the second number is drawn.) Table 7.5 also lists the means of these samples. The probability distribution for these means and the extensions are given in Table 7.6, along with the calculation of the mean and the standard error of the mean for the sampling distribution. The histogram for the sampling distribution of sample means is shown in Figure 7.7.

Let's now check the truth of the three facts about the sampling distribution of sample means:

1. The mean $\mu_{\bar{x}}$ of the sampling distribution will equal the mean μ of the population: Both μ and $\mu_{\bar{x}}$ have the value **4.0.**
2. The standard error of the mean $\sigma_{\bar{x}}$ for the sampling distribution will equal the standard deviation σ of the population divided by the square root of the sample size, n: $\sigma_{\bar{x}} = \mathbf{1.15}$ and $\sigma = 1.63$, $n = 2$, $\frac{\sigma}{\sqrt{n}} = \frac{1.63}{\sqrt{2}} = \mathbf{1.15}$**;** they are equal: $\sigma_{\bar{x}} = \frac{\sigma}{\sqrt{n}}$.
3. The distribution will become approximately normally distributed: The histogram in Figure 7.7 very strongly suggests normality. ■

Illustration 7.3, a theoretical situation, suggests that all three facts appear to hold true. Do these three facts hold when actual data are collected? Let's look back at Illustration 7.2 (p. 314) and see if all three facts are supported by the empirical sampling distribution there.

First, let's look at the population—the theoretical probability distribution from which the samples in Illustration 7.2 were taken. Figure 7.3 is a histogram showing the probability distribution for randomly selected data from the population of equally likely integers 1, 2, 3, 4, 5. The population mean μ equals 3.0. The population standard deviation σ is $\sqrt{2}$, or 1.41. The population has a uniform distribution.

Now let's look at the empirical distribution of the 30 sample means found in Illustration 7.2. From the 30 values of $\bar{x}$ in Table 7.3, the observed mean of the $\bar{x}$'s, $\bar{\bar{x}}$, is 2.98 and the observed standard error of the mean, $s_{\bar{x}}$, is 0.638. The histogram of the sampling distribution in Figure 7.4 appears to be mounded, approximately symmetrical, and centered near the value 3.0.

Now let's check the truth of the three specific properties:

1. $\mu_{\bar{x}}$ and μ will be equal: The mean of the population μ is 3.0, and the observed sampling distribution mean $\bar{\bar{x}}$ is 2.98; they are very close in value.
2. $\sigma_{\bar{x}}$ will equal $\frac{\sigma}{\sqrt{n}}$: $\sigma = 1.41$ and $n = 5$; therefore, $\frac{\sigma}{\sqrt{n}} = \frac{1.41}{\sqrt{5}} = \mathbf{0.632}$, and $s_{\bar{x}} = \mathbf{0.638}$**;** they are very close in value. (Remember that we have taken only 30 samples, not all possible samples, of size 5.)
3. The sampling distribution of $\bar{x}$ will be approximately normally distributed. Even though the population has a rectangular distribution, the histogram in Figure 7.4 suggests that the $\bar{x}$ distribution has some of the properties of normality (mounded, symmetric).

FIGURE 7.6
Population

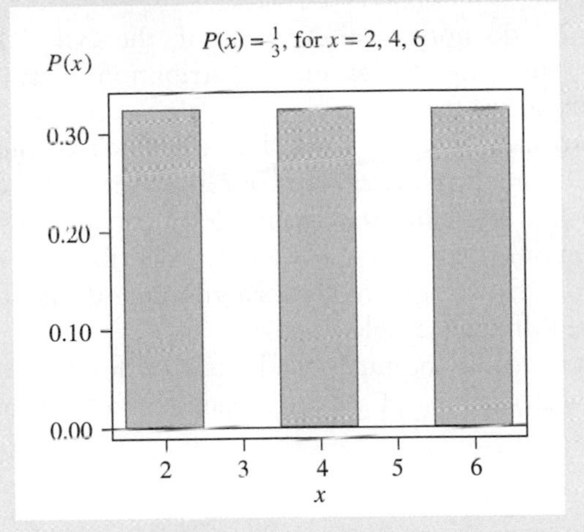

TABLE 7.4 **Extensions Table for x**

x	$P(x)$	$xP(x)$	$x^2P(x)$
2	$\frac{1}{3}$	$\frac{2}{3}$	$\frac{4}{3}$
4	$\frac{1}{3}$	$\frac{4}{3}$	$\frac{16}{3}$
6	$\frac{1}{3}$	$\frac{6}{3}$	$\frac{36}{3}$
Σ	$\frac{3}{3}$	$\frac{12}{3}$	$\frac{56}{3}$
	1.0	4.0	18.66

$\mu = \mathbf{4.0}$

$\sigma = \sqrt{18.6\overline{6} - (4.0)^2} = \sqrt{2.6\overline{6}} = \mathbf{1.63}$

draw samples

TABLE 7.5 **All Nine Possible Samples of Size 2**

Sample	$\bar{x}$	Sample	$\bar{x}$	Sample	$\bar{x}$
2, 2	2	4, 2	3	6, 2	4
2, 4	3	4, 4	4	6, 4	5
2, 6	4	4, 6	5	6, 6	6

TABLE 7.6 **Extensions Table for $\bar{x}$**

$\bar{x}$	$P(\bar{x})$	$\bar{x}P(\bar{x})$	$\bar{x}^2P(\bar{x})$
2	$\frac{1}{9}$	$\frac{2}{9}$	$\frac{4}{9}$
3	$\frac{2}{9}$	$\frac{6}{9}$	$\frac{18}{9}$
4	$\frac{3}{9}$	$\frac{12}{9}$	$\frac{48}{9}$
5	$\frac{2}{9}$	$\frac{10}{9}$	$\frac{50}{9}$
6	$\frac{1}{9}$	$\frac{6}{9}$	$\frac{36}{9}$
Σ	$\frac{9}{9}$	$\frac{36}{9}$	$\frac{156}{9}$
	1.0	4.0	$17.3\overline{3}$

$\mu_{\bar{x}} = \mathbf{4.0}$

$\sigma_{\bar{x}} = \sqrt{17.3\overline{3} - (4.0)^2} = \sqrt{1.3\overline{3}} = \mathbf{1.15}$

FIGURE 7.7
Sampling Distribution of Sample Means

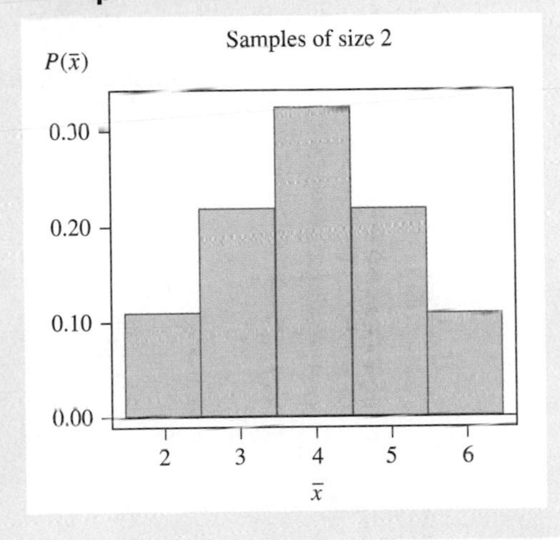

ANSWER NOW 7.16

If a population has a standard deviation σ of 25 units, what is the standard error of the mean if samples of size 16 are selected? Samples of size 36? Samples of size 100?

Although Illustrations 7.2 and 7.3 do not constitute a proof, the evidence seems to strongly suggest that both statements, the sampling distribution of sample means and the central limit theorem, are true.

Having taken a look at these two specific illustrations, let's now look at four graphic illustrations that present the sampling distribution information and the CLT in a slightly different form. Each of these illustrations has four distributions. The first graph shows the distribution of the parent population, the distribution of the individual x values. Each of the other three graphs shows a sampling distribution of sample means, $\bar{x}$'s, using three different sample sizes.

In Figure 7.8 we have a uniform distribution, much like Figure 7.3 for the integer illustration, and the resulting distributions of sample means for samples of sizes 2, 5, and 30.

FIGURE 7.8
Uniform Distribution

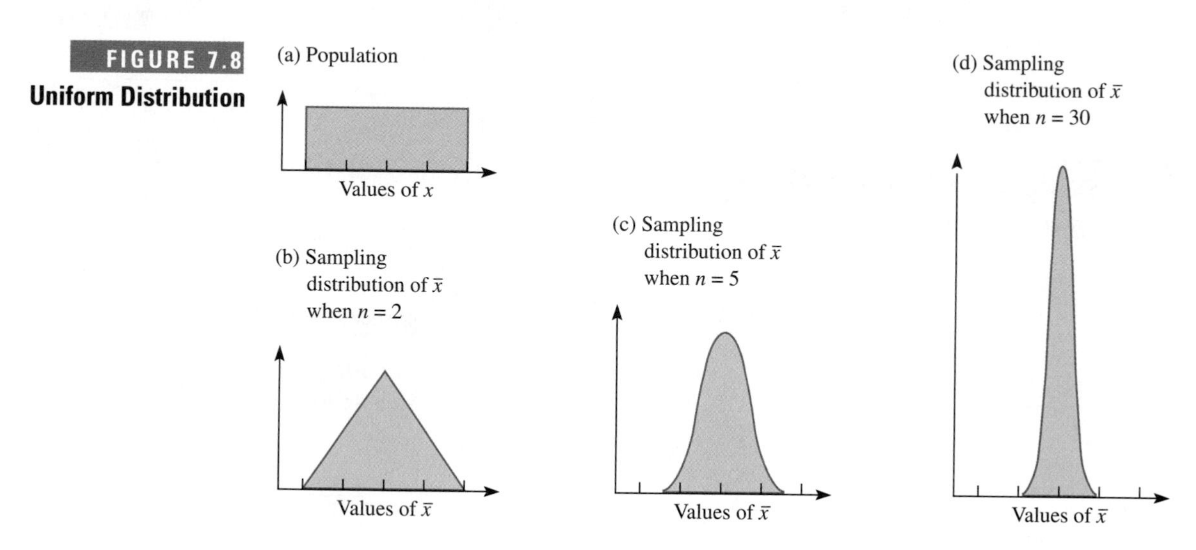

Figure 7.9 shows a U-shaped population and the three sampling distributions.

FIGURE 7.9
U-Shaped Distribution

(a) Population

Values of x

(b) Sampling distribution of $\bar{x}$ when $n = 2$

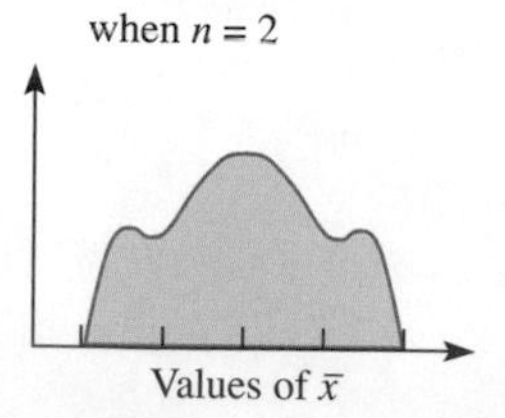

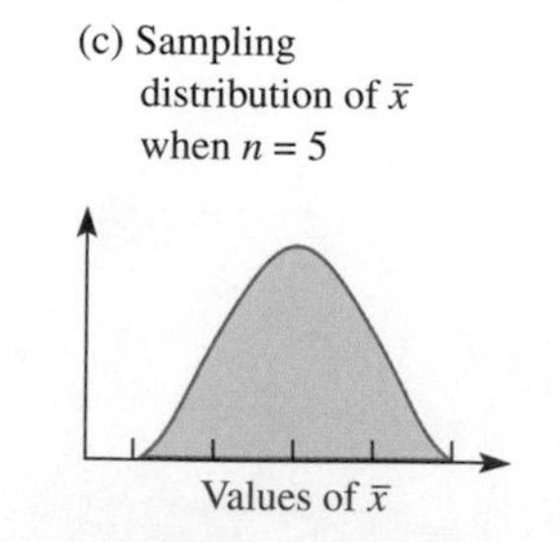

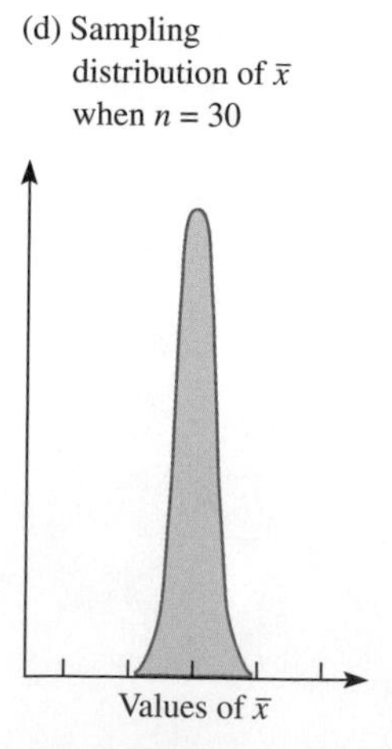

Figure 7.10 shows a J-shaped population and the three sampling distributions.

FIGURE 7.10
J-Shaped Distribution

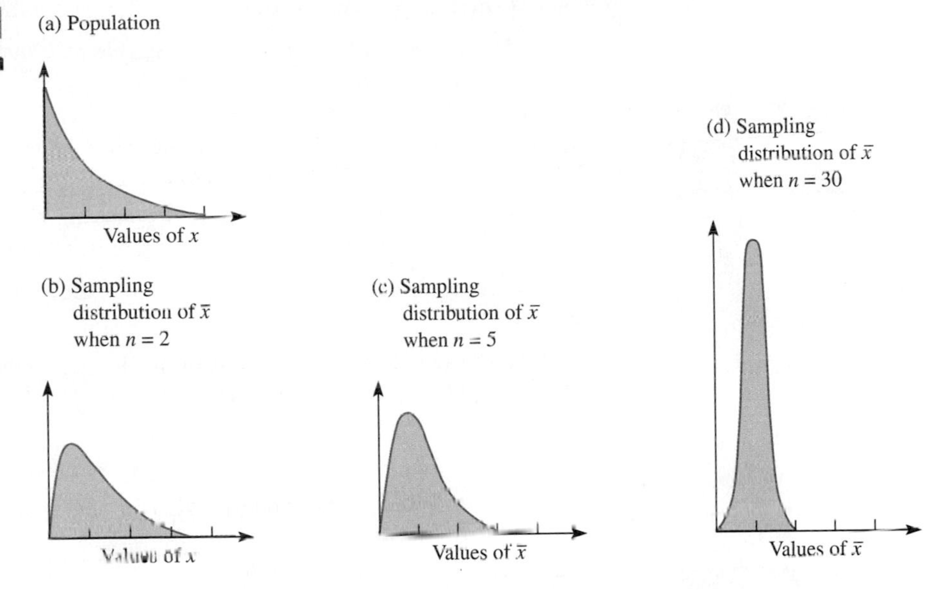

All three nonnormal distributions seem to verify the CLT; the sampling distributions of sample means appear to be approximately normal for all three when samples of size 30 were used. With the normal population (see Figure 7.11), the sampling distributions for all sample sizes appear to be normal. Thus, you have seen an amazing phenomenon: No matter what the shape of a population, the sampling distribution of sample means either is normal or becomes approximately normal when n becomes sufficiently large.

Figure 7.11 shows a normally distributed population and the three sampling distributions.

FIGURE 7.11
Normal Distribution

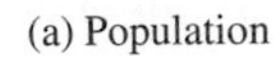

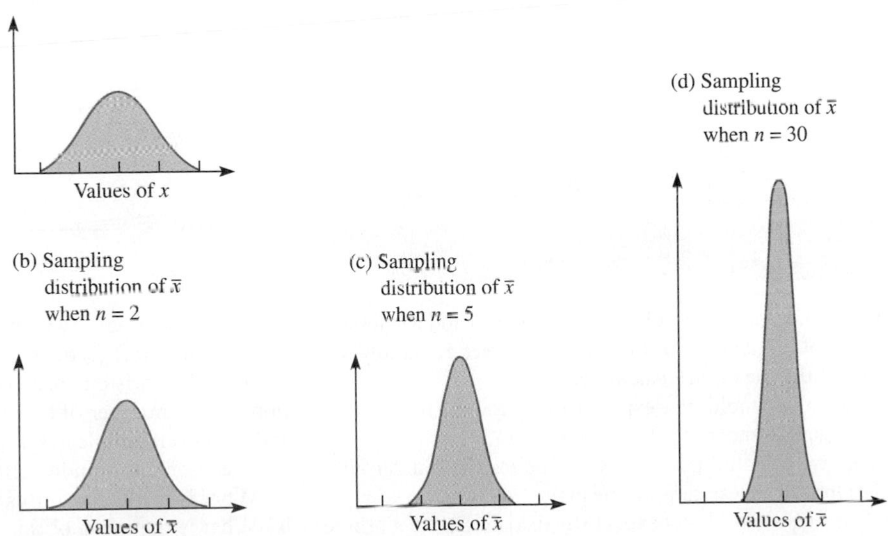

You should notice one other point: The sample mean becomes less variable as the sample size increases. Notice that as n increases from 2 to 30, all the distributions become narrower and taller.

ANSWER NOW 7.17

a. **What is the total measure of the area for any probability distribution?**

b. **Justify the statement "$\bar{x}$ becomes less variable as n increases."**

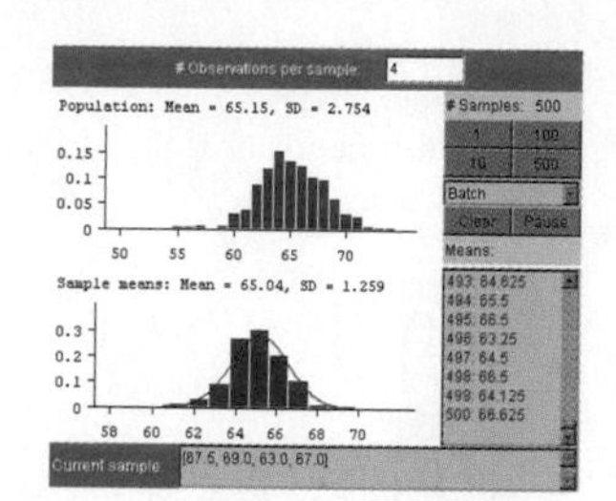

ANSWER NOW 7.18 INTERACTIVITY

Interactivity 7-A simulates taking samples of size 4 from an approximately normal population, where $\mu = 65.15$ and $\sigma = 2.754$.

a. **Click "1" for "# Samples." Note the four pieces of data and their mean. Change "slow" to "batch" and take at least 1000 samples using "500" for "# Samples."**

b. **What is the mean for the 1001 sample means? How close is it to the population mean, μ?**

c. **Compare the sample standard deviation to the population standard deviation, σ. What is happening to the sample standard deviation? Compare it to $\sigma/\sqrt{n}$, which is $2.754/\sqrt{4}$.**

d. **Does the histogram of sample means have an approximately normal shape?**

e. **Relate your findings to the sampling distribution of sample means.**

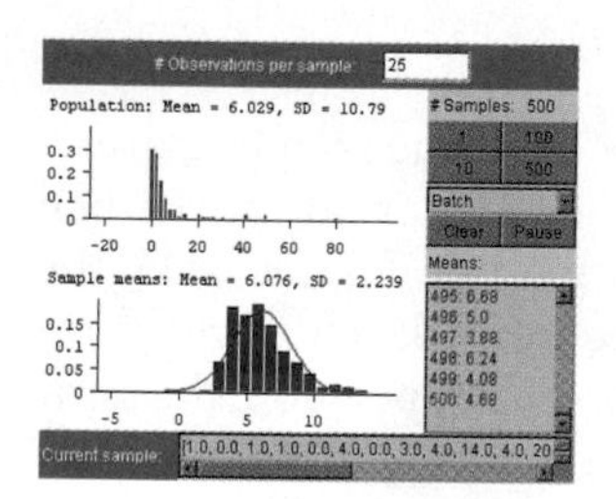

ANSWER NOW 7.19 INTERACTIVITY

Interactivity 7-B simulates sampling from a skewed population, where $\mu = 6.029$ and $\sigma = 10.79$.

a. **Change the "# Observations per sample" to "4." Using batch and 500, take 1000 samples of size 4.**

b. **Compare the mean and standard deviation for the sample means to μ and σ. Compare the sample standard deviation to $\sigma/\sqrt{n}$, which is $10.79/\sqrt{4}$. Does the histogram have an approximately normal shape? What shape is it?**

c. **Using the "clear" button each time, repeat the directions in parts a and b for samples of size 25, 100, and 1000. Table your findings for each sample size.**

d. **Relate your findings to the sampling distribution of sample means and the central limit theorem.**

Exercises

7.20 A certain population has a mean of 500 and a standard deviation of 30. Many samples of size 36 are randomly selected and the means are calculated.

a. What value would you expect to find for the mean of all these sample means?

b. What value would you expect to find for the standard deviation of all these sample means?

c. What shape would you expect the distribution of all these sample means to have?

7.21 "21st Century Consumer," an article in the January 2001 *Consumer Reports*, stated that Americans have an average of 2.4 televisions per household. If the standard deviation for the number of televisions in a U.S. household is 1.2 and a random sample of 80 American households is selected, the mean of this sample belongs to a sampling distribution.

a. What is the shape of this sampling distribution?

b. What is the mean of this sampling distribution?

c. What is the standard deviation of this sampling distribution?

7.22 According to the 2001 *World Factbook*, the 2001 total fertility rate (estimated mean number of children born per woman) for Madagascar is 5.8. Suppose that the standard deviation of the total fertility rate is 2.6. The mean number of children for a random sample of 200 women is one value of many that form the sampling distribution of sample means.

a. What is the mean value for this sampling distribution?

b. What is the standard deviation of this sampling distribution?

c. Describe the shape of this sampling distribution.

7.23 For the last 20 years the price that turkey farmers receive for turkeys has been relatively stable. According to *The World Almanac and Book of Facts 1998*, turkey farmers received an average of 43.3 cents per pound for their birds in 1996. Suppose the standard deviation of all prices received by turkey farmers is 7.5 cents per pound. The mean price received by a random sample of 150 turkey farmers in 1996 is one value of many that form the sampling distribution of sample means.

a. What is the mean value for this sampling distribution?

b. What is the standard deviation of this sampling distribution?

c. Describe the shape of this sampling distribution.

7.24 A researcher wants to take a simple random sample of about 5% of the student body at each of two schools. The university has approximately 20,000 students and the college has about 5,000 students. Which of the following statements is correct?

a. The sampling variability is the same for both schools.

b. The sampling variability for the university is higher than that for the college.

c. The sampling variability for the university is lower than that for the college.

d. No conclusion about the sampling variability can be stated without knowing the results of the study.

7.25

a. Use a computer to select 100 random samples of size 6 from a normal population with a mean $\mu = 20$ and standard deviation $\sigma = 4.5$.

b. Find the mean $\bar{x}$ for each of the 100 samples.

c. Using the 100 sample means, construct a histogram, find the mean $\bar{\bar{x}}$, and find the standard deviation $s_{\bar{x}}$.

d. Compare the results of part c with the three statements made for the SDSM (p. 318).

MINITAB (Release 13)

a. Use the Normal RANDOM DATA commands on page 285, replacing generate with 100, store in with C1-C6, mean with 20, and standard deviation with 4.5.

b. Use the ROW STATISTICS commands on page 317, replacing input variables with C1-C6 and store result in with C7.

c. Use the HISTOGRAM commands on page 58 for the data in C7, entering cutpoints from 12.8 to 27.2 in increments of 1.8. Use the MEAN and STANDARD DEVIATION commands on page 69 and 82 for the data in C7.

Excel XP

a. Use the Normal RANDOM NUMBER GENERATION commands on page 285, replacing number of variables with 6, number of random numbers with 100, mean with 20, and standard deviation with 4.5.

b. Activate cell G1:

```
Choose:   Insert function, fx > Statistical
          > AVERAGE > OK
Enter:    Number1: (A1:F1 or select cells)
Drag:     Bottom right corner of average value
          box down to give other averages
```

c. Use the RANDOM NUMBER GENERATION Patterned Distribution commands in Exercise 6.58a on page 292, replacing the first value with 12.8, the last value with 27.2, the steps with 1.8, and the output range with H1.
Use the HISTOGRAM commands on page 59, with column G as the input range and column H as the bin range.
Use the MEAN and STANDARD DEVIATION commands on pages 69 and 82 for the data in column G.

TI-83 Plus

a. Use the Integer RANDOM DATA and STO commands on page 286, replacing the Enter with 20, 4.5, 100). Repeat the above commands five more times, storing data in L2, L3, L4, L5, L6, respectively.

```
b.  Enter:    (L1 + L2 + L3 + L4 + L5 + L6)/6
    Choose:   STO→ L7 (use ALPHA key for the 'L'
              or use 'MEAN')
c.  Choose:   2nd > STAT PLOT > 1:Plot1
```

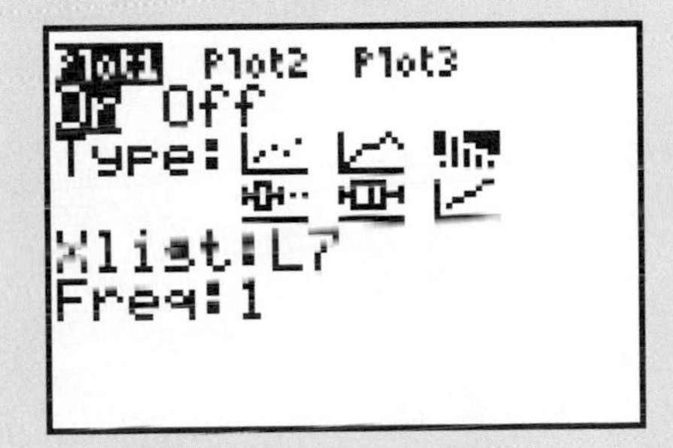


```
Choose:   Window
Enter:    12.8, 27.2, 1.8, 0, 40, 5, 1
Choose:   Trace >>>
Choose:   STAT > CALC > 1:1-VAR STATS > 2nd > LIST
Select:   L7
```

7.26

a. Use a computer to select 200 random samples of size 24 from a normal population with mean $\mu = 20$ and standard deviation $\sigma = 4.5$.

b. Find the mean $\bar{x}$ for each of the 200 samples.

c. Using the 200 sample means, construct a histogram, find the mean $\bar{\bar{x}}$, and find the standard deviation $s_{\bar{x}}$.

d. Compare the results of part c with the three statements made in the SDSM (p. 318).

e. Compare these results to the results obtained in Exercise 7.25. Specifically, what effect did the increase in sample size from 6 to 24 have? What effect did the increase from 100 samples to 200 samples have?

FYI

If you use a computer, see Exercise 7.25.

7.3 Application of the Sampling Distribution of Sample Means

When the sampling distribution of sample means is normally distributed, or approximately normally distributed, we will be able to answer probability questions with the aid of the standard normal distribution (Table 3 of Appendix B).

Illustration 7.4 Converting $\bar{x}$ Information into z-Scores

Consider a normal population with $\mu = 100$ and $\sigma = 20$. If a random sample of size 16 is selected, what is the probability that this sample will have a mean value between 90 and 110? That is, what is $P(90 < \bar{x} < 110)$?

SOLUTION

Since the population is normally distributed, the sampling distribution of $\bar{x}$'s is normally distributed. To determine probabilities associated with a normal distribution, we will need to convert the statement $P(90 < \bar{x} < 110)$ to a probability statement involving the **z-score** in order to use Table 3 in Appendix B, the standard normal distribution table. The sampling distribution is shown in the figure, where the shaded area represents $P(90 < \bar{x} < 110)$.

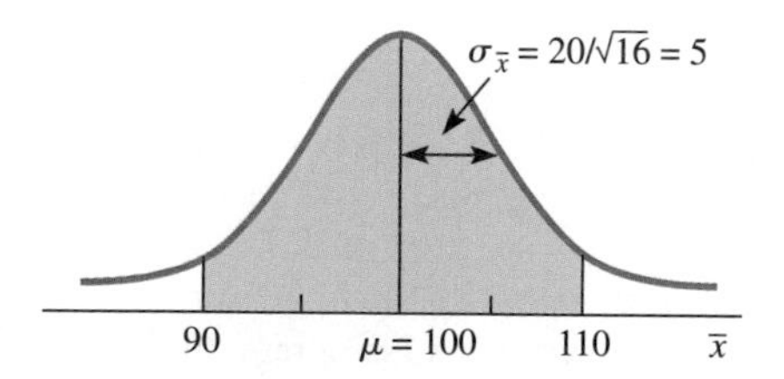

The formula for finding the z-score corresponding to a known value of $\bar{x}$ is

$$z = \frac{\bar{x} - \mu_{\bar{x}}}{\sigma_{\bar{x}}}. \quad (7.1)$$

The mean and standard error of the mean are $\mu_{\bar{x}} = \mu$ and $\sigma_{\bar{x}} = \frac{\sigma}{\sqrt{n}}$. Therefore, we will rewrite formula (7.1) in terms of μ, σ, and n:

$$z = \frac{\bar{x} - \mu}{\sigma/\sqrt{n}} \quad (7.2)$$

Returning to the illustration and applying formula (7.2), we find:

$$z\text{-score for } \bar{x} = 90: \quad z = \frac{\bar{x} - \mu}{\sigma/\sqrt{n}} = \frac{90 - 100}{20/\sqrt{16}} = \frac{-10}{5} = -2.00$$

$$z\text{-score for } \bar{x} = 110: \quad z = \frac{\bar{x} - \mu}{\sigma/\sqrt{n}} = \frac{110 - 100}{20/\sqrt{16}} = \frac{10}{5} = 2.00$$

Therefore,

$$P(90 < \bar{x} < 110) = P(-2.00 < z < 2.00) = 2(0.4772) = \mathbf{0.9544}$$

ANSWER NOW 7.27
Explain how 0.4772 was obtained and what it is.

Before we look at more illustrations, let's consider what is implied by $\sigma_{\bar{x}} = \frac{\sigma}{\sqrt{n}}$. To demonstrate, let's suppose that $\sigma = 20$ and let's use a sampling distribution of samples of size 4. Now $\sigma_{\bar{x}}$ is $20/\sqrt{4}$, or 10, and approximately 95% (0.9544) of all such sample means should be within the interval from 20 below to 20 above the population mean (within two standard deviations of the population mean). However, if the sample size is increased to 16, $\sigma_{\bar{x}}$ becomes $20/\sqrt{16} = 5$ and approximately 95% of the sampling distribution should be within 10 units of the mean, and so on. As the sample size increases, the size of $\sigma_{\bar{x}}$ becomes smaller so that the distribution of sample means becomes much narrower. Figure 7.12 illustrates what happens to the distribution of $\bar{x}$'s as the size of the individual samples increases.

FIGURE 7.12
Distributions of Sample Means

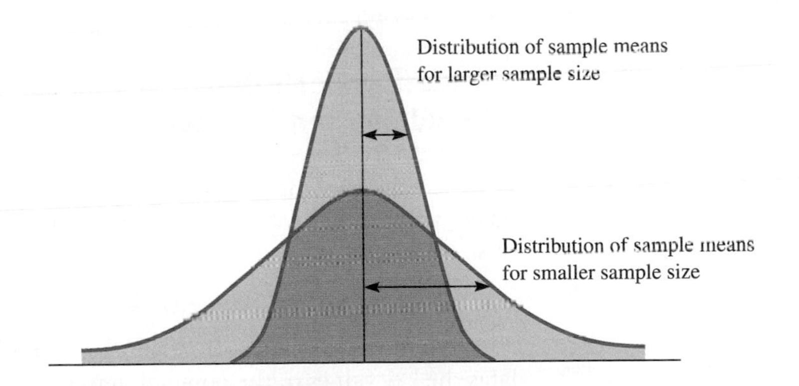

Recall that the area (probability) under the normal curve is always exactly one. So as the width of the curve narrows, the height has to increase in order to maintain this area.

Illustration 7.5 Calculating Probabilities for the Mean Height of Kindergarten Children

Kindergarten children have heights that are approximately normally distributed about a mean of 39 in. and a standard deviation of 2 in. A random sample of size 25 is taken and the mean $\bar{x}$ is calculated. What is the probability that this mean value will be between 38.5 and 40.0 in.?

SOLUTION

We want to find $P(38.5 < \bar{x} < 40.0)$. The values of $\bar{x}$, 38.5 and 40.0, must be converted to z-scores (necessary for use of Table 3) using $z = \frac{\bar{x} - \mu}{\sigma/\sqrt{n}}$:

$$\bar{x} = 38.5: \qquad z = \frac{\bar{x} - \mu}{\sigma/\sqrt{n}} = \frac{38.5 - 39.0}{2/\sqrt{25}} = \frac{-0.5}{0.4} = \mathbf{-1.25}$$

$$\bar{x} = 40.0: \qquad z = \frac{\bar{x} - \mu}{\sigma/\sqrt{n}} = \frac{40.0 - 39.0}{2/\sqrt{25}} = \frac{1.0}{0.4} = \mathbf{2.50}$$

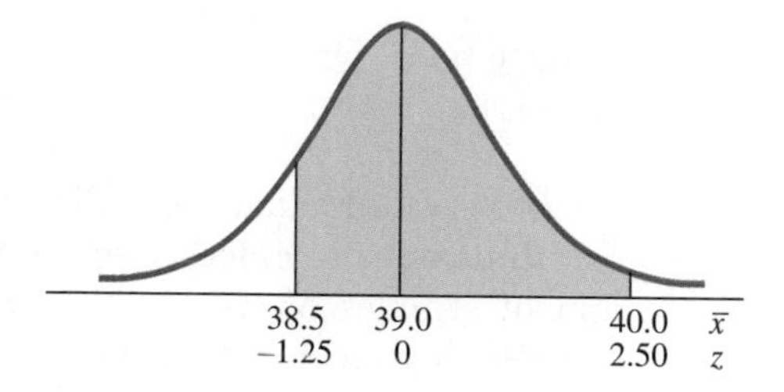

Therefore,

$$P(38.5 < \bar{x} < 40.0) = P(-1.25 < z < 2.50) = 0.3944 + 0.4938 = \mathbf{0.8882}$$

ANSWER NOW 7.28

What is the probability that the sample of kindergarten children have a mean height less than 39.75 in.?

■

Illustration 7.6 Calculating Mean Height Limits for the Middle 90% of Kindergarten Children

Use the heights of kindergarten children given in Illustration 7.5. Within what limits does the middle 90% of the sampling distribution of sample means for samples of size 100 fall?

SOLUTION

The two tools we have to work with are formula (7.2) and Table 3. The formula relates the key values of the population to the key values of the sampling distribution, and Table 3 relates areas to z-scores. First, using Table 3, we find that the middle 0.9000 is bounded by $z = \pm 1.65$.

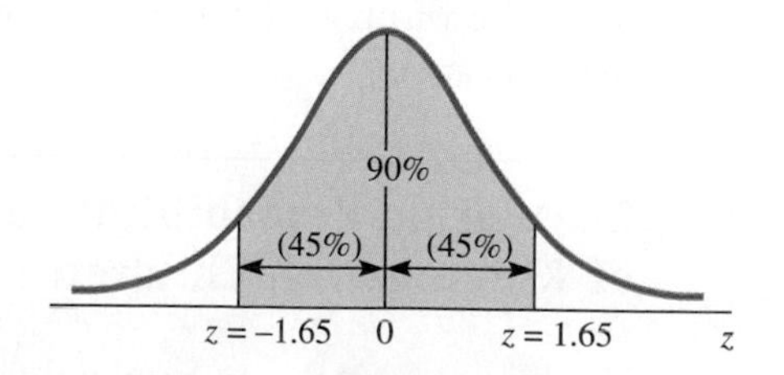

FYI

Remember: If value is exactly halfway, use the larger z.

z	...	0.04		0.05	...
:			↑		
1.6	...	0.4495	*0.4500*	0.4505	

Second, we use formula (7.2), $z = \frac{\bar{x} - \mu}{\sigma/\sqrt{n}}$:

$$z = -1.65: \quad -1.65 = \frac{\bar{x} - 39.0}{2/\sqrt{100}} \qquad z = 1.65: \quad 1.65 = \frac{\bar{x} - 39.0}{2/\sqrt{100}}$$

$$\bar{x} - 39 = (-1.65)(0.2) \qquad \bar{x} - 39 = (1.65)(0.2)$$
$$\bar{x} = 39 - 0.33 \qquad \bar{x} = 39 + 0.33$$
$$= \mathbf{38.67} \qquad = \mathbf{39.33}$$

Thus,

$$\mathbf{P(38.67 < \bar{x} < 39.33) = 0.90}$$

Therefore, 38.67 in. and 39.33 in. are the limits that capture the middle 90% of the sample means.

ANSWER NOW 7.29
What height bounds the lower 25% of all samples of size 25? ■

Exercises

7.30 A random sample of size 36 is to be selected from a population that has a mean μ of 50 and a standard deviation σ of 10.

a. This sample of 36 has a mean value of $\bar{x}$ that belongs to a sampling distribution. Find the shape of this sampling distribution.
b. Find the mean of this sampling distribution.
c. Find the standard error of this sampling distribution.
d. What is the probability that this sample mean will be between 45 and 55?
e. What is the probability that the sample mean will have a value greater than 48?
f. What is the probability that the sample mean will be within 3 units of the mean?

7.31 Consider the approximately normal population of heights of male college students with a mean μ of 69 in. and a standard deviation σ of 4 in. A random sample of 16 heights is obtained.

a. Describe the distribution of x, height of male college student.
b. Find the proportion of male college students whose height is greater than 70 in.
c. Describe the distribution of $\bar{x}$, the mean of samples of size 16.
d. Find the mean and standard error of the $\bar{x}$ distribution.
e. Find $P(\bar{x} > 70)$.
f. Find $P(\bar{x} < 67)$.

7.32 The amount of fill (weight of contents) put into a glass jar of spaghetti sauce is normally distributed with mean $\mu = 850$ g and standard deviation $\sigma = 8$ g.

a. Describe the distribution of x, the amount of fill per jar.
b. Find the probability that one jar selected at random contains between 848 and 855 g.
c. Describe the distribution of $\bar{x}$, the mean weight for a sample of 24 such jars of sauce.
d. Find the probability that a random sample of 24 jars has a mean weight between 848 and 855 g.

7.33 The heights of the kindergarten children mentioned in Illustration 7.5 (p. 327) are approximately normally distributed with $\mu = 39$ and $\sigma = 2$.

a. If an individual kindergarten child is selected at random, what is the probability that he or she has a height between 38 and 40 in.?
b. A classroom of 30 of these children is used as a sample. What is the probability that the class mean $\bar{x}$ is between 38 and 40 in.?
c. If an individual kindergarten child is selected at random, what is the probability that he or she is taller than 40 in.?
d. A classroom of 30 of these kindergarten children is used as a sample. What is the probability that the class mean $\bar{x}$ is greater than 40 in.?

7.34 Compensations to chief executive officers of corporations vary substantially from one company to the next, both in amount and in form. A study published in *The Accounting Review* ("Accounting Transactions and CEO Cash Compensations," Vol. 73, No. 2, p. 248) showed that the mean annual salary and bonus compensation to CEOs in 1997 was $634,961 with a standard deviation of $441,690. Assume that annual salaries and bonus compensations are normally distributed.

a. What is the probability that a randomly selected CEO received more than $750,000 in 1997?
b. Explain why the assumption of a normal distribution may be very unlikely.
c. A sample of 20 CEOs is taken and salaries and bonus compensations are reported. What is the probability that the sample mean salary and bonus compensation falls between $600,000 and $700,000?
d. Explain why the standard normal distribution can be used to answer part c with more confidence than part a.

7.35 Based on 52 years of data compiled by the National Climatic Data Center (http://lwf.ncdc.noaa.gov/oa/climate/online/ccd/avgwind.html), the average speed of winds in Honolulu is 11.3 miles per hour as of April 2002. Assume that wind speeds are approximately normally distributed with a standard deviation of 3.5 miles per hour.
a. Find the probability that the wind speed on any one reading will exceed 13.5 mph.
b. Find the probability that the mean speed of a random sample of nine readings exceeds 13.5 mph.
c. Do you think the assumption of normality is reasonable? Explain.
d. What effect do you think the assumption of normality had on the answers to parts a and b? Explain.

7.36 TIMSS 1999, successor to the 1995 Third International Mathematics and Science Study, focused on the mathematics and science achievement of eighth-grade students throughout the world. A total of 38 countries (including the United States) participated in the study. The mean score for U.S. students was 502 with a standard deviation of 88. Assume the scores are normally distributed and a sample of 150 students was taken.
a. Find the probability that the mean TIMSS score for a randomly selected group of eighth-grade students is between 495 and 510.
b. Find the probability that the mean TIMSS score for a randomly selected group of eighth-grade students is less than 520.
c. Do you think the assumption of normality is reasonable? Explain.

7.37 According to the State of Utah Online Services (http://occ.dws.state.ut.us/Advisory/Minutes/StateOfChildCare.asp), the average annual child-care cost for a 12-month-old was $4651 in December 1997. Assume that costs are normally distributed with a standard deviation of $310.
a. Find the probability that the mean child-care cost for a 12-month-old at 50 randomly selected daycare centers in Utah is between $4500 and $4750.
b. Find the probability that the mean child-care cost for a 12-month-old at 50 randomly selected daycare centers in Utah is greater than $4800.
c. Do you think the assumption of normality is reasonable? Explain.

7.38 S Wageweb (www.wageweb.com/health1.htm) provides compensation information and services on more than 160 positions. As of October 1, 2001, the national average salary for a registered nurse was $44,902. Suppose the standard deviation is $7750. Find the following probabilities for the mean of a random sample of 100 such nurses:
a. the probability that the mean of the sample is less than $44,000
b. the probability that the sample mean is between $43,000 and $46,000
c. the probability that the sample mean is greater than $45,000
d. Explain why the assumption of normality about the distribution of wages was not involved in the solution to parts a–c.

7.39
a. Find $P(4 < \bar{x} < 6)$ for a random sample of size 4 drawn from a normal population with mean $\mu = 5$ and standard deviation $\sigma = 2$.
b. Use a computer to generate 100 random samples, each of size 4, from a normal probability distribution with mean $\mu = 5$ and standard deviation $\sigma = 2$, and calculate the mean, $\bar{x}$, for each sample.
c. How many of the sample means in part b have values between 4 and 6? What percentage is that?
d. Compare the answers to parts a and c, and explain any differences that occurred.

MINITAB (Release 13)
a. Input the numbers 4 and 6 into C1. Use the CUMULATIVE NORMAL PROBABILITY DISTRIBUTION commands on page 287, replacing the mean with 5, the standard deviation with 1 ($2/\sqrt{4}$), the input column with C1, and the optional storage with C2. Find CDF(6)–CDF(4).
b. Use the Normal RANDOM DATA commands on page 285, replacing generate with 100, store in with C3–C6, mean with 5, and standard deviation with 2. Use the ROW STATISTICS commands on page 317, replacing input variables with C3–C6 and store result in with C7.
c. Use the HISTOGRAM commands on page 58 for the data in C7, entering cutpoints from 1 to 9 in increments of 1 and selecting data labels.

Excel XP
a. Input the numbers 4 and 6 into column A. Activate cell B1. Use the CUMULATIVE NORMAL DISTRIBUTION commands on page 288, replacing X with A1:A2. Find CDF(6)–CDF(4).
b. Use the Normal RANDOM NUMBER GENERATION commands on page 285, replacing number of variables with 4, number of random numbers with 100, mean with 5, standard deviation with 2, and output range with C1. Activate cell G1. Use the AVERAGE INSERT FUNCTION commands in Exercise 7.14b on page 317, replacing Number1 with C1:F1.
c. Use the RANDOM NUMBER GENERATION Patterned Distribution commands in Exercise 6.58a on page 292, replacing the first value with 0, the last value with 9, the steps with 1, and the output range with H1.

Excel XP (continued)

Use the HISTOGRAM commands on page 58 with column G as the input range, column H as the bin range, and column I as the output range.

TI-83 Plus

a. Use the CUMULATIVE NORMAL PROBABILITY commands on page 288, replacing the Enter with 4,6,5,1) [The standard deviation is 1; from $2/\sqrt{4}$.]

b. Use the Normal RANDOM DATA and STO commands on page 286, replacing the Enter with 5,2,100). Repeat these commands three more times, storing data in L2, L3, L4, respectively.

```
Choose:      STAT > EDIT > 1:Edit
Highlight:   L5 (column heading)
Enter:       (L1+L2+L3+L4)/4
```

c. Use the HISTOGRAM and TRACE commands on page 59 to count. Enter 0,9,1,0,45,1 for the Window.

7.40

a. Find $P(46 < \bar{x} < 55)$ for a random sample of size 16 drawn from a normal population with mean $\mu = 50$ and standard deviation $\sigma = 10$.

b. Use a computer to generate 200 random samples, each of size 16, from a normal probability distribution with mean $\mu = 50$ and standard deviation $\sigma = 10$, and calculate the mean, $\bar{x}$, for each sample.

c. How many of the sample means in part b have values between 46 and 55? What percentage is that?

d. Compare the answers to parts a and c, and explain any differences that occurred.

FYI
If you use a computer, see Exercise 7.39.

CHAPTER SUMMARY

RETURN TO CHAPTER CASE STUDY

© Michael Dwyer/Stock Boston

The U.S. Census and Sampling It

A second sample of 100 ages has been collected from the U.S. 2000 census data and is listed here:

CS07Re

14	6	59	64	39	12	8	34	27	4
16	18	17	33	56	60	65	73	53	43
26	42	60	87	58	42	82	21	35	64
58	53	36	66	63	66	39	62	58	49
31	27	39	35	12	28	28	20	3	54
41	41	63	39	37	23	79	43	28	17
12	45	52	10	11	32	32	23	86	61
50	27	19	15	3	51	5	36	83	39
35	44	59	30	31	69	40	16	40	66
15	55	32	4	43	41	23	46	61	30

FIRST THOUGHTS 7.41

a. Describe these sample data. Be sure to include both visual and numerical measures.

b. How well did this sample describe the population? Explain.

c. Using the graph that you constructed in part a, describe the shape of this distribution of sample data.

d. How do the statistics of this sample compare to the statistics of the sample given in the Chapter Case Study? Be specific.

e. How do the graphs compare?

PUTTING CHAPTER 7 TO WORK 7.42

a. Is the distribution of ages for this population normal? Approximately normal?

b. Will the sampling distribution of sample means (SDSM) apply to samples taken from this population? Explain.

c. Will the central limit theorem (CLT) apply to samples taken from this population? Explain.
d. Describe the sampling distribution of sample means for samples of size 100. Be sure to include center, spread, and shape.
e. Describe the sampling distribution of sample means for samples of size 30. Be sure to include center, spread, and shape.
f. Describe the sampling distribution of sample means for samples of size 1000. Be sure to include center, spread, and shape.
g. Relate your findings in parts d, e, and f to the sampling distribution of sample means (SDSM) and the central limit theorem (CLT).

YOUR OWN MINI-STUDY 7.43

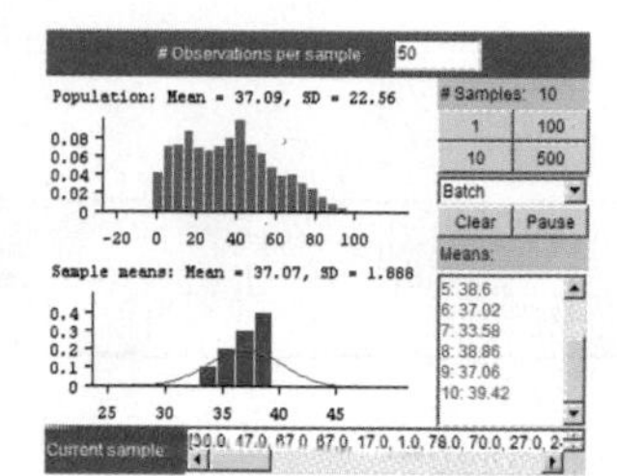

Interactivity 7-C simulates taking samples of size 50 from the population of American ages from the 2000 census, where $\mu = 37.09$ and $\sigma = 22.56$ and the shape is skewed right.

a. Click "1" for number of samples. Note the 50 pieces of data and their mean. Change "slow" to "batch" and take at least 1000 samples of size 50.
b. What is the mean of the sample means? How close is it to the population mean?
c. What is the standard deviation of the sample means?
d. Based on the sampling distribution of sample means (as described in Section 7.3), what should you expect for the standard deviation of sample means? How close was your standard deviation from part c?
e. What shape is the histogram of the 1000 means?
f. Relate your findings to the sampling distribution of sample means and the central limit theorem.

In Retrospect

In Chapters 6 and 7 we have learned to use the standard normal probability distribution. We now have two formulas for calculating a z-score:

$$z = \frac{x - \mu}{\sigma} \quad \text{and} \quad z = \frac{\bar{x} - \mu}{\sigma/\sqrt{n}}$$

You must be careful to distinguish between these two formulas. The first gives the standard score when we have individual values from a normal distribution (x values). The second formula deals with a sample mean ($\bar{x}$ value). The key to distinguishing between the formulas is to decide whether the problem deals with an individual x or a sample mean $\bar{x}$. If it deals with the individual values of x, we use the first formula, as presented in Chapter 6. If the problem deals with a sample mean, $\bar{x}$, we use the second formula and proceed as illustrated in this chapter.

The basic purpose for considering what happens when a population is repeatedly sampled, as discussed in this chapter, is to form sampling distributions. The sampling distribution is then used to describe the variability that occurs from one sample to the next. Once this pattern of variability is known and understood for a specific sample statistic, we are able to make predictions about the corresponding population parameter with a measure of how accurate the prediction is. The central limit theorem helps describe the distribution for sample means. We will begin to make inferences about population means in Chapter 8.

There are other reasons for repeated sampling. Repeated samples are commonly used in the field of production control, in which samples are taken to determine whether a product is of the proper size or quantity. When the sample statistic does not fit the standards, a mechanical adjustment of the machinery is necessary. The adjustment is then followed by another sampling to be sure the production process is in control.

The "standard error of the _____" is the name used for the standard deviation of the sampling distribution for whatever statistic is named in the blank. In this chapter we have been concerned with the standard error of the mean. However, we could also work with the standard error of the proportion, median, or any other statistic.

You should now be familiar with the concept of a sampling distribution and, in particular, with the sampling distribution of sample means. In Chapter 8 we will begin to make predictions about the values of population parameters.

Chapter Exercises

7.44 The diameters of Red Delicious apples in a certain orchard are normally distributed with a mean of 2.63 in. and a standard deviation of 0.25 in.

a. What percentage of the apples in this orchard have diameters less than 2.25 in.?
b. What percentage of the apples in this orchard are larger than 2.56 in. in diameter?
c. A random sample of 100 apples is gathered and their mean diameter is $\bar{x} = 2.56$. If another sample of size 100 is taken, what is the probability that its sample mean will be greater than 2.56 in.?
d. Why is the z-score used in answering parts a–c?
e. Why is the formula for z used in part c different from that used in parts a and b?

7.45

a. Find a value for e such that 95% of the apples in Exercise 7.44 are within e units of the mean 2.63; that is, find e such that $P(2.63 - e < x < 2.63 + e) = 0.95$.
b. Find a value for E such that 95% of the samples of 100 apples taken from the orchard in Exercise 7.44 will have mean values within E units of the mean 2.63; that is, find E such that $P(2.63 - E < \bar{x} < 2.63 + E) = 0.95$.

7.46 Americans spend $2.8 billion on veterinary care each year, and the health care services offered to animals rival those provided to humans. The average dog owner in 1997 spent $275 per year on vet care. In addition, about 20% of all dog owners have indicated that they would be willing to spend more than $5000 to extend their pet's life (*Life*, "Animal E.R.," July 1998). Assume that annual dog owner expenditures on vet care are normally distributed with a mean of $275 and a standard deviation of $95.

a. What is the probability that a dog owner, randomly selected from the population, spent more than $300 on vet care in 1997?
b. Suppose a survey of 300 dog owners is conducted and each owner is asked to report the total of vet care bill for the year. What is the probability that the mean annual expenditure of this sample falls between $268 and $280?
c. Assume that life-extending offers are normally distributed with a standard deviation of $2000. What is the mean amount that dog owners are willing to spend to extend their pet's life?

7.47 According to an article in *Newsweek* ("China by the Numbers: Portrait of a Nation," June 29, 1998), the rate of water pollution in China is more than twice that measured in the United States and more than three times the amount measured for Japan. The mean emission of organic pollutants is 11.7 million pounds per day in China. Assume that water pollution in China is normally distributed throughout the year with a standard deviation of 2.8 million pounds of organic emissions per day.

a. What is the probability that on any given day the water pollution in China exceeds 15 million pounds per day?
b. If 20 days of water pollution readings in China are taken, what is the probability that the mean of this sample is less than 10 million pounds of organic emissions?

7.48 A recent study from the University of Michigan, as noted in *Newsweek* (March 25, 2002), stated that today men average 16 hours of housework each week (up from an average of 12 hours in 1965). If we assume that the number of hours in which men engage in housework each week is normally distributed with a standard deviation of 5.4 hours, what is the probability that the mean number of housework hours for a sample of 20 randomly selected men is between 15 and 18?

7.49 A shipment of steel bars will be accepted if the mean breaking strength of a random sample of ten steel bars is greater than 250 pounds per square inch. In the past, the breaking strength of such bars has had a mean of 235 and a variance of 400.

a. What is the probability, assuming that the breaking strengths are normally distributed, that one randomly selected steel bar will have a breaking strength in the range of 245 to 255?
b. What is the probability that the shipment will be accepted?

7.50 A report in *Time* magazine (April 15, 2002) stated that the average age for women to marry in the United States is now 25. If the standard deviation is assumed to be 3.2 years, find the probability that a random sample of 40 U.S. women would show a mean age at marriage of less than or equal to 24 years.

7.51 A manufacturer of light bulbs says that its light bulbs have a mean lifetime of 700 hr and a standard deviation of 120 hr. You purchased 144 of these bulbs with the idea that you would purchase more if the mean lifetime of your sample was longer than 680 hr. What is the probability that you will not buy again from this manufacturer?

7.52 A tire manufacturer claims (based on years of experience with its tires) that the tires' mean mileage is 35,000 mi and the standard deviation is 5000 mi. A consumer agency randomly selects 100 of these tires and finds a sample mean of 31,000 mi. Should the consumer agency doubt the manufacturer's claim?

7.53 For large samples, the sample sum (Σx) has an approximately normal distribution. The mean of the sample sum is $n \cdot \mu$ and the standard deviation is $\sqrt{n} \cdot \sigma$. The distribution of savings per account for a savings and loan institution has a mean equal to $750 and a standard deviation equal to $25. For a sample of 50 such accounts, find the probability that the sum in the 50 accounts exceeds $38,000.

7.54 The baggage weights for passengers using a particular airline are normally distributed with a mean of 20 lb and a standard deviation of 4 lb. If the limit on total luggage weight is 2125 lb, what is the probability that the limit will be exceeded for 100 passengers?

7.55 A trucking firm delivers appliances for a large retail operation. The packages (or crates) have a mean weight of 300 lb and a variance of 2500 lb.

a. If a truck can carry 4000 lb and 25 appliances need to be picked up, what is the probability that the 25 appliances will have an aggregate weight greater than the truck's capacity? (Assume that the 25 appliances represent a random sample.)
b. If the truck has a capacity of 8000 lb, what is the probability that it will be able to carry the entire lot of 25 appliances?

7.56 A pop-music record firm wants the distribution of lengths of cuts on its records to have an average of 2 min 15 sec (135 sec) and a standard deviation of 10 sec, so that disc jockeys will have plenty of time for commercials within each 5-minute period. The population of times for cuts is approximately normally distributed with only a negligible skew to the right. You have just timed the cuts on a new release and have found that the ten cuts average 140 sec.

a. What percentage of the time will the average be 140 sec or longer, if the new release is randomly selected?
b. If the music firm wants ten cuts to average 140 sec less than 5% of the time, what must the population mean be, given that the standard deviation remains at 10 sec?

7.57 Let's simulate the sampling distribution related to the disc jockey's concern for "length of cut" in Exercise 7.56.

a. Use a computer to generate 50 random samples, each of size 10, from a normal distribution with mean 135 and standard deviation 10. Find the sample total and the sample mean for each sample.
b. Using the 50 sample means, construct a histogram and find their mean and standard deviation.
c. Using the 50 sample totals, construct a histogram and find their mean and standard deviation.
d. Compare the results obtained in parts b and c. Explain any similarities and any differences observed.

MINITAB (Release 13)

a. Use the Normal RANDOM DATA commands on page 285, replacing generate with 50, store in with C1–C10, mean with 135, and standard deviation with 10. Use the ROW STATISTICS commands on page 317, with selecting Sum, replacing input variables with C1–C10, and store result in with C11. Use the ROW STATISTICS commands, again with selecting Mean and then replacing input variables with C1-C10, and store result in C12.
b. Use the HISTOGRAM commands on page 58 for the data in C12, selecting midpoints and automatic. Use the MEAN and STANDARD DEVIATION commands on pages 69 and 82 for the data in C12.
c. Use the HISTOGRAM commands on page 58 for the data in C11, selecting midpoints and automatic. Use the MEAN and STANDARD DEVIATION commands on pages 69 and 82 for the data in C11.
d. Use the DISPLAY DESCRIPTIVE STATISTICS commands on page 100 for the data in C11 and C12.

Excel XP

a. Use the Normal RANDOM NUMBER GENERATION commands on page 285, replacing number of variables with 10, number of random numbers with 50, mean with 135, and standard deviation with 10.
Activate cell K1:

```
Choose:    Insert function, fx > All > SUM
           > OK
Enter:     Number1: (A1:J1 or select cells)
Drag:      Bottom right corner of sum value box
           down to give other sums
```

Activate cell L1. Use the AVERAGE INSERT FUNCTION commands in Exercise 7.14b on page 317, replacing Number1 with A1:J1.
b. Use the RANDOM NUMBER GENERATION Patterned Distribution commands in Exercise 6.58a on page 292, replacing the first value with 125.4, the last value with 147.8, the steps with 3.2, and the output range with M1. Use the HISTOGRAM commands on page 58 with column L as the input range and column M as the bin range. Use the MEAN and STANDARD DEVIATION commands on pages 69 and 82 for the data in column L.
c. Use the RANDOM NUMBER GENERATION Patterned Distribution commands in Exercise 6.58a on page 292, replacing the first value with 1254, the last value with 1478, the steps with 32, and the output range with M20. Use the HISTOGRAM commands on page 58, with column L as the input range and cells M20–M26 as the bin range. Use the MEAN and STANDARD DEVIATION commands on pages 69 and 82 for the data in column K.
d. Use the DESCRIPTIVE STATISTICS commands on page 100 for the data in columns K and L.

7.58

a. Find the mean and standard deviation of x for a binomial probability distribution with $n = 16$ and $p = 0.5$.
b. Use a computer to construct the probability distribution and histogram for the binomial probability experiment with $n = 16$ and $p = 0.5$.
c. Use a computer to generate 200 random samples of size 25 from a binomial probability distribution with $n = 16$ and $p = 0.5$, and calculate the mean of each sample.
d. Construct a histogram and find the mean and standard deviation of the 200 sample means.
e. Compare the probability distribution of x found in part b and the frequency distribution of $\bar{x}$ in part d. Does your information support the CLT? Explain.

MINITAB (Release 13)

b. Use the MAKE PATTERNED DATA commands in Exercise 6.58a on page 292, replacing the first value with 0, the last value with 16, and the steps with 1. Use the BINOMIAL PROBABILITY DISTRIBUTIONS commands on page 255, replacing n with 16, p with 0.5, input column with C1, and optional storage with C2. Use the PLOT commands on page 139, replacing Y with C2, X with C1, and data display with area.
c. Use the BINOMIAL RANDOM DATA commands on page 263, replacing generate with 200, store in with C3–C27, number of trials with 16, and probability with 0.5. Use the ROW STATISTICS commands for a mean on page 317, replacing input variables with C3–C27 and store result in with C28.
d. Use the HISTOGRAM commands on page 58 for the data in C28, selecting midpoints and automatic. Use the MEAN and STANDARD DEVIATION commands on pages 69 and 82 for the data in C28.

Excel XP

b. Input 0 through 16 into column A. Continue with the binomial probability commands on page 255, using $n = 16$ and $p = 0.5$. Activate columns A and B, then continue with:

```
Choose:   Chart Wizard > Column > 1st picture > Next > Series
Choose:   Series 1 > Remove
Enter:    Category (x)axis labels: (A1:A17 or
          select 'x value' cells)
Choose:   Next > Finish
```

c. Use the Binomial RANDOM NUMBER GENERATION commands from Exercise 5.92 on page 264, replacing number of variables with 25, number of random numbers with 200, p-value with 0.5, number of trials with 16, and output range with C1. Activate cell BB1. Use the AVERAGE INSERT FUNCTION commands in Exercise 7.14b on page 317, replacing Number1 with C1:AA1.
d. Use the RANDOM NUMBER GENERATION Patterned Distribution commands in Exercise 6.58a on page 292, replacing the first value with 6.8, the last value with 9.6, the steps with 0.4, and the output range with AC1. Use the HISTOGRAM commands on page 58, with column AB as the input range and column AC as the bin range. Use the MEAN and STANDARD DEVIATION commands on pages 69 and 82 for the data in column AB.

7.59

a. Find the mean and standard deviation of x for a binomial probability distribution with $n = 200$ and $p = 0.3$.
b. Use a computer to construct the probability distribution and histogram for the random variable x of the binomial probability experiment with $n = 200$ and $p = 0.3$.
c. Use a computer to generate 200 random samples of size 25 from a binomial probability distribution with $n = 200$ and $p = 0.3$ and calculate the mean $\bar{x}$ of each sample.
d. Construct a histogram and find the mean and standard deviation of the 200 sample means.
e. Compare the probability distribution of x found in part b and the frequency distribution of $\bar{x}$ in part d. Does your information support the CLT? Explain.

FYI

Use the commands in Exercise 7.58, making the necessary adjustments.

7.60 A random sample of 144 values is selected from a population with mean μ equal to 45 and standard deviation σ equal to 18.

a. Determine the interval (smallest value to largest value) within which you would expect a sample mean to lie.
b. What is the amount of deviation from the mean for a sample mean of 45.3?
c. What is the maximum deviation you have allowed for in your answer to part a?
d. How is this maximum deviation related to the standard error of the mean?

Vocabulary and Key Concepts

Be able to define each term. Pay special attention to the key terms, which are printed in **red.** In addition, describe each term in your own words and give an example of each. Your examples should not be ones given in class or in the textbook. The bracketed numbers indicate the chapter in which the term first appeared, but you should define the terms again to show increased understanding of their meaning. Page numbers indicate the first appearance of the term in Chapter 7.

central limit theorem (p. 319)
frequency distribution [2] (p. 314)
probability distribution [5] (p. 313)
random sample [2] (p. 315)
repeated sampling (p. 312)
sampling distribution (p. 312)
sampling distribution of sample means (pp. 313, 318)
standard error of the mean (p. 319)
z-score [2, 6] (p. 326)

Chapter Practice Test

PART 1: KNOWING THE DEFINITIONS

Answer "True" if the statement is always true. If the statement is not always true, replace the words shown in bold with words that make the statement always true.

7.1 A sampling distribution **is** a distribution listing all the sample statistics that describe a particular sample.

7.2 The histograms of **all** sampling distributions are symmetric.

7.3 The mean of the sampling distribution of $\bar{x}$'s is equal to the mean of the **sample.**

7.4 The standard error of the mean is the standard deviation of the population **from which the samples have been taken.**

7.5 The standard error of the mean **increases** as the sample size increases.

7.6 The shape of the distribution of sample means is always that of a **normal** distribution.

7.7 A **probability** distribution of a sample statistic is a distribution of all the values of that statistic that were obtained from all possible samples.

7.8 The sampling distribution of sample means provides us with a description of the three characteristics of a sampling distribution of sample **medians.**

7.9 A **frequency** sample is obtained in such a way that all possible samples of a given size have an equal chance of being selected.

7.10 We **do not need** to take repeated samples in order to use the concept of the sampling distribution.

PART II: APPLYING THE CONCEPTS

7.11 The lengths of the lake trout in Conesus Lake are believed to have a normal distribution with a mean of 15.6 in. and a standard deviation of 3.8 in.

a. Kevin is going fishing at Conesus Lake tomorrow. If he catches one lake trout, what is the probability that it is less than 15.0 in. long?

b. If Captain Brian's fishing boat takes ten people fishing on Conesus Lake tomorrow and they catch a random sample of 16 lake trout, what is the probability that the mean length of their total catch is less than 15 in.?

7.12 Cigarette lighters manufactured by EasyVice Company are claimed to have a mean lifetime of 20 months with a standard deviation of 6 months. The money-back guarantee allows you to return the lighter if it does not last at least 12 months from the date of purchase.

a. If the lifetimes of these lighters are normally distributed, what percentage of the lighters will be returned to the company?

b. If a random sample of 25 lighters is tested, what is the probability the sample mean lifetime will be more than 18 months?

7.13 Aluminum rivets produced by Rivets Forever, Inc., are believed to have shearing strengths that are distributed about a mean of 13.75 with a standard deviation of 2.4. If this information is true and a sample of 64 such rivets is tested for shear strength, what is the probability that the mean strength will be between 13.6 and 14.2?

PART III: UNDERSTANDING THE CONCEPTS

7.14 "Two heads are better than one." If that's true, then how good would several heads be? To find out, a statistics instructor drew a line across the chalkboard and asked her class to estimate its length to the nearest inch. She collected their estimates, which ranged from 33 to 61 in., and calculated the mean value. She reported that the mean was 42.25 in. She then measured the line and found it to be 41.75 in. long. Does this show that "several heads are better than one"? What statistical theory supports this occurrence? Explain how.

7.15 The sampling distribution of sample means is more than just a distribution of the mean values that occur from many repeated samples taken from the same population. Describe what other specific condition must be met in order to have a sampling distribution of sample means.

7.16 Student A states, "A sampling distribution of the standard deviations tells you how the standard deviation varies from sample to sample." Student B argues, "A population distribution tells you that." Who is right? Justify your answer.

7.17 Student A says it is the "size of each sample used" and Student B says it is the "number of samples used" that determines the spread of an empirical sampling distribution. Who is right? Justify your choice.

WORKING WITH YOUR OWN DATA

© StockTrek/Getty Images

The sampling distribution of sample means and the central limit theorem are very important to the development of the rest of this course. The proof, which requires the use of calculus, is not included in this textbook. However, the truth of the SDSM and the CLT can be demonstrated both theoretically and by experimentation. The following activities will help to verify both statements.

A THE POPULATION

Consider the theoretical population that contains the three numbers 0, 3, and 6 in equal proportions.

1. a. Construct the theoretical probability distribution for the drawing of a single number, with replacement, from this population.
   b. Draw a histogram of this probability distribution.
   c. Calculate the mean μ and the standard deviation σ for this population.

B THE SAMPLING DISTRIBUTION, THEORETICALLY

Let's study the theoretical sampling distribution formed by the means of all possible samples of size 3 that can be drawn from the given population.

2. Construct a list showing all the possible samples of size 3 that could be drawn from this population. (There are 27 possibilities.)
3. Find the mean for each of the 27 possible samples listed in answer to question 2.
4. Construct the probability distribution (the theoretical sampling distribution of sample means) for these 27 sample means.
5. Construct a histogram for this sampling distribution of sample means.
6. Calculate the mean $\mu_{\bar{x}}$ and the standard error of the mean $\sigma_{\bar{x}}$ using the probability distribution found in question 4.
7. Show that the results found in questions 1c, 5, and 6 support the three claims made by the sampling distribution of sample means and the central limit theorem. Cite specific values to support your conclusions.

C THE SAMPLING DISTRIBUTION, EMPIRICALLY

Let's now see whether the sampling distribution of sample means and the central limit theorem can be verified empirically; that is, does it hold when the sampling distribution is formed by the sample means that result from several random samples?

8. Draw a random sample of size 3 from the given population. List your sample of three numbers and calculate the mean for this sample.

You may use a computer to generate your samples. You may take three identical "tags" numbered 0, 3, and 6, put them in a "hat," and draw your sample using replacement between each drawing. Or you may use dice; let 0 be represented by 1 and 2, let 3 be represented by 3 and 4, and 6 by 5 and 6. You may also use random numbers to simulate the drawing of your samples. Or you may draw your sample from the list of random samples at the end of this section. Describe the method you decide to use. (Ask your instructor for guidance.)

9. Repeat question 8 forty-nine (49) more times so that you have a total of 50 sample means that have resulted from samples of size 3.

10. Construct a frequency distribution of the 50 sample means found in questions 8 and 9.
11. Construct a histogram of the frequency distribution of observed sample means.
12. Calculate the mean $\bar{\bar{x}}$ and standard deviation $s_{\bar{x}}$ of the frequency distribution formed by the 50 sample means.
13. Compare the observed values of $\bar{\bar{x}}$ and $s_{\bar{x}}$ with the values of $\mu_{\bar{x}}$ and $\sigma_{\bar{x}}$. Do they agree? Does the empirical distribution of $\bar{x}$ look like the theoretical one?

Here are 100 random samples of size 3 that were generated by computer:

6 3 0	0 3 0	6 6 0	3 3 6	6 6 3	6 3 3
0 0 3	3 0 6	3 3 0	3 6 6	0 3 0	6 6 3
6 6 6	0 3 0	6 3 6	0 6 3	6 0 3	6 3 3
6 0 0	3 0 6	6 3 3	3 3 0	3 3 0	3 3 3
3 3 3	3 0 0	6 6 6	3 3 6	0 0 6	0 6 3
6 6 6	0 0 6	3 3 0	0 6 6	0 0 3	6 6 3
0 0 6	0 0 6	6 6 6	6 3 6	6 6 0	3 0 0
3 6 6	6 3 0	3 6 3	3 0 0	3 3 6	0 6 0
3 0 0	0 3 6	6 3 3	6 0 6	3 3 6	6 0 3
0 3 6	3 6 3	6 6 3	6 6 0	3 3 3	3 0 0
6 3 0	6 6 0	0 3 0	6 6 0	3 6 6	0 3 6
6 3 3	0 3 0	6 6 0	6 6 3	6 6 0	3 0 3
3 6 3	3 6 0	0 0 6	0 3 3	3 6 6	0 3 6
0 6 0	6 0 0	0 6 0	0 6 6	0 3 3	0 3 6
3 3 6	3 3 3	3 3 6	6 3 6	3 3 3	3 6 6
6 3 3	3 0 0	3 0 6	6 0 3	3 6 6	6 0 3
0 3 3	6 3 0	0 3 6	0 3 6		

PART

3 Inferential Statistics

The central limit theorem gave us some very important information about the sampling distribution of sample means (SDSM). Specifically, it stated that in many realistic cases (when the random sample is large enough), a distribution of sample means is normally or approximately normally distributed about the mean of the population. With this information we were able to make probability statements about the likelihood of certain sample mean values occurring when samples are drawn from a population with a known mean and a known standard deviation. We are now ready to turn this situation around to the case in which the population mean is not known. We will draw one sample, calculate its mean value, and then make an inference about the value of the population mean based on the sample's mean value.

The objective of inferential statistics is to use the information contained in the sample data to increase our knowledge of the sampled population. In this part of the textbook we will learn about making two types of inferences: (1) estimating the value of a population parameter and (2) testing a hypothesis. Specifically, we will learn about making these two types of inferences for the mean μ of a normal population, for the standard deviation σ of a normal population, and for the probability parameter p of a binomial population.

The sampling distribution of sample means (SDSM) is the key to making these inferences, as shown in the figure.

The Statistical Process

Population Being Studied
Parameter of interest, μ
Collect a random sample.
Sample
Data collected
Analyze the sample data.
Sample Statistics
Graphic:
Numeric: $\bar{x} = 107.2$, $s = 7.93$
Use the sample statistic $\bar{x}$ (and the sampling distribution) to make an inference about the population mean, μ.

Where the Sampling Distribution Fits into the Statistical Process

William Gosset

William Gosset ("Student"), a British industrial statistician, was born in Canterbury, England, on June 13, 1876, to Frederick and Agnes (Vidal) Gosset. William's educational background included studies at Winchester College, New College, and Oxford University. Gosset was employed by the Arthur Guinness & Son Brewing Company as a brewer. In 1906 Gosset married Marjory Surtees Philpotts, and they became the parents of two daughters and a son. William Gosset died in Beaconsfield, England, on October 16, 1937.

Guinness liked its employees to use pen names when publishing papers, so in 1908 Gosset adopted the pen name "Student" under which he published what was probably his most noted contribution to statistics, "The Probable Error of a Mean." Guinness sent Gosset to work under Karl Pearson at the University of London, and eventually Gosset took charge of the new Guinness Brewery in London.

In his paper "The Probable Error of a Mean," Student set out to find the distribution of the amount of error in the sample mean, $(\bar{x} - \mu)$, when divided by s, where s was the estimate of σ from a sample of any known size. The probable error of a mean, $\bar{x}$, could be calculated for any size sample by using this distribution of $(\bar{x} - \mu)/(s/\sqrt{n})$. Even though Student was well aware of the insufficiency of a small sample to determine the form of the distribution of $\bar{x}$, he chose the normal distribution for simplicity, stating his opinion: "It appears probable that the deviation from normality must be very severe to lead to serious error."

Student's t-distribution did not immediately gain popularity. In September 1922, even 14 years after its publication, Student wrote to Fisher: "I am sending you a copy of Student's Tables as you are the only man that's ever likely to use them!" Today, Student's t-distribution is widely used and respected in statistical research.

CHAPTER

8 Introduction to Statistical Inferences

Chapter Outline and Objectives

CHAPTER CASE STUDY: WERE THEY SHORTER BACK THEN?
Comparing today's women to 17th- and 18th-century women

8.1 THE NATURE OF ESTIMATION
Estimations occur in two forms: a **point estimate** and an **interval estimate.**

8.2 ESTIMATION OF MEAN μ (σ KNOWN)
Information from the sampling distribution of sample means and the central limit theorem is used in **estimating** the value of an unknown **population mean.**

8.3 THE NATURE OF HYPOTHESIS TESTING
To test a claim we must formulate a **null hypothesis** and an **alternative hypothesis.**

8.4 HYPOTHESIS TEST OF MEAN μ (σ KNOWN): A PROBABILITY-VALUE APPROACH
The **probability-value approach** makes full use of the computer's capability in doing the work of the decision-making process.

8.5 HYPOTHESIS TEST OF MEAN μ (σ KNOWN): A CLASSICAL APPROACH
The **classical approach** uses critical values in doing the work of the decision-making process.

RETURN TO CHAPTER CASE STUDY

We deal with questions about the population mean using two methods that assume the value of the population standard deviation is a known quantity. This assumption is seldom realized in real-life problems, but it will make our first look at the techniques of inference much simpler.

Chapter Case Study

Were They Shorter Back Then?

Were They Shorter Back Then?
The average height for an early 17th-century English man was approximately 5′6″. For 17th-century English women, it was about 5′1/2″. While average heights in England remained virtually unchanged in the 17th and 18th centuries, American colonists grew taller. Averages for modern Americans are just over 5′9″ for men and about 5′3 3/4″ for women. The main reasons for this difference are improved nutrition, notably increased consumption of meat and milk, and antibiotics. Source: http://www.plimoth.org/Library/l-short.htm

The National Center for Health Statistics (NCHS) provides statistical information that will guide actions and policies to improve the health of the American people. Recent data from NCHS give the average height of women in the United States as 63.7 inches with a standard deviation of 2.75 inches.

A random sample of 50 women from the health profession yielded these height data: CS08

65.0	66.0	64.0	67.0	59.0	69.0	66.0	69.0	64.0	61.5
63.0	62.0	63.0	64.0	72.0	66.0	65.0	64.0	67.0	68.0
70.0	63.0	63.0	68.0	58.0	60.0	63.5	66.0	64.0	62.0
64.5	69.0	63.5	69.0	62.0	58.0	66.0	68.0	59.0	56.0
64.0	66.0	65.0	69.0	67.0	66.5	67.5	62.0	70.0	62.0

ANSWER NOW 8.1

a. What population was sampled to obtain the data?

b. Describe the sample data using the mean and standard deviation plus any other statistic(s) that helps describe the sample. Construct a histogram and comment on the shape of the distribution.

c. How is the distribution of the sample data related to the distribution of the population? To the sampling distribution of sample means?

d. Using the techniques of Chapter 7, find the limits that would bound the middle 90% of the sampling distribution of sample means for random samples of size 50 selected from the population of female heights with a known mean of 63.7 inches and a standard deviation of 2.75 inches.

e. On the histogram drawn in part b draw a vertical line at the sample mean and then draw a horizontal line segment showing the interval found in part d. Does the sample mean fall within the interval? Explain what this means.

f. Using the techniques of Chapter 7, find $P(\bar{x} \geq 64.72)$ for a random sample of 50 drawn from a population with a known mean of 63.7 inches and a standard deviation of 2.75 inches. Explain what the resulting value means.

g. Does this sample of 50 data appear to belong to the population described by the NCHS? Explain.

After completing Chapter 8, further investigate the Chapter Case Study in the Return to Chapter Case Study section with Exercises 8.131, 8.132, and 8.133 (p. 402).

8.1 The Nature of Estimation

A company manufactures rivets for use in building aircraft. One characteristic of extreme importance is the "shearing strength" of each rivet. The company's engineers must monitor production to be certain that the shearing strength of the rivets meets the required specs. To accomplish this, they take a sample and determine the mean shearing strength of the sample. Based on this sample information, the company can estimate the mean shearing strength for all the rivets it is manufacturing.

A random sample of 36 rivets is selected, and each rivet is tested for shearing strength. The resulting sample mean is $\bar{x} = 924.23$ lb. Based on this sample, we say, "We believe the mean shearing strength of all such rivets is 924.23 lb."

NOTE 1 Shearing strength is the force required to break a material in a "cutting" action. Obviously, the manufacturer is not going to test all rivets because the test destroys each rivet tested. Therefore, samples are tested and the information about the sample must be used to make inferences about the population of all such rivets.

NOTE 2 Throughout Chapter 8 we will treat the standard deviation σ as a known, or given, quantity and concentrate on learning the procedures for making statistical inferences about the population mean μ. Therefore, to continue the explanation of statistical inferences, we will assume $\sigma = 18$ for the specific rivets described in our example.

POINT ESTIMATE FOR A PARAMETER
A single number designed to estimate a quantitative parameter of a population, usually the value of the corresponding **sample statistic.**

That is, the sample mean, $\bar{x}$, is the point estimate (single number value) for the mean μ of the sampled population. For our rivet example, 924.23 is the point estimate for μ, the mean shearing strength of all rivets.

The quality of this point estimate should be questioned. Is the estimate exact? Is the estimate likely to be high? Or low? Would another sample yield the same result? Would another sample yield an estimate of nearly the same value? Or a value that is very different? How is "nearly the same" or "very different" measured? The quality of an estimation procedure (or method) is greatly enhanced if the sample statistic is both *less variable* and *unbiased.* The variability of a statistic is measured by the standard error of its sampling distribution. The sample mean can be made less variable by reducing its standard error, $\sigma/\sqrt{n}$. That requires using a larger sample because as n increases, the standard error decreases.

UNBIASED STATISTIC
A sample statistic whose sampling distribution has a mean value equal to the value of the population parameter being estimated. A statistic that is not unbiased is a **biased statistic.**

ANSWER NOW 8.2

Using a tremendously large sample does not solve the question of quality for an estimator. What problems do you anticipate with very large samples?

Figure 8.1 illustrates the concept of being unbiased and the effect of variability on the point estimate. The value A is the parameter being estimated, and the dots represent possible sample statistic values from the sampling distribution of the statistic. If A represents the true population mean μ, then the dots represent possible sample means from the $\bar{x}$ sampling distribution.

FIGURE 8.1

Effects of Variability and Bias

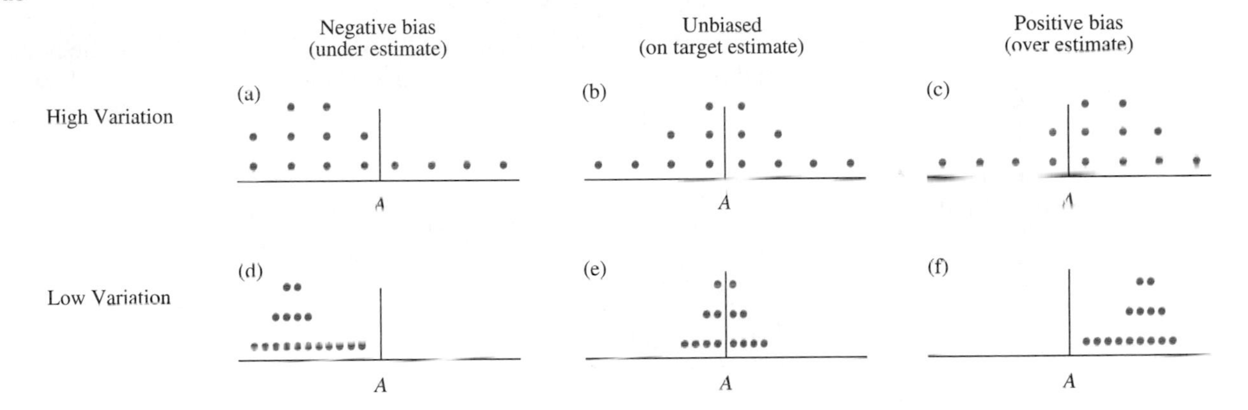

Figure 8.1(a), (c), (d), and (f) show biased statistics; (a) and (d) show sampling distributions whose mean values are less than the value of the parameter, while (c) and (f) show sampling distributions whose mean values are greater than the parameter. Figure 8.1(b) and (e) show sampling distributions that appear to have a mean value equal to the value of the parameter; therefore, they are unbiased. Figure 8.1 (a), (b), and (c) show more variability, while (d), (e), and (f) show less variability in the sampling distributions. Diagram (e) represents the best situation, an estimator that is unbiased (on target) and has low variability (all values close to the target).

The sample mean, $\bar{x}$, is an unbiased statistic because the mean value of the sampling distribution of sample means, $\mu_{\bar{x}}$, is equal to the population mean, μ. (Recall that the sampling distribution of sample means has a mean $\mu_{\bar{x}} = \mu$.) Therefore, the sample statistic $\bar{x} = 924.23$ is an unbiased point estimate for the mean strength of all rivets being manufactured in our example.

Sample means vary in value and form a sampling distribution in which not all samples result in $\bar{x}$ values equal to the population mean. Therefore, we should not expect this sample of 36 rivets to produce a point estimate (sample mean) that is exactly equal to the mean μ of the sampled population. We should, however, expect the point estimate to be fairly close in value to the population mean. The sampling distribution of sample means (SDSM) and the central limit theorem (CLT) provide the information needed to describe how close the point estimate, $\bar{x}$, is expected to be to the population mean, μ.

Recall that approximately 95% of a normal distribution is within two standard deviations of the mean and that the central limit theorem describes the sampling distribution of sample means as being nearly normal when samples are large enough. Samples of size 36 from populations of variables like rivet strength are generally considered large enough. Therefore, we should anticipate that 95% of all random samples selected from a population with unknown mean μ and standard deviation $\sigma = 18$ will have means $\bar{x}$ between

$$\mu - 2(\sigma_{\bar{x}}) \quad \text{and} \quad \mu + 2(\sigma_{\bar{x}})$$

$$\mu - 2\left(\frac{\sigma}{\sqrt{n}}\right) \quad \text{and} \quad \mu + 2\left(\frac{\sigma}{\sqrt{n}}\right)$$

$$\mu - 2\left(\frac{18}{\sqrt{36}}\right) \quad \text{and} \quad \mu + 2\left(\frac{18}{\sqrt{36}}\right)$$

$$\mu - 6 \quad \text{and} \quad \mu + 6$$

This suggests that 95% of all random samples of size 36 selected from the population of rivets should have a mean $\bar{x}$ between $\mu - 6$ and $\mu + 6$. Figure 8.2 shows the middle 95% of the distribution, the bounds of the interval covering the 95%, and the mean μ.

FIGURE 8.2

Sampling Distribution of $\bar{x}$'s, Unknown μ

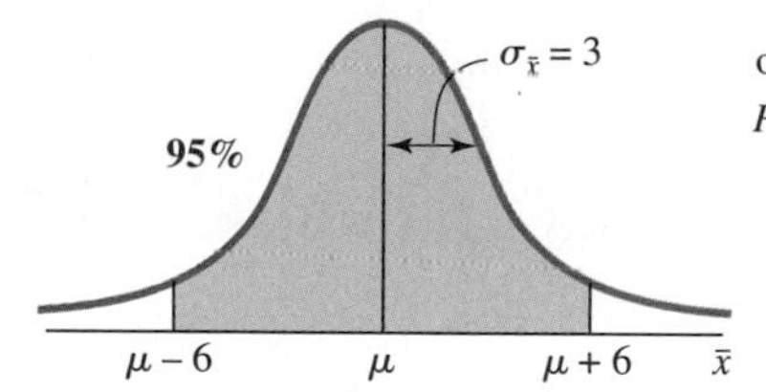

or expressed algebraically:
$P(\mu - 6 < \bar{x} < \mu + 6) = 0.95$

ANSWER NOW 8.3

Explain why the standard error of sample means is 3 for the rivet example.

Now let's put all of this information together in the form of a *confidence interval.*

INTERVAL ESTIMATE

An interval bounded by two values and used to estimate the value of a population parameter. The values that bound this interval are statistics calculated from the sample that is being used as the basis for the estimation.

LEVEL OF CONFIDENCE $1 - \alpha$

The proportion of all interval estimates that include the parameter being estimated.

CONFIDENCE INTERVAL

An interval estimate with a specified level of confidence.

To construct the confidence interval, we will use the point estimate $\bar{x}$ as the central value of an interval in much the same way as we used the mean μ as the central value to find the interval that captures the middle 95% of the $\bar{x}$ distribution in Figure 8.2.

For our rivet example, we can find the bounds to an interval centered at $\bar{x}$:

$$\mu - 2(\sigma_{\bar{x}}) \quad \text{to} \quad \mu + 2(\sigma_{\bar{x}})$$

$$924.23 - 6 \quad \text{to} \quad 924.23 + 6$$

The resulting interval is 918.23 to 930.23

The level of confidence assigned to this interval is approximately 95%, or 0.95. The bounds of the interval are 2 multiples ($z = 2.0$) of the standard error from the sample mean, and by looking at Table 3 in Appendix B, we can more accurately determine the level of confidence as 0.9544. Putting all of this information together, we express the estimate as a confidence interval: **918.23 to 930.23** *is the 95.44% confidence interval for the mean shear strength of the rivets.* Or in an abbreviated form: **918.23 to 930.23**, *the 95.44% confidence interval for* μ.

ANSWER NOW 8.4

Verify that the level of confidence for a two-standard-deviation interval is 95.44%.

Application 8.1 Yellowstone Park's Old Faithful

Courtesy of National Park Service, Yellowstone National Park

Welcome to the Old Faithful WebCam

Predictions for the time of the next eruption of Old Faithful are made by the rangers using a formula that takes into account the length of the previous eruption. The formula used has proven to be accurate, plus or minus 10 minutes, 90% of the time. At 9:15 A.M. on October 27, 2002, the posted prediction time of the next eruption was:

Next Prediction: 10:24 A.M. +/- 10 min.

Note the time at which the picture was recorded, 10:18:02.

http://www.nps.gov/yell/oldfaithfulcam.htm

FYI
Visit the Old Faithful Webcam. When is the next eruption predicted to occur?

ANSWER NOW 8.5

a. **What does "10:24 A.M. +/− 10 min." mean? Explain.**

b. **Did this eruption occur during the predicted time interval?**

c. **What does "90% of the time" mean? Explain.**

Exercises

8.6 Explain the difference between a point estimate and an interval estimate.

8.7 Identify each numerical value by "name" (mean, variance, etc.) and by symbol ($\bar{x}$, etc.):

a. The mean height of 24 junior high school girls is 4′11″.
b. The standard deviation for IQ scores is 16.
c. The variance among the test scores on last week's exam was 190.
d. The mean height of all cadets who have ever entered West Point is 69 inches.

8.8 Ex08-008 A random sample of taxi fares from downtown to the airport was obtained:

15	19	17	23	21	17	16	18
12	18	20	22	15	18	20	

Use the data to find a point estimate for each of these parameters:

a. mean b. variance c. standard deviation

Figure for Exercise 8.9

a. b. c.

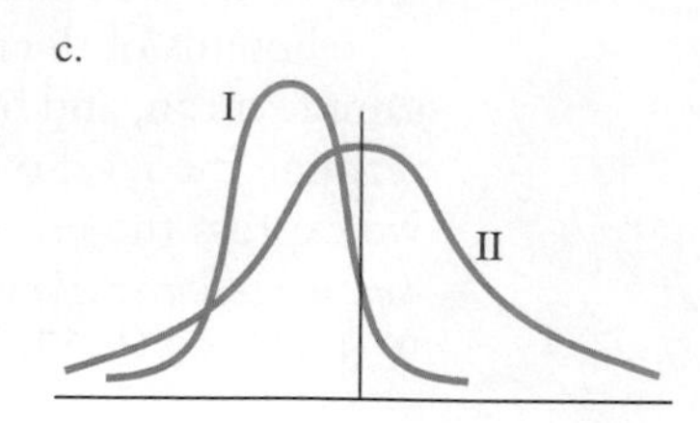

8.9 In each diagram above, I and II represent the sampling distributions of two statistics that might be used to estimate a parameter. In each case, identify the statistic that you think would be the better estimator and describe why it is your choice.

8.10 Suppose there are two statistics that will serve as an estimator for the same parameter. One of them is biased and the other unbiased.

a. Everything else being equal, explain why you usually would prefer an unbiased estimator to a biased estimator.
b. If a statistic is unbiased, does that ensure it is a good estimator? Why or why not? What other considerations must be taken into account?
c. Describe a situation in which the biased statistic might be a better choice as an estimator than the unbiased statistic.

8.11 Being unbiased and having a small variability are two desirable characteristics of a statistic if it is going to be used as an estimator. Describe how the central limit theorem addresses both of these properties when estimating the mean of a population.

8.12 Find the level of confidence assigned to an interval estimate of the mean formed using each interval:

a. $\bar{x} - 1.3(\sigma_{\bar{x}})$ to $\bar{x} + 1.3(\sigma_{\bar{x}})$
b. $\bar{x} - 1.65(\sigma_{\bar{x}})$ to $\bar{x} + 1.65(\sigma_{\bar{x}})$
c. $\bar{x} - 1.96(\sigma_{\bar{x}})$ to $\bar{x} + 1.96(\sigma_{\bar{x}})$
d. $\bar{x} - 2.3(\sigma_{\bar{x}})$ to $\bar{x} + 2.3(\sigma_{\bar{x}})$
e. $\bar{x} - 2.6(\sigma_{\bar{x}})$ to $\bar{x} + 2.6(\sigma_{\bar{x}})$

8.13 *Accounting Horizons* (Vol. 12, No. 2, June 1998) reported the results of a study concerning the KPMG Peat Marwick's "Research Opportunities in Auditing" (ROA) program that supported audit research projects with data from actual audits. A total of 174 projects were funded between 1977 and 1993. A sample of 25 of these projects revealed that 19 were valued at $17,320 each and 6 were valued at $20,200 each. From the sample data, estimate the total value of the funding for all the projects.

8.14 A stamp dealer wishes to purchase a stamp collection that is believed to contain approximately 7000 individual stamps and approximately 4000 first-day covers. Devise a plan that she might use to estimate the collection's worth.

8.2 Estimation of Mean μ (σ Known)

In Section 8.1 we surveyed the basic ideas of estimation: point estimate, interval estimate, level of confidence, and confidence interval. These basic ideas are interrelated and used throughout statistics when an inference calls for an estimate. In this section we formalize the interval estimation process as it applies to estimating the population mean μ based on a random sample under the restriction that the population standard deviation σ is a known value.

The sampling distribution of sample means and the central limit theorem provide us with the information we need to ensure that the necessary *assumptions* are satisfied.

ASSUMPTIONS
The conditions that need to exist in order to correctly apply a statistical procedure.

NOTE The word *assumptions* is somewhat of a misnomer. It does not mean that we "assume" something to be the situation and continue, but that we must be sure the conditions expressed by the assumptions do exist before we apply a particular statistical method.

THE ASSUMPTION FOR ESTIMATING MEAN μ USING A KNOWN σ

The sampling distribution of $\bar{x}$ has a normal distribution.

The information needed to ensure that this assumption (or condition) is satisfied is contained in the sampling distribution of sample means (SDSM) and in the central limit theorem (see Chapter 7, p. 318–319):

> The sampling distribution of sample means $\bar{x}$ is distributed about a mean equal to μ with a standard error equal to $\sigma/\sqrt{n}$; and (1) if the randomly sampled population is normally distributed, then $\bar{x}$ is normally distributed for all sample sizes, or (2) if the randomly sampled population is not normally distributed, then $\bar{x}$ is approximately normally distributed for sufficiently large sample sizes.

Therefore, we can satisfy the required assumption by either (1) knowing that the sampled population is normally distributed or (2) using a random sample that contains a sufficiently large number of data. The first possibility is obvious. We either know enough about the population to know that it is normally distributed or we don't. The second way to satisfy the assumption is by applying the CLT. Inspection of various graphic displays of the sample data should yield an indication of the type of distribution the population possesses. The CLT can be applied to smaller samples (say, $n = 15$ or larger) when the data provide a strong indication of a unimodal distribution that is approximately symmetric. If there is evidence of some skewness in the data, then the sample size needs to be much larger (perhaps $n \geq 50$). If the data provide evidence of an extremely skewed or J-shaped distribution, the CLT will still apply if the sample is large enough. In extreme cases, "large enough" may be unrealistically or impracticably large.

FYI
The help of a professional statistician should be sought when treating extremely skewed data.

NOTE: There is no hard and fast rule defining "large enough"; the sample size that is "large enough" varies greatly according to the distribution of the population.

ANSWER NOW 8.15

Determining the emissions of an engine before it is manufactured is the subject of the "Estimation" video on your Student Suite CD (found inside the back cover of this textbook). View the video clip and answer this question: How does GM determine whether a prototype engine will perform within legal limits when it is mass produced?

The $1 - \alpha$ confidence interval for the estimation of mean μ is found using the formula

$$\bar{x} - z(\alpha/2)\left(\frac{\sigma}{\sqrt{n}}\right) \quad \text{to} \quad \bar{x} + z(\alpha/2)\left(\frac{\sigma}{\sqrt{n}}\right) \tag{8.1}$$

Here are the parts of the confidence interval formula:

1. $\bar{x}$ is the point estimate and the center point of the confidence interval.
2. $z(\alpha/2)$ is the **confidence coefficient.** It is the number of multiples of the standard error needed to formulate an interval estimate of the correct width to have a level of confidence of $1 - \alpha$. Figure 8.3 shows the relationship among:

FIGURE 8.3
Confidence Coefficient, $z(\alpha/2)$

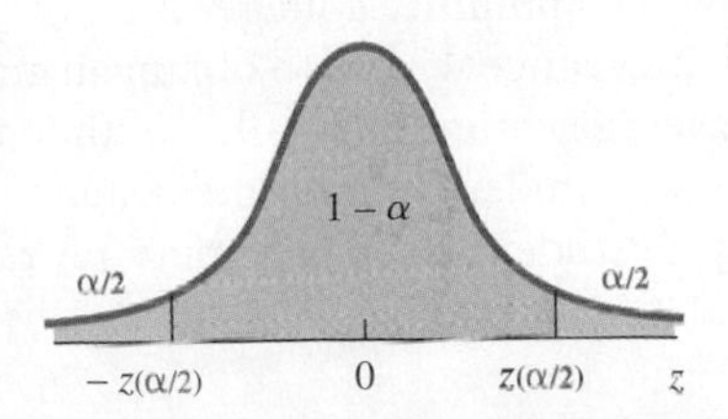

the level of confidence $1 - \alpha$ (the middle portion of the distribution), $\alpha/2$ (the "area to the right" used with the critical-value notation), and the confidence coefficient $z(\alpha/2)$ (whose value is found using Table 4B of Appendix B).

3. $\sigma/\sqrt{n}$ is the **standard error of the mean,** or the standard deviation of the sampling distribution of sample means.
4. $z(\alpha/2)\left(\frac{\sigma}{\sqrt{n}}\right)$ is one-half the width of the confidence interval (the product of the confidence coefficient and the standard error) and is called the **maximum error of estimate, *E*.**
5. $\bar{x} - z(\alpha/2)\left(\frac{\sigma}{\sqrt{n}}\right)$ is called the **lower confidence limit** (LCL), and $\bar{x} + z(\alpha/2)\left(\frac{\sigma}{\sqrt{n}}\right)$ is called the **upper confidence limit** (UCL) for the confidence interval.

FYI
Basically, the confidence interval is "point estimate ± maximum error."

The estimation procedure is organized into a five-step process that will take into account all of the above information and produce both the point estimate and the confidence interval.

THE CONFIDENCE INTERVAL: A FIVE-STEP PROCEDURE

Step 1 **The Set-Up:**
Describe the population parameter of interest.

Step 2 **The Confidence Interval Criteria:**
a. Check the assumptions.
b. Identify the probability distribution and the formula to be used.
c. State the level of confidence, $1 - \alpha$.

Step 3 **The Sample Evidence:**
Collect the sample information.

Step 4 **The Confidence Interval:**
a. Determine the confidence coefficient.
b. Find the maximum error of estimate.
c. Find the lower and upper confidence limits.

Step 5 **The Results:**
State the confidence interval.

Illustration 8.1 Constructing a Confidence Interval for the Mean One-Way Commute Distance

The student body at many community colleges is considered a "commuter population." The student activities office wishes to obtain an answer to the question: How far (one way) does the average community college student commute to college each day? (Typically the "average student's commute distance" is meant to be the "mean distance" commuted by all students who commute.) A random sample of 100 commuting students was identified, and the one-way distance each commuted was obtained. The resulting sample mean distance was 10.22 miles.

Estimate the mean one-way distance commuted by all commuting students using: (a) a point estimate and (b) a 95% confidence interval. (Use $\sigma = 6$ miles.)

SOLUTION

(a) The point estimate for the mean one-way distance is **10.22** miles (the sample mean).

(b) We use the five-step procedure to find the 95% confidence interval.

STEP 1 **The Set-Up:**
Describe the population parameter of interest.
The mean μ of the one-way distances commuted by all commuting community college students is the parameter of interest.

STEP 2 **The Confidence Interval Criteria:**

a. Check the assumptions.
σ is known. The variable "distance commuted" most likely has a skewed distribution because the vast majority of the students will commute between 0 and 25 miles, with fewer commuting more than 25 miles. A sample size of 100 should be large enough for the CLT to satisfy the assumption; the x sampling distribution is approximately normal.

b. Identify the probability distribution and the formula to be used.
The standard normal distribution, z, will be used to determine the confidence coefficient, and formula (8.1) with $\sigma = 6$.

c. State the level of confidence, $1 - \alpha$.
The question asks for 95% confidence, or $1 - \alpha = 0.95$.

STEP 3 **The Sample Evidence:**
Collect the sample information.
The sample information is given in the statement of the problem: $n = 100$, $\bar{x} = 10.22$.

STEP 4 **The Confidence Interval:**

a. Determine the confidence coefficient.
The confidence coefficient is found using Table 4B:

A Portion of Table 4B

α	...	0.05
$z(\alpha/2)$	...	**1.96**
$1 - \alpha$	...	0.95

Level of confidence: $1 - \alpha = 0.95 \rightarrow$ (row $1 - \alpha$)

$\rightarrow$ Confidence coefficient: $z(\alpha/2) = 1.96$

b. Find the maximum error of estimate.
Use the maximum error part of formula (8.1):

$$E = z(\alpha/2)\left(\frac{\sigma}{\sqrt{n}}\right) = 1.96\left(\frac{6}{\sqrt{100}}\right) = (1.96)(0.6) = 1.176$$

c. **Find the lower and upper confidence limits.**
Using the point estimate, $\bar{x}$, from Step 3 and the maximum error, E, from Step 4b, we find the confidence interval limits:

$$\bar{x} - z(\alpha/2)\left(\frac{\sigma}{\sqrt{n}}\right) \quad \text{to} \quad \bar{x} + z(\alpha/2)\left(\frac{\sigma}{\sqrt{n}}\right)$$

$$10.22 - 1.176 \quad \text{to} \quad 10.22 + 1.176$$

$$9.044 \quad \text{to} \quad 11.396$$

$$9.04 \quad \text{to} \quad 11.40$$

STEP 5 **The Results:**
State the confidence interval.
9.04 to 11.40, the 95% confidence interval for μ. That is, with 95% confidence we can say, "The mean one-way distance is between 9.04 and 11.40 miles." ■

ANSWER NOW 8.16

In your own words, describe the relationships among the point estimate, the level of confidence, the maximum error, and the confidence interval.

Let's look at another illustration of the estimation procedure.

Illustration 8.2 Constructing a Confidence Interval for the Mean Particle Size

"Particle size" is an important property of latex paint and is monitored during production as part of the quality-control process. Thirteen particle-size measurements were taken using the Dwight P. Joyce Disc, and the sample mean was 3978.1 angstroms [where 1 angstrom (1 Å) = 10^{-8} cm]. The particle size, x, is normally distributed with a standard deviation $\sigma = 200$ angstroms. Find the 98% confidence interval for the mean particle size for this batch of paint.

SOLUTION

STEP 1 **The Set-Up:**
Describe the population parameter of interest.
The mean particle size, μ, for the batch of paint from which the sample was drawn.

STEP 2 **The Confidence Interval Criteria:**

a. **Check the assumptions.**
σ is known. The variable "particle size" is normally distributed; therefore, the sampling distribution of sample means is normal for all sample sizes.

b. **Identify the probability distribution and the formula to be used.**
The standard normal variable z, and formula (8.1) with $\sigma = 200$.

c. **State the level of confidence, $1 - \alpha$.**
98%, or $1 - \alpha = 0.98$.

STEP 3 **The Sample Evidence:**
Collect the sample information: $n = 13$ and $\bar{x} = 3978.1$.

STEP 4 **The Confidence Interval:**

a. Determine the confidence coefficient.

The confidence coefficient is found using Table 4B:
$z(\alpha/2) = z(0.01) = 2.33$.

A Portion of Table 4B

α	...	0.02
$z(\alpha/2)$	...	**2.33**
$1 - \alpha$	...	0.98

Level of confidence: $1 - \alpha = 0.98 \rightarrow$ ($1 - \alpha$ row)

$\rightarrow$ Confidence coefficient: $z(\alpha/2) = \mathbf{2.33}$

b. Find the maximum error of estimate.

$$E = z(\alpha/2)\left(\frac{\sigma}{\sqrt{n}}\right) = 2.33\left(\frac{200}{\sqrt{13}}\right) = (2.33)(55.47) = 129.2$$

c. Find the lower and upper confidence limits.

Using the point estimate, $\bar{x}$, from Step 3 and the maximum error, E, from Step 4b, we find the confidence interval limits:

$$\bar{x} - z(\alpha/2)\left(\frac{\sigma}{\sqrt{n}}\right) \quad \text{to} \quad \bar{x} + z(\alpha/2)\left(\frac{\sigma}{\sqrt{n}}\right)$$

$$3978.1 - 129.2 = 3848.9 \quad \text{to} \quad 3978.1 + 129.2 = 4107.3$$

STEP 5 **The Results:**

State the confidence interval.

3848.9 to 4107.3, the 98% confidence interval for μ. With 98% confidence we can say, "The mean particle size is between 3848.9 and 4107.3 angstroms." ■

ANSWER NOW 8.17

A machine produces parts whose lengths are normally distributed with $\sigma = 0.5$ inch. A sample of ten parts has a mean length of 75.92 inches.

a. **Find the point estimate for μ.**

b. **Find the 98% confidence maximum error of estimate for μ.**

c. **Find the 98% confidence interval for μ.**

Let's take another look at the concept "level of confidence." It was defined to be the probability that the sample to be selected will produce interval bounds that contain the parameter.

Illustration 8.3 Demonstrating the Meaning of a Confidence Interval

Single-digit random numbers, like the ones in Table 1 in Appendix B, have a mean value $\mu = 4.5$ and a standard deviation $\sigma = 2.87$ (see Exercise 5.33, p. 246). Draw a sample of 40 single-digit numbers from Table 1 and construct the 90% confidence interval for the mean. Does the resulting interval contain the expected value of μ, 4.5? If we were to select another sample of 40 single-digit numbers from

Table 1, would we get the same result? What might happen if we selected a total of 15 different samples and constructed the 90% confidence interval for each? Would the expected value for μ— namely, 4.5—be contained in all of them? Should we expect all 15 confidence intervals to contain 4.5? Think about the definition of "level of confidence"; it says that in the long run 90% of the samples will result in bounds that contain μ. In other words, 10% of the samples will not contain μ. Let's see what happens.

First we need to address the assumptions; if the assumptions are not satisfied, we cannot expect the 90% and the 10% to occur. We know: (1) the distribution of

Application 8.2 Rockies Snow Brings Little Water

When snow melts it becomes water, sometimes more water than at other times. This newspaper article compares the water contents of snow from two areas in the United States that typically get about the same amount of snow annually. However, the water contents are very different. Several point estimates for the average are included in the *USA Today* article.

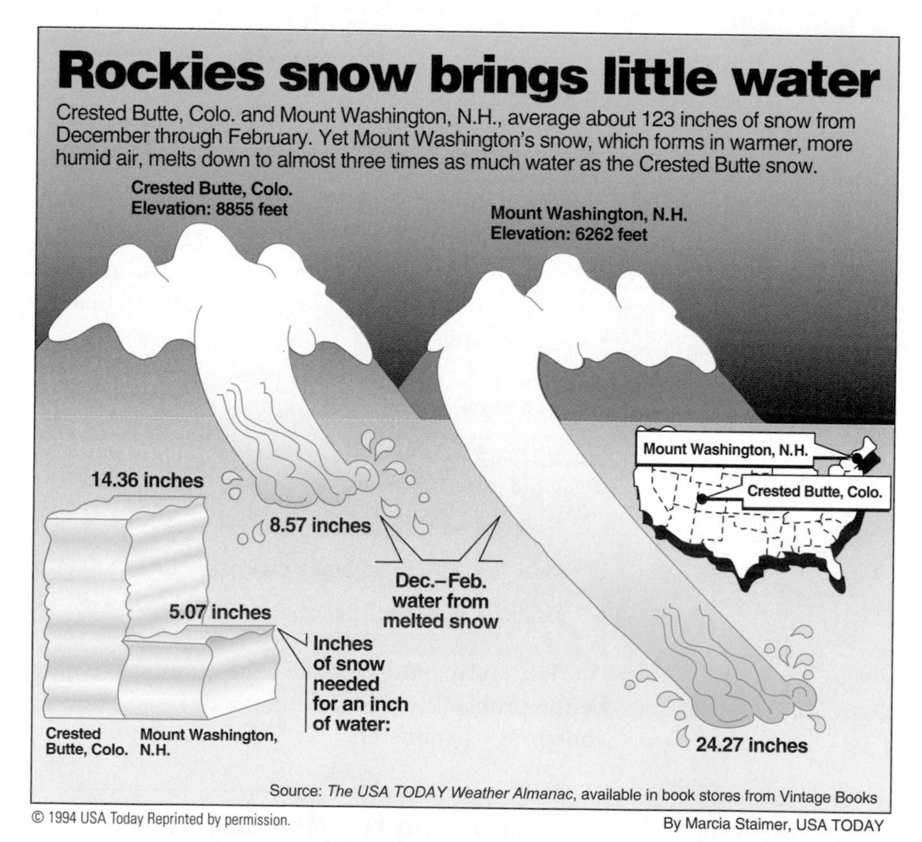

By Marcia Staimer, USA TODAY

ANSWER NOW 8.18

"Rockies snow brings little water" lists "14.36 inches" and "5.07 inches" as statistics and uses them as point estimates. Describe why these numbers are statistics and why they are also point estimates.

single-digit random numbers is rectangular (definitely not normal), (2) the distribution of single-digit random numbers is symmetric about their mean, (3) the $\bar{x}$ distribution for very small samples ($n = 5$) in Illustration 7.2 (p. 314) displayed a distribution that appeared to be approximately normal, and (4) there should be no skewness involved. Therefore, it seems reasonable to assume that $n = 40$ is large enough for the CLT to apply.

The first random sample was drawn from Table 1 in Appendix B: II08-03

2	8	2	1	5	5	4	0	9	1
0	4	6	1	5	1	1	3	8	0
3	6	8	4	8	6	8	9	5	0
1	4	1	2	1	7	1	7	9	3

The sample statistics are $n = 40$, $\sum x = 159$, and $\bar{x} = 3.975$. Here is the resulting 90% confidence interval:

$$\bar{x} \pm z(\alpha/2)\left(\frac{\sigma}{\sqrt{n}}\right): \quad 3.975 \pm 1.65\left(\frac{2.87}{\sqrt{40}}\right)$$

$$3.975 \pm (1.65)(0.454)$$

$$3.975 \pm 0.749$$

$$3.975 - 0.749 = 3.23 \quad \text{to} \quad 3.975 + 0.749 = 4.72$$

3.23 to 4.72, the 90% confidence interval for μ

Figure 8.4 shows this interval estimate, its bounds, and the expected mean μ.

FIGURE 8.4
The 90% Confidence Interval

With 90% confidence, we think μ is somewhere within this interval

3.23 $\mu = 4.50$ 4.72 $\bar{x}$

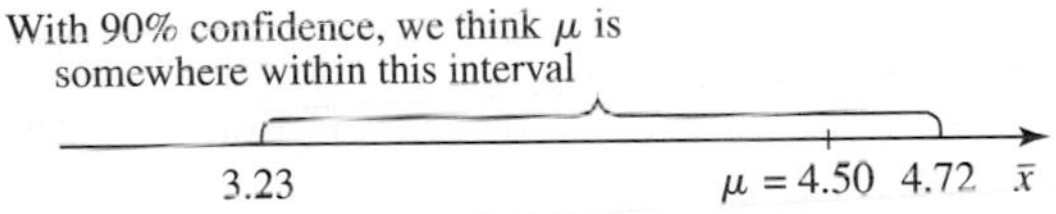

The expected value for the mean, 4.5, does fall within the bounds of the confidence interval for this sample. Let's now select 14 more random samples from Table 1 in Appendix B, each of size 40.

Table 8.1 lists the mean from the first sample and the means obtained from the 14 additional random samples of size 40. The 90% confidence intervals for the estimation of μ based on each of the 15 samples are listed in Table 8.1 and shown in Figure 8.5 on page 356.

TABLE 8.1 **Fifteen Samples of Size 40**

TaU8-01

Sample Number	Sample Mean, $\bar{x}$	90% Confidence Interval Estimate for μ	Sample Number	Sample Mean, $\bar{x}$	90% Confidence Interval Estimate for μ
1	3.975	3.23 to 4.72	9	4.08	3.33 to 4.83
2	4.64	3.89 to 5.39	10	5.20	4.45 to 5.95
3	4.56	3.81 to 5.31	11	4.88	4.13 to 5.63
4	3.96	3.21 to 4.71	12	5.36	4.61 to 6.11
5	5.12	4.37 to 5.87	13	4.18	3.43 to 4.93
6	4.24	3.49 to 4.99	14	4.90	4.15 to 5.65
7	3.44	2.69 to 4.19	15	4.48	3.73 to 5.23
8	4.60	3.85 to 5.35			

FIGURE 8.5
Confidence Intervals from Table 8.1

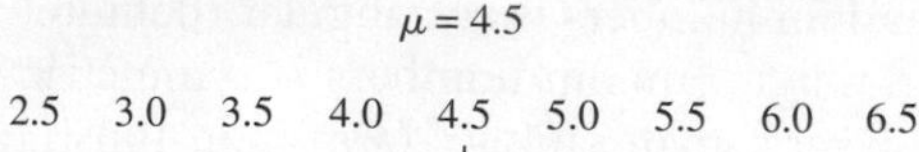

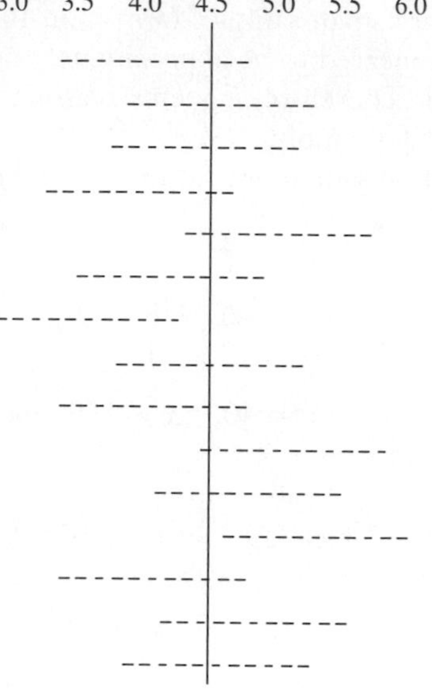

We see that 86.7% (13 of the 15) of the intervals contain μ and two of the 15 samples (sample 7 and sample 12) do not contain μ. The results here are "typical"; repeated experimentation might result in any number of intervals that contain 4.5. However, in the long run we should expect approximately $1 - \alpha = 0.90$ (or 90%) of the samples to result in bounds that contain 4.5 and approximately 10% that do not contain 4.5. ■

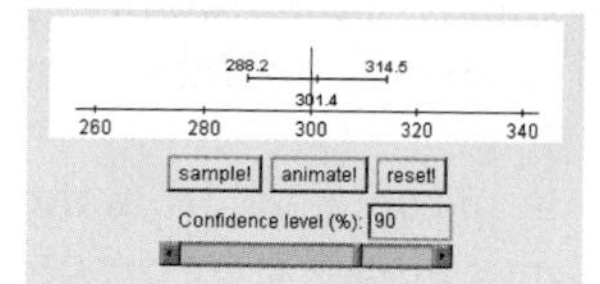

ANSWER NOW 8.19 INTERACTIVITY

Interactivity 8-A demonstrates the effect that the level of confidence $(1 - \alpha)$ has on the width of a confidence interval. Consider sampling from a population where $\mu = 300$ and $\sigma = 80$.

a. Set the slider for level of confidence to 68%. Click "sample!" to construct one 68% confidence interval. Note the upper and lower confidence limits and calculate the width of the interval. Using "animate!" construct many samples and note the percentage of the intervals that contain the true mean of 300. Click "stop" and "reset".

b. Set the slider for level of confidence to 95%. Click "sample!" to construct one 95% confidence interval. Note the upper and lower confidence limits and calculate the width of the interval. Using "animate!" construct many samples and note the percentage of the intervals that contain the true mean of 300. Click "stop" and "reset".

c. Set the slider for level of confidence to 99%. Click "sample!" to construct one 99% confidence interval. Note the upper and lower confidence limits and calculate the width of the interval. Using "animate!" construct many samples and note the percentage of the intervals that contain the true mean of 300. Click "stop".

d. From the information collected in parts a, b, and c, what effect does the level of confidence have on the width of the interval? Why is this happening?

Technology Instructions: Confidence Interval for Mean μ with a Given σ

MINITAB (Release 13)

Input the data into C1; then continue with:

Choose:	Stat > Basic Statistics > 1-Sample Z
Enter:	Variables: C1
	Sigma: σ
Select:	Options
Enter:	Confidence Level: $1-\alpha$ (ex.: 0.95 or 95.0)
Select:	Alternative: not equal

Excel XP

Input the data into column A; then continue with:

Choose:	Tools > Data Analysis Plus > Z-Test: Mean > OK
Enter:	Input Range: (A1:A20 or select cells) > OK
	Standard Deviation (SIGMA): σ > OK
	Alpha: α (ex.: 0.05) > OK

TI-83 Plus

Input the data into L1; then continue with the following, entering the appropriate values and highlighting Calculate:

Choose:	STAT > TESTS > 7:ZInterval

```
ZInterval
 Inpt:Data Stats
 σ:0
 List:L1
 Freq:1
 C-Level:.95
 Calculate
```

ANSWER NOW 8.20

Using a computer or calculator, select a random sample of 40 single-digit numbers and find the 90% confidence interval for μ. Repeat several times, observing whether or not 4.5 is in the interval each time. Describe your results.

(Use the commands for generating integer data on p. 183 and then continue with the confidence interval commands above.)

SAMPLE SIZE

The confidence interval has two basic characteristics that determine its quality: its level of confidence and its width. It is preferred that the interval have a high level of confidence and be precise (narrow) at the same time. The higher the level of confidence, the more likely the interval is to contain the parameter, and the narrower the interval, the more precise the estimation. However, these two properties seem to work against each other, since it would seem that a narrower interval would tend to have a lower probability and a wider interval would be less precise. The maximum error part of the confidence interval formula specifies the relationship involved.

MAXIMUM ERROR OF ESTIMATE

$$E = z(\alpha/2)\left(\frac{\sigma}{\sqrt{n}}\right) \quad (8.2)$$

This formula has four components: (1) the maximum error E, half of the width of the confidence interval; (2) the confidence coefficient, $z(\alpha/2)$, which is determined by the level of confidence; (3) the sample size, n; and (4) the standard deviation, σ. The standard deviation σ is not a concern in this discussion because it is a constant (the standard deviation of a population does not change in value). That leaves three factors. Inspection of formula (8.2) indicates the following: Increasing the level of confidence will make the confidence coefficient larger and thereby require either the maximum error to increase or the sample size to increase; decreasing the maximum error will require the level of confidence to decrease or the sample size to increase; and decreasing the sample size will force the maximum error to become larger or the level of confidence to decrease. We have a "three-way tug of war," as pictured in Figure 8.6. An increase or decrease to any one of the three factors has an effect on one or both of the other two factors. The statistician's job is to "balance" the level of confidence, the sample size, and the maximum error so that an acceptable interval results.

FYI
When the denominator increases, the value of the fraction decreases.

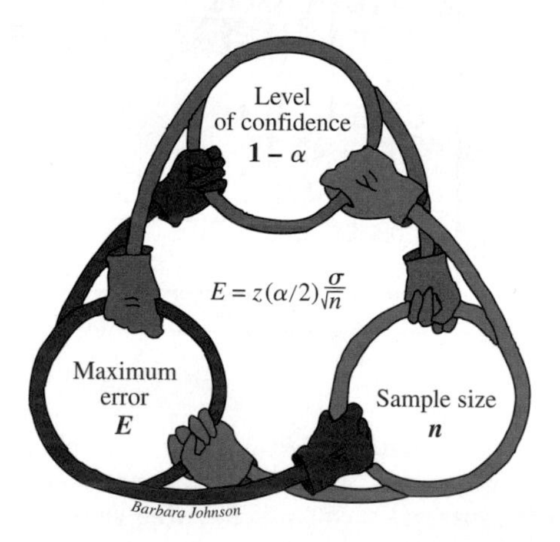

FIGURE 8.6
The "Three-Way Tug of War" Between $1 - \alpha$, n, and E

Let's look at an illustration of this relationship in action.

Illustration 8.4 Determining the Sample Size for a Confidence Interval

Determine the size sample needed to estimate the mean weight of all second-grade boys if we want to be accurate within 1 lb with 95% confidence. Assume a normal distribution and that the standard deviation of the boys' weights is 3 lb.

FYI
Instructions for using Table 4B are given on page 351.

SOLUTION
The desired level of confidence determines the confidence coefficient: The confidence coefficient is found using Table 4B: $z(\alpha/2) = z(0.025) = \mathbf{1.96.}$

The desired maximum error is $E = 1.0$. Now we are ready to use the maximum error formula:

$$E = z(\alpha/2)\left(\frac{\sigma}{\sqrt{n}}\right): \quad 1.0 = 1.96\left(\frac{3}{\sqrt{n}}\right)$$

$$\text{Solve for } n: \quad 1.0 = \frac{5.88}{\sqrt{n}}$$

$$\sqrt{n} = 5.88$$

$$n = (5.88)^2 = 34.57 = \mathbf{35}$$

Therefore, $\boldsymbol{n = 35}$ is the sample size needed if you want a 95% confidence interval with a maximum error no greater than 1 lb. ■

NOTE When we solve for the sample size n, it is customary to round up to the next larger integer, no matter what fraction (or decimal) results.

Using the maximum error formula (8.2) can be made a little easier by rewriting the formula in a form that expresses n in terms of the other values.

FYI
Complete Answer Now 8.21 to see the effect that increasing the level of confidence has on the sample size when the maximum error is kept the same.

SAMPLE SIZE

$$n = \left(\frac{z(\alpha/2) \cdot \sigma}{E}\right)^2 \tag{8.3}$$

ANSWER NOW 8.21
Find the sample size needed to estimate μ for a normal population with $\sigma = 3$ to within 1 unit at the 98% level of confidence.

If the maximum error is expressed as a multiple of the standard deviation σ, then the actual value of σ is not needed in order to calculate the sample size.

Illustration 8.5 Determining the Sample Size Without a Known Value of Sigma (σ)

Find the sample size needed to estimate the population mean to within $\frac{1}{5}$ of a standard deviation with 99% confidence.

SOLUTION
Determine the confidence coefficient (using Table 4B): $1 - \alpha = 0.99$, $z(\alpha/2) = 2.58$. The desired maximum error is $E = \frac{\sigma}{5}$. Now we are ready to use the sample size formula (8.3):

$$n = \left(\frac{z(\alpha/2) \cdot \sigma}{E}\right)^2: \quad n = \left(\frac{(2.58) \cdot \sigma}{\sigma/5}\right)^2 = \left(\frac{(2.58\sigma)(5)}{\sigma}\right)^2 = [(2.58)(5)]^2$$

$$= (12.90)^2 = 166.41 = \mathbf{167}$$

■

Exercises

8.22 Discuss the effect each of the following has on the confidence interval:
a. point estimate
b. level of confidence
c. sample size
d. variability of the characteristic being measured

8.23 Discuss the conditions that must exist before we can estimate the population mean using the interval techniques of formula (8.1).

8.24 Determine the value of the confidence coefficient $z(\alpha/2)$ for each situation:
a. $1 - \alpha = 0.90$
b. $1 - \alpha = 0.95$
c. 98% confidence
d. 99% confidence

8.25 Given the information: The sampled population is normally distributed, $n = 16$, $\bar{x} = 28.7$, and $\sigma = 6$.
a. Find the 0.95 confidence interval for μ.
b. Are the assumptions satisfied? Explain why.

8.26 Given the information: The sampled population is normally distributed, $n = 55$, $\bar{x} = 78.2$, and $\sigma = 12$.
a. Find the 0.98 confidence interval for μ.
b. Are the assumptions met? Explain.

8.27 Given the information: $n = 86$, $\bar{x} = 128.5$, and $\sigma = 16.4$.
a. Find the 0.90 confidence interval for μ.
b. Are the assumptions satisfied? Explain why.

8.28 Given the information: $n = 22$, $\bar{x} = 72.3$, and $\sigma = 6.4$.
a. Find the 0.99 confidence interval for μ.
b. Are the assumptions satisfied? Explain why.

8.29 The Channel Tunnel train that connects England with France carries up to 650 passengers, and peak speeds greater than 190 mph are occasionally obtained. Recently Prince Harry paid $206 for his ticket to take the two-hour trip, reportedly whisking along at 186 mph (*People*, "Match Point," July 13, 1998). Assume the standard deviation is 19 mph in the course of all the journeys back and forth and that the train's speed is normally distributed. Suppose speed readings are made during the next 20 trips of the Channel Tunnel train, and the mean speed of these measurements is 184 mph.
a. What is the variable being studied?
b. Find the 90% confidence interval estimate for the mean speed.
c. Find the 95% confidence interval estimate for the mean speed.

8.30 Ex08-030 A certain adjustment to a machine will change the length of the parts it is making but will not affect the standard deviation. The length of the parts is normally distributed, and the standard deviation is 0.5 mm. After an adjustment is made, a random sample is taken to determine the mean length of parts now being produced. The resulting lengths are:

75.3	76.0	75.0	77.0	75.4
76.3	77.0	74.9	76.5	75.8

a. What is the parameter of interest?
b. Find the point estimate for the mean length of all parts now being produced.
c. Find the 0.99 confidence interval for μ.

8.31 A sample of 60 night school students' ages is obtained in order to estimate the mean age of night school students. The result is $\bar{x} = 25.3$ years. The population variance is 16.
a. Give a point estimate for μ.
b. Find the 95% confidence interval for μ.
c. Find the 99% confidence interval for μ.

8.32 The lengths of 200 fish caught in Cayuga Lake had a mean of 14.3 in. The population standard deviation is 2.5 in.
a. Find the 90% confidence interval for the population mean length.
b. Find the 98% confidence interval for the population mean length.

8.33 The Third International Mathematics and Science Study (TIMSS) in 1999 examined eighth-graders' proficiency in math and science. The mean geometry score for the sample of eighth-grade students in the United States was 473 with a standard error of 4.4. Construct a 95% confidence interval for the mean geometry score for all eighth-grade students in the United States.

8.34 About 67% of married adults say they consult with their spouse before spending $352, the average amount for which married adults say they consult (Yankelovich Partner for Lutheran Brotherhood).
a. Based on this information, what can you conclude about the variable "amount spent before consulting with spouse"? What is the $352?

A survey of 500 married adults was taken from a nearby neighborhood and gave a sample mean of $289.75.
b. Construct a 0.98 confidence interval for the mean amount for all married adults. Use $\sigma = \$600$.
c. Based on the above answers, what can you conclude about the mean amount spent before consulting with spouse for adults in the sampled neighborhood compared to the general population?

8.35 Ex08-035 The atomic weight of a reference sample of silver was measured at NIST (National Institute of Standards and Technology) using two nearly identical mass spectrometers. This project was undertaken in conjunction with the redetermination of the Faraday constant. Here are 48 observations:

107.8681568	107.8681465	107.8681572	107.8681785	107.8681446	107.8681903
107.8681526	107.8681494	107.8681616	107.8681587	107.8681519	107.8681486
107.8681419	107.8681569	107.8681508	107.8681672	107.8681385	107.8681518
107.8681662	107.8681424	107.8681360	107.8681333	107.8681610	107.8681477
107.8681079	107.8681344	107.8681513	107.8681197	107.8681604	107.8681385
107.8681642	107.8681365	107.8681151	107.8681082	107.8681517	107.8681448
107.8681198	107.8681482	107.8681334	107.8681609	107.8681101	107.8681512
107.8681469	107.8681360	107.8681254	107.8681261	107.8681450	107.8681368

Source: StatLib http://lib.stat.cmu.edu/datasets/

a. Notice that the data differ only in the fifth, sixth and seventh decimal places. Most computers will round the data and their calculated results, so the variation is seemingly lost. The statistics can be found by using just the last three digits of each data (i.e., 107.8681568 becomes 568). Algebraically the coding looks like this:
   Atomic weight coded = (Atomic weight − 107.8681000) × 10,000,000.
b. Construct a graph of the coded data. How does the coding show on the graph?
c. Find the mean and standard deviation of the coded data.
d. Convert the answers found in part c to original units.
e. Determine whether the data have an approximately normal distribution. Present your case.
f. Do the SDSM and CLT apply? Explain.
g. Is sigma known?
h. If the goal is to find a 95% confidence interval for the mean value of all observations, what would you do?
i. Find the 95% confidence interval for the mean value of all such observations. Justify your method.

8.36 How large a sample should be taken if the population mean is to be estimated with 99% confidence to within \$75? The population has a standard deviation of \$900.

8.37 A high-tech company wants to estimate the mean number of years of college education its employees have completed. A good estimate of the standard deviation for the number of years of college is 1.0. How large a sample needs to be taken to estimate μ to within 0.5 year with 99% confidence?

8.38 By measuring the amount of time it takes a component of a product to move from one work station to the next, an engineer has estimated that the standard deviation is 5 sec.

a. How many measurements should be made in order to be 95% certain that the maximum error of estimation will not exceed 1 sec?
b. What sample size is required for a maximum error of 2 sec?

8.39 The new mini-laptop computers can deliver as much computing power as machines ten times their size, but they weigh in at less than 3 lb. Experts predict they will soon replace their traditional 6- to 8-lb older laptop brothers as the computer industry continues its relentless quest for smallness (*Fortune*, "Time to Dump Heavyweights," May 25, 1998). If the standard deviation of the weight of all mini-laptops is 0.4 lb, how large a sample would be needed to estimate the population mean weight if the maximum error of estimate is to be 0.15 lb with 95% confidence?

8.40 According to *USA Today* (October 7, 1998), adults visit a public library an average of seven times per year. A random sample of the adults in a metropolitan area is to be commissioned by the library's planning board. The board is interested in estimating the mean number of visits made to the library by the adults in the community. How large will the sample need to be if the board wants to estimate the mean within 0.3 of one standard deviation with 0.98 confidence?

8.3 The Nature of Hypothesis Testing

We all make decisions every day of our lives. Some of these decisions are of major importance; others are seemingly insignificant. All decisions follow the same basic pattern. We weigh the alternatives; then, based on our beliefs and preferences and whatever evidence is available, we arrive at a decision and take the appropriate action. The statistical hypothesis test follows much the same process, except that it involves statistical information. In this section we develop many of the concepts and attitudes of the hypothesis test while looking at several decision-making situations without using any statistics.

A friend is having a party (Super Bowl party, home-from-college party—you know the situation, any excuse will do), and you have been invited. You must make a decision: attend or not attend. That's simple; well maybe, except that you want to go only if you can be convinced the party is going to be more fun than your friend's typical party; furthermore, you definitely do not want to go if the party is going to be just another dud. You have taken the position that "the party will be a dud" and you will not go unless you become convinced otherwise. Your friend assures you, "Guaranteed, the party will be a great time!" Do you go or not?

The decision-making process starts by identifying **something of concern** and then formulating **two hypotheses** about it.

HYPOTHESIS
A statement that something is true.

Your friend's statement, "The party will be a great time," is a hypothesis. Your position, "The party will be a dud," is also a hypothesis.

STATISTICAL HYPOTHESIS TEST
A process by which a decision is made between two opposing hypotheses. The two opposing hypotheses are formulated so that each hypothesis is the negation of the other. (That way one of them is always true, and the other one is always false.) Then one hypothesis is tested in hopes that it can be shown to be a very improbable occurrence, thereby implying the other hypothesis is likely the truth.

The two hypotheses involved in making a decision are known as the *null hypothesis* and the *alternative hypothesis.*

NULL HYPOTHESIS,* H_o
The hypothesis we will test. Generally this is a statement that a population parameter has a specific value. The null hypothesis is so named because it is the "starting point" for the investigation. (The phrase "there is no difference" is often used in its interpretation.)

ALTERNATIVE HYPOTHESIS, H_a
A statement about the same population parameter that is used in the null hypothesis. Generally this is a statement that specifies the population parameter has a value different, in some way, from the value given in the null hypothesis. The rejection of the null hypothesis will imply the likely truth of this alternative hypothesis.

With regard to your friend's party, the two opposing viewpoints or hypotheses are: "The party will be a great time" and "The party will be a dud." Which statement becomes the null hypothesis, and which becomes the alternative hypothesis?

---

*We use the notation H_o for the null hypothesis to contrast it with H_a for the alternative hypothesis. Other texts may use H_0 (subscript zero) in place of H_o and H_1 in place of H_a.

Determining the statement of the null hypothesis and the statement of the alternative hypothesis is a very important step. The *basic idea* of the hypothesis test is for the evidence to have a chance to "disprove" the null hypothesis. The null hypothesis is the statement that the evidence might disprove. *Your concern* (belief or desired outcome), as the person doing the testing, is expressed in the alternative hypothesis. As the person making the decision, you believe that the evidence will demonstrate the feasibility of your "theory" by demonstrating the *unlikeliness* of the truth of the null hypothesis. The alternative hypothesis is sometimes referred to as the *research hypothesis*, since it represents what the researcher hopes will be found to be "true." (If so, he or she will get a paper out of the research.)

Since the "evidence" (who's going to the party, what is going to be served, and so on) can only demonstrate the unlikeliness of the party being a dud, your initial position, "The party will be a dud," becomes the null hypothesis. Your friend's claim, "The party will be a great time," then becomes the alternative hypothesis.

H_o: "Party will be a dud" vs. H_a: "Party will be a great time"

ANSWER NOW 8.41
Behavior patterns of lobsters are the subject of the "Formulating a Hypothesis" video on your Student Suite CD (found inside the back cover of this textbook). View the video clip and then formulate some hypotheses that you could test by observing lobsters.

Illustration 8.6

Writing Hypotheses

You are testing a new design for airbags used in automobiles, and you are concerned that they might not open properly. State the null and alternative hypotheses.

SOLUTION
The two opposing possibilities are "Bags open properly" and "Bags do not open properly." Testing can only produce evidence that discredits the hypothesis "Bags open properly." Therefore, the null hypothesis is "Bags open properly" and the alternative hypothesis is "Bags do not open properly." ■

The alternative hypothesis can be the statement the experimenter wants to show to be true.

Illustration 8.7

Writing Hypotheses

An engineer wishes to show that the new formula that was just developed results in a quicker-drying paint. State the null and alternative hypotheses.

SOLUTION
The two opposing possibilities are "does dry quicker" and "does not dry quicker." Since the engineer wishes to show "does dry quicker," the alternative hypothesis is "Paint made with the new formula does dry quicker" and the null hypothesis is "Paint made with the new formula does not dry quicker." ■

Occasionally it might be reasonable to hope that the evidence does not lead to a rejection of the null hypothesis. Such is the case in Illustration 8.8.

Illustration 8.8 Writing Hypotheses

You suspect that a brand-name detergent outperforms the store's brand of detergent, and you wish to test the two detergents because you would prefer to buy the cheaper store brand. State the null and alternative hypotheses.

SOLUTION

Your suspicion, "the brand-name detergent outperforms the store brand," is the reason for the test and therefore becomes the alternative hypothesis.

H_o: "There is no difference in detergent performance."
H_a: "The brand-name detergent performs better than the store brand."

However, as a consumer, you are hoping not to reject the null hypothesis for budgetary reasons. ■

ANSWER NOW 8.42
You are testing a new detonating system for explosives and you are concerned that the system is not reliable. State the null and alternative hypotheses.

Before returning to our example about the party, we need to look at the four possible outcomes that could result from the null hypothesis being either true or false and the decision being either to "reject H_o" or to "fail to reject H_o." Table 8.2 shows these four possible outcomes.

A **type A correct decision** occurs when the null hypothesis is true, and we decide in its favor. A **type B correct decision** occurs when the null hypothesis is false, and the decision is in opposition to the null hypothesis. A **type I error** is committed when a true null hypothesis is rejected—that is, when the null hypoth-

Application 8.3 Evaluation of Teaching Techniques

Abstract: This study tests the effect of homework collection and quizzes on exam scores.

The hypothesis for this study is that an instructor can improve a student's performance (exam scores) through influencing the student's perceived effort-reward probability. An instructor accomplishes this by assigning tasks (teaching techniques) which are a part of a student's grade and are perceived by the student as a means of improving his or her grade in the class. The student is motivated to increase effort to complete those tasks which should also improve understanding of course material. The expected final result is improved exam scores. The null hypothesis for this study is:

H_o: Teaching techniques have no significant effect on students' exam scores. . . .

SOURCE: David R. Vruwink and Janon R. Otto, *The Accounting Review*, Vol. LXII, No. 2, April 1987. Reprinted by permission.

ANSWER NOW 8.43
State the instructor's hypothesis, the alternative hypothesis.

esis is true but we decide against it. A **type II error** is committed when we decide in favor of a null hypothesis that is actually false.

TABLE 8.2 Four Possible Outcomes in a Hypothesis Test

	Null Hypothesis	
Decision	**True**	**False**
Fail to reject H_o	Type A correct decision	Type II error
Reject H_o	Type I error	Type B correct decision

Illustration 8.9

Describing the Possible Outcomes and Resulting Actions (on Hypothesis Tests)

Describe the four possible outcomes and the resulting actions that would occur for the hypothesis test in Illustration 8.8.

SOLUTION

Recall: H_o: "There is no difference in detergent performance."
H_a: "The brand-name detergent performs better than the store brand."

	Null Hypothesis Is True	Null Hypothesis Is False
Fail to Reject H_o	**Type A Correct Decision** Truth of situation: There is no difference between the detergents. Conclusion: It was determined that there was no difference. Action: The consumer bought the cheaper detergent, saving money and getting the same results.	**Type II Error** Truth of situation: The brand-name detergent is better. Conclusion: It was determined that there was no difference. Action: The consumer bought the cheaper detergent, saving money and getting inferior results.
Reject H_o	**Type I Error** Truth of situation: There is no difference between the detergents. Conclusion: It was determined that the brand-name detergent was better. Action: The consumer bought the brand-name detergent, spending extra money to attain no better results.	**Type B Correct Decision** Truth of situation: The brand-name detergent is better. Conclusion: It was determined that the brand name detergent was better. Action: The consumer bought the brand-name detergent, spending more and getting better results. ■

NOTE The type II error often results in what represents a "lost opportunity"; lost in this situation is the chance to use a product that yields better results.

ANSWER NOW 8.44
Describe the four possible decisions and the resulting actions with regard to your friend's party.

ANSWER NOW 8.45
Describe how the type II error in the party example represents a "lost opportunity."

When a decision is made, it would be nice to always make the correct decision. This, however, is not possible in statistics because we make our decisions on the basis of sample information. The best we can hope for is to control the probability with which an error occurs. The probability assigned to the type I error is **α** (called **"alpha"**; α is the first letter of the Greek alphabet). The probability of the type II error is **β** (called **"beta"**; β is the second letter of the Greek alphabet). See Table 8.3.

TABLE 8.3 Probability with Which Decisions Occur

Error in Decision	Type	Probability	Correct Decision	Type	Probability
Rejection of a true H_o	I	α	Failure to reject a true H_o	A	$1 - \alpha$
Failure to reject a false H_o	II	β	Rejection of a false H_o	B	$1 - \beta$

ANSWER NOW 8.46

Explain why α is not always the probability of rejecting the null hypothesis.

To control these errors we will assign a small probability to each of them. The most frequently used probability values for α and β are 0.01 and 0.05. The probability assigned to each error depends on its seriousness. The more serious the error, the less willing we are to have it occur, and therefore a smaller probability will be assigned. α and β are probabilities of errors, each under separate conditions, and they cannot be combined. Therefore, we cannot determine a single probability for making an incorrect decision. Likewise, the two correct decisions are distinctly separate and each has its own probability; $1 - \alpha$ is the probability of a correct decision when the null hypothesis is true, and $1 - \beta$ is the probability of a correct decision when the null hypothesis is false. $1 - \beta$ is called the *power of the statistical test*, since it is the measure of the ability of a hypothesis test to reject a false null hypothesis, a very important characteristic.

NOTE: Regardless of the outcome of a hypothesis test, you are never certain that a correct decision has been reached.

ANSWER NOW 8.47

Explain how assigning a small probability to an error controls the likeliness of its occurrence.

Let's look back at the two possible errors in decision that could occur in Illustration 8.9. Most people would become upset if they found out they were spending extra money for a detergent that performed no better than the cheaper brand. Likewise, most people would become upset if they found out they could have been buying a better detergent. Evaluating the relative seriousness of these errors requires knowing whether this is your personal laundry or a professional laundry business, how much extra the brand-name detergent costs, and so on.

There is an interrelationship among the probability of the type I error (α), the probability of the type II error (β), and the sample size (n). This is very much like the interrelationship among level of confidence, maximum error, and sample size discussed on page 358. Figure 8.7 shows the "three-way tug of war" among α, β, and n. If any one of the three is increased or decreased, it has an effect on one or both of the others. The statistician's job is to "balance" the three values of α, β, and n to achieve an acceptable testing situation.

FIGURE 8.7
The "Three-Way Tug of War" Between α, β, and n

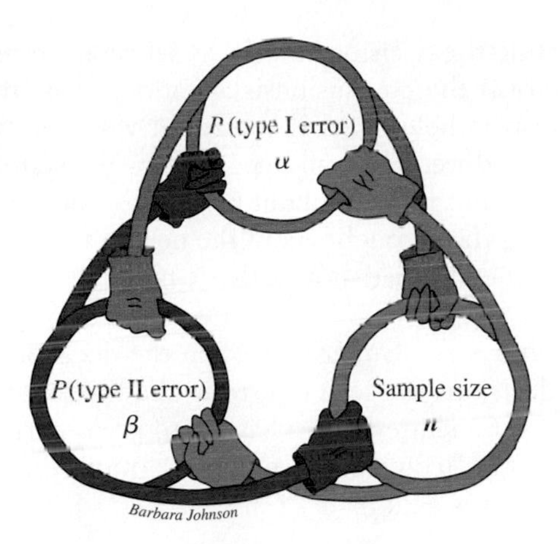

If α is reduced, then either β must increase or n must be increased; if β is decreased, then either α increases or n must be increased; if n is decreased, then either α increases or β increases. The choices for α, β, and n are definitely not arbitrary. At this time in our study of statistics, α will be given in the statement of the problem, as will the sample size n. Further discussion on the role of β, *P*(type II error), is left for another time.

LEVEL OF SIGNIFICANCE α
The probability of committing the type I error.

Establishing the level of significance can be thought of as a "managerial decision." Typically, someone in charge determines the level of probability with which he or she is willing to risk a type I error.

At this point in the hypothesis test procedure, the evidence is collected and summarized and the value of a *test statistic* is calculated.

TEST STATISTIC
A random variable whose value is calculated from the sample data and is used in making the decision "fail to reject H_o" or "reject H_o."

The value of the calculated test statistic is used in conjunction with a decision rule to determine either "reject H_o" or "fail to reject H_o." This **decision rule** must be established prior to collecting the data; it specifies how you will reach the decision.

Back to your friend's party: You have to weigh the history of your friend's parties, the time and place, others going, and so on, against your own criteria and then make your decision. As a result of the decision about the null hypothesis ("The party will be a dud"), you will take the appropriate action; you will either go to or not go to the party.

To complete a hypothesis test, you will need to write a conclusion that carefully describes the meaning of the decision relative to the intent of the hypothesis test.

THE CONCLUSION

a. If the decision is "reject H_o," then the conclusion should be worded something like, "There is sufficient evidence at the α level of significance to show that . . . (the meaning of the alternative hypothesis)."
b. If the decision is "fail to reject H_o," then the conclusion should be worded something like, "There is not sufficient evidence at the α level of significance to show that . . . (the meaning of the alternative hypothesis)."

When writing the decision and the conclusion, remember that (1) the decision is about H_o and (2) the conclusion is a statement about whether or not the contention of H_a was upheld. This is consistent with the "attitude" of the whole hypothesis test procedure. The null hypothesis is the statement that is "on trial," and therefore the decision must be about it. The contention of the alternative hypothesis is the thought that brought about the need for a decision. Therefore, the question that led to the alternative hypothesis must be answered when the conclusion is written.

We must always remember that when the decision is made, nothing has been proved. Both decisions can lead to errors: "Fail to reject H_o" could be a type II error (the lack of sufficient evidence has led to great parties being missed more than once), and "reject H_o" could be a type I error (more than one person has decided to go to a party that was a dud).

ANSWER NOW 8.48

Two methods of preparing for SAT exams are the subject of the "Hypothesis Testing" video on your Student Suite CD (found inside the back cover of this textbook). View the video clip and answer these questions:

a. What two situations are being compared?

b. What claim is made in the comparison of the two groups?

Exercises

8.49 State the null and alternative hypotheses for each of the following:

a. You are investigating a complaint that "special delivery mail takes too much time" to be delivered.
b. You want to show that people find the new design for a recliner chair more comfortable than the old design.
c. You are trying to show that cigarette smoke has an effect on the quality of a person's life.
d. You are testing a new formula for hair conditioner, hoping to show it is effective on "split ends."

8.50 When a parachute is inspected, the inspector is looking for anything to indicate that the parachute might not open.

a. State the null and alternative hypotheses.
b. Describe the four possible outcomes that can result depending on the truth of the null hypothesis and the decision reached.
c. Describe the seriousness of the two possible errors.

8.51 When a medic at the scene of a serious accident inspects the victims, she administers the appropriate medical assistance to each unless she is certain the victim is dead.

a. State the null and alternative hypotheses.
b. Describe the four possible outcomes that can result depending on the truth of the null hypothesis and the decision reached.
c. Describe the seriousness of the two possible errors.

8.52 A supplier of highway materials claims he can supply an asphalt mixture that will make roads paved with his materials less slippery when wet. A general contractor who builds roads wishes to test the supplier's claim. The null hypothesis is "Roads paved with this asphalt mixture are no less slippery than roads paved with other asphalt." The alternative hypothesis is "Roads paved with this asphalt mixture are less slippery than roads paved with other asphalt."

a. Describe the meaning of the two possible types of errors that can occur in the decision when this hypothesis test is completed.
b. Describe how the null hypothesis, as stated above, is a "starting point" for the decision to be made about the asphalt.

8.53 Describe the action that would result in a type I error and a type II error if each of the following null hypotheses were tested. (Remember, the alternative hypothesis is the negation of the null hypothesis.)

a. H_o: The majority of Americans favor laws against assault weapons.
b. H_o: This fast-food menu is not low in salt.
c. H_o: This building must not be demolished.
d. H_o: There is no waste in government spending.

8.54 Describe the action that would result in a correct decision type A and a correct decision type B, if each of the null hypotheses in Exercise 8.53 were tested.

8.55
a. If the null hypothesis is true, what decision error could be made?
b. If the null hypothesis is false, what decision error could be made?
c. If the decision "reject H_o" is made, what decision error could have been made?
d. If the decision "fail to reject H_o" is made, what decision error could have been made?

8.56 The director of an advertising agency is concerned about the effectiveness of a television commercial.
a. What null hypothesis is she testing if she commits a type I error when she erroneously says that the commercial is effective?
b. What null hypothesis is she testing if she commits a type II error when she erroneously says that the commercial is effective?

8.57
a. If α is assigned the value 0.001, what are we saying about the type I error?
b. If α is assigned the value 0.05, what are we saying about the type I error?
c. If α is assigned the value 0.10, what are we saying about the type I error?

8.58
a. If β is assigned the value 0.001, what are we saying about the type II error?
b. If β is assigned the value 0.05, what are we saying about the type II error?
c. If β is assigned the value 0.10, what are we saying about the type II error?

8.59
a. If the null hypothesis is true, the probability of a decision error is identified by what name?
b. If the null hypothesis is false, the probability of a decision error is identified by what name?

8.60 Suppose that a hypothesis test is to be carried out by using $\alpha = 0.05$. What is the probability of committing a type I error?

8.61 The conclusion is part of the hypothesis test that communicates the findings of the test to the reader. As such, it needs special attention so that the reader receives an accurate picture of the findings.
a. Carefully describe the "attitude" of the statistician and the statement of the conclusion when the decision is "reject H_o."
b. Carefully describe the "attitude" and the statement of the conclusion when the decision is "fail to reject H_o."

8.62 Find the power of a test when the probability of the type II error is:
a. 0.01 b. 0.05 c. 0.10

8.63 A normally distributed population is known to have a standard deviation of 5, but its mean is in question. It has been argued to be either $\mu = 80$ or $\mu = 90$, and the following hypothesis test has been devised to settle the argument. The null hypothesis, H_o: $\mu = 80$, will be tested using one randomly selected data and comparing it to the critical value 86. If the data is greater than or equal to 86, the null hypothesis will be rejected.
a. Find α, the probability of the type I error.
b. Find β, the probability of the type II error.

8.64 Suppose the argument in Exercise 8.63 is to be settled using a sample of size 4; find α and β.

8.4 Hypothesis Test of Mean μ (σ Known): A Probability-Value Approach

In Section 8.3 we surveyed the concepts and much of the reasoning behind a hypothesis test while looking at nonstatistical illustrations. In this section we are going to formalize the hypothesis test procedure as it applies to statements concerning the mean μ of a population under the restriction that σ, the population standard deviation, is a known value.

THE ASSUMPTION FOR HYPOTHESIS TESTS ABOUT MEAN μ USING A KNOWN σ

The sampling distribution of $\bar{x}$ has a normal distribution.

The information we need to ensure that this assumption is satisfied is contained in the sampling distribution of sample means and in the central limit theorem (see Chapter 7, pp. 318–319):

The sampling distribution of sample means $\bar{x}$ is distributed about a mean equal to μ with a standard error equal to $\sigma/\sqrt{n}$; and (1) if the randomly sampled population is normally distributed, then $\bar{x}$ is normally distributed for all sample sizes, or (2) if the randomly sampled population is not normally distributed, then $\bar{x}$ is approximately normally distributed for sufficiently large sample sizes.

The hypothesis test is a well-organized, step-by-step procedure used to make a decision. Two different formats are commonly used for hypothesis testing. The *probability-value approach*, or simply *p-value approach*, is the hypothesis test process that has gained popularity in recent years, largely as a result of the convenience and the "number crunching" ability of the computer. This approach is organized as a five-step procedure.

THE PROBABILITY-VALUE HYPOTHESIS TEST: A FIVE-STEP PROCEDURE

Step 1 **The Set-Up:**
a. Describe the population parameter of interest.
b. State the null hypothesis (H_o) and the alternative hypothesis (H_a).

Step 2 **The Hypothesis Test Criteria:**
a. Check the assumptions.
b. Identify the probability distribution and the test statistic to be used.
c. Determine the level of significance, α.

Step 3 **The Sample Evidence:**
a. Collect the sample information.
b. Calculate the value of the test statistic.

Step 4 **The Probability Distribution:**
a. Calculate the p-value for the test statistic.
b. Determine whether or not the p-value is smaller than α.

Step 5 **The Results:**
a. State the decision about H_o.
b. State the conclusion about H_a.

A commercial aircraft manufacturer buys rivets to use in assembling airliners. Each rivet supplier that wants to sell rivets to the aircraft manufacturer must demonstrate that its rivets meet the required specifications. One of the specs is: "The mean shearing strength of all such rivets, μ, is at least 925 lb." Each time the aircraft manufacturer buys rivets, it is concerned that the mean strength might be less than the 925-lb specification.

FYI
Think about the consequences of using weak rivets.

NOTE 1 Each individual rivet has a shearing strength, which is determined by measuring the force required to shear ("break") the rivet. Clearly, not all the rivets can be tested. Therefore, a sample of rivets will be tested, and a decision about the mean strength of all the untested rivets will be based on the mean from those sampled and tested.

ANSWER NOW 8.65
In this example, the aircraft builder, the buyer of the rivets, is concerned that the rivets might not meet the mean-strength specification. State the aircraft manufacturer's null and alternative hypotheses.

NOTE 2 We will use $\sigma = 18$ for our rivet example.

STEP 1 **The Set-Up:**

a. Describe the population parameter of interest.
The population parameter of interest is the mean μ, the mean shearing strength (or mean force required to shear) of the rivets being considered for purchase.

b. State the null hypothesis (H_o) and the alternative hypothesis (H_a).
The null hypothesis and the alternative hypothesis are formulated by inspecting the problem or statement to be investigated and first formulating two opposing statements about the mean μ. For our example, these two opposing statements are: (A) "The mean shearing strength is less than 925" ($\mu < 925$, the aircraft manufacturer's concern), and (B) "The mean shearing strength is at least 925" ($\mu = 925$, the rivet supplier's claim and the aircraft manufacturer's spec).

FYI
More specific instructions are given on pages 362–363.

NOTE The trichotomy law from algebra states that two numerical values must be related in exactly one of three possible relationships: <, =, or >. All three of these possibilities must be accounted for in the two opposing hypotheses in order for the two hypotheses to be negations of each other. The three possible combinations of signs and hypotheses are shown in Table 8.4. Recall that the null hypothesis assigns a specific value to the parameter in question, and therefore "equals" will always be part of the null hypothesis.

TABLE 8.4 The Three Possible Statements of Null and Alternative Hypotheses

Null Hypothesis	Alternative Hypothesis
1. greater than or equal to (≥)	less than (<)
2. less than or equal to (≤)	greater than (>)
3. equal to (=)	not equal to (≠)

The parameter of interest, the population mean μ, is related to the value 925. Statement (A) becomes the alternative hypothesis:

$$H_a\colon \mu < 925 \text{ (the mean is less than 925)}$$

This statement represents the aircraft manufacturer's concern and says, "The rivets do not meet the required specs." Statement (B) becomes the null hypothesis:

$$H_o\colon \mu = 925\ (\geq) \text{ (the mean is at least 925)}$$

This hypothesis represents the negation of the aircraft manufacturer's concern and says, "The rivets do meet the required specs."

NOTE We will write the null hypothesis with just the equal sign, thereby stating the exact value assigned. When "equal" is paired with "less than" or paired with "greater than," the combined symbol is written beside the null hypothesis as a reminder that all three signs have been accounted for in these two opposing statements.

Before continuing with our example, let's look at three illustrations that demonstrate formulating the statistical null and alternative hypotheses involving the population mean μ. Illustrations 8.10 and 8.11 each demonstrate a "one-tailed" alternative hypothesis.

Illustration 8.10 Writing Null and Alternative Hypotheses (One-Tailed Situation)

Suppose the EPA was suing the city of Rochester for noncompliance with carbon monoxide standards. Specifically, the EPA would want to show that the mean level of carbon monoxide in downtown Rochester's air is dangerously high, higher than 4.9 parts per million. State the null and alternative hypotheses.

SOLUTION

To state the two hypotheses, we first need to identify the population parameter in question: the "mean level of carbon monoxide in Rochester." The parameter μ is being compared to the value 4.9 parts per million, the specific value of interest. The EPA is questioning the value of μ and wishes to show it is higher than 4.9 (that is, $\mu > 4.9$). The three possible relationships—(1) $\mu < 4.9$, (2) $\mu = 4.9$, and (3) $\mu > 4.9$—must be arranged to form two opposing statements: One states the EPA's position, "The mean level is higher than 4.9 ($\mu > 4.9$)," and the other states the negation, "The mean level is not higher than 4.9 ($\mu \leq 4.9$)." One of these two statements will become the null hypothesis H_o, and the other will become the alternative hypothesis H_a.

NOTE: Recall that there are two rules for forming the hypotheses: (1) The null hypothesis states that the parameter in question has a specified value ("H_o must contain the equal sign"), and (2) the EPA's contention becomes the alternative hypothesis ("higher than"). Both rules indicate:

$$H_o\text{: } \mu = 4.9\ (\leq) \quad \text{and} \quad H_a\text{: } \mu > 4.9$$

■

Illustration 8.11 Writing Null and Alternative Hypotheses (One-Tailed Situation)

An engineer wants to show that applications of paint made with the new formula dry and are ready for the next coat in a mean time of less than 30 minutes. State the null and alternative hypotheses for this test situation.

SOLUTION

The parameter of interest is the mean drying time per application, and 30 minutes is the specified value. $\mu < 30$ corresponds to "The mean time is less than 30," whereas $\mu \geq 30$ corresponds to the negation, "The mean time is not less than 30." Therefore, the hypotheses are

$$H_o\text{: } \mu = 30\ (\geq) \quad \text{and} \quad H_a\text{: } \mu < 30$$

■

Illustration 8.12 demonstrates a "two-tailed" alternative hypothesis.

Illustration 8.12 Writing Null and Alternative Hypotheses (Two-Tailed Situation)

Job satisfaction is very important to worker productivity. A standard job-satisfaction questionnaire was administered by union officers to a sample of assembly-line workers in a large plant in hopes of showing that the assembly workers' mean score on this questionnaire would be different from the established mean of 68. State the null and alternative hypotheses.

SOLUTION

Either the mean job satisfaction score is different from 68 ($\mu \neq 68$) or the mean is equal to 68 ($\mu = 68$). Therefore,

$$H_o\colon \mu = 68 \quad \text{and} \quad H_a\colon \mu \neq 68$$

■

NOTE 1 The alternative hypothesis is referred to as being "two-tailed" when H_a is "not equal."

NOTE 2 When "less than" is combined with "greater than," they become "not equal to."

The viewpoint of the experimenter greatly affects the way the hypotheses are formed. Generally, the experimenter is trying to show that the parameter value is different from the value specified. Thus, the experimenter is often hoping to be able to reject the null hypothesis so that the experimenter's theory has been substantiated. Illustrations 8.10, 8.11, and 8.12 also represent the three possible arrangements for the $<$, $=$, and $>$ relationships between the parameter μ and a specified value.

ANSWER NOW 8.66
Professor Hart does not believe a statement he heard: "The mean weight of college women is 54.4 kg." State the null and alternative hypotheses he would use to challenge this statement.

Table 8.5 lists some additional common phrases used in claims and indicates their negations and the hypothesis in which each phrase will be used. Again, notice that "equals" is always in the null hypothesis. Also notice that the negation of "less than" is "greater than or equal to." Think of negation as "all the others" from the set of three signs.

TABLE 8.5 Common Phrases and Their Negations

H_o: ($\geq$)	H_a: ($<$)	H_o: ($\leq$)	H_a: ($>$)	H_o: ($=$)	H_a: ($\neq$)
at least	less than	at most	more than	is	is not
no less than	less than	no more than	more than	not different from	different from
not less than	less than	not greater than	greater than	same as	not same as

After the null and alternative hypotheses are established, we will work under the assumption that the null hypothesis is a true statement until there is sufficient evidence to reject it. This situation might be compared to a courtroom trial, where the accused is assumed to be innocent (H_o: Defendant is innocent vs. H_a: Defendant is not innocent) until sufficient evidence has been presented to show that innocence is totally unbelievable ("beyond reasonable doubt"). At the conclusion of the hypothesis test, we will make one of two possible decisions. We will decide in opposition to the null hypothesis and say that we "reject H_o" (this corresponds to "conviction" of the accused in a trial), or we will decide in agreement with the null hypothesis and say that we "fail to reject H_o" (this corresponds to "fail to convict" or an "acquittal" of the accused in a trial).

ANSWER NOW 8.67
State the null and alternative hypotheses used to test each claim:

a. **The mean reaction time is greater than 1.25 seconds.**

b. **The mean score on that qualifying exam is less than 335.**

c. **The mean selling price of homes in the area is not $230,000.**

Let's return to the rivet example we interrupted on page 371 and continue with Step 2. Recall that

$$H_o: \mu = 925 \ (\geq) \text{ (at least 925)} \qquad H_a: \mu < 925 \text{ (less than 925)}$$

STEP 2 The Hypothesis Test Criteria:

a. **Check the assumptions.**
σ is known. Variables like shearing strength typically have a mounded distribution; therefore, a sample of size 50 should be large enough for the CLT to apply and ensure that the sampling distribution of sample means will be normally distributed.

b. **Identify the probability distribution and the test statistic to be used.**
The standard normal probability distribution is used because $\bar{x}$ is expected to have a normal distribution.

For a hypothesis test of μ, we want to compare the value of the sample mean to the value of the population mean as stated in the null hypothesis. This comparison is accomplished using the test statistic in formula (8.4):

$$z\bigstar = \frac{\bar{x} - \mu}{\sigma/\sqrt{n}} \qquad (8.4)$$

The resulting calculated value is identified as $z\bigstar$ ("z star") because it is expected to have a standard normal distribution when the null hypothesis is true and the assumptions have been satisfied. The ★ ("star") is to remind us that this is the calculated value of the test statistic.

$$\text{The test statistic to be used is } z\bigstar = \frac{\bar{x} - \mu}{\sigma/\sqrt{n}} \quad \text{with } \sigma = 18$$

c. **Determine the level of significance, α**

Setting α was described as a managerial decision in Section 8.3. To see what is involved in determining α, the probability of the type I error, for our rivet example, we start by identifying the four possible outcomes, their meaning, and the action related to each.

FYI
Do Answer Now 8.68 before continuing.

ANSWER NOW 8.68
Identify the four possible outcomes and describe the situation involved with each outcome with regard to the aircraft manufacturer testing and buying rivets. Which is the more serious error: type I or type II? Explain.

The type I error occurs when a true null hypothesis is rejected. This would occur when the manufacturer tested rivets that in truth did meet the specs, and rejected them. Undoubtedly this would lead to the rivets not being purchased even though they did meet the specs. In order for the manager to set a level of significance, related information is needed—namely, how soon is the new supply of rivets needed? If they are needed tomorrow and this is the only vendor with an available supply, waiting a week to find acceptable rivets could be very expensive; therefore, rejecting good rivets could be considered a serious error. On the other hand, if the rivets are not needed until next month, then this error may not be very serious. Only the manager will know all the ramifications, and therefore the manager's input is important here.

FYI
There is more to this scenario, but we hope you get the idea.

FYI
α will be assigned in the statement of exercises.

After much consideration, the manager assigns the level of significance: $\alpha = 0.05$.

STEP 3 **The Sample Evidence:**

a. **Collect the sample information.**
We are ready for the data. The sample must be a random sample drawn from the population whose mean μ is being questioned. A random sample of 50 rivets is selected, each rivet is tested, and the sample mean shearing strength is calculated: $\bar{x} = 921.18$ and $n = 50$.

b. **Calculate the value of the test statistic.**
The sample evidence ($\bar{x}$ and n found in Step 3b) is next converted into the **calculated value of the test statistic, $z\bigstar$,** using formula (8.4). (μ is 925 from H_o; $\sigma = 18$ is the known quantity, given in Note 2 on p. 371.) We have

$$z\bigstar = \frac{\bar{x} - \mu}{\sigma/\sqrt{n}}: \quad z\bigstar = \frac{921.18 - 925.0}{18/\sqrt{50}} = \frac{-3.82}{2.5456} = -1.50$$

ANSWER NOW 8.69
Calculate the test statistic $z\bigstar$, given H_o: $\mu = 56$, $\sigma = 7$, $\bar{x} = 54.3$, and $n = 36$.

STEP 4 **The Probability Distribution:**

a. **Calculate the p-value for the test statistic.**

PROBABILITY VALUE, OR p-VALUE
The probability that the test statistic could be the value it is or a more extreme value (in the direction of the alternative hypothesis) when the null hypothesis is true. (*Note:* The symbol **P** will be used to represent the p-value, especially in algebraic situations.)

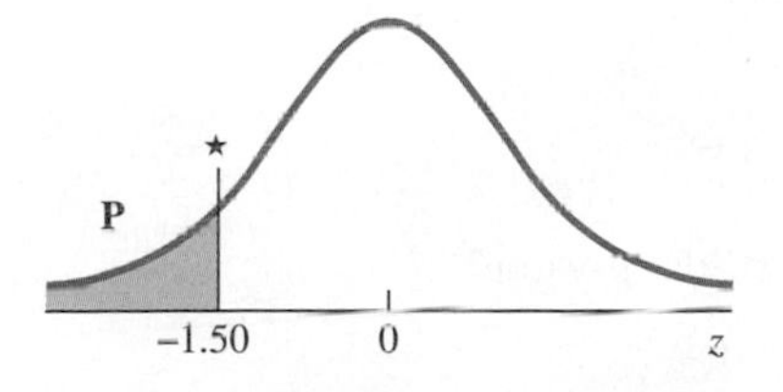

Draw a sketch of the standard normal distribution and locate $z\bigstar$(found in Step 3b) on it. To identify the area that represents the p-value, look at the sign in the alternative hypothesis. For this test, the alternative hypothesis indicates that we are interested in that part of the sampling distribution that is "*less than*" $z\bigstar$. Therefore, the p-value is the area that lies to the *left* of $z\bigstar$. Shade this area.

To find the p-value, you may use any one of three methods:

Method 1: Use Table 3 in Appendix B to determine the tabled area related to $z = 1.50$; then calculate the p-value by subtracting from 0.5000:

FYI
Complete instructions for using Table 3 are given on page 274.

$$p\text{-value} = P(z < z\star) = P(z < -1.50) = P(z > 1.50) = 0.5000 - 0.4332 = \mathbf{0.0668}$$

Method 2: Use Table 5 in Appendix B and the symmetry property: Table 5 is set up to allow you to read the p-value directly from the table. Since $P(z < -1.50) = P(z > 1.50)$, simply locate $z\star = 1.50$ on Table 5 and read the p-value:

FYI
Let the tables do the work the computer will typically do.

$$P(z < -1.50) = \mathbf{0.0668}$$

Method 3: Use the cumulative probability function on a computer or calculator to find the p-value:

FYI
Instructions for using this command are given on page 287.

$$P(z < -1.50) = \mathbf{0.0668}$$

FYI
Try it! See if you get the same answer.

ANSWER NOW 8.70
The meaningfulness of research results is the subject of the "The p-value" video on your Student Suite CD (found inside the back cover of this textbook). View the video clip and answer these questions:

a. **What question must a researcher ask?**

b. **How is its likeliness measured?**

ANSWER NOW 8.71

a. **Calculate the p-value, given H_a: $\mu < 45$ and $z\star = -2.3$.**

b. **Calculate the p-value, given H_a: $\mu > 58$ and $z\star = 1.8$.**

b. Determine whether or not the p-value is smaller than α.
The p-value (0.0668) is not smaller than α (0.05).

STEP 5 The Results:

a. State the decision about H_o.
Is the p-value small enough to indicate that the sample evidence is highly unlikely in the event that the null hypothesis is true? In order to make the decision, we need to know the *decision rule*.

DECISION RULE

a. If the p-value is *less than or equal to* the level of significance α, then the decision must be **reject H_o.**
b. If the p-value is *greater than* the level of significance α, then the decision must be **fail to reject H_o.**

Decision about H_o: Fail to reject H_o.

ANSWER NOW 8.72

a. **What decision is reached when the p-value is greater than α?**

b. **What decision is reached when α is greater than the p-value?**

FYI
Specific information about writing the conclusion is given on page 367.

b. State the conclusion about H_a.
There is not sufficient evidence at the 0.05 level of significance to show that the mean shearing strength of the rivets is less than 925. "We failed to convict" the null hypothesis. In other words, a sample mean as small as 921.18 is likely to occur (as defined by α) when the true population mean value is 925.0 and $\bar{x}$ is normally distributed. The resulting action by the manager would be to buy the rivets.

NOTE When the decision reached is "fail to reject H_o" (or "accept H_o," as many say improperly), it simply means "for the lack of better information, act as if the null hypothesis is true."

Before looking at another example, let's look at the procedures for finding the p-value. The p-value is represented by the area under the curve of the probability distribution for the test statistic that is more extreme than the calculated value of the test statistic. There are three separate cases, and the direction (or sign) of the alternative hypothesis is the key. Table 8.6 outlines the procedure for all three cases.

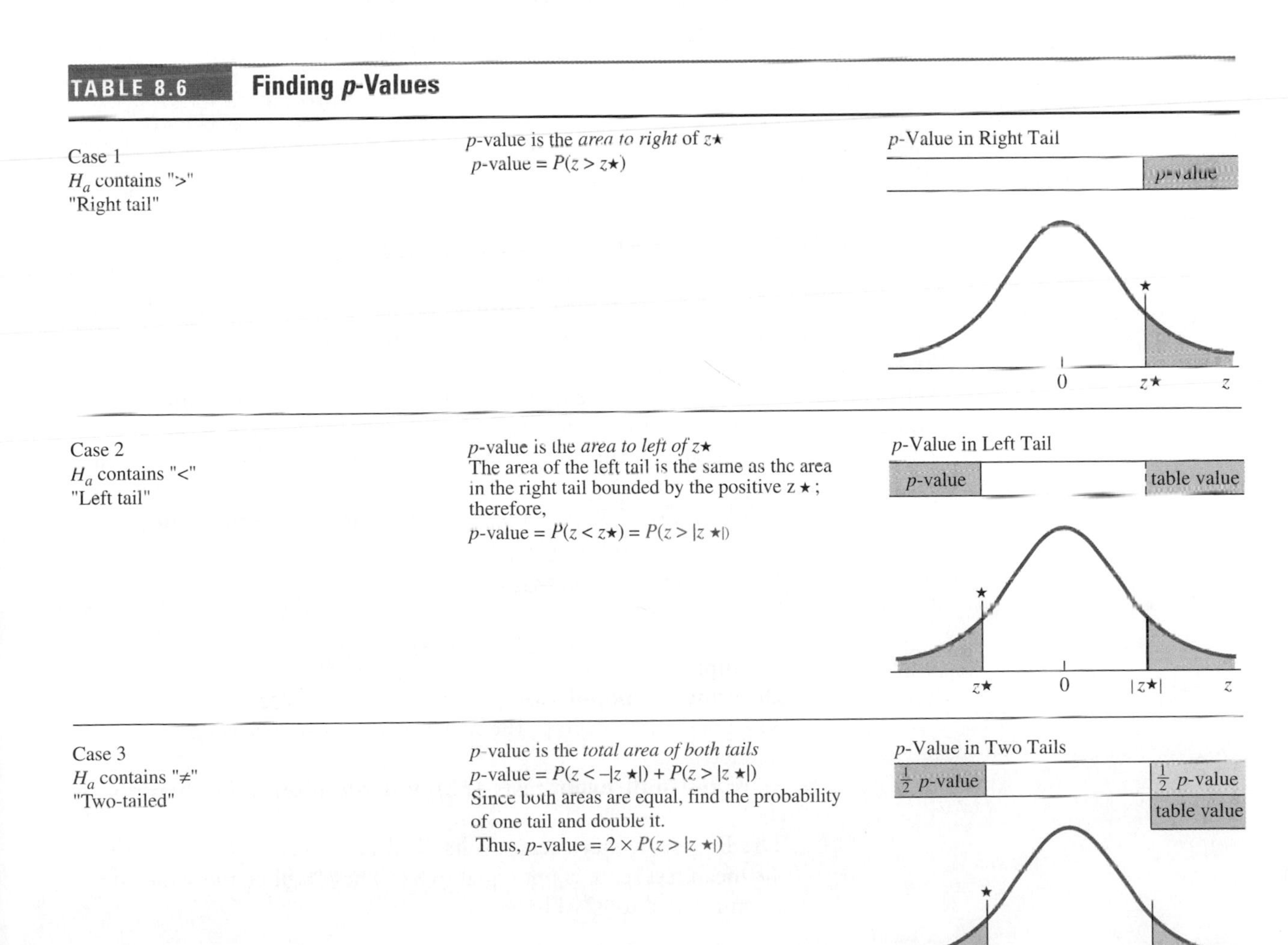

TABLE 8.6 Finding p-Values

Case 1 H_a contains ">" "Right tail"	p-value is the *area to right* of $z\star$ p-value $= P(z > z\star)$	p-Value in Right Tail p-value 0, $z\star$, z
Case 2 H_a contains "<" "Left tail"	p-value is the *area to left of* $z\star$ The area of the left tail is the same as the area in the right tail bounded by the positive $z\star$; therefore, p-value $= P(z < z\star) = P(z > \lvert z\star \rvert)$	p-Value in Left Tail p-value, table value $z\star$, 0, $\lvert z\star \rvert$, z
Case 3 H_a contains "≠" "Two-tailed"	p-value is the *total area of both tails* p-value $= P(z < -\lvert z\star \rvert) + P(z > \lvert z\star \rvert)$ Since both areas are equal, find the probability of one tail and double it. Thus, p-value $= 2 \times P(z > \lvert z\star \rvert)$	p-Value in Two Tails $\frac{1}{2}$ p-value, $\frac{1}{2}$ p-value, table value $-\lvert z\star \rvert$, 0, $\lvert z\star \rvert$, z

ANSWER NOW 8.73

Find the test statistic $z\bigstar$ and the p-value for each situation:

a. H_o: $\mu = 22.5$ vs. H_a: $\mu > 22.5$, $\bar{x} = 24.5$, $\sigma = 6$, $n = 36$

b. H_o: $\mu = 200$ vs. H_a: $\mu < 200$, $\bar{x} = 192.5$, $\sigma = 40$, $n = 50$

c. H_o: $\mu = 12.4$ vs. H_a: $\mu \neq 12.4$, $\bar{x} = 11.52$, $\sigma = 2.2$, $n = 16$

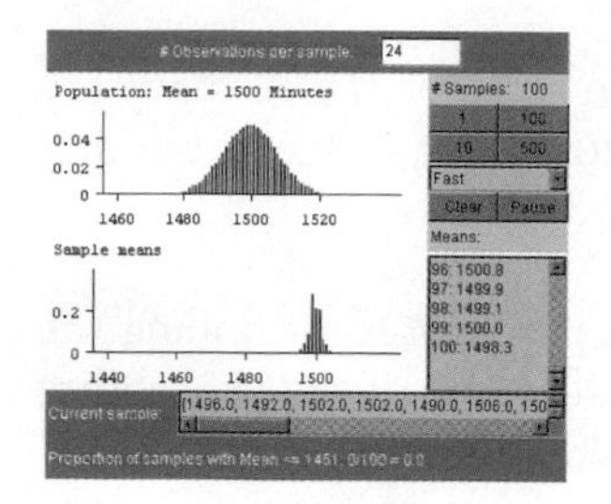

ANSWER NOW 8.74 INTERACTIVITY

Interactivity 8-B estimates the p-value for a one-tailed hypothesis test by simulating the taking of many samples. The given hypothesis test is for H_o: $\mu = 1500$ versus H_a: $\mu < 1500$. A sample of 24 has been taken and the sample mean is 1451. To determine the likelihood that $\bar{x} = 1451$ came from a population with $\mu = 1500$:

a. **Click "10" for "# of samples". Note the sample means and the probability of being less than 1451 if the true mean is really 1500.**

b. **Change to "Batch" and simulate 1000 more samples. What is the probability of being less than 1451? This is your estimated p-value.**

c. **How does your estimated p-value show on the histogram formed from the taking of many samples? Explain what this p-value means with respect to the test.**

d. **If the level of significance was 0.01, what would your decision be?**

Let's look at an illustration involving the two-tailed procedure.

Illustration 8.13 Two-Tailed Hypothesis Test

Many large companies in a certain city have for years used the Kelley Employment Agency for testing prospective employees. The employment selection test used has historically resulted in scores normally distributed about a mean of 82 and a standard deviation of 8. The Brown Agency has developed a new test that is quicker and easier to administer and therefore less expensive. Brown claims that its test results are the same as those obtained on the Kelley test. Many of the companies are considering a change from the Kelley Agency to the Brown Agency in order to cut costs. However, they are unwilling to make the change if the Brown test results have a different mean value. An independent testing firm tested 36 prospective employees with the Brown test. A sample mean of 79 resulted. Determine the p-value associated with this hypothesis test. (Assume $\sigma = 8$.)

SOLUTION

STEP 1 The Set-Up:

a. **Describe the population parameter of interest.**

The population mean μ, the mean of all test scores using the Brown Agency test.

b. **State the null hypothesis (H_o) and the alternative hypothesis (H_a).**

The Brown Agency's test results "will be different" (the concern) if the mean test score is not equal to 82. They "will be the same" if the mean is equal to 82. Therefore,

H_o: $\mu = 82$ (test results have the same mean)
H_a: $\mu \neq 82$ (test results have a different mean)

STEP 2 The Hypothesis Test Criteria:

a. **Check the assumptions.**
σ is known. If the Brown test scores are distributed the same as the Kelley test scores, they will be normally distributed and the sampling distribution will be normal for all sample sizes.

b. **Identify the probability distribution and the test statistic to be used.**
The standard normal probability distribution and the test statistic $z\star = \dfrac{\bar{x} - \mu}{\sigma/\sqrt{n}}$ will be used with $\sigma = 8$.

c. **Determine the level of significance, α.**
The level of significance is omitted because the question asks for the *p*-value and not a decision.

STEP 3 The Sample Evidence:

a. **Collect the sample information:** $n = 36$, $\bar{x} = 79$.

b. **Calculate the value of the test statistic.**
μ is 82 from H_o; $\sigma = 8$ is the known quantity. We have

$$z\star = \frac{\bar{x} - \mu}{\sigma/\sqrt{n}}: \qquad z\star = \frac{79 - 82}{8/\sqrt{36}} = \frac{-3}{1.3333} = -2.25$$

STEP 4 The Probability Distribution:

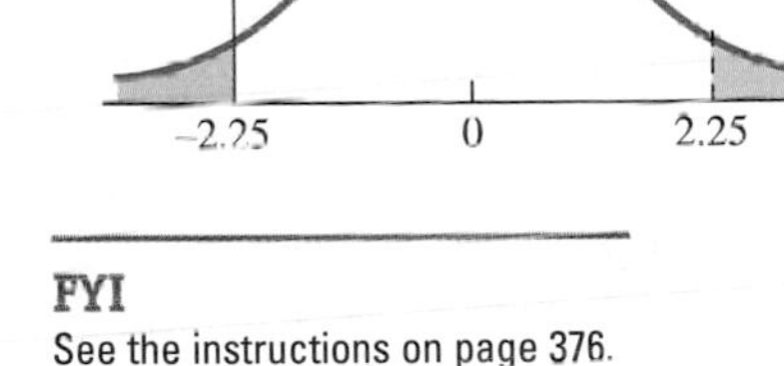

a. **Calculate the *p*-value for the test statistic.**
Since the alternative hypothesis indicates a two-tailed test, we must find the probability associated with both tails. The *p*-value is found by doubling the area of one tail (see Table 8.6, p. 377). Since $z\star = -2.25$, the value of $|z\star| = 2.25$.
The *p*-value $= 2 \times P(z > |z\star|) = 2 \times P(z > 2.25)$.
From Table 3: *p*-value $= 2 \times P(z > 2.25) = 2 \times (0.5000 - 0.4878) = 2(0.0122) = 0.0244$.
or
From Table 5: *p*-value $= 2 \times P(z > 2.25) = 2(0.0122) =$ **0.0244.**
or
Use the cumulative probability function on a computer or calculator.

b. **Determine whether or not the *p*-value is smaller than α.**
A comparison is not possible; no α value was given in the statement of the question.

FYI
See the instructions on page 376.

STEP 5 The Results:
The *p*-value for this hypothesis test is 0.0244. Each individual company now will decide whether to continue to use the Kelley Agency's services or change to the Brown Agency. Each will need to establish the level of significance that best fits its own situation and then make a decision using the decision rule described previously. ■

The *fundamental idea of the p-value* is to express the degree of belief in the null hypothesis:

- When the *p*-value is minuscule (something like 0.0003), the null hypothesis would be rejected by everybody because the sample results are very unlikely for a true H_o.

- When the p-value is fairly small (like 0.012), the evidence against H_o is quite strong and H_o will be rejected by many.
- When the p-value begins to get larger (say, 0.02 to 0.08), there is too much probability that data like the sample involved could have occurred even if H_o were true, and the rejection of H_o is not an easy decision.
- When the p-value gets large (like 0.15 or more), the data are not at all unlikely if the H_o is true, and no one will reject H_o.

The *advantages of the p-value approach* are: (1) The results of the test procedure are expressed in terms of a continuous probability scale from 0.0 to 1.0, rather than simply on a "reject" or "fail to reject" basis. (2) A p-value can be reported and the user of the information can decide on the strength of the evidence as it applies to his or her own situation. (3) Computers can do all the calculations and report the p-value, thus eliminating the need for tables.

FYI
Do your opponents show you their poker hands before you bet?

The *disadvantage of the p-value approach* is the tendency for people to put off determining the level of significance. This should not be allowed to happen, as it is then possible for someone to set the level of significance after the fact, leaving open the possibility that the "preferred" decision will result. This is probably important only when the reported p-value falls in the "hard choice" range (say, 0.02 to 0.08), as described above.

Illustration 8.14 Two-Tailed Hypothesis Test with Sample Data

According to the results of Exercise 5.33 (p. 246), the mean of single-digit random numbers is 4.5 and the standard deviation is $\sigma = 2.87$. Draw a random sample of 40 single-digit numbers from Table 1 in Appendix B and test the hypothesis, "The mean of the single-digit numbers in Table 1 is 4.5." Use $\alpha = 0.10$.

SOLUTION

STEP 1 The Set-Up:

a. **Describe the population parameter of interest.**
The population parameter of interest is the mean μ of the population of single-digit numbers in Table 1 of Appendix B.

b. **State the null hypothesis (H_o) and the alternative hypothesis (H_a).**

$$H_o: \mu = 4.5 \text{ (mean is 4.5)}$$
$$H_a: \mu \neq 4.5 \text{ (mean is not 4.5)}$$

STEP 2 The Hypothesis Test Criteria:

a. **Check the assumptions.**
σ is known. Samples of size 40 should be large enough to satisfy the CLT; see the discussion of this issue on page 370.

b. **Identify the probability distribution and the test statistic to be used.**
We use the standard normal probability distribution, and the test statistic is $z\bigstar = \dfrac{\bar{x} - \mu}{\sigma/\sqrt{n}}$; $\sigma = 2.87$.

c. **Determine the level of significance, α.**
$\alpha = 0.10$ (given in the statement of the problem).

STEP 3 **The Sample Evidence:**

a. Collect the sample information.

This random sample was drawn from Table 1 in Appendix B:

II08-14

2	8	2	1	5	5	4	0	9	1
0	4	6	1	5	1	1	3	8	0
3	6	8	4	8	6	8	9	5	0
1	4	1	2	1	7	1	7	9	3

From the sample: $\bar{x} = 3.975$ and $n = 40$.

b. Calculate the value of the test statistic.

We use formula (8.4), and μ is 4.5 from H_o; $\sigma = 2.87$:

$$z\bigstar = \frac{\bar{x} - \mu}{\sigma/\sqrt{n}}; \qquad z\bigstar = \frac{3.975 - 4.50}{2.87/\sqrt{40}} = \frac{-0.525}{0.454} = -1.156 = -1.16$$

STEP 4 **The Probability Distribution:**

$\frac{1}{2}$ p-value | table value

-1.16 0 1.16 z

a. Calculate the p-value for the test statistic.

Since the alternative hypothesis indicates a two-tailed test, we must find the probability associated with both tails. The p-value is found by doubling the area of one tail. Since $z\bigstar = -1.16$, the value of $|z\bigstar| = 1.16$.

The p-value $= 2 \times P(z > |z\bigstar|)$:

$$\mathbf{P} = 2 \times P(z > 1.16) = 2 \times (0.5000 - 0.3770)$$
$$= 2(0.1230) = 0.2460$$

b. Determine whether or not the p-value is smaller than α.

The p-value (0.2460) is greater than α (0.10).

STEP 5 **The Results:**

a. State the decision about H_o: Fail to reject H_o.

b. State the conclusion about H_a.

The observed sample mean is not significantly different from 4.5 at the 0.10 level of significance. ■

ANSWER NOW 8.75

Calculate the p-value, given H_a: $\mu \neq 245$ and $z\bigstar = 1.1$.

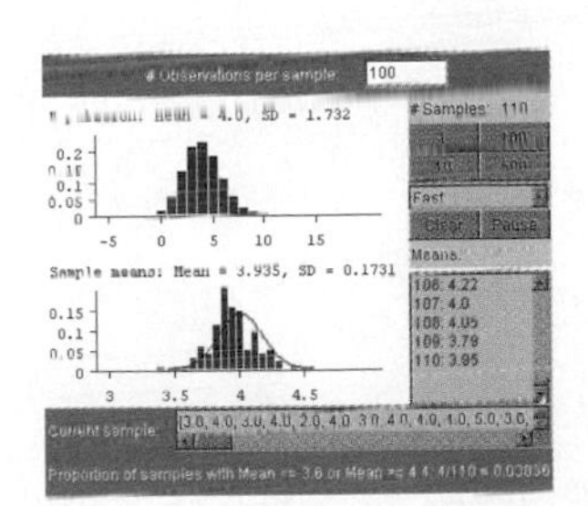

ANSWER NOW 8.76 INTERACTIVITY

Interactivity 8-C estimates the p-value for a two-tailed hypothesis test by simulating the taking of many samples. The given hypothesis test is H_o: $\mu = 4$ versus H_a: $\mu \neq 4$. A sample of 100 has been taken and the sample mean is 3.6. To determine the likelihood that $\bar{x} = 3.6$ came from a population with $\mu = 4$:

a. **Click "10" for "# of samples". Note the sample means and the probability of being less than 3.6 and greater than 4.4. Why are we including greater than 4.4?**

b. **Change to "Batch" and simulate 1000 more samples. What is the probability of being less than 3.6 and greater than 4.4? This is your estimated p-value.**

c. **How does your estimated p-value show on the histogram formed from the taking of many samples? Explain what this p-value means with respect to the test.**

d. **If the level of significance was 0.05, what would your decision be?**

Suppose we were to take another sample of size 40 from Table 1. Would we obtain the same results? Suppose we took a third sample and a fourth. What results might we expect? What does the p-value in Illustration 8.14 measure? Table 8.7 lists (a) the means obtained from 50 different random samples of size 40 that were taken from Table 1 in Appendix B, (b) the 50 values of $z\bigstar$ corresponding to the 50 $\bar{x}$'s, and (c) their 50 corresponding p-values. Figure 8.8 shows a histogram of the 50 $z\bigstar$-values.

TABLE 8.7

a. The Means of 50 Random Samples Taken from Table 1 in Appendix B Ta08-07

3.850	5.075	4.375	4.675	5.200	4.250	3.775	4.075	5.800	4.975
4.225	4.125	4.350	4.925	5.100	4.175	4.300	4.400	4.775	4.525
4.225	5.075	4.325	5.025	4.725	4.600	4.525	4.800	4.550	3.875
4.750	4.675	4.700	4.400	5.150	4.725	4.350	3.950	4.300	4.725
4.975	4.325	4.700	4.325	4.175	3.800	3.775	4.525	5.375	4.225

b. The $z\bigstar$-Values Corresponding to the 50 Means

−1.432	1.267	−0.275	0.386	1.543	−0.551	−1.598	−0.937	2.865	1.047
−0.606	−0.826	−0.331	0.937	1.322	−0.716	−0.441	−0.220	0.606	0.055
−0.606	1.267	−0.386	1.157	0.496	0.220	0.055	0.661	0.110	−1.377
0.551	0.386	0.441	−0.220	1.432	0.496	−0.331	−1.212	−0.441	0.496
1.047	−0.386	0.441	−0.386	−0.716	−1.543	−1.598	0.055	1.928	−0.606

c. The p-Values Corresponding to the 50 Means

0.152	0.205	0.783	0.700	0.123	0.582	0.110	0.349	0.004	0.295
0.545	0.409	0.741	0.349	0.186	0.474	0.659	0.826	0.545	0.956
0.545	0.205	0.700	0.247	0.620	0.826	0.956	0.509	0.912	0.168
0.582	0.700	0.659	0.826	0.152	0.620	0.741	0.226	0.659	0.620
0.295	0.700	0.659	0.700	0.474	0.123	0.110	0.956	0.054	0.545

FIGURE 8.8

The 50 Values of $z\bigstar$ from Table 8.7

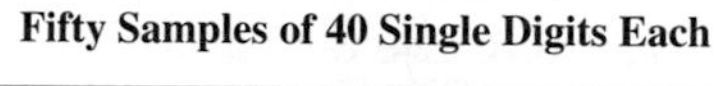

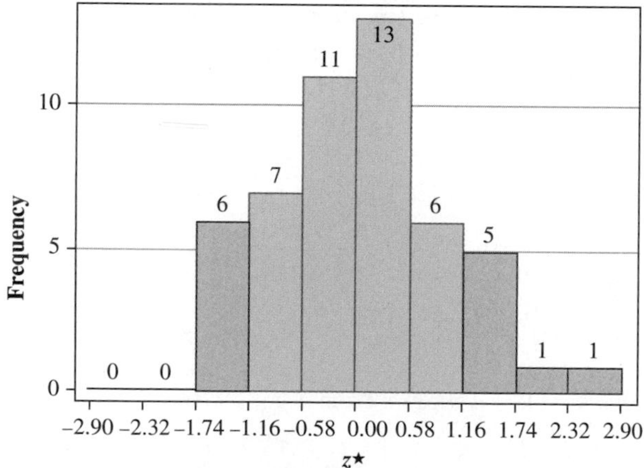

The histogram shows that 6 values of $z\bigstar$ were less than -1.16 and 7 values were greater than 1.16. That means 13 of the 50 samples, or 26%, have mean values more extreme than the mean ($\bar{x} = 3.975$) in Illustration 8.14. This observed relative frequency of 0.26 represents an empirical look at the p-value. Notice that the empirical value for the p-value (0.26) is very similar to the calculated p-value of 0.2460. Check the list of p-values; do you find that 13 of the 50 p-values are less than 0.2460? Which samples resulted in $|z\bigstar| = 1.16$? Which samples resulted in a p-value greater than 0.2460? How do they compare?

ANSWER NOW 8.77
Describe in your own words what the p-value measures.

Technology Instructions: Hypothesis Test for Mean μ with a Given σ

MINITAB (Release 13)

Input the data into C1; then continue with:

```
Choose:   Stat > Basic Statistics > 1-Sample Z
Enter:    Variables: C1
          Sigma: σ
          Test mean: μ
Select:   Options
Choose:   Alternative: less than or not equal or greater
          than > OK > OK
```

Excel XP

Input the data into column A; then continue with:

```
Choose:   Tools > Data Analysis Plus > Z-Test: Mean > OK
Enter:    Input Range: (A1:A20 or select cells)
          Hypothesized Mean: μ
          Standard Deviation (SIGMA): σ > OK
Gives p-values for both one-tailed and two-tailed tests.
```

TI-83 Plus

Input the data into L1; then continue with the following, entering the appropriate values and highlighting Calculate:

```
Choose:   STAT > TESTS > 1:Z-Test
```

```
Z-Test
 Inpt:Data Stats
 μ0:0
 σ:0
 List:L1
 Freq:1
 μ:≠μ0 <μ0 >μ0
 Calculate Draw
```

The MINITAB solution to the rivet example, used in this section (pp. 370–371, 374–377), is shown here:

```
Z-Test
TEST OF MU = 925.00 VS MU < 925.00
THE ASSUMED SIGMA = 18.0
N     MEAN      STDEV     SE MEAN     Z        P VALUE
50    921.18    17.58     2.546       -1.50    0.0668
```

FYI
The p-value approach was "made" for the computer!

When the computer is used, all that is left is for you is to make the decision and to write the conclusion.

ANSWER NOW 8.78
Describe how MINITAB found each of the six numerical values it reported as results.

ANSWER NOW 8.79
Use a computer or calculator to select 40 random single-digit numbers. Find the sample mean, $z\bigstar$, and the p-value. Repeat several times as in Table 8.7. Describe your findings.

(Use the commands for generating integer data on page 183; then continue with the hypothesis test commands on page 383.)

Exercises

8.80 State the null hypothesis H_o and the alternative hypothesis H_a that would be used for a hypothesis test related to each of these statements:

a. The mean age of the students enrolled in evening classes at a certain college is greater than 26 years.
b. The mean weight of packages shipped on Air Express during the past month was less than 36.7 lb.
c. The mean lifetime of fluorescent light bulbs is at least 1600 hr.

8.81 State the null hypothesis H_o and the alternative hypothesis H_a that would be used for a hypothesis test related to each of these statements:

a. The mean weight of college football players is no more than 210 lb.
b. The mean strength of welds by a new process is different from 570 lb per unit area, the mean strength of welds by the old process.
c. The mean hourly wage for a child-care giver is at most $9.00.

8.82 A manufacturer wishes to test the hypothesis that by changing the formula of its toothpaste, it will give its users improved protection. The null hypothesis represents "The change will not improve the protection," and the alternative represents "The change will improve the protection." Describe the meaning of the two possible types of errors that can occur in the decision when this test of the hypothesis is conducted.

8.83 Suppose we want to test the hypothesis that the mean hourly charge for automobile repairs is at least $60 at the repair shops in a nearby city. Explain the conditions that would exist if we make an error in decision by committing:

a. a type I error b. a type II error

8.84 Describe how the null hypothesis, as stated in Illustration 8.11 (p. 372), is a "starting point" for the decision to be made about the drying time for paint made with the new formula.

8.85 Assume that z is the test statistic, and calculate the value of $z\bigstar$ for each of the following:

a. H_o: $\mu = 10$, $\sigma = 3$, $n = 40$, $\bar{x} = 10.6$
b. H_o: $\mu = 120$, $\sigma = 23$, $n = 25$, $\bar{x} = 126.2$
c. H_o: $\mu = 18.2$, $\sigma = 3.7$, $n = 140$, $\bar{x} = 18.93$
d. H_o: $\mu = 81$, $\sigma = 13.3$, $n = 50$, $\bar{x} = 79.6$

8.86 Assume that z is the test statistic, and calculate the value of $z\bigstar$ for each of the following:

a. H_o: $\mu = 51$, $\sigma = 4.5$, $n = 40$, $\bar{x} = 49.6$
b. H_o: $\mu = 20$, $\sigma = 4.3$, $n = 75$, $\bar{x} = 21.2$
c. H_o: $\mu = 138.5$, $\sigma = 3.7$, $n = 14$, $\bar{x} = 142.93$
d. H_o: $\mu = 815$, $\sigma = 43.3$, $n = 60$, $\bar{x} = 799.6$

8.87 There are only two possible decisions as a result of a hypothesis test.

a. State the two possible decisions.
b. Describe the conditions that lead to each of the two decisions identified in part a.

8.88 For each pair of values, state the decision that will occur and state why:

a. p-value $= 0.014$, $\alpha = 0.02$
b. p-value $= 0.118$, $\alpha = 0.05$
c. p-value $= 0.048$, $\alpha = 0.05$
d. p-value $= 0.064$, $\alpha = 0.10$

8.89 The calculated p-value for a hypothesis test is 0.084. What decision about the null hypothesis would occur if:

a. the hypothesis test is completed at the 0.05 level of significance
b. the hypothesis test is completed at the 0.10 level of significance

8.90

a. A one-tailed hypothesis test is to be completed at the 0.05 level of significance. What calculated values of p will cause a rejection of H_o?
b. A two-tailed hypothesis test is to be completed at the 0.02 level of significance. What calculated values of p will cause a "fail to reject H_o" decision?

8.91 Calculate the p value for each of the following:
a. H_o: $\mu = 10$ vs. H_a: $\mu > 10$, $z★ = 1.48$
b. H_o: $\mu = 105$ vs. H_a: $\mu < 105$, $z★ = -0.85$
c. H_o: $\mu = 13.4$ vs. H_a: $\mu \neq 13.4$, $z★ = 1.17$
d. H_o: $\mu = 8.56$ vs. H_a: $\mu < 8.56$, $z★ = -2.11$
e. H_o: $\mu = 110$ vs. H_a: $\mu \neq 110$, $z★ = -0.93$

8.92 Calculate the p-value for each of the following:
a. H_o: $\mu = 20$ vs. H_a: $\mu < 20$, $\bar{x} = 17.8$, $\sigma = 9$, $n = 36$
b. H_o: $\mu = 78.5$ vs. H_a: $\mu > 78.5$, $\bar{x} = 79.8$, $\sigma = 15$, $n = 100$
c. H_o: $\mu = 1.587$ vs. H_a: $\mu \neq 1.587$, $\bar{x} = 1.602$, $\sigma = 0.15$, $n = 50$

8.93 Find the value of $z★$ for each of the following:
a. H_o: $\mu = 35$ vs. H_a: $\mu > 35$, p-value $= 0.0582$
b. H_o: $\mu = 35$ vs. H_a: $\mu < 35$, p-value $= 0.0166$
c. H_o: $\mu = 35$ vs. H_a: $\mu \neq 35$, p-value $= 0.0042$

8.94 The null hypothesis, H_o: $\mu = 48$, was tested against the alternative hypothesis, H_a: $\mu > 48$. A sample of 75 resulted in a calculated p-value of 0.102. If $\sigma = 3.5$, find the value of the sample mean, $\bar{x}$.

8.95 This computer output was used to complete a hypothesis test:

```
TEST OF MU = 525.00 VS MU < 525.00
THE ASSUMED SIGMA = 60.0
N    MEAN     STDEV    SE MEAN     Z       P VALUE
38   512.14   64.78    9.733      -1.32    0.093
```

a. State the null and alternative hypotheses.
b. If the test is completed using $\alpha = 0.05$, what decision and conclusion are reached?
c. Verify the value of the standard error of the mean.

8.96 This computer output was used to complete a hypothesis test:

```
TEST OF MU = 6.250 VS MU not = 6.250
THE ASSUMED SIGMA = 1.40
N    MEAN     STDEV    SE MEAN     Z       P VALUE
78   6.596    1.273    0.1585      2.18    0.029
```

a. State the null and alternative hypotheses.
b. If the test is completed using $\alpha = 0.05$, what decision and conclusion are reached?
c. Verify the value of the standard error of the mean.
d. Find the values for $\sum x$ and $\sum x^2$.

8.97 An article titled "Comparisons of Mathematical Competencies and Attitudes of Elementary Education Majors with Established Norms of a General College Population" (*School Science and Mathematics*, Vol. 93, No. 3, March 1993) reported the mean score on a test of mathematical competency for 165 elementary education majors to be 32.63. Test the null hypothesis that μ, the mean score for the population of elementary education majors, is 35.70 (the established norm of the general college population) versus the alternative hypothesis that $\mu < 35.70$. Assume $\sigma = 6.73$.
a. Describe the parameter of interest.
b. State the null and alternative hypotheses.
c. Calculate the value for $z★$ and find the p-value.
d. State your decision and conclusion using $\alpha = 0.001$.

8.98 When the workers for a major automobile manufacturer go on strike, there are repercussions throughout the rest of the economy; in particular, the dealers who sell the cars and trucks feel the pinch. Dealers like to maintain a two-month supply to give their customers adequate selection, but when the manufacturer cannot deliver the vehicles, inventories dwindle. The June 1997 days' supply of Chevrolet S-10 pickup trucks was 106 vehicles, but shortly after the United Auto Workers strike in Flint, Michigan, the June 1998 days' supply of these trucks had dropped to a mean of 38 and a standard deviation of 16 (*Newsweek*, "Big, Empty Lots," July 27, 1998). Suppose one month after the strike was settled, 150 dealers are sampled and the S-10 inventories yielded a mean days' supply of 41 trucks. Based on this new evidence, complete the hypothesis test of H_o: $\mu = 38$ versus H_a: $\mu > 38$ at the 0.01 level of significance using the probability-value approach.
a. Define the parameter.
b. State the null and alternative hypotheses.
c. Specify the hypothesis test criteria.
d. Present the sample evidence.
e. Find the probability distribution information.
f. Determine the results.

8.99 Who says that the more you spend on a wristwatch, the more accurate the watch will be? Some say that nowadays you can buy a quartz watch for less than $25 that runs just as accurately as watches that cost four times as much. Suppose the average accuracy for all watches being sold today, irrespective of price, is within 19.8 seconds per month with a standard deviation of 9.1 seconds. A random sample of 36 quartz watches priced at less than $25 is taken, and their accuracy check reveals a sample mean error of 22.7 seconds per month. Based on this evidence, complete the hypothesis test of H_o: $\mu = 20$ versus H_a: $\mu > 20$ at the 0.05 level of significance using the probability-value approach.
a. Define the parameter.
b. State the null and alternative hypotheses.
c. Specify the hypothesis test criteria.
d. Present the sample evidence.
e. Find the probability distribution information.
f. Determine the results.

8.100 Ex08-100 The National Health and Nutrition Examination Survey (NHANES) in 1999 indicates that more U.S. adults are becoming either overweight or obese, defined as having a body mass index (BMI) of 25 or greater. Data from the National Centers for Disease Control and Prevention (CDC) indicate that for women aged 35 to 55, the mean BMI is 25.12 with a standard deviation of 5.3. In a similar study of female cardiovascular technologists registered in the United States within the same age range, these BMI scores resulted:

22	28	26	19	26	23	25	21	21	31
23	21	25	39	25	23	18	25	23	29
22	27	24	19	26	21	30	19	23	27
21	23	19	22	21	28	29	22	28	19
24	22	28	22	21	16	22	33	22	34
23	22	22	23	21	28	22	19	39	25
21	24	30	23	38	26	24	27	20	28
37	29	25	18	27	20	18	25	19	31
26	20	26	22	26	28	21	18	38	22
28	22	18	22	18	22	33	25	22	29

Source: "An Assessment of Cardiovascular Risk Behaviors of Registered Cardiovascular Technologists," dissertation by Dr. Suzanne Wambold, University of Toledo, 2002. Reprinted by permission.

Test the claim that the cardiovascular technologists have a lower average BMI than the general population. Use $\alpha = 0.05$.

a. Describe the parameter of interest.
b. State the null and alternative hypotheses.
c. Calculate the value for $z\star$ and find the p-value.
d. State your decision and conclusion using $\alpha = 0.05$.

8.5 Hypothesis Test of Mean μ (σ Known): A Classical Approach

In Section 8.3 we surveyed the concepts and much of the reasoning behind a hypothesis test while looking at nonstatistical illustrations. In this section we are going to formalize the hypothesis test procedure as it applies to statements concerning the mean μ of a population under the restriction that σ, the population standard deviation, is a known value.

THE ASSUMPTION FOR HYPOTHESIS TESTS ABOUT MEAN μ USING A KNOWN σ

The sampling distribution of $\bar{x}$ has a normal distribution.

The information we need to ensure that this assumption is satisfied is contained in the sampling distribution of sample means and in the central limit theorem (see Chapter 7, p. 318).

The sampling distribution of sample means $\bar{x}$ is distributed about a mean equal to μ with a standard error equal to $\sigma/\sqrt{n}$; and (1) if the randomly sampled population is normally distributed, then $\bar{x}$ is normally distributed for all sample sizes, or (2) if the randomly sampled population is not normally distributed, then $\bar{x}$ is approximately normally distributed for sufficiently large sample sizes.

The hypothesis test is a well-organized, step-by-step procedure used to make a decision. Two different formats are commonly used for hypothesis testing. The *classical approach* is the hypothesis test process that has enjoyed popularity for many years. This approach is organized as a five-step procedure.

THE CLASSICAL HYPOTHESIS TEST: A FIVE-STEP PROCEDURE

Step 1 **The Set-Up:**

a. Describe the population parameter of interest.
b. State the null hypothesis (H_o) and the alternative hypothesis (H_a).

Step 2 **The Hypothesis Test Criteria:**
a. Check the assumptions.
b. Identify the probability distribution and the test statistic to be used.
c. Determine the level of significance, α.

Step 3 **The Sample Evidence:**
a. Collect the sample information.
b. Calculate the value of the test statistic.

Step 4 **The Probability Distribution:**
a. Determine the critical region and critical value(s).
b. Determine whether or not the calculated test statistic is in the critical region.

Step 5 **The Results:**
a. State the decision about H_o.
b. State the conclusion about H_a.

A commercial aircraft manufacturer buys rivets to use in assembling airliners. Each rivet supplier that wants to sell rivets to the aircraft manufacturer must demonstrate that its rivets meet the required specifications. One of the specs is: "The mean shearing strength of all such rivets, μ, is at least 925 lb." Each time the aircraft manufacturer buys rivets, it is concerned that the mean strength might be less than the 925-lb specification.

FYI
Using weak rivets could have terrible consequences.

NOTE 1 Each individual rivet has a shearing strength, which is determined by measuring the force required to shear ("break") the rivet. Clearly, not all the rivets can be tested. Therefore, a sample of rivets will be tested, and a decision about the mean strength of all the untested rivets will be based on the mean from those sampled and tested.

ANSWER NOW 8.101
In this example, the aircraft builder, the buyer of the rivets, is concerned that the rivets might not meet the mean-strength specs. State the aircraft manufacturer's null and alternative hypotheses.

NOTE 2 We will use $\sigma = 18$ for our rivet example.

STEP 1 **The Set-Up:**
a. **Describe the population parameter of interest.**
The population parameter of interest is the mean μ, the mean shearing strength (or mean force required to shear) of the rivets being considered for purchase.
b. **State the null hypothesis (H_o) and the alternative hypothesis (H_a).**
The null hypothesis and the alternative hypothesis are formulated by inspecting the problem or statement to be investigated and first formulating two opposing statements about the mean μ. For our example, these two opposing statements are: (A) "The mean shearing strength is less than 925" ($\mu < 925$, the aircraft manufacturer's concern), and (B) "The mean shearing strength is at least 925" ($\mu = 925$, the rivet supplier's claim and the aircraft manufacturer's spec).

NOTE The trichotomy law from algebra states that two numerical values must be related in exactly one of three possible relationships: <, =, or >. All three of these possibilities must be accounted for in the two opposing hypotheses in order for the two hypotheses to be negations of each other. The three possible combinations of signs and hypotheses are shown in Table 8.8. Recall that the null hypothesis assigns a specific value to the parameter in question, and therefore "equals" will always be part of the null hypothesis.

FYI
More specific instructions are given on pages 362–363.

TABLE 8.8 The Three Possible Statements of Null and Alternative Hypotheses

Null Hypothesis	Alternative Hypothesis
1. greater than or equal to (≥)	less than (<)
2. less than or equal to (≤)	greater than (>)
3. equal to (=)	not equal to (≠)

The parameter of interest, the population mean μ, is related to the value 925. Statement (A) becomes the alternative hypothesis:

$$H_a\text{: } \mu < 925 \text{ (the mean is less than 925)}$$

This statement represents the aircraft manufacturer's concern and says, "The rivets do not meet the required specs." Statement (B) becomes the null hypothesis:

$$H_o\text{: } \mu = 925\ (\geq) \text{ (the mean is at least 925)}$$

This hypothesis represents the negation of the aircraft manufacturer's concern and says, "The rivets do meet the required specs."

NOTE We will write the null hypothesis with just the equal sign, thereby stating the exact value assigned. When "equal" is paired with "less than" or paired with "greater than," the combined symbol is written beside the null hypothesis as a reminder that all three signs have been accounted for in these two opposing statements.

Before continuing with our example, let's look at three illustrations that demonstrate formulating the statistical null and alternative hypotheses involving population mean μ. Illustrations 8.15 and 8.16 each demonstrate a "one-tailed" alternative hypothesis.

Illustration 8.15 Writing Null and Alternative Hypotheses (One-Tailed Situation)

A consumer advocate group would like to disprove a car manufacturer's claim that a specific model will average 24 miles per gallon of gasoline. Specifically, the group would like to show that the mean miles per gallon is considerably less than 24. State the null and alternative hypotheses.

SOLUTION

To state the two hypotheses, we first need to identify the population parameter in question: the "mean mileage attained by this car model." The parameter μ is being compared to the value 24 miles per gallon, the specific value of interest. The advocates are questioning the value of μ and wish to show it to be less than 24 (that is, $\mu < 24$). There are three possible relationships: (1) $\mu < 24$, (2) $\mu = 24$, and (3) $\mu > 24$. These three cases must be arranged to form two opposing statements: One states what the advocates are trying to show, "The mean level is less than 24 ($\mu < 24$)," whereas the "negation" is "The mean level is not less than 24 ($\mu \geq 24$)." One of these two statements will become the null hypothesis H_o, and the other will become the alternative hypothesis H_a.

NOTE: Recall that there are two rules for forming the hypotheses: (1) The null hypothesis states that the parameter in question has a specified value ("H_o must contain the equal sign"), and (2) the consumer advocate group's contention becomes the alternative hypothesis ("less than"). Both rules indicate:

$$H_o\text{: } \mu = 24\ (\geq) \quad \text{and} \quad H_a\text{: } \mu < 24$$

■

Illustration 8.16 Writing Null and Alternative Hypotheses (One-Tailed Situation)

Suppose the EPA is suing a large manufacturing company for not meeting federal emissions guidelines. Specifically, the EPA is claiming that the mean amount of sulfur dioxide in the air is dangerously high, higher than 0.09 part per million. State the null and alternative hypotheses for this test situation.

SOLUTION

The parameter of interest is the mean amount of sulfur dioxide in the air, and 0.09 part per million is the specified value. $\mu > 0.09$ corresponds to "The mean amount is greater than 0.09," whereas $\mu \leq 0.09$ corresponds to the negation, "The mean amount is not greater than 0.09." Therefore, the hypotheses are

$$H_o\text{: } \mu = 0.09\ (\leq) \quad \text{and} \quad H_a\text{: } \mu > 0.09$$

■

Illustration 8.17 demonstrates a "two-tailed" alternative hypothesis.

Illustration 8.17 Writing Null and Alternative Hypotheses (Two-Tailed Situation)

Job satisfaction is very important to worker productivity. A standard job-satisfaction questionnaire was administered by union officers to a sample of assembly-line workers in a large plant in hopes of showing that the assembly workers' mean score on this questionnaire would be different from the established mean of 68. State the null and alternative hypotheses.

SOLUTION

Either the mean job satisfaction score is different from 68 ($\mu \neq 68$) or the mean score is equal to 68 ($\mu = 68$). Therefore,

$$H_o\text{: } \mu = 68 \quad \text{and} \quad H_a\text{: } \mu \neq 68$$

■

NOTE 1 When "less than" is combined with "greater than," they become "not equal to."

NOTE 2 The alternative hypothesis is referred to as being "two-tailed" when H_a is "not equal."

The viewpoint of the experimenter greatly affects the way the hypotheses are formed. Generally, the experimenter is trying to show that the parameter value is different from the value specified. Thus, the experimenter is often hoping to be able to reject the null hypothesis so that the experimenter's theory has been substantiated. Illustrations 8.15, 8.16, and 8.17 also represent the three possible arrangements for the $<$, $=$, and $>$ relationships between the parameter μ and a specified value.

ANSWER NOW 8.102
Professor Hart does not believe the statement "The mean distance commuted daily by the nonresident students at our college is no more than 9 miles." State the null and alternative hypotheses he would use to challenge this statement.

Table 8.9 lists some additional common phrases used in claims and indicates the phrase of its negation and the hypothesis in which each phrase will be used. Again, notice that "equals" is always in the null hypothesis. Also notice that the negation of "less than" is "not less than," which is equivalent to "greater than or equal to." Think of negation of one sign as the other two signs combined.

TABLE 8.9 Common Phrases and Their Negations

H_o: ($\geq$)	H_a: ($<$)	H_o: ($\leq$)	H_a: ($>$)	H_o: ($=$)	H_a: ($\neq$)
at least	less than	at most	more than	is	is not
no less than	less than	no more than	more than	not different from	different from
not less than	less than	not greater than	greater than	same as	not same as

After the null and alternative hypotheses are established, we will work under the assumption that the null hypothesis is a true statement until there is sufficient evidence to reject it. This situation might be compared to a courtroom trial, where the accused is assumed to be innocent (H_o: Defendant is innocent vs. H_a: Defendant is not innocent) until sufficient evidence has been presented to show that innocence is totally unbelievable ("beyond reasonable doubt"). At the conclusion of the hypothesis test, we will make one of two possible decisions. We will decide in opposition to the null hypothesis and say that we "reject H_o" (this corresponds to "conviction" of the accused in a trial), or we will decide in agreement with the null hypothesis and say that we "fail to reject H_o" (this corresponds to "fail to convict" or an "acquittal" of the accused in a trial).

ANSWER NOW 8.103
State the null and alternative hypotheses used to test each of the following claims:

a. **The mean reaction time is less than 1.25 seconds.**

b. **The mean score on that qualifying exam is different from 335.**

c. **The mean selling price of homes in the area is no more than $230,000.**

Let's return to the rivet example we interrupted on page 387 and continue with Step 2. Recall that

$$H_o\text{: } \mu = 925\ (\geq)\ \text{(at least 925)} \qquad H_a\text{: } \mu < 925\ \text{(less than 925)}$$

STEP 2 **The Hypothesis Test Criteria:**

a. Check the assumptions.

σ is known. Variables like shearing strength typically have a mounded distribution; therefore, a sample of size 50 should be large enough for the CLT to satisfy the assumption; the sampling distribution of sample means is normally distributed.

b. Identify the probability distribution and the test statistic to be used.

The standard normal probability distribution is used because $\bar{x}$ is expected to have a normal or approximately normal distribution.

For a hypothesis test of μ, we want to compare the value of the sample mean to the value of the population mean as stated in the null hypothesis. This comparison is accomplished using the test statistic in formula (8.4):

$$z\bigstar = \frac{\bar{x} - \mu}{\sigma/\sqrt{n}} \qquad (8.4)$$

The resulting calculated value is identified as $z\bigstar$ ("z star") because it is expected to have a standard normal distribution when the null hypothesis is true and the assumptions have been satisfied. The ★ ("star") is to remind us that this is the calculated value of the test statistic.

The test statistic to be used is $z\bigstar = \dfrac{\bar{x} - \mu}{\sigma/\sqrt{n}}$.

c. Determine the level of significance, α.

FYI
Do Answer Now 8.104 before continuing.

Setting α was described as a managerial decision in Section 8.3. To see what is involved in determining α, the probability of the type I error, for our rivet example, we start by identifying the four possible outcomes, their meaning, and the action related to each.

ANSWER NOW 8.104
Identify the four possible outcomes and describe the situation involved with each outcome with regard to the aircraft manufacturer testing and buying rivets. Which is the more serious error: type I or type II? Explain.

The type I error occurs when a true null hypothesis is rejected. This would occur when the manufacturer tested rivets that in truth did meet the specs, and rejected them. Undoubtedly this would lead to the rivets not being purchased even though the manufacturer did meet the specs. In order for the manager to set a level of significance, related information is needed—namely, how soon is the new supply of rivets needed? If they are needed tomorrow and this is the only vendor with an available supply, waiting a week to find acceptable rivets could be very expensive; therefore, rejecting good rivets could be considered a serious error. On the other hand, if the rivets are not needed until next month, then this error may not be very serious. Only the manager will know all the ramifications, and therefore the manager's input is important here.

FYI
There is more to this scenario, but we hope you get the idea.

FYI
α will be assigned in the statement of exercises.

After much consideration, the manager assigns the level of significance: $\alpha = 0.05$.

STEP 3 **The Sample Evidence:**

a. Collect the sample information.

We are ready for the data. The sample must be a random sample drawn from the population whose mean μ is being questioned. A random sample of 50 rivets is selected, each rivet is tested, and the sample mean shearing strength is calculated: $\bar{x} = 921.18$ and $n = 50$.

b. Calculate the value of the test statistic.

The sample evidence (x and n found in Step 3a) is next converted into the calculated value of the test statistic, $z\bigstar$, using formula (8.4). (μ is 925 from H_o; $\sigma = 18$ is the known quantity, given in Note 2 on p. 387.) We have

$$z\bigstar = \frac{\bar{x} - \mu}{\sigma/\sqrt{n}}: \qquad z\bigstar = \frac{921.18 - 925.0}{18/\sqrt{50}} = \frac{-3.82}{2.5456} = -1.50$$

STEP 4 **The Probability Distribution:**

a. Determine the critical region and critical value(s).

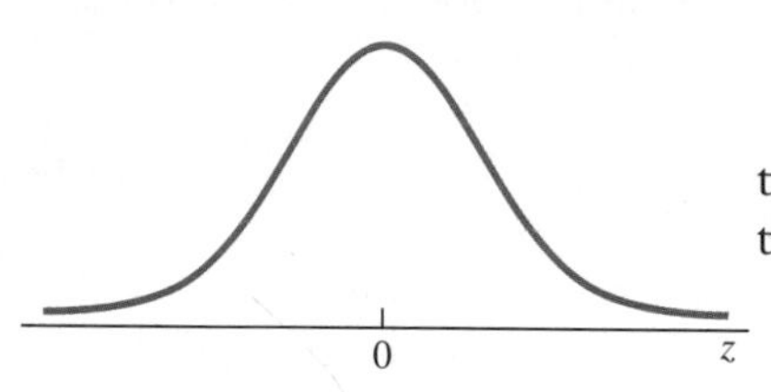

The standard normal variable z is our test statistic for this hypothesis test; therefore, we draw a sketch of the standard normal distribution, label the scale as z, and locate its mean value, 0.

CRITICAL REGION

The set of values for the test statistic that will cause us to reject the null hypothesis. The set of values that are not in the critical region is called the **noncritical region** (sometimes called the *acceptance region*).

Recall that we are working under the assumption that the null hypothesis is true. Thus, we are assuming that the mean shearing strength of all rivets in the sampled population is 925. If this is the case, then when we select a random sample of 50 rivets, we can expect this sample mean, $\bar{x}$, to be part of a normal distribution that is centered at 925 and to have a standard error of $\sigma/\sqrt{n} = 18/\sqrt{50}$, or approximately 2.55. Approximately 95% of the sample mean values will be greater than 920.8 (a value 1.65 standard errors below the mean: $925 - (1.65)(2.55) = 920.8$). Thus, if H_o is true and $\mu = 925$, then we expect $\bar{x}$ to be greater than 920.8 approximately 95% of the time and less than 920.8 only 5% of the time.

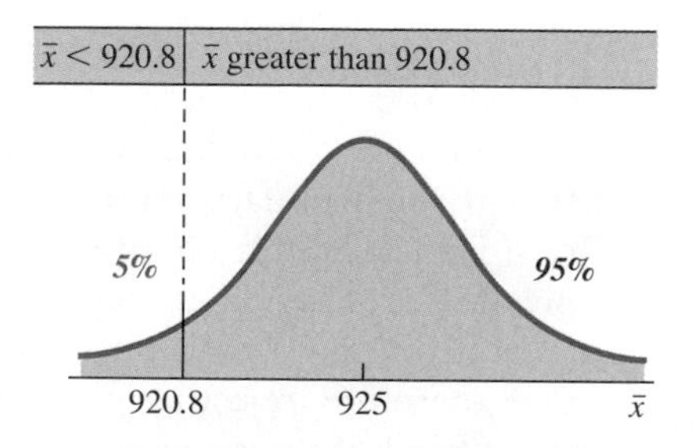

If, however, the value of $\bar{x}$ that we obtain from our sample is less than 920.8—say, 919.5—we will have to make a choice. It could be that either: (A) such an $\bar{x}$ value (919.5) is a member of the sampling distribution with mean 925 although it has a very low probability of occurrence (less than 0.05), or (B) $\bar{x} = 919.5$ is a member of a sampling distribution whose mean is less than 925, which would make it a value that is more likely to occur.

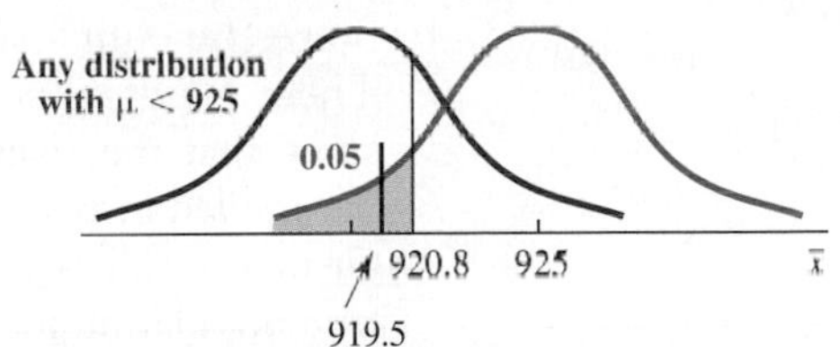

In statistics, we "bet" on the "more probable to occur" and consider the second choice (B) to be the right one. Thus, the left-hand tail of the z-distribution becomes the critical region. And the level of significance α becomes the measure of its area.

CRITICAL VALUE(S)
The "first" or "boundary" value(s) of the critical region(s).

FYI
Information about the critical value notation, $z(\alpha)$, is given on pages 293–295.

FYI
Shading will be used to identify the critical region.

The critical value for our illustration is $-z(0.05)$ and has the value of -1.65, as found in Table 4A in Appendix B.

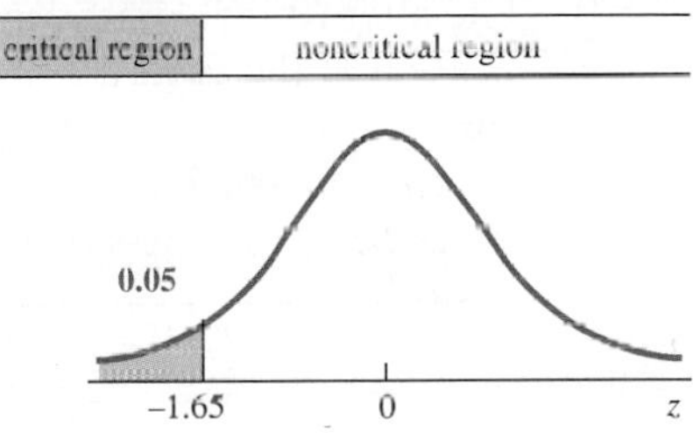

b. **Determine whether or not the calculated test statistic is in the critical region.**
Graphically this determination is shown by locating the value for $z\bigstar$ on the sketch in Step 4a.
The calculated value of z, $z\bigstar = -1.50$, is **not in the critical region** (it is in the unshaded portion of the figure).

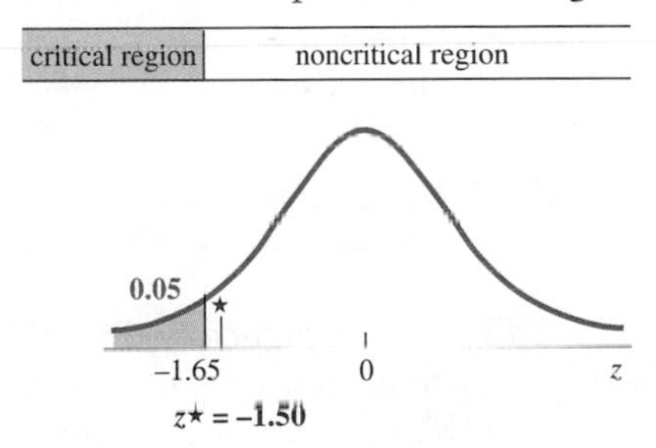

ANSWER NOW 8.105
Calculate the test statistic $z\bigstar$, given H_o: $\mu = 356$, $\sigma = 17$, $\bar{x} = 354.3$, and $n = 120$.

STEP 5 **The Result:**
a. **State the decision about H_o.**
In order to make the decision, we need to know the *decision rule.*

DECISION RULE

a. If the test statistic falls *within the critical region,* then the decision must be **reject H_o.** (The critical value is part of the critical region.)
b. If the test statistic is *not in the critical region,* then the decision must be **fail to reject H_o.**

The decision is: Fail to reject H_o.

FYI
Specific information about writing the conclusion is given on pages 367–368.

b. State the conclusion about H_a.
There is not sufficient evidence at the 0.05 level of significance to show that the rivets have a mean shearing strength less than 925. "We failed to convict" the null hypothesis. In other words, a sample mean as small as 921.18 is likely to occur (as defined by α) when the true population mean value is 925.0. Therefore, the resulting action would be to buy the rivets.

ANSWER NOW 8.106

a. **What decision is reached when the test statistic falls in the critical region?**

b. **What decision is reached when the test statistic falls in the noncritical region?**

Before we look at another illustration, let's summarize briefly some of the details we have seen thus far:

1. The null hypothesis specifies a particular value of a population parameter.
2. The alternative hypothesis can take three forms. Each form dictates a specific location of the critical region(s), as shown in the following table.
3. For many hypothesis tests, the sign in the alternative hypothesis "points" in the direction in which the critical region is located. [Think of the not equal to sign ($\neq$) as being both less than ($<$) and greater than ($>$), thus pointing in both directions.]

Sign in the Alternative Hypothesis

	$<$	$\neq$	$>$
Critical Region	One region Left side **One-tailed test**	Two regions Half on each side **Two-tailed test**	One region Right side **One-tailed test**

ANSWER NOW 8.107
Find the critical region and value(s) for H_a: $\mu < 19$ and $\alpha = 0.01$.

ANSWER NOW 8.108
Find the critical region and value(s) for H_a: $\mu > 34$ and $\alpha = 0.02$.

The value assigned to α is called the *significance level* of the hypothesis test. Alpha cannot be interpreted to be anything other than the risk (or probability) of rejecting the null hypothesis when it is actually true. We will seldom be able to determine whether the null hypothesis is true or false; we will decide only to "reject H_o" or to "fail to reject H_o." The relative frequency with which we reject a true hypothesis is α, but we will never know the relative frequency with which we make an error in decision. The two ideas are quite different; that is, a type I error and an error in decision are two different things altogether. Remember that there are two types of errors: type I and type II.

Let's look at another hypothesis test, one involving the two-tailed procedure.

Illustration 8.18 Two-Tailed Hypothesis Test

It has been claimed that the mean weight of women students at a college is 54.4 kg. Professor Hart does not believe the claim and sets out to show that the mean weight is not 54.4 kg. To test the claim he collects a random sample of 100 weights from among the women students. A sample mean of 53.75 kg results. Is this sufficient evidence for Professor Hart to reject the statement? Use $\alpha = 0.05$ and $\sigma = 5.4$ kg.

ANSWER NOW 8.109
How many pounds is 54.4 kilograms?

SOLUTION

STEP 1 **The Set-Up:**

a. **Describe the population parameter of interest.**
The population parameter of interest is the mean μ, the mean weight of all women students at the college.

b. **State the null hypothesis (H_o) and the alternative hypothesis (H_a).**
The mean weight is equal to 54.4 kg, or the mean weight is not equal to 54.4 kg.

H_o: $\mu = 54.4$ (mean weight is 54.4)

H_a: $\mu \neq 54.4$ (mean weight is not 54.4)
(Remember: $\neq$ is $<$ and $>$ together.)

STEP 2 **The Hypothesis Test Criteria:**

a. **Check the assumptions.**
σ is known. The weights of an adult group of women are generally approximately normally distributed; therefore, a sample of $n = 100$ is large enough to allow the CLT to apply.

b. **Identify the probability distribution and the test statistic to be used.**
The standard normal probability distribution and the test statistic

$$z\bigstar = \frac{\bar{x} - \mu}{\sigma/\sqrt{n}} \text{ will be used; } \sigma = 5.4.$$

c. **Determine the level of significance, α.**
$\alpha = 0.05$ (given in the statement of problem).

STEP 3 **The Sample Evidence:**

a. **Collect the sample information:** $\bar{x} = 53.75$ and $n = 100$.

b. **Calculate the value of the test statistic.**
Use formula (8.4), information from H_o: $\mu = 54.4$, and $\sigma = 5.4$ (known):

$$z\bigstar = \frac{\bar{x} - \mu}{\sigma/\sqrt{n}}: \quad z\bigstar = \frac{53.75 - 54.4}{5.4/\sqrt{100}} = \frac{-0.65}{0.54} = -1.204 = -1.20$$

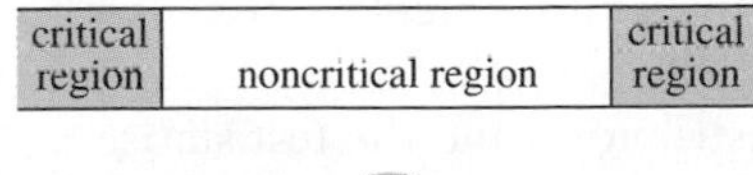

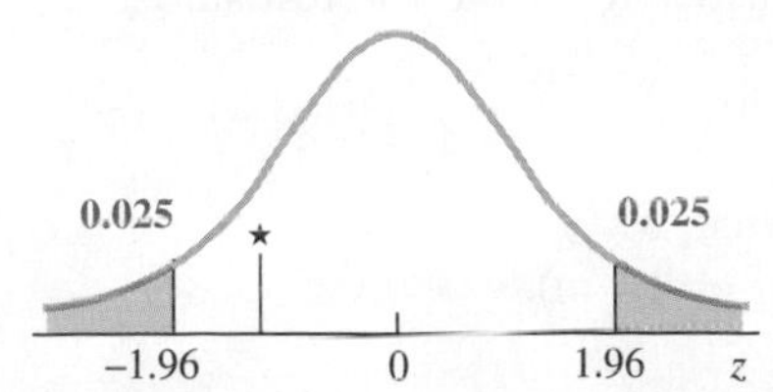

STEP 4 **The Probability Distribution:**

a. **Determine the critical region and critical value(s).**
The critical region is both the left tail and the right tail because both smaller and larger values of the sample mean suggest that the null hypothesis is wrong. The level of significance will be split in half, with 0.025 being the measure of each tail. The critical values are found in Table 4B in Appendix B: $\pm z(0.025) = \pm 1.96$. (Table 4B instructions are on page 351.)

b. Determine whether or not the calculated test statistic is in the critical region.
The calculated value of z, $z\star = -1.20$, is not in the critical region (shown in red on the previous figure).

STEP 5 **The Results:**

a. State the decision about H_o: Fail to reject H_o.

b. State the conclusion about H_a.
There is not sufficient evidence at the 0.05 level of significance to show that the women students have a mean weight different from the 54.4 kg claimed. In other words, there is no statistical evidence to support Professor Hart's contentions. ■

Illustration 8.19 One-Tailed Hypothesis Test

The student body at many community colleges is considered a "commuter population." The following question was asked of the student affairs office at one such college: How far (one way) does the average community college student commute to college daily? The office answered: No more than 9.0 mi. The inquirer was not convinced and decided to test the statement. She took a random sample of 50 students and found a mean commuting distance of 10.22 mi. Test the hypothesis stated above at a significance level of $\alpha = 0.05$, using $\sigma = 5$ mi.

SOLUTION

STEP 1 **The Set-Up:**

a. Describe the population parameter of interest.
The population parameter of interest is the mean μ, the mean one-way distance traveled by all commuting students.

b. State the null hypothesis (H_o) and the alternative hypothesis (H_a).
The claim "no more than 9.0 mi" implies that the three possible relationships should be grouped "no more than 9.0" ($\leq$) versus "more than 9.0" ($>$).

$$H_o\text{: } \mu = 9.0 \ (\leq) \text{ (no more than 9.0 mi)}$$

$$H_a\text{: } \mu > 9.0 \text{ (more than 9.0 mi)}$$

STEP 2 **The Hypothesis Test Criteria:**

a. Check the assumptions.
σ is known. Commuting distance is a mounded and skewed distribution, but with a sample size of $n = 50$, the CLT will hold; therefore, the assumptions are satisfied.

b. Identify the probability distribution and the test statistic to be used.
The standard normal probability distribution and the test statistic

$z\star = \dfrac{\bar{x} - \mu}{\sigma/\sqrt{n}}$ will be used; $\sigma = 5$.

c. Determine the level of significance, α.
$\alpha = 0.05$ (given in the statement of problem).

STEP 3 **The Sample Evidence:**

a. **Collect the sample information:** $\bar{x} = 10.22$ and $n = 50$.

b. **Calculate the value of the test statistic.**

Use formula (8.4), information from H_o: $\mu = 9.0$, and $\sigma = 5.0$:

$$z\bigstar = \frac{\bar{x} - \mu}{\sigma/\sqrt{n}}: \qquad z\bigstar = \frac{10.22 - 9.0}{5.0/\sqrt{50}} = \frac{1.22}{0.707} = 1.73$$

STEP 4 **The Probability Distribution:**

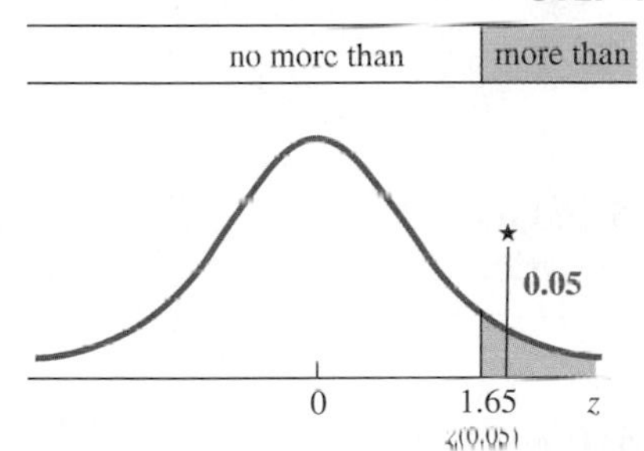

a. **Determine the critical region and critical value(s).**

The critical region is the right tail because the alternative hypothesis is "greater than." The critical value is $z(0.05) = 1.65$.

b. **Determine whether or not the calculated test statistic is in the critical region.**

The calculated value of z, $z\bigstar = 1.73$, is in the critical region (shown in red on the figure at the left).

STEP 5 **The Results:**

a. **State the decision about H_o:** Reject H_o.

b. **State the conclusion about H_a.**

There is sufficient evidence at the 0.05 level of significance to conclude that the average commuting community college student probably travels more than 9.0 miles one way to college.

■

Illustration 8.20 Two-Tailed Hypothesis Test with Sample Data

According to the results of Exercise 5.33 (p. 246), the mean of single-digit random numbers is 4.5 and the standard deviation is $\sigma = 2.87$. Draw a random sample of 40 single-digit numbers from Table 1 in Appendix B and test the hypothesis, "The mean of the single-digit numbers in Table 1 is 4.5." Use $\alpha = 0.10$.

SOLUTION

STEP 1 **The Set-Up:**

a. **Describe the population parameter of interest.**

The parameter of interest is the mean μ of the population of single-digit numbers in Table 1 of Appendix B.

b. **State the null hypothesis (H_o) and the alternative hypothesis (H_a).**

H_o: $\mu = 4.5$ (mean is 4.5)
H_a: $\mu \neq 4.5$ (mean is not 4.5)

STEP 2 **The Hypothesis Test Criteria:**

a. **Check the assumptions.**

σ is known. Samples of size 40 should be large enough to satisfy the CLT; see the discussion of this issue on page 386.

b. **Identify the probability distribution and the test statistic to be used.**

We use the standard normal probability distribution and the test statistic $z\bigstar = \dfrac{\bar{x} - \mu}{\sigma/\sqrt{n}}$; $\sigma = 2.87$.

c. **Determine the level of significance, α.**
$\alpha = 0.10$ (given in the statement of problem).

STEP 3 The Sample Evidence:

a. **Collect the sample information.**

This random sample was drawn from Table 1 in Appendix B: II08-20

2	8	2	1	5	5	4	0	9	1
0	4	6	1	5	1	1	3	8	0
3	6	8	4	8	6	8	9	5	0
1	4	1	2	1	7	1	7	9	3

The sample statistics are $\bar{x} = 3.975$ and $n = 40$.

b. **Calculate the value of the test statistic.**
Use formula (8.4), information from H_o: $\mu = 4.5$, and $\sigma = 2.87$:

$$z\bigstar = \frac{\bar{x} - \mu}{\sigma/\sqrt{n}}: \qquad z\bigstar = \frac{3.975 - 4.50}{2.87/\sqrt{40}} = \frac{-0.525}{0.454} = -1.156 = -1.16$$

STEP 4 The Probability Distribution:

is not | is | is not

0.05 0.05

−1.65 0 1.65 z

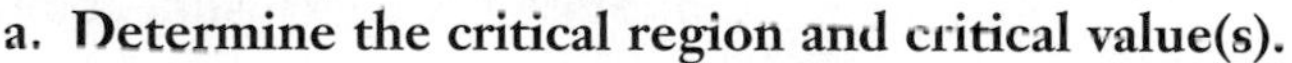

a. **Determine the critical region and critical value(s).**
A two-tailed critical region will be used, and 0.05 will be the area in each tail. The critical values are $\pm z(0.05) = \pm 1.65$.

b. **Determine whether or not the calculated test statistic is in the critical region.**
The calculated value of z, $z\bigstar = -1.16$, is not in the critical region (shown in red on the figure).

STEP 5 The Result:

a. **State the decision about H_o:** Fail to reject H_o.

b. **State the conclusion about H_a.**
The observed sample mean is not significantly different from 4.5 at the 0.10 significance level. ■

Suppose we were to take another sample of size 40 from Table 1. Would we obtain the same results? Suppose we took a third sample and a fourth. What results might we expect? What is the level of significance? Yes, its value is 0.10, but what does it measure? Table 8.10 lists the means obtained from 20 different random samples of size 40 that were taken from Table 1 in Appendix B. The calculated value of $z\bigstar$ that corresponds to each $\bar{x}$ and the decision each would dictate are also listed. The 20 calculated z-scores are shown in Figure 8.9. Note that 3 of the 20 samples (or 15%) caused us to reject the null hypothesis, even though we know the null hypothesis is true for this situation. Can you explain this?

NOTE: Remember that α is the probability that we "reject H_o" when it is actually a true statement. Therefore, we can anticipate that a type I error will occur α of the time when testing a true null hypothesis. In the above empirical situation, we observed a 15% rejection rate. If we were to repeat this experiment many times, the proportion of samples that would lead to a rejection would vary, but the observed relative frequency of rejection should be approximately α or 10%.

TABLE 8.10 **Twenty Random Samples of Size 40 Taken from Table 1 in Appendix B** Ta08-10

Sample Number	Sample Mean, $\bar{x}$	Calculated z, $z\bigstar$	Decision Reached
1	4.62	+0.26	Fail to reject H_o
2	4.55	+0.11	Fail to reject H_o
3	4.08	−0.93	Fail to reject H_o
4	5.00	+1.10	Fail to reject H_o
5	4.30	−0.44	Fail to reject H_o
6	3.65	−1.87	Reject H_o
7	4.60	+0.22	Fail to reject H_o
8	4.15	−0.77	Fail to reject H_o
9	5.05	+1.21	Fail to reject H_o
10	4.80	+0.66	Fail to reject H_o
11	4.70	+0.44	Fail to reject H_o
12	4.88	+0.83	Fail to reject H_o
13	4.45	−0.11	Fail to reject H_o
14	3.93	−1.27	Fail to reject H_o
15	5.28	+1.71	Reject H_o
16	4.20	−0.66	Fail to reject H_o
17	3.48	−2.26	Reject H_o
18	4.78	+0.61	Fail to reject H_o
19	4.28	−0.50	Fail to reject H_o
20	4.23	−0.61	Fail to reject H_o

FIGURE 8.9

z-Scores from Table 8.10

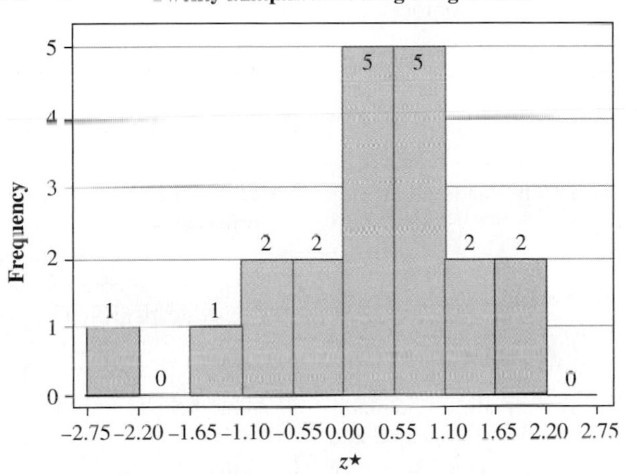

ANSWER NOW 8.110

Use a computer or calculator to select 40 random single-digit numbers. Repeat the selection several times as in Table 8.10. Find the sample mean and $z\bigstar$. Describe your findings after several tries. (Use the commands for generating integer data on pages 183–184; then continue with the hypothesis test commands on page 383.)

Exercises

8.111 State the null hypothesis, H_o, and the alternative hypothesis, H_a, that would be used for a hypothesis test related to each of these statements:

a. The mean age of the youths who hang out at the mall is less than 16 years.
b. The mean height of professional basketball players is more than 6 ft 6 in.
c. The mean elevation drop for ski trails at eastern ski centers is at least 285 ft.
d. The mean diameter of the rivets is no more than 0.375 in.
e. The mean cholesterol level of male college students is different from 200 units.

8.112 Suppose you want to test the hypothesis that the mean salt content of frozen "lite" dinners is more than 350 mg per serving. An average of 350 mg is an acceptable amount of salt per serving; therefore, you use it as the standard. The null hypothesis is "The average content is not more than 350 mg" ($\mu = 350$). The alternative hypothesis is "The average content is more than 350 mg" ($\mu > 350$).

a. Describe the conditions that would exist if your decision results in a type I error.
b. Describe the conditions that would exist if your decision results in a type II error.

8.113 Suppose you wanted to test the hypothesis that the mean minimum home service call charge for plumbers is at most \$85 in your area. Explain the conditions that would exist if you make an error in decision by committing:

a. a type I error b. a type II error

8.114 Describe how the null hypothesis in Illustration 8.19 is a "starting point" for the decision to be made about the mean one-way commuting distance.

8.115

a. What is the critical region?
b. What is the critical value?

8.116 Since the size of the type I error can always be made smaller by reducing the size of the critical region, why don't we always choose critical regions that make α extremely small?

8.117 Determine the critical region and critical values for z that would be used to test the null hypothesis at the given level of significance, as described in each of the following:

a. H_o: $\mu = 20$ vs. H_a: $\mu \neq 20$, $\alpha = 0.10$
b. H_o: $\mu = 24$ ($\leq$) vs. H_a: $\mu > 24$, $\alpha = 0.01$
c. H_o: $\mu = 10.5$ ($\geq$) vs. H_a: $\mu < 10.5$, $\alpha = 0.05$
d. H_o: $\mu = 35$ vs. H_a: $\mu \neq 35$, $\alpha = 0.01$

8.118 Determine the critical region and the critical values used to test the following null hypotheses:

a. H_o: $\mu = 55$ ($\geq$) vs. H_a: $\mu < 55$, $\alpha = 0.02$
b. H_o: $\mu = -86$ ($\geq$) vs. H_a: $\mu < -86$, $\alpha = 0.01$
c. H_o: $\mu = 107$ vs. H_a: $\mu \neq 107$, $\alpha = 0.05$
d. H_o: $\mu = 17.4$ ($\leq$) vs. H_a: $\mu > 17.4$, $\alpha = 0.10$

8.119 The null hypothesis, H_o: $\mu = 250$, was tested against the alternative hypothesis, H_a: $\mu < 250$. A sample of $n = 85$ resulted in a calculated test statistic of $z\bigstar = -1.18$. If $\sigma = 22.6$, find the value of the sample mean, $\bar{x}$. Find the sum of the sample data, $\sum x$.

8.120 Find the value of $\bar{x}$ for each of the following:

a. H_o: $\mu = 580$, $z\bigstar = 2.10$, $\sigma = 26$, $n = 55$
b. H_o: $\mu = 75$, $z\bigstar = -0.87$, $\sigma = 9.2$, $n = 35$

8.121 The calculated value of the test statistic is actually the number of standard errors that the sample mean differs from the hypothesized value of μ in the null hypothesis. Suppose that the null hypothesis is H_o: $\mu = 4.5$, σ is known to be 1.0, and a sample of size 100 results in $\bar{x} = 4.8$.

a. How many standard errors is $\bar{x}$ above 4.5?
b. If the alternative hypothesis is H_a: $\mu > 4.5$ and $\alpha = 0.01$, would you reject H_o?

8.122 Consider the hypothesis test where the hypotheses are H_o: $\mu = 26.4$ and H_a: $\mu < 26.4$. A random sample of size 64 is selected and yields a sample mean of 23.6.

a. If it is known that $\sigma = 12$, how many standard errors below $\mu = 26.4$ is the sample mean $\bar{x} = 23.6$?
b. If $\alpha = 0.05$, would you reject H_o? Explain.

8.123 There are only two possible decisions as a result of a hypothesis test.

a. State the two possible decisions.
b. Describe the conditions that will lead to each of the two decisions identified in part a.

8.124

a. What proportion of the probability distribution is in the critical region, provided the null hypothesis is correct?
b. What error could be made if the test statistic falls in the critical region?
c. What proportion of the probability distribution is in the noncritical region, provided the null hypothesis is not correct?
d. What error could be made if the test statistic falls in the noncritical region?

8.125 This computer output was used to complete a hypothesis test:

```
TEST OF MU = 15.000 VS MU not = 15.0000
THE ASSUMED SIGMA = 0.50
N       MEAN       STDEV      SE MEAN      Z
30      15.6333    0.4270     0.0913       6.94
```

a. State the null and alternative hypotheses.
b. If the test is completed using $\alpha = 0.01$, what decision and conclusion are reached?
c. Verify the value of the standard error of the mean.

8.126 This computer output was used to complete a hypothesis test:

```
TEST OF MU = 72.00 VS MU > 72.00
THE ASSUMED SIGMA = 12.0
N       MEAN       STDEV      SE MEAN      Z
36      75.2       11.87      2.00         1.60
```

a. State the null and alternative hypotheses.
b. If the test is completed using $\alpha = 0.05$, what decision and conclusion are reached?
c. Verify the value of the standard error of the mean.

8.127 The Texas Department of Health published the statewide results from the Emergency Medical Services Certification Examination. Those who took the paramedic exam for the first time in October 2001 got an average score of 79.68 (out of a possible 100) with a standard deviation of 9.06. Suppose a random sample of 50 individuals who took the exam yielded a mean score of 81.05. Is there sufficient evidence to conclude that the population from which this random sample was taken, on the average, scored higher than the state average? Use $\alpha = 0.05$.

8.128 The dollar value of a college education is often measured by comparing the average annual pay for workers who graduated from college to the pay earned by those who never attended college after receiving a high school diploma. *USA Today* (September 11, 1998) listed Census Bureau data that showed average earnings in 1996 of male workers who possessed a bachelor's degree as \$53,102 and earnings of male high school graduates with no college experience as \$34,034. Suppose a random sample of 150 male college graduates with bachelor's degrees is taken in 1998 to check for any possible salary increase and a sample mean of \$53,500 is obtained. Test the hypothesis that no increase in average salary occurred during the two-year period at the 0.05 level of significance. Assume that σ = \$3,900 per year. Use the classical approach.

a. Define the parameter.
b. State the null and alternative hypotheses.
c. Specify the hypothesis test criteria.
d. Present the sample evidence.
e. Find the probability distribution information.
f. Determine the results.

8.129 S Ex08-129 The manager at Air Express thinks that packages shipped recently weigh less than packages in the past. Records show that in the past packages have had a mean weight of 36.5 lb and a standard deviation of 14.2 lb. A random sample of last month's shipping records yielded these 64 weights:

32.1	41.5	16.1	8.9	36.2	12.3	28.4	40.4
45.5	15.2	26.5	13.3	23.5	33.7	18.3	16.3
15.4	39.7	50.3	14.8	44.4	47.7	45.8	52.3
48.4	10.4	59.9	5.5	6.7	17.1	20.0	28.1
48.1	29.5	22.9	47.8	24.8	20.1	40.1	12.6
24.3	43.3	32.4	57.7	42.9	36.7	15.5	46.4
51.3	38.6	39.4	27.1	55.7	37.7	39.4	55.5
26.9	15.7	32.3	47.8	33.2	29.1	31.1	34.5

Is this sufficient evidence to reject the null hypothesis in favor of the manager's claim? Use $\alpha = 0.01$.

8.130 S A fire insurance company decided that the mean distance from a home to the nearest fire department in a suburb of Chicago was at least 4.7 mi. It set its fire insurance rates accordingly. Members of the community set out to show that the mean distance was less than 4.7 mi. This, they thought, would convince the insurance company to lower its rates. They randomly identified 64 homes and measured the distance to the nearest fire department for each. The resulting sample mean was 4.4. If σ = 2.4 mi, does the sample provide sufficient evidence to support the community's claim at the $\alpha = 0.05$ level of significance?

CHAPTER SUMMARY

RETURN TO CHAPTER CASE STUDY

© Joe Sohm/Stock Boston

Were They Shorter Back Then?

Data from the National Center for Health Statistics (NCHS) indicate that the average height of a woman in the United States is 63.7 inches with a standard deviation of 2.75 inches. Use these data from the Chapter Case Study (p. 343) to answer the questions that follow: CS08

65.0	66.0	64.0	67.0	59.0	69.0	66.0	69.0	64.0	61.5
63.0	62.0	63.0	64.0	72.0	66.0	65.0	64.0	67.0	68.0
70.0	63.0	63.0	68.0	58.0	60.0	63.5	66.0	64.0	62.0
64.5	69.0	63.5	69.0	62.0	58.0	66.0	68.0	59.0	56.0
64.0	66.0	65.0	69.0	67.0	66.5	67.5	62.0	70.0	62.0

FIRST THOUGHTS 8.131

a. Do you expect the mean of a random sample to be exactly equal to the mean of the population from which the sample was drawn? What do you expect? Explain.
b. Draw a histogram and calculate the mean height for the sample of 50 heights (see Answer Now 8.1b, p. 343).
c. How is the distribution of the sample data related to the distribution of the population? To the sampling distribution of sample means?
d. Using the techniques of Chapter 7, find the limits that bound the middle 95% of the sampling distribution of sample means for random samples of size 50 selected from the population of women's heights with a known mean of 63.7 inches and a standard deviation of 2.75 inches.
e. On the histogram drawn in part b draw a vertical line at the sample mean, and then draw a horizontal line segment showing the interval found in part d. Does the sample mean fall in the interval? Explain what this means.

PUTTING CHAPTER 8 TO WORK 8.132

a. Are the assumptions of the confidence interval and hypothesis test methods of this chapter satisfied? Explain.
b. Using the sample data and a 95% level of confidence, estimate the mean height of women in the health profession. Use the given population standard deviation of 2.75 inches.
c. Test the claim that the mean height of women in the health profession is different from 63.7 inches, the mean height for all American women. Use a 0.05 level of significance.
d. On the same histogram used in 8.131e, draw a vertical line at the hypothesized population mean value, 63.7. Then draw a horizontal line segment showing the 95% confidence interval found in part b. Does the mean $\mu = 63.7$ fall in the interval? Explain what this means.
e. Describe the relationship between the two lines drawn on your graph in 8.131e and the two lines drawn for 8.132d.
f. Based on the results obtained above, does it appear that the women in this study, on the average, are the same height as all women in the United States as reported by the NCHS? Explain.

YOUR OWN MINI-STUDY 8.133

a. Design your own study of women's heights. Define a specific population that you will sample, describe your sampling plan, collect your data, and answer questions 8.131(b) and 8.132(a, b, c, and f), replacing health profession with your particular population.
b. Discuss the differences and similarities between your sample and the population and between your sample and the sample of 50 female health professionals.

In Retrospect

Two forms of inference were presented in this chapter: estimation and hypothesis testing. They may be, and often are, used separately. It seems natural, however, for the rejection of a null hypothesis to be followed by a confidence interval. (If the value claimed is wrong, we often want an estimate for the true value.)

These two forms of inference are quite different, but they are related. There is a certain amount of crossover between the use of the two inferences. For example, suppose that you had sampled and calculated a 90% confidence interval for the mean of a population. The interval was 10.5 to 15.6. Then someone claims that the true mean is 15.2. Your confidence interval can be compared to this claim. If the claimed value falls within your interval estimate, you would fail to reject the null hypothesis that $\mu = 15.2$ at a 10% level of significance in a two-tailed test. If the claimed value (say, 16.0) falls outside the interval, you would reject the null hypothesis that $\mu = 16.0$ at $\alpha = 0.10$ in a two-tailed test. If a one-tailed test is required, or if you prefer a different value of α, a separate hypothesis test must be used.

Many users of statistics (especially those marketing a product) will claim that their statistical results prove that their product is superior. But remember, the hypothesis test does not *prove* or *disprove* anything. The decision reached in a hypothesis test has probabilities associated with the four various situations. If "fail to reject H_o" is the decision, it is possible that an error has occurred. Furthermore, if "reject H_o" is the decision reached, it is possible for this to be an error. Both errors have probabilities greater than zero.

In this chapter we have restricted our discussion of inferences to the mean of a population for which the standard deviation is known. In Chapters 9 and 10 we will discuss inferences about the population mean and remove the restriction about the known value for standard deviation. We will also look at inferences about the parameters proportion, variance and standard deviation.

Chapter Exercises

8.134 A sample of 64 measurements is taken from a continuous population, and the sample mean is found to be 32.0. The standard deviation of the population is known to be 2.4. An interval estimation is to be made of the mean with a level of confidence of 90%. State or calculate these values:

a. $\bar{x}$ b. σ c. n d. $1 - \alpha$ e. $z(\alpha/2)$ f. $\sigma_{\bar{x}}$
g. E (maximum error of estimate)
h. upper confidence limit
i. lower confidence limit

8.135 Suppose a confidence interval is assigned a level of confidence of $1 - \alpha = 95\%$.

a. How is the 95% used in constructing the confidence interval?
b. If $1 - \alpha$ is changed to 90%, what effect does this have on the confidence interval?

8.136 Suppose a hypothesis test is assigned a level of significance of $\alpha = 0.01$.

a. How is the 0.01 used in completing the hypothesis test?
b. If α is changed to 0.05, what effect does this have on the test procedure?

8.137 The expected mean of a continuous population is 100, and its standard deviation is 12. A sample of 50 measurements gives a sample mean of 96. At the 0.01 level of significance, a test is to be made to decide between "The population mean is 100" and "The population mean is different from 100." State or find each of these:

a. H_o b. H_a c. α d. μ (based on H_o)
e. $\bar{x}$ f. σ g. $\sigma_{\bar{x}}$ h. $z\bigstar$, z-score for $\bar{x}$
i. p-value j. decision
k. Sketch the standard normal curve and locate $z\bigstar$, the p-value, and α.

8.138 The expected mean of a continuous population is 200, and its standard deviation is 15. A sample of 80 measurements gives a sample mean of 205. At the 0.01 level of significance, a test is to be made to decide between "The population mean is 200" and "The population mean is different from 200." State or find each of the following:

a. H_o b. H_a c. α d. $z(\alpha/2)$
e. μ (based on H_o) f. $\bar{x}$ g. σ
h. $\sigma_{\bar{x}}$ i. $z\bigstar$, z-score for $\bar{x}$ j. decision
k. Sketch the standard normal curve and locate $\alpha/2$, $z(\alpha/2)$, the critical region, and $z\bigstar$.

8.139 From a population of unknown mean μ and standard deviation $\sigma = 5.0$, a sample of $n = 100$ is selected and the sample mean 40.6 is found. Complete this exercise to compare the concepts of estimation and hypothesis testing.

a. Determine the 95% confidence interval for μ.
b. Complete the hypothesis test involving H_a: $\mu \neq 40$ using the p-value approach and $\alpha = 0.05$.

(continued)

c. Complete the hypothesis test involving H_a: $\mu \neq 40$ using the classical approach and $\alpha = 0.05$.
d. On one sketch of the standard normal curve, locate: the interval that represents the confidence interval from part a; $z\bigstar$, the p-value, and α from part b; and $z\bigstar$ and the critical regions from part c. Describe the relationship between these three separate procedures.

8.140 From a population of unknown mean μ and standard deviation $\sigma = 5.0$, a sample of $n = 100$ is selected and the sample mean 41.5 is found. Compare the concepts of estimation and hypothesis testing by completing the following:
a. Determine the 95% confidence interval for μ.
b. Complete the hypothesis test involving H_a: $\mu \neq 40$ using the p-value approach and $\alpha = 0.05$.
c. Complete the hypothesis test involving H_a: $\mu \neq 40$ using the classical approach and $\alpha = 0.05$.
d. On one sketch of the standard normal curve, locate: the interval that represents the confidence interval from part a; $z\bigstar$, the p-value, and α from part b; and $z\bigstar$ and the critical regions from part c. Describe the relationship between these three separate procedures.

8.141 From a population of unknown mean μ and standard deviation $\sigma = 5.0$, a sample of $n = 100$ is selected and the sample mean 40.9 is found. Compare the concepts of estimation and hypothesis testing by completing the following:
a. Determine the 95% confidence interval for μ.
b. Complete the hypothesis test involving H_a: $\mu > 40$ using the p-value approach and $\alpha = 0.05$.
c. Complete the hypothesis test involving H_a: $\mu > 40$ using the classical approach and $\alpha = 0.05$.
d. On one sketch of the standard normal curve, locate: the interval that represents the confidence interval from part a; $z\bigstar$, the p-value, and α from part b; and $z\bigstar$ and the critical regions from part c. Describe the relationship between these three separate procedures.

8.142 The standard deviation of a normally distributed population is equal to 10. A sample of 25 is selected, and its mean is found to be 95.
a. Find an 80% confidence interval for μ.
b. If the sample size were 100, what would be the 80% confidence interval?
c. If the sample size were 25 but the standard deviation were 5 (instead of 10), what would be the 80% confidence interval?

8.143 The weights of full boxes of a certain kind of cereal are normally distributed with a standard deviation of 0.27 oz. A random sample of 18 boxes produced a mean weight of 9.87 oz.
a. Find the 95% confidence interval for the true mean weight of a box of this cereal.
b. Find the 99% confidence interval for the true mean weight of a box of this cereal.
c. What effect did the increase in the level of confidence have on the width of the confidence interval?

8.144 Waiting times (in hours) at a popular restaurant are believed to be approximately normally distributed with a variance of 2.25 during busy periods.
a. A sample of 20 customers revealed a mean waiting time of 1.52 hr. Construct the 95% confidence interval for the population mean.
b. Suppose that the mean of 1.52 hr had resulted from a sample of 32 customers. Find the 95% confidence interval.
c. What effect does a larger sample size have on the confidence interval?

8.145 A random sample of the scores of 100 applicants for clerk-typist positions at a large insurance company showed a mean score of 72.6. The preparer of the test maintained that qualified applicants should average 75.0.
a. Determine the 99% confidence interval for the mean score of all applicants at the insurance company. Assume that the standard deviation of test scores is 10.5.
b. Can the insurance company conclude that it is getting qualified applicants (as measured by this test)?

8.146 Over the years, the general trend has been that farms have been increasing in size, while the rural population has declined. The National Agricultural Statistics Service of the U.S. Department of Agriculture reported in 1998, however, that the average acreage per farm was 470 in both 1996 and 1997 for the 2.06 million or so farms in the nation. Total farm acreage declined slightly, but so did the number of farms. Therefore, the average farm size did not change. A random sample of 100 farms in rural America in 1998 reveals an average size of 495 acres. Find the 95% confidence interval for the mean farm size in 1998. Assume the standard deviation is 190 acres per farm.

8.147 An article titled "Evaluation of a Self-Efficacy Scale for Preoperative Patients" (*AORN Journal*, Vol. 60, No. 1, July 1994) describes a 32-item rating scale used to determine efficacy (measure of effectiveness) expectations as well as outcome expectations. The 32-item scale was administered to 200 preoperative patients. The mean efficacy expectation score for the ambulating item was 4.00. Construct the 0.95 confidence interval for the mean of all such preoperative patients. Use $\sigma = 0.94$.

8.148 An automobile manufacturer wants to estimate the mean gasoline mileage that its customers will obtain with its new compact model. How many sample runs must be performed so that the estimate will be accurate to within 0.3 mpg at 95% confidence? (Assume that $\sigma = 1.5$.)

8.149 A fish hatchery manager wants to estimate the mean length of her three-year-old hatchery-raised trout. She wants to make a 99% confidence interval accurate to within $\frac{1}{3}$ of a standard deviation. How large a sample does she need to take?

8.150 We are interested in estimating the mean life of a new product. How large a sample do we need to take in order to estimate the mean to within $\frac{1}{10}$ of a standard deviation with 90% confidence?

8.151 According to an article in *Health* magazine (March 1991) supplementation with potassium reduced the blood pressure readings in a group of mild hypertensive patients from an average of $\frac{158}{100}$ to $\frac{143}{84.5}$. Consider a study involving the use of potassium supplementation to reduce the systolic blood pressure for mild hypertensive patients. Suppose 75 patients with mild hypertension were placed on potassium for six weeks. The response measured was the systolic reading at the beginning of the study minus the systolic reading at the end of the study. The mean drop in the systolic reading was 12.5 units. Assume the population standard deviation, σ, is 7.5 units. Calculate the value of the test statistic $z\bigstar$ and the p-value for testing H_o: $\mu = 10.0$ versus H_a: $\mu > 10.0$.

8.152 The college bookstore tells prospective students that the average cost of its textbooks is \$32 per book with a standard deviation of \$4.50. The engineering science students think that the average cost of their books is higher than the average for all students. To test the bookstore's claim against their alternative, the engineering students collect a random sample of size 45.

a. If they use $\alpha = 0.05$, what is the critical value of the test statistic?
b. The engineering students' sample data are summarized by $n = 45$ and $\sum x = 1470.25$. Is this sufficient evidence to support their contention?

8.153 A manufacturing process produces ball bearings with diameters that have a normal distribution and a standard deviation of $\sigma = 0.04$ cm. Ball bearings with diameters that are too small or too large are undesirable. To test the null hypothesis that $\mu = 0.50$ cm, a random sample of 25 is selected and the sample mean is found to be 0.51.

a. Design null and alternative hypotheses such that rejection of the null hypothesis will imply that the ball bearings are undesirable.
b. Using the decision rule established in part a, what is the p-value for the sample results?
c. If the decision rule in part a is used with $\alpha = 0.02$, what is the critical value for the test statistic?

8.154 A rope manufacturer, after conducting a large number of tests over a long period of time, has found that the rope has a mean breaking strength of 300 lb and a standard deviation of 24 lb. Assume that these values are μ and σ. It is believed that by using a recently developed high-speed process, the mean breaking strength has been decreased.

a. Design null and alternative hypotheses such that rejection of the null hypothesis will imply that the mean breaking strength has decreased.
b. Using the decision rule established in part a, what is the p-value associated with rejecting the null hypothesis when 45 tests result in a sample mean of 295?
c. If the decision rule in part a is used with $\alpha = 0.01$, what is the critical value for the test statistic and what value of $\bar{x}$ corresponds to it if a sample of size 45 is used?

8.155 Worker honeybees leave the hive on a regular basis and travel to flowers and other sources of pollen and nectar before returning to the hive to deliver their cargo. They repeat the process several times each day in order to feed younger bees and support the hive's production of honey and wax. The worker bee can carry an average of 0.0113 g of pollen and nectar per trip, with a standard deviation of 0.0063 g. Fuzzy Drone is entering the honey and beeswax business with a new strain of Italian bees that are reportedly capable of carrying larger loads of pollen and nectar than the typical honeybee can. After installing three hives, Fuzzy isolated 200 bees before and after their return trip and carefully weighed their cargoes. The sample mean weight of the pollen and nectar was 0.0124 g. Can Fuzzy's bees carry a larger load of pollen and nectar than the rest of the honeybee population? Complete the appropriate hypothesis test at the 0.01 level of significance.

a. Solve using the p-value approach.
b. Solve using the classical approach.

8.156 For a population of humans to sustain itself, there must be an average of just over two births for each woman of reproductive age. This fertility rate varies substantially from nation to nation. Peter Drucker pointed out in *Fortune* magazine ("Peter Drucker Takes the Long View," September 28, 1998, p. 164) that the average for Japan dropped to 1.5 births for each woman of reproductive age in 1997, which could reduce Japan's population from 135 million in 1997 to 50 million by the end of the 21st century. Suppose a random sample of 300 Japanese women of reproductive age is taken in 1998, and the sample mean fertility rate is measured at 1.45. Assume the standard deviation is 0.75. Did the rate decline? Complete the appropriate hypothesis test at the 0.05 level of significance.

a. Solve using the p-value approach.
b. Solve using the classical approach.

8.157 In a large supermarket the customers' waiting times to check out are approximately normally distributed with a standard deviation of 2.5 min. A sample of 24 customer waiting times produced a mean of 10.6 min. Is this evidence sufficient to reject the supermarket's claim that its customer checkout times average no more than 9 min? Complete this hypothesis test using the 0.02 level of significance.

a. Solve using the p-value approach.
b. Solve using the classical approach.

8.158 At a very large firm, the clerk-typists were sampled to see whether the salaries differed among departments for workers in similar categories. For a sample of 50 of the firm's accounting clerks, the average annual salary was $16,010. The firm's personnel office insists that the average salary paid to all clerk-typists in the firm is $15,650 and that the standard deviation is $1800. At the 0.05 level of significance, can we conclude that the accounting clerks receive, on the average, a salary different from that of the clerk-typists?

a. Solve using the p-value approach.
b. Solve using the classical approach.

8.159 Jack Williams is vice president of marketing for one of the largest natural gas companies in the nation. During the past four years, he has watched two major factors erode away the profits and sales of the company. First, the average price of crude oil has been virtually flat, and many of his industrial customers are burning heavy oil rather than natural gas to fire their furnaces, irrespective of added smokestack emissions. Second, both residential and commercial customers are still pursuing energy conservation techniques (e.g., adding extra insulation, installing clock-drive thermostats, and sealing cracks around doors and windows to eliminate cold air infiltration). In 1997 residential customers bought an average of 129.2 mcf (thousands of cubic feet) of natural gas from Jack's company ($\sigma = 18$ mcf), based on internal company billing records, but environmentalists have claimed that conservation is cutting fuel consumption up to 3% per year. Jack has commissioned you to conduct a spot check to see whether annual usage has changed before his next meeting with the officers of the corporation. A random sample of 300 customers selected from the billing records reveals an average of 127.1 mcf during the past 12 months. Did consumption decline?

a. Complete the appropriate hypothesis test at the 0.01 level of significance using the probability-value approach so that you can properly advise Jack prior to his meeting.
b. Since you are Jack's assistant, why is it best for you to use the p-value approach?

8.160 A manufacturer of automobile tires believes it has developed a new rubber compound that has superior wearing qualities. It produced a test run of tires made with this new compound and had them road tested. The data recorded were the amounts of tread wear per 10,000 miles. In the past, the mean amount of tread wear per 10,000 miles, for tires of this quality, has been 0.0625 inch. The null hypothesis to be tested here is: "The mean amount of wear on the tires made with the new compound is the same mean amount of wear as with the old compound, 0.0625 inch per 10,000 miles"; H_o: $\mu = 0.0625$. Three possible alternative hypotheses could be used: H_a: $\mu < 0.0625$, H_a: $\mu \neq 0.0625$, and H_a: $\mu > 0.0625$.

a. Explain the meaning of each of these three alternatives.
b. Which one of the possible alternative hypotheses should the manufacturer use if it hopes to conclude that use of the new compound does yield superior wear?

8.161 All drugs must be approved by the Food and Drug Administration (FDA) before they can be marketed by a drug company. The FDA must weigh the error of marketing an ineffective drug, with the usual risks of side effects, against the consequences of not allowing an effective drug to be sold. Suppose, using standard medical treatment, that the mortality rate, r, of a certain disease is known to be A. A manufacturer submits for approval a drug that is supposed to treat this disease. The FDA sets up the hypothesis to test the mortality rate for the drug as (1) H_o: $r = A$ vs. H_a: $r < A$, $\alpha = 0.005$; or (2) H_o: $r = A$ vs. H_a: $r > A$, $\alpha = 0.005$.

a. If $A = 0.95$, which test do you think the FDA should use? Explain.
b. If $A = 0.05$, which test do you think the FDA should use? Explain.

8.162 The drug manufacturer in Exercise 8.161 has a different viewpoint on the matter. It wants to market the new drug starting as soon as possible so that it can beat its competitors to the marketplace and make lots of money. Its position is: Market the drug unless the drug is totally ineffective.

a. How would the drug company set up the alternative hypothesis if it were doing the testing: H_a: $r < A$, H_a: $r \neq A$, or H_a: $r > A$? Explain why.
b. Does the mortality rate ($A = 0.95$ or $A = 0.05$) of the existing treatment affect the alternative? Explain.

8.163 Ex08-163 This computer output shows a simulated sample of size 25 randomly generated from a normal population with $\mu = 130$ and $\sigma = 10$. A confidence interval command was then used to set a 95% confidence interval for μ.

```
116.187   119.832   121.782   122.320   141.436
129.197   119.172   120.713   135.765   131.153
122.307   126.155   137.545   141.154   123.405
143.331   121.767   109.742   140.524   150.600
121.655   127.992   136.434   139.768   125.594

N    MEAN     STDEV   SE MEAN   95.0 PERCENT C.I.
25   129.02   10.18    2.00     (125.10, 132.95)
```

a. State the confidence interval that resulted.
b. Verify the values reported for the standard error of mean and the interval bounds.

8.164 Use a computer and generate 50 random samples, each of size $n = 25$, from a normal probability distribution with $\mu = 130$ and standard deviation $\sigma = 10$.

a. Calculate the 95% confidence interval based on each sample mean.

b. What proportion of these confidence intervals contain $\mu = 130$?

c. Explain what the proportion found in part b represents.

8.165 Ex08-165 This computer output shows a simulated sample of size 28 randomly generated from a normal population with $\mu = 18$ and $\sigma = 4$. Computer commands were then used to complete a hypothesis test for $\mu = 18$ against a two-tailed alternative.

```
18.7734   21.4352   15.5438   20.2764   23.2434
15.7222   13.9368   14.4112   15.7403   19.0970
19.0032   20.0688   12.2466   10.4158    8.9755
18.0094   20.0112   23.2721   16.6458   24.6140
17.8078   16.5922   16.1385   12.3115   12.5674
18.9141   22.9315   13.3658

TEST OF MU = 18.000 VS MU not = 18.000
THE ASSUMED SIGMA = 4.00

N      MEAN     STDEV    SE MEAN      Z       P VALUE
28    17.217    4.053     0.756     -1.04      0.30
```

a. State the alternative hypothesis, the decision, and the conclusion that resulted.

b. Verify the values reported for the standard error of mean, $z\bigstar$, and the p-value.

8.166 Use a computer and generate 50 random samples, each of size $n = 28$, from a normal probability distribution with $\mu = 18$ and standard deviation $\sigma = 4$.

a. Calculate the $z\bigstar$ corresponding to each sample mean.

b. In regard to the p-value approach, find the proportion of 50 $z\bigstar$-values that are "more extreme" than the $z = -1.04$ used in Exercise 8.165 (H_a: $\mu \neq 18$). Explain what this proportion represents.

c. In regard to the classical approach, find the critical values for a two-tailed test using $\alpha = 0.01$; find the proportion of 50 $z\bigstar$-values that fall in the critical region. Explain what this proportion represents.

8.167 Use a computer and generate 50 random samples, each of size $n = 28$, from a normal probability distribution with $\mu = 19$ and standard deviation $\sigma = 4$.

a. Calculate the $z\bigstar$ corresponding to each sample mean that would result when testing the null hypothesis $\mu = 18$.

b. In regard to the p-value approach, find the proportion of 50 $z\bigstar$-values that are "more extreme" than the $z = -1.04$ used in Exercise 8.165 (H_a: $\mu \neq 18$). Explain what this proportion represents.

c. In regard to the classical approach, find the critical values for a two-tailed test using $\alpha = 0.01$; find the proportion of 50 $z\bigstar$-values that fall in the noncritical region. Explain what this proportion represents.

Vocabulary and Key Concepts

Be able to define each term. Pay special attention to the key terms, which are printed in **red**. In addition, describe each term in your own words and give an example of each. Your examples should not be ones given in class or in the textbook. The bracketed numbers indicate the chapters in which the terms first appeared, but you should define the terms again to show increased understanding of their meaning. Page numbers indicate the first appearance of the term in Chapter 8.

alpha (α) (p. 366)
alternative hypothesis (pp. 362, 371, 387)
assumptions (pp. 348, 349, 368, 386)
beta (β) (p. 366)
biased statistics (p. 344)
calculated value ($z\bigstar$) (pp. 375, 392)
conclusion (pp. 367, 377, 394)
confidence coefficient (p. 349)
confidence interval (p. 346)
confidence interval procedure (p. 350)
critical region (p. 392)
critical value (pp. 392, 393)
decision rule (pp. 367, 376, 393)
estimation (p. 344)
hypothesis (p. 362)
hypothesis test (p. 362)
hypothesis test, classical procedure (p. 386)
hypothesis test, p-value procedure (p. 370)
interval estimate (p. 346)
level of confidence (p. 346)
level of significance (p. 367)
lower confidence limit (p. 350)
maximum error of estimate (pp. 350, 358)
noncritical region (p. 392)
null hypothesis (pp. 362, 371, 387)
parameter [1] (p. 344)
point estimate (p. 344)
p-value (p. 375)
sample size (pp. 357, 359)
sample statistic [1, 2] (p. 344)
standard error of mean [7] (p. 350)
test criteria (pp. 374, 391)
test statistic (pp. 367, 374, 391)
type A correct decision (p. 364)
type B correct decision (p. 364)
type I error (p. 364)
type II error (p. 365)
unbiased statistic (p. 344)
upper confidence limit (p. 350)
$z(\alpha)$ [6] (pp. 349, 393)

Chapter Practice Test

PART I: KNOWING THE DEFINITIONS

Answer "True" if the statement is always true. If the statement is not always true, replace the words shown in bold with words that make the statement always true.

8.1 **Beta** is the probability of a type I error.

8.2 **$1 - \alpha$** is known as the level of significance of a hypothesis test.

8.3 The standard error of the mean is the standard deviation of the **sample selected.**

8.4 The maximum error of estimate is controlled by three factors: **level of confidence, sample size,** and **standard deviation.**

8.5 Alpha is the measure of the area under the curve of the standard score that lies in the **rejection region** for H_o.

8.6 The risk of making a **type I error** is directly controlled in a hypothesis test by establishing a level for α.

8.7 Failing to reject the null hypothesis when it is false is a **correct decision.**

8.8 If the noncritical region in a hypothesis test is made wider (assuming σ and n remain fixed), α becomes larger.

8.9 Rejection of a null hypothesis that is false is a **type II error.**

8.10 To conclude that the mean is higher (or lower) than a claimed value, the value of the test statistic must fall in the **acceptance region.**

PART II: APPLYING THE CONCEPTS

Answer all questions, showing all formulas, substitutions, and work.

8.11 An unhappy post office customer is frustrated with the waiting time to buy stamps. Upon registering his complaint, he was told, "The average waiting time in the past has been about 4 minutes with a standard deviation of 2 minutes." The customer collected a sample of $n = 45$ customers and found the mean wait was 5.3 min. Find the 95% confidence interval for the mean waiting time.

8.12 State the null (H_o) and the alternative (H_a) hypotheses that would be used to test each of these claims:
a. The mean weight of professional football players is more than 245 lb.
b. The mean monthly amount of rainfall in Monroe County is less than 4.5 in.
c. The mean weight of the baseball bats used by major league players is not equal to 35 oz.

8.13 Determine the level of significance, test statistic, critical region, and critical value(s) that would be used in completing each hypothesis test using $\alpha = 0.05$:
a. H_o: $\mu = 43$; H_a: $\mu < 43$ (given $\sigma = 6$)
b. H_o: $\mu = 0.80$; H_a: $\mu > 0.80$ (given $\sigma = 0.13$)
c. H_o: $\mu = 95$; H_a: $\mu \neq 95$ (given $\sigma = 12$)

8.14 Find each value:
a. $z(0.05)$ b. $z(0.01)$ c. $z(0.12)$

8.15 In the past, the grapefruits grown in a particular orchard have had a mean diameter of 5.50 in. and a standard deviation of 0.6 in. The owner believes this year's crop is larger than in the past. He collected a random sample of 100 grapefruits and found a sample mean diameter of 5.65 in.
a. Find the value of the test statistic, $z\bigstar$, that corresponds to $\bar{x} = 5.65$.
b. Calculate the p-value for the owner's hypothesis.

8.16 A manufacturer claims that its light bulbs have a mean lifetime of 1520 hr with a standard deviation of 85 hr. A random sample of 40 such bulbs is selected for testing. If the sample produces a mean value of 1498.3 hr, is there sufficient evidence to claim that the mean lifetime is less than the manufacturer claimed? Use $\alpha = 0.01$.

PART III: UNDERSTANDING THE CONCEPTS

8.17 Sugar Creek Convenience Stores has commissioned a statistics firm to survey its customers in order to estimate the mean amount spent per customer. From previous records the standard deviation is believed to be $\sigma = \$5$. In its proposal to Sugar Creek, the statistics firm states that it plans to base the estimate for the mean amount spent on a sample of size 100 and use the 95% confidence level. Sugar Creek's president has suggested that the sample size be increased to 400. If nothing else changes, what effect will this increase in the sample size have on:

a. the point estimate for the mean
b. the maximum error of estimation
c. the confidence interval

The CEO wants the level of confidence increased to 99%. If nothing else changes, what effect will this increase in level of confidence have on:

d. the point estimate for the mean
e. the maximum error of estimation
f. the confidence interval

8.18 The noise level in a hospital may be a critical factor influencing a patient's speed of recovery. Suppose for the sake of discussion that a research commission has recommended a maximum mean noise level of 30 decibels (db) with a standard deviation of 10 db. The staff of a hospital intends to sample one of its wards to determine whether the noise level is significantly higher than the recommended level. The following hypothesis test will be completed:

H_o: $\mu = 30$ ($\leq$) vs. H_a: $\mu > 30$, $\alpha = 0.05$

a. Identify the correct interpretation for each hypothesis with regard to the recommendation and justify your choice.

H_o: (1) Noise level is not significantly higher than the recommended level, or (2) Noise level is significantly higher than the recommended level

H_a: (1) Noise level is not significantly higher than the recommended level, or (2) Noise level is significantly higher than the recommended level

b. Which statement best describes the type I error?
1. Decision reached was that noise level is within the recommended level, when in fact it actually was within.
2. Decision reached was that noise level is within the recommended level, when in fact it actually exceeded it.
3. Decision reached was that noise level exceeds the recommended level, when in fact it actually was within.
4. Decision reached was that noise level is within the recommended level, when in fact it actually exceeded it.

c. Which statement in part b best describes the type II error?

d. If α were changed from 0.05 to 0.01, identify and justify the effect (increases, decreases, or remains the same) on P(type I error) and on P(type II error).

8.19 The alternative hypothesis is sometimes called the "research hypothesis." The conclusion is a statement written about the alternative hypothesis. Explain why these two statements are compatible.

CHAPTER

9 INFERENCES INVOLVING ONE POPULATION

Chapter Outline and Objectives

CHAPTER CASE STUDY: GET ENOUGH DAILY EXERCISE?

Inferences about the mean time we exercise.

9.1 INFERENCES ABOUT THE MEAN μ (σ UNKNOWN)

The standard deviation of a population is seldom known in real-world applications. When the standard deviation of the population is unknown, the **Student's *t*-distribution** is used to make inferences about the population mean μ.

9.2 INFERENCES ABOUT THE BINOMIAL PROBABILITY OF SUCCESS

The observed sample proportion p' is **approximately normally distributed** under certain conditions and is the basis for inference when our concern is about the population parameter p, the binomial probability of success.

9.3 INFERENCES ABOUT THE VARIANCE AND STANDARD DEVIATION

The **chi-square distribution** is employed when testing hypotheses concerning variance, σ^2, or standard deviation, σ.

RETURN TO CHAPTER CASE STUDY

Chapter Case Study

© Journal Courier/ Steve Warmowski/ The Image Works

Get Enough Daily Exercise?

Do we get enough exercise from our daily activities?

Most Americans get little vigorous exercise at work or during leisure hours. Today, only a few jobs require vigorous physical activity. People usually ride in cars or buses and watch TV during their free time rather than be physically active. Activities like golfing and bowling provide people with some benefit. But they do not provide the same benefits as regular, more vigorous exercise.

Evidence suggests that even low- to moderate-intensity activities can have both short- and long-term benefits. If done daily, they help lower your risk of heart disease. Such activities include pleasure walking, stair climbing, gardening, yard work, moderate to heavy housework, dancing, and home exercise. More vigorous exercise can help improve fitness of the heart and lungs, which can provide even more consistent benefits for lowering heart disease risk.

Today, many people are rediscovering the benefits of regular, vigorous exercise—activities like swimming, brisk walking, running, or jumping rope. These kinds of activities are sometimes called "aerobic"—meaning the body uses oxygen to produce the energy needed for the activity. Aerobic exercises can condition your heart and lungs if performed at the proper intensity for at least 30 minutes, 3–4 times a week.

Source: U.S. Department of Health and Human Services, NIH Publication No. 93-1677, Revised August 1993

The article recommends that adults should exercise at least 90 minutes a week to lower their risk of heart disease. The following data are from a survey of cardiovascular technicians (individuals who perform various cardiovascular diagnostic procedures) asked about their own physical exercise, measured in minutes per week:

CS09

60	40	50	30	60	50	90	30	60	60
60	80	90	90	60	30	20	120	60	50
20	60	30	120	50	30	90	20	30	40
50	40	30	40	20	30	60	50	60	80

ANSWER NOW 9.1

a. **Find the mean and standard deviation for the amounts of time the cardiovascular technicians exercised per week.**

b. **How would you estimate the mean amount of time exercised per week by all cardiovascular technicians?**

c. **Does it appear that cardiovascular technicians exercise at least 90 minutes per week? Justify your answer.**

After completing Chapter 9, further investigate the Chapter Case Study in the Return to Chapter Case Study section with Exercises 9.123, 9.124, and 9.125 (p. 461).

9.1 Inferences About the Mean μ (σ Unknown)

Inferences about the population mean μ are based on the sample mean $\bar{x}$ and information obtained from the sampling distribution of sample means. Recall that the sampling distribution of sample means has a mean μ and a **standard error** of $\sigma/\sqrt{n}$ for all samples of size n, and it is normally distributed when the sampled population has a normal distribution or approximately normally distributed when the **sample size** is sufficiently large. This means the test statistic $z\bigstar = \frac{\bar{x} - \mu}{\sigma/\sqrt{n}}$ has a standard normal distribution. However, when **σ is unknown,** the standard error $\sigma/\sqrt{n}$ is also unknown. Therefore, the sample standard deviation s will be used as the point estimate for σ. As a result, an estimated standard error of the mean, $s/\sqrt{n}$, will be used and our test statistic will become $\frac{\bar{x} - \mu}{s/\sqrt{n}}$.

When a **known σ** is being used to make an inference about the mean μ, a sample provides one value for use in the formulas; that one value is $\bar{x}$. When the sample standard deviation s is also used, the sample provides two values: the sample mean $\bar{x}$ and the estimated standard error $s/\sqrt{n}$. As a result, the z-statistic will be replaced with a statistic that accounts for the use of an estimated standard error. This new statistic is known as the **Student's t-statistic.**

In 1908 W. S. Gosset, an Irish brewery employee, published a paper about this t-distribution under the pseudonym "Student." In deriving the t-distribution, Gosset assumed that the samples were taken from normal populations. Although this might seem to be restrictive, satisfactory results are obtained when large samples are selected from many nonnormal populations.

Figure 9.1 presents a diagrammatic organization for the inferences about the population mean as discussed in Chapter 8 and in this first section of Chapter 9. Two situations exist: σ is known, or σ is unknown. As stated before, σ is almost never a known quantity in real-world problems; therefore, the standard error will almost always be estimated by $s/\sqrt{n}$. The use of an estimated standard error of the mean requires the use of the t-distribution. Almost all real-world inferences about the population mean will be made with the Student's t-statistic.

The t-distribution has the following properties (see also Figure 9.2):

PROPERTIES OF THE t-DISTRIBUTION (df $>$ 2)*

1. t is distributed with a mean of zero.
2. t is distributed symmetrically about its mean.
3. t is distributed so as to form a family of distributions, a separate distribution for each different number of degrees of freedom (df $\geq$ 1).
4. The t-distribution approaches the **standard normal distribution** as the number of degrees of freedom increases.
5. t is distributed with a variance greater than 1, but as the degrees of freedom increases, the variance approaches 1.
6. t is distributed so as to be less peaked at the mean and thicker at the tails than is the normal distribution.

*Not all of the properties hold for df = 1 and df = 2. Since we will not encounter situations where df = 1 or 2, these special cases are not discussed further.

FYI
Explore Interactivity 9A on your CD.

FIGURE 9.1

Do I Use the *z*-Statistic or the *t*-Statistic?

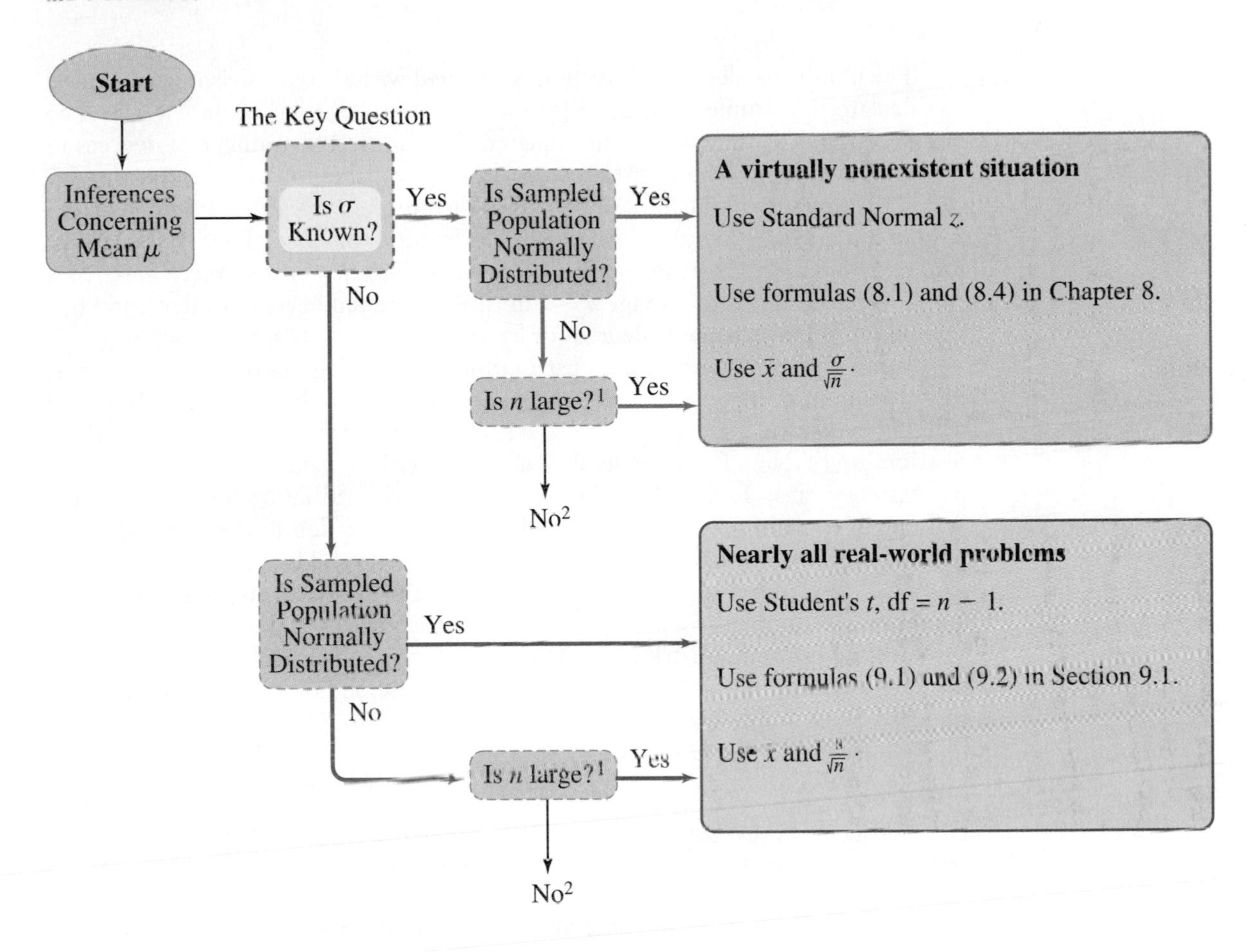

1. Is n large? Samples as small as $n = 15$ or 20 may be considered large enough for the central limit theorem to hold if the sample data are unimodal, nearly symmetric, short-tailed, and without outliers. Samples that are not symmetric require larger sample sizes, with 50 sufficing except for extremely skewed samples. See the discussion on page 349.

2. Requires the use of a nonparametric technique; see Chapter 14.

FIGURE 9.2

Student's *t*-Distributions

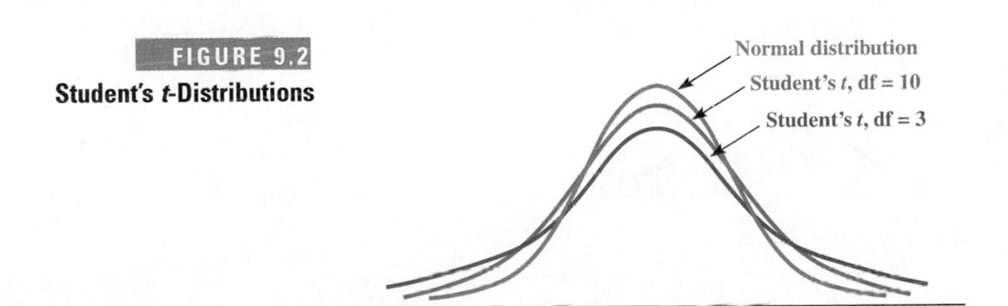

DEGREES OF FREEDOM, df

A **parameter** that identifies each different distribution of Student's *t*-distribution. For the methods presented in this chapter, the value of df will be the sample size minus 1: df $= n - 1$.

The number of degrees of freedom associated with s^2 is the divisor $(n - 1)$ used to calculate the sample variance s^2 [formula (2.6), p. 80]; that is, df $= n - 1$. The sample variance is the mean of the squared deviations. The number of degrees of freedom is the "number of unrelated deviations" available for use in estimating σ^2. Recall that the sum of the deviations, $\Sigma(x - \bar{x})$, must be zero. From a sample of size n, only the first $n - 1$ of these deviations has freedom of value. That is, the last, or nth, value of $(x - \bar{x})$ must make the sum of the n deviations total exactly zero. As a result, variance is said to average $n - 1$ unrelated squared deviation values, and this number, $n - 1$, was named "degrees of freedom."

Although there is a separate *t*-distribution for each degrees of freedom, df = 1, df = 2, . . . , df = 20, . . . , df = 40, and so on, only certain key **critical values of *t*** will be necessary for our work. Consequently, the table for the Student's *t*-distribution (Table 6 in Appendix B) is a table of critical values rather than a complete table, such as Table 3 is for the standard normal distribution for z. As you look at Table 6, you will note that the left side of the table is identified by "df," degrees of freedom. This left-hand column starts at 3 at the top and lists consecutive df values to 30, then jumps to 35, . . . , to "df > 100" at the bottom. As we stated, as the degrees of freedom increases, the *t*-distribution approaches the characteristics of the standard normal *z*-distribution. Once df is "greater than 100," the critical values of the *t*-distribution are the same as the corresponding critical values of the standard normal distribution as given in Table 4A in Appendix B.

ANSWER NOW 9.2

List four numbers that total zero. How many numbers were you able to pick without restriction? Explain how this demonstrates "degrees of freedom."

The critical values of the Student's *t*-distribution that are to be used both for constructing a confidence interval and for hypothesis testing will be obtained from Table 6 in Appendix B. To find the value of t, you will need to know two identifying values: (1) df, the number of degrees of freedom (identifying the distribution of interest), and (2) α, the area under the curve to the right of the right-hand critical value. A notation much like that used with z will be used to identify a critical value. $t(\text{df}, \alpha)$, read as "t of df, α," is the symbol for the value of t with df degrees of freedom and an area of α in the right-hand tail, as shown in Figure 9.3.

FIGURE 9.3
***t*-Distribution Showing $t(\text{df}, \alpha)$**

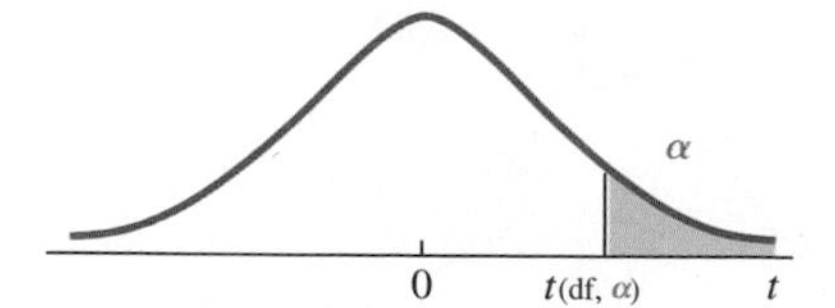

ANSWER NOW 9.3

Explain the relationship between the critical values found in the bottom row of Table 6 and the critical values of z given in Table 4A.

Illustration 9.1 *t* on the Right Side of the Mean

Find the value of $t(10, 0.05)$ (see the diagram).

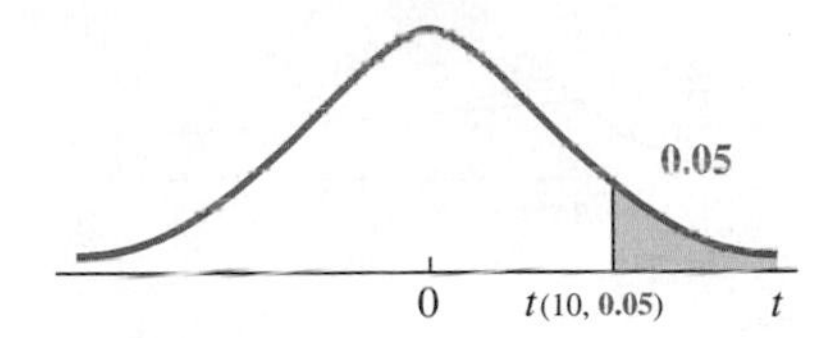

SOLUTION

There are 10 degrees of freedom, and 0.05 is the area to the right of the critical value. In Table 6 of Appendix B, we look for the row df = 10 and the column marked "Amount of α in One Tail," α = 0.05. At their intersection, we see that $t(10, 0.05) = \mathbf{1.81}$.

Portion of Table 6

	Amount of α in One Tail		
df	...	0.05	...
⋮			
10		1.81	⟶ $t(10, 0.05) = \mathbf{1.81}$

■

ANSWER NOW 9.4

Find the values: a. $t(12, 0.01)$ b. $t(22, 0.025)$

For the values of *t* on the left side of the mean, we can use one of two notations. The *t*-value shown in Figure 9.4 could be named $t(\text{df}, 0.95)$, since the area to the right of it is 0.95, or it could be identified by $-t(\text{df}, 0.05)$, since the *t*-distribution is symmetric about its mean, zero.

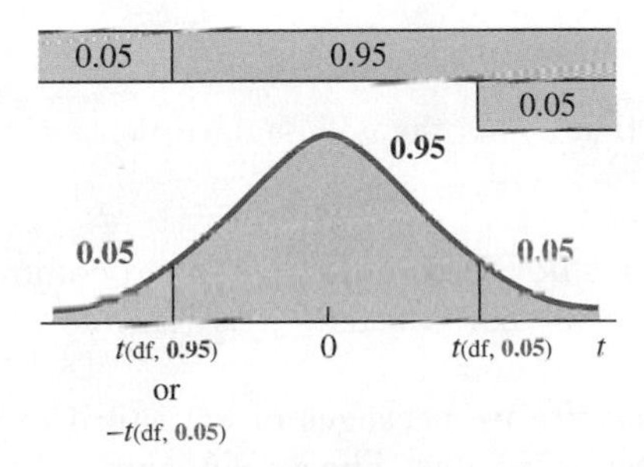

FIGURE 9.4
***t*-Value on Left Side**

Illustration 9.2 *t* on the Left Side of the Mean

Find the value of $t(15, 0.95)$.

SOLUTION

There are 15 degrees of freedom. In Table 6 we look for the column marked α = 0.05 (one tail) and its intersection with the row df = 15. The table gives us

$t(15, 0.05) = 1.75$; therefore, $t(15, 0.95) = -t(15, 0.05) = -1.75$. The value is negative because it is to the left of the mean; see the figure.

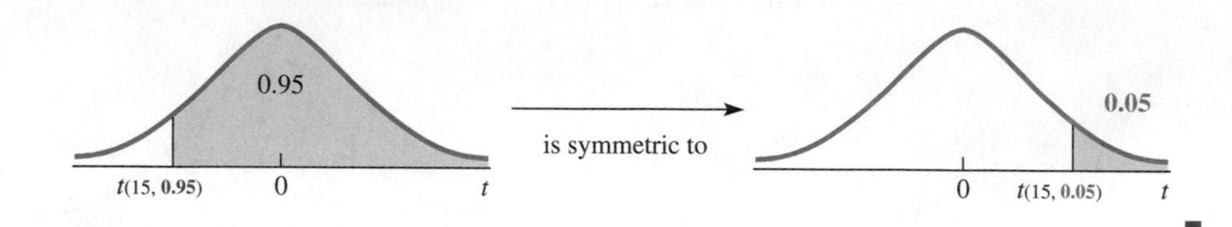

ANSWER NOW 9.5

Find the values: a. $t(18, 0.90)$ b. $t(9, 0.99)$

Illustration 9.3 *t*-Values that Bound a Middle Percentage

Find the values of the *t*-distribution that bound the middle 0.90 of the area under the curve for the distribution with df = 17.

SOLUTION

The middle 0.90 leaves 0.05 for the area of each tail. The value of *t* that bounds the right-hand tail is $t(17, 0.05) =$ **1.74,** as found in Table 6. The value that bounds the left-hand tail is **−1.74** because the *t*-distribution is symmetric about its mean, zero.

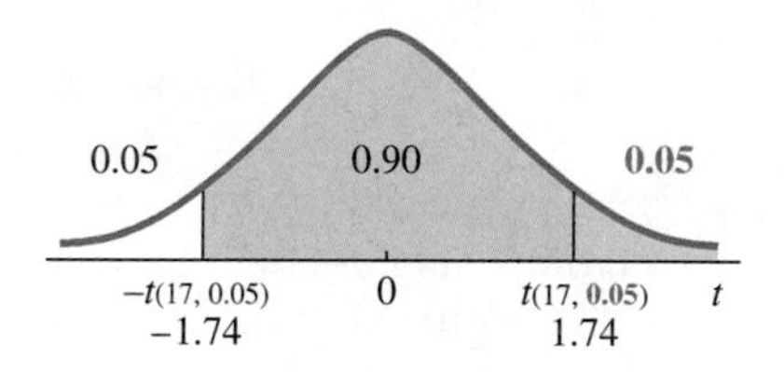

ANSWER NOW 9.6

Find the values of *t* that bound the middle 0.95 of the distribution for df = 12.

If the df needed is not listed in the left-hand column of Table 6, then use the next smaller value of df that is listed. For example, $t(72, 0.05)$ is estimated using $t(70, 0.05) = 1.67$.

Most computer software packages or statistical calculators will calculate the area related to a specified *t*-value. The accompanying figure shows the relationship between the cumulative probability distribution and a specific *t*-value for a *t*-distribution with df degrees of freedom.

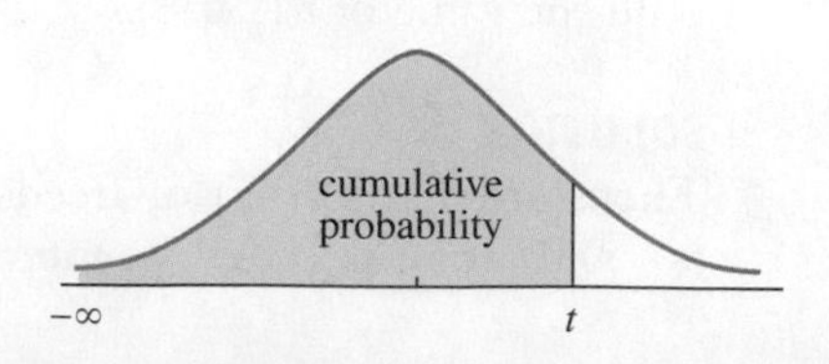

Technology Instructions: Probability Associated With a Specified Value of *t*

MINITAB (Release 13)

Cumulative probability for a specified value of *t*:

```
Choose:   Calc > Probability Distribution > t
Select:   Cumulative Probability
          Noncentrality parameter: 0.0
Enter:    Degrees of freedom: df
Select:   Input constant*
Enter:    t-value (ex. 1.74) > OK
```

*Select Input column if several *t*-values are stored in C1. Use C2 for optional storage. If the area in the right tail is needed, subtract the calculated probability from 1.

Excel XP

Probability in one or two tails for a given *t*-value:

If several *t*-values (nonnegative) are to be used, input the values into column A and activate B1; then continue with:

```
Choose:   Insert function fx > Statistical > TDIST > OK
Enter:    X: individual t-value or (A1:A5 or select
          't-value' cells)*
          Deg_freedom: df
          Tails: 1 or 2 (one or two-tailed distributions) > OK
Drag*:    Bottom right corner of the B1 cell down to give
          other probabilities
```

To find the probability within the two tails or the cumulative probability for one tail, subtract the calculated probability from 1.

TI-83 Plus

Cumulative probability for a specified value of *t*:

```
Choose:   2nd > DISTR > 5:tcdf(
Enter:    -1EE99, t-value, df)
```

NOTE: To find the probability between two *t*-values, enter the two values in place of −1EE99 and *t*-value.

If the area in the right tail is needed, subtract the calculated probability from 1.

ANSWER NOW 9.7

Use a computer or calculator to find the area to the left of $t = -2.12$ with df = 18. Draw a sketch showing the question with the answer.

ANSWER NOW 9.8

Use a computer or calculator to find the area to the right of $t = 1.12$ with df = 15. Draw a sketch showing the question with the answer.

We are now ready to make inferences about the population mean μ using the sample standard deviation. As we mentioned earlier, use of the *t*-distribution has a condition.

THE ASSUMPTION FOR INFERENCES ABOUT THE MEAN μ WHEN σ IS UNKNOWN

The sampled population is normally distributed.

CONFIDENCE INTERVAL PROCEDURE

The procedure to make confidence intervals using the sample standard deviation is very similar to that used when σ is known (see pp. 348–353). The difference is the use of the Student's t in place of the standard normal z and the use of s, the sample standard deviation, as an estimate of σ. The central limit theorem implies that this technique can also be applied to nonnormal populations when the sample size is sufficiently large.

The formula for the $1 - \alpha$ confidence interval for μ is:

$$\bar{x} - t(\text{df}, \alpha/2)\left(\frac{s}{\sqrt{n}}\right) \quad \text{to} \quad \bar{x} + t(\text{df}, \alpha/2)\left(\frac{s}{\sqrt{n}}\right), \quad \text{with df} = n - 1 \qquad (9.1)$$

Illustration 9.4 Confidence Interval for μ With σ Unknown

A random sample of 20 weights is taken from babies born at Northside Hospital. A mean of 6.87 lb and a standard deviation of 1.76 lb were found for the sample. Estimate, with 95% confidence, the mean weight of all babies born in this hospital. Based on past information, it is assumed that weights of newborns are normally distributed.

SOLUTION

FYI
The five-step confidence interval procedure is given on page 350.

Step 1 **The Set-Up:**
Describe the population parameter of interest.
μ, the mean weight of newborns at Northside Hospital.

Step 2 **The Confidence Interval Criteria:**

a. **Check the assumptions.**
σ is unknown, and past information indicates that the sampled population is normal.

b. **Identify the probability distribution and the formula to be used.**
The Student's t-distribution will be used with formula (9.1).

c. **State the level of confidence:** $1 - \alpha = 0.95$.

Step 3 **The Sample Evidence:**
Collect the sample information: $n = 20$, $\bar{x} = 6.87$, and $s = 1.76$.

FYI
Recall that confidence intervals are two-tailed situations.

Step 4 **The Confidence Interval:**

a. **Determine the confidence coefficients.**
Since $1 - \alpha = 0.95$, $\alpha = 0.05$, and therefore $\alpha/2 = 0.025$. Also, since $n = 20$, df = 19. At the intersection of row df = 19 and one-tailed column $\alpha = 0.025$ in Table 6, we find $t(\text{df}, \alpha/2) = t(19, 0.025) = 2.09$. See the figure.

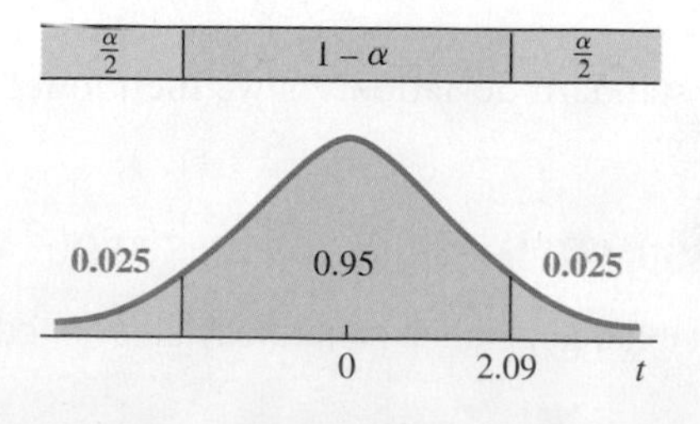

FYI
df is used to find the confidence coefficient in Table 6; n is used in the formula.

Information about the confidence coefficient and using Table 6 is on pages 414–416.

b. **Find the maximum error of estimate.**

$$E = t(\text{df}, \alpha/2)\left(\frac{s}{\sqrt{n}}\right): \ E = t(19, 0.025)\left(\frac{s}{\sqrt{n}}\right)$$

$$= 2.09\left(\frac{1.76}{\sqrt{20}}\right) = (2.09)(0.394) = 0.82$$

c. **Find the lower and upper confidence limits.**

$$\bar{x} - E \quad \text{to} \quad \bar{x} + E$$

$$6.87 - 0.82 \quad \text{to} \quad 6.87 + 0.82$$

$$6.05 \quad \text{to} \quad 7.69$$

Step 5 **The Results:**
State the confidence interval.
6.05 to 7.69, the 95% confidence interval for μ. That is, with 95% confidence we estimate the mean weight to be between 6.05 and 7.69 lb. ■

ANSWER NOW 9.9

Construct a 95% confidence interval estimate for the mean μ using the sample information $n = 24$, $\bar{x} = 16.7$, and $s = 2.6$.

Technology Instructions: $1 - \alpha$ Confidence Interval for Mean μ with σ Unknown

Minitab (Release 13)

Input the data into C1; then continue with:

```
Choose:   Stat > Basic Statistics > 1-Sample t
Enter:    Variables: C1
Select:   Options
Enter:    Confidence level: 1 − α   (ex. 95.0)
Choose:   Alternative: not equal
```

Excel XP

Input the data into column A; then continue with:

```
Choose:   Tools > Data Analysis Plus > t-Estimate: Mean > OK
Enter:    Input Range: (A1:A20 or select cells)
Enter:    Alpha: α (ex. 0.05) > OK
```

TI-83 Plus

Input the data into L1; then continue with the following, entering the appropriate values and highlighting Calculate:

```
Choose:   STAT > TESTS >
          8:TInterval
```

```
TInterval
 Inpt:Data Stats
 List:L1
 Freq:1
 C-Level:.95
 Calculate
```

FYI
Compare the MINITAB output to the solution of Illustration 9.4.

The MINITAB solution to Illustration 9.4 looks like this:

```
Confidence Interval
Variable   N    Mean   StDev   SE Mean       95% CI
C1        20   6.870   1.760    0.394    (6.047, 7.693)
```

ANSWER NOW 9.10

Use a computer or calculator to construct a 0.98 confidence interval using these sample data:

6 7 12 9 10 8 5 9 7 9 6 5

HYPOTHESIS-TESTING PROCEDURE

The t-statistic is used to complete a hypothesis test about the population mean μ in much the same manner z was used in Chapter 8. In hypothesis-testing situations, we use formula (9.2) to calculate the value of the **test statistic $t\bigstar$**:

$$t\bigstar = \frac{\bar{x} - \mu}{s/\sqrt{n}} \quad \text{with df} = n - 1 \tag{9.2}$$

The **calculated t** is the number of estimated standard errors $\bar{x}$ is from the hypothesized mean μ. As with confidence intervals, the central limit theorem indicates that the t-distribution can also be applied to nonnormal populations when the **sample size** is sufficiently large.

ILLUSTRATION 9.5 One-Tailed Hypothesis Test for μ with σ Unknown

Let's return to the hypothesis of Illustration 8.10 (p. 372) where the EPA wanted to show that the mean carbon monoxide level is higher than 4.9 parts per million. Does a random sample of 22 readings (sample results: $\bar{x} = 5.1$ and $s = 1.17$) present sufficient evidence to support the EPA's claim? Use $\alpha = 0.05$. Previous studies have indicated that such readings have an approximately normal distribution.

FYI
The five-step p-value hypothesis test procedure is given on page 370.

SOLUTION

Step 1 **The Set-Up:**

a. **Describe the population parameter of interest.**
μ, the mean carbon monoxide level of air in downtown Rochester.

b. **State the null hypothesis (H_o) and the alternative hypothesis (H_a).**

FYI
Procedures for writing H_o and H_a are discussed on pages 371–374.

H_o: $\mu = 4.9$ ($\leq$) (no higher than)

H_a: $\mu > 4.9$ (higher than)

Step 2 **The Hypothesis Test Criteria:**

a. **Check the assumptions.**
The assumptions are satisfied because the sampled population is approximately normal and the sample size is large enough for the CLT to apply (see p. 412); σ is unknown.

b. Identify the probability distribution and the test statistic to be used.
The t-distribution with df $= n - 1 = 21$, and the test statistic is $t\bigstar$, formula (9.2).

c. Determine the level of significance: $\alpha = 0.05$.

Step 3 **The Sample Evidence:**

a. Collect the sample information: $n = 22$, $\bar{x} = 5.1$, and $s = 1.17$.

b. Calculate the value of the test statistic.
Use formula (9.2):

$$t\bigstar = \frac{\bar{x} - \mu}{s/\sqrt{n}}: \quad t\bigstar = \frac{5.1 - 4.9}{1.17/\sqrt{22}} = \frac{0.20}{0.2494} = 0.8018 = 0.80$$ ■

ANSWER NOW 9.11

Calculate the value of $t\bigstar$ for the hypothesis test: H_o: $\mu = 32$ vs. H_a: $\mu > 32$, $n = 16$, $\bar{x} = 32.93$, $s = 3.1$.

Step 4 **The Probability Distribution:**

Using the p-value procedure:

a. Calculate the p-value for the test statistic.
Use the right-hand tail because H_a expresses concern for values related to "higher than."
$\mathbf{P} = P(t\bigstar > 0.80$, with df $= 21)$ as shown on the figure.

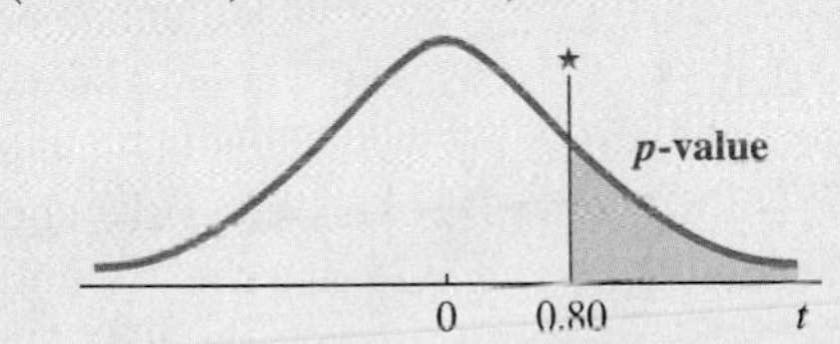

To find the p-value, use one of three methods:

1. Use Table 6 in Appendix B to place bounds on the p-value: $0.10 < \mathbf{P} < 0.25$.
2. Use Table 7 in Appendix B to read the value directly: $\mathbf{P} = 0.216$.
3. Use a computer or calculator to calculate the p-value: $\mathbf{P} = 0.2163$.

Specific details follow this illustration.

b. Determine whether or not the p-value is smaller than α.
The p-value is not smaller than α, the level of significance.

OR

Using the classical procedure:

a. Determine the critical region and critical value(s).
The critical region is the right-hand tail because H_a expresses concern for values related to "higher than."
The critical value is found at the intersection of the df $= 21$ row and the one-tailed 0.05 column of Table 6: $t(21, 0.05) = 1.72$.

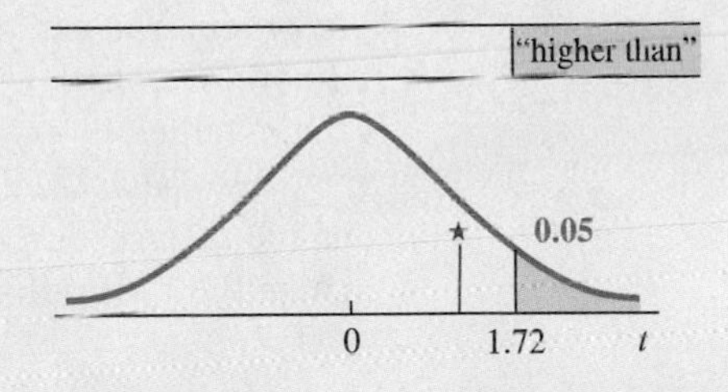

Specific instructions are given on pages 414–415.

b. Determine whether or not the calculated test statistic is in the critical region.
$t\bigstar$ is not in the critical region, as shown in **red** on the figure above.

Step 5 **The Results:**

a. State the decision about H_o: Fail to reject H_o.

b. State the conclusion about H_a.
At the 0.05 level of significance, the EPA does not have sufficient evidence to show that the mean carbon monoxide level is higher than 4.9. ■

ANSWER NOW 9.12

Find the value of P and state the decision for the hypothesis test in Answer Now 9.11 using $\alpha = 0.05$.

ANSWER NOW 9.13

Find the critical region and the critical value and state the decision for the hypothesis test in Answer Now 9.11 using $\alpha = 0.05$.

Calculating the *p*-value when using the *t*-distribution

Method 1: Use Table 6 in Appendix B to place bounds on the p-value. By inspecting the df = 21 row of Table 6, you can determine an interval within which the *p*-value lies. Locate $t\bigstar$ along the row labeled df = 21. If $t\bigstar$ is not listed, locate the two table values it falls between, and read the bounds for the *p*-value from the top of the table. In this case, $t\bigstar = 0.80$ is between 0.686 and 1.32; therefore, **P** is between 0.10 and 0.25. Use the one-tailed heading, since H_a is one-tailed in this illustration. (Use the two-tailed heading when H_a is two-tailed.)

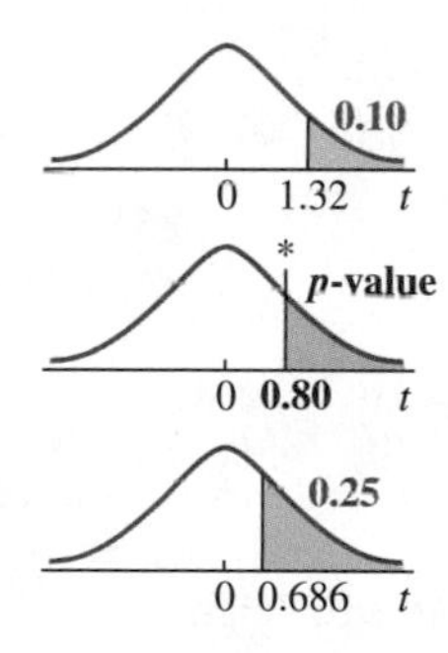

	Portion of Table 6		
	Amount of α in one-tail		
df	0.25	P	0.10
⋮	⋮	↑	⋮
21	0.686	0.80	1.32

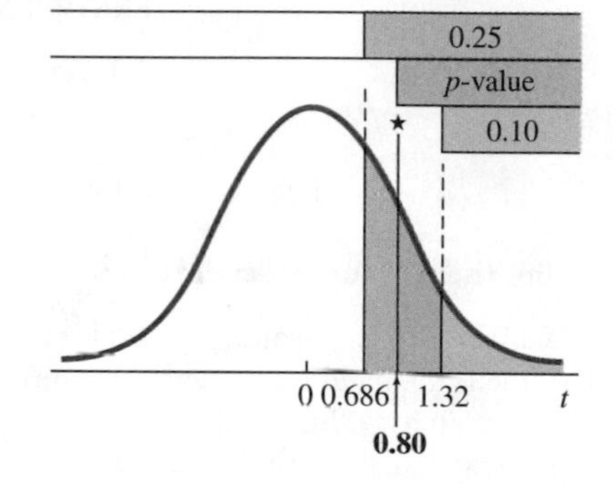

The 0.686 entry in the table tells us that $P(t > 0.686) = 0.25$, as shown on the figure in blue. The 1.32 entry in the table tells us that $P(t > 1.32) = 0.10$, as shown in green. You can see that the *p*-value **P** (shown in red) is between 0.10 and 0.25. Therefore, $0.10 < \mathbf{P} < 0.25$, and we say that 0.10 and 0.25 are the "bounds" for the *p*-value.

Method 2: Use Table 7 in Appendix B to read the p-value directly. Locate the *p*-value at the intersection of the $t\bigstar = 0.80$ row and the df = 21 column. The *p*-value for $t\bigstar = 0.80$ with df = 21 is **0.216**.

Portion of Table 7

t★	df	...	21
⋮			
0.80			0.216

→ **P** = $P(t^* > 0.80$, with df $= 21) =$ **0.216**

Method 3: If you are doing the hypothesis test with the aid of a computer or calculator, most likely it will calculate the *p*-value for you, or you may use the cumulative probability distribution commands described on page 417.

Let's look at a two-tailed hypothesis-testing situation.

Illustration 9.6 **Two-Tailed Hypothesis Test for μ with σ Unknown**

On a popular self-image test that results in normally distributed scores, the mean score for public-assistance recipients is expected to be 65. A random sample of 28 public-assistance recipients in Emerson County is given the test. They achieve a mean score of 62.1, and their scores have a standard deviation of 5.83. Do the

Emerson County public-assistance recipients test differently, on the average, than what is expected, at the 0.02 level of significance?

SOLUTION

Step 1 **The Set-Up:**

a. **Describe the population parameter of interest.**

μ, the mean self-image test score for all Emerson County public-assistance recipients.

b. **State the null hypothesis (H_o) and the alternative hypothesis (H_a).**

H_o: $\mu = 65$ (mean is 65)

H_a: $\mu \neq 65$ (mean is different from 65)

Step 2 **The Hypothesis Test Criteria:**

a. **Check the assumptions.**

The test is expected to produce normally distributed scores; therefore, the assumption has been satisfied; σ is unknown.

b. **Identify the probability distribution and the test statistic to be used.**

The t-distribution with df $= n - 1 = 27$, and the test statistic is $t\bigstar$, formula (9.2).

c. **Determine the level of significance:** $\alpha = 0.02$ (given in statement of problem).

Step 3 **The Sample Evidence:**

a. **Collect the sample information:** $n = 28$, $\bar{x} = 62.1$, and $s = 5.83$.

b. **Calculate the value of the test statistic.**

Use formula (9.2):

$$t\bigstar = \frac{\bar{x} - \mu}{s/\sqrt{n}}: \qquad t\bigstar = \frac{62.1 - 65.0}{5.83/\sqrt{28}} = \frac{-2.9}{1.1018} = -2.632 = -2.63$$

■

ANSWER NOW 9.14

Calculate the value of $t\bigstar$ for this hypothesis test: H_o: $\mu = 73$ vs. H_a: $\mu \neq 73$, $\alpha = 0.05$, $n = 12$, $\bar{x} = 71.46$, $s = 4.1$.

Step 4 **The Probability Distribution:**

Using the p-value procedure:

a. **Calculate the p-value for the test statistic.**

Use both tails because H_a expresses concern for values related to "different from."

$\mathbf{P} = P(t < -2.63) + P(t > 2.63) = 2 \cdot P(t > 2.63)$, with df $= 27$ as shown on the figure.

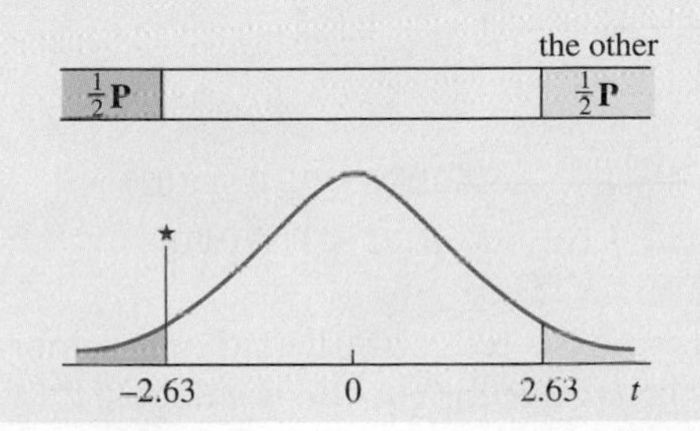

OR

Using the classical procedure:

a. **Determine the critical region and critical value(s).**

The critical region is both tails because H_a expresses concern for values related to "different from." The critical value is found at the intersection of the df = 27 row and the one-tailed 0.01 column of Table 6: $t(27, 0.01) = 2.47$.

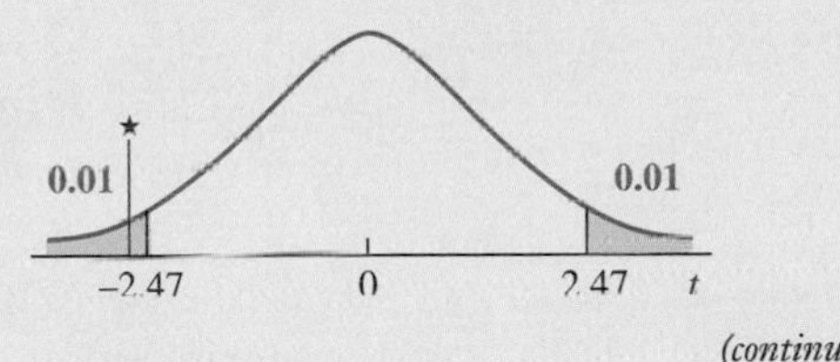

(continued)

To find the p-value, use one of three methods:

1. Use Table 6 in Appendix B to place bounds on the p-value: $0.01 < \mathbf{P} < 0.02$.
2. Use Table 7 in Appendix B to place bounds on the p-value: $0.012 < \mathbf{P} < 0.016$.
3. Use a computer or calculator to calculate the p-value: $\mathbf{P} = 0.0140$.

Specific details follow this illustration.

b. Determine whether or not the p-value is smaller than α.

The p-value is smaller than the level of significance, α.

b. Determine whether or not the calculated test statistic is in the critical region.

$t\bigstar$ is in the critical region, as shown in **red** on the figure above.

Step 5 The Results:

a. State the decision about H_o: Reject H_o.

b. State the conclusion about H_a.

At the 0.02 level of significance, we do have sufficient evidence to conclude that the Emerson County assistance recipients test significantly different, on the average, from the expected 65. ■

ANSWER NOW 9.15

Use Table 6 or Table 7 to find the value of P for the hypothesis test in Answer Now 9.14; state the decision using $\alpha = 0.05$.

ANSWER NOW 9.16

Find the critical region and critical values for the hypothesis test in Answer Now 9.14; state the decision using $\alpha = 0.05$.

Calculating the p-value when using the t-distribution

Method 1: Using Table 6, find 2.63 between two entries in the df = 27 row and read the bounds for **P** from the two-tailed heading at the top of the table:

$$\mathbf{0.01 < P < 0.02.}$$

Method 2: Generally, bounds found using Table 7 will be narrower than bounds found using Table 6. The table below shows you how to read the bounds from Table 7; find $t\bigstar = 2.63$ between two rows and df = 27 between two columns, and locate the four intersections of these columns and rows. The value of $^1/_2\mathbf{P}$ is bounded by the upper left and the lower right of these table entries.

Portion of Table 7

	Degrees of Freedom		
$t\bigstar$	25	**27**	29
⋮			⋮
2.6	0.008	↓	0.007
2.63	→	$^1/_2$ P	→ $0.006 < {}^1/_2\,\mathrm{P} < 0.008$
2.7	0.006		0.006

$\mathbf{0.012 < P < 0.016}$

Method 3: If you are doing the hypothesis test with the aid of a computer or calculator, most likely it will calculate the p-value for you (do not double it). Or you may use the cumulative probability distribution commands described on page 417.

ANSWER NOW 9.17

Use a computer or calculator to find the p-value for Answer Now 9.14.

Technology Instructions: Hypothesis Test for Mean μ When σ Unknown

MINITAB (Release 13)

Input the data into C1; then continue with:

Choose:	**Stat > Basic Statistics > 1-Sample t**
Enter:	Variables: **C1**
	Test mean: μ (value in H_o)
Select:	**Options**
Choose:	Alternative: **less than** or **not equal** or **greater than**

Excel XP

Input the data into column A; then continue with:

Choose:	**Tools > Data Analysis Plus > t-Test: Mean > OK**
Enter:	Input Range: **(A1:A20 or select cells) > OK**
	Hypothesized Mean: μ
	Alpha: α (ex. 0.05) > OK
	Gives p-values and critical values for both one-tailed and two-tailed tests.

TI-83 Plus

Input the data into L1; then continue with the following, entering the appropriate values and highlighting Calculate:

Choose: **STAT > TESTS > 2:T-Test**

```
T-Test
 Inpt:Data Stats
 μ0:0
 List:L1
 Freq:1
 μ:≠μ0 <μ0 >μ0
 Calculate Draw
```

FYI
Compare the MINITAB results to the solution found in Illustration 9.6.

Here is the MINITAB solution to Illustration 9.6:

```
T-Test of the Mean
Test of mu = 65 vs mu ≠ 65
Variable  N    Mean   StDev   SE Mean     T      P-Value
C1        28   62.1   5.83    1.102     -2.63    0.0140
```

ANSWER NOW 9.18 Ex09-018

Use a computer or calculator to complete the hypothesis test H_o: $\mu = 52$ vs. H_a: $\mu < 52$, $\alpha = 0.01$ using these data:

45 47 46 58 59 49 46 54 53 52 47 41.

Application 9.1 Mothers' Use of Personal Pronouns When Talking With Toddlers

The calculated t-value and the probability value for five different hypothesis tests are given in the following article. The expression $t(44) = 1.92$ means $t\bigstar = 1.92$ with df = 44 and is significant with p-value < 0.05. Can you verify the p-values? Explain.

Abstract. The verbal interaction of 2-year-old children ($N = 46$; 16 girls, 30 boys) and their mothers was audiotaped, transcribed, and analyzed for the use of personal pronouns, the total number of utterances, the child's mean length of utterance, and the mother's responsiveness to her child's utterances. Mothers' use of the personal pronoun "we" was significantly related to their children's performance on the Stanford-Binet at age 5 and the Wechsler Intelligence Scale for Children at age 8. Mothers' use of "we" in social-vocal interchange, indicating a system for establishing a shared relationship with the child, was closely connected with their verbal responsiveness to their children. The total amount of maternal talking, the number of personal pronouns used by mothers, and their verbal responsiveness to their children were not related to mothers' social class or years of education.

Mothers tended to use more first person singular pronouns (I and me), $t(44) = 1.81$, $p < .10$, and used significantly more first person plural pronouns (we), $t(44) = 1.92$, $p < .05$, with female children than with male children. The mothers also were more verbally responsive to their female children, $t(44) = 2.0$, $p < .06$.

In general, mothers talked more to their first born children, $t(44) = 3.41$, $p < .001$, and were more responsive to their first born children, $t(44) = 3.71$, $p < .001$. Yet, the proportion of personal pronouns used when speaking to first born children was not different from that used when speaking to later born children.

Source: Dan R. Laks, Leila Beckwith, and Sarale E. Cohen, *The Journal of Genetic Psychology,* 151(1), 25–32, 1990. Reprinted with permission of the Helen Dwight Reid Educational Foundation. Published by Heldref Publications, 1319 Eighteenth St. N.W., Washington, D.C., 20036-1802.

ANSWER NOW 9.19

a. **Verify that $t(44) = 1.92$ is significant at the 0.05 level.**

b. **Verify that $t(44) = 3.41$ is significant at the 0.01 level.**

c. **Explain why $t(44) = 1.81$, $p < 0.10$, makes sense only if the hypothesis test is two-tailed. If the test is one-tailed, what level would be reported?**

Exercises

9.20 Find these critical values using Table 6 in Appendix B:
a. $t(25, 0.05)$ b. $t(10, 0.10)$
c. $t(15, 0.01)$ d. $t(21, 0.025)$

9.21 Find these critical values using Table 6 in Appendix B:
a. $t(21, 0.95)$ b. $t(26, 0.975)$
c. $t(27, 0.99)$ d. $t(60, 0.025)$

9.22 Using the notation of Exercise 9.20, name and find the following critical values of t:

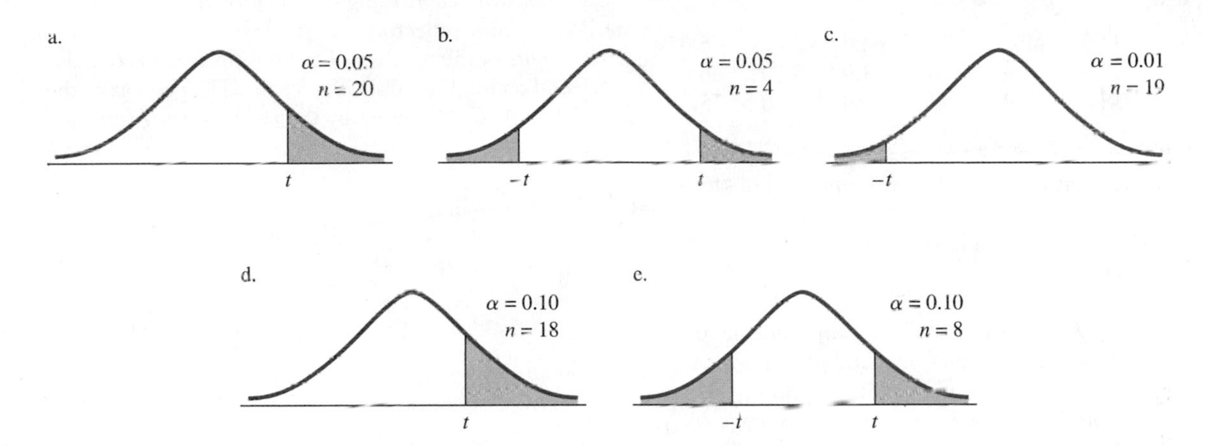

9.23 Ninety percent of the Student's t-distribution lies between $t = -1.89$ and $t = 1.89$ for how many degrees of freedom?

9.24 Ninety percent of the Student's t-distribution lies to the right of $t = -1.37$ for how many degrees of freedom?

9.25

a. Find the first percentile of the Student's t-distribution with 24 degrees of freedom.
b. Find the 95th percentile of the Student's t-distribution with 24 degrees of freedom.
c. Find the first quartile of the Student's t-distribution with 24 degrees of freedom.

9.26 Find the percentage of the Student's t-distribution that lies between the following values:

a. df = 12 and t ranges from -1.36 to 2.68
b. df = 15 and t ranges from -1.75 to 2.95

9.27

a. State two ways in which the standard normal distribution and the Student's t-distribution are alike.
b. State two ways in which they are different.

9.28 The variance for each of the Student's t-distributions is equal to df/(df − 2). Find the standard deviation for the Student's t-distribution with each of the following degrees of freedom:

a. 10 b. 20 c. 30
d. Explain how this verifies property 5 of the t-distributions listed on page 412.

9.29 The data from a study reported in "White-Collar Crime and Criminal Careers: Some Preliminary Findings" (*Crime and Delinquency*, July 1990) indicate that white-collar criminals are likely to be older and to show a lower frequency of offending than are street criminals. For example, the mean age of onset of offending for those convicted of antitrust offenses was 54 years with $n = 35$. If the standard deviation is estimated to be 7.5 years, find the 90% confidence interval for the true mean age.

9.30 The Robertson square drive screw was invented in 1908, but it has gained popularity with American woodworkers and home craftsmen only within the last ten years. The advantages of square drives over conventional screws is indeed remarkable—most notably, greater strength, increased holding power, and reduced driving resistance and "cam-out." Strength test results published in McFeely's 1998 catalog (Catalog 98D, Fall 1998) revealed that #8 square drive screws fail only after an average of 48 inch-pounds of torque is applied, nearly 50% more strength than the more common slotted- or Phillips-head wood screw. Suppose an independent testing laboratory selects a random sample of 22 square drive screws from a box of 1000 screws, and obtains a mean failure torque of 48.2 inch-pounds and a standard deviation of 5.1 inch-pounds. Estimate with 95% confidence the mean failure torque of the #8 wood screws based on the study by the independent laboratory. Specify the population parameter of interest, the criteria, the sample evidence, and the interval limits.

9.31 While doing an article on the high cost of college education, a reporter took a random sample of the cost of new textbooks for a semester. The random variable x is the cost of one book. Her sample data can be summarized by $n = 41$, $\sum x = 1962.26$, and $\sum (x - \bar{x})^2 = 8202.496$.

a. Find the sample mean, $\bar{x}$.
b. Find the sample standard deviation, s.
c. Find the 90% confidence interval to estimate the true mean textbook cost for the semester based on this sample.

9.32 Ex09-032 Twenty-four oat-producing counties were randomly identified from across the United States for the purpose of obtaining data to estimate the 2001 national mean county yield. For each county identified, the 2001 oat crop yield, in bushels per acre, was obtained. Here are the resulting data:

37.0 30.0 42.9 60.0 37.0 64.0 68.0 76.0
65.7 73.0 50.0 64.4 71.7 43.0 57.0 66.0
69.0 77.9 38.0 72.9 56.0 95.0 38.0 75.7

Source: http://www.usda.gov/nass/graphics/county01/data/ot01.csv

a. Show evidence that the normality assumptions are satisfied.
b. Find the 95% confidence interval for "mean oat yield, bushels per acre."

9.33 Ex09-033 The National Adoption Information Clearinghouse, at naic@calib.com, tracks and posts information about child adoptions in the United States. Twenty U.S. states were randomly identified and the percentage of change in the number of adoptions per year from 1993 to 1997 was recorded:

6 8 −17 18 8 11 2 13 14 22
−5 −11 0 −20 −23 12 −1 32 5 5

Source: http://www.calib.com/naic/pubs/s_flang3.htm

Find the 95% confidence interval for the true mean percentage change in the number of adoptions from 1993 to 1997.

9.34 Ex02-167 The addition of a new accelerator is claimed to decrease the drying time of latex paint by more than 4%. Eight test samples were conducted, with the following percentage decreases in drying time:

5.2 6.4 3.8 6.3 4.1 2.8 3.2 4.7

We assume that the percentage decreases in drying time are normally distributed.

a. Find the 95% confidence interval for the true mean decrease in drying time based on this sample. (The sample mean and standard deviation were found in Exercise 2.167.)
b. Did the interval estimate in part a result in the same conclusion as you expressed in answering part c of Exercise 2.167 for these same data?

9.35 Ex09-035 The pulse rates for 13 adult women were

83 58 70 56 76 64 80
76 70 97 68 78 108

Verify the results shown on the last line of this MINITAB output:

```
MTB > TINTERVAL 90 PERCENT CONFIDENCE INTERVAL
FOR DATA IN C1
      N   MEAN    STDEV   SE MEAN       90% CI
C1   13  75.69   14.54     4.03    (68.50, 82.88)
```

9.36 Ex09-036 James Short (1708–1768), a Scottish optician, constructed the highest quality reflectors of his time. With these reflectors Short obtained the following measurements of the parallax of the sun (in seconds of a degree), based on the 1761 transit of Venus. The parallax of the sun is the angle α subtended by the earth, as seen from the surface of the sun. (See the diagram.)

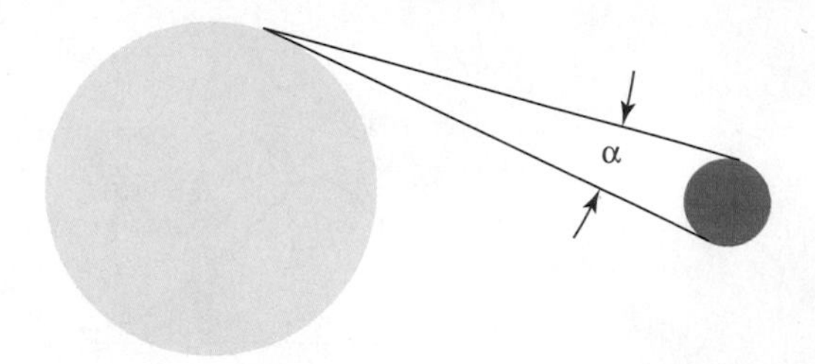

8.50 8.50 7.33 8.64 9.27 9.06 9.25 9.09 8.50
8.06 8.43 8.44 8.14 7.68 10.34 8.07 8.36 9.71
8.65 8.35 8.71 8.31 8.36 8.58 7.80 7.71 8.30
9.71 8.50 8.28 9.87 8.86 5.76 8.44 8.23 8.50
8.80 8.40 8.82 9.02 10.57 9.11 8.66 8.34 8.60
7.99 8.58 8.34 9.64 8.34 8.55 9.54 9.07

Source: The data and descriptive information are based on material from Stephen M. Stigler, "Do robust estimators work with real data?" *Annals of Statistics* 5 (1977), 1055–1098.

a. Determine whether an assumption of normality is reasonable. Explain.
b. Construct a 95% confidence interval for the estimate of the mean parallax of the sun.
c. If the true value is 8.798 seconds of a degree, what does the confidence interval suggest about Short's measurements?

9.37 State the null hypothesis, H_o, and the alternative hypothesis, H_a, that would be used to test each of these claims:

a. The mean weight of honeybees is at least 11 g.
b. The mean age of patients at Memorial Hospital is no more than 54 years.
c. The mean amount of salt in granola snack bars is different from 75 mg.

9.38 Determine the p-value for the following hypothesis tests involving the Student's t-distribution with 10 degrees of freedom:

a. H_o: $\mu = 15.5$ vs. H_a: $\mu < 15.5$, $t\star = -2.01$
b. H_o: $\mu = 15.5$ vs. H_a: $\mu > 15.5$, $t\star = 2.01$
c. H_o: $\mu = 15.5$ vs. H_a: $\mu \neq 15.5$, $t\star = 2.01$
d. H_o: $\mu = 15.5$ vs. H_a: $\mu \neq 15.5$, $t\star = -2.01$

9.39 Determine the critical region and critical value(s) that would be used in the classical approach to test these null hypotheses:

a. H_o: $\mu = 10$ vs. H_a: $\mu \neq 10$, $\alpha = 0.05$, $n = 15$
b. H_o: $\mu = 37.2$ vs. H_a: $\mu > 37.2$, $\alpha = 0.01$, $n = 25$
c. H_o: $\mu = -20.5$ vs. H_a: $\mu < -20.5$, $\alpha = 0.05$, $n = 18$
d. H_o: $\mu = 32.0$ vs. H_a: $\mu > 32.0$, $\alpha = 0.01$, $n = 42$

9.40 Compare the p-value and decision of the p-value approach to the critical values and decision of the classical approach for each of the following situations. Use $\alpha = 0.05$.
a. H_o: $\mu = 128$ vs. H_a: $\mu \neq 128$, $n = 15$, $t\bigstar = 1.60$
b. H_o: $\mu = 18$ vs. H_a: $\mu > 18$, $n = 25$, $t\bigstar = 2.16$
c. H_o: $\mu = 38$ vs. H_a: $\mu < 38$, $n = 45$, $t\bigstar = -1.73$
d. Compare the results of the two techniques for each case.

9.41 A student group maintains that the average student must travel for at least 25 minutes to reach college each day. The college admissions office obtained a random sample of 22 one-way travel times from students. The sample had a mean of 19.4 min and a standard deviation of 9.6 min. Does the admissions office have sufficient evidence to reject the students' claim? Use $\alpha = 0.01$.
a. Solve using the p-value approach.
b. Solve using the classical approach.

9.42 Homes in a nearby college town have a mean value of $88,950. It is assumed that homes in the vicinity of the college have a higher value. To test this theory, a random sample of 12 homes is chosen from the college area. Their mean valuation is $92,460 and the standard deviation is $5,200. Complete a hypothesis test using $\alpha = 0.05$. Assume home prices are normally distributed.
a. Solve using the p-value approach.
b. Solve using the classical approach.

9.43 An article in the *American Journal of Public Health* (March 1994) describes a study involving 20,143 individuals. The article states that the mean percentage intake of kilocalories from fat was 38.4% with a range from 6% to 71.6%. A small-sample study was conducted at a university hospital to determine whether the mean intake at that hospital was different from 38.4%. A sample of 15 patients had a mean intake of 40.5% with a standard deviation equal to 7.5%. Assuming that the sample is from a normally distributed population, test the hypothesis of "different from" using a level of significance equal to 0.05.
a. What evidence do you have that the assumption of normality is reasonable? Explain.
b. Complete the test using the p-value approach. Include $t\bigstar$, the p-value, and your conclusion.
c. Complete the test using the classical approach. Include the critical values, $t\bigstar$, and your conclusion.

9.44 Consumers enjoy the selection of merchandise made possible by specialty stores that sacrifice breadth for greater depth. Consider stores that carry only Levi Strauss pants. The company reports that a fully stocked Levi's store carries 130 ready-to-wear pairs of jeans for any given waist and inseam, and the company is phasing in two more lines of pants (Personal Pair and Original Spin) that it claims will eventually quadruple that number (*Fortune*, "The Customized, Digitized, Have-It-Your-Way Economy," September 28, 1998). Suppose a random sample of 24 Levi stores is chosen two months after the two new lines have been launched, and inventories are taken at each of the stores in the sample for all sizes of jeans. The sample mean number of choices for any given size is 141.3, and the standard deviation is 36.2. Does this sample of stores carry a greater selection of jeans, on the average, than what is expected, at the 0.01 level of significance?
a. Solve using the p-value approach.
b. Solve using the classical approach.

9.45 It is claimed that the students at a certain university will score an average of 35 on a given test. Is the claim reasonable if a random sample of test scores from this university are 33, 42, 38, 37, 30, 42? Complete a hypothesis test using $\alpha = 0.05$. Assume the test results are normally distributed.
a. Solve using the p-value approach.
b. Solve using the classical approach.

9.46 Ex02-168 Gasoline pumped from a supplier's pipeline is supposed to have an octane rating of 87.5. On 13 consecutive days a sample was taken and analyzed, with the following results:

88.6	86.4	87.2	88.4	87.2	87.6	86.8
86.1	87.4	87.3	86.4	86.6	87.1	

a. If the octane ratings have a normal distribution, is there sufficient evidence to show that these octane readings were taken from gasoline with a mean octane rating significantly less than 87.5 at the 0.05 level? (The sample mean and standard deviation were found in Exercise 2.168.)
b. Was the statistical decision reached in part a the same conclusion you expressed in part c of Exercise 2.168 for these same data?

9.47 Ex09-047 To test the null hypothesis "The mean weight of adult men equals 160 lb" against the alternative "The mean weight of adult men exceeds 160 lb," the weights of 16 men were found:

173	178	145	146	157	175	173	137
152	171	163	170	135	159	199	131

Assume normality and verify the results shown on the following MINITAB analysis by calculating the values yourself.

```
TEST OF MU = 160.00 VS MU > 160.00
      N    MEAN     STDEV    SE MEAN    T      P
C1   16   160.25    18.49    4.62      0.05   0.48
```

9.48 Ex09-048 The density of the earth relative to the density of water is known to be 5.517 g/cm³. Henry Cavendish (1731–1810), an English chemist and physicist, was the first scientist to accurately measure the density of the earth. These 29 measurements were taken by Cavendish in 1798 using a torsion balance:

5.50	5.61	4.88	5.07	5.26	5.55	5.36	5.29	5.58	5.65
5.57	5.53	5.62	5.29	5.44	5.34	5.79	5.10	5.27	5.39
5.42	5.47	5.63	5.34	5.46	5.30	5.75	5.68	5.85	

Source: The data and descriptive information are based on material from Stephen M. Stigler, "Do robust estimators work with real data?" *Annals of Statistics* 5 (1977), 1055–1098.

a. What evidence do you have that the assumption of normality is reasonable? Explain.
b. Is the mean of Cavendish's data significantly less than today's recognized standard? Use a 0.05 level of significance.

9.49 Ex09-049 Use a computer or calculator to complete the calculations and the hypothesis test for this exercise. Delco Products, a division of General Motors, produces commutators designed to be 18.810 mm in overall length. (A commutator is a device used in the electrical system of an automobile.) The following sample of 35 commutators was taken while monitoring the manufacturing process:

18.802	18.810	18.780	18.757	18.824	18.827
18.825	18.809	18.794	18.787	18.844	18.824
18.829	18.817	18.785	18.747	18.802	18.826
18.810	18.802	18.780	18.830	18.874	18.836
18.758	18.813	18.844	18.861	18.824	18.835
18.794	18.853	18.823	18.863	18.808	

Source: With permission of Delco Products Division, GMC

Is there sufficient evidence to reject the claim that these parts meet the design requirements "mean length is 18.810" at the $\alpha = 0.01$ level of significance?

9.50 How important is the assumption that the sampled population is normally distributed to the use of the Student's t-distribution? Using a computer, simulate drawing 100 random samples of size 10 from each of three different types of population distributions: a normal, a uniform, and an exponential. First generate 1000 data from the population and construct a histogram to see what the population looks like. Then generate 100 samples of size 10 from the same population; each row represents a sample. Calculate the mean and standard deviation for each of the 100 samples. Calculate $t\bigstar$ for each of the 100 samples. Construct histograms of the 100 sample means and the 100 $t\bigstar$-values. (Additional details can be found in the *Statistical Tutor*.)

For the samples from the normal population:

a. Does the $\bar{x}$-distribution appear to be normal? Find percentages for intervals and compare to the normal distribution.
b. Does the distribution of $t\bigstar$ appear to have a t-distribution with df = 9? Find percentages for intervals and compare them to the t-distribution.

For the samples from the rectangular or uniform population:

c. Does the $\bar{x}$-distribution appear to be normal? Find percentages for intervals and compare them to the normal distribution.
d. Does the distribution of $t\bigstar$ appear to have a t-distribution with df = 9? Find percentages for intervals and compare them to the t-distribution.

For the samples from the skewed (exponential) population:

e. Does the $\bar{x}$-distribution appear to be normal? Find percentages for intervals and compare them to the normal distribution.
f. Does the distribution of $t\bigstar$ appear to have a t-distribution with df = 9? Find percentages for intervals and compare them to the t-distribution.

In summary:

g. In each of the three situations, the sampling distribution of $\bar{x}$ appears to be slightly different from the distribution of $t\bigstar$. Explain why.
h. Does the normality condition appear to be necessary in order for the calculated test statistic $t\bigstar$ to have the Student's t-distribution? Explain.

9.2 Inferences About the Binomial Probability of Success

Perhaps the most common inference involves the **binomial parameter p,** the "probability of success." Yes, every one of us uses this inference, even if only casually. In thousands of situations we are concerned about something either "happening" or "not happening." There are only two possible outcomes of concern, and that is the fundamental property of a **binomial experiment.** The other necessary ingredient is multiple independent trials. Asking five people whether they are "for" or "against" some issue can create five independent trials; if 200 people are asked the same question, 200 independent trials may be involved; if 30 items are inspected to see if each "exhibits a particular property" or "not," there will be 30 repeated trials; these are the makings of a binomial inference.

The binomial parameter p is defined to be the probability of success on a single trial in a binomial experiment. We define p', the **observed** or **sample binomial probability,** as

$$p' = \frac{x}{n} \tag{9.3}$$

where the **random variable** x represents the number of successes that occur in a sample consisting of n trials. Recall that the mean and standard deviation of the binomial random variable x are found by using formula (5.7), $\mu = np$, and formula (5.8), $\sigma = \sqrt{npq}$, where $q = 1 - p$. The distribution of x is considered to be approximately normal if n is greater than 20 and if np and nq are both greater than 5. This commonly accepted *rule of thumb* allows us to use the **standard normal distribution** to estimate probabilities for the binomial random variable x, the number of successes in n trials, and to make inferences concerning the binomial parameter p, the probability of success on an individual trial.

FYI
Complete details about binomial experimentation can be found on pages 250–253.

Generally, it is easier and more meaningful to work with the distribution of p' (the observed probability of occurrence) than with x (the number of occurrences). Consequently, we will convert formulas (5.7) and (5.8) from units of x (integers) to units of proportions (percentages expressed as decimals) by dividing each formula by n, as shown in Table 9.1.

TABLE 9.1 Formulas (9.4) and (9.5)

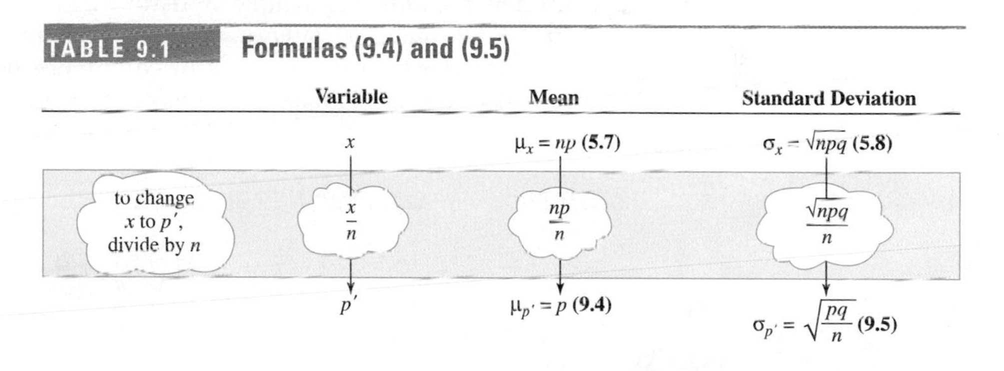

Variable
Mean
Standard Deviation
x
$\mu_x = np$ (5.7)
$\sigma_x = \sqrt{npq}$ (5.8)
to change x to p', divide by n
$\frac{x}{n}$
$\frac{np}{n}$
$\frac{\sqrt{npq}}{n}$
p'
$\mu_{p'} = p$ (9.4)
$\sigma_{p'} = \sqrt{\frac{pq}{n}}$ (9.5)

Recall that $\mu_{p'} = p$ and that the *sample statistic* p' is an **unbiased estimator for** p**.** Therefore, the information about the sampling distribution of p' is summarized as follows:

If a random sample of size n is selected from a large population with $p = P(\text{success})$, then the sampling distribution of p' has:

1. a mean $\mu_{p'}$ equal to p,
2. a standard error $\sigma_{p'}$ equal to $\sqrt{\frac{pq}{n}}$, and
3. an approximately normal distribution if n is sufficiently large.

ANSWER NOW 9.51

a. **Does it seem reasonable that the mean of the sampling distribution of observed values of p' should be p, the true proportion? Explain.**

b. **Explain why p' is an unbiased estimator for the population p.**

ANSWER NOW 9.52

Show that $\frac{\sqrt{npq}}{n}$ **simplifies to** $\sqrt{\frac{pq}{n}}$.

In practice, using these guidelines will ensure normality:

1. The sample size is greater than 20.
2. The products np and nq are both greater than 5.
3. The sample consists of less than 10% of the population.

THE ASSUMPTIONS FOR INFERENCES ABOUT THE BINOMIAL PARAMETER *p*

The n random observations that form the sample are selected independently from a population that is not changing during the sampling.

CONFIDENCE INTERVAL PROCEDURE

FYI
The standard deviation of a sampling distribution is called the "standard error."

Inferences concerning the population binomial parameter p, P(success), are made using procedures that closely parallel the inference procedures used for the population mean μ. When we estimate the **population proportion *p*,** we will base our estimations on the **unbiased sample statistic *p*′.** The point estimate, p', becomes the center of the confidence interval, and the maximum error of estimate is a multiple of the **standard error.** The **level of confidence** determines the confidence coefficient, the number of multiples of the standard error.

CONFIDENCE INTERVAL FOR A PROPORTION

$$p' - z(\alpha/2)\left(\sqrt{\frac{p'q'}{n}}\right) \quad \text{to} \quad p' + z(\alpha/2)\left(\sqrt{\frac{p'q'}{n}}\right) \tag{9.6}$$

where $p' = \frac{x}{n}$ and $q' = 1 - p'$. Notice that the standard error, $\sqrt{\frac{pq}{n}}$, has been replaced by $\sqrt{\frac{p'q'}{n}}$. Since we are estimating p, we do not know its value and therefore we must use the best replacement available. That replacement is p', the observed value or the point estimate for p. This replacement will cause little change in the standard error or the width of our confidence interval provided n is sufficiently large.

Illustration 9.7 Confidence Interval for p

In a discussion about the cars that fellow students drive, several statements were made about types, ages, makes, colors, and so on. Dana decided he wanted to estimate the proportion of convertibles students drive, so he randomly identified 200 cars in the student parking lot and found 17 to be convertibles. Find the 90% confidence interval for the proportion of convertibles driven by students.

FYI
The five-step confidence interval procedure is given on page 350.

SOLUTION

Step 1 **The Set-Up:**
Describe the population parameter of interest.
p, the proportion (percentage) of convertibles driven by students.

Step 2 **The Confidence Interval Criteria:**

a. **Check the assumptions.**
The sample was randomly selected, and each student's response is independent of those of the others surveyed.

b. **Identify the probability distribution and the formula to be used.**
The standard normal distribution will be used with formula (9.6) as the test statistic. p' is expected to be approximately normal because: (1) $n = 200$ is greater than 20, and
(2) both np [approximated by $np' = 200(17/200) = 17$] and nq [approximated by $nq' = 200(183/200) = 183$] are greater than 5

c. **State the level of confidence:** $1 - \alpha = 0.90$.

Step 3 **The Sample Evidence:**
Collect the sample information.
$n = 200$ cars were identified, and $x = 17$ were convertibles:

$$p' = \frac{x}{n} = \frac{17}{200} = 0.085$$

Step 4 **The Confidence Interval:**

a. **Determine the confidence coefficient.**
This is the z-score [$z(\alpha/2)$, "z of one-half of alpha"] identifying the number of standard errors needed to attain the level of confidence and is found using Table 4 in Appendix B; $z(\alpha/2) = z(0.05) = 1.65$ (see the diagram).

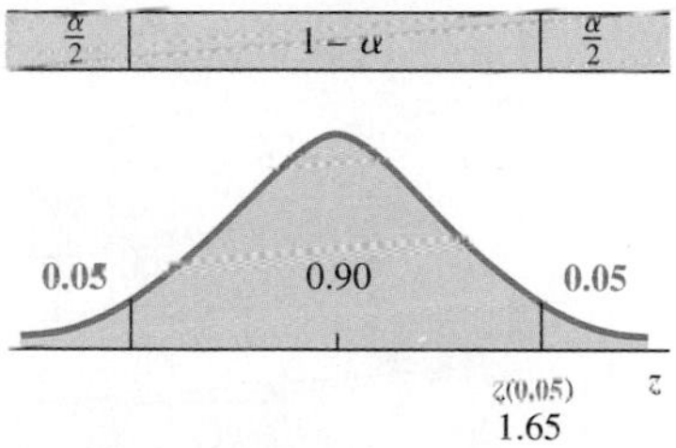

FYI
Specific instructions are given on page 349–350.

$\frac{\alpha}{2}$	$1-\alpha$	$\frac{\alpha}{2}$

0.05 0.90 0.05
$z(0.05)$ 1.65 z

b. **Find the maximum error of estimate.**
Use the maximum error part of formula (9.6):

$$E = z(\alpha/2)\left(\sqrt{\frac{p'q'}{n}}\right) = 1.65\left(\sqrt{\frac{(0.085)(0.915)}{200}}\right)$$
$$= (1.65)\sqrt{0.000389} = (1.65)(0.020) = \mathbf{0.033}$$

c. **Find the lower and upper confidence limits.**

$$p' - E \quad \text{to} \quad p' + E$$
$$0.085 - 0.033 \quad \text{to} \quad 0.085 + 0.033$$
$$0.052 \quad \text{to} \quad 0.118$$

Step 5 **The Results:**
State the confidence interval.
0.052 to 0.118 is the 90% confidence interval for $p = P$(drives convertible).
That is, the true proportion of students who drive convertibles is between 0.052 and 0.118, with 90% confidence. ■

ANSWER NOW 9.53

Another sample is taken to estimate the proportion of convertibles (see Illustration 9.7). The results are $n = 400$ and $x = 92$.

a. **Find the estimate for the standard error.**

b. **Find the 95% confidence interval.**

Technology Instructions: $1 - \alpha$ Confidence Interval for a Proportion p

MINITAB (Release 13)

```
Choose:   Stat > Basic Statistics > 1 Proportion
Select:   Summarized Data
Enter:    Number of trials: n
          Number of successes: x
Select:   Options
Enter:    Confidence level: 1 − α (ex. 95.0)
Select:   Alternative: not equal
          Use test and interval based on normal distribu-
          tion.
```

Excel XP

Input the data into column A using 0's for failures (or no's) and 1's for successes (or yes's); then continue with:

```
Choose:   Tools > Data Analysis Plus > Z-Estimate: Proportion
Enter:    Input Range: (A2:A20 or select cells) > OK
          Code for success: 1
          Alpha: α (ex. 0.05) > OK
```

TI-83 Plus

```
Choose:   STAT > TESTS > A:1-PropZint
          Enter the appropriate values and highlight Calcu-
          late.
```

```
1-PropZInt
 x:0
 n:0
 C-Level:.95
 Calculate
```

Application 9.2 Myth and Reality in Reporting Sampling Error

On almost every occasion when we release a new survey, someone in the media will ask, "What is the margin of error for this survey?" When the media print sentences such as "the margin of error is plus or minus three percentage points," they strongly suggest that the results are accurate to within the percentage stated. They want to warn people about sampling error. But they might be better off assuming that all surveys, all opinion polls are estimates, which may be wrong.

In the real world, "random sampling error"—or the likelihood that a pure probability sample would produce replies within a certain band of percentages only because of the sample size—is one of the least of our measurement problems.

For this reason, we (Harris) include a strong warning in all of the surveys that we publish. Typically, it goes as follows: In theory, with a sample of this size, one can say with 95 percent certainty that the results have a statistical precision of plus or minus __ percentage points of what they would be if the entire adult population had been polled with complete accuracy. Unfortunately, there are several other possible sources of error in all polls or surveys that are probably more serious than theoretical calculations of sampling error. They include refusals to be interviewed (non-response), question wording and question order, interviewer bias, weighting by demographic control data, and screening. It is difficult or impossible to quantify the errors that may result from these factors.

If journalists are the least bit interested in all of this they may well ask, "If there are so many sources of error in surveys, why should we bother to read or report any poll results?" To which I normally give two replies:

1. Well-designed, well-conducted surveys work. Their record overall is pretty good. Most social, and marketing, researchers would be very happy with the average forecasting errors of the polls. However, there are enough disasters in the history of election predictions for readers to be cautious about interpreting the results.
2. (And this is more effective.) I re-word Winston Churchill's famous remarks about democracy and say, "Polls are the worst way of measuring public opinion and public behavior, or of predicting elections—except for all of the others."

The Polling Report, May 4, 1998, by Humphrey Taylor, Chairman, Louis Harris & Assoc., Inc.
http://www.pollingreport.com/sampling.htm

ANSWER NOW 9.54

a. **If x successes result from a binomial experiment with $n = 1000$ and $p = P(\text{success})$, and the 95% confidence interval for the true probability of success is determined, what is the maximum value possible for the "maximum error of estimate"?**

b. **Compare the numerical value of the "maximum error of estimate" found in part a to the "margin of error" discussed in Application 9.2.**

c. **Under what conditions are they the same? Not the same?**

d. **Explain how the results of national polls, like those of Harris and Gallup, are related (similarities and differences) to the confidence interval technique studied in this section.**

e. **The theoretical sampling error with a level of confidence can be calculated, but the polls typically report only a "margin of error" with no probability (level of confidence). Why is that?**

Sample Size

By using the maximum error part of the confidence interval formula, it is possible to determine the **size of the sample** that must be taken in order to estimate p with a desired accuracy. Here is the formula for the **maximum error of estimate for a proportion:**

$$E = z(\alpha/2)\left(\sqrt{\frac{pq}{n}}\right) \tag{9.7}$$

In order to determine the sample size from this formula, we must decide on the quality we want for our final confidence interval. This quality is measured in two ways: the level of confidence and the preciseness (narrowness) of the interval. The level of confidence we establish will in turn determine the confidence coefficient, $z(\alpha/2)$. The desired preciseness will determine the maximum error of estimate, E. (Remember that we are estimating p, the binomial probability; therefore, E will typically be expressed in hundredths.)

For ease of use, we can solve formula (9.7) for n as follows:

Sample Size for $1 - \alpha$ Confidence Interval of p

$$n = \frac{[z(\alpha/2)]^2 \cdot p^* \cdot q^*}{E^2} \tag{9.8}$$

where p^* and q^* are provisional values of p and q used for planning.

FYI
Remember that $q = 1 - p$.

By inspecting formula (9.8), we can observe that three components determine the sample size:

1. the level of confidence ($1 - \alpha$, which in turn determines the confidence coefficient)
2. the provisional value of p (p^* determines the value of q^*)
3. the maximum error, E

An increase or decrease in one of these three components affects the sample size. If the level of confidence is increased or decreased (while the other components are held constant), then the sample size will increase or decrease, respectively. If the product of p^* and q^* is increased or decreased (with other components held constant), then the sample size will increase or decrease, respectively. (The product $p^* \cdot q^*$ is largest when $p^* = 0.5$ and decreases as the value of p^* becomes further from 0.5.) An increase or decrease in the desired maximum error will have the opposite effect on the sample size, since E appears in the denominator of the formula. If no provisional values for p and q are available, then use $p^* = 0.5$ and $q^* = 0.5$. Using $p^* = 0.5$ is safe because it gives the largest sample size of any possible value of p. Using $p^* = 0.5$ works reasonably well when the true value is "near 0.5" (say, between 0.3 and 0.7); however, as p gets nearer to either zero or one, a sizable overestimate in sample size will occur.

Illustration 9.8 Sample Size for Estimating p

Determine the sample size that is required to estimate the true proportion of blue-eyed community college students if you want your estimate to be within 0.02 with 90% confidence.

SOLUTION

STEP 1 The level of confidence is $1 - \alpha = 0.90$; therefore, the confidence coefficient is $z(\alpha/2) = z(0.05) = 1.65$ from Table 4 in Appendix B; see the diagram.

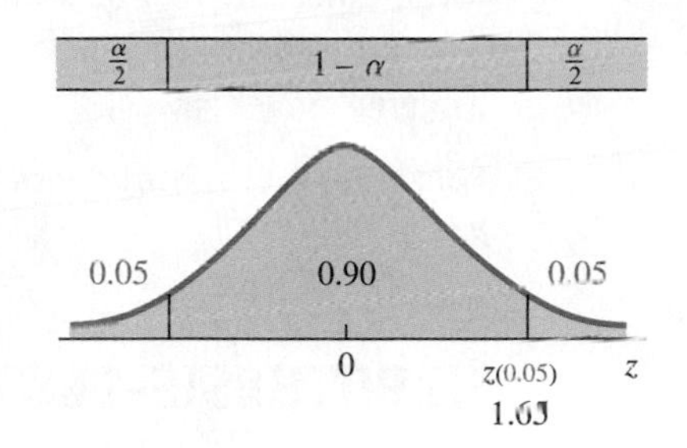

STEP 2 The desired maximum error is $E = 0.02$.

STEP 3 Since no estimate was given for p, use $p^* = 0.5$ and $q^* = 1 - p^* = 0.5$.

STEP 4 Use formula (9.8) to find n:

$$n = \frac{[z(\alpha/2)]^2 \cdot p^* \cdot q^*}{E^2}: \quad n = \frac{(1.65)^2 \cdot 0.5 \cdot 0.5}{(0.02)^2} = \frac{0.680625}{0.0004} = 1701.56 = \mathbf{1702}$$

■

ANSWER NOW 9.55

Find the sample size n needed for a 95% interval estimate in Illustration 9.8.

Illustration 9.9 Sample Size for Estimating p (Prior Information)

An automobile manufacturer purchases bolts from a supplier who claims the bolts are approximately 5% defective. Determine the sample size that will be required to estimate the true proportion of defective bolts if we want our estimate to be within ± 0.02 with 90% confidence.

SOLUTION

STEP 1 The level of confidence is $1 - \alpha = 0.90$; the confidence coefficient is $z(\alpha/2) = z(0.05) = 1.65$.

STEP 2 The desired maximum error is $E = 0.02$.

STEP 3 Since there is an estimate for p (supplier's claim is "5% defective"), use $p^* = 0.05$ and $q^* = 1 - p^* = 0.95$.

STEP 4 Use formula (9.8) to find n:

$$n = \frac{[z(\alpha/2)]^2 \cdot p^* \cdot q^*}{E^2}: \quad n = \frac{(1.65)^2 \cdot 0.05 \cdot 0.95}{(0.02)^2} = \frac{0.12931875}{0.0004} = 323.3 = \mathbf{324}$$

■

FYI
When finding the sample size n, always round up to the next larger integer, no matter how small the decimal.

Notice the difference in the sample sizes required in Illustrations 9.8 and 9.9. The only mathematical difference between the problems is the value used for p^*. In Illustration 9.8 we used $p^* = 0.5$, and in Illustration 9.9 we used $p^* = 0.05$. Recall that the use of the provisional value $p^* = 0.5$ gives the maximum sample size. As you can see, it will be an advantage to have some indication of the value expected for p, especially as p becomes increasingly further from 0.5.

ANSWER NOW 9.56

Find n for a 90% confidence interval for p with $E = 0.02$ using an estimate of $p = 0.25$.

HYPOTHESIS-TESTING PROCEDURE

When the binomial parameter p is to be tested using a hypothesis-testing procedure, we will use a test statistic that represents the difference between the observed proportion and the hypothesized proportion, divided by the standard error. This test statistic is assumed to be normally distributed when the null hypothesis is true, when the assumptions for the test have been satisfied, and when n is sufficiently large ($n > 20$, $np > 5$, and $nq > 5$).

FYI

p' is from the sample, p is from H_o, and $q = 1 - p$.

The value of the **test statistic $z\bigstar$** is calculated using formula (9.9):

$$z\bigstar = \frac{p' - p}{\sqrt{\frac{pq}{n}}} \quad \text{with } p' = \frac{x}{n} \tag{9.9}$$

Illustration 9.10 One-Tailed Hypothesis Test for Proportion p

Many people sleep-in on the weekends to make up for "short nights" during the workweek. The Better Sleep Council reports that 61% of us get more than seven hours of sleep per night on the weekend. A random sample of 350 adults found that 235 had more than seven hours of sleep each night last weekend. At the 0.05 level of significance, does this evidence show that more than 61% sleep seven or more hours per night on the weekend?

SOLUTION

Step 1 **The Set-Up:**

a. **Describe the population parameter of interest.**
   p, the proportion of adults who get more than seven hours of sleep per night on weekends.

b. **State the null hypothesis (H_o) and the alternative hypothesis (H_a).**

   H_o: $p = P(7+ \text{ hours of sleep}) = 0.61$ ($\leq$) (no more than 61%)

   H_a: $p > 0.61$ (more than 61%)

Step 2 **The Hypothesis Test Criteria:**

a. **Check the assumptions.**
   The random sample of 350 adults was independently surveyed.

b. Identify the probability distribution and the test statistic to be used.
The standard normal z will be used with formula (9.9). Since $n = 350$ is greater than 20 and both $np = (350)(0.61) = 213.5$ and $nq = (350)(0.39) = 136.5$ are greater than 5, p' is expected to be approximately normally distributed.

c. Determine the level of significance: $\alpha = 0.05$.

Step 3 **The Sample Evidence:**

a. Collect the sample information: $n = 350$ and $x = 235$:

$$p' = \frac{x}{n} = \frac{235}{350} = 0.671$$

b. Calculate the value of the test statistic.
Use formula (9.9):

$$z\bigstar = \frac{p' - p}{\sqrt{\frac{pq}{n}}}: \quad z\bigstar = \frac{0.671 - 0.61}{\sqrt{\frac{(0.61)(0.39)}{350}}} = \frac{0.061}{\sqrt{0.0006797}} = \frac{0.061}{0.0261} = 2.34$$

ANSWER NOW 9.57

Calculate the test statistic $z\bigstar$ used in testing H_o: $p = 0.70$ vs. H_a: $p > 0.70$, with the sample statistics $n = 300$ and $x = 224$.

Step 4 **The Probability Distribution:**

Using the p-value procedure:

a. Calculate the p-value for the test statistic.
Use the right-hand tail because H_a expresses concern for values related to "more than."
$\mathbf{P} = p\text{-value} = P(z > 2.34)$ as shown on the figure.

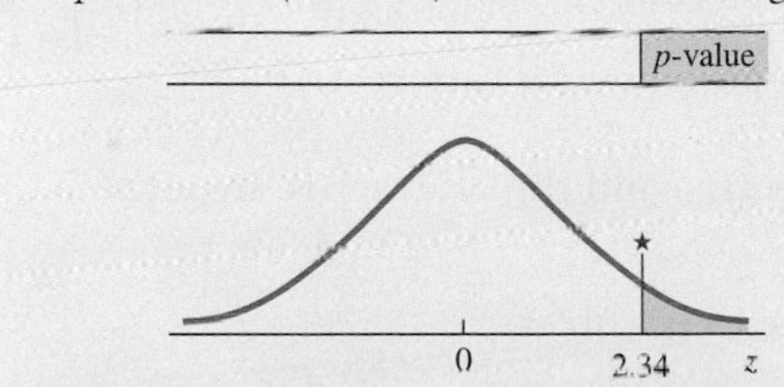

To find the p-value, use one of three methods:

1. Use Table 3 in Appendix B to calculate the p-value: $\mathbf{P} = 0.5000 - 0.4904 = \mathbf{0.0096}$.
2. Use Table 5 in Appendix B to place bounds on the p-value: $0.0094 < \mathbf{P} < 0.0107$.
3. Use a computer or calculator to calculate the p-value: $\mathbf{P} = 0.0096$.

For specific instructions, see Method 3 on page 440.

b. Determine whether or not the p-value is smaller than α.
The p-value is smaller than α.

OR

Using the classical procedure:

a. Determine the critical region and critical value(s).
The critical region is the right-hand tail because H_a expresses concern for values related to "more than." The critical value is obtained from Table 4A:
$z(0.05) = \mathbf{1.65}$.

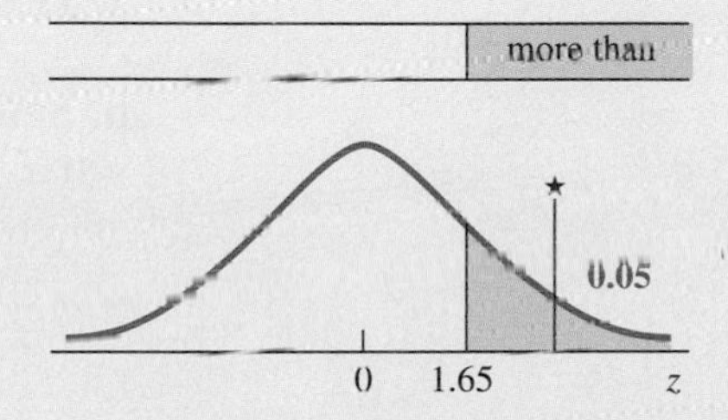

Specific instructions for finding critical values are given on pages 392–393.

b. Determine whether or not the calculated test statistic is in the critical region.
$z\bigstar$ is in the critical region, as shown in **red** on the figure above.

Step 5 **The Results:**

a. **State the decision about H_o:** Reject H_o.

b. **State the conclusion about H_a.**
There is sufficient reason to conclude that the proportion of adults in the sampled population who are getting more than seven hours of sleep nightly on weekends is significantly higher than 61% at the 0.05 level of significance. ■

Method 3: If you are doing the hypothesis test with the aid of a computer or calculator, most likely it will calculate the p-value for you, or you may use the cumulative probability distribution commands described on pages 287–288.

ANSWER NOW 9.58

Find the value of P for the hypothesis test in Answer Now 9.57; state the decision using $\alpha = 0.05$.

ANSWER NOW 9.59

Find the critical region and critical values for the hypothesis test in Answer Now 9.57; state the decision using $\alpha = 0.05$.

Illustration 9.11 Two-Tailed Hypothesis Test for Proportion p

While talking about the cars that fellow students drive (see Illustration 9.7, p. 432), Tom claimed that 15% of the students drive convertibles. Jody finds this hard to believe, and she wants to check the validity of Tom's claim using Dana's random sample. At a level of significance of 0.10, is there sufficient evidence to reject Tom's claim if there were 17 convertibles in his sample of 200 cars?

SOLUTION

Step 1 **The Set-Up:**

a. **Describe the population parameter of interest.**
$p = P(\text{student drives convertible})$.

b. **State the null hypothesis (H_o) and the alternative hypothesis (H_a).**

H_o: $p = 0.15$ (15% do drive convertibles)

H_a: $p \neq 0.15$ (the percentage is different from 15)

Step 2 **The Hypothesis Test Criteria:**

a. **Check the assumptions.**
The sample was randomly selected, and each subject's response is independent of other responses.

b. **Identify the probability distribution and the test statistic to be used.**
The standard normal z and formula (9.9) will be used. Since $n = 200$ is greater than 20 and both np and nq are greater than 5, p' is expected to be approximately normally distributed.

c. **Determine the level of significance:** $\alpha = 0.10$.

Step 3 **The Sample Evidence:**

a. Collect the sample information: $n = 200$ and $x = 17$:

$$p' = \frac{x}{n} = \frac{17}{200} = 0.085$$

b. Calculate the value of the test statistic.

Use formula (9.9):

$$z\bigstar = \frac{p' - p}{\sqrt{\frac{pq}{n}}}: \qquad z\bigstar = \frac{0.085 - 0.150}{\sqrt{\frac{(0.15)(0.85)}{200}}}$$

$$= \frac{-0.065}{\sqrt{0.00064}} = \frac{0.065}{0.022525} = -2.57$$

Step 4 **The Probability Distribution:**

Using the *p*-value procedure:

a. Calculate the *p*-value for the test statistic.

Use both tails because H_a expresses concern for values related to "different from."

$$\mathbf{P} = p\text{-value} = P(z < -2.57) + P(z > 2.57)$$

$$= 2 \times P(|z| > 2.57) \quad \text{as shown on the figure.}$$

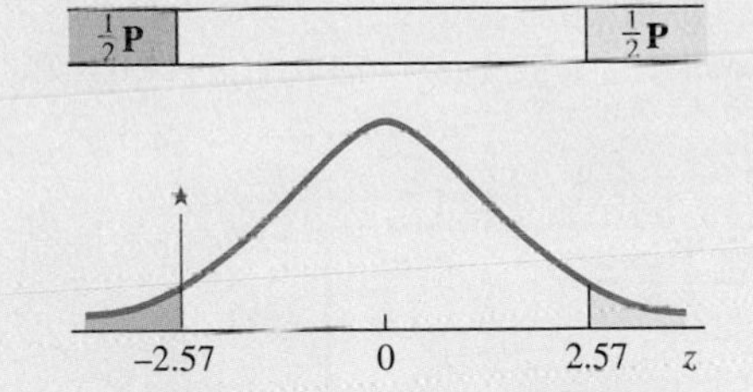

To find the *p*-value, use one of three methods:

1. Use Table 3 in Appendix B to calculate the *p*-value: $\mathbf{P} = 2 \times (0.5000 - 0.4949) = 0.0102$.
2. Use Table 5 in Appendix B to place bounds on the *p*-value. $0.0091 < \mathbf{P} < 0.0108$.
3. Use a computer or calculator to calculate the *p*-value: $\mathbf{P} = 0.0102$.

For specific instructions, see page 376.

b. Determine whether or not the *p*-value is smaller than α.

The *p*-value is smaller than α.

OR

Using the classical procedure:

a. Determine the critical region and critical value(s).

The critical region is two-tailed because H_a expresses concern for values related to "different from." The critical value is obtained from Table 4B: $z(0.05) = 1.65$.

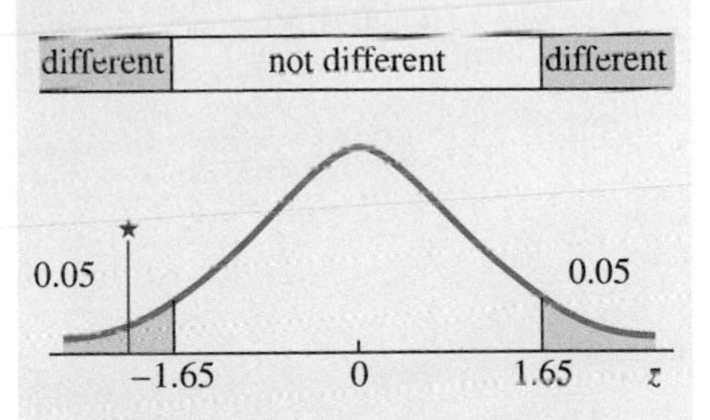

For specific instructions, see page 392–393.

b. Determine whether or not the calculated test statistic is in the critical region.

$z\bigstar$ is in the critical region, as shown in **red** on the figure above.

Step 5 **The Results:**

a. State the decision about H_o: Reject H_o.

b. State the conclusion about H_a.

There is sufficient evidence to reject Tom's claim and conclude that the percentage of students who drive convertibles is different from 15% at the 0.10 level of significance. ■

Technology Instructions: Hypothesis Test for a Proportion p

MINITAB (Release 13)

```
Choose:   Stat > Basic Statistics > 1 Proportion
Select:   Summarized Data
Enter:    Number of trials: n
          Number of successes: x
Select:   Options
Enter:    Test proportion: p
Select:   Alternative: less than or not equal or greater
          than
          Use test and interval based on normal distribu-
          tion.
```

Excel XP

Input the data into column A using 0's for failures (or no's) and 1's for successes (or yes's); then continue with:

```
Choose:   Tools > Data Analysis Plus > Z-Test: Proportion
Enter:    Input Range: (A2:A20 or select cells) > OK
          Code for success: 1
          Hypothesized Proportion: p
          Alpha: α (ex. 0.05)
Choose:   Alternative: less than or not equal or greater
          than > OK
          Gives p-values and critical values for both one-
          tailed and two-tailed tests.
```

TI-83 Plus

```
Choose:   STAT > TESTS > 5:1-PropZTest
          Enter the appropriate values and highlight Calcu-
          late.
```

```
1-PropZTest
 p0:0
 x:0
 n:0
 prop≠p0 <p0 >p0
 Calculate Draw
```

There is a relationship between confidence intervals and two-tailed hypothesis tests when the level of confidence and the level of significance add up to one. The confidence coefficients and the critical values are the same, which means the width of the confidence interval and the width of the noncritical region are the same. The point estimate is the center of the confidence interval, and the hypothesized mean is the center of the noncritical region. Therefore, if the hypothesized value of p is contained in the confidence interval, then the test statistic will be in the noncritical region (see Figure 9.5). Furthermore, if the hypothesized probability p does not fall within the confidence interval, then the test statistic will be in the critical region (see Figure 9.6). This comparison should be used only when the hypothesis test is two-tailed and when the same value of α is used in both procedures.

FIGURE 9.5
Confidence Interval Contains p

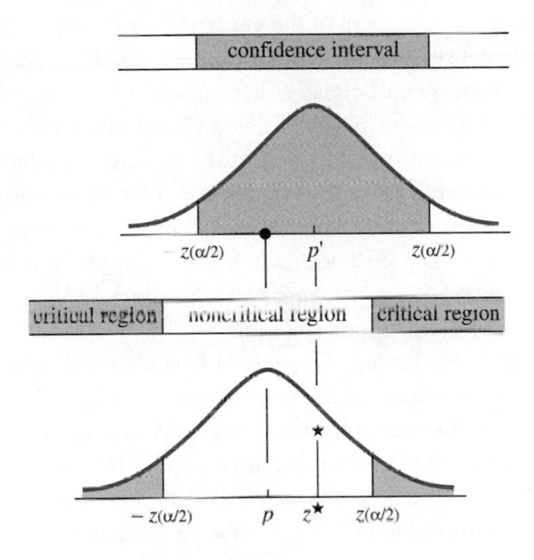

FIGURE 9.6
Confidence Interval Does Not Contain p

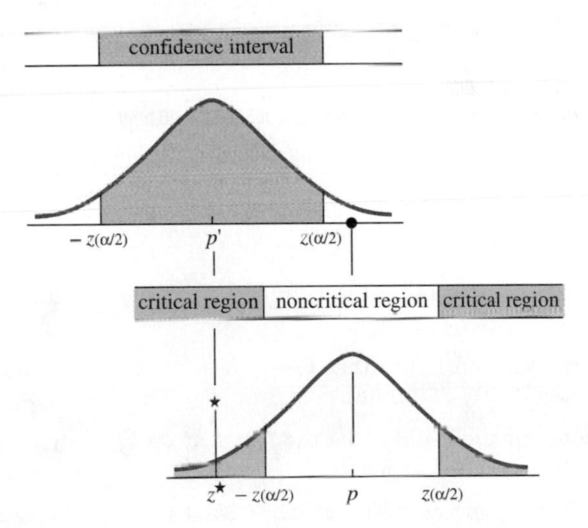

FYI
Explore Interactivity 9B on your CD.

ANSWER NOW 9.60

Show that the hypothesis test completed as Illustration 9.11 was unnecessary because the confidence interval had already been found in Illustration 9.7.

Application 9.3 Heads or Tails?

europa.eu.int/euro

Heads, Belgium wins—and wins

Memo to all teams playing Belgium in the World Cup this year: "Don't let them use their own coins for the toss."

Mathematicians say the coins issued in the eurozone's administrative heartland are more likely to land heads up than down. While the notes which began circulating in the 12 members of the eurozone on January 1 are all the same, the coins show national symbols on one side and a map of Europe on the other. King Albert, who appears on Belgian coins, appears to be a bit of a lightweight, according to Polish mathematicians Tomasz Gliszczynski and Waclaw Zawadowski. The two professors and their students at the Podlaska Academy in Siedlce spun a Belgian one euro coin 250 times, and found it landed heads up 140 times.

"The euro is struck asymmetrically," Prof Gliszczynski, who teaches statistics, told Germany's *Die Welt* newspaper. The head of the mint said yesterday that the Polish mathematicians' findings were "just luck." "When the coins were made they were struck in exactly the same way on all sides and the metal was evenly distributed," said Romain Coenen. "I haven't heard of any problems with the coins." But a variation of the experiment at the Guardian office suggested that the Polish mathematicians may be right. When tossed 250 times, the one euro coin came up heads 139 times and tails 111. "It looks very suspicious to me," said Barry Blight, a statistics lecturer at the London School of Economics. "If the coin were unbiased the chance of getting a result as extreme as that would be less than 7%."

Charlotte Denny and Sarah Dennis
The Guardian
Friday, January 4, 2002
http://www.guardian.co.uk/euro/story/0,11306,627496,00.html

ANSWER NOW 9.61

a. **Find the probability that 139 or more heads result when a balanced coin is tossed fairly 250 times.**

b. **Find the probability that 110 or fewer heads result when a balanced coin is tossed fairly 250 times.**

c. **Find the probability that the result of fairly tossing a balanced coin 250 times is as extreme as 139 heads.**

d. **What is Romain Coenen's claim? What is it that Barry Blight says is "suspicious"? How do these two statements form the opposing sides of a hypothesis test? State the null and alternative hypotheses.**

e. **Barry Blight's statement, "If the coin were unbiased the chance of getting a result as extreme as that would be less than 7%," is a statement of a p-value for a two-tailed hypothesis test. Explain why.**

f. **If the sample results had been used to estimate the probability that the euro coin lands heads up, what would have been the 0.95 confidence interval?**

g. **If the media had reported these results as 56% with a $\pm 6\%$ margin of error, how would this be similar to and different from a national opinion poll that resulted in 56% $\pm$ 6%?**

Exercises

9.62 Forty-five of the 150 elements in a random sample are classified as "success."

a. Explain why x and n are assigned the values 45 and 150, respectively.

b. Determine the value of p'. Explain how p' is found and the meaning of p'.

For each of the following cases, find p':

c. $x = 24$ and $n = 250$

d. $x = 640$ and $n = 2050$

e. 892 of 1280 responded "yes"

9.63

a. What is the relationship between $p = P(\text{success})$ and $q = P(\text{failure})$? Explain.

b. Explain why the relationship between p and q can be expressed by the formula $q = 1 - p$.

c. If $p = 0.6$, what is the value of q?

d. If $q' = 0.273$, what is the value of p'?

9.64 Find (1) α, (2) the area of one tail, and (3) the confidence coefficients of z that are used with each of the following levels of confidence:

a. $1 - \alpha = 0.90$ b. $1 - \alpha = 0.95$

c. $1 - \alpha = 0.98$ d. $1 - \alpha = 0.99$

9.65 "You say tomato; burger lovers say ketchup!" According to a recent T.G.I. Friday's® restaurants random survey of 1027 Americans, half (47%) say that ketchup is their preferred burger condiment. The survey quoted a margin of error of plus or minus 3.1%. (Source: Harris Interactive/Yankelovich Partners for T.G.I. Friday's restaurants: http://www.knoxville3.com/fridays/News/burger.htm)

a. Describe how this survey of 1027 Americans fits the properties of a binomial experiment. Specifically identify n, a trial, success, p, and x.

b. What is the point estimate for the proportion of all Americans who prefer ketchup on their burger? Is it a parameter or a statistic?

c. Calculate the 95% confidence maximum error of estimate for a binomial experiment of 1027 trials that results in an observed proportion of 0.47.

d. How is the maximum error found in part c related to the margin of error ±3.1% quoted in the survey report?

e. Find the 95% confidence interval for the true proportion p based on a binomial experiment of 1027 trials that results in an observed proportion of 0.47.

9.66 A bank randomly selected 250 checking-account customers and found that 110 of them also had savings accounts at this same bank. Construct a 95% confidence interval for the true proportion of checking-account customers who also have savings accounts.

9.67 In a sample of 60 randomly selected students, only 22 favored the amount budgeted for next year's intramural and interscholastic sports. Construct the 99% confidence interval for the proportion of all students who support the proposed budget amount.

9.68 *USA Today* (February 2, 1995) reported on a poll of 750 children and teenagers aged 10 to 16. Two of the findings reported were: "4 out of 5 kids say Entertainment television should teach youngsters right from wrong," and "2 out of 3 say shows like *The Simpsons* and *Married . . . With Children* encourage kids to disrespect their parents." The poll (margin of error: plus/minus 3%) was sponsored by Children Now, an advocacy group.

a. Find the point estimate, the maximum error of estimate, and the 95% confidence interval that results from a "4 out of 5" binomial experiment with $n = 750$.

b. Find the point estimate, the maximum error of estimate, and the 95% confidence interval that results from a "2 out of 3" binomial experiment with $n = 750$.

c. Compare the maximum error calculated in parts a and b to the margin of error mentioned in the article. What level of confidence do you believe the margin of error "plus/minus 3%" is based on? Explain.

9.69 Of the 1742 managers and professionals polled by Management Recruiters International in May 2002, 27.8% work late five days a week on average. Using a 99% confidence interval for the true binomial proportion based on a random sample of 1742 binomial trials and an observed proportion of 0.278, estimate the proportion of managers and professionals who work late five days a week.

9.70 Adverse drug reactions to legally prescribed medicines are among the leading causes of death in the United States. According to a recent study published in the *Journal of the American Medical Association*, just over 100,000 people die due to complications resulting from legally prescribed drugs, whereas 5,000 to 10,000 die from using illicit drugs (*People*, "Poison Pills," September 7, 1998). You decide to investigate drug-related deaths in your city by monitoring the next 250 incidences. From among these cases, 223 were caused by legally prescribed drugs and the rest were the result of illicit drug use. Find the 95% confidence interval for the proportion of drug-related deaths that are caused by legally prescribed drugs.

9.71 A nationwide telephone survey of 1000 people by the Cambridge Consumer Credit Index found that most Americans are not easily swayed by the lure of reward points or rebates when deciding to use a credit card or pay by cash or check. The survey found that two out of three consumers don't even have credit cards that offer reward points or rebates. Explain why you would be reluctant to use this information to construct a confidence interval to estimate the true proportion of consumers who do not have credit cards offering reward points or rebates.

9.72 Construct 90% confidence intervals for the binomial parameter p for each of the following pairs of values. Write your answers on the chart.

Observed Proportion $p' = x/n$	Sample Size	Lower Limit	Upper Limit
a. $p' = 0.3$	$n = 30$		
b. $p' = 0.7$	$n = 30$		
c. $p' = 0.5$	$n = 10$		
d. $p' = 0.5$	$n = 100$		
e. $p' = 0.5$	$n = 1000$		

f. Explain the relationship between answers a and b.
g. Explain the relationship between answers c, d, and e.

9.73 CNN/*USA Today*/Gallup Poll reported the following three nationwide poll results:

June 21–23, 2002. $N = 1{,}020$ adults nationwide. MoE ± 3.
"Do you think school districts should or should not be allowed to test public school students for illegal drugs before those students can participate in non-athletic activities?"
Should be allowed 70%—Should not be allowed 29%—No opinion 1%

June 28–30, 2002. $N = 1{,}019$ adults nationwide. MoE ± 3.
"A proposal has been made that would allow people to put a portion of their Social Security payroll taxes into personal retirement accounts that would be invested in private stocks and bonds. Do you favor or oppose this proposal?"
Favor 57%—Oppose 39%—No opinion 4%

June 23–25, 2000. $N = 1{,}020$ adults nationwide. MoE ± 3.
"How often do you watch *Monday Night Football:* regularly, occasionally, or never?"
Regularly 24%—Occasionally 32%—Never 44%

Each of the polls is based on approximately 1020 randomly selected adults.

a. Calculate the 95% confidence maximum error of estimate for the true binomial proportion based on binomial experiments with the same sample size and observed proportion as listed first in each article.
b. Explain what caused the values of the maximum errors to vary.
c. The margin of error reported is typically the value of the maximum error rounded to the next larger whole percentage. Do your results in part a verify this?
d. Explain why the round-up practice is considered conservative.

(continued)

e. What value of p should be used to calculate the standard error if the most conservative margin of error is desired?

9.74 Karl Pearson once tossed a coin 24,000 times and recorded 12,012 heads.

a. Calculate the point estimate for $p = P(\text{head})$ based on Pearson's results.
b. Determine the standard error of proportion.
c. Determine the 95% confidence interval estimate for $p = P(\text{head})$.
d. It must have taken Mr. Pearson many hours to toss a coin 24,000 times. You can simulate 24,000 coin tosses using the computer and calculator commands listed below.

(*Note:* A Bernoulli experiment is like a "single" trial binomial experiment. That is, one toss of a coin is one Bernoulli experiment with $p = 0.5$, and 24,000 tosses of a coin either is a binomial experiment with $n = 24{,}000$ or is 24,000 Bernoulli experiments. Code 0 = tail and 1 = head. The sum of the 1's will be the number of heads in the 24,000 tosses.)

MINITAB (Release 13)

Choose Calc > Random Data > Bernoulli, entering 24000 for generate, C1 for Store in column(s), and 0.5 for Probability of success. Sum the data and divide by 24,000.

Excel XP

Choose Tools > Data Analysis > Random Number Generation > Bernoulli, entering 1 for Number of Variables, 24000 for Number of Random Numbers, and 0.5 for p Value. Sum the data and divide by 24,000.

TI-83 Plus

Choose MATH > PRB > 5:randInt; then enter 0, 1, number of trials. The maximum number of elements (trials) in a list is 999. (slow process for large n's) Sum the data and divide by n.

e. How do your simulated results compare to Pearson's?
f. Use these commands to generate another set of 24,000 coin tosses. Compare these results to those above. Explain what you can conclude from these results.

9.75 The rule of thumb stated on page 431 indicated that we would expect the sampling distribution of p' to be approximately normal when $n > 20$ and both np and nq are greater than 5. What happens when these guidelines are not followed?

a. Use the following set of computer and calculator commands to see what happens. Try $n = 15$ and $p = 0.1$. Do the distributions look normal? Explain what causes the "gaps." Why do the histograms look alike? Try some different combinations of n and p.

MINITAB (Release 13)

Choose Calc > Random Data > Binomial to simulate 1000 trials for an n of 15 and a p of 0.5. Divide each generated value by n, forming a column of sample p's. Calculate a z-value for each sample p by using $z = (p' - p)/\sqrt{p(1-p)/n}$. Construct a histogram for the sample p's and another histogram for the z's.

Excel XP

Choose Tools > Data Analysis > Random Number Generation > Binomial to simulate 1000 trials for an n of 15 and a p of 0.5. Divide each generated value by n, forming a column of sample p's. Calculate a z-value for each sample p by using $z = (p' - p)/\sqrt{p(1-p)/n}$. Construct a histogram for the sample p's and another histogram for the z's.

TI-83 Plus

Choose MATH > PRB > 7:randBin; then enter n, p, number of trials. The maximum number of elements (trials) in a list is 999. (slow process for large n's) Divide each generated value by n, forming a list of sample p's. Calculate a z-value for each sample p by using $z = (p' - p)/\sqrt{p(1-p)/n}$. Construct a histogram for the sample p's and another histogram for the z's.

b. Try $n = 15$ and $p = 0.01$.
c. Try $n = 50$ and $p = 0.03$.
d. Try $n = 20$ and $p = 0.2$.
e. Try $n = 20$ and $p = 0.8$.
f. What happens when the rule of thumb is not followed?

9.76 Has the law requiring bike helmet use failed or what? During August 1998, Yankelovich Partners conducted a survey of bicycle riders in the United States. Only 60% of the nationally representative sample of 1020 bike riders reported owning a bike helmet (http://www.cpsc.gov/library/helmet.html).

a. Find the 95% confidence interval for the true proportion p for a binomial experiment of 1020 trials that resulted in an observed proportion of 0.60. Use this to estimate the percentage of bike riders who report owning a helmet.
b. Do you think the law is working?

Suppose you wish to conduct a survey in your city to determine what percentage of bicyclists own helmets. Use the national figure of 60% for your initial estimate of p.

c. Find the sample size if you want your estimate to be within 0.02 with 95% confidence.
d. Find the sample size if you want your estimate to be within 0.04 with 95% confidence.
e. Find the sample size if you want your estimate to be within 0.02 with 90% confidence.
f. What effect does changing the maximum error have on the sample size? Explain.
g. What effect does changing the level of confidence have on the sample size? Explain.

9.77 According to *USA Today* (May 14, 2002), 81% of all drivers use their seat belts. You wish to conduct a survey in your city to determine what percentage of the drivers use seat belts. Use the national figure of 81% for your initial estimate of p.

a. Find the sample size if you want your estimate to be within 0.02 with 90% confidence.
b. Find the sample size if you want your estimate to be within 0.04 with 90% confidence.
c. Find the sample size if you want your estimate to be within 0.02 with 98% confidence.
d. What effect does changing the maximum error have on the sample size? Explain.
e. What effect does changing the level of confidence have on the sample size? Explain.

9.78 Lung cancer is the leading cause of cancer deaths in both women and men in the United States, Canada, and China. In several other countries, lung cancer is the number one cause of cancer deaths in men, and the second or third cause among women. Only about 14% of all people who develop lung cancer survive for five years (*eMedicine Consumer Journal*, Vol. 3, No. 6, June 20 2002) Suppose you wanted to see if this survival rate were still true. How large a sample would you need to estimate the true proportion surviving for five years after diagnosis to within 1% with 95% confidence? (Use 14% as the value of p.)

9.79 State the null hypothesis, H_o, and the alternative hypothesis, H_a, that would be used to test these claims:

a. More than 60% of all students at our college work part-time jobs during the academic year.
b. The probability of our team winning tonight is less than 0.50.
c. No more than one-third of cigarette smokers are interested in quitting.
d. At least 50% of all parents believe in spanking their children when appropriate.
e. A majority of the voters will vote for the school budget this year.
f. At least three-quarters of the trees in our county were seriously damaged by the storm.
g. The results show the coin was not tossed fairly.
h. The single-digit numbers generated by the computer do not seem to be equally likely with regard to being odd or even.

9.80 Determine the test criteria that would be used to test the following hypotheses when z is used as the test statistic with the classical approach:

a. H_o: $p = 0.5$ vs. H_a: $p > 0.5$, $\alpha = 0.05$
b. H_o: $p = 0.5$ vs. H_a: $p \neq 0.5$, $\alpha = 0.05$
c. H_o: $p = 0.4$ vs. H_a: $p < 0.4$, $\alpha = 0.10$
d. H_o: $p = 0.7$ vs. H_a: $p > 0.7$, $\alpha = 0.01$

9.81 Determine the p-value for each of these hypothesis-testing situations:

a. H_o: $p = 0.5$ vs. H_a: $p \neq 0.5$, $z\bigstar = 1.48$
b. H_o: $p = 0.7$ vs. H_a: $p \neq 0.7$, $z\bigstar = -2.26$
c. H_o: $p = 0.4$ vs. H_a: $p > 0.4$, $z\bigstar = 0.98$
d. H_o: $p = 0.2$ vs. H_a: $p < 0.2$, $z\bigstar = -1.59$

9.82 The binomial random variable x may be used as the test statistic when testing hypotheses about the binomial parameter p when n is small (say, 15 or less). Use Table 2 in Appendix B to determine the p-value for each of these situations:

a. H_o: $p = 0.5$ vs. H_a: $p \neq 0.5$, $n = 15$, $x = 12$
b. H_o: $p = 0.8$ vs. H_a: $p \neq 0.8$, $n = 12$, $x = 4$
c. H_o: $p = 0.3$ vs. H_a: $p > 0.3$, $n = 14$, $x = 7$
d. H_o: $p = 0.9$ vs. H_a: $p < 0.9$, $n = 13$, $x = 9$

9.83 The binomial random variable x may be used as the test statistic when testing hypotheses about the binomial parameter p. When n is small (say, 15 or less), Table 2 in Appendix B provides the probabilities for each value of x separately, thereby making it unnecessary to estimate probabilities of the discrete binomial random variable with the continuous standard normal variable z. Use Table 2 and determine the value of α for each of the following:

a. H_o: $p = 0.5$ vs. H_a: $p > 0.5$, where $n = 15$ and the critical region is $x = 12, 13, 14, 15$
b. H_o: $p = 0.3$ vs. H_a: $p < 0.3$, where $n = 12$ and the critical region is $x = 0, 1$
c. H_o: $p = 0.6$ vs. H_a: $p \neq 0.6$, where $n = 10$ and the critical region is $x = 0, 1, 2, 3, 9, 10$
d. H_o: $p = 0.05$ vs. H_a: $p > 0.05$, where $n = 14$ and the critical region is $x = 4, 5, 6, 7, \ldots, 14$

9.84 Use Table 2 in Appendix B and determine the critical region used in testing each of the following hypotheses. (*Note:* Since x is discrete, choose critical regions that do not exceed the value of α given.)

a. H_o: $p = 0.5$ vs. H_a: $p > 0.5$, $n = 15$, $\alpha = 0.05$
b. H_o: $p = 0.5$ vs. H_a: $p \neq 0.5$, $n = 14$, $\alpha = 0.05$
c. H_o: $p = 0.4$ vs. H_a: $p < 0.4$, $n = 10$, $\alpha = 0.10$
d. H_o: $p = 0.7$ vs. H_a: $p > 0.7$, $n = 13$, $\alpha = 0.01$

9.85 You are testing the hypothesis $p = 0.7$ and have decided to reject this hypothesis if after 15 trials you observe 14 or more successes.

a. If the null hypothesis is true and you observe 13 successes, then which of the following will you do: correctly fail to reject H_o, correctly reject H_o, commit a type I error, or commit a type II error?
b. Find the significance level of your test.
c. If the true probability of success is $\frac{1}{2}$ and you observe 13 successes, then which of the following will you do: correctly fail to reject H_o, correctly reject H_o, commit a type I error, or commit a type II error?
d. Calculate the p-value for your hypothesis test after 13 successes are observed.

9.86 You are testing the null hypothesis $p = 0.4$ and will reject this hypothesis if $z\bigstar$ is less than -2.05.

a. If the null hypothesis is true and you observe $z\bigstar$ equal to -2.12, then which of the following results: correctly fail to reject H_o, correctly reject H_o, commit a type I error, or commit a type II error?
b. What is the significance level for this test?
c. What is the p-value for $z\bigstar = -2.12$?

9.87 An insurance company states that 90% of its claims are settled within 30 days. A consumer group selected a random sample of 75 of the company's claims to test this statement. If the consumer group found that 55 of the claims were settled within 30 days, do they have sufficient reason to support their contention that fewer than 90% of the claims are settled within 30 days? Use $\alpha = 0.05$.

a. Solve using the p-value approach.
b. Solve using the classical approach.

9.88 A recent survey conducted by ZOOM and Applied Research & Consulting LLC reported that the events of September 11, 2001, have motivated kids to volunteer and that more than 80% volunteer. A disbeliever of this information took a separate random sample of 500 kids in an attempt to show that the true percentage of kids who volunteer is less than 80%.

a. Find the p-value if 384 of the surveyed kids said they do volunteer work.
b. Explain why it is important for the level of significance to be established before the sample results are known.

9.89 A politician claims that she will receive 60% of the vote in an upcoming election. The results of a properly designed random sample of 100 voters showed that 50 of those sampled will vote for her. Is it likely that her assertion is correct at the 0.05 level of significance?

a. Solve using the p-value approach.
b. Solve using the classical approach.

9.90 The full-time student body of a college is 50% men and 50% women. Does a random sample of students (30 men, 20 women) from an introductory chemistry course show sufficient evidence to reject the hypothesis that the proportions of male and female students who take this course are the same as those in the whole student body? Use $\alpha = 0.05$.

a. Solve using the p-value approach.
b. Solve using the classical approach.

9.91 The first baby ever conceived through in vitro fertilization (IVF) was born in England in 1978. In the 20 years that followed, 10 million women received such care for infertility. The procedure, which can cost upward of $12,000, has had an average success rate of 22.5% nationwide, but that rate is increasing with advances in technology (*Family Circle*, "Moms at Last," September 15, 1998). Suppose a recent study of 200 women who are attempting to overcome infertility using the IVF procedure shows that 61 were actually successful in becoming pregnant. Do the results show a higher success rate for the sample than expected based on the historical success rate? Use $\alpha = 0.05$.

a. Solve using the p-value approach.
b. Solve using the classical approach.

9.92 The popularity of personal watercraft (PWCs, also known as jet skis) continues to increase, despite the apparent danger associated with their use. In fact, a sample of 54 watercraft accidents reported to the Nebraska Game and Parks Commission in 1997 revealed that 85% of them involved PWCs even though only 8% of the motorized boats registered in the state are PWCs (*Nebraskaland*, "Officer's Notebook: The Personal Problem," June 1998). Suppose the national average proportion of watercraft accidents in 1997 involving PWCs was 78%. Does the watercraft accident rate for PWCs in Nebraska exceed the rate in the nation as a whole? Use a 0.01 level of significance.

a. Solve using the p-value approach.
b. Solve using the classical approach.

9.93 *USA Today* ("Facing a crowd isn't easy," May 30, 2002) reported that 35% of the country's professional women fear public speaking. Suppose you conduct a survey of 1000 randomly chosen professional women to test H_o: $p = 0.35$ versus H_a: $p < 0.35$, where p represents the proportion who fear public speaking. Of the 1000 sampled, 324 feared public speaking. Use $\alpha = 0.01$.

a. Calculate the value of the test statistic.
b. Solve using the p-value approach.
c. Solve using the classical approach.

9.3 Inferences About the Variance and Standard Deviation

Problems often arise that require us to make inferences about variability. For example, a soft-drink bottling company has a machine that fills 16-oz bottles. The company needs to control the standard deviation σ (or variance σ^2) in the amount of soft drink, x, put into each bottle. The mean amount placed in each bottle is important, but a correct mean amount does not ensure that the filling machine is working correctly. If the variance is too large, many bottles will be overfilled and many underfilled. Thus, the bottling company wants to maintain as small a standard deviation (or variance) as possible.

When discussing inferences about the spread of data, we usually talk about variance instead of standard deviation because the techniques (the formulas used) employ the sample variance rather than the standard deviation. However, remember that the standard deviation is the positive square root of the variance; thus, talking about the variance of a population is comparable to talking about the standard deviation.

Inferences about the variance of a normally distributed population use the **chi-square,** χ^2, distributions ("*ki-square*": that's "*ki*" as in "*kite*" and χ is the Greek lowercase letter chi). The chi-square distributions, like the Student t-distributions, are a family of probability distributions, each one identified by the **parameter** number of **degrees of freedom.** In order to use the chi-square distribution, we must be aware of its properties (see Figure 9.7 on page 450).

PROPERTIES OF THE CHI-SQUARE DISTRIBUTION

1. χ^2 is nonnegative in value; it is zero or positively valued.
2. χ^2 is not symmetrical; it is skewed to the right.
3. χ^2 is distributed so as to form a family of distributions, a separate distribution for each different number of degrees of freedom.

FIGURE 9.7
Various Chi-Square Distributions

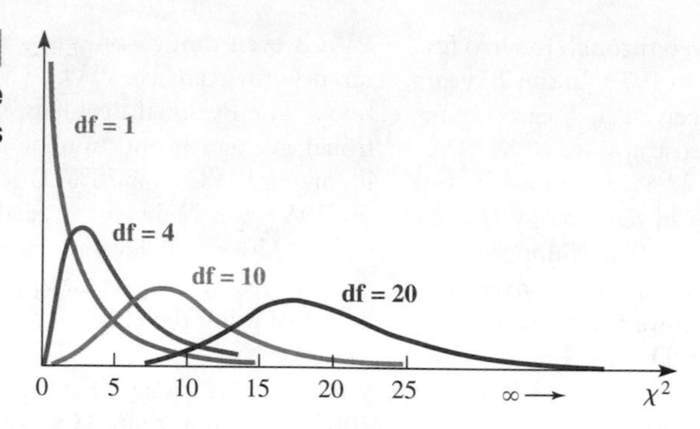

The **critical values for chi-square** are obtained from Table 8 in Appendix B. Each critical value is identified by two pieces of information: degrees of freedom (df) and area under the curve to the right of the critical value being sought. Thus, $\chi^2(\text{df}, \alpha)$ (read "*chi-square of df, alpha*") is the symbol used to identify the critical value of chi-square with df degrees of freedom and with α area to the right, as shown in Figure 9.8. Since the chi-square distribution is not symmetrical, the critical values associated with the right and left tails are given separately in Table 8.

FIGURE 9.8
Chi-Square Distribution Showing $\chi^2(\text{df}, \alpha)$

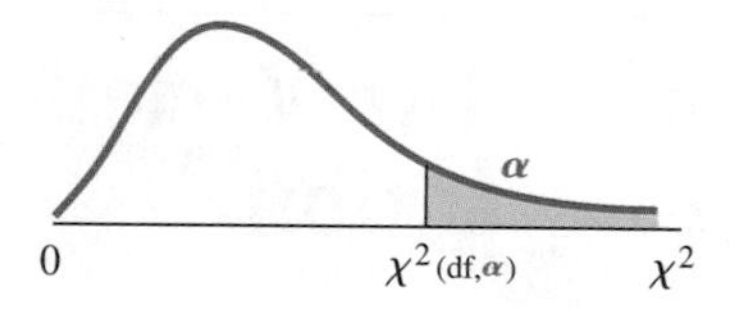

Illustration 9.12 χ^2 Associated with the Right Tail

Find $\chi^2(20, 0.05)$.

SOLUTION
See the figure. Use Table 8 in Appendix B to find the value of $\chi^2(20, 0.05)$ at the intersection of row df = 20 and column $\alpha = 0.05$, as shown in the portion of the table below:

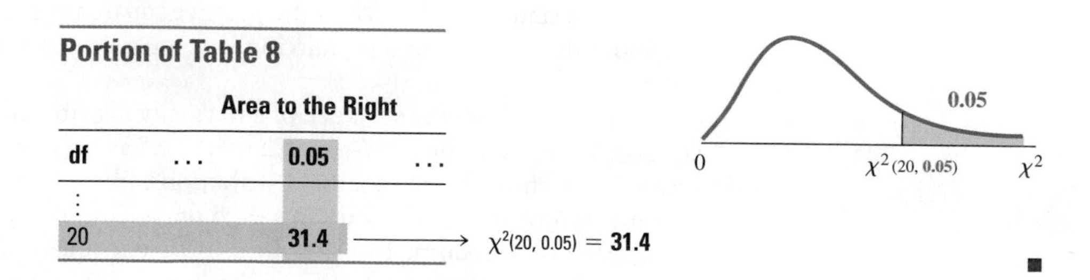

Portion of Table 8

		Area to the Right	
df	. . .	0.05	. . .
⋮			
20		31.4	

⟶ $\chi^2(20, 0.05) = \mathbf{31.4}$

■

ANSWER NOW 9.94

Find: a. $\chi^2(10, 0.01)$ **b.** $\chi^2(12, 0.025)$

NOTE When df > 2, the mean value of the chi-square distribution is df. The mean is located to the right of the mode (the value where the curve reaches its high point) and just to the right of the median (the value that splits the distribution, 50% on either side). By locating zero at the left extreme and the value of df on your sketch of the χ^2 distribution, you will establish an approximate scale so that other values can be located in their respective positions. See Figure 9.9.

FIGURE 9.9
Location of Mean, Median, and Mode for χ^2 Distribution

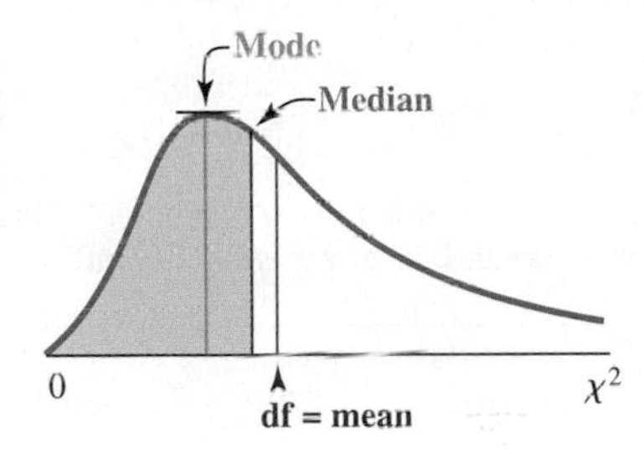

Illustration 9.13 χ^2 Associated with the Left Tail

Find $\chi^2(14, 0.90)$.

SOLUTION
See the figure. Use Table 8 in Appendix B to find the value of $\chi^2(14, 0.90)$ at the intersection of row df = 14 and column α = 0.90, as shown in the portion of the table below:

Portion of Table 8

	Area to the Right		
df	. . .	0.90	. . .
⋮			
14		7.79	

→ $\chi^2(14, 0.90) = \mathbf{7.79}$ ■

ANSWER NOW 9.95
Find: a. $\chi^2(10, 0.95)$ b. $\chi^2(22, 0.995)$

Most computer software packages or statistical calculators will calculate the area related to a specified χ^2-value. The accompanying figure shows the relationship between the cumulative probability distribution and a specific χ^2-value for a χ^2-distribution with df degrees of freedom.

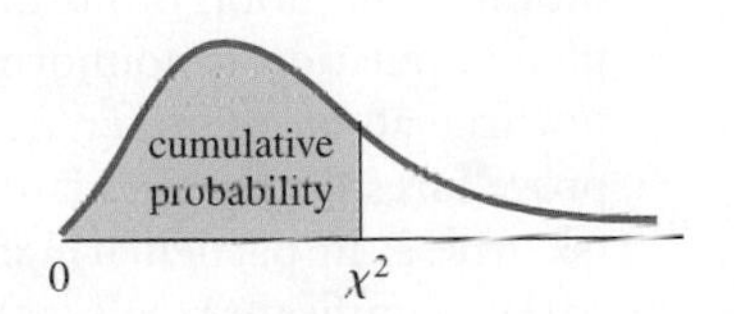

FYI
Explore Interactivity 9C on your CD.

Technology Instructions: Cumulative Probabilities for χ^2

MINITAB (Release 13)

Input the data into C1; then continue with:

```
Choose:   Calc > Probability Distributions > Chi-Square
Select:   Cumulative Probability
          Noncentrality Parameter: 0.0
Enter:    Degrees of freedom: df
Select:   Input constant*
Enter:    χ²-value (ex. 47.25)
```

*Select Input column if several χ^2-values are stored in C1. Use C2 for optional storage. If the area in the right tail is needed, subtract the calculated probability from one.

Excel XP

If several χ^2-values are to be used, input the values into column A and activate B1; then continue with:

```
Choose:   Insert function fx > Statistical > CHIDIST > OK
Enter:    X: individual χ²-value or (A1:A5 or select
          "χ²-value" cells)*
          Deg_freedom: df > OK
Drag*:    Bottom right corner of the B1 cell down to give
          other probabilities
```

TI-83 Plus

```
Choose:   2nd > DISTR > 7: χ²cdf(
Enter:    0, χ²-value, df)
```

If the area in the right tail is needed, subtract the calculated probability from one.

ANSWER NOW 9.96

Use a computer or calculator to find the area (a) to the left and (b) to the right of $\chi^2\bigstar = 20.2$ with df = 15.

We are now ready to use chi-square to make inferences about the population variance or standard deviation.

THE ASSUMPTIONS FOR INFERENCES ABOUT THE VARIANCE σ^2 OR STANDARD DEVIATION σ

The sampled population is normally distributed.

The t procedures for inferences about the mean (see Section 9.1) were based on the assumption of normality, but they are generally useful even when the sampled population is nonnormal, especially for larger samples. However, the same is not true about the inference procedures for the standard deviation. The statistical procedures for the standard deviation are very sensitive to nonnormal distributions (skewness, in particular), and this makes it difficult to determine whether an apparent significant result is the result of the sample evidence or a violation of the assumptions. Therefore, the only inference procedure to be presented here is the hypothesis test for the standard deviation of a normal population.

The **test statistic** that will be used in testing hypotheses about the population variance or standard deviation is obtained by using the formula

$$\chi^2\bigstar = \frac{(n-1)s^2}{\sigma^2}, \qquad \text{with df} = n - 1 \tag{9.10}$$

When random samples are drawn from a normal population with a known variance σ^2, the quantity $\frac{(n-1)s^2}{\sigma^2}$ possesses a probability distribution that is known as the chi-square distribution with $n - 1$ degrees of freedom.

HYPOTHESIS-TESTING PROCEDURE

Let's return to the illustration about the bottling company that wishes to detect when the variability in the amount of soft drink placed into each bottle gets out of control. A variance of 0.0004 is considered acceptable, and the company wants to adjust the bottle-filling machine when the variance, σ^2, becomes larger than this value. The decision will be made using the hypothesis-testing procedure.

Illustration 9.14 One-Tailed Hypothesis Test for Variance, σ^2

The soft-drink bottling company wants to control the variability in the amount of fill by not allowing the variance to exceed 0.0004. Does a sample of size 28 with a variance of 0.0007 indicate that the bottling process is out of control (with regard to variance) at the 0.05 level of significance?

SOLUTION

Step 1 The Set-Up:

a. **Describe the population parameter of interest.**
σ^2, the variance in the amount of fill of a soft drink during a bottling process.

b. **State the null hypothesis (H_o) and the alternative hypothesis (H_a).**

H_o: $\sigma^2 = 0.0004$ ($\leq$) (variance is not larger than 0.0004)

H_a: $\sigma^2 > 0.0004$ (variance is larger than 0.0004)

Step 2 The Hypothesis Test Criteria:

a. **Check the assumptions.**
The amount of fill put into a bottle is generally normally distributed. By checking the distribution of the sample, we could verify this.

b. **Identify the probability distribution and the test statistic to be used.**
The chi-square distribution will be used and formula (9.10), with df $= n - 1 = 28 - 1 = 27$.

c. **Determine the level of significance:** $\alpha = 0.05$.

Step 3 The Sample Evidence:

a. **Collect the sample information:** $n = 28$ and $s^2 = 0.0007$.

b. Calculate the value of the test statistic.
Use formula (9.10):

$$\chi^2\bigstar = \frac{(n-1)s^2}{\sigma^2}: \qquad \chi^2\bigstar = \frac{(28-1)(0.0007)}{0.0004} = \frac{(27)(0.0007)}{0.0004} = 47.25$$

ANSWER NOW 9.97

Find the test statistic for the hypothesis test H_o: $\sigma^2 = 532$ versus H_a: $\sigma^2 > 532$ using the sample information $n = 18$ and $s^2 = 785$.

Step 4 The Probability Distribution:

Using the *p*-value procedure:

a. Calculate the *p*-value for the test statistic.
Use the right-hand tail because H_a expresses concern for values related to "larger than."
$\mathbf{P} = P(\chi^2\bigstar > 47.25$, with df $= 27)$ as shown on the figure.

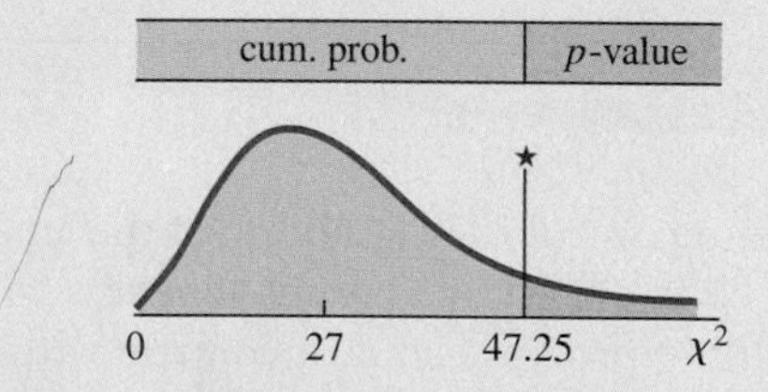

To find the *p*-value, use one of two methods:

1. Use Table 8 in Appendix B to place bounds on the *p*-value: $0.005 < \mathbf{P} < 0.01$.
2. Use a computer or calculator to calculate the *p*-value: $\mathbf{P} = 0.0093$.

Specific instructions follow this illustration.

b. Determine whether or not the *p*-value is smaller than α.
The *p*-value is smaller than the level of significance, α (0.05).

OR

Using the classical procedure:

a. Determine the critical region and critical value(s).
The critical region is the right-hand tail because H_a expresses concern for values related to "larger than." The critical value is obtained from Table 8, at the intersection of row df = 27 and column α = 0.05: $\chi^2(27, 0.05) = 40.1$.

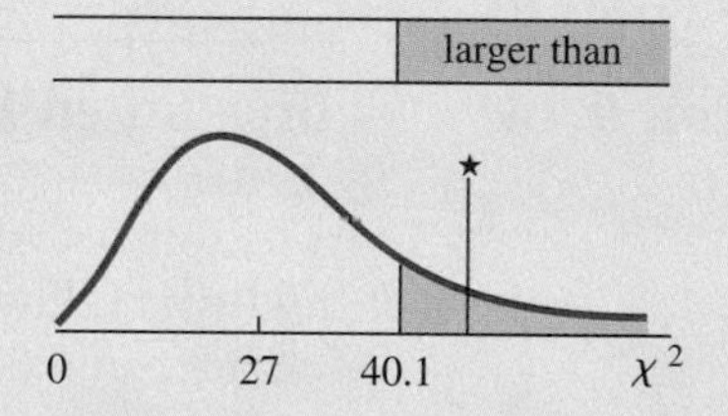

For specific instructions, see page 450.

b. Determine whether or not the calculated test statistic is in the critical region.
$\chi^2\bigstar$ is in the critical region, as shown in **red** on the figure above.

Step 5 The Results:

a. State the decision about H_o: Reject H_o.

b. State the conclusion about H_a.
At the 0.05 level of significance, we conclude that the bottling process is out of control with regard to the variance. ■

ANSWER NOW 9.98

Complete the hypothesis test in Answer Now 9.97 using the *p*-value method and α = 0.01.

ANSWER NOW 9.99

Complete the hypothesis test in Answer Now 9.97 using the classical method and α = 0.01.

Calculating the *p*-value when using the χ^2-distribution
Method 1: Use Table 8 in Appendix B to place bounds on the p-value. By inspecting the df = 27 row of Table 8, you can determine an interval within which the *p*-value lies. Locate $\chi^2\bigstar$ along the row labeled df = 27. If $\chi^2\bigstar$ is not listed, locate the two values that $\chi^2\bigstar$ falls between, and then read the bounds for the *p*-value from the top of the table. In this case, $\chi^2\bigstar = 47.25$ is between 47.0 and 49.6; therefore, **P** is between 0.005 and 0.01.

Portion of Table 8

		Area in Right-Hand Tail			
df	. . .	0.01	P	0.005	→ 0.005 < P < 0.01
⋮		↑		↑	
27		47.0	47.25	49.6	

Method 2: Use a computer or calculator. Use the χ^2 probability distribution commands on page 452 to find the *p*-value associated with $\chi^2\bigstar = 47.25$.

Illustration 9.15 One-Tailed *p*-Value Hypothesis Test for Variance, σ^2

Find the *p*-value for this hypothesis test:

$$H_o: \sigma^2 = 12$$
$$H_a: \sigma^2 < 12 \quad \text{with df} = 15 \text{ and } \chi^2\bigstar = 7.88$$

SOLUTION
Since the concern is for "smaller" values (the alternative hypothesis is "less than"), the *p*-value is the area to the left of $\chi^2\bigstar = 7.88$, as shown in the figure:

$$\mathbf{P} = P(\chi^2\bigstar < 7.88 \text{ with df} = 15)$$

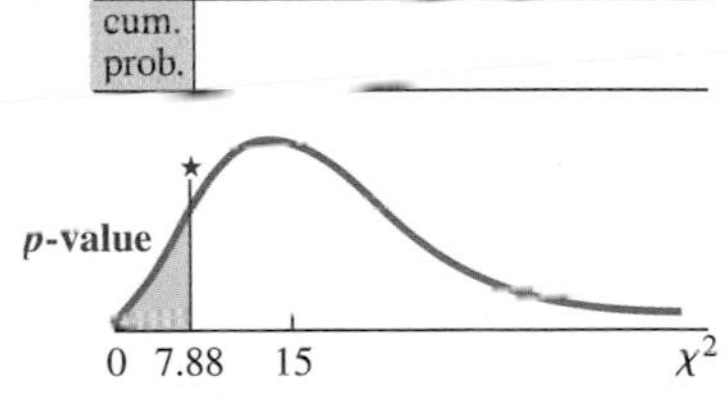

To find the *p*-value, use one of two methods:
Method 1: Use Table 8 in Appendix B to place bounds on the p-value. Inspect the df = 15 row to find $\chi^2\bigstar = 7.88$. The $\chi^2\bigstar$ value is between entries, so the interval that bounds **P** is read from the Left-Hand Tail heading at the top of the table.

Portion of Table 8

		Area in Left-Hand Tail			
df	. . .	0.05	P	0.10	→ 0.05 < P < 0.10
⋮		↑		↑	
15		7.26	7.88	8.55	

Method 2: Use a computer or calculator. Use the χ^2 probability distribution commands on page 452 to find the *p*-value associated with $\chi^2\bigstar = 7.88$. ■

Illustration 9.16 Two-Tailed Hypothesis Test for Standard Deviation, σ

The manufacturer claims that a photographic chemical has a shelf life that is normally distributed about a mean of 180 days with a standard deviation of no more than 10 days. As a user of this chemical, Fast Photo is concerned that the standard deviation might be different from 10 days; otherwise, it will buy a larger quantity while the chemical is part of a special promotion. Twelve random samples were selected and tested, with a standard deviation of 14 days resulting. At the 0.05 level of significance, does this sample present sufficient evidence to show that the standard deviation is different from 10 days?

SOLUTION

Step 1 **The Set-Up:**

a. **Describe the population parameter of interest.**
σ, the standard deviation for the shelf life of the chemical.

b. **State the null hypothesis (H_o) and the alternative hypothesis (H_a).**

H_o: $\sigma = 10$ (standard deviation is 10 days)

H_a: $\sigma \neq 10$ (standard deviation is different from 10 days)

Step 2 **The Hypothesis Test Criteria:**

a. **Check the assumptions.**
The manufacturer claims shelf life is normally distributed; this could be verified by checking the distribution of the sample.

b. **Identify the probability distribution and the test statistic to be used.**
The chi-square distribution will be used and formula (9.10), with $\text{df} = n - 1 = 12 - 1 = 11$.

c. **Determine the level of significance:** $\alpha = 0.05$.

Step 3 **The Sample Evidence:**

a. **Collect the sample information:** $n = 12$ and $s = 14$.

b. **Calculate the value of the test statistic.**
Use formula (9.10):

$$\chi^2\bigstar = \frac{(n-1)s^2}{\sigma^2}: \quad \chi^2\bigstar = \frac{(12-1)(14)^2}{(10)^2} = \frac{2156}{100} = 21.56$$

ANSWER NOW 9.100

Find the test statistic for the hypothesis test H_o: $\sigma^2 = 52$ versus H_a: $\sigma^2 \neq 52$ using the sample information $n = 41$ and $s^2 = 78.2$.

Step 4 **The Probability Distribution:**

Using the p-value procedure:

a. **Calculate the p-value for the test statistic.**
Since the concern is for values "different from" 10, the p-value is the area of both tails. The area of each tail will represent $\frac{1}{2}$**P**. Since

OR

Using the classical procedure:

a. **Determine the critical region and critical value(s).**
The critical region is split into two equal

$\chi^2\bigstar = 21.56$ is in the right tail, the area of the right tail is $\frac{1}{2}$**P**:
$\frac{1}{2}\mathbf{P} = P(\chi^2 > 21.56$, with df $= 11)$ as shown on the figure.

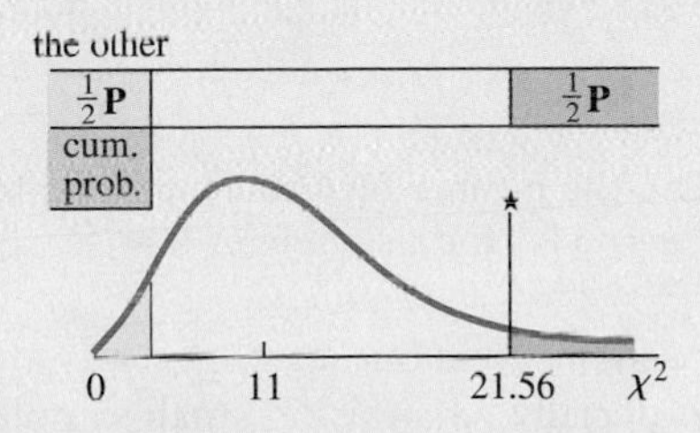

To find $\frac{1}{2}$**P**, use one of two methods:

1. Use Table 8 in Appendix B to place bounds on $\frac{1}{2}$**P**: $0.025 < \frac{1}{2}\mathbf{P} < 0.05$. Double both bounds to find the bounds for **P**: $2 \times (0.025 < \frac{1}{2}\mathbf{P} < 0.05)$ becomes $0.05 < \mathbf{P} < 0.10$.
2. Use a computer or calculator to find $\frac{1}{2}$**P**: $\frac{1}{2}\mathbf{P} = 0.0280$; therefore, $\mathbf{P} = 0.0560$.

Specific instructions follow this illustration.

b. Determine whether or not the *p*-value is smaller than α.

The *p*-value is not smaller than the level of significance, α (0.05).

parts because H_a expresses concern for values related to "different from." The critical values are obtained from Table 8 at the intersections of row df = 11 with columns α = 0.975 and 0.025 (area to right):
$\chi^2(11, 0.975) = 3.82$ and
$\chi^2(11, 0.025) = 21.9$.

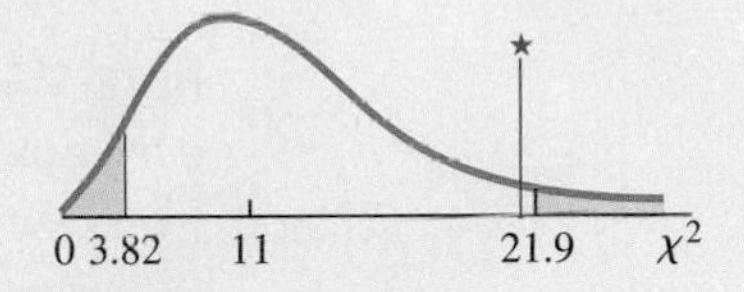

For specific instructions, see pages 450–451.

b. Determine whether or not the calculated test statistic is in the critical region.

$\chi^2\bigstar$ is not in the critical region; see the figure above.

Step 5 **The Results:**

a. State the decision about H_o: Fail to reject H_o.

b. State the conclusion about H_a.

There is not sufficient evidence at the 0.05 significance level to conclude that the shelf life of this chemical has a standard deviation different from 10 days. Therefore, Fast Photo should purchase the chemical accordingly.

ANSWER NOW 9.101

Complete the hypothesis test in Answer Now 9.100 using the *p*-value method and α = 0.05.

ANSWER NOW 9.102

Complete the hypothesis test in Answer Now 9.100 using the classical method and α = 0.05.

Calculating the *p*-value when using the χ^2-distribution

Method 1: Use Table 8 in Appendix B to place bounds on the p-value. Inspect the df = 11 row to locate $\chi^2\bigstar = 21.56$. Notice that 21.56 is between two table entries. The bounds for $\frac{1}{2}$**P** are read from the Right-Hand Tail heading at the top of the table.

Portion of Table 8

	Area in Right-Hand Tail				
df	...	**0.05**	**½P**	**0.025**	⟶ $0.025 < \frac{1}{2}\mathbf{P} < 0.05$
⋮		↑		↑	
11		19.7	**21.56**	21.9	

Double both bounds to find the bounds for **P**: $2 \times (0.025 < \frac{1}{2}\mathbf{P} < 0.05)$ becomes $\mathbf{0.05 < P < 0.10}$.

Method 2: Use a computer or calculator. Use the χ^2 probability distribution commands on page 452 to find the p-value associated with $\chi^2\bigstar = 21.56$. Remember to double the probability. ■

ANSWER NOW 9.103

Use a computer or calculator to find the p-value for the hypothesis test H_o: $\sigma^2 = 7$ versus H_a: $\sigma^2 \neq 7$, if $\chi^2\bigstar = 6.87$ for a sample of $n = 15$.

NOTE When sample data are skewed, just one outlier can greatly affect the standard deviation. It is very important, especially when using small samples, that the sampled population be normal; otherwise, the procedures are not reliable.

Exercises

9.104

a. Calculate the standard deviation for each set:
A: 5, 6, 7, 7, 8, 10 and **B:** 5, 6, 7, 7, 8, 15.

b. What effect did the largest value changing from 10 to 15 have on the standard deviation?

c. Why do you think 15 might be called an outlier?

9.105 Find these critical values using Table 8 in Appendix B:

a. $\chi^2(18, 0.01)$ b. $\chi^2(16, 0.025)$ c. $\chi^2(8, 0.10)$
d. $\chi^2(28, 0.01)$ e. $\chi^2(22, 0.95)$ f. $\chi^2(10, 0.975)$
g. $\chi^2(50, 0.90)$ h. $\chi^2(24, 0.99)$

9.106 Using the notation of Exercise 9.105, name and find the critical values of χ^2 shown below.

Figures for Exercise 9.106

a.

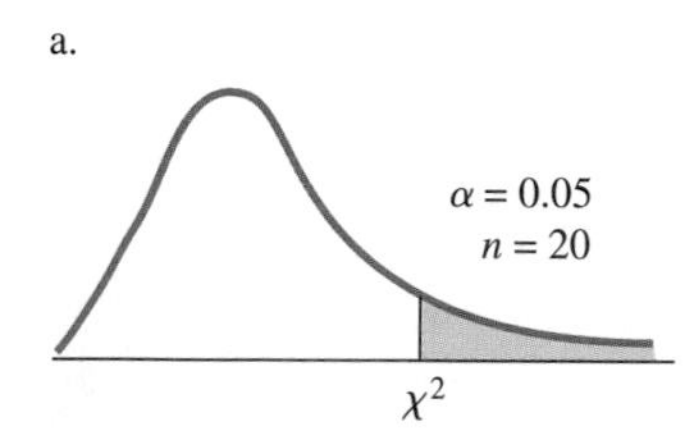

b.

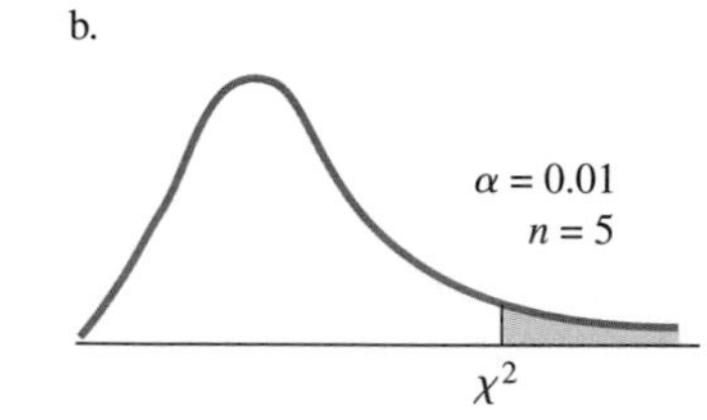

c.

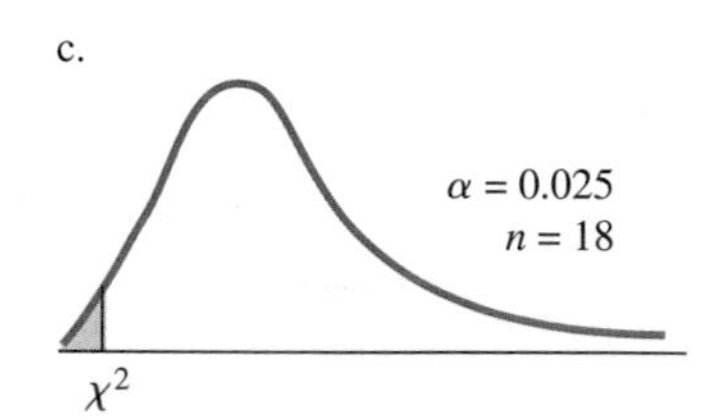

d.

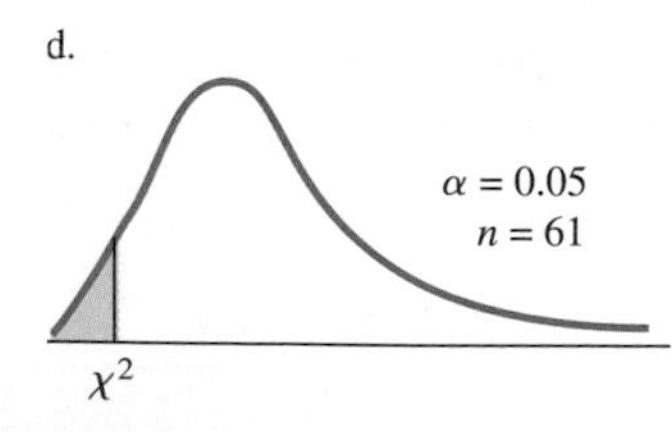

e.

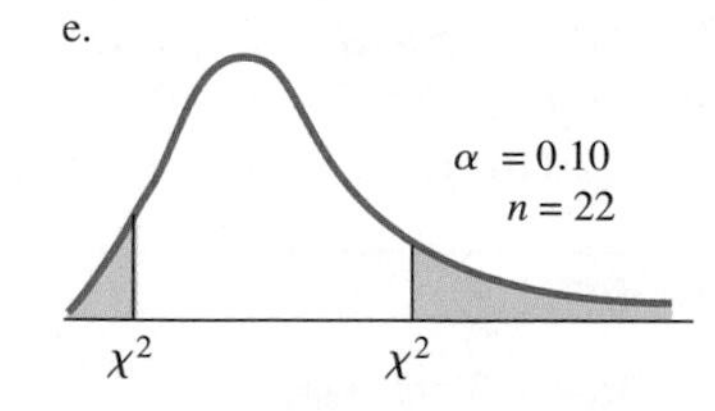

f.

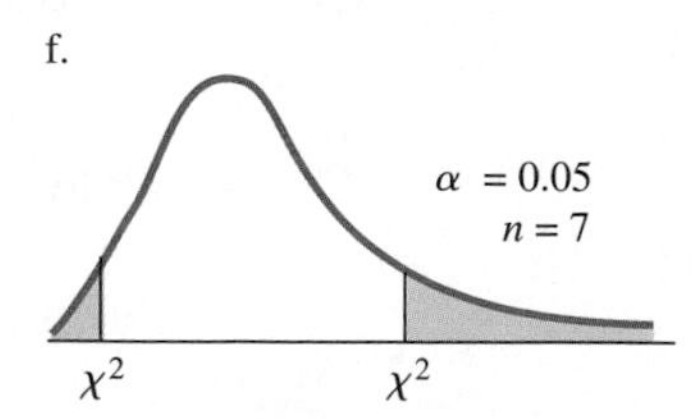

g.

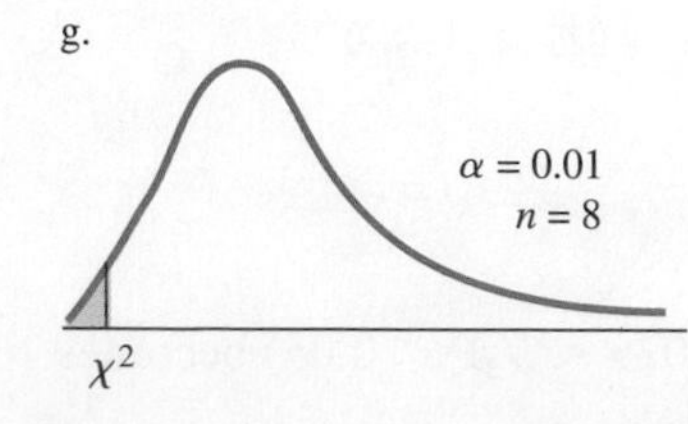

h.

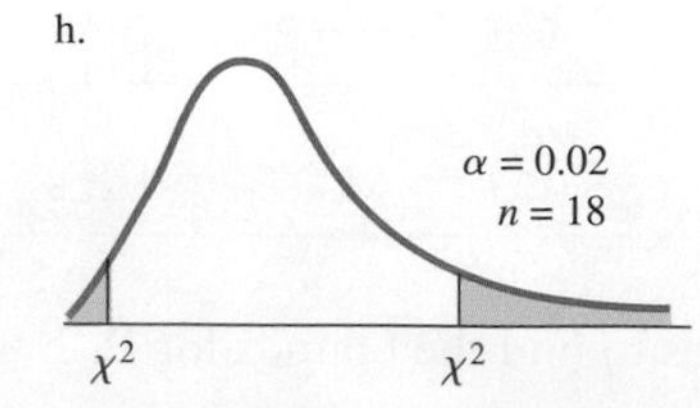

9.107

a. What value of chi-square for 5 degrees of freedom subdivides the area under the distribution curve such that 5% is to the right and 95% is to the left?
b. What is the value of the 95th percentile for the chi-square distribution with 5 degrees of freedom?
c. What is the value of the 90th percentile for the chi-square distribution with 5 degrees of freedom?

9.108

a. The central 90% of the chi-square distribution with 11 degrees of freedom lies between what values?
b. The central 95% of the chi-square distribution with 11 degrees of freedom lies between what values?
c. The central 99% of the chi-square distribution with 11 degrees of freedom lies between what values?

9.109 For a chi-square distribution with 12 degrees of freedom, find the area under the curve for chi-square values ranging from 3.57 to 21.0.

9.110 For a chi-square distribution with 35 degrees of freedom, find the area under the curve between $\chi^2(35, 0.96)$ and $\chi^2(35, 0.15)$.

9.111 State the null hypothesis, H_o, and the alternative hypothesis, H_a, that would be used to test these claims:

a. The standard deviation has increased from its previous value of 24.
b. The standard deviation is no larger than 0.5 oz.
c. The standard deviation is not equal to 10.
d. The variance is no less than 18.
e. The variance is different from the value of 0.025, the value called for in the specs.
f. The variance has increased from 34.5.

9.112 Calculate the value of the test statistic, $\chi^2\bigstar$, for each of these situations:

a. H_o: $\sigma^2 = 20$, $n = 15$, $s^2 = 17.8$
b. H_o: $\sigma^2 = 30$, $n = 18$, $s = 5.7$
c. H_o: $\sigma^2 = 42$, $n = 25$, $s = 37.8$
d. H_o: $\sigma^2 = 12$, $n = 37$, $s^2 = 163$

9.113 Calculate the p-value for each of these hypothesis tests:

a. H_a: $\sigma^2 \neq 20$, $n = 15$, $\chi^2\bigstar = 27.8$
b. H_a: $\sigma^2 > 30$, $n = 18$, $\chi^2\bigstar = 33.4$
c. H_a: $\sigma^2 \neq 42$, df $= 25$, $\chi^2\bigstar = 37.9$
d. H_a: $\sigma^2 < 12$, df $= 40$, $\chi^2\bigstar = 26.3$

9.114 Determine the critical region and critical value(s) that would be used to test the following using the classical approach:

a. H_o: $\sigma = 0.5$ vs. H_a: $\sigma > 0.5$, $n = 18$, $\alpha = 0.05$
b. H_o: $\sigma^2 = 8.5$ vs. H_a: $\sigma^2 < 8.5$, $n = 15$, $\alpha = 0.01$
c. H_o: $\sigma = 20.3$ vs. H_a: $\sigma \neq 20.3$, $n = 10$, $\alpha = 0.10$
d. H_o: $\sigma^2 = 0.05$ vs. H_a: $\sigma^2 \neq 0.05$, $n = 8$, $\alpha = 0.02$
e. H_o: $\sigma = 0.5$ vs. H_a: $\sigma < 0.5$, $n = 12$, $\alpha = 0.10$

9.115 A random sample of 51 observations was selected from a normally distributed population. The sample mean was $\bar{x} = 98.2$, and the sample variance was $s^2 = 37.5$. Does this sample provide sufficient reason to conclude that the population standard deviation is not equal to 8 at the 0.05 level of significance?

9.116 In the past the standard deviation of weights of certain 32.0-oz packages filled by a machine was 0.25 oz. A random sample of 20 packages showed a standard deviation of 0.35 oz. Is the apparent increase in variability significant at the 0.10 level of significance? Assume package weights are normally distributed.

a. Solve using the p-value approach.
b. Solve using the classical approach.

9.117 In the United States, 36% of all people have a medically treatable foot problem. Among patients who require surgery, 80% to 90% are women. The problem is compounded by mail-order sales of shoes that do not fit, but people wear them anyway rather than send them back. Although many experts have blamed high-heeled shoes for most of the troubles, a recent study of 368 women complaining of foot problems showed that 88% were wearing shoes that were too small (*Ladies Home Journal*, "My Aching Feet," June 1998). Suppose the standard deviation of shoe sizes for all manufacturers is 0.32. A separate study is conducted of 27 mail-order sellers of women's shoes, and a standard deviation of 0.51 is obtained from the sample. Do mail-order distributors sell shoes that vary more in size than shoes sold by all manufacturers, at the 0.01 level of significance?

a. What role does the assumption of normality play in this solution? Explain.
b. Describe how you might attempt to determine whether or not it is realistic to assume that shoe sizes are normally distributed.
c. Solve using the p-value approach.
d. Solve using the classical approach.

9.118 A commercial farmer harvests his entire field of a vegetable crop at one time. Therefore, he would like to plant a variety of green beans that mature all at one time (small standard deviation for the maturity times of individual plants). A seed company has developed a new hybrid strain of green beans that it believes is better for the commercial farmer. The maturity times of the standard variety have an average of 50 days and a standard deviation of 2.1 days. A random sample of 30 plants of the new hybrid showed a standard deviation of 1.65 days. Does this sample show a lower standard deviation at the 0.05 level of significance? Assume that maturity times are normally distributed.

a. Solve using the *p*-value approach.
b. Solve using the classical approach.

9.119 Ex09-119 A car manufacturer claims that the miles per gallon for a certain model have a mean of 40.5 mi with a standard deviation equal to 3.5 mi. Use the following data, obtained from a random sample of 15 such cars, to test the hypothesis that the standard deviation differs from 3.5. Use $\alpha = 0.05$. Assume normality.

37.0	38.0	42.5	45.0	34.0	32.0	36.0	35.5
38.0	42.5	40.0	42.5	35.0	30.0	37.5	

a. Solve using the *p*-value approach.
b. Solve using the classical approach.

9.120 Farm real estate values in rural America fluctuate substantially from state to state and county to county, thus making it difficult for buyers or landowners to know precisely what a property is actually worth. For example, the average value of ranch land in Missouri in 1998 was $548 per acre, whereas the same average in three nearby states (Kansas, Nebraska, and Oklahoma) was more than $200 less (*Regional Economic Digest*, "Survey of Agricultural Credit Conditions," First Quarter 1998). This discrepancy could be caused by an exaggerated variability in the value of ranch land acreage in Missouri. Assume that the combined four-state region yielded a standard deviation of $85 per acre. Suppose a random sample of 31 landowners in Missouri who recently sold their property resulted in a sample standard deviation of $125 per acre. Is the variability in ranch land value in Missouri, at the 0.05 level of significance, greater than the variability for the region as a whole?

a. Solve using the *p*-value approach.
b. Solve using the classical approach.

9.121 The chi-square distribution was described on pages 449–450 as a family of distributions. Let's investigate these distributions and observe some of their properties.

a. Use the MINITAB commands to generate several large random samples of data from various chi-square distributions. Use df values of 1, 2, 3, 5, 10, 20, and 80 (others if you wish).

```
Choose: Calc > Random Data > ChiSquare
Enter:  Generate: 1000 rows of data
        Store in column(s): C1
        Degrees of freedom: df
Use Stat > Basic Statistics > Display Descrip-
tive Statistics to calculate the mean and
median of the data in C1. Use Graph >
Histogram to construct a histogram of the
data in C1.
```

b. What appears to be the relationship between the mean of the sample and the number of degrees of freedom?
c. How do the values of the mean, median, and mode appear to be related? Do your results agree with the information on page 451?
d. Have the computer generate samples for two additional degrees of freedom, df = 120 and 150. Describe how these distributions seem to be changing as df increases.

9.122 How important is the assumption "the sampled population is normally distributed" for the use of the chi-square distributions? Use a computer and the two sets of MINITAB commands found in the *Statistical Tutor* to simulate drawing 200 samples of size 10 from each of two different types of population distributions. The first commands will generate 2000 data and construct a histogram so that you can see what the population looks like. The next commands will generate 200 samples of size 10 from the same population; each row represents a sample. The following commands will calculate the standard deviation and $\chi^2\bigstar$ for each of the 200 samples. The last commands will construct histograms of the 200 sample standard deviations and the 200 $\chi^2\bigstar$-values. (Additional details can be found in the *Statistical Tutor.*)

For the samples from the normal population:

a. Does the sampling distribution of sample standard deviations appear to be normal? Describe the distribution.
b. Does the χ^2-distribution appear to have a chi-square distribution with df = 9? Find percentages for intervals (less than 2, less than 4, . . . , more than 15, more than 20, etc.), and compare them to the percentages expected as estimated using Table 8 in Appendix B.

For the samples from the skewed population:

c. Does the sampling distribution of sample standard deviations appear to be normal? Describe the distribution.
d. Does the χ^2-distribution appear to have a chi-square distribution with df = 9? Find percentages for intervals (less than 2, less than 4, . . . , more than 15, more than 20, etc.), and compare them to the percentages expected as estimated using Table 8.

In summary:

e. Does the normality condition appear to be necessary in order for the calculated test statistic $\chi^2\bigstar$ to have a χ^2-distribution? Explain.

CHAPTER SUMMARY

RETURN TO CHAPTER CASE STUDY

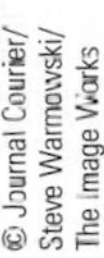
© Journal Courier/ Steve Warmowski/ The Image Works

Many studies have reached the conclusion that we need to exercise to lower health risks such as high blood pressure, heart disease, and high cholesterol levels. But knowing and doing are not the same thing. People in the health profession should be even more aware of the need for exercise. The data below are from the Chapter Case Study (p. 411), a survey of cardiovascular technicians (individuals who perform various cardiovascular diagnostic procedures) asked about their own physical exercise, measured in minutes per week: CS09

60	40	50	30	60	50	90	30	60	60
60	80	90	90	60	30	20	120	60	50
20	60	30	120	50	30	90	20	30	40
50	40	30	40	20	30	60	50	60	80

FIRST THOUGHTS 9.123

a. Describe the population of interest.
b. Describe the amount of physical exercise done by the technicians in the sample using one graph, the mean, and the standard deviation.

PUTTING CHAPTER 9 TO WORK 9.124

a. What evidence do you have that the assumption of normality is reasonable? Explain.
b. Estimate the mean amount of weekly exercise time for all cardiovascular technicians using a point estimate and a 95% confidence interval.
c. The article in the case study says that people should exercise at least 90 minutes a week. Based on the data in the case study, determine whether the technicians exercise at least 90 minutes a week. Use a 0.05 level of significance.

YOUR OWN MINI-STUDY 9.125

a. Define the population whose amount of exercise time per week you are interested in investigating.
b. Collect the "times" from a sample of 40 members of your population.
c. Find the mean and standard deviation for the amounts of time exercised per week by the members of your sample.
d. Construct a graph displaying the distribution of your data.
e. Estimate the mean amount of weekly exercise time for your population using a point estimate and a 95% confidence interval.
f. The article in the case study says that people should exercise at least 90 minutes a week. Does it appear that the members of your sample exercise at least 90 minutes per week? Use a 0.05 level of significance. Justify your answer.
g. Did your data satisfy the assumptions? Explain.

In Retrospect

We have been studying inferences, both confidence intervals and hypothesis tests, for the three basic population parameters (mean μ, proportion p, and standard deviation σ) of a single population. Most inferences about a single population are concerned with one of these three parameters. Figure 9.10 (p. 462) presents a visual organization of the techniques presented in Chapters 8 and 9 along with the key questions that you must ask as you are deciding which test statistic and formula to use.

(continues on p. 463)

FIGURE 9.10 Choosing the Right Inference Technique

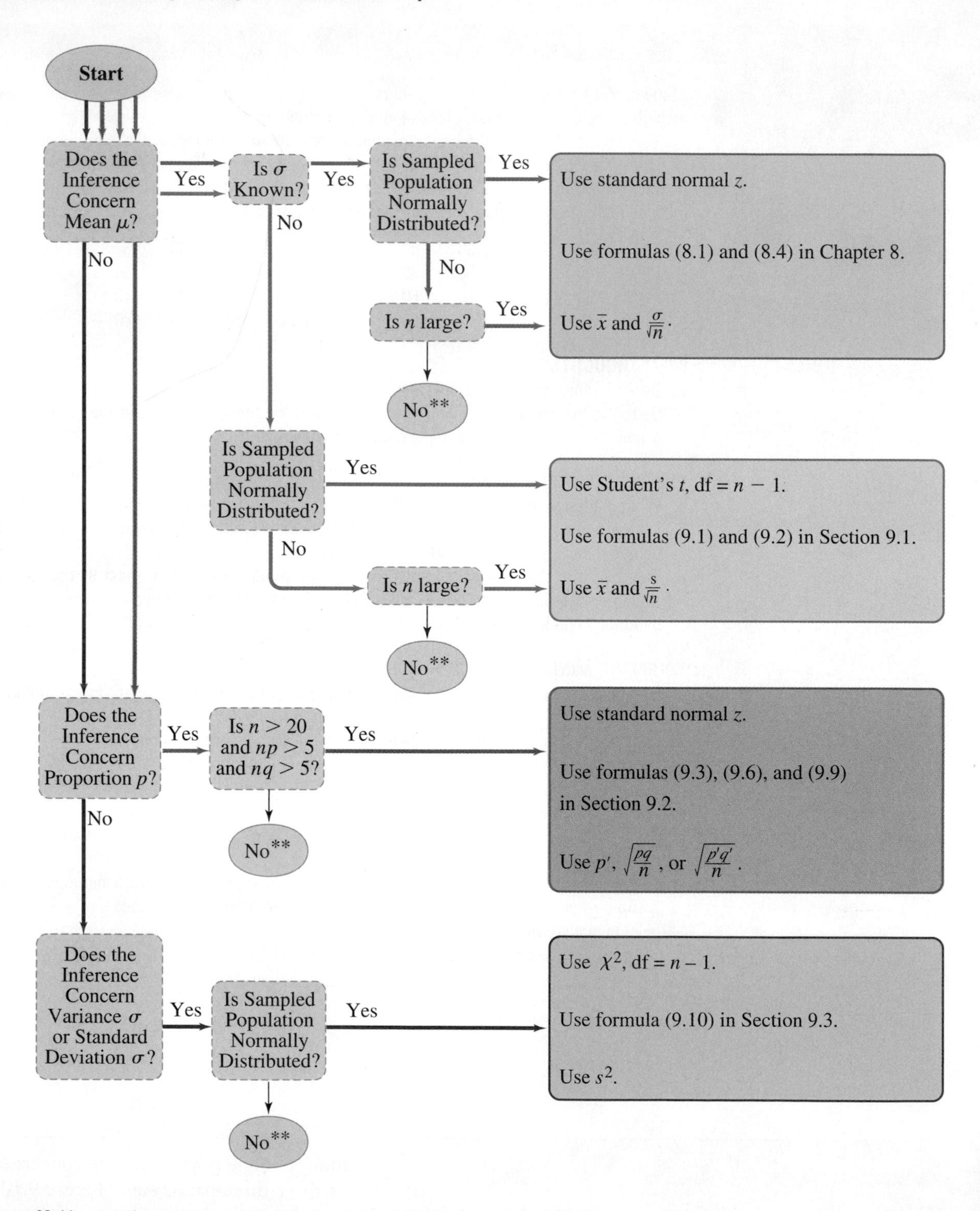

No** means that a nonparametric technique (normal distribution not required) is used: see Chaper 14.

In this chapter we also used the maximum error of estimate, formula (9.7), to determine the size of the sample required to make estimations about the population proportion with the desired accuracy. In Application 9.2 the margin of error reported by the media is described, and its relationship to the maximum error of estimate, as presented in this chapter, is discussed. By combining the reported point estimate and the sample size, we can determine the corresponding binomial proportion maximum error of estimate. Most polls and surveys use the 95% confidence level and then use the maximum error as an estimate for margin of error and do not report a level of confidence, as Humphrey Taylor explained.

In the next chapter we will discuss inferences about two populations whose respective means, proportions, and standard deviations are to be compared.

Chapter Exercises

9.126 A natural-gas utility is considering a contract for purchasing tires for its fleet of service trucks. The decision will be based on expected mileage. For a random sample of 100 tires tested, the mean mileage was 36,000 and the standard deviation was 2000 miles. Estimate the mean mileage that the utility should expect from these tires using a 98% confidence interval.

9.127 One objective of a large medical study was to estimate the mean physician fee for cataract removal. For 25 randomly selected cataract cases the mean fee was found to be $1550 with a standard deviation of $125. Set a 99% confidence interval on μ, the mean fee for all physicians. Assume fees are normally distributed.

9.128 Oranges are selected at random from a large shipment that just arrived. The sample was taken to estimate the size (circumference, in inches) of the oranges. The sample data are summarized as $n = 100$, $\Sigma x = 878.2$, and $\Sigma(x - \bar{x})^2 = 49.91$.

a. Determine the sample mean and standard deviation.
b. What is the point estimate for μ, the mean circumference of all oranges in the shipment?
c. Find the 95% confidence interval for μ.

9.129 Ex09-129 Molds are used in the manufacture of contact lenses so that the lens material, on proper preparation and curing, will be consistent and meet designated dimensional criteria. Molds were fabricated and a critical dimension measured for 15 randomly selected molds. (Data have been doubly coded to ensure propriety.) (Courtesy of Bausch & Lomb)

140	130	15	180	95	135	220	105
195	110	150	150	130	120	120	

a. Construct a histogram and find the mean and standard deviation.
b. Demonstrate how this set of data satisfies the assumptions for inference.
c. Find the 95% confidence interval for μ.
d. Interpret the meaning of the confidence interval.

9.130 A company claims that its battery lasts no less than 42.5 hours in continuous use in a specified toy. A simple random sample of batteries yields a sample mean lifetime of 41.89 hr with a standard deviation of 4.75 hr. A computer calculates a test statistic of $t = -1.09$ and a p-value of 0.139. If the test utilizes 71 degrees of freedom, then what is the best estimate of the sample size?

9.131 A manufacturer of television sets claims that the maintenance expenditures for its product will average no more than $50 during the first year following the expiration of the warranty. A consumer group has asked you to substantiate or discredit the claim. The results of a random sample of 50 owners of such television sets showed that the mean expenditure was $61.60 and the standard deviation was $32.46. At the 0.01 level of significance, should you conclude that the producer's claim is true or not likely to be true?

9.132 Ex09-132 In a large cherry orchard the average yield has been 4.35 tons per acre for the last several years. A new fertilizer was tested on 15 randomly selected one-acre plots. The yields from these plots follow:

3.56	5.00	4.88	4.93	3.92	4.25	5.12	5.13
5.35	4.81	3.48	4.45	4.72	4.79	4.45	

At the 0.05 level of significance, do we have sufficient evidence to claim that there has been a significant increase in production? Assume yields per acre are normally distributed.

9.133 Ex09-133 The water pollution readings at State Park Beach seem to be lower than last year. A random sample of 12 readings was selected from this year's daily readings:

3.5	3.9	2.8	3.1	3.1	3.4
4.8	3.2	2.5	3.5	4.4	3.1

Does this sample provide sufficient evidence to conclude that the mean of this year's pollution readings is significantly lower than last year's mean of 3.8 at the 0.05 level? Assume that all such readings have a normal distribution.

9.134 Ex09-134 It has been suggested that more abnormal male children tend to be born to older-than-average parents. Case histories of 20 abnormal males were obtained, along with the ages of their 20 mothers:

31	21	29	28	34	45	21	41	27	31
43	21	39	38	32	28	37	28	16	39

The mean age at which mothers in the general population give birth is 28.0 years.

a. Calculate the sample mean and standard deviation.
b. Does the sample give sufficient evidence to support the claim that more abnormal male children have older-than-average mothers? Use $\alpha = 0.05$. Assume ages have a normal distribution.

9.135 Ex09-135 Twenty-four oat-producing counties were randomly identified from across the United States for the purpose of testing the claim: The 2001 oat crop mean yield is less than 60 bushels per acre. For each county identified, the yield was obtained, in bushels of oats per harvested acre:

44.0	65.0	78.5	50.0	52.5	51.0	47.5	67.5
76.7	33.3	73.6	20.0	57.0	52.0	68.6	42.9
63.0	80.0	37.0	70.0	43.0	30.0	60.0	67.5

Source: http://www.usda.gov/nass/graphics/county01/data/ot01.csv

a. Are the test assumptions satisfied? Explain.
b. Complete the test using $\alpha = 0.05$.

9.136 Ex09-136 Presented here are 100 measurements of the velocity of light in air (km/sec) made by Albert Michelson, an American physicist, from June 5 to July 2, 1879. The measurements have had 299,000 subtracted from them and then been adjusted for corrections used by Michelson. In this form, the true constant value for the velocity of light in air becomes 734.5 km/sec. Do Michelson's measurements support the true value that he was trying to measure? Use a 0.01 level of significance.

850	740	900	1070	930	850	950	980
980	880	1000	980	930	650	760	810
1000	1000	960	960	960	940	960	940
880	800	850	880	900	840	830	790
810	880	880	830	800	790	760	800
880	880	880	860	720	720	620	860
970	950	880	910	850	870	840	840
850	840	840	840	890	810	810	820
800	770	760	740	750	760	910	920
890	860	880	720	840	850	850	780
890	840	780	810	760	810	790	810
820	850	870	870	810	740	810	940
950	800	810	870				

Source: http://lib.stat.cmu.edu/DASL/Stories/SpeedofLight.html

Note: The currently accepted "true" value is 299,792.5 km/sec with no adjustments.

9.137 Even with a heightened awareness of beef quality, 82% of Americans indicated their recent burger-eating behavior has remained the same, according to a recent T.G.I. Friday's® restaurants random survey of 1027 Americans. In fact, half of Americans eat at least one beef burger each week. That's a minimum of 52 burgers each year. (Source: Harris Interactive/Yankelovich Partners for T.G.I. Friday's restaurants: http://www.knoxville3.com/fridays/News/burger.htm.)

a. What is the point estimate for the proportion of all Americans who eat at least one beef burger per week?
b. Find the 98% confidence interval for the true proportion p in the binomial situation where $n = 1027$ and the observed proportion is 1/2.
c. Use the results of part b to estimate the percentage of all Americans who eat at least one beef burger per week.

9.138 The marketing research department of an instant-coffee company conducted a survey of married men to determine the proportion who preferred their brand. Twenty of the 100 men in the random sample preferred the company's brand. Use a 95% confidence interval to estimate the proportion of all married men who prefer this company's brand of instant coffee. Interpret your answer.

9.139 A company is drafting an advertising campaign that will involve endorsements by noted athletes. In order for the campaign to succeed, the endorser must be both highly respected and easily recognized. A random sample of 100 prospective customers are shown photos of various athletes. If the customer recognizes an athlete, then the customer is asked whether he or she respects the athlete. In the case of a top woman golfer, 16 of the 100 respondents recognized her picture and indicated that they also respected her. At the 95% level of confidence, what is the true proportion for which this woman golfer is both recognized and respected?

9.140 A local auto dealership advertises that 90% of customers whose autos were serviced by their service department are pleased with the results. As a researcher, you take exception to this statement because you are aware that many people are reluctant to express dissatisfaction even if they are not pleased. A research experiment was set up in which those in the sample had received service by this dealer within the past two weeks. During the interview, the individuals were led to believe that the interviewer was new in town and was considering taking his car to this dealer's service department. Of the 60 sampled, 14 said that they were dissatisfied and would not recommend the department.

a. Estimate the proportion of dissatisfied customers using a 95% confidence interval.
b. Given your answer to part a, what can you conclude about the dealer's claim?

9.141 In obtaining the sample size to estimate a proportion, the formula $n = \frac{[z(\alpha/2)]^2 pq}{E^2}$ is used. If a reasonable estimate of p is not available, then it is suggested that $p = 0.5$ be used because this will give the maximum value for n. Calculate the value of $pq = p(1 - p)$ for $p = 0.1, 0.2, 0.3, \ldots, 0.8, 0.9$ to obtain some idea about the behavior of the quantity pq.

9.142 S The so-called "glass ceiling" and numerous other reasons have prevented women from reaching the top of the corporate employment ladder compared to men. *Fortune* ("The Global Glass Ceiling", October 12, 1998) reported that women make up 11% of corporate directors in the *Fortune* 500 companies, even though women represent a much higher percentage (40%) of the total work force employed in management positions in America. The percentage of female corporate directors and officers, however, has been rising steadily, and power appears to be shifting to people who are not in traditional corporate America. You wish to conduct a study to estimate the percentage of female corporate directors in the companies with headquarters in your state. Assume the population proportion is 11% as reported by *Fortune*. What sample size must you use if you want your estimate to be within:

a. 0.03 with 90% confidence
b. 0.06 with 95% confidence
c. 0.09 with 99% confidence

9.143 S The CEO of a small business wishes to hire your consulting firm to conduct a simple random sample of its customers. She wants to determine the proportion of her customers who consider her company their primary source of her product. She requests the margin of error in the proportion be no greater than 3% with 95% confidence. Earlier studies have indicated that the approximate proportion is 37%.

a. What is the minimum size of the sample that you would recommend to meet the requirements of your client if you use the earlier results?
b. What is the minimum size of the sample that you would recommend to meet the requirements of your client if you ignore the earlier results?
c. Is the earlier approximate proportion of value in conducting the survey? Explain.

9.144 The article "Bringing Up Adultolescents" in the March 25, 2002, issue of *Newsweek* quoted an online survey by *Monster-TRAK.com*. The survey found that 60% of college students planned to live at home after graduation. How large a sample size would you need to estimate the true proportion of students who plan on living at home after graduation to within 2% with 98% confidence?

9.145 A machine is considered to be operating in an acceptable manner if it produces 0.5% or fewer defective parts. It is not performing in an acceptable manner if more than 0.5% of its production is defective. The null hypothesis H_o: $p = 0.005$ is tested against the alternative hypothesis H_a: $p > 0.005$ by taking a random sample of 50 parts produced by the machine. The null hypothesis is rejected if two or more defective parts are found in the sample. Find the probability of the type I error.

9.146 You are interested in comparing the hypothesis $p = 0.8$ against the alternative $p < 0.8$. In 100 trials you observe 73 successes. Calculate the p-value associated with this result.

9.147 An instructor asks each of the 54 members of his class to write down "at random" one of the numbers 1, 2, 3, . . . , 13, 14, 15. Since the instructor believes that students like to gamble, he considers that 7 and 11 are lucky numbers. He counts the number of students, x, who selected 7 or 11. How large must x be before the hypothesis of randomness can be rejected at the 0.05 level?

9.148 Today's newspapers and magazines often report the findings of surveys about various aspects of life. *The American Gender Evolution* (November 1990) reports "62% of the men believe both partners should earn a living." Other bits of information given in the article are: "telephone survey of 1201 adults" and "has a sampling error of ±3 percent." Relate this information to the statistical inferences you have been studying in this chapter.

a. Is a percentage of people a population parameter, and if so, how is it related to any of the parameters that we have studied?
b. Based on the information given, find the 95% confidence interval for the true proportion of men who believe both partners should earn a living.
c. Explain how the terms "point estimate," "level of confidence," "maximum error of estimate," and "confidence interval" relate to the values reported in the article and to your answers in part b.

9.149 To test the hypothesis that the standard deviation of test scores is 12, a random sample of 40 students was tested. The sample variance of their scores was found to be 155. Does this sample provide sufficient evidence to show that the standard deviation differs from 12 at the 0.05 level of significance?

9.150 Bright-Lite claims that its 60-watt light bulb burns with a length of life that is approximately normally distributed with a standard deviation of 81 hours. A sample of 101 bulbs had a variance of 8075. Is this sufficient evidence to reject Bright-Lite's claim in favor of the alternative, "The standard deviation is longer than 81 hours," at the 0.05 level of significance?

9.151 A production process is considered out of control if the produced parts have a mean length that is different from 27.5 mm or a standard deviation that is greater than 0.5 mm. A random sample of 30 parts yields a sample mean of 27.63 mm and a sample standard deviation of 0.87 mm. If we assume part length is a normally distributed variable, does this sample indicate that the process should be adjusted in order to correct the standard deviation of the product? Use $\alpha = 0.05$.

9.152 Julia Jackson operates a franchised restaurant that specializes in soft ice cream cones and sundaes. Recently she received a letter from corporate headquarters warning her that her shop was in danger of losing its franchise because the average sales per customer had dropped "substantially below the average for the rest of the corporation." The statement may be true, but Julie is convinced that such a statement is invalid as a justification for threatening a closing. The variation in sales at her restaurant is bound to be larger than most, primarily because she serves more children, elderly, and single adults than the large families that run up big bills at the other restaurants. Therefore, her average ticket is likely to be smaller and exhibit greater variability. To prove her point, Julie obtained the sales records from the whole company and found that the standard deviation was $2.45 per sales ticket. She then conducted a study of the last 71 sales at her store and found a standard deviation of $2.95 per ticket. Is the variability in sales at Julie's franchise, at the 0.05 level of significance, greater than the variability for the company?

Vocabulary and Key Concepts

Be able to define each term. Pay special attention to the key terms, which are printed in **red.** In addition, describe each term in your own words and give an example of each. Your examples should not be ones given in class or in the textbook. The bracketed numbers indicate the chapter(s) in which the term appeared previously, but you should define the terms again to show increased understanding of their meaning. Page numbers indicate the first appearance of the term in Chapter 9.

assumptions [8] (pp. 417, 432, 452)
binomial experiment [5] (p. 430)
calculated value [8] (pp. 421, 439, 454)
chi-square (p. 449)
conclusion [8] (pp. 421, 440, 454)
confidence interval [8] (pp. 418, 432)
critical region [8] (pp. 421, 439, 454)
critical value [8] (pp. 414, 439, 454)
decision [8] (pp. 421, 440, 454)
degrees of freedom (pp. 414, 449)
hypothesis test [8] (pp. 420, 438, 453)
inference [8] (pp. 412, 430, 449)
level of confidence [8] (pp. 418, 432)
level of significance [8] (pp. 421, 439, 453)
maximum error of estimate [8] (pp. 419, 433, 436)
observed binomial probability, p' (p. 431)
p-value [8] (pp. 421, 439, 455)
parameter [1, 8] (pp. 414, 430, 449)
proportion [6] (p. 432)
random variable [5, 6] (p. 431)
rule of thumb (p. 431)
sample size [8] (pp. 420, 436)
sample statistic [1, 2] (p. 431)
σ known[8] (p. 412)
σ unknown (p. 412)
standard error [7, 8] (pp. 412, 432)
standard normal, z [2, 6, 8] (pp. 412, 432)
Student's t (p. 412)
test statistic [8] (pp. 420, 438, 453)
unbiased estimator [8] (p. 431)

Chapter Practice Test

PART I: KNOWING THE DEFINITIONS

Answer "True" if the statement is always true. If the statement is not always true, replace the words shown in bold with words that make the statement always true.

9.1 The Student's t-distributions have an approximately normal distribution but are **more** dispersed than the standard normal distribution.

9.2 The **chi-square** distribution is used for inferences about the mean when σ is unknown.

9.3 The **Student's t**-distribution is used for all inferences about a population's variance.

9.4 If the test statistic falls in the critical region, the null hypothesis has **been proven true.**

9.5 When the test statistic is t and the number of degrees of freedom gets very large, the critical value of t is very close to that of the **standard normal z.**

9.6 When making inferences about one mean when the value of σ is not known, the **z-score** is the test statistic.

9.7 The chi-square distribution is a skewed distribution whose mean value is **2** for df > 2.

9.8 Often the concern with testing the variance (or standard deviation) is to keep its size under control or relatively small. Therefore, many of the hypothesis tests with chi-square are **one-tailed.**

9.9 $\sqrt{npq}$ is the standard error of proportion.

9.10 The sampling distribution of p' is distributed approximately as a **Student's t**-distribution.

PART II: APPLYING THE CONCEPTS

Answer all questions, showing all formulas, substitutions, and work.

9.11 Find each value:
a. $z(0.02)$ b. $t(18, 0.95)$ c. $\chi^2(25, 0.95)$

9.12 A random sample of 25 data was selected from a normally distributed population for the purpose of estimating the population mean, μ. The sample statistics are $n = 25$, $\bar{x} = 28.6$, and $s = 3.50$.
a. Find the point estimate for μ.
b. Find the maximum error of estimate for the 0.95 confidence interval estimate.
c. Find the lower confidence limit (LCL) and the upper confidence limit (UCL) for the 0.95 confidence interval estimate for μ.

9.13 Thousands of area elementary school students were recently given a nationwide standardized exam to test their composition skills. If 64 of a random sample of 100 students passed this exam, construct the 0.98 confidence interval estimate for the true proportion of all area students who passed the exam.

9.14 State the null (H_o) and the alternative (H_a) hypotheses that would be used to test each of these claims:
a. The mean weight of professional basketball players is no more than 225 lb.
b. Approximately 40% of daytime students own their own car.
d. The standard deviation for the monthly amounts of rainfall in Monroe County is less than 3.7 in.

9.15 Determine the level of significance, test statistic, critical region, and critical values(s) that would be used in completing each hypothesis test using the classical approach with $\alpha = 0.05$.
a. H_o: $\mu = 43$ vs. H_a: $\mu < 43$, $\sigma = 6$
b. H_o: $\mu = 95$ vs. H_a: $\mu \neq 95$, σ unknown, $n = 22$
c. H_o: $p = 0.80$ vs. H_a: $p > 0.80$
d. H_o: $\sigma = 12$ vs. H_a: $\sigma \neq 12$, $n = 28$

9.16 The automobile manufacturer of the Alero claims that the typical Alero will average 32 mpg of gasoline. An independent consumer group is somewhat skeptical of this claim and thinks the mean gas mileage is less than the 32 claimed. A sample of 24 randomly selected Aleros produced these sample statistics: mean 30.15 and standard deviation 4.87. At the 0.05 level of significance, does the consumer group have sufficient evidence to refute the manufacturer's claim?

9.17 A coffee machine is supposed to dispense 6 fluid ounces of coffee into a paper cup. In reality, the amount dispensed varies from cup to cup. However, if the machine is operating properly, the standard deviation of the amounts dispensed should be 0.1 oz or less. A random sample of 15 cups produced a standard deviation of 0.13 oz. Does this represent sufficient evidence, at the 0.10 level of significance, to conclude that the machine is not operating properly?

9.18 An unhappy customer is frustrated with the waiting time at the post office when buying stamps. Upon registering his complaint, he was told, "You wait more than 1 minute for service no more than half of the time when you only buy stamps." Not believing this to be the case, the customer collected some data from people who had just purchased stamps only. The sample statistics are $n = 60$ and $x = n$ (wait more than 1 minute) $= 35$. At the 0.02 level of significance, does our unhappy customer have sufficient evidence to refute the post office's claim?

PART III: UNDERSTANDING THE CONCEPTS

9.19 Student B says the range of a set of data may be used to obtain a crude estimate for the standard deviation of a population. Student A is not sure. How will student B correctly explain how and under what circumstances his statement is true?

9.20 Is it the null hypothesis or the alternative hypothesis that the researcher usually believes to be true? Explain.

9.21 When you reject a null hypothesis, student A says that you are expressing disbelief in the value of the parameter as claimed in the null hypothesis. Student B says that instead you are expressing the belief that the sample statistic came from a population other than the one related to the parameter claimed in the null hypothesis. Who is correct? Explain.

9.22 "The Student t-distribution must be used when making inferences about the population mean, μ, when the population standard deviation, σ, is not known" is a true statement. Student A states that the z-score sometimes plays a role when the t-distribution is used. Explain the conditions that exist and the role played by z that make student A's statement correct.

9.23 Student A says that the percentage of the sample means that fall outside the critical values of the sampling distribution determined by a true null hypothesis is the p-value for the test. Student B says that the percentage student A is describing is the level of significance. Who is correct? Explain.

9.24 Student A carries out a study in which she is willing to run a 1% risk of making a type I error. She rejects the null hypothesis and claims that her statistic is significant at the 99% level of confidence. Student B argues that student A's claim is not properly worded. Who is correct? Explain.

9.25 Student A claims that when you employ a 95% confidence interval to determine an estimation, you do not know for sure whether or not your inference is correct (the parameter is contained within the interval). Student B claims that you do know; you have shown that the parameter cannot be less than the lower limit or greater than the upper limit of the interval. Who is right? Explain.

9.26 Student A says that the best way to improve a confidence interval estimate is to increase the level of confidence. Student B argues that using a high confidence level does not really improve the resulting interval estimate. Who is right? Explain.

CHAPTER

10 Inferences Involving Two Populations

Chapter Outline and Objectives

CHAPTER CASE STUDY

A statistical look at students, credit cards, and debt.

10.1 DEPENDENT AND INDEPENDENT SAMPLES

Dependent samples result from using **paired** subjects; **independent** samples are obtained by using **unrelated sets** of subjects.

10.2 INFERENCES CONCERNING THE MEAN DIFFERENCE USING TWO DEPENDENT SAMPLES

Using dependent samples helps **control** otherwise **untested factors.**

10.3 INFERENCES CONCERNING THE DIFFERENCE BETWEEN MEANS USING TWO INDEPENDENT SAMPLES

The comparison of the **mean values of two populations** is a common practice.

10.4 INFERENCES CONCERNING THE DIFFERENCE BETWEEN PROPORTIONS USING TWO INDEPENDENT SAMPLES

Questions comparing **two binomial probabilities** are answered using the standard normal variable z.

10.5 INFERENCES CONCERNING THE RATIO OF VARIANCES USING TWO INDEPENDENT SAMPLES

To investigate the relationship between the **variances of two populations,** we use the ***F*-distribution.**

RETURN TO CHAPTER CASE STUDY

Since we will be studying several different situations, Figure 10.1 (p. 473) is offered as a "road map" to help you organize these various inference techniques.

Chapter Case Study

Students, Credit Cards, and Debt

© Wayne Scarberry/The Image Works

We all know that college is expensive and credit cards are readily available. We also know and believe that young adults need experience at handling their own finances. But two old adages, "Buyers beware" and "Know the facts," probably should be needed in a college student's approach to credit card use. Here is a portion of the report Nellie Mae published in 2002.

Undergraduate Students and Credit Cards

Not surprisingly, the freshman population has a lower overall percentage of credit card holders and lower debt levels on their cards than upperclassmen. However, more than half of all freshmen (54%) had at least one credit card with the average number of cards being 2.5, and, among those who have cards, 26% have four or more. Freshman debt levels are also lower than the overall counts in all categories. Their median debt amount is $901, lower than the overall median of $1,770; their average balance is $1,533 vs. $2,327 overall; those with balances exceeding $7,000 are 4% as opposed to 6% overall; and those with high-level balances between $3,000 and $7,000 are 8% compared to 21% overall.

Credit Card Usage by Grade Level	01Fresh	02Soph	03Jr	04/05Sr/+
Percentage who have credit cards	54%	92%	87%	96%
Average number of credit cards	2.5	3.67	4.5	6.13
Percentage who have 4 or more cards	26%	44%	50%	66%
Average credit card debt	$1,533	$1,825	$2,705	$3,262
Median credit card debt	$901	$1,564	$1,872	$2,185
Percentage with balances between $3,000-$7,000	8%	18%	24%	31%
Percentage with balances exceeding $7,000	4%	4%	7%	9%

As students progress through their four (or more) years in college, there is a steady increase in usage rates and balances each year from first to final year. By graduation, most students have more than doubled their average debt, and almost tripled the number of cards they hold. Most dramatic, however, is the 70% jump occurring between freshman and sophomore year in the percentage of students with at least one card—from 54% to 92% of the total population. Once freshmen arrive on campus, there are many tempting incentives to sign up for new credit cards and many opportunities to use them. The fact that the average number of cards per student continues to increase is not surprising. The proliferation of on-campus, mail, and Internet offers of free gifts, bonus airline miles, and low introductory rates for each new card is difficult for students to resist.

Random samples of 200 freshman and 200 sophomore college students were asked, "Do you have your own credit card?" Ninety-seven freshmen and 187 sophomores answered that they had one or more credit cards with their name on the cards. The first 40 freshmen and the first 44 sophomores who answered yes were then asked for their current total credit card debt balance. The total credit card debt balances are listed here.

Freshman Debt Balance, $n = 40$ CS10

1011.97	3998.72	2447.93	2457.39	855.63	1602.74	912.39	2478.49
1014.39	444.48	1293.36	1065.82	989.56	412.53	321.85	2578.39
2103.35	2917.65	3218.54	1384.34	4368.28	244.33	190.24	2778.17
1702.65	616.31	491.73	2205.95	1130.09	2402.92	767.42	657.83
1150.78	1102.28	154.11	1494.48	1324.01	2054.76	1762.31	644.31

Sophomore Debt Balance, $n = 44$

690.08	595.04	2983.50	1761.21	1020.91	2143.18	3048.87	1314.36
1378.99	1456.10	1893.37	1287.47	284.93	7135.64	3194.07	2565.71
3298.15	2747.14	839.57	393.20	1422.73	1652.03	2214.77	1126.76
3433.80	3962.25	1849.23	3037.52	328.29	3074.19	1194.87	889.40
1480.94	486.22	1688.81	1317.27	2624.01	2286.74	5341.94	633.37
873.18	3601.18	2023.29	4898.46				

ANSWER NOW 10.1

a. **What percentage of each group has their own credit card? How does this compare with the findings reported by the Nellie Mae organization?**

b. **Describe the shape of the distribution you believe total credit card debt will display. Explain.**

c. **Construct a histogram of total credit card debt for each class. Use the same class intervals for both histograms. Compare your findings to your thoughts in part b.**

d. **Find the mean and standard deviation for each data set.**

Compare your findings to Nellie Mae's.

e. **Estimate the mean credit card debt for freshmen with a 95% confidence interval.**

f. **Does the sampled population of sophomores have a significantly higher credit card debt than the national average reported by Nellie Mae using $\alpha = 0.05$?**

After completing Chapter 10, further investigate the Chapter Case Study in the Return to Chapter Case Study with Exercises 10.109, 10.110, and 10.111 (p. 526).

10.1 Dependent and Independent Samples

In this chapter we are going to study the procedures for making inferences about two populations. When comparing two populations, we need two samples, one from each population. Two basic kinds of samples can be used: independent and dependent. The dependence or independence of two samples is determined by the sources of the data. A **source** can be a person, an object, or anything that yields a piece of data. If the same set of sources or related sets are used to obtain the data representing both populations, we have **dependent samples.** If two unrelated sets of sources are used, one set from each population, we have **independent samples.** The following illustrations should clarify these ideas.

FIGURE 10.1

"Road Map" to Two Population Inferences

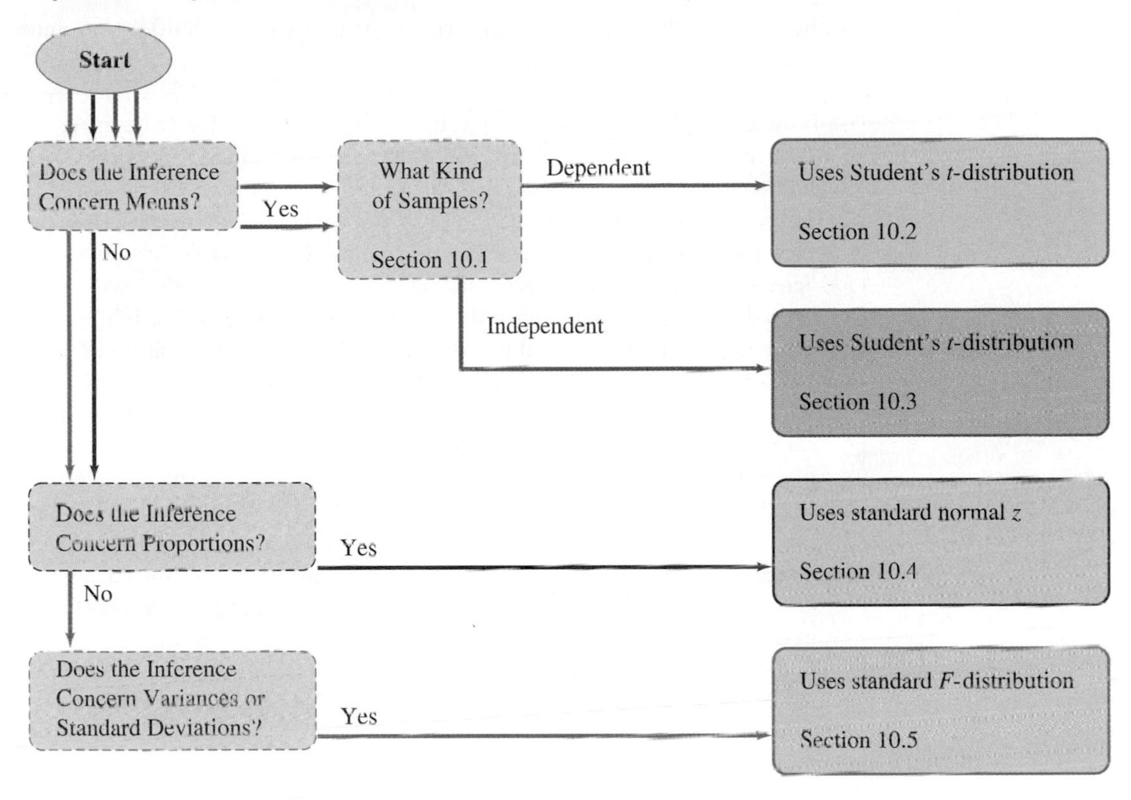

Illustration 10.1 Dependent Versus Independent Samples

A test will be conducted to see whether the participants in a physical fitness class actually improve in their level of fitness. It is anticipated that approximately 500 people will sign up for this course. The instructor decides that she will give 50 of the participants a set of tests before the course begins (a pretest), and then she will give another set of tests to 50 participants at the end of the course (a posttest). Two sampling procedures are proposed:

Plan A: Randomly select 50 participants from the list of those enrolled and give them the pretest. At the end of the course, make a second random selection of size 50 and give them the posttest.

Plan B: Randomly select 50 participants and give them the pretest; give the same set of 50 the posttest when they complete the course.

Plan A illustrates independent sampling; the sources (the class participants) used for each sample (pretest and posttest) were selected separately. Plan B illustrates dependent sampling; the sources used for both samples (pretest and posttest) are the same. ■

Typically, when both a pretest and a posttest are used, the same subjects participate in the study. Thus, pretest versus posttest (before versus after) studies usually use dependent samples.

Application 10.1 Exploring the Traits of Twins

Studies that involve identical twins are a natural for the dependent sampling technique discussed in this section.

A New Study Shows That Key Characteristics May Be Inherited

Like many identical twins reared apart, Jim Lewis and Jim Springer found they had been leading eerily similar lives. Separated four weeks after birth in 1940, the Jim twins grew up 45 miles apart in Ohio and were reunited in 1979. Eventually they discovered that both drove the same model blue Chevrolet, chain-smoked Salems, chewed their fingernails, and owned dogs named Toy. Each had spent a good deal of time vacationing at the same three-block strip of beach in Florida. More important, when tested for such personality traits as flexibility, self-control, and sociability, the twins responded almost exactly alike.

The project is considered the most comprehensive of its kind. The Minnesota researchers report the results of six-day tests of their subjects, including 44 pairs of identical twins who were brought up apart. Well-being, alienation, aggression, and the shunning of risk or danger were found to owe as much or more to nature as to nurture. Of eleven key traits or clusters of traits analyzed in the study, researchers estimated that a high of 61 percent of what they call "society potency" (a tendency toward leadership or dominance) is inherited, while "social closeness" (the need for intimacy, comfort, and help) was lowest, at 33 percent.

ANSWER NOW 10.2
Explain why studies involving identical twins result in dependent samples of data.

Illustration 10.2 Dependent Versus Independent Samples

A test is being designed to compare the wearing quality of two brands of automobile tires. The automobiles will be selected and equipped with the new tires and then driven under "normal" conditions for one month. Then a measurement will be taken to determine how much wear took place. Two plans are proposed:

Plan C: A sample of cars will be selected randomly, equipped with brand A tires, and driven for the month. Another sample of cars will be selected, equipped with brand B tires, and driven for the month.

Plan D: A sample of cars will be selected randomly, equipped with one tire of brand A and one tire of brand B (the other two tires are not part of the test), and driven for the month.

We suspect that many other factors must be taken into account when testing automobile tires—such as age, weight, and mechanical condition of the car; driving habits of drivers; location of the tire on the car; and where and how much the car is driven. However, at this time we are trying only to illustrate dependent and independent samples. Plan C is independent (unrelated sources), and plan D is dependent (common sources). ■

ANSWER NOW 10.3

a. **Describe how you could select two independent samples from among your classmates to compare the heights of female and male students.**

b. **Describe how you could select two dependent samples from among your classmates to compare their heights as they entered high school to when they entered college.**

Independent and dependent samples each have their advantages; these will be emphasized later. Both methods of sampling are often used.

Exercises

10.4 In trying to estimate the amount of growth that took place in the trees planted by the County Parks Commission recently, 36 trees were randomly selected from the 4000 planted. The heights of these trees were measured and recorded. One year later another set of 42 trees was randomly selected and measured. Do the two sets of data (36 heights, 42 heights) represent dependent or independent samples? Explain.

10.5 Twenty people were selected to participate in a psychology experiment. They answered a short multiple-choice quiz about their attitudes on a particular subject and then viewed a 45-minute film. The following day the same 20 people were asked to answer a follow-up questionnaire about their attitudes. At the completion of the experiment, the experimenter will have two sets of scores. Are these two samples dependent or independent? Explain.

10.6 An experiment is designed to study the effect diet has on the uric acid level. Twenty white rats are used for the study. Ten rats are randomly selected and given a junk-food diet. The other ten receive a high-fiber, low-fat diet. The uric acid levels of the two groups are determined. Do the resulting sets of data represent dependent or independent samples? Explain.

10.7 Two different types of disc centrifuges are used to measure the particle size in latex paint. A gallon of paint is randomly selected, and ten specimens are taken from it for testing on each of the centrifuges. There will be two sets of data, ten data each, as a result of the testing. Do the two sets of data represent dependent or independent samples? Explain.

10.8 An insurance company is concerned that garage A charges more for repair work than garage B charges. It plans to send 25 cars to each garage and obtain separate estimates for the repairs needed for each car.

a. How can the company do this and obtain independent samples? Explain in detail.

b. How can the company do this and obtain dependent samples? Explain in detail.

10.9 A study is being designed to determine the reasons adults choose to follow a healthy diet plan. One thousand men and 1000 women will be surveyed. Upon completion, the reasons that men choose a healthy diet will be compared to the reasons that women choose a healthy diet.

a. How can the data be collected if independent samples are to be obtained? Explain in detail.

b. How can the data be collected if dependent samples are to be obtained? Explain in detail.

10.10 Suppose that 400 students in a certain college are taking elementary statistics this semester. Two samples of size 25 are needed in order to test some precourse skill against the same skill after the students complete the course.

a. Describe how you would obtain your samples if you were to use dependent samples.

b. Describe how you would obtain your samples if you were to use independent samples.

10.2 Inferences Concerning the Mean Difference Using Two Dependent Samples

The procedures for comparing two population means are based on the relationship between two sets of sample data, one sample from each population. When dependent samples are involved, the data are thought of as "paired data." The data may be paired as a result of being obtained from "before" and "after" studies; from pairs of identical twins as in Application 10.1; from a "common" source, as with the amounts of tire wear for each brand in plan D of Illustration 10.2; or from matching two subjects with similar traits to form "matched pairs." The pairs of data values are compared directly to each other by using the difference in their numerical values. The resulting difference is called a **paired difference.**

PAIRED DIFFERENCE

$$d = x_1 - x_2 \tag{10.1}$$

Using paired data this way has a built-in ability to remove the effect of otherwise uncontrolled factors. The tire-wear problem in Illustration 10.2 is an excellent example of such additional factors. The wearing ability of a tire is greatly affected by a multitude of factors: the size, weight, age, and condition of the car, the driving habits of the driver, the number of miles driven, the condition and types of roads driven on, the quality of the material used to make the tire, and so on. We create paired data by mounting one tire from each brand on the same car. Since one tire of each brand will be tested under the same conditions, same car, same driver, and so on, the extraneous causes of wear are neutralized.

A test was conducted to compare the wearing quality of the tires produced by two tire companies using plan D, as described in Illustration 10.2. All the aforementioned factors will have an equal effect on both brands of tires, car by car. One tire of each brand was placed on each of six test cars. The position (left or right side, front or back) was determined with the aid of a random-number table. Table 10.1 lists the amounts of wear (in thousandths of an inch) that resulted from the test.

TABLE 10.1 Amount of Tire Wear Ta10-01

Car	1	2	3	4	5	6
Brand A	125	64	94	38	90	106
Brand B	133	65	103	37	102	115

Since the various cars, drivers, and conditions are the same for each tire of a paired set of data, it makes sense to use a third variable, the paired difference ***d***. Our two dependent samples of data may be combined into one set of d values, where $d = \text{B} - \text{A}$.

Car	1	2	3	4	5	6
$d = \text{B} - \text{A}$	8	1	9	−1	12	9

The sample statistics needed will be the mean of the sample differences, $\bar{d}$:

$$\bar{d} = \frac{\sum d}{n} \tag{10.2}$$

and the standard deviation of the sample differences, s_d:

$$s_d = \sqrt{\frac{\sum d^2 - \left[\frac{(\sum d)^2}{n}\right]}{n - 1}} \tag{10.3}$$

FYI
Formulas (10.2) and (10.3) are adaptations of formulas (2.1) and (2.10).

Illustration 10.3 Calculating the Mean and Standard Deviation for Paired Differences

Find the mean and standard deviation of the paired differences in Table 10.1.

SOLUTION

The summary of data: $n = 6$, $\sum d = 38$, and $\sum d^2 = 372$.

FYI
n = number of dependent pairs

The mean:

$$\bar{d} = \frac{\sum d}{n}: \qquad \bar{d} = \frac{38}{6} = 6.333 = \mathbf{6.3}$$

The standard deviation:

$$s_d = \sqrt{\frac{\sum d^2 - \left[\frac{(\sum d)^2}{n}\right]}{n-1}}: \qquad s_d = \sqrt{\frac{372 - \left[\frac{(38)^2}{6}\right]}{6-1}} = \sqrt{26.27} = 5.13 = \mathbf{5.1}$$

■

ANSWER NOW 10.11
Consider this set of paired data:

Pairs	1	2	3	4	5
Sample A	3	6	1	4	7
Sample B	2	5	1	2	8

a. **Find the paired differences, d = A − B.**

b. **Find the mean $\bar{d}$ of the paired differences.**

c. **Calculate the standard deviation s_d of the paired differences.**

The difference between the two population means, when dependent samples are used (often called **"dependent means"**), is equivalent to the **mean of the paired differences.** Therefore, when an inference is to be made about the difference of two means and paired differences are used, the inference will in fact be about the mean of the paired differences. The sample mean of the paired differences will be used as the point estimate for these inferences.

In order to make inferences about the mean of all possible paired differences μ_d, we need to know about the *sampling distribution* of $\bar{d}$.

When paired observations are randomly selected from normal populations, the paired difference, $d = x_1 - x_2$, will be approximately normally distributed about a mean μ_d with a standard deviation of σ_d.

This is another situation in which the t-test for one mean is applied; namely, we wish to make inferences about an unknown mean (μ_d) where the random variable (d) involved has an approximately normal distribution with an unknown standard deviation (σ_d).

Inferences about the mean of all possible paired differences μ_d are based on samples of n dependent pairs of data and the **t-distribution** with $n - 1$ degrees of freedom, under the following assumption:

ASSUMPTION FOR INFERENCES ABOUT THE MEAN OF PAIRED DIFFERENCES μ_d

The paired data are randomly selected from normally distributed populations.

CONFIDENCE INTERVAL PROCEDURE

The $1 - \alpha$ **confidence interval for estimating the mean difference μ_d** is found using this formula:

FYI
Formula (10.4) is an adaptation of formula (9.1).

CONFIDENCE INTERVAL FOR μ_d

$$\bar{d} - t(\text{df}, \alpha/2) \cdot \frac{s_d}{\sqrt{n}} \quad \text{to} \quad \bar{d} + t(\text{df}, \alpha/2) \cdot \frac{s_d}{\sqrt{n}}, \qquad \text{where df} = n - 1 \qquad (10.4)$$

Illustration 10.4 Constructing a Confidence Interval for μ_d

Construct the 95% confidence interval for the mean difference in the paired data on tire wear, as reported in Table 10.1. The sample information is $n = 6$ pieces of paired data, $\bar{d} = 6.3$, and $s_d = 5.1$ (calculated in Illustration 10.3). Assume the amounts of wear are approximately normally distributed for both brands of tires.

SOLUTION

Step 1 **The Set-Up:**
Describe the population parameter of interest.
μ_d, the mean difference in the amounts of wear between the two brands of tires.

Step 2 **The Confidence Interval Criteria:**

a. **Check the assumptions.**
Both sampled populations are approximately normal.

b. **Identify the probability distribution and the formula to be used.**
The t-distribution with df $= 6 - 1 = 5$ and formula (10.4) will be used.

c. **State the level of confidence:** $1 - \alpha = 0.95$.

Step 3 **The Sample Evidence:**
Collect the sample information: $n = 6$, $\bar{d} = 6.3$, and $s_d = 5.1$.

Step 4 **The Confidence Interval:**

a. **Determine the confidence coefficient.**
This is a two-tailed situation with $\alpha/2 = 0.025$ in one tail. From Table 6 in Appendix B, $t(\text{df}, \alpha/2) = t(5, 0.025) = \mathbf{2.57}$.

For specific instructions about confidence coefficients and Table 6, see page 418.

ANSWER NOW 10.12

Find $t(15, 0.025)$. Describe the role this number plays in the confidence interval.

b. Find the maximum error of estimate.

Using the maximum error part of formula (10.4), we have

$$E = t(\text{df}, \alpha/2) \cdot \frac{s_d}{\sqrt{n}}: \qquad E = 2.57 \cdot \left(\frac{5.1}{\sqrt{6}}\right) = (2.57)(2.082) = 5.351 = \mathbf{5.4}$$

c. Find the lower and upper confidence limits.

$$\bar{d} \pm E$$

$$6.3 \pm 5.4$$

$$6.3 - 5.4 = \mathbf{0.9} \quad \text{to} \quad 6.3 + 5.4 = \mathbf{11.7}$$

Step 5 **The Results:**

State the confidence interval.

0.9 to 11.7 is the 95% confidence interval for μ_d. That is, with 95% confidence we can say that the mean difference in the amounts of wear is between 0.9 and 11.7 thousandths of an inch. ■

NOTE This confidence interval is quite wide, in part because of the small sample size. Recall from the central limit theorem that as the sample size increases, the standard error (estimated by $s_d/\sqrt{n}$) decreases.

ANSWER NOW 10.13

a. **Find the 95% confidence interval for μ_d given $n = 26$, $\bar{d} = 6.3$, and $s_d = 5.1$.**

b. **Compare your interval to the interval found in Illustration 10.4.**

Technology Instructions: $1 - \alpha$ Confidence Interval for Mean μ_d with Unknown Standard Deviation for Two Dependent Sets of Sample Data

MINITAB (Release 13)

Input the paired data into C1 and C2; then continue with:

```
Choose:   Stat > Basic Statistics > Paired t
Enter:    First sample: C1*
          Second sample: C2
Select:   Options
Enter:    Confidence level: 1 − α (ex. 0.95 or 95.0)
```

*Paired t evaluates the first sample minus the second sample.

Excel XP

Input the paired data into columns A and B; activate C1 or C2 (depending on whether column headings are used or not); then continue with:

```
Enter:    = A2 − B2* (if column headings are used)
Drag:     Bottom right corner of C2 down to give other dif-
          ferences
```

(continued)

Technology Instructions: $1 - \alpha$ Confidence Interval for Mean μ_d with Unknown Standard Deviation for Two Dependent Sets of Sample Data (Continued)

Excel XP (continued)

```
Choose:    Tools > Data Analysis Plus
                 > t-Estimate: Mean
Enter:     Input range: (C2:C20 or select cells)
Select:    Labels (if necessary)
Enter:     Alpha: α (ex. 0.05) > OK
```

*Enter the expression in the order that is needed: A2 − B2 or B2 − A2.

TI-83 Plus

Input the paired data into L1 and L2; then continue with the following, entering the appropriate values and highlighting Calculate:

```
Highlight:   L3
Enter:       L3= L1 − L2*
Choose:      STAT > TESTS > 8:Tinterval
```

```
TInterval
 Inpt:Data Stats
 List:L3
 Freq:1
 C-Level:.95
 Calculate
```

*Enter the expression in the order that is needed: L1 − L2 or L2 − L1.

The solution to Illustration 10.4 looks like this when solved in MINITAB:

```
Paired T for Brand B - Brand A
              N        Mean      StDev     SE Mean
Brand B       6        92.5      35.2      14.4
Brand A       6        86.2      30.9      12.6
Difference    6        6.33      5.13      2.09
95% CI for mean difference: (0.95, 11.71)
```

ANSWER NOW 10.14 Ex10-014

Use a computer or calculator to find the 95% confidence interval for estimating μ_d based on these paired data:

Before	75	68	40	30	43	65
After	70	69	32	30	39	63

HYPOTHESIS-TESTING PROCEDURE

When we test a null **hypothesis about the mean difference,** the test statistic used will be the difference between the sample mean $\bar{d}$ and the hypothesized value of μ_d, divided by the estimated **standard error.** This statistic is assumed to have a t-distribution when the null hypothesis is true and the assumptions for the test are satisfied. The value of the **test statistic $t\bigstar$** is calculated as follows:

FYI
Formula (10.5) is an adaptation of formula (9.2).

TEST STATISTIC FOR μ_d

$$t\bigstar = \frac{\bar{d} - \mu_d}{s_d/\sqrt{n}}, \quad \text{where df} = n - 1 \qquad (10.5)$$

NOTE A hypothesized mean difference, μ_d, can be any specified value. The most common value specified is zero; however, the difference can be nonzero.

Illustration 10.5

One-Tailed Hypothesis Test for μ_d

In a study on high blood pressure and the drugs used to control it, the effect of calcium channel blockers on pulse rate was one of many specific concerns. Twenty-six patients were randomly selected from a large pool of potential subjects, and their pulse rates were recorded. A calcium channel blocker was administered to each patient for a fixed period of time, and then each patient's pulse rate was again determined. The two resulting sets of data appeared to have approximately normal distributions, and the statistics were $\bar{d} = 1.07$ and $s_d = 1.74$ (d = before − after). Does the sample information provide sufficient evidence to show that this calcium channel blocker lowered the pulse rate? Use $\alpha = 0.05$.

FYI
"Lower rate" means "after" is less than "before" and "before − after" is positive.

SOLUTION

Step 1 **The Set-Up:**

a. **Describe the population parameter of interest.**
μ_d, the mean difference (reduction) in pulse rate from before to after using the calcium channel blocker for the time period of the test.

b. **State the null hypothesis (H_o) and the alternative hypothesis (H_a).**

H_o: $\mu_d = 0$ ($\leq$) (did not lower rate)   Remember: d = before − after.
H_a: $\mu_d > 0$ (did lower rate)

Step 2 **The Hypothesis Test Criteria:**

a. **Check the assumptions.**
Since the data in both sets are approximately normal, it seems reasonable to assume that the two populations are approximately normally distributed.

b. **Identify the probability distribution and the test statistic to be used.**
The t-distribution with df $= n - 1 = 25$, and the test statistic is $t\bigstar$ from formula (10.5).

c. **Determine the level of significance, α.**
$\alpha = 0.05$.

Step 3 **The Sample Evidence:**

a. **Collect the sample information:** $n = 26$, $\bar{d} = 1.07$, and $s_d = 1.74$.

b. **Calculate the value of the test statistic.**

$$t\bigstar = \frac{\bar{d} - \mu_d}{s_d/\sqrt{n}}: \qquad t\bigstar = \frac{1.07 - 0.0}{1.74/\sqrt{26}} = \frac{1.07}{0.34} = \mathbf{3.14}$$

Step 4 **The Probability Distribution:**

Using the p-value Procedure

a. Calculate the p-value for the test statistic.
Use the right-hand tail because H_a expresses concern for values related to "greater than." **P** = $P(t\bigstar > 3.14$, with df = 25) as shown on the figure.

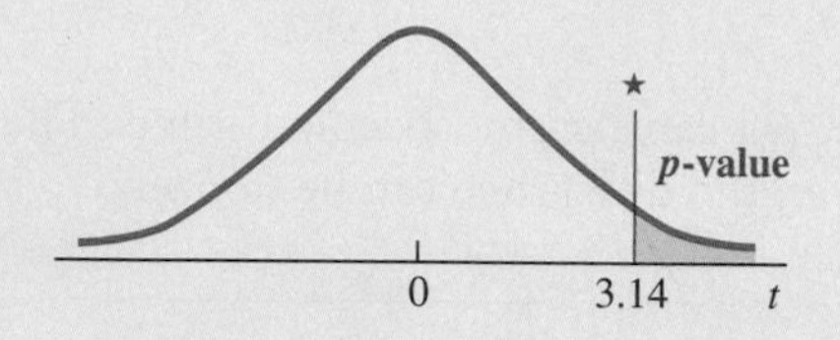

To find the p-value, use one of three methods:

1. Use Table 6 in Appendix B to place bounds on the p-value: **P < 0.005.**
2. Use Table 7 in Appendix B to read the value directly: **P = 0.002.**
3. Use a computer or calculator to find the p-value: **P = 0.0022.**

Specific instructions are on page 421.

b. Determine whether or not the p-value is smaller than α.
The p-value is smaller than the level of significance, α.

OR

Using the Classical Procedure

a. Determine the critical region and critical value(s).
The critical region is the right-hand tail because H_a expresses concern for values related to "greater than." The critical value is obtained from Table 6: $t(25, 0.05)$ = **1.71.**

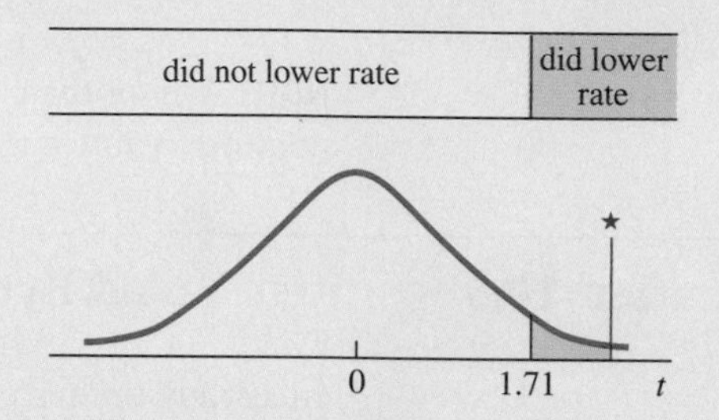

Specific instructions are on pages 414–415.

b. Determine whether or not the calculated test statistic is in the critical region.
$t\bigstar$ is in the critical region, as shown in **red** on the figure.

Step 5 **The Results:**

a. State the decision about H_o: Reject H_o.

b. State the conclusion about H_a.
At the 0.05 level of significance, we can conclude that the calcium channel blocker does lower the pulse rate. ■

ANSWER NOW 10.15 Ex10-015
Complete the hypothesis test with the alternative hypothesis $\mu_d > 0$ based on the paired data listed here and d = B − A. Use α = 0.05. Assume normality.

A	700	830	860	1080	930
B	720	820	890	1100	960

a. **Use the p-value approach.** b. **Use the classical approach.**

"*Statistical significance*" does not always have the same meaning when the "practical" application of the results is considered. In the preceding detailed hypothesis test, the results showed a statistical significance with a p-value of 0.002—that is, 2 chances in 1000. However, a more practical question might be: Is lowering the pulse rate by this small average amount, estimated to be 1.07 beats per minute, worth the risks of possible side effects of this medication? Actually the whole issue is much broader than just this one issue of pulse rate.

Technology Instructions: Hypothesis Test for the Mean μ_d with Unknown Standard Deviation for Two Dependent Sets of Sample Data.

MINITAB (Release 13)

Input the paired data into C1 and C2; then continue with:

```
Choose:   Stat > Basic Statistics > Paired t
Enter:    First sample: C1*
          Second sample: C2
Select:   Options
Enter:    Test mean: 0.0 or μd
Choose:   Alternative: less than or not equal or greater
          than
```

*Paired t evaluates the first sample minus the second sample.

Excel XP

Input the paired data into columns A and B; then continue with:

```
Choose:   Tools > Data Analysis > t-Test: Paired Two Sample
          for Means
Enter:    Variable 1 Range: (A1:A20 or select cells)
          Variable 2 Range: (B1:B20 or select cells)
                    (subtracts: Var1 - Var2)
          Hypothesized Mean Difference: μd (usually 0)
Select:   Labels (if necessary)
Enter:    α (ex. 0.05)
Select:   Output Range
Enter:    (C1 or select cell) > OK
```

Use Format > Column > AutoFit Selection to make the output more readable. The output shows p-values and critical values for one- and two-tailed tests. The hypothesis test may also be done by first subtracting the two columns and then using the inference about a mean (sigma unknown) commands on page 425 on the differences.

TI-83 Plus

Input the paired data into L1 and L2; then continue with the following, entering the appropriate values, and highlighting Calculate:

```
Highlight:   L3
Enter:       L3= L1 - L2*
Choose:      STAT > TESTS > 2:T-Test...
```

```
T-Test
 Inpt:Data Stats
 μ0:0
 List:L3
 Freq:1
 μ:≠μ0 <μ0 >μ0
 Calculate Draw
```

*Enter the expression in the order that is needed: L1 − L2 or L2 − L1.

The solution to Illustration 10.5 looks like this when solved in MINITAB:

```
Paired T for Before - After

                 N        Mean       StDev      SE Mean
Difference      26        1.07        1.74         0.34

T-Test of mean difference = 0 (vs > 0): T-Value = 3.14
P-Value = 0.002
```

ANSWER NOW 10.16
Use a computer or calculator to complete the hypothesis test with the alternative hypothesis $\mu_d < 0$ based on the paired data listed here and $d = M - N$. Use $\alpha = 0.02$. Assume normality.

M	58	78	45	38	49	62
N	62	86	42	39	47	68

Illustration 10.6 Two-Tailed Hypothesis Test for μ_d

Suppose the sample data in Table 10.1 (p. 476) were collected with the hope of showing that the two brands do not wear equally. Do the data provide sufficient evidence for us to conclude that the two brands show unequal wear, at the 0.05 level of significance? Assume the amounts of wear are approximately normally distributed for both brands of tires.

SOLUTION

Step 1 **The Set-Up:**

a. **Describe the population parameter of interest.**
μ_d, the mean difference in the amounts of wear between the two brands.

b. **State the null hypothesis (H_o) and the alternative hypothesis (H_a).**

H_o: $\mu_d = 0$ (no difference) Remember: $d = \text{B} - \text{A}$.

H_a: $\mu_d \neq 0$ (difference)

Step 2 **The Hypothesis Test Criteria:**

a. **Check the assumptions.**
The assumption of normality is included in the statement of this problem.

b. **Identify the probability distribution and the test statistic to be used.**
The t-distribution with df $= n - 1 = 6 - 1 = 5$, and $t\bigstar = \dfrac{\bar{d} - \mu_d}{s_d/\sqrt{n}}$.

c. **Determine the level of significance:** $\alpha = 0.05$.

Step 3 **The Sample Evidence:**

a. **Collect the sample information:** $n = 6$, $\bar{d} = 6.3$, and $s_d = 5.1$.

b. **Calculate the value of the test statistic.**

$$t\bigstar = \frac{\bar{d} - \mu_d}{s_d/\sqrt{n}}: \qquad t\bigstar = \frac{6.3 - 0.0}{5.1/\sqrt{6}} = \frac{6.3}{2.08} = \mathbf{3.03}$$

Step 4 The Probability Distribution:

Using the p-Value Procedure

a. Calculate the p-value for the test statistic.

Use both tails because H_a expresses concern for values related to "different from."

$$\mathbf{P} = p\text{-value} = P(t\bigstar < -3.03) + P(t\bigstar > 3.03)$$

$$= 2 \times P(|t\bigstar| > 3.03) \quad \text{as shown on the figure}$$

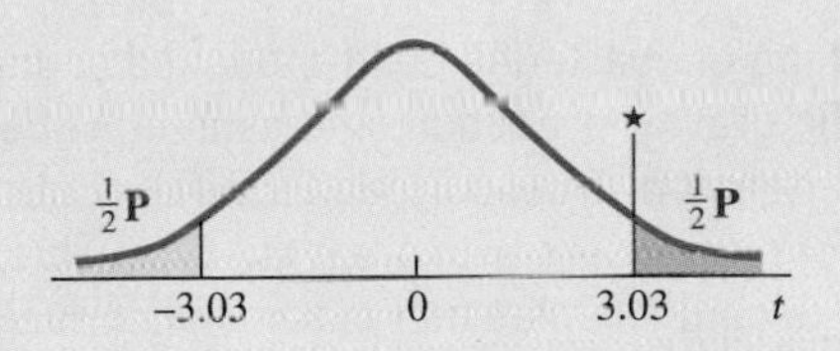

To find the p-value, you have three options:

1. Use Table 6 in Appendix B: $\mathbf{0.02 < P < 0.05}$.
2. Use Table 7 in Appendix B to place bounds on the p-value: $\mathbf{0.026 < P < 0.030}$.
3. Use a computer or calculator to find the p-value: $\mathbf{P} = 2 \times 0.0145 = \mathbf{0.0290}$.

For specific instructions, see pages 421–422.

b. Determine whether or not the p-value is smaller than α.

The p-value is smaller than α.

OR

Using the Classical Procedure

a. Determine the critical region and critical value(s).

The critical region is two-tailed because H_a expresses concern for values related to "different than." The critical value is obtained from Table 6: $t(5, 0.025) = \mathbf{2.57}$.

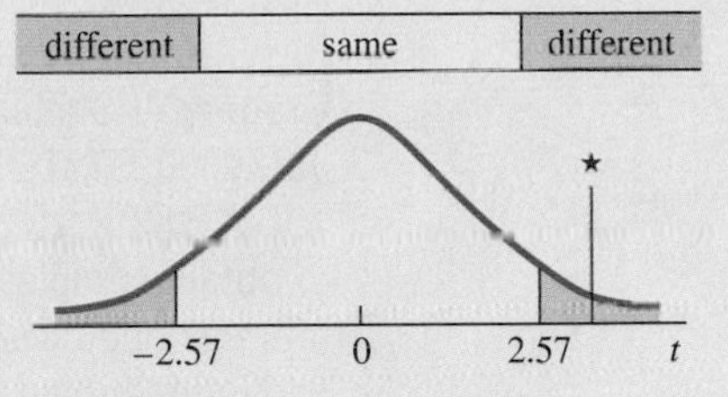

For specific instructions, see pages 414–415.

b. Determine whether or not the calculated test statistic is in the critical region.

$t\bigstar$ is in the critical region, as shown in **red** on the figure.

Step 5 The Results:

a. State the decision about H_o: Reject H_o.

b. State the conclusion about H_a.

There is a significant difference in the mean amounts of wear at the 0.05 level of significance. ■

ANSWER NOW 10.17

Complete the hypothesis test with the alternative hypothesis $\mu_d \neq 0$ based on the paired data listed here and d = oldest − youngest. Use $\alpha = 0.01$. Assume normality.

Oldest	100	162	174	159	173
Youngest	194	162	167	156	176

a. Use the p-value approach. b. Use the classical approach.

Application 10.2 Testing Asphalt Sampling Procedures

This application is an excerpt from a Florida Department of Transportation research report.

COMPARISON OF THE SCOOPING VS. QUARTERING METHODS FOR OBTAINING ASPHALT MIXTURE SAMPLES
Research Report FL/DOT/SMO/00-441
Gregory A. Sholar, James A. Musselman, Gale C. Page
July 2000, STATE MATERIALS OFFICE

ABSTRACT The standard method of quartering plant-produced asphalt mix to obtain samples for maximum specific gravity, gradation, and asphalt binder content has been used by the Florida Department of Transportation (FDOT), contractors, and independent testing laboratories for many years with great success. This report examines an alternative method for obtaining samples that is somewhat easier and less time-consuming than the traditional quartering method. This method, hereafter referred to as the "scooping" method, involves some of the same procedures and techniques that are used with the quartering method. The principal difference is that samples are scooped from the pile of asphalt mix until the desired sample weight is obtained instead of quartering the pile down until the desired sample weight is obtained. Twelve different mixtures were sampled for this study and the following mixture properties were compared for the two different sampling methods: bulk density, maximum specific gravity, % air voids, asphalt binder content, and gradation. Analysis of the data indicates that the two sampling methods provide statistically equivalent results for the aforementioned mixture properties. Included in this report is a new version of FM 1-T 168, "Sampling Bituminous Paving Mixtures," which encompasses this new method for sampling asphalt mixtures.

DATA ANALYSIS Theoretically, if the two sampling methods were identical, then the average difference between values obtained for any asphalt property (e.g., asphalt binder content) for a particular mix would be zero. A paired-difference analysis was performed for each property measured. A paired-difference analysis is a *t*-test performed on the differences between each sampling method. A 95% confidence interval was used (i.e., $\alpha = 0.05$) to calculate the two-sided *t*-critical value. The null hypothesis is that the average difference is zero. If *t*-calculated is less than *t*-critical, then the null hypothesis cannot be rejected. In the *t*-test summaries, the important values are the "*t*-calculated" and the "*t*-critical" values. For simplicity, all of these "*t*" values have been summarized in Table 14. Examination of the statistical results indicates that for all of the properties measured, except for % passing the No. 4 sieve, the null hypothesis cannot be rejected. This indicates that the two methods are statistically equivalent. The one exception is for the % passing the No. 4 sieve. The *t*-calculated and *t*-critical values were nearly identical (2.224 vs. 2.228).

TABLE 14 Summary of Paired Difference Analysis

Asphalt Mixture Property	Absolute Value		
	t-calculated	*t*-critical	*t*-calc. < *t*-crit. ?
Gmb (Nmax)	1.442	2.306	Yes
Gmm	0.802	2.201	Yes
% air voids	1.719	2.306	Yes
% AC (ignition)	0.534	2.201	Yes
Sieve size			
1/2″	0.672	2.228	Yes
3/8″	0.783	2.228	Yes
No. 4	2.224	2.228	Equal
No. 8	1.819	2.228	Yes
No. 16	1.047	2.228	Yes
No. 30	0.814	2.228	Yes
No. 50	0.753	2.228	Yes
No. 100	0.387	2.228	Yes
No. 200	0.305	2.228	Yes

CONCLUSION Based on the statistical analysis of the data, the two methods of sampling are equivalent with respect to Gmb, Gmm, asphalt binder content, and gradation. Since the scooping method is easier and faster, it is recommended that the revised Florida method for sampling (FM 1-T 168) be accepted and implemented statewide.

ANSWER NOW 10.18

a. What null hypothesis is being tested in each of these 13 tests?

b. Why are the "*t*-calculated" and the "*t*-critical" values the important values?

c. Why is it correct to report the absolute values for both *t*-values in Table 14?

d. What decision is reached for each of these 13 hypothesis tests?

e. What conclusion is reached as a result of these tests?

f. What action is recommended to the State of Florida as a result of the conclusion?

Exercises

10.19 Ex10-019 Salt-free diets are often prescribed to people with high blood pressure. The following data were obtained from an experiment designed to estimate the reduction in diastolic blood pressure as a result of following a salt-free diet for two weeks. Assume the diastolic readings are normally distributed.

Before	93	106	87	92	102	95	88	110
After	92	102	89	92	101	96	88	105

a. What is the point estimate for the mean reduction in the diastolic reading after two weeks on this diet?
b. Find the 98% confidence interval for the mean reduction.

10.20 Ex10-020 All students who enroll in a certain memory course are given a pretest before the course begins. A random sample of ten students who completed the course were given a posttest, and their scores are listed here.

Student	1	2	3	4	5	6	7	8	9	10
Before	93	86	72	54	92	65	80	81	62	73
After	98	92	80	62	91	78	89	78	71	80

MINITAB was used to find the 95% confidence interval for the mean improvement in memory resulting from taking the memory course, as measured by the difference in test scores (d = after − before). Verify the results shown on the output by calculating the values yourself.

```
Confidence Intervals
Variable  N   Mean  StDev  SE Mean  95% C.I.
C3        10  6.10  4.79   1.52     (2.67, 9.53)
```

10.21 Ex10-021 An experiment was designed to estimate the mean difference in weight gain for pigs fed ration A as compared to those fed ration B. Eight pairs of pigs were used. The pigs within each pair were littermates. The rations were assigned at random to the two animals within each pair. The gains (in pounds) after 45 days are shown in the table.

Litter	1	2	3	4	5	6	7	8
Ration A	65	37	40	47	49	65	53	59
Ration B	58	39	31	45	47	55	59	51

Assuming weight gain is normal, find the 95% confidence interval estimate for the mean of the differences μ_d, where d = ration A − ration B.

10.22 Ex10-022 A sociologist is studying the effects of viewing a certain motion picture on the attitudes of black men toward white men. Twelve black men were randomly selected and asked to fill out a questionnaire before and after viewing the film. The scores received by the 12 men are listed in the table. Assuming the questionnaire scores are normal, construct a 95% confidence interval for the mean shift in score that takes place when this film is viewed.

Before	10	13	18	12	9	8	14	12	17	20	7	11
After	5	9	13	17	4	5	11	14	13	18	7	12

10.23 Ex10-023 Two men, A and B, who usually commute to work together, decide to conduct an experiment to see whether one route is faster than the other. The men think that their driving habits are approximately the same, and therefore they decide on the following procedure: Each morning for two weeks A will drive to work on one route and

Table for Exercise 10.23

	Day									
Route	**M**	**Tu**	**W**	**Th**	**F**	**M**	**Tu**	**W**	**Th**	**F**
I	29	26	25	25	25	24	26	26	30	31
II	25	26	25	25	24	23	27	25	29	30

B will use the other route. On the first morning A will toss a coin. If heads appear, he will use route I; if tails appear, he will use route II. On the second morning B will toss the coin: heads, route I; tails, route II. The times, recorded to the nearest minute, are given in the table. Assume commute times are normal, and estimate the mean of the differences with a 95% confidence interval.

10.24 Ex10-024 In evaluating different measuring instruments one must first determine whether there is a systematic difference between the instruments. Lenses of different powers were measured once each by two different instruments. The measurement differences (instrument A − instrument B) were recorded. The units have been coded for proprietary reasons. (Courtesy: Bausch & Lomb)

4	5	−2	−3	−7
10	11	−1	3	7
−5	3	−4	−5	−7
4	−1	−18	0	−17
12	9	4	17	−2

Does there appear to be a systematic difference between the two instruments?

a. Describe the data using a histogram and one other graph.
b. Find the mean and the standard deviation.
c. Are the assumptions for inferences satisfied? Explain.
d. Using a 95% confidence interval, estimate the mean of the differences.
e. Is there any evidence of a difference? Explain.

10.25 State the null hypothesis, H_o, and the alternative hypothesis, H_a, that would be used to test these claims:

a. The mean of the differences between the posttest and the pretest scores shows an improvement.
b. As a result of a special training session, it is believed that the mean of the differences in performance scores will not be zero.
c. The mean weight gain, due to the change in diet for the laboratory animals, is at least 10 oz.
d. The mean weight loss experienced by people on a new diet plan was no less than 12 lb.

10.26 Determine the test criteria that would be used with the classical approach to test the following hypotheses when t is used as the test statistic.

a. H_o: $\mu_d = 0$ vs. H_a: $\mu_d > 0$ with $n = 15$ and $\alpha = 0.05$
b. H_o: $\mu_d = 0$ vs. H_a: $\mu_d \neq 0$ with $n = 25$ and $\alpha = 0.05$
c. H_o: $\mu_d = 1.45$ vs. H_a: $\mu_d < 1.45$ with $n = 12$ and $\alpha = 0.10$
d. H_o: $\mu_d = 0.75$ vs. H_a: $\mu_d > 0.75$ with $n = 18$ and $\alpha = 0.01$

10.27 The corrosive effects of various soils on coated and uncoated steel pipe were tested by using a dependent sampling plan. The data collected are summarized by $n = 40$, $\sum d = 220$, and $\sum d^2 = 6222$, where d is the amount of corrosion on the coated portion subtracted from the amount of corrosion on the uncoated portion. Does this sample provide sufficient evidence to conclude that the coating is beneficial? Use $\alpha = 0.01$.

a. Solve using the p-value approach.
b. Solve using the classical approach.

10.28 Ex10-028 To test the effect of a physical fitness course on one's physical ability, the number of sit-ups that a person could do in 1 minute was recorded both before and after the course. Ten randomly selected participants scored as shown in the table. Can you conclude that a significant amount of improvement took place? Use $\alpha = 0.01$.

Before	29	22	25	29	26	24	31	46	34	28
After	30	26	25	35	33	36	32	54	50	43

a. Solve using the p-value approach.
b. Solve using the classical approach.

10.29 Ex10-029 Ten recently diagnosed diabetics were tested to determine whether an educational program was effective in increasing their knowledge of diabetes. They were given a test before and after the educational program concerning self-care aspects of diabetes. The scores on the test were as follows:

Patient	1	2	3	4	5	6	7	8	9	10
Before	75	62	67	70	55	59	60	64	72	59
After	77	65	68	72	62	61	60	67	75	68

The MINITAB output may be used to determine whether the scores improved as a result of the program. Verify the values shown on the output [mean difference (MEAN), standard deviation (STDEV), standard error of the difference (SE MEAN), $t\star$ (T), and p-value] by calculating them yourself.

```
TEST OF MU = 0.000 VS MU G.T. 0.000
     N    MEAN    STDEV   SE MEAN    T      P VALUE
C3   10   3.200   2.741   0.867      3.69   0.0025
```

10.30 Ex10-030 As metal parts experience wear, the metal is displaced. The table lists displacement measurements (in mm) on metal parts that have undergone durability cycling for the equivalent of 100,000-plus miles. The first column is the serial number for the part, the second column lists the before-test (BT) displacement measurements of the new part, the third column lists the end-of-test (EOT) measurements, and the fourth column lists the changes (i.e., wear) in the parts.

	Serial Displacement (mm)		
Number	BT	EOT	Difference
1	4.609	4.604	−0.005
2	5.227	5.208	−0.019
3	5.255	5.193	−0.062
4	4.622	4.601	−0.021
5	4.630	4.589	−0.041
6	5.207	5.100	0.019
7	5.239	5.198	−0.041
8	4.605	4.596	−0.009
9	4.622	4.576	−0.046
10	4.753	4.736	−0.017
11	5.226	5.218	−0.008
12	5.094	5.057	−0.037
13	4.702	4.683	−0.019
14	5.152	5.111	−0.041

Source: Problem data provided by AC Rochester Division, General Motors, Rochester, NY

a. Does it seem right that all the difference values are negative? Explain.

Use a computer or calculator to complete the following parts:

b. Find the sample mean, variance, and standard deviation for the before-test (BT) data.

c. Find the sample mean, variance, and standard deviation for the end-of-test (EOT) data.

d. Find the sample mean, variance, and standard deviation for the difference data.

e. How is the sample mean difference related to the means of BT and EOT? Are the variances and standard deviations related in the same way?

f. At the 0.01 level, do the data show that a significant amount of wear took place?

10.31 Ex10-031 The amount of general anesthetic a patient should receive prior to surgery has received considerable public attention. According to the American Society of Anesthesiologists (*People*, "Wake-Up Call," June 29, 1998), every year about 40,000 (some researchers have put the figure closer to 200,000) of the 28 million patients who undergo general anesthesia experience limited awareness during surgery because of resistance to the medication or too small a dose. Patients commonly report overhearing doctors conversing with nurses and assistants during operations.

Suppose a study is conducted using 20 patients who are having eye surgery performed on both eyes, with two weeks separating the treatments on each eye. Ten of the patients are given a lighter dose of general anesthetic prior to surgery on the first eye, and the other ten patients are given a heavier dose. The following week the procedure is reversed. Two days after each surgery is performed, the patients are asked to rate the amount of pain and discomfort they experienced on a scale from 0 (none) to 10 (unbearable).

Subject	Light Dose	Heavy Dose	Subject	Light Dose	Heavy Dose
1	4	3	11	6	7
2	6	5	12	7	5
3	5	8	13	10	7
4	8	4	14	3	2
5	4	5	15	1	0
6	9	6	16	5	6
7	3	2	17	6	3
8	7	8	18	8	5
9	8	5	19	4	2
10	9	7	20	2	0

Can you conclude that the heavier dose of anesthetic resulted in the patient's experiencing less pain and discomfort after the eye surgery? Use the 0.01 level of significance.

a. Solve using the p-value approach.

b. Solve using the classical approach.

10.32 Ex10-032 A research project was undertaken to evaluate two focimeters. Each of 20 lenses of varying powers was read once on each focimeter. The measurement differences were then calculated, where each difference is focimeter A − focimeter B. (Courtesy: Bausch & Lomb)

−0.016	0.013	0.009	0.000	−0.005
−0.015	−0.006	−0.016	−0.022	−0.006
−0.020	0.015	−0.017	−0.010	−0.003
0.011	−0.012	0.008	−0.005	−0.009

a. Using a t-test on these paired differences and $\alpha = 0.01$, determine whether the resulting differences are significantly different from zero.

b. Construct a 99% confidence interval for these differences.

c. Explain what both evaluation methods indicate about the differences.

d. If an enterprising experimenter performed this same test using $\alpha = 0.10$, what would the outcome be? Offer comments about proceeding using these ground rules.

10.3 Inferences Concerning the Difference Between Means Using Two Independent Samples

When comparing the means of two populations, we typically consider the difference between their means, $\mu_1 - \mu_2$ (often called **"independent means"**). The inferences about $\mu_1 - \mu_2$ will be based on the difference between the observed sample means, $\bar{x}_1 - \bar{x}_2$. This observed difference, $\bar{x}_1 - \bar{x}_2$, belongs to a sampling distribution with the characteristics described in the following statement.

If independent samples of sizes n_1 and n_2 are drawn randomly from large populations with means μ_1 and μ_2 and variances σ_1^2 and σ_2^2, respectively, then the sampling distribution of $\bar{x}_1 - \bar{x}_2$, the difference between the sample means, has

1. mean $\mu_{\bar{x}_1 - \bar{x}_2} = \mu_1 - \mu_2$ and
2. standard error $\sigma_{\bar{x}_1 - \bar{x}_2} = \sqrt{\left(\frac{\sigma_1^2}{n_1}\right) + \left(\frac{\sigma_2^2}{n_2}\right)}$. (10.6)

If both populations have normal distributions, then the sampling distribution of $\bar{x}_1 - \bar{x}_2$ will also be normally distributed.

FYI

Why is $\bar{x}_1 - \bar{x}_2$ an *unbiased* estimator of $\mu_1 - \mu_2$?

The preceding statement is true for all sample sizes given that the populations involved are normal and the population variances σ_1^2 and σ_2^2 are known quantities. However, as with inferences about one mean, the variance of a population is generally an unknown quantity. Therefore, it will be necessary to estimate the standard error by replacing the variances, σ_1^2 and σ_2^2, in formula (10.6) with the best estimates available—namely, the sample variances, s_1^2 and s_2^2. The *estimated standard error* will be found using the following formula:

$$\text{estimated standard error} = \sqrt{\left(\frac{s_1^2}{n_1}\right) + \left(\frac{s_2^2}{n_2}\right)} \qquad (10.7)$$

ANSWER NOW 10.33

Two independent random samples resulted in these values:

Sample 1: $n_1 = 12$, $s_1^2 = 190$
Sample 2: $n_2 = 18$, $s_2^2 = 150$

Find the estimate for the standard error for the difference between two means.

Inferences about the difference between two population means, $\mu_1 - \mu_2$, will be based on the following assumptions.

ASSUMPTIONS FOR INFERENCES ABOUT THE DIFFERENCE BETWEEN TWO MEANS, $\mu_1 - \mu_2$:

The samples are randomly selected from normally distributed populations, and the samples are selected in an independent manner.

NO ASSUMPTIONS ARE MADE ABOUT THE POPULATION VARIANCES.

Since the samples provide the information for determining the standard error, the ***t*-distribution** will be used as the test statistic. The inferences are divided into two cases.

Case 1: The t-distribution will be used, and the number of degrees of freedom will be calculated.

Case 2: The t-distribution will be used, and the number of degrees of freedom will be approximated.

Case 1 will occur when you are completing the inference *using a computer or statistical calculator and the statistical software or program calculates the number of degrees of freedom* for you. The calculated value for df is a function of both sample sizes and their relative sizes, and both sample variances and their relative sizes. The value of df will be a number between the smaller of $df_1 = n_1 - 1$ or $df_2 = n_2 - 1$, and the sum of the degrees of freedom, $df_1 + df_2 = [(n_1 - 1) + (n_2 - 1)] = n_1 + n_2 - 2$.

Case 2 will occur when you are completing the inference *without the aid of a computer or calculator and its statistical software package.* Use of the t-distribution with the smaller of $df_1 = n_1 - 1$ or $df_2 = n_2 - 1$ will give conservative results. Because of this approximation, the true level of confidence for an interval estimate will be slightly higher than the reported level of confidence; or the true p-value and the true level of significance for a hypothesis test will be slightly less than reported. The gap between these reported values and the true values will be quite small, unless the sample sizes are quite small and unequal or the sample variances are very different. The gap will decrease as the samples increase in size or as the sample variances are more alike.

ANSWER NOW 10.34

Two independent random samples of sizes 18 and 24 were obtained to make inferences about the difference between two means. What is the number of degrees of freedom? Discuss both cases.

Since the only difference between the two cases is the number of degrees of freedom used to identify the t-distribution involved, we will study case 2 first.

FYI

Would you say the difference between 5 and 8 is −3? How would you express the difference? Explain.

NOTE $A > B$ ("A is greater than B") is equivalent to $B < A$ ("B is less than A"). When the difference between A and B is being discussed, it is customary to express the difference as "larger − smaller" so that the resulting difference is positive: $A - B > 0$. To express the difference as "smaller − larger" results in $B - A < 0$ (the difference is negative) and is at best clever and is usually unnecessarily confusing. Therefore, it is recommended that the difference be expressed as "larger − smaller."

CONFIDENCE INTERVAL PROCEDURE

We will use the following formula for calculating the endpoints of the **$1 - \alpha$ confidence interval.**

CONFIDENCE INTERVAL FOR THE DIFFERENCE BETWEEN TWO MEANS (INDEPENDENT SAMPLES)

$$(\bar{x}_1 - \bar{x}_2) - t(\text{df}, \alpha/2) \cdot \sqrt{\left(\frac{s_1^2}{n_1}\right) + \left(\frac{s_2^2}{n_2}\right)} \quad \text{to} \quad (\bar{x}_1 - \bar{x}_2) + t(\text{df}, \alpha/2) \cdot \sqrt{\left(\frac{s_1^2}{n_1}\right) + \left(\frac{s_2^2}{n_2}\right)} \tag{10.8}$$

where df equals the smaller of df_1 or df_2 when the confidence interval is calculated without the aid of a computer and its statistical software.

Illustration 10.7 Constructing a Confidence Interval for the Difference Between Two Means

The heights (in inches) of 20 randomly selected women and 30 randomly selected men were independently obtained from the student body of a certain college in order to estimate the difference in their mean heights. The sample information is given in Table 10.2. Assume that heights are approximately normally distributed for both populations.

TABLE 10.2 Sample Information on Student Heights

Sample	Number	Mean	Standard Deviation
Female (*f*)	20	63.8	2.18
Male (*m*)	30	69.8	1.92

Find the 95% confidence interval for the difference between the mean heights, $\mu_m - \mu_f$.

SOLUTION

Step 1 **The Set-Up:**
Describe the population parameter of interest.
$\mu_m - \mu_f$, the difference between the mean height of male students and the mean height of female students.

Step 2 **The Confidence Interval Criteria:**
a. **Check the assumptions.**
Both populations are approximately normal, and the samples were random and independently selected.
b. **Identify the probability distribution and the formula to be used.**
The t-distribution with df = 19, the smaller of $n_m - 1 = 30 - 1 = 29$ or $n_f - 1 = 20 - 1 = 19$, and formula (10.8).
c. **State the level of confidence:** $1 - \alpha = 0.95$.

Step 3 **The Sample Evidence:**
Collect the sample information. See Table 10.2.

Step 4 **The Confidence Interval:**
a. **Determine the confidence coefficient.**
We have a two-tailed situation with $\alpha/2 = 0.025$ in one tail and df = 19. From Table 6 in Appendix B, $t(\text{df},\alpha/2) = t(19, 0.025) = 2.09$. See the figure.

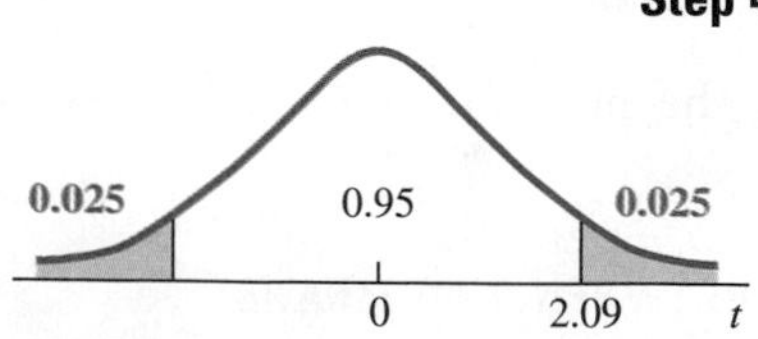

See pages 414–416 for instructions on using Table 6.

b. **Find the maximum error of estimate.**
Using the maximum error part of formula (10.8), we have

$$E = t(\text{df}, \alpha/2) \cdot \sqrt{\left(\frac{s_1^2}{n_1}\right) + \left(\frac{s_2^2}{n_2}\right)}: \qquad E = 2.09 \cdot \sqrt{\left(\frac{1.92^2}{30}\right) + \left(\frac{2.18^2}{20}\right)}$$

$$= (2.09)(0.60) = \mathbf{1.25}$$

c. Find the lower and upper confidence limits.

$$(\bar{x}_1 - \bar{x}_2) \pm E$$

$$6.00 \pm 1.25$$

$$6.00 - 1.25 = \mathbf{4.75} \quad \text{to} \quad 6.00 + 1.25 = \mathbf{7.25}$$

Step 5 **The Results:**
State the confidence interval.
4.75 to 7.25 is the 95% confidence interval for $\mu_m - \mu_f$. That is, with 95% confidence, we can say that the difference between the mean heights of the male and female students is between 4.75 and 7.25 inches. ■

ANSWER NOW 10.35
Find the 90% confidence interval for the difference between two means based on this information about two samples. Assume independent samples from normal populations.

Sample	Number	Mean	Standard Deviation
1	20	35	22
2	15	30	16

HYPOTHESIS-TESTING PROCEDURE

When we test a null **hypothesis about the difference between two population means,** the test statistic used will be the difference between the observed difference of the sample means and the hypothesized difference of the population means, divided by the estimated standard error. The test statistic is assumed to have approximately a t-distribution when the null hypothesis is true and the normality assumption has been satisfied. The calculated value of the **test statistic** is found using this formula:

TEST STATISTIC FOR THE DIFFERENCE BETWEEN TWO MEANS

$$t\star = \frac{(\bar{x}_1 - \bar{x}_2) - (\mu_1 - \mu_2)}{\sqrt{\left(\frac{s_1^2}{n_1}\right) + \left(\frac{s_2^2}{n_2}\right)}} \qquad (10.9)$$

where df is the smaller of df_1 or df_2 when $t\star$ is calculated *without* the aid of a computer or statistical calculator and its statistical software or programs.

NOTE A hypothesized difference between the two population means, $\mu_1 - \mu_2$, can be any specified value. The most common value specified is zero; however, the difference can be nonzero.

Illustration 10.8 One-Tailed Hypothesis Test for the Difference Between Two Means

Suppose that we are interested in comparing the academic success of college students who belong to fraternal organizations with the academic success of those who do not belong to fraternal organizations. The reason for the comparison is the recent concern that fraternity members, on the average, are achieving at a lower academic level than nonfraternal students achieve. (Cumulative grade-point average is used to measure academic success.) Random samples of size 40 are taken from each population. The sample results are listed in Table 10.3.

TABLE 10.3 Sample Information on Academic Success

Sample	Number	Mean	Standard Deviation
Fraternity members (*f*)	40	2.03	0.68
Nonmembers (*n*)	40	2.21	0.59

Complete a hypothesis test using $\alpha = 0.05$. Assume that the grade-point averages for both groups are approximately normally distributed.

SOLUTION

Step 1 **The Set-Up:**

a. **Describe the population parameter of interest.**
$\mu_n - \mu_f$, the difference between the mean grade-point averages for the nonfraternity members and the fraternity members.

b. **State the null hypothesis (H_o) and the alternative hypothesis (H_a).**

FYI
Remember: "Larger − smaller" results in a positive difference.

H_o: $\mu_n - \mu_f = 0$ ($\leq$) (fraternity averages are no lower)
H_a: $\mu_n - \mu_f > 0$ (fraternity averages are lower)

Step 2 **The Hypothesis Test Criteria:**

a. **Check the assumptions.**
Both populations are approximately normal, and random samples were selected. Since the two populations are separate, the samples are independent.

b. **Identify the probability distribution and the test statistic to be used.**
The t-distribution with df = the smaller of df_n or df_f; since both n's are 40, df = 40 − 1 = **39**; and $t\bigstar$ is calculated using formula (10.9).

c. **Determine the level of significance:** $\alpha = 0.05$.

Step 3 **The Sample Evidence:**

a. **Collect the sample information:** See Table 10.3.

b. **Calculate the value of the test statistic.**

$$t\bigstar = \frac{(\bar{x}_1 - \bar{x}_2) - (\mu_1 - \mu_2)}{\sqrt{\left(\frac{s_1^2}{n_1}\right) + \left(\frac{s_2^2}{n_2}\right)}}: \qquad t\bigstar = \frac{(2.21 - 2.03) - (0.00)}{\sqrt{\left(\frac{0.59^2}{40}\right) + \left(\frac{0.68^2}{40}\right)}}$$

$$= \frac{0.18}{\sqrt{0.00870 + 0.01156}} = \frac{0.18}{0.1423} = \mathbf{1.26}$$

ANSWER NOW 10.36

Find the value of $t\star$ for the difference between two means based on this information about two samples:

Sample	Number	Mean	Standard Deviation
1	18	38.2	14.2
2	25	43.1	10.6

FYI
When df is not in the table, use the next smaller df value.

Step 4 **The Probability Distribution:**

Using the *p*-Value Procedure

a. Calculate the *p*-value for the test statistic.
Use the right-hand tail because H_a expresses concern for values related to "greater than." **P** = $P(t\star > 1.26$, with df $= 39)$ as shown on the figure.

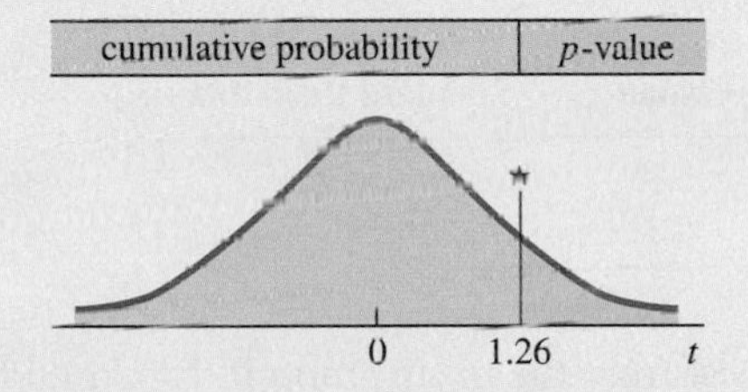

To find the *p*-value, use one of three methods:

1. Use Table 6 in Appendix B to place bounds on the *p*-value: $\mathbf{0.10 < P < 0.25}$.
2. Use Table 7 in Appendix B to place bounds on the *p*-value: $\mathbf{0.100 < P < 0.119}$.
3. Use a computer or calculator to find the *p*-value: $\mathbf{P = 0.1076}$.

Specific details follow this illustration.

b. Determine whether or not the *p*-value is smaller than α.
The *p*-value is not smaller than α.

OR

Using the Classical Procedure

a. Determine the critical region and critical value(s).
The critical region is the right-hand tail because H_a expresses concern for values related to "greater than." The critical value is obtained from Table 6: $t(39, 0.05) = \mathbf{1.69}$.

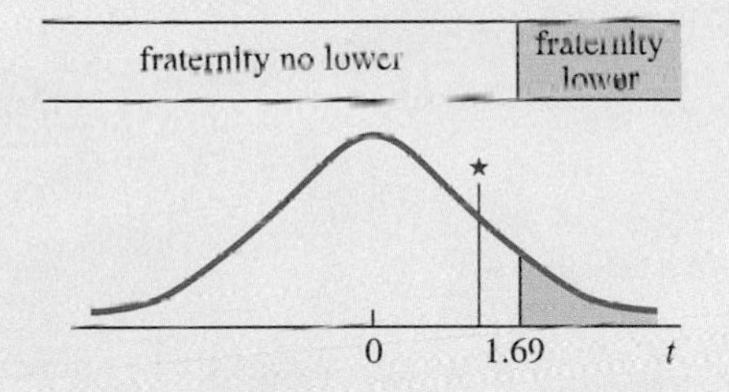

See pages 414–415 for information about critical values.

b. Determine whether or not the calculated test statistic is in the critical region.
$t\star$ is not in the critical region, as shown in **red** on the figure.

Step 5 **The Results:**

a. State the decision about H_o: Fail to reject H_o.

b. State the conclusion about H_a.
At the 0.05 level of significance, the claim that the fraternity members achieve at a lower level than nonmembers is not supported by the sample data.

To find the *p*-value for Illustration 10.8, use one of three methods:

Method 1: Use Table 6. Find 1.26 between two entries in the df = 39 row and read the bounds for **P** from the one-tail heading at the top of the table: $\mathbf{0.10 < P < 0.25}$.

Method 2: Use Table 7. Find $t\star = 1.26$ between two rows and df = 39 between two columns; read the bounds for $P(t\star > 1.26 \mid df = 39)$; $\mathbf{0.100 < P < 0.119}$.

Method 3: If you are doing the hypothesis test with the aid of a computer or calculator, most likely it will calculate the p-value for you (see p. 491), or you may use the cumulative probability distribution commands described in Chapter 9 (p. 417). ■

Illustration 10.9 Two-Tailed Hypothesis for the Difference Between Two Means

Many students have complained that the soft-drink vending machine A (in the student recreation room) dispenses a different amount of drink than machine B (in the faculty lounge). To test this belief, a student randomly sampled several servings from each machine and carefully measured them, with the results shown in Table 10.4.

TABLE 10.4 Sample Information on Vending Machines

Machine	Number	Mean	Standard Deviation
A	10	5.38	1.59
B	12	5.92	0.83

Does this evidence support the hypothesis that the mean amount dispensed by machine A is different from the amount dispensed by machine B? Assume the amounts dispensed by both machines are normally distributed, and complete the test using $\alpha = 0.10$.

SOLUTION

Step 1 **The Set-Up:**

a. **Describe the population parameter of interest.**
$\mu_B - \mu_A$, the difference between the mean amount dispensed by machine B and the mean amount dispensed by machine A.

b. **State the null hypothesis (H_o) and the alternative hypothesis (H_a).**

H_o: $\mu_B - \mu_A = 0$ (A dispenses the same amount as B)

H_a: $\mu_B - \mu_A \neq 0$ (A dispenses a different amount than B)

FYI
"Larger − smaller" results in a positive difference.

Step 2 **The Hypothesis Test Criteria:**

a. **Check the assumptions.**
Both populations are assumed to be approximately normal, and the samples were random and independently selected.

b. **Identify the probability distribution and the test statistic to be used.**
The t-distribution with df = the smaller of $n_A - 1 = 10 - 1 = 9$ or $n_B - 1 = 12 - 1 = 11$, df = 9, and $t\star$ calculated using formula (10.9).

c. **Determine the level of significance:** $\alpha = 0.10$.

Step 3 **The Sample Evidence:**

a. **Collect the sample information:** See Table 10.4.

b. Calculate the value of the test statistic.

$$t\bigstar = \frac{(\bar{x}_B - \bar{x}_A) - (\mu_B - \mu_A)}{\sqrt{\left(\frac{s_B^2}{n_B}\right) + \left(\frac{s_A^2}{n_A}\right)}} : \quad t\bigstar = \frac{(5.92 - 5.38) - (0.00)}{\sqrt{\left(\frac{0.83^2}{12}\right) + \left(\frac{1.59^2}{10}\right)}}$$

$$= \frac{0.54}{\sqrt{0.0574 + 0.2528}} = \frac{0.54}{0.557} = \mathbf{0.97}$$

Step 4 **The Probability Distribution:**

Using the p-Value Procedure

a. Calculate the p-value for the test statistic.

Use both tails because H_a expresses concern for values related to "different than."

$\mathbf{P} = p\text{-value} = P(t\bigstar < -0.97) + P(t\bigstar > 0.97)$

$= 2 \times P(|t\bigstar| > 0.97 | \text{df} = 9)$ as in the figure.

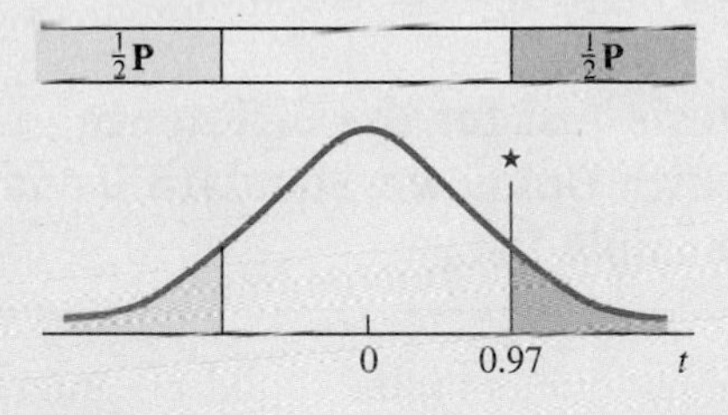

To find the p-value, you have three options:

1. Use Table 6 in Appendix B: $\mathbf{0.20 < P < 0.50}$.
2. Use Table 7 in Appendix B to place bounds on the p value: $\mathbf{0.340 < P < 0.394}$.
3. Use a computer or calculator to find the p-value: $\mathbf{P} = 2 \times 0.1787 = \mathbf{0.3574}$.

For specific instructions, see below.

b. Determine whether or not the p-value is smaller than α.

The p-value is not smaller than α.

OR

Using the Classical Procedure

a. Determine the critical region and critical value(s).

The critical region is two-tailed because H_a expresses concern for values related to "different than." The right-hand critical value is obtained from Table 6: $t(9, 0.05) = \mathbf{1.83}$. See the figure.

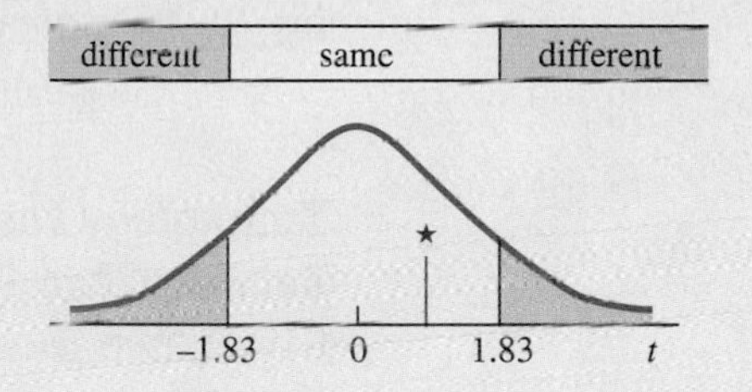

For specific instructions, see pages 414–415.

b. Determine whether or not the calculated test statistic is in the critical region.

$t\bigstar$ is not in the critical region as shown in **red** on the figure.

Step 5 **The Results:**

a. State the decision about H_o: Fail to reject H_o.

b. State the conclusion about H_a.

The evidence is not sufficient to show that machine A dispenses a different amount of soft drink than machine B, at the 0.10 level of significance. Thus, for lack of evidence we will proceed as though the two machines dispense, on average, the same amount.

To find the p-value for Illustration 10.9, use one of three methods:

Method 1: Use Table 6. Find 0.97 between two entries in the df = 9 row and read the bounds for **P** from the two-tail heading at the top of the table: $\mathbf{0.20 < P < 0.50}$.

Method 2: *Use Table* 7. Find $t\bigstar = 0.97$ between two rows and df = 9 between two columns; read the bounds for $P(t\bigstar > 0.97 \mid df = 9)$: $0.170 < {}^1/_2\mathbf{P} < 0.197$; therefore, $\mathbf{0.340 < P < 0.394}$.

Method 3: If you are doing the hypothesis test with the aid of a computer or calculator, most likely it will calculate the *p*-value (do not double) for you (see p. 491), or you may use the cumulative probability distribution commands described in Chapter 9 (p. 417). ■

Most computer or calculator statistical packages will complete the inferences for the difference between two means by calculating the number of degrees of freedom.

ANSWER NOW 10.37

Suppose the calculated $t\bigstar$ had been 1.80 in Illustration 10.9. Using df = 9 or df = 20 results in different answers. Explain how the word *conservative* (p. 491) applies here.

ANSWER NOW 10.38

The hypothesis test involves H_a: $\mu_B - \mu_A \neq 0$ with df = 18 and $t\bigstar = 1.3$.

a. Find the *p*-value. b. Find the critical values given $\alpha = 0.05$.

Technology Instructions: Hypothesis Test for the Difference Between Two Population Means with Unknown Standard Deviation Given Two Independent Sets of Sample Data

MINITAB (Release 13)

MINITAB's TWOSAMPLE command performs both the confidence interval and the hypothesis test at the same time.

Input the two independent sets of data into C1 and C2; then continue with:

```
Choose:   Stat > Basic Statistics > 2-Sample t
Select:   Samples in different columns
Enter:    First: C1
          Second: C2
Select:   Assume equal variances (if known)
Select:   Options
Enter:    Confidence level: 1 − α (ex. 0.95 or 95.0)
          Test mean: 0.0
Choose:   Alternative: less than or not equal or greater
          than
```

Excel XP

Input the two independent sets of data into columns A and B; then continue with:

```
Choose:   Tools > Data Analysis
                > t-Test: Two-Sample Assuming Unequal Vari-
                ances
Enter:    Variable 1 Range: (A1:A20 or select cells)
          Variable 2 Range: (B1:B20 or select cells)
          Hypothesized Mean Difference: μB − μA (usually 0)
Select:   Labels (if necessary)
Enter:    α (ex. 0.05)
Select:   Output Range
Enter:    (C1 or select cell) > OK
```

(continued)

Excel XP (continued)

Use Format > Column > AutoFit Selection to make the output more readable. The output shows *p*-values and critical values for one- and two-tailed tests.

TI-83 Plus

Input the two independent sets of data into L1 and L2.*

To construct a $1 - \alpha$ confidence interval for the mean difference, continue with the following, entering the appropriate values and highlighting Calculate:

```
Choose:   STAT > TESTS > 0:2-SampTInt...
```

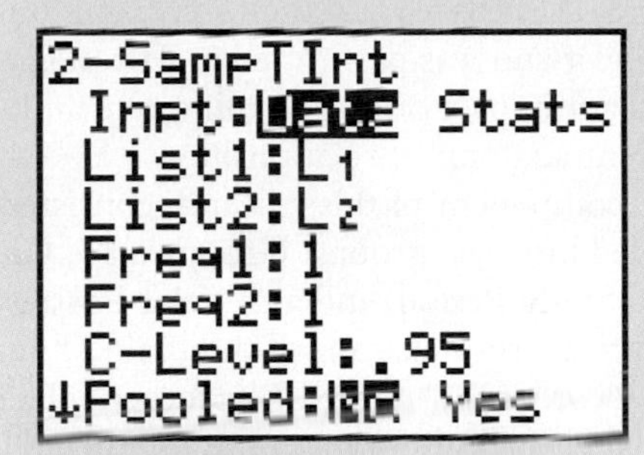


```
2-SampTInt
↑List1:L1
 List2:L2
 Freq1:1
 Freq2:1
 C-Level:.95
 Pooled:No Yes
 Calculate
```

To complete a hypothesis test for the mean difference, continue with the following, entering the appropriate values and highlighting Calculate:

```
Choose:   STAT > TESTS > 4:2-SampTTest...
```

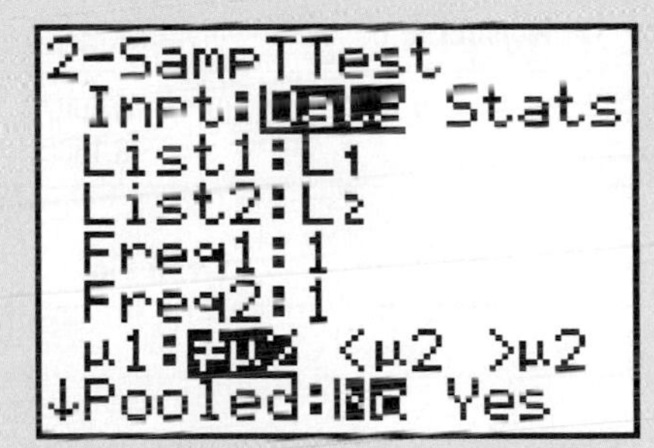


```
2-SampTTest
↑List1:L1
 List2:L2
 Freq1:1
 Freq2:1
 μ1:≠μ2 <μ2 >μ2
 Pooled:No Yes
 Calculate Draw
```

*Enter the data in the order that is needed; the program subtracts as L1 − L2.

Highlight No for Pooled if there are no assumptions about the equality of variances.

Illustration 10.8 was solved using MINITAB. With 40 cumulative grade-point averages for nonmembers in C1 and 40 averages for fraternity members in C2, the above commands resulted in the output shown here. Compare these results to the solution of Illustration 10.8. Notice the difference in **P** and df values. Explain.

```
Two Sample T-Test and Confidence Interval
Twosample T for C1 vs C2
       N      Mean          StDev          SE Mean
C1     40     2.21          0.59           0.09
C2     40     2.03          0.68           0.11
95% C.I. for mu C1 − mu C2: (−0.10, 0.46)
T-Test mu C1 = mu C2 (vs >): T = 1.26 P = 0.106 DF = 75
```

Application 10.3 Sectional Anatomy: Strategy for Mastery

A strategy for mastery of sectional anatomy must contain research to determine whether or not the methodology is sound. The research-based approach presented below demonstrates the effectiveness of a specific strategy. One application that deals with a comprehensive understanding of human anatomical features and their adjacent structures is a prescribed, sequenced labeling method (PSLM). The method contrasts the modern convention of random labeled human anatomical sections. The PSLM is based on studies utilizing images acquired from The Visible Human Project and was performed at Triton College. These studies have shown a significant impact on the learning rate and the comprehension of structures and adjacent anatomical relationships.

The initial component of this research consisted of a group of 28 students who were divided into two groups: Group A: 15, Group B: 13. Both groups were given a Pretest, Study Period, and a singular Posttest. Both images presented for study were identical cross sections of the brain. The Group A image for study was labeled in accordance with PSLM protocol. The Group B image for study was randomly labeled. Both groups were instructed to "list and recognize the parts/layers of a transverse brain section, from superficial to deep" by writing their answers in spaces provided as the only posttest in this preliminary study.

TABLE First Set of PSLM Studies

Variable	Number of Cases	Mean	SD	SE of Mean
Group A	15	9.6667	1.589	0.410
Group B	13	3.1538	3.023	0.839

Variances	*t*-value	df	2-Tail Sig	SE of Difference	95% CI of Diff
Unequal	6.98	17.57	.000	0.933	(4.548, 8.477)

Group A's average score was 9.6667 of a total possible right of 11 with a standard deviation of 1.589. Group B scored on average 3.1538 out of 11 possible points. The mean difference between Group A and Group B is 6.5129, which is highly significant.

Alexander Lane, Ph.D.
Triton College, River Grove, Illinois
http://www.nlm.nih.gov/research/visible/vhpconf98/AUTHORS/LANE/LANE.HTM

ANSWER NOW 10.39

a. **The "Variances—Unequal" is equivalent to making no assumption about the variances. Verify the value reported for the estimated standard error, "SE of Difference."**

b. **Verify the *t*-value.**

c. **What is the range of possible values for df for this study?**

d. **Explain how the df of 17.57 was obtained.**

e. **What df value would you use based on the material presented in Section 10.3?**

f. **Explain why the *t*-value of 6.98 is said to be "highly significant."**

Exercises

10.40 Find the confidence coefficient, $t(\text{df}, \alpha/2)$, that would be used to find the maximum error for each of the following situations when estimating the difference between two means, $\mu_1 - \mu_2$.

a. $1 - \alpha = 0.95$, $n_1 = 25$, $n_2 = 15$
b. $1 - \alpha = 0.98$, $n_1 = 43$, $n_2 = 32$
c. $1 - \alpha = 0.99$, $n_1 = 19$, $n_2 = 45$

10.41 An experiment was conducted to compare the mean absorptions of two drugs in specimens of muscle tissue. Seventy-two tissue specimens were randomly divided into two equal groups. Each group was tested with one of the two drugs. The sample results were $\bar{x}_A = 7.9$, $\bar{x}_B = 8.5$, $s_A = 0.11$, and $s_B = 0.10$. Assume both populations are normal. Construct the 98% confidence interval for the difference in the mean absorption rates.

10.42 A study comparing attitudes toward death was conducted in which organ donors (individuals who had signed organ donor cards) were compared with nondonors. The study is reported in the journal *Death Studies* (Vol. 14, No. 3, 1990). Templer's Death Anxiety Scale (DAS) was administered to both groups. On this scale, high scores indicate high anxiety concerning death. These results were reported:

	n	Mean	Standard Deviation
Organ Donors	25	5.36	2.91
Nonorgan Donors	69	7.62	3.45

Construct the 95% confidence interval for the difference between the means, $\mu_{non} - \mu_{donor}$.

10.43 *USA Today* (October 12, 1994) reported the longest average workweeks for nonsupervisory employees in private industry to be in mining (45.4 hours) and manufacturing (42.3 hours). The same article reported the shortest average workweeks to be in retail trade (29 hours) and services (32.4 hours). A study conducted in Missouri found these results for a similar study:

Industry	n	Average Hours/ Week	Standard Deviation
Mining	15	47.5	5.5
Manufacturing	10	43.5	4.9

Set a 95% confidence interval on the difference in average length of workweek between mining and manufacturing. Assume that the sampled populations were normal and the samples were selected randomly.

10.44 Experimentation with a new rocket nozzle has led to two slightly different designs. These data summaries resulted from testing the two designs:

	n	$\sum x$	$\sum x^2$
Design 1	36	278.4	2163.76
Design 2	42	310.8	2332.26

Determine the 99% confidence interval for the difference in the means for these two rocket nozzles.

10.45 Ex10-045 A study was designed to estimate the difference in diastolic blood pressures between men and women. MINITAB was used to construct a 99% confidence interval for the difference between the means based on the following sample data:

Men

76	76	74	70	80	68	90	70
76	80	68	72	96	80	90	72

Women

76	70	82	90	68	60	62
60	62	72	68	80	74	

```
Twosample T for Males vs Females
          N        Mean       StDev     SE Mean
Males     16       77.37      8.35      2.1
Females   13       71.08      9.22      2.6
99% C.I. for mu male - mu female: (-2.9, 15.5)
```

Verify the results (the two sample means and standard deviations, and the confidence interval bounds) by calculating the values yourself.

10.46 Ex10-046 Is the length of a steel bar affected by the heat treatment technique used? This question was being tested when the following lengths (to the nearest inch) were collected:

Treatment 1

156	159	151	153	157	159	155	155	151	152	158
154	156	156	157	155	156	159	153	157	157	159
158	155	159	152	150	154	156	156	157	160	

Treatment 2

154	156	150	151	156	155	153	154	149	150	150
151	154	155	155	154	154	156	150	151	156	154
153	154	149	150	150	151	154	148	155	158	

a. Find the means and standard deviations for the two sets of data.

b. Find evidence about the sample data (both graphic and numerical) that supports the assumption of normality for the two sampled populations.
c. Find the 95% confidence interval for $\mu_1 - \mu_2$.

10.47 Ex10-047 Sucrose, ordinary table sugar, is probably the single most abundant pure organic chemical in the world and the one most widely known to nonchemists. Whether from sugar cane (20% by weight) or sugar beets (15% by weight), and whether raw or refined, common sugar is still sucrose. Fifteen U.S. sugar-beet–producing counties were randomly selected and their 2001 sucrose percentages recorded. Similarly, 12 U.S. sugar-cane–producing counties were randomly selected and their 2001 sucrose percentages recorded.

Sugar beet sucrose

17.30	16.46	16.20	17.53	17.00	18.53	16.77	16.11
15.30	17.90	15.98	17.30	17.94	17.30	16.60	

Sugar cane sucrose

14.1	13.5	15.2	15.0	13.6	13.6
14.8	13.7	11.7	14.3	13.8	13.8

Source: http://www.usda.gov/nass/graphics/county01/data/

Find the 95% confidence interval for the difference between the mean sucrose percentages for all U.S. sugar-beet–producing counties and sugar-cane–producing counties.

10.48 Ex10-048 At a large university, a mathematics placement exam is administered to all students. Samples of 36 males and 30 females are randomly selected from this year's student body and these scores recorded:

Men

72	68	75	82	81	60	75	85	80
70	71	84	68	85	82	80	54	81
86	79	99	90	68	82	60	63	67
72	77	51	61	71	81	74	79	76

Women

81	76	94	89	83	78	85	91	83	83
84	80	84	88	77	74	63	69	80	82
89	69	74	97	73	79	55	76	78	81

a. Describe each set of data with a histogram (use the same class intervals on both histograms) and with the mean and standard deviation.
b. Construct a 95% confidence interval for the mean score for all male students. For all female students.
c. Do the results in part b show that the mean scores for men and women could be the same? Justify your answer. Be careful!
d. Construct a 95% confidence interval for the difference between the mean scores for male and female students.
e. Do the results in part d show that the mean scores for men and women could be the same? Explain.

(continued)

f. Explain why the results in part b cannot be used to draw conclusions about the difference between the two means.

10.49 State the null and alternative hypotheses that would be used to test the following claims:
a. There is a difference between the mean ages of employees at two different large companies.
b. The mean of population 1 is greater than the mean of population 2.
c. The difference between the means of the two populations is more than 20 pounds.
d. The mean of population A is less than 50 more than the mean of population B.

10.50 Calculate the estimate for the standard error of the difference between two independent means for each of these cases:
a. $s_1^2 = 12$, $s_2^2 = 15$, $n_1 = 16$, and $n_2 = 21$
b. $s_1^2 = 0.054$, $s_2^2 = 0.087$, $n_1 = 8$, and $n_2 = 10$
c. $s_1 = 2.8$, $s_2 = 6.4$, $n_1 = 16$, and $n_2 = 21$

10.51 Determine the p-value for each hypothesis test for the difference between two means with population variances unknown.
a. H_a: $\mu_1 - \mu_2 > 0$, $n_1 = 6$, $n_2 = 10$, $t\bigstar = 1.3$
b. H_a: $\mu_1 - \mu_2 < 0$, $n_1 = 16$, $n_2 = 9$, $t\bigstar = -2.8$
c. H_a: $\mu_1 - \mu_2 \neq 0$, $n_1 = 26$, $n_2 = 16$, $t\bigstar = 1.8$
d. H_a: $\mu_1 - \mu_2 \neq 5$, $n_1 = 26$, $n_2 = 35$, $t\bigstar = -1.8$

10.52 Determine the critical values that would be used for the following hypothesis tests (using the classical approach) about the difference between two means with population variances unknown.
a. H_a: $\mu_1 - \mu_2 \neq 0$, $n_1 = 26$, $n_2 = 16$, $\alpha = 0.05$
b. H_a: $\mu_1 - \mu_2 < 0$, $n_1 = 36$, $n_2 = 27$, $\alpha = 0.01$
c. H_a: $\mu_1 - \mu_2 > 0$, $n_1 = 8$, $n_2 = 11$, $\alpha = 0.10$
d. H_a: $\mu_1 - \mu_2 \neq 10$, $n_1 = 14$, $n_2 = 15$, $\alpha = 0.05$

10.53 Many cheeses are produced in the shape of a wheel. Because of the differences in consistency between these different types of cheese, the amount of cheese (measured by weight) varies from wheel to wheel. Heidi Cembert wishes to determine whether there is a significant difference, at the 10% level, between the weights per wheel of gouda and brie. She randomly samples 16 wheels of gouda and finds the mean is 1.2 pounds with a standard deviation of 0.32 pound and 14 wheels of Brie and finds a mean of 1.05 pounds and a standard deviation of 0.25 pound. What is the p-value for Heidi's hypothesis of equality?

10.54 S If a random sample of 18 homes south of Center Street in Provo has a mean selling price of \$125,000 and a standard deviation of \$2,400, and a random sample of 18 homes north of Center Street has a mean selling price of \$126,000 and a standard deviation of \$4,800, can you conclude that there is a significant difference between the selling price of homes in these two areas of Provo at the 0.05 level?

a. Solve using the p-value approach.
b. Solve using the classical approach.

10.55 The computer age has enabled teachers to use electronic tutorials to motivate their students to learn. *Issues in Accounting Education* (Vol. 13, No. 2, May 1998) published the results of a 1998 study revealing that an electronic tutorial, along with intentionally induced peer pressure, was effective in enhancing preclass preparation and in improving class attendance, test scores, and course evaluations when used by students studying tax accounting.

Suppose a similar study is conducted at your school using an electronic study guide (ESG) as a tutor for students in accounting principles. For one course section, the students were required to use a new ESG computer program that generated and scored chapter review quizzes and practice examinations, presented textbook chapter reviews, and tracked progress. Students could use the computer to build, take, and score their own simulated tests and review materials at their own pace before they took their formal in-class quizzes and exams composed of different questions. The same instructor taught the other course section, used the same textbook, and gave the same daily assignments, but he did not require the students to use the ESG. Identical tests were administered to both sections, and the mean scores of all tests and assignments at the end of the year were tabulated.

Section	n	Mean Score	Standard Deviation
ESG (1)	38	79.6	6.9
No ESG (2)	36	72.8	7.6

Do these results show that the mean score of tests and assignments for students taking accounting principles with an ESG to help them is significantly higher than the mean for those not using an ESG? Use the 0.01 level of significance.

a. Solve using the p-value approach.
b. Solve using the classical approach.

10.56 The purchasing department for a regional supermarket chain is considering two sources from which to purchase 10-lb bags of potatoes. A random sample taken from each source gives these results:

	Idaho Supers	Idaho Best
Number of bags weighed	100	100
Mean weight	10.2 lb	10.4 lb
Sample variance	0.36 lb	0.25 lb

At the 0.05 level of significance, is there a difference between the mean weights of the 10-lb bags of potatoes?

a. Solve using the p-value approach.
b. Solve using the classical approach.

10.57 Some 20 million Americans visit chiropractors annually, and the number of practitioners has nearly doubled in the last two decades, to 55,000 in the United States, according to the American Chiropractic Association. The *New England Journal of Medicine* (Fall 1998) released a report showing the results of a study that compared chiropractic spinal manipulation (CSM) with physical therapy for treating acute lower back pain. After two years of treatment, CSM was found to be no more effective at either reducing missed work or preventing a relapse.

Suppose a similar study of 60 patients is made by dividing the sample into two groups. One group is given CSM and the other receives physical therapy for one year. During the period, the numbers of missed days at work due to lower back pain were recorded:

Group	n	Mean	Standard Deviation
CSM (1)	32	10.6	4.8
Therapy (2)	28	12.5	6.3

Do these results show that the mean number of missed days of work for people suffering from acute back pain is significantly less for those receiving CSM than for those undergoing physical therapy? Use the 0.01 level of significance.

a. Solve using the p-value approach.
b. Solve using the classical approach.

10.58 A study was conducted to assess the safety and efficiency of receiving nitroglycerin from a transdermal system (i.e., a patch worn on the skin), which intermittently delivers the medication, versus oral medication (pills). Twenty patients who suffer from angina (chest pain) due to physical effort were enrolled in trials. All received patches; some ($n = 8$) contained nitroglycerin and the others ($n = 12$) contained a placebo. The resulting "times to angina" (in seconds) are summarized in the table.

	Active	Placebo	Difference	Standard Error	p-Value*
Day 1 A.M.	320.00	287.00	33.00	9.68	0.0029
Day 7 P.M.	314.00	285.25	28.75	13.74	0.0500

*For treatment difference

a. Determine the value of t for the difference between two independent means given the difference and the standard error for the day 1 A.M. data.
b. Verify the p-value.
c. Determine the value of t for the difference between two independent means given the difference and the standard error for the day 7 P.M. data.
d. Verify the p-value.

10.59 Ex10-059 MINITAB was used to complete a t-test of the difference between the two means using these two independent samples:

Sample 1

33.7	21.6	32.1	38.2	33.2	35.9	34.1
23.5	21.2	23.3	18.9	30.3	39.8	

Sample 2

28.0	59.9	22.3	43.3	43.6	24.1	6.9	14.1
30.2	3.1	13.9	19.7	16.6	13.8	62.1	28.1

```
Twosample T for sample 1 vs sample 2
          N       Mean     StDev      SE Mean
sample1   13      29.68    7.07       2.0
sample2   16      26.9     17.4       4.4
T-Test mu sample1 = mu sample2 (vs not =):
T=0.59 P=0.56 DF=20
```

a. Verify the results (two sample means and standard deviations, and the calculated $t\bigstar$) by calculating the values yourself.
b. Use Table 7 in Appendix B to verify the p-value based on the calculated df.
c. Find the p-value using the smaller number of degrees of freedom. Compare the two p-values.

10.60 Ex10-060 The quality of latex paint is monitored by measuring different characteristics of the paint. One characteristic of interest is the particle size. Two different types of disc centrifuges (JLDC, Joyce Loebl Disc Centrifuge, and the DPJ, Dwight P. Joyce disc) are used to measure the particle size. It is thought that these two methods yield different measurements. Thirteen readings were taken from the same batch of latex paint using both the JLDC and the DPJ discs.

JLDC	4714	4601	4696	4896	4905	4870	4987
	5144	3962	4006	4561	4626	4924	
DPJ	4295	4271	4326	4530	4618	4779	4752
	4744	3764	3797	4401	4339	4700	

Source: With permission of SCM Corporation

Assume the particle sizes are normally distributed.
a. Determine whether there is a significant difference between the readings at the 0.10 level of significance.
b. What is your estimate for the difference between the two readings?

10.61 Ex10-061 Twenty laboratory mice were randomly divided into two groups of 10. Each group was fed a prescribed diet. At the end of three weeks, the weight gained by each animal was recorded. Do the data in the table justify the conclusion that the mean weight gained on diet B was greater than the mean weight gained on diet A, at the $\alpha = 0.05$ level of significance?

Diet A	5	14	7	9	11	7	13	14	12	8
Diet B	5	21	16	23	4	16	13	19	9	21

a. Solve using the p-value approach.
b. Solve using the classical approach.

10.62 Ex10-062 In evaluating different measuring instruments one must first determine whether there is a systematic difference between the instruments. Lenses from two different groups (A and B) were measured once each by two different instruments. The measurement differences (instrument A − instrument B) were recorded. The units have been coded for propriety reasons. (Courtesy: Bausch & Lomb)

Group A

4	5	−2	−3	−7	10	11	−1	3
7	−5	3	−4	−5	−7	4	−1	
−18	0	−17	12	9	4	17	−2	

Group B

−13	−12	−5	11	15	7
−33	−10	−6	−2	−16	2
0	−19	6	−17	−4	−19
−22	−4	8	10	−6	

Does there appear to be a systematic difference between the two instruments?
a. Describe each set of data separately using a histogram and comparatively using one side-by-side graph.
b. Find the mean and the standard deviation for each set of data.
c. Are the assumptions satisfied? Explain.
d. Test the hypothesis that there is no difference between the means of the two differences. Use $\alpha = 0.05$.
e. Is there any evidence of a difference between the two instruments? Explain.

10.63 Use a computer to demonstrate the truth of the statement describing the sampling distribution of $\bar{x}_1 - \bar{x}_2$. Use two theoretical normal populations for the simulation: $N_1(100, 20)$ and $N_2(120, 20)$.
a. To get acquainted with the two theoretical populations, randomly select a very large sample from each. Generate 2000 data values, calculate the mean and standard deviation, and construct a histogram using class boundaries that are multiples of one-half of a standard deviation (10) starting at the mean for each population.
b. If samples of size 8 are randomly selected from each population, what do you expect the distribution of $\bar{x}_1 - \bar{x}_2$ to be like (shape of distribution, mean, standard error)?
c. Randomly draw a sample of size 8 from each population, and find the mean of each sample. Find the difference between the sample means. Repeat 99 more times.
d. The set of 100 $(\bar{x}_1 - \bar{x}_2)$-values form an empirical sampling distribution of $\bar{x}_1 - \bar{x}_2$. Describe the empirical distribution: shape (histogram), mean, and standard error. (Use class boundaries that are multiples of the standard

error from the mean for easy comparison to the expected.)

e. Using the information you have found, verify the statement about the $\bar{x}_1 - \bar{x}_2$ sampling distribution made on page 490.

f. Repeat the experiment a few times and compare the results.

FYI
See the *Statistical Tutor* for additional information about commands.

10.64 One reason for being conservative when determining the number of degrees of freedom to use with the t-distribution was the possibility that the population variances might be unequal. Extremely different values cause a lower df to be used. Repeat Exercise 10.63 using theoretical normal distributions of $N(100, 9)$ and $N(120, 27)$ and both sample sizes of 8. Check all three properties of the sampling distribution: normality, mean value, and standard error. Describe in detail what you discover. Do you think we should be concerned about the choice of df? Explain.

10.65 Unbalanced sample sizes are a factor in determining the number of degrees of freedom for inferences concerning the difference between two means. Repeat Exercise 10.63 using theoretical normal distributions of $N(100, 20)$ and $N(120, 20)$ and sample sizes of 5 and 20. Check all three properties of the sampling distribution: normality, mean value, and standard error. Describe in detail what you discover. Do you think we should be concerned when using unbalanced sample sizes? Explain.

10.66 One of the assumptions for the two-sample t-test is that the sampled populations are to be normally distributed. What happens when they are not normally distributed? Repeat Exercise 10.63 using two theoretical populations that are not normal and using samples of size 10. The exponential distribution uses a continuous random variable, it has a J-shaped distribution, and its mean and standard deviation are the same value. Use the two exponential distributions with means of 50 and 80: exp(50) and exp(80). Check all three properties of the sampling distribution: normality, mean value, and standard error. Describe in detail what you discover. Do you think we should be concerned when sampling nonnormal populations? Explain.

10.4 Inferences Concerning the Difference Between Proportions Using Two Independent Samples

We are often interested in making statistical comparisons between the **proportions, percentages, or probabilities** associated with two populations. These questions ask for such comparisons: Is the proportion of homeowners who favor a certain tax proposal different from the proportion of renters who favor it? Did a larger percentage of this semester's class than of last semester's class pass statistics? Is the probability of a Democratic candidate winning in New York greater than the probability of a Republican candidate winning in Texas? Do students' opinions about the new code of conduct differ from those of the faculty? You have probably asked similar questions.

FYI
The 3 "p" words (*proportion, percentage, probability*) are all the binomial parameter p, P(success).

NOTE These are the properties of a **binomial experiment:**

FYI
Binomial experiments are defined in more detail on page 250.

1. The observed probability is $p' = x/n$, where x is the number of observed successes in n trials.
2. $q' = 1 - p'$.
3. p is the probability of success on an individual trial in a binomial probability experiment of n repeated independent trials.

ANSWER NOW 10.67
Only 75 of 250 people interviewed were able to name the vice president of the United States. Find the values for x, n, p', and q'.

In this section, we will compare two population proportions by using the difference between the observed proportions, $p'_1 - p'_2$, of two independent samples. The observed difference, $p'_1 - p'_2$, belongs to a sampling distribution with the characteristics described in the following statement.

If independent samples of sizes n_1 and n_2 are drawn randomly from large populations with $p_1 = P_1$ (success) and $p_2 = P_2$ (success), respectively, then the sampling distribution of $p'_1 - p'_2$ has these properties:

1. mean $\mu_{p'_1 - p'_2} = p_1 - p_2$,
2. standard error $\sigma_{p'_1 - p'_2} = \sqrt{\left(\frac{p_1 q_1}{n_1}\right) + \left(\frac{p_2 q_2}{n_2}\right)}$, and (10.10)
3. an approximately normal distribution if n_1 and n_2 are sufficiently large.

In practice, we use the following *guidelines to ensure normality*:

1. The sample sizes are both larger than 20.
2. The products n_1p_1, n_1q_1, n_2p_2, and n_2q_2 are all larger than 5.
3. The samples consist of less than 10% of their respective populations.

NOTE p_1 and p_2 are unknown; therefore, the products mentioned in guideline 2 will be estimated by $n_1p'_1$, $n_1q'_1$, $n_2p'_2$, and $n_2q'_2$.

ANSWER NOW 10.68
Assume $n_1 = 40$, $p'_1 = 0.9$, $n_2 = 50$, and $p'_2 = 0.9$.

a. **Find the estimated values for both *np*'s and both *nq*'s.**

b. **Does this situation satisfy the guidelines for approximately normal? Explain.**

Inferences about the difference between two population proportions, $p_1 - p_2$, will be based on the following assumptions.

ASSUMPTIONS FOR INFERENCES ABOUT THE DIFFERENCE BETWEEN TWO PROPORTIONS $p_1 - p_2$

The n_1 random observations and the n_2 random observations that form the two samples are selected independently from two populations that are not changing during the sampling.

CONFIDENCE INTERVAL PROCEDURE

When we estimate the **difference between two proportions,** $p_1 - p_2$, we will base our estimates on the **unbiased sample statistic** $p'_1 - p'_2$. The point estimate, $p'_1 - p'_2$, becomes the center of the confidence interval and the confidence interval limits are found using the following formula:

CONFIDENCE INTERVAL FOR THE DIFFERENCE BETWEEN TWO PROPORTIONS

$$(p'_1 - p'_2) - z(\alpha/2) \cdot \sqrt{\left(\frac{p'_1 q'_1}{n_1}\right) + \left(\frac{p'_2 q'_2}{n_2}\right)} \quad \text{to} \quad (p'_1 - p'_2) + z(\alpha/2) \cdot \sqrt{\left(\frac{p'_1 q'_1}{n_1}\right) + \left(\frac{p'_2 q'_2}{n_2}\right)}$$

(10.11)

Illustration 10.10 Constructing a Confidence Interval for the Difference Between Two Proportions

In studying his campaign plans, Mr. Morris wishes to estimate the difference between men's and women's views regarding his appeal as a candidate. He asks his campaign manager to take two random independent samples and find the 99% confidence interval for the difference. A sample of 1000 voters was taken from each population, with 388 men and 459 women favoring Mr. Morris.

SOLUTION

FYI
It is customary to place the larger value first; that way, the point estimate for the difference is a positive value.

Step 1 **The Set-Up:**
Describe the population parameter of interest.
$p_w - p_m$, the difference between the proportion of women voters and the proportion of men voters who plan to vote for Mr. Morris.

Step 2 **The Confidence Interval Criteria:**

a. **Check the assumptions.**
The samples are randomly and independently selected.

b. **Identify the probability distribution and the formula to be used.**
The standard normal distribution. The populations are large (all voters); the sample sizes are larger than 20; and the estimated values for $n_m p_m$, $n_m q_m$, $n_w p_w$, and $n_w q_w$ are all larger than 5. Therefore, the sampling distribution of $p'_w - p'_m$ should have an approximately normal distribution. $z\bigstar$ will be calculated using formula (10.11).

c. **State the level of confidence:** $1 - \alpha = 0.99$.

Step 3 **The Sample Evidence:**
Collect the sample information.
We have $n_m = 1000$, $x_m = 388$, $n_w = 1000$, and $x_w = 459$.

$$p'_m = \frac{x_m}{n_m} = \frac{388}{1000} = \mathbf{0.388} \qquad q'_m = 1 - 0.388 = \mathbf{0.612}$$

$$p'_w = \frac{x_w}{n_w} = \frac{459}{1000} = \mathbf{0.459} \qquad q'_w = 1 - 0.459 = \mathbf{0.541}$$

Step 4 **The Confidence Interval:**

a. **Determine the confidence coefficient.**
This is a two-tailed situation, with $\alpha/2$ in each tail. From Table 4B, $z(\alpha/2) = z(0.005) = 2.58$.

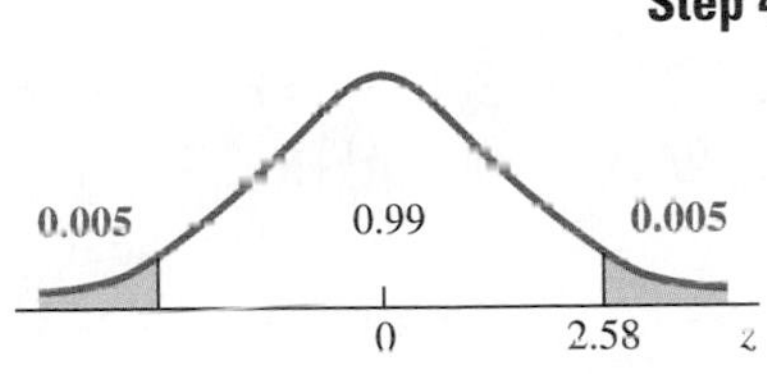

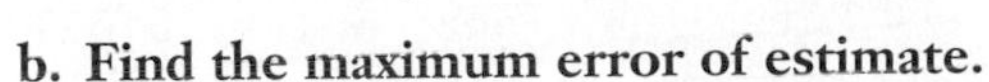

Instructions for using Table 4B are on page 349.

b. **Find the maximum error of estimate.**
Using the maximum error part of formula (10.11), we have

$$E = z(\alpha/2) \cdot \sqrt{\left(\frac{p'_w q'_w}{n_w}\right) + \left(\frac{p'_m q'_m}{n_m}\right)}:$$

$$E = 2.58 \cdot \sqrt{\left(\frac{(0.459)(0.541)}{1000}\right) + \left(\frac{(0.388)(0.612)}{1000}\right)}$$

$$= 2.58\sqrt{0.000248 + 0.000237} = (2.58)(0.022) = \mathbf{0.057}$$

c. Find the lower and upper confidence limits.

$$(p'_w - p'_m) \pm E$$

$$0.071 \pm 0.057$$

$$0.071 - 0.057 = \mathbf{0.014} \quad \text{to} \quad 0.071 + 0.057 = \mathbf{0.128}$$

Step 5 **The Results:**
State the confidence interval.
0.014 to 0.128 is the 99% confidence interval for $p_w - p_m$. With 99% confidence, we can say that there is a difference of from 1.4% to 12.8% in Mr. Morris's voter appeal. That is, a larger proportion of women than men favor Mr. Morris, and the difference in the proportions is between 1.4% and 12.8%. ■

ANSWER NOW 10.69
Find the 95% confidence interval for $p_A - p_B$.

Sample	n	x
A	125	45
B	150	48

Technology Instructions: Confidence Intervals for the Difference Between Two Proportions Given Two Independent Sets of Sample Data

MINITAB (Release 13)

```
Choose:   Stat > Basic Statistics > 2 Proportions
Select:   Summarized data:
Enter:    First Sample:    n (trials)    x (successes)
          Second Sample:   n (trials)    x (successes)
Select:   Options
Enter:    Confidence level: 1 − α (ex. 0.95 or 95.0)
```

Excel XP

Input the data for the first sample into column A using 0's for failures (or no's) and 1's for successes (or yes's); then repeat the same procedure for the second sample in column B; then continue with:

```
Choose:   Tools > Data Analysis Plus > Z-Estimate: Two Pro-
          portions
Enter:    Variable 1 Range: (A2:A20 or select cells)
          Variable 2 Range: (B1:B20 or select cells)
          Code for success: 1
Select:   Labels (if necessary)
Enter:    Alpha: α (ex. 0.05) > OK
```

TI-83 Plus

```
Choose:   STAT > TESTS >
          B:2-PropZint
```

Enter the appropriate values and highlight Calculate.

```
2-PropZInt
 x1:0
 n1:0
 x2:0
 n2:0
 C-Level:.95
 Calculate
```

HYPOTHESIS-TESTING PROCEDURE

When the null **hypothesis, There is no difference between two proportions,** is being tested, the **test statistic** will be the difference between the observed proportions divided by the **standard error;** it is found with the following formula:

TEST STATISTIC FOR THE DIFFERENCE BETWEEN TWO PROPORTIONS—POPULATION PROPORTION KNOWN

$$z\star = \frac{p'_1 - p'_2}{\sqrt{pq\left[\left(\frac{1}{n_1}\right) + \left(\frac{1}{n_2}\right)\right]}} \tag{10.12}$$

NOTES

1. The null hypothesis is $p_1 = p_2$ or $p_1 - p_2 = 0$ (the difference is zero).
2. Nonzero differences between proportions are not discussed in this section.
3. The numerator of formula (10.12) could be written as $(p'_1 - p'_2) - (p_1 - p_2)$, but since the null hypothesis is assumed to be true during the test, $p_1 - p_2 = 0$. By substitution, the numerator becomes simply $p'_1 - p'_2$.
4. Since the null hypothesis is $p_1 = p_2$, the standard error of $p'_1 - p'_2$, $\sqrt{\left(\frac{p_1 q_1}{n_1}\right) + \left(\frac{p_2 q_2}{n_2}\right)}$, can be written as $\sqrt{pq\left[\left(\frac{1}{n_1}\right) + \left(\frac{1}{n_2}\right)\right]}$, where $p = p_1 = p_2$ and $q = 1 - p$.
5. When the null hypothesis states $p_1 = p_2$ and does not specify the value of either p_1 or p_2, the two sets of sample data will be pooled to obtain the estimate for p. This pooled probability (known as p'_p) is the total number of successes divided by the total number of observations with the two samples combined; it is found using the next formula:

$$p'_p = \frac{x_1 + x_2}{n_1 + n_2} \tag{10.13}$$

and q'_p is its complement,

$$q'_p = 1 - p'_p \tag{10.14}$$

ANSWER NOW 10.70
Find the values of p'_p and q'_p for these samples:

Sample	x	n
E	15	250
R	25	275

When the pooled estimate, p'_p, is being used, formula (10.12) becomes formula (10.15):

TEST STATISTIC FOR THE DIFFERENCE BETWEEN TWO PROPORTIONS—POPULATION PROPORTION UNKNOWN

$$z\star = \frac{p'_1 - p'_2}{\sqrt{(p'_p)(q'_p)\left[\left(\frac{1}{n_1}\right) + \left(\frac{1}{n_2}\right)\right]}} \tag{10.15}$$

Illustration 10.11 One-Tailed Hypothesis Test for the Difference Between Two Proportions

A salesman for a new manufacturer of cellular phones claims not only that they cost the retailer less but also that the percentage of defective cellular phones found among his products will be no higher than the percentage of defectives found in a competitor's line. To test this statement, the retailer took random samples of each manufacturer's product. The sample summaries are given in Table 10.5. Can we reject the salesman's claim at the 0.05 level of significance?

TABLE 10.5 Cellular Phone Sample Information

Product	Number Defective	Number Checked
Salesman's	15	150
Competitor's	6	150

SOLUTION

Step 1 **The Set-Up:**

a. **Describe the population parameter of interest.**
$p_s - p_c$, the difference between the proportion of defectives in the salesman's product and the proportion of defectives in the competitor's product.

b. **State the null hypothesis (H_o) and the alternative hypothesis (H_a).**
The concern of the retailer is that the salesman's less expensive product may be of a poorer quality, meaning a greater proportion of defectives. If we use the difference "suspected larger proportion − smaller proportion," then the alternative hypothesis is "The difference is positive (greater than zero)."

H_o: $p_s - p_c = 0$ (≤) (salesman's defective rate is no higher than competitor's)
H_a: $p_s - p_c > 0$ (salesman's defective rate is higher than competitor's)

Step 2 **The Hypothesis Test Criteria:**

a. **Check the assumptions.**
Random samples were selected from the products of two different manufacturers.

b. **Identify the probability distribution and the test statistic to be used.**
The standard normal distribution. Populations are very large (all cellular phones produced); the samples are larger than 20; and the estimated products $n_s p'_s$, $n_s q'_s$, $n_c p'_c$, and $n_c q'_c$ are all larger than 5. Therefore, the sampling distribution should have an approximately normal distribution. $z\bigstar$ will be calculated using formula (10.15).

c. **Determine the level of significance:** $\alpha = 0.05$.

Step 3 **The Sample Evidence:**

a. **Collect the sample information.**

$$p'_s = \frac{x_s}{n_s} = \frac{15}{150} = \mathbf{0.10} \qquad p'_c = \frac{x_c}{n_c} = \frac{6}{150} = \mathbf{0.04}$$

$$p'_p = \frac{x_1 + x_2}{n_1 + n_2} = \frac{15 + 6}{150 + 150} = \frac{21}{300} = \mathbf{0.07} \qquad q'_p = 1 - p'_p = 1 - 0.07 = \mathbf{0.93}$$

b. Calculate the value of the test statistic.

$$z\bigstar = \frac{p'_s - p'_c}{\sqrt{(p'_p)(q'_p)\left[\left(\frac{1}{n_s}\right) + \left(\frac{1}{n_c}\right)\right]}}: \quad z\bigstar = \frac{0.10 - 0.04}{\sqrt{(0.07)(0.93)\left[\left(\frac{1}{150}\right) + \left(\frac{1}{150}\right)\right]}}$$

$$= \frac{0.06}{\sqrt{0.000868}} = \frac{0.06}{0.02946} = \mathbf{2.04}$$

ANSWER NOW 10.71
Find the value of $z\bigstar$ that would be used to test the difference between the proportions.

Sample	n	x
G	380	323
H	420	332

Step 4 **The Probability Distribution:**

Using the p-Value Procedure

a. Calculate the p-value for the test statistic.
Use the right-hand tail because H_a expresses concern for values related to "higher than." $\mathbf{P} = p\text{-value} = P(z\bigstar > 2.04)$ as shown on the figure.

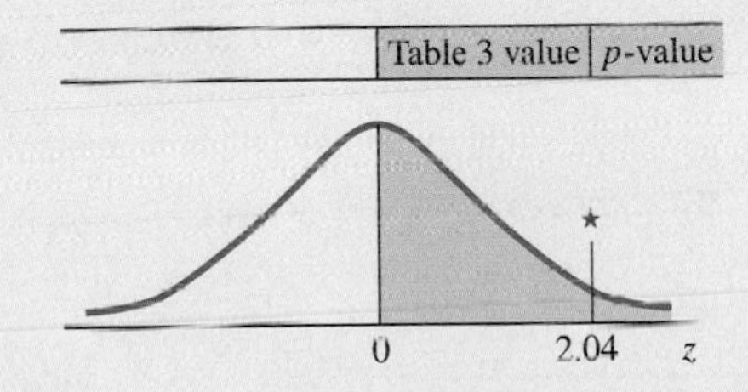

To find the p-value, you have three options:

1. Use Table 3 in Appendix B to calculate the p-value: $\mathbf{P} = 0.5000 - 0.4793 = \mathbf{0.0207}$.
2. Use Table 5 in Appendix B to place bounds on the p-value: $\mathbf{0.0202 < P < 0.0228}$.
3. Use a computer or calculator: $\mathbf{P = 0.0207}$.

For specific instructions, see pages 375–376.

b. Determine whether or not the p-value is smaller than α.
The p-value is smaller than α.

OR

Using the Classical Procedure

a. Determine the critical region and critical value(s).
The critical region is the right-hand tail because H_a expresses concern for values related to "higher than." The critical value is obtained from Table 4A: $z(0.05) = \mathbf{1.65}$.

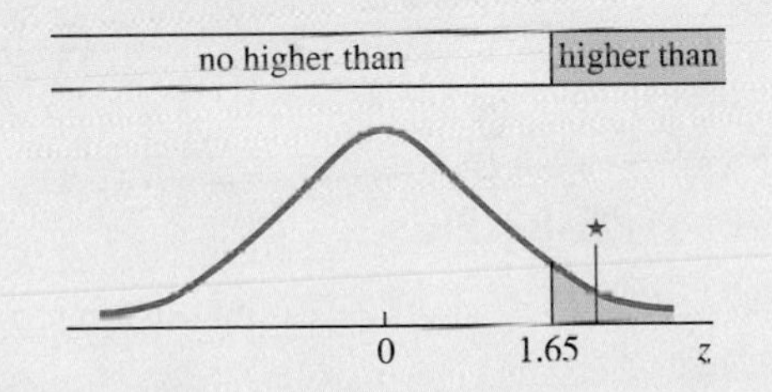

For specific instructions, see pages 392–393.

b. Determine whether or not the calculated test statistic is in the critical region.
$z\bigstar$ is in the critical region, as shown in **red** on the figure.

Step 5 **The Results:**

a. State the decision about H_o: Reject H_o.

b. State the conclusion about H_a.
At the 0.05 level of significance, there is sufficient evidence to reject the salesman's claim; the proportion of his company's cellular phones that are defective is higher than the proportion of his competitor's cellular phones that are defective. ■

ANSWER NOW 10.72
Find the p-value for the test with the alternative hypothesis $p_E < p_R$ using the data in Answer Now 10.70.

Technology Instructions: Hypothesis Test for the Difference Between Two Proportions, $p_1 - p_2$, for Two Independent Sets of Sample Data

MINITAB (Release 13)

```
Choose:  Stat > Basic Statistics > 2 Proportions
Select:  Summarized data:
Enter:   First Sample:    n (trials)    x (successes)
         Second Sample:   n (trials)    x (successes)
Select:  Options
Enter:   Test difference: 0.0
Choose:  Alternative: less than or not equal or greater
         than
Select:  Use pooled estimate of p for test
```

Excel XP

Input the data for the first sample into column A using 0's for failures (or no's) and 1's for successes (or yes's); then repeat the same procedure for the second sample in column B; then continue with:

```
Choose:  Tools > Data Analysis Plus > Z-Test: Two Propor-
         tions
Enter:   Variable 1 Range: (A1:A20 or select cells)
         Variable 2 Range: (B1:B20 or select cells)
         Code for success: 1
         Hypothesized difference: 0
Select:  Labels (if necessary)
Enter:   Alpha: α (ex. 0.05)
```

TI-83 Plus

```
Choose:  STAT > TESTS > 6:2-PropZTest...
```

Enter the appropriate values and highlight Calculate.

```
2-PropZTest
 x1:0
 n1:0
 x2:0
 n2:0
 p1:≠p2 <p2 >p2
 Calculate Draw
```

Application 10.4 Cadaver Kidneys Are Good for Transplants

In a discovery that could ease the severe shortage of donor organs, Swiss researchers found that kidneys transplanted from cadavers keep working just as long as those from a patient whose heart is still beating. Most transplant organs are taken from brain-dead patients whose hearts have not stopped because doctors have long believed that if they wait until the heart stops, the organs will become damaged from lack of oxygen.

But in the first long-term study comparing the two approaches, doctors at University Hospital Zurich followed nearly 250 transplant patients for up to 15 years and found nearly identical survival rates. At 10 years, 79 percent of patients whose kidney came from a donor with no heartbeat were alive, as were 77 percent of patients whose organ came from a brain-dead donor whose heart was beating. The study, published in Thursday's *New England Journal of Medicine*, could prove especially influential because it was a head-to-head comparison of the two approaches and was the first to follow patients for many years.

Doctors believe similar results may be found for transplants of the liver, pancreas, and lungs. By using organs from "cardiac death" donors, the number of kidneys available could increase up to 30 percent, meaning some 1000 or more extra U.S. donors a year, experts estimate.

ANSWER NOW 10.73

a. **79% versus 77% hardly seems like much of a difference. Assuming the 250 patients were equally divided between the two groups, is the difference of 2% significant?**

b. **What additional information is needed to test the statistical significance of the difference?**

c. **Is the fact that there is such a small difference of major importance to organ recipients? Explain.**

Exercises

10.74 In a random sample of 40 brown-haired individuals, 22 indicated that they used hair coloring. In another random sample of 40 blonde individuals, 26 indicated that they used hair coloring. Use a 92% confidence interval to estimate the difference in the proportions of these groups that use hair coloring.

10.75 An article titled "Nurse Executive Turnover" (*Nursing Economics*, January/February 1993) compared two groups of nurse executives. One group had participated in a unique program for nurse executives called the Wharton Fellows Program, and the other group had not participated in the program. Eighty-seven of 341 Wharton Fellows had experienced one change in position, and 9 of 40 non-Wharton Fellows had experienced one change in position. Find a 99% confidence interval for the difference in population proportions.

10.76 In a survey of 300 people from city A, 128 preferred New Spring soap to all other brands of deodorant soap. In city B, 149 of 400 people preferred New Spring. Find the 98% confidence interval for the difference in the two proportions.

10.77 The proportions of defective parts produced by two machines were compared, and these data were collected:

Machine 1: $n = 150$, number of defective parts = 12
Machine 2: $n = 150$, number of defective parts = 6

Determine a 90% confidence interval for $p_1 - p_2$.

10.78 Show that the standard error of $p'_1 - p'_2$, which is $\sqrt{\left(\frac{p_1q_1}{n_1}\right) + \left(\frac{p_2q_2}{n_2}\right)}$, reduces to $\sqrt{pq\left[\left(\frac{1}{n_1}\right) + \left(\frac{1}{n_2}\right)\right]}$, when $p_1 = p_2 = p$.

10.79 State the null hypothesis, H_o, and the alternative hypothesis, H_a, that would be used to test these claims:

a. There is no difference between the proportions of men and women who will vote for the incumbent in next month's election.
b. The percentage of boys who cut classes is greater than the percentage of girls who cut classes.
c. The percentage of college students who drive old cars is higher than the percentage of noncollege people of the same age who drive old cars.

10.80 Determine the critical region and critical value(s) that would be used to test (classical procedure) the following hypotheses when z is used as the test statistic.

a. H_o: $p_1 = p_2$ vs. H_a: $p_1 > p_2$, with $\alpha = 0.05$
b. H_o: $p_A = p_B$ vs. H_a: $p_A \neq p_B$, with $\alpha = 0.05$
c. H_o: $p_1 - p_2 = 0$ vs. H_a: $p_1 - p_2 < 0$, with $\alpha = 0.04$
d. H_o: $p_m - p_f = 0$ vs. H_a: $p_m - p_f > 0$, with $\alpha = 0.01$

10.81 PC users are often victimized by hardware problems. A 1998 study (*PC World*, "Which PC Makers Can You Trust?," November 1998) revealed that hardware problems reported to manufacturers could not be fixed by one in three owners of personal computers. Home PC owners fared even worse than those with work PCs, facing longer waits for service and getting even fewer problems resolved. Relatively few owners gave service technicians high marks for having adequate knowledge or for exerting sincere efforts to help solve the problems with the hardware.

Suppose a study is conducted to compare the service provided by manufacturers to both home PC owners and work PC owners. Of 220 home PC owners who had trouble, 98 reported that their problem was not resolved satisfactorily. When the same question was asked of 180 work PC owners who experienced difficulty, 52 reported that the problem was not resolved. Did the home PC owners experience more problems that could not be solved with help from the manufacturer? Use the 0.05 level of significance to answer the question.

a. Solve using the p-value approach.
b. Solve using the classical approach.

10.82 In a survey of working parents (both parents working), one of the questions asked was: Have you refused a job, promotion, or transfer because it would mean less time with your family? Two hundred men and 200 women were asked this question. Twenty-nine percent of the men and 24% of the women responded yes. Based on this survey, can we conclude that there is a difference in the proportion of men and women who respond yes at the 0.05 level of significance?

10.83 Two randomly selected groups of citizens were exposed to different media campaigns that dealt with the image of a political candidate. One week later the citizen groups were surveyed to see whether they would vote for the candidate. The results were as given in the table.

	Exposed to Conservative Image	Exposed to Moderate Image
Number in sample	100	100
Proportion for the candidate	0.40	0.50

Is there sufficient evidence to show a difference in the effectiveness of the two image campaigns at the 0.05 level of significance?

a. Solve using the p-value approach.
b. Solve using the classical approach.

10.84 U.S. military active duty personnel and their dependents are provided with free medical care. A survey was conducted to compare obstetrical care of military and civilian (pay for medical care) families (*Public Health Reports*, May/June 1989). The numbers of women who began prenatal care by the second trimester were reported as follows:

	Military	Civilian
Prenatal care by second trimester	358	6786
Total sample	407	7363

Is there a significant difference between the proportions of military and civilian women who begin prenatal care by the second trimester? Use $\alpha = 0.02$.

10.85 The July 28, 1990, issue of *Science News* reported that smoking boosts the death risk for diabetics. The death risk is increased more for women than for men. Suppose as a follow-up study we investigated the smoking rates for male and female diabetics and obtained these data:

Gender	n	Number Who Smoke
Male	500	215
Female	500	170

a. Test the research hypothesis that the smoking rate (proportion of smokers) is higher for men than for women. Calculate the p-value.
b. What decision and conclusion would be reached at the 0.05 level of significance?

10.86 The guidelines to ensure that the sampling distribution of $p'_1 - p'_2$ is normal include several conditions about the size of several values (see page 506). The two binomial distributions $B(100, 0.3)$ and $B(100, 0.4)$ satisfy all of those guidelines.

a. Verify that $B(100, 0.3)$ and $B(100, 0.4)$ satisfy all the guidelines.

b. Use a computer to generate 200 random samples from each of the binomial populations. Find the observed proportion for each sample and the value of the 200 differences between two proportions.
c. Describe the observed sampling distribution using both graphic and numerical statistics.
d. Does the empirical sampling distribution appear to have an approximately normal distribution? Explain.

FYI
See the *Statistical Tutor* for additional information about commands.

10.5 Inferences Concerning the Ratio of Variances Using Two Independent Samples

When comparing two populations, we naturally compare their two most fundamental distribution characteristics, their "center" and their "spread," by comparing their means and standard deviations. We have learned, in two of the previous sections, how to use the t-distribution to make inferences comparing two population means with either dependent or independent samples. These procedures were intended to be used with normal populations, but they work quite well even when the populations are not exactly normally distributed.

The next logical step in comparing two populations is to compare their standard deviations, the most often used measure of spread. However, sampling distributions that deal with sample standard deviations (or variances) are very sensitive to slight departures from the assumptions. Therefore, the only inference procedure to be presented here will be the **hypothesis test for the equality of standard deviations (or variances)** for two normal populations.

The soft-drink bottling company discussed in Section 9.3 (p. 449) is trying to decide whether to install a modern, high-speed bottling machine. There are, of course, many concerns in making this decision, and one of them is that the increased speed may result in increased variability in the amount of fill placed in each bottle; such an increase would not be acceptable. To this concern, the manufacturer of the new system responded that the variance in fills will be no greater with the new machine than with the old. (The new system will fill several bottles in the same amount of time as the old system fills one bottle; this is the reason the change is being considered.) A test is set up to statistically test the bottling company's concern, "Standard deviation of new machine is greater than standard deviation of old," against the manufacturer's claim, "Standard deviation of new is no greater than standard deviation of old."

Illustration 10.12 Writing Hypotheses for the Equality of Variances

State the null and alternative hypotheses to be used for comparing the variances of the two soft-drink bottling machines.

SOLUTION
There are several equivalent ways to express the null and alternative hypotheses, but since the test procedure uses the ratio of variances, the recommended conven-

tion is to express the null and alternative hypotheses as ratios of the population variances. Furthermore, it is recommended that the "larger" or "expected to be larger" variance be the numerator. The concern of the soft-drink company is that the new modern machine (m) will result in a larger standard deviation in the amounts of fill than its present machine (p); $\sigma_m > \sigma_p$ or equivalently $\sigma_m^2 > \sigma_p^2$, which becomes $\frac{\sigma_m^2}{\sigma_p^2} > 1$. We want to test the manufacturer's claim (the null hypothesis) against the company's concern (the alternative hypothesis):

$$H_o: \frac{\sigma_m^2}{\sigma_p^2} = 1 \quad (m \text{ is no more variable})$$

$$H_a: \frac{\sigma_m^2}{\sigma_p^2} > 1 \quad (m \text{ is more variable})$$

■

ANSWER NOW 10.87
Explain why the inequality $\sigma_m^2 > \sigma_p^2$ is equivalent to $\frac{\sigma_m^2}{\sigma_p^2} > 1$.

Inferences about the ratio of variances for two normally distributed populations use the ***F*-distribution.** The F-distribution, similar to the Student's t-distribution and the χ^2-distribution, is a family of probability distributions. Each F-distribution is identified by two numbers of degrees of freedom, one for each of the two samples involved.

Before continuing with the details of the hypothesis-testing procedure, let's learn about the F-distribution.

PROPERTIES OF THE *F*-DISTRIBUTION

1. F is nonnegative; it is zero or positive.
2. F is nonsymmetrical; it is skewed to the right.
3. F is distributed so as to form a family of distributions; there is a separate distribution for each pair of numbers of degrees of freedom.

For inferences discussed in this section, the number of degrees of freedom for each sample is $df_1 = n_1 - 1$ and $df_2 = n_2 - 1$. Each different combination of degrees of freedom results in a different F-distribution, and each F-distribution looks approximately like the distribution shown in Figure 10.2.

FIGURE 10.2
***F*-Distribution**

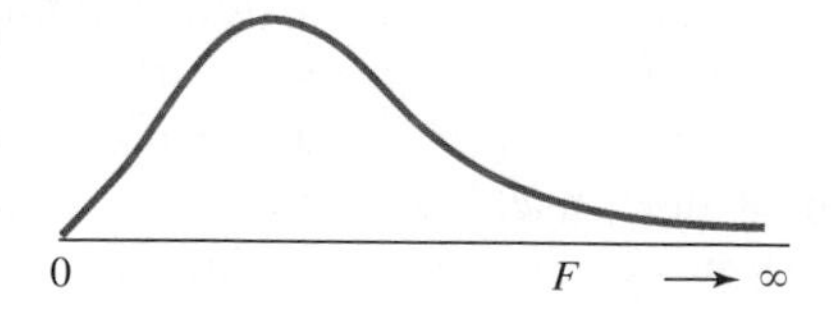

FYI
Explore Interactivity 10A on your CD.

The critical values for the F-distribution are identified using three values:

df_n, the degrees of freedom associated with the sample whose variance is in the numerator of the calculated F,

df_d, the degrees of freedom associated with the sample whose variance is in the denominator, and

α, the area under the distribution curve to the right of the critical value being sought.

Therefore, the symbolic name for a critical value of F will be $F(df_n, df_d, \alpha)$, as shown in Figure 10.3.

FIGURE 10.3
A Critical Value of F

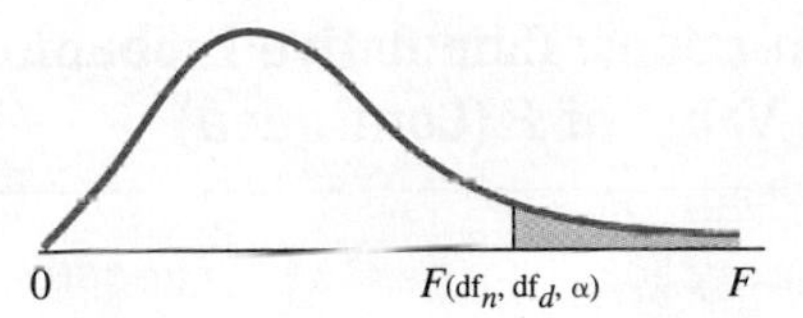

Since it takes three values to identify a single critical value of F, making tables for F is not as simple as with previously studied distributions. The tables presented in this textbook are organized so as to have a different table for each different value of α, the "area to the right." Table 9a in Appendix B shows the critical values for $F(\text{df}_n, \text{df}_d, \alpha)$, when $\alpha = 0.05$; Table 9b gives the critical values when $\alpha = 0.025$; Table 9c gives the values when $\alpha = 0.01$.

Illustration 10.13 Finding Critical F-Values

Find $F(5, 8, 0.05)$, the critical F-value for samples of size 6 and size 9 with 5% of the area in the right-hand tail.

SOLUTION

Using Table 9a ($\alpha = 0.05$), find the intersection of column df = 5 (for the numerator) and row df = 8 (for the denominator) and read the value: $F(5, 8, 0.05) = \mathbf{3.69}$. See the accompanying partial table.

Portion of Table 9a ($\alpha = 0.05$)

df for Denominator		df for Numerator				
		...	5	...	8	...
	⋮					
	5				4.82 ← $F(8, 5, 0.05)$	
	⋮					
	8		3.69 ← $F(5, 8, 0.05)$			
	⋮					

FYI
Explore Interactivity 10B on your CD.

Notice that $F(8, 5, 0.05)$ is 4.82. The degrees of freedom associated with the numerator and with the denominator must be kept in the correct order; 3.69 is different from 4.82. Check some other pairs to verify that interchanging the degrees of freedom numbers will result in different F-values. ■

ANSWER NOW 10.88
Find the values of $F(12, 24, 0.01)$ and $F(24, 12, 0.01)$.

Technology Instructions: Cumulative Probability Associated with a Specified Value of F

MINITAB (Release 13)

```
Choose:   Calc > Probability Distribution > F
Select:   Cumulative Probability
```

(continued)

Technology Instructions: Cumulative Probability Associated with a Specified Value of F (Continued)

MINITAB (Release 13) (continued)

```
Enter:    Numerator degrees of freedom: df_n
          Denominator degrees of freedom: df_d
Select:   Input constant*
Enter:    F-value (ex. 1.74)
```

*Select the Input column if several F-values are stored in C1. Use C2 for optional storage. If the area in the right tail is needed, subtract the calculated probability from one.

Excel XP

If several F-values are to be used, input the values into column A and activate B1; then continue with:

```
Choose:   Insert function f_x > Statistical > FDIST > OK
Enter:    X: individual F-value or (A1:A5 or select
          'F-value' cells)*
          Deg_freedom 1: df_n
          Deg_freedom 2: df_d > OK
*Drag:    Bottom right corner of the B1 cell down to give
          other probabilities
```

To find the probability for the left tail (the cumulative probability up to the F-value), subtract the calculated probability from one.

TI-83 Plus

```
Choose:   2nd > DISTR > 9:Fcdf(
Enter:    0, F-value, df_n, df_d)
```

Note: To find the probability between two F-values, enter the two values in place of 0 and the F-value.
If the area in the right tail is needed, subtract the calculated probability from one.

We are ready to use F to complete a hypothesis test about the ratio of two population variances.

ASSUMPTIONS FOR INFERENCES ABOUT THE RATIO OF TWO VARIANCES

The samples are randomly selected from normally distributed populations, and the two samples are selected in an independent manner.

Illustration 10.14 One-Tailed Hypothesis Test for the Equality of Variances

Recall that our soft-drink bottling company was to make a decision about the equality of the variances of amounts of fill between its present machine and a modern high-speed outfit. Does the sample information in Table 10.6 present sufficient evidence to reject the null hypothesis (the manufacturer's claim) that the modern high-speed bottle-filling machine fills bottles with no greater variance than the company's present machine? Assume the amounts of fill are normally distributed for both machines, and complete the test using $\alpha = 0.01$.

TABLE 10.6 Sample Information on Variances of Fills

Sample	n	s^2
Present machine (p)	22	0.0008
Modern high-speed machine (m)	25	0.0018

SOLUTION

Step 1 **The Set-Up:**

a. Describe the population parameter of interest.

$\frac{\sigma_m^2}{\sigma_p^2}$, the ratio of the variances in the amounts of fill placed in bottles for the modern machine versus the company's present machine.

b. State the null hypothesis (H_o) and the alternative hypothesis (H_a).

The hypotheses were established in Illustration 10.12 (pp. 515–516):

$$H_o: \frac{\sigma_m^2}{\sigma_p^2} = 1 \quad (m \text{ is no more variable})$$

$$H_a: \frac{\sigma_m^2}{\sigma_p^2} > 1 \quad (m \text{ is more variable})$$

NOTE When the "expected to be larger" variance is in the numerator for a one-tailed test, the alternative hypothesis states "The ratio of the variances is greater than one."

ANSWER NOW 10.89
Express the H_o and H_a of Illustration 10.14 equivalently in terms of standard deviations.

Step 2 **The Hypothesis Test Criteria:**

a. Check the assumptions.

The sampled populations are normally distributed (given in the statement of the problem), and the samples are independently selected (drawn from two separate populations).

b. Identify the probability distribution and the test statistic to be used.

The F-distribution with the ratio of the sample variances and formula (10.16):

TEST STATISTIC FOR EQUALITY OF VARIANCES

$$F\bigstar = \frac{s_m^2}{s_p^2}, \quad \text{with } df_m = n_m - 1 \text{ and } df_p = n_p - 1 \tag{10.16}$$

The sample variances are assigned to the numerator and denominator in the order established by the null and alternative hypotheses for one-tailed tests. The calculated ratio, $F\bigstar$, will have an F-distribution with $df_n = n_n - 1$ (numerator) and $df_d = n_d - 1$ (denominator) when the assumptions are met and the null hypothesis is true.

c. Determine the level of significance: $\alpha = 0.01$.

Step 3 **The Sample Evidence:**

a. Collect the sample information: See Table 10.6.

b. Calculate the value of the test statistic.
Using formula (10.16), we have

$$F\bigstar = \frac{s_m^2}{s_p^2}: \quad F\bigstar = \frac{0.0018}{0.0008} = \mathbf{2.25}$$

ANSWER NOW 10.90
Calculate $F\bigstar$ given $s_1 = 3.2$ and $s_2 = 2.6$.

The number of degrees of freedom for the numerator is $df_n = 24$ (or $25 - 1$) because the sample from the modern high-speed machine is associated with the numerator, as specified by the null hypothesis. Also, $df_d = 21$ because the sample associated with the denominator has size 22.

Step 4 **The Probability Distribution:**

Using the *p*-Value Procedure
a. Calculate the *p*-value for the test statistic.
Use the right-hand tail because H_a expresses concern for values related to "more than." $\mathbf{P} = P(F\bigstar > 2.25$, with $df_n = 24$ and $df_d = 21)$ as shown on the figure.

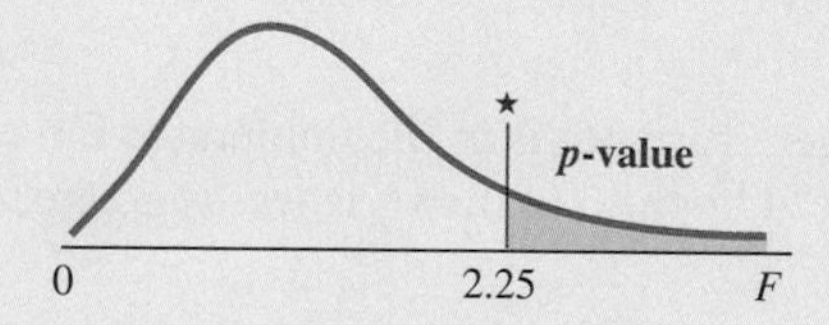

To find the *p*-value, you have two options:

1. Use Tables 9a and 9b in Appendix B to place bounds on the *p*-value: $\mathbf{0.025 < P < 0.05}$.
2. Use a computer or calculator to find the *p*-value: $\mathbf{P = 0.0323}$.

Specific instructions follow this illustration.

b. Determine whether or not the *p*-value is smaller than α.
The *p*-value is not smaller than the level of significance, α (0.01).

OR

Using the Classical Procedure
a. Determine the critical region and critical value(s).
The critical region is the right-hand tail because H_a expresses concern for values related to "more than." $df_n = 24$ and $df_d = 21$. The critical value is obtained from Table 9c: $F(24,21,0.01) = \mathbf{2.80}$.

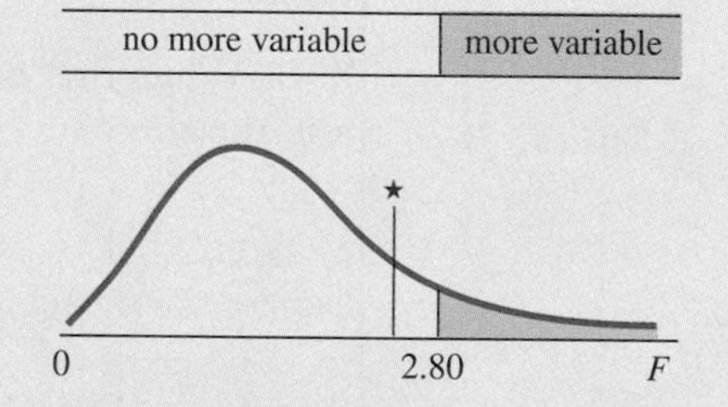

For additional instructions, see page 517.

b. Determine whether or not the calculated test statistic is in the critical region.
$F\bigstar$ is not in the critical region, as shown in **red** on the figure.

Step 5 **The Results:**

a. State the decision about H_o: Fail to reject H_o.

b. State the conclusion about H_a.
At the 0.01 level of significance, the samples do not present sufficient evidence to indicate an increase in variance.

Calculating the *p*-value when using the *F*-distribution
Method 1: Use Table 9 in Appendix B to place bounds on the p-value. Using Tables 9a, 9b, and 9c in Appendix B to estimate the *p*-value is very limited. However, for Illustration 10.14, the *p*-value can be estimated. By inspecting Tables 9a and 9b, you

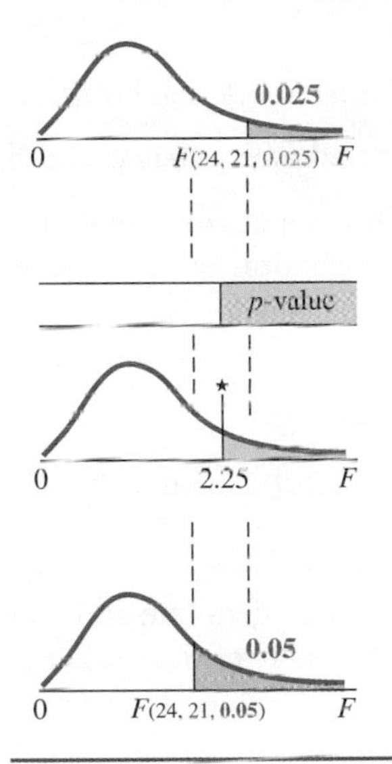

will find that $F(24,21,0.025) = 2.37$ and $F(24,21,0.05) = 2.05$. $F\star = 2.25$ is between the values 2.37 and 2.05; therefore, the p-value is between 0.025 and 0.05: $\mathbf{0.025 < P < 0.05}$. (See figure in margin.)

Method 2: If you are doing the hypothesis test with the aid of a computer or calculator, most likely it will calculate the p-value for you, or you may use the cumulative probability distribution commands described on pages 517–518. ■

The tables of critical values for the F-distribution give only the right-hand critical values. This will not be a problem because the right-hand critical value is the only critical value that will be needed. You can adjust the numerator–denominator order so that all the "activity" is in the right-hand tail. There are two cases: one-tailed tests and two-tailed tests.

One-tailed tests: Arrange the null and alternative hypotheses so that the alternative is always "greater than." The $F\star$-value is calculated using the same order as specified in the null hypothesis (as in Illustration 10.14; also see Illustration 10.15).

Two-tailed tests: When the value of $F\star$ is calculated, always use the sample with the larger variance for the numerator; this will make $F\star$ greater than one and place it in the right-hand tail of the distribution. Thus, you will need only the critical value for the right-hand tail (see Illustration 10.16).

FYI

α must still be split between the two tails for a two-tailed H_a.

ANSWER NOW 10.91

Find the critical value for the hypothesis test with H_a: $\sigma_1 > \sigma_2$, $n_1 = 7$, $n_2 = 10$, using the classical approach and $\alpha = 0.05$.

Illustration 10.15 Format for Writing Hypotheses for the Equality of Variances

Reorganize the alternative hypothesis so that the critical region will be the right-hand tail:

$$H_a: \sigma_1^2 < \sigma_2^2 \quad \text{or} \quad \frac{\sigma_1^2}{\sigma_2^2} < 1 \quad \text{(population 1 is less variable)}$$

SOLUTION

Reverse the direction of the inequality, and reverse the roles of the numerator and denominator.

$$H_a: \sigma_2^2 > \sigma_1^2 \quad \text{or} \quad \frac{\sigma_2^2}{\sigma_1^2} > 1 \quad \text{(population 2 is more variable)}$$

The calculated test statistic $F\star$ will be $\frac{s_2^2}{s_1^2}$. ■

Illustration 10.16 Two-Tailed Hypothesis Test for the Equality of Variances

Find $F\star$ and the critical values for the following hypothesis test so that only the right-hand critical value is needed. Use $\alpha = 0.05$ and the sample information $n_1 = 10$, $n_2 = 8$, $s_1 = 5.4$, and $s_2 = 3.8$.

$$H_o: \sigma_2^2 = \sigma_1^2 \quad \text{or} \quad \frac{\sigma_2^2}{\sigma_1^2} = 1$$

$$H_a: \sigma_2^2 \neq \sigma_1^2 \quad \text{or} \quad \frac{\sigma_2^2}{\sigma_1^2} \neq 1$$

SOLUTION

When the alternative hypothesis is two-tailed ($\neq$), the calculated $F\bigstar$ can be either $F\bigstar = \frac{s_1^2}{s_2^2}$ or $F\bigstar = \frac{s_2^2}{s_1^2}$. The choice is ours; we only need to make sure that we keep df_n and df_d in the correct order. We make the choice by looking at the sample information and using the sample with the larger standard deviation or variance as the numerator. Therefore, in this illustration,

$$F\bigstar = \frac{s_1^2}{s_2^2} = \frac{5.4^2}{3.8^2} = \frac{29.16}{14.44} = \mathbf{2.02}$$

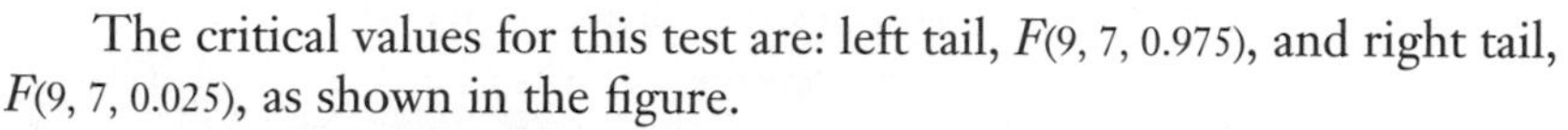

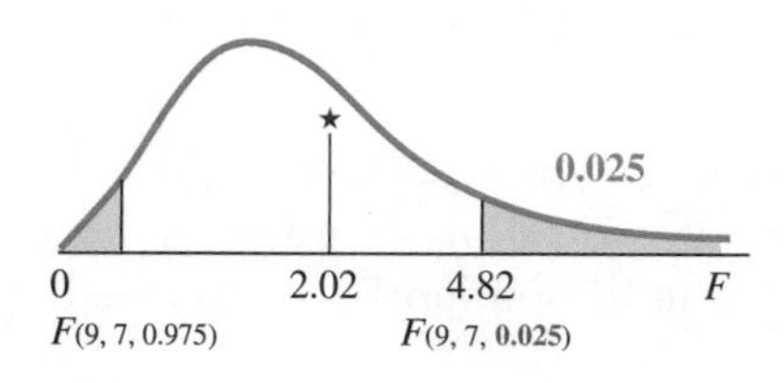

The critical values for this test are: left tail, $F(9, 7, 0.975)$, and right tail, $F(9, 7, 0.025)$, as shown in the figure.

Since we chose the sample with the larger standard deviation (or variance) for the numerator, the value of $F\bigstar$ will be greater than one and will be in the right-hand tail; therefore, only the right-hand critical value is needed. (All critical values for left-hand tails will be values between 0 and 1.) ■

ANSWER NOW 10.92

What would the value of $F\bigstar$ in Illustration 10.16 be if $F\bigstar = \frac{s_2^2}{s_1^2}$ were used? Why is it less than one?

Technology Instructions: Hypothesis Test for the Ratio Between Two Population Variances, σ_1^2/σ_2^2, for Two Independent Sets of Sample Data

MINITAB (Release 13)

```
Choose:   Stat > Basic Statistics > 2 Variances
Select:   Samples in different columns:
Enter:    First:       C1*
          Second:      C2
Select:   Storage
          Standard Deviations
```

*The 2 Variances procedure evaluates the first sample divided by the second sample.

Excel XP

Input the data for the numerator (larger spread) into column A and the data for the denominator (smaller spread) into column B; then continue with:

```
Choose:   Tools > Data Analysis > F-Test: Two-Sample for
          Variances
Enter:    Variable 1 Range: (A1:A20 or select cells)
          Variable 2 Range: (B1:B20 or select cells)
Select:   Labels (if necessary)
Enter:    α (ex. 0.05)
Select:   Output Range
Enter:    (C1 or select cell) > OK
```

Use Format > Column > AutoFit Selection to make the output more readable. The output shows the p value and the critical value for a one-tailed test.

ANSWER NOW 10.93
When a hypothesis test is two-tailed and Excel is used to calculate the *p*-value, as shown above, what additional step must be taken?

TI-83 Plus

Input the data for the numerator (larger spread) into L1 and the data for the denominator (smaller spread) into L2; then continue with the following, entering the appropriate values and highlighting Calculate:

Choose: STAT > TESTS > D:2-SampFTest...

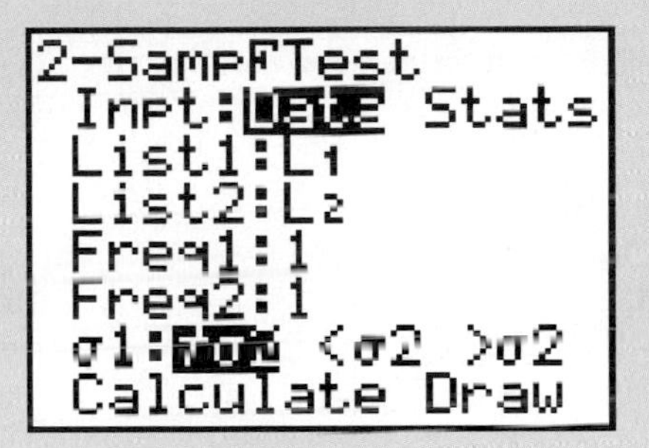

Application 10.5 Personality Characteristics of Police Academy Applicants

Bruce N. Carpenter and Susan M. Raza concluded that "police applicants are somewhat more like each other than are those in the normative population" when the *F*-test of homogeneity of variance resulted in a *p*-value of less than 0.005. *Homogeneity* means that the group's scores are less variable than the scores for the normative population.

Comparisons Across Subgroups and with Other Populations

To determine whether police applicants are a more homogeneous group than the normative population, the *F*-test of homogeneity of variance was used. With the exception of scales *F*, *K*, and 6, where the differences are nonsignificant, the results indicate that the police applicants form a somewhat more homogeneous group than the normative population [$F(237, 305) = 1.36, p < 0.005$]. Thus, police applicants are somewhat more like each other than are individuals in the normative population.

Source: Reproduced from the *Journal of Police Science and Administration*, Vol. 15, no. 1, pp. 10–17, with permission of the International Association of Chiefs of Police, PO Box 6010, 13 Firstfield Road, Gaithersburg, MD 20878.

ANSWER NOW 10.94

a. **What null and alternative hypotheses did Carpenter and Raza test?**

b. **What does "$p < 0.005$" mean?**

c. **Use a computer or calculator to find the *p*-value for $F(237, 305) = 1.36$.**

Exercises

10.95 State the null hypothesis, H_o, and the alternative hypothesis, H_a, that would be used to test these claims:

a. The variances of populations A and B are not equal.
b. The standard deviation of population I is larger than the standard deviation of population II.
c. The ratio of the variances for populations A and B is different from one.
d. The variability within population C is less than the variability within population D.

10.96 Using the $F(df_1, df_2, \alpha)$ notation, name each of the critical values shown on the figures.

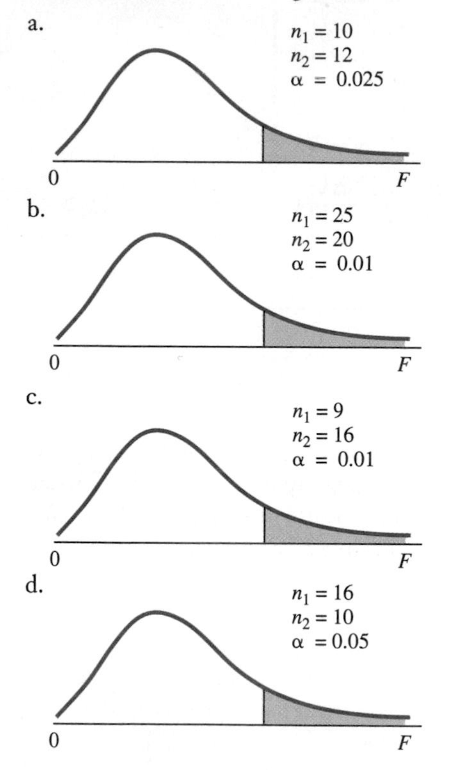

10.97 Find the following critical values for F from Tables 9a, 9b, and 9c in Appendix B.

a. $F(24, 12, 0.05)$
b. $F(30, 40, 0.01)$
c. $F(12, 10, 0.05)$
d. $F(5, 20, 0.01)$
e. $F(15, 18, 0.025)$
f. $F(15, 9, 0.025)$
g. $F(40, 30, 0.05)$
h. $F(8, 40, 0.01)$

10.98 Determine the critical region and critical value(s) that would be used to test these hypotheses using the classical model when $F\bigstar$ is used as the test statistic.

a. H_o: $\sigma_1^2 = \sigma_2^2$ vs. H_a: $\sigma_1^2 > \sigma_2^2$ with $n_1 = 10$, $n_2 = 16$, and $\alpha = 0.05$

b. H_o: $\frac{\sigma_1^2}{\sigma_2^2} = 1$ vs. H_a: $\frac{\sigma_1^2}{\sigma_2^2} \neq 1$ with $n_1 = 25$, $n_2 = 31$, and $\alpha = 0.05$

c. H_o: $\frac{\sigma_1^2}{\sigma_2^2} = 1$ vs. H_a: $\frac{\sigma_1^2}{\sigma_2^2} > 1$ with $n_1 = 10$, $n_2 = 10$, and $\alpha = 0.01$

d. H_o: $\sigma_1 = \sigma_2$ vs. H_a: $\sigma_1 < \sigma_2$ with $n_1 = 25$, $n_2 = 16$, and $\alpha = 0.01$

10.99

a. Two independent samples, each of size 3, are drawn from a normally distributed population. Find the probability that one of the sample variances is at least 19 times larger than the other one.
b. Two independent samples, each of size 6, are drawn from a normally distributed population. Find the probability that one of the sample variances is no more than 11 times larger than the other one.

10.100 A study in *Pediatric Emergency Care* (June 1994) compared the injury severity of younger and older children. One measure reported was the Injury Severity Score (ISS). The standard deviation of ISS scores for 37 children 8 years or younger was 23.9, and the standard deviation for 36 children older than 8 was 6.8. Assume that ISS scores are normally distributed for both age groups. At the 0.01 level of significance, is there sufficient reason to conclude that the standard deviation of ISS scores for younger children is larger than the standard deviation of ISS scores for older children?

10.101 A bakery is considering buying one of two gas ovens. The bakery requires that the temperature remain constant during a baking operation. A study was conducted to measure the variance in temperatures of the ovens during the baking process. The variance in temperatures before the thermostat restarted the flame for the Monarch oven was 2.4 for 16 measurements. The variance for the Kraft oven was 3.2 for 12 measurements. Does this test provide sufficient evidence to conclude that there is a difference in the variances for the two ovens? Assume measurements are normally distributed, and use a 0.02 level of significance.

10.102 Sucrose, ordinary table sugar, is probably the single most abundant pure organic chemical in the world and the one most widely known to nonchemists. Whether from sugar cane (20% by weight) or sugar beets (15% by weight), and whether raw or refined, common sugar is still sucrose. Fifteen U.S. sugar-beet–producing counties were randomly selected, and their 2001 sucrose percentages had a standard deviation of 0.862. Similarly, 12 U.S. sugar-cane–producing counties were randomly selected, and their 2001 sucrose percentages had a standard deviation of 0.912. At the 0.05 level, is there significant difference between the standard deviations for sugar beet and sugar cane sucrose percentages?

10.103 Television viewing time appears to differ from one age group to the next. According to Nielsen Media Research (*The World Almanac and Book of Facts 1998*), in 1997 adults over age 54 watched TV for an average of 11.2 hours per week between 8:00 and 11:00 P.M. In contrast, adults between the ages of 25 and 54 averaged 8.5 hours of viewing time per week. Media experts claim that such data are subject to interpretation because younger adults are more unsettled and exhibit a greater variety of lifestyles than older adults, many of whom are retired and display more stability. A recent study is conducted to check the results of the national survey, and the sampling data are shown in the table.

Viewing Age Group	n	Weekly Mean Viewing Time (hours)	Standard Deviation
25 to 54 (y)	27	8.8	2.9
55 and over (o)	28	11.6	1.6

Test the null hypothesis of equal variances against the alternative hypothesis of unequal variances. Use the 0.05 level of significance.

a. Solve using the p-value approach.
b. Solve using the classical approach.

10.104 Ex10-104 A study was conducted to determine whether or not there was equal variability in male and female systolic blood pressures. Random samples of 16 men and 13 women were used to test the experimenter's claim that the variances were unequal. MINITAB was used to calculate the standard deviations, $F\bigstar$, and the p-value.

Men

120	120	118	112	120	114	130	114
130	100	120	108	112	122	124	125

Women

122	102	118	126	108	130	104
120	118	130	116	102	122	

```
Standard deviation of Men = 7.8864
Standard deviation of Women = 9.9176
F-Test (normal distribution)
Test Statistic: 1.581
P-Value : 0.398
```

Verify these results by calculating the values yourself.

10.105 Ex10-105 The quality of the end product is somewhat determined by the quality of the materials used. Textile mills monitor the tensile strength of the fibers used in weaving their yard goods. These data are the tensile strengths of cotton fibers from two suppliers:

Supplier A

78	82	85	83	77	84	90	82
80	82	77	80	80	93	82	

Supplier B

76	79	83	78	72	73	69	80
78	78	73	76	78	79	74	77

Calculate the observed value of F, $F\bigstar$, for comparing the variances of these two sets of data.

10.106 Ex10-106 Several counties in Minnesota and Wisconsin were randomly selected, and information about the 2001 sweet corn crop was collected. These yield rates (in tons of sweet corn per acre harvested) resulted:

Yield, MN	5.90	6.50	6.20	5.90	6.20	5.70
	5.91	5.90	6.00	5.60	6.51	6.30
Yield, WI	7.80	6.80	7.00	5.30	6.50	6.90
	6.60	5.70	6.60	6.00	6.20	

Source: http://www.usda.gov/nass/graphics/county01/data/vsc01.csv

a. Is there a difference in the variability of county yields as measured by the standard deviation of the yield rates? Use $\alpha = 0.05$.
b. Is the mean yield rate in Wisconsin significantly higher than the mean yield rate in Minnesota? Use $\alpha = 0.05$.

10.107 Use a computer to demonstrate the truth of the theory presented in this section.

a. The underlying assumption is that the populations are normally distributed. While a hypothesis test is conducted for the equality of two standard deviations, it is also assumed that the standard deviations are equal. Generate very large samples of two theoretical populations: $N(100, 20)$ and $N(120, 20)$. Find graphic and numerical evidence that the populations satisfy the assumptions.
b. Randomly select 100 samples, each of size 8, from both populations and find the standard deviation of each sample.
c. Using the first sample drawn from each population as a pair, calculate the $F\bigstar$-statistic. Repeat for all samples. Describe the sampling distribution of the 100 $F\bigstar$-values using both graphic and numerical statistics.
d. Generate the probability distribution for $F(7, 7)$, and compare it to the observed distribution of $F\bigstar$. Do the two graphs agree? Explain.

FYI
See the *Statistical Tutor* for additional information about commands.

10.108 It was stated in this section that the F-test is very sensitive to minor digressions from the assumptions. Repeat Exercise 10.107 using $N(100, 20)$ and $N(120, 30)$. Notice that the only change from Exercise 10.107 is the seemingly slight increase in the standard deviation of the second population. Answer the same questions using the same kind of information, and you will see very different results.

CHAPTER SUMMARY

RETURN TO CHAPTER CASE STUDY

Students, Credit Cards, and Debt

© Wayne Scarberry/The Image Works

Let's return to the Chapter Case Study (p. 471) and continue our investigation of students, credit cards, and debt.

FIRST THOUGHTS 10.109

a. Find the proportion of each sample that have at least one credit card.
b. Find the point estimate for the difference between the two proportions.
c. Find the mean credit card debt for the freshmen and sophomores in the samples.
d. Draw dotplots for the amounts of debt for both groups using a common scale.
e. Find the point estimate for the difference between the two means.

PUTTING CHAPTER 10 TO WORK 10.110

How do credit card debts for freshmen and sophomores compare?

a. Check the assumptions of normality for both sets of credit card debt. Verify.
b. Is the difference between the two means observed in 10.109 significantly greater than the $292 difference reported by Nellie Mae? Use $\alpha = 0.05$.
c. Are the assumptions for the difference between two proportions satisfied? Explain.
d. Find the 95% confidence interval for the difference between the proportions of sophomore and freshman students who have their own credit cards.

YOUR OWN MINI-STUDY 10.111

Design your own study involving two populations.

a. Determine a set of questions that compare the means, proportions, or variances of two populations of interest to you. You might consider two different classes as the two populations and questions similar to: Is there is a difference between their mean costs of books and supplies for a semester? Is there is a difference between the proportions of those with credit cards of their own? Is there is a difference between the proportions of those with four or more credit cards of their own?
b. Define two specific populations that you will sample, describe your sampling plan, and collect the data needed to answer your questions.
c. Discuss any differences and similarities between your study and the Chapter Case Study.

In Retrospect

In this chapter we began the comparisons of two populations by distinguishing between independent and dependent samples, which are statistically important and useful sampling procedures. We then proceeded to examine the inferences concerning the comparison of means, proportions, and variances for two populations.

Confidence intervals and hypothesis tests can sometimes be interchanged; that is, a confidence interval can be used in place of a hypothesis test. For example, Illustration 10.10 (p. 507) called for a confidence interval. Now suppose that Mr. Morris asked, "Is there a difference in my voter appeal to men voters as opposed to women voters?" To answer his question, you would not need to complete a hypothesis test if you chose to test at $\alpha = 0.01$ using a two-tailed test. "No difference" would mean a difference of zero, which is not included in the interval from 0.014 to 0.128 (the interval determined in Illustration 10.10). Therefore, a null hypothesis of "no difference" would be rejected, thereby substantiating the conclusion that a significant difference exists in voter appeal between the two groups.

We are always making comparisons between two groups. We compare means and we compare proportions. In this chapter we have learned how to statistically compare two populations by making inferences about their means, proportions, or variances. For convenience, Table 10.7 identifies the formulas to use when making inferences about comparisons between two populations.

In Chapters 8, 9, and 10 we have learned how to use confidence intervals and hypothesis tests to answer questions about means, proportions, and standard deviations for one or two populations. From here we can expand our techniques to include inferences about more than two populations as well as inferences of different types.

TABLE 10.7 Formulas to Use for Inferences Involving Two Populations

Situations	Test Statistic	Formula to Be Used: Confidence Interval	Formula to Be Used: Hypothesis Test
Difference between two means			
Dependent samples	t	Formula (10.4) (p. 478)	Formula (10.5) (p. 481)
Independent samples	t	Formula (10.8) (p. 491)	Formula (10.9) (p. 493)
Difference between two proportions	z	Formula (10.11) (p. 506)	Formula (10.12) (p. 509)
Difference between two variances	F		Formula (10.16) (p. 519)

Chapter Exercises

10.112 Ex10-112 The diastolic blood pressures for 15 patients were determined using two techniques: the standard method used by medical personnel and a method using an electronic device with a digital readout. The results were as follows:

Patient	1	2	3	4	5	6	7	8
Standard	72	80	88	80	80	75	92	77
Digital	70	76	87	77	81	75	90	75

Patient	9	10	11	12	13	14	15
Standard	80	65	69	96	77	75	60
Digital	82	64	72	95	80	70	61

Assuming blood pressures are normally distributed, determine the 90% confidence interval for the mean difference in the two readings, where d = standard method − digital method.

10.113 Ex10-113 Using a 95% confidence interval, estimate the difference in IQ between the oldest and the youngest members (brothers and sisters) of a family based on the following random sample of IQs. Assume normality.

Oldest	145	133	116	128	85	100
	105	150	97	110	120	130
Youngest	131	119	103	93	108	100
	111	130	135	113	108	125

10.114 Ex10-114 We want to know which of two types of filters we should use. A test was designed in which the strength of a signal could be varied from zero to the point where the operator first detects the image. At this point, the intensity setting is recorded. Lower settings are better. Twenty operators were asked to make one reading for each filter.

Assuming the intensity readings are normally distributed, estimate the mean difference between the two readings using a 90% confidence interval.

Table for Exercise 10.114

Operator	1	2	3	4	5	6	7	8	9	10
Filter 1	96	83	97	93	99	95	97	91	100	92
Filter 2	92	84	92	90	93	91	92	90	93	90
Operator	11	12	13	14	15	16	17	18	19	20
Filter 1	88	89	85	94	90	92	91	78	77	93
Filter 2	88	89	86	91	89	90	90	80	80	90

Table for Exercise 10.115

Time of Competition	Recruit 1	2	3	4	5	6	7	8	9	10
First day	72	29	62	60	68	59	61	73	38	48
One week later	75	43	63	63	61	72	73	82	47	43

10.115 Ex10-115 Ten new recruits participated in a rifle-shooting competition at the end of their first day at training camp. The same ten competed again at the end of a full week of training and practice. Their resulting scores are shown in the above table.

Does this set of ten pairs of data show that there was a significant improvement in the recruits' shooting abilities during the week? Use $\alpha = 0.05$.

10.116 Ex10-116 Immediate-release medications deliver their drug content quickly, with the maximum concentration reached in a short time, while sustained-release medications take longer to reach maximum concentration. As part of a study, immediate-release codeine (irc) was compared with sustained-release codeine (src) on 13 healthy patients. They were randomly assigned to one of the two types of codeine and treated for 2.5 days; then after a 7-day wash-out period, each patient was given the other type of codeine. Thus, each patient received both types. The total amounts (A) of drug available over the life of the treatment [in (ng · mL)/hr] are given here.

Patient	1	2	3	4	5	6	7
Airc	1091.3	1064.5	1281.1	1921.4	1649.9	1423.6	1308.4
Asrc	1308.5	1494.2	1382.2	1978.3	2004.6	—	1211.1

Patient	8	9	10	11	12	13
Airc	1192.1	766.2	978.6	1618.9	582.9	972.1
Asrc	1002.4	866.6	1345.8	979.2	576.3	999.1

Source: http://exploringdata.cqu.edu.au/ws_coedn.htm

a. Explain why this is a paired-difference design.
b. What adjustment is needed since there is no Asrc for patient 6?

Is there is a significant difference in the total amount of drug available over the life of the treatment?

c. Check the test assumptions and describe your findings.
d. Test the claim using $\alpha = 0.05$.

10.117 A test that measures math anxiety was given to 50 male and 50 female students. The results follow:

Males: $\bar{x} = 70.5$, $s = 13.2$
Females: $\bar{x} = 75.7$, $s = 13.6$

Construct a 95% confidence interval for the difference between the mean anxiety scores.

10.118 The same achievement test is given to soldiers selected at random from two units. The scores they attained are summarized here:

Unit 1: $n_1 = 70, \bar{x}_1 = 73.2, s_1 = 6.1$
Unit 2: $n_2 = 60, \bar{x}_2 = 70.5, s_2 = 5.5$

Construct a 90% confidence interval for the difference in the mean levels of the two units.

10.119 Ex10-119 Ten soldiers were selected at random from each of two companies to participate in a rifle-shooting competition. Their scores are listed in the table.

Company A	72	29	62	60	68
	59	61	73	38	48
Company B	75	43	63	63	61
	72	73	82	47	43

Construct a 95% confidence interval for the difference between the mean scores for the two companies.

10.120 Ex10-120 An achievement test in a beginning computer science course was administered to two groups. One group had a previous computer science course in high school; the other group did not. The test results are given here. Assuming the test scores are normal, construct a 98% confidence interval for the difference between the two means.

Group 1	17	18	27	19	24	36	27	26
(high school course)	29	26	33	35	22	18	29	
Group 2	19	25	28	27	21	24	18	14
(no high school	28	21	22	21	14	29	28	25
course)	17	20	28	31	27	20		

10.121 Ex10-121 Two methods were used to study the latent heat of ice fusion. Both method A (an electrical method) and method B (a method of mixtures) were conducted with the specimens cooled to −0.72°C. The data in the table are the changes in total heat from 0.72°C to water at 0°C (in calories per gram of mass).

Method A

79.98	80.04	80.02	80.04	80.03	80.03	80.04
80.02	80.00	80.02	80.05	80.03	79.97	

Method B

80.02	79.94	79.98	79.97
79.97	80.03	79.95	79.97

Construct a 95% confidence interval for the difference between the means.

10.122 A test concerning some of the fundamental facts about AIDS was administered to two groups: one consisting of college graduates and the other high school graduates. A summary of the test results follows:

College graduates: $n = 75$, $\bar{x} = 77.5$, $s = 6.2$
High school graduates: $n = 75$, $\bar{x} = 50.4$, $s = 9.4$

Do these data show that the college graduates, on the average, score significantly higher on the test? Use $\alpha = 0.05$.

10.123 George Johnson is the head coach of a college football team that trains and competes at home on artificial turf. George is concerned that the 40-yard sprint time recorded by his players and others increases substantially on natural turf as opposed to artificial turf. George's next opponent plays on grass, so he surveyed all the starters in the next game and obtained their best 40-yard sprint times. He then compared them to the best times turned in by his own players. The results are shown in the table.

	n	Mean (seconds)	Standard Deviation
Artificial turf	22	4.85	0.31
Grass	22	4.96	0.42

Do Coach Johnson's players have a lower mean sprint time? Test at the 0.05 level of significance to advise Coach Johnson.

a. Solve using the p-value approach.
b. Solve using the classical approach.

10.124 To compare the merits of two short-range rockets, eight of the first kind and ten of the second kind are fired at a target. If the first kind has a mean target error of 36 ft and a standard deviation of 15 ft, while the second kind has a mean target error of 52 ft and a standard deviation of 18 ft, does this indicate that the second kind of rocket is less accurate than the first? Use $\alpha = 0.01$ and assume a normal distribution for target error.

10.125 Ex10-125 The material used in making parts affects not only how long the part lasts but also how difficult it is to take them apart to repair. The measurements are the screw removal torques (in Newton-meters) for a specific screw after several uses. The first row (see table below) lists the part number, the second row gives the screw torque measurements for assemblies made with material A, and the third row lists the screw torque measurements for assemblies made with material B. Assume the torque measurements are normally distributed.

a. Find the sample mean, variance, and standard deviation for the material A data.
b. Find the sample mean, variance, and standard deviation for the material B data.
c. At the 0.01 level, do these data show a significant difference in the mean torque required to remove the screws made with the two different materials?

10.126 Ex10-126 A group of 17 students participated in an evaluation of a special training session that claimed to improve memory. The students were randomly assigned to two groups: group A, the test group, and group B, the control group. All 17 students were tested for their ability to remember certain material. Group A was given the special training; group B was not. After one month both groups were tested again, with the results shown in the table at the bottom of the page. Do these data support the alternative hypothesis that the special training is effective at the $\alpha = 0.01$ level of significance?

10.127 Ex10-048 At a large university, a mathematics placement exam is administered to all students. This exam has a history of producing scores with a mean of 77. Samples of 36 male and 30 female students are randomly

Table for Exercise 10.125

Part Number	1	2	3	4	5	6	7	8	9	10	11	12	13	14	15
Material A	16	14	13	17	18	15	17	16	14	16	15	17	14	16	15
Material B	11	14	13	13	10	15	14	12	11	14	13	12	11	13	12

Source: Problem data provided by AC Rochester Division, General Motors, Rochester, NY.

Table for Exercise 10.126

	Group A Students									Group B Students							
	1	2	3	4	5	6	7	8	9	10	11	12	13	14	15	16	17
Before	23	22	20	21	23	18	17	20	23	22	20	23	17	21	19	20	20
After	28	29	26	23	31	25	22	26	26	23	25	26	18	21	17	18	20

selected from this year's student body and the following scores recorded:

Men

72	68	75	82	81	60	75	85	80
70	71	84	68	85	82	80	54	81
86	79	99	90	68	82	60	63	67
72	77	51	61	71	81	74	79	76

Women

81	76	94	89	83	78	85	91	83	83
84	80	84	88	77	74	63	69	80	82
89	69	74	97	73	79	55	76	78	81

a. Describe each set of data with a histogram (use the same class intervals on both histograms) and with the mean and standard deviation.
b. Test the hypotheses: "Mean score for all men is 77" and "Mean score for all women is 77" using $\alpha = 0.05$.
c. Do your results show that the mean scores for men and women are the same? Justify your answer. Be careful!
d. Test the hypothesis "There is no difference between the mean scores for male and female students" using $\alpha = 0.05$.
e. Do the results in part d show that the mean scores for men and women are the same? Explain.
f. Explain why the results in part b cannot be used to conclude that the two means are the same.

10.128 A survey was conducted to determine the proportion of Democrats as well as Republicans who support a "get tough" policy in South America. Here are the results of the survey:

Democrats: $n = 250$, number in support = 120
Republicans: $n = 200$, number in support = 105

Construct the 98% confidence interval for the difference between the proportions of support.

10.129 Of a random sample of 100 stocks on the New York Stock Exchange, 32 made a gain today. A random sample of 100 stocks on the NASDAQ showed 27 stocks making a gain.

a. Construct a 99% confidence interval to estimate the difference in the proportions of stocks making a gain.
b. Does the answer to part a suggest that there is a significant difference between the proportions of stocks making gains on the two stock exchanges?

10.130 According to a report in *Science News* (Vol. 137, No. 8), the percentage of seniors who used an illicit drug during the previous month was 19.7% in 1989. The figure was 21.3% in 1988. The annual survey of 17,000 seniors is conducted by researchers at the University of Michigan in Ann Arbor.

a. Set a 95% confidence interval for the true decrease in usage.
b. Does the interval in part a suggest that there has been a significant decrease in the usage of illicit drugs by seniors? Explain.

10.131 A consumer group compared the reliability of two comparable microcomputers from two different manufacturers. The proportion that required service within the first year after purchase was determined for samples from each of two manufacturers.

Manufacturer	Sample Size	Proportion Needing Service
1	75	0.15
2	75	0.09

Find a 0.95 confidence interval for $p_1 - p_2$.

10.132 In determining the "goodness" of a test question, a teacher will often compare the percentage of the better students who answer it correctly to the percentage of the poorer students who answer it correctly. One expects that the better students will answer the question correctly more frequently than the poorer students. On the last test, 35 of the students with the top 60 grades and 27 with the bottom 60 answered a certain question correctly. Did the students with the top grades do significantly better on this question? Use $\alpha = 0.05$.

a. Solve using the p-value approach.
b. Solve using the classical approach.

10.133 According to *USA Today* (December 24, 1997), 65% of adult men and 55% of adult women have suffered from holiday depression. Give details to support each of your answers.

a. If these statistics came from samples of 100 men and 100 women, is the difference significant?
b. If these statistics came from samples of 150 men and 150 women, is the difference significant?
c. If these statistics came from samples of 200 men and 200 women, is the difference significant?

10.134 The April 4, 1991, issue of *USA Today* reported results from the *New England Journal of Medicine.* In a study of 987 deaths in southern California, the average right-hander died at age 75 and the average left-hander died at age 66. In addition, it was found that 7.9% of the lefties died from accident-related injuries, excluding vehicles, versus 1.5% for the right-handers; and 5.3% of the left-handers died while driving vehicles versus 1.4% of the right-handers.

Suppose you examine 1000 randomly selected death certificates of which 100 were left-handers and 900 were right-handers. If you found that 5 of the left-handers and 18 of the right-handers died while driving a vehicle, would you have evidence to show that the proportion of left-handers who die at the wheel is significantly higher than the proportion of right-handers? Calculate the p-value and interpret its meaning.

10.135 Who wins disputed cases when a change is made to the tax laws: the taxpayer or the Internal Revenue Service? The latest trend indicates that the burden of proof in all court cases has shifted from the taxpayers to the IRS, which tax experts predict could set off more intrusive questioning. Of the accountants, lawyers, and other tax professionals surveyed by RIA Group, a tax-information publisher, 55% expect at least a slight increase in taxpayer wins. (*Fortune*, "Tax Reform?" September 28, 1998, p. 322).

Suppose samples of 175 accountants and 165 lawyers are asked: Do you expect taxpayers to win more court cases because of the new burden of proof rules? Of those surveyed, 101 accountants replied yes and 84 lawyers said yes. Do the two expert groups differ in their opinions? Use a 0.01 level of significance to answer the question.

a. Solve using the p-value approach.
b. Solve using the classical approach.

10.136 "It's a draw, according to two Australian researchers. By age 25, up to 29% of all men and up to 34% of all women have some gray hair, but this difference is so small that it's considered insignificant" (*Family Circle*, June 26, 1990). If 1000 men and 1000 women were involved in this research, would the 5% difference mentioned be significant at the 0.01 level? Explain, including details to support your answer.

10.137 A manufacturer designed an experiment to compare men and women with respect to the times they require to assemble a product. Fifteen men and 15 women were tested to determine the time they required, on the average, to assemble the product. The times required by the men had a standard deviation of 4.5 min, and the times required by the women had a standard deviation of 2.8 min. Do these data show that the amount of time needed by men is more variable than the time needed by the women? Use $\alpha = 0.05$ and assume the times are approximately normally distributed.

a. Solve using the p-value approach.
b. Solve using the classical approach.

10.138 A soft-drink distributor is considering two new models of dispensing machines. Both the Harvard Company machine and the Fizzit machine can be adjusted to fill cups to a certain mean amount. However, the variation in the amounts dispensed from cup to cup is a primary concern. Ten cups dispensed from the Harvard machine showed a variance of 0.065, whereas 15 cups dispensed from the Fizzit machine showed a variance of 0.033. The factory representative from the Harvard Company maintains that his machine had no more variability than the Fizzit machine. Assume the amounts dispensed are normally distributed. At the 0.05 level of significance, does the sample refute the representative's assertion?

a. Solve using the p-value approach.
b. Solve using the classical approach.

10.139 Mindy Fernandez is in charge of production at the new sport utility vehicle (SUV) assembly plant that just opened in her town. Lately she has been concerned that the wheel studs don't match the chrome lug nuts closely enough to keep the wheel assemblies operating smoothly. Workers are complaining that cross-threading is happening so often that threads are being stripped by the air wrenches and torque settings also have to be adjusted downward to prevent stripped threads even if the parts match up. In an effort to determine whether the fault lies with the lug nuts or the studs, Mindy has decided to ask the quality control department to test a random sample of 60 lug nuts and 40 studs to see if the variances in threads are the same for both parts. The report from the technician indicated that the thread variance of the sampled lug nuts was 0.00213 and that the thread variance for the sampled studs was 0.00166. What can Mindy conclude about the equality of the variances at the 0.05 level of significance?

a. Solve using the p-value approach.
b. Solve using the classical approach.

10.140 Ex10-140 Random samples of counties in North Dakota and South Dakota were selected from the USDA-NASS Web site for the purpose of estimating the difference between the mean 2001 yield rates for oat production for the two states.

Yield, ND

66.0	73.8	51.9	61.3	67.4	54.0	71.4
66.7	64.4	75.7	74.4	58.0	56.2	40.0

Yield, SD

65.0	62.5	30.0	90.0	70.0	62.7	42.4	47.1
56.1	50.0	65.6	45.5	79.2	59.3	62.2	76.2

Source: http://www.usda.gov/nass/graphics/county01/data/ot01.csv

a. Are the assumptions of normality satisfied? Explain.
b. Is there sufficient evidence to reject the hypothesis of equal variances for the oat production yield rates for these two states? Use $\alpha = 0.05$.
c. Is there sufficient evidence to reject the hypothesis that there is no difference between the mean oat production yield rates for these two states? Use $\alpha = 0.05$.

FYI

When using the two-sample t-test, select "assume equal variances" according to the result in part b.

10.141 Ex10-141 A research project was undertaken to evaluate the amount of force needed to elicit a designated response on equipment made from two distinct designs: the existing design and an improved design. The expectation was that improved equipment would require less force than the current equipment. Fifty units of each design were tested and the required force recorded. A lower force level and reduced variability are both considered desirable.

Control or Existing

0.003562	0.005216	0.005710	0.004985	0.005852
0.003542	0.005453	0.006380	0.005644	0.005767
0.005711	0.005157	0.005413	0.004762	0.004346
0.004194	0.006363	0.004697	0.005302	0.005863
0.005606	0.004988	0.006509	0.005790	0.005515
0.006340	0.005179	0.005657	0.004937	0.005752
0.005612	0.005096	0.005198	0.005785	0.005280
0.005468	0.005517	0.005053	0.005645	0.005720
0.005677	0.006553	0.004770	0.005459	0.005728
0.004489	0.005273	0.008243	0.006207	0.005965

Test or New Design

0.002477	0.002725	0.002170	0.002559	0.002348
0.002524	0.002812	0.002734	0.002516	0.002731
0.003429	0.003243	0.003649	0.003112	0.003491
0.003108	0.003593	0.003716	0.002852	0.003491
0.003342	0.003429	0.003799	0.003579	0.003877
0.003211	0.003553	0.003117	0.003490	0.003702
0.003382	0.004878	0.003840	0.003818	0.004005
0.004224	0.003605	0.003467	0.003894	0.004309
0.004563	0.003323	0.003574	0.005654	0.004191
0.004383	0.003834	0.003590	0.003828	0.004593

a. Describe both sets of data using means, standard deviations, and histograms.
b. Check the assumptions for comparing the variances and means of two independent samples. Describe your findings.
c. Will one- or two-tailed tests be appropriate for testing the expectations for the new design? Why?
d. Is there significant evidence to show that the new design has reduced the variability in the required force? Use $\alpha = 0.05$.
e. Is there significant evidence to show that the new design has reduced the mean amount of force? Use $\alpha = 0.05$.
f. Does the new design live up to expectations? Explain.

Vocabulary and Key Concepts

Be able to define each term. Pay special attention to the key terms, which are printed in **red**. In addition, describe each term in your own words and give an example of each. Your examples should not be ones given in class or in the textbook. The bracketed numbers indicate the chapters in which the term previously appeared, but you should define the terms again to show increased understanding of their meaning. Page numbers indicate the first appearance of the term in Chapter 10.

assumptions [8, 9] (pp. 478, 490, 506, 518)
binomial experiment [5] (p. 505)
binomial p [5, 9] (p. 505)
confidence interval [8, 9] (pp. 478, 491, 506)
dependent means (p. 477)
dependent samples (pp. 472, 475)
F-distribution (p. 516)
***F*-statistic** (p. 519)
hypothesis test [8, 9] (pp. 480, 493, 509, 515, 518)
independent means (p. 490)
independent samples (pp. 472, 490, 505, 515)
mean difference (p. 477)
paired difference (p. 476)
percentage [5] (p. 505)
pooled observed probability (p. 509)
probability [5] (p. 505)
proportion [5] (p. 505)
p-value [8, 9] (pp. 482, 495, 511, 520)
source (of data) (p. 472)
standard error [8, 9] (pp. 481, 490, 506, 509)
t-distribution [9] (pp. 478, 490)
test statistic (pp. 481, 493, 509, 520)
***t*-statistic** [9] (pp. 481, 494)
***z*-statistic** [8, 9] (p. 511)

Chapter Practice Test

PART I: KNOWING THE DEFINITIONS

Answer "True" if the statement is always true. If the statement is not always true, replace the words shown in bold with words that make the statement always true.

10.1 When the means of two unrelated samples are used to compare two populations, we are dealing with **two dependent means.**

10.2 The use of **paired data (dependent means)** often allows for the control of unmeasurable or confounding variables because each pair is subjected to these confounding effects equally.

10.3 The **chi-square distribution** is used for making inferences about the ratio of the variances of two populations.

10.4 The **z-distribution** is used when two dependent means are to be compared.

10.5 In comparing two independent means when the σ's are unknown, we need to use the **standard normal** distribution.

10.6 The **standard normal score** is used for all inferences concerning population proportions.

10.7 The F-distribution is a **symmetric** distribution.

10.8 The number of degrees of freedom for the critical value of t is equal to **the smaller of $n_1 - 1$ or $n_2 - 1$** when inferences are made about the difference between two independent means in the case when the degrees of freedom are estimated.

10.9 In a confidence interval for the mean difference in paired data, the interval **increases** in width when the sample size is increased.

10.10 A **pooled estimate** for any statistic in a problem dealing with two populations is a value arrived at by combining the two separate sample statistics so as to achieve the best possible point estimate.

PART II: APPLYING THE CONCEPTS

Answer all questions, showing all formulas, substitutions, and work.

10.11 State the null (H_o) and the alternative (H_a) hypotheses that would be used to test each of these claims.

a. There is no significant difference in the mean batting averages for the baseball players of the two major leagues.

b. The standard deviation for the monthly amounts of rainfall in Monroe County is less than the standard deviation for the monthly amounts of rainfall in Orange County.

c. There is a significant difference between the percentages of male and female college students who own their own car.

10.12 Determine the test statistic, critical region, and critical value(s) that would be used in completing each hypothesis test using the classical procedure with $\alpha = 0.05$.

a. H_o: $p_1 - p_2 = 0$
H_a: $p_1 - p_2 \neq 0$

b. H_o: $\mu_d = 12$
H_a: $\mu_d \neq 12$
$(n = 28)$

c. H_o: $\mu_1 - \mu_2 = 17$
H_a: $\mu_1 - \mu_2 > 17$
$(n_1 = 8, n_2 = 10)$

d. H_o: $\mu_1 - \mu_2 = 37$
H_a: $\mu_1 - \mu_2 < 37$
$(n_1 = 38, n_2 = 50)$

(continued)

e. H_o: $\sigma_m^2 = \sigma_p^2$
H_a: $\sigma_m^2 > \sigma_p^2$
$(n_m = 16, n_p = 25)$

10.13 Find each of the following:

a. $z(0.02)$ b. $t(15, 0.025)$ c. $F(24, 12, 0.05)$
d. $F(12, 24, 0.05)$ e. $z(0.04)$
f. $t(38, 0.05)$ g. $t(23, 0.99)$ h. $z(0.90)$

10.14 Pt10-14 Twenty college freshmen were randomly divided into two groups. Members of one group were assigned to a statistics section that used programmed materials only. Members of the other group were assigned to a section in which the professor lectured. At the end of the semester, all were given the same final exam. Here are the results:

Programmed	76	60	85	58	91
	44	82	64	79	88
Lecture	81	62	87	70	86
	77	90	63	85	83

At the 5% level of significance, do these data provide sufficient evidence to conclude that on the average the students in the lecture sections performed significantly better on the final exam?

10.15 Pt10-15 The weights of eight people before they stopped smoking and five weeks after they stopped smoking are as follows:

	1	2	3	4	5	6	7	8
Before	148	176	153	116	129	128	120	132
After	154	179	151	121	130	136	125	128

At the 0.05 level of significance, does this sample present enough evidence to justify the conclusion that weight increases if one quits smoking?

10.16 In a nationwide sample of 600 school-age boys and 500 school-age girls, 288 boys and 175 girls admitted to having committed a destruction-of-property offense. Use this sample data to construct a 95% confidence interval for the difference between the proportions of boys and girls who have committed this offense.

PART III: UNDERSTANDING THE CONCEPTS

10.17 To compare the accuracy of two short-range missiles, eight of the first kind and ten of the second kind are fired at a target. Let x be the distance by which the missile missed the target. Do these two sets of data (eight distances and ten distances) represent dependent or independent samples? Explain.

10.18 Let's assume that 400 students in our college are taking elementary statistics this semester. Describe how you could obtain two dependent samples of size 20 from these students in order to test some precourse skill against the same skill after completing the course. Be very specific.

10.19 Student A says, "I don't see what all the fuss is about the difference between independent and dependent means; the results are almost the same regardless of the method used." Professor C suggests student A should compare the procedures a bit more carefully. Help student A discover that there is a substantial difference between the procedures.

10.20 Suppose you are testing H_o: $\mu_d = 0$ versus H_a: $\mu_d < 0$ and the sample paired differences are all negative. Does this mean there is sufficient evidence to reject the null hypothesis? How can it not be significant? Explain.

10.21 Truancy is very disruptive to the educational system. A group of high school teachers and counselors have developed a group counseling program that they hope will help improve the truancy situation in their school. They have selected the 80 students in their school with the worst truancy records and have randomly assigned half of them to the group counseling program. At the end of the school year, the 80 students will be rated with regard to their truancy. When the scores have been collected, they will be turned over to you for evaluation. Explain what you will do to complete the study.

10.22 You wish to estimate and compare the proportion of Catholic families whose children attend a private school to the proportion of non-Catholic families whose children attend private schools. How would you go about estimating the two proportions and the difference between them?

WORKING WITH YOUR OWN DATA

© StockTrek/Getty Images

History contains many stories about consumers and the various products they purchase. An exhibit at the Boston Museum of Science tells such a mathematician–baker story. A man named Poincaré bought one loaf of bread daily from his local baker, a loaf that was supposed to weigh 1 kilogram. After a year of weighing and recording the weight of each loaf, Poincaré found a normal distribution with a mean of 950 grams. The police were called and the baker was told to behave himself; however, a year later Poincaré reported that the baker had not reformed and the police confronted the baker again. The baker questioned, "How could Poincaré have known that we always gave him the largest loaf?" Poincaré then showed the police the second year of his record, a bell-shaped curve with a mean of 950 grams but truncated on the left side.

As consumers, we all purchase many bottled, boxed, canned, and packaged products. Seldom, if ever, do any of us question whether or not the content is really the amount stated on the container. Here are some content listings found on containers we purchase:

28 FL OZ (1 PT 12 OZ)	750 ml
5 FL OZ (148 ml)	32 FL OZ (1 QT) 0.95 l
NET WT 10 OZ 283 GRAMS	NET WT $3^3/_4$ OZ 106 g—48 tea bags
140 1-PLY NAPKINS	77 SQ FT—92 TWO-PLY SHEETS—11 × 11 IN.

Have you ever wondered, "Am I getting the amount that I am paying for?" And if this thought did cross your mind, did you attempt to check the validity of the content claim? The following article appeared in the *Times Union* of Rochester, New York, in 1972.

Milk Firm Accused of Short Measure*

The processing manager of Dairylea Cooperative, Inc., has been named in a warrant charging that the cooperative is distributing cartons of milk in the Rochester area containing less than the quantity represented.

. . . an investigator found shortages in four quarts of Dairylea milk purchased Friday.

Asst. Dist. Atty. Howard R. Relin, who issued the warrant, said the shortages ranged from $1\frac{1}{8}$ to $1\frac{1}{4}$ ounces per quart. A quart of milk contains 32 fluid ounces.

. . . . the state Agriculture and Markets Law . . . provides that a seller of a commodity shall not sell or deliver less of the commodity than the quantity represented to be sold.

> . . . the purpose of the law under which . . . the dairy is charged is to ensure honest, accurate, and fair dealing with the public. There is no requirement that intent to violate the law be proved, he said.
>
> *From *The Times-Union,* Rochester, NY, February 16, 1972

This situation poses a very interesting legal problem: There is no need to show intent to "short the customer." If caught, violators are fined automatically and the fines are often quite severe.

A A HIGH-SPEED FILLING OPERATION

A high-speed piston-type machine used to fill cans with hot tomato juice was sold to a canning company. The guarantee stated that the machine would fill 48-oz cans with a mean amount of 49.5 oz, a standard deviation of 0.072 oz, and a maximum spread of 0.282 oz while operating at a rate of filling 150 to 170 cans per minute. On August 12, 1994, a sample of 42 cans was gathered and the following weights were recorded. The weights, measured to the nearest $\frac{1}{8}$ oz, are recorded as variations from 49.5 oz.

$$\begin{array}{cccccccccccccc} -\frac{1}{8} & 0 & -\frac{1}{8} & 0 & 0 & 0 & -\frac{1}{8} & 0 & 0 & 0 & 0 & 0 & -\frac{1}{8} & 0 \\ \frac{1}{8} & 0 & \frac{1}{8} & 0 & \frac{1}{8} & 0 & 0 & 0 & -\frac{1}{8} & 0 & 0 & 0 & 0 & -\frac{1}{8} \\ 0 & 0 & 0 & 0 & 0 & 0 & 0 & 0 & 0 & 0 & 0 & 0 & 0 & 0 \end{array}$$

1. Calculate the mean $\bar{x}$, the standard deviation s, and the range of the sample data.
2. Construct a histogram picturing the sample data.
3. Does the amount of fill differ from the prescribed 49.5 oz at the $\alpha = 0.05$ level? Test the hypothesis that $\mu = 49.5$ against an appropriate alternative.
4. Does the amount of variation, as measured by the range, satisfy the guarantee?
5. Assuming that the filling machine continues to fill cans with an amount of tomato juice that is distributed normally and the mean and standard deviation are equal to the values found in question 1, what is the probability that a randomly selected can will contain less than the 48 oz claimed on the label?
6. If the amount of fill per can is normally distributed and the standard deviation can be maintained, find the setting for the mean value that would allow only 1 can in every 10,000 to contain less than 48 oz.

B YOUR OWN INVESTIGATION

Select a packaged product that has a quantity of fill per package that you can and would like to investigate.

1. Describe your selected product, including the quantity per package, and describe how you plan to obtain your data.
2. Collect your sample of data. (Consult your instructor for advice on size of sample.)
3. Calculate the mean $\bar{x}$ and the standard deviation s for your sample data.
4. Construct a histogram or stem-and-leaf diagram picturing the sample data.
5. Does the mean amount of fill agree with the amount given on the label? Test using $\alpha = 0.05$.
6. Assume that the item you selected is filled continually. The amount of fill is normally distributed, and the mean and standard deviation are equal to the values found in question 3. What is the probability that one randomly selected package contains less than the precribed amount?

PART

4 More Inferential Statistics

© StockTrek/Getty Images

In Part Three we studied the two inferential statistical techniques: confidence intervals and hypothesis tests, for one and two populations, and the three parameters: mean, μ; proportion, p; and standard deviation, σ. In Part Four we will learn how to use the confidence interval and the hypothesis-testing techniques in other situations. Some of these situations will be extensions of previous methods. (For example, analysis of variance will be used to deal with more than two populations when the problem involves the mean, and the binomial experiment will be expanded to a multinomial experiment.) Other situations will be alternatives to methods previously studied (such as the nonparametric techniques) or will deal with new inferences (such as the nonparametric methods and the regression and correlation inference techniques).

Sir Ronald A. Fisher

SIR RONALD A. FISHER, British statistician, was born in London on February 17, 1890. In 1912, after earning a B.A. from Gonville and Caius College in Cambridge, Fisher worked as a statistician for an investment company until 1915, at which time he became a public school teacher. In 1917 he married Ruth Eillean Gralton Guiness with whom he had eight children. Fisher received his M.A. in 1920 and his Sc.D. in 1926. Many years later, he retired to Australia, where he died at the age of 72 on July 29, 1962.

In 1919 Fisher was hired by Rothamsted Experimental Station to do statistical work with its plant breeding experiments. It was there that he pioneered the applications of statistical procedures to the design of scientific experiments. During his employment at Rothamsted, Fisher introduced the principle of randomization and the concept of analysis of variance (an even more important achievement). In 1925 Fisher wrote "Statistical Methods for Research," a work that remained in print for more than 50 years. Fisher has played a major role in the development and application of statistics.

CHAPTER

11 Applications of Chi-Square

Chapter Outline and Objectives

CHAPTER CASE STUDY

A statistical look at cooling methods after one eats hot, spicy foods.

11.1 CHI-SQUARE STATISTIC

The chi-square distribution will be used to test hypotheses concerning **enumerative data.**

11.2 INFERENCES CONCERNING MULTINOMIAL EXPERIMENTS

A multinomial experiment differs from a binomial experiment in that **each trial has many outcomes** rather than only two outcomes.

11.3 INFERENCES CONCERNING CONTINGENCY TABLES

Contingency tables are tabular representations of frequency counts for data in a **two-way classification.**

RETURN TO CHAPTER CASE STUDY

Chapter Case Study

Cooling Your Mouth After a Great Hot Taste

© Dion Ogust/The Image Works

If you like hot foods, you probably have a preferred way to "cool" your mouth after eating a delicious spicy favorite. Some of the more common methods used by people are drinking water, milk, soda, or beer, and eating bread or other food. There are even a few who prefer not to cool their mouth on such occasions and therefore do nothing. The USA Snapshot shown here appeared in *USA Today* on March 6, 1998, and shows the top six ways adults say they cool their mouths after eating hot sauce.

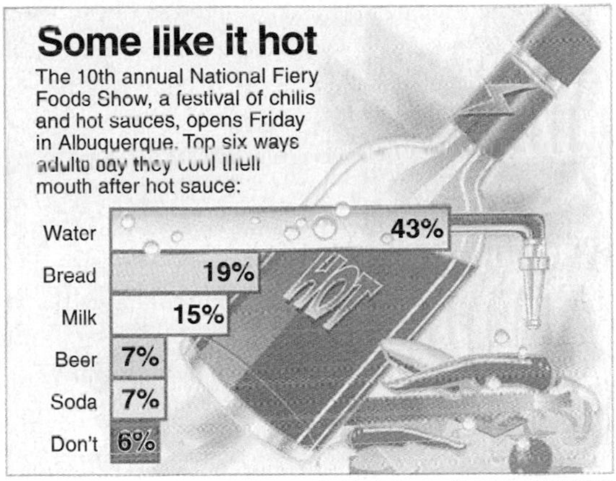
USA SNAPSHOTS®
A look at statistics that shape our lives
Some like it hot
The 10th annual National Fiery Foods Show, a festival of chilis and hot sauces, opens Friday in Albuquerque. Top six ways adults say they cool their mouth after hot sauce:
Water 43%
Bread 19%
Milk 15%
Beer 7%
Soda 7%
Don't 6%
HOT
Source: Cholula Hot Sauce
By Anne R. Carey and Suzy Parker, USA TODAY
© 1998 USA Today. Reprinted by permission.

Two hundred adults who professed to love hot, spicy food were asked to name their favorite way to cool their mouth after eating food with hot sauce. Here is the summary of the sample:

Method	Water	Milk	Soda	Beer	Bread	Other	Nothing
Number	73	35	20	19	29	11	13

ANSWER NOW 11.1 CS11

How similar is the distribution in the sample to the distribution of percentages in the USA Snapshot?

a. **Construct a horizontal bar graph of the 200 adults, using relative frequency for the horizontal scale.**

b. **Superimpose the bar graph from "Some like it hot" on the bar graph in part a.**

c. **Does the sample's distribution looks similar to or quite different from the distribution shown in the "Some like it hot" graph? Explain your answer.**

After completing Chapter 11, further investigate the Chapter Case Study in the Return to Chapter Case Study section with Exercises 11.37, 11.38, and 11.39 (p. 564).

11.1 Chi-Square Statistic

There are many problems for which **enumerative** data are categorized and the results shown by way of counts. For example, a set of final exam scores can be displayed as a frequency distribution. These frequency numbers are counts, the number of data that fall in each cell. A survey asks voters whether they are registered as Republican, Democrat, or other, and whether or not they support a particular candidate. The results are usually displayed on a chart that shows the number of voters in each possible category. Numerous illustrations of this way of presenting data have been given throughout the previous ten chapters.

Suppose that we have a number of **cells** into which n observations have been sorted. (The term *cell* is synonymous with the term *class;* the terms *class* and *frequency* were defined and first used in earlier chapters. Before you continue, a brief review of Sections 2.1, 2.2, and 3.1 might be beneficial.) The **observed frequencies** in each cell are denoted by $O_1, O_2, O_3, \ldots, O_k$ (see Table 11.1). Note that the sum of all the observed frequencies is

$$O_1 + O_2 + \cdots + O_k = n$$

where n is the sample size. What we would like to do is compare the observed frequencies with some **expected,** or theoretical **frequencies,** denoted by $E_1, E_2, E_3, \ldots, E_k$ (see Table 11.1), for each of these cells. Again, the sum of these expected frequencies must be exactly n:

$$E_1 + E_2 + \cdots + E_k = n$$

TABLE 11.1 Observed Frequencies

	k Categories					
	1st	2nd	3rd	···	*k*th	Total
Observed frequencies	O_1	O_2	O_3	···	O_k	n
Expected frequencies	E_1	E_2	E_3	···	E_k	n

We will then decide whether the observed frequencies seem to agree or disagree with the expected frequencies. We will do this by using a **hypothesis test** with **chi-square, χ^2** ("*ki-square*"; that's "*ki*" as in *kite;* χ is the Greek lowercase letter chi).

The calculated value of the test statistic will be $\chi^2\star$:

$$\chi^2\star = \sum_{\text{all cells}} \frac{(O - E)^2}{E} \tag{11.1}$$

This calculated value for chi-square is the sum of several nonnegative numbers, one from each cell (or category). The numerator of each term in the formula for $\chi^2\star$ is the square of the difference between the values of the observed and the expected frequencies. The closer together these values are, the smaller the value of $(O - E)^2$; the farther apart, the larger the value of $(O - E)^2$. The denominator for each cell puts the size of the numerator into perspective; that is, a difference $(O - E)$ of 10 resulting from frequencies of 110 (O) and 100 (E) is quite different from a difference of 10 resulting from 15 (O) and 5 (E).

These ideas suggest that small values of chi-square indicate agreement between the two sets of frequencies, whereas larger values indicate disagreement. Therefore, it is customary for these tests to be one-tailed, with the critical region on the right.

In repeated sampling, the calculated value of $\chi^2\bigstar$ in formula (11.1) will have a sampling distribution that can be approximated by the chi-square probability distribution when n is large. This approximation is generally considered adequate when all the expected frequencies are equal to or greater than 5. Recall that the chi-square distributions, like the Student t-distributions, are a family of probability distributions, each one being identified by the parameter number of **degrees of freedom,** df. The appropriate value of df will be described with each specific test. In order to use the chi-square distribution, we must be aware of its properties, which were listed in Section 9.3 on page 449. (Also see Figure 9.7.) The critical values for chi-square are obtained from Table 8 in Appendix B. (Specific instructions were given in Section 9.3.)

ANSWER NOW 11.2

Find:

a. $\chi^2(10, 0.01)$ b. $\chi^2(12, 0.025)$ c. $\chi^2(10, 0.95)$ d. $\chi^2(22, 0.995)$

ASSUMPTION FOR USING CHI-SQUARE TO MAKE INFERENCES BASED ON ENUMERATIVE DATA

The sample information is obtained using a random sample drawn from a population in which each individual is classified according to the categorical variable(s) involved in the test.

A *categorical variable* is a variable that classifies or categorizes each individual into exactly one of several cells or classes; these cells or classes are all-inclusive and mutually exclusive. The side facing up on a rolled die is a categorical variable: The list of outcomes {1, 2, 3, 4, 5, 6} is a set of all-inclusive and mutually exclusive categories.

In this chapter we permit a certain amount of "liberalization" with respect to the null hypothesis and its testing. In previous chapters the null hypothesis was always a statement about a population parameter (μ, σ, or p). However, there are other types of hypotheses that can be tested, such as "This die is fair" or "The height and weight of individuals are independent." Notice that these hypotheses are not claims about a parameter, although sometimes they could be stated with parameter values specified.

Suppose that I claim "This die is fair," $p = P(\text{any one number}) = \frac{1}{6}$, and you want to test the claim. What would you do? Was your answer something like: Roll this die many times and record the results? Suppose that you decide to roll the die 60 times. If the die is fair, what do you expect will happen? Each number (1, 2, . . . , 6) should appear approximately $\frac{1}{6}$ of the time (that is, 10 times). If it happens that approximately 10 of each number occur, you will certainly accept the claim of fairness ($p = \frac{1}{6}$ for each value). If it happens that the die seems to favor some particular numbers, you will reject the claim. (The test statistic $\chi^2\bigstar$ will have a large value in this case, as we will soon see.)

Exercises

11.3 Find these critical values by using Table 8 of Appendix B:

a. $\chi^2(18, 0.01)$
b. $\chi^2(16, 0.025)$
c. $\chi^2(40, 0.10)$
d. $\chi^2(45, 0.01)$

Using the notation in parts a–d, name and find the critical values of χ^2.

e.
f.

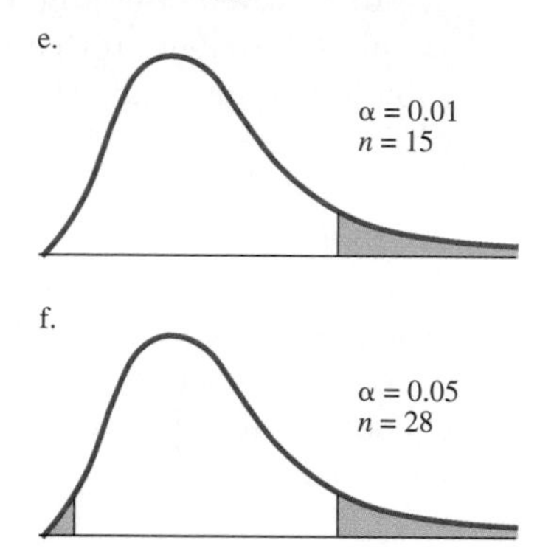

11.2 Inferences Concerning Multinomial Experiments

The preceding die problem is a good illustration of a **multinomial experiment.** Let's consider this problem again. Suppose that we want to test this die (at $\alpha = 0.05$) and decide whether to fail to reject or reject the claim "This die is fair." (The probability of each number is $\frac{1}{6}$.) The die is rolled from a cup onto a smooth, flat surface 60 times, with the following observed frequencies:

Number	1	2	3	4	5	6
Observed frequency	7	12	10	12	8	11

The null hypothesis that the die is fair is assumed to be true. This allows us to calculate the expected frequencies. If the die is fair, we certainly expect 10 occurrences of each number.

Now let's calculate an observed value of χ^2. These calculations are shown in Table 11.2. The calculated value is $\chi^2\bigstar = 2.2$.

TABLE 11.2 Computations for Calculating χ^2

Number	Observed (O)	Expected (E)	$O - E$	$(O - E)^2$	$\frac{(O-E)^2}{E}$
1	7	10	−3	9	0.9
2	12	10	2	4	0.4
3	10	10	0	0	0.0
4	12	10	2	4	0.4
5	8	10	−2	4	0.4
6	11	10	1	1	0.1
Total	60	60	0 ck		**2.2**

NOTE $\sum(O - E)$ must equal zero because $\sum O = \sum E = n$. You can use this fact as a check, as shown in Table 11.2.

Now let's use our familiar hypothesis-testing format.

Step 1 **The Set-Up:**

a. **Describe the population parameter of interest.**
The probability with which each side faces up: $P(1)$, $P(2)$, $P(3)$, $P(4)$, $P(5)$, $P(6)$.

b. **State the null hypothesis (H_o) and the alternative hypothesis (H_a).**

H_o: The die is fair (each $p = \frac{1}{6}$).

H_a: The die is not fair (at least one p is different from the others).

Step 2 **The Hypothesis Test Criteria:**

a. **Check the assumptions.**
The data were collected in a random manner, and each outcome is one of the six numbers.

b. **Identify the probability distribution and the test statistic to be used.**
The chi-square distribution and formula (11.1), with df $= k - 1 = 6 - 1 = 5$.

In a multinomial experiment, df $= k - 1$, where k is the number of cells.

c. **Determine the level of significance, α.**
$\alpha = 0.05$.

Step 3 **The Sample Evidence:**

a. **Collect the sample information:** See Table 11.2.

b. **Calculate the value of the test statistic.**
Using formula (11.1), we have

$$\chi^2\bigstar = \sum_{\text{all cells}} \frac{(O-E)^2}{E}: \quad \chi^2\bigstar = \mathbf{2.2} \text{ (calculations are shown in Table 11.2)}$$

Step 4 **The Probability Distribution:**

Using the p-value procedure

a. **Calculate the p-value for the test statistic.**
Use the right-hand tail because "larger" values of chi-square disagree with the null hypothesis:
$\mathbf{P} = P(\chi^2\bigstar > 2.2 \mid \text{df} = 5)$ as shown on the figure.

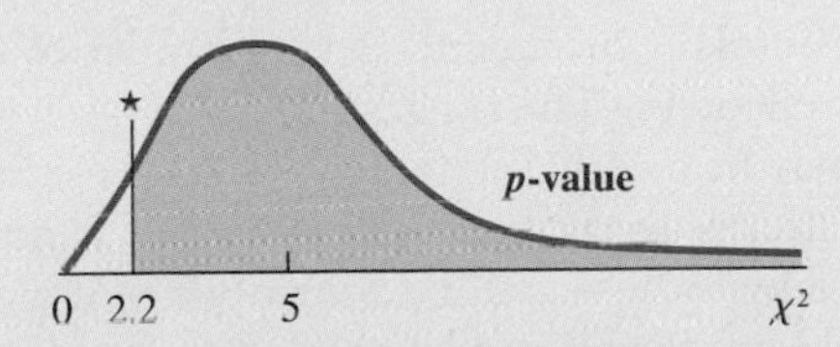

To find the p-value, you have two options:

1. Use Table 8 in Appendix B to place bounds on the p-value: $\mathbf{0.75 < P < 0.90}$.
2. Use a computer or calculator to find the p-value: $\mathbf{P = 0.821}$.

For specific instructions, see page 455. *(continued)*

OR

Using the classical procedure

a. **Determine the critical region and critical value(s).**
The critical region is the right-hand tail because "larger" values of chi-square disagree with the null hypothesis. The critical value is obtained from Table 8, at the intersection of row df = 5 and column $\alpha = 0.05$:

$$\chi^2(5, 0.05) = \mathbf{11.1}$$

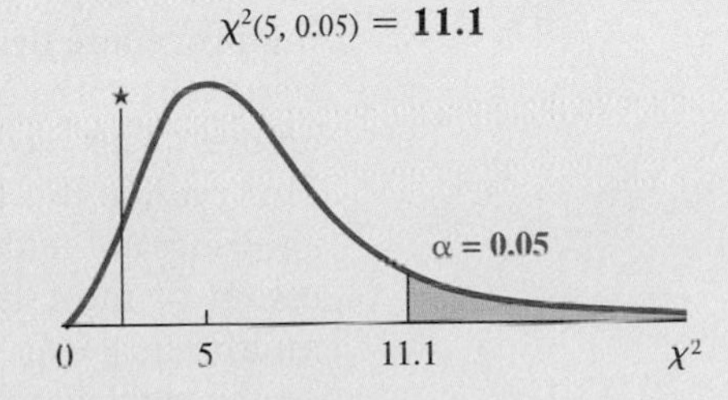

For specific instructions, see page 450. *(continued)*

b. Determine whether or not the p-value is smaller than α.
The p-value is not smaller than the level of significance, α.

b. Determine whether or not the calculated test statistic is in the critical region.
$\chi^2\bigstar$ is not in the critical region, as shown in **red** on the figure.

FYI
Computer and calculator commands to find the probability associated with a specified chi-square value can be found in Chapter 9 (p. 452).

Step 5 **The Results:**
a. State the decision about H_o: Fail to reject H_o.
b. State the conclusion about H_a.
At the 0.05 level of significance, the observed frequencies are not significantly different from those expected of a fair die.

Before we look at other illustrations, we must define the term *multinomial experiment* and state the guidelines for completing the chi-square test for it.

MULTINOMIAL EXPERIMENT

A multinomial experiment has the following characteristics:
1. It consists of n identical independent trials.
2. The outcome of each trial fits into exactly one of k possible cells.
3. There is a probability associated with each particular cell, and these individual probabilities remain constant during the experiment. (It must be the case that $p_1 + p_2 + \cdots + p_k = 1$.)
4. The experiment will result in a set of k observed frequencies, $O_1, O_2, \ldots, O_k$, where each O_i is the number of times a trial outcome falls into that particular cell. (It must be the case that $O_1 + O_2 + \cdots + O_k = n$.)

The die example meets the definition of a multinomial experiment because it has all four of the characteristics described in the definition.

1. The die was rolled n (60) times in an identical fashion, and these trials were independent of each other. (The result of each trial was unaffected by the results of other trials.)
2. Each time the die was rolled, one of six numbers resulted, and each number was associated with a cell.
3. The probability associated with each cell was $\frac{1}{6}$, and this was constant from trial to trial. (Six values of $\frac{1}{6}$ sum to 1.0.)
4. When the experiment was complete, we had a list of six frequencies (7, 12, 10, 12, 8, and 11) that summed to 60, indicating that each of the outcomes was taken into account.

The testing procedure for multinomial experiments is very similar to the testing procedure described in previous chapters. The biggest change comes with the statement of the null hypothesis. It may be a verbal statement, such as in the die example: "This die is fair." Often the alternative to the null hypothesis is not stated. However, in this book the alternative hypothesis will be shown, since it aids in organizing and understanding the problem. It will not be used to determine the location of the critical region, though, as was the case in previous chapters. For multinomial experiments we will always use a one-tailed critical region, and it will be the right-hand tail of the χ^2-distribution because larger deviations (positive or negative) from the expected values lead to an increase in the calculated $\chi^2\bigstar$ value.

The critical value will be determined by the level of significance assigned (α) and the number of degrees of freedom. The number of degrees of freedom (df) will be 1 less than the number of cells (k) into which the data are divided:

$$\text{df} = k - 1 \tag{11.2}$$

Each expected frequency, E_i, will be determined by multiplying the total number of trials n by the corresponding probability (p_i) for that cell; that is,

$$E_i = n \cdot p_i \tag{11.3}$$

One guideline should be met to ensure a good approximation to the chi-square distribution: Each expected frequency should be at least 5 (that is, each $E_i \geq 5$). Sometimes it is possible to combine "smaller" cells to meet this guideline. If this guideline cannot be met, then corrective measures to ensure a good approximation should be used. These corrective measures are not covered in this book but are discussed in many other sources.

Illustration 11.1 A Multinomial Hypothesis Test with Equal Expected Frequencies

College students have regularly insisted on freedom of choice when they register for courses. This semester there were seven sections of a particular mathematics course. The sections were scheduled to meet at various times with a variety of instructors. Table 11.3 shows the number of students who selected each of the seven sections. Do the data indicate that the students had a preference for certain sections, or do they indicate that each section was equally likely to be chosen?

TABLE 11.3 Data on Section Enrollments

	Section							
	1	2	3	4	5	6	7	Total
Number of students	18	12	25	23	8	19	14	119

SOLUTION

If no preference were shown in the selection of sections, then we would expect the 119 students to be equally distributed among the seven classes. We would expect 17 students to register for each section. The hypothesis test is completed at the 5% level of significance.

Step 1 **The Set-Up:**

a. Describe the population parameter of interest.

Preference for each section, the probability that a particular section is selected at registration.

b. State the null hypothesis (H_o) and the alternative hypothesis (H_a).

H_o: There was no preference shown (equally distributed).
H_a: There was a preference shown (not equally distributed).

Step 2 **The Hypothesis Test Criteria:**

a. Check the assumptions.
The 119 students represent a random sample of the population of all students who register for this particular course. Since no new regulations were introduced in the selection of courses and registration seemed to proceed in its usual pattern, there is no reason to believe this is other than a random sample.

b. Identify the probability distribution and the test statistic to be used.
The chi-square distribution and formula (11.1), with df = 6.

c. Determine the level of significance, α.
$\alpha = 0.05$.

Step 3 **The Sample Evidence:**

a. Collect the sample information: See Table 11.3.

b. Calculate the value of the test statistic.
Using formula (11.1), we have

$$\chi^2\bigstar = \sum_{\text{all cells}} \frac{(O-E)^2}{E}: \quad \chi^2\bigstar = \frac{(18-17)^2}{17} + \frac{(12-17)^2}{17} + \frac{(25-17)^2}{17} + \frac{(23-17)^2}{17} + \frac{(8-17)^2}{17} + \frac{(19-17)^2}{17} + \frac{(14-17)^2}{17}$$

$$= \frac{(1)^2 + (-5)^2 + (8)^2 + (6)^2 + (-9)^2 + (2)^2 + (-3)^2}{17}$$

$$= \frac{1 + 25 + 64 + 36 + 81 + 4 + 9}{17} = \frac{220}{17} = 12.9411$$

$$= \mathbf{12.94}$$

Step 4 **The Probability Distribution:**

Using the *p*-Value Procedure

a. Calculate the *p*-value for the test statistic.
Use the right-hand tail because "larger" values of chi-square disagree with the null hypothesis:
$\mathbf{P} = P(\chi^2\bigstar > 12.94 \mid df = 6)$ as shown on the figure:

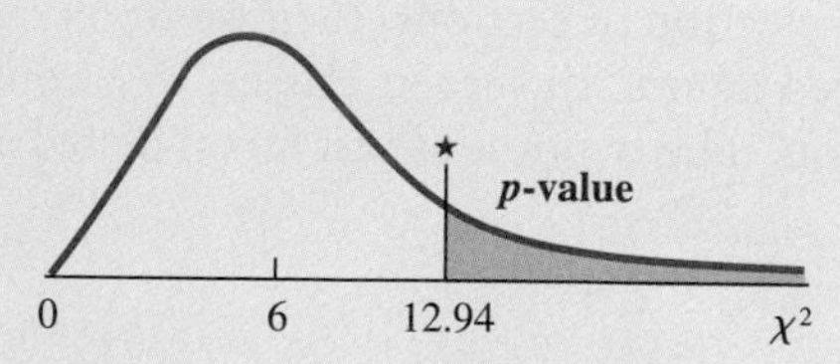

To find the *p*-value, you have two options:

1. Use Table 8 in Appendix B to place bounds on the *p*-value: $\mathbf{0.025 < P < 0.05}$.
2. Use a computer or calculator to find the *p*-value: $\mathbf{P = 0.044}$.

For specific instructions, see page 455.

b. Determine whether or not the *p*-value is smaller than α.
The *p*-value is smaller than the level of significance, α.

OR

Using the Classical Procedure

a. Determine the critical region and critical value(s).
The critical region is the right-hand tail because "larger" values of chi-square disagree with the null hypothesis. The critical value is obtained from Table 8, at the intersection of row df = 6 and column $\alpha = 0.05$:

$$\chi^2(6, 0.05) = \mathbf{12.6}$$

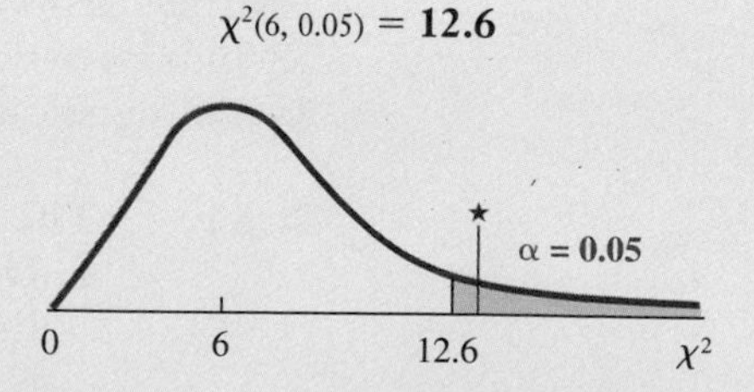

For specific instructions, see page 450.

b. Determine whether or not the calculated test statistic is in the critical region.
$\chi^2\bigstar$ is in the critical region, as shown in **red** on the figure.

Step 5 **The Results:**

a. **State the decision about H_o.** Reject H_o.

b. **State the conclusion about H_a.**

At the 0.05 level of significance, there does seem to be a preference shown. We cannot determine from the given information what the preference is. It could be teacher preference, time preference, or a schedule conflict. ■

Conclusions must be worded carefully to avoid suggesting conclusions that the data cannot support.

Not all multinomial experiments result in equal expected frequencies, as we will see in Illustration 11.2.

Illustration 11.2 A Multinomial Hypothesis Test with Unequal Expected Frequencies

The Mendelian theory of inheritance claims that the frequencies of round and yellow, wrinkled and yellow, round and green, and wrinkled and green peas will occur in the ratio 9:3:3:1 when two specific varieties of peas are crossed. In testing this theory, Mendel obtained frequencies of 315, 101, 108, and 32, respectively. Do these sample data provide sufficient evidence to reject the theory at the 0.05 level of significance?

SOLUTION

Step 1 **The Set-Up:**

a. **Describe the population parameter of interest.**

The proportions: *P*(round and yellow), *P*(wrinkled and yellow), *P*(round and green), *P*(wrinkled and green).

b. **State the null hypothesis (H_o) and the alternative hypothesis (H_a).**

H_o: 9:3:3:1 is the ratio of inheritance.

H_a: 9:3:3:1 is not the ratio of inheritance.

Step 2 **The Hypothesis Test Criteria:**

a. **Check the assumptions.**

We will assume that Mendel's results form a random sample.

b. **Identify the probability distribution and the test statistic to be used.**

The chi-square distribution and formula (11.1), with df = 3.

c. **Determine the level of significance, α.**

$\alpha = 0.05$.

Step 3 **The Sample Evidence:**

a. **Collect the sample information.**

The observed frequencies were: 315, 101, 108, and 32.

b. **Calculate the value of the test statistic.**

The ratio 9:3:3:1 indicates probabilities of $\frac{9}{16}$, $\frac{3}{16}$, $\frac{3}{16}$, and $\frac{1}{16}$.

ANSWER NOW 11.4
Explain how 9:3:3:1 becomes $\frac{9}{16}, \frac{3}{16}, \frac{3}{16}$, and $\frac{1}{16}$.

Therefore, the expected frequencies are $\frac{9n}{16}, \frac{3n}{16}, \frac{3n}{16}$, and $\frac{1n}{16}$. We have

$$n = \sum O_i = 315 + 101 + 108 + 32 = 556$$

The computations for calculating $\chi^2\bigstar$ are shown in Table 11.4.

TABLE 11.4 Computations Needed to Calculate $\chi^2\bigstar$

O	E	$O - E$	$\frac{(O-E)^2}{E}$
315	312.75	2.25	0.0162
101	104.25	−3.25	0.1013
108	104.25	3.75	0.1349
32	34.75	−2.75	0.2176
556	556.00	0 ⓒⓚ	**0.4700**

$$\rightarrow \chi^2\bigstar = \sum_{\text{all cells}} \frac{(O-E)^2}{E} = \mathbf{0.47}$$

ANSWER NOW 11.5
Explain how 312.75, 2.25, and 0.0162 were obtained in Table 11.4.

Step 4 The Probability Distribution:

Using the *p*-Value Procedure

a. Calculate the *p*-value for the test statistic.

Use the right-hand tail because "larger" values of chi-square disagree with the null hypothesis:

$\mathbf{P = P(\chi^2\bigstar > 0.47 \mid df = 3)}$ as shown on the figure.

To find the *p*-value, you have two options:

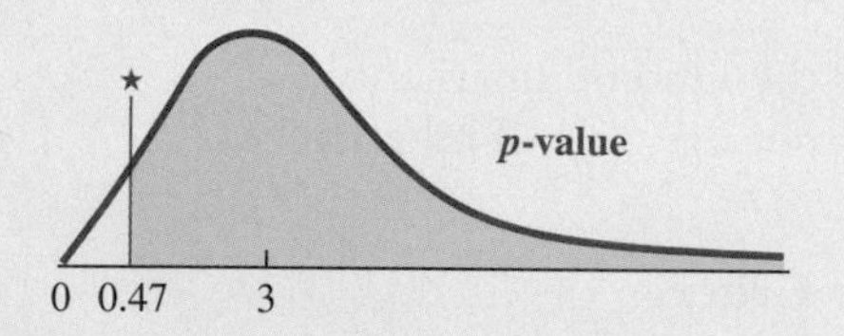

1. Use Table 8 in Appendix B to place bounds on the *p*-value: $\mathbf{0.90 < P < 0.95}$.
2. Use a computer or calculator to find the *p*-value: $\mathbf{P = 0.925}$.

For specific instructions, see page 455.

b. Determine whether or not the *p*-value is smaller than α.

The *p*-value is not smaller than the level of significance, α.

OR

Using the Classical Procedure

a. Determine the critical region and critical value(s).

The critical region is the right-hand tail because "larger" values of chi-square disagree with the null hypothesis. The critical value is obtained from Table 8, at the intersection of row df = 3 and column α = 0.05:

$$\chi^2(3, 0.05) = \mathbf{7.82}$$

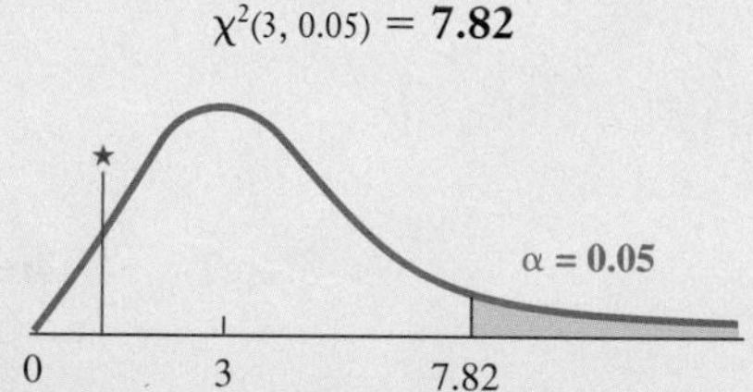

For specific instructions, see page 450.

b. Determine whether or not the calculated test statistic is in the critical region.

$\chi^2\bigstar$ is not in the critical region, as shown in **red** on the figure.

Step 5 The Results:

a. State the decision about H_o: Fail to reject H_o.

b. State the conclusion about H_a.

At the 0.05 level of significance, there is not sufficient evidence to reject Mendel's theory. ■

Application 11.1 Money Lessons

The data collected from a survey of 1000 people and used to create "Personal finance lessons" are actually those of a multinomial experiment.

USA SNAPSHOTS®

Personal finance lessons

When people were asked which of the following taught them the most about personal money management, here's what they said:

Mistakes 64%

School 31%

Unsure 5%

Note: Based on a poll of 1,000 people

Source: VISA USA By Shannon Reilly and Suzy Parker, USA TODAY

ANSWER NOW 11.6

Verify that "Personal finance lessons" is a multinomial experiment. Be specific.

a. What is one trial?

b. What is the variable?

c. What are the many levels of results from each trial?

Application 11.2 Time for Chocolate

USA SNAPSHOTS®

A look at statistics that shape our lives

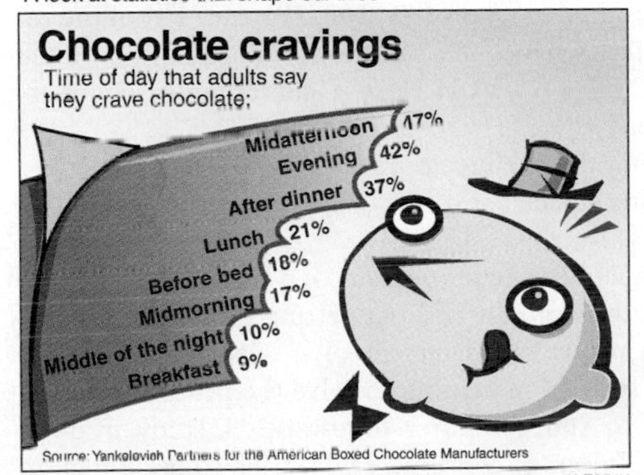

By Cindy Hall and Alejandro Gonzalez, USA TODAY

"Chocolate cravings" displays the results of surveying adults about the time of day they say they crave chocolate.

ANSWER NOW 11.7

Why is the information shown in "Chocolate cravings" not that of a multinomial experiment? Be specific.

Exercises

11.8 State the null hypothesis H_o and the alternative hypothesis H_a that would be used to test each statement:
a. The five numbers 1, 2, 3, 4, and 5 are equally likely to be drawn.
b. That multiple-choice question has a history of students selecting answers in the ratio of 2:3:2:1.
c. The poll will show a distribution of 16%, 38%, 41%, and 5% for the possible ratings of excellent, good, fair, and poor on that issue.

11.9 Determine the p-value for each hypothesis test involving the χ^2-distribution with 12 degrees of freedom.
a. H_o: $P(1) = P(2) = P(3) = P(4) = 0.25$, with $\chi^2\bigstar = 12.25$
b. H_o: $P(\text{I}) = 0.25$, $P(\text{II}) = 0.40$, $P(\text{III}) = 0.35$, with $\chi^2\bigstar = 5.98$

11.10 Determine the critical value and critical region that would be used in the classical approach to test the null hypothesis for each of these multinomial experiments:
a. H_o: $P(1) = P(2) = P(3) = P(4) = 0.25$, with $\alpha = 0.05$
b. H_o: $P(\text{I}) = 0.25$, $P(\text{II}) = 0.40$, $P(\text{III}) = 0.35$, with $\alpha = 0.01$

11.11 A manufacturer of floor polish conducted a consumer-preference experiment to determine which of five different floor polishes was the most appealing in appearance. A sample of 100 consumers viewed five patches of flooring that had each received one of the five polishes. Each consumer indicated the patch he or she preferred. The lighting and background were approximately the same for all patches. Here are the results:

Polish	A	B	C	D	E	Total
Frequency	27	17	15	22	19	100

a. State the hypothesis for "no preference" in statistical terminology.
b. What test statistic will be used in testing this null hypothesis?
c. Complete the hypothesis test using $\alpha = 0.10$.
   (1) Solve using the p-value approach.
   (2) Solve using the classical approach.

11.12 A certain type of flower seed will produce magenta, chartreuse, and ochre flowers in the ratio 6:3:1 (one flower per seed). A total of 100 seeds are planted and all germinate, yielding these results:

Magenta	Chartreuse	Ochre
52	36	12

a. If the null hypothesis (6:3:1) is true, what is the expected number of magenta flowers?
b. How many degrees of freedom are associated with chi-square?

(continued)

c. Complete the hypothesis test using $\alpha = 0.10$.
   (1) Solve using the p-value approach.
   (2) Solve using the classical approach.

11.13 Ex11-13 Over the years, African-American actors in major cinema releases are more likely to have major roles in comedies than are white actors. The table shows the percentages of all roles by type of picture.

Type of Picture	Percentage of roles
Action and adventure	13.2
Comedy	31.9
Drama	23.0
Horror and suspense	12.5
Romantic comedy	8.2
Other	11.2

The next table shows the numbers of leading roles played by African-Americans in each type of film.

Type of Picture	Number of roles
Action and adventure	9
Comedy	40
Drama	17
Horror and suspense	11
Romantic comedy	5
Other	7

Does the distribution of African/American roles differ from the overall distribution of roles? Use $\alpha = 0.05$.
a. Solve using the p-value approach.
b. Solve using the classical approach.

11.14 A large supermarket carries four qualities of ground beef. Customers are believed to purchase these four varieties with probabilities of 0.10, 0.30, 0.35, and 0.25, respectively, from the least to most expensive. A sample of 500 purchases resulted in sales of 46, 162, 191, and 101 of the respective qualities. Does this sample contradict the expected proportions? Use $\alpha = 0.05$. Use a computer to complete this exercise.
a. Solve using the p-value approach.
b. Solve using the classical approach.

11.15 One of the major benefits of e-mail is that the user can communicate rapidly without getting a busy signal or no answer, two major criticisms of voice telephone calls. But does e-mail succeed in helping to solve the problems people have trying to run computer hardware? A study in 1998 polled the opinions of consumers who tried to use e-mail to get help by posting a message online to their PC manufacturer or authorized representative. Results are shown in the table.

Result of Online Query	Percentage
Never got a response	14
Got a response, but it didn't help	30
Response helped but didn't solve the problem	34
Response solved the problem	22

Source: *PC World*, "PC World's 1998 Reliability and Service Survey," November 1998

As marketing manager for a large PC manufacturer, you decide to conduct a survey of your customers, comparing your e-mail records with the published results. In order to make a fair comparison, you elect to use the same questionnaire and examine the returns from 500 customers who attempted to use e-mail to get help from your technical support staff. Here are the results:

Result of Online Query	Number Responding
Never got a response	35
Got a response, but it didn't help	102
Response helped but didn't solve the problem	125
Response solved the problem	238
Total	500

Does your distribution of responses differ from the distribution obtained in the published survey? Test at the 0.01 level of significance.

a. Solve using the p-value approach.
b. Solve using the classical approach.

11.16 The 1993 edition of the *Digest of Educational Statistics* gives the following distribution of the ages of persons 18 and over who hold a bachelor's or higher degree:

Age	18–24	25–34	35–44	45–54	55–64	65 or over
Percentage	5	29	30	16	10	10

A survey of 500 randomly chosen persons age 18 and over who hold a bachelor's or higher degree in Alaska gave the following distribution:

Age	18–24	25–34	35–44	45–54	55–64	65 or over
Number	30	150	155	75	35	55

Test the null hypothesis that the age distribution is the same in Alaska as it is nationally at a level of significance equal to 0.05. Include in your answer the calculated $\chi^2\bigstar$, p-value or critical value, decision, and conclusion.

11.17 A program for generating random numbers on a computer is to be tested. The program is instructed to generate 100 single-digit integers between 0 and 9. The frequencies of the observed integers were as follows:

Integer	0	1	2	3	4	5	6	7	8	9
Frequency	11	8	7	7	10	10	8	11	14	14

At the 0.05 level of significance, is there sufficient reason to believe that the integers are not being generated uniformly?

a. Solve using the p-value approach.
b. Solve using the classical approach.

11.18 *Nursing Magazine* (March 1991) reported the results of a survey of more than 1800 nurses across the country concerning job satisfaction and retention. Nurses from magnet hospitals (hospitals that successfully attract and retain nurses) describe the staffing situation in their units as follows:

Staffing Situation	Percentage
1. Desperately short of help—patient care has suffered	12
2. Short, but patient care hasn't suffered	32
3. Adequate	38
4. More than adequate	12
5. Excellent	6

A survey of 500 nurses from nonmagnet hospitals gave these responses to the staffing situation:

Staffing situation	1	2	3	4	5
Number	165	140	125	50	20

Do the data indicate that the nurses from the nonmagnet hospitals have a different distribution of opinions? Use $\alpha = 0.05$.

a. Solve using the p-value approach.
b. Solve using the classical approach.

11.19 Why is the chi-square test typically a one-tailed test with the critical region in the right tail?

a. What kind of value would result if the observed frequencies and the expected frequencies were very close in value? Explain how you would interpret this situation.
b. Suppose you had to roll a die 60 times as an experiment to test the fairness of the die as discussed in the example on page 542, but instead of rolling the die yourself, you paid your little brother $1 to roll it 60 times and keep a tally of the numbers. He agreed to perform this task and ran off to his room with the die, returning in a few minutes with his frequencies. He demanded his $1. You, of course, pay him before he hands over his results, which are 10, 10, 10, 10, 10, and 10. The observed results are exactly what you had "expected." Right? Explain your reactions. What value of $\chi^2\bigstar$ will result? What do you think happened? What do you demand of your little brother and why? What possible role might the left tail have in the hypothesis test?
c. Why is the left tail not typically of concern?

11.20 Ex11.20 According to the Harris poll released on May 30, 2001, the proportions of all adults who live in households with rifles (29%), shotguns (29%), or pistols (23%) have not changed significantly since 1996. However, today more people live in households with no guns (61%). The 1014 adults surveyed gave these results:

	All Adults (%)	All Gun Owners (%)
Have rifle, shotgun, and pistol (3 out of 3)	16	41
Have 2 out of 3 (rifle, shotgun, or pistol)	11	27
Have 1 out of 3 (rifle, shotgun, or pistol)	11	29
Decline to answer/ Not sure	1	3
Total	39	100

In a survey of 2000 adults in Memphis who said they own guns, 780 said they owned all three types, 550 said they owned 2 of the 3, 560 said they owned 1 of the 3 types, and 110 declined to specify what types of guns they owned.

a. Test the null hypothesis that the distribution of the number of types owned is the same in Memphis as it is nationally as reported by the Harris poll. Use a 0.05 level of significance.

b. What caused the calculated value of $\chi^2\bigstar$ to be so large? Does it seem right that one cell should have this much effect on the results? How could this test be completed differently (hopefully, more meaningfully) so that the results might not be affected as they were in part a? Be specific.

11.3 Inferences Concerning Contingency Tables

A **contingency table** is an arrangement of data in a two-way classification. The data are sorted into cells, and the number of data in each cell is reported. The contingency table involves two factors (or variables), and a common question concerning such tables is whether the data indicate that the two variables are independent or dependent (see pp. 130–133).

Two different tests use the contingency table format. The first one we will look at is the *test of independence.*

TEST OF INDEPENDENCE

To illustrate a test of independence, let's consider a random sample that shows the gender of liberal arts college students and their favorite academic area.

Illustration 11.3

Hypothesis Test for Independence

Each person in a group of 300 students was identified as male or female and then asked whether he or she preferred taking liberal arts courses in the area of math–science, social science, or humanities. Table 11.5 is a contingency table that shows the frequencies found for these categories. Does this sample present sufficient evidence to reject the null hypothesis: "Preference for math–science, social science, or humanities is independent of the gender of a college student"? Complete the **hypothesis test** using the 0.05 level of significance.

TABLE 11.5 Sample Results for Gender and Subject Preference

	Favorite Subject Area			
Gender	Math–Science (MS)	Social Science (SS)	Humanities (H)	Total
Male (M)	37	41	44	122
Female (F)	35	72	71	178
Total	72	113	115	300

SOLUTION

Step 1 **The Set-Up:**

a. **Describe the population parameter of interest.**
Determining the independence of the variables "gender" and "favorite subject area" requires us to discuss the probability of the various cases and the effect that answers about one variable have on the probability of answers about the other variable. Independence, as defined in Chapter 4, requires $P(\text{MS} \mid \text{M}) = P(\text{MS} \mid \text{F}) = P(\text{MS})$; that is, gender has no effect on the probability of a person's choice of subject area.

b. **State the null hypothesis (H_o) and the alternative hypothesis (H_a).**

H_o: Preference for math–science, social science, or humanities is independent of the gender of a college student.

H_a: Subject area preference is not independent of the gender of the student.

Step 2 **The Hypothesis Test Criteria:**

a. **Check the assumptions.**
The sample information is obtained using one random sample drawn from one population, with each individual then classified according to gender and favorite subject area.

b. **Identify the probability distribution and the test statistic to be used.**

In the case of contingency tables, the number of degrees of freedom is exactly the same as the number of cells in the table that may be filled in freely when you are given the *marginal totals*. The totals in this illustration are shown in the following table:

			122
			178
72	113	115	300

Given these totals, you can fill in only two cells before the others are all determined. (The totals must, of course, remain the same.) For example, once we pick two arbitrary values (say, 50 and 60) for the first two cells of the first row, the other four cell values are fixed (see the following table):

50	60	C	122
D	E	F	178
72	113	115	300

The values have to be $C = 12$, $D = 22$, $E = 53$, and $F = 103$. Otherwise, the totals will not be correct. Therefore, for this problem there are two free choices. Each free choice corresponds to 1 degree of freedom. Hence, the number of degrees of freedom for our example is 2 (df = 2).

The chi-square distribution will be used along with formula (11.1), with df = 2.

c. **Determine the level of significance, α.**
$\alpha = 0.05$.

Step 3 **The Sample Evidence:**

a. **Collect the sample information:** See Table 11.5.

b. **Calculate the value of the test statistic.**

Before we can calculate the value of chi-square, we need to determine the expected values, *E*, for each cell. To do this we must recall the null hypothesis, which asserts that these factors are independent. Therefore, we would expect the values to be distributed in proportion to the marginal totals. There are 122 males; we would expect them to be distributed among MS, SS, and H proportionally to the 72, 113, and 115 totals. Thus, the expected cell counts for males are

$$\frac{72}{300} \cdot 122 \qquad \frac{113}{300} \cdot 122 \qquad \frac{115}{300} \cdot 122$$

Similarly, we would expect for the females

$$\frac{72}{300} \cdot 178 \qquad \frac{113}{300} \cdot 178 \qquad \frac{115}{300} \cdot 178$$

Thus, the expected values are as shown in Table 11.6. Always check the marginal totals for the expected values against the marginal totals for the observed values.

TABLE 11.6 Expected Values

	MS	SS	H	Total
Male	29.28	45.95	46.77	122.00
Female	42.72	67.05	68.23	178.00
Total	72.00	113.00	115.00	300.00

NOTE We can think of the computation of the expected values in a second way. Recall that we assume the null hypothesis to be true until there is evidence to reject it. Having made this assumption in our example, we are saying in effect that the event that a student picked at random is male and the event that a student picked at random prefers math–science courses are independent. Our point estimate for the probability that a student is male is $\frac{122}{300}$, and the point estimate for the probability that the student prefers math–science courses is $\frac{72}{300}$. Therefore, the probability that both events occur is the product of the probabilities. [Refer to formula (4.8a), p. 212.] Thus, $\left(\frac{122}{300}\right)\left(\frac{72}{300}\right)$ is the probability of a selected student being male and preferring math–science. The number of students out of 300 that are expected to be male and prefer math–science is found by multiplying the probability (or proportion) by the total number of students (300). Thus, the expected number of males who prefer math–science is $\left(\frac{122}{300}\right)\left(\frac{72}{300}\right)(300) = \left(\frac{122}{300}\right)(72) = 29.28$. The other expected values can be determined in the same manner.

ANSWER NOW 11.21
Find the expected value for the cell shown.

☐ ···	50
40	200

Typically the contingency table is written so that it contains all this information (see Table 11.7).

TABLE 11.7 Contingency Table Showing Sample Results and Expected Values

	Favorite Subject Area			
Gender	**Math–Science**	**Social Science**	**Humanities**	**Total**
Male	37 (29.28)	41 (45.95)	44 (46.77)	122
Female	35 (42.72)	72 (67.05)	71 (68.23)	178
Total	72	113	116	300

The calculated chi-square is

$$\chi^2\bigstar = \sum_{\text{all cells}} \frac{(O-E)^2}{E}: \quad \chi^2\bigstar = \frac{(37-29.28)^2}{29.28} + \frac{(41-45.95)^2}{45.95} + \frac{(44-46.77)^2}{46.77}$$

$$+ \frac{(35-42.72)^2}{42.72} + \frac{(72-67.05)^2}{67.05} + \frac{(71-68.23)^2}{68.23}$$

$$= 2.035 + 0.533 + 0.164 + 1.395 + 0.365 + 0.112$$

$$= \mathbf{4.604}$$

Step 4 The Probability Distribution:

Using the *p*-Value Procedure

a. Calculate the *p*-value for the test statistic.

Use the right-hand tail because "larger" values of chi-square disagree with the null hypothesis:

$\mathbf{P = P(\chi^2\bigstar > 4.604 \mid df = 2)}$ as shown on the figure.

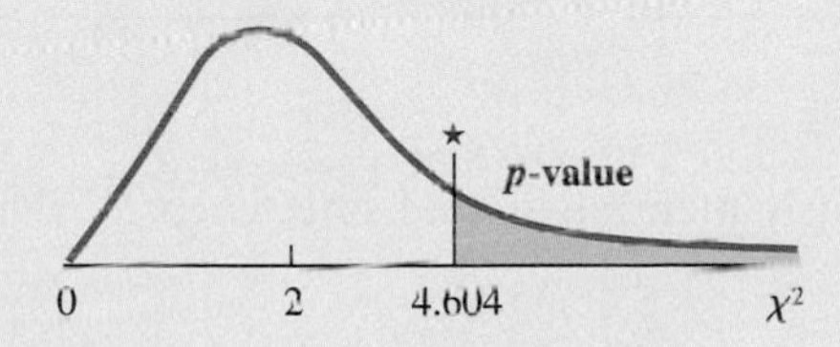

To find the *p*-value, you have two options:

1. Use Table 8 in Appendix B to place bounds on the *p*-value: $\mathbf{0.10 < P < 0.25}$.
2. Use a computer or calculator to find the *p*-value: $\mathbf{P = 0.1001}$.

For specific instructions, see page 455.

b. Determine whether or not the *p*-value is smaller than α.

The *p*-value is not smaller than α.

OR

Using the Classical Procedure

a. Determine the critical region and critical value(s).

The critical region is the right-hand tail because "larger" values of chi-square disagree with the null hypothesis. The critical value is obtained from Table 8, at the intersection of row df = 2 and column α = 0.05:

$$\chi^2(2, 0.05) = \mathbf{5.99}$$

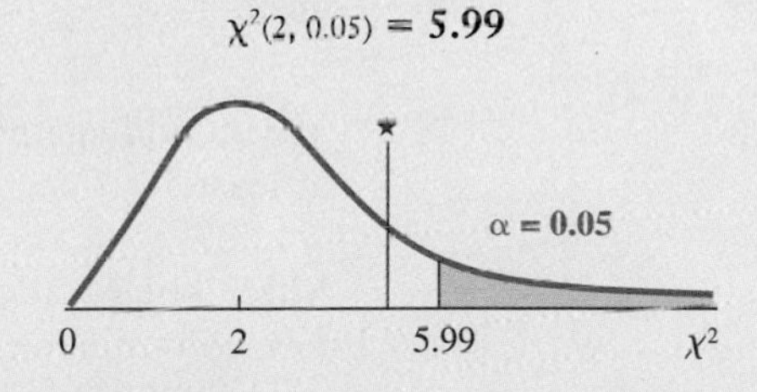

For specific instructions, see page 450.

b. Determine whether or not the calculated test statistic is in the critical region.

$\chi^2\bigstar$ is not in the critical region, as shown in **red** on the figure.

Step 5 **The Results:**

a. **State the decision about H_o:** Fail to reject H_o.

b. **State the conclusion about H_a.**

At the 0.05 level of significance, the evidence does not allow us to reject independence between the gender of a student and the student's preferred academic subject area. ■

In general, the $r \times c$ **contingency table** (r is the number of **rows;** c is the number of **columns**) is used to test the independence of the row factor and the column factor. The number of **degrees of freedom** is determined by

$$\text{df} = (r - 1) \cdot (c - 1) \tag{11.4}$$

where r and c are both greater than 1. (This value for df should agree with the number of cells counted according to the general description on page 553.)

The **expected frequencies** for an $r \times c$ contingency table are found by means of the formulas given in each cell in Table 11.8, where n = grand total. In general, the expected frequency at the intersection of the ith row and the jth column is given by

$$E_{i,j} = \frac{\text{row total} \times \text{column total}}{\text{grand total}} = \frac{R_i \times C_j}{n} \tag{11.5}$$

TABLE 11.8 **Expected Frequencies for an $r \times c$ Contingency Table**

Row	Column 1	2	. . .	jth column	. . .	c	Total
1	$\frac{R_1 \times C_1}{n}$	$\frac{R_1 \times C_2}{n}$	. . .	$\frac{R_1 \times C_j}{n}$	. . .	$\frac{R_1 \times C_c}{n}$	R_1
2	$\frac{R_2 \times C_1}{n}$						R_2
⋮	⋮			⋮			⋮
ith row	$\frac{R_i \times C_1}{n}$		. . .	$\frac{R_i \times C_j}{n}$	. . .		R_i
⋮	⋮			⋮			⋮
r	$\frac{R_r \times C_1}{n}$						
Total	C_1	C_2	. . .	C_j	. . .	. . .	n

We should again observe the previously mentioned guideline: Each $E_{i,j}$ should be at least 5.

NOTE The notation used in Table 11.8 and formula (11.5) may be unfamiliar to you. For convenience in referring to cells or entries in a table, we use $E_{i,j}$ to denote the entry in the ith row and the jth column. That is, the first letter in the subscript corresponds to the row number and the second letter corresponds to the column number. Thus, $E_{1,2}$ is the entry in the first row, second column, and $E_{2,1}$ is the entry in the second row, first column. In Table 11.6 (p. 554), $E_{1,2}$ is 45.95 and $E_{2,1}$ is 42.72. The notation used in Table 11.8 is interpreted in a similar manner; that is, R_1 corresponds to the total from row 1, and C_1 corresponds to the total from column 1.

ANSWER NOW 11.22
Identify these values from Table 11.7:
a. C_2 b. R_1 c. n d. $E_{2,3}$

TEST OF HOMOGENEITY

The second type of contingency table problem is called a *test of homogeneity*. This test is used when one of the two variables is controlled by the experimenter so that the row or column totals are predetermined.

For example, suppose that we want to poll registered voters about a piece of legislation proposed by the governor. In the poll, 200 urban, 200 suburban, and 100 rural residents are randomly selected and asked whether they favor or oppose the governor's proposal. That is, a simple random sample is taken for each of these three groups. A total of 500 voters are polled. But notice that it has been predetermined (before the sample is taken) just how many are to fall within each row category, as shown in Table 11.9, and each category is sampled separately.

TABLE 11.9 Registered Voter Poll with Predetermined Row Totals

	Governor's Proposal		
Residence	Favor	Oppose	Total
Urban			200
Suburban			200
Rural			100
Total			500

In a test of this nature, we are actually testing the hypothesis: The distribution of proportions within the rows is the same for all rows. That is, the distribution of proportions in row 1 is the same as in row 2, is the same as in row 3, and so on. The alternative is: The distribution of proportions within the rows is not the same for all rows. This type of example may be thought of as a comparison of several multinomial experiments.

Beyond this conceptual difference, the actual testing for independence and homogeneity with contingency tables is the same. Let's demonstrate this **hypothesis test** by completing the polling illustration.

Illustration 11.4

Hypothesis Test for Homogeneity

Each person in a random sample of 500 registered voters (200 urban, 200 suburban, and 100 rural residents) was asked his or her opinion about the governor's proposed legislation. Does the sample evidence shown in Table 11.10 (p. 558) support the hypothesis: "Voters within the different residence groups have different opinions about the governor's proposal"? Use $\alpha = 0.05$.

TABLE 11.10 Sample Results for Residence and Opinion

Residence	Governor's Proposal		Total
	Favor	Oppose	
Urban	143	57	200
Suburban	98	102	200
Rural	13	87	100
Total	254	246	500

SOLUTION

Step 1 **The Set-Up:**

a. **Describe the population parameter of interest.**
The proportion of voters who favor or oppose (that is, the proportion of urban voters who favor, the proportion of suburban voters who favor, the proportion of rural voters who favor, and the proportion of all three groups, separately, who oppose).

b. **State the null hypothesis (H_o) and the alternative hypothesis (H_a).**

H_o: The proportion of voters who favor the proposed legislation is the same in all three residence groups.

H_a: The proportion of voters who favor the proposed legislation is not the same in all three groups. (That is, in at least one group the proportion is different from the others.)

Step 2 **The Hypothesis Test Criteria:**

a. **Check the assumptions.**
The sample information is obtained using three random samples drawn from three separate populations in which each individual is classified according to his or her opinion.

b. **Identify the probability distribution and the test statistic to be used.**
The chi-square distribution and formula (11.1),
with df $= (r - 1)(c - 1) = (3 - 1)(2 - 1) = 2$

c. **Determine the level of significance, α.**
$\alpha = 0.05$.

Step 3 **The Sample Evidence:**

a. **Collect the sample information:** See Table 11.10.

b. **Calculate the value of the test statistic.**
The expected values are found by using formula (11.5) (p. 556) and are given in Table 11.11.

TABLE 11.11 Sample Results and Expected Values

Residence	Governor's Proposal		Total
	Favor	Oppose	
Urban	143 (101.6)	57 (98.4)	200
Suburban	98 (101.6)	102 (98.4)	200
Rural	13 (50.8)	87 (49.2)	100
Total	254	246	500

NOTE Each expected value is used twice in the calculation of $\chi^2\bigstar$; therefore, it is a good idea to keep extra decimal places while doing the calculations.

The calculated chi-square is

$$\chi^2\bigstar = \sum_{\text{all cells}} \frac{(O-E)^2}{E}: \quad \chi^2\bigstar = \frac{(143-101.6)^2}{101.6} + \frac{(57-98.4)^2}{98.4} + \frac{(98-101.6)^2}{101.6}$$

$$+ \frac{(102-98.4)^2}{98.4} + \frac{(13-50.8)^2}{50.8} + \frac{(87-49.2)^2}{49.2}$$

$$= 16.87 + 17.42 + 0.13 + 0.13 + 28.13 + 29.04$$

$$= \mathbf{91.72}$$

Step 4 **The Probability Distribution:**

Using the *p*-Value Procedure

a. Calculate the *p*-value for the test statistic.

Use the right-hand tail because "larger" values of chi-square disagree with the null hypothesis:

$\mathbf{P} = P(\chi^2\bigstar > 91.72 \mid df = 2)$ as shown on the figure.

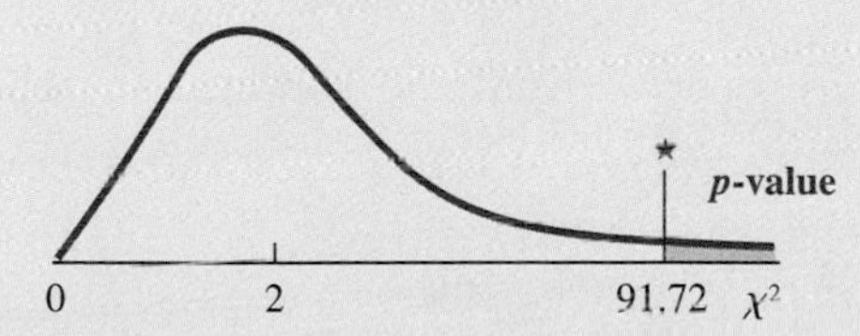

To find the p-value, you have two options:

1. Use Table 8 in Appendix B to place bounds on the p-value: $\mathbf{P < 0.005}$.
2. Use a computer or calculator to find the p-value: $\mathbf{P = 0.000+}$.

For specific instructions, see page 455.

b. Determine whether or not the *p*-value is smaller than α.

The p-value is smaller than α.

OR

Using the Classical Procedure

a. Determine the critical region and critical value(s).

The critical region is the right-hand tail because "larger" values of chi-square disagree with the null hypothesis. The critical value is obtained from Table 8, at the intersection of row df = 2 and column $\alpha = 0.05$:

$$\chi^2(2, 0.05) = \mathbf{5.99}$$

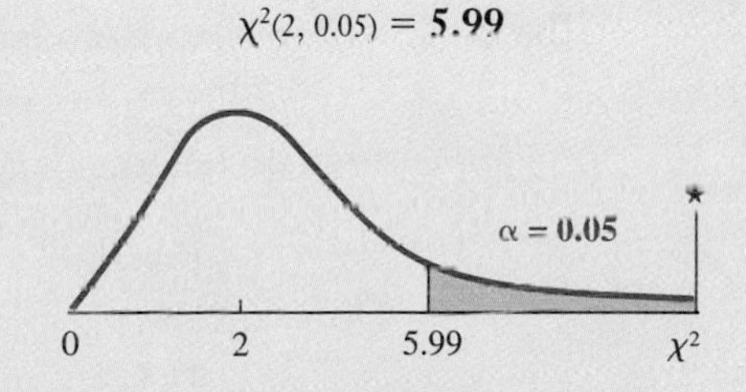

For specific instructions, see page 450.

b. Determine whether or not the calculated test statistic is in the critical region.

$\chi^2\bigstar$ is in the critical region, as shown in **red** on the figure.

Step 5 **The Results:**

a. State the decision about H_o: Reject H_o.

b. State the conclusion about H_a.

The three groups of voters do not all have the same proportions favoring the proposed legislation. ■

Technology Instructions: Hypothesis Test of Independence or Homogeneity

MINITAB (Release 13)

Input each column of observed frequencies from the contingency table into C1, C2, . . . ; then continue with:

```
Choose:   Stat > Tables > Chi-Square Test
Enter:    Columns containing the table: C1 C2
```

```
COMPUTER SOLUTION MINITAB Printout for Illustration 11.4:
     C1       C2
    143       57
     98      102
     13       87
Expected counts are printed below observed counts
          favor    oppose     Total
1           143        57       200
         101.60     98.40
2            98       102       200
         101.60     98.40
3            13        87       100
          50.80     49.20
Total       254       246       500
Chi-Sq = 16.870 + 17.418 +
          0.128 +  0.132 +
         28.127 + 29.041 = 91.715
DF = 2, P-Value = 0.000
```

Excel XP

Input each column of observed frequencies from the contingency table into columns A, B, . . . ; then continue with:

```
Choose:   Tools  > Data Analysis Plus
                 > Contingency Table > OK
Enter:    Input range: (A1:B4 or select cells)
Select:   Labels (if necessary)
Enter:    Alpha: α (ex. 0.05)
```

TI-83 Plus

Input the observed frequencies from the $r \times c$ contingency table into an $r \times c$ matrix A. Set up matrix B as an empty $r \times c$ matrix for the expected frequencies.

```
Choose:   MATRX > EDIT > 1:[A]
Enter:    r > ENTER > c > ENTER
          Each observed frequency with an ENTER afterwards
```

Then continue with:

```
Choose:     MATRX > EDIT > 2[B]
Enter:      r > ENTER > c > ENTER
Choose:     STAT > TESTS > C:χ² Test...
Enter:          Observed: [A] or wherever the contingency
                table is located
                Expected: [B] place for expected frequencies
Highlight:      Calculate > ENTER
```

Application 11.3 Westerners Bake Their Spuds

The USA Snapshot reports the percentage of Americans who prefer to eat baked potatoes by region as well as for the whole country. If the actual number of people in each category were given, we would have a contingency table and we would be able to complete a hypothesis test about the homogeneity of the four regions.

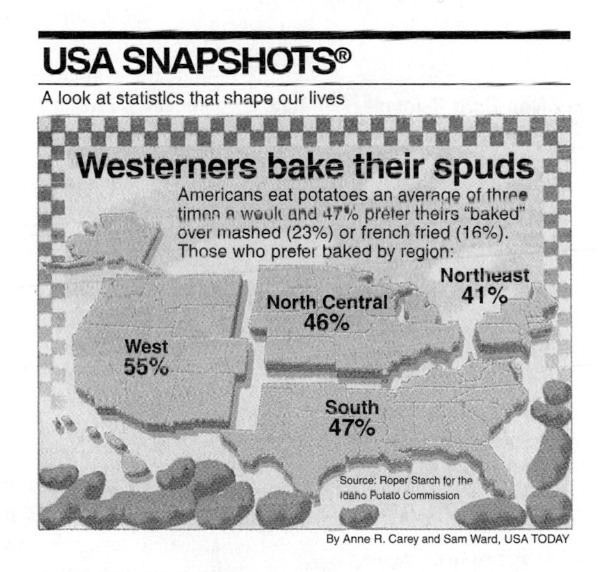

ANSWER NOW 11.23

a. **Express the percentages of Americans who prefer baked by region as a 2×4 contingency table.**

b. **Explain why the following question could be tested using the chi-square statistic: Is the preference for baked the same in all four regions of the United States?**

c. **Explain why this is a test of homogeneity.**

Exercises

11.24 State the null hypothesis H_o and the alternative hypothesis H_a that would be used to test each statement:

a. The voters expressed preferences that were not independent of their party affiliations.
b. The distribution of opinions is the same for all three communities.
c. The proportion of yes responses was the same for all categories surveyed.

11.25 The test of independence and the test of homogeneity are completed in identical fashion, using the contingency table to display and organize the calculations. Explain how these two hypothesis tests differ.

11.26 Many articles have been published about the record number of boat-related manatee deaths. An article in *The News-Press* (December 30, 2001), "Lee sets new manatee death mark," noted that the number of deaths had increased. The table compares two adjacent counties in Florida and the number of boat-related deaths compared to all deaths in 2001.

County	Boat-Related Deaths	Non-Boat-Related Deaths	Total Deaths
Lee	23	25	48
Collier	8	23	31

Is the proportion of boat-related deaths independent of county? Use $\alpha = 0.05$.

11.27 A random sample of 500 married men was taken. Each person was cross-classified as to the size community that he was presently residing in and the size community that he was reared in. The results are listed in the table.

Size of Community Reared in	Size of Community Residing in: Under 10,000	10,000 to 49,999	50,000 or over	Total
Under 10,000	24	45	45	114
10,000 to 49,999	18	64	70	152
50,000 or over	21	54	159	234
Total	63	163	274	500

Does this sample contradict the claim of independence at the 0.01 level of significance?

a. Solve using the p-value approach.
b. Solve using the classical approach.

11.28 A survey of randomly selected travelers who visited the service station restrooms of a large U.S. petroleum distributor showed the following results:

Gender of Respondent	Quality of Restroom Facilities: Above Average	Average	Below Average	Total
Female	7	24	28	59
Male	8	26	7	41
Total	15	50	35	100

Using $\alpha = 0.05$, does the sample present sufficient evidence to reject the hypothesis: "Quality of facilities is independent of gender of the respondent"?

a. Solve using the p-value approach.
b. Solve using the classical approach.

11.29 A survey of employees at an insurance firm was concerned with worker–supervisor relationships. One statement for evaluation was "I am not sure what my supervisor expects." The answers found were as listed in the contingency table.

Years of Employment	Answer: True	Not True	Total
Less than 1 year	18	13	31
1 to 3 years	20	8	28
3 to 10 years	28	9	37
10 years or more	26	8	34
Total	92	38	130

Can we reject the hypothesis: "The responses to the statement and the years of employment are independent," at the 0.10 level of significance?

a. Solve using the p-value approach.
b. Solve using the classical approach.

11.30 Ex11-30 The table is from the July 1993 issue of *Vital and Health Statistics* from the Centers for Disease Control and Prevention/National Center for Health Statistics. The individuals surveyed have an eye irritation, or a nose irritation, or a throat irritation; each person has only one of the three.

Type of Irritation	Age: 18–29	30–44	45–64	65 and over
Eye	440	567	349	59
Nose	924	1311	794	102
Throat	253	311	157	19

Is there sufficient evidence to reject the hypothesis that the type of irritation is independent of age at a level of significance equal to 0.05?

a. Solve using the p-value approach.
b. Solve using the classical approach.

11.31 The manager of an assembly process wants to determine whether the number of defective articles manufactured depends on the day of the week the articles are produced. She collected this information:

	Mon.	Tues.	Wed.	Thurs.	Fri.
Nondefective	85	90	95	95	90
Defective	15	10	5	5	10

Is there sufficient evidence to reject the hypothesis that the number of defective articles is independent of the day of the week on which they are produced? Use $\alpha = 0.05$.

a. Solve using the p-value approach.
b. Solve using the classical approach.

11.32 A study of the Harvard Business School conducted in 1998 concentrated on the career paths and lifestyles of the women who were graduates of the program. The focus was on the class of 1973 and the class of 1983. The class of 1973 was the first to include "a solid number of women." But did things change ten years later? Consider this table:

Class Year	Men	Women	Total
1973	742	34	776
1983	538	189	727
Total	1280	223	1503

Source: *Fortune*, "Tales of the Trailblazers," October 12, 1998

At the 0.01 level of significance, did the distribution of men and women who completed the program change significantly between 1973 and 1983?

a. Solve using the p-value approach.
b. Solve using the classical approach.

11.33 Students use many criteria when selecting courses. "Teacher who is a very easy grader" is often one criterion. Three teachers are scheduled to teach statistics next semester. Here is a sample of the grade distributions given by these three professors:

	Professor		
Grade	1	2	3
A	12	11	27
B	16	29	25
C	35	30	15
Other	27	40	23

At the 0.01 level of significance, is there sufficient evidence to conclude: "The distribution of grades is not the same for all three professors"?

a. Solve using the p-value approach.
b. Solve using the classical approach.
c. Which professor is the easiest grader? Explain, citing specific supporting evidence.

11.34 Fear of darkness is a common emotion. The following data were obtained by asking 200 individuals in each age group whether they had serious fears of darkness. At $\alpha = 0.01$, do we have sufficient evidence to reject the hypothesis: "The same proportion of each age group has serious fears of darkness"? (*Hint:* The contingency table must account for all 1000 people.)

Age Group	Elementary	Jr. High	Sr. High	College	Adult
Number Who Fear Darkness	83	72	49	36	114

a. Solve using the p-value approach.
b. Solve using the classical approach.

11.35 Ex11-35 Every two years the Josephson Institute of Ethics conducts a survey of American teenage ethics. One of the objectives of the study is to determine the propensity of teenagers to lie, cheat, and steal. The 2000 report split the sample of 15,877 respondents into two major student groups for comparison: high school and middle school. The students admitted to performing the various acts during the past year. The table summarizes two portions of the results:

	Took a Gun to School		
	At Least Once	Never	Total
Middle School	663	5,633	6,296
High School	1,265	7,523	8,788
Total	1,928	13,156	15,084

	Hit a Person in Anger		
	At Least Once	Never	Total
Middle School	4,379	1,899	6,278
High School	5,954	2,798	8,752
Total	10,333	4,697	15,030

Source: http://www.josephsoninstitute.org/Survey2000/violence2000-commentary.htm.

Does the sample evidence show that students in high school and middle school have a different tendency to take guns to school? Hit someone because they are angry? Use the 0.01 level of significance in each case.

a. Solve using the p-value approach.
b. Solve using the classical approach.

11.36 Application 11.3 (p. 561) reports percentages describing people's preferences on how their potatoes are prepared. Do you believe there is a significant difference between the four regions of the United States with regard to the percentage who prefer baked? Notice that the article does not mention the sample size.

a. Assume the percentages reported were based on four samples of size 100 from each region and calculate $\chi^2\bigstar$ and its p-value.
b. Repeat part a using sample sizes of 200 and 300.
c. Are the four percentages reported in the USA Snapshot who prefer baked potatoes significantly different? Describe in detail the circumstances for which they are significantly different.

CHAPTER SUMMARY

RETURN TO CHAPTER CASE STUDY

© Dion Ogust/The Image Works

Cooling Your Mouth After a Great Hot Taste

Two hundred adults who professed to love hot, spicy food were asked to name their favorite way to cool their mouth after eating food with hot sauce. Here is the summary of the resulting sample:

Method	Water	Milk	Soda	Beer	Bread	Other	Nothing
Number	73	35	20	19	29	11	13

FIRST THOUGHTS 11.37 CS11

a. What information was collected from each adult in the sample?
b. Define the population and the variable involved in the sample.
c. Construct a graph that you believe best displays the distribution of the data and compares it to the distribution of percentages in "Some like it hot" on page 539.
d. Do the data appear to have a different distribution than the distribution of the percentages?

PUTTING CHAPTER 11 TO WORK 11.38

a. Does the sample show a distribution that is significantly different from the distribution shown in the "Some like it hot" graph (p. 539)? Use $\alpha = 0.05$.
b. Write a paragraph (50+ words) describing why the statistical method used in part a is appropriate for this set of data.
c. Write a paragraph (50+ words) describing the meaning of the assumptions and the results of the above statistical procedure.

YOUR OWN MINI-STUDY 11.39

Design your own study for the favorite way people cool their mouth after eating something hot.

a. Define a specific population that you will sample, describe your sampling plan, and collect a random sample of at least 100 data.
b. Make a descriptive presentation of your sample data, using a chart, at least one graph, and a descriptive paragraph.
c. Does your sample show a distribution that is significantly different from the distribution shown in "Some like it hot" (p. 539)? Use $\alpha = 0.05$.
d. Discuss the differences and similarities between your sample and the distribution shown in "Some like it hot."

In Retrospect

In this chapter we have been concerned with tests of hypotheses using chi-square, with the cell probabilities associated with the multinominal experiment, and with the simple contingency table. In each case the basic assumptions are that a large number of observations have been made and that the resulting test statistic, $\sum \frac{(O-E)^2}{E}$, is approximately distributed as chi-square. In general, if n is large and the minimum allowable expected cell size is 5, then this assumption is satisfied.

The contingency table can be used to test independence and homogeneity. The test for homogeneity and the test for independence look very similar and, in fact, are carried out in exactly the same way. The concepts being tested, however—same distributions and independence—are quite different. The two tests are easily distinguished because the test of homogeneity has predetermined marginal totals in one direction in the table. That is, before the data are collected, the experimenter determines how many subjects will be observed in each category. The only predetermined number in the test of independence is the grand total.

A few words of caution: The correct number of degrees of freedom is critical if the test results are to be meaningful. The degrees of freedom determine, in part, the critical region, and its size is important. As in other tests of hypothesis, failure to reject H_o does not mean outright acceptance of the null hypothesis.

Chapter Exercises

11.40 The psychology department at a certain college claims that the grades in its introductory course are distributed as follows: 10% A's, 20% B's, 40% C's, 20% D's, and 10% F's. In a poll of 200 randomly selected students who had completed this course, it was found that 16 had received A's, 43 B's, 65 C's, 48 D's, and 28 F's. Does this sample contradict the department's claim at the 0.05 level?

a. Solve using the p-value approach.
b. Solve using the classical approach.

11.41 When interbreeding two strains of roses, we expect the hybrid to appear in three genetic classes in the ratio 1:3:4. If the results of an experiment yield 80 hybrids of the first type, 340 of the second type, and 380 of the third type, do we have sufficient evidence to reject the hypothesized genetic ratio at the 0.05 level of significance?

a. Solve using the p-value approach.
b. Solve using the classical approach.

11.42 A sample of 200 individuals are tested for their blood type, and the results are used to test the hypothesized distribution of blood types:

Blood Type	A	B	O	AB
Percent	0.41	0.09	0.46	0.04

The observed results were as follows:

Blood Type	A	B	O	AB
Number	75	20	95	10

At the 0.05 level of significance, is there sufficient evidence to show that the stated distribution is incorrect?

a. Solve using the p-value approach.
b. Solve using the classical approach.

11.43 Ex11-43 As reported in *USA Today* (October 6, 1997) about 8.9 million families sent students to college that year and more than half lived away from home. The table shows where students lived:

Parent or guardian's home	46%
Campus housing	26%
Off-campus rental	18%
Own off-campus housing	9%
Other arrangements	2%

(The total exceeds 100% due to rounding error.) A random sample of 1000 college students resulted in the following information:

Parent or guardian's home	484
Campus housing	230
Off-campus rental	168
Own off-campus housing	96
Other arrangements	22

Is the distribution of this sample significantly different from the distribution reported in the newspaper? Use $\alpha = 0.05$. (To adjust for the rounding error, subtract 2 from each expected frequency.)

a. Solve using the p-value approach.
b. Solve using the classical approach.

11.44 Ex11-44 How often do you review your pay stub to check that the correct taxes are being withheld? *USA Today* (June 8, 1998) reported that American adults check as follows:

Always	53%
Most of the time	12%
Occasionally	14%
Never	10%
Don't get a paycheck	10%
Not sure	1%

A random sample of 650 workers resulted in the following frequencies:

Always	342
Most of the time	94
Occasionally	68
Never	73
Don't get a paycheck	60
Not sure	13

Is the distribution of this sample significantly different from the distribution reported in the newspaper? Use $\alpha = 0.05$.

a. Solve using the p-value approach.
b. Solve using the classical approach.

11.45 Ex11-45 Most golfers are probably happy to play 18 holes of golf whenever they get a chance. In 1998, Ben Winter, a club professional, played 306 holes in one day at a charity golf marathon in Stevens, Pennsylvania. A nationwide survey conducted by *Golf* magazine over the Internet revealed the following frequency distribution of the most holes ever played by the respondents in one day:

Most Holes Played in One Day	Percentage
18	5
19 to 27	12
28 to 36	28
37 to 45	20
46 to 54	18
55 or more	17

Source: *Golf*, "18 Is Not Enough," September 1998

Suppose one of your local public golf courses asks the next 200 golfers who tee off to answer the same question. This table summarizes their responses:

Most Holes Played in One Day	Number
18	12
19 to 27	35
28 to 36	60
37 to 45	44
46 to 54	35
55 or more	14

Does the distribution of "marathon golfers" at your public course differ from the distribution compiled by *Golf* magazine using responses polled on the Internet? Test at the 0.01 level of significance.

a. Solve using the p-value approach.
b. Solve using the classical approach.

11.46 Ex11-46 The weights of 300 adult men were determined and used to test the hypothesis that the weights were normally distributed with a mean of 160 lb and a standard deviation of 15 lb. The data were grouped into the following classes:

Weight (x)	Observed Frequency
$x < 130$	7
$130 \le x < 145$	38
$145 \le x < 160$	100
$160 \le x < 175$	102
$175 \le x < 190$	40
190 and over	13

From the normal tables, the percentages for the classes are 2.28%, 13.59%, 34.13%, 34.13%, 13.59%, and 2.28%, respectively. Do the observed data provide significant reason to discredit the hypothesis that the weights are normally distributed with a mean of 160 lb and a standard deviation of 15 lb? Use $\alpha = 0.05$.

a. Verify the percentages for the classes.
b. Solve using the p-value approach.
c. Solve using the classical approach.

11.47 Ex11-47 The table gives the color counts for a sample of 30 bags (47.9-gram size) of M&M's.

Case	Red	Green	Blue	Orange	Yellow	Brown
1	15	9	3	3	9	19
2	9	17	19	3	3	8
3	14	8	6	8	19	4
4	15	7	3	8	16	8
5	10	3	7	9	22	4
6	12	7	6	5	17	11
7	6	7	3	6	26	10
8	14	11	4	1	14	17
9	4	2	10	6	18	18
10	9	9	3	9	8	15
11	9	11	13	0	7	18
12	8	8	6	5	11	20
13	12	9	13	2	6	13
14	9	7	7	2	18	7
15	6	6	6	4	21	13
16	4	6	9	4	12	20
17	3	5	11	12	11	16
18	14	5	6	6	21	6
19	5	5	16	12	7	12
20	8	9	13	4	15	11
21	8	7	7	13	7	18
22	9	8	3	8	23	8

(continued)

Case	Red	Green	Blue	Orange	Yellow	Brown
23	20	2	7	5	13	9
24	12	6	1	12	6	19
25	8	9	4	6	21	7
26	4	6	7	6	14	19
27	10	12	11	6	11	7
28	5	4	2	9	18	16
29	15	11	4	13	7	8
30	11	6	7	12	12	13

Source: http://www.math.uah.edu/stat/
Christine Nickel and Jason York, ST 687 project, Fall 1998

Before the Global Color Vote of 2002, the target percentages were: brown—30%; red and yellow—20%; blue, green, and orange—10%. Use $\alpha = 0.05$.

a. Does case 1 have a significantly different distribution of colors from the target distribution?
b. Combine cases 1 and 2. Does the total show a significantly different distribution of colors from the target distribution?
c. Combine the results of all 30 cases. Does the total of all 30 show a significantly different distribution of colors from the target distribution?
d. Discuss your findings.

11.48 Ex11-48 Based on data from the U.S. Bureau of the Census, the National Association of Home Builders forecasts a rise in homeownership rates for this decade. Part of the forecast is to predict new housing starts by region. The table shows the forecasts.

Region	Average Housing Starts 1996–2000	2001–2005	2006–2010
Northeast	145	161	170
South	710	687	688
Midwest	331	314	313
West	302	385	373

Do the data present sufficient evidence to reject the hypothesis that the distribution of housing starts across the regions is the same for all years? (Use $\alpha = 0.05$.)

a. Solve using the p-value approach.
b. Solve using the classical approach.

11.49 Does test failure reduce academic aspirations and thereby contribute to the decision to drop out of school? These were the concerns of a study titled "Standards and School Dropouts: A National Study of Tests Required for High School Graduation." The table gives the responses of 283 students selected from schools with low graduation rates to the question "Do tests required for graduation discourage some students from staying in school?"

	Urban	Suburban	Rural	Total
Yes	57	27	47	131
No	23	16	12	51
Unsure	45	25	31	101
Total	125	68	90	283

Source: *American Journal of Education* (November 1989), University of Chicago Press © 1989 by The University of Chicago. All rights reserved.

Does there appear to be a relationship at the 0.05 level of significance between a student's response and the school's location?

a. Solve using the p-value approach.
b. Solve using the classical approach.

11.50 Ex11-50 The table shows the numbers of reported crimes committed last year in the inner part of a large city. Each crime was classified according to type and the district of the inner city where it occurred. Do these data show sufficient evidence to reject the hypothesis that the type of crime and the district in which it occurred are independent? Use $\alpha = 0.01$.

District	Robbery	Assault	Burglary	Larceny	Stolen Vehicle
1	54	331	227	1090	41
2	42	274	220	488	71
3	50	306	206	422	83
4	48	184	148	480	42
5	31	102	94	596	56
6	10	53	92	236	45

a. Solve using the p-value approach.
b. Solve using the classical approach.

11.51 Based on the results of a survey questionnaire, 400 individuals were classified as politically conservative, moderate, or liberal. In addition, each was classified by age, as shown in the table.

	Age 20–35	36–50	Over 50	Total
Conservative	20	40	20	80
Moderate	80	85	45	210
Liberal	40	25	45	110
Total	140	150	110	400

Is there sufficient evidence to reject the hypothesis that political preference is independent of age? Use $\alpha = 0.01$.

a. Solve using the p-value approach.
b. Solve using the classical approach.

11.52 "Cramped quarters" is a common complaint of airline travelers. A random sample of 150 business travelers and 150 leisure travelers were asked where they would most like more space. Their answers are summarized in the table.

Place	Business	Leisure
Overhead space on plane	15	9
Hotel room	29	49
Leg room on plane	91	66
Rental car size	10	20
Other	5	6

Does this sample information present sufficient evidence to conclude that the business traveler and the leisure traveler differ in where they would most like additional space? Use $\alpha = 0.05$.

a. Solve using the p-value approach.
b. Solve using the classical approach.

11.53 Four brands of popcorn were tested for popping. One hundred kernels of each brand were popped, and the number of kernels not popped was recorded in each test (see the table). Can we reject the null hypothesis that all four brands pop equally? Test at $\alpha = 0.05$.

Brand	A	B	C	D
Number Not Popped	14	8	11	15

a. Solve using the p-value approach.
b. Solve using the classical approach.

11.54 Ex11-54 An average of two players per boys' or girls' high school basketball team are injured during a season. The table shows the distribution of injuries for a random sample of 1000 girls and 1000 boys taken from the 1996–1997 season records of all reported injuries.

Injury	Girls	Boys
Ankle/foot	360	383
Hip/thigh/leg	166	147
Knee	130	103
Forearm/wrist/hand	112	115
Face/scalp	88	122
All others	144	130

Does this sample information present sufficient evidence to conclude that the distribution of injuries is different for girls than for boys? Use $\alpha = 0.05$.

a. Solve using the p-value approach.
b. Solve using the classical approach.

11.55 Ex11-55 The Centers for Disease Control and Prevention (CDC) is a federal agency that protects the health and safety of people by collecting information and working closely with various health and community organizations. While they collect data every year, data are not collected on all topics each year. In 1999 CDC collected information about cholesterol awareness, as given in the table. The table also includes data from a survey of 100 nurses who were asked whether they had been told that they had high blood cholesterol.

	Male		Female	
	High	Not High	High	Not High
CDC	29.7%	70.3%	29.7%	70.3%
Nurses	14	20	17	49

Source: CDC Media Relations: Press Release Fact Sheet, September 7, 2000

Is the likeliness of nurses having high cholesterol independent of gender? Use $\alpha = 0.05$.

a. Solve using the p-value approach.
b. Solve using the classical approach.

11.56 Last year's record of absenteeism in each of four categories for 100 randomly selected employees is given in the table. Do these data provide sufficient evidence to reject the hypothesis that the rate of absenteeism is the same for all categories of employees? Use $\alpha = 0.01$ and 240 workdays for the year.

	Married Male	Single Male	Married Female	Single Female
Number of employees	40	14	16	30
Days absent	180	110	75	135

a. Solve using the p-value approach.
b. Solve using the classical approach.

11.57 If you were to roll a die 600 times, how different from 100 could the observed frequencies for each face be before the results would become significantly different from equally likely at the 0.05 level?

11.58 Consider this set of data:

	Response		
	Yes	No	Total
Group 1	75	25	100
Group 2	70	30	100
Total	145	55	200

a. Compute the value of the test statistic $z\bigstar$ that would be used to test the null hypothesis that $p_1 = p_2$, where p_1 and p_2 are the proportions of yes responses in the respective groups.
b. Compute the value of the test statistic $\chi^2\bigstar$ that would be used to test the hypothesis that response is independent of group.
c. Show that $\chi^2\bigstar = (z\bigstar)^2$.

Vocabulary and Key Concepts

Be able to define each term. Pay special attention to the key terms, which are printed in **red.** In addition, describe each term in your own words and give an example of each. Your examples should not be ones given in class or in the textbook. The bracketed numbers indicate the chapters in which the term previously appeared, but you should define the terms again to show increased understanding of their meaning. Page numbers indicate the first appearance of the term in Chapter 11.

assumptions [8, 9, 10] (p. 541)
cell (p. 540)
chi-square [9] (p. 540)
column [3] (p. 556)
contingency table (pp. 552, 556)
degrees of freedom [9, 10] (pp. 541, 556)
enumerative data (p. 540)
expected frequency (pp. 540, 556)
homogeneity (p. 557)
hypothesis test [8, 9, 10] (pp. 540, 543, 552, 557)
independence [4] (p. 552)
marginal totals [3] (p. 553)
multinomial experiment (pp. 542, 544)
observed frequency [2, 4] (pp. 540)
$r \times c$ contingency table (p. 556)
rows [3] (p. 556)
statistic [1, 2, 8] (p. 540)

Chapter Practice Test

PART I: KNOWING THE DEFINITIONS

Answer "True" if the statement is always true. If the statement is not always true, replace the words shown in bold with words that make the statement always true.

11.1 The number of degrees of freedom for a test of a multinomial experiment is **equal to** the number of cells in the experimental data.

11.2 The **expected frequency** in a chi-square test is found by multiplying the hypothesized probability of a cell by the number of pieces of data in the sample.

11.3 The **observed** frequency of a cell should not be allowed to be smaller than 5 when a chi-square test is being conducted.

11.4 In a **multinomial experiment** we have $(r - 1)(c - 1)$ degrees of freedom (r is the number of rows, and c is the number of columns).

11.5 A multinomial experiment consists of ***n* identical independent trials.**

11.6 A **multinomial experiment** arranges the data in a two-way classification such that the totals in one direction are predetermined.

11.7 The charts for both the multinomial experiment and the contingency table **must** be set in such a way that each piece of data will fall into exactly one of the categories.

11.8 The test statistic $\sum\frac{(O - E)^2}{E}$ has a distribution that is **approximately normal.**

11.9 The data used in a chi-square multinomial test are always **enumerative.**

11.10 The null hypothesis being tested by a test of **homogeneity** is that the distribution of proportions is the same for each of the subpopulations.

PART II: APPLYING THE CONCEPTS

Answer all questions. Show formulas, substitutions, and work.

11.11 State the null and alternative hypotheses that would be used to test each of these claims:

a. The single-digit numerals generated by a certain random-number generator were not equally likely.
b. The results of the last election in our city suggest that the votes cast were not independent of the voter's registered party.
c. The distributions of types of crimes committed against society are the same in the four largest U.S. cities.

11.12 Find each value:

a. $\chi^2(12, 0.975)$ b. $\chi^2(17, 0.005)$

11.13 Three hundred consumers were asked to identify which one of three different items they found to be the most appealing. The table shows the number that preferred each item.

Item	1	2	3
Number	85	103	
112			

Do these data present sufficient evidence at the 0.05 level of significance to indicate that the three items are not equally preferred?

11.14 To study the effect of the type of soil on the amount of growth attained by a new hybrid plant, saplings were planted in three different types of soil and their subsequent amounts of growth classified into three categories:

	Soil Type		
Growth	**Clay**	**Sand**	**Loam**
Poor	16	8	14
Average	31	16	21
Good	18	36	25
Total	65	60	60

Does the quality of growth appear to be distributed differently for the tested soil types at the 0.05 level?

a. State the null and alternative hypotheses.
b. Find the expected value for the cell containing 36.
c. Calculate the value of chi-square for these data.
d. Find the p-value.
e. Find the test criteria [level of significance, test statistic, its distribution, critical region, and critical value(s)].
f. State the decision and the conclusion for this hypothesis test.

PART III: UNDERSTANDING THE CONCEPTS

11.15 Explain how a multinomial experiment and a binomial experiment are similar and also how they are different.

11.16 Explain the distinction between a test for independence and a test for homogeneity.

11.17 Student A says that tests for independence and homogeneity are the same, and student B says that they are not at all alike because they are tests of different concepts. Both students are partially right and partially wrong. Explain.

11.18 You are interpreting the results of an opinion poll on the role of recycling in your town. A random sample of 400 people were asked to respond strongly in favor, slightly in favor, neutral, slightly against, or strongly against on each of several questions. There are four key questions that concern you, and you plan to analyze their results.

a. How do you calculate the expected probabilities for each answer?
b. How would you decide whether the four questions were answered the same?

CHAPTER

12 Analysis of Variance

Chapter Outline and Objectives

CHAPTER CASE STUDY

Does your age influence how much time you spend reading the newspaper?

12.1 INTRODUCTION TO THE ANALYSIS OF VARIANCE TECHNIQUE

Analysis of variance (ANOVA) is used to test a hypothesis about **several population means.**

12.2 THE LOGIC BEHIND ANOVA

Between-sample variance and **within-sample variation** are compared in an ANOVA test.

12.3 APPLICATIONS OF SINGLE-FACTOR ANOVA

Notational considerations and a **mathematical model** explaining the composition of each piece of data are presented.

RETURN TO CHAPTER CASE STUDY

We have tested hypotheses about two means. In this chapter we are concerned with testing a hypothesis about several means. The analysis of variance (ANOVA) technique, which we are about to explore, will be used to test a hypothesis about several means—for example,

$$H_o\text{: } \mu_1 = \mu_2 = \mu_3 = \mu_4 = \mu_5$$

By using our former technique for testing hypotheses about two means, we could test several hypotheses if each stated a comparison of two means. For example, we could test

$$\begin{array}{lllll} H_1\text{: } \mu_1 = \mu_2 & H_2\text{: } \mu_1 = \mu_3 & H_3\text{: } \mu_1 = \mu_4 & H_4\text{: } \mu_1 = \mu_5 & H_5\text{: } \mu_2 = \mu_3 \\ H_6\text{: } \mu_2 = \mu_4 & H_7\text{: } \mu_2 = \mu_5 & H_8\text{: } \mu_3 = \mu_4 & H_9\text{: } \mu_3 = \mu_5 & H_{10}\text{: } \mu_4 = \mu_5 \end{array}$$

In order to test the null hypothesis, H_o, that all five means are equal, we would have to test each of these ten hypotheses using our former technique. Rejection of any one of the ten hypotheses about two means would cause us to reject the null hypothesis that all five means are equal. If we failed to reject all ten hypotheses, we would fail to reject the main null hypothesis. By testing in this manner, the overall type I error rate would become much larger than the value of α associated with a single test. The ANOVA techniques allow us to test the null hypothesis (all means are equal) against the alternative hypothesis (at least one mean value is different) with a specified value of α.

In this chapter we introduce ANOVA. ANOVA experiments can be very complex, depending on the situation. We will restrict our discussion to the most basic experimental design, the single-factor ANOVA.

Chapter Case Study

Time Spent Reading the Newspaper

© Kent Meireis/The Image Works

How much time did you spend reading the newspaper yesterday? How much time did your parents spend reading the newspaper yesterday? Does everybody spend the same amount of time? What was the mean amount of time spent reading the newspaper yesterday by people of your age? What was the mean amount of time spent reading the newspaper yesterday by people of your parents' age? Do you think that a person's age has any effect on the amount of time spent reading the newspaper yesterday? The USA Snapshot "Speed readers" that appeared in USA Today on July 14, 1998, seems to suggest that older adults spend more time reading the newspaper than younger adults do.

USA SNAPSHOTS®
A look at statistics that shape the nation
Speed readers
The average adult surveyed who said they read a newspaper the previous day spent about 18 minutes doing so. Minutes spent by age:
18-24
9
25-29
11
11
35-49
16
50-64
21
65-up
33
Source: Pew Research Center for the People and the Press
By Anne R. Carey and Web Bryant, USA TODAY

© 1998 USA Today. Reprinted with permission.

Twenty-five newspaper readers were asked how much time, estimated to the nearest minute, they spent reading yesterday's newspaper. Their times are listed by the age category of the reader.

Age CS12

	18–24	25–29	30–34	35–49	50–64	65 and up
x,	10	13	14	20	20	8
time spent	5	8	7	12	12	30
reading	15	20	20	5	18	35
newspaper	4	10		28	15	20
yesterday				15	30	

ANSWER NOW 12.1

a. **Construct a graphic representation of the data using six side-by-side dotplots. Locate the mean for each age group with an X.**

b. **Does it appear that a person's age has an effect on the average amount of time spent daily reading the newspaper? Explain.**

After completing Chapter 12, further investigate the Chapter Case Study in the Return to Chapter Case Study section with Exercises 12.26, 12.27, and 12.28 (p. 594).

12.1 Introduction to the Analysis of Variance Technique

We will begin our discussion of the analysis of variance technique by looking at an illustration.

Illustration 12.1 Hypothesis Test for Several Means

The temperature at which a manufacturing plant is maintained is believed to affect the rate of production in the plant. The data in Table 12.1 are the number, x, of units produced in one hour for randomly selected one-hour periods when the production process in the plant was operating at each of three temperature *levels*. The data values from repeated samplings are called **replicates.** Four replicates, or data values, were obtained for two of the temperatures and five were obtained for the third temperature. Do these data suggest that temperature has a significant effect on the production level at $\alpha = 0.05$?

TABLE 12.1 Sample Results on Temperature and Production

	Temperature Levels		
	Sample from 68°F ($i = 1$)	Sample from 72°F ($i = 2$)	Sample from 76°F ($i = 3$)
	10	7	3
	12	6	3
	10	7	5
	9	8	4
		7	
Column totals	$C_1 = 41$ $\bar{x}_1 = 10.25$	$C_2 = 35$ $\bar{x}_2 = 7.0$	$C_3 = 15$ $\bar{x}_3 = 3.75$

The level of production is measured by the mean value; $\bar{x}_i$ indicates the observed production mean at level i, where $i = 1$, 2, and 3 corresponds to temperatures of 68°F, 72°F, and 76°F, respectively. There is a certain amount of variation among these means. Since sample means are not necessarily the same when repeated samples are taken from a population, some variation can be expected, even if all three population means are equal. We will next pursue the question: Is this variation among the $\bar{x}$'s due to chance, or is it due to the effect that temperature has on the production rate?

ANSWER NOW 12.2

Draw a dotplot of the data in Table 12.1. Represent the data using integers 1, 2, and 3 to indicate the factor level the data are from. Do you see a difference between the levels?

SOLUTION

STEP 1 The Set-Up:

a. Describe the population parameter of interest.

The "mean" at each *level of the test factor* is of interest: the mean production rate at 68°F, μ_{68}; the mean production rate at 72°F, μ_{72}; and the mean production rate at 76°F, μ_{76}. The factor being tested, plant temperature, has three levels: 68°F, 72°F, or 76°F.

b. State the null hypothesis (H_o) and the alternative hypothesis (H_a).

$$H_o: \mu_{68} = \mu_{72} = \mu_{76}$$

That is, the true production mean is the same at each temperature level tested. In other words, the temperature does not have a significant effect on the production rate. The alternative to the null hypothesis is

H_a: Not all temperature level means are equal.

Thus, we will want to reject the null hypothesis if the data show that one or more of the means are significantly different from the others.

ANSWER NOW 12.3
Each department at the large industrial plant is rated weekly. State the hypotheses used to test: The mean weekly ratings are the same in three departments.

STEP 2 **The Hypothesis Test Criteria:**

a. Check the assumptions.
The data were randomly collected and are independent of each other. The effects due to chance and untested factors are assumed to be normally distributed. (See p. 582 for further discussion.)

b. Identify the probability distribution and the test statistic to be used.
We will make the decision to reject H_o or fail to reject H_o by using the F-distribution and an F-test statistic.

c. Determine the level of significance, α.
$\alpha = 0.05$ (given in the statement of the problem).

STEP 3 **The Sample Evidence:**

a. Collect the sample information: See Table 12.1.

b. Calculate the value of the test statistic.

Recall from Chapter 10 that the calculated value of F is the ratio of two variances. The analysis of variance procedure will separate the variation among the entire set of data into two categories. To accomplish this separation, we first work with the numerator of the fraction used to define **sample variance,** formula (2.6):

$$s^2 = \frac{\sum(x - \bar{x})^2}{n - 1}$$

The numerator of this fraction is called the **sum of squares:**

TOTAL SUM OF SQUARES

$$\text{sum of squares} = \sum(x - \bar{x})^2 \qquad (12.1)$$

We calculate the **total sum of squares, SS(total),** for the total set of data by using a formula that is equivalent to formula (12.1) but does not require the use of $\bar{x}$. This equivalent formula is

SHORTCUT FOR TOTAL SUM OF SQUARES

$$\text{SS(total)} = \sum(x^2) - \frac{(\sum x)^2}{n} \tag{12.2}$$

Now we can find SS(total) for our illustration by using formula (12.2). First,

$$\sum(x^2) = 10^2 + 12^2 + 10^2 + 9^2 + 7^2 + 6^2 + 7^2 + 8^2 + 7^2 + 3^2 + 3^2 + 5^2 + 4^2 = 731$$

$$\sum x = 10 + 12 + 10 + 9 + 7 + 6 + 7 + 8 + 7 + 3 + 3 + 5 + 4 = 91$$

Then, using formula (12.2), we have

$$\text{SS(total)} = \sum(x^2) - \frac{(\sum x)^2}{n}: \qquad \text{SS(total)} = 731 - \frac{(91)^2}{13} = 731 - 637 = \mathbf{94}$$

Next, 94, SS(total), must be separated into two parts: the sum of squares due to temperature levels, SS(temperature), and the sum of squares due to experimental error of replication, SS(error). This splitting is often called **partitioning,** since SS(temperature) + SS(error) = SS(total); that is, in our illustration SS(temperature) + SS(error) = 94. The sum of squares, **SS(factor)** [SS(temperature) for our illustration], that measures the **variation between the factor levels** (temperatures) is found by using formula (12.3):

SUM OF SQUARES DUE TO FACTOR

$$\text{SS(factor)} = \left(\frac{C_1^2}{k_1} + \frac{C_2^2}{k_2} + \frac{C_3^2}{k_3} + \cdots\right) - \frac{(\sum x)^2}{n} \tag{12.3}$$

where C_i represents the column total, k_i represents the number of replicates at each level of the factor, and n represents the total sample size ($n = \sum k_i$).

NOTE The data have been arranged so that each column represents a different level of the factor being tested.

Now we can find SS(temperature) for our illustration by using formula (12.3):

$$\text{SS(factor)} = \left(\frac{C_1^2}{k_1} + \frac{C_2^2}{k_2} + \frac{C_3^2}{k_3} + \cdots\right) - \frac{(\sum x)^2}{n}:$$

$$\text{SS(temperature)} = \left(\frac{41^2}{4} + \frac{35^2}{5} + \frac{15^2}{4}\right) - \frac{(91)^2}{13}$$

$$= (420.25 + 245.00 + 56.25) - 637.0 = 721.5 - 637.0 = \mathbf{84.5}$$

The sum of squares, **SS(error),** that measures the **variation within the rows** is found by using formula (12.4):

SUM OF SQUARES DUE TO ERROR

$$\text{SS(error)} = \sum(x^2) - \left(\frac{C_1^2}{k_1} + \frac{C_2^2}{k_2} + \frac{C_3^2}{k_3} + \cdots\right) \tag{12.4}$$

The SS(error) for our illustration can now be found. First,

$$\sum(x^2) = 731 \quad \text{(found previously)}$$

$$\left(\frac{C_1^2}{k_1} + \frac{C_2^2}{k_2} + \frac{C_3^2}{k_3} + \cdots\right) = 721.5 \quad \text{(found previously)}$$

Then, using formula (12.4), we have

$$\text{SS(error)} = \sum(x^2) - \left(\frac{C_1^2}{k_1} + \frac{C_2^2}{k_2} + \frac{C_3^2}{k_3} + \cdots\right): \quad \text{SS(error)} = 731.0 - 721.5 = \mathbf{9.5}$$

NOTE SS(total) = SS(factor) + SS(error). Inspection of formulas (12.2), (12.3), and (12.4) will verify this.

For convenience we will use an ANOVA table to record the sums of squares and to organize the rest of the calculations. The format of an ANOVA table is shown in Table 12.2.

TABLE 12.2 Format for ANOVA Table

Source	df	SS	MS
Factor		84.5	
Error		9.5	
Total		94.0	

We have calculated the three sums of squares for our illustration. The degrees of freedom, df, associated with each of the three sources are determined as follows:

1. df(factor) is 1 less than the number of levels (columns) for which the factor is tested:

DEGREES OF FREEDOM FOR FACTOR

$$\text{df(factor)} = c - 1 \tag{12.5}$$

where c is the number of *levels for which the factor is being tested* (number of columns on the data table).

2. df(total) is 1 less than the total number of data:

DEGREES OF FREEDOM FOR TOTAL

$$\text{df(total)} = n - 1 \tag{12.6}$$

where n is the number of data in the total sample (that is, $n = k_1 + k_2 + k_3 + \cdots$, where k_i is the number of replicates at each level tested).

3. df(error) is the sum of the degrees of freedom for all the levels tested (columns in the data table). Each column has $k_i - 1$ degrees of freedom; therefore,

$$\text{df(error)} = (k_1 - 1) + (k_2 - 1) + (k_3 - 1) + \cdots$$

ANSWER NOW 12.6
Consider the following table for a single-factor ANOVA. Find these values:

a. $x_{1,2}$ b. $x_{2,1}$ c. C_1 d. $\sum x$ e. $\sum (C_i)^2$

	Level of Factor		
Replicates	1	2	3
1	3	2	7
2	0	5	4
3	1	4	5

A **mathematical model** (equation) is often used to express a particular situation. In Chapter 3 we used a mathematical model to help explain the relationship between the values of bivariate data. The equation $\hat{y} = b_0 + b_1x$ served as the model when we believed that a straight-line relationship existed. The probability functions studied in Chapter 5 are also examples of mathematical models. For the single-factor ANOVA, the mathematical model, formula (12.13), is an expression of the composition of each piece of data entered in our data table:

$$x_{c,k} = \mu + F_c + \epsilon_{k(c)} \tag{12.13}$$

We interpret each term of this model as follows:

$x_{c,k}$ is the value of the variable at the kth replicate of level c.
μ is the mean value for all the data without respect to the test factor.
F_c is the effect that the factor being tested has on the response variable at each different level c.
$\epsilon_{k(c)}$ (ϵ is the lowercase Greek letter epsilon) is the *experimental error* that occurs among the k replicates in each of the c columns.

Let's look at another hypothesis test using an analysis of variance.

Illustration 12.4 Hypothesis Test for the Equality of Several Means

TABLE 12.7

Sample Results on Target Shooting

Ta12-7

Method of Sighting		
Right Eye	Left Eye	Both Eyes
12	10	16
10	17	14
18	16	16
12	13	11
14		20
		21

A rifle club performed an experiment on a randomly selected group of first-time shooters. The purpose of the experiment was to determine whether shooting accuracy is affected by the method of sighting used: only the right eye open, only the left eye open, or both eyes open. Fifteen first-time shooters were selected and divided into three groups. Each group experienced the same training and practicing procedures with one exception: the method of sighting used. After completing training, each student was given the same number of rounds and asked to shoot at a target. Their scores are listed in Table 12.7.

At the 0.05 level of significance, is there sufficient evidence to reject the claim that the three methods of sighting are equally effective?

SOLUTION
In this experiment the factor is method of sighting and the levels are the three different methods of sighting (right eye, left eye, and both eyes open). The replicates are the scores received by the students in each group. The null hypothesis to be tested is: the three methods of sighting are equally effective, or the mean scores attained using each of the three methods are the same.

Step 1 **The Set-Up:**

a. **Describe the population parameter of interest.**
The "mean" at each level of the test factor is of interest: the mean score using the right eye, μ_R; the mean score using the left eye, μ_L; and the mean score using both eyes, μ_B. The factor being tested, "method of sighting," has three levels: right, left, and both.

b. **State the null hypothesis (H_o) and the alternative hypothesis (H_a).**
H_o: $\mu_R = \mu_L = \mu_B$
H_a: The means are not all equal (that is, at least one mean is different).

Step 2 **The Hypothesis Test Criteria:**

a. **Check the assumptions.**
The shooters were randomly assigned to the method, and their scores are independent of each other. The effects due to chance and untested factors are assumed to be normally distributed.

b. **Identify the probability distribution and the test statistic to be used.**
The F-distribution and formula (12.12) will be used with df(numerator) = df(method) = 2 and df(denominator) = df(error) = 12.

c. **Determine the level of significance, α.**
$\alpha = 0.05$.

Step 3 **The Sample Evidence:**

a. **Collect the sample information.** See Table 12.8.

b. **Calculate the value of the test statistic.**
The test statistic is $F\bigstar$: Table 12.8 is used to find the column totals.

TABLE 12.8 Sample Results for Target Shooting

	Factor Levels: Method of Sighting		
Replicates	Right Eye	Left Eye	Both Eyes
$k = 1$	12	10	16
$k - 2$	10	17	14
$k = 3$	18	16	16
$k = 4$	12	13	11
$k = 5$	14		20
$k = 6$			21
Totals	$C_R = 66$	$C_L = 56$	$C_B = 98$

First, the summations $\sum x$ and $\sum x^2$ need to be calculated:

$$\sum x = 12 + 10 + 18 + 12 + 14 + 10 + 17 + \cdots + 21 = \mathbf{220}$$
$$(\text{Or } 66 + 56 + 98 = 220 \text{ ck})$$

$$\sum x^2 = 12^2 + 10^2 + 18^2 + 12^2 + 14^2 + 10^2 + \cdots + 21^2 = \mathbf{3392}$$

Using formula (12.2), we find

$$\text{SS(total)} = \sum(x^2) - \frac{(\sum x)^2}{n}: \quad \text{SS(total)} = 3392 - \frac{(220)^2}{15}$$

$$= 3392 - 3226.67 = \mathbf{165.33}$$

Using formula (12.3), we find

$$\text{SS(method)} = \left(\frac{C_1^2}{k_1} + \frac{C_2^2}{k_2} + \frac{C_3^2}{k_3} + \cdots\right) - \frac{(\sum x)^2}{n}:$$

$$\text{SS(method)} = \left(\frac{66^2}{5} + \frac{56^2}{4} + \frac{98^2}{6}\right) - \frac{(220)^2}{15}$$

$$= (871.2 + 784 + 1600.67) - 3226.67 = 3255.87 - 3226.67 = \mathbf{29.20}$$

To find SS(error) we need first:

$$\sum(x^2) = 3392 \quad \text{(found previously)}$$

$$\left(\frac{C_1^2}{k_1} + \frac{C_2^2}{k_2} + \frac{C_3^2}{k_3} + \cdots\right) = 3255.87 \quad \text{(found previously)}$$

Then using formula (12.4), we have

$$\text{SS(error)} = \sum(x^2) - \left(\frac{C_1^2}{k_1} + \frac{C_2^2}{k_2} + \frac{C_3^2}{k_3} + \cdots\right):$$

$$\text{SS(error)} = 3392 - 3255.87 = \mathbf{136.13}$$

We use formula (12.8) to check the sum of squares:

$$\text{SS(method)} + \text{SS(error)} = \text{SS(total)}: \quad 29.20 + 136.13 = 165.33$$

The degrees of freedom are found using formulas (12.5), (12.6), and (12.7):

$$\text{df(method)} = c - 1 = 3 - 1 = \mathbf{2}$$

$$\text{df(total)} = n - 1 = 15 - 1 = \mathbf{14}$$

$$\text{df(error)} = n - c = 15 - 3 = \mathbf{12}$$

Using formulas (12.10) and (12.11), we find

$$\text{MS(method)} = \frac{\text{SS(method)}}{\text{df(method)}}: \quad \text{MS(method)} = \frac{29.20}{2} = \mathbf{14.60}$$

$$\text{MS(error)} = \frac{\text{SS(error)}}{\text{df(error)}}: \quad \text{MS(error)} = \frac{136.13}{12} = \mathbf{11.34}$$

The results of these computations are recorded in the ANOVA table in Table 12.9.

TABLE 12.9 ANOVA Table for Illustration 12.4

Source	df	SS	MS
Method	2	29.20	14.60
Error	12	136.13	11.34
Total	14	165.33	

The calculated value of the test statistic is then found using formula (12.12):

$$F\bigstar = \frac{MS(\text{factor})}{MS(\text{error})}: \qquad F\bigstar = \frac{MS(\text{method})}{MS(\text{error})} = \frac{14.60}{11.34} = \mathbf{1.287}$$

Step 4 **The Probability Distribution:**

Using the p-Value Procedure

a. Calculate the p-value for the test statistic.

Use the right-hand tail: **P** = $P(F\bigstar > 1.287$, with $df_n = 2$ and $df_d = 12)$ as shown on the figure.

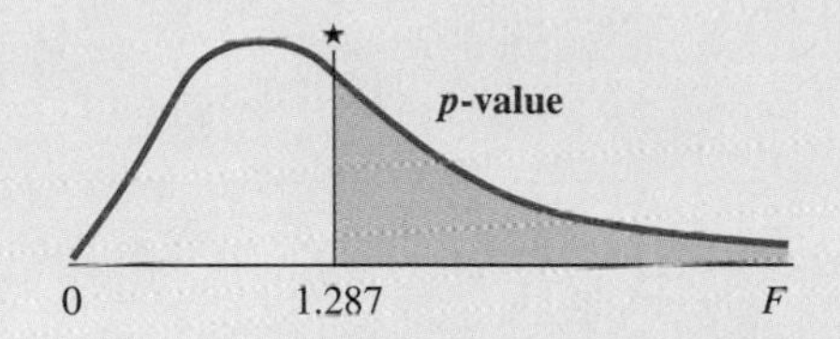

To find the p-value, you have two options:

1. Use Table 9a in Appendix B to place bounds on the p-value: **P > 0.05.**
2. Use a computer or calculator to find the p-value: **P = 0.312.**

For additional instructions, see pages 520–521.

b. Determine whether or not the p-value is smaller than α.

The p-value is not smaller than the level of significance, α (0.05).

OR

Using the Classical Procedure

a. Determine the critical region and critical value(s).

The critical region is the right-hand tail; the critical value is obtained from Table 9a:

$$F(2, 12, 0.05) = 3.89$$

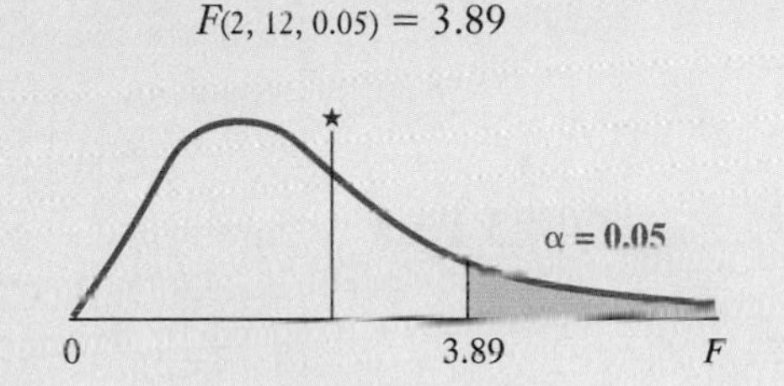

For additional instructions, see page 517.

b. Determine whether or not the calculated test statistic is in the critical region.

$\chi^2\bigstar$ is not in the critical region, as shown in **red** on the figure.

Step 5 **The Results:**

a. State the decision about H_o: Fail to reject H_o.

b. State the conclusion about H_a.

The data show no evidence to reject the null hypothesis that the three methods are equally effective. ■

Technology Instructions: One-Way Analysis of Variance

MINITAB (Release 13)

Input the data for each level into columns C1, C2, . . . ; then continue with:

```
Choose:   Stat > ANOVA > Oneway (unstacked)
Enter:    Responses: C1 C2 ...*
```

OR

Input all of the data into C1 with the corresponding levels of factors into C2; then continue with:

```
Choose:   Stat > ANOVA > Oneway
Enter:    Response: C1
          Factor: C2*
```

*Optional for either method:

Choose: **Graphs...**

Select: **Dotplots of data** and/or **Boxplots of data > OK**

Technology Instructions: One-Way Analysis of Variance (Continued)

Excel XP

Input the data for each level into columns A, B, . . . ; then continue with:

```
Choose:   Tools > Data Analysis > Anova: Single Factor
Enter:    Input Range: (A1:C4 or select cells)
Select:   Grouped By: Columns
          Labels in First Row (if necessary)
Enter:    Alpha: α
Select:   Output Range:
Enter:    (D1 or select cell)
```

To make the output more readable, continue with: Format > Column > Autofit Selection.

TI-83 Plus

Input the data for each level into lists L1, L2, . . . ; then continue with:

```
Choose:   STAT > TESTS > F:ANOVA(
Enter:    L1, L2, ...)
```

NOTE Side-by-side dotplots are very useful in visualizing the within-sample variation, the between-sample variation, and the relationship between them. Commands for side-by-side dotplots can be found in Chapter 2, p. 47–49, 136.

Computer Solution MINITAB Printout for Illustration 12.4:

Information given to computer →

Row	Right eye	Left eye	Both eyes
1	12	10	16
2	10	17	14
3	18	16	16
4	12	13	11
5	14		20
6			21

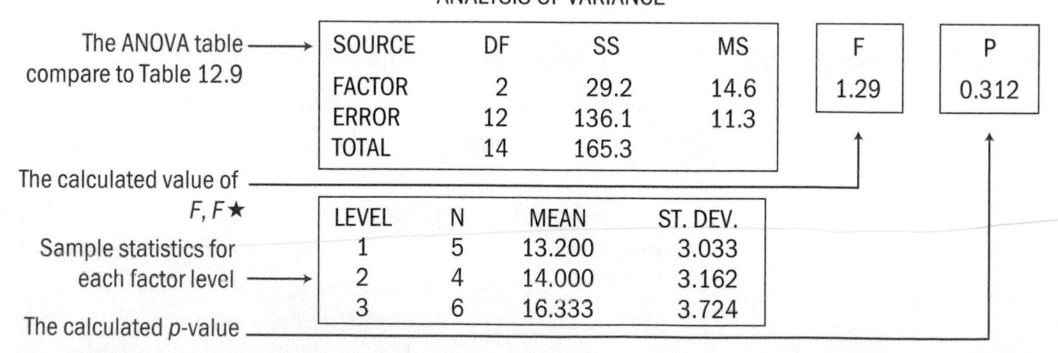

ANALYSIS OF VARIANCE

SOURCE	DF	SS	MS	F	P
FACTOR	2	29.2	14.6	1.29	0.312
ERROR	12	136.1	11.3		
TOTAL	14	165.3			

LEVEL	N	MEAN	ST. DEV.
1	5	13.200	3.033
2	4	14.000	3.162
3	6	16.333	3.724

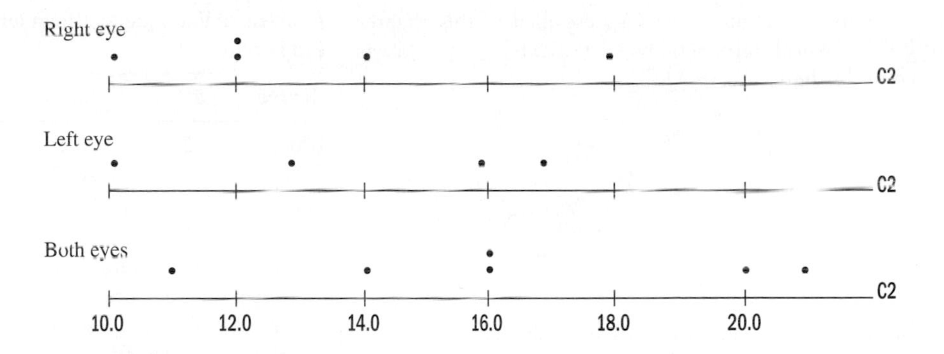

Recall the null hypothesis: There is no difference between the levels of the factor being tested. A "fail to reject H_o" decision must be interpreted as the conclusion that there is no evidence of a difference due to the levels of the tested factor, whereas the rejection of H_o implies that there is a difference between the levels. That is, at least one level is different from the others. If there is a difference, the next problem is to locate the level or levels that are different. Locating this difference may be the main objective of the analysis. In order to find the difference, the only method that is appropriate at this stage is to inspect the data. It may be obvious which level(s) caused the rejection of H_o. In Illustration 12.1 it seems quite obvious that at least one of the levels [level 1 (68°F) or level 3 (76°F) because they have the largest and the smallest sample means] is different from the other two. If the higher values are more desirable for finding the "best" level to use, we would choose that corresponding level of the factor.

Thus far we have discussed analysis of variance for data dealing with one factor. It is not unusual for problems to have several factors of interest. The ANOVA techniques presented in this chapter can be developed further and applied to more complex cases.

Exercises

12.7 State the null hypothesis H_o and the alternative hypothesis H_a that would be used to test these statements:

a. The mean value of x is the same at all five levels of the experiment.
b. The scores are the same at all four locations.
c. The four levels of the test factor do not significantly affect the data.
d. The three different methods of treatment do affect the variable.

12.8 Find the p-value for each situation:

a. $F\bigstar = 3.852$, df(factor) = 3, df(error) = 12
b. $F\bigstar = 4.152$, df(factor) = 5, df(error) = 18
c. $F\bigstar = 4.572$, df(factor) = 5, df(error) = 22

12.9 Determine the critical region(s) and critical value(s) that would be used to test, using the classical approach, the null hypothesis for these multinomial experiments:

a. H_o: $\mu_1 = \mu_2 = \mu_3 = \mu_4$, with $n = 18$ and $\alpha = 0.05$
b. H_o: $\mu_1 = \mu_2 = \mu_3 = \mu_4 = \mu_5$, with $n = 15$ and $\alpha = 0.01$
c. H_o: $\mu_1 = \mu_2 = \mu_3$, with $n = 25$ and $\alpha = 0.05$

12.10 Why does df(factor), the number of degrees of freedom associated with the factor, always appear first in the critical value notation F[df(factor), df(error), α]?

12.11 Suppose that an F-test (as described in this chapter using the p-value approach) has a p-value of 0.04.

a. What is the interpretation of p-value = 0.04?
b. What is the interpretation of the situation if you had previously decided on a 0.05 level of significance?
c. What is the interpretation of the situation if you had previously decided on a 0.02 level of significance?

12.12 Suppose that an F-test (as described in this chapter using the classical approach) has a critical value of 2.2, as shown in this figure:

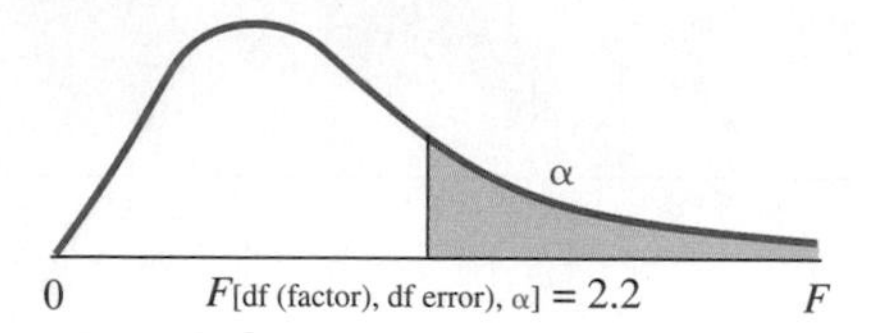

a. What is the interpretation of a calculated value of F larger than 2.2?
b. What is the interpretation of a calculated value of F smaller than 2.2?
c. What is the interpretation if the calculated F were 0.1? 0.01?

12.13
a. State the null hypothesis, in a general form, for the one-way ANOVA.
b. State the alternative hypothesis, in a general form, for the one-way ANOVA.
c. What must happen in order to "reject H_o"? (If using the p-value approach. The classical approach.)
d. How would a decision of "reject H_o" be interpreted?
e. What must happen in order to "fail to reject H_o"? (If using the p-value approach. The classical approach.)
f. How would a decision of "fail to reject H_o" be interpreted?

12.14 The table of data is to be used for single-factor ANOVA. Find each of the following:
a. $x_{3,2}$ b. $x_{4,3}$ c. C_3 d. $\sum x$ e. $\sum(C_i)^2$

	Level of Factor			
Replicates	**1**	**2**	**3**	**4**
1	13	12	16	14
2	17	8	18	11
3	9	15	10	19

12.15 The article "An Investigation of High School Preparation as Predictors of the Cultural Literacy of Developmental, Nondevelopmental and ESL College Students" (*RTDE*, Fall 1990) reported on a study that examined the cultural literacy of developmental, nondevelopmental, and ESL (English as a second language) college freshmen.

Analysis of Variance by Group for Total Score

Source	df	SS	MS	F	P
Group	2	4062.06	2031.03	14.49	0.0001
Error	117	16394.53	140.12		
Total	119	20456.59			

Analysis of Variance by Group for Foreign Language Preparation

Source	df	SS	MS	F	P
Group	2	0.95	0.475	1.93	0.1493
Error	117	28.75	0.246		
Total	119	29.70			

a. How many student scores were in the samples?
b. The students were divided into how many groups?
c. Given the sum of squares (SS) and degree of freedom (df) values, verify the mean square (MS), the calculated F-value, and the p-value for each figure.
d. Do the statistics in the first table show that the total scores were different for the groups involved? Explain.
e. Do the statistics in the second table show that the foreign language preparation scores were different for the groups involved? Explain.

12.16 An article titled "The Effectiveness of Biofeedback and Home Relaxation Training on Reduction of Borderline Hypertension" (*Health Education*, October/November 1988) compared different methods of reducing blood pressure. Biofeedback ($n = 13$ subjects), biofeedback/relaxation ($n = 15$), and relaxation ($n = 14$) were the three methods compared. There were no differences among the three groups on pretest diastolic or systolic blood pressures. There was a significant posttest difference between groups on the systolic measure, $F(2, 39) = 4.14$, $\mathbf{P} < 0.025$, and the diastolic measure, $F(2, 39) = 5.56$, $\mathbf{P} < 0.008$.
a. Verify that df(method) = 2 and df(error) = 39.
b. Use Tables 9a, 9b, and 9c in Appendix B to verify that for systolic, $\mathbf{P} < 0.025$, and for diastolic, $\mathbf{P} < 0.008$.

12.17 Ex12-17 A new operator was recently assigned to a crew of workers who perform a certain job. From the records of the number of units of work completed by each worker each day last month, a sample of size five was randomly selected for each of the two experienced workers and the new worker. At the 0.05 level of significance, does the evidence provide sufficient reason to reject the claim that there is no difference in the amount of work done by the three workers?

	Workers		
	New	**A**	**B**
Units of work (replicates)	8	11	10
	10	12	13
	9	10	9
	11	12	12
	8	13	13

a. Solve using the p-value approach.
b. Solve using the classical approach.

12.18 Ex12-18 An employment agency wants to see which of three types of ads in the help-wanted section of local newspapers is the most effective. Three types of ads (big headline, straightforward, and bold print) were randomly alternated over several weeks, and the numbers of people responding to the ads were noted each week. Do these data support the null hypothesis that there is no difference in the effectiveness of the ads, as measured by the mean number responding, at the 0.01 level of significance?

	Type of Advertisement		
	Big Headline	Straightforward	Bold Print
Number of responses (replicates)	23	19	28
	42	31	33
	36	18	46
	48	24	29
	33	26	34
	26		34

a. Solve using the p-value approach.
b. Solve using the classical approach.

12.19 Ex12-19 Cities across the United States have restaurants that offer themes associated with foreign countries. The U.S. embargo of Cuba keeps most Americans away from that nation, but domestic Cuban restaurants serve authentic samples of the island's food and beverages. The following ratings, based on three categorical judgments of food quality, décor, and service, were collected from different restaurants in different cities. The ratings were made on the same scale from 0 to 30 (the higher, the better).

Restaurant Rating Category		
Food Quality	Décor	Service
20	16	18
21	12	16
19	18	18
24	21	21
20	17	19
22	18	21

Source: *Fortune*, "Feed Your Face Like Fidel," July 6, 1998, p. 32

Is there any significant difference in the ratings given to the Cuban restaurants in each category? Construct a one-way ANOVA table and test for the difference at the 0.05 level of significance.
a. Solve using the p-value approach.
b. Solve using the classical approach.

12.20 Ex12-20 Thirty-nine counties from the six-state Upper Midwest were randomly selected from the USDA-NASS Web site, and the following 2001 oat-production yield per acre data were obtained:

County	IA	MN	ND	NE	SD	WI
1	76.2	53.0	71.4	60.0	76.5	52.0
2	65.3	70.0	64.3	37.0	50.0	53.0
3	86.0	71.0	66.7	53.0	42.0	72.0
4	73.6	54.0	61.4	50.0	62.5	81.0
5	61.3	64.0	66.0	56.0	55.7	57.0
6	74.3	40.0		58.0	59.1	64.0
7	58.3				59.3	
8	56.0					
9	61.4					

Source: http://www.usda.gov/nass/graphics/county01/data/ot01.csv

a. Do the data show a significant difference in the mean yield rates for the six states? Use $\alpha = 0.05$.
b. Draw a graph that demonstrates the results found in part a.
c. Explain the meaning of the results, including an explanation of how the graph portrays the results.

12.21 Ex12-21 A number of sports enthusiasts have argued that major league baseball players from teams in the Central Division have an unfair advantage over coastal players in the Western and Eastern Divisions. This is because the impact of time differences is likely to be greater when playing on the road (i.e., games away from home). Players from teams on the coasts could gain (going west) or lose (going east) up to three hours, whereas Central Division players would seldom gain or lose more than one hour. The following data are the won/loss percentages for games played on the road for all three divisions at the end of the 1997 season:

Major League Division		
Eastern	Central	Western
63.0	46.9	51.9
49.4	44.4	50.6
46.9	44.4	44.4
40.7	39.5	45.7
37.0	32.1	55.5
64.2	51.9	46.9
60.5	43.2	46.9
45.7	38.3	37.0
48.1	40.7	
42.0	42.0	

Source: *The World Almanac and Book of Facts 1998*, "Sports—Baseball," pp. 943, 946

Complete an ANOVA table for won/loss percentages by teams in each division. Test the null hypothesis that when teams play on the road, the mean won/loss percentages are the same for the three divisions. Use the 0.05 level of significance.

a. Solve using the p-value approach.
b. Solve using the classical approach.

12.22 Ex12-22 A study was conducted to assess the effectiveness of treating vertigo (motion sickness) with the transdermal therapeutic system (TTS, a patch worn on the skin). Two other treatments were both oral: one pill containing a drug and one a placebo. The age and the gender (m or f) of the patients for each treatment are listed in the table.

TTS		Antivert		Placebo	
47-f	53-m	51-f	43-f	67-f	38-m
41-f	58-f	53-f	56-f	52-m	59-m
63-m	62-f	27-m	48-m	47-m	33-f
59-f	34-f	29-f	52-f	35-f	32-f
62-f	47-f	31-f	19-f	37-f	26-f
24-m	35-f	25-f	31-f	40-f	37-m
43-m	34-f	52-f	48-f	31-f	49-f
20-m	63-m	55-f	53-m	45-f	49-m
55-f	46-f	32-f	63-m	41-f	38-f
		51-f	54-m	49-m	
		21-f			

Is there a significant difference between the mean ages of the three test groups? Use $\alpha = 0.05$. Use a computer or calculator to complete this exercise.

a. Solve using the p-value approach.
b. Solve using the classical approach.

12.23 Ex12-23 Snacks are a concern of people who are trying not to eat too much between meals. "The trick is to give your body what it's asking for, but without sacrificing good nutrition," according to an article written by a nutritionist for *Woman's Day*. The author assembled a list of 50 snacks that contain 100 calories or less, together with the fat content. The list was then divided into four categories: Crunchy Choices, Salty Sensations, Creamy Concoctions, and Sinless Sweets. The table below summarizes the nutritional data.

Complete an ANOVA table for calories and another table for fat content. Test the null hypotheses that the calorie content and the fat content are the same for each of the four categories of snacks. Use the 0.01 level of significance.

a. Solve using the p-value approach.
b. Solve using the classical approach.

Table for Exercise 12.23

Crunchy Choices		Salty Sensations		Creamy Concoctions		Sinless Sweets	
Calories	Fat (grams)	Calories	Fat (grams)	Calories	Fat (grams)	Calories	Fat (grams)
89	1.5	100	0.0	99	5.0	90	1.0
99	1.3	32	0.0	97	0.2	99	0.5
91	0.5	60	0.0	65	5.0	94	0.0
76	0.6	65	7.0	79	2.0	91	1.0
90	2.0	100	1.3	99	0.0	100	0.0
90	2.0	70	0.8	90	0.0	54	2.5
97	8.5	5	0.0	90	1.0	87	0.5
52	0.9	14	0.1	100	5.0	88	0.0
93	5.0	52	0.4	50	0.0	96	1.5
80	1.5	59	2.8	94	8.0	50	0.4
82	0.5			100	1.5	70	1.0
90	1.0			70	0.5	90	0.0
						75	0.4
						100	0.3
						70	0.0
						88	3.0

Source: *Woman's Day*, "50 Snacks Under 100 Calories," November 1, 1998

12.24 Ex12-24 Albert Michelson, the first American citizen to be awarded a Nobel Prize in physics, conducted many experiments to determine the velocity of light in air. The table shows five trials of 20 measurements each taken by Michelson from June 5 to July 2, 1879. The measurements have had 299,000 subtracted from them.

Trial 1		Trial 2		Trial 3		Trial 4		Trial 5	
850	1000	960	830	880	880	890	910	890	870
740	980	940	790	880	910	810	920	840	870
900	930	960	810	880	850	810	890	780	810
1070	650	940	880	860	870	820	860	810	740
930	760	880	880	720	840	800	880	760	810
850	810	800	830	720	840	770	720	810	940
950	1000	850	800	620	850	760	840	790	950
980	1000	880	790	860	840	740	850	810	800
980	960	900	760	070	840	750	850	820	810
880	960	840	800	950	840	760	780	850	870

Source: http://lib.stat.cmu.edu/DASL/Stories/SpeedofLight.html

a. Construct a boxplot showing the five trials side by side. What does this graph suggest?

b. Using a one-way ANOVA, test the claim that the five population means are not all the same. Use a level of significance of 0.05.

c. What conclusions can you draw from the results of the hypothesis test?

12.25 Ex12-25 The Bureau of Labor Statistics posts a table of average hourly wages paid to production or nonsupervisory workers on private nonfarm payrolls by major industry. The following average hourly earnings were taken from the Web site on October 15, 2002.

a. At the 0.05 level of significance, does the evidence provide sufficient reason to reject the claim that there is no difference in the mean hourly wages paid by month? Show graphic evidence to visually support your conclusion.

b. At the 0.05 level of significance, does the evidence provide sufficient reason to reject the claim that there is no difference in the mean hourly wages paid each year? Show graphic evidence to visually support your conclusion.

Table for Exercise 12.25

Year	Jan.	Feb.	Mar.	Apr.	May	June	July	Aug.	Sep.	Oct.	Nov.	Dec.
1992	10.45	10.47	10.50	10.52	10.54	10.57	10.59	10.62	10.62	10.65	10.67	10.69
1993	10.72	10.73	10.78	10.78	10.80	10.82	10.84	10.86	10.89	10.91	10.93	10.96
1994	10.99	11.02	11.03	11.05	11.07	11.09	11.12	11.14	11.17	11.21	11.22	11.25
1995	11.27	11.30	11.33	11.35	11.37	11.41	11.45	11.47	11.50	11.54	11.56	11.59
1996	11.63	11.65	11.67	11.72	11.74	11.81	11.83	11.86	11.90	11.93	11.98	12.02
1997	12.06	12.09	12.14	12.16	12.20	12.24	12.26	12.33	12.37	12.43	12.48	12.51
1998	12.54	12.60	12.64	12.69	12.73	12.77	12.79	12.85	12.08	12.92	12.95	12.99
1999	13.04	13.06	13.10	13.15	13.19	13.24	13.28	13.30	13.35	13.38	13.40	13.44
2000	13.51	13.55	13.59	13.64	13.67	13.73	13.76	13.81	13.86	13.93	13.97	14.03
2001	14.05	14.12	14.17	14.21	14.24	14.29	14.33	14.38	14.43	14.46	14.52	14.56
2002	14.59	14.62	14.65	14.68	14.70	14.75	14.78	14.82	14.87			

Source: http://www.bls.gov/

CHAPTER SUMMARY

RETURN TO CHAPTER CASE STUDY

Time Spent Reading the Newspaper

How much time did you spend reading the newspaper yesterday? Do you think that a person's age has any effect on the amount of time spent reading the newspaper yesterday? The USA Snapshot "Speed readers" that appeared in USA Today on July 14, 1998, seems to suggest that older adults spend more time reading the newspaper than younger adults do (see p. 573).

© Kent Meireis/The Image Works

FIRST THOUGHTS 12.26

Twenty-five newspaper readers were asked how much time, estimated to the nearest minute, they spent reading yesterday's newspaper. Their times are listed by the age category of the reader.

	Age CS12					
	18–24	**25–29**	**30–34**	**35–49**	**50–64**	**65 and up**
x,	10	13	14	20	20	8
time spent	5	8	7	12	12	30
reading	15	20	20	5	18	35
newspaper	4	10		28	15	20
yesterday				15	30	

a. Construct a scatter diagram to display the data, locating age along the horizontal axis (use the center of each age group as the age value: 21, 27, 32, 42, 57, 70) and reading time along the vertical axis. Plot the mean reading time for each age group on the diagram with an X.
b. Does your graph show visual evidence to suggest that the average amount of time spent daily reading the newspaper increases with the age of the reader? Justify your answer.
c. Does the sample show that a person's age has an effect on the ***amount of time*** spent daily reading the newspaper? Does the sample show that a person's age has an effect on the ***average amount of time*** spent daily reading the newspaper? Are these different questions? Explain.

PUTTING CHAPTER 12 TO WORK 12.27

a. With the ANOVA technique learned in this chapter, do these data show sufficient evidence to claim that a person's age has an effect on the average amount of time spent daily reading the newspaper? Use $\alpha = 0.05$.
b. Does the statistical answer found in part a agree with your response in 12.26, part b? Explain why your answers agree or disagree, citing statistical information learned in this chapter.

YOUR OWN MINI-STUDY 12.28

Design your own "Speed readers" study.

a. Define a specific population that you will sample, describe your sampling plan, and collect a random sample of at least 25 data.
b. Make a descriptive presentation of your sample data, using a chart or graph and a descriptive paragraph.
c. With the ANOVA technique learned in this chapter, do these data show sufficient evidence to claim that a person's age has an effect on the average amount of time spent daily reading the newspaper? Use $\alpha = 0.05$.
d. Discuss the similarities and differences between the results of your study and the results found in 12.27.

In Retrospect

In this chapter we have presented an introduction to the statistical techniques known as analysis of variance. The techniques studied here were restricted to the test of a hypothesis that dealt with questions about the means from several populations. We were restricted to normal populations and populations with homogeneous (equal) variances. The test of multiple means is done by partitioning the sum of squares into two segments: (1) the sum of squares due to variation between the levels of the factor being tested and (2) the sum of squares due to variation between the replicates within each level. The null hypothesis about means is then tested by using the appropriate variance measurements.

Note that we restricted our development to one-factor experiments. This one-factor technique represents only a beginning to the study of analysis of variance techniques.

Chapter Exercises

12.29 Ex12-29 Samples of peanut butter produced by three different manufacturers were tested for salt content, with the following results:

Brand 1	2.5	8.3	3.1	4.7	7.5	6.3
Brand 2	4.5	3.8	5.6	7.2	3.2	2.7
Brand 3	5.3	3.5	2.4	6.8	4.2	3.0

Is there a significant difference in the mean amounts of salt in these samples? Use $\alpha = 0.05$.

a. State the null and alternative hypotheses.
b. Determine the test criteria: assumptions, level of significance, test statistic.
c. Using the information on the computer printout, state the decision and conclusion to the hypothesis test.
d. What does the p-value tell you? Explain.

(*Hint:* Each level of data is entered as a separate column.)

```
Analysis of Variance
Source    DF     SS         MS        F         P
Factor     2     4.68       2.34      0.64      0.541
Error     15    54.88       3.66
Total     17    59.56

                               Individual 95% CIs
                               For Mean Based on
                               Pooled StDev
Level     N    Mean    StDev   ---+----+---+---+-
Brand1    6    5.400   2.359        (------*------)
Brand2    6    4.500   1.669      (----*----)
Brand3    6    4.700   1.621     (----*----)
                               ----+---+---+---+-
Pooled StDev = 1.913              3.0  4.5  6.0 7.5
```

12.30 Ex12-30 A new all-purpose cleaner is being test-marketed with sales displays placed in three different locations in various supermarkets. The numbers of bottles sold from each location in each of the supermarkets tested are reported here:

Location I	40	35	44	38
Location II	32	38	30	35
Location III	45	48	50	52

a. State the null and alternative hypotheses for testing: The location of the sales display had no effect on the number of bottles sold.
b. Using $\alpha = 0.01$, determine the test criteria: assumptions, level of significance, test statistic.
c. Using the information on the computer printout shown below, state the decision and conclusion to the hypothesis test.
d. What does the p-value tell you? Explain.

12.31 An experiment was designed to compare the lengths of time that four different drugs provided pain relief following surgery. The results (in hours) are given in the table:

Drug			
A	**B**	**C**	**D**
8	6	8	4
6	6	10	4
4	4	10	2
2	4	10	
		12	

Printout for Exercise 12.30 (*Hint:* Each level of data is entered as a separate column.)

```
Analysis of Variance
Source      DF      SS        MS        F        P
Factor       2      460.7     230.3     19.51    0.001
Error        9      106.2      11.8
Total       11      566.9

                                          Individual 95% CIs For Mean
                                          Based on Pooled StDev
Level        N     Mean      StDev   -------+--------+--------+--------
Location     4     39.250    3.775          (----*----)
Location     4     33.750    3.500    (----*----)
Location     4     48.750    2.986                        (---*---)
                                     -------+--------+--------+--------
Pooled StDev = 3.436                       35.0     42.0     49.0
```

Is there enough evidence to reject the null hypothesis that there is no significant difference in the lengths of pain relief for the four drugs at $\alpha = 0.05$?
a. Solve using the p-value approach.
b. Solve using the classical approach.

12.32 Ex12-32 The distance required to stop a vehicle on wet pavement was measured to compare the stopping power of four major brands of tires. A tire of each brand was tested on the same vehicle on a controlled wet pavement. The resulting distances (in feet) are shown in the table. At $\alpha = 0.05$, is there sufficient evidence to conclude that there is a difference in the mean stopping distances?

Brand A	Brand B	Brand C	Brand D
37	37	33	41
34	40	34	41
38	37	38	40
36	42	35	39
40	38	42	41
32		34	43

a. Solve using the p-value approach.
b. Solve using the classical approach.

12.33 Ex12-33 A certain vending company's soft-drink dispensing machines are supposed to serve 6 ounces of beverage. Various machines were sampled and the resulting amounts of dispensed drink were recorded, as given in the table. Does this sample evidence provide sufficient reason to reject the null hypothesis that all five machines dispense the same average amount of soft drink? Use $\alpha = 0.01$.

Machine				
A	B	C	D	E
3.8	6.8	4.4	6.5	6.2
4.2	7.1	4.1	6.4	4.5
4.1	6.7	3.9	6.2	5.3
4.4		4.5		5.8

a. Solve using the p-value approach.
b. Solve using the classical approach.

12.34 Ex12-34 Suburbs, each with its own attributes, are located around every metropolitan area. There is always the "rich" one (the most expensive one), the least expensive one, and so on. Does the suburb affect the transfer value of its homes? The table lists transfer values, the amounts on which county transfer taxes are paid.
a. Do the sample data show sufficient evidence to conclude that the suburbs represented do have a significant effect on the transfer value of their homes? Use $\alpha = 0.01$.
b. Construct a graph that demonstrates the conclusion reached in part a.

Suburb A	Suburb B	Suburb C	Suburb D	Suburb E
105	101	95	74	79
114	88	107	135	89
85	105	101	165	140
177	100	92	114	114
104	161	91	80	80
135	113	89	115	86
	94			94
				102

12.35 Ex12-35 The question arises every year when the playoffs begin in the National Football League: Which division teams are the toughest, Eastern, Central or Western? Two ways to measure the strength of the football teams are the number of points they score and the number of points their opponents score. The final results for the 16 games played in the 2001 season are shown in the table:

Eastern		Central		Western	
Points	Opp. Points	Points	Opp. Points	Points	Opp. Points
371	272	352	212	399	327
344	290	303	265	301	324
308	295	285	319	340	339
413	486	336	388	320	344
265	420	294	286	332	321
343	208	226	309	503	273
256	303	338	203	409	282
294	321	390	266	333	409
295	343	324	280	291	377
246	338	290	390	253	410
		270	424		

Source: http://www.nfl.com/standings/2001/regular

Complete an ANOVA table for points scored and another table for points scored by opposing teams. In each case, test the null hypothesis that the mean points scored is the same for each of the three divisions. Use the 0.05 level of significance.
a. Solve using the p-value approach.
b. Solve using the classical approach.

12.36 Ex12-36 To compare the effectiveness of three different methods of teaching reading, 26 children of equal reading aptitude were divided into three groups. Each group was instructed for a given period of time using one of the three methods. After completing the instruction period, all students were tested. The test scores are shown in the table. Is the evidence sufficient to reject the hypothesis that all three instruction methods are equally effective? Use $\alpha = 0.05$.

Method I	Method II	Method III
45	45	44
51	44	50
48	46	45
50	44	55
46	41	51
48	43	51
45	46	45
48	49	47
47	44	

a. Solve using the p-value approach.
b. Solve using the classical approach.

12.37 Ex12-37 The 1997 *County and City Data Book* gave the median family incomes for three counties in Nebraska: Lancaster—\$39,478, Hall—\$35,764, and Sarpy—\$49,644. The following data are the family incomes (in thousands) for nine randomly selected individuals from each of the three counties.

Lancaster	Hall	Sarpy
\$48.1	\$36.9	\$51.3
42.6	34.9	53.3
45.1	41.9	56.3
38.1	39.9	50.8
43.1	33.4	51.3
40.1	42.4	49.3
47.1	35.9	62.3
51.6	42.5	58.8
53.1	29.9	52.3

Complete an ANOVA for the data, and test the null hypothesis that the mean family income is the same for each of the three counties. Use a 0.05 level of significance.
a. Solve using the p-value approach.
b. Solve using the classical approach.

12.38 Ex12-38 The table gives the numbers of arrests made last year for violations of the narcotic drug laws in 24 communities. The data are rates of arrest per 10,000 inhabitants. At $\alpha = 0.05$, is there sufficient evidence to reject the hypothesis that the mean rates of arrests are the same in all four sizes of communities?

Cities over 250,000	Cities under 250,000	Suburban Communities	Rural Communities
45	23	25	8
34	18	17	16
41	27	19	14
42	21	28	17
37	26	31	10
28	34	37	23

a. Solve using the p-value approach.
b. Solve using the classical approach.

12.39 Ex12-39 Seven golf balls from each of six manufacturers were randomly selected and tested for durability. Each ball was hit 300 times or until failure occurred, whichever came first. Do these sample data show sufficient reason to reject the null hypothesis that the six different brands tested withstood the durability test equally well? Use $\alpha = 0.05$.

Manufacturer					
A	B	C	D	E	F
300	190	228	276	162	264
300	164	300	296	175	168
300	238	268	62	157	254
260	200	280	300	262	216
300	221	300	230	200	257
261	132	300	175	256	183
300	156	300	211	92	93

a. Solve using the p-value approach.
b. Solve using the classical approach.

12.40 Ex12-40 The Bureau of Labor Statistics posts a table of average weekly hours worked by production or

Table for Exercise 12.40

Year	Jan.	Feb.	Mar.	Apr.	May.	June	July	Aug.	Sep.	Oct.	Nov.	Dec.
1992	32.5	32.5	32.5	32.7	32.7	32.5	32.6	32.7	32.7	32.6	32.7	32.6
1993	32.7	32.7	32.5	32.7	32.7	32.7	32.8	32.7	32.7	32.8	32.7	32.8
1994	32.8	32.7	32.8	32.8	32.8	32.8	32.8	32.7	32.7	32.8	32.7	32.8
1995	32.7	32.6	32.6	32.7	32.6	32.7	32.6	32.6	32.6	32.6	32.7	32.6
1996	32.4	32.6	32.7	32.6	32.7	32.7	32.6	32.7	32.8	32.7	32.7	32.8
1997	32.7	32.8	32.8	32.8	32.8	32.7	32.8	32.9	32.9	32.9	32.9	32.9
1998	33.0	32.9	32.8	32.9	33.0	32.8	32.9	32.9	32.9	32.9	32.9	32.9
1999	32.9	32.9	32.8	32.8	32.8	32.8	32.9	32.9	32.7	32.9	32.8	32.9
2000	32.8	32.8	32.8	32.8	32.8	32.8	32.7	32.8	32.8	32.8	32.8	32.7
2001	32.8	32.7	32.7	32.7	32.7	32.7	32.7	32.7	32.7	32.6	32.6	32.7
2002	32.7	32.7	32.8	32.7	32.8	32.8	32.6	32.7	32.9			

Source: http://www.bls.gov/

nonsupervisory workers on private nonfarm payrolls by major industry. The following average weekly hours were taken from the Web site on October 15, 2002.

a. At the 0.05 level of significance, does the evidence provide sufficient reason to reject the claim that there is no difference in the mean weekly hours worked by month? Show graphic evidence to visually support your conclusion.
b. At the 0.05 level of significance, does the evidence provide sufficient reason to reject the claim that there is no difference in the mean weekly hours worked each year? Show graphic evidence to visually support your conclusion.

12.41 Ex12-41 Ronald Fisher, an English statistician (1890–1962), collected measurements for a sample of 150 irises. Of concern were the variables species, petal width (PW), petal length (PL), sepal width, and sepal length (all in millimeters). (Sepals are the outermost leaves that encase the flower before it has opened.) The goal of Fisher's experiment was to produce a simple function that could be used to classify flowers correctly. A sample of his data is listed here.

Species	PW	SW	Species	PW	SW
0	2	35	1	24	28
2	18	32	1	19	25
1	19	27	0	1	31
0	3	35	1	23	32
0	3	38	2	13	23
2	12	26	2	15	30
1	20	38	1	25	33
2	15	31	1	21	33
2	15	29	0	2	37
2	12	27	1	18	27
1	22	28	1	17	25
1	13	30	1	24	34
0	2	29	0	2	36
2	16	27	2	10	22
0	5	33	0	2	32

a. Is there a significant difference in the mean petal width for the three species? Use a 0.05 level of significance.
b. Is there a significant difference in the mean sepal width for the three species? Use a 0.05 level of significance.
c. How could Fisher use these outcomes to help him classify irises into the correct species?

12.42 Ex12-42 Cicadas are flying, plant-eating insects. The particular variety 13-year cicadas *(Magicicada)* spend five juvenile stages in underground burrows. During the 13 years underground, the cicadas grow from approximately the size of a small ant to nearly the size of an adult cicada. Every 13 years, these cicadas then emerge from their burrows as adults. Three different species of these 13-year cicadas are listed with their adult body weights (BW) in grams and their body lengths (BL) in millimeters.

BW	BL	Species	BW	BL	Species
0.15	22	*tredecula*	0.18	24	*tredecula*
0.29	26	*tredecim*	0.21	20	*tredecassini*
0.17	24	*tredecim*	0.15	24	*tredecula*
0.18	23	*tredecula*	0.17	23	*tredecula*
0.39	32	*tredecim*	0.13	22	*tredecassini*
0.26	27	*tredecim*	0.17	23	*tredecassini*
0.17	24	*tredecassini*	0.23	25	*tredecassini*
0.16	24	*tredecassini*	0.12	24	*tredecim*
0.14	25	*tredecassini*	0.26	26	*tredecula*
0.14	25	*tredecassini*	0.19	25	*tredecula*
0.28	27	*tredecassini*	0.20	23	*tredecassini*
0.12	29	*tredecim*	0.14	22	*tredecula*

a. Is there any significant difference in the body weights of adult cicadas with respect to species? Construct a one-way ANOVA table and test for the difference at the 0.01 level of significance.
b. Is there any significant difference in the body lengths of adult cicadas with respect to species? Construct a one-way ANOVA table and test for the difference at the 0.01 level of significance.

12.43 For the following data, find SS(error) and show that

$$\text{SS(error)} = [(k_1 - 1)s_1^2 + (k_2 - 1)s_2^2 + (k_3 - 1)s_3^2]$$

where s_i^2 is the variance for the ith factor level.

Factor Level

1	2	3
8	6	10
4	6	12
2	4	14

12.44 For the following data, show that

$$\text{SS(factor)} = k_1(\bar{x}_1 - \bar{x})^2 + k_2(\bar{x}_2 - \bar{x})^2 + k_3(\bar{x}_3 - \bar{x})^2$$

where $\bar{x}_1$, $\bar{x}_2$, and $\bar{x}_3$ are the means for the three factor levels and $\bar{x}$ is the overall mean.

Factor Level

1	2	3
6	13	9
8	12	11
10	14	7

12.45 An article in the *Journal of Pharmaceutical Sciences* (December 1987) discusses the plasma protein binding of diazepam at various concentrations of imipramine. Suppose the results were reported as follows:

Diazepam Alone (1.25 mg/mL)	Diazepam with Imipramine 1.25	2.50	5.00
97.99	97.68	96.29	93.92

The values represent mean plasma protein binding, and $n = 8$ for each of the four groups. Find the sum of squares among the four groups.

12.46 A study reported in the *Journal of Research and Development in Education* (Summer 1989) evaluates the effectiveness of social skills training and cross-age tutoring for improving academic skills and social communication behaviors among boys with learning disabilities. Twenty boys were divided into three groups, and their scores on the Test of Written Spelling (TWS) are summarized as follows:

Group	n	TWS Mean	Standard Deviation
Social skills training and tutoring	7	21.43	9.48
Social skills training only	7	20.00	8.91
Neither component	6	20.83	9.06

Calculate the entries of the ANOVA table using these results.

Vocabulary and Key Concepts

Be able to define each term. Pay special attention to the key terms, which are printed in **red.** In addition, describe each term in your own words and give an example of each. Your examples should not be ones given in class or in the textbook. The bracketed numbers indicate the chapters in which the term previously appeared, but you should define the terms again to show increased understanding of their meaning. Page numbers indicate the first appearance of the term in Chapter 12.

analysis of variance (ANOVA) (pp. 572, 580)
assumptions [8, 9, 10, 11] (p. 582)
between-sample variation (p. 580)
degrees of freedom [9, 10, 11] (p. 577)
experimental error (p. 584)
levels of the tested factor (pp. 574, 577, 580)
mathematical model (p. 584)
mean square, MS(factor), MS(error) (p. 578)
partitioning (p. 576)
randomize [2] (p. 582)
replicate (p. 574)
response variable [1] (p. 580)
sum of squares (p. 575)
test statistic, $F\star$ (p. 579)
total sum of squares, SS(total) (p. 576)
variance [2, 9, 10] (p. 575)
variation between levels, MS(factor) (pp. 576, 578, 580)
variation within a level, MS(error) (pp. 576, 578, 580)
within-sample variation (pp. 576, 580)

Chapter Practice Test

PART I: KNOWING THE DEFINITIONS

Answer "True" if the statement is always true. If the statement is not always true, replace the words shown in bold with words that make the statement always true.

12.1 To partition the sum of squares for the total is to separate the numerical value of SS(total) into two values such that the **sum** of these two values is equal to SS(total).

12.2 A **sum of squares** is actually a measure of variance.

12.3 **Experimental error** is the name given to the variability that takes place between the levels of the test factor.

12.4 **Experimental error** is the name given to the variability that takes place among the replicates of an experiment as it is repeated under constant conditions.

12.5 **Fail to reject H_o** is the desired decision when the means for the levels of the factor being tested are all different.

12.6 The **mathematical model** for a particular problem is an equational statement showing the anticipated makeup of an individual piece of data.

12.7 The degrees of freedom for the factor are equal to the **number of factors tested.**

12.8 The measure of a specific level of a factor being tested in an ANOVA is the **variance** of that factor level.

12.9 We **need not** assume that the observations are independent to do analysis of variance.

12.10 The rejection of H_o **indicates** that you have identified the level(s) of the factor that is (are) different from the others.

PART II: APPLYING THE CONCEPTS

12.11 Determine the truth (T/F) of each statement with regard to the one-factor analysis of variance technique.

_____ a. The mean squares are measures of variance.

_____ b. "There is no difference between the mean values of the random variable at the various levels of the test factor" is a possible interpretation of the null hypothesis.

_____ c. "The factor being tested has no effect on the random variable x" is a possible interpretation of the alternative hypothesis.

_____ d. "There is no variance among the mean values of x for each of the different factor levels" is a possible interpretation of the null hypothesis.

_____ e. The "partitioning" of the variance occurs when SS(total) is separated into SS(factor) and SS(error).

_____ f. We will reject the null hypothesis and conclude that the factor has an effect on the variable when the amount of variance assigned to the factor is significantly larger than the variance assigned to error.

_____ g. In order to apply the F-test, the sample size from each factor level must be the same.

_____ h. In order to apply the F-test, the sample standard deviation from each factor level must be the same.

_____ i. If 20 is subtracted from every data value, then the calculated value of the $F\bigstar$ statistic is also reduced by 20.

When the calculated value of F, $F\bigstar$, is greater than the table value for F,

_____ j. the decision will be: Fail to reject H_0.

_____ k. the conclusion will be: The factor being tested does have an effect on the variable.

Independent samples were collected in order to test the effect a factor had on a variable. The data are summarized in this ANOVA table:

	SS	df
Factor	810	2
Error	720	8
Total	1530	10

Is there sufficient evidence to reject the null hypothesis that all levels of the test factor have the same effect on the variable?

_____ l. The null hypothesis could be: $\mu_A = \mu_B = \mu_C = \mu_D$.

_____ m. The calculated value of F is 1.125.

_____ n. The critical value of F for $\alpha = 0.05$ is 6.06.

_____ o. The null hypothesis can be rejected at $\alpha = 0.05$.

12.12 Consider this table:

	SS	df	MS	$F\bigstar$
Factor	***A***	4	18	***E***
Error	***B***	18	***D***	
Total	144	***C***		

Find the values:

a. A b. B c. C d. D e. E

PART III: UNDERSTANDING THE CONCEPTS

12.13 In 50 words or less, explain what a single-factor ANOVA experiment is.

12.14 A state environmental agency tested three different scrubbers used to reduce the resulting air pollution in the generation of electricity. The primary concern was the emission of particulate matter. Several trials were run with each scrubber. The amount of particulate emission was recorded for each trial.

PT12-14

	Amounts of Emission					
Scrubber I	11	10	12	9	13	12
Scrubber II	12	10	12	8	9	
Scrubber III	9	11	10	7	8	

a. State the mathematical model for this experiment.

b. State the null and alternative hypotheses.

c. Calculate and form the ANOVA table.

d. Complete the testing of H_o using a 0.05 level of significance. State the decision and conclusion clearly.

e. Construct a graph representing the data that is helpful in picturing the results of the hypothesis test.

CHAPTER 13

Linear Correlation and Regression Analysis

Chapter Outline and Objectives

CHAPTER CASE STUDY

The relationship between acres planted and acres harvested can be important information in the agriculture industry.

13.1 LINEAR CORRELATION ANALYSIS

Analysis of linear dependency uses two measures: **covariance** and the **coefficient of linear correlation.**

13.2 INFERENCES ABOUT THE LINEAR CORRELATION COEFFICIENT

We **question and interpret** the correlation coefficient once we have obtained it.

13.3 LINEAR REGRESSION ANALYSIS

To analyze two related variables, we use the **line of best fit.**

13.4 INFERENCES CONCERNING THE SLOPE OF THE REGRESSION LINE

We judge the **usefulness of the line of best fit equation** before predicting the value of one variable given the value of another variable.

13.5 CONFIDENCE INTERVALS FOR REGRESSION

If the line of best fit is usable, we can establish confidence interval estimates.

13.6 UNDERSTANDING THE RELATIONSHIP BETWEEN CORRELATION AND REGRESSION

Linear correlation analysis tells us whether or not we are dealing with a **linear relationship;** regression analysis tells us **what the relationship is.**

RETURN TO CHAPTER CASE STUDY

The basic ideas of regression and linear correlation analysis were introduced in Chapter 3. (If these concepts are not fresh in your mind, review Chapter 3 before beginning this chapter.) Chapter 3 was only a first look: a presentation of the basic graphic (the scatter diagram) and descriptive statistical aspects of linear correlation and regression analysis. In this chapter we take a second, more detailed look at linear correlation and regression analysis.

Chapter Case Study

© Francis de Richemond/The Image Works

Wheat! Beautiful Golden Wheat!

Wheat is a common name for cereal grass of a genus of the grass family. It has been cultivated for food since prehistoric times and is one of our most important grain crops. The common types of wheat grown in the United States are spring wheat, planted in the spring for fall harvest, and winter wheat, planted in the fall for spring harvest. The main use of wheat is in flour for bread and other food products. And it's also used to a limited extent in the making of beer, whiskey, industrial alcohol, and other products.

The U.S. Wheat Crop for 2002 is Expected to Be the Smallest in a Quarter Century

July 12, 2002 (EIRNS)—The area harvested in the United States this year for winter wheat (the predominant wheat variety in U.S. latitudes) is estimated to be only 29.8 million acres (12.06 million hectares)—the same as in 1917! (The United States harvested winter wheat area in recent years has been between 35 and over 40 million acres.) Farmers have abandoned large amounts of sown land because of drought and related pests and disease. Estimates now put the total U.S. wheat harvest (all types) this year at around 1.79 billion bushels (48.9 million metric tons), about the same as in 1974, and way down from the 64 million ton levels of recent yearly harvests. Western Canada could potentially harvest 19.7 million metric tons of wheat, down from the five-year average of 23.3 million tons, which itself has been declining.

In terms of world trade in basic foodstuffs, the United States and Canada are a major source of world wheat supplies—now severely contracted. Australia's wheat output next season is expected to drop. Argentina is in turmoil. Only Europe (principally France) expects a good harvest. World wheat stocks are way down.

Source: http://committeerepubliccanada.ca/English/News/Slug012.htm

The decisions a grain farmer makes as he runs his business are not as simple as the statistical relationship between the four variables listed in the table (p. 604). However, an understanding of the relationship between these variables is an important component of what he needs to know in order to make decisions about how many acres to plant, what kind of grain to plant, and other issues.

Twenty randomly selected wheat-producing counties in Kansas were identified and data were collected for these variables:

Planted = 1000s of acres planted with winter wheat

Harvested = 1000s of acres harvested (not all planted acres are harvested for a variety of reasons)

Yield = bushels of wheat harvested per acre

Production = 1000s of bushels of wheat harvested

CS13

County	Planted	Harvested	Yield	Production
Allen	32.0	29.5	49	1,451
Chautauqua	5.5	4.8	43	206
Cherokee	64.4	61.1	46	2,791
Coffey	26.4	25.1	51	1,281
Gove	118.0	79.9	34	2,739
Gray	127.0	103.9	36	3,689
Greenwood	6.1	5.6	43	242
Johnson	5.5	5.0	52	261
Linn	14.6	14.3	54	774
Logan	145.0	84.2	32	2,706
McPherson	208.0	192.3	41	7,942
Miami	10.0	9.0	55	491
Nemaha	27.7	24.8	46	1,134
Neosho	37.2	35.3	43	1,523
Sherman	185.0	171.8	35	5,984
Stafford	141.0	127.4	38	4,781
Sumner	350.0	317.8	40	12,726
Thomas	196.0	166.9	44	7,400
Washington	88.0	80.3	43	3,433
Wilson	41.5	39.1	49	1,932

Source: http://www.usda.gov/nass/graphics/county01/data/ww01.csv

ANSWER NOW 13.1

a. **Construct a scatter diagram and find the linear correlation coefficient and the equation of the line of best fit for Planted (x) and Harvested (y).**

b. **Construct a scatter diagram of Harvested (x) and Production (y).**

c. **Construct a histogram of Yield.**

d. **Explain what you have learned in parts a–c.**

After completing Chapter 13, further investigate the Chapter Case Study in the Return to Chapter Case Study section with Exercises 13.63, 13.64, and 13.65 (p. 646).

13.1 Linear Correlation Analysis

In Chapter 3 the linear correlation coefficient was presented as a quantity that measures the strength of a linear relationship (dependency). Now let's take a second look at this concept and see how r, the coefficient of linear correlation, works. Intuitively, we want to think about how to measure the mathematical linear dependency of one variable on another. As x increases, does y tend to increase or decrease? How strong (consistent) is this tendency? We are going to use two measures of dependence—covariance and the coefficient of linear correlation—to measure the relationship between two variables. We'll begin our discussion by examining a set of bivariate data and identifying some related facts as we prepare to define covariance.

Illustration 13.1 Understanding and Calculating Covariance

Let's consider the sample of six pieces of bivariate data: (2, 1), (3, 5), (6, 3), (8, 2), (11, 6), (12, 1). See Figure 13.1. The mean of the six x values (2, 3, 6, 8, 11, 12) is $\bar{x} = 7$. The mean of the six y values (1, 5, 3, 2, 6, 1) is $\bar{y} = 3$.

FIGURE 13.1
Graph of Bivariate Data

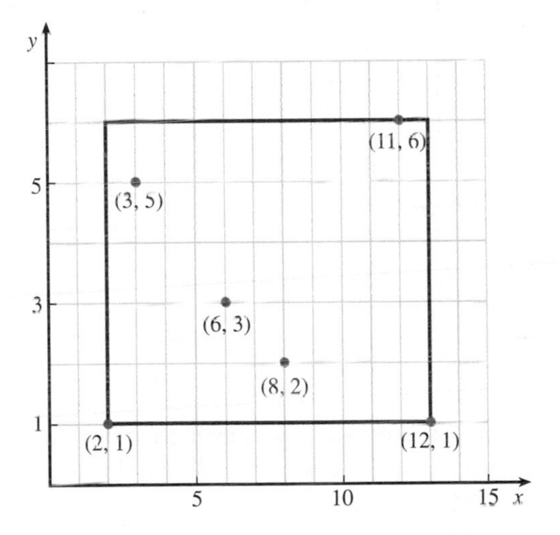

The point $(\bar{x}, \bar{y})$, which is (7, 3), is located as shown on the graph of the sample points in Figure 13.2 (p. 606). The point $(\bar{x}, \bar{y})$ is called the **centroid** of the data. A vertical and a horizontal line drawn through the centroid divide the graph into four sections, as shown in Figure 13.2. Each point (x, y) lies a certain distance from each of these two lines: $(x - \bar{x})$ is the horizontal distance from (x, y) to the vertical line that passes through the centroid and $(y - \bar{y})$ is the vertical distance from (x, y) to the horizontal line that passes through the centroid. Both the horizontal and vertical distances of each data point from the centroid can be measured, as shown in Figure 13.3 (p. 606). The distances may be positive, negative, or zero, depending on the position of the point (x, y) in relation to $(\bar{x}, \bar{y})$. [Figure 13.3 shows $(x - \bar{x})$ and $(y - \bar{y})$ represented by braces, with positive or negative signs.]

FIGURE 13.2
The Point (7, 3) Is the Centroid

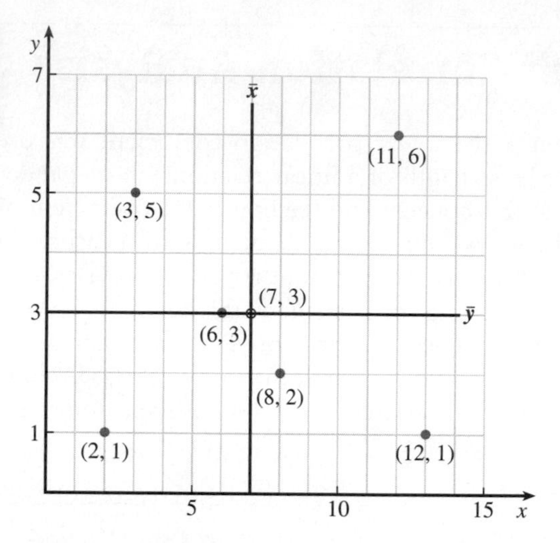

FIGURE 13.3
Measuring the Distance of Each Data Point from the Centroid

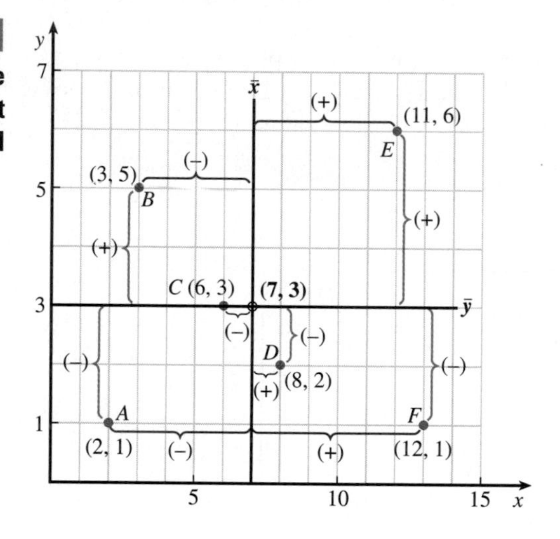

■

One measure of linear dependency is the covariance. The **covariance of x and y** is defined as the sum of the products of the distances of all values of x and y from the centroid,

$$\sum[(x - \bar{x})(y - \bar{y})], \text{ divided by } n - 1:$$

COVARIANCE OF X AND Y

$$\text{covar}(x, y) = \frac{\sum_{i=1}^{n} (x_i - \bar{x})(y_i - \bar{y})}{n - 1} \qquad (13.1)$$

Calculations for the covariance for the data in Illustration 13.1 are given in Table 13.1. The covariance, written as covar(x, y), of the data is $\frac{3}{5}$ = **0.6.**

NOTE 1 $\sum(x - \bar{x}) = 0$ and $\sum(y - \bar{y}) = 0$. This will always happen. Why? (See pp. 78–80.)

NOTE 2 Even though the variance of a single set of data is always positive, the covariance of bivariate data can be negative.

TABLE 13.1 Calculations for Finding covar(x, y) for the Data of Illustration 13.1

Points	$x - \bar{x}$	$y - \bar{y}$	$(x - \bar{x})(y - \bar{y})$
(2, 1)	−5	−2	10
(3, 5)	−4	2	8
(6, 3)	−1	0	0
(8, 2)	1	−1	−1
(11, 6)	4	3	12
(12, 1)	5	−2	−10
Total	0 ck	0 ck	3

The covariance is positive if the graph is dominated by points to the upper right and to the lower left of the centroid. The products of $(x - \bar{x})$ and $(y - \bar{y})$ are positive in these two sections. If the majority of the points are to the upper left and the lower right of the centroid, then the sum of the products is negative. Figure 13.4 shows data that represent a positive dependency (a), a negative dependency (b), and little or no dependency (c). The covariances for these three situations would definitely be positive in part a, negative in b, and near zero in c. (The sign of the covariance is always the same as the sign of the slope of the regression line.)

FIGURE 13.4
Data and Covariance

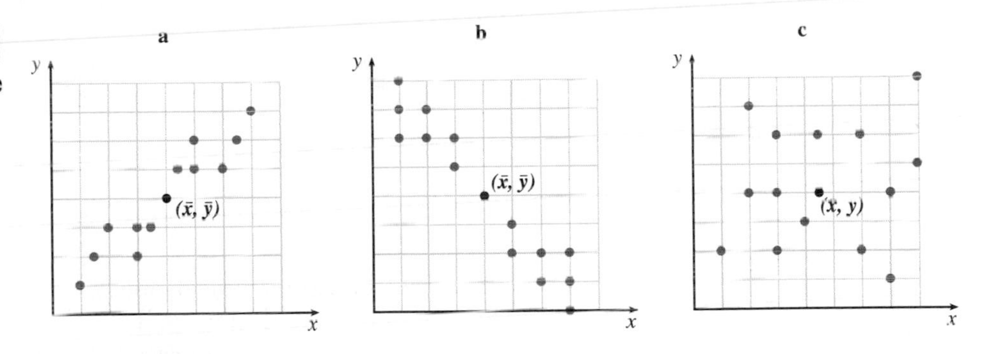

The biggest disadvantage of covariance as a measure of linear dependency is that it does not have a standardized unit of measure. One reason for this is that the spread of the data is a strong factor in the size of the covariance. For example, if we multiply each data point in Illustration 13.1 by 10, we have (20, 10), (30, 50), (60, 30), (80, 20), (110, 60), and (120, 10). The relationship of the points to each other is changed only in that they are much more spread out. However, the covariance for this new set of data is 60. Does this mean that the dependency between the x and y variables is stronger than in the original case? No, it does not; the relationship is the same, even though each data value has been multiplied by 10. This is the trouble with covariance as a measure. We must find some way to eliminate the effect of the spread of the data when we measure dependency.

FYI
This calculation is assigned in Exercise 13.11 (p. 611).

If we standardize x and y by dividing the distance of each from the respective mean by the respective standard deviation:

$$x' = \frac{x - \bar{x}}{s_x} \quad \text{and} \quad y' = \frac{y - \bar{y}}{s_y}$$

and then compute the covariance of x' and y', we will have a covariance that is not affected by the spread of the data. This is exactly what is accomplished by the linear correlation coefficient. It divides the covariance of x and y by a measure of the spread of x and by a measure of the spread of y (the standard deviations of x and of y are used as measures of spread). Therefore, by definition, the **coefficient of linear correlation** is:

COEFFICIENT OF LINEAR CORRELATION

$$r = \text{covar}(x', y') = \frac{\text{covar}(x, y)}{s_x \cdot s_y} \qquad (13.2)$$

The coefficient of linear correlation standardizes the measure of dependency and allows us to compare the relative strengths of dependency of different sets of data. [Formula (13.2) for linear correlation is also commonly referred to as **Pearson's product moment, *r*.**]

We can find the value of r, the coefficient of linear correlation, for the data in Illustration 13.1 by calculating the two standard deviations and then dividing:

$$s_x = 4.099 \quad \text{and} \quad s_y = 2.098$$

$$r = \frac{\text{covar}(x, y)}{s_x \cdot s_y}: \quad r = \frac{0.6}{(4.099)(2.098)} = \mathbf{0.07}$$

Finding the correlation coefficient using formula (13.2) can be a very tedious arithmetic process. We can write the formula in a more workable form, however, as it was in Chapter 3:

FYI
Refer to Chapter 3 (pp. 146–147) for an illustration of the use of this formula.

SHORTCUT FOR COEFFICIENT OF LINEAR CORRELATION

$$r = \frac{\text{covar}(x, y)}{s_x \cdot s_y} = \frac{\dfrac{\sum[(x-\bar{x})(y-\bar{y})]}{n-1}}{s_x \cdot s_y} = \frac{\text{SS}(xy)}{\sqrt{\text{SS}(x) \cdot \text{SS}(y)}} \qquad (13.3)$$

FYI
Computer and calculator commands to find the correlation coefficient were presented in Chapter 3 (p. 148).

Formula (13.3) avoids the separate calculations of $\bar{x}$, $\bar{y}$, s_x, and s_y as well as the calculations of the deviations from the means. Therefore, formula (13.3) is much easier to use and, more important, it is more accurate when decimals are involved because it minimizes round-off error.

Exercises

13.2 Consider a set of paired bivariate data.

a. Explain why $\sum(x - \bar{x}) = 0$ and $\sum(y - \bar{y}) = 0$.

b. Describe the effect that the lines $x = \bar{x}$ and $y = \bar{y}$ have on the graph of these points.

c. Describe the relationship of the ordered pairs that will cause $\sum[(x - \bar{x})(y - \bar{y})]$ to be positive. Negative. Near zero.

13.3 Ex13-03 These data are from a random sample of 40 college students, showing their gender, ACT composite scores, and grade-point averages (GPA) after their first term in college.

Data for Females, Exercise 13.3

ACT	GPA	ACT	GPA
23	1.833	15	3.000
28	4.000	22	3.600
22	3.057	20	2.665
20	4.000	17	2.934
23	3.550	21	3.422
19	2.583	18	3.002
20	3.165	17	3.000
29	3.398	25	4.000
27	3.868	25	3.472
18	2.918	25	3.550
17	2.360		

Data for Males, Exercise 13.3

ACT	GPA	ACT	GPA
33	3.333	16	3.313
17	2.835	13	3.053
26	3.249	16	2.600
25	2.290	27	2.000
20	2.178	19	2.500
23	2.835	22	4.000
19	2.364	33	2.833
21	3.000	17	3.438
22	3.934	26	2.418
29	3.533		

Source: http://www.act.org/research/briefs/97-2.html

a. Construct a scatter diagram of the data with ACT scores on the horizontal axis and GPAs on the vertical axis. Be sure to identify the male and female students.
b. Do the patterns for males and females appear to be the same or are they different? Identify specific similarities and differences.
c. Given that a student had an ACT score of 25, what would you predict for that student's GPA at the end of the first term in college?
d. Does there appear to be any relationship between ACT scores and first-term GPAs?

13.4 Ex13-04 Aerial photographs are one of many techniques used to monitor wildlife populations. Knowing the numbers of animals and their locations relative to areas inhabited by the human population is very useful. It is also important to monitor physical characteristics of the animals. Is it possible to use the length of a bear, as estimated from an aerial photograph, to estimate the bear's age, weight, or both? [That would be much safer than asking the bear to stand on a scale!] The data in the table are bears' ages in months, genders (1 = male, 2 = female), lengths in inches, and weights in pounds.

Age	Gender	Length	Weight
19	1	45.0	65
29	2	62.0	121
20	1	56.0	100
67	1	78.0	371
81	1	72.0	416
140	1	75.0	386
104	2	62.0	166
100	2	70.0	220
70	1	78.0	334
56	1	73.5	262
51	1	68.5	360
57	2	64.0	204
53	2	58.0	144
68	1	73.0	332
8	1	37.0	34
44	2	63.0	140
32	1	67.0	180
20	2	52.0	105
32	1	59.0	166
56	1	66.0	250
9	2	36.0	26
30	1	66.5	210
177	1	72.0	436
69	2	61.0	176
84	2	57.0	180
23	1	57.8	140
9	1	40.0	40
45	1	63.0	220
18	1	45.0	60
33	1	66.5	154
57	2	60.5	116
45	2	60.0	182
22	1	65.0	180
22	1	63.0	172
82	2	64.0	356
70	2	65.0	316
31	1	69.0	289
10	1	47.0	86
43	1	64.0	166
34	1	72.0	270
34	1	65.0	202
58	2	63.0	202
58	1	70.5	365
11	1	48.0	79
17	1	50.5	90
23	1	50.0	148
70	1	76.5	446
11	2	46.0	62
83	2	61.5	236
35	1	63.5	212
16	1	48.0	60
16	1	41.0	64
17	1	53.0	114
17	2	52.5	76
17	2	46.0	48
8	2	43.5	29
83	1	75.0	514
18	1	57.3	140

Source: MINITAB's *Bears.mtw*

a. Investigate the relationship between the lengths and ages of the bears. Be sure to include the gender variable.
b. Does there seem to be a predictable pattern for the relationship between length and age? How does gender of the bear affect the relationship? Explain. Describe the pattern.
c. Investigate the relationship between the lengths and weights of the bears. Be sure to include the gender variable.
d. Does there seem to be a predictable pattern for the relationship between length and weight? How does gender of the bear affect the relationship? Explain. Describe the pattern.

(continued)

e. If the gender of a smaller or younger bear cannot be determined, how will this affect the estimate for age or weight? Explain.

13.5

a. Construct a scatter diagram of the bivariate data.

Point	*A*	*B*	*C*	*D*	*E*	*F*	*G*	*H*	*I*	*J*
x	1	1	3	3	5	5	7	7	9	9
y	1	2	2	3	3	4	4	5	5	6

b. Calculate the covariance.
c. Calculate s_x and s_y.
d. Calculate r using formula (13.2).
e. Calculate r using formula (13.3).

13.6 Consider these bivariate data:

Point	*A*	*B*	*C*	*D*	*E*	*F*	*G*	*H*	*I*	*J*
x	0	1	1	2	3	4	5	6	6	7
y	6	6	7	4	5	2	3	0	1	1

a. Draw a scatter diagram for the data.
b. Calculate the covariance.
c. Calculate s_x and s_y.
d. Calculate r by formula (13.2).
e. Calculate r by formula (13.3).

13.7 Ex13-07 The MINITAB output shows the preliminary calculations for the table of bivariate data. Form the extensions table; calculate the summations $\sum x$, $\sum y$, $\sum x^2$, $\sum xy$, and $\sum y^2$; and find SS(x), SS(y), and SS(xy). Verify the results by calculating the values yourself.

x	45	52	49	60	67	61
y	22	26	21	28	33	32

```
Row      X      Y      XSQ      XY     YSQ
 1      45     22     2025     990     484
 2      52     26     2704    1352     676
 3      49     21     2401    1029     441
 4      60     28     3600    1680     784
 5      67     33     4489    2211    1089
 6      61     32     3721    1952    1024

Row     C6     C7       C8      C9     C10
 1     334    162    18940    9214    4498
SS(X)   347.333
SS(Y)   124.000
SS(XY)  196.000
```

13.8 Ex13-08 Use a computer to form the extensions table; calculate the summations $\sum x$, $\sum y$, $\sum x^2$, $\sum xy$, and $\sum y^2$; and find SS(x), SS(y), and SS(xy) for this set of bivariate data:

x	11.4	9.4	6.5	7.3	7.9	9.0	9.3	10.6
y	8.1	8.2	5.8	6.4	5.9	6.5	7.1	7.8

13.9 Ex13-09 NFL football enthusiasts often look at a team's total points scored for (Pts) and total points scored against (PtsA) as a way of comparing the strengths of teams. The 2001 season totals for the 31 teams in the NFL are presented here:

Pts	PtsA	Pts	PtsA	Pts	PtsA
371	272	352	212	399	327
344	290	303	265	301	324
308	295	285	319	340	339
413	486	336	388	320	344
265	420	294	286	332	321
343	208	226	309	503	273
256	303	338	203	409	282
294	321	390	266	333	409
295	343	324	280	291	377
246	338	290	390	253	410
		270	424		

a. Calculate the linear correlation coefficient (Pearson's product moment, r) for the points scored for and against.
b. What conclusion can you draw from the answer in part a?
c. Construct the scatter diagram and comment on how it supports, or disagrees with, your comments in part b.

FYI

See page 148 for information about using MINITAB, Excel, and TI-83 to find the correlation coefficient.

13.10 Ex13-10 The performances of personal computers are rated on many different dimensions and applications. Three of the most popular are multimedia, word processing, and spreadsheet. *Windows*® magazine's technical staff conducted benchmark tests in the summer of 1998 on a collection of 12 machines representing both desktop and notebook designs. Results of the tests are shown in the table of scores.

Multimedia	Word Processing	Spreadsheet
121	115	112
115	83	116
100	95	92
78	87	86
110	72	51
80	79	74
76	76	68
83	45	49
57	50	42
131	123	120
120	118	115
119	115	115

Source: *Windows*®, "Winscore 2.0 Results," September 1998

a. Calculate the linear correlation coefficient (Pearson's product moment, r) between (1) multimedia and word processing, (2) multimedia and spreadsheet, and (3) word processing and spreadsheet.
b. What conclusions might you draw from your answers in part a?

13.11
a. Calculate the covariance of the set of data (20, 10), (30, 50), (60, 30), (80, 20), (110, 60), and (120, 10).
b. Calculate the standard deviation of the six x values and the standard deviation of the six y values.
c. Calculate r, the coefficient of linear correlation, for the data in part a.
d. Compare these results to those found in the text for Illustration 13.1 (pp. 606–607).

13.12 A formula that is sometimes given for computing the correlation coefficient is

$$r = \frac{n(\sum xy) - (\sum x)(\sum y)}{\sqrt{n(\sum x^2) - (\sum x)^2}\sqrt{n(\sum y^2) - (\sum y)^2}}$$

Use this expression as well as the formula

$$r = \frac{SS(xy)}{\sqrt{SS(x) \cdot SS(y)}}$$

to compute r for the data in the table.

x	2	4	3	4	0
y	6	7	5	6	3

13.2 Inferences About the Linear Correlation Coefficient

In Section 13.1 we learned that covariance is a measure of linear dependency. Also noted was the fact that its value is affected by the spread of the data; therefore, we standardize the covariance by dividing it by the standard deviations of both x and y. This standardized form is known as r, the coefficient of linear correlation. Standardizing enables us to compare different sets of data, thereby allowing r to play a role much like z or t does for $\bar{x}$. The calculated r value becomes $r\bigstar$, the test statistic for inferences about ρ, the population correlation coefficient. (ρ is the lowercase Greek letter "*rho*.")

ASSUMPTIONS FOR INFERENCES ABOUT THE LINEAR CORRELATION COEFFICIENT

The set of (x, y) ordered pairs forms a random sample, and the y values at each x have a normal distribution. Inferences use the t-distribution with $n - 2$ degrees of freedom.

CAUTION

Inferences about the linear correlation coefficient are about the pattern of behavior of the two variables involved and the usefulness of one variable in predicting the other. *Significance of the linear correlation coefficient does not mean that you have established a cause-and-effect relationship.* Cause and effect is a separate issue. (See the causation discussion on pages 150–151.)

CONFIDENCE INTERVAL PROCEDURE

As with other parameters, a **confidence interval** may be used to estimate the value of ρ, the linear correlation coefficient of the population. Usually this is accomplished by using a table that shows **confidence belts.** Table 10 in Appendix B gives confidence belts for 95% confidence intervals. This table is a bit tricky to read and utilizes n, the sample size; so be extra careful when you use it. The next illustration demonstrates the procedure for estimating ρ.

Illustration 13.2

Constructing a Confidence Interval for the Population Correlation Coefficient

A random sample of 15 ordered pairs of data has a calculated r value of 0.35. Find the 95% confidence interval for ρ, the population linear correlation coefficient.

SOLUTION

Step 1 **The Set-Up:**
Describe the population parameter of interest.
The linear correlation coefficient for the population, ρ.

Step 2 **The Confidence Interval Criteria:**
a. **Check the assumptions.**
The ordered pairs form a random sample, and we will assume that the y values at each x have a normal distribution.
b. **Identify the formula to be used.**
The calculated linear correlation coefficient, r.
c. **State the level of confidence:** $1 - \alpha = 0.95$.

Step 3 **The Sample Evidence:**
Collect the sample information: $n = 15$ and $r = 0.35$.

Step 4 **The Confidence Interval:**
The confidence interval is read from Table 10 in Appendix B. Find $r = 0.35$ at the bottom of Table 10. (See the arrow on Figure 13.5.) Visualize a vertical line drawn through that point. Find the two points where the belts marked for the correct sample size cross the vertical line. The sample size is 15. These two points are circled in Figure 13.5. Now look horizontally from the two circled points to the vertical scale on the left and read the confidence interval. The values are **−0.20** and **0.72.**

Step 5 **The Results:**
State the confidence interval.
The 95% confidence interval for ρ, the population coefficient of linear correlation, is −0.20 to 0.72. ■

ANSWER NOW 13.13
Find the 95% interval when a sample of $n = 25$ results in $r = 0.35$.

HYPOTHESIS-TESTING PROCEDURE

After the linear correlation coefficient, r, has been calculated for the sample data, it seems necessary to ask this question: Does the value of r indicate that there is a linear dependency between the two variables in the population from which the

FIGURE 13.5
Using Table 10 of Appendix B, Confidence Belts for the Correlation Coefficient

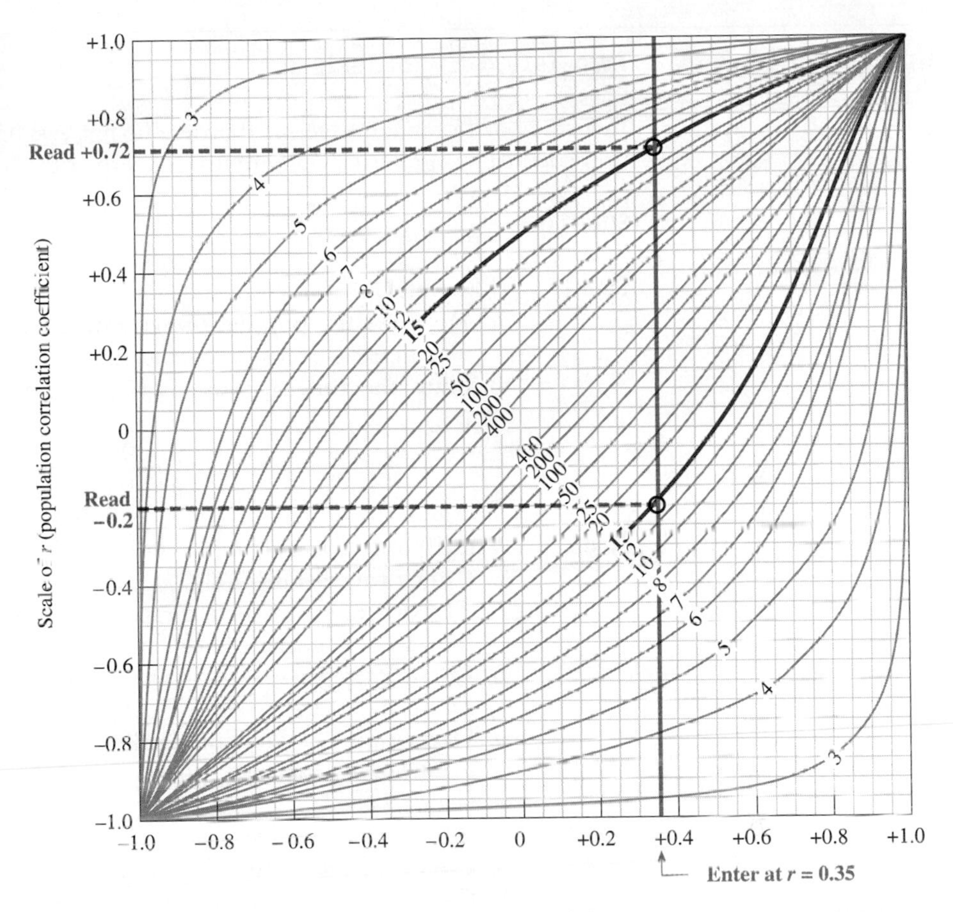

sample was drawn? To answer this question we can perform a **hypothesis test.** The null hypothesis is: The two variables are linearly unrelated ($\rho = 0$), where ρ is the linear correlation coefficient for the population. The alternative hypothesis may be either one-tailed or two-tailed. Most frequently it is two-tailed, $\rho \neq 0$. However, when we suspect that there is only a positive or only a negative correlation, we should use a one-tailed test. The alternative hypothesis of a one-tailed test is $\rho > 0$ or $\rho < 0$.

The area that represents the *p*-value or the critical region for the test is on the right when a positive correlation is expected and on the left when a negative correlation is expected. The test statistic used to test the null hypothesis is the calculated value of *r* from the sample data. Probability bounds for the *p*-value or critical values for *r* are found in Table 11 of Appendix B. The number of degrees of freedom for the *r* statistic is 2 less than the sample size, $\text{df} = n - 2$. Specific details for using Table 11 follow Illustration 13.3.

Rejection of the null hypothesis means that there is evidence of a linear relationship between the two variables in the population. Failure to reject the null hypothesis is interpreted as meaning that a linear relationship between the two variables in the population has not been shown.

Now let's look at an illustration of a hypothesis test.

Illustration 13.3 Two-Tailed Hypothesis Test

In a study of 15 randomly selected ordered pairs, $r = 0.548$. Is this linear correlation coefficient significantly different from zero at the 0.02 level of significance?

SOLUTION

Step 1 **The Set-Up:**

a. **Describe the population parameter of interest.**
The linear correlation coefficient for the population, ρ.

b. **State the null hypothesis (H_o) and the alternative hypothesis (H_a).**

$$H_o\colon \rho = 0$$
$$H_a\colon \rho \neq 0$$

Step 2 **The Hypothesis Test Criteria:**

a. **Check the assumptions.**
The ordered pairs form a random sample, and we will assume that the y values at each x have a normal distribution.

b. **Identify the test statistic to be used.**
$r\bigstar$, formula (13.3), with df $= n - 2 = 15 - 2 = 13$.

c. **Determine the level of significance:** $\alpha = 0.02$ (given in the statement of the problem).

Step 3 **The Sample Evidence:**

a. **Collect the sample information:** $n = 15$ and $r = 0.548$

b. **Calculate the value of the test statistic:**
The calculated sample linear correlation coefficient is the test statistic: $r\bigstar = 0.548$

Step 4 **The Probability Distribution:**

Using the *p*-Value Procedure

a. **Calculate the *p*-value for the test statistic.**
Use both tails because H_a expresses concern for values related to "different from."
$\mathbf{P} = P(r < -0.548) + P(r > 0.548) = 2 \cdot P(r > 0.548)$, with df $= 13$ as shown on the figure.

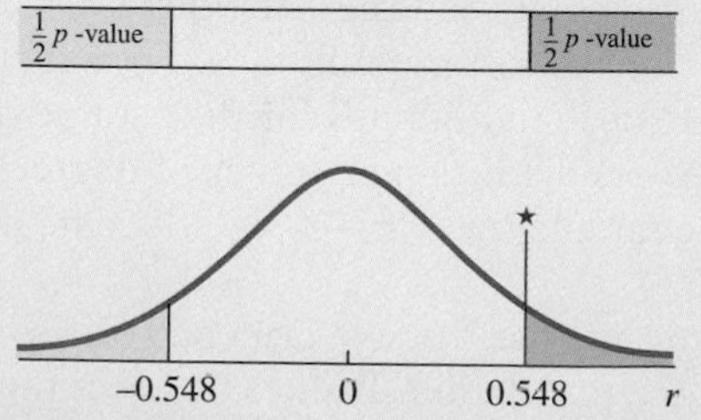

Use Table 11 in Appendix B to place bounds on the *p*-value: $\mathbf{0.02 < P < 0.05}$.
Specific details follow this illustration.

b. **Determine whether or not the *p*-value is smaller than α.**
The *p*-value is not smaller than the level of significance, α.

OR

Using the Classical Procedure

a. **Determine the critical region and critical value(s).**
The critical region is both tails because H_a expresses concern for values related to "different from." The critical value is found at the intersection of the df $= 13$ row and the two-tailed 0.02 column of Table 11: **0.592.**

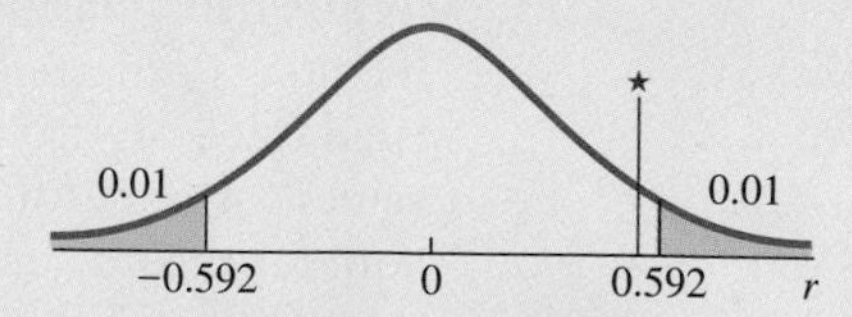

Specific details follow this illustration.

b. **Determine whether or not the calculated test statistic is in the critical region.**
$r\bigstar$ is not in the critical region, as shown in **red** on the figure.

Step 5 **The Results:**

a. State the decision about H_o: Fail to reject H_o.

b. State the conclusion about H_a.

At the 0.02 level of significance, we have failed to show that x and y are correlated.

Calculating the p-value

Use Table 11 in Appendix B to "place bounds" on the p-value. By inspecting the df = 13 row of Table 11, you can determine an interval within which the p-value lies. Locate $r\bigstar$ along the row labeled df = 13. If $r\bigstar$ is not listed, locate the two table values it falls between, and read the bounds for the p-value from the top of the table. In this case, $r\bigstar = 0.548$ is between 0.514 and 0.592; therefore, **P** is between 0.02 and 0.05. Table 11 shows only two-tailed values and the alternative hypothesis is two-tailed; therefore, the bounds for the p-value are read directly from the table.

Portion of Table 11

	Amount of α in two tails				
df	...	0.05	P	0.02	...
⋮		⋮	↑	⋮	
13		0.514	0.548	0.592	

→ $0.02 < P < 0.05$

NOTE: When H_a is one-tailed, divide the column headings by 2 to place bounds on the p-value.

ANSWER NOW 13.14

Place bounds on the p-value that results from a sample with $n = 18$ and $r = 0.444$:

a. if H_a is two-tailed

b. if H_a is one-tailed

Use Table 11 in Appendix B to find the critical values. The critical value is at the intersection of the df = 13 and the two-tailed $\alpha = 0.02$ column. Table 11 shows only two-tailed values and the alternative hypothesis is two-tailed; therefore, the critical values are read directly from the table.

Portion of Table 11

	Amount of α in two tails		
df	...	0.02	...
⋮		⋮	
13		0.592	

→ Critical values: ± 0.592

NOTE: When H_a is one-tailed, divide the column headings by 2. ■

ANSWER NOW 13.15

What are the critical values of r for $\alpha = 0.05$ and $n = 20$?

a. if H_a is two-tailed

b. if H_a is one-tailed

Application 13.1 Use of Correlation in a Medical Study

Correlation of Activated Clotting Time and Activated Partial Thromboplastin Time to Plasma Heparin Concentration

Study Objective: Determine the correlation between activated clotting time (ACT) or activated partial thromboplastin time (aPTT) and plasma heparin concentration

Design: Two-phase prospective study

Patients: Thirty patients receiving continuous-infusion intravenous heparin

Interventions: Measurement of ACT, aPTT, and plasma heparin concentrations

Heparin has been administered for over 50 years as an anticoagulant and is known to have a narrow therapeutic range. Underdosing of heparin is associated with recurrent thromboembolism, whereas excessive dosing may increase the risk of hemorrhagic complications. Several clotting time tests are available to monitor heparin, including whole blood clotting time, activated partial thromboplastin time (aPTT), and activated clotting time (ACT).

The study was conducted in two phases. In phase 1 (intraperson phase), sequential blood draws from five patients were evaluated. The goal was to determine if there was a significant relationship between plasma heparin concentrations and clotting time tests within an individual. In phase 2 (interperson phase), single random blood draws from 25 additional patients were evaluated with the same collection technique and analysis as in phase 1. Blood draws were performed within 48 hours after the start of heparin therapy. The goal of phase 2 was to determine the quantitative relationship between ACT or aPTT and plasma heparin concentration between individuals.

For both phases, correlations between ACT or aPTT results and plasma heparin concentrations were performed using the Pearson moment R correlation test. Phase 1: Linear correlation coefficients (r) for the five patients were 0.93 ($p = 0.02$), 0.99 ($p = 0.009$), 0.89 ($p = 0.12$), 0.96 ($p = 0.04$), and 0.90 ($p = 0.10$). Phase 2: Correlation coefficient for these data was 0.58 (linear, $p = 0.008$). The linear regression line formula is 137 + (52.9)(plasma heparin concentration), which, for a therapeutic heparin range of 0.3–0.7 U/ml (by antifactor Xa), equates to an ACT range of 153–174 seconds. Linear regression lines for aPTT versus plasma heparin concentration are shown in Figure 7. Correlation coefficient for these data was 0.89 (linear, $p = 0.0001$). The linear regression line formula was 14.4 + (135.4)(plasma heparin concentration), which, for the same therapeutic heparin range, equates to an aPTT range of 55–109 seconds.

FIGURE 7 **Linear aPTT versus plasma heparin concentration for phase 2 (interperson correlation and regression). Vertical dashed lines indicate the therapeutic range for plasma heparin concentration by antifactor Xa.**

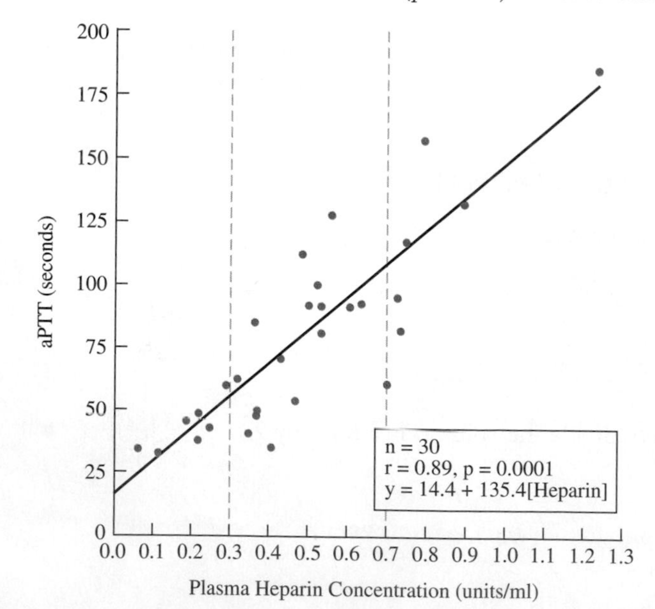

The decision analysis results indicate that a standard clotting time test therapeutic range (not derived from heparin concentration) often results in incorrect patient management decisions. The ACT based on a standard therapeutic range may result in dosage adjustment decisions that may increase the risk of bleeding (in 43% of patients). The aPTT based on a standard therapeutic range may result in dosage adjustment decisions that may increase the risk of thrombosis (in 37% of patients). A larger study in 200 patients is under way to confirm these results using heparin concentration-derived therapeutic ranges for both aPTT and ACT.

John M. Koerber, B.S., Maureen A. Smythe, Pharm.D., Robert L. Begle, M.D., Joan C. Mattson, M.D., Beverly P. Kershaw, M.S., and Susan J. Westley, M.T. (ASCP)
Pharmacotherapy 19(8):922–931, 1999.
http://www.medscape.com/viewarticle/418017_3.

ANSWER NOW 13.16

a. **Explain the meaning of "Correlation coefficient for these data was 0.58 (linear, $p = 0.008$)" as reported for Phase 2.**

b. **What is the critical value for a two-tailed test of $\rho = 0.00$ at the $\alpha = 0.01$ level?**

c. **Is $r = 0.58$ significant?**

Exercises

13.17 Using graphs to illustrate, explain the meaning of a correlation coefficient whose value is:
a. -1.0 b. 0.0 c. $+1.0$ d. $+0.5$ e. 0.6

13.18 Use Table 10 of Appendix B to determine a 95% confidence interval for the true population linear correlation coefficient based on these sample statistics:
a. $n = 8, r = 0.20$ b. $n = 100, r = -0.40$
c. $n = 25, r = +0.65$ d. $n = 15, r = 0.23$

13.19 Ex13-19 The test–retest method is one way of establishing the reliability of a test. The test is administered and then, at a later date, the same test is re-administered to the same individuals. The correlation coefficient is computed between the two sets of scores. The following test scores were obtained in a test–retest situation:

First Score	75	87	60	75	98	80	68	84	47	72
Second Score	72	90	52	75	94	78	72	80	53	70

Find r and set a 95% confidence interval for ρ.

13.20 Ex13-20 Three measures of a company's performance are its net income (earnings), revenues (sales), and the market value of its stock. But are these related? The table shows these data for 15 of the top 100 fastest growing companies listed by *Fortune* in mid-1998 ($ millions):

Company	Net Income	Revenues	Market Value of Stock
Noble Drilling	175.2	764.8	1,986
Funco	0.1	172.2	94
Marine Drilling	71.7	230.6	645
Vitesse Semiconductor	46.5	151.9	2,360
Central Garden & Pet	33.4	1,181.3	641
Jabil Circuit	74.5	1,265.0	1,178
Cliff's Drilling	54.9	316.7	325
Pairgain Technologies	48.5	291.2	895
RMI Titanium	82.5	355.4	423
Sanmina	59.8	642.6	1,717
UTI Energy	15.6	202.4	165
Veritas DGC	62.4	497.3	611
Fidelity National Financial	73.1	935.2	827
Waste Management	365.9	3,148.5	29,447
TJX	376.4	7,770.6	8,425

Source: *Fortune*, "America's Fastest Growing Companies," September 28, 1998

Calculate the correlation coefficient and use it and Table 10 of Appendix B to determine a 95% confidence interval for ρ in each case:
a. net income and revenues
b. net income and market value of stock
c. revenues and market value of stock

13.21 Ex13-21 The *State and Metropolitan Area Data Book—1991* gives the 1988 rates per 100,000 population for physicians and nurses for the West North Central states as follows:

State	Physicians	Nurses
Minnesota	212	787
Iowa	145	805
Missouri	190	745
North Dakota	167	935
South Dakota	138	809
Nebraska	168	725
Kansas	169	676

a. Calculate r. b. Set a 95% confidence interval for ρ.
c. Describe the meaning of the answer in part b.
d. Explain the meaning of the width of the interval in part b.

13.22 Ex13-22 State finances have always fascinated economists, and political leaders seeking office have often used state financial statistics to raise debatable issues at election time. Do states with higher taxes spend more? Do states with higher taxes also acquire more debt? Do states with higher expenditures acquire more debt? The data in the table below (all per capita) from the 50 states in fiscal year 1995 are offered to help study and provide an insight to these questions. Calculate the correlation coefficient and use it and Table 10 of Appendix B to determine a 95% confidence interval for ρ in each case:

a. taxes and debt
b. debt and expenditures
c. taxes and expenditures

13.23 State the null hypothesis, H_o, and the alternative hypothesis, H_a, that would be used to test each of these statements:

a. The linear correlation coefficient is positive.
b. There is no linear correlation.
c. There is evidence of negative correlation.
d. There is positive linear relationship.

13.24 Determine the critical values that would be used to test each set of hypotheses using the classical approach.

a. H_o: $\rho = 0$ vs. H_a: $\rho \neq 0$, with $n = 18$ and $\alpha = 0.05$
b. H_o: $\rho = 0$ vs. H_a: $\rho > 0$, with $n = 32$ and $\alpha = 0.01$
c. H_o: $\rho = 0$ vs. H_a: $\rho < 0$, with $n = 16$ and $\alpha = 0.05$

Table for Exercise 13.22

State	Debt	Taxes	Expenditures	State	Debt	Taxes	Expenditures
AL	$ 884	$1194	$2714	MT	$2540	$1396	$3434
AK	5351	3183	9270	NE	836	1356	2596
AZ	720	1475	2646	NV	1305	1764	2994
AR	798	1365	2663	NH	5306	800	2697
CA	1526	1686	3458	NJ	3066	1713	4104
CO	899	1209	2616	NM	1083	1688	3776
CT	4719	2282	4145	NY	3775	1891	4487
DE	4916	2224	4156	NC	632	1588	2840
FL	1085	1311	2453	ND	1334	1496	3452
GA	781	1317	2660	OH	1103	1362	3138
HI	4377	2422	5067	OK	1140	1347	2742
ID	1120	1490	2889	OR	1745	1365	3512
IL	1855	1402	2789	PA	1184	1513	3263
IN	940	1386	2634	RI	5571	1505	4308
IA	743	1549	3021	SC	1367	1297	3164
KS	447	1468	2774	SD	2282	952	2578
KY	1839	1628	2952	TN	537	1124	2556
LA	1962	1077	3331	TX	530	1084	2384
ME	2451	1461	3368	UT	1056	1371	2963
MD	1872	1599	2989	VT	2851	1370	3442
MA	4566	1910	3998	VA	1317	1327	2575
MI	1313	1856	3631	WA	1624	1877	3904
MN	975	2023	3553	WV	1415	1494	3425
MS	713	1335	2749	WI	1608	1763	3182
MO	1261	1268	2344	WY	1642	1389	4261

Source: Census Bureau, U.S. Department of Commerce, 1996

13.25 A sample of 20 pieces of bivariate data has a linear correlation coefficient of $r = 0.43$. Does this provide sufficient evidence to reject the null hypothesis that $\rho = 0$ in favor of a two-sided alternative? Use $\alpha = 0.10$.

13.26 If a sample of size 18 has a linear correlation coefficient of -0.50, is there significant reason to conclude that the linear correlation coefficient of the population is negative? Use $\alpha = 0.01$.

13.27 In a study involving 24 coastal drainage basins in the Mendocino triple junction region of northern California (*Geological Society of America Bulletin*, November 1989), it is reported that the Pearson correlation coefficient between the uplift rate and the length of the drainage basin equals 0.16942. Use Table 11 in Appendix B to determine whether these data provide evidence sufficient to reject H_o: $\rho = 0$ in favor of H_a: $\rho \neq 0$ at $\alpha = 0.05$.

13.28 Is a value of $r = +0.24$ significant in showing that ρ is greater than zero for a sample of 62 data at the 0.05 level of significance?

13.29 Ex13-29 The population (in millions) and the violent crime rate (per 1000) were recorded for ten metropolitan areas.

Population	10.0	1.3	2.1	7.0	4.4	0.3	0.3	0.2	0.2	0.4
Crime Rate	12.0	9.5	9.2	8.4	8.2	7.3	7.1	7.0	6.9	6.9

Do these data provide evidence to reject the null hypothesis that $\rho = 0$ in favor of $\rho \neq 0$ at $\alpha = 0.05$?

13.30 Ex13-30 Going to work for a particular company is a major decision in anyone's career path. The popularity of any company may be indicated by both the number of job applications it considers each year and the number of people it already employs. The following table was extracted from *Fortune's* 100 best companies to work for, published in early 1998:

Company	U.S. Employees	Applicants
Southwest Airlines	24,757	150,000
Kingston Technology	552	4,000
SAS Institute	3,154	12,000
W.L. Gore	4,118	23,717
Microsoft	14,936	150,000
Merck	31,767	165,000
Hewlett-Packard	66,300	255,000
Goldman Sachs	6,546	8,000
Corning	8,127	60,000
Harley-Davidson	5,288	8,000

Source: *Fortune*, "The 100 Best Companies to Work for in America," January 12, 1998

a. Do these data provide evidence to reject the null hypothesis that $\rho = 0$ in favor of $\rho > 0$ at $\alpha = 0.01$?
b. Explain the meaning of the apparent positive correlation.

13.31 Ex13-31 Two indicators of the level of economic activity in a given geographical area are its total personal income and the value of the construction contracts. The table lists the data for seven states during the third quarter of 1997:

State	Personal Income ($ millions)	Construction Contracts (Index, 1980 = 100)
Colorado	93,706	274.0
Kansas	56,092	202.4
Missouri	115,588	239.3
Nebraska	35,588	263.9
New Mexico	30,254	162.0
Oklahoma	60,819	125.1
Wyoming	9,678	88.8

Source: *Regional Economic Digest*, "Economic Indicators," first quarter 1998

a. Calculate the correlation coefficient between the two variables.
b. Test for a significant correlation at the 0.05 level of significance and draw your conclusion.

13.32 Ex13-32 Sugar beet growers are interested in realizing higher yields and higher sucrose percentages from their crops. But do the two go together? The data listed are for Minnesota's 2001 sugar beet crop: Values listed are by county, Yield is tons per acre, and Sucrose is percentage of sucrose.

County	Yield	Sucrose
Becker	18.0	17.8
Chippewa	19.3	16.6
Clay	17.9	18.1
Grant	18.3	17.0
Kandiyohi	21.3	16.6
Kittson	11.2	17.9
McLeod	17.9	16.8
Mahnomen	18.8	17.9
Marshall	15.7	19.0
Meeker	19.2	17.0
Norman	19.1	17.6
Otter Tail	19.8	17.3
Polk	18.7	17.9
Pope	21.5	17.1
Red Lake	19.9	16.9
Redwood	20.0	16.3
Renville	20.6	16.5
Sibley	19.7	16.2
Stearns	20.2	17.3

(continued)

County	Yield	Sucrose
Stevens	21.3	16.8
Swift	18.6	17.0
Traverse	16.6	17.3
Wilkin	17.7	17.6
Yellow Medicine	19.0	16.7

a. What, if any, relationship do you expect to find between the yield per acre and the sucrose percentage for sugar beets?
b. Draw the scatter diagram for yield in tons per acre (x) and sucrose percentage (y) for the Minnesota data. Describe the relationship as seen on the scatter diagram. Is it what you anticipated?
c. Find the linear correlation coefficient.
d. Is the linear correlation coefficient significantly different from zero at the 0.05 level?
e. Two of the ordered pairs appear to be outside of the pattern created by the other 22 ordered pairs. What effect do you think removing these two pairs from the data would have on the appearance of the scatter diagram? On the linear correlation coefficient? On the answer in part d?
f. Remove Kittson and Marshall Counties from the data and redo parts b, c, and d. Compare the results with your answers to part e.

13.3 Linear Regression Analysis

Recall that the **line of best fit** results from an analysis of two (or more) related quantitative variables. (We will restrict our work to two variables.) When two variables are studied jointly, we often would like to control one variable by controlling the other. Or we might want to predict the value of a variable based on knowledge about another variable. In both cases we want to find the line of best fit, provided one exists, that will best predict the value of the dependent, or output, variable from a value of the independent, or input, variable. Recall that the variable we know or can control is called the *independent*, or input, variable; the variable that results from using the equation of the line of best fit is called the *dependent*, or predicted, variable.

In Chapter 3 we developed the method of least squares. From this concept, formulas (3.7) and (3.6) were obtained and used to calculate b_0 (the **y-intercept**) and b_1 (the **slope of the line of best fit**):

$$b_0 = \frac{1}{n}(\sum y - b_1 \cdot \sum x) \tag{3.7}$$

$$b_1 = \frac{SS(xy)}{SS(x)} \tag{3.6}$$

Then these two coefficients are used to write the equation of the line of best fit in the form

$$\hat{y} = b_0 + b_1 x$$

When the line of best fit is plotted, it does more than just show us a pictorial representation of the line. It tells us two things: (1) whether or not there really is a linear relationship between the two variables and (2) the quantitative (equation) relationship between the two variables. When there is no relationship between the variables, a horizontal line of best fit will result. A horizontal line has a slope of zero, which implies that the value of the input variable has no effect on the output variable. (This idea will be amplified later in this chapter.)

The result of regression analysis is the mathematical equation of the line of best fit. We will, as mentioned before, restrict our work to the **simple linear** case—

that is, one input variable and one output variable where the line of best fit is straight. However, you should be aware that not all relationships are of this nature. If the scatter diagram suggests something other than a straight line, the relationship may be **curvilinear regression.** In cases of this type we must introduce terms to higher powers, x^2, x^3, and so on, or other functions, e^x, $\log x$, and so on; or we must introduce other input variables. Maybe two or three input variables would improve the usefulness of our regression equation. These possibilities are examples of curvilinear regression and **multiple regression.**

The linear model used to explain the behavior of linear bivariate data in the population is

$$\hat{y} = \beta_0 + \beta_1 x + \epsilon \tag{13.4}$$

This equation represents the linear relationship between the two variables in a population. β_0 is the y-intercept and β_1 is the slope. ϵ (lowercase Greek letter "epsilon") is the random **experimental error** in the observed value of y at a given value of x.

The **regression line** from the sample data gives us b_0, which is our estimate of β_0, and b_1, our estimate of β_1. The error ϵ is approximated by $e = y - \hat{y}$, the difference between the observed value of y and the predicted value of y, $\hat{y}$, at a given value of x:

Estimate of the Experimental error

$$e = y - \hat{y} \tag{13.5}$$

FYI
Explore Interactivity 13A on your CD.

The random variable e (also known as the "residual") is positive when the observed value of y is larger than the predicted value, $\hat{y}$; e is negative when y is less than $\hat{y}$. The sum of the errors (residuals) for all values of y for a given value of x is exactly zero. (This is part of the least squares criteria.) Thus the mean value of the experimental error is zero; its variance is σ_ϵ^2. Our next goal is to estimate this **variance of the experimental error.**

Before we estimate the variance of ϵ, let's try to understand exactly what the error represents: ϵ is the amount of error in our observed value of y. That is, it is the difference between the observed value of y and the mean value of y at that particular value of x. Since we do not know the mean value of y, we will use the regression equation and estimate it with $\hat{y}$, the predicted value of y at this same value of x. Thus the best estimate that we have for ϵ is $e = y - \hat{y}$, as shown in Figure 13.6.

The Error e Is $y - \hat{y}$

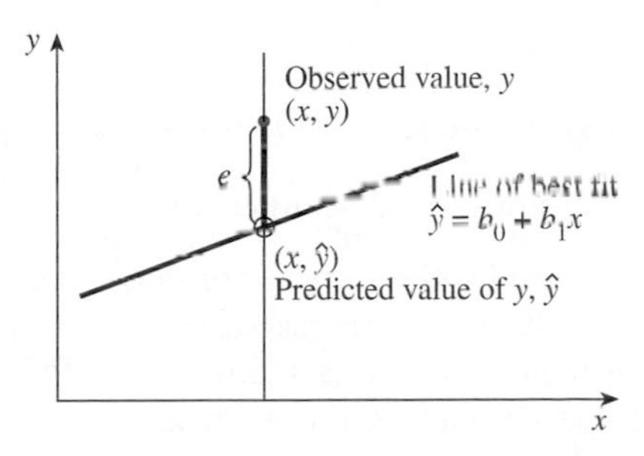

NOTE e is the observed error in measuring y at a specified value of x.

If we were to observe several values of y at a given value of x, we could plot a distribution of y values about the line of best fit (about $\hat{y}$, in particular). Figure 13.7 (p. 622) shows a sample of bivariate values that share a common x value. Figure 13.8 shows the theoretical distribution of all possible y values at a given x value. A

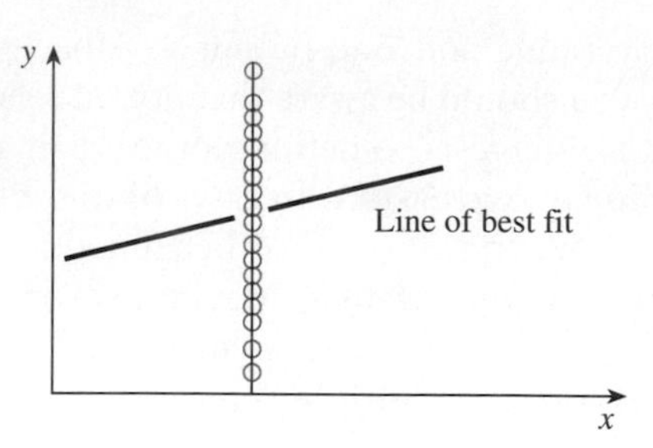

FIGURE 13.7
Sample of y Values at a Given x

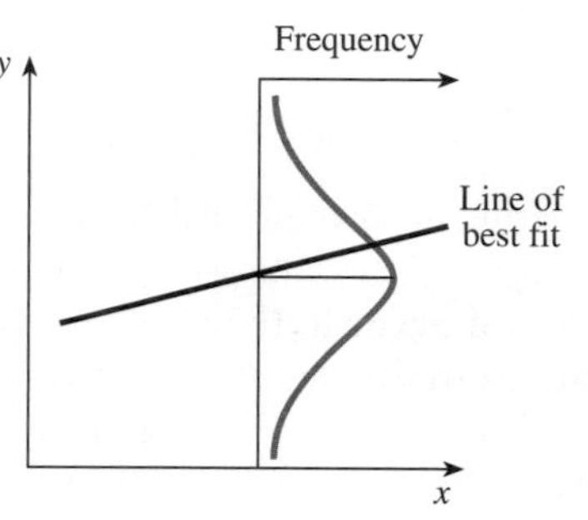

FIGURE 13.8
Theoretical Distribution of y Values for a Given x

similar distribution occurs at each different value of x. The mean of the observed y's at a given value of x varies, but it can be estimated by $\hat{y}$.

Before we can make any inferences about a regression line, we must assume that the distribution of y's is approximately normal and that the variances of the distributions of y at all values of x are the same. That is, the standard deviation of the distribution of y about $\hat{y}$ is the same for all values of x, as shown in Figure 13.9.

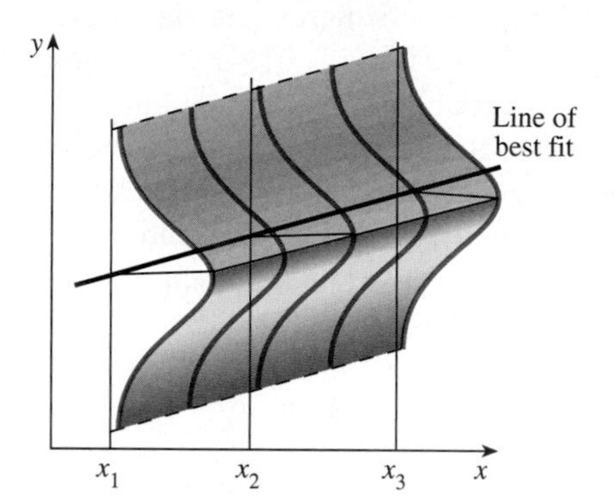

FIGURE 13.9
Standard Deviation of the Distribution of y Values Is the Same for All x

Before we look at the variance of e, let's review the definition of sample variance. The sample variance, s^2, is defined as $\frac{\sum(x - \bar{x})^2}{n - 1}$, the sum of the squares of each deviation divided by the number of degrees of freedom, $n - 1$, associated with a sample of size n. The variance of y involves an additional complication: There is a different mean for y at each value of x. (Notice the many distributions in Figure 13.9.) However, each of these "means" is actually the predicted value, $\hat{y}$, that corresponds to the x that fixes the distribution. So the variance of the error e is estimated by the formula

$$s_e^2 = \frac{\sum(y - \hat{y})^2}{n - 2} \tag{13.6}$$

where $n - 2$ is the number of degrees of freedom.

NOTE The variance of y about the line of best fit is the same as the variance of the error e. Recall that $e = y - \hat{y}$.

Formula (13.6) can be rewritten by substituting $b_0 + b_1x$ for $\hat{y}$. Since $\hat{y} = b_0 + b_1x$, we have

$$s_e^2 = \frac{\sum(y - b_0 - b_1x)^2}{n - 2} \tag{13.7}$$

With some algebra and some patience, this formula can be rewritten once again into a more workable form. The form we will use is

VARIANCE OF THE ERROR *e*

$$s_e^2 = \frac{(\sum y^2) - (b_0)(\sum y) - (b_1)(\sum xy)}{n - 2} \tag{13.8}$$

For ease of discussion let's agree to call the numerator of formulas (13.6), (13.7), and (13.8) the **sum of squares for error** (SSE).

Now let's see how we can use all of this information.

Illustration 13.4 Determining the Variance of *y* About the Regression Line

Suppose that you move to a new city and take a job. You will, of course, be concerned about the problems you will face commuting to and from work. For example, you would like to know how long it will take you to drive to work each morning. Let's use "one-way distance to work" as a measure of where you live. You live x miles away from work and want to know how long it will take you to commute each day. Your new employer, foreseeing this question, has already collected a random sample of data to be used in answering your question. Fifteen of your co-workers were asked to give their one-way travel times and distances to work. The resulting data are shown in Table 13.2. (For convenience the data have been arranged so that the x values are in numerical order.) Find the line of best fit and the variance of y about the line of best fit, s_e^2.

TABLE 13.2 Data on Commute Distances and Times Ta13-2

Co-worker	Miles (x)	Minutes (y)	x^2	xy	y^2
1	3	7	9	21	49
2	5	20	25	100	400
3	7	20	49	140	400
4	8	15	64	120	225
5	10	25	100	250	625
6	11	17	121	187	289
7	12	20	144	240	400
8	12	35	144	420	1225
9	13	26	169	338	676
10	15	25	225	375	625
11	15	35	225	525	1225
12	16	32	256	512	1024
13	18	44	324	792	1936
14	19	37	361	703	1369
15	20	45	400	900	2025
Total	184	403	2616	5623	12,493

SOLUTION

The extensions and summations needed for this problem are shown in Table 13.2. The line of best fit can now be calculated using formulas (2.9), (3.4), (3.6), and (3.7). From formula (2.9):

$$SS(x) = \sum x^2 - \frac{(\sum x)^2}{n}: \qquad SS(x) = 2616 - \frac{(184)^2}{15} = 358.9333$$

From formula (3.4):

$$SS(xy) = \sum xy - \frac{\sum x \cdot \sum y}{n}: \qquad SS(xy) = 5623 - \frac{(184)(403)}{15} = 679.5333$$

We use formula (3.6) for the slope:

$$b_1 = \frac{SS(xy)}{SS(x)}: \qquad b_1 = \frac{679.5333}{358.9333} = 1.893202 = \mathbf{1.89}$$

We use formula (3.7) for the y-intercept:

FYI
Use extra decimal place during these calculations.

$$b_0 = \frac{\sum y - (b_1 \cdot \sum x)}{n}: \qquad b_0 = \frac{403 - (1.893202)(184)}{15} = 3.643387 = \mathbf{3.64}$$

Therefore, the equation of the line of best fit is

$$\hat{y} = 3.64 + 1.89x$$

The variance of y about the regression line is calculated by using formula (13.8):

$$s_e^2 = \frac{(\sum y^2) - (b_0)(\sum y) - (b_1)(\sum xy)}{n - 2}:$$

$$s_e^2 = \frac{(12{,}493) - (3.643387)(403) - (1.893202)(5623)}{15 - 2} = \frac{379.2402}{13} = \mathbf{29.17}$$

$s_e^2 = 29.17$ is the variance of the 15 e's. In Figure 13.10 the 15 e's are shown as vertical line segments.

NOTE Extra decimal places are often needed for this type of calculation. Notice that b_1 (1.893202) was multiplied by 5623. If 1.89 had been used instead, that one product would have changed the numerator by approximately 18. That, in turn, would have changed the final answer by almost 1.4—a sizable round-off error.

FIGURE 13.10
The 15 Random Errors as Line Segments

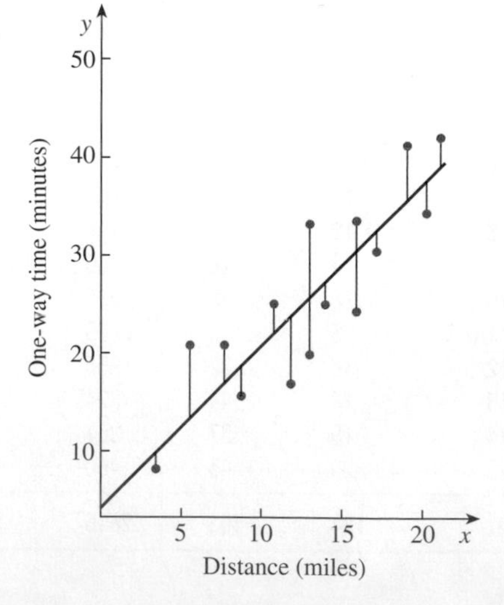

■

FYI
Computer and calculator commands to find the regression line for a set of bivariate data can be found in Chapter 3 (pp. 161–162).

In the sections that follow, we will use the variance of e in much the same way as the variance of x (as calculated in Chapter 2) was used in Chapters 8, 9, and 10 to complete the statistical inferences studied there.

Exercises

13.33 Ex13-33 Ten salespeople were surveyed, and the average number of client contacts per month, x, and the sales volume, y (in thousands), were recorded for each:

x	12	14	16	20	23	46	50	48	50	55
y	15	25	30	30	30	80	90	95	110	130

Refer to the computer output and verify that the equation of the line of best fit is $\hat{y} = 13.4 + 2.3x$ and that $s_e = 10.17$ by calculating these values yourself.

```
The regression equation is y = -13.4 + 2.30 x
Predictor        Coef
Constant      -13.414
x              2.3028
s = 10.17
```

13.34 Ex13-34 Twelve of Wisconsin's sweet-corn–producing counties were randomly selected and the following information about their 2001 crop was recorded: acres planted (100s of acres) and total production (100s of tons).

County	Acres Planted	Production
Jefferson	15	89
Walworth	21	146
Dane	12	78
Ozaukee	15	102
Racine	12	64
Juneau	32	212
Green Lake	51	332
Winnebago	36	209
Sheboygan	32	169
Adams	110	765
Green	10	57
St. Croix	20	105

Source: http://www.usda.gov/nass/graphics/county01/data/vsc01.csv

a. Investigate the relationship between the number of acres planted and the total tons of sweet corn produced. Include a scatter diagram, the linear correlation coefficient and line of best fit, and a statement about their meaning.

b. If you were advising a Wisconsin sweet corn grower, based on your information, how many tons of sweet corn, on the average, can he expect to produce for each acre planted?

13.35 Ex13-35 Fourteen of Minnesota's sweet-corn–producing counties were randomly selected and the following information about their 2001 crop was recorded: acres planted (100s of acres) and yield rate (tons of sweet corn produced per acre harvested).

County	Acres Planted	Yield Rate
Waseca	60	5.90
Freeborn	70	6.41
Martin	20	5.90
Dakota	52	5.00
McLeod	22	5.50
Redwood	62	5.90
Stearns	10	6.30
Dodge	46	5.80
Kandiyohi	20	6.50
Olmsted	87	5.60
Goodhue	54	5.70
Meeker	15	6.00
Nicollet	35	5.91
Sherburne	10	7.00

Source: http://www.usda.gov/nass/graphics/county01/data/vsc01.csv

Investigate the relationship between the number of acres planted and the yield rate. Include a scatter diagram, the linear correlation coefficient and line of best fit, and a statement about their meaning.

13.36 Ex13-36 Diamonds are often thought of as a cherished item with a personal value well in excess of their monetary value. The monetary value of a diamond is determined by its exact quality as defined by the four C's: cut, color, clarity, and carat weight. The price and the carat weight of a diamond are its two most known characteristics. In order to understand the role carat weight has in determining the price of a diamond, the carat weight and price of 20 loose round diamonds, all of color D and clarity VS1, were obtained October 19, 2002, on the Internet. (See table at the top of p. 626.)

a. Draw a scatter diagram of the data: carat weight (x) and price (y).

b. Do the data suggest a linear relationship for the domain 0.50 to 0.66 carat? Discuss your findings in part a.

Carat Weight	Price	Carat Weight	Price
0.58	$2791	0.55	$2739
0.64	2803	0.59	2467
0.50	1974	0.52	2595
0.53	2559	0.56	2317
0.50	2553	0.52	2623
0.50	2109	0.60	2979
0.56	2348	0.52	2651
0.51	2011	0.50	2499
0.55	2278	0.66	2909
0.50	1997	0.53	2226

Source: http://www.overnightdiamonds.com/diamondlist1.htm

c. Diamonds smaller than 0.50 carat and diamonds larger than 0.66 carat may not fit the linear pattern demonstrated by these data. Explain.
d. Find the equation of the line of best fit.
e. According to this information, what would be a typical price for a 0.50-carat loose diamond of this quality?
f. On the average, by how much does the price increase for each extra 0.01 carat in weight? Within what interval of x values would you expect this to be true?
g. Find the variance of y about the regression line. What characteristics in the scatter diagram support this large value?

13.37 Ex13-37 There's an old adage that higher fat content in foods means more calories. A list of restaurant foods appeared in the December 1997 *Nutrition Action Healthletter* that gave both the calories and fat contents of foods that people typically consume at restaurants.

Restaurant Foods	Calories	Fat (grams)
Kung Pao chicken with rice	1620	76
Chicken burrito with extras	1530	68
Spaghetti with meatballs	1155	39
Porterhouse steak (20 oz.)	1100	82
Tuna salad sandwich with mayo	835	56
Turkey club sandwich	735	34
Pizza Hut Pan Pizza	700	34
Movie theater popcorn with butter	630	50
Baked potato with sour cream	620	31
Meat loaf	570	38
Burger King BK Broiler	550	29
Ham and cheese omelette	510	39
Grilled cheese sandwich	510	33
AuBon Pain Blueberry Muffin	430	18
McDonald's Cheese Danish	410	22
KFC Chicken Breast	400	24
McDonald's Grilled Chicken Salad	350	23

(continued)

Source: *Nutrition Action Healthletter*, Vol. 24, No. 10, December 1997

a. Draw a scatter diagram of the data. Use calories as the dependent variable (y) and fat content as the independent variable (x).
b. Find the equation of the line of best fit and draw it on the scatter diagram.
c. What is the average increase in calories for each additional gram of fat in the food?
d. Find the variance of y about the regression line.

13.38 Ex13-38 The computer-science aptitude score, x, and the achievement score, y (measured by a comprehensive final), were recorded (see table at bottom) for 20 students in a beginning computer-science course. Find the equation of the line of best fit and s_e^2.

13.39

a. Using the ten points given in the table, find the equation of the line of best fit, $\hat{y} = b_0 + b_1x$, and draw it on a scatter diagram.

Point	A	B	C	D	E	F	G	H	I	J
x	1	1	3	3	5	5	7	7	9	9
y	1	2	2	3	3	4	4	5	5	6

b. Find the ordinates $\hat{y}$ of the points on the line of best fit whose abscissas are $x = 1, 3, 5, 7$, and 9.
c. Find the value of e for each of the points in the given data ($e = y - \hat{y}$).
d. Find the variance s_e^2 of those points about the line of best fit by using formula (13.6).
e. Find the variance s_e^2 by using formula (13.8). (Your answers to parts d and e should be the same.)

13.40 Ex13-40 The following data show the number of hours studied for an exam, x, and the grade received on the exam, y (y is measured in 10's; that is, $y = 8$ means that the grade, rounded to the nearest 10 points, is 80).

x	2	3	3	4	4	5	5	6	6	6	7	7	7	8	8
y	5	5	7	5	7	7	8	6	9	8	7	9	10	8	9

a. Draw a scatter diagram of the data.

Table for Exercise 13.38

x	4	16	20	13	22	21	15	20	19	16	18	17	8	6	5	20	18	11	19	14
y	19	19	24	36	27	26	25	28	17	27	21	24	18	18	14	28	21	22	20	21

b. Find the equation of the line of best fit and draw it on the scatter diagram.
c. Find the ordinates $\hat{y}$ that correspond to $x = 2, 3, 4, 5, 6, 7$, and 8.
d. Find the five values of e that are associated with the points where $x = 3$ and $x = 6$.
e. Find the variance s_e^2 of all the points about the line of best fit.

13.4 Inferences Concerning the Slope of the Regression Line

Now that the equation of the line of best fit has been found and the linear model has been verified (by inspection of the scatter diagram), we are ready to determine whether we can use the equation to predict y. We will test the null hypothesis: The equation of the line of best fit is of no value in predicting y given x. That is, the null hypothesis to be tested is: β_1 (the slope of the relationship in the population) is zero. If $\beta_1 = 0$, then the linear equation will be of no real use in predicting y.

Before we look at the confidence interval or the hypothesis test, let's discuss the **sampling distribution** of the slope. If random samples of size n are repeatedly taken from a bivariate population, then the calculated slopes, the b_1's, will form a sampling distribution that is normally distributed with a mean of β_1, the population value of the slope, and with a variance of $\sigma_{b_1}^2$, where

$$\sigma_{b_1}^2 = \frac{\sigma_\epsilon^2}{\sum(x - \bar{x})^2} \tag{13.9}$$

provided there is no lack of fit. An appropriate estimator for $\sigma_{b_1}^2$ is obtained by replacing σ_ϵ^2 by s_e^2, the estimate of the variance of the error about the regression line:

$$s_{b_1}^2 = \frac{s_e^2}{\sum(x - \bar{x})^2} \tag{13.10}$$

This formula may be rewritten in the following, more managcable form:

$$s_{b_1}^2 = \frac{s_e^2}{\sum x^2 - \dfrac{(\sum x)^2}{n}} \tag{13.11}$$

NOTE The "**standard error** of __" is the standard deviation of the sampling distribution of __. Therefore, the *standard error of regression* (slope) is σ_{b_1} and is estimated by s_{b_1}.

FYI
Recall that we found SS(x) with formula (2.9).

In our illustration of commute times and distances, the variance among the b_1's is estimated by using formula (13.11):

$$s_{b_1}^2 = \frac{s_e^2}{\sum x^2 - \dfrac{(\sum x)^2}{n}}: \quad s_{b_1}^2 = \frac{29.1723}{358.9333} = 0.081275 = \mathbf{0.0813}$$

ASSUMPTIONS FOR INFERENCES ABOUT LINEAR REGRESSION

The set of (x, y) ordered pairs forms a random sample, and the y values at each x have a normal distribution. Since the population standard deviation is unknown and replaced with the sample standard deviation, the t-distribution will be used with $n - 2$ degrees of freedom.

CONFIDENCE INTERVAL PROCEDURE

The slope β_1 of the regression line of the population can be estimated by means of a confidence interval. The confidence interval is determined by

$$b_1 \pm t(n-2,\alpha/2) \cdot s_{b_1} \tag{13.12}$$

Illustration 13.5 Constructing a Confidence Interval for β_1, the Population Slope of the Line of Best Fit

Find the 95% confidence interval for the population's slope, β_1, for Illustration 13.4.

SOLUTION

Step 1 **The Set-Up:**
Describe the population parameter of interest.
The slope, β_1, of the line of best fit for the population.

Step 2 **The Confidence Interval Criteria:**

a. **Check the assumptions.**
The ordered pairs form a random sample, and we will assume that the y values (minutes) at each x (miles) have a normal distribution.

b. **Identify the probability distribution and the formula to be used.**
The Student's t-distribution and formula (13.12).

c. **State the level of confidence:** $1 - \alpha = 0.95$

Step 3 **The Sample Evidence:**
Collect the sample information: $n = 15$, $b_1 = 1.89$, and $s^2_{b_1} = 0.0813$.

Step 4 **The Confidence Interval:**

a. **Determine the confidence coefficients.**
From Table 6 in Appendix B, we find $t(\text{df},\alpha/2) = t(13, 0.025) = 2.16$.

b. **Find the maximum error of estimate.**
We use formula (13.12) to find

$$E = t(n-2,\alpha/2) \cdot s_{b_1}: \qquad E = (2.16) \cdot \sqrt{0.0813} = 0.6159$$

c. **Find the lower and upper confidence limits.**

$$b_1 - E \quad \text{to} \quad b_1 + E$$

$$1.89 - 0.62 \quad \text{to} \quad 1.89 + 0.62$$

Thus, 1.27 to 2.51 is the 95% confidence interval for β_1.

Step 5 **The Results:**
State the confidence interval.
We can say that the slope of the line of best fit of the population from which the sample was drawn is between 1.27 and 2.51 with 95% confidence. ■

Application 13.2 Reexamining the Use of Seriousness Weights in an Index of Crime

Regression of the Arizona UCR Index on the average seriousness index produces the linear relationship depicted in the figure. Also shown is the 95% confidence interval (3.001, 3.262), which is based upon a standard error of 0.065 on the estimate of the slope. The regression equation for this relationship is

$$S_t = -3953.85 + 3.13A_t$$

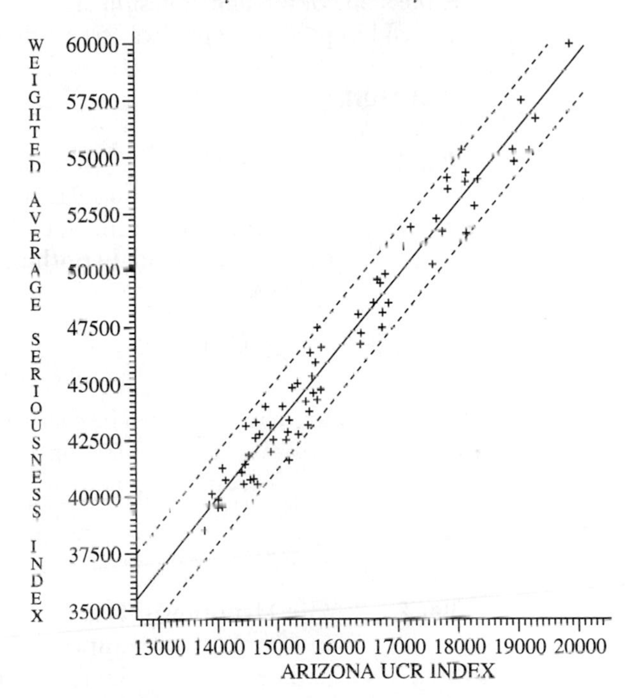

Source: Reprinted with permission from the *Journal of Criminal Justice*, Volume 17, Thomas Epperlein and Barbara C. Nienstedt, "Reexamining the Use of Seriousness Weights in an Index of Crime," 1989, Pergamon Press, Inc.

ANSWER NOW 13.41

a. The vertical scale on the graph is drawn at $A_t = 12{,}600$, and the line of best fit appears to intersect the vertical scale at approximately 35,500. Verify the coordinates of this point of intersection.

b. The article gives an interval estimate of (3.001, 3.262). Verify this 95% interval using the information given in the article.

HYPOTHESIS-TESTING PROCEDURE

We are now ready to test the hypothesis $\beta_1 = 0$. That is, we want to determine whether the equation of the line of best fit is of any real value in predicting y. For this hypothesis test, the null hypothesis is always H_o: $\beta_1 = 0$. It will be tested using the Student's t-distribution with df $= n - 2$ degrees of freedom and the test statistic $t\bigstar$ found using formula (13.13):

$$t\bigstar = \frac{b_1 - \beta_1}{s_{b_1}} \qquad (13.13)$$

Illustration 13.6 One-Tailed Hypothesis Test for the Slope of the Regression Line

Is the slope of the line of best fit significant enough to show that one-way distance is useful in predicting one-way travel time in Illustration 13.4? Use $\alpha = 0.05$.

SOLUTION

Step 1 **The Set-Up:**

a. **Describe the population parameter of interest.**
β_1, the slope of the line of best fit for the population.

b. **State the null hypothesis (H_o) and the alternative hypothesis (H_a).**

H_o: $\beta_1 = 0$ (This implies that x is of no use in predicting y; that is, $\hat{y} = \bar{y}$ would be as effective.)

The alternative hypothesis can be either one-tailed or two-tailed. If we suspect that the slope is positive, as in Illustration 13.4, a one-tailed test is appropriate.

H_a: $\beta_1 > 0$ (We expect travel time y to increase as the distance x increases.)

Step 2 **The Hypothesis Test Criteria:**

a. **Check the assumptions.**
The ordered pairs form a random sample, and we will assume that the y values (minutes) at each x (miles) have a normal distribution.

b. **Identify the probability distribution and the test statistic to be used.**
The t-distribution with df $= n - 2 = 13$, and the test statistic $t\bigstar$ from formula (13.13).

c. **Determine the level of significance:** $\alpha = 0.05$.

Step 3 **The Sample Evidence:**

a. **Collect the sample information:** $n = 15$, $b_1 = 1.89$, and $s_{b_1}^2 = 0.0813$.

b. **Calculate the value of the test statistic.**
Using formula (13.13), we find the observed value of t:

$$t\bigstar = \frac{b_1 - \beta_1}{s_{b_1}}: \qquad t\bigstar = \frac{1.89 - 0.0}{\sqrt{0.0813}} = 6.629 = \mathbf{6.63}$$

Step 4 **The Probability Distribution:**

Using the *p*-Value Procedure

a. Calculate the *p*-value for the test statistic.

Use the right-hand tail because H_a expresses concern for values related to "positive."

P = $P(t\star > 6.63$, with df = 13) as shown on the figure.

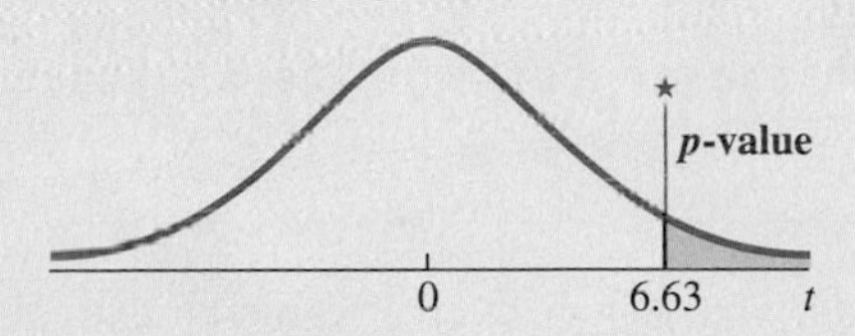

To find the *p*-value, use one of three methods:

1. Use Table 6 in Appendix B to place bounds on the *p*-value: **P < 0.005.**
2. Use Table 7 in Appendix B to place bounds on the *p*-value: **P < 0.001.**
3. Use a computer or calculator to find the *p*-value: **P = 0.0000082.**

Specific details are on page 422.

b. Determine whether or not the *p*-value is smaller than α.

The *p*-value is smaller than the level of significance, α.

OR

Using the Classical Procedure

a. Determine the critical region and critical value(s).

The critical region is the right-hand tail because H_a expresses concern for values related to "positive." The critical value is found in Table 6:

$t(13, 0.05) = \mathbf{1.77}$

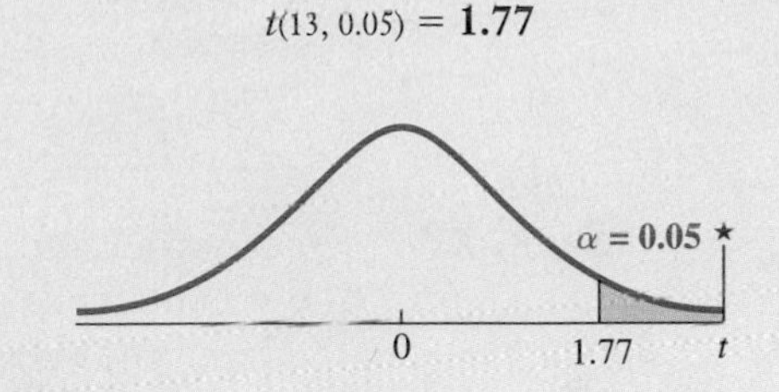

Specific instructions are on pages 414–415.

b. Determine whether or not the calculated test statistic is in the critical region.

$t\star$ is in the critical region, as shown in **red** on the figure.

Step 5 **The Results:**

a. **State the decision about H_o:** Reject H_o.

b. **State the conclusion about H_a.**

At the 0.05 level of significance, we conclude that the slope of the line of best fit in the population is greater than zero. The evidence indicates that there is a linear relationship and that the one-way distance (x) is useful in predicting the travel time to work (y). ■

Technology Instructions: Regression Analysis

Output includes the equation for the regression line, information for a *t*-test concerning the slope of the regression line, the standard deviation of error, r and/or r^2, and a scatter diagram showing the regression line.

MINITAB (Release 13)

MINITAB output also includes the predicted y values for given x values and residuals.

Input the x-variable data into C1 and the corresponding y-variable data into C2; then continue with:

```
Choose:   Stat > Regression > Regression...
Enter:    Response (y): C2
          Predictors (x): C1
Select:   Type of Regression: Linear
Choose:   Storage
Select:   Residuals
          Fits > OK > OK
```

(continued)

Technology Instructions: Regression Analysis (continued)

MINITAB (Release 13) (continued)

```
Choose:   Graph > Plot
Enter:    Graph 1: Y: C2 X: C1
Choose:   Annotation > Title
Enter:    your title
Choose:   Annotation > Line
Enter:    Points: C1 C3(whichever column Fits is located)
          Type: Solid
```

Excel XP

Excel output also includes predicted y values for given x values, residuals, and a $1 - \alpha$ confidence interval for the slope.

Input the x-variable data into column A and the corresponding y-variable data into column B; then continue with:

```
Choose:   Tools > Data Analysis > Regression
Enter:    Input Y Range: (B1:B10 or select cells)
          Input X Range: (A1:A10 or select cells)
Select:   Labels (if necessary)
          Confidence Level:
Enter:    95% (desired level)
Select:   Output Range:
Enter:    (C1 or select cell)
Select:   Line Fits Plots > OK
```

To make the output more readable, continue with: Format > Column > Autofit Selection.

FYI
Additional commands to adjust the window can be found on pages 139–140.

TI-83 Plus

Input the x-variable data into L1 and the corresponding y-variable data into L2; then continue with the following, entering the apppropriate values and highlighting Calculate:

```
Choose:   STAT > TESTS >
          E:LinRegTTest
(To enter Y1, use: VARS > YVARS >
1:Function... > 1:Y1.)
```

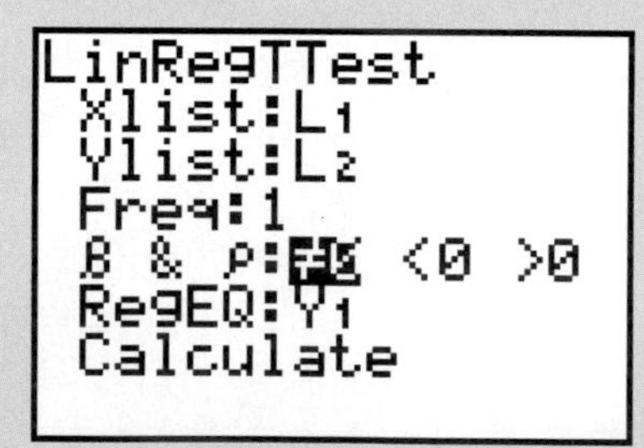

Enter the following to obtain a scatter diagram with regression line:

```
Choose:   2nd > STATPLOT >
          1:Plot1...On

Choose:   ZOOM > 9:ZoomStat >
          Trace
```

Here is the MINITAB printout with explanations for parts of Illustration 13.4.

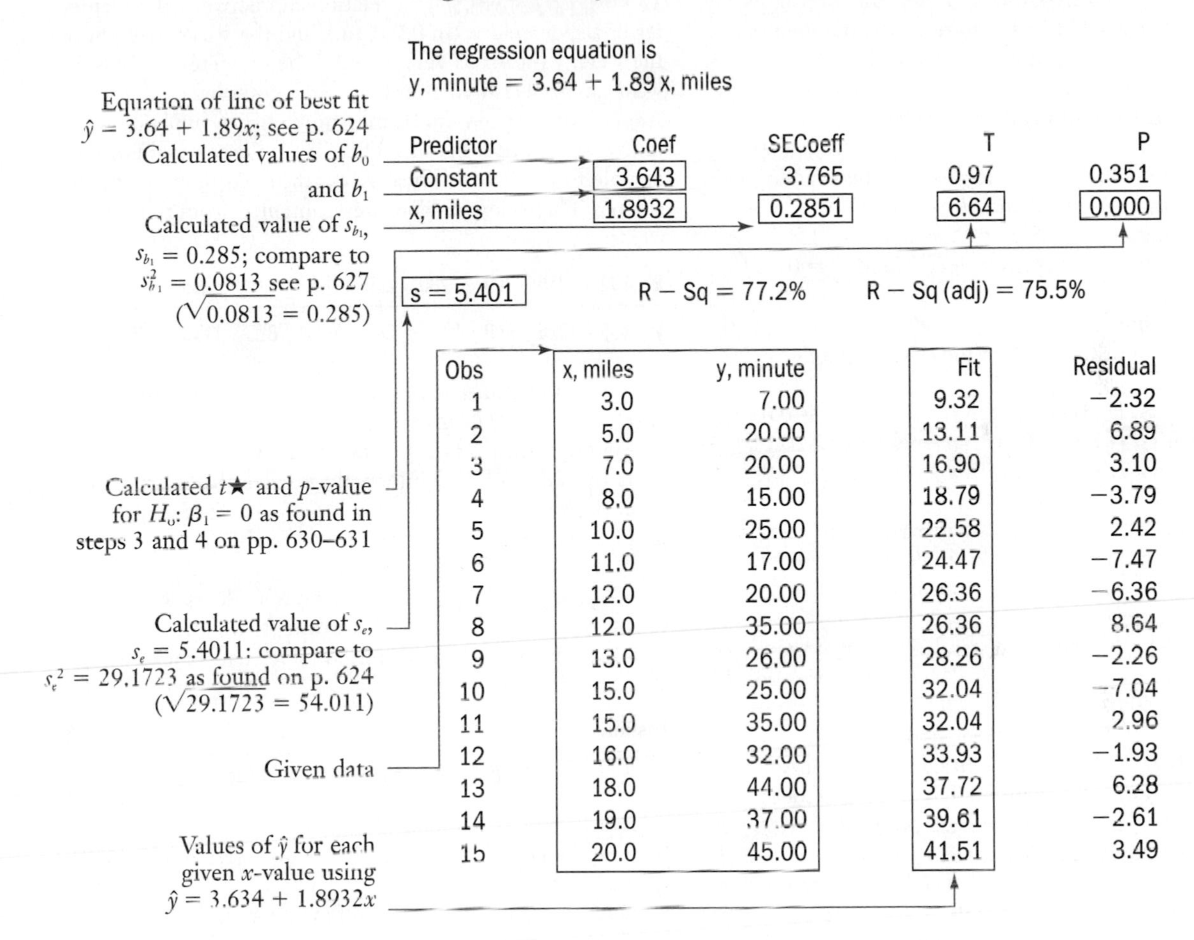

Regression Analysis

The regression equation is
y, minute = 3.64 + 1.89 x, miles

Predictor	Coef	SECoeff	T	P
Constant	3.643	3.765	0.97	0.351
x, miles	1.8932	0.2851	6.64	0.000

s = 5.401 R − Sq = 77.2% R − Sq (adj) = 75.5%

Obs	x, miles	y, minute	Fit	Residual
1	3.0	7.00	9.32	−2.32
2	5.0	20.00	13.11	6.89
3	7.0	20.00	16.90	3.10
4	8.0	15.00	18.79	−3.79
5	10.0	25.00	22.58	2.42
6	11.0	17.00	24.47	−7.47
7	12.0	20.00	26.36	−6.36
8	12.0	35.00	26.36	8.64
9	13.0	26.00	28.26	−2.26
10	15.0	25.00	32.04	−7.04
11	15.0	35.00	32.04	2.96
12	16.0	32.00	33.93	−1.93
13	18.0	44.00	37.72	6.28
14	19.0	37.00	39.61	−2.61
15	20.0	45.00	41.51	3.49

Equation of line of best fit $\hat{y} = 3.64 + 1.89x$; see p. 624

Calculated values of b_0 and b_1

Calculated value of s_{b_1}, $s_{b_1} = 0.285$; compare to $s^2_{b_1} = 0.0813$ see p. 627 $(\sqrt{0.0813} = 0.285)$

Calculated $t\bigstar$ and p-value for H_o: $\beta_1 = 0$ as found in steps 3 and 4 on pp. 630–631

Calculated value of s_e, $s_e = 5.4011$: compare to $s_e^2 = 29.1723$ as found on p. 624 $(\sqrt{29.1723} = 54.011)$

Given data

Values of $\hat{y}$ for each given x-value using $\hat{y} = 3.634 + 1.8932x$

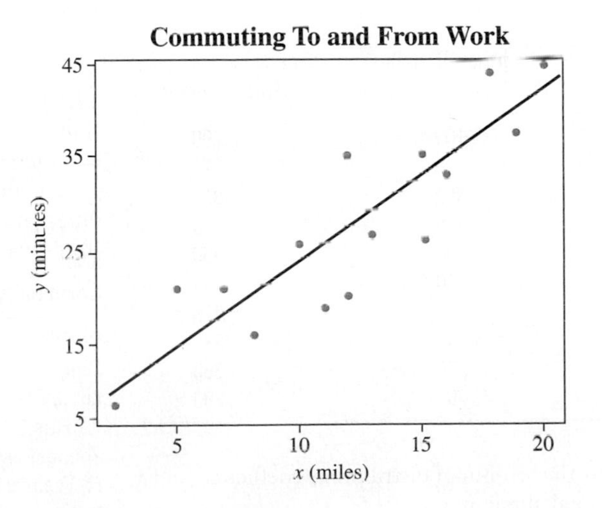

Exercises

13.42 State the null hypothesis H_o and the alternative hypothesis H_a that would be used to test each statement.
a. The slope of the line of best fit is positive.
b. There is no regression.
c. There is evidence of negative regression.

13.43 Determine the p-value for each of these situations:
a. H_a: $\beta_1 > 0$, with $n = 18$ and $t\bigstar = 2.4$
b. H_a: $\beta_1 \neq 0$, with $n = 15$, $b_1 = 0.16$, and $s_{b_1} = 0.08$
c. H_a: $\beta_1 < 0$, with $n = 24$, $b_1 = -1.29$, and $s_{b_1} = 0.82$

13.44 Determine the critical value(s) and regions that would be used in testing each of the following null hypotheses using the classical approach:
a. H_o: $\beta_1 = 0$ vs. H_a: $\beta_1 \neq 0$, with $n = 18$ and $\alpha = 0.05$
b. H_o: $\beta_1 = 0$ vs. H_a: $\beta_1 > 0$, with $n = 28$ and $\alpha = 0.01$
c. H_o: $\beta_1 = 0$ vs. H_a: $\beta_1 < 0$, with $n = 16$ and $\alpha = 0.05$

13.45 Calculate the estimated standard error of regression, s_{b_1}, for the computer-science aptitude score–achievement score relationship in Exercise 13.38 (p. 626).

13.46 Calculate the estimated standard error of regression, s_{b_1}, for the number of hours studied–exam grade relationship in Exercise 13.40 (p. 626).

13.47 Ex13-47 An article titled "Statistical Approach for the Estimation of Strontium Distribution Coefficient" (*Environmental Science & Technology*, November 1993) reports a linear correlation coefficient of 0.55 between the strontium distribution coefficient (mL/g) and the total aluminum (mmol/100 g-soil) for soils collected from the surface throughout Japan. Consider the data for ten such samples:

Soil Sample	Strontium Distribution Coefficient	Total Aluminum
1	100	200
2	120	225
3	300	325
4	250	310
5	400	350
6	500	400
7	450	375
8	445	385
9	310	350
10	200	290

Let y represent the strontium distribution coefficient and x represent the total aluminum.
a. Find the equation of the line of best fit.
b. Find a 95% confidence interval for β_1.
c. Explain the meaning of the interval in part b.

13.48 Ex13-48 The relationship between the diameter of a spot weld, x (in 0.001 in.), and the shear strength of the weld, y (in lb), is very useful. The diameter of the spot weld can be measured after the weld is completed. The shear strength of the weld can be measured only by applying force to the weld until it breaks. Thus, it would be very useful to be able to predict the shear strength based only on the diameter. The following data were obtained from ten sample welds:

x	190	215	200	230	209	250	215	265	215	250
y	680	1025	800	1100	780	1030	885	1175	975	1300

Complete these questions with the aid of a computer.
a. Draw a scatter diagram.
b. Find the equation of the line of best fit.
c. Is the value of b_1 significantly greater than zero at the 0.05 level?
d. Find the 95% confidence interval for β_1.

13.49 Ex13-49 A sample of ten students were asked for the distance (in miles) and the time (in minutes) required to commute to college yesterday. The data collected are shown in the table.

Distance	1	3	5	5	7	7	8	10	10	12
Time	5	10	15	20	15	25	20	25	35	35

a. Draw a scatter diagram of these data.
b. Find the equation that describes the regression line for these data.
c. Does the value of b_1 show sufficient strength to conclude that β_1 is greater than zero at the $\alpha = 0.05$ level?
d. Find the 98% confidence interval for the estimation of β_1. (Retain these answers to use in Exercise 13.54.)

13.50 Ex13-50 The National Association of Home Builders periodically publishes the median family incomes and the median sales prices of homes for hundreds of U.S. communities. Listed here are 26 randomly selected communities and their 2001 median family incomes ($1000s) and third-quarter median sales prices ($1000s) for homes.

Community	Income	Price
Rockford, IL	57.1	99
Binghamton, NY	44.7	72
Lima, OH	49.6	97
Kansas City, MO	62.2	117
Youngstown, OH	44.3	85
Ft. Walton Beach, FL	48.9	118
Duluth, MN	49.8	109
Greensboro, NC	53.1	125
Nashville, TN	60.7	137
Saginaw, MI	53.1	95

(continued)

Community	Income	Price
Rochester, NY	52.9	107
Buffalo, NY	48.4	92
Tulsa, OK	46.8	109
Tampa–St. Pete, Fl	47.7	116
Springfield, MA	49.7	127
Milwaukee, WI	63.5	133
Biloxi, MS	44.4	104
Reno, NV	50.4	167
Las Vegas, NV	52.1	150
Detroit, MI	66.5	152
Waterbury, CT	60.7	165
Austin, TX	64.7	173
Worcester, MA	57.0	175
New York, NY	59.1	185
Charleston, SC	46.3	162
Jersey City, NJ	54.8	200

Source: National Association of Home Builders, third quarter 2001

a. Construct a scatter diagram for median family income (x) versus median selling price of homes (y).
b. Find the equation of the line of best fit.
c. Is the slope significant at the $\alpha = 0.05$ level of significance?
d. Explain what the answer in part c means in terms of the usefulness of median income in predicting median selling price of homes.

13.51 Ex13-51 The September 1994 issue of *Popular Mechanics* gives specifications and dimensions for various jet boats. The table summarizes some of this information.

Model	Base Price	Engine Horsepower
Baja Blast	$8,395	120
Bayliner Jazz	8,495	90
Boston Whaler Rage 15	11,495	115
Dynasty Jet Storm	8,495	90
Four Winds Fling	9,568	115
Regal Rush	9,995	90
Sea-Doo Speedster	11,499	160
Sea Ray Sea Rayder	8,495	90
Seaswirl Squirt	8,495	115
Suga Sand Mirage	0,395	120

a. Find the equation of the line of best fit. Let x equal the horsepower, and let y equal the base price.
b. Find the standard deviation along the line of best fit and the standard error of the slope.
c. Describe the meaning of the two answers in part b.
d. Is horsepower an effective predictor of base price? Explain your answer using statistical evidence.

13.52 Ex13-52 Grain farmers are interested in the relationship between what they plant and what they harvest. The process starts with acres of land being planted and ends when most of these acres are harvested (some acres are lost due to weather and other factors). The harvest results in bushels of grain produced, and the quality of the harvest is measured using a yield rate (bushels of grain per acre harvested). A random sample of 15 oat-producing U.S. counties was selected from the USDA-NASS Web site and the following 2001 oat-production data recorded:

State	County	Acres Planted (100s)	Bushels Produced (100s)	Yield/Acre (bushels)
GA	Miller	46	1620	65.0
IA	Sac	18	740	74.0
ID	Gem	16	250	83.3
IL	Henry	40	2813	97.0
MI	Mecosta	11	430	48.0
MN	Marshall	39	2145	66.0
NC	Stanly	23	660	78.0
NC	Union	30	1040	77.0
NY	Clinton	7	240	60.0
OH	Trumbull	24	1375	62.5
SD	Union	25	890	80.9
TX	Gonzales	10	35	35.0
TX	Wichita	15	28	28.0
UT	Beaver	21	140	70.0
WI	La Crosse	20	850	57.0

Source: http://www.usda.gov/nass/graphics/county01/data/ot01.csv

a. Investigate the relationship between the variables number of acres planted (x) and number of bushels of oats harvested (y). Include a scatter diagram and line of best fit.
b. Test the significance of the results (slope of the line of best fit) found in part a. Use $\alpha = 0.05$.
c. How many bushels of oats, on the average, can a farmer expect to produce for each acre planted?
d. If the farmer plants more acres, can he expect to increase his total harvest? Explain.
e. Investigate the relationship between the variables number of acres planted (x) and yield rate (y) in bushels of oats per harvested acre. Include a scatter diagram and line of best fit.
f. Test the significance of the results (slope of the line of best fit) found in part e. Use $\alpha = 0.05$.
g. If a farmer plants more acres, can he expect an improvement in the yield rate? Explain.
h. Explain how the two investigations (a–d and e–g) are asking for different information.
i. Explain why the results should have been anticipated.

13.5 Confidence Intervals for Regression

Once the equation of the line of best fit has been obtained and determined usable, we are ready to use the equation to make predictions. We can estimate two different quantities: (1) the mean of the population y values at a given value of x, written $\mu_{y|x_0}$, and (2) the individual y value selected at random that will occur at a given value of x, written y_{x_0}. The best point estimate, or **prediction, for both $\mu_{y|x_0}$ and y_{x_0}** is $\hat{y}$. This is the y value obtained when an x value is substituted into the equation of the line of best fit. Like other point estimates, it is seldom correct. The calculated value of $\hat{y}$ will vary above and below the actual values for both $\mu_{y|x_0}$ and y_{x_0}.

Before we develop interval estimates of $\mu_{y|x_0}$ and y_{x_0}, recall the development of confidence intervals for the population mean μ in Chapter 8 when the variance was known and in Chapter 9 when the variance was estimated. The sample mean, $\bar{x}$, was the best point estimate of μ. We used the fact that $\bar{x}$ is normally distributed, or approximately normally distributed, with a standard deviation of $\frac{\sigma}{\sqrt{n}}$ to construct formula (8.1) for the confidence interval for μ. When σ has to be estimated, we used formula (9.1) for the confidence interval.

The **confidence interval for $\mu_{y|x_0}$** and the **prediction interval for y_{x_0}** are constructed in a similar fashion with $\hat{y}$ replacing $\bar{x}$ as our point estimate. If we were to randomly select several samples from the population, construct the line of best fit for each sample, calculate $\hat{y}$ for a given x using each regression line, and plot the various $\hat{y}$ values (they would vary since each sample would yield a slightly different regression line), we would find that the $\hat{y}$ values form a normal distribution. That is, the **sampling distribution of $\hat{y}$** is normal, just as the sampling distribution of $\bar{x}$ is normal. What about the appropriate standard deviation of $\hat{y}$? The standard deviation in both cases ($\mu_{y|x_0}$ and y_{x_0}) is calculated by multiplying the square root of the variance of the error by an appropriate correction factor. Recall that the variance of the error, s_e^2, is calculated by means of formula (13.8).

Before we look at the correction factors for the two cases, let's see why they are necessary. Recall that the line of best fit passes through the point $(\bar{x}, \bar{y})$, the centroid. In Section 13.3 we formed a confidence interval for the slope β_1 (see Illustration 13.4) by using formula (13.12). If we draw lines with slopes equal to the extremes of that confidence interval, 1.27 to 2.51, through the point $(\bar{x}, \bar{y})$ [which is (12.3, 26.9)] on the scatter diagram, we will see that the value for $\hat{y}$ fluctuates considerably for different values of x (Figure 13.11). Therefore, we should suspect a need for a wider confidence interval as we select values of x that are farther away from $\bar{x}$. Hence we need a correction factor to adjust for the distance between x_0 and $\bar{x}$. This factor must also adjust for the variation of the y values about $\hat{y}$.

First, let's estimate the mean value of y at a given value of x, $\mu_{y|x_0}$. The confidence interval formula is:

$$\hat{y} \pm t(n-2, \alpha/2) \cdot s_e \cdot \sqrt{\frac{1}{n} + \frac{(x_0 - \bar{x})^2}{\sum(x - \bar{x})^2}} \qquad (13.14)$$

NOTE The numerator of the second term under the radical sign is the square of the distance of x_0 from $\bar{x}$. The denominator is closely related to the variance of x and has a "standardizing" effect on this term.

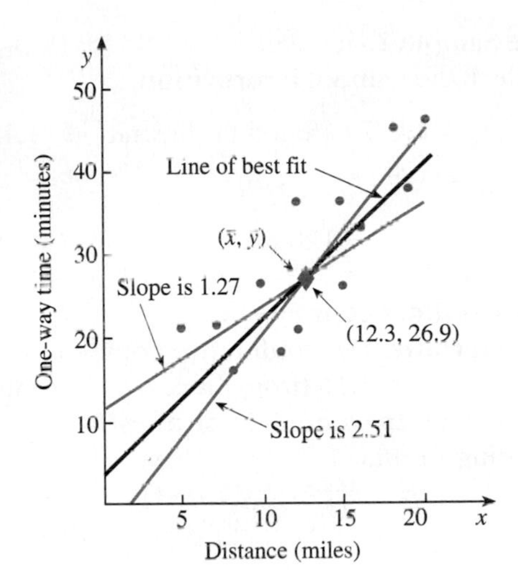

FIGURE 13.11
Lines Representing the Confidence Interval for Slope

Formula (13.14) can be modified to avoid having $\bar{x}$ in the denominator. Here is the new form:

CONFIDENCE INTERVAL FOR $\mu_{y|x_0}$

$$\hat{y} \pm t(n-2, \alpha/2) \cdot s_e \cdot \sqrt{\frac{1}{n} + \frac{(x_0 - \bar{x})^2}{\text{SS}(x)}} \quad (13.15)$$

Let's compare formula (13.14) with formula (9.1): $\hat{y}$ replaces $\bar{x}$, and

$$s_e \cdot \sqrt{\frac{1}{n} + \frac{(x_0 - \bar{x})^2}{\sum(x - \bar{x})^2}} \quad \text{(the \textbf{standard error of} } \hat{y}\text{)}$$

the estimated standard deviation of $\hat{y}$ in estimating $\mu_{y|x_0}$, replaces $\frac{s}{\sqrt{n}}$, the standard deviation of $\bar{x}$. The degrees of freedom are now $n - 2$ instead of $n - 1$ as before.

These ideas are explored in the next illustration.

Illustration 13.7 Constructing a Confidence Interval for $\mu_{y|x_0}$

Construct a 95% confidence interval for the mean travel time for the co-workers who travel 7 miles to work (refer to Illustration 13.4).

SOLUTION

Step 1 **The Set-Up:**
Describe the population parameter of interest.
$\mu_{y|x=7}$, the mean travel time for co-workers who travel 7 miles to work.

Step 2 **The Confidence Interval Criteria:**

a. **Check the assumptions.**
The ordered pairs form a random sample, and we will assume that the y values (minutes) at each x (miles) have a normal distribution.

b. **Identify the probability distribution and the formula to be used.**
The Student's t-distribution and formula (13.15).

c. **State the level of confidence:** $1 - \alpha = 0.95$.

Step 3 **The Sample Evidence:**
Collect the sample information.

$s_e^2 = 29.17$ (found in Illustration 13.4)

$s_e = \sqrt{29.17} = 5.40$

$\hat{y} = 3.64 + 1.89x = 3.64 + 1.89(7) = 16.87$

Step 4 **The Confidence Interval:**

a. Determine the confidence coefficient.
$t(13, 0.025) = 2.16$ (from Table 6 in Appendix B).

b. Find the maximum error of estimate.
Using formula (13.15), we have

$$E = t(n-2, \alpha/2) \cdot s_e \cdot \sqrt{\frac{1}{n} + \frac{(x_0 - \bar{x})^2}{\text{SS}(x)}}: \quad E = (2.16)(5.40)\sqrt{\frac{1}{15} + \frac{(7 - 12.27)^2}{358.933}}$$

$$= (2.16)(5.40)\sqrt{0.06667 + 0.07738}$$

$$= (2.16)(5.40)(0.38) = 4.43$$

c. Find the lower and upper confidence limits.

$$\hat{y} - E \quad \text{to} \quad \hat{y} + E$$

$$16.87 - 4.43 \quad \text{to} \quad 16.87 + 4.43$$

Thus, **12.44 to 21.30** is the 95% confidence interval for $\mu_{y|x=7}$.

This confidence interval is shown in Figure 13.12 by the dark red vertical line. The confidence belt showing the upper and lower boundaries of all intervals at

FIGURE 13.12
Confidence Belts for $\mu_{y|x_0}$

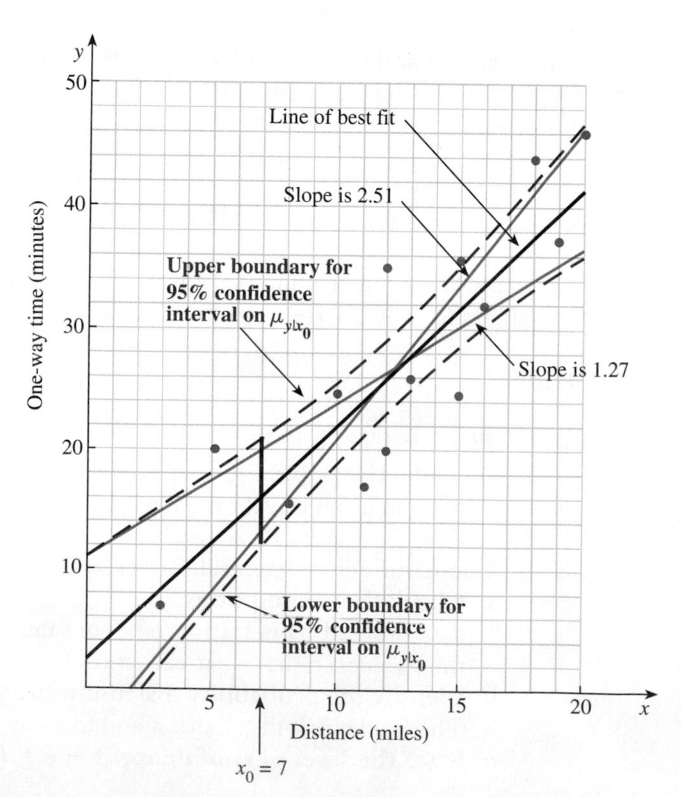

95% confidence is also shown in red. Notice that the boundary lines for the x values far away from $\bar{x}$ become close to the two lines that represent the equations with slopes equal to the extreme values of the 95% confidence interval for the slope (see Figure 13.12). ■

Often we want to predict the value of an individual y. For example, you live 7 miles from your place of business and you are interested in an estimate of how long it will take you to get to work. You are somewhat less interested in the average time for all of those who live 7 miles away. The formula for the prediction interval of the value of a single randomly selected y is

PREDICTION INTERVAL FOR $y_{x=x_0}$

$$\hat{y} \pm t(n-2, \alpha/2) \cdot s_e \cdot \sqrt{1 + \frac{1}{n} + \frac{(x_0 - \bar{x})^2}{SS(x)}} \qquad (13.16)$$

Illustration 13.8 Constructing a Prediction Interval for $y_{x=x_0}$

What is the 95% prediction interval for the time it will take you to commute to work if you live 7 miles away?

SOLUTION

Step 1 **The Set-Up:**
Describe the population parameter of interest.
$y_{x=7}$, the travel time for one co-worker who travels 7 miles to work.

Step 2 **The Confidence Interval Criteria:**
a. **Check the assumptions.**
The ordered pairs form a random sample, and we will assume that the y values (minutes) at each x (miles) have a normal distribution.
b. **Identify the probability distribution and the formula to be used.**
The Student's t-distribution and formula (13.16).
c. **State the level of confidence.** $1 - \alpha = 0.95$.

Step 3 **The Sample Evidence:**
Collect the sample information: $s_e = 5.40$ and $\hat{y}_{x=7} = 16.87$ (from Illustration 13.7)

Step 4 **The Confidence Interval:**
a. **Determine the confidence coefficient.**
$t(13, 0.025) = 2.16$ (from Table 6 in Appendix B).

b. Find the maximum error of estimate.
Using formula (13.15), we have

$$E = t(n-2, \alpha/2) \cdot s_e \cdot \sqrt{1 + \frac{1}{n} + \frac{(x_0 - \bar{x})^2}{SS(x)}}:$$

$$E = (2.16)(5.40)\sqrt{1 + \frac{1}{15} + \frac{(7 - 12.27)^2}{358.933}}$$

$$= (2.16)(5.40)\sqrt{1 + 0.06667 + 0.07738}$$

$$= (2.16)(5.40)\sqrt{1.14405}$$

$$= (2.16)(5.40)(1.0696) = 12.48$$

c. Find the lower and upper confidence limits.

$$\hat{y} - E \quad \text{to} \quad \hat{y} + E$$

$$16.87 - 12.48 \quad \text{to} \quad 16.87 + 12.48$$

Thus, **4.39 to 29.35** is the 95% prediction interval for $y_{x=7}$.

The prediction interval is shown in Figure 13.13 as the blue vertical line segment at $x_0 = 7$. Notice that it is much longer than the confidence interval for $\mu_{y|x=7}$. The dashed blue lines represent the prediction belts, the upper and lower boundaries of the prediction intervals for individual y values for all given x values.

FIGURE 13.13
Prediction Belts for y_{x_0}

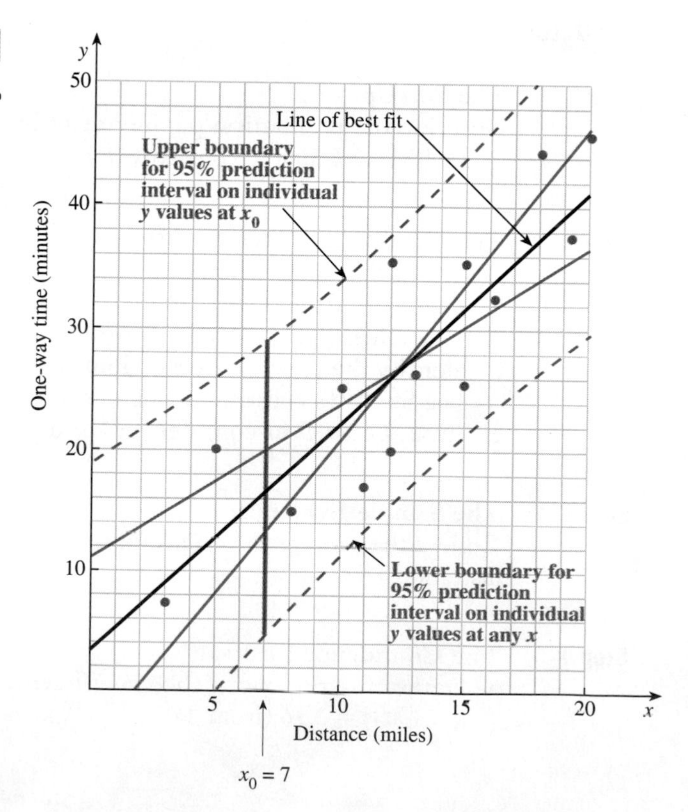

Can you justify the fact that the prediction interval for individual values of y is wider than the confidence interval for the mean values? Think about "individual values" and "mean values" and study Figure 13.14.

FIGURE 13.14

Confidence Belts for the Mean Value of y and Prediction Belts for Individual y's

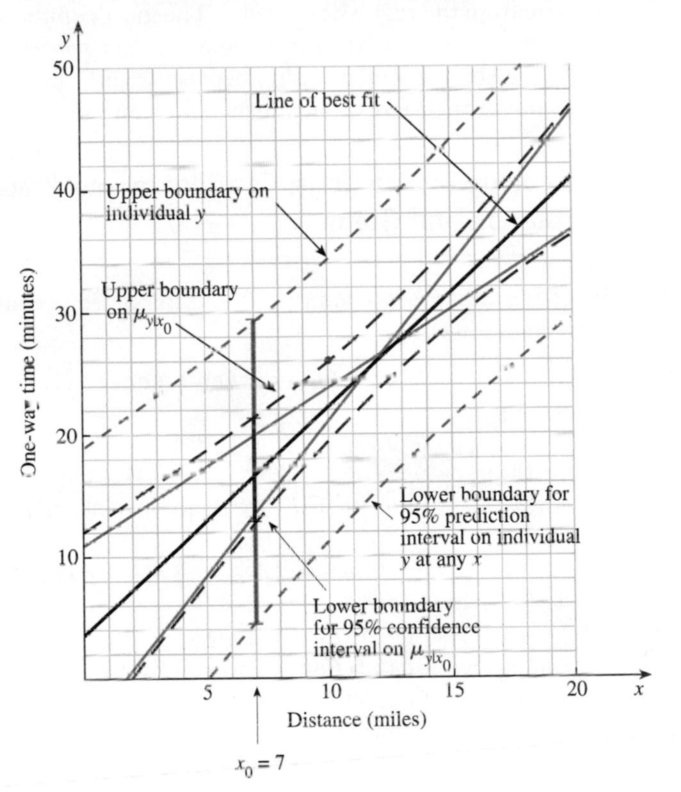

There are three basic precautions that you need to be aware of as you work with regression analysis:

1. Remember that the regression equation is meaningful only in the domain of the x variable studied. Estimation outside this domain is extremely dangerous; it requires that we know or assume that the relationship between x and y remains the same outside the domain of the sample data. For example, Joe says that he lives 75 miles from work, and he wants to know how long it will take him to commute. We certainly can use $x = 75$ in all the formulas, but we do not expect the answers to have the confidence or validity of the values of x between 3 and 20, which were in the sample. The 75 miles may represent a distance to the heart of a nearby major city. Do you think the estimated times, which were based on local distances of 3 to 20 miles, would be good predictors in this situation? Also, at $x = 0$ the equation has no real meaning. However, although projections outside the interval may be somewhat dangerous, they may be the best predictors available.
2. Don't get caught by the common fallacy of applying the regression results inappropriately. For example, this fallacy would include applying the results of Illustration 13.4 to another company. But suppose that the second company had a city location, whereas the first company had a rural location, or vice versa. Do you think the results for a rural location would be valid for a city location? Basically, the results of one sample should not be used to make inferences about a population other than the one from which the sample was drawn.
3. Don't jump to the conclusion that the results of the regression prove that x *causes* y to change. (This is perhaps the most common fallacy.) Regressions only

measure movement between x and y; they never prove causation. (See pp. 150–151 for a discussion of causation.) A judgment of causation can be made only when it is based on theory or knowledge of the relationship separate from the regression results. The most common difficulty in this regard occurs because of what is called the *missing variable*, or *third-variable, effect.* That is, we observe a relationship between x and y because a third variable, one that is not in the regression, affects both x and y.

Technology Instructions: Confidence and Prediction Intervals Calculation and Graph

MINITAB (Release 13)

Input the x-variable data into C1 and the corresponding y-variable data into C2; then continue with:

```
Choose:   Stat > Regression > Regression...
Enter:    Response (y): C2
          Predictors (x): C1
Select:   Options
Enter:    Prediction intervals for new observations:
                 x-value or C1 (C1-list of x values)
          Confidence level: 1 − α (ex. 95.0)
Select:   Confidence limits
          Prediction limits             > OK > OK
Choose:   Stat > Regression > Fitted Line Plot
Enter:    Response (y): C2
          Predictors (x): C1
Select:   Type of Regression: Linear
Choose:   Options
Select:   Display options:     Display confidence bands
                               Display prediction bands
Enter:    Confidence level: 1 − α (ex. 95.0)
Choose:   Storage
Select:   Residuals
          Fits
```

Here is the MINITAB printout for parts of Illustrations 13.4, 13.5, and 13.6.

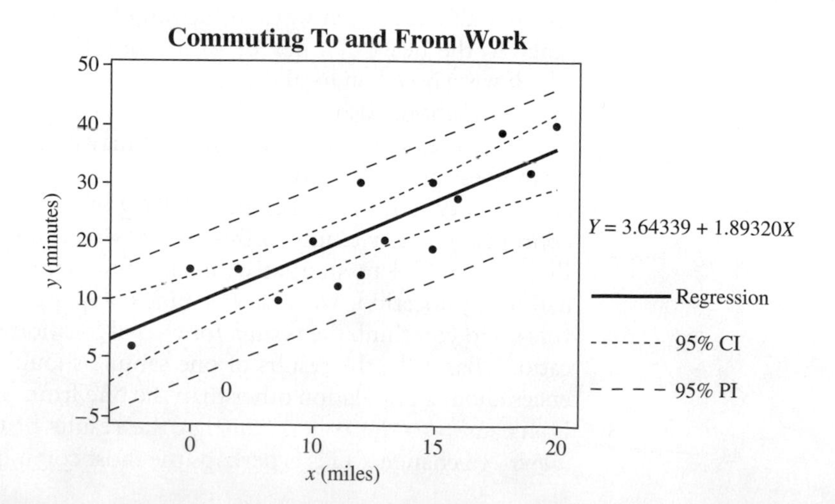

Application 13.3 Using Regression Confidence Intervals in an Environmental Study

Much time, money, and effort are spent studying our environmental problems so that effective and appropriate management practices might be implemented. Here are excerpts from a study in South Florida in which linear regression analysis was an important tool.

Methodology for Estimating Nutrient Loads Discharged from the East Coast Canals to Biscayne Bay, Miami–Dade County, Florida

U.S. Geological Survey
Water-Resources Investigations Report 99-4094
by A. C. Lietz

A major concern in many coastal areas across the nation is the ecological health of bays and estuaries. One common problem in many of these areas is nutrient enrichment as a result of agricultural and urban activities. Nutrients are essential compounds for the growth and maintenance of all organisms and especially for the productivity of aquatic environments. Nitrogen and phosphorus compounds are especially important to seagrass, macroalgae, and phytoplankton. However, heavy nutrient loads transported to bays and estuaries can result in conditions conducive to eutrophication and the attendant problems of algal blooms and high phytoplankton productivity. Additionally, reduced light penetration in the water column because of phytoplankton blooms can adversely affect seagrasses, which many commercial and sport fish rely on for their habitat.

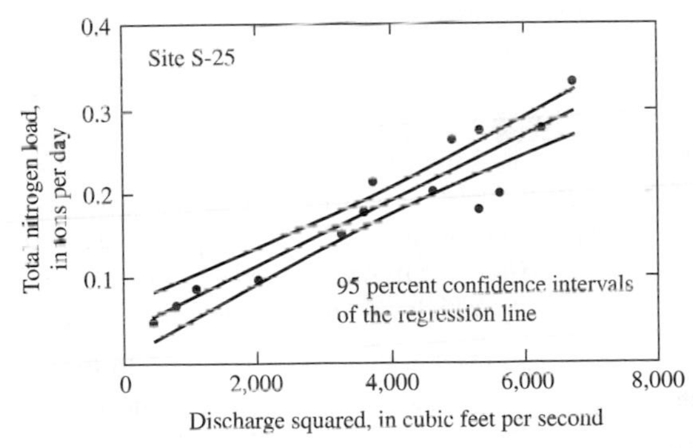

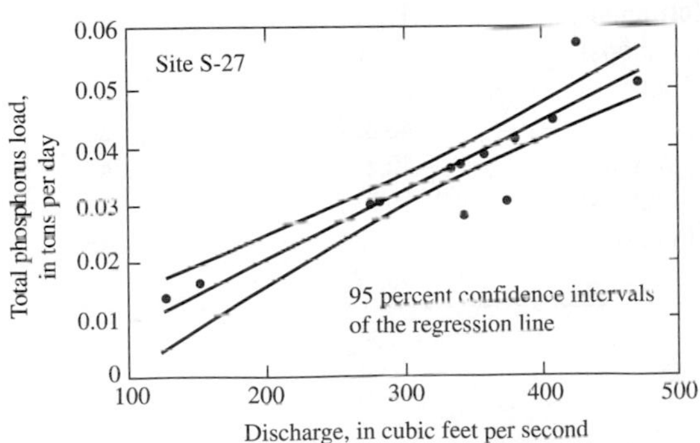

The purpose of this report is to present methodology that can be used to estimate nutrient loads discharged from the east coast canals into Biscayne Bay in southeastern Florida. Water samples were collected from the gated control structures at the east coast canal sites in Miami–Dade County for the purpose of developing models that could be used to estimate nitrogen and phosphorus loads.

An ordinary least-squares regression technique was used to develop predictive equations for the purpose of estimating total nitrogen and total phosphorus loads discharged from the east coast canals to Biscayne Bay. The predictive equations can be used to estimate the value of a dependent variable from observations on a related or independent variable. In this study, load was used as the dependent or response variable and discharge as the independent or explanatory variable. All of the total nitrogen load models had p-values less than 0.05, indicating they were statistically significant at an alpha level of 0.05. Plots showing total nitrogen load as a function of discharge at the east coast canal sites are shown in figure 17. [Sites S25 and S27 from figure 17 are shown here.]

ANSWER NOW 13.53

The graphs for Site S-25 and Site S-27 display 95% confidence intervals of the regression line. What distinguishing feature would 95% prediction intervals have with respect to these graphs? Explain the difference between confidence intervals and prediction intervals.

Exercises

13.54 Use the data and the answers from Exercise 13.49 (p. 634) to make the following estimates:

a. Give a point estimate for the mean time required to commute 4 miles.
b. Find a 90% confidence interval for the mean travel time required to commute 4 miles.
c. Find a 90% prediction interval for the travel time required for one person to commute 4 miles.
d. Do parts a, b, and c for $x = 9$.

13.55 A study in *Physical Therapy* (April 1991) reports on seven different methods used to determine crutch length plus two new techniques utilizing linear regression. One of the regression techniques uses the patient's reported height. The study considered 107 individuals, and the mean of the self-reported heights was 68.84 in. The regression equation determined was $\hat{y} = 0.68x + 4.8$, where y = crutch length and x = self-reported height. The MSE (s_e^2) was reported to be 0.50. In addition, the standard deviation of the self-reported heights was 7.35 in. Use this information to determine a 95% confidence interval estimate for the mean crutch length for individuals who say they are 70 in. tall.

13.56 Ex13-56 Cicadas are flying, plant-eating insects. The 13-year cicadas *(Magicicada)* spend five juvenile stages in underground burrows. During their 13 years underground, the cicadas grow from approximately the size of a small ant to nearly the size of an adult cicada. Every 13 years, the cicadas emerge from their burrows as adults. The table lists three different species of these 13-year cicadas along with their adult body weights (BW), in grams, and their wing lengths (WL), in millimeters.

BW	WL	Species	BW	WL	Species
0.15	28	*tredecula*	0.18	29	*tredecula*
0.29	32	*tredecim*	0.21	27	*tredecassini*
0.17	27	*tredecim*	0.15	30	*tredecula*
0.18	30	*tredecula*	0.17	27	*tredecula*
0.39	35	*tredecim*	0.13	27	*tredecassini*
0.26	31	*tredecim*	0.17	29	*tredecassini*
0.17	29	*tredecassini*	0.23	30	*tredecassini*
0.16	28	*tredecassini*	0.12	22	*tredecim*
0.14	25	*tredecassini*	0.26	30	*tredecula*
0.14	28	*tredecassini*	0.19	30	*tredecula*
0.28	25	*tredecassini*	0.20	30	*tredecassini*
0.12	28	*tredecim*	0.14	23	*tredecula*

a. Draw a scatter diagram with body weight as the independent variable and wing length as the dependent variable. Find the equation of the line of best fit.
b. Is body weight an effective predictor of wing length for a 13-year cicada? Use the 0.05 level of significance.
c. Give a 90% confidence interval for the mean wing length for all cicadas with 0.20-gm body weight.

13.57 Ex13-57 An experiment was conducted to study the effect of a new drug in lowering the heart rate in adults. The data are shown in the table below.

a. Find the 95% confidence interval for the mean heart-rate reduction for a dose of 2.00 mg.
b. Find the 95% prediction interval for the heart-rate reduction expected for an individual who received a dose of 2.00 mg.

13.58 Ex13-58 The relationship between the strength, x, and fineness, y, of cotton fibers was the subject of a study that producedthe data below:

a. Draw a scatter diagram.
b. Find the 99% confidence interval for the mean measurement of fineness for fibers with a strength of 80.
c. Find the 99% prediction interval for an individual measurement of fineness for fibers with a strength of 75.

13.59 Ex13-59 An article titled "Ailing and Well Babies: 'Gap Is Striking'" appeared in the September 8, 1994, issue of the Omaha *World-Herald.* The article gave weighted median household incomes in 1989 and percentages of households with up-to-date immunizations for children age 2 for Douglas County, which was divided into eight sections.

Section	Median Household Income	Up-to-Date Immunization
East/Northeast	$17,723	43%
West/Northeast	27,005	51
North/Central	33,424	62
Northwest	43,337	66
East/Southeast	19,226	46
West/Southeast	29,775	59
South/Central	40,607	65
Southwest	45,496	62

Table for Exercise 13.57

x (drug dose in mg)	0.50	0.75	1.00	1.25	1.50	1.75	2.00	2.25	2.50	2.75
y (heart-rate reduction)	10	7	15	12	15	14	20	20	18	21

Table for Exercise 13.58

x	76	69	71	76	83	72	78	74	80	82	90	81	78	80	81	78
y	4.4	4.6	4.6	4.1	4.0	4.1	4.9	4.8	4.2	4.4	3.8	4.1	3.8	4.2	3.8	4.2

Let y represent the percentage with up-to-date immunizations and x represent the median household income.

a. Find the equation of the line of best fit.
b. Find a 95% confidence interval for the mean percentage with up-to-date immunizations for families with a median household income equal to $40,000.
c. Find a 95% prediction interval for the probability that a family will have up-to-date immunizations if the median household income equals $40,000.
d. Explain the meaning of the intervals found in parts b and c.

13.60 Ex13-60 People not only live longer today but also are living independently longer; even if an individual becomes temporarily dependent at some age, he or she still may enjoy years of independent living during the remaining life. The May/June 1989 issue of *Public Health Reports* included an article titled "A Multistate Analysis of Active Life Expectancy." Two of the variables studied were the age at which people became dependent, x, and the number of independent years they had remaining, y. The data follow:

x	65	66	67	68	70	72	74	76	78	80	83	85
y	11.1	10.0	10.4	9.3	8.2	6.8	6.8	4.4	5.4	2.5	2.7	0.9

a. Draw a scatter diagram of the data.
b. Calculate the equation of the line of best fit.
c. Draw the line of best fit on the scatter diagram.
d. For a person who becomes dependent at age 80, how many more years of independent living can be expected? Find the answer in two different ways: Use the equation from part b and use the line on the scatter diagram from part c.
e. Construct a 99% prediction interval for the number of years of independent living remaining for a person who becomes dependent at age 80.
f. Draw a vertical line segment on the scatter diagram representing the interval found in part e.

13.61 Explain why a 95% confidence interval for the mean value of y at a particular x is much narrower than a 95% prediction interval for an individual y value at the same value of x.

13.62 When $x_0 = \bar{x}$, is the formula for the standard error of $\hat{y}_{x_0}$ what you expected it to be, $s \cdot \frac{1}{\sqrt{n}}$? Explain.

13.6 Understanding the Relationship Between Correlation and Regression

Now that we have taken a closer look at both correlation and regression analysis, it is necessary to decide when to use them. Do you see any duplication of work?

The primary use of the linear correlation coefficient is in answering the question: Are these two variables linearly related? Other words may be used to ask this basic question—for example: Is there a linear correlation between the annual consumption of alcoholic beverages and the salary paid to firemen?

The linear correlation coefficient can be used to indicate the usefulness of x as a predictor of y in the case where the linear model is appropriate. The test concerning the slope of the regression line (H_o: $\beta_1 = 0$) tests this same basic concept. Either one of the two is sufficient to determine the answer to this query.

The choice of mathematical model can be tested statistically (called a "lack of fit" test); however, these procedures are beyond the scope of this text. We do perform this test informally, or subjectively, when we view the scatter diagram and use the presence of a linear pattern as our reason for using the linear model.

The concepts of linear correlation and regression are quite different because each measures different characteristics. It is possible to have data that yield a strong linear correlation coefficient and have the wrong model. For example, the straight line can be used to approximate almost any curved line if the domain is restricted sufficiently. In such a case the linear correlation coefficient can become quite high, but the curve will still not be a straight line. Figure 13.15 illustrates one interval where r could be significant but the scatter diagram does not suggest a straight line.

FIGURE 13.15

The Value of r Is High but the Relationship Is Not Linear

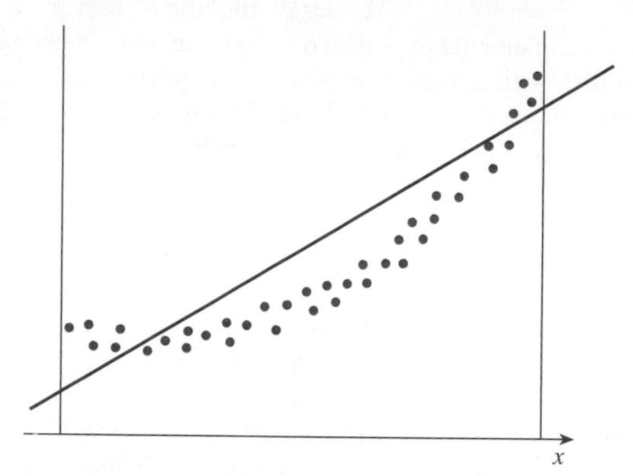

Regression analysis should be used to answer questions about the relationship between two variables. Such questions as What is the relationship? and How are two variables related? require this regression analysis.

CHAPTER SUMMARY

RETURN TO CHAPTER CASE STUDY

© Francis de Richemond/The Image Works

Wheat! Beautiful Golden Wheat!

Let's return to the Chapter Case Study (p. 604) and continue our investigation of the U.S. Wheat crop. CS13

FIRST THOUGHTS 13.63

a. Construct a scatter diagram, and find the linear correlation coefficient and the equation of the line of best fit for Planted (x) and Production (y).
b. Construct a scatter diagram, and find the linear correlation coefficient and the equation of the line of best fit for Planted (x) and Yield (y).

PUTTING CHAPTER 13 TO WORK 13.64

Use the results from 13.63.

a. Find the 95% confidence interval for the true linear correlation coefficient between Planted and Production.
b. Calculate the 95% confidence interval for the slope β_1 for the line of best fit relating Planted and Production.
c. Predict the total production for a county with 125 thousand acres planted using a 95% prediction interval.
d. Test the significance of the correlation between Planted and Yield. Use $\alpha = 0.05$.
e. Test the significance of the slope β_1 for the line of best fit relating Planted and Yield. Use $\alpha = 0.05$.
f. Estimate the mean yield for all counties with 100 thousand acres planted with a 95% confidence interval.

YOUR OWN MINI-STUDY 13.65

Design your own study to investigate another grain. Answer questions similar to those in 13.63 and 13.64. Use the Internet and the U.S. Department of Agriculture's National Agricultural Statistics Service Web site: http://www.usda.gov/nass/ to find your data. The data sets posted will contain extra information that makes the data in its original format unusable. However, this unwanted information can easily be removed by deleting the unwanted rows and columns.

In Retrospect

In this chapter we have made a more thorough inspection of the linear relationship between two variables. Although the curvilinear and multiple regression situations were mentioned only in passing, the basic techniques and concepts have been explored. We would only have to modify our mathematical model and our formulas if we wanted to deal with these other relationships.

Although it was not directly emphasized, we have applied many of the topics of earlier chapters in this chapter. The ideas of confidence interval and hypothesis testing were applied to the regression problem. Reference was made to the sampling distribution of the sample slope b_1. This allowed us to make inferences about β_1, the slope of the population from which the sample was drawn. We estimated the mean value of y at a fixed value of x by pooling the variance for the slope with the variance of the y's. This was allowable because they are independent. Recall that in Chapter 10 we presented formulas for combining the variances of independent samples. The idea here is much the same. Finally, we added a measure of variance for individual values of y and made estimates for these individual values of y at fixed values of x.

Application 13.2 presents the results of regression analysis on data collected to compare two crime-reporting indices. (Take another look at Application 13.2, p. 629.) The scatter diagram very convincingly shows that the two crime indices being compared are related to each other in a very strong and predictable pattern. Thus, as stated in the original article, "the weighted index contributed no further information" since the two indices are basically the same. Thus, the introduction of the weighted index seems unnecessary since the Uniform Crime Reports index is a recognized standard.

As this chapter ends, you should be aware of the basic concepts of regression analysis and correlation analysis. You should now be able to collect the data for, and do a complete analysis on, any two-variable linear relationship.

Chapter Exercises

13.66 Answer the following as "sometimes," "always," or "never." Explain each "never" and "sometimes" response.

a. The correlation coefficient has the same sign as the slope of the least squares line fitted to the same data.
b. A correlation coefficient of 0.99 indicates a strong causal relationship between the variables under consideration.
c. An r value greater than zero indicates that ordered pairs with high x values will have low y values.
d. The two coefficients for the line of best fit have the same sign.
e. If x and y are independent, then the population correlation coefficient equals zero.

13.67 A study in the *Journal of Range Management* (September 1990) examines the relationships between elements in Russian wild rye. The correlation coefficient between magnesium and calcium was reported to be 0.69 for a sample of size 45. Is there a significant correlation between magnesium and calcium contents in Russian wild rye (i.e., is $\rho = 0$)?

13.68 A study concerning the plasma concentration of the drug Ranitidine was reported in the *Journal of Pharmaceutical Sciences* (December 1989). The drug was administered (coded I), and the plasma concentration of Ranitidine was followed for 12 hours. The time to the first peak in concentration was called T_{max1}. The same experiment was repeated one week later (coded II). Twelve subjects participated in the study. The correlation coefficient between T_{max1}, I and T_{max1}, II was reported to be 0.818. Use Table 11 in Appendix B to determine bounds on the p-value for the hypothesis test of H_o: $\rho = 0$ versus H_a: $\rho \neq 0$.

13.69 Ex13-69 Shopping for a new personal computer may be one of the most difficult tasks facing consumers today. A higher price usually means higher quality. But is that true for PCs? The staff of *PC World* published the results of their study of the top 20 power desktops and top 20 budget desktops in late 1998. The street prices of each machine were listed along with their overall rating.

Power Desktops

Street Price	Overall Rating	Street Price	Overall Rating
$2579	87	2499	82
2599	86	2750	80
2704	85	2798	80
2299	85	2499	80
2250	85	2823	80
2199	83	2168	79
2783	83	2609	78
2704	83	2297	77
2395	83	2890	77
2675	82	2099	77

(continued)

Budget Desktops

Street Price	Overall Rating	Street Price	Overall Rating
1999	82	1299	79
1499	81	999	79
1199	81	1699	78
1599	80	1649	78
1748	80	1397	78
1599	80	1925	77
1797	80	1499	77
1849	80	1879	77
1599	79	1749	76
1649	79	1100	76

Source: *PC World,* "Top 100," November 1998

a. Calculate the linear correlation coefficient (Pearson's product moment, r) between street price and overall rating for power desktops and for budget desktops.
b. What conclusions might you draw from your answers in part a?

13.70 Ex13-70 About 2527 athletes competed in the 2002 Olympic Winter Games held in Salt Lake City, Utah, for medals in 78 events. Athletes from 78 nations and territories participated. The table shows the distribution of gold, silver, and bronze medals awarded to athletes representing the 18 nations that won the most.

Nation	Gold	Silver	Bronze
Germany	12	16	7
United States	10	13	11
Norway	11	7	6
Canada	6	3	8
Austria	2	4	10
Russian Federation	6	6	4
Italy	4	4	4
France	4	5	2
Switzerland	3	2	6
China	2	2	4
Netherlands	3	5	0
Finland	4	2	1
Sweden	0	2	4
Croatia	3	1	0
Korea	2	2	0
Bulgaria	0	1	2
Estonia	1	1	1
Great Britain	1	0	2

Source: http://www.wikipedia.org/wiki/2002_Winter_Olympic_Games

Calculate the correlation coefficient and use it along with Table 10 of Appendix B to determine a 95% confidence interval on ρ for each of these cases:

a. gold and silver
b. gold and bronze
c. silver and bronze

13.71 The use of electrical stimulation (ES) to increase muscular strength is discussed in the *Journal of Orthopedic and Sports Physical Therapy* (September 1990). Seventeen healthy volunteers were used in the experiment. Muscular strength, y, was measured as a torque in foot-pounds, and electrical stimulation, x, was measured in mA (micro-amps). The equation for the line of best fit is given as $\hat{y} = 1.8x + 28.7$ and the Pearson correlation coefficient as 0.61.

a. Was the correlation coefficient significantly different from zero? Use $\alpha = 0.05$.
b. Predict the torque for a current of 50 mA.

13.72 Ex13-72 Innovative companies continue their search for new products to market. One measure of a company's attempts in pioneering new products is the number of U.S. patents it receives. But are the patented products successful in generating earnings for the company after they are launched? The table lists 13 corporations that received patents in 1996, together with the 12-month total return to shareholders of their stock as of October 1998.

Company	Number of Patents	12-month Total Return
IBM	1867	22.2%
Canon K.K.	1541	−28.5
Motorola	1064	−38.9
Hitachi	963	−50.2
Sony	855	−26.5
Matsushita	841	−24.3
General Electric	819	18.7
Eastman Kodak	768	21.5
Xerox	703	2.2
Texas Instruments	600	−20.4
MMM	537	−18.2
AT&T	510	35.1
Hewlett-Packard	501	−23.2

Sources: *Technology Assessment and Forecast Report,* U.S. Patent and Trademark Office, 1997; and *Stock Guide,* Standard & Poor's, October 1998

a. Calculate the correlation coefficient between the two variables.
b. Test for a significant correlation at the 0.05 level of significance and draw your conclusion.

13.73 An article in *Geology* (September 1989) gives the following equation relating pressure, P, and total aluminum content, AL, for 12 hornblende rims:

$$P = -3.46(+0.24) + 4.23(+0.13)AL.$$

The quantities shown in parentheses are standard errors for the y-intercept and slope estimates. Find a 95% confidence interval for the slope, β_1.

13.74 Ex13-74 The tobacco settlement negotiated by an eight-member team of attorneys-general on behalf of 41 states resulted in $206 billion to be paid by the tobacco industry to recoup Medicaid costs the states incurred while treating sick smokers. Payments are to be made in annual increments over a 25-year span, starting in 1998. The table shows the populations (in millions) and the settlements (in $billions) awarded to 46 states, the District of Columbia, and Puerto Rico.

State	Settlement	Population	State	Settlement	Population
AL	3.17	4.27	NE	1.17	1.65
AK	0.67	0.61	NV	1.19	1.60
AZ	2.89	4.43	NH	1.30	1.16
AR	1.62	2.61	NJ	7.58	7.99
CA	25.00	31.88	NM	1.17	1.71
CO	2.69	3.82	NY	25.00	18.18
CT	3.64	3.27	NC	4.57	7.32
DE	7.75	0.72	ND	0.72	0.64
DC	1.19	0.54	OH	9.87	11.17
GA	4.81	7.35	OK	2.03	3.30
HI	1.18	1.18	OR	2.25	3.20
ID	0.71	1.19	PA	11.30	12.06
IL	9.12	11.85	PR	2.20	3.78
IN	4.00	5.84	RI	1.41	0.99
IA	1.70	2.85	SC	2.30	0.70
KS	1.60	2.57	SD	0.68	0.73
KY	3.45	3.88	TN	4.78	5.32
LA	4.42	4.35	UT	0.87	2.00
ME	1.51	1.24	VT	0.81	0.59
MD	4.43	5.07	VA	4.01	6.68
MA	7.91	6.09	WA	4.02	5.53
MI	8.53	9.59	WV	1.74	1.83
MO	4.46	5.36	WI	4.06	5.16
MT	0.83	0.88	WY	0.49	0.48

Sources: Washington State Attorney General Office, and Bureau of the Census, U.S. Department of Commerce

a. Draw a scatter diagram of these data with tobacco settlement as the dependent variable, y, and population as the predictor variable, x.
b. Calculate the regression equation and draw the regression line on the scatter diagram.
c. If your state's population were 11.5 million people, what would you estimate the tobacco settlement to be? Make your estimate based on the equation, and then draw a line on the scatter diagram to illustrate it.
d. Construct a 95% prediction interval for the estimate you obtained in part c.

13.75 Ex13-75 The following data resulted from an experiment performed for the purpose of regression analysis. The input variable, x, was set at five different levels, and observations were made at each level.

x	0.5	1.0	2.0	3.0	4.0
y	3.8	3.2	2.9	2.4	2.3
	3.5	3.4	2.6	2.5	2.2
	3.8	3.3	2.7	2.7	2.3
		3.6	3.2	2.3	

a. Draw a scatter diagram.
b. Draw the regression line by eye.
c. Place a star, ★, at each level approximately where the mean of the observed y values is located. Does your regression line look like the line of best fit for these five mean values?
d. Calculate the equation of the regression line.
e. Find the standard deviation of y about the regression line.
f. Construct a 95% confidence interval for the true value of β_1.
g. Construct a 95% confidence interval for the mean value of y at $x = 3.0$. At $x = 3.5$.
h. Construct a 95% prediction interval for an individual value of y at $x = 3.0$. At $x = 3.5$.

13.76 Ex13-76 The set of 25 scores was randomly selected from a teacher's class list. Let x be the prefinal average and y the final examination score. (The final examination had a maximum of 75 points.)

Student	x	y	Student	x	y
1	75	64	14	73	62
2	86	65	15	78	66
3	68	57	16	71	62
4	83	59	17	86	71
5	57	63	18	71	55
6	66	61	19	96	72
7	55	48	20	96	75
8	84	67	21	59	49
9	61	59	22	81	71
10	68	56	23	58	58
11	64	52	24	90	67
12	76	63	25	92	75
13	71	66			

a. Draw a scatter diagram for these data.
b. Draw the regression line (by eye) and estimate its equation.
c. Estimate the value of the coefficient of linear correlation.
d. Calculate the equation of the line of best fit.
e. Draw the line of best fit on your graph. How does it compare with your estimate?
f. Calculate the linear correlation coefficient. How does it compare with your estimate?
g. Test the significance of r at $\alpha = 0.10$.
h. Find the 95% confidence interval for the true value of ρ.
i. Find the standard deviation of the y values about the regression line.
j. Calculate a 95% confidence interval for the true value of the slope β_1.
k. Test the significance of the slope at $\alpha = 0.05$.
l. Estimate the mean final-exam grade that all students with an 85 prefinal average will obtain (95% confidence interval).

(continued)

m. Using the 95% prediction interval, predict the grade that John Henry will receive on his final, knowing that his prefinal average is 78.

13.77 Ex13-77 Twenty-one mature flowers of a particular species were dissected, and the numbers of stamens (x) and carpels (y) present in each flower were counted.

x	y	x	y	x	y
52	20	65	30	45	27
68	31	43	19	72	21
70	28	37	25	59	35
38	20	36	22	60	27
61	19	74	29	73	33
51	29	38	28	76	35
56	30	35	25	68	34

a. Is there sufficient evidence to claim a linear relationship between these two variables at $\alpha = 0.05$?
b. What is the relationship between the numbers of stamens and carpels in this variety of flower?
c. Is the slope of the regression line significant at $\alpha = 0.05$?
d. Give the 95% prediction interval for the number of carpels that one would expect to find in a mature flower of this variety if the number of stamens were 64.

13.78 Ex13-78 It is believed that the amount of nitrogen fertilizer used per acre has a direct effect on the amount of wheat produced. The data are the amount of nitrogen fertilizer used per test plot and the amount of wheat harvested per test plot.

x, Pounds of Fertilizer	y, 100 Pounds of Wheat	x, Pounds of Fertilizer	y, 100 Pounds of Wheat
30	5	40	14
30	9	40	18
30	14	50	12
40	6	50	14

(continued)

x, Pounds of Fertilizer	y, 100 Pounds of Wheat	x, Pounds of Fertilizer	y, 100 Pounds of Wheat
50	23	80	32
60	18	80	35
60	24	90	27
60	28	90	32
70	19	90	38
70	23	100	34
70	31	100	35
80	24	100	39

a. Is there sufficient reason to conclude that the use of more fertilizer results in a higher yield? Use $\alpha = 0.05$.
b. Estimate, with a 98% confidence interval, the mean yield that could be expected if 50 pounds of fertilizer were used per plot.
c. Estimate, with a 98% confidence interval, the mean yield that could be expected if 75 pounds of fertilizer were used per plot.

13.79 The correlation coefficient, r, is related to the slope of the line of best fit, b_1, by the equation

$$r = b_1\sqrt{\frac{SS(x)}{SS(y)}}$$

Verify the equation using the following data:

x	1	2	3	4	6
y	4	6	7	9	12

13.80 The following equation is known to be true for any set of data:

$$\sum(y - \bar{y})^2 = \sum(y - \hat{y})^2 + \sum(\hat{y} - \bar{y})^2$$

Verify the equation with these data:

x	0	1	2
y	1	3	2

Vocabulary and Key Concepts

Be able to define each term. Pay special attention to the key terms, which are printed in **red**. In addition, describe each term in your own words and give an example of each. Your examples should not be ones given in class or in the textbook. The bracketed numbers indicate the chapters in which the term previously appeared, but you should define the terms again to show increased understanding of their meaning. Page numbers indicate the first appearance of the term in Chapter 13.

assumptions [8, 9, 10, 11, 12] (pp. 611, 627)
bivariate data [3] (p. 605)
centroid (p. 605)
coefficient of linear correlation [3] (p. 608)
confidence belts (p. 611)
confidence interval [8, 9, 10] (pp. 611, 628, 636, 637)
covariance (p. 606)
curvilinear regression (p. 621)
experimental error (ϵ or e) (p. 621)
hypothesis tests [8, 9, 10, 11, 12] (pp. 612, 629)
intercept (b_0 or β_0) [3] (p. 620)
line of best fit [3] (p. 620)
linear correlation (p. 605)
linear regression [3] (p. 620)
multiple regression (p. 621)
Pearson's product moment, r (p. 608)
predicted value of μ_y (p. 636)
predicted value of y ($\hat{y}$) (p. 636)

Chapter Practice Test

PART I: KNOWING THE DEFINITIONS

Answer "True" if the statement is always true. If the statement is not always true, replace the words shown in bold with words that make the statement always true.

13.1 The error **must be** normally distributed if inferences are to be made.

13.2 Both x and y **must be** normally distributed.

13.3 A high correlation between x and y **proves** that x causes y.

13.4 The value of the input variable **must be** randomly selected to achieve valid results.

13.5 The output variable must be **normally distributed** about the regression line for each value of x.

13.6 **Covariance** measures the strength of the linear relationship and is a standardized measure.

13.7 The **sum of squares for error** is the name given to the numerator of the formula used to calculate the variance of y about the line of regression.

13.8 **Correlation** analysis attempts to find the equation of the line of best fit for two variables.

13.9 There are $\boldsymbol{n-3}$ degrees of freedom involved with the inferences about the regression line.

13.10 $\hat{y}$ serves as the **point estimate** for both $\mu_{y|x_0}$ and y_{x_0}.

PART II: APPLYING THE CONCEPTS

Answer all questions, showing formulas and work.

It is believed that the amount of nitrogen fertilizer used per acre has a direct effect on the amount of wheat produced. The data show the amount of nitrogen fertilizer used per test plot and the amount of wheat harvested per test plot. All test plots were the same size.

Pt13-11

x, Pounds of Fertilizer	y, 100 Pounds of Wheat
30	9
30	11
30	14
50	12
50	14
50	23
70	19
70	22
70	31
90	29
90	33
90	35

13.11 Draw a scatter diagram of the data (use graph paper and a straightedge). Be sure to label completely.

13.12 Complete an extensions table.

13.13 Calculate SS(x), SS(xy), and SS(y).

13.14 Calculate the linear correlation coefficient, r.

13.15 Determine the 95% confidence interval estimate for the population linear correlation coefficient.

13.16 Calculate the equation of the line of best fit.

13.17 Draw the line of best fit on the scatter diagram (in red ink).

13.18 Calculate the standard deviation of the y values about the line of best fit.

13.19 Does the value of b_1 show strength significant enough to conclude that the slope is greater than zero at the 0.05 level?

13.20 Determine the 0.95 confidence interval for the mean yield when 85 pounds of fertilizer are used per plot.

13.21 Draw a line on the scatter diagram representing the 95% confidence interval found in question 13.20 (in blue ink).

PART III: UNDERSTANDING THE CONCEPTS

13.22 "There is a high correlation between how frequently skiers have their bindings tested and the incidence of lower-leg injuries, according to researchers at the Rochester Institute of Technology. To make sure your bindings release properly when you begin to fall, you should have them serviced by a ski mechanic every 15 to 30 ski days or at least at the start of each ski season" (University of California, Berkeley, "Wellness Letter," February 1991). Explain what two variables are discussed in this statement, and interpret the "high correlation" mentioned.

13.23 Describe why the method used to define the correlation coefficient is referred to as "a product moment."

13.24 If you know that the value of r is very close to zero, what value would you anticipate for b_1? Explain why.

13.25 Describe why the method used to find the line of best fit is referred to as "the method of least squares."

13.26 You wish to study the relationship between the amount of sugar in a child's breakfast and the child's hyperactivity in school during the four hours after breakfast. You ask 200 mothers of fifth-grade children to keep a careful record of what the children eat and drink each morning. Each parent's report is analyzed and the sugar consumption is determined. During the same time period, data on hyperactivity are collected at school. What statistic will measure the strength and kind of relationship that exists between the amount of sugar and the amount of hyperactivity? Explain why the statistic you selected is appropriate and what value you expect this statistic might have.

13.27 You are interested in studying the relationship between the length of time a person has been supported by welfare and self-esteem. You believe that the longer a person is supported, the lower the self-esteem. What data would you need to collect and what statistics would you calculate if you wish to predict a person's level of self-esteem after having been on welfare for a certain period of time? Explain in detail.

CHAPTER

14

Elements of Nonparametric Statistics

Chapter Outline and Objective

CHAPTER CASE STUDY

The statistics suggest that teenagers' attitudes toward moral and social values are much more conventional than widely believed.

14.1 NONPARAMETRIC STATISTICS

Distribution-free, or **nonparametric,** methods provide test statistics for an unspecified distribution.

14.2 COMPARING STATISTICAL TESTS

When choosing between parametric and nonparametric tests, we are interested primarily in the **control of error,** the relative **power** of the test, and **efficiency.**

14.3 THE SIGN TEST

A simple **count** of plus and minus signs tells us whether or not to reject the null hypothesis.

14.4 THE MANN–WHITNEY *U* TEST

The **rank number** for each piece of data is used to compare **two independent samples.**

14.5 THE RUNS TEST

A **sequence of data** that possesses a common property, a **run,** is used to test the question of randomness.

14.6 RANK CORRELATION

The linear correlation coefficient's **nonparametric alternative,** rank correlation, uses only rankings to determine whether or not to reject the null hypothesis.

RETURN TO CHAPTER CASE STUDY

This chapter presents only a sampling of the many nonparametric tests. The selections presented demonstrate their ease of application and variety of technique.

Chapter Case Study

Teenagers' Attitudes

© Peter Hvizdak/The Image Works

A national survey revealed that teenagers' attitudes toward moral and social values are much more conventional than widely believed.

How Teenagers See Things

One could think of them as Generation "V"—for values. According to The Mood of American Youth study, today's teens are neither as rebellious as ado lescents in the 1970s nor as materialistic as those of the 1980s. What they want is not to change the world or to own a chunk of it, but to be happy. Among the teens' greatest concerns: the decline in moral and social values.

The study, conducted by NFO Research, Inc., included 938 young people aged 13 to 17 who are representative of America's adolescent population as a whole. Of those polled, nine in 10 say they don't drink or smoke. Seven in 10 say religion is important in their lives. Most of the respondents respect their parents, get along well with them, and consider their rules strict but fair.

Teens' Views on Contemporary Issues

Families and Children	Agree
Teenagers are not prepared to have babies.	91%
A single parent can raise a family.	75%
I am very likely to raise my children differently than I was raised.	55%
In the Schools	
Local school officials should be able to censor the books and materials used in their schools.	47%
School prayer should not be permitted.	32%
Social Concerns	
The "V" chip will unfairly censor what teens can watch on TV.	67%
It is important to control information on the Internet.	60%
. . . And What About the Government?	
Government spending on AIDS research should be increased.	83%
Adequate health care for all should be provided through a national health plan.	81%
Being rich is necessary to get elected to high office.	55%

Part of the survey asked the teens to indicate the one thing they wanted most from life. The following table (p. 656) lists the choices and the percentage of teens selecting each.

The One Thing Teens Want Most from Life

CS14

Choices	All	Boys	Girls
Happiness	28%	23%	32%
Long, enjoyable life	16	18	14
Marriage and family	9	8	11
Financial success	8	11	4
Career success	8	9	6
Religious satisfaction	8	7	7
Love	7	6	7
Personal success	6	6	5
Personal contribution to society	2	3	2
Friends	2	3	1
Health	2	2	2
Education	2	1	2

*Some teens didn't respond, so figures don't total 100%.

Dianne Hales
Parade Magazine, August 18, 1996.

ANSWER NOW 14.1

a. **Do teenage boys and girls agree on what they want most from life?**

b. **Present statistical evidence to support your answer in part a.**

After completing Chapter 14, further investigate the Chapter Case Study in the Return to Chapter Case Study section with Exercises 14.58, 14.59, and 14.60 (p. 699).

14.1 Nonparametric Statistics

Most of the statistical procedures we have studied in this book are known as **parametric methods.** For a statistical procedure to be parametric, either we assume that the parent population is at least approximately normally distributed or we rely on the central limit theorem to give us a normal approximation. This is particularly true of the statistical methods studied in Chapters 8, 9, and 10.

The **nonparametric methods,** or **distribution-free methods** as they are also known, do not depend on the distribution of the population being sampled. The nonparametric statistics are usually subject to much less confining restrictions than are their parametric counterparts. Some, for example, require only that the parent population be continuous.

The recent popularity of nonparametric statistics can be attributed to the following characteristics:

1. Nonparametric methods require few assumptions about the parent population.
2. Nonparametric methods are generally easier to apply than their parametric counterparts.
3. Nonparametric methods are relatively easy to understand.
4. Nonparametric methods can be used in situations where the normality assumptions cannot be made.
5. Nonparametric methods are generally only slightly less efficient than their parametric counterparts.

14.2 Comparing Statistical Tests

Only four nonparametric tests are presented in this chapter. They represent a very small sampling of the many different nonparametric tests that exist. Many of the nonparametric tests can be used in place of certain parametric tests. The question is, then: Which statistical test do we use, the parametric or the nonparametric? Sometimes there is more than one nonparametric test to choose from.

The decision about which test to use must be based on the answer to the question: Which test will do the job best? First, let's agree that when we compare two or more tests, they must be equally qualified for use. That is, each test has a set of assumptions that must be satisfied before it can be applied. From this starting point we will attempt to define "best" to mean the test that is best able to control the risks of error and at the same time keep the size of the sample to a number that is reasonable to work with. (Sample size means cost—cost to you or your employer.)

Let's look first at the ability to control the risk of error. The risk associated with a type I error is controlled directly by the level of significance α. Recall that $P(\text{type I error}) = \alpha$ and $P(\text{type II error}) = \beta$. Therefore, it is β that we must control. Statisticians like to talk about *power* (as do others), and the **power of a statistical test** is defined to be $1 - \beta$. Thus, the power of a test, $1 - \beta$, is the probability that we reject the null hypothesis when we should have rejected it. If two tests with the same α are equal candidates for use, then the one with the greater power is the one you would want to choose.

The other factor is the sample size required to do a job. Suppose that you set the levels of risk you can tolerate, α and β, and then you are able to determine the sample size it would take to meet your specified challenge. The test that required the smaller sample size would seem to have the edge. Statisticians usually use the term *efficiency* to talk about this concept. *Efficiency* is the ratio of the sample size of the best parametric test to the sample size of the best nonparametric test when compared under a fixed set of risk values. For example, the efficiency rating for the sign test is approximately 0.63. This means that a sample of size 63 with a parametric test will do the same job as a sample of size 100 will do with the sign test.

The power and the efficiency of a test cannot be used alone to determine the choice of test. Sometimes you will be forced to use a certain test because of the data you are given. When there is a decision to be made, the final decision rests in a trade-off of three factors: (1) the power of the test, (2) the efficiency of the test, and (3) the data (and the number of data) available. Table 14.1 shows how the nonparametric tests discussed in this chapter compare with the parametric tests covered in previous chapters.

TABLE 14.1 Comparison of Parametric and Nonparametric Tests

Test Situation	Parametric Test	Nonparametric Test	Efficiency of Nonparametric Test
One mean	*t*-test (p. 412)	Sign test (p. 658)	0.63
Two independent means	*t*-test (p. 490)	*U* test (p. 670)	0.95
Two dependent means	*t*-test (p. 475)	Sign test (p. 662)	0.63
Correlation	Pearson's (p. 611)	Spearman test (p. 689)	0.91
Randomness		Runs test (p. 601)	Not meaningful; there is no parametric test for comparison

14.3 The Sign Test

The sign test is a versatile and exceptionally easy-to-apply nonparametric method that uses only plus and minus signs. Three sign test applications are presented here: (1) a confidence interval for the median of one population, (2) a hypothesis test concerning the value of the median for one population, and (3) a hypothesis test concerning the median difference (paired difference) for two **dependent samples.** These sign tests are carried out using the same basic confidence interval and hypothesis test procedures as described in earlier chapters. They are the nonparametric alternatives to the *t*-tests used for one mean (see Section 9.1) and the difference between two dependent means (see Section 10.2).

ASSUMPTIONS FOR INFERENCES ABOUT THE POPULATION MEDIAN USING THE SIGN TEST

The *n* random observations that form the sample are selected independently, and the population is continuous in the vicinity of the median *M*.

SINGLE-SAMPLE CONFIDENCE INTERVAL PROCEDURE

The sign test can be applied to obtain a confidence interval for the unknown **population median, *M*.** To accomplish this we will need to arrange the sample data in ascending order (smallest to largest). The data are identified as x_1 (smallest), x_2, $x_3, \ldots, x_n$ (largest). The critical value, k (known as the "maximum allowable number of signs"), is obtained from Table 12 in Appendix B, and tells us the number of positions to be dropped from each end of the ordered data. The remaining extreme values become the bounds of the $1 - \alpha$ confidence interval. That is, the lower boundary for the confidence interval is x_{k+1}, the $(k + 1)$th piece of data; the upper boundary is x_{n-k}, the $(n - k)$th piece of data. The next illustration will clarify this procedure.

Illustration 14.1 Constructing a Confidence Interval for a Population Median

Suppose that we have 12 pieces of data in ascending order ($x_1, x_2, x_3, \ldots, x_{12}$) and we wish to form a 95% confidence interval for the population median. Table 12 shows a critical value of 2 ($k = 2$) for $n = 12$ and $\alpha = 0.05$ for a hypothesis test. This means that we drop the last two values on each end (x_1 and x_2 on the left; x_{11} and x_{12} on the right). The confidence interval is bounded by x_3 and x_{10}, inclusively. That is, the 95% confidence interval is x_3 to x_{10} and is expressed as:

$$\mathbf{x_3 \text{ to } x_{10}}, \quad 95\% \text{ confidence interval for } M$$

In general, the two pieces of data that bound the confidence interval occupy positions $k + 1$ and $n - k$, where k is the critical value read from Table 12. Thus,

$$\mathbf{x_{k+1} \text{ to } x_{n-k}}, \quad 1 - \alpha \text{ confidence interval for } M$$

■

ANSWER NOW 14.2 Ex14-02

Ten randomly selected shut-ins were asked how many hours of television they watched last week. The results are listed here:

82 66 90 84 75 88 80 94 110 91

Determine the 90% confidence interval estimate for the median number of hours of television watched per week by shut-ins.

SINGLE-SAMPLE HYPOTHESIS-TESTING PROCEDURE

The sign test can be used when the null hypothesis to be tested concerns the value of the population median *M*. The test may be either one- or two-tailed. This test procedure is presented in the following illustration.

Illustration 14.2 Two-Tailed Hypothesis Test

A random sample of 75 students was selected, and each student was asked to carefully measure the amount of time it takes to commute from his or her front door to the college parking lot. The data collected were used to test the hypothesis: The median time required for students to commute is 15 minutes, against the alternative that the median is unequal to 15 minutes. The 75 pieces of data were summarized as follows:

Under 15:	18
15:	12
Over 15:	45

Use the sign test to test the null hypothesis against the alternative hypothesis.

SOLUTION

The data are converted to + and − signs according to whether the data is more or less than 15. A plus sign will be assigned to each piece of data larger than 15, a minus sign to each piece of data smaller than 15, and a zero to those data equal to 15. The sign test uses only the plus and minus signs; therefore, the zeros are discarded and the usable sample size becomes 63. That is, $n(+) = 45$, $n(-) = 18$, and $n = n(+) + n(-) = 45 + 18 = 63$.

Step 1 **The Set-Up:**

a. **Describe the population parameter of interest.**
   M, the population median time to commute.

b. **State the null hypothesis (H_o) and the alternative hypothesis (H_a).**

$$H_o\text{: } M = 15$$

$$H_a\text{: } M \neq 15$$

Step 2 **The Hypothesis Test Criteria:**

a. **Check the assumptions.**
   The 75 observations were randomly selected, and the variable commute time is continuous.

b. **Identify the test statistic to be used.**
The test statistic that will be used is the number of the less frequent sign: the smaller of $n(+)$ and $n(-)$, which is $n(-)$ for our illustration. We will want to reject the null hypothesis whenever the number of the less frequent sign is extremely small. Table 12 in Appendix B gives the maximum allowable number of the less frequent sign, k, that will allow us to reject the null hypothesis. That is, if the number of the less frequent sign is less than or equal to the critical value in the table, we will reject H_o. If the observed value of the less frequent sign is larger than the table value, we will fail to reject H_o. In the table, n is the total number of signs, not including zeros. Therefore, the test statistic = $\boldsymbol{x\star = n(-)}$.

c. **Determine the level of significance;** $\alpha = 0.05$ for a two-tailed test.

Step 3 **The Sample Evidence:**

a. **Collect the sample information.**
$n = 63$; the observed value of the test statistic is $x = n(-) = 18$.

b. **Calculate the value of the test statistic:** $x\star = n(-) = \mathbf{18}$.

Step 4 **The Probability Distribution:**

Using the *p*-Value Procedure

a. **Calculate the *p*-value for the test statistic.**
Since the concern is for values "not equal to," the p-value is the area of both tails. We will find the left tail and double it: $\mathbf{P} = 2 \times P(x \leq 18, \text{ for } n = 63)$

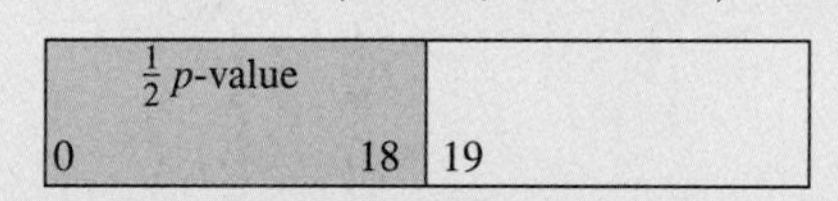

Number of less frequent sign

To find the p-value, you have two options:

1. Use Table 12 in Appendix B to place bounds on the p-value. Table 12 lists only two-tailed values (do not double): $\mathbf{P < 0.01}$.
2. Use a computer or calculator to find the p-value: $\mathbf{P = 0.0011}$.

Specific instructions follow this illustration.

b. **Determine whether or not the *p*-value is smaller than α.**
The p-value is smaller than α.

OR

Using the Classical Procedure

a. **Determine the critical region and critical value(s).**
The critical region is split into two equal parts because H_a expresses concern for values related to "not equal to." Since the table is for two-tailed tests, the critical value is located at the intersection of the $\alpha = 0.05$ column and the $n = 63$ row of Table 12: **23.**

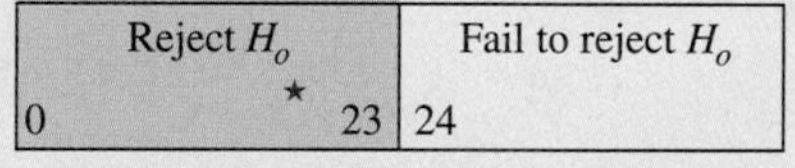

Number of less frequent sign

b. **Determine whether or not the calculated test statistic is in the critical region.**
$x\star$ is in the critical region, as shown in the figure.

Step 5 **The Results:**

a. **State the decision about H_o:** Reject H_o.

b. **State the conclusion about H_a.**
The sample shows sufficient evidence at the 0.05 level to conclude that the median commute time is not equal to 15 minutes.

Calculating the *p*-value when using the sign test

Method 1: Use Table 12 in Appendix B to place bounds on the p-value. By inspecting the $n = 63$ row of Table 12, you can determine an interval within which the p-value lies. Locate the value of x along the $n = 63$ row and read the bounds from the top of the table. Table 12 lists only two-tailed values (therefore, do not double): **$P < 0.01$.**

Method 2: If you are doing the hypothesis test with the aid of a computer or graphing calculator, most likely it will calculate the p-value for you. Specific instructions are described below. ■

Technology Instructions: Sign Test for a Single-Sample Hypothesis Test of the Median

MINITAB (Release 13)

Input the set of data into C1; then continue with:

Choose:	Stat > Nonparametrics > 1-Sample Sign
Enter:	Variables: C1
Select:	Test median:*
Enter:	M (hypothesized median value)
	Alternative; less than or not equal or greater than

*A confidence interval may also be selected.

(If original data are not given, just the number of plus and minus signs, then input data values above and below the median that will compute into the correct number of each sign.)

Excel XP

The following Excel commands will compute the differences between the data values and the hypothesized median. The data will then be sorted so that the number of + and − signs can be easily counted.

Input the data into column A and select cell B1; then continue with:

Choose:	Insert function f_x > All > SIGN > OK
Enter:	Number: A1 − hypothesized median value > OK
Drag:	Bottom right corner of the B1 cell down to give other differences

Select the data in columns A and B; then continue with:

Choose:	Data > Sort
Enter:	Sort by: Column B
Select:	Ascending > OK

TI-83 Plus

Input the data into L1; then continue with:

Choose:	PRGM > EXEC > SIGNTEST*
Select:	PROCEDURE: 3: HYP TEST
	INPUT? 2:DATA: 1 LIST
Enter:	DATA: L1
	MED0: hypothesized median value
Select:	ALT HYP? 1: > or 2: < or 3: ≠

*Program SIGNTEST is one of many programs that are available for downloading from www.duxbury.com. See page 40 for specific directions.

TWO-SAMPLE HYPOTHESIS-TESTING PROCEDURE

The sign test may also be applied to a hypothesis test dealing with the median difference between **paired data** that result from **two dependent samples.** A familiar application is the use of before-and-after testing to determine the effectiveness of some activity. In a test of this nature, the signs of the differences are used to carry out the test. Again, zeros are disregarded.

ASSUMPTIONS FOR INFERENCES ABOUT THE MEDIAN OF PAIRED DIFFERENCES USING THE SIGN TEST

The paired data are selected independently, and the variables are ordinal or numerical.

The following illustration shows this procedure.

Illustration 14.3 One-Tailed Hypothesis Test for the Median of Paired Differences

A new no-exercise, no-starve weight-reducing plan has been developed and advertised. To test the claim that "you will lose weight within two weeks or . . . ," a local statistician obtained the before-and-after weights of 18 people who had used this plan. Table 14.2 lists the people, their weights, and a minus (−) for those who lost weight during the two weeks, a 0 for those who remained the same, and a plus (+) for those who actually gained weight.

TABLE 14.2 Sample Results on Weight-Reducing Plan Ta14-02

Person	Weight		Sign of Difference,
	Before	After	After − Before
Mrs. Smith	146	142	−
Mr. Brown	175	178	+
Mrs. White	150	147	−
Mr. Collins	190	187	−
Mr. Gray	220	212	−
Ms. Collins	157	160	+
Mrs. Allen	136	135	−
Mrs. Noss	146	138	−
Ms. Wagner	128	132	+
Mr. Carroll	187	187	0
Mrs. Black	172	171	−
Mrs. McDonald	138	135	−
Ms. Henry	150	151	+
Ms. Greene	124	126	+
Mr. Tyler	210	208	−
Mrs. Williams	148	148	0
Mrs. Moore	141	138	−
Mrs. Sweeney	164	159	−

The claim being tested is that people lose weight. The null hypothesis that will be tested is: There is no weight loss (or the median weight loss is zero), meaning that only a rejection of the null hypothesis will allow us to conclude in favor of the advertised claim. Actually we will be testing to see whether there are significantly more minus signs than plus signs. If the weight-reducing plan is of absolutely no value, we would expect to find an equal number of plus and minus signs. If it works, there should be significantly more minus signs than plus signs. Thus, the test performed here will be a one-tailed test. (We want to reject the null hypothesis in favor of the advertised claim if there are "many" minus signs.)

SOLUTION

Step 1 **The Set-Up:**

a. Describe the population parameter of interest.
M, the median weight loss.

b. State the null hypothesis (H_o) and the alternative hypothesis (H_a).

$$H_o: M = 0 \quad \text{(no weight loss)}$$

$$H_a: M < 0 \quad \text{(weight loss)}$$

Step 2 **The Hypothesis Test Criteria:**

a. Check the assumptions.
The 18 observations were randomly selected, and the variables, weight before and weight after, are both continuous.

b. Identify the test statistic to be used.
The number of the less frequent sign: the test statistic = $x\bigstar = n(+)$.

c. Determine the level of significance: $\alpha = 0.05$ for a one-tailed test.

Step 3 **The Sample Evidence:**

a. Collect the sample information.
$n = 16$ [$n(+) = 5$, $n(-) = 11$]; the observed value of the test statistic is $x\bigstar = n(+) = 5$.

b. Calculate the value of the test statistic: $x\bigstar = n(+) = 5$.

Step 4 **The Probability Distribution:**

Using the p-Value Procedure:

a. Calculate the p-value for the test statistic.
Since the concern is for values "less than," the p-value is the area to the left: $\mathbf{P} = P(x \leq 5, \text{ for } n = 16)$

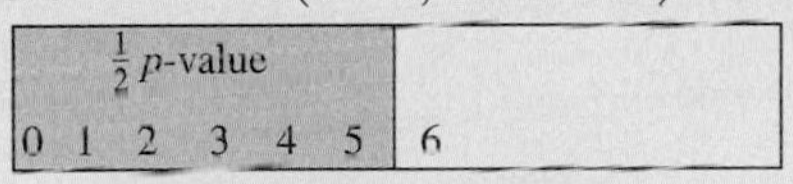

Number of less frequent sign

To find the p-value, you have two options:

1. Use Table 12 in Appendix B to estimate the p-value. Table 12 lists only two-tailed α (this is one-tailed, so divide α by two): $\mathbf{P} \approx \mathbf{0.125}$.
2. Use a computer or calculator to find the p-value: $\mathbf{P} = \mathbf{0.1051}$.

For specific instructions, see pp. 660–661. *(continued)*

OR

Using the Classical Procedure:

a. Determine the critical region and critical value(s).
The critical region is one-tailed because H_a expresses concern for values related to "less than." Since the table is for two-tailed tests, the critical value is located at the intersection of the $\alpha = 0.10$ column ($\alpha = 0.05$ in each tail) and the $n = 16$ row of Table 12:

$$k = 4.$$

Reject H_o | Fail to reject H_o
0 1 2 3 4 | ★ 5

Number of less frequent sign

(continued)

b. Determine whether or not the p-value is smaller than α.
The p-value is not smaller than α.

b. Determine whether or not the calculated test statistic is in the critical region.
$x\bigstar$ is not in the critical region, as shown in the figure.

Step 5 **The Results:**

a. State the decision about H_o: Fail to reject H_o.

b. State the conclusion about H_a.
The evidence observed is not sufficient to allow us to reject the no-weight-loss null hypothesis at the 0.05 level of significance. ■

Technology Instructions: Sign Test for the Median of Paired Differences

MINITAB (Release 13)

Input the paired set of data into C1 and C2; then continue with:

```
Choose:   Calc > Calculator
Enter:    Store result in: C3
          Expression: C1-C2 (whichever order is needed,
          based on Ha)
Choose:   Stat > Nonparametrics > 1-Sample Sign...
Enter:    Variables: C3
Select:   Test median:*
Enter:    0 (hypothesized median value)
          Alternative: less than or not equal or greater
          than
```

*As before, the confidence interval may be selected.

Excel XP

Input the paired data into columns A and B; then continue with:

```
Choose:   Tools > Data Analysis Plus > Sign Test > OK
Enter:    Variable 1 Range: (A1:A20 or select cells)
          Variable 2 Range: (B1:B20 or select cells)
Select:   Labels (if necessary)
Enter:    Alpha: α (ex. 0.05)
```

TI-83 Plus

Input the paired data into L1 and L2; then continue with:

```
Highlight:  L3
Enter:      L1-L2 (whichever order is needed, based on Ha)
Choose:     PRGM > EXEC > SIGNTEST*
```

(continued)

```
Select:   PROCEDURE: 3: HYP TEST
          INPUT? 2:DATA: 1 LIST
Enter:    DATA: L3
          MEDO: hypothesized median value
Select:   ALT HYP? 1:>or 2:<or 3:≠
```

*Program SIGNTEST is one of many programs that are available for downloading from www.duxbury.com. See page 40 for specific directions.

NORMAL APPROXIMATION

The sign test may be carried out by means of a normal approximation using the standard normal variable z. The normal approximation will be used if Table 12 does not show the particular levels of significance desired or if n is large.

NOTES

1. x may be the number of the less frequent sign or the more frequent sign. You will have to determine this in such a way that the direction is consistent with the interpretation of the situation.
2. x is really a **binomial random variable,** where $p = 0.5$. The sign test statistic satisfies the properties of a binomial experiment (see p. 250). Each sign is the result of an independent trial. There are n trials, and each trial has two possible outcomes (+ or −). Since the median is used, the probabilities for each outcome are both 0.5. Therefore, the mean, μ_x, is equal to

$$\mu_x = \frac{n}{2} \qquad \left[\mu = np = n \cdot \frac{1}{2} = \frac{n}{2}\right]$$

   and the standard deviation, σ_x, is equal to

$$\sigma_x = \frac{1}{2}\sqrt{n} \qquad \left[\sigma = \sqrt{npq} = \sqrt{n \cdot \frac{1}{2} \cdot \frac{1}{2}} = \frac{1}{2}\sqrt{n}\right]$$

3. x is a discrete variable. But recall that the normal distribution must be used only with continuous variables. However, although the binomial random variable is discrete, it does become approximately normally distributed for large n. Nevertheless, when using the normal distribution for testing, we should make an adjustment in the variable so that the approximation is more accurate. (See Section 6.5, p. 298, on the normal approximation.) This adjustment is illustrated in Figure 14.1 and is called a **continuity correction.** For this discrete variable the area that represents the probability is a rectangular bar. Its width is 1 unit wide,

FIGURE 14.1
Continuity Correction

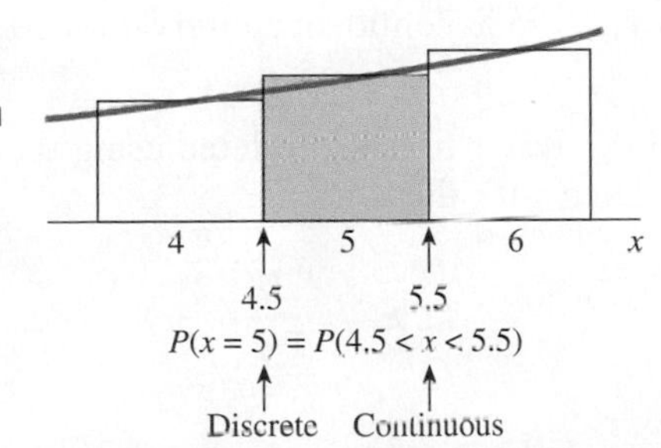

from $\frac{1}{2}$ unit below to $\frac{1}{2}$ unit above the value of interest. Therefore, when z is to be used, we will need to make a $\frac{1}{2}$-unit adjustment before calculating the observed value of z. So x' will be the adjusted value for x. If x is larger than $\frac{n}{2}$, then $x' = x - \frac{1}{2}$. If x is smaller than $\frac{n}{2}$, then $x' = x + \frac{1}{2}$. The test is then completed by the usual procedure, using x'.

If the normal approximation is to be used (including the continuity correction), the position numbers for a $1 - \alpha$ confidence interval for M are found using the formula:

$$\frac{1}{2}(n) \pm \left(\frac{1}{2} + \frac{1}{2} \cdot z(\alpha/2) \cdot \sqrt{n}\right) \tag{14.1}$$

The interval is

$$x_L \text{ to } x_U, \quad 1 - \alpha \text{ confidence interval for } M \text{ (median)}$$

where

$$L = \frac{n}{2} - \frac{1}{2} - z(\alpha/2) \cdot \sqrt{n} \quad \text{and} \quad U = \frac{n}{2} + \frac{1}{2} + z(\alpha/2) \cdot \sqrt{n}$$

NOTE L should be rounded down and U should be rounded up to be sure that the level of confidence is at least $1 - \alpha$.

Illustration 14.4 Constructing a Confidence Interval for a Population Median

Estimate the population median with a 95% confidence interval for a given set of 60 pieces of data: $x_1, x_2, x_3, \ldots, x_{59}, x_{60}$.

SOLUTION

When we use formula (14.1), the position numbers L and U are

$$\frac{1}{2}(n) \pm \left(\frac{1}{2} + \frac{1}{2} \cdot z(\alpha/2) \cdot \sqrt{n}\right): \quad \frac{1}{2}(60) \pm \left(\frac{1}{2} + \frac{1}{2} \cdot 1.96 \cdot \sqrt{60}\right)$$

$$30 \pm (0.50 + 7.59)$$

$$30 \pm 8.09$$

Thus,

$L = 30 - 8.09 = 21.91$, rounded down becomes 21 (21st piece of data)

$U = 30 + 8.09 = 38.09$, rounded up becomes 39 (39th piece of data)

Therefore,

$$\boldsymbol{x_{21}} \textbf{ to } \boldsymbol{x_{39}}, \quad 95\% \text{ confidence interval for } M \text{ (median)}$$

■

When a hypothesis test is to be completed using the standard normal distribution, z will be calculated with the formula:

$$z\bigstar = \frac{x' - \frac{n}{2}}{\frac{1}{2} \cdot \sqrt{n}} \tag{14.2}$$

(See Note 3 on p. 665 with regard to x'.)

Illustration 14.5 One-Tailed Hypothesis Test

Use the sign test to test the hypothesis that the median number of hours, M, worked by students of a certain college is at least 15 hours per week. A survey of 120 students was taken; a plus sign was recorded if the number of hours the student worked last week was equal to or greater than 15, and a minus sign was recorded if the number of hours was less than 15. Totals showed 80 minus signs and 40 plus signs.

SOLUTION

Step 1 The Set-Up:

a. **Describe the population parameter of interest.**
M, the median number of hours worked by students.

b. **State the null hypothesis (H_o) and the alternative hypothesis (H_a).**

H_o: $M = 15$ ($\geq$) (at least as many plus signs as minus signs)

H_a: $M < 15$ (fewer plus signs than minus signs)

Step 2 The Hypothesis Test Criteria:

a. **Check the assumptions.**
The random sample of 120 adults was independently surveyed, and the variable, hours worked, is continuous.

b. **Identify the probability distribution and the test statistic to be used.**
The standard normal z and formula (14.2).

c. **Determine the level of significance:** $\alpha = 0.05$.

Step 3 The Sample Evidence:

a. **Collect the sample information.**
$n(+) = 40$ and $n(-) = 80$; therefore, $n = 120$ and x is the number of plus signs; $x = 40$.

b. **Calculate the value of the test statistic.**
Using formula (14.2), we have

$$z\bigstar = \frac{x' - \frac{n}{2}}{\frac{1}{2} \cdot \sqrt{n}}:$$

$$z\bigstar = \frac{40.5 - \frac{120}{2}}{\frac{1}{2} \cdot \sqrt{120}} = \frac{40.5 - 60}{\frac{1}{2} \cdot (10.95)} = \frac{-19.5}{5.475}$$

$$= -3.562 = -3.56$$

Step 4 **The Probability Distribution:**

Using the *p*-Value Procedure:

a. Calculate the *p*-value for the test statistic.

Use the left-hand tail because H_a expresses concern for values related to "fewer than." **P** = $P(z < -3.56)$ as shown on the figure.

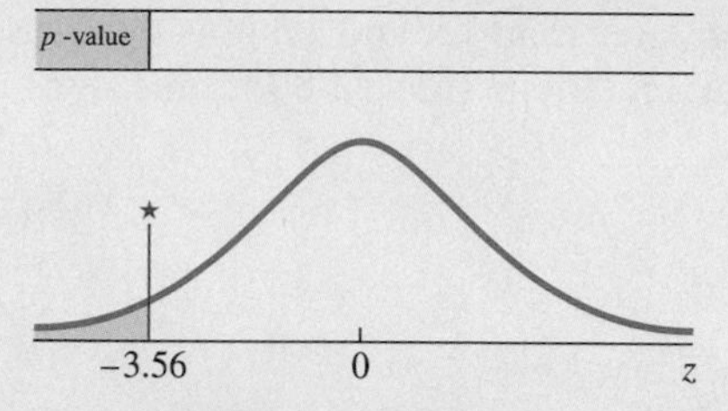

To find the *p*-value, you have three options:

1. Use Table 3 in Appendix B to calculate the *p*-value: **P** = 0.5000 − 0.4998 = **0.0002.**
2. Use Table 5 in Appendix B to place bounds on the *p*-value: **P = 0.0002.**
3. Use a computer or calculator to find the *p*-value: **P = 0.0002.**

For specific instructions, see pp. 375–376.

b. Determine whether or not the *p*-value is smaller than α.

The *p*-value is smaller than α.

OR

Using the Classical Procedure:

a. Determine the critical region and critical value(s).

The critical region is the left-hand tail because H_a expresses concern for values related to "fewer than." The critical value is obtained from Table 4A:

$$-z(0.05) = -1.65$$

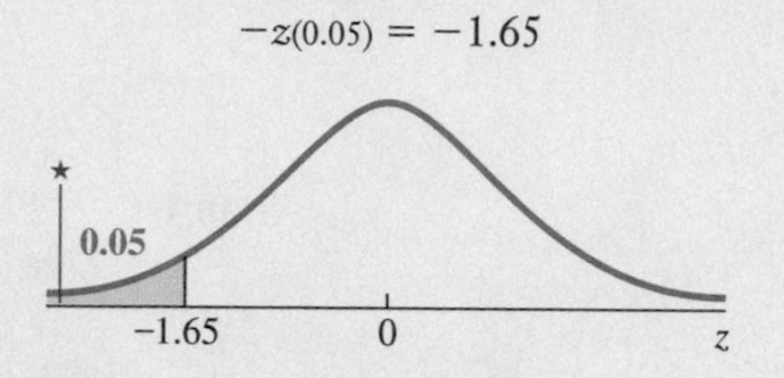

Specific instructions for finding critical values are on pages 392–393.

b. Determine whether or not the calculated test statistic is in the critical region.

$z★$ is in the critical region, as shown in **red** on the figure.

FYI
See pages 664 and 665 for computer and calculator commands.

Step 5 **The Results:**

a. State the decision about H_o: Reject H_o.

b. State the conclusion about H_a.

At the 0.05 level, there are significantly more minus signs than plus signs, thereby implying that the median is less than the claimed 15 hours. ■

Exercises

14.3 Ex14-03 The following daily high temperatures were recorded in the city of Rochester, New York, on 20 randomly selected December days.

47	46	40	40	46	35	34	59	54	33
65	39	48	47	46	46	42	36	45	38

Use the sign test to determine the 95% confidence interval for the median daily high temperature in Rochester during December.

14.4 Ex14-04 Fifteen North Carolina peanut-producing counties were randomly identified and the 2001 peanut yield rates (in pounds of peanuts harvested per acre) were recorded.

3460	2635	2570	3935	2975	2965	2580	4965
2345	2890	3390	2650	2590	2655	2700	

Source: http://www.usda.gov/nass/graphics/county01/data/pe01.csv

Use the sign test to determine the 95% confidence interval for the median yield rate for peanuts.

14.5 Ex14-05 Every year sixth-grade students in Ohio schools take proficiency tests. The following list is sixth-grade reading score changes from the prior year. Negative values indicate a decrease in score, positive values show an increase, and a zero shows no change from the prior year.

−4	−4	−10	−9	−30	6	18	−3	2
−5	−6	−12	−9	1	−1	−2	19	6
−1	−14	−13	5	12	−8	6	−3	−8
−14	−16	−6	2	0	16	−7	6	−11
6	−8	−4	13	9	−12	12	−10	

Construct a 95% confidence interval for the median change in reading scores.

14.6 Ex14-06 A sample of the daily rental-car rates for a compact car was collected in order to estimate the average daily cost of renting a compact car.

39.93	41.00	42.99	38.99	42.93	35.00	40.95	29.99
49.93	50.95	34.95	28.99	43.93	43.00	41.99	42.99
36.93	34.95	35.99	31.99	45.93	46.50	34.90	29.80

(continued)

32.93 29.70 32.99 27.94 53.93 46.00 35.94 34.99
29.93 28.70 34.99 31.48 37.93 37.90 37.92 35.99

Find the 99% confidence interval for the median daily rental cost.

14.7 State the null hypothesis, H_o, and the alternative hypothesis, H_a, that would be used to test each statement:

a. The median value is at least 32.
b. People prefer the taste of the bread made with the new recipe.
c. There is no change in weight from weigh-in until after two weeks of the diet.

14.8 Determine the critical value that would be used to test the null hypothesis for the following situations using the classical approach and the sign test:

a. H_o: $P(+) = 0.5$ vs. H_a: $P(+) \neq 0.5$, with $n = 18$ and $\alpha = 0.05$
b. H_o: $P(+) = 0.5$ vs. H_a: $P(+) > 0.5$, with $n = 78$ and $\alpha = 0.05$
c. H_o: $P(+) = 0.5$ vs. H_a: $P(+) < 0.5$, with $n = 38$ and $\alpha = 0.05$
d. H_o: $P(+) = 0.5$ vs. H_a: $P(+) \neq 0.5$, with $n = 148$ and $\alpha = 0.05$

14.9 An article titled "Venocclusive Disease of the Liver: Development of a Model for Predicting Fatal Outcome After Marrow Transplantation" (*Journal of Clinical Oncology*, September 1993) gives the median age of 355 patients who underwent marrow transplantation at the Fred Hutchinson Cancer Research Center as 30 years. A sample of 100 marrow transplantation patients was recently selected for a study, and it was found that 40 of the patients were over 30 and 60 were under 30 years of age. Test the null hypothesis that the median age of the population from which the 100 patients were selected equals 30 years versus the alternative that the median does not equal 30 years. Use $\alpha = 0.05$.

14.10 Ex14-05 Every year sixth-grade students in Ohio schools take proficiency tests. The list of sixth-grade reading score changes from the prior year is given in Exercise 14.5. Negative values indicate a decrease in score, positive values show an increase, and a zero shows no change from the prior year. Use the sign test to test the hypothesis: On the average, reading scores have decreased from the prior year. Use $\alpha = 0.05$.

14.11 According to an article in a *Newsweek* special issue (Fall/Winter 1990), 51.1% of 17-year-olds answered the following question correctly:

If $7X + 4 = 5X + 8$, then $X =$ ___. (a) 1 (b) 2 (c) 4 (d) 6

Suppose we wished to test the null hypothesis that one-half of all 16-year-olds could solve the problem against the alternative hypothesis: The proportion who can solve differs from one-half. Furthermore, suppose we asked 75 randomly selected 17-year-olds to solve the problem. Let + represent a correct solution and − represent an incorrect solution. Do we have sufficient evidence to show that the proportion who can solve the problem is different from one-half? Explain.

a. if we obtain 20 + signs and 55 − signs
b. if we obtain 27 + signs and 48 − signs
c. if we obtain 30 + signs and 45 − signs
d. if we obtain 33 + signs and 42 − signs

14.12 Ex14-12 Part of the results from the Third International Mathematics and Science Study was a comparison of eighth-grade science achievement by nation from 1995 to 1999. The table gives the nations with their 1995 and 1999 average scores.

Nation	1995	1999
Australia	527	540
Belgium–Flemish	533	535
Bulgaria	545	518
Canada	514	533
Cyprus	452	460
Czech Republic	555	539
England	533	538
Hong Kong	510	530
Hungary	537	552
Iran, Islamic Republic of	463	448
Italy	497	498
Japan	554	550
Korea, Republic of	546	549
Latvia	476	503
Lithuania	464	488
Netherlands	541	545
New Zealand	511	510
Romania	471	472
Russian Federation	523	529
Singapore	580	568
Slovak Republic	532	535
Slovenia	541	533
United States	533	515

Source: http://nces.ed.gov/quicktables/detail.asp?Key=479

a. Construct a table showing the sign of the difference between the years for each country.
b. Using $\alpha = 0.05$, has there been a significant improvement in science scores?

14.13 Ex14-13 According to "The Annual Report on the Economic Status of the Profession, 2001–2002" done by the American Association of University Professors, the mean salary of a full professor was $83,282. The table lists the average salaries for a random sample of institutions in Colorado.

54,500	63,000	83,600	67,000	49,700
60,800	47,700	82,200	86,800	73,900
57,700	58,200	62,200	82,000	78,500
70,000	96,100	89,700	57,200	55,400

Test the claim that the median salary of full professors in Colorado is lower than the mean for the whole country. Use $\alpha = 0.05$.

a. Solve using the p-value approach.
b. Solve using the classical approach.

14.14 An article titled "Naturally Occurring Anticoagulants and Bone Marrow Transplantation: Plasma Protein C Predicts the Development of Venocclusive Disease of the Liver" (*Blood*, June 1993) compared baseline values for antithrombin III with antithrombin II values seven days after a bone marrow transplant for 45 patients. The differences were found to be nonsignificant. Suppose 17 of the differences were positive and 28 were negative. The null hypothesis is that the median difference is zero, and the alternative hypothesis is that the median difference is not zero. Use the 0.05 level of significance. Perform the test and carefully state your conclusion.

14.15 A blind taste test was conducted to determine people's preference for the taste of the "classic" cola and "new" cola. The results were that 645 preferred the new, 583 preferred the old, and 272 had no preference. Is the preference for the taste of the new cola significantly greater than one-half? Use $\alpha = 0.01$.

14.16 A taste test was conducted with a regular beef pizza. Each of 133 individuals was given two pieces of pizza, one with a whole-wheat crust and the other with a white crust. Each person was then asked whether she or he preferred whole-wheat or white crust. The results were that 65 preferred whole-wheat, 53 preferred white, and 15 had no preference. Is there sufficient evidence to verify the hypothesis that whole-wheat crust is preferred to white crust at the $\alpha = 0.05$ level of significance?

14.17 According to an article in *USA Today* (June 7, 1991), "only 46% of high school seniors can solve problems involving fractions, decimals, and percentages." Suppose we wish to test the null hypothesis: One-half of all seniors can solve problems involving fractions, decimals, and percentages, against an alternative that the proportion who can solve differs from one-half. Let + represent passed and − represent failed the test on fractions, decimals, and percentages. If a random sample of 1500 students is tested, what value of x, the number of the less frequent sign, will be the critical value at the 0.05 level of significance?

14.4 The Mann–Whitney *U* Test

The Mann–Whitney *U* test is a nonparametric alternative for the t-test for the difference between two independent means. The usual two-sample situation occurs when the experimenter wants to see whether the difference between the two samples is sufficient to reject the null hypothesis that the two sampled populations are identical.

ASSUMPTIONS FOR INFERENCES ABOUT TWO POPULATIONS USING THE MANN–WHITNEY *U* TEST

The two **independent random samples** are independent within each sample as well as between samples, and the random variables are ordinal or numerical.

This test is often used in situations in which the two samples are drawn from the same population of subjects but different "treatments" are used on each set. We will demonstrate the procedure in the next illustration.

HYPOTHESIS-TESTING PROCEDURE

Illustration 14.6 **Two-Tailed Hypothesis Test**

In a large lecture class, when the instructor gives a one-hour exam, she gives two "equivalent" examinations. It is reasonable to ask: Are these two different exams equivalent? Students in even-numbered seats take exam A, and those in the odd-numbered seats take exam B. To test this "equivalent" hypothesis, two random samples were taken. Table 14.3 lists the exam scores of the two samples.

TABLE 14.3 Data on Exam Scores Ta14-03

Exam A	52	78	56	90	65	86	64	90	49	78
Exam B	72	62	91	88	90	74	98	80	81	71

If we assume that the odd- or even-numbered seats had no effect, does the sample present sufficient evidence to reject the hypothesis: The exam forms yielded scores that had identical distributions? Test using $\alpha = 0.05$.

SOLUTION

Step 1 **The Set-Up:**

a. Describe the population parameter of interest.
The distribution of scores for each version of the exam.

b. State the null hypothesis (H_o) and the alternative hypothesis (H_a).

H_o: Exam A and exam B have test scores with identical distributions.

H_a: The two distributions are not the same.

Step 2 **The Hypothesis Test Criteria:**

a. Check the assumptions.
The two samples are independent, and the random variable, exam score, is numerical.

b. Identify the test statistic to be used.
The Mann–Whitney U statistic.

c. Determine the level of significance: $\alpha = 0.05$.

Step 3 **The Sample Evidence:**

a. Collect the sample information.
The sample data are listed in Table 14.3.

b. Calculate the value of the test statistic.

The size of the individual samples will be called n_a and n_b; actually, it makes no difference which way these are assigned. In our illustration they both have the value 10. The two samples are combined into one sample (all $n_a + n_b$ pieces of data) and ordered from smallest to largest:

49	52	56	62	64	65	71	72	74	78
78	80	81	86	88	90	90	90	91	98

Each piece of data is then assigned a **rank** number. The smallest (49) is assigned rank 1, the next smallest (52) is assigned rank 2, and so on, up to the largest, which is assigned rank $n_a + n_b$ (20). Ties are handled by assigning to each of the tied observations the mean rank of those rank positions that they occupy. For example, in our illustration there are two 78s; they are the 10th and 11th pieces of data. The mean rank for each is then $\frac{10 + 11}{2} = 10.5$. In the case of the three 90s—the 16th,

17th, and 18th pieces of data—each is assigned 17 because $\frac{16 + 17 + 18}{3} = 17$. The rankings are shown in Table 14.4.

TABLE 14.4 Ranked Exam Score Data

Ranked Data	Rank	Source	Ranked Data	Rank	Source
49	1	A	78	10.5	A
52	2	A	80	12	B
56	3	A	81	13	B
62	4	B	86	14	A
64	5	A	88	15	B
65	6	A	90	17	A
71	7	B	90	17	A
72	8	B	90	17	B
74	9	B	91	19	B
78	10.5	A	98	20	B

Figure 14.2 shows the relationship between the two sets of data, first by using the data values and second by comparing the rank numbers for the data.

FIGURE 14.2
Comparing the Data of Two Samples

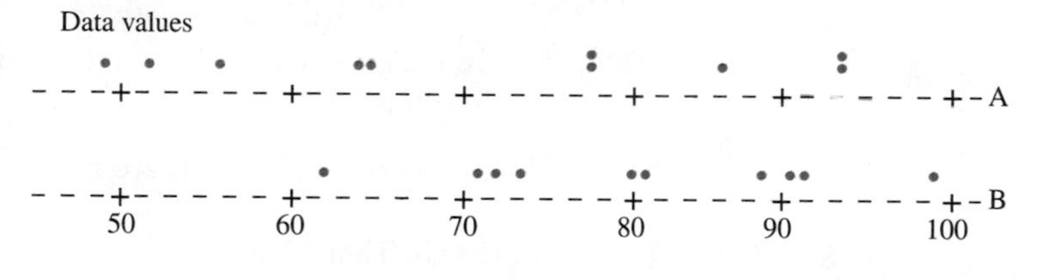

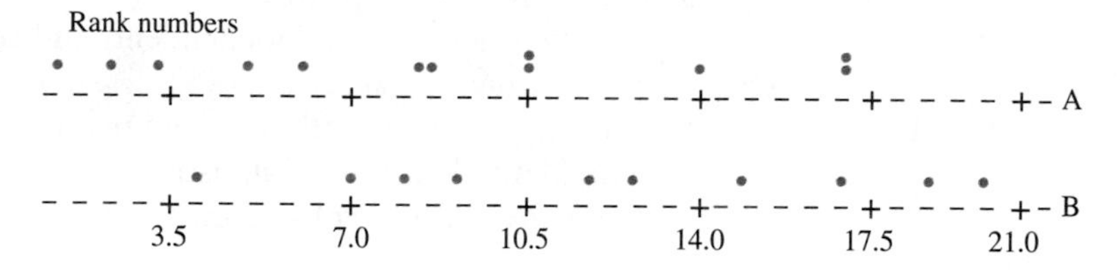

ANSWER NOW 14.18
Do you see a different relationship between the two sets of data? Explain.

The calculation of the **test statistic** U is a two-step procedure. We first determine the sum of the ranks for each of the two samples. Then, using the two sums of ranks, we calculate a U score for each sample. The smaller U score is the test statistic.

The sum of ranks R_a for sample A is computed as

$$R_a = 1 + 2 + 3 + 5 + 6 + 10.5 + 10.5 + 14 + 17 + 17 = \mathbf{86}$$

The sum of ranks R_b for sample B is

$$R_b = 4 + 7 + 8 + 9 + 12 + 13 + 15 + 17 + 19 + 20 = \mathbf{124}$$

The U score for each sample is obtained by using the following pair of formulas:

MANN–WHITNEY U TEST STATISTIC

$$U_a = n_a \cdot n_b + \frac{(n_b)(n_b + 1)}{2} - R_b \qquad (14.3)$$

$$U_b = n_a \cdot n_b + \frac{(n_a)(n_a + 1)}{2} - R_a \qquad (14.4)$$

$U★$, the test statistic, is the smaller of U_a and U_b.

For our illustration, we obtain

$$U_a = (10)(10) + \frac{(10)(10 + 1)}{2} - 124 = 31$$

$$U_b = (10)(10) + \frac{(10)(10 + 1)}{2} - 86 = 69$$

Therefore, $U★ = \mathbf{31}$.

Before we carry out the test for this illustration, let's try to understand some of the underlying possibilities. Recall that the null hypothesis is that the distributions are the same and that we will most likely want to conclude from this that the averages are approximately equal. Suppose for a moment that the distributions are indeed quite different; say, all of one sample comes before the smallest piece of data in the second sample when they are ranked together. This would certainly mean that we want to reject the null hypothesis. What kind of a value can we expect for U in this case? Suppose that the ten A values had ranks 1 through 10 and the ten B values had ranks 11 through 20. Then we would obtain

$$R_a = 55 \quad \text{and} \quad R_b = 155$$

$$U_a = (10)(10) + \frac{(10)(10 + 1)}{2} - 155 = 0$$

$$U_b = (10)(10) + \frac{(10)(10 + 1)}{2} - 55 = 100$$

Therefore, $U★ = \mathbf{0}$.

If this were the case, we certainly would want to reach the decision: Reject the null hypothesis.

Suppose, on the other hand, that both samples were perfectly matched; that is, a score in each set is identical to one in the other.

54	54	62	62	71	71	72	72	. . .
A	B	A	B	A	B	A	B	. . .
1.5	1.5	3.5	3.5	5.5	5.5	7.5	7.5	. . .

Now what would happen?

$$R_a = R_b = 105$$

$$U_a = U_b = (10)(10) + \frac{(10)(10 + 1)}{2} - 105 = 50$$

Therefore, $U★ = \mathbf{50}$. If this were the case, we certainly would want to reach the decision: Fail to reject the null hypothesis.

NOTE The sum of the two U's ($U_a + U_b$) will always be equal to the product of the two sample sizes ($n_a \cdot n_b$). For this reason we need only to concern ourselves with the smaller U value.

Now, let's return to the solution of Illustration 14.6.

Step 4 The Probability Distribution:

Using the *p*-Value Procedure

a. Calculate the *p*-value for the test statistic.

Since the concern is for values related to "not the same," the p-value is the probability of both tails. It will be found by finding the probability of the left tail and doubling:

$$\mathbf{P} = 2 \times P(U \leq 31 \text{ for } n_1 = 10 \text{ and } n_2 = 10)$$

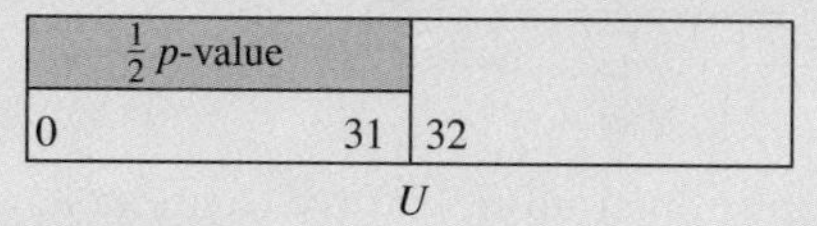

To find the p-value, you have two options:

1. Use Table 13 in Appendix B to place bounds on the p-value: $\mathbf{P > 0.10}$.
2. Use a computer or calculator to find the p-value: $\mathbf{P = 0.1612}$.

Specific instructions follow this illustration.

b. Determine whether or not the *p*-value is smaller than α.

The p-value is not smaller than α.

OR

Using the Classical Procedure

a. Determine the critical region and critical value(s).

The critical region is two-tailed because H_a expresses concern for values related to "not the same." Use Table 13A for two-tailed $\alpha = 0.05$. The critical value is at the intersection of column $n_1 = 10$ and row $n_2 = 10$: **23.** The critical region is $U \leq 23$.

Reject H_o | Fail to reject H_o ★
0 23 | 24

b. Determine whether or not the calculated test statistic is in the critical region.

U★ is not in the critical region, as shown in the figure.

Step 5 The Results:

a. State the decision about H_o: Fail to reject H_o.

b. State the conclusion about H_a.

We do not have sufficient evidence to reject the "equivalent" hypothesis.

Calculating the *p*-value when using the Mann–Whitney test

Method 1: Use Table 13 in Appendix B to place bounds on the p-value. By inspecting Table 13A and B at the intersection of column $n_1 = 10$ and row $n_2 = 10$, you can determine that the p-value is greater than 0.10; the larger two-tailed value of α is 0.10 in Table 13B.

Method 2: If you are doing the hypothesis test with the aid of a computer or graphing calculator, most likely it will calculate the p-value for you. Specific instructions are described on pages 677–678. ■

NORMAL APPROXIMATION

If the samples are larger than size 20, we may make the test decision with the aid of the standard normal variable, z. This is possible because the distribution of U is approximately normal with a mean

$$\mu_U = \frac{n_a \cdot n_b}{2} \tag{14.5}$$

and a standard deviation

$$\sigma_u = \sqrt{\frac{n_a \cdot n_b \cdot (n_a + n_b + 1)}{12}} \qquad (14.7)$$

The hypothesis test is then completed using the **test statistic $z\bigstar$:**

$$z\bigstar = \frac{U\bigstar - \mu_U}{\sigma_U} \qquad (14.7)$$

The standard normal distribution may be used whenever n_a and n_b are both greater than 10.

Illustration 14.7 One-Tailed Hypothesis Test

A dog-obedience trainer is training 27 dogs to obey a certain command. The trainer is using two different training techniques: (I) the reward-and-encouragement method and (II) the no-reward method. Table 14.5 shows the numbers of obedience sessions that were necessary before the dogs would obey the command. Does the trainer have sufficient evidence to claim that the reward method will, on average, require less training time ($\alpha = 0.05$)?

TABLE 14.5 Data on Dog Training Ta14-05

Method I	29	27	32	25	27	28	23	31	37	28	22	24	28	31	34
Method II	40	44	33	26	31	29	34	31	38	33	42	35			

SOLUTION

Step 1 **The Set-Up:**

a. **Describe the population parameter of interest.**
The distribution of training times for each technique.

b. **State the null hypothesis (H_o) and the alternative hypothesis (H_a).**

H_o: The distributions of the training times required are the same for both methods.

H_a: The reward method, on the average, requires less time.

Step 2 **The Test Criteria:**

a. **Check the assumptions.**
The two samples are independent, and the random variable, training time, is continuous.

b. **Identify the test statistic to be used.**
The Mann–Whitney U statistic.

c. **Determine the level of significance:** $\alpha = 0.05$.

Step 3 **The Sample Evidence:**

a. **Collect the sample information.**
The sample data are listed in Table 14.5.

b. **Calculate the value of the test statistic.**
The two sets of data are ranked jointly and ranks are assigned as shown in Table 14.6 (p. 676).

TABLE 14.6 Rankings for Training Methods

Number of Sessions	Group	Rank		Number of Sessions	Group	Rank	
22	I	1		31	II	15	14.5
23	I	2		31	II	16	14.5
24	I	3		32	I	17	
25	I	4		33	II	18	18.5
26	II	5		33	II	19	18.5
27	I	6	6.5	34	I	20	20.5
27	I	7	6.5	34	II	21	20.5
28	I	8	9	35	II	22	
28	I	9	9	37	I	23	
28	I	10	9	38	II	24	
29	I	11	11.5	40	II	25	
29	II	12	11.5	42	II	26	
31	I	13	14.5	44	II	27	
31	I	14	14.5				

The sums are:

$$R_{\text{I}} = 1 + 2 + 3 + 4 + 6.5 + \cdots + 20.5 + 23 = 151.0$$

$$R_{\text{II}} = 5 + 11.5 + 14.5 + \cdots + 26 + 27 = 227.0$$

The U scores are found using formulas (14.3) and (14.4):

$$U_{\text{I}} = (15)(12) + \frac{(12)(12+1)}{2} - 227 = 180 + 78 - 227 = 31$$

$$U_{\text{II}} = (15)(12) + \frac{(15)(15+1)}{2} - 151 = 180 + 120 - 151 = 149$$

Therefore, $U\bigstar = \mathbf{31}$. Now we use formulas (14.5), (14.6), and (14.7) to determine the z statistic:

$$\mu_U = \frac{n_a \cdot n_b}{2}: \qquad \mu_U = \frac{12 \cdot 15}{2} = 90$$

$$\sigma_U = \sqrt{\frac{n_a \cdot n_b \cdot (n_a + n_b + 1)}{12}}: \qquad \sigma_u = \sqrt{\frac{12 \cdot 15 \cdot (12 + 15 + 1)}{12}}$$

$$= \sqrt{\frac{(180)(28)}{12}} = \sqrt{420} = 20.49$$

$$z\bigstar = \frac{U\bigstar - \mu_U}{\sigma_U}: \qquad z\bigstar = \frac{31 - 90}{20.49} = \frac{-59}{20.49} = -2.879 = \mathbf{-2.88}$$

Step 4 The Probability Distribution:

Using the *p*-Value Procedure

a. Calculate the *p*-value for the test statistic.

Use the left-hand tail because H_a expresses concern for values related to "less than." $\mathbf{P} = P(z < -2.88)$ as shown on the figure.

(continued)

OR

Using the Classical Procedure

a. Determine the critical region and critical value(s).

The critical region is the left-hand tail because H_a expresses concern for values related to "less than." The critical value is obtained from Table 4A:

$$-z(0.05) = \mathbf{-1.65}$$

(continued)

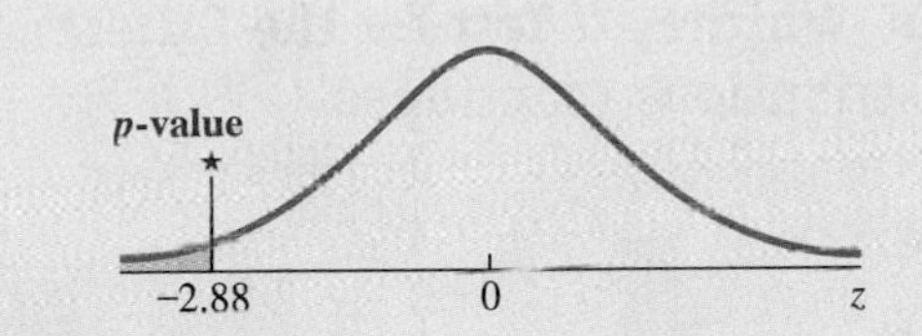

To find the p-value, you have three options:

1. Use Table 3 in Appendix B to calculate the p-value: **P = 0.5000 − 0.4980 = 0.0020.**
2. Use Table 5 in Appendix B to place bounds on the p-value: **0.0019 < P < 0.0022.**
3. Use a computer or calculator to find the p-value: **P = 0.0020.**

For specific instructions, see pp. 375–376.

b. Determine whether or not the p-value is smaller than α.
The p-value is smaller than α.

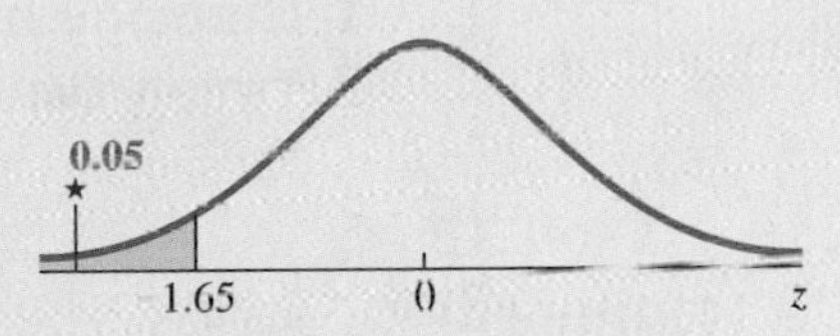

Specific instructions for finding critical values are on pages 392–393.

b. Determine whether or not the calculated test statistic is in the critical region.
$z\star$ is in the critical region, as shown in red on the figure.

Step 5 **The Results:**

a. State the decision about H_o: Reject H_o.

b. State the conclusion about H_a.
At the 0.05 level of significance, the data show sufficient evidence to conclude that the reward method does, on the average, require less training time. ■

Technology Instructions: Mann–Whitney *U* Test for the Difference Between Two Independent Distributions

MINITAB (Release 13) Input the two independent sets of data into C1 and C2; then continue with:

```
Choose:   Stat > Nonparametrics > Mann-Whitney
Enter:    First Sample: C1
          Second Sample: C2
          Alternative: less than or not equal or greater
          than
```

With respect to the p-value approach, the p-value is given. With respect to the classical approach, just the sum of the ranks for one of the samples is given. Use this to find U for that one sample. The U for the other sample is found by subtracting U from the product of n_1 and n_2.

Excel XP Input the two independent sets of data into column A and column B; then continue with:

```
Choose:   Tools > Data Analysis Plus > Wilcoxon Rank Sum
          Test*
Enter:    Variable 1 Range: (A1:A20 or select cells)
          Variable 2 Range: (B1:B20 or select cells)
```

*The Wilcoxon rank sum test is equivalent to the Mann–Whitney test.

(continued)

Technology Instructions: Mann–Whitney U Test for the Difference Between Two Independent Distributions (continued)

Excel XP (continued)

```
Select:   Labels (if necessary)
Enter:    Alpha: α (ex. 0.05)
```

The sum of the ranks is given for both samples and also the p-value.

TI–83 Plus

Input the two independent sets of data into L1 and L2; then continue with:

```
Choose:   PRGM > EXEC > MANNWHIT
Enter:    XLIST: L1
          YLIST: L2
          NULL HYPOTHESIS D0 = difference amount (ex. 0)
Select:   ALT HYP? 1:U1-U2 > D0 or 2:U1-U2 < D0 or 3:U1-U2 ≠ D0
```

*Program MANNWHIT is one of many programs that are available for downloading from www.duxbury.com. See page 40 for specific directions.

Application 14.1 Sea Otters

Quantitative Assessment of Sea Otter Benthic Prey Communities within the Olympic Coast National Marine Sanctuary: 1999 Re-survey of 1995 and 1985 Monitoring Stations

This report summarizes the changes in the distribution and abundance of selected benthic species within sea otter prey communities along the Washington State Olympic coast between 1987 and 1999. During this 12-year period, the Washington otter population has undergone a dramatic increase in both numbers and range, now occupying habitats that were otter free when first sampled in 1987. Invertebrate prey such as commercially harvested sea urchins that were abundant just outside the boundaries of the 1987 sea otter range are now virtually absent along the entire outer rocky coast. Understory foliose red, coralline, and brown algal cover have also undergone changes as otters removed large invertebrate grazers from the newly occupied habitats. In 1995 a test comparison was conducted at Chibahdehl Rocks to compare invertebrate size and abundance data collected using both methods. Results showed no significant difference (t-tests, $p = 0.32$ and 0.24 for abundance and size, respectively).

Hypotheses

H_1: As the Washington State sea otter population continues to grow, it will expand north, drawn by and depleting the rich prey resources found there.

H_2: If sea otters move into northern habitats, significant changes in benthic algal cover will occur with reduced abundance of sea urchins and other invertebrate grazers.

H_3: Sea otters will be slower to colonize areas with higher water velocities, resulting in a higher prey biomass in those areas.

Results

For 1999, there was no significant difference in prey abundance between sites. Foliose red, coralline, and brown algal cover were followed at three sites, Neah Bay, Anderson Pt. and Cape Alava for all years. The only significant difference in foliose red cover between 1995 and 1999 was the decline at Anderson Pt. (Mann–Whitney U test $p < 0.0001$). Coralline cover continued to drop dramatically and significantly at Neah Bay (100%, 44%, 1%) (Mann–Whitney U test $p < 0.0001$) and at Anderson Pt. (18%, 17%, 6%) (Mann–Whitney U test $p < 0.0001$), while fluctuating slightly but significantly at Cape Alava (Mann–Whitney U test $p = 0.0006$). Brown algae has increased steadily and significantly from 0% to 33% at Neah Bay since 1987 (Mann–Whitney U test $p = 0.009$), fluctuated significantly between 4% and 34% at Anderson Point (Mann–Whitney U test $p < 0.0001$), and did not change significantly at Cape Alava (Mann–Whitney U test $p = 0.20$).

Conclusions Otter numbers have increased within their range since 1987, and their range has expanded to the north as predicted (H1). Prey abundance and biomass have declined by an order of magnitude to very low levels at newly otter-occupied sites on either side of Cape Flattery by 1995, also as predicted (H2). By 1999, the high prey numbers and biomass found at Cape Flattery and Tatoosh Island in 1995 had also dropped to levels comparable with the other monitoring site, refuting the high current prey refuge hypothesis (H3). The removal of urchin grazers by sea otters was most likely responsible for the rise in cover of more palatable algae at the recently occupied Neah Bay and Anderson Pt sites. The most dramatic change in algal cover occurred at Neah Bay, the site that experienced the greatest decline in urchin abundance following the movement of sea otters into the area.

Rikk Kvitek, Pat Iampietro, and Kate Thomas
California State University Monterey Bay, Seaside, CA
http://seafloor.csumb.edu/publications/posters/OCNMS.pdf
Reprinted with permission

ANSWER NOW 14.19

Of the seven Mann–Whitney *U* test *p*-values given, six are all less than 0.001 and the seventh is 0.20. Explain how these *p*-values relate to statements containing phrases like: significant, drop dramatically, increased steadily, and did not change significantly.

Exercises

14.20 State the null hypothesis H_o and the alternative hypothesis H_a that would be used to test each statement.

a. There is a difference in the values of the variable between the two groups of subjects.
b. The average value is not the same for both groups.
c. The blood pressure for group A is higher than for group B.

14.21 Determine the critical value that would be used to test these hypotheses for experiments involving two independent samples, using the classical method:

a. H_o: Average(A) = Average(B) vs. H_a: Average(A) > Average(B), with $n_A = 18$, $n_B = 15$, and $\alpha = 0.05$
b. H_o: The average score is the same for both groups, vs. H_a: Group I average scores are less than those for group II, with $n_I = 78$, $n_{II} = 45$, and $\alpha = 0.05$

14.22 The July 4, 1994, issue of *Newsweek* quotes several tobacco executives. Cigarette makers point out that some brands have more nicotine and tar than others do. Consider a study designed to compare the nicotine contents of two different brands of cigarettes. The nicotine content was determined for 25 cigarettes of brand A and 25 cigarettes of brand B. The sum of the ranks for brand A equals 688, and the sum of the ranks for brand B equals 587. Use the Mann–Whitney U statistic to test the null hypothesis that the average nicotine content is the same for the two brands versus the alternative that the average nicotine content differs. Use $\alpha = 0.01$.

14.23 Ex14-23 Pulse rates were recorded for 16 men and 13 women. Here are the results:

Men	61	73	58	64	70	64	72	60
	65	80	55	72	56	56	74	65
Women	83	58	70	56	76	64	80	
	68	78	108	76	70	97		

These data were used to test the hypothesis that the distribution of pulse rates differs for men and women. The MINITAB output gives the sum of ranks for men (W = 192.0) and the p-value of 0.0373. Verify these two values by calculating them yourself.

```
Mann-Whitney Confidence Interval and Test

Males   N = 16 Median = 64.50
Females N = 13 Median = 76.00
W = 192.0
Test of ETA1 = ETA2 vs ETA1 not = ETA2 is
significant at 0.0373
```

14.24 Ex14-24 An article in the *International Journal of Sports Medicine* (July 1994) discusses the use of the Mann–Whitney U test to compare the total cholesterol (mg/dL) of 35 adipose (obese) boys with that of 27 adipose girls. No significant difference was found between the two groups with respect to total cholesterol. A similar study involving six adipose boys and eight adipose girls gave the following total-cholesterol values:

Adipose Boys	175	185	160	200	170	150		
Adipose Girl	160	190	175	190	185	150	140	195

Use the Mann–Whitney U test to test the research hypothesis that the total-cholesterol values differ for the two groups, using the 0.05 level of significance.

14.25 A study titled "Textbook Pictures and First-Grade Children's Perception of Mathematical Relationships" by Patricia Campbell (*Journal for Research in Mathematics Education*, November 1978) investigated the influence of artistic style and the number of pictures on first-grade children's perception of mathematics textbook pictures. Analysis with the Mann–Whitney U test indicated that students who initially viewed and described sequences of pictures had significantly higher story-response scores than students who viewed only single pictures. Consider the following data from two such groups. Group 1 viewed sequences of pictures, and group 2 viewed only single pictures.

Group 1	30	35	40	42	45	36
Group 2	25	32	27	39	30	

Using the Mann–Whitney U test, determine whether the group 1 scores are significantly higher than the group 2 scores. Use $\alpha = 0.05$.

14.26 Ex14-26 Eighteen Texas and 12 Oklahoma peanut-producing counties were randomly identified and the 2001 peanut yield rates (in pounds of peanuts harvested per acre) were recorded.

Texas		Oklahoma	
County	**Yield**	**County**	**Yield**
Stonewall	290	Pittsburg	2340
Cottle	1485	Johnston	2135
Lee	1285	Harmon	3000
Eastland	1385	Grady	2465
Donley	3600	Washita	2775
Andrews	4035	Caddo	3035
Gaines	3820	Kiowa	2770
Motley	2415	Greer	1885
Briscoe	3675	Bryan	1865
Wilson	2555	Atoka	1555
Hall	1275	McClain	1300
Terry	3210	Love	3000
Dawson	3800		
Comanche	1475		
Atascosa	3000		
Childress	1875		
Frio	3330		
Erath	1305		

Source: http://www.usda.gov/nass/graphics/county01/data/pe01.csv

Use the Mann–Whitney U statistic to test the hypothesis that the average yields differ for the two states. Use $\alpha = 0.05$.

14.27 Ex14-27 As part of a study to determine whether cloud seeding increased rainfall, clouds were randomly seeded or not seeded with silver nitrate. The amounts of rainfall (in inches) that followed are listed here.

Unseeded

4.9	41.1	21.7	372.4	26.3	17.3	36.6	26.1
47.3	95.0	147.8	321.2	11.5	68.5	29.0	24.4
1202.6	87.0	28.6	830.1	81.2	4.9	163.0	345.5
244.3							

Seeded

129.6	334.1	274.7	198.6	430.0	274.7	31.4	115.3
1656.0	118.3	489.1	302.8	255.0	32.7	119.0	17.5
242.5	2745.6	7.7	40.6	978.0	200.7	703.4	92.4
1697.8							

Do the data show that cloud seeding will significantly increase the average amount of rainfall? Use $\alpha = 0.05$.

14.28 Ex14-28 The 2001 Ohio State Proficiency tests result for Toledo fourth-grade students was the highest recorded since the start of the statewide proficiency testing. Though the results were an improvement districtwide, in some subjects there was not as much improvement as in other subjects. The results show the changes in the reading and writing scores. Change in scores is indicated by positive for improvement, negative for lower scores, and zero for no change.

Writing

2	0	3	30	10	25	7	17	2
6	15	−9	−2	6	13	−5	−5	10
24	6	29	4	27	16	1	−4	−8
−6	13	8	5	−23	3	14	−1	7
16	−12	10	42	−2	4	8	38	24

Reading

23	25	2	6	40	3	3	32	−2
8	28	−1	8	5	34	−6	7	6
34	6	19	27	23	6	46	23	35
−4	10	11	31	−13	10	20	10	−10
−5	17	22	20	19	11	13	3	21

Test the claim that there was equal improvement in fourth-grade writing and reading scores. Use $\alpha = 0.05$.

14.5 The Runs Test

The runs test is used most frequently to test the **randomness** of data (or lack of randomness). A **run** is a sequence of data that possesses a common property. One run ends and another starts when a piece of data does not display the property in question. The **test statistic** in this test is V, the number of runs observed.

Illustration 14.8 Determining the Number of Runs

To illustrate the idea of runs, let's draw a sample of ten single-digit numbers from the telephone book, listing the next-to-last digit from each of the selected telephone numbers:

Sample: 2 3 1 1 4 2 6 6 6 7

Let's consider the property of "odd" (o) or "even" (e). The sample, as it was drawn, becomes e, o, o, o, e, e, e, e, e, o, which displays four runs:

e o o o *e e e e e* o

Thus, $V\bigstar = 4$. ■

In Illustration 14.8, if the sample contained no randomness, there would be only two runs—all the evens, then all the odds, or the other way around. We would also not expect to see them alternate—odd, even, odd, even. The maximum number of possible runs would be $n_1 + n_2$ or less (provided n_1 and n_2 are not equal), where n_1 and n_2 are the numbers of data that have each of the two properties being identified.

ASSUMPTION FOR INFERENCES ABOUT RANDOMNESS USING THE RUNS TEST

Each sample data can be classified into one of two categories.

The runs test is generally a two-tailed test. We will reject the hypothesis when there are too few runs because this indicates that the data are "separated" according to the two properties. We will also reject when there are too many runs because that indicates that the data alternate between the two properties too often to be random. For example, if the data alternated all the way down the line, we might suspect that the data had been tampered with. There are many aspects to the concept of randomness. The occurrence of odd and even as discussed in Illustration 14.8 is one aspect. Another aspect of randomness that we might wish to check is the ordering of fluctuations of the data above or below the mean or median of the sample.

Illustration 14.9 Hypothesis Test for Randomness

Consider the following sample and determine whether the data points form a random sequence with regard to being above or below the median value.

2	5	3	8	4	2	9	3	2	3	7	1	7	3	3
6	3	4	1	9	5	2	5	5	2	4	3	4	0	4

Test the null hypothesis that this sequence is random. Use $\alpha = 0.05$.

SOLUTION

Step 1 **The Set-Up:**

a. **Describe the population parameter of interest.**
   Randomness of the values above or below the median.

b. **State the null hypothesis (H_o) and the alternative hypothesis (H_a).**

   H_o: The numbers in the sample form a random sequence with respect to the two properties "above" and "below" the median value.

   H_a: The sequence is not random.

Step 2 **The Hypothesis Test Criteria:**

a. **Check the assumptions.**
   Each sample data can be classified as "above" or "below" the median.

b. **Identify the test statistic to be used.**
   V, the number of runs in the sample data.

c. **Determine the level of significance:** $\alpha = 0.05$

Step 3 **The Sample Evidence:**

a. **Collect the sample information.**
   The sample data are listed at the beginning of the illustration.

b. **Calculate the value of the test statistic.**
   First we must rank the data and find the median. The ranked data are

0	1	1	2	2	2	2	2	3	3	3	3	3	3	3
4	4	4	4	4	5	5	5	5	6	7	7	8	9	9

Since there are 30 pieces of data, the depth of the median is at the $d(\tilde{x}) = 15.5$ position. Thus, $\tilde{x} = \dfrac{3 + 4}{2} = 3.5$. By comparing each number in the original sample to the value of the median, we obtain the following sequence of **a**'s (above) and *b*'s (below):

$$b\,\mathbf{a}\,b\,\mathbf{a}\,\mathbf{a}\,b\,\mathbf{a}\,b\,b\,b\,\mathbf{a}\,b\,\mathbf{a}\,b\,b\,\mathbf{a}\,b\,\mathbf{a}\,b\,\mathbf{a}\,\mathbf{a}\,b\,\mathbf{a}\,\mathbf{a}\,b\,\mathbf{a}\,b\,\mathbf{a}\,b\,\mathbf{a}$$

We observe $n_a = 15$, $n_b = 15$, and 24 runs. So $V\bigstar = 24$.

If n_1 and n_2 are both less than or equal to 20 and a two-tailed test at $\alpha = 0.05$ is desired, then Table 14 in Appendix B is used to complete the hypothesis test.

Step 4 **The Probability Distribution:**

Using the *p*-Value Procedure

a. Calculate the *p*-value for the test statistic.

Since the concern is for values related to "not random," the test is two-tailed. The *p*-value is found by finding the probability of the right tail and doubling:

$$\mathbf{P} = 2 \times P(V \geq 24 \text{ for } n_a = 15 \text{ and } n_b = 15)$$

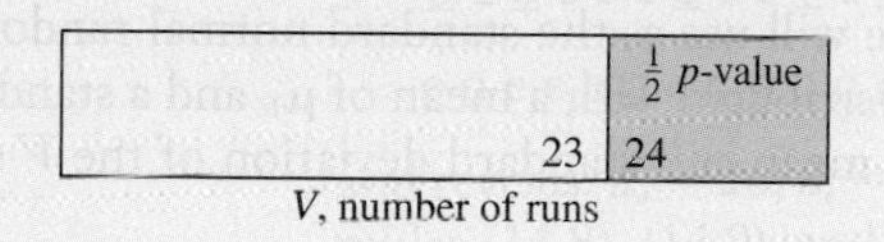

To find the *p*-value, you have two options:

1. Use Table 14 in Appendix B to place bounds on the *p*-value: **P < 0.05.**
2. Use a computer or calculator to find the *p*-value: **P = 0.003.**

Specific instructions follow this illustration.

b. Determine whether or not the *p*-value is smaller than α.

The *p*-value is smaller than α.

OR

Using the Classical Procedure

a. Determine the critical region and critical value(s).

Since the concern is for values related to "not random," the test is two-tailed. Use Table 14 for two-tailed α = 0.05. The critical values are at the intersection of column $n_1 = 15$ and row $n_2 = 15$: 10 and 22. The critical region is $V \leq 10$ or $V \geq 22$.

Reject H_o | Fail to reject H_0 | Reject H_o ★

10 11 21 22

V, number of runs

Specific instructions follow this illustration.

b. Determine whether or not the calculated test statistic is in the critical region.

$V\bigstar$ is in the critical region, as shown in the figure.

Step 5 **The Results:**

a. State the decision about H_o: Reject H_o.

b. State the conclusion about H_a.

We are able to reject the hypothesis of randomness at the 0.05 level of significance and conclude that the sequence is not random with regard to above and below the median.

Calculating the *p*-value when using the runs test

Method 1: Use Table 14 in Appendix B to place bounds on the p-value. By inspecting Table 14 at the intersection of column $n_1 = 15$ and row $n_2 = 15$, you can determine that the *p*-value is less than 0.05; the observed value of $V\bigstar = 24$ is larger than the larger critical value listed.

Method 2: If you are doing the hypothesis test with the aid of a computer or graphing calculator, most likely it will calculate the *p*-value for you. Specific instructions are given on page 686. ■

FYI
Try this link to an online random number generator. Use the Runs test to check randomness.

ANSWER NOW 14.29

Research Randomizer is a free service offered to students and researchers interested in conducting random assignment and random sampling. Although every effort has been made to develop a useful means of generating random numbers, Research Randomizer and its staff do not guarantee the quality or randomness of numbers generated by Research Randomizer. Any use to which these numbers are put remains the sole responsibility of the user who generated them.

a. **Go to the Web site http://www.randomizer.org/about.htm and generate one set of ten random numbers from 1 to 9, where each number can repeat (select "No" for**

Step 5 **The Results:**

a. State the decision about H_o: Fail to reject H_o.

b. State the conclusion about H_a.

At the 0.10 level of significance, we are unable to reject the hypothesis of randomness and conclude that these data are a random sequence. ■

Technology Instructions: Runs Test for Testing Randomness Above and Below the Median

MINITAB (Release 13)

Input the set of data into C1; then continue with:

```
Choose:   Stat > Nonparametrics > Runs
Enter:    Variable: C1
Select:   Above and below mean
          or
          Above and below:
Enter:    Median value
```

Excel XP

The following commands will compute only the differences between the data values and the median. Then to complete the runs test you will need to count the number of runs created by the sequence of + and − signs.
Input the data into column A; select cell B1; then continue with:

```
Choose:   Edit Formula (=)
Enter:    A1 -median(
          (A1:A20 or select cells) > OK
Drag:     Bottom right corner of the B1 cell down to give
          other differences
```

TI-83 Plus

Input the data into L1; then continue with:

```
Highlight:   L2
Enter:       L1 - median*(L1)        (*2nd LIST > MATH >
             4:median( )
Choose:      PRGM > EXEC > RUNSTEST*
Enter:       n1 = # of observations with particular
             characteristic (ex. below median)
             n2 = # of observations with other
             characteristic (ex. above median)
             V = # of runs
```

*Program RUNSTEST is one of many programs that are available for downloading from www.duxbury.com. See page 40 for specific directions.

Illustrati

Application 14.2 Casino Gaming Rules

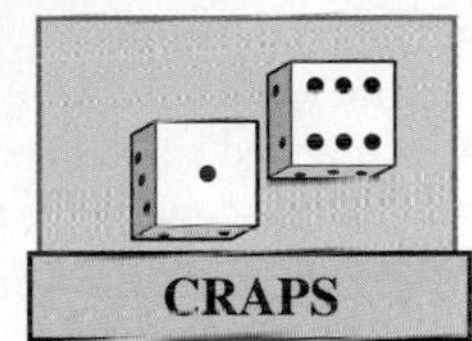

Many casino games rely on electronically generated random numbers for "fair" play. Here is a sample of the rules governing these casino games.

Requirements Relating to Electronic Gaming Devices in International Casinos

These conditions are drafted pursuant to the Casino Act (Fi1999: 355). The purpose of the conditions is to guarantee the player security in relation to casinos and manufacturers of games, mainly as regards cheating through manipulation of gaming devices. Electronic gaming devices used in a casino must meet the specifications set forth in this rule.

The following conditions apply to randomness events and randomness testing:

(a) A random event has a given set of possible outcomes that has a given probability of occurrence.

(b) Two events are called independent if both of the following conditions exist:
  - (i) The outcome of one event does not have an influence on the outcome of the other event.
  - (ii) The outcome of one event does not affect the probability of occurrence of the other event.

(c) An electronic gaming device shall be equipped with a random number generator to make the selection process. A selection process is considered random if all of the following specifications are met:
  - (i) The random number generator satisfies not less than a 99% confidence level using chi-square tests.
  - (ii) The random number generator does not produce a statistic with regard to producing patterns of occurrences. Each reel position is considered random if it meets not less than 99% confidence level with regard to the runs test or any similar pattern testing statistic.
  - (iii) The random number generator produces numbers that are independently chosen without regard to any other symbol produced during that play. This test is the correlation test. Each pair of reels is considered random if the pair of reels meet not less than 99% confidence level using standard correlation analysis.

http://216.239.37.100/search?q=cache:KD66UcpLAzEaC:www.kemaquality.com/en/pages/services/gaming/Downloads/SW/Electronic%2520Gaming%2520Devices.pdf+testing+randomness,+%22runs+test%22&hl=en&ie=UTF-8

ANSWER NOW 14.30

a. **What aspect of randomness will be tested using the chi-square test mentioned in part (i) of rule (c)? Describe how it will be used.**

b. **What aspect of randomness will be tested using the runs test mentioned in part (ii) of rule (c)? Describe how it will be used.**

c. **What aspect of randomness will be tested using the correlation analysis mentioned in part (iii) of rule (c)? Describe how it will be used.**

d. **These gaming rules are written using the phrase "99% confidence level" instead of "0.01 level of significance" as hypothesis tests typically use. Explain why this seems appropriate.**

Exercises

14.31 State the null hypothesis, H_o, and the alternative hypothesis, H_a, that would be used to test each statement.

a. The data did not occur in a random order about the median.
b. The sequence of odd and even is not random.
c. The gender of customers entering a grocery store was recorded; the entry is not random in order.

14.32 Determine the critical values that would be used to complete these hypothesis tests for the runs tests using the classical approach:

a. H_o: The results collected occurred in random order above and below the median, vs. H_a: The results were not random, with $n(\text{A}) = 14$, $n(\text{B}) = 15$, and $\alpha = 0.05$
b. H_o: The two properties alternated randomly, vs. H_a: The two properties didn't occur in random fashion, with $n(\text{I}) = 78$, $n(\text{II}) = 45$, and $\alpha = 0.05$

14.33 A manufacturing firm hires both men and women. The following sequence shows the genders of the last 20 individuals hired (M = male, F = female), in order of occurrence:

M	M	F	M	F	F	M	M	M	M
M	M	F	M	M	F	M	M	M	M

At the $\alpha = 0.05$ level of significance, are we correct in concluding that this sequence is not random?

14.34 A student was asked to perform an experiment that involved tossing a coin 25 times. After each toss, the student recorded the results. These data were reported (H = heads, T = tails), in order of occurrence:

H	T	H	T	H	T	H	T	H	H
T	T	H	H	T	T	H	T	H	T
H	T	H	T	H					

Use the runs test at a 5% level of significance to test the student's claim that the results reported are random.

14.35 Ex14-35 The following data were collected in an attempt to show that the number of minutes the city bus is late is steadily growing larger. The numbers of minutes are in order of occurrence.

6	1	3	9	10	10	2	5	5	6
12	3	7	8	9	4	5	8	11	14

At $\alpha = 0.05$, do these data show sufficient lack of randomness to support the claim?

14.36 In an attempt to answer the question: Does the husband (h) or wife (w) do the family banking?, the results of a sample of 28 married customers doing the family banking show the following sequence of arrivals at the bank:

w	w	w	w	h	w	h	h	h	h	w	w	w	w
w	h	h	w	w	w	h	h	h	h	w	h	h	w

Do these data show lack of randomness with regard to whether the husband or wife does the family banking? Use $\alpha = 0.05$.

14.37 A USA Snapshot (*USA Today*, October 14, 1994) titled "School buildings aging" gives these average ages for schools in five cities: Washington, D.C.—75 years, St. Louis—74 years, San Diego—30 years, Baltimore—30 years, and Fresno, California—30 years. The following ages of school buildings were collected in the sequence given for Spokane:

5	13	25	45	15	17	22	35	16	23	36	22	35	35

a. Determine the median age and the number of runs above and below the median for Spokane.
b. Use the runs test to test these data for randomness about the median. Use $\alpha = 0.05$.

14.38 Ex14-38 The article "Water Boosts Hemoglobin's Love for Oxygen" (*Science News*, March 30, 1991) discusses the ability of the iron-rich protein pigment hemoglobin to carry oxygen throughout the body. The article states that the protein's conversion to an oxygen-loving state involves between 60 and 80 water molecules. Suppose 20 different determinations resulted in the following numbers of water molecules needed:

79	75	69	70	70	65	75	75	65	70
60	62	63	63	67	65	70	60	65	62

a. Determine the median and the number of runs above and below the median.
b. Use the runs test to test these data for randomness about the median.
c. State your conclusion.

14.39 Ex14-39 The following are 24 consecutive downtimes (in minutes) of a particular machine:

20	33	33	35	36	36	22	22	25	27	30	30
30	31	31	32	32	36	40	40	50	45	45	40

The null hypothesis of randomness is to be tested against the alternative that there is a trend. Here is a MINITAB analysis of the number of runs above and below the median:

```
Runs Test
  Median of Downtime K = 32.5000
  The observed number of runs = 4
  The expected number of runs = 13.0000
  12 Observations above K                    12 below
    The test is significant at 0.0002
```

a. Confirm the values reported for the median and the number of runs by calculating them yourself.
b. Compute the value of $z\bigstar$ and the p-value.
c. Would you reject the hypothesis of randomness? Explain.
d. Construct a graph that displays the sample data that visually supports the answer in part c.

14.40 Posted on the Economic Statistics Briefing Room page of the White House's Web site on January 17, 2003, was the statement, "Median household income in 2001 in the United States was \$42,228." A random sample of 250 incomes has a median value different from any of the 250 incomes in the sample. The data contain 105 runs above and below the median. Test the null hypothesis that the incomes in the sample form a random sequence with respect to the two properties above and below the median value versus the alternative that the sequence is not random at $\alpha = 0.05$.

14.41 Ex14-41 The numbers of absences recorded at a lecture that met at 8 A.M. on Mondays and Thursdays last semester are given in order of occurrence:

5	16	6	9	18	11	16	21	14	17	12	14	10
6	8	12	13	4	5	5	6	1	7	18	26	6

Do these data show a randomness about the median value at $\alpha = 0.05$? Complete this test by using (a) critical values from Table 14 in Appendix B and (b) the standard normal distribution.

14.42 Research Randomizer is a free service offered to students and researchers interested in conducting random assignment and random sampling. Although every effort has been made to develop a useful means of generating random numbers, Research Randomizer and its staff do not guarantee the quality or randomness of numbers generated by Research Randomizer. Any use to which these numbers are put remains the sole responsibility of the user who generated them.

a. Go to the Web site http://www.randomizer.org/about.htm and generate one set of 20 random numbers from 1 to 9, where each number can repeat (select "No" for each number to be unique). (Use your computer, calculator, or Table 1 in Appendix B if you do not have a Web connection.)
b. Test your set for randomness above and below the median value of 5. Use $\alpha = 0.05$.
c. Conduct the test again with the same parameters.
d. Test your new set for randomness. Use $\alpha = 0.05$. Did you get the same results?
e. Solve part d using the standard normal distribution. Did you reach the same conclusion?
f. Compare your results from samples of size 20 with the results from samples of size 10 taken in Answer Now 14.29 (p. 683).

14.6 Rank Correlation

Charles Spearman developed the rank correlation coefficient in the early 1900s. It is a nonparametric alternative to the linear correlation coefficient (Pearson's product moment, r) that was discussed in Chapters 3 and 13.

The **Spearman rank correlation coefficient,** r_s, is found by using this formula:

SPEARMAN RANK CORRELATION COEFFICIENT

$$r_s = 1 - \frac{6\sum(d_i)^2}{n(n^2 - 1)} \quad (14.11)$$

where d_i is the difference in the **paired rankings** and n is the number of pairs of data. The value of r_s will range from -1 to $+1$ and will be used in much the same manner as Pearson's linear correlation coefficient r was used.

FYI
The subscript s is used in honor of Spearman, the originator.

The Spearman rank coefficient is defined by using formula (3.1) with data rankings substituted for quantitative x and y values. The original data may be rankings or, if the data are quantitative, each variable must be ranked separately; then the rankings are used as pairs. If there are no ties in the rankings, formula (14.11) is equivalent to formula (3.1). Formula (14.11) provides us with an easier procedure to use for calculating the r_s statistic.

ASSUMPTIONS FOR INFERENCES ABOUT RANK CORRELATION

The n ordered pairs of data form a random sample, and the variables are ordinal or numerical.

The null hypothesis that we will be testing is: There is no correlation between the two rankings. The alternative hypothesis may be either two-tailed, there is correlation, or one-tailed if we anticipate either positive or negative correlation. The critical region will be on the side(s) corresponding to the specific alternative that is expected. For example, if we suspect negative correlation, then the critical region will be in the left-hand tail.

Illustration 14.11 Calculating the Spearman Rank Correlation Coefficient

Let's consider a hypothetical situation in which four judges rank five contestants in a contest. Let's identify the judges as A, B, C, and D and the contestants as a, b, c, d, and e. Table 14.7 lists the awarded rankings.

TABLE 14.7 Rankings for Five Contestants

	Judge			
Contestant	**A**	**B**	**C**	**D**
a	1	5	1	5
b	2	4	2	2
c	3	3	3	1
d	4	2	4	4
e	5	1	5	3

When we compare judges A and B, we see that they ranked the contestants in exactly the opposite order: perfect disagreement (see Table 14.8). From our previous work with correlation, we expect the calculated value for r_s to be exactly -1 for these data. We have:

TABLE 14.8 Rankings of A and B

Contestant	A	B	$d_i = \text{A} - \text{B}$	$(d_i)^2$
a	1	5	−4	16
b	2	4	−2	4
c	3	3	0	0
d	4	2	2	4
e	5	1	4	16
				40

$$r_s = 1 - \frac{6\sum(d_i)^2}{n(n^2-1)}: \quad r_s = 1 - \frac{(6)(40)}{5(5^2-1)} = 1 - \frac{240}{120} = 1 - 2 = \mathbf{-1}$$

When judges A and C are compared, we see that their rankings of the contestants are identical (see Table 14.9). We would expect to find a calculated correlation coefficient of +1 for these data:

TABLE 14.9 Rankings of A and C

Contestant	A	C	$d_i = A - C$	$(d_i)^2$
a	1	1	0	0
b	2	2	0	0
c	3	3	0	0
d	4	4	0	0
e	5	5	0	0
				0

$$r_s = 1 - \frac{6\sum(d_i)^2}{n(n^2 - 1)}: \qquad r_s = 1 - \frac{(6)(0)}{5(5^2 - 1)} = 1 - \frac{0}{120} = 1 - 0 = \mathbf{1}$$

By comparing the rankings of judge A with those of judge B and then with those of judge C, we have seen the extremes: total agreement and total disagreement. Now let's compare the rankings of judge A with those of judge D (see Table 14.10). There seems to be no real agreement or disagreement here. Let's compute r_s:

TABLE 14.10 Rankings of A and D

Contestant	A	D	$d_i = A - D$	$(d_i)^2$
a	1	5	−4	16
b	2	2	0	0
c	3	1	2	4
d	4	4	0	0
e	5	3	2	4
				24

$$r_s = 1 - \frac{6\sum(d_i)^2}{n(n^2 - 1)}: \qquad r_s = 1 - \frac{(6)(24)}{5(5^2 - 1)} = 1 - \frac{144}{120} = 1 - 1.2 = \mathbf{-0.2}$$

The result is fairly close to zero, which is what we should have suspected, since there was no real agreement or disagreement. ■

The test of significance will result in a failure to reject the null hypothesis when r_s is close to zero; the test will result in a rejection of the null hypothesis when r_s is found to be close to +1 or −1. The critical values in Table 15 in Appendix B are the positive critical values only. Since the null hypothesis is: The population correlation coefficient is zero (that is, $\rho_s = 0$), we have a symmetric test statistic. Hence we need only add a plus or minus sign to the value found in the table, as appropriate. The sign is determined by the specific alternative that we have in mind.

When there are only a few ties, it is common practice to use formula (14.11). Even though the resulting value of r_s is not exactly equal to the value that would occur if formula (3.1) were used, it is generally considered to be an acceptable estimate. Illustration 14.12 shows the procedure for handling ties and uses formula (14.11) for the calculation of r_s.

When ties occur in either set of the ordered pairs of rankings, assign each tied observation the mean of the ranks that would have been assigned had there been no ties, as was done for the Mann–Whitney U test (see pp. 671–672).

Illustration 14.12 One-Tailed Hypothesis Test

Students who finish exams more quickly than the rest of the class are often thought to be smarter. Table 14.11 presents the scores and order of finish for 12 students on a recent one-hour exam. At the 0.01 level, do these data support the alternative hypothesis that the first students to complete an exam have higher grades?

TABLE 14.11 Data on Exam Scores Ta14-11

Order of Finish	1	2	3	4	5	6	7	8	9	10	11	12
Exam Score	90	78	76	60	92	86	74	60	78	70	68	64

SOLUTION

Step 1 **The Set-Up:**

a. **Describe the population parameter of interest.**
The rank correlation coefficient between score and order of finish, ρ_s.

b. **State the null hypothesis (H_o) and the alternative hypothesis (H_a).**

H_o: Order of finish has no relationship to exam score.

H_a: The first to finish tend to have higher grades.

Step 2 **The Hypothesis Test Criteria:**

a. **Check the assumptions.**
The 12 ordered pairs of data form a random sample; order of finish is an ordinal variable and test score is numerical.

b. **Identify the test statistic to be used.**
The Spearman rank correlation coefficient, r_s.

c. **Determine the level of significance:** $\alpha = 0.01$ for a one-tailed test.

Step 3 **The Sample Evidence:**

a. **Collect the sample information.**
The data are given in Table 14.11.

b. **Calculate the value of the test statistic.**
Rank the scores from highest to lowest, assigning the highest score the rank number 1, as shown. (Order of finish is already ranked.)

92	90	86	78	78	76	74	70	68	64	60	60
1	2	3	4	5	6	7	8	9	10	11	12
			4.5	4.5						11.5	11.5

The rankings and preliminary calculations are shown in Table 14.12.

TABLE 14.12 Rankings of Test Scores and Differences

Order of Finish	Test Score Rank	Difference (d_i)	$(d_i)^2$
1	2	−1	1.00
2	4.5	−2.5	6.25
3	6	−3	9.00
4	11.5	−7.5	56.25
5	1	4	16.00
6	3	3	9.00
7	7	0	0.00
8	11.5	−3.5	12.25
9	4.5	4.5	20.25
10	8	2	4.00
11	9	2	4.00
12	10	2	4.00
			142.00

FYI
For comparison, Exercise 14.56 (p. 698) asks you to calculate r_s using formula (3.2).

Using formula (14.11), we obtain

$$r_s = 1 - \frac{6\sum(d_i)^2}{n(n^2-1)}: \quad r_s = 1 - \frac{(6)(142.0)}{12(12^2-1)} = 1 - \frac{852}{1716} = 1 - 0.4965 = 0.503$$

Thus, $r_s\bigstar = 0.503$.

Step 4 The Probability Distribution:

Using the p-Value Procedure

a. Calculate the p-value for the test statistic.

Since the concern is for values "positive," the p-value is the area to the right:

$$\mathbf{P} = P(r_s \geq 0.503 \text{ for } n = 12)$$

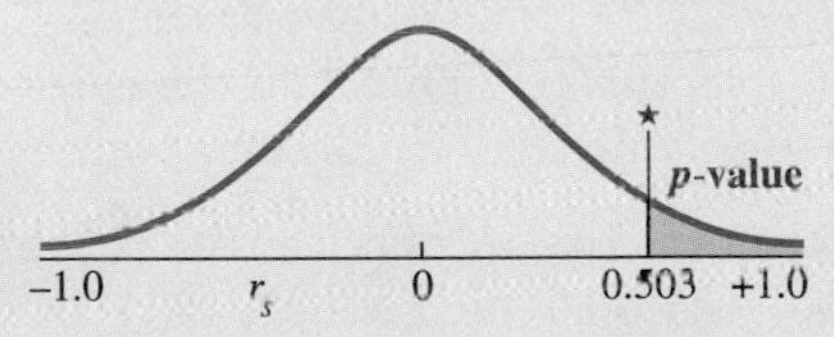

To find the p-value, you have two options:

1. Use Table 15 in Appendix B to place bounds on the p-value. Table 15 lists only two-tailed α (this test is one-tailed, so divide the column heading by 2): $\mathbf{0.025 < P < 0.05}$.
2. Use a computer or calculator to find the p-value: $\mathbf{P = 0.048}$.

Specific instructions follow this illustration.

b. Determine whether or not the p-value is smaller than α.

The p-value is not smaller than α.

OR

Using the Classical Procedure

a. Determine the critical region and critical value(s).

The critical region is one-tailed because H_a expresses concern for values related to "positive." Since the table is for two-tailed, the critical value is located at the intersection of the α = 0.02 column (α = 0.01 in each tail) and the $n = 12$ row of Table 15: **0.703**.

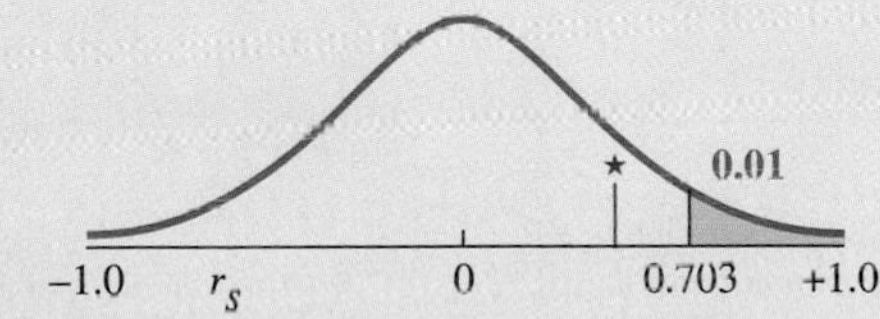

Specific instructions follow this illustration.

f. Determine whether or not the calculated test statistic is in the critical region.

$r_s\bigstar$ is not in the critical region, as shown in the figure.

Step 5 **The Results:**

a. State the decision about H_o: Fail to reject H_o.

b. State the conclusion about H_a.
These sample data do not provide sufficient evidence to enable us to conclude that the first students to finish have higher grades, at the 0.01 level of significance.

Calculating the p-value for the Spearman rank correlation test

Method 1: *Use Table 15 in Appendix B to place bounds on the p-value.* By inspecting the $n = 12$ row of Table 15, you can determine an interval within which the p-value lies. Locate the value of r_s along the $n = 12$ row and read the bounds from the top of the table. Table 15 lists only two-tailed values (therefore, you must divide by 2 for a one-tailed test). We find $\mathbf{0.025 < P < 0.05}$.

Method 2: If you are doing the hypothesis test with the aid of a computer or graphing calculator, most likely it will calculate the p-value for you. Specific instructions are described below. MINITAB and Excel calculate a two-tailed p-value; therefore, you must divide by 2 when the test is one-tailed. ■

Technology Instructions: Spearman's Rank Correlation Coefficient

MINITAB (Release 13)

Input the set of data for the first variable into C1 and the corresponding data values for the second variable into C2; then continue with:

```
Choose:   Manip > Rank...
Enter:    Rank data in: C1
          Store ranks in: C3
```

Repeat the above commands for the data in C2 and store in C4.

```
Choose:   Stat > Basic Statistics > Correlation
Enter:    Variables: C3 C4
```

Excel XP

Input the set of data for the first variable into column A and the corresponding data values for the second variable into column B; then continue with:

```
Choose:   Tools > Data Analysis Plus > Correlation (Spearman)
Enter:    Variable 1 range: (A1:A10 or select cells)
          Variable 2 range: (B1:B10 or select cells)
Select:   Labels (if necessary)
Enter:    Alpha: α (ex. 0.05)
```

FYI
Both Excel and TI-83 Plus use the normal approximation to complete the Spearman rank correlation test.

TI-83 Plus

Input the set of data for the first variable into L1 and the corresponding data values for the second variable into L2; then continue with:

```
Choose:   PRGM > EXEC > SPEARMAN*
Enter:    XLIST: L1
          YLIST: L2
Select:   DATA? 1:UNRANKED
          ALT HYP? 1:RHO > 0 or 2:RHO < 0 or 3:RHO ≠ 0
```

*Program SPEARMAN is one of many programs that are available for downloading from www.duxbury.com. See page 40 for specific directions.

Application 14.3 Calculus or Tartar

Consistency of Calculus Formation in Controlled Clinical Studies

ABSTRACT

Methods for demonstrating clinical efficacy of anti-tartar agents have been well established. These clinical models typically select high tartar-forming populations from short-term pre-test periods. This paper evaluates the consistency of tartar formation in pre-test periods in a common population. Two randomized, controlled 6-month tartar control studies were conducted over a 4-year period at a single center using a common design except for use of different examiners. Both studies used a 2-month pre-test period to evaluate calculus formation of participating subjects with a non-tartar control dentifrice. At the beginning of this period, all subjects received a prophylaxis, and were provided a regular dentifrice and a toothbrush and instructed to brush their teeth twice daily. After 8 weeks, subjects were assessed for the accumulation of supragingival calculus on the lingual surfaces of 6 mandibular anterior teeth using the Volpe-Manhold Calculus Index (VMI). The two studies have a total of 58 common subjects which completed the pre-test phase. In the first study, VMI scores at the end of pre-test phase ranged from 5.0–40.5 with a mean of 17.0 ± 7.8, and in the second study, the scores ranged from 0.5–45.5 with a mean of 9.7 ± 8.7. While the mean scores are significantly different ($p = 0.0001$), Pearson correlation analysis shows that VMI scores in the two studies were significantly correlated ($r = 0.60$, $p = 0.0001$). Spearman's rank correlation coefficient was also highly significant ($r = 0.57$, $p = 0.0001$). The data demonstrate that VMI scores in pre-test phases can be strongly related within subjects over several years and different examiners.

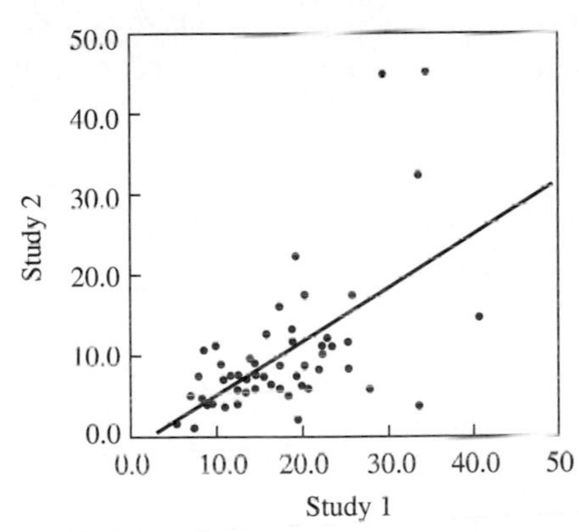

H. Liu and others
Procter & Gamble Company, Mason, OH

Research presented at the 30th Annual Meeting of the AADR, March 7–10, 2001

ANSWER NOW 14.43

a. Explain the meaning of "the mean scores are significantly different ($p = 0.0001$)." What means? What methodology might have been used to establish this significance?

b. Explain the meaning of "Pearson correlation analysis shows that VMI scores in the two studies were significantly correlated ($r = 0.60$, $p = 0.0001$)."

c. Explain the meaning of "Spearman's rank correlation coefficient was also highly significant ($r = 0.57$, $p = 0.0001$)."

d. What is the relationship between the Pearson correlation and Spearman correlation analysis?

Exercises

14.44 State the null hypothesis H_o and the alternative hypothesis H_a that would be used to test each statement:
a. There is no relationship between the two rankings.
b. The two variables are unrelated.
c. There is a positive correlation between the two variables.
d. Age has a decreasing effect on monetary value.

14.45 Determine the test criteria that would be used to test the null hypothesis for these Spearman rank correlation experiments:
a. H_o: No relationship between the two variables, vs. H_a: There is a relationship, with $n = 14$ and $\alpha = 0.05$
b. H_o: No correlation, vs. H_a: Positively correlated, with $n = 27$ and $\alpha = 0.05$
c. H_o: Variable A has no effect on variable B, vs. H_a: Variable B decreases as A increases, with $n = 18$ and $\alpha = 0.01$

14.46 Ex14-46 When it comes to getting workers to produce, money is not everything; feeling appreciated is more important. Do the rankings assigned by workers and the boss show a significant difference in what each thinks is important? Test using $\alpha = 0.05$.

Component of Job Satisfaction	Worker Ranking	Boss Ranking
Full appreciation of work done	1	8
Feeling of being in on things	2	10
Sympathetic help on personal problems	3	9
Job security	4	2
Good wages	5	1
Interesting work	6	5
Promotion and growth in the organization	7	3
Personal loyalty to employees	8	6
Good working conditions	9	4
Tactful discipline	10	7

Source: *Philadelphia Inquirer*, December 29, 1976

14.47 Ex14-47 Consumer product testing groups commonly supply ratings of all sorts of products to consumers in an effort to assist them in their purchase decisions. Different manufacturers' products are usually tested for their performance and then given an overall rating. *PC World* ranked the top ten 17-inch computer monitors in late 1998 and also supplied the street prices. The ranks of each are shown in the table, with the highest priced monitor given a rank of 1 and the lowest a 10.

Overall Rating	Street Price
1	3
2	4
3	6.5
4	8.5
5	5
6	2
7	8.5
8	6.5
9	10
10	1

Source: *PC World*, "Top 10 Monitors," December 1998

a. Compute the Spearman rank correlation coefficient for the overall rating and the street price of the 17-inch monitors.
b. Does a higher price yield a higher rating? Test the null hypothesis that there is no relationship between the overall ratings of the monitors and their street prices versus the alternative that there is a relationship between them. Use $\alpha = 0.05$.

14.48 Ex14-48 An article in *Self* magazine (February 1991) discusses the relationship between the pace of life and the coronary-heart-disease death rate. New York City, for example, was ranked as only the third-fastest-paced city but was number one for deadly heart attacks. Suppose the data from another such study involving eight cities were as follows:

City	Rank for Pace of Life	Rank for Heart-Disease Death Rate
Salt Lake City	4	7
Buffalo	2	2
Columbus	5	8
Worcester	6	4
Boston	1	6
Providence	8	3
New York	3	1
Paterson	7	5

Find the Spearman rank correlation coefficient.

14.49 Ex14-49 The data are the ages (x) of 12 subjects and the mineral concentration (y, in parts per million) in their tissue samples.

x	82	83	64	53	47	50	70	62	34	27	75	28
y	170	40	64	5	15	5	48	34	3	7	50	10

Refer to the MINITAB output and verify that the Spearman rank correlation coefficient equals 0.753 by calculating it yourself.

```
Correlations (Pearson)
Correlation of Rank, x and Rank, y = 0.753,
P-Value = 0.005
```

14.50 Ex14-50 Many people are concerned about eating foods that have a high sodium content. They are also advised of the benefits of obtaining sufficient fiber in their diets. Do foods high in fiber tend to have more sodium? The table was obtained by selecting 11 soups from a list published in *Nutrition Action Healthletter*. Both the sodium content and fiber were measured.

Soup	Sodium (mg)	Fiber (g)
A	480	12
B	830	0
C	510	1
D	460	5
F	490	3
F	580	7
G	420	2
H	290	4
I	450	10
J	430	6
K	390	9

Source: *Nutrition Action Healthletter*, Vol. 24, No. 7, December 1997

a. Rank the soups in ascending order based on their sodium content and on their fiber content, and show your results in a table.
b. Compute the Spearman rank correlation coefficient for the two rankings.
c. Does higher sodium content accompany foods that are higher in fiber? Test the null hypothesis that there is no relationship between the fiber and sodium contents of the soups versus the alternative that there is a positive relationship between them. Use $\alpha = 0.05$.

14.51 Ex14-51 An article titled "The Graduate Record Examination as an Admission Requirement for the Graduate Nursing Program" (*Journal of Professional Nursing*, October 1994) reported a significant correlation between undergraduate GPA and GPA at graduation from a graduate nursing program. The data below were collected on ten nursing students who graduated from a graduate nursing program. Compute the Spearman rank correlation coefficient and test the null hypothesis of no relationship versus a positive relationship. Use a level of significance equal to 0.05.

14.52 Ex14-52 The Aviation Consumer Protection Division of the U.S. Department of Transportation tracks and reports a wide variety of information regarding airlines. One of the reports is about the number of passengers denied boarding. The table shows the numbers of passengers involuntary denied boarding per 10,000 passengers for the ten major airlines in the United States for the indicated periods.

Airline	Jan.–Mar. 2002	Jan.–Mar. 2001
Alaska Airlines	2.21	0.76
American Airlines	0.28	0.38
American Eagle	0.18	0.02
America West	0.33	0.49
Continental	1.85	1.30
Delta Airline	0.89	0.41
Northwest	0.73	0.52
Southwest	1.14	1.57
United Airlines	0.65	0.82
US Airways	0.38	0.53

At the 0.05 level of significance, test the claim there is no correlation between the 2002 and 2001 numbers of passengers involuntarily denied boarding.

14.53 Ex14-53 Most people in the United States work in either trade or service jobs. "Trade" is defined as businesses like retail stores, car dealers, and restaurants—places where goods are sold. The "service" area includes such jobs as those in health care, the hotel industry, and cleaning services—jobs where people perform services rather than sell items. The table shows nine job classifications and compares Lee County, Florida, with all of Florida and the United States.

Job Category	Lee County	Florida	United States
Agriculture	2.0%	2.0%	2.6%
Construction	8.0	5.0	6.5
Manufacturing	5.0	8.0	16.0
Transportation, communications, and public utilities	5.0	5.0	7.1
Retail trade	25.0	21.0	16.8
Wholesale trade	3.0	5.0	7.1
Finance, insurance, and real estate	6.0	6.0	6.5
Services	30.0	34.0	35.9
Government	16.0	15.0	NA

Sources: Florida Department of Labor; U.S. Department of Labor

(continued)

Table for Exercise 14.51

Undergraduate GPA	3.5	3.1	2.7	3.7	2.5	3.3	3.0	2.9	3.8	3.2
GPA at Graduation	3.4	3.2	3.0	3.6	3.1	3.4	3.0	3.4	3.7	3.8

a. Construct a new table ranking the percentages for Lee County, Florida, and the United States separately.
b. Using Spearman's rank correlation and a 0.05 level of significance, determine whether there is a relationship between Lee County and all of Florida.
c. Using Spearman's rank correlation and a 0.05 level of significance, determine whether there is a relationship between Lee County and all of the United States.
d. Using Spearman's rank correlation and a 0.05 level of significance, determine whether there is a relationship between all of Florida and all of the United States.
e. Review the results of parts b, c, and d, and comment on your combined findings.

14.54 S Ex14-54 "The NAHB's 2000 Survey of Home Buyer Preferences" was conducted by the National Association of Home Builders to determine the features that homebuyers really want. Respondents were to rate each feature as desirable as well as essential. The following table shows the percentage of respondents rating each feature as desirable or essential.

Feature	Desirable	Essential
Laundry room	40	52
Linen closet	56	32
Exhaust fan	44	42
Dining room	43	36
Walk-in pantry	59	19
Island work area	55	16
Separate shower enclosure	49	20
Temperature control faucets	49	18
Whirlpool tub	46	12
White bathroom fixtures	40	16
Ceramic wall tiles	43	12
Solid-surface countertops	48	7
Den/library	43	11
Wood-burning fireplace	39	15
Special use storage	47	6

It is not surprising that the ratings in the Desirable column are considerably higher than the ratings in the Essential column. There is no question about there being a difference in the ratings; however, an appropriate question is: Do the items on the list appear in the same order of preference in both columns?

a. Use the Mann–Whitney U test to test the hypothesis that the items follow essentially the same distribution using $\alpha = 0.05$.
b. Use the Spearman rank correlation coefficient to test the hypothesis that the rankings of the items are not correlated using $\alpha = 0.05$.
c. State your conclusion.

14.55 Ex14-55 As this chart shows, what's "good enough" to qualify as "proficient" varies widely from state to state. *Education Week* (February 2, 2002) compared the percentages of students who scored at or above proficient on the National Assessment of Educational Progress (NAEP) and on state assessments in mathematics.

State	Percentage of Students at or Above Proficient Level: Statewide Assessment	NAEP Assessment
Arkansas	41	13
Connecticut	30	32
Georgia	62	18
Idaho	16	21
Kansas	39	30
Louisiana	12	14
Massachusetts	40	33
Michigan	75	29
Missouri	37	23
New York	65	22
North Carolina	84	28
North Dakota	15	25
Rhode Island	28	23
South Carolina	24	18
Texas	43	27
Vermont	38	29
Wyoming	27	25

Source: Education Week on the Web (www.edweek.com)

a. Present the information on a bar graph to visualize any relationship between the two different assessments. Does there appear to be a relationship? Explain.
b. Find the rank numbers for each set of percentages separately.
c. Present the information on a scatter diagram to visualize any relationship between the two different assessments. Does there appear to be any relationship? Explain.
d. Use the Spearman rank correlation coefficient to test the hypothesis that there is no correlation between the two sets of percentages using $\alpha = 0.05$.

14.56 Use formula (3.2) to calculate the Spearman rank correlation coefficient for the data in Illustration 14.12 (p. 692). Recall that formula (3.2) is equivalent to the definition formula (3.1) and that rank numbers must be used with this formula in order for the resulting statistic to be the Spearman r_s.

14.57 Refer to these bivariate data:

x	−2	−1	1	2
y	4	1	1	4

a. Construct a scatter diagram.
b. Calculate the Spearman rank correlation coefficient, r_s, using formula (14.11).
c. Calculate the Pearson correlation coefficient, r, using formula (3.2).
d. Compare the results from parts b and c. Do the two measures of correlation measure the same thing?

CHAPTER SUMMARY

RETURN TO CHAPTER CASE STUDY

Teenagers' Attitudes

Let's return to the Chapter Case Study (p. 655) and continue our investigation of teenagers' attitudes. CS14

FIRST THOUGHTS 14.58

a. Are the responses received from boys and girls the same?
b. Construct a graph that shows the two sets of responses side by side.
c. Explain why the chi-square test of homogeneity studied in Chapter 11 cannot be completed using this information.

PUTTING CHAPTER 14 TO WORK 14.59

a. Rank the choices for the boys and for the girls separately.
b. Do the responses from the boys and the girls have the same distribution? Use the Mann–Whitney *U* test and $\alpha = 0.05$.
c. Do the boys' preferences correlate to the girls' preferences? Use Spearman's rank correlation coefficient to test at the 0.05 level of significance.
d. Compare the results obtained in parts b and c.

YOUR OWN MINI-STUDY 14.60

Define a population of your choice and randomly sample variable 1: gender or class level (freshmen, sophomore, upper classmen, . . .), and variable 2: "the one thing I want most from life"—choose from: happiness, long enjoyable life, marriage and family, financial success, career success, religious satisfaction, love, personal success, personal contribution to society, friends, health, or education.

a. Rank the choices for each level of variable 1.
b. Do the responses from the levels of variable 1 have the same distribution? Use the Mann–Whitney *U* test and $\alpha = 0.05$.
c. Do the levels of the responses correlate to each other? Use Spearman's rank correlation coefficient to test at the 0.05 level of significance.
d. Write a paragraph comparing and contrasting the results obtained in parts b and c.

In Retrospect

In this chapter you have become acquainted with some of the basic concepts of nonparametric statistics. While learning about the use of nonparametric methods and specific nonparametric tests of significance, you should have also come to realize and understand some of the basic assumptions that are needed when the parametric techniques of the earlier chapters are encountered. You now have seen a variety of tests, many of which somewhat duplicate the job done by others. Keep in mind that you should use the best test for your particular needs. The power of the test and the cost of sampling, as related to the size and availability of the desired response variable, will play important roles in determining the specific test to be used.

Chapter Exercises

14.61 Ex14-61 "Because regional-scale atmospheric deposition data in the Rocky Mountains are sparse, a program was designed by the U.S. Geological Survey to more thoroughly determine the quality of precipitation and to identify sources of atmospherically deposited pollution in a network of high-elevation sites. Depth-integrated samples of seasonal snowpacks at 52 sampling sites, in a network from New Mexico to Montana, were collected and analyzed each year since 1993." One of a number of chemical characteristics sampled was hydrogen. Following are the five-year average results from each of the 52 sites.

7.3	3.3	4.3	4.8	4.5	5.1
5.0	6.7	3.5	6.1	5.1	8.3
4.1	9.7	5.4	3.8	5.8	6.1
4.5	8.8	5.2	7.2	4.9	2.0
3.6	6.3	7.8	5.5	11.1	9.4
5.1	15.2	5.0	9.9	3.8	5.4
7.8	9.4	4.5	10.6	3.6	2.7
10.5	12.4	3.1	2.8	8.7	4.3
8.3	5.9	4.6	6.1		

Source: U.S. Geological Survey, 2002

a. Construct a stem-and-leaf plot of the data.
b. Describe the pattern you see in the stem-and-leaf plot.
c. Using the stem-and-leaf plot and the sign test, find the 95% confidence interval for the population median.

14.62 Ex14-62 Research about the health practices of cardiovascular technologists compared the body mass index (BMI) of the technologists with the BMI of the general population. Weight classification by BMI is: underweight if less than 19, normal from 19 to 24, overweight 25 to 29, and obese if 30 and higher. The list is the BMI values for a sample of 30 technologists:

16	50	39	33	33	25	29	30	39	23
21	24	19	28	26	34	19	20	18	21
24	24	20	18	26	22	24	18	25	25

a. Construct a stem-and-leaf plot for the BMIs of the cardiovascular technologists.
b. Describe the sample of technologists as underweight, normal, overweight, or obese.
c. Find the 95% confidence interval for the median BMI.

14.63 Ex14-63 A sample of 32 students received the following scores on an exam:

41	42	48	46	50	54	51	42
51	50	45	42	32	45	43	56
55	47	45	51	60	44	57	57
47	28	41	42	54	48	47	32

a. Does this sample show that the median score for the exam differs from 50? Use $\alpha = 0.05$.
b. Does this sample show that the median score for the exam is less than 50? Use $\alpha = 0.05$.

14.64 Ex14-64 Is the absentee rate in the 8 A.M. statistics class the same as in the 11 A.M. statistics class? The following sample of the daily number of absences was taken from the attendance records of the two classes.

	Day											
Class	**1**	**2**	**3**	**4**	**5**	**6**	**7**	**8**	**9**	**10**	**11**	**12**
8 A.M.	0	1	3	1	0	2	4	1	3	5	3	2
11 A.M.	1	0	1	0	1	2	3	0	1	3	2	1

Is there sufficient reason to conclude that there is more absence in the 8 A.M. class? Use $\alpha = 0.05$.

14.65 Ex14-65 Track coaches, runners, and fans talk a lot about the "speed of the track." The surface of the track is believed to have a direct effect on the amount of time that it takes a runner to cover the required distance. To test this effect, ten runners were asked to run a 220-yard sprint on each of two tracks. Track A is a cinder track, and track B is made of a new synthetic material. The running times (in seconds) are given in the table. Test the claim that the surface on track B is conducive to faster running times.

	Runner									
Track	**1**	**2**	**3**	**4**	**5**	**6**	**7**	**8**	**9**	**10**
A	27.7	26.8	27.0	25.5	26.6	27.4	27.2	27.4	25.8	25.1
B	27.0	26.7	25.3	26.0	26.1	25.3	26.7	27.1	24.8	27.1

a. State the null and alternative hypotheses being tested. Complete the test using $\alpha = 0.05$.
b. State your conclusions.

14.66 A candy company has developed two new chocolate-covered candy bars. Six randomly selected people all preferred candy bar I. Is this statistical evidence, at $\alpha = 0.05$, that the general public will prefer candy bar I?

14.67 An article in *Sedimentary Geology* (Vol. 57, 1988) compares a measure called the "roughness coefficient" for translucent and opaque quartz sand grains. If you measured the roughness coefficient for 20 sand grains of each type (translucent and opaque), for what values of the Mann–Whitney U statistic would you reject the null hypothesis in a two-tailed test with alpha equal to 0.05?

14.68 Ex14-68 Twenty students were randomly divided into two equal groups. Group 1 was taught an anatomy course using a standard lecture approach. Group 2 was taught using a computer-assisted approach. The test scores on a comprehensive final exam were as follows:

Group 1	75	83	60	89	77	92	88	90	55	70
Group 2	77	92	90	85	72	59	65	92	90	79

Test the claim that a computer-assisted approach produces higher achievement (as measured by final exam scores) in anatomy courses than does a lecture approach. Use $\alpha = 0.05$.

14.69 Ex14-69 The use of nuclear magnetic resonance (NMR) spectroscopy for detection of malignancy is discussed in *Clinical Chemistry* (Vol. 34, No. 3, 1988). The line width at the half height of peaks in the NMR spectra is measured. The spectrum is produced from assaying plasma from an individual. Suppose the following line widths were obtained from a normal group and a group known to have malignancies. Would you reject a two tailed research hypothesis at the 0.05 level of significance?

Normal Group	35.1	32.9	30.6	30.5	30.9
Malignancy Group	28.5	29.5	30.7	27.5	28.0

14.70 Ex14-70 A firm is currently testing two different procedures for adjusting the cutting machines used in the production of greeting cards. The results of two samples show the following recorded adjustment times (in minutes):

Method 1	17	15	14	18	16	15	17	
	18	15	14	14	16	15		
Method 2	14	14	13	13	15	12	16	14
	16	13	14	13	12	15	17	13

Is there sufficient reason to conclude that method 2 requires less time (on the average) than method 1 at the 0.05 level of significance?

14.71 Ex14-71 Two statistics that baseball enthusiasts use to compare the strengths of teams are team batting (the higher the batting average, the better) and team pitching (the lower the earned run average—ERA, the better). The 2002 results for the National and American Leagues follow:

NL Team	Batting Average	ERA	AL Team	Batting Average	ERA
Colorado	0.274	5.20	Boston	0.277	3.75
San Diego	0.253	4.62	New York	0.275	3.87
Atlanta	0.260	3.13	Cleveland	0.249	4.91
Los Angeles	0.263	3.69	Seattle	0.275	4.07
Chicago	0.246	4.29	Texas	0.269	5.15
New York	0.256	3.89	Chicago	0.270	4.55
Pittsburgh	0.244	4.23	Anaheim	0.282	3.69
Florida	0.261	4.36	Minnesota	0.272	4.12
Houston	0.262	4.00	Baltimore	0.246	4.46
Montreal	0.261	3.97	Kansas City	0.256	5.21
San Francisco	0.267	3.54	Tampa Bay	0.253	5.29
Philadelphia	0.259	4.17	Oakland	0.261	3.68
St. Louis	0.268	3.70	Detroit	0.248	4.92
Cincinnati	0.253	4.27	Toronto	0.261	4.80
Milwaukee	0.253	4.72			
Arizona	0.267	3.92			

(continued)

Source: *MLB.com*

a. Convert the table to ranks of the (1) batting averages and (2) earned run averages for the National League and the American League, showing the league (A or N) represented by a team's rank.
b. Use the Mann–Whitney U test to test these hypotheses: (1) The batting average of the American League is higher, and (2) The earned run average of the National League is lower. Use the 0.05 level of significance.

14.72 Ex14-72 Two table-tennis-ball manufacturers have agreed that the quality of their products can be measured by the height to which the balls rebound. A test is arranged, the balls are dropped from a constant height, and the rebound heights are measured. The results (in inches) are shown in the table. Manufacturer A claims, "The results show my product to be superior." Manufacturer B replies, "I know of no statistical test that supports this claim." Can you find a test that supports A's claim?

A	14.0	12.5	11.5	12.2	12.4	12.3	11.8	11.9	13.7	13.2
B	12.0	12.5	11.6	13.3	13.0	13.0	12.1	12.8	12.2	12.6

a. Does the appropriate parametric test show that A's product is superior? What parametric test (or tests) is appropriate, and what exactly does it show?
b. Does the appropriate nonparametric test show that A's product is superior?

14.73 Consider this sequence of defective parts (d) and nondefective parts (n) produced by a machine:

n n n d n n n n n d n n n n n n n d n d n n n n

Can we reject the hypothesis of randomness at $\alpha = 0.05$?

14.74 Ex14-74 A patient was given two different types of vitamin pills, one containing iron and one iron-free. The patient was instructed to take the pills on alternate days. To free himself from remembering which pill he needed to take, he mixed all of the pills together in a large bottle. Each morning he took the first pill that came out of the bottle. To see whether this was a random process, for

25 days he recorded an "I" each morning that he took a vitamin with iron and an "N" for no iron. Here are his results:

Day	1	2	3	4	5	6	7	8	9	10	11	12	13
Type	I	I	N	I	I	N	N	I	N	N	N	N	N

Day	14	15	16	17	18	19	20	21	22	23	24	25
Type	I	I	I	N	I	I	I	I	N	I	I	N

Is there sufficient reason to reject the null hypothesis that the vitamins were taken in random order at the 0.05 level of significance?

14.75 Ex14-75 What makes one company more attractive to work for than another? One possibility is the growth in new jobs. In 1998, the editors of *Fortune* developed a list of the top 100 companies to work for in America. Included in the list was the percentage change in full-time positions of each company during the past two years. The top 20 companies are shown in the table.

Company	Job Growth	Company	Job Growth
1	26	11	23
2	54	12	13
3	34	13	17
4	10	14	23
5	31	15	9
6	48	16	3
7	26	17	15
8	22	18	11
9	24	19	1
10	10	20	122

Source: *Fortune*, "The 100 Best Companies to Work for in America," January 12, 1998

a. Determine the median job growth percentage and the number of runs above and below the median.

b. Use the runs test to test whether the growth rates are listed in a random sequence about the median.

c. Do companies ranked higher also have higher job growth rates? State your conclusion.

14.76 Ex14-76 In a study to see whether spouses are consistent in their preferences for television programs, a market research firm asked several married couples to rank a list of 12 programs (1 represents the highest score; 12 represents the lowest). The average ranks for the programs, rounded to the nearest integer, were as follows:

	Program											
Rank	1	2	3	4	5	6	7	8	9	10	11	12
Husbands	12	2	6	10	3	11	7	1	9	5	8	4
Wives	5	4	1	9	3	12	2	8	6	10	7	11

Is there significant evidence of negative correlation at the 0.01 level of significance?

14.77 Ex14-77 Can today's high temperature be accurately predicted from yesterday's high? Pairs of yesterday's and today's high temperatures were randomly selected. The results are shown in the table. Do the data present sufficient evidence to justify the statement: "Today's high temperature tends to correlate with yesterday's high temperature"? Use $\alpha = 0.05$.

Yesterday	40	58	46	33	40	51	55	81	85
Today	40	56	34	59	46	51	74	77	83

Yesterday	83	89	64	73	63	46	58	28	69
Today	84	85	68	65	60	54	62	34	66

14.78 Ex14-78 U.S. commercial radio stations are classified by the primary format of their broadcasts. As people change their listening preferences, the stations are likely to react to the change by adjusting their formats. The table shows the percentages of radio stations in 1991, 1997, and 2002 broken down by their primary format:

Primary Format	1991	1997	2002
Country	25.61	24.26	20.18
Adult contemporary (AC)	21.76	14.75	14.34
News, talk, business, sports	5.49	12.73	14.73
Religion (teaching and music)	8.33	10.22	9.90
Rock	5.51	9.13	8.51
Oldies	7.34	7.30	7.70
Spanish and ethnic	3.86	5.32	6.57
Adult standards	4.25	5.20	5.29
Urban, black, urban AC	3.24	3.47	3.69
Top-40	7.04	3.40	4.47
Easy listening	2.19	0.84	2.61
Variety	0.84	0.49	0.40
Jazz	0.55	0.48	0.78
Classical, fine arts	0.53	0.45	0.30
Pre-teen	0.04	0.35	0.47
All other	3.42	1.61	0.06

Source: M Street Corporation, Nashville, TN, 1997.

http://www.rab.com/station/marketing_guide/rmfb0203.pdf

a. Construct a table that lists the ranks of the relative frequency of stations within each format for 1991, 1997, and 2002.

b. Use the Spearman rank correlation coefficient to test at the 0.01 level of significance the hypothesis that there is no correlation between the ranks of the formats offered by radio stations in 1991 and 1997, 1991 and 2002, 1997 and 2002.

c. Has the distribution of primary formats changed? Describe how the results in part b support your answer.

14.79 Ex14-79 Every year *Sports Illustrated* presents its college football preview and ranks the top 25 teams before the season starts based primarily on scouting reports. As the season progresses, other college football polls provide a weekly ranking of the teams, evaluations that are largely influenced by how well the teams are playing and who they play against. The table shows the ranks of the top 25 college teams by *Sports Illustrated* prior to the first snap of the 2002 season and the ranks bestowed by *USA Today*/ESPN and the AP Top 25 after the regular season of play had transpired.

Team	*Sports Illustrated* (Preseason)	*USA Today*/ ESPN (After Regular Season)	AP Top 25 (After Regular Season)
Oklahoma	1	8	8
Miami	2	1	1
Tennessee	3	29.5	30
Texas	4	9	9
Florida State	5	16	16
Colorado	6	14	14
Washington State	7	7	7
Florida	8	20	22
Oregon	9	29.5	30
Georgia	10	4	4
Washington	11	29.5	30
Ohio State	12	2	2
LSU	13	25	30
Nebraska	14	29.5	30
Virginia Tech	15	19	21
Maryland	16	18	20
Marshall	17	24	27
South Carolina	18	29.5	30
Louisville	19	29.5	30
Michigan	20	11	12
USC	21	5	5
Texas A&M	22	29.5	30
Auburn	23	22	19
Illinois	24	29.5	30
Penn State	25	10	10

Sources: http://sportsillustrated.cnn.com, http://sports.espn.go.com, http://www.sltrib.com

NOTE: The ranks given to teams that were ranked in the preseason poll but no longer ranked after the season were obtained by $[26 + 27 + 28 + \cdots + (25 + n)] \div n$, where n is the number of teams no longer ranked in the top 25.

a. Compute the Spearman rank correlation coefficient for the *SI* preseason poll and the *USA Today*/ESPN poll; the *SI* preseason poll and the AP Top 25; and the *USA Today*/ESPN poll and the AP Top 25.
b. Test the null hypothesis that there is no relationship between the polls versus the alternative that there is a relationship for each of the three possible paired comparisons. Use $\alpha = 0.05$.

14.80 Nonparametric tests are also called distribution-free tests. However, the normal distributions are used in the inference-making procedures.
a. To what does the term *distribution-free* apply? (The population? The sample? The sampling distribution?) Explain.
b. What is it that has the normal distribution? Explain.

Vocabulary and Key Concepts

Be able to define each term. Pay special attention to the key terms, which are printed in **red**. In addition, describe each term in your own words and give an example of each. Your examples should not be ones given in class or in the textbook. The bracketed numbers indicate the chapters in which the term previously appeared, but you should define the terms again to show increased understanding of their meaning. Page numbers indicate the first appearance of the term in Chapter 14.

assumptions [8–13] (pp. 658, 662, 670, 681, 690)
binomial random variable [5, 9] (p. 665)
continuity correction [6] (p. 665)
correlation [3, 13] (p. 689)
dependent sample [10] (p. 662)
distribution-free test (p. 656)
efficiency (p. 657)
independent sample [10] (p. 670)
Mann–Whitney U test (p. 670)
median, M [2] (p. 658)
nonparametric test (p. 656)
normal approximation [6] (pp. 665, 674, 684)
paired data [3, 10, 13] (p. 662)
parametric test (p. 656)
power (p. 657)
randomness [2, 7] (p. 681)
rank (p. 671)
run (p. 681)
runs test (p. 681)
sign test (p. 658)
Spearman rank correlation coefficient (p. 689)
test statistic [8–13] (pp. 660, 666, 673, 675, 681, 684, 689)

Chapter Practice Test

PART I: KNOWING THE DEFINITIONS

Answer "True" if the statement is always true. If the statement is not always true, replace the words shown in bold with words that make the statement always true.

14.1 One of the advantages of the nonparametric tests is the necessity for **less restrictive** assumptions.

14.2 The sign test is a possible replacement for the ***F*-test.**

14.3 The **sign test** can be used to test the randomness of a set of data.

14.4 If a tie occurs in a set of ranked data, the data that form the tie are **removed from the set.**

14.5 Two dependent **means** can be compared nonparametrically by using the sign test.

14.6 The sign test is a possible alternative to the Student's t-test for **one mean value.**

14.7 The **runs test** is a nonparametric alternative to the difference between two independent means.

14.8 The **confidence level** of a statistical hypothesis test is measured by $1 - \beta$.

14.9 Spearman's rank correlation coefficient is an alternative to using the **linear correlation coefficient.**

14.10 The **efficiency** of a nonparametric test is the probability that a false null hypothesis is rejected.

PART II: APPLYING THE CONCEPTS

14.11 The weights (in pounds) of nine people before they stopped smoking and five weeks after they stopped smoking are listed here: PT14-11

Person	1	2	3	4	5	6	7	8	9
Before	148	176	153	116	128	129	120	132	154
After	155	178	151	120	130	136	126	128	158

Find the 95% confidence interval estimate for the average weight change.

14.12 The following data show the weight gains (in ounces) for 20 laboratory mice, half of which were fed one diet and half a different diet. Test to determine whether the difference in weight gain is significant at $\alpha = 0.05$. PT14-12

Diet A	41	40	36	43	36	43	39	36	24	41
Diet B	35	34	27	39	31	41	37	34	42	38

14.13 A large textbook publishing company hired nine new sales representatives three years ago. At the time of hire, the nine were ranked according to their potential. Now three years later the company president wants to know how well their potential ranks correlate with their sales totals for the three years. PT14-13

Sales Representative	a	b	c	d	e	f	g	h	I
Potential	2	5	6	1	4	3	9	8	7
Sales Total	450	410	350	345	330	400	250	310	270

Is there significant correlation at the 0.05 level?

14.14 The new school principal thought there might be a pattern to the order in which discipline problems arrived at his office. He had his secretary record the grade levels of the students as they arrived. PT14-14

9	10	11	9	12	11	9	10	10	11
10	11	10	10	11	12	12	9	9	11
12	10	9	12	10	11	12	11	10	10

At the 0.05 level, is there significant evidence of randomness?

PART III: UNDERSTANDING THE CONCEPTS

14.15 What advantages do nonparametric statistics have over parametric methods?

14.16 Explain how the sign test is based on the binomial distribution and is often approximated by the normal distribution.

14.17 Why does the sign test use a null hypothesis about the median instead of the mean like a t-test uses?

14.18 Explain why a nonparametric test is not as sensitive to an extreme data as a parametric test might be.

14.19 A restaurant has collected data on which of two seating arrangements its customers prefer. In a sign test to determine whether one seating arrangement is significantly preferred, which null hypothesis would be used?

a. $M = 0$ b. $M = 0.5$ c. $p = 0$ d. $p = 0.5$

Explain your choice.

Working With Your Own Data

© StockTrek/Getty Images

The existence of bivariate data is commonplace in everyday life and there are multiple options for analyzing the relationship between the two variables. In Chapter 10, the relationship between the paired data was analyzed as paired differences. In Chapter 12, the analysis of variance methods tested for an effect that one variable might have on a second variable. In Chapter 13, the methods of correlation and regression were used to investigate the relationship between the variables in order to determine whether they have a mathematical relationship that can be approximated by means of a straight line. The following illustrates such a situation.

A THE AGE AND VALUE OF PEGGY'S CAR

Peggy would like to sell her 1994 Corvette, and she wants to determine an asking price for it in order to advertise. Her Corvette features the typical Corvette equipment with no customizing and is in average condition for a well-cared-for 1994 Corvette. She wants to advertise using an average asking price and expects to get an average price for it (average for a Corvette!) when it sells. Presently she must answer the question, What is an average asking price for a 1994 Corvette?

Inspection of many classified sections of newspapers turned up only three advertisements for 1994 Corvettes. The prices listed varied a great deal, and Peggy needed more information in order to determine her asking price. She decided to use the Internet and search for prices. She restricted her search to used Chevrolet Corvettes, 1988 to present, that were in good repair, not customized or show cars, and were being sold by their owner, not a dealer or an auction. Peggy collected the following data on January 8, 2003. DS-4

Year	Asking Price	Year	Asking Price	Year	Asking Price	Year	Asking Price	Year	Asking Price
2001	$35,800	1989	$15,700	1989	$18,200	1998	$26,500	1995	$18,900
1995	$19,900	2002	$35,500	1992	$17,900	2001	$36,000	2000	$33,000
1996	$26,700	2001	$35,700	2002	$35,900	1997	$21,000	1994	$23,400
1994	$19,000	1996	$21,900	1997	$28,900	1991	$11,500	1999	$31,500
1991	$16,700	1999	$33,000	1998	$28,600	1994	$16,000	1999	$31,900
1991	$17,500	1992	$15,500	1997	$30,000	1995	$24,000	1998	$30,000
1997	$30,600	1999	$29,000	2000	$30,500	1995	$21,700	1988	$14,400
2000	$32,000	1988	$ 9,500	1993	$17,500	1989	$10,500	1990	$16,000
1998	$33,000	1993	$16,500	1991	$15,500	1990	$17,500	1996	$21,000
1994	$16,000	1992	$16,000	2000	$35,000	1990	$14,800	2001	$34,000
1996	$25,800	1993	$16,300	1997	$25,000	1992	$19,800		
1988	$10,500	2002	$35,000	1993	$14,000	1993	$19,500		

Peggy knows that the price she should ask, and the amount she receives, for her Corvette is affected by its age. The general questions are: Is the effect of age on price predictable? and, Can a meaningful relationship between the age and the typical asking price for used Corvettes be established?

Independent variable, x: The age of the car as measured in years and defined by

$$x = (\text{present calendar year}) - (\text{year of manufacture}) + 1$$

Example: During 2003, Peggy's 1994 Corvette is considered to be 10 years old.

$$x = (2003 - 1994) + 1 = 9 + 1 = 10$$

Dependent variable, y: The advertised asking price

1. Discuss why age should be used instead of manufacture year.
2. Convert manufacture year to the variable age, x, using the formula above.
3. Construct side-by-side vertical dotplots of the asking price for each year of age.
4. Does it appear that the asking price for a Corvette is affected by its age? Describe the effect as pictured on the graph shown for question 3.
5. Complete a one-way ANOVA and test the hypothesis that age has no effect on the asking price. Use $\alpha = 0.05$.
6. What effect would using year of manufacture instead of age have on the above analysis?
7. Discuss why using age instead of manufacture year results in a scatter diagram that is more representative of behavior of the price or value of a used car as it ages. Describe how a scatter diagram using year of manufacture as x would differ from one using age as x.
8. Construct and label a scatter diagram of Peggy's data.
9. Discuss the relationship between the side-by-side dotplots drawn in question 3 and the scatter diagram drawn in question 8.
10. Determine the equation for the line of best fit.
11. Draw the line of best fit on the scatter diagram.
12. Test the question of the line of best fit to see whether the linear model is appropriate for the data. Use $\alpha = 0.05$.
13. Construct a 95% confidence interval for the mean advertised price for 1994 Corvettes.
14. Draw a line segment on the scatter diagram that represents the interval estimate found for question 13.
15. What does the value of the slope, b_1, represent? Explain.
16. What does the value of the y-intercept, b_0, represent? Explain.
17. Write a meaningful paragraph answering the general questions of concern:
    a. Is the effect of age on price predictable?
    b. Can a meaningful relationship between the age and the typical asking price for used Corvettes be established?
    c. What is an average asking price for a 1994 Corvette that is for sale?

B YOUR OWN INVESTIGATION

Identify a situation of interest to you that can be investigated statistically using bivariate data. (Consult your instructor for specific guidance.)

1. Define the population, the independent variable, the dependent variable, and the purpose for studying these two variables as a regression analysis.
2. Collect 15 to 20 ordered pairs of data.
3. Partition the independent variable values into three or more categories that are meaningful or appropriate for your data.
4. Complete a one-way ANOVA and test the hypothesis that the independent variable has no effect on the dependent variable.
5. Construct and label a scatter diagram of your data.
6. Determine the equation for the line of best fit.
7. Draw the line of best fit on the scatter diagram.

8. Test the equation of the line of best fit to see whether the linear model is appropriate for the data. Use $\alpha = 0.05$.
9. Construct a 95% confidence interval for the mean value of the dependent variable at the following value of x: Let x be equal to one-third the sum of the lowest value of x in your sample and twice the largest value; that is,

$$x = \frac{L + 2H}{3}$$

10. Draw a line segment on the scatter diagram that represents the interval estimate found for question 9.
11. What does the value of the slope, b_1, represent? Explain.
12. What does the value of the y-intercept, b_0, represent? Explain.
13. Write a meaningful paragraph comparing and/or contrasting the results from your ANOVA test and the test for the appropriateness of a linear model. What conclusions can you reach based on the results of these tests?

APPENDIX

A Basic Principles of Counting

Appendix A is available on the Student Suite CD.

APPENDIX

B Tables

TABLE 1 Random Numbers

10 09 73 25 33	76 52 01 35 86	34 67 35 48 76	80 95 90 91 17	39 29 27 49 45
37 54 20 48 05	64 89 47 42 96	24 80 52 40 37	20 63 61 04 02	00 82 29 16 65
08 42 26 89 53	19 64 50 93 03	23 20 90 25 60	15 95 33 43 64	35 08 03 36 06
99 01 90 25 29	09 37 67 07 15	38 31 13 11 65	88 67 67 43 97	04 43 62 76 59
12 80 79 99 70	80 15 73 61 47	64 03 23 66 53	98 95 11 68 77	12 17 17 68 33
66 06 57 47 17	34 07 27 68 50	36 69 73 61 70	65 81 33 98 85	11 19 92 91 70
31 06 01 08 05	45 57 18 24 06	35 30 34 26 14	86 79 90 74 39	23 40 30 97 32
05 26 97 76 02	02 05 16 56 92	68 66 57 48 18	73 05 38 52 47	18 02 38 85 79
63 57 33 21 35	05 32 54 70 48	90 55 35 75 48	28 46 82 87 09	83 49 12 56 24
73 79 64 57 53	03 52 96 47 78	35 80 83 42 82	60 93 52 03 44	35 27 38 84 35
98 52 01 77 67	14 90 56 86 07	22 10 94 05 58	60 97 09 34 33	50 50 07 39 98
11 80 50 54 31	39 80 82 77 32	50 72 56 82 48	29 40 52 42 01	52 77 56 78 51
83 45 29 96 34	06 28 89 80 83	13 74 67 00 78	18 47 54 06 10	68 71 17 78 17
88 68 54 02 00	86 50 75 84 01	36 76 66 79 51	90 36 47 64 93	29 60 91 10 62
99 59 46 73 48	87 51 76 49 69	91 82 60 89 28	93 78 56 13 68	23 47 83 41 13
65 48 11 76 74	17 46 85 09 50	58 04 77 69 74	73 03 95 71 86	40 21 81 65 44
80 12 43 56 35	17 72 70 80 15	45 31 82 23 74	21 11 57 82 53	14 38 55 37 63
74 35 09 98 17	77 40 27 72 14	43 23 60 02 10	45 52 16 42 37	96 28 60 26 55
69 91 62 68 03	66 25 22 91 48	36 93 68 72 03	76 62 11 39 90	94 40 05 64 18
09 89 32 05 05	14 22 56 85 14	46 42 75 67 88	96 29 77 88 22	54 38 21 45 98
91 49 91 45 23	68 47 92 76 86	46 16 28 35 54	94 75 08 99 23	37 08 92 00 48
80 33 69 45 98	26 94 03 68 58	70 29 73 41 35	54 14 03 33 40	42 05 08 23 41
44 10 48 19 49	85 15 74 79 54	32 97 92 65 75	57 60 04 08 81	22 22 20 64 13
12 55 07 37 42	11 10 00 20 40	12 86 07 46 97	96 64 48 94 39	28 70 72 58 15
63 60 64 93 29	16 50 53 44 84	40 21 95 25 63	43 65 17 70 82	07 20 73 17 90
61 19 69 04 46	26 45 74 77 74	51 92 43 37 29	65 39 45 95 93	42 58 26 05 27
15 47 44 52 66	95 27 07 99 53	59 36 78 38 48	82 39 61 01 18	33 21 15 94 66
94 55 72 85 73	67 89 75 43 87	54 62 24 44 31	91 19 04 25 92	92 92 74 59 73
42 48 11 62 13	97 34 40 87 21	16 86 84 87 67	03 07 11 20 59	25 70 14 66 70
23 52 37 83 17	73 20 88 98 37	68 93 59 14 16	26 25 22 96 63	05 52 28 25 62
04 49 35 24 94	75 24 63 38 24	45 86 25 10 25	61 96 27 93 35	65 33 71 24 72
00 54 99 76 54	64 05 18 81 59	96 11 96 38 96	54 69 28 23 91	23 28 72 95 29
35 96 31 53 07	26 89 80 93 54	33 35 13 54 62	77 97 45 00 24	90 10 33 93 33
59 80 80 83 91	45 42 72 68 42	83 60 94 97 00	13 02 12 48 92	78 56 52 01 06
46 05 88 52 36	01 39 09 22 86	77 28 14 40 77	93 91 08 36 47	70 61 74 29 41
32 17 90 05 97	87 37 92 52 41	05 56 70 70 07	86 74 31 71 57	85 39 41 18 38
69 23 46 14 06	20 11 74 52 04	15 95 66 00 00	18 74 39 24 23	97 11 89 63 38
19 56 54 14 30	01 75 87 53 79	40 41 92 15 85	66 67 43 68 06	84 96 28 52 07
45 15 51 49 38	19 47 60 72 46	43 66 79 45 43	59 04 79 00 33	20 82 66 95 41
94 86 43 19 94	36 16 81 08 51	34 88 88 15 53	01 54 03 54 56	05 01 45 11 76
98 08 62 48 26	45 24 02 84 04	44 99 90 88 96	39 09 47 34 07	35 44 13 18 80
33 18 51 62 32	41 94 15 09 49	89 43 54 85 81	88 69 54 19 94	37 54 87 30 43
80 95 10 04 06	96 38 27 07 74	20 15 12 33 87	25 01 62 52 98	94 62 46 11 71
79 75 24 91 40	71 96 12 82 96	69 86 10 25 91	74 85 22 05 39	00 38 75 95 79
18 63 33 25 37	98 14 50 65 71	31 01 02 46 74	05 45 56 14 27	77 93 89 19 36

For specific details about using this table, see page 24 or the *Statistical Tutor.*

TABLE 1 Random Numbers (continued)

74 02 94 39 02	77 55 73 22 70	97 79 01 71 19	52 52 75 80 21	80 81 45 17 48
54 17 84 56 11	80 99 33 71 43	05 33 51 29 69	56 12 71 92 55	36 04 09 03 24
11 66 44 98 83	52 07 98 48 27	59 38 17 15 39	09 97 33 34 40	88 46 12 33 56
48 32 47 79 28	31 24 96 47 10	02 29 53 68 70	32 30 75 75 46	15 02 00 99 94
69 07 49 41 38	87 63 79 19 76	35 58 40 44 01	10 51 82 16 15	01 84 87 69 38
09 18 82 00 97	32 82 53 95 27	04 22 08 63 04	83 38 98 73 74	64 27 85 80 44
90 04 58 54 97	51 98 15 06 54	94 93 88 19 97	91 87 07 61 50	68 47 66 46 59
73 18 95 02 07	47 67 72 62 69	62 29 06 44 64	27 12 46 70 18	41 36 18 27 60
75 76 87 64 90	20 97 18 17 49	90 42 91 22 72	95 37 50 58 71	93 82 34 31 78
54 01 64 40 56	66 28 13 10 03	00 68 22 73 98	20 71 45 32 95	07 70 61 78 13
08 35 86 99 10	78 54 24 27 85	13 66 15 88 73	04 61 89 75 53	31 22 30 84 20
28 30 60 32 64	81 33 31 05 91	40 51 00 78 93	32 60 46 04 75	94 11 90 18 40
53 84 08 62 33	81 59 41 36 28	51 21 59 02 90	28 46 66 87 95	77 76 22 07 91
91 75 75 37 41	61 61 36 22 69	50 26 39 02 12	55 78 17 65 14	83 48 34 70 55
89 41 59 26 94	00 39 75 83 91	12 60 71 76 46	48 94 97 23 06	94 54 13 74 08
77 51 30 38 20	86 83 42 99 01	68 41 48 27 74	51 90 81 39 80	72 89 35 55 07
19 50 23 71 74	69 97 92 02 88	55 21 02 97 73	74 28 77 52 51	65 34 46 74 15
21 81 85 93 13	93 27 88 17 57	05 68 67 31 56	07 08 28 50 46	31 85 33 84 52
51 47 46 64 99	68 10 72 36 21	94 04 99 13 45	42 83 60 91 91	08 00 74 54 49
99 55 96 83 31	62 53 52 41 70	69 77 71 28 30	74 81 97 81 42	43 86 07 28 34
33 71 34 80 07	93 58 47 28 69	51 92 66 47 21	58 30 32 98 22	93 17 49 39 72
85 27 48 68 93	11 30 32 92 70	28 83 43 41 37	73 51 59 04 00	71 14 84 36 43
84 13 38 96 40	44 03 55 21 66	73 85 27 00 91	61 22 26 05 61	62 32 71 84 23
56 73 21 62 34	17 39 59 61 31	10 12 39 16 22	85 49 65 75 60	81 60 41 88 80
65 13 85 68 06	87 60 88 52 61	34 31 36 58 61	45 87 52 10 69	85 64 44 72 77
38 00 10 21 76	81 71 91 17 11	71 60 29 29 37	74 21 96 40 49	65 58 44 96 98
37 40 29 63 97	01 30 47 75 86	56 27 11 00 86	47 32 46 26 05	40 03 03 74 38
97 12 54 03 48	87 08 33 14 17	21 81 53 92 50	75 23 76 20 47	15 50 12 95 78
21 82 64 11 34	47 14 33 40 72	64 63 88 59 02	49 13 90 64 41	03 85 65 45 52
73 13 54 27 42	95 71 90 90 35	85 79 47 42 96	08 78 98 81 56	64 69 11 92 02
07 63 87 79 29	03 06 11 80 72	96 20 74 41 56	23 82 19 95 38	04 71 36 69 94
60 52 88 34 41	07 95 41 98 14	59 17 52 06 95	05 53 35 21 39	61 21 20 64 55
83 59 63 56 55	06 95 89 29 83	05 12 80 97 19	77 43 35 37 83	92 30 15 04 98
10 85 06 27 46	99 59 91 05 07	13 49 90 63 19	53 07 57 18 39	06 41 01 93 62
39 82 09 89 52	43 62 26 31 47	64 42 18 08 14	43 80 00 93 51	31 02 47 31 67
59 58 00 64 78	75 56 97 88 00	88 83 55 44 86	23 76 80 61 56	04 11 10 84 08
38 50 80 73 41	23 79 34 87 63	90 82 29 70 22	17 71 90 42 07	95 95 44 99 53
30 69 27 06 68	94 68 81 61 27	56 19 68 00 91	82 06 76 34 00	05 46 26 92 00
65 44 39 56 59	18 28 82 74 37	49 63 22 40 41	08 33 76 56 76	96 29 99 08 36
27 26 75 02 64	13 19 27 22 94	07 47 74 46 06	17 98 54 89 11	97 34 13 03 58
91 30 70 69 91	19 07 22 42 10	36 69 95 37 28	28 82 53 57 93	28 97 66 62 52
68 43 49 46 88	84 47 31 36 22	62 12 69 84 08	12 84 38 25 90	09 81 59 31 46
48 90 81 58 77	54 74 52 45 91	35 70 00 47 54	83 82 45 26 92	54 13 05 51 60
06 91 34 51 97	42 67 27 86 01	11 88 30 95 28	63 01 19 89 01	14 97 44 03 44
10 45 51 60 19	14 21 03 37 12	91 34 23 78 21	88 32 58 08 51	43 66 77 08 83
12 88 39 73 43	65 02 76 11 84	04 28 50 13 92	17 97 41 50 77	90 71 22 67 69
21 77 83 09 76	38 80 73 69 61	31 64 94 20 96	63 28 10 20 23	08 81 64 74 49
19 52 35 95 15	65 12 25 96 59	86 28 36 82 58	69 57 21 37 98	16 43 59 15 29
67 24 55 26 70	35 58 31 65 63	79 24 68 66 86	76 46 33 42 22	26 65 59 08 02
60 58 44 73 77	07 50 03 79 92	45 13 42 65 29	26 76 08 36 37	41 32 64 43 44
53 85 34 13 77	36 06 69 48 50	58 83 87 38 59	49 36 47 33 31	96 24 04 36 42
24 63 73 97 36	74 38 48 93 42	52 62 30 79 92	12 36 91 86 01	03 74 28 38 73
83 08 01 24 51	38 99 22 28 15	07 75 95 17 77	97 37 72 75 85	51 97 23 78 67
16 44 42 43 34	36 15 19 90 73	27 49 37 09 39	85 13 03 25 52	54 84 65 47 59
60 79 01 81 57	57 17 86 57 62	11 16 17 85 76	45 81 95 29 79	65 13 00 48 60

From tables of the RAND Corporation. Reprinted from Wilfred J. Dixon and Frank J. Massey, Jr., *Introduction to Statistical Analysis.* 3rd ed. (New York: McGraw-Hill, 1969), pp. 446–447. Reprinted by permission of the RAND Corporation.

TABLE 2 Binomial Probabilities $[\binom{n}{x} \cdot p^x \cdot q^{n-x}]$

n	x	0.01	0.05	0.10	0.20	0.30	0.40	0.50	0.60	0.70	0.80	0.90	0.95	0.99	x
2	0	.980	.902	.810	.640	.490	.360	.250	.160	.090	.040	.010	.002	0+	0
	1	.020	.095	.180	.320	.420	.400	.500	.480	.420	.320	.180	.095	.020	1
	2	0+	.002	.010	.040	.090	.160	.250	.360	.490	.640	.810	.902	.980	2
3	0	.970	.857	.729	.512	.343	.216	.125	.064	.027	.008	.001	0+	0+	0
	1	.029	.135	.243	.384	.441	.432	.375	.288	.189	.096	.027	.007	0+	1
	2	0+	.007	.027	.096	.189	.288	.375	.432	.441	.384	.243	.135	.029	2
	3	0+	0+	.001	.008	.027	.064	.125	.216	.343	.512	.729	.857	.970	3
4	0	.961	.815	.656	.410	.240	.130	.062	.026	.008	.002	0+	0+	0+	0
	1	.039	.171	.292	.410	.412	.346	.250	.154	.076	.026	.004	0+	0+	1
	2	.001	.014	.049	.154	.265	.346	.375	.346	.265	.154	.049	.014	.001	2
	3	0+	0+	.004	.026	.076	.154	.250	.346	.412	.410	.292	.171	.039	3
	4	0+	0+	0+	.002	.008	.026	.062	.130	.240	.410	.656	.815	.961	4
5	0	.951	.774	.590	.328	.168	.078	.031	.010	.002	0+	0+	0+	0+	0
	1	.048	.204	.328	.410	.360	.259	.156	.077	.028	.006	0+	0+	0+	1
	2	.001	.021	.073	.205	.309	.346	.312	.230	.132	.051	.008	.001	0+	2
	3	0+	.001	.008	.051	.132	.230	.312	.346	.309	.205	.073	.021	.001	3
	4	0+	0+	0+	.006	.028	.077	.156	.259	.360	.410	.328	.204	.048	4
	5	0+	0+	0+	0+	.002	.010	.031	.078	.168	.328	.590	.774	.951	5
6	0	.941	.735	.531	.262	.118	.047	.016	.004	.001	0+	0+	0+	0+	0
	1	.057	.232	.354	.393	.303	.187	.094	.037	.010	.002	0+	0+	0+	1
	2	.001	.031	.098	.246	.324	.311	.234	.138	.060	.015	.001	0+	0+	2
	3	0+	.002	.015	.082	.185	.276	.312	.276	.185	.082	.015	.002	0+	3
	4	0+	0+	.001	.015	.060	.138	.234	.311	.324	.246	.098	.031	.001	4
	5	0+	0+	0+	.002	.010	.037	.094	.187	.303	.393	.354	.232	.057	5
	6	0+	0+	0+	0+	.001	.004	.016	.047	.118	.262	.531	.735	.941	6
7	0	.932	.698	.478	.210	.082	.028	.008	.002	0+	0+	0+	0+	0+	0
	1	.066	.257	.372	.367	.247	.131	.055	.017	.004	0+	0+	0+	0+	1
	2	.002	.041	.124	.275	.318	.261	.164	.077	.025	.004	0+	0+	0+	2
	3	0+	.004	.023	.115	.227	.290	.273	.194	.097	.029	.003	0+	0+	3
	4	0+	0+	.003	.029	.097	.194	.273	.290	.227	.115	.023	.004	0+	4
	5	0+	0+	0+	.004	.025	.077	.164	.261	.318	.275	.124	.041	.002	5
	6	0+	0+	0+	0+	.004	.017	.055	.131	.247	.367	.372	.257	.066	6
	7	0+	0+	0+	0+	0+	.002	.008	.028	.002	.210	.470	.608	.032	7
8	0	.923	.663	.430	.168	.058	.017	.004	.001	0+	0+	0+	0+	0+	0
	1	.075	.279	.383	.336	.198	.090	.031	.008	.001	0+	0+	0+	0+	1
	2	.003	.051	.149	.294	.296	.209	.109	.041	.010	.001	0+	0+	0+	2
	3	0+	.005	.033	.147	.254	.279	.219	.124	.047	.009	0+	0+	0+	3
	4	0+	0+	.005	.046	.136	.232	.273	.232	.136	.046	.005	0+	0+	4
	5	0+	0+	0+	.009	.047	.124	.219	.279	.254	.147	.033	.005	0+	5
	6	0+	0+	0+	.001	.010	.041	.109	.209	.296	.294	.149	.051	.003	6
	7	0+	0+	0+	0+	.001	.008	.031	.090	.198	.336	.383	.279	.075	7
	8	0+	0+	0+	0+	0+	.001	.004	.017	.058	.168	.430	.663	.923	8

For specific details about using this table, see page 254.

TABLE 2 **Binomial Probabilities [$\binom{n}{x} \cdot p^{x} \cdot q^{n-x}$] (continued)**

		P													
n	*x*	0.01	0.05	0.10	0.20	0.30	0.40	0.50	0.60	0.70	0.80	0.90	0.95	0.99	*x*
9	0	.914	.630	.387	.134	.040	.010	.002	0+	0+	0+	0+	0+	0+	0
	1	.083	.299	.387	.302	.156	.060	.018	.004	0+	0+	0+	0+	0+	1
	2	.003	.063	.172	.302	.267	.161	.070	.021	.004	0+	0+	0+	0+	2
	3	0+	.008	.045	.176	.267	.251	.164	.074	.021	.003	0+	0+	0+	3
	4	0+	.001	.007	.066	.172	.251	.246	.167	.074	.017	.001	0+	0+	4
	5	0+	0+	.001	.017	.074	.167	.246	.251	.172	.066	.007	.001	0+	5
	6	0+	0+	0+	.003	.021	.074	.164	.251	.267	.176	.045	.008	0+	6
	7	0+	0+	0+	0+	.004	.021	.070	.161	.267	.302	.172	.063	.003	7
	8	0+	0+	0+	0+	0+	.004	.018	.060	.156	.302	.387	.299	.083	8
	9	0+	0+	0+	0+	0+	0+	.002	.010	.040	.134	.387	.630	.914	9
10	0	.904	.599	.349	.107	.028	.006	.001	0+	0+	0+	0+	0+	0+	0
	1	.091	.315	.387	.268	.121	.040	.010	.002	0+	0+	0+	0+	0+	1
	2	.004	.075	.194	.302	.233	.121	.044	.011	.001	0+	0+	0+	0+	2
	3	0+	.010	.057	.201	.267	.215	.117	.042	.009	.001	0+	0+	0+	3
	4	0+	.001	.011	.088	.200	.251	.205	.111	.037	.006	0+	0+	0+	4
	5	0+	0+	.001	.026	.103	.201	.246	.201	.103	.026	.001	0+	0+	5
	6	0+	0+	0+	.006	.037	.111	.205	.251	.200	.088	.011	.001	0+	6
	7	0+	0+	0+	.001	.009	.042	.117	.215	.267	.201	.057	.010	0+	7
	8	0+	0+	0+	0+	.001	.011	.044	.121	.233	.302	.194	.075	.004	8
	9	0+	0+	0+	0+	0+	.002	.010	.040	.121	.268	.387	.315	.091	9
	10	0+	0+	0+	0+	0+	0+	.001	.006	.028	.107	.349	.599	.904	10
11	0	.895	.569	.314	.086	.020	.004	0+	0+	0+	0+	0+	0+	0+	0
	1	.099	.329	.384	.236	.093	.027	.005	.001	0+	0+	0+	0+	0+	1
	2	.005	.087	.213	.295	.200	.089	.027	.005	.001	0+	0+	0+	0+	1
	3	0+	.014	.071	.221	.257	.177	.081	.023	.004	0+	0+	0+	0+	3
	4	0+	.001	.016	.111	.220	.236	.161	.070	.017	.002	0+	0+	0+	4
	5	0+	0+	.002	.039	.132	.221	.226	.147	.057	.010	0+	0+	0+	5
	6	0+	0+	0+	.010	.057	.147	.226	.221	.132	.039	.002	0+	0+	6
	7	0+	0+	0+	.002	.017	.070	.161	.236	.220	.111	.016	.001	0+	7
	8	0+	0+	0+	0+	.004	.023	.081	.177	.257	.221	.071	.014	0+	8
	9	0+	0+	0+	0+	.001	.005	.027	.089	.200	.295	.213	.087	.005	9
	10	0+	0+	0+	0+	0+	.001	.005	.027	.093	.236	.384	.329	.099	10
	11	0+	0+	0+	0+	0+	0+	0+	.004	.020	.086	.314	.569	.895	11
12	0	.886	.540	.282	.069	.014	.002	0+	0+	0+	0+	0+	0+	0+	0
	1	.107	.341	.377	.206	.071	.017	.003	0+	0+	0+	0+	0+	0+	1
	2	.006	.099	.230	.283	.168	.064	.016	.002	0+	0+	0+	0+	0+	2
	3	0+	.017	.085	.236	.240	.142	.054	.012	.001	0+	0+	0+	0+	3
	4	0+	.002	.021	.133	.231	.213	.121	.042	.008	.001	0+	0+	0+	4
	5	0+	0+	.004	.053	.158	.227	.193	.101	.029	.003	0+	0+	0+	5
	6	0+	0+	0+	.016	.079	.177	.226	.177	.079	.016	0+	0+	0+	6
	7	0+	0+	0+	.003	.029	.101	.193	.227	.158	.053	.004	0+	0+	7
	8	0+	0+	0+	.001	.008	.042	.121	.213	.231	.133	.021	.002	0+	8
	9	0+	0+	0+	0+	.001	.012	.054	.142	.240	.236	.085	.017	0+	9
	10	0+	0+	0+	0+	0+	.002	.016	.064	.168	.283	.230	.099	.006	10
	11	0+	0+	0+	0+	0+	0+	.003	.017	.071	.206	.377	.341	.107	11
	12	0+	0+	0+	0+	0+	0+	0+	.002	.014	.069	.282	.540	.886	12

TABLE 2 **Binomial Probabilities $[\binom{n}{x} \cdot p^x \cdot q^{n-x}]$ (continued)**

		p													
n	*x*	0.01	0.05	0.10	0.20	0.30	0.40	0.50	0.60	0.70	0.80	0.90	0.95	0.99	*x*
13	0	.878	.513	.254	.055	.010	.001	0+	0+	0+	0+	0+	0+	0+	0
	1	.115	.351	.367	.179	.054	.011	.002	0+	0+	0+	0+	0+	0+	1
	2	.007	.111	.245	.268	.139	.045	.010	.001	0+	0+	0+	0+	0+	2
	3	0+	.021	.100	.246	.218	.111	.035	.006	.001	0+	0+	0+	0+	3
	4	0+	.003	.028	.154	.234	.184	.087	.024	.003	0+	0+	0+	0+	4
	5	0+	0+	.006	.069	.180	.221	.157	.066	.014	.001	0+	0+	0+	5
	6	0+	0+	.001	.023	.103	.197	.209	.131	.044	.006	0+	0+	0+	6
	7	0+	0+	0+	.006	.044	.131	.209	.197	.103	.023	.001	0+	0+	7
	8	0+	0+	0+	.001	.014	.066	.157	.221	.180	.069	.006	0+	0+	8
	9	0+	0+	0+	0+	.003	.024	.087	.184	.234	.154	.028	.003	0+	9
	10	0+	0+	0+	0+	.001	.006	.035	.111	.218	.246	.100	.021	0+	10
	11	0+	0+	0+	0+	0+	.001	.010	.045	.139	.268	.245	.111	.007	11
	12	0+	0+	0+	0+	0+	0+	.002	.011	.054	.179	.367	.351	.115	12
	13	0+	0+	0+	0+	0+	0+	0+	.001	.010	.055	.254	.513	.878	13
14	0	.869	.488	.229	.044	.007	.001	0+	0+	0+	0+	0+	0+	0+	0
	1	.123	.359	.356	.154	.041	.007	.001	0+	0+	0+	0+	0+	0+	1
	2	.008	.123	.257	.250	.113	.032	.006	.001	0+	0+	0+	0+	0+	2
	3	0+	.026	.114	.250	.194	.085	.022	.003	0+	0+	0+	0+	0+	3
	4	0+	.004	.035	.172	.229	.155	.061	.014	.001	0+	0+	0+	0+	4
	5	0+	0+	.008	.086	.196	.207	.122	.041	.007	0+	0+	0+	0+	5
	6	0+	0+	.001	.032	.126	.207	.183	.002	.023	.002	0+	0+	0+	6
	7	0+	0+	0+	.009	.062	.157	.209	.157	.062	.009	0+	0+	0+	7
	8	0+	0+	0+	.002	.023	.092	.183	.207	.126	.032	.001	0+	0+	8
	9	0+	0+	0+	0+	.007	.041	.122	.207	.196	.086	.008	0+	0+	9
	10	0+	0+	0+	0+	.001	.014	.061	.155	.229	.172	.035	.004	0+	10
	11	0+	0+	0+	0+	0+	.003	.022	.085	.194	.250	.114	.026	0+	11
	12	0+	0+	0+	0+	0+	.001	.006	.032	.113	.250	.257	.123	.008	12
	13	0+	0+	0+	0+	0+	0+	.001	.007	.041	.154	.356	.359	.123	13
	14	0+	0+	0+	0+	0+	0+	0+	.001	.007	.044	.229	.488	.869	14
15	0	.860	.463	.206	.035	.005	0+	0+	0+	0+	0+	0+	0+	0+	0
	1	.130	.366	.343	.132	.031	.005	0+	0+	0+	0+	0+	0+	0+	1
	2	.009	.135	.267	.231	.092	.022	.003	0+	0+	0+	0+	0+	0+	2
	3	0+	.031	.129	.250	.170	.063	.014	.002	0+	0+	0+	0+	0+	3
	4	0+	.005	.043	.188	.219	.127	.042	.007	.001	0+	0+	0+	0+	4
	5	0+	.001	.010	.103	.206	.186	.092	.024	.003	0+	0+	0+	0+	5
	6	0+	0+	.002	.043	.147	.207	.153	.061	.012	.001	0+	0+	0+	6
	7	0+	0+	0+	.014	.081	.177	.196	.118	.035	.003	0+	0+	0+	7
	8	0+	0+	0+	.003	.035	.118	.196	.177	.081	.014	0+	0+	0+	8
	9	0+	0+	0+	.001	.012	.061	.153	.207	.147	.043	.002	0+	0+	9
	10	0+	0+	0+	0+	.003	.024	.092	.186	.206	.103	.010	.001	0+	10
	11	0+	0+	0+	0+	.001	.007	.042	.127	.219	.188	.043	0.05	0+	11
	12	0+	0+	0+	0+	0+	.002	.014	.063	.170	.250	.129	.031	0+	12
	13	0+	0+	0+	0+	0+	0+	.003	.022	.092	.231	.267	.135	.009	13
	14	0+	0+	0+	0+	0+	0+	0+	.005	.031	.132	.343	.366	.130	14
	15	0+	0+	0+	0+	0+	0+	0+	0+	.005	.035	.206	.463	.860	15

TABLE 3 Areas of the Standard Normal Distribution

The entries in this table are the probabilities that a random variable with a standard normal distribution assumes a value between 0 and z; the probability is represented by the shaded area under the curve in the accompanying figure. Areas for negative values of z are obtained by symmetry.

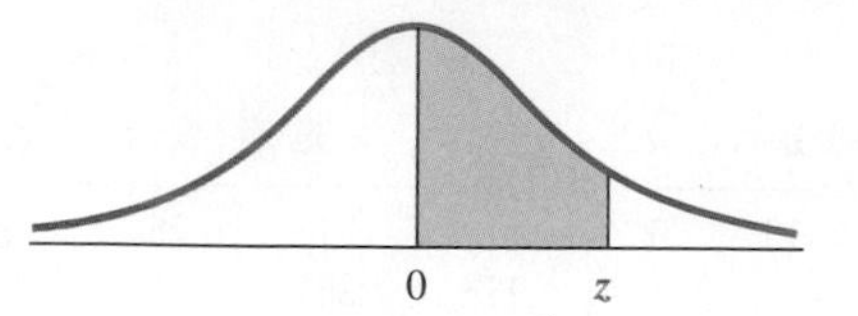

Second Decimal Place in z

z	0.00	0.01	0.02	0.03	0.04	0.05	0.06	0.07	0.08	0.09
0.0	0.0000	0.0040	0.0080	0.0120	0.0160	0.0199	0.0239	0.0279	0.0319	0.0359
0.1	0.0398	0.0438	0.0478	0.0517	0.0557	0.0596	0.0636	0.0675	0.0714	0.0753
0.2	0.0793	0.0832	0.0871	0.0910	0.0948	0.0987	0.1026	0.1064	0.1103	0.1141
0.3	0.1179	0.1217	0.1255	0.1293	0.1331	0.1368	0.1406	0.1443	0.1480	0.1517
0.4	0.1554	0.1591	0.1628	0.1664	0.1700	0.1736	0.1772	0.1808	0.1844	0.1879
0.5	0.1915	0.1950	0.1985	0.2019	0.2054	0.2088	0.2123	0.2157	0.2190	0.2224
0.6	0.2257	0.2291	0.2324	0.2357	0.2389	0.2422	0.2454	0.2486	0.2517	0.2549
0.7	0.2580	0.2611	0.2642	0.2673	0.2704	0.2734	0.2764	0.2794	0.2823	0.2852
0.8	0.2881	0.2910	0.2939	0.2967	0.2995	0.3023	0.3051	0.3078	0.3106	0.3133
0.9	0.3159	0.3186	0.3212	0.3238	0.3264	0.3289	0.3315	0.3340	0.3365	0.3389
1.0	0.3413	0.3438	0.3461	0.3485	0.3508	0.3531	0.3554	0.3577	0.3599	0.3621
1.1	0.3643	0.3665	0.3686	0.3708	0.3729	0.3749	0.3770	0.3790	0.3810	0.3830
1.2	0.3849	0.3869	0.3888	0.3907	0.3925	0.3944	0.3962	0.3980	0.3997	0.4015
1.3	0.4032	0.4049	0.4066	0.4082	0.4099	0.4115	0.4131	0.4147	0.4162	0.4177
1.4	0.4192	0.4207	0.4222	0.4236	0.4251	0.4265	0.4279	0.4292	0.4306	0.4319
1.5	0.4332	0.4345	0.4357	0.4370	0.4382	0.4394	0.4406	0.4418	0.4429	0.4441
1.6	0.4452	0.4463	0.4474	0.4484	0.4495	0.4505	0.4515	0.4525	0.4535	0.4545
1.7	0.4554	0.4564	0.4573	0.4582	0.4591	0.4599	0.4608	0.4616	0.4625	0.4633
1.8	0.4641	0.4649	0.4656	0.4664	0.4671	0.4678	0.4686	0.4693	0.4699	0.4706
1.9	0.4713	0.4719	0.4726	0.4732	0.4738	0.4744	0.4750	0.4756	0.4761	0.4767
2.0	0.4772	0.4778	0.4783	0.4788	0.4793	0.4798	0.4803	0.4808	0.4812	0.4817
2.1	0.4821	0.4826	0.4830	0.4834	0.4838	0.4842	0.4846	0.4850	0.4854	0.4857
2.2	0.4861	0.4864	0.4868	0.4871	0.4875	0.4878	0.4881	0.4884	0.4887	0.4890
2.3	0.4893	0.4896	0.4898	0.4901	0.4904	0.4906	0.4909	0.4911	0.4913	0.4916
2.4	0.4918	0.4920	0.4922	0.4925	0.4927	0.4929	0.4931	0.4932	0.4934	0.4936
2.5	0.4938	0.4940	0.4941	0.4943	0.4945	0.4946	0.4948	0.4949	0.4951	0.4952
2.6	0.4953	0.4955	0.4956	0.4957	0.4959	0.4960	0.4961	0.4962	0.4963	0.4964
2.7	0.4965	0.4966	0.4967	0.4968	0.4969	0.4970	0.4971	0.4972	0.4973	0.4974
2.8	0.4974	0.4975	0.4976	0.4977	0.4977	0.4978	0.4979	0.4979	0.4980	0.4981
2.9	0.4981	0.4982	0.4982	0.4983	0.4984	0.4984	0.4985	0.4985	0.4986	0.4986
3.0	0.4987	0.4987	0.4987	0.4988	0.4988	0.4989	0.4989	0.4989	0.4990	0.4990
3.1	0.4990	0.4991	0.4991	0.4991	0.4992	0.4992	0.4992	0.4992	0.4993	0.4993
3.2	0.4993	0.4993	0.4994	0.4994	0.4994	0.4994	0.4994	0.4995	0.4995	0.4995
3.3	0.4995	0.4995	0.4995	0.4996	0.4996	0.4996	0.4996	0.4996	0.4996	0.4997
3.4	0.4997	0.4997	0.4997	0.4997	0.4997	0.4997	0.4997	0.4997	0.4997	0.4998
3.5	0.4998	0.4998	0.4998	0.4998	0.4998	0.4998	0.4998	0.4998	0.4998	0.4998
3.6	0.4998	0.4998	0.4999	0.4999	0.4999	0.4999	0.4999	0.4999	0.4999	0.4999
3.7	0.4999									
4.0	0.49997									
4.5	0.499997									
5.0	0.4999997									

For specific details about using this table to find: probabilities, see page 274; confidence coefficients, page 351; *p*-values, pages 376, 377, 379; critical values, page 293.

TABLE 4 Critical Values of Standard Normal Distribution

A ONE-TAILED SITUATIONS

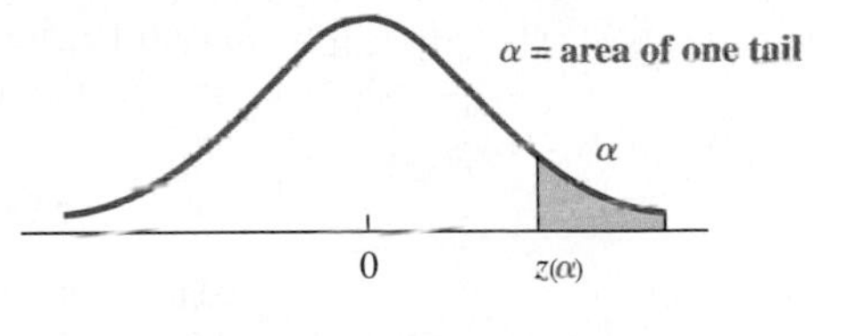

The entries in this table are the critical values for z for which the area under the curve representing α is in the right-hand tail. Critical values for the left-hand tail are found by symmetry.

	Amount of α in one tail						
α	0.25	0.10	0.05	0.025	0.02	0.01	0.005
$z(\alpha)$	0.67	1.28	1.65	1.96	2.05	2.33	2.58

One-tailed example:
$\alpha = 0.05$
$z(\alpha) = z(0.05) = 1.65$

B TWO-TAILED SITUATIONS

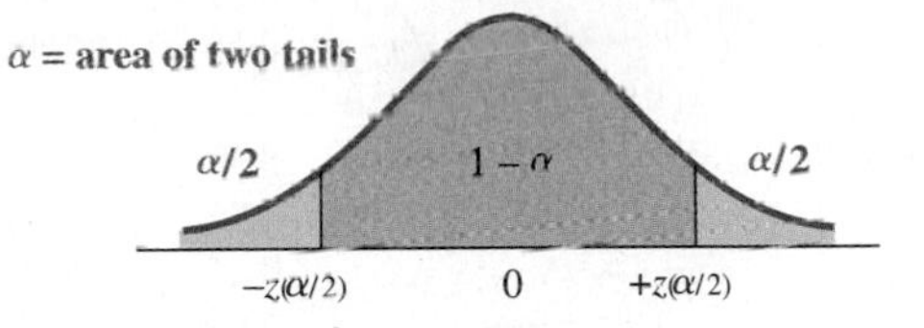

The entries in this table are the critical values for z for which the area under the curve representing α is split equally between the two tails.

	Amount of α in two-tails					
α	0.25	0.20	0.10	0.05	0.02	0.01
$z(\alpha/2)$	1.15	1.28	1.65	1.96	2.33	2.58
$1-\alpha$	0.75	0.80	0.90	0.95	0.98	0.99
	Area in the "center"					

Two-tailed example:
$\alpha = 0.05$ or $1 - \alpha = 0.95$
$\alpha/2 = 0.025$
$z(\alpha/2) = z(0.025) = 1.96$

For specific details about using this table to find: confidence coefficients, see page 351; critical values, pages 393, 395.

TABLE 5 *p*-Values for Standard Normal Distribution

The entries in this table are the *p*-values related to the right-hand tail for the calculated $z\star$ for the standard normal distribution.

$z\star$	*p*-value	$z\star$	*p*-value	$z\star$	*p*-value	$z\star$	*p*-value
0.00	0.5000	**1.00**	0.1587	**2.00**	0.0228	**3.00**	0.0013
0.05	0.4801	**1.05**	0.1469	**2.05**	0.0202	**3.05**	0.0011
0.10	0.4602	**1.10**	0.1357	**2.10**	0.0179	**3.10**	0.0010
0.15	0.4404	**1.15**	0.1251	**2.15**	0.0158	**3.15**	0.0008
0.20	0.4207	**1.20**	0.1151	**2.20**	0.0139	**3.20**	0.0007
0.25	0.4013	**1.25**	0.1056	**2.25**	0.0122	**3.25**	0.0006
0.30	0.3821	**1.30**	0.0968	**2.30**	0.0107	**3.30**	0.0005
0.35	0.3632	**1.35**	0.0885	**2.35**	0.0094	**3.35**	0.0004
0.40	0.3446	**1.40**	0.0808	**2.40**	0.0082	**3.40**	0.0003
0.45	0.3264	**1.45**	0.0735	**2.45**	0.0071	**3.45**	0.0003
0.50	0.3085	**1.50**	0.0668	**2.50**	0.0062	**3.50**	0.0002
0.55	0.2912	**1.55**	0.0606	**2.55**	0.0054	**3.55**	0.0002
0.60	0.2743	**1.60**	0.0548	**2.60**	0.0047	**3.60**	0.0002
0.65	0.2578	**1.65**	0.0495	**2.65**	0.0040	**3.65**	0.0001
0.70	0.2420	**1.70**	0.0446	**2.70**	0.0035	**3.70**	0.0001
0.75	0.2266	**1.75**	0.0401	**2.75**	0.0030	**3.75**	0.0001
0.80	0.2119	**1.80**	0.0359	**2.80**	0.0026	**3.80**	0.0001
0.85	0.1977	**1.85**	0.0322	**2.85**	0.0022	**3.85**	0.0001
0.90	0.1841	**1.90**	0.0287	**2.90**	0.0019	**3.90**	0+
0.95	0.1711	**1.95**	0.0256	**2.95**	0.0016	**3.95**	0+

For specific details about using this table to find *p*-values, see pages 376, 377, 379.

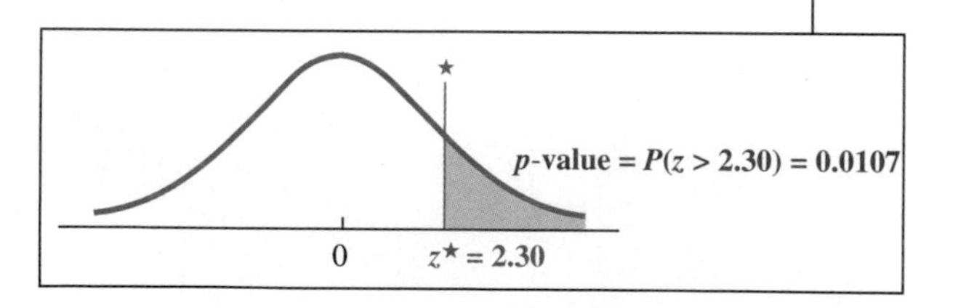

TABLE 6 Critical Values of Student's *t*-Distribution

The entries in this table, $t(\text{df}, \alpha)$, are the critical values for Student's t-distribution for which the area under the curve in the right-hand tail is α. Critical values for the left-hand tail are found by symmetry.

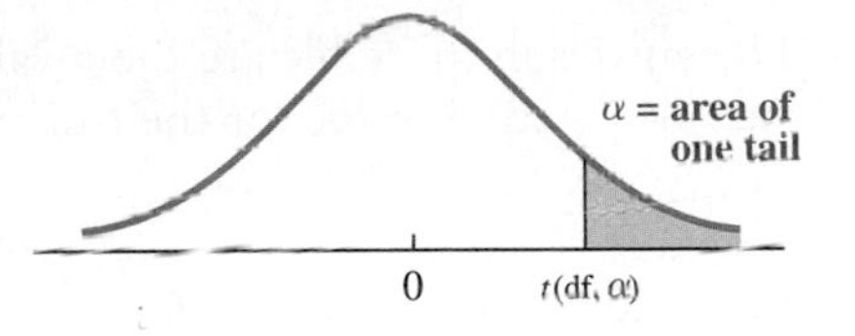

	Amount of α in One Tail					
	0.25	0.10	0.05	0.025	0.01	0.005
	Amount of α in Two Tails					
df	0.50	0.20	0.10	0.05	0.02	0.01
3	0.765	1.64	2.35	3.18	4.54	5.84
4	0.741	1.53	2.13	2.78	3.75	4.60
5	0.729	1.48	2.02	2.57	3.37	4.03
6	0.718	1.44	1.94	2.45	3.14	3.71
7	0.711	1.42	1.89	2.36	3.00	3.50
8	0.706	1.40	1.86	2.31	2.90	3.36
9	0.703	1.38	1.83	2.26	2.82	3.25
10	0.700	1.37	1.81	2.23	2.76	3.17
11	0.697	1.36	1.80	2.20	2.72	3.11
12	0.696	1.36	1.78	2.18	2.68	3.05
13	0.694	1.35	1.77	2.16	2.65	3.01
14	0.692	1.35	1.76	2.14	2.62	2.98
15	0.691	1.34	1.75	2.13	2.60	2.95
16	0.690	1.34	1.75	2.12	2.58	2.92
17	0.689	1.33	1.74	2.11	2.57	2.90
18	0.688	1.33	1.73	2.10	2.55	2.88
19	0.688	1.33	1.73	2.09	2.54	2.86
20	0.687	1.33	1.72	2.09	2.53	2.85
21	0.686	1.32	1.72	2.08	2.52	2.83
22	0.686	1.32	1.72	2.07	2.51	2.82
23	0.685	1.32	1.71	2.07	2.50	2.81
24	0.685	1.32	1.71	2.06	2.49	2.80
25	0.684	1.32	1.71	2.06	2.49	2.79
26	0.684	1.32	1.71	2.06	2.48	2.78
27	0.684	1.31	1.70	2.05	2.47	2.77
28	0.683	1.31	1.70	2.05	2.47	2.76
29	0.683	1.31	1.70	2.05	2.46	2.76
30	0.683	1.31	1.70	2.04	2.46	2.75
35	0.682	1.31	1.69	2.03	2.44	2.73
40	0.681	1.30	1.68	2.02	2.42	2.70
50	0.679	1.30	1.68	2.01	2.40	2.68
70	0.678	1.29	1.67	1.99	2.38	2.65
100	0.677	1.29	1.66	1.98	2.36	2.63
df > 100	0.675	1.28	1.65	1.96	2.33	2.58

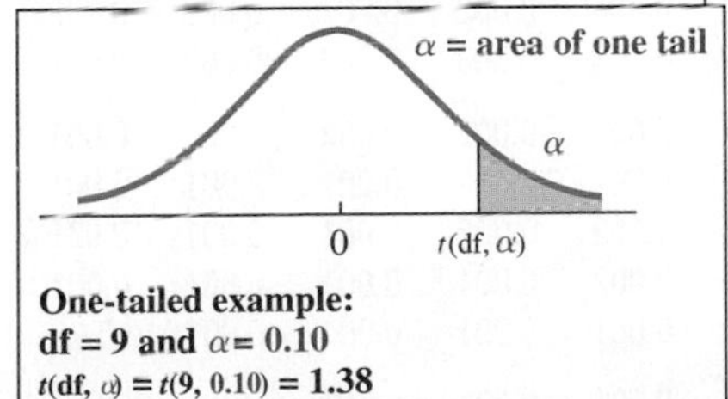

One-tailed example:
df = 9 and α = 0.10
$t(\text{df}, \alpha) = t(9, 0.10) = 1.38$

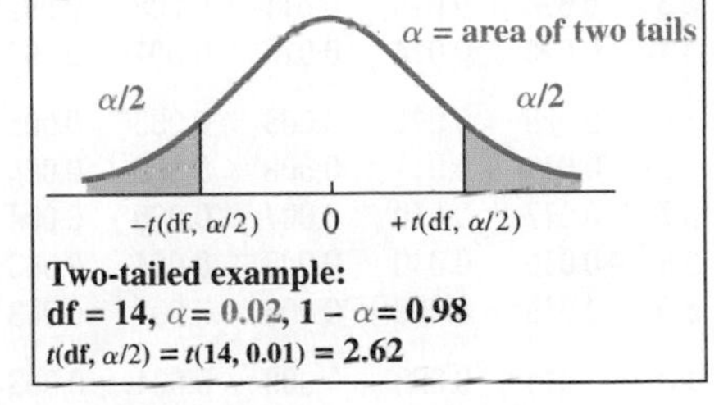

Two-tailed example:
df = 14, α = 0.02, $1 - \alpha$ = 0.98
$t(\text{df}, \alpha/2) = t(14, 0.01) = 2.62$

For specific details about using this table to find: confidence coefficients, see page 418; *p*-values, pages 422, 424; critical values, page 414.

TABLE 7 Probability-Values for Student's t-distribution

The entries in this table are the p-values related to the right-hand tail for the calculated $t★$ value for the t-distribution of df degrees of freedom.

	Degrees of Freedom														
t★	3	4	5	6	7	8	10	12	15	18	21	25	29	35	df ≥ 45
0.0	0.500	0.500	0.500	0.500	0.500	0.500	0.500	0.500	0.500	0.500	0.500	0.500	0.500	0.500	0.500
0.1	0.463	0.463	0.462	0.462	0.462	0.461	0.461	0.461	0.461	0.461	0.461	0.461	0.461	0.460	0.460
0.2	0.427	0.426	0.425	0.424	0.424	0.423	0.423	0.422	0.422	0.422	0.422	0.422	0.421	0.421	0.421
0.3	0.392	0.390	0.388	0.387	0.386	0.386	0.385	0.385	0.384	0.384	0.384	0.383	0.383	0.383	0.383
0.4	0.358	0.355	0.353	0.352	0.351	0.350	0.349	0.348	0.347	0.347	0.347	0.346	0.346	0.346	0.346
0.5	0.326	0.322	0.319	0.317	0.316	0.315	0.314	0.313	0.312	0.312	0.311	0.311	0.310	0.310	0.310
0.6	0.295	0.290	0.287	0.285	0.284	0.283	0.281	0.280	0.279	0.278	0.277	0.277	0.277	0.276	0.276
0.7	0.267	0.261	0.258	0.255	0.253	0.252	0.250	0.249	0.247	0.246	0.246	0.245	0.245	0.244	0.244
0.8	0.241	0.234	0.230	0.227	0.225	0.223	0.221	0.220	0.218	0.217	0.216	0.216	0.215	0.215	0.214
0.9	0.217	0.210	0.205	0.201	0.199	0.197	0.195	0.193	0.191	0.190	0.189	0.188	0.188	0.187	0.186
1.0	0.196	0.187	0.182	0.178	0.175	0.173	0.170	0.169	0.167	0.165	0.164	0.163	0.163	0.162	0.161
1.1	0.176	0.167	0.161	0.157	0.154	0.152	0.149	0.146	0.144	0.143	0.142	0.141	0.140	0.139	0.139
1.2	0.158	0.148	0.142	0.138	0.135	0.132	0.129	0.127	0.124	0.123	0.122	0.121	0.120	0.119	0.118
1.3	0.142	0.132	0.125	0.121	0.117	0.115	0.111	0.109	0.107	0.105	0.104	0.103	0.102	0.101	0.100
1.4	0.128	0.117	0.110	0.106	0.102	0.100	0.096	0.093	0.091	0.089	0.088	0.087	0.086	0.085	0.084
1.5	0.115	0.104	0.097	0.092	0.089	0.086	0.082	0.080	0.077	0.075	0.074	0.073	0.072	0.071	0.070
1.6	0.104	0.092	0.085	0.080	0.077	0.074	0.070	0.068	0.065	0.064	0.062	0.061	0.060	0.059	0.058
1.7	0.094	0.082	0.075	0.070	0.066	0.064	0.060	0.057	0.055	0.053	0.052	0.051	0.050	0.049	0.048
1.8	0.085	0.073	0.066	0.061	0.057	0.055	0.051	0.049	0.046	0.044	0.043	0.042	0.041	0.040	0.039
1.9	0.077	0.065	0.058	0.053	0.050	0.047	0.043	0.041	0.038	0.037	0.036	0.035	0.034	0.033	0.032
2.0	0.070	0.058	0.051	0.046	0.043	0.040	0.037	0.034	0.032	0.030	0.029	0.028	0.027	0.027	0.026
2.1	0.063	0.052	0.045	0.040	0.037	0.034	0.031	0.029	0.027	0.025	0.024	0.023	0.022	0.022	0.021
2.2	0.058	0.046	0.040	0.035	0.032	0.029	0.026	0.024	0.022	0.021	0.020	0.019	0.018	0.017	0.016
2.3	0.052	0.041	0.035	0.031	0.027	0.025	0.022	0.020	0.018	0.017	0.016	0.015	0.014	0.014	0.013
2.4	0.048	0.037	0.031	0.027	0.024	0.022	0.019	0.017	0.015	0.014	0.013	0.012	0.012	0.011	0.010
2.5	0.044	0.033	0.027	0.023	0.020	0.018	0.016	0.014	0.012	0.011	0.010	0.010	0.009	0.009	0.008
2.6	0.040	0.030	0.024	0.020	0.018	0.016	0.013	0.012	0.010	0.009	0.008	0.008	0.007	0.007	0.006
2.7	0.037	0.027	0.021	0.018	0.015	0.014	0.011	0.010	0.008	0.007	0.007	0.006	0.006	0.005	0.005
2.8	0.034	0.024	0.019	0.016	0.013	0.012	0.009	0.008	0.007	0.006	0.005	0.005	0.005	0.004	0.004
2.9	0.031	0.022	0.017	0.014	0.011	0.010	0.008	0.007	0.005	0.005	0.004	0.004	0.004	0.003	0.003
3.0	0.029	0.020	0.015	0.012	0.010	0.009	0.007	0.006	0.004	0.004	0.003	0.003	0.003	0.002	0.002
3.1	0.027	0.018	0.013	0.011	0.009	0.007	0.006	0.005	0.004	0.003	0.003	0.002	0.002	0.002	0.002
3.2	0.025	0.016	0.012	0.009	0.008	0.006	0.005	0.004	0.003	0.002	0.002	0.002	0.002	0.001	0.001
3.3	0.023	0.015	0.011	0.008	0.007	0.005	0.004	0.003	0.002	0.002	0.002	0.001	0.001	0.001	0.001
3.4	0.021	0.014	0.010	0.007	0.006	0.005	0.003	0.003	0.002	0.002	0.001	0.001	0.001	0.001	0.001
3.5	0.020	0.012	0.009	0.006	0.005	0.004	0.003	0.002	0.002	0.001	0.001	0.001	0.001	0.001	0.001
3.6	0.018	0.011	0.008	0.006	0.004	0.004	0.002	0.002	0.001	0.001	0.001	0.001	0.001	0+	0+
3.7	0.017	0.010	0.007	0.005	0.004	0.003	0.002	0.002	0.001	0.001	0.001	0.001	0+	0+	0+
3.8	0.016	0.010	0.006	0.004	0.003	0.003	0.002	0.001	0.001	0.001	0.001	0+	0+	0+	0+
3.9	0.015	0.009	0.006	0.004	0.003	0.002	0.001	0.001	0.001	0.001	0+	0+	0+	0+	0+
4.0	0.014	0.008	0.005	0.004	0.003	0.002	0.001	0.001	0.001	0+	0+	0+	0+	0+	0+

For specific details about using this table to find p-values, see pages 422, 424.

TABLE 8 Critical Values of χ^2 ("Chi-Square") Distribution

The entries in this table, χ^2 (df, α), are the critical values for the χ^2 distribution for which the area under the curve to the right is α.

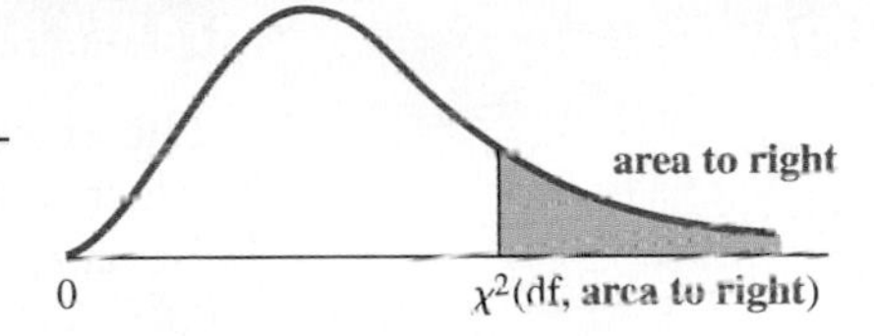

	Area to the Right												
	0.995	0.99	0.975	0.95	0.90	0.75	0.50	0.25	0.10	0.05	0.025	0.01	0.005
	Area in Left-hand Tail						Median		Area in Right-hand Tail				
df	0.005	0.01	0.025	0.05	0.10	0.25	0.50	0.25	0.10	0.05	0.025	0.01	0.005
1	0.0000393	0.000157	0.000982	0.00393	0.0158	0.101	0.455	1.32	2.71	3.84	5.02	6.63	7.88
2	0.0100	0.0201	0.0506	0.103	0.211	0.575	1.39	2.77	4.61	5.99	7.38	9.21	10.6
3	0.0717	0.115	0.216	0.352	0.584	1.21	2.37	4.11	6.25	7.82	9.35	11.3	12.8
4	0.207	0.297	0.484	0.711	1.06	1.92	3.36	5.39	7.78	9.49	11.1	13.3	14.9
5	0.412	0.554	0.831	1.15	1.61	2.67	4.35	6.63	9.24	11.1	12.8	15.1	16.8
6	0.676	0.872	1.24	1.64	2.20	3.45	5.35	7.84	10.6	12.6	14.5	16.8	18.6
7	0.990	1.24	1.69	2.17	2.83	4.25	6.35	9.04	12.0	14.1	16.0	18.5	20.3
8	1.34	1.65	2.18	2.73	3.49	5.07	7.34	10.2	13.4	15.5	17.5	20.1	22.0
9	1.73	2.09	2.70	3.33	4.17	5.90	8.04	11.4	14.7	16.9	19.0	21.7	23.6
10	2.16	2.56	3.25	3.94	4.87	6.74	9.34	12.5	16.0	18.3	20.5	23.2	25.2
11	2.60	3.05	3.82	4.57	5.58	7.58	10.34	13.7	17.3	19.7	21.9	24.7	26.8
12	3.07	3.57	4.40	5.23	6.30	8.44	11.34	14.8	18.5	21.0	23.3	26.2	28.3
13	3.57	4.11	5.01	5.89	7.04	9.30	12.34	16.0	19.8	22.4	24.7	27.7	29.8
14	4.07	4.66	5.63	6.57	7.79	10.2	13.34	17.1	21.1	23.7	26.1	29.1	31.3
15	4.60	5.23	6.26	7.26	8.55	11.0	14.34	18.2	22.3	25.0	27.5	30.6	32.8
16	5.14	5.81	6.91	7.96	9.31	11.9	15.34	19.4	23.5	26.3	28.8	32.0	34.3
17	5.70	6.41	7.56	8.67	10.1	12.8	16.34	20.5	24.8	27.6	30.2	33.4	35.7
18	6.26	7.01	8.23	9.39	10.9	13.7	17.34	21.6	26.0	28.9	31.5	34.8	37.2
19	6.84	7.63	8.91	10.1	11.7	14.6	18.34	22.7	27.2	30.1	32.9	36.2	38.6
20	7.43	8.26	9.59	10.9	12.4	15.5	19.34	23.8	28.4	31.4	34.2	37.6	40.0
21	8.03	8.90	10.3	11.6	13.2	16.3	20.34	24.9	29.6	32.7	35.5	38.9	41.4
22	8.64	9.54	11.0	12.3	14.0	17.2	21.34	26.0	30.8	33.9	36.8	40.3	42.8
23	9.26	10.2	11.7	13.1	14.8	18.1	22.34	27.1	32.0	35.2	38.1	41.6	44.2
24	9.89	10.9	12.4	13.8	15.7	19.0	23.34	28.2	33.2	36.4	39.4	43.0	45.6
25	10.5	11.5	13.1	14.6	16.5	19.9	24.34	29.3	34.4	37.7	40.6	44.3	46.9
26	11.2	12.2	13.8	15.4	17.3	20.8	25.34	30.4	35.6	38.9	41.9	45.6	48.3
27	11.8	12.9	14.6	16.2	18.1	21.7	26.34	31.5	36.7	40.1	43.2	47.0	49.6
28	12.5	13.6	15.3	16.9	18.9	22.7	27.34	32.6	37.9	41.3	44.5	48.3	51.0
29	13.1	14.3	16.0	17.7	19.8	23.6	28.34	33.7	39.1	42.6	45.7	49.6	52.3
30	13.8	15.0	16.8	18.5	20.6	24.5	29.34	34.8	40.3	43.8	47.0	50.9	53.7
40	20.7	22.2	24.4	26.5	29.1	33.7	39.34	45.6	51.8	55.8	59.3	63.7	66.8
50	28.0	29.7	32.4	34.8	37.7	42.9	49.33	56.3	63.2	67.5	71.4	76.2	79.5
60	35.5	37.5	40.5	43.2	46.5	52.3	59.33	67.0	74.4	79.1	83.3	88.4	92.0
70	43.3	45.4	48.8	51.7	55.3	61.7	69.33	77.6	85.5	90.5	95.0	100.0	104.0
80	51.2	53.5	57.2	60.4	64.3	71.1	79.33	88.1	96.6	102.0	107.0	112.0	116.0
90	59.2	61.8	65.6	69.1	73.3	80.6	89.33	98.6	108.0	113.0	118.0	124.0	128.0
100	67.3	70.1	74.2	77.9	82.4	90.1	99.33	109.0	118.0	124.0	130.0	136.0	140.0

Left-tail example:
Find χ^2 with df = 28; area in left-tail = 0.10.

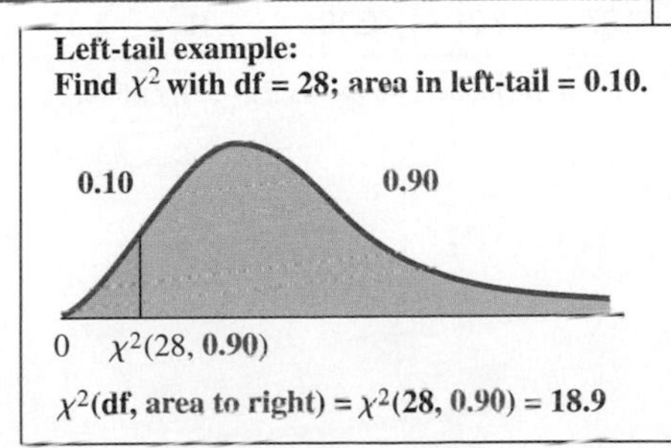

χ^2(df, area to right) = χ^2(28, 0.90) = 18.9

Right-tail example:
Find χ^2 with df = 23; area in right-tail = 0.025

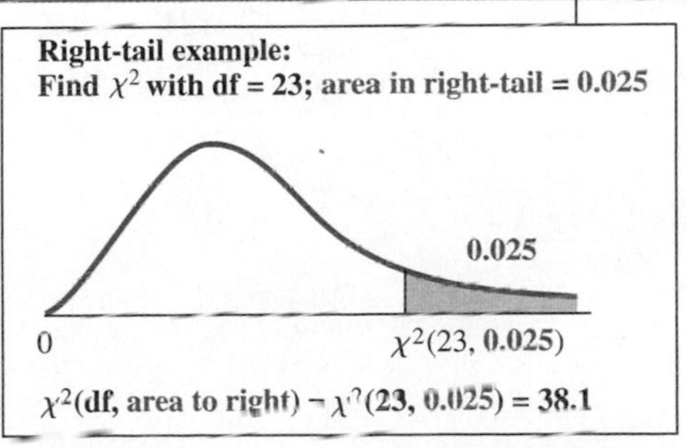

χ^2(df, area to right) = χ^2(23, 0.025) = 38.1

For specific details about using this table to find: p-values, see page 455; critical values, page 450.

TABLE 9A Critical Values of the F Distribution ($\alpha = 0.05$)

The entries in this table are critical values of F for which the area under the curve to the right is equal to 0.05.

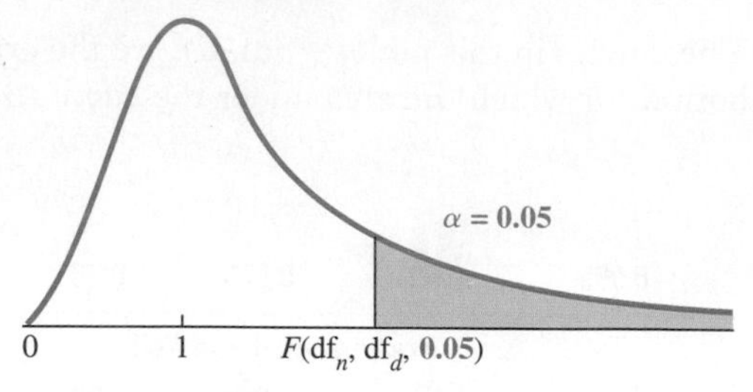

Degrees of Freedom for Numerator

Degrees of Freedom for Denominator	1	2	3	4	5	6	7	8	9	10
1	161.	200.	216.	225.	230.	234.	237.	239.	241.	242.
2	18.5	19.0	19.2	19.2	19.3	19.3	19.4	19.4	19.4	19.4
3	10.1	9.55	9.28	9.12	9.01	8.94	8.89	8.85	8.81	8.79
4	7.71	6.94	6.59	6.39	6.26	6.16	6.09	6.04	6.00	5.96
5	6.61	5.79	5.41	5.19	5.05	4.95	4.88	4.82	4.77	4.74
6	5.99	5.14	4.76	4.53	4.39	4.28	4.21	4.15	4.10	4.06
7	5.59	4.74	4.35	4.12	3.97	3.87	3.79	3.73	3.68	3.64
8	5.32	4.46	4.07	3.84	3.69	3.58	3.50	3.44	3.39	3.35
9	5.12	4.26	3.86	3.63	3.48	3.37	3.29	3.23	3.18	3.14
10	4.96	4.10	3.71	3.48	3.33	3.22	3.14	3.07	3.02	2.98
11	4.84	3.98	3.59	3.36	3.20	3.09	3.01	2.95	2.90	2.85
12	4.75	3.89	3.49	3.26	3.11	3.00	2.91	2.85	2.80	2.75
13	4.67	3.81	3.41	3.18	3.03	2.92	2.83	2.77	2.71	2.67
14	4.60	3.74	3.34	3.11	2.96	2.85	2.76	2.70	2.65	2.60
15	4.54	3.68	3.29	3.06	2.90	2.79	2.71	2.64	2.59	2.54
16	4.49	3.63	3.24	3.01	2.85	2.74	2.66	2.59	2.54	2.49
17	4.45	3.59	3.20	2.96	2.81	2.70	2.61	2.55	2.49	2.45
18	4.41	3.55	3.16	2.93	2.77	2.66	2.58	2.51	2.46	2.41
19	4.38	3.52	3.13	2.90	2.74	2.63	2.54	2.48	2.42	2.38
20	4.35	3.49	3.10	2.87	2.71	2.60	2.51	2.45	2.39	2.35
21	4.32	3.47	3.07	2.84	2.68	2.57	2.49	2.42	2.37	2.32
22	4.30	3.44	3.05	2.82	2.66	2.55	2.46	2.40	2.34	2.30
23	4.28	3.42	3.03	2.80	2.64	2.53	2.44	2.37	2.32	2.27
24	4.26	3.40	3.01	2.78	2.62	2.51	2.42	2.36	2.30	2.25
25	4.24	3.39	2.99	2.76	2.60	2.49	2.40	2.34	2.28	2.24
30	4.17	3.32	2.92	2.69	2.53	2.42	2.33	2.27	2.21	2.16
40	4.08	3.23	2.84	2.61	2.45	2.34	2.25	2.18	2.12	2.08
60	4.00	3.15	2.76	2.53	2.37	2.25	2.17	2.10	2.04	1.99
120	3.92	3.07	2.68	2.45	2.29	2.18	2.09	2.02	1.96	1.91
∞	3.84	3.00	2.60	2.37	2.21	2.10	2.01	1.94	1.88	1.83

For specific details about using this table to find: *p*-values, see pages 520–521; critical values, page 517.

TABLE 9A (Continued)

Degrees of Freedom for Numerator

Degrees of Freedom for Denominator	12	15	20	24	30	40	60	120	∞
1	244.	246.	248.	249.	250.	251.	252.	253.	254.
2	19.4	19.4	19.4	19.5	19.5	19.5	19.5	19.5	19.5
3	8.74	8.70	8.66	8.64	8.62	8.59	8.57	8.55	8.53
4	5.91	5.86	5.80	5.77	5.75	5.72	5.69	5.66	5.63
5	4.68	4.62	4.56	4.53	4.50	4.46	4.43	4.40	4.37
6	4.00	3.94	3.87	3.84	3.81	3.77	3.74	3.70	3.67
7	3.57	3.51	3.44	3.41	3.38	3.34	3.30	3.27	3.23
8	3.28	3.22	3.15	3.12	3.08	3.04	3.01	2.97	2.93
9	3.07	3.01	2.94	2.90	2.86	2.83	2.79	2.75	2.71
10	2.91	2.85	2.77	2.74	2.70	2.66	2.62	2.58	2.54
11	2.79	2.72	2.65	2.61	2.57	2.53	2.49	2.45	2.40
12	2.69	2.62	2.54	2.51	2.47	2.43	2.38	2.34	2.30
13	2.60	2.53	2.46	2.42	2.38	2.34	2.30	2.25	2.21
14	2.53	2.46	2.39	2.35	2.31	2.27	2.22	2.18	2.13
15	2.48	2.40	2.33	2.29	2.25	2.20	2.16	2.11	2.07
16	2.42	2.35	2.28	2.24	2.19	2.15	2.11	2.06	2.01
17	2.38	2.31	2.23	2.19	2.15	2.10	2.06	2.01	1.96
18	2.34	2.27	2.19	2.15	2.11	2.06	2.02	1.97	1.92
19	2.31	2.23	2.16	2.11	2.07	2.03	1.98	1.93	1.88
20	2.28	2.20	2.12	2.08	2.04	1.99	1.95	1.90	1.84
21	2.25	2.18	2.10	2.05	2.01	1.96	1.92	1.87	1.81
22	2.23	2.15	2.07	2.03	1.98	1.94	1.89	1.84	1.78
23	2.20	2.13	2.05	2.01	1.96	1.91	1.86	1.81	1.76
24	2.18	2.11	2.03	1.98	1.94	1.89	1.84	1.79	1.73
25	2.16	2.09	2.01	1.96	1.92	1.87	1.82	1.77	1.71
30	2.09	2.01	1.93	1.89	1.84	1.79	1.74	1.68	1.62
40	2.00	1.92	1.84	1.79	1.74	1.69	1.64	1.58	1.51
60	1.92	1.84	1.75	1.70	1.65	1.59	1.53	1.47	1.39
120	1.83	1.75	1.66	1.61	1.55	1.50	1.43	1.35	1.25
∞	1.75	1.67	1.57	1.52	1.46	1.39	1.32	1.22	1.00

TABLE 9B Critical Values of the *F* Distribution ($\alpha = 0.025$)

The entries in this table are critical values of *F* for which the area under the curve to the right is equal to 0.025.

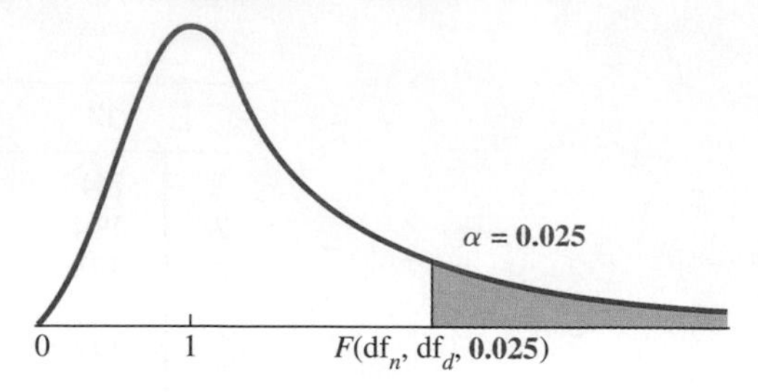

Degrees of Freedom for Numerator

Degrees of Freedom for Denominator	1	2	3	4	5	6	7	8	9	10
1	648.	800.	864.	900.	922.	937.	948.	957.	963.	969.
2	38.5	39.0	39.2	39.2	39.3	39.3	39.4	39.4	39.4	39.4
3	17.4	16.0	15.4	15.1	14.9	14.7	14.6	14.5	14.5	14.4
4	12.2	10.6	9.98	9.60	9.36	9.20	9.07	8.98	8.90	8.84
5	10.0	8.43	7.76	7.39	7.15	6.98	6.85	6.76	6.68	6.62
6	8.81	7.26	6.60	6.23	5.99	5.82	5.70	5.60	5.52	5.46
7	8.07	6.54	5.89	5.52	5.29	5.12	4.99	4.90	4.82	4.76
8	7.57	6.06	5.42	5.05	4.82	4.65	4.53	4.43	4.36	4.30
9	7.21	5.71	5.08	4.72	4.48	4.32	4.20	4.10	4.03	3.96
10	6.94	5.46	4.83	4.47	4.24	4.07	3.95	3.85	3.78	3.72
11	6.72	5.26	4.63	4.28	4.04	3.88	3.76	3.66	3.59	3.53
12	6.55	5.10	4.47	4.12	3.89	3.73	3.61	3.51	3.44	3.37
13	6.41	4.97	4.35	4.00	3.77	3.60	3.48	3.39	3.31	3.25
14	6.30	4.86	4.24	3.89	3.66	3.50	3.38	3.28	3.21	3.15
15	6.20	4.77	4.15	3.80	3.58	3.41	3.29	3.20	3.12	3.06
16	6.12	4.69	4.08	3.73	3.50	3.34	3.22	3.12	3.05	2.99
17	6.04	4.62	4.01	3.66	3.44	3.28	3.16	3.06	2.98	2.92
18	5.98	4.56	3.95	3.61	3.38	3.22	3.10	3.01	2.93	2.87
19	5.92	4.51	3.90	3.56	3.33	3.17	3.05	2.96	2.88	2.82
20	5.87	4.46	3.86	3.51	3.29	3.13	3.01	2.91	2.84	2.77
21	5.83	4.42	3.82	3.48	3.25	3.09	2.97	2.87	2.80	2.73
22	5.79	4.38	3.78	3.44	3.22	3.05	2.93	2.84	2.76	2.70
23	5.75	4.35	3.75	3.41	3.18	3.02	2.90	2.81	2.73	2.67
24	5.72	4.32	3.72	3.38	3.15	2.99	2.87	2.78	2.70	2.64
25	5.69	4.29	3.69	3.35	3.13	2.97	2.85	2.75	2.68	2.61
30	5.57	4.18	3.59	3.25	3.03	2.87	2.75	2.65	2.57	2.51
40	5.42	4.05	3.46	3.13	2.90	2.74	2.62	2.53	2.45	2.39
60	5.29	3.93	3.34	3.01	2.79	2.63	2.51	2.41	2.33	2.27
120	5.15	3.80	3.23	2.89	2.67	2.52	2.39	2.30	2.22	2.16
∞	5.02	3.69	3.12	2.79	2.57	2.41	2.29	2.19	2.11	2.05

For specific details about using this table to find: *p*-values, see pages 520–521; critical values, page 517.

TABLE 9B (Continued)

Degrees of Freedom for Numerator

Degrees of Freedom for Denominator	12	15	20	24	30	40	60	120	∞
1	977.	985.	993.	997.	1001.	1006.	1010.	1014.	1018.
2	39.4	39.4	39.4	39.5	39.5	39.5	39.5	39.5	39.5
3	14.3	14.3	14.2	14.1	14.1	14.0	14.0	13.9	13.9
4	8.75	8.66	8.56	8.51	8.46	8.41	8.36	8.31	8.26
5	6.52	6.43	6.33	6.28	6.23	6.18	6.12	6.07	6.02
6	5.37	5.27	5.17	5.12	5.07	5.01	4.96	4.90	4.85
7	4.67	4.57	4.47	4.42	4.36	4.31	4.25	4.20	4.14
8	4.20	4.10	4.00	3.95	3.89	3.84	3.78	3.73	3.67
9	3.87	3.77	3.67	3.61	3.56	3.51	3.45	3.39	3.33
10	3.62	3.52	3.42	3.37	3.31	3.26	3.20	3.14	3.08
11	3.43	3.33	3.23	3.17	3.12	3.06	3.00	2.94	2.88
12	3.28	3.18	3.07	3.02	2.96	2.91	2.85	2.79	2.72
13	3.15	3.05	2.95	2.89	2.84	2.78	2.72	2.66	2.60
14	3.05	2.95	2.84	2.79	2.73	2.67	2.61	2.55	2.49
15	2.96	2.86	2.76	2.70	2.64	2.59	2.52	2.46	2.40
16	2.89	2.79	2.68	2.63	2.57	2.51	2.45	2.38	2.32
17	2.82	2.72	2.62	2.56	2.50	2.44	2.38	2.32	2.25
18	2.77	2.67	2.56	2.50	2.44	2.38	2.32	2.26	2.19
19	2.72	2.62	2.51	2.45	2.39	2.33	2.27	2.20	2.13
20	2.68	2.57	2.46	2.41	2.35	2.29	2.22	2.16	2.09
21	2.64	2.53	2.42	2.37	2.31	2.25	2.18	2.11	2.04
22	2.60	2.50	2.39	2.33	2.27	2.21	2.14	2.08	2.00
23	2.57	2.47	2.36	2.30	2.24	2.18	2.11	2.04	1.97
24	2.54	2.44	2.33	2.27	2.21	2.15	2.08	2.01	1.94
25	2.51	2.41	2.30	2.24	2.18	2.12	2.05	1.98	1.91
30	2.41	2.31	2.20	2.14	2.07	2.01	1.94	1.87	1.79
40	2.29	2.18	2.07	2.01	1.94	1.88	1.80	1.72	1.64
60	2.17	2.06	1.94	1.88	1.82	1.74	1.67	1.58	1.48
120	2.05	1.95	1.82	1.76	1.69	1.61	1.53	1.43	1.31
∞	1.94	1.83	1.71	1.64	1.57	1.48	1.39	1.27	1.00

TABLE 9C Critical Values of the F Distribution ($\alpha = 0.01$)

The entries in the table are critical values of F for which the area under the curve to the right is equal to 0.01

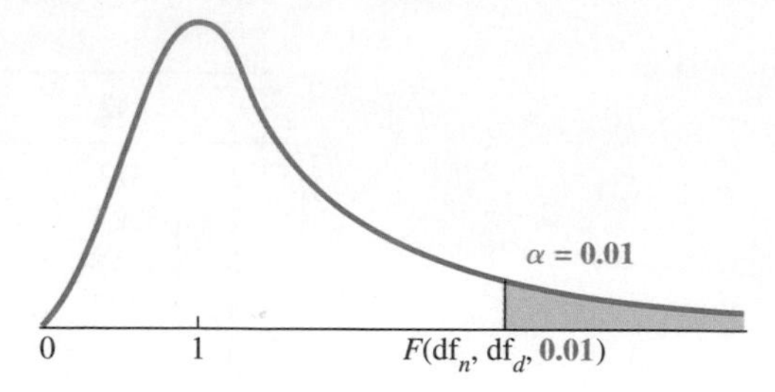

Degrees of Freedom for Numerator

Degrees of Freedom for Denominator	1	2	3	4	5	6	7	8	9	10
1	4052.	5000.	5403.	5625.	5764.	5859.	5928.	5982.	6024.	6056.
2	98.5	99.0	99.2	99.2	99.3	99.3	99.4	99.4	99.4	99.4
3	34.1	30.8	29.5	28.7	28.2	27.9	27.7	27.5	27.3	27.2
4	21.2	18.0	16.7	16.0	15.5	15.2	15.0	14.8	14.7	14.5
5	16.3	13.3	12.1	11.4	11.0	10.7	10.5	10.3	10.2	10.1
6	13.7	10.9	9.78	9.15	8.75	8.47	8.26	8.10	7.98	7.87
7	12.2	9.55	8.45	7.85	7.46	7.19	6.99	6.84	6.72	6.62
8	11.3	8.65	7.59	7.01	6.63	6.37	6.18	6.03	5.91	5.81
9	10.6	8.02	6.99	6.42	6.06	5.80	5.61	5.47	5.35	5.26
10	10.0	7.56	6.55	5.99	5.64	5.39	5.20	5.06	4.94	4.85
11	9.65	7.21	6.22	5.67	5.32	5.07	4.89	4.74	4.63	4.54
12	9.33	6.93	5.95	5.41	5.06	4.82	4.64	4.50	4.39	4.30
13	9.07	6.70	5.74	5.21	4.86	4.62	4.44	4.30	4.19	4.10
14	8.86	6.51	5.56	5.04	4.70	4.46	4.28	4.14	4.03	3.94
15	8.68	6.36	5.42	4.89	4.56	4.32	4.14	4.00	3.89	3.80
16	8.53	6.23	5.29	4.77	4.44	4.20	4.03	3.89	3.78	3.69
17	8.40	6.11	5.19	4.67	4.34	4.10	3.93	3.79	3.68	3.59
18	8.29	6.01	5.09	4.58	4.25	4.01	3.84	3.71	3.60	3.51
19	8.19	5.93	5.01	4.50	4.17	3.94	3.77	3.63	3.52	3.43
20	8.10	5.85	4.94	4.43	4.10	3.87	3.70	3.56	3.46	3.37
21	8.02	5.78	4.87	4.37	4.04	3.81	3.64	3.51	3.40	3.31
22	7.95	5.72	4.82	4.31	3.99	3.76	3.59	3.45	3.35	3.26
23	7.88	5.66	4.76	4.26	3.94	3.71	3.54	3.41	3.30	3.21
24	7.82	5.61	4.72	4.22	3.90	3.67	3.50	3.36	3.26	3.17
25	7.77	5.57	4.68	4.18	3.86	3.63	3.46	3.32	3.22	3.13
30	7.56	5.39	4.51	4.02	3.70	3.47	3.30	3.17	3.07	2.98
40	7.31	5.18	4.31	3.83	3.51	3.29	3.12	2.99	2.89	2.80
60	7.08	4.98	4.13	3.65	3.34	3.12	2.95	2.82	2.72	2.63
120	6.85	4.79	3.95	3.48	3.17	2.96	2.79	2.66	2.56	2.47
∞	6.63	4.61	3.78	3.32	3.02	2.80	2.64	2.51	2.41	2.32

For specific details about using this table to find: p-values, see pages 520–521; critical values, page 517.

TABLE 9C (Continued)

Degrees of Freedom for Denominator	Degrees of Freedom for Numerator								
	12	15	20	24	30	40	60	120	∞
1	6106.	6157.	6209.	6235.	6261.	6287.	6313.	6339.	6366.
2	99.4	99.4	99.4	99.5	99.5	99.5	99.5	99.5	99.5
3	27.1	26.9	26.7	26.6	26.5	26.4	26.3	26.2	26.1
4	14.4	14.2	14.0	13.9	13.8	13.7	13.7	13.6	13.5
5	9.89	9.72	9.55	9.47	9.38	9.29	9.20	9.11	9.02
6	7.72	7.56	7.40	7.31	7.23	7.14	7.06	6.97	6.88
7	6.47	6.31	6.16	6.07	5.99	5.91	5.82	5.74	5.65
8	5.67	5.52	5.36	5.28	5.20	5.12	5.03	4.95	4.86
9	5.11	4.96	4.81	4.73	4.65	4.57	4.48	4.40	4.31
10	4.71	4.56	4.41	4.33	4.25	4.17	4.08	4.00	3.91
11	4.40	4.25	4.10	4.02	3.94	3.86	3.78	3.69	3.60
12	4.16	4.01	3.86	3.78	3.70	3.62	3.54	3.45	3.36
13	3.96	3.82	3.66	3.59	3.51	3.43	3.34	3.25	3.17
14	3.80	3.66	3.51	3.43	3.35	3.27	3.18	3.09	3.00
15	3.67	3.52	3.37	3.29	3.21	3.13	3.05	2.96	2.87
16	3.55	3.41	3.26	3.18	3.10	3.02	2.93	2.84	2.75
17	3.46	3.31	3.16	3.08	3.00	2.92	2.83	2.75	2.65
18	3.37	3.23	3.08	3.00	2.92	2.84	2.75	2.66	2.57
19	3.30	3.15	3.00	2.92	2.84	2.76	2.67	2.58	2.49
20	3.23	3.09	2.94	2.86	2.78	2.69	2.61	2.52	2.42
21	3.17	3.03	2.88	2.80	2.72	2.64	2.55	2.46	2.36
22	3.12	2.98	2.83	2.75	2.67	2.58	2.50	2.40	2.31
23	3.07	2.93	2.78	2.70	2.62	2.54	2.45	2.35	2.26
24	3.03	2.89	2.74	2.66	2.58	2.49	2.40	2.31	2.21
25	2.99	2.85	2.70	2.62	2.53	2.45	2.36	2.27	2.17
30	2.84	2.70	2.55	2.47	2.39	2.30	2.21	2.11	2.01
40	2.66	2.52	2.37	2.29	2.20	2.11	2.02	1.92	1.80
60	2.50	2.35	2.20	2.12	2.03	1.94	1.84	1.73	1.60
120	2.34	2.19	2.03	1.95	1.86	1.76	1.66	1.53	1.38
∞	2.18	2.04	1.88	1.79	1.70	1.59	1.47	1.32	1.00

From E. S. Pearson and H. O. Hartley, *Biometrika Tables for Statisticians*, vol. I (1958), pp. 159-163. Reprinted by permission of the Biometrika Trustees.

TABLE 10 Confidence Belts for the Correlation Coefficient $(1 - \alpha) = 0.95$

The numbers on the curves are sample sizes.

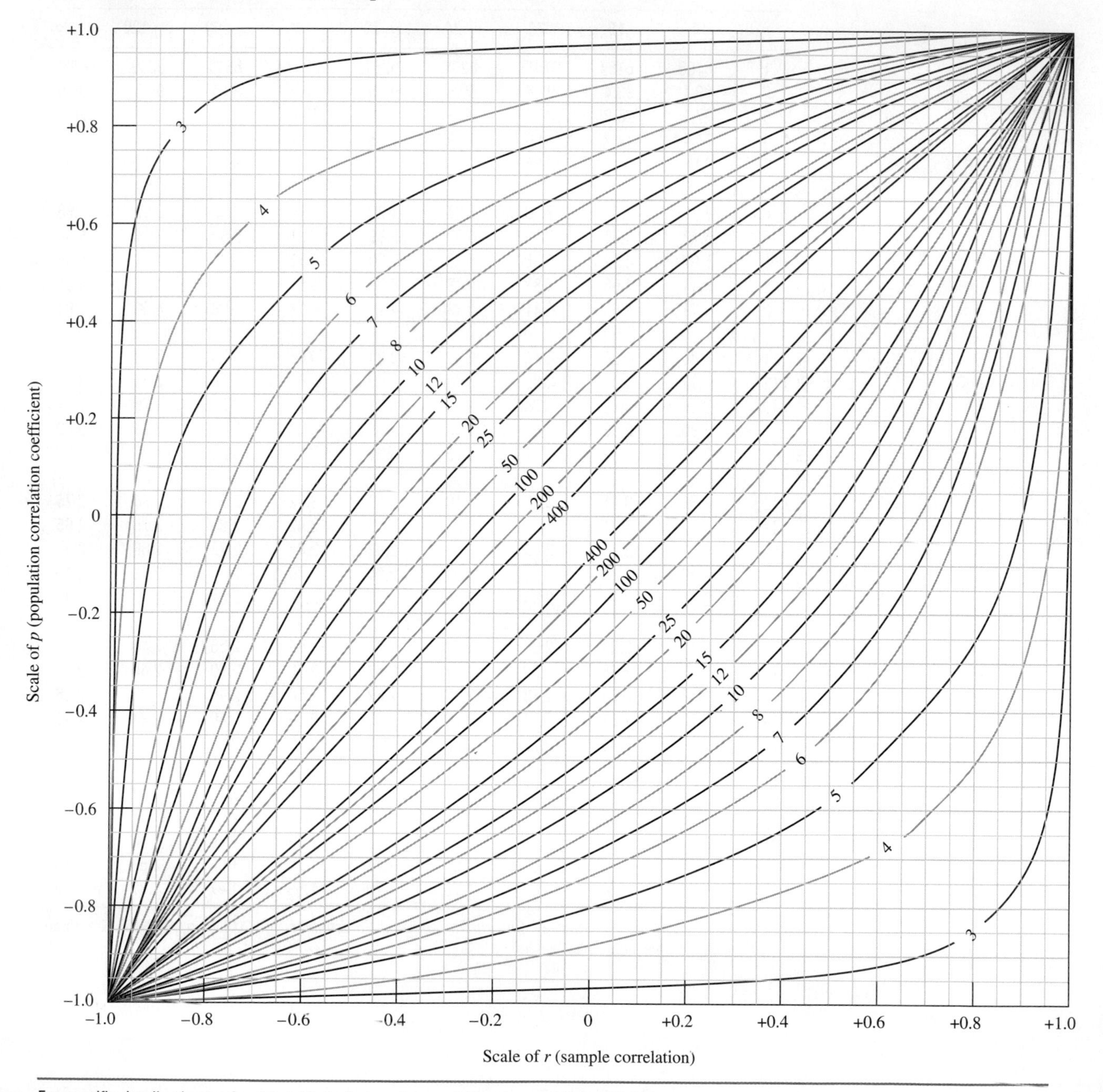

For specific details about using this table to find confidence intervals, see pages 612–613.

TABLE 11 Critical Values of r When $\rho = 0$

The entries in this table are the critical values of r for a two-tailed test at α. For simple correlation, df $= n - 2$, where n is the number of pairs of data in the sample. For a one-tailed test, the value of α shown at the top of the table is double the value of α being used in the hypothesis test.

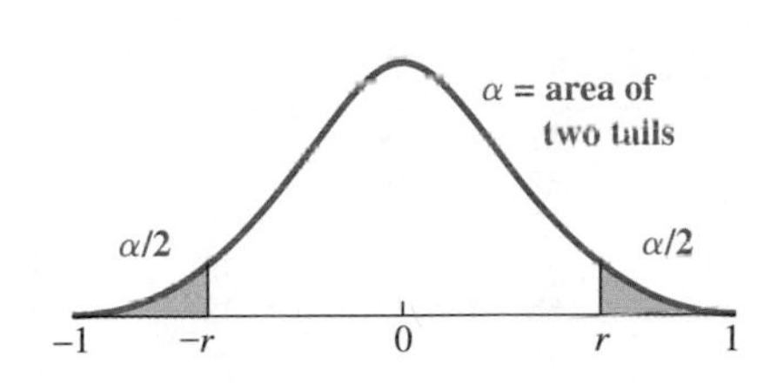

df \ α	0.10	0.05	0.02	0.01
1	0.988	0.997	1.000	1.000
2	0.900	0.950	0.980	0.990
3	0.805	0.878	0.934	0.959
4	0.729	0.811	0.882	0.917
5	0.669	0.754	0.833	0.874
6	0.621	0.707	0.789	0.834
7	0.582	0.666	0.750	0.798
8	0.549	0.632	0.716	0.765
9	0.521	0.602	0.685	0.735
10	0.497	0.576	0.658	0.708
11	0.476	0.553	0.634	0.684
12	0.458	0.532	0.612	0.661
13	0.441	0.514	0.592	0.641
14	0.426	0.497	0.574	0.623
15	0.412	0.482	0.558	0.606
16	0.400	0.468	0.542	0.590
17	0.389	0.456	0.528	0.575
18	0.378	0.444	0.516	0.561
19	0.369	0.433	0.503	0.549
20	0.360	0.423	0.492	0.537
25	0.323	0.381	0.445	0.487
30	0.296	0.349	0.409	0.449
35	0.275	0.325	0.381	0.418
40	0.257	0.304	0.358	0.393
45	0.243	0.288	0.338	0.372
50	0.231	0.273	0.322	0.354
60	0.211	0.250	0.295	0.325
70	0.195	0.232	0.274	0.302
80	0.183	0.217	0.256	0.283
90	0.173	0.205	0.242	0.267
100	0.164	0.195	0.230	0.254

From E. S. Pearson and H. O. Hartley, *Biometrika Tables for Statisticians,* vol. 1 (1962), p. 138. Reprinted by permission of the Biometrika Trustees.

For specific details about using this table to find: p-values, see page 615; critical values, page 615.

TABLE 12 Critical Values of the Sign Test

The entries in this table are the critical values for the number of the least frequent sign for a two-tailed test at α for the binomial $p = 0.5$. For a one-tailed test, the value of α shown at the top of the table is double the value of α being used in the hypothesis test.

	α					α			
n	0.01	0.05	0.10	0.25	n	0.01	0.05	0.10	0.25
1					51	15	18	19	20
2					52	16	18	19	21
3				0	53	16	18	20	21
4				0	54	17	19	20	22
5			0	0	55	17	19	20	22
6		0	0	1	56	17	20	21	23
7		0	0	1	57	18	20	21	23
8	0	0	1	1	58	18	21	22	24
9	0	1	1	2	59	19	21	22	24
10	0	1	1	2	60	19	21	23	25
11	0	1	2	3	61	20	22	23	25
12	1	2	2	3	62	20	22	24	25
13	1	2	3	3	63	20	23	24	26
14	1	2	3	4	64	21	23	24	26
15	2	3	3	4	65	21	24	25	27
16	2	3	4	5	66	22	24	25	27
17	2	4	4	5	67	22	25	26	28
18	3	4	5	6	68	22	25	26	28
19	3	4	5	6	69	23	25	27	29
20	3	5	5	6	70	23	26	27	29
21	4	5	6	7	71	24	26	28	30
22	4	5	6	7	72	24	27	28	30
23	4	6	7	8	73	25	27	28	31
24	5	6	7	8	74	25	28	29	31
25	5	7	7	9	75	25	28	29	32
26	6	7	8	9	76	26	28	30	32
27	6	7	8	10	77	26	29	30	32
28	6	8	9	10	78	27	29	31	33
29	7	8	9	10	79	27	30	31	33
30	7	9	10	11	80	28	30	32	34
31	7	9	10	11	81	28	31	32	34
32	8	9	10	12	82	28	31	33	35
33	8	10	11	12	83	29	32	33	35
34	9	10	11	13	84	29	32	33	36
35	9	11	12	13	85	30	32	34	36
36	9	11	12	14	86	30	33	34	37
37	10	12	13	14	87	31	33	35	37
38	10	12	13	14	88	31	34	35	38
39	11	12	13	15	89	31	34	36	38
40	11	13	14	15	90	32	35	36	39
41	11	13	14	16	91	32	35	37	39
42	12	14	15	16	92	33	36	37	39
43	12	14	15	17	93	33	36	38	40
44	13	15	16	17	94	34	37	38	40
45	13	15	16	18	95	34	37	38	41
46	13	15	16	18	96	34	37	39	41
47	14	16	17	19	97	35	38	39	42
48	14	16	17	19	98	35	38	40	42
49	15	17	18	19	99	36	39	40	43
50	15	17	18	20	100	36	39	41	43

From Wilfred J. Dixon and Frank J. Massey, Jr., *Introduction to Statistical Analysis*, 3d ed. (New York: McGraw-Hill, 1969), p. 509. Reprinted by permission.

For specific details about using this table to find: *p*-values, see pages 660–661; critical values, page 658.

TABLE 13 Critical Values of U in the Mann-Whitney Test

A. The entries are the critical values of U for a one-tailed test at 0.025 or for a two-tailed test at 0.05.

n_2 \ n_1	1	2	3	4	5	6	7	8	9	10	11	12	13	14	15	16	17	18	19	20
1																				
2								0	0	0	0	1	1	1	1	1	2	2	2	2
3					0	1	1	2	2	3	3	4	4	5	5	6	6	7	7	8
4				0	1	2	3	4	4	5	6	7	8	9	10	11	11	12	13	13
5			0	1	2	3	5	6	7	8	9	11	12	13	14	15	17	18	19	20
6			1	2	3	5	6	8	10	11	13	14	16	17	19	21	22	24	25	27
7			1	3	5	6	8	10	12	14	16	18	20	22	24	26	28	30	32	34
8		0	2	4	6	8	10	13	15	17	19	22	24	26	29	31	34	36	38	41
9		0	2	4	7	10	12	15	17	20	23	26	28	31	34	37	39	42	45	48
10		0	3	5	8	11	14	17	20	23	26	29	33	36	39	42	45	48	52	55
11		0	3	6	9	13	16	19	23	26	30	33	37	40	44	47	51	55	58	62
12		1	4	7	11	14	18	22	26	29	33	37	41	45	49	53	57	61	65	69
13		1	4	8	12	16	20	24	28	33	37	41	45	50	54	59	63	67	72	70
14		1	5	9	13	17	22	26	31	36	40	45	50	55	59	64	67	74	78	83
15		1	5	10	14	19	24	29	34	39	44	49	54	59	64	70	75	80	85	90
16		1	6	11	15	21	26	31	37	42	47	53	59	64	70	75	81	86	92	98
17		2	6	11	17	22	28	34	39	45	51	57	63	67	75	81	87	93	99	105
18		2	7	12	18	24	30	36	42	48	55	61	67	74	80	86	93	99	106	112
19		2	7	13	19	25	32	38	45	52	58	65	72	78	85	92	99	106	113	119
20		2	8	13	20	27	34	41	48	55	62	69	76	83	90	98	105	112	119	127

B. The entries are the critical values of U for a one-tailed test at 0.05 or for a two-tailed test at 0.10.

n_2 \ n_1	1	2	3	4	5	6	7	8	9	10	11	12	13	14	15	16	17	18	19	20
1																			0	0
2					0	0	0	1	1	1	1	2	2	2	3	3	3	4	4	4
3			0	0	1	2	2	3	3	4	5	5	6	7	7	8	9	9	10	11
4			0	1	2	3	4	5	6	7	8	9	10	11	12	14	15	16	17	18
5		0	1	2	4	5	6	8	9	11	12	13	15	16	18	19	20	22	23	25
6		0	2	3	5	7	8	10	12	14	16	17	19	21	23	25	26	28	30	32
7		0	2	4	6	8	11	13	15	17	19	21	24	26	28	30	33	35	37	39
8		1	3	5	8	10	13	15	18	20	23	26	28	31	33	36	39	41	44	47
9		1	3	6	9	12	15	18	21	24	27	30	33	36	39	42	45	48	51	54
10		1	4	7	11	14	17	20	24	27	31	34	37	41	44	48	51	55	58	62
11		1	5	8	12	16	19	23	27	31	34	38	42	46	50	54	57	61	65	69
12		2	5	9	13	17	21	26	30	34	38	42	47	51	55	60	64	68	72	77
13		2	6	10	15	19	24	28	33	37	42	47	51	56	61	65	70	75	80	84
14		2	7	11	16	21	26	31	36	41	46	51	56	61	66	71	77	82	87	92
15		3	7	12	18	23	28	33	39	44	50	55	61	66	72	77	83	88	94	100
16		3	8	14	19	25	30	36	42	48	54	60	65	71	77	83	89	95	101	107
17		3	9	15	20	26	33	39	45	51	57	64	70	77	83	89	96	102	109	115
18		4	9	16	22	28	35	41	48	55	61	68	75	82	88	95	102	109	116	123
19	0	4	10	17	23	30	37	44	51	58	65	72	80	87	94	101	109	116	123	130
20	0	4	11	18	25	32	39	47	54	62	69	77	84	92	100	107	115	123	130	138

For specific details about using this table to find: p-values, see page 674, critical values, page 674.

TABLE 14 Critical Values for Total Number of Runs (V)

The entries in this table are the critical values for a two-tailed test using $\alpha = 0.05$. For a one-tailed test, $\alpha = 0.025$ and use only one of the critical values: the smaller critical value for a left-hand critical region, the larger for a right-hand critical region.

The smaller of n_1 and n_2	The larger of n_1 and n_2: 5	6	7	8	9	10	11	12	13	14	15	16	17	18	19	20
2								2	2	2	2	2	2	2	2	2
								6	6	6	6	6	6	6	6	6
3		2	2	2	2	2	2	2	2	2	3	3	3	3	3	3
		8	8	8	8	8	8	8	8	8	8	8	8	8	8	8
4	2	2	2	3	3	3	3	3	3	3	3	4	4	4	4	4
	9	9	10	10	10	10	10	10	10	10	10	10	10	10	10	10
5	2	3	3	3	3	3	4	4	4	4	4	4	4	5	5	5
	10	10	11	11	12	12	12	12	12	12	12	12	12	12	12	12
6		3	3	3	4	4	4	4	5	5	5	5	5	5	6	6
		11	12	12	13	13	13	13	14	14	14	14	14	14	14	14
7			3	4	4	5	5	5	5	5	6	6	6	6	6	6
			13	13	14	14	14	14	15	15	15	16	16	16	16	16
8				4	5	5	5	6	6	6	6	6	7	7	7	7
				14	14	15	15	16	16	16	16	17	17	17	17	17
9					5	5	6	6	6	7	7	7	7	8	8	8
					15	16	16	16	17	17	18	18	18	18	18	18
10						6	6	7	7	7	7	8	8	8	8	9
						16	17	17	18	18	18	19	19	19	20	20
11							7	7	7	8	8	8	9	9	9	9
							17	18	19	19	19	20	20	20	21	21
12								7	8	8	8	9	9	9	10	10
								19	19	20	20	21	21	21	22	22
13									8	9	9	9	10	10	10	10
									20	20	21	21	22	22	23	23
14										9	9	10	10	10	11	11
										21	22	22	23	23	23	24
15											10	10	11	11	11	12
											22	23	23	24	24	25
16												11	11	11	12	12
												23	24	25	25	25
17													11	12	12	13
													25	25	26	26
18														12	13	13
														26	26	27
19															13	13
															27	27
20																14
																28

From C. Eisenhart and F. Swed, "Tables for testing randomness of grouping in a sequence of alternatives," *Annals of Statistics*, vol. 14 (1943): 66–87. Reprinted by permission.

For specific details about using this table to find: p-values, see page 683; critical values, page 683.

TABLE 15 Critical Values of Spearman's Rank Correlation Coefficient

The entries in this table are the critical values of r_s for a two-tailed test at α. For a one-tailed test, the value of α shown at the top of the table is double the value of α being used in the hypothesis test.

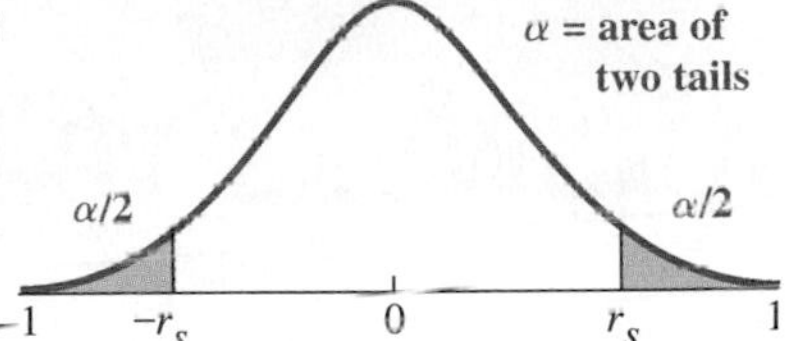

n	$\alpha = 0.10$	$\alpha = 0.05$	$\alpha = 0.02$	$\alpha = 0.01$
5	0.900	—	—	—
6	0.829	0.886	0.943	—
7	0.714	0.786	0.893	—
8	0.643	0.738	0.833	0.881
9	0.600	0.700	0.783	0.833
10	0.564	0.648	0.745	0.794
11	0.536	0.618	0.736	0.818
12	0.497	0.591	0.703	0.780
13	0.475	0.566	0.673	0.745
14	0.457	0.545	0.646	0.716
15	0.441	0.525	0.623	0.689
16	0.425	0.507	0.601	0.666
17	0.412	0.490	0.582	0.645
18	0.399	0.476	0.564	0.625
19	0.388	0.462	0.549	0.608
20	0.377	0.450	0.534	0.591
21	0.368	0.438	0.521	0.576
22	0.359	0.428	0.508	0.562
23	0.351	0.418	0.496	0.549
24	0.343	0.409	0.485	0.537
25	0.336	0.400	0.475	0.526
26	0.329	0.392	0.465	0.515
27	0.323	0.385	0.456	0.505
28	0.317	0.377	0.448	0.496
29	0.311	0.370	0.440	0.487
30	0.305	0.364	0.432	0.478

From E. G. Olds, "Distribution of sums of squares of rank differences for small numbers of individuals," *Annals of Statistics*, vol. 9 (1938), pp. 138–148, and amended, vol. 20 (1949), pp. 117–118. Reprinted by permission.

For specific details about using this table to find: p-values, see page 694; critical values, page 693.

ANSWERS TO SELECTED EXERCISES

Chapter 1

1.1 **a.** Americans **b, c, d.** Answers will vary
1.2 **a.** university students **b.** 1,746
c. 82% like to snack at home **d.** Multiple answers
1.3 **a.** employed American adults
b. U.S. Labor Department
c. number of cashiers surveyed in 1996; median hourly pay rate
1.5 **a.** American travelers on trips of 100 miles or more
b. 38-year-old male, income $50,000 or more, travels by car between 100 and 300 miles, usually to visit friends or relatives and stays at their homes
1.6 **a.** American adults **b.** within ±4%
c. Actual percentage could range from 50% to 58%
1.9 **a.** descriptive **b.** inferential
1.11 **a.** U.S.A. drivers **b.** 837
c. has driver ever had to suddenly swerve, inadvertently speeded up, know someone who had crash while talking on cellphone
d. 41% of those surveyed said they had inadvertently speeded up
e. 0.41(837) = 343
1.13 **a.** 45% **b.** Percentages are from different groups
1.15 **a.** took samples from various spots in the bay
b. new pollution laws enacted
1.16 population: all U.S. adults; sample: 1200 adults; variable: "allergy status"; one data: yes, dust; data: set of yes/no responses or allergies; experiment: method to select the adults; parameter: percent of all U.S. adults with an allergy, 36%; statistic: 33.2% based on sample
1.17 Parameter has one specific value. Statistics vary by sample.
1.18 **c.** Sample averages vary less for samples of size 10—the larger sample size
1.19 ZIP code, gender, highest level of education
1.20 annual income, age, distance to store
1.21 marital status, ZIP code
1.22 highest level of education, rating for first impression
1.23 **a.** Scores are counted. **b.** Time is measured.
1.24 **a.** gender, nominal **b.** height, continuous
1.25 **a.** residents of Lee County, Florida
b. households of Lee County, residents, registered adults
c. income, age, political affiliation
d. income, continuous; age, continuous; political affiliation, nominal
e. Collected data were counted by categories.
1.27 **a.** severity of side-effects **b.** ordinal
1.29 **a.** all individuals who have hypertension and use prescription drugs to control it
b. 5,000 in study
c. proportion of population for which drug is effective
d. proportion of the sample for which the drug is effective, 80%
e. no
1.31 **a.** all assembled parts from the assembly line
b. infinite **c.** parts checked
d. attribute, attribute, numerical
1.33 **a.** all people suffering from migraine headaches
b. 2,633 people given the drug
c. amount of drug dosage, the side effects encountered
d. dosage, quantitative; type of side effects, qualitative
1.35 **a.** numerical **b.** attribute
c. numerical **d.** attribute
e. numerical **f.** numerical
1.37 **a.** The population contains all objects of interest; sample contains only those actually studied.
b. convenience, availability, practicality
1.39 football players, cover a wider range of values
1.41 Price/standard unit makes price the only variable.
1.43 **c.** Sample size of 4 shows more variability.
d. Sample size of 10, less variability
1.44 Answers include reference to an economic bias.
1.45 volunteer; yes
1.46 observational
1.47 **a.** biased, the results were not representative
b. experiment
1.48 volunteer, bias
1.49 U.S. Senate
1.50 U.S. House of Representatives
1.51 Each precinct is a cluster, not all precincts are sampled.
1.52 **a.** American children
b. 16,202 children of participants in Nurses' Health Study II
c. age, frequency of family dinner, home food, fast food, physical activity, team sports, television
d. percent of family dinners for each age
e. no **f.** convenient
1.55 **a.** observational or volunteer **b.** yes

1.57 **a.** (list) from which a sample is actually drawn
b. a computer list of the full-time students
c. random number table, student numbered 1288
1.59 judgment sampling
1.61 probability samples
1.63 convenience sampling
1.65 A proportional sample would work best.
1.67 Not all adults are registered voters.
1.68 **a.** probability **b.** statistics
1.69 **a.** statistics **b.** probability
c. statistics **d.** probability
1.71 perform many of the computations and test, quickly and easily
1.73 Computers cannot determine whether or not a study has been conducted properly.
1.74 **a.** Americans, yes
b. opinion on the current state of movies
c. 53% of those people sampled said they used "pens, pencils and paper" during meetings.
1.77 **a.** color of hair, major, gender
b. number of courses taken, height, distance from hometown to college
1.79 **a.** piece of data
b. What is the average number of times per week the people in the sample went shopping?
c. What is the average number of times per week that people (all people) go shopping?
1.81 **a.** all people in the U.S. who died in 1997
b. death from heart disease, state of residence, age, obesity, inactivity
c. mortality rate per 100,000 people in the U.S.
d. death from heart disease—attribute
state of residence—attribute
age at death—numerical
obesity—attribute
inactivity—attribute
1.83 **a.** young men who do not go to college and take restaurant jobs
b. the 2,000 individuals in the study
c. most likely, judgment sample

Chapter 2

2.2

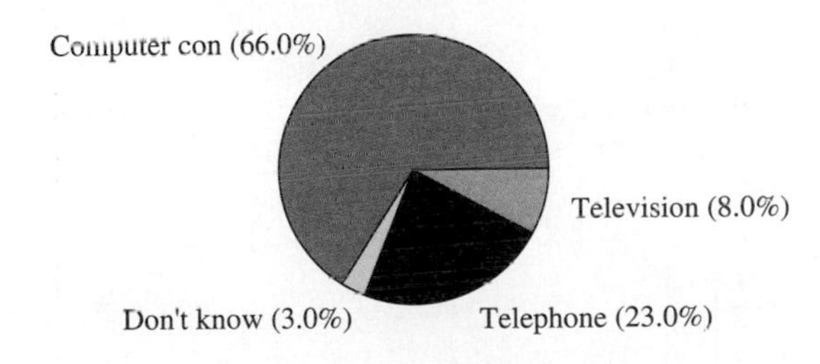

2.3

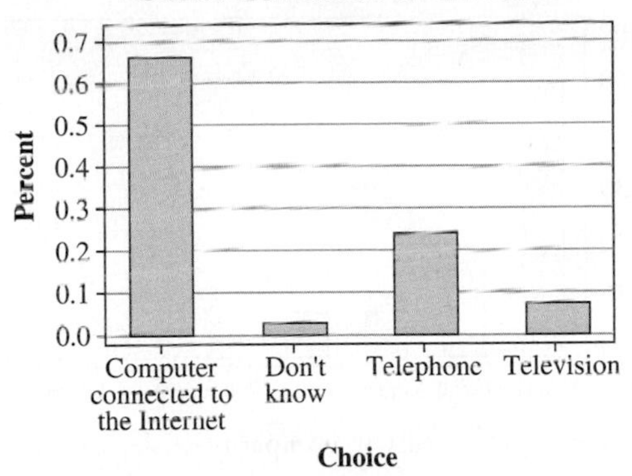

2.4 Circle graph represents the relative proportions as a whole; bar graph represents the relative proportions between the individual answers.

2.5

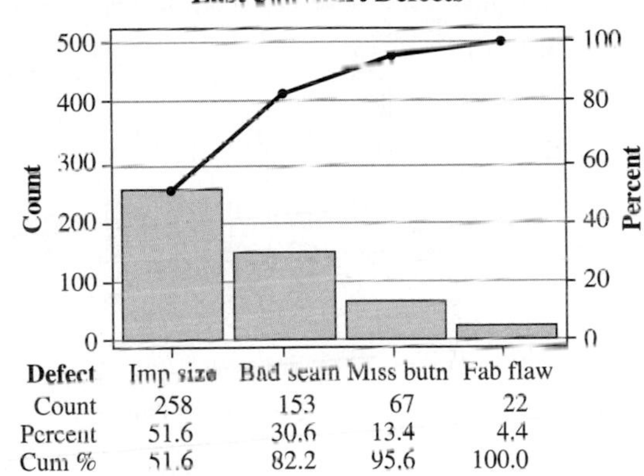

2.6

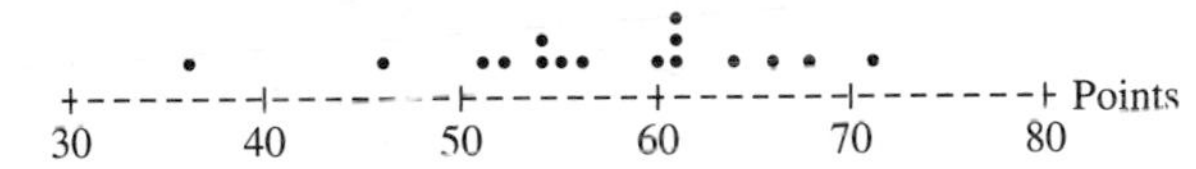

2.7 96: leaf 6 on 9 stem,
66: leaf 6 on 6 stem
2.8 Points scored per game

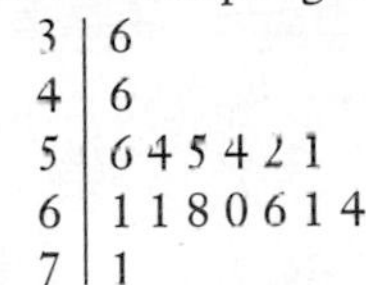

3	6
4	6
5	6 4 5 4 2 1
6	1 1 8 0 6 1 4
7	1

2.9 Leaf is 1.0 unit wide.
2.11 **a.**

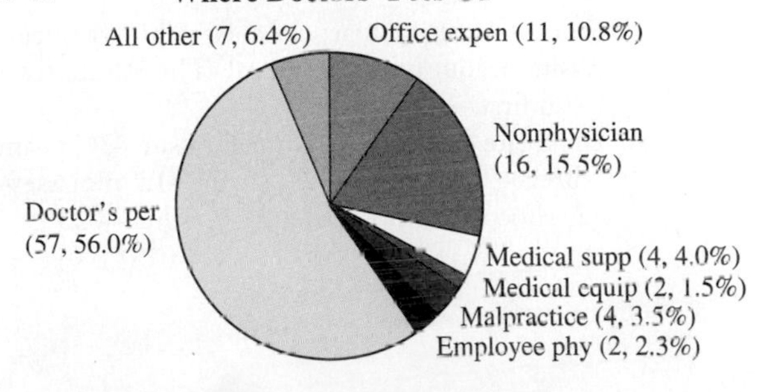

b.

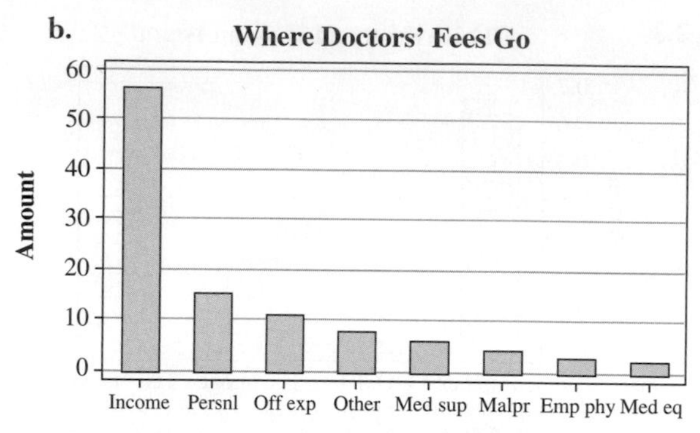

c. circle graph, easy to compare the size of each part to the whole; bar graph, easy to compare the sizes of the parts to each other.

2.13 a.

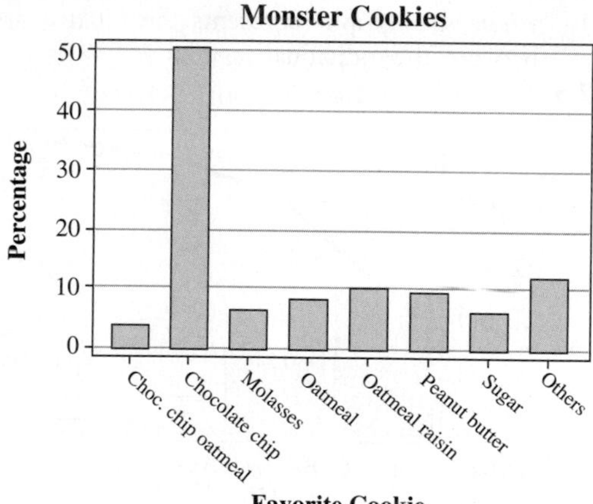

b.

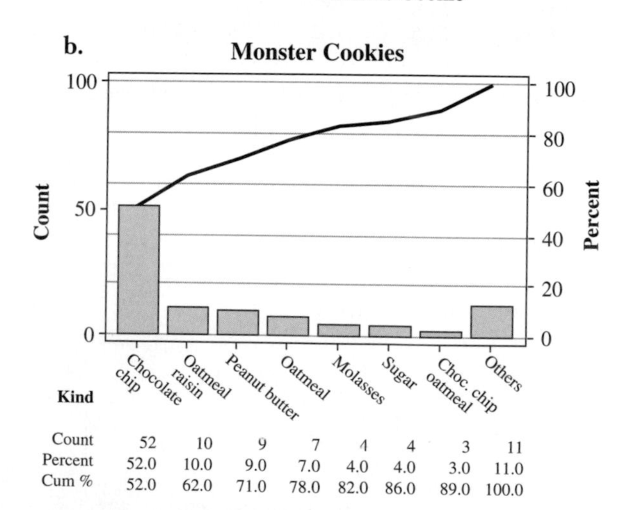

c. Stock 4 most popular: chocolate chip, oatmeal raisin, peanut butter, oatmeal. The Pareto lists them first.

d. chocolate chip—156, oatmeal raisin—30, peanut butter—27, oatmeal—21, sugar—12, molasses—12, chocolate chip oatmeal—9, others—33

2.15 a.

Major Chores Mothers Would Like Family Help With

Defect	Cleaning	Laundry	Other	Cooking	Dishes
Count	53	18	12	9	8
Percent	53.0	18.0	12.0	9.0	8.0
Cum %	53.0	71.0	83.0	92.0	100.0

b. It is a collection of several answers, needs to be broken down.

2.17 a.

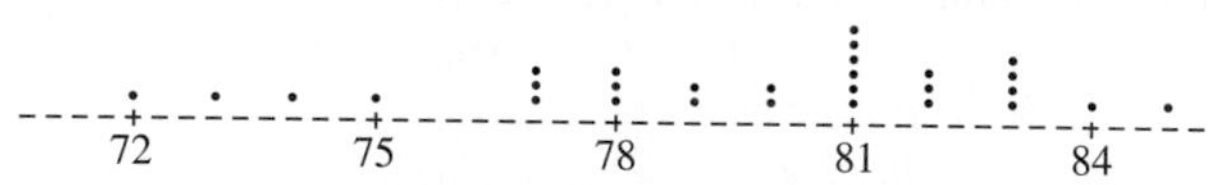

b. shortest—72 inches, tallest—85

c. 81 inches, six players

d. "tallest" stack of points

2.19 a. 15 **b.** 11.2, 11.2, 11.3, 11.4, 11.7

c. 15.6 **d.** 13.7; 3

2.21 a. Profitability last 12 mos (pct)

10	4
11	37
12	25589
13	4689
14	3688
15	17
16	
17	9
18	247
19	
20	2

b. skewed to the right; larger cluster centered around 12, 13 and 14 percent; smaller cluster centered at 18 percent.

2.23 a. Place value of the leaves is in the hundredths place.

b. 16 **c.** 5.97, 6.01, 6.04, 6.08

d. cumulative frequencies starting at the top and the bottom

2.25

x	f
0	2
1	5
2	3
3	0
4	2
	12

2.26 **a.** f is frequency; values of 70 or more but less than 80 occurred 8 times.
b. 19 **c.** number of data, or sample size

2.27 **a.** sectors of the circle
b. angle forming the sector **c.** $2 \le x < 4$
d.

Class boundaries	Relative frequency
$0 \le x < 1$	0.05
$1 \le x < 2$	0.20
$2 \le x < 4$	0.33
$4 \le x$	0.39
don't know	0.03

2.28 **a.** $65 \le x < 75$
b. values greater than or equal to 65 and also less than 75
c. difference between upper and lower class boundaries

2.29

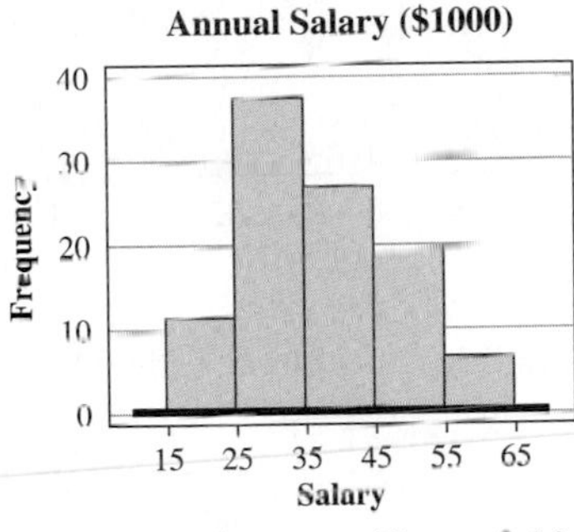

2.30 Shapes are the same; Figure 2.10 uses class mid points, frequency; Figure 2.11 uses the class boundaries, percentages.

2.31 same classes of data, same shape; vertical vs. horizontal, individual data points vs. grouped

2.32 symmetric: strength of string
uniform: rolling a die several hundred times
skewed right: salaries
skewed left: hour exam scores
bimodal: weights for groups containing both male and female
J-shaped: amount of television watched per day

2.33 **a.** uniform **b.** J-shaped **c.** skewed right

2.34

Class Boundaries	Cumulative Frequency
$15 \le x < 25$	12
$25 \le x < 35$	49
$35 \le x < 45$	75
$45 \le x < 55$	94
$55 \le x \le 65$	100

2.35

Class Boundaries	Cum. Rel. Frequency
$15 \le x < 25$	0.12
$25 \le x < 35$	0.49
$35 \le x < 45$	0.75
$45 \le x < 55$	0.94
$55 \le x \le 65$	1.00

2.36

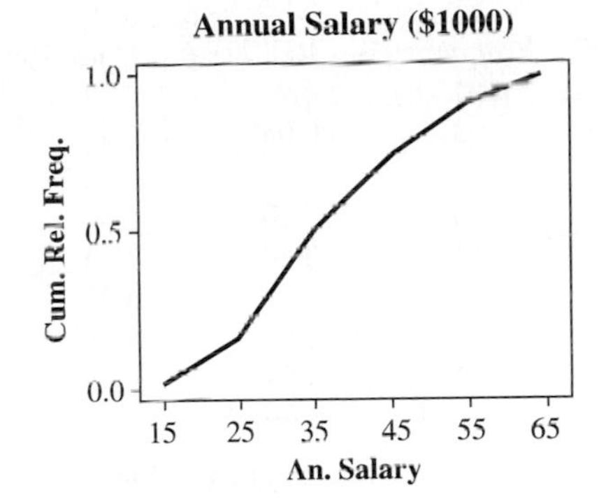

2.37 **a.** lower left, upper right
b. upper left, lower right

2.39 **a.** & **c.**

AP Score	Frequency	Rel. Freq.
1	3	0.081
2	18	0.487
3	9	0.243
4	3	0.081
5	4	0.108

b.

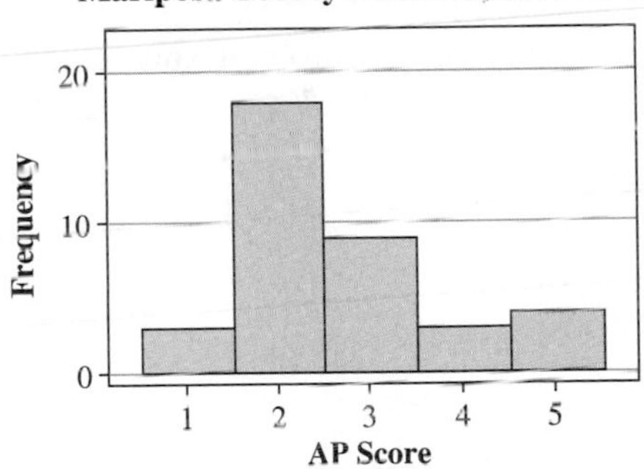

d. 43.2%

2.41 **a.**

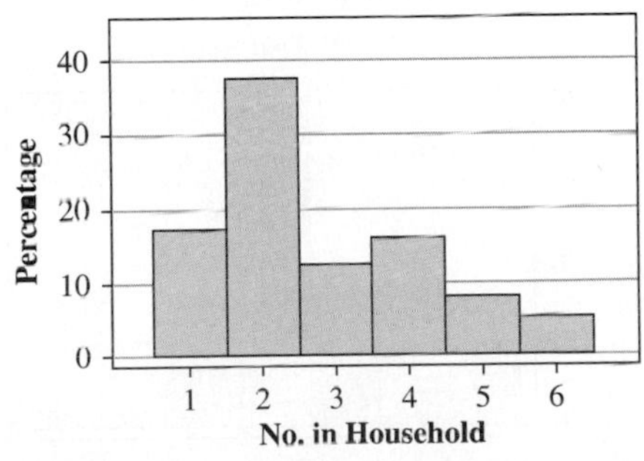

b. mounded, skewed right
c. Two is the most common number; majority between 1 and 4 people.

2.43 **a.**, **b.**, **d.**

Age	Frequency	Rel. Freq.	Cum. Rel. Freq.
17	1	0.02	0.02
18	3	0.06	0.08
19	16	0.32	0.40
20	10	0.20	0.60
21	12	0.24	0.84
22	5	0.10	0.94
23	1	0.02	0.96
24	2	0.04	1.00
	50	1.00	

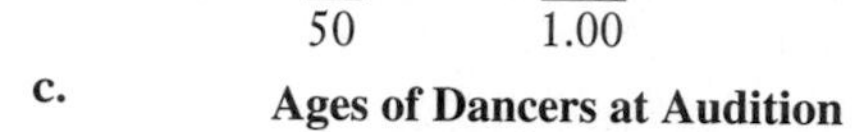

c. Ages of Dancers at Audition

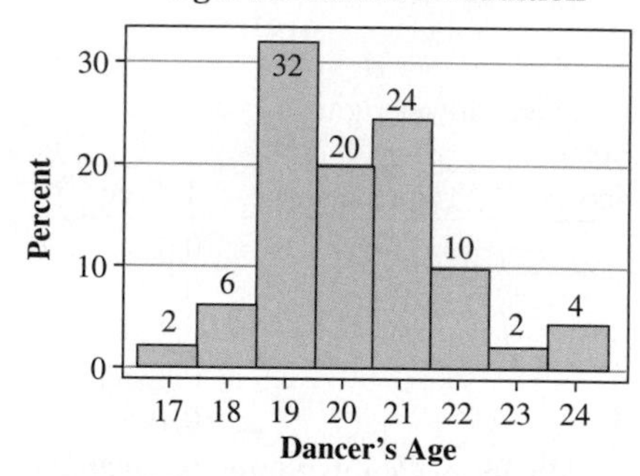

e. Ages of Dancers at Audition

Cum. Rel. Freq.

Dancer's Age

2.45 **a.** 12 and 16 **b.** 2, 6, 10, 14, 18, 22, 26 **c.** 4.0
d. 0.08, 0.16, 0.16, 0.40, 0.12, 0.06, 0.02
e.

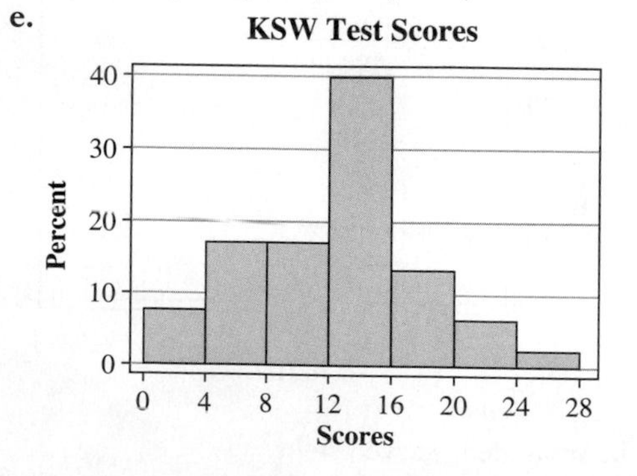

2.47 **a.**

Class Limits	Frequency
12–18	1
18–24	14
24–30	22
30–36	8
36–42	5
42–48	3
48–54	2

b. 6 **c.** 27, 24, 30
d.

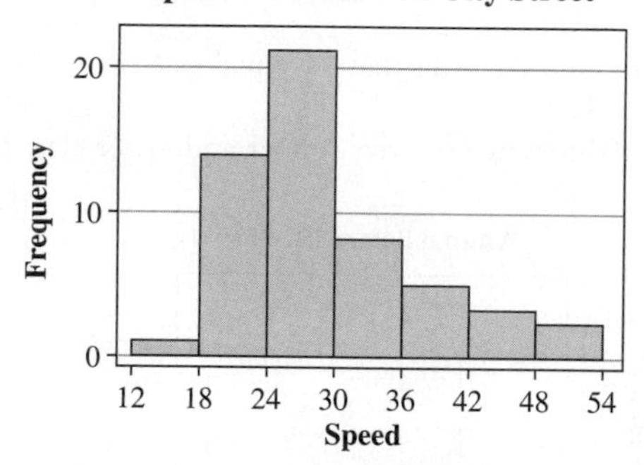

2.49 **a.**

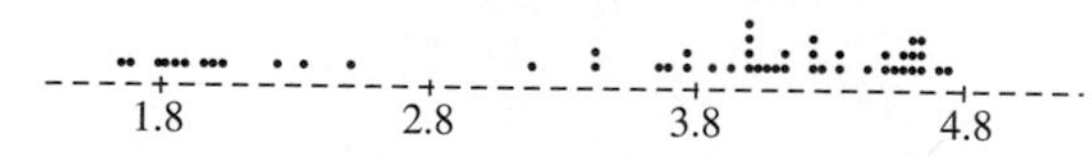

b.

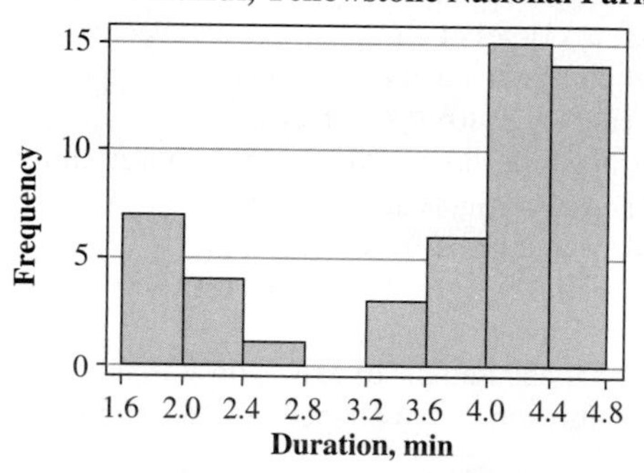

c. and **d.** The histograms will vary as class boundaries and class widths are changed.

e.

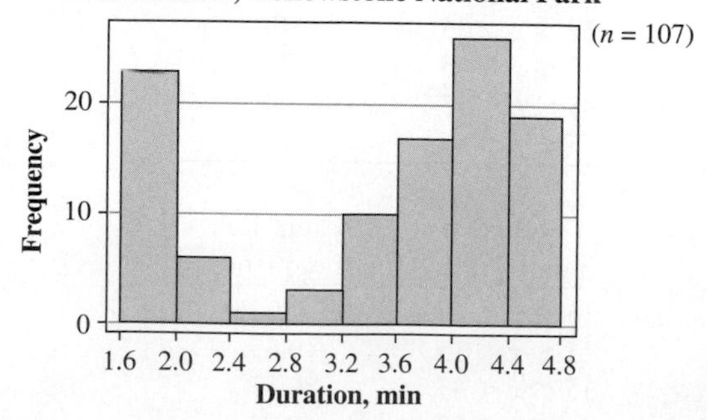

f. Answers will vary. Histogram is easier to read.

2.51 **a.**

AP Score	Cum. Rel. Freq.
1	0.081
2	0.568
3	0.811
4	0.892
5	1.000

b. **Advanced Placement Tests**

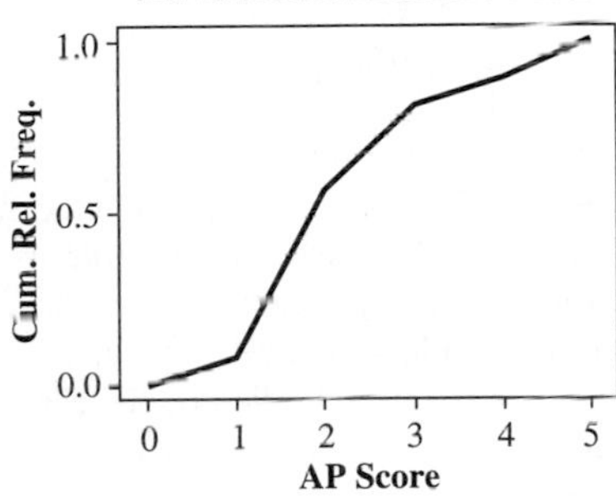

2.53 $\bar{x} = 18/8 = 2.25$

2.54 **a.** 9 **b.** 0

2.55 70, 72, 73, 74, 76
$d(\tilde{x}) = 3\text{rd}$; $\tilde{x} = 73$

2.56 4.15, 4.25, 4.25, 4.50, 4.60, 4.60, 4.75, 4.90;
$d(\tilde{x}) = 4.5\text{th}$; $\tilde{x} = 4.55$

2.57 **a.** mean—larger, median—same
b. mean—smaller, median—same
c. median

2.58 2

2.59 $5.48

2.60 $\bar{x} = 49/6 = 8.2$
6, 7, 8, 9, 9, 10
$d(\tilde{x}) = 3.5\text{th}$; $\tilde{x} = 8.5$
mode = 9
midrange = 8.0

2.61 **a.** 25,500 31,500 31,500 31,500 31,500
35,250 36,750 37,500 39,000 54,000
$\bar{x} = 354{,}000/10 = 35{,}400.00$
$d(\tilde{x}) = 5.5\text{th}$; $\tilde{x} = 33{,}375$
mode = 31,500
midrange = 39,750
Values all agree with those in the article.
b. The large value of 54,000 is pulling the mean and midrange toward the larger data value.
c. The higher figures exceed the mean by a total of $25,650. The lower figures fall short of the mean by a total of $25,650.

2.62 **a.** time of strike
b. one-hour interval
c. Most strikes occur in early afternoon.
d. height of graph for those intervals

2.63 Data resulting from a quantitative variable are numbers with which arithmetic can be performed; for a qualitative variable it is not possible to add.

2.65 **a.** $\bar{x} = 36/6 = 6.0$
b. $d(\tilde{x}) = 3.5\text{th}$; $\tilde{x} = 6.5$
c. 7 **d.** 5.5

2.67 **a.** $\sum x = 423.00$, $\bar{x} = 32.5$ hours
b. 24.6 26.9 28.4 29.3 30.5 32.7 33.3
33.4 33.5 34.7 34.9 36.4 44.4
$d(\tilde{x}) = 7^{\text{th}}$; $\tilde{x} = 33.3$ hours
c. 34.5 **d.** no mode

2.69 **a.** midrange = $2,192
b. only given the two extreme values
c. most likely skewed to the right

2.71 **a.**

Third Graders at Roth Elementary School

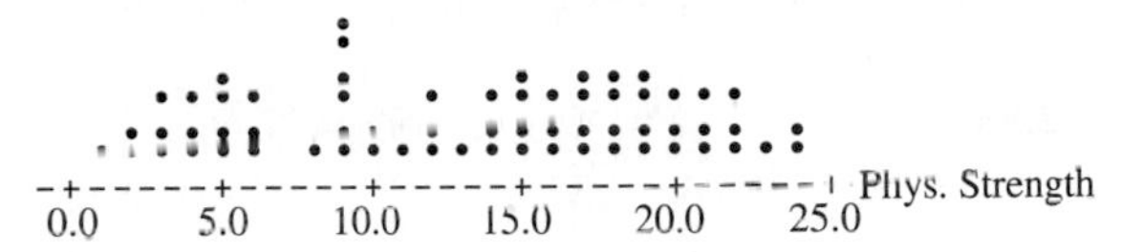

b. mode = 9

c. **Third Graders Physical Strength Test**

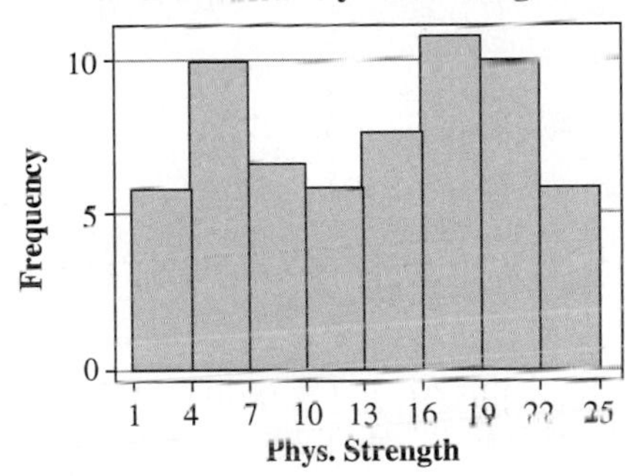

d. Bimodal; modal classes are 4–7 and 16–19.
e. Dotplot shows mode to be 9, which is in the 7–10 class; two modal classes are 4–7 and 16–19. The mode is not in either modal class.
f. no, only one numerical value per class
g. The mode is simply the single data value that occurs most often, while a modal class results from data tending to bunch up.

2.73 **a.** & **b.**

	Runs at Home	Runs Away	Difference
Mean	9.77	9.80	−0.03
Median	9.65	9.70	−0.06
Maximum	13.65	11.06	4.89
Minimum	7.64	8.67	−1.74
Midrange	10.65	9.87	1.58

c. At Coors Field, the maximum number of runs scored (13.65); away, only 8.76 runs, which ranked second from the bottom.

2.75 **a.** $\sum x$ needs to be 500
b. need two numbers smaller than 70 and one larger
c. need multiple 87's
d. need any two numbers that total 140
e. need two numbers smaller than 70
f. need two numbers of 87 and a third number so total of all five is 500

g. total 500 and total of L and H to be 140; impossible

h. two 87's; impossible

2.77 \$1800

2.78 **a.** 12 units above the mean
b. 20 units below the mean

2.79 $\sum(x - \bar{x}) = \sum x - n\bar{x} = \sum x - n \cdot (\sum x/n) = \sum x - \sum x = 0$

2.80 $n = 5, \sum x = 25, \sum(x - \bar{x})^2 = 46$; 11.5

2.81 $n = 5, \sum x = 25, \sum x^2 = 171$; 11.5, same

2.83 **a.** 7 **b.** $n = 5, \sum x = 30, \sum(x - \bar{x})^2 = 34$; 8.5
c. 2.9

2.85 **a.** $n = 10, \sum x = 72, \sum(x - \bar{x})^2 = 73.60$; 8.2
b. $n = 10, \sum x = 72, \sum x^2 = 592$; 8.2
c. 2.9

2.87 **a.** 19.800
b. $n = 13, \sum x = 423.00, \sum x^2 = 14058$; 24.51923
c. 4.96

2.89 **a.**

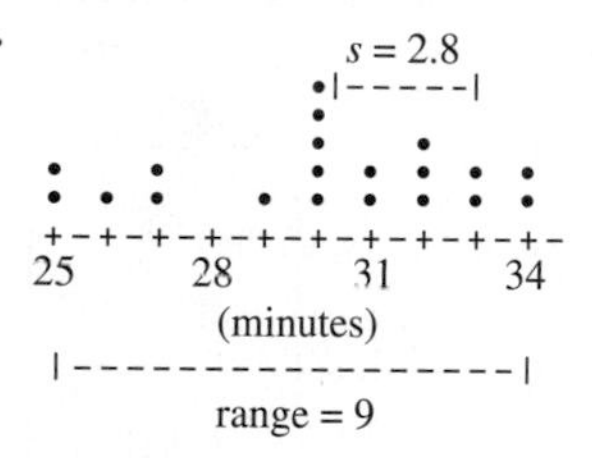

$n = 20, \sum x = 601, \sum x^2 = 18{,}209$
b. 30.05 **c.** 9 **d.** 7.8 **e.** 2.8
g. Except for the value $x = 30$, the distribution looks rectangular.

2.91 **a.** range = 25.100,
$n = 25, \sum x = 2002.0, \sum x^2 = 160{,}955; s = 5.14$
b.

```
  1   6  4  ------------------- (range top)
  1   6
  2   6  9
  2   7
  2   7
  3   7  4
  6   7  667        -+-                range
 10   7  8899        | St. dev.
(4)   8  0111       -+-
 11   8  2222222
  4   8  44
  2   8  6
  1   8  9  ------------------- (range bottom)
```

c. Standard deviation is approximately one-fifth the range.

2.93

	$\sum x$	$\sum\lvert x - \bar{x}\rvert$	$\sum(x - \bar{x})^2$	Range
Set 1:	250	14	54	9
Set 2:	250	46	668	35

2.95 **a.** All data must be the same value.
b. 99, 99.5, 100.5, 100 [100]; $s = 0.57$
c. 107, 95, 94, 108, [100]; $s = 6.53$
d. 75, 78, 123, 124, [100]; $s = 23.53$

2.98 **a.**

x	f	xf	x^2f
0	1	0	0
1	3	3	3
2	8	16	32
3	5	15	45
4	3	12	48
$\sum$	20	46	128

b. $\sum f = 20$; $\sum xf = 46$;
$\sum x^2f = 128$
c. 4 is one of the possible data values
8 is the number of times an x value occurred
$\sum f$: sum of the frequencies = sample size
$\sum xf$: the sum of the data

2.99 **a.** no meaning unless each value occurred only once
b. Each data value is multiplied by times it occurred. Summing these products will give the same sum as if all data values were listed individually.

2.100 $n = 20, \sum xf = 46$, 2.3

2.101 $n = 20, \sum xf = 46, \sum x^2f = 128$; 1.2

2.102 1.1

2.103 $n = 40, \sum xf = 516, \sum x^2f = 7504; \bar{x} = 12.9$,
$s^2 = 21.7, s = 4.7$

2.105 $n = 73, \sum xf = 134, \sum x^2f = 378$;
$\bar{x} = 1.8, s^2 = 1.8, s = 1.4$

2.107 $n = 40, \sum xf = 447.0, \sum x^2f = 5472.00$; 11.2, 12.2, 3.5

2.109 $n = 125, \sum xf = 1100, \sum x^2f = 10{,}376$; 8.8, 2.37

2.111 $n = 100, \sum xf = 4210, \sum x^2f = 198100$; 42.1, 14.5

2.113 **a.** 23 is the number of students tested at Caruthers Unified, 441 is the average verbal score for those 23 students.
b. $23 \times 441 = 10143$
c. 3199
d. 1,508,912
e. 471.7

2.115 **a.** $n = 997, \sum xf = 63{,}665, \sum x^2f = 4{,}302{,}725$; 63.9, 65, 65, 55
b. 238.25, 15.4

2.117 **a.** and **b.**
Using class midpoints 10, 12, . . . ;
$\sum f = 50, \sum xf = 1406.0, \sum x^2f = 41{,}748$
$\bar{x} = 28.12$
$s^2 = 45.128, s = 6.7177$
Ungrouped: using 50 data:
$n = 50, \sum x = 1393.0, \sum x^2 = 41{,}057$
$\bar{x} = 27.86$
$s^2 = 45.878, s = 6.773$
c. $(28.12 - 27.86)/27.86 = 0.9\%$
$(45.128 - 45.877959)/45.877959 = -1.6\%$
$(6.7177 - 6.773)/6.773 = -0.8\%$

2.118 44th position from L; 7th position from H

2.119 $nk/100 = 10.0$, $d(P_{20}) = 10.5^{\text{th}}$, $P_{20} = 64$
$nk/100 = 17.5$, $d(P_{35}) = 18\text{th}$, $P_{35} = 70$

2.120 $nk/100 = 10.0$, $d(P_{80}) = 10.5\text{th}$
$P_{80} = 88.5$
$nk/100 = 2.5$, $d(P_{95}) = 3\text{rd}$
$P_{95} = 95$

2.121 symmetric about the mean

2.122

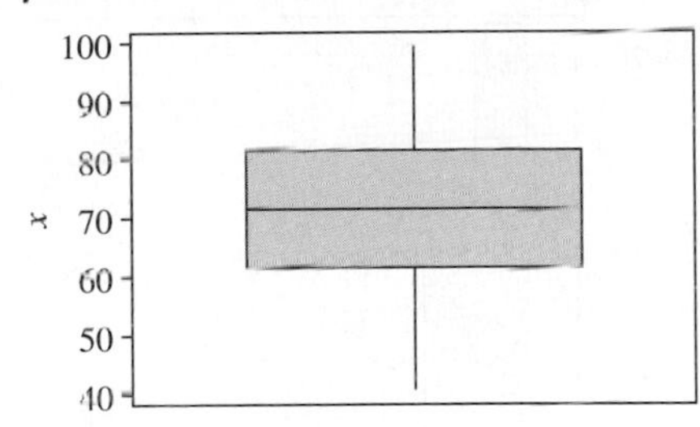

2.123 1.67, −0.75

2.125

2.6 2.7 3.4 3.6 3.7 3.9 4.0 4.4 4.8 4.8
4.8 5.0 5.1 5.6 5.6 5.6 5.8 6.8 7.0 7.0

a. $nk/100 = 5.0$, $d(P_{25}) = 5.5^{th}$, $Q_1 = 3.8$
$nk/100 = 15.0$, $d(P_{75}) = 15.5^{th}$, $Q_3 = 5.6$

b. 4.7

c. $nk/100 = 3.0$, $d(P_{15}) = 3.5^{th}$, $P_{15} = 3.5$
$nk/100 = 6.6$, $d(P_{33}) = 7^{th}$, $P_{33} = 4.0$
$nk/100 = 18.0$, $d(P_{90}) = 18.5^{th}$, $P_{90} = 6.9$

2.127 **a.** $nk/100 = 13$, $d(Q_1) = 13.5^{th}$, $Q_1 = 3.05$

b. $d(\text{median}) = 26.5^{th}$, $Q_2 = 4.0$

c. $Q_3 = 4.65$ **d.** 3.85

e. $nk/100 = 15.6$, $d(P_{30}) = 16^{th}$, $P_{30} = 3.2$

f. 1.4, 3.05, 4.0, 4.65, 13.3

g. **U.S. Geological Survey, Rocky Mountains**

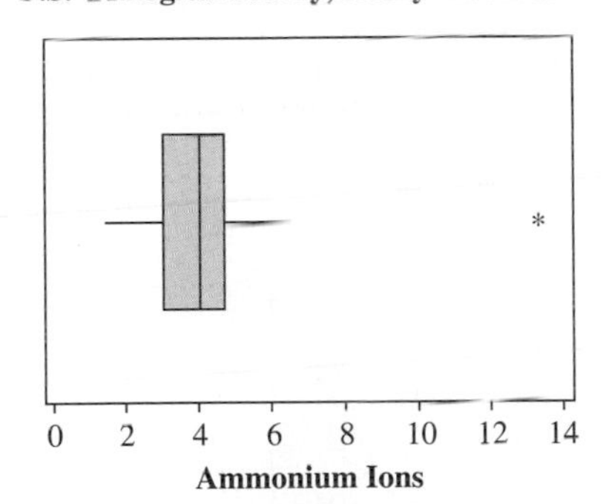

2.129 **a.** 1.0
b. 0.0
c. 2.25
d. −1.25

2.131 680

2.133 **a.** one and one-half standard deviations above the mean
b. 2.1 standard deviations below the mean
c. The number of standard deviations from the mean

2.135 1.625, 1.2, A has the higher relative position

2.137 from 175 through 225 words, inclusive

2.138 **a.** 50%, **b.** ≈68%, **c.** ≈84%

2.139 at least 93.75%

2.141 99.7%, lies within 3 standard deviations of the mean

2.143 **a.** 2.5%
b. 70.4 to 97.6 hours

2.145 **a.** range equal to 6 times the standard deviation
b. dividing the range by 6

2.147 **a.** at most 11% **b.** at most 6.25%

2.149 $n = 50$, $\sum x = 5{,}917{,}489$, $\sum x^2 = 810{,}941{,}217{,}297$
a. \$118,350, \$47,511
b. \$70,839 and \$165,861
c. 40/50 = 80%
d. \$23,328 and \$213,372
e. 46/50 = 92%
f. −\$24,183 and \$260,883
g. 49/50 = 98%
h. Both percentages exceed the values cited.
i. 80%, 92% and 98% as a set are not close to the 68%, 95% and 99.7% cited by the empirical rule, not approximately normal.

2.153 yes

2.155 **a.** Answers will vary.
b. **Service Complaints**

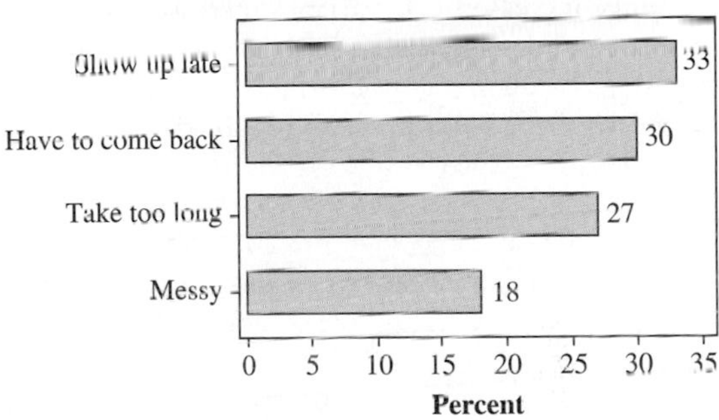

c. "Messy" is substantial.

2.156 **a.** **Overweight Adult Males**

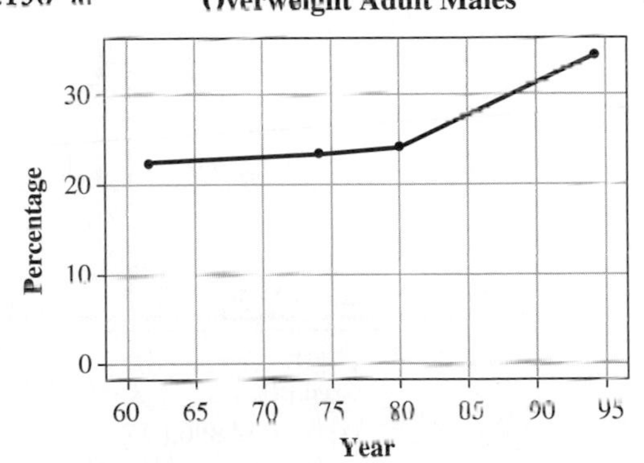

b. The percentage has increased greatly; however, the graph in the application seems to overstate the increase.

2.157 The class width is not uniform.

2.163 **a.** .15, .34, .18, .24, .09

b.

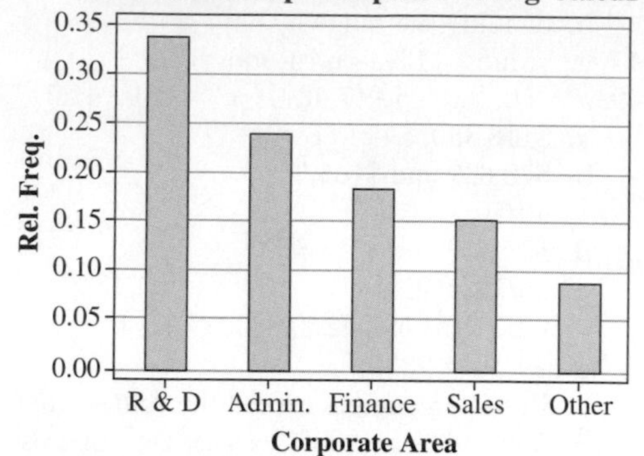

c. The bars are touching in a histogram, emphasizing the sequence of the data.

2.165 **a.** mean increased **b.** median unchanged **c.** mode unchanged **d.** midrange increased **e.** range increased **f.** variance increased **g.** standard deviation increased

2.167 $n = 8, \sum x = 36.5, \sum x^2 = 179.11$
a. 4.56 **b.** 1.34 **c.** very close to 4%

2.169 $n = 118, \sum x = 2364$
a. 20.0 **b.** 17 **c.** 16 **d.** 15, 21 **e.** 14, 43

2.171 $n = 25, \sum x = 1{,}997; \sum x^2 = 163{,}205$; 79.9, 12.4

2.173 **a.** Population, all airline passengers; variable, passenger luggage status, lost or not. Other variables: airline, date of flight.
b. statistics **c.** statistic **d.** no

2.175 **a.**

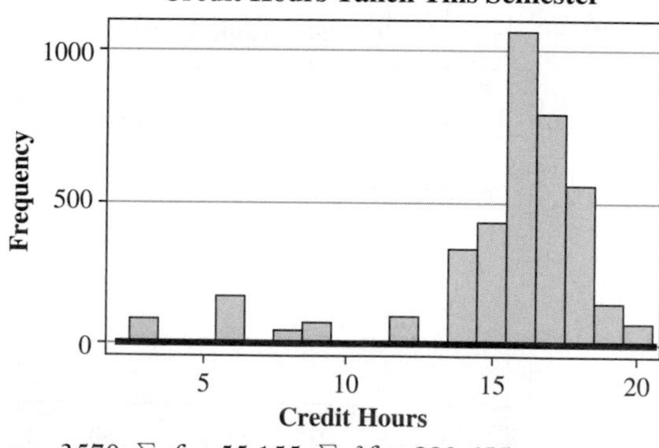

$n = 3570; \sum xf = 55{,}155; \sum x^2f = 890{,}655$
b. $\bar{x} = 15.4$, $\tilde{x} = 16$, mode = 16, midrange = 11.5, midquartile = 16
c. $Q_1 = 15, Q_3 = 17$
d. $P_{15} = 14, P_{12} = 14$
e. Range = 17, $s^2 = 10.7967$, $s = 3.3$

2.177 **a.**

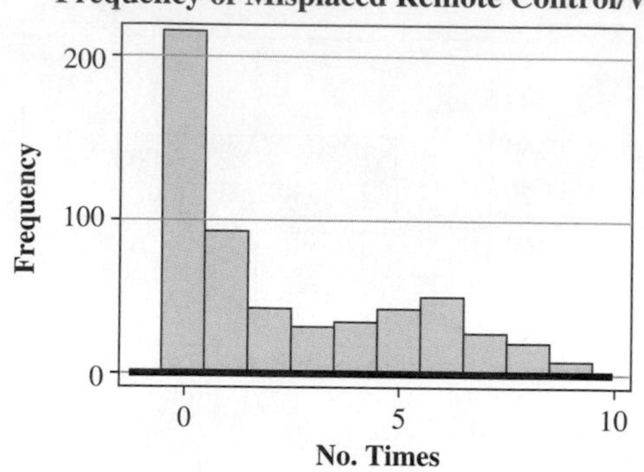

b. $n = \sum f = 500, \sum xf = 994, \sum x^2f = 5200$;
$\bar{x} = 1.988$, $\tilde{x} = 1$, mode = 0, midrange = 4.5
c. 6.46, 2.5 **d.** $Q_1 = 0, Q_3 = 4, P_{90} = 6$
e. 2 **f.** 5-number summary: 0, 0, 1, 4, 9

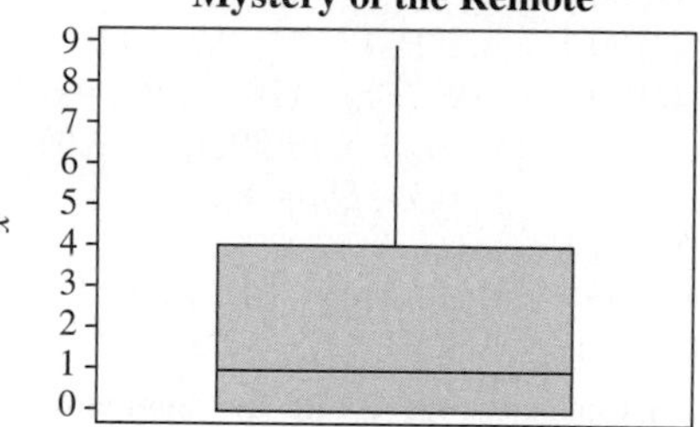

2.179 **a.**

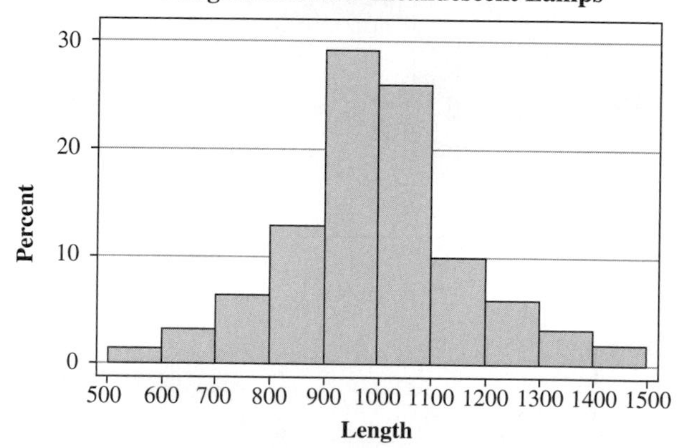

$n = 220; \sum xf = 219{,}100; \sum x^2f = 224{,}470{,}000$
b. 995.9
c. 169.2

2.181 a.

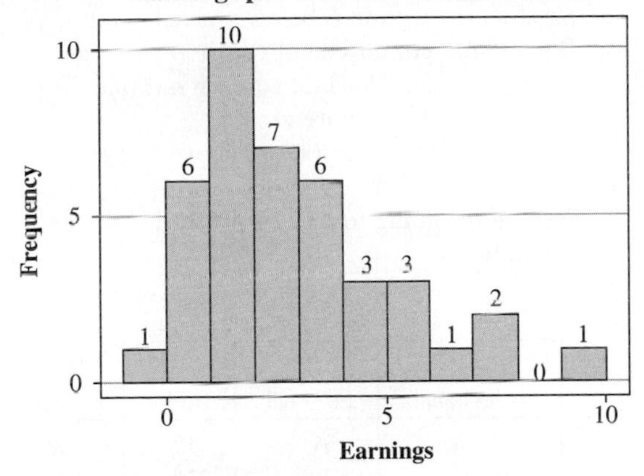

b. Median is in the class \$2.00–\$3.00.

2.183 a. Hourly earnings are increasing annually.
b. same observations

2.185 a. P_{98} **b.** P_{16}
c.

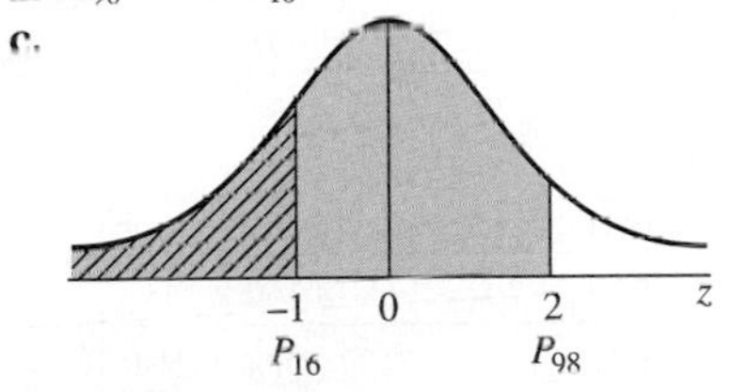

2.187 58, 0, 9.8, 6.9, 181

2.189 $n = 8, \sum x = 31{,}825, \sum x^2 = 126{,}894{,}839$
a. 3978.1
b. 203.9
c. 3570.3 to 4385.9

2.191 a. Calories: $n = 50$; $\sum x = 5805$; $\sum x^2 = 781{,}425$
$\bar{x} = 116.1, s = 46.83$
Sodium: $n = 50$; $\sum x = 28{,}990$;
$\sum x^2 = 19{,}880{,}100$
$\bar{x} = 579.8, s = 250.4$
b. Calories: between 22.44 and 209.76
Sodium: between 79.04 and 1,080.56 mg
yes, 94%, 98%
c. 329.42 and 830.18 mg; 64%, 98%; approximately normal

2.193 $n = 40$; $\sum x = 103{,}545$; $\sum x^2 = 281{,}671{,}245$
a. 2588.6, 591.22
b. 1393, 2127.5, 2621.5 3131, 3439
c. 1406.2 and 3771; yes, 97.5%
d. 1997.4 and 3179.8; no, 57.5%
e. The points distribution for the top 40 is part of a skewed left distribution.

2.195 Many possible answers, only one is shown.
a. 70, 77.5, 77.5, 77.5, 85 yields $s = 5.30$
b. 70, 76, 85, 89, 95 yields $s = 10.02$
c. 70, 85, 90, 99, 110 yields $s = 15.02$
d. Data had to become more dispersed.

2.197 a.

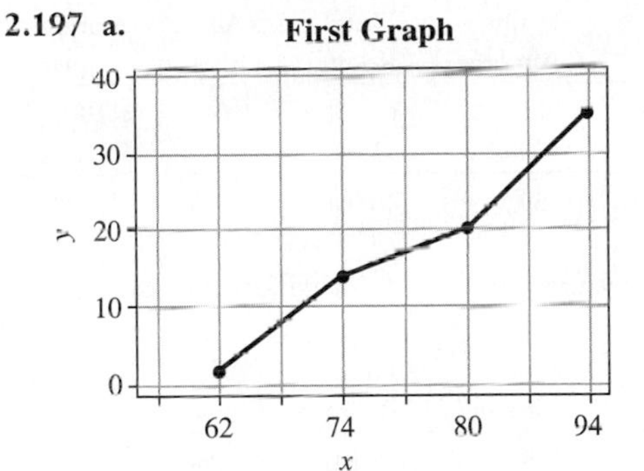

b.

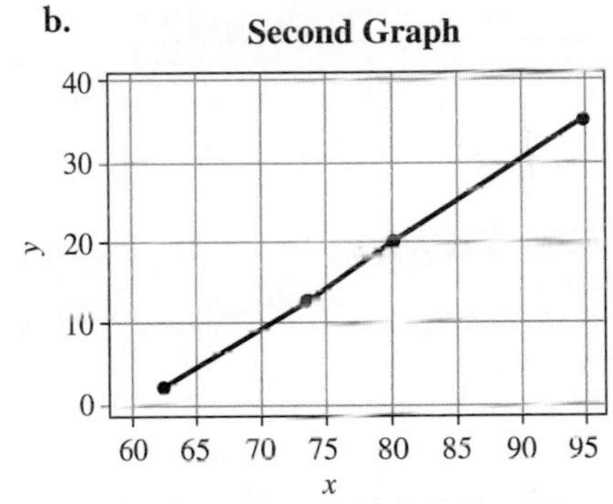

c. (a) suggests an accelerated rate from 1980 to 1995. (b) suggests rate of increase has been constant from 1962 to 1995.
d. By adjusting the horizontal intervals, it is possible to cause the line to show different slopes in those different intervals.

2.201 Samples of size 30 usually demonstrated some of the properties of the population. As the sample size was increased, more of the properties of the population were shown. The suggested distributions in this exercise seem to require sample sizes greater than 30 for a closer match to the population.

Chapter 3

3.1 a. Yes. A relationship seems to exist, the higher number of personal fouls go with the higher values of points scored per game.
b. somewhat; explanations will vary

3.2

	On Airplane	Hotel Room	All Other	Marginal Total
Business	35.5%	9.5%	5.0%	50%
Leisure	25.0%	16.5%	8.5%	50%
Marginal Total	60.5%	26.0%	13.5%	100%

3.3

	On Airplane	Hotel Room	All Other	Marginal Total
Business	71.0%	19.0%	10.0%	100%
Leisure	50.0%	33.0%	17.0%	100%
Marginal Total	60.5%	26.0%	13.5%	100%

Business and leisure are separate distributions.

3.4

	On Airplane	Hotel Room	All Other	Marginal Total
Business	58.7%	36.5%	37.0%	50%
Leisure	41.3%	63.5%	63.0%	50%
Marginal Total	100%	100%	100%	100%

Each of the categories is a separate distribution.

3.5

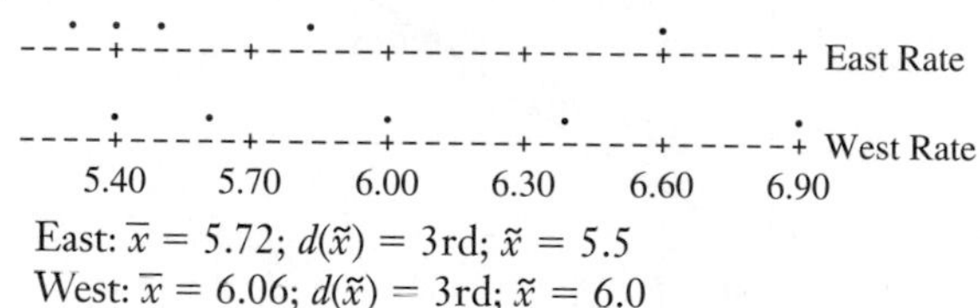

East: $\bar{x} = 5.72$; $d(\tilde{x}) = 3\text{rd}$; $\tilde{x} = 5.5$
West: $\bar{x} = 6.06$; $d(\tilde{x}) = 3\text{rd}$; $\tilde{x} = 6.0$

3.6 height, weight is often predicted

3.7

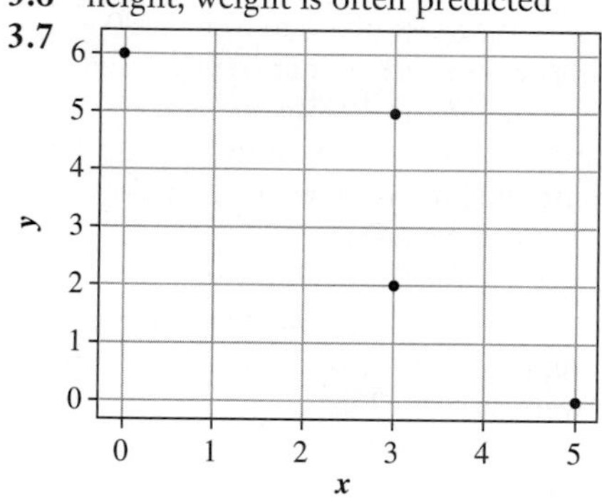

3.8 **a.**

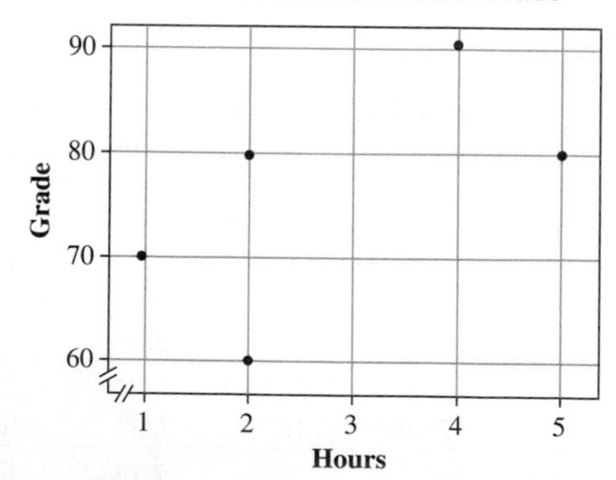

b. As hours studied increased, it seems that exam grades increased.

3.9 **a.** school's poverty level, passage rate
b. Yes, there is a pattern.
c. As poverty level increased, passage rate decreased.

3.10 **a.** age, height
b. age = 3 yr, height = 87 cm

c. Sarah's growth was well below normal.
d. Sarah's information is below normal band.

3.11 **a.** Population: employed college graduates age 30–55 who had been out of college 10 or more years.
Variables: Type of worker: tech worker or other worker
Reason for going to school: professional, personal, both

b.

	Reason for Going to School			
Worker	Professional	Personal	Both	Marginal Total
Tech	28%	31%	41%	100%
Other	47%	20%	33%	100%

3.13 **a.**

	Interstate Highway Speed Limits (mph)					
Vehicle Type	75	70	65	60	55	Row Totals
Cars	10	16	22	0	2	50
Trucks	9	11	20	3	7	50
Column Totals	19	27	42	3	9	100

b.

	Interstate Highway Speed Limits (mph)					
Vehicle Type	75	70	65	60	55	Row Totals
Cars	10%	16%	22%	0%	2%	50%
Trucks	9%	11%	20%	3%	7%	50%
Column Totals	19%	27%	42%	3%	9%	100%

c.

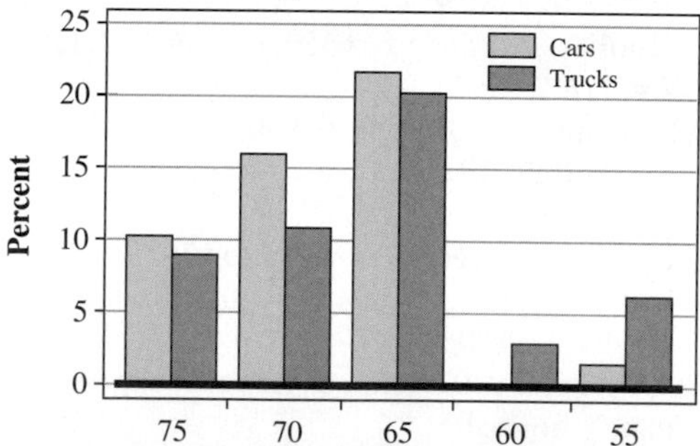

3.15 **a.** 3000
b. Two variables, political affiliation and news information preferred, are paired together. Both variables are qualitative.
c. 950 **d.** 50% [1500/3000] **e.** 25% [200/800]

3.17 **a.**

Three-Month CDs

Minimum deposit ($100): 25, 10, 5

Annual Percent of Return: 1.0, 1.5, 2.0

b.

	Low	Q_1	Median	Q_3	High
5:	1.19	1.265	1.545	2.068	2.16
10:	0.98	1.35	1.73	2.13	2.23
25:	1.00	1.35	1.490	1.990	2.00

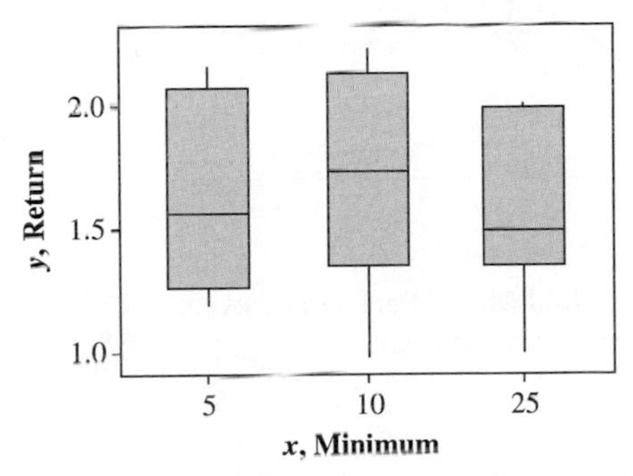

c. 1. The $500 minimum does not have any of the very low return rates like the other two have.
2. All three have similar interquartile range values.
3. The $2500 minimum does not have any of the higher return rates.

3.19 a.

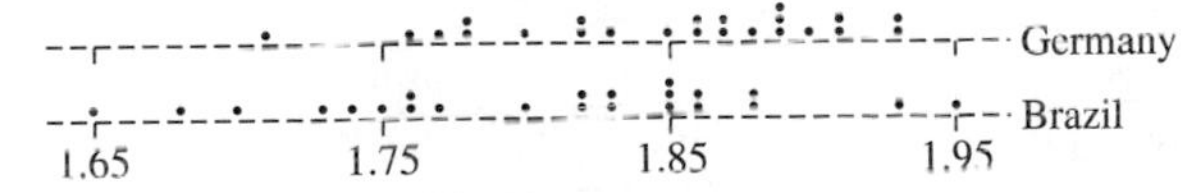

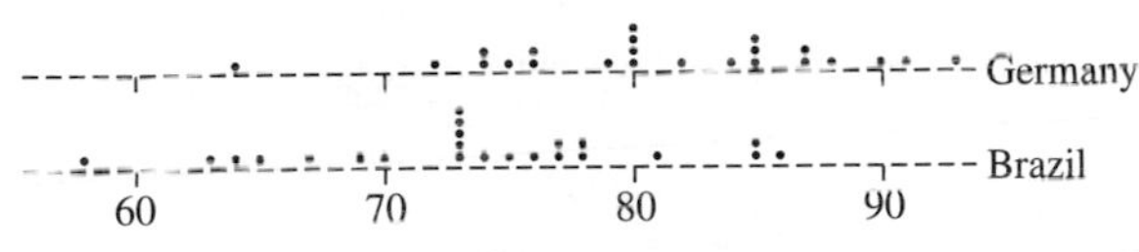

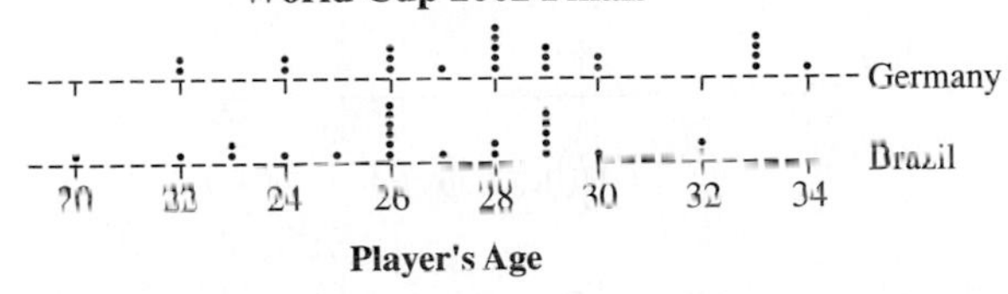

b. German team is slightly taller and slightly heavier, and a bit older than the Brazilian team.

c. Each variable was used separately, players on each team are different—no pairing took place.

3.21

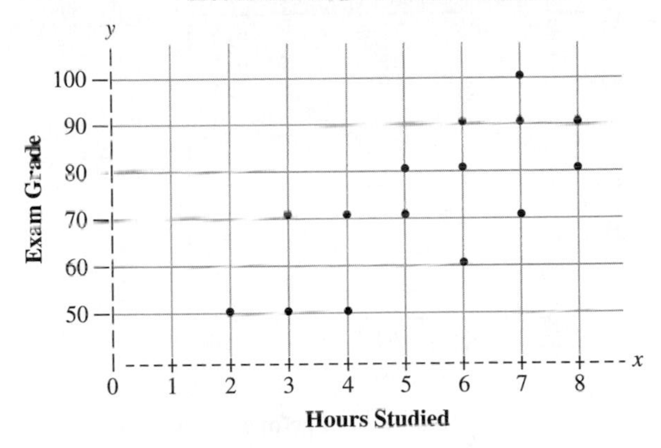

3.23

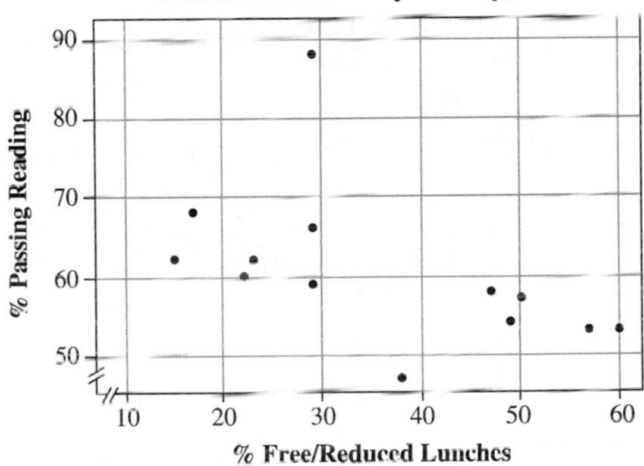

3.25 a.

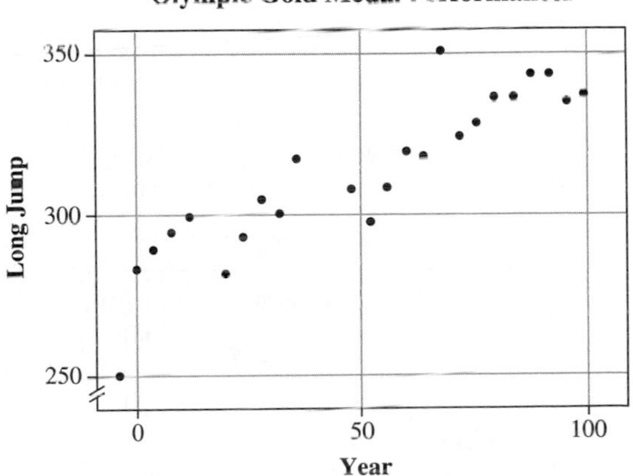

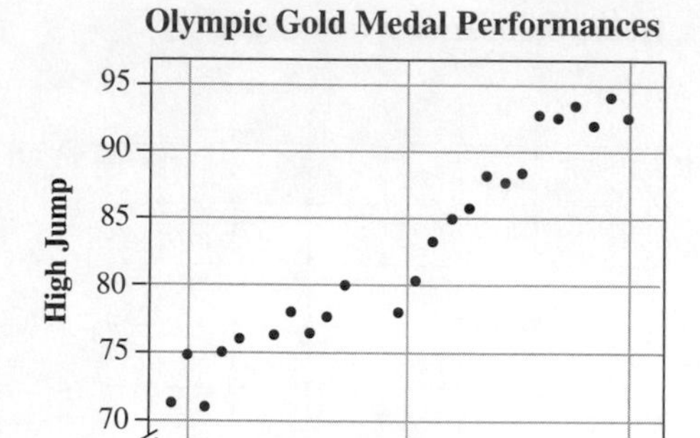

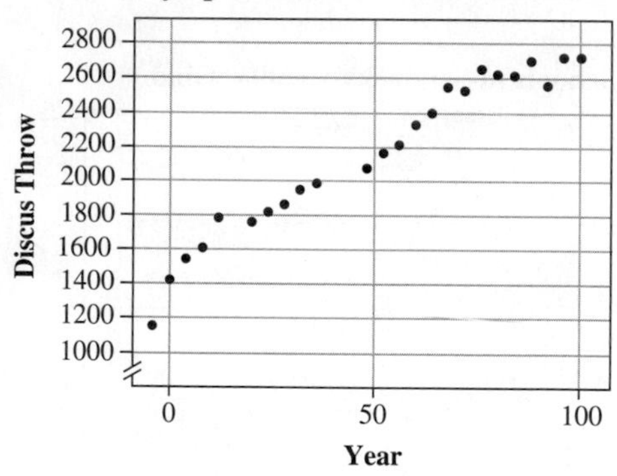

b. There is a very strong increasing pattern in all three, and even though it is not a perfectly straight line, they all seem to follow a fairly straight pattern.

c. These three scatter diagrams indicate that our Olympic athletes, at least the gold-medal-winning ones, jump higher and longer and throw the discus further. All of these are strength skills and indicate that these athletes are stronger today. It is reasonable to anticipate that the general population will follow a similar pattern.

d.

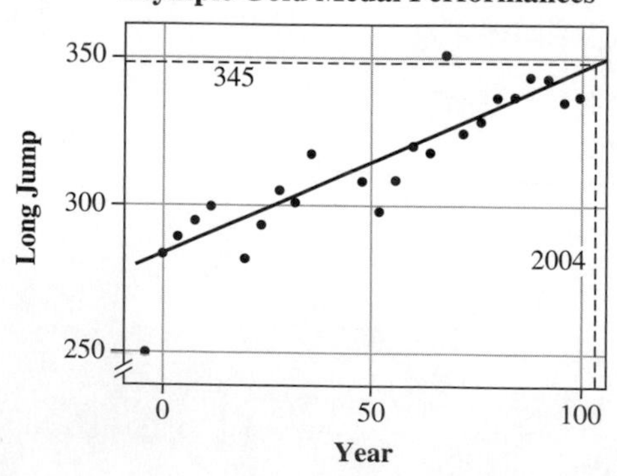

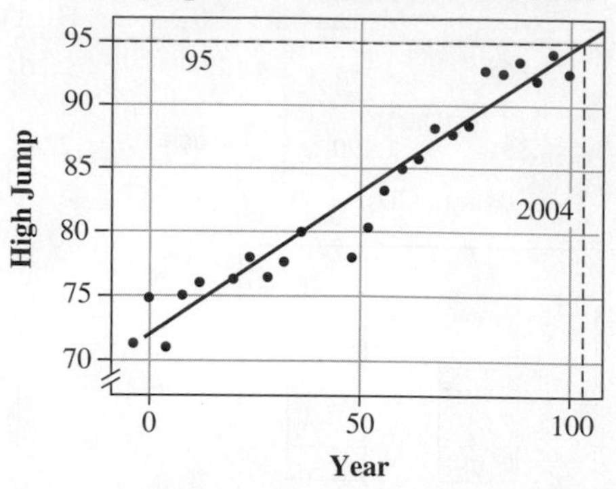

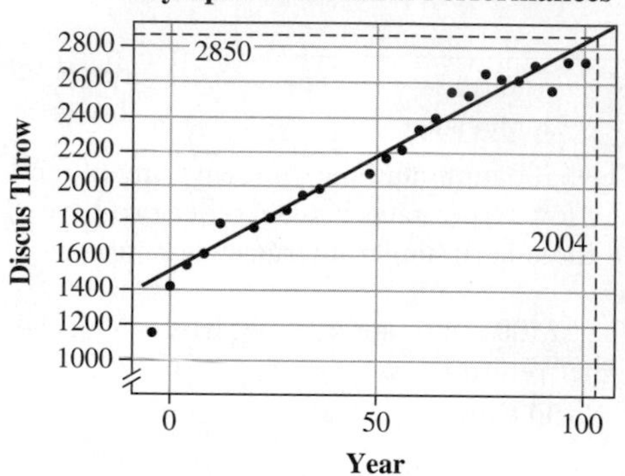

Predictions based on the scatter diagrams (in inches): Long jump—345, high jump—95, discus throw—2850.

e.

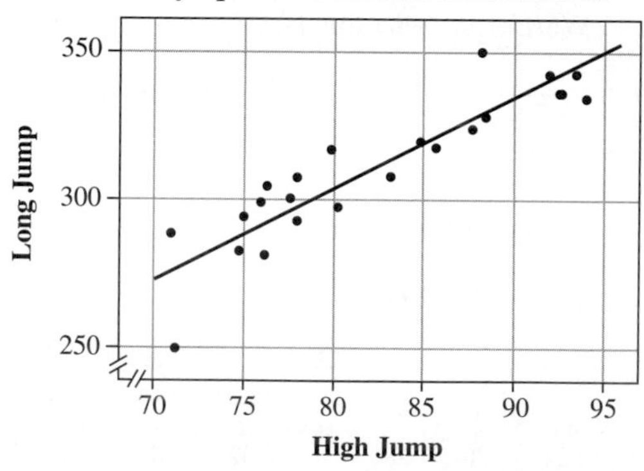

They seem to display a linear relationship. This should not be surprising.

3.27 **a.** closer to a straight line with a positive slope
b. closer to a straight line with a negative slope

3.28 **a.** $n = 5$, $\sum x = 14$, $\sum y = 380$, $\sum x^2 = 50$, $\sum xy = 1110$, $\sum y^2 = 29{,}400$;
$\text{SS}(x) = 10.8$, $\text{SS}(y) = 520$, $\text{SS}(xy) = 46$
b. 0.61

3.29 Answers will vary.

3.30 **a.** $r = 1$ or $r = -1$, two points make a straight line
b. Answers will vary.

3.31 −0.75; 0.00; +0.75

3.32 **a.** manatees, powerboats
b. number of registrations, manatee deaths
c. As one increases other does also

3.33 **a.** $\sum x = 315$; $\sum y = 294.7$; $\sum x^2 = 14{,}875$; $\sum xy = 15{,}274$; $\sum y^2 = 19{,}177$; 0.92
b.

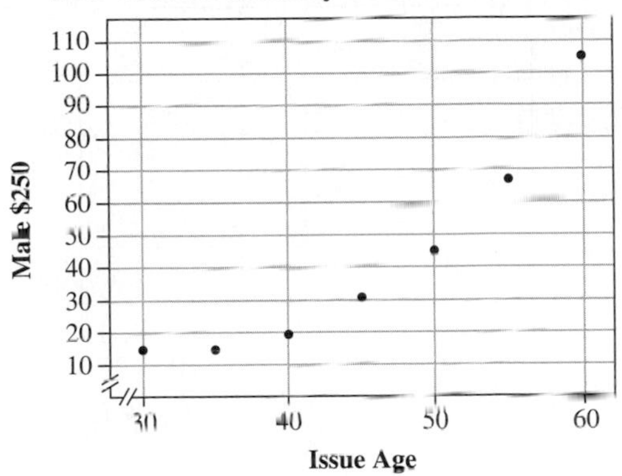

c. elongated pattern
d. Rate for insurance increases (accelerates) as the insured person's issue age increases, thus the "upward-bending" pattern.

3.34 **a.** number of drownings, ice cream sales
b. no, both increase during hot weather

3.35 **a.** Impossible. There must be a calculation or typographical error.
b. very little or no linear correlation

3.37 $n = 13$; $\sum x = 465$; $\sum y = 787$; $\sum x^2 = 19{,}453$; $\sum xy = 27{,}218$, $\sum y^2 = 48{,}849$
a. 2820.307692
b. 1205.23
c. −932.385
d. −0.506

3.39 **a.** 0.7
b. $\sum x = 81$, $\sum y = 110$, $\sum x^2 = 487$, $\sum xy = 625$, $\sum y^2 = 842$; 0.74

3.41 **a.**

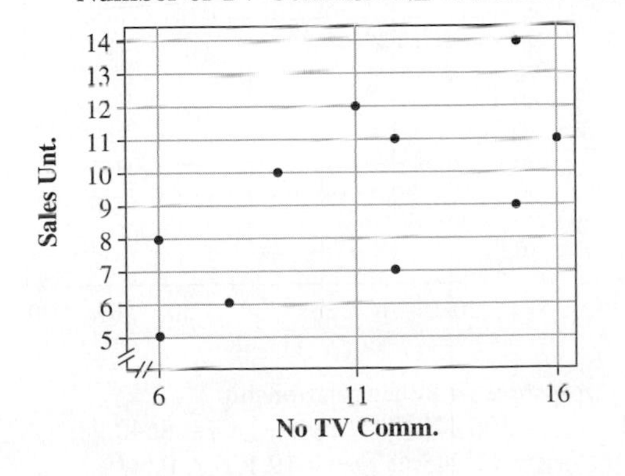

b. from 1/2 to 2/3
c. $\sum x = 110$, $\sum y = 93$, $\sum x^2 = 1332$, $\sum xy = 1085$, $\sum y^2 = 937$; 0.66

3.43

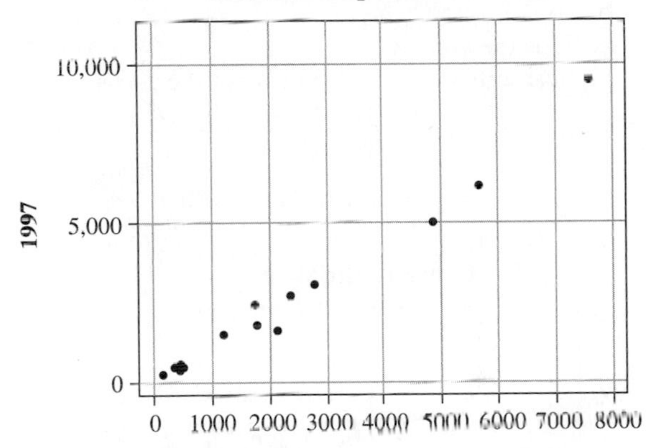

$\sum x = 33{,}019.0$; $\sum y = 36{,}413.0$; $\sum x^2 = 140{,}541{,}807$; $\sum xy = 159{,}275{,}750$; $\sum y^2 = 182{,}712{,}391$; $r = 0.990$; definitely appears to be a linear relationship

3.45 No, both increase during months of warmer weather and decrease during months of cooler weather.

3.47 $b_0 = \dfrac{\sum y - (b_1 \cdot \sum x)}{n} = \dfrac{\sum y}{n} - \dfrac{(b_1 \cdot \sum x)}{n} = \bar{y} - b_1\bar{x}$

3.48 **a.** $\sum x^2 = 13{,}717$; $SS(x) = 1396.9$
$\sum y^2 = 15{,}298$; $SS(y) = 858.0$
$\sum xy = 14{,}257$; $SS(xy) = 919.0$
b. $\sum$'s are sums of data values; SS( ) are parts of complex formulas.

3.49 **a.** 28.1, 47.9 **b.** yes

3.50 **a.** $\sum x = 14$, $\sum y = 380$, $\sum x^2 = 50$, $\sum xy = 1110$; $\hat{y} = 64.1 + 4.26x$
b. use (1, 68.4) and (3, 76.9)

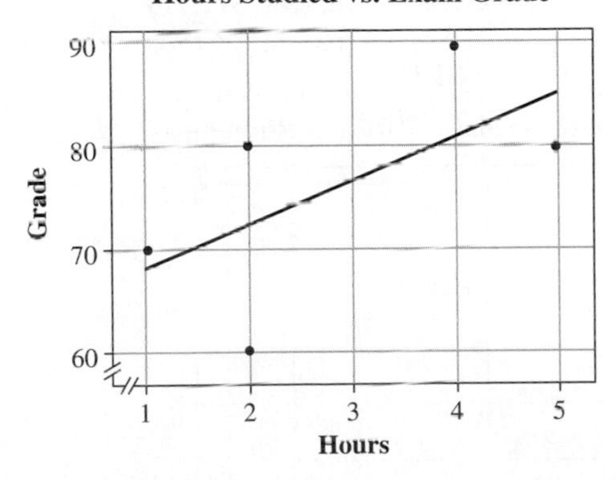

c. Yes, as hours studied increased, exam grades increased.

3.51 **a.** As height increases by one inch, weight increases by 4.71 pounds.
b. The scale for the y-axis starts at $y = 95$ and the scale for the x-axis starts at $x = 60$.

3.52 **a.** $\hat{y} = 14.9 + 0.66(40) = 41.3 = 41$
b. no
c. 41 is the average

3.53 Vertical scale is at $x = 58$ and is not the y-axis.

3.54 5.83; −260.61; $b_1 \approx \dfrac{y_2 - y_1}{x_2 - x_1} = \dfrac{130 - 95}{67 - 61} = 5.83;$

$b_0 = y - b_1 x = 130 - 5.83(67) = -260.61$

3.55 **a.**

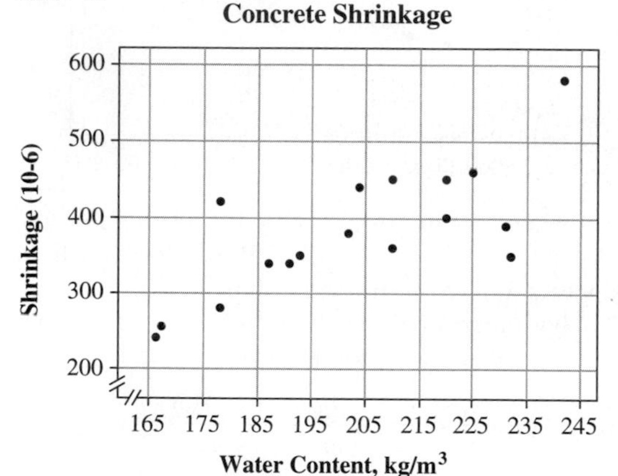

b. yes; an elongated pattern from lower left to upper right
c. $\sum x = 3456$; $\sum y = 6485$; $\sum x^2 = 711{,}306$; $\sum xy = 1{,}341{,}865$; $\sum y^2 = 2{,}586{,}325$; $\hat{y} = -166.4 + 2.69x$

3.57 **a.** When no long distance calls are made there is still the monthly phone charge of \$23.65.
b. \$1.28 is the rate at which the total phone bill will increase for each additional long distance call.

3.59 **a.** $\hat{y} = 185.7 - 21.52(3) = 121.14$, \$12,114
b. $\hat{y} = 185.7 - 21.52(6) = 56.58$, \$5,658
c. 21.52(\$100) = \$2,152

3.61 **a.**

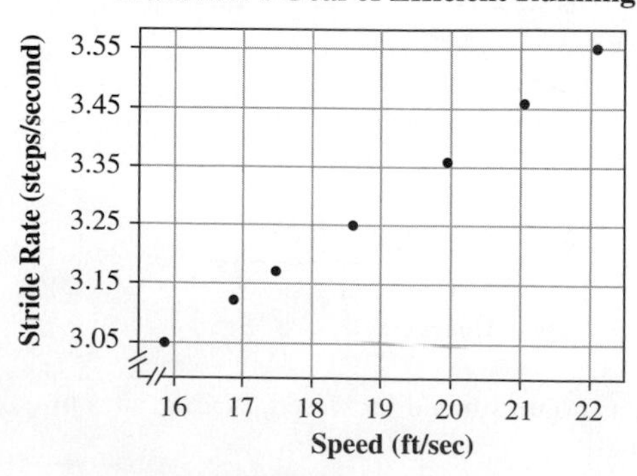

b. yes, linear pattern
c. $\sum x = 132.00$, $\sum y = 22.96$, $\sum x^2 = 2520.61$, $\sum xy = 435.486$; $\hat{y} = 1.77 + 0.0803x$
d. For each one foot per second increase in speed, the stride will increase by 0.0803 of a step per second.
e.

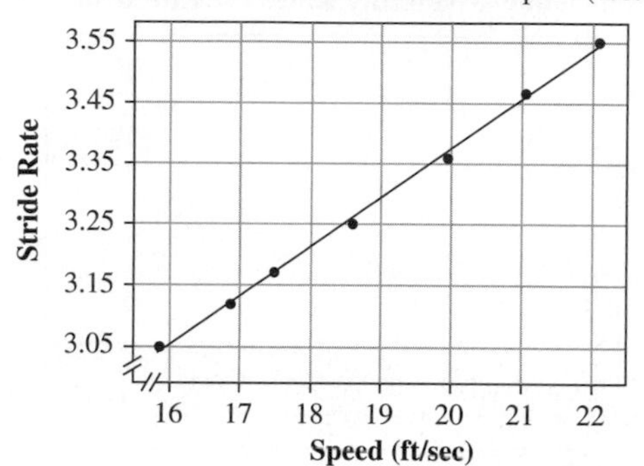

f.

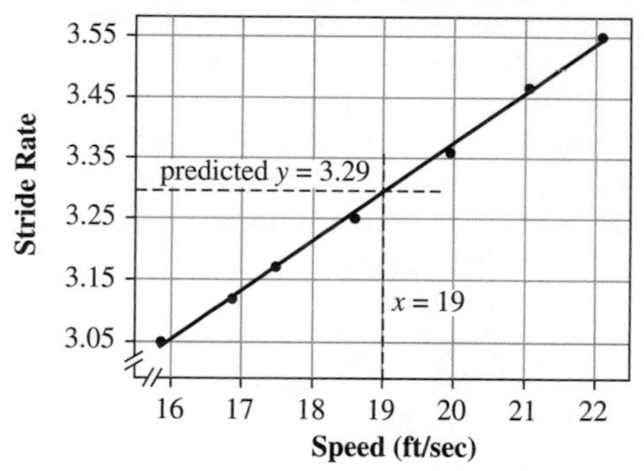

g. domain of the study is speed 16 to 22 feet per second; a running speed and the regression methods do not apply outside the domain

3.63 **a.**

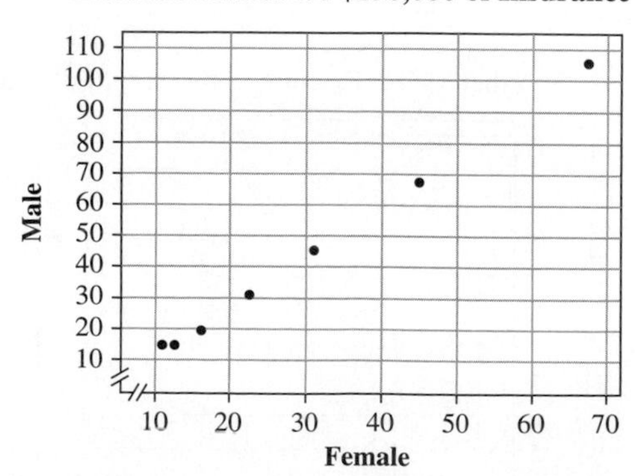

yes, shows a linear relationship
b. $\sum x = 206.47$; $\sum y = 294.7$; $\sum x^2 = 8640.98$; $\sum xy = 12{,}845.6$; $\sum y^2 = 19{,}176.8$; 0.999

c. $\hat{y} = -5.92 + 1.63x$
d. $18.53
e. Males pay $1.63 for every $1.00 that females pay

3.67 a. A relationship seems to exist; as the number of personal fouls committed increases, so does the number of points scored per game.
b. It seems contrary to common sense, but if there is a predictable pattern, then it might be possible to make such a prediction.

3.71 a. Population: workers ages 25–64
Variables: gender, nominal; savings set aside, ordinal
b.

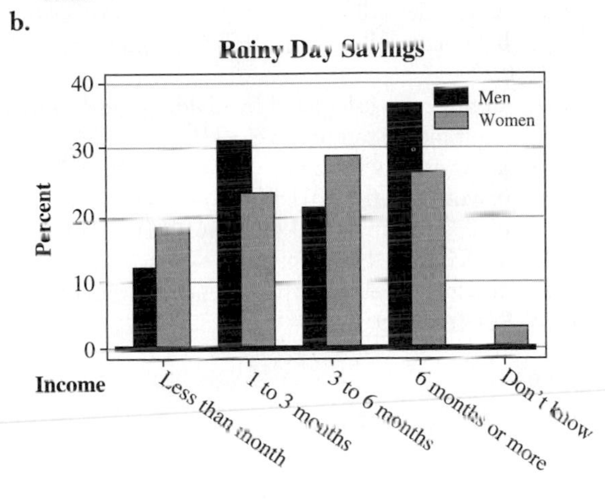

c. no

3.73 a.

	Less Than 6 mo	6 mo–1 yr	More Than 1 yr	Total
Under 28	413	192	295	900
28–40	574	208	218	1000
Over 40	653	288	259	1200
Total	1640	688	772	3100

b.

	Less Than 6 mo	6 mo–1 yr	More Than 1 yr	Total
Under 28	13.3%	6.2%	9.5%	29.0%
28–40	18.5%	6.7%	7.0%	32.2%
Over 40	21.1%	9.3%	8.4%	38.8%
Total	52.9%	22.2%	24.9%	100%

c.

	Less Than 6 mo	6 mo–1 yr	More Than 1 yr	Total
Under 28	45.9%	21.3%	32.8%	100%
28–40	57.4%	20.8%	21.8%	100%
Over 40	54.4%	24.0%	21.6%	100%
Total	52.9%	22.2%	24.9%	100%

d.

	Less Than 6 mo	6 mo–1 yr	More Than 1 yr	Total
Under 28	25.2%	27.9%	38.2%	29.0%
28–40	35.0%	30.2%	28.2%	32.3%
Over 40	39.8%	41.9%	33.6%	38.7%
Total	100%	100%	100%	100%

e.

Last Saw Your Doctor When?

3.75 a. correlation **b.** regression **c.** correlation
d. regression **e.** correlation

3.77

Scatterplot

$n = 5$, $\sum x = 10$, $\sum y = 25$, $\sum x^2 = 30$, $\sum xy = 70$, $\sum y^2 = 165$; $r = 1.00$; $\hat{y} = 1.0 + 2.0x$

3.79 Answers will vary.

3.81 Answers will vary.

3.83 a.

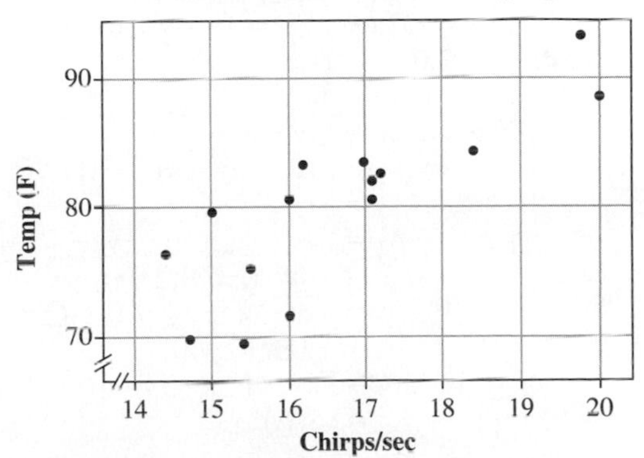

b. linearly increasing pattern
c. $n = 15, \sum x = 249.8, \sum y = 1200.60,$
$\sum x^2 = 4200.56, \sum xy = 20{,}127.5; \hat{y} = 25.2 + 3.29x$
d. $x = 14$, 71F; $x = 20$, 91F
e. seems reasonable
f. $x = 16$, 78F

3.85 a.

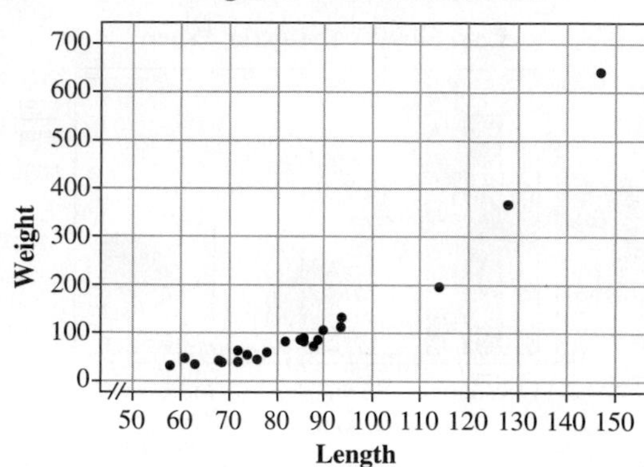

b. yes, follows a tight pattern, just not a straight line
c. no
d. not a straight line
e. $n = 25; \sum x = 2124; \sum y = 2705; \sum x^2 = 190{,}282;$
$\sum xy = 287{,}819; \sum y^2 = 702{,}127; 0.914$
f. Data are very elongated.

3.87 a.

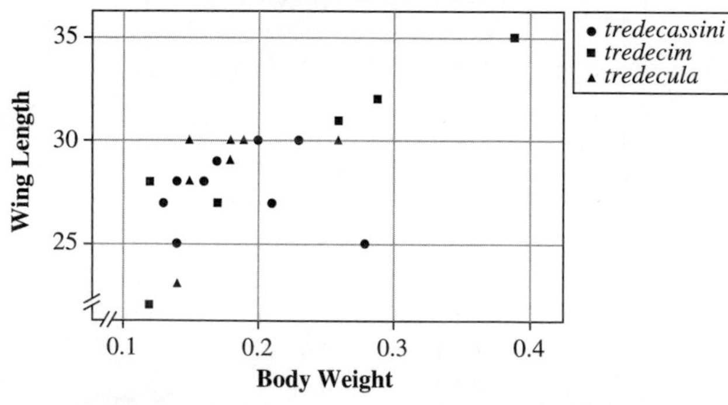

b. linear, moderately strong, increasing, all species are intermingled
c. $n = 24, \sum x = 4.6, \sum y = 680, \sum x^2 = 0.978400,$
$\sum xy = 133.050, \sum y^2 = 19448.0; 0.649$
d. $\hat{y} = 23.0 + 28.1x$
e. 28.62 mm

3.89 a. Numerator $= \sum(x - \bar{x})(y - \bar{y})$
$= \sum[xy - \bar{x}y - x\bar{y} + \bar{x}\bar{y}]$
$= \sum xy - \bar{x} \cdot \sum y - \bar{y} \cdot \sum x + n\bar{x}\bar{y}$
$= \sum xy - [(\sum x/n) \cdot \sum y] - [(\sum y/n) \cdot \sum x]$
$+ [n \cdot (\sum x/n)(\sum y/n)]$
$= \sum xy - [(\sum x \cdot \sum y/n) - (\sum x \cdot \sum y/n)$
$+ (\sum x \cdot \sum y/n)]$
$= \sum xy - [(\sum x \cdot \sum y)/n]$
$= \mathrm{SS}(xy)$

Denominator of formula (3.1):
Denominator $= (n - 1)s_x s_y$
$= (n - 1) \cdot \sqrt{SS(x)/(n - 1)} \cdot \sqrt{SS(y)/(n - 1)}$
$= \sqrt{SS(x) \cdot SS(y)}$
Therefore, formula (3.1) is equivalent to formula (3.2).

Chapter 4

4.1 **a.** red, yellow, and purple; blue, brown, green, and orange
b. not exactly, but similar
4.3 Results will vary but have denominators of 25.
4.5 **a.** Results will vary but have denominators of 50.
b. Results will vary but have denominators of 100.
4.7 0.225
4.8 All three are calculated by dividing the experimental count by the sample size.
4.9 **a.** Answers will vary.
b. yes, $\frac{1}{2} = 0.5$, $P(\text{red}) = \frac{1}{2}$, $P(\text{black}) = \frac{1}{2}$
c. Probability is substantially greater than 0.5.
4.11 **a.** 0.5397 **b.** 0.0849
4.13 Results will vary but have denominators of 50.
4.15 Results will vary.
4.17 {0, 1, 2, 3, 4, 5, 6, 7, 8, 9}
4.18

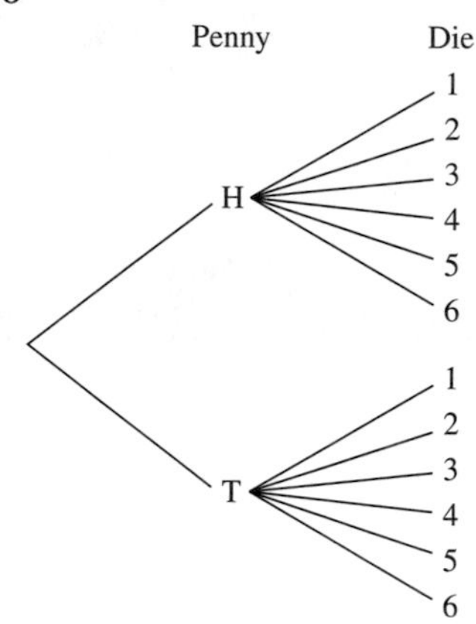

4.19

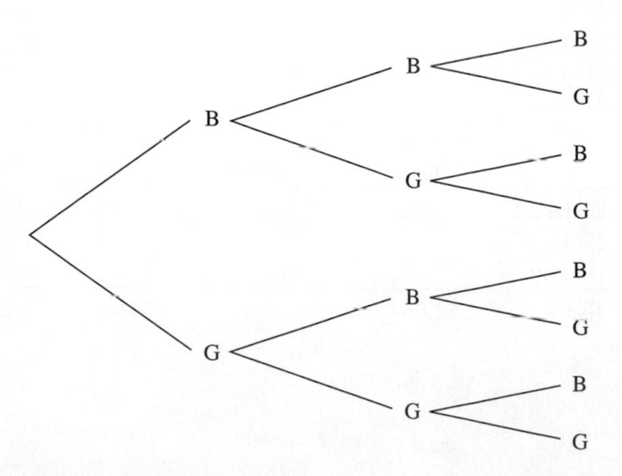

4.20 **a.** $S = \{\$1, \$5, \$10, \$20\}$
b.

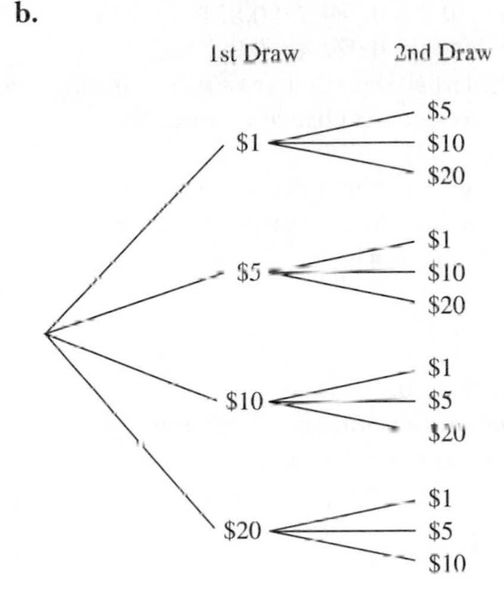

c.

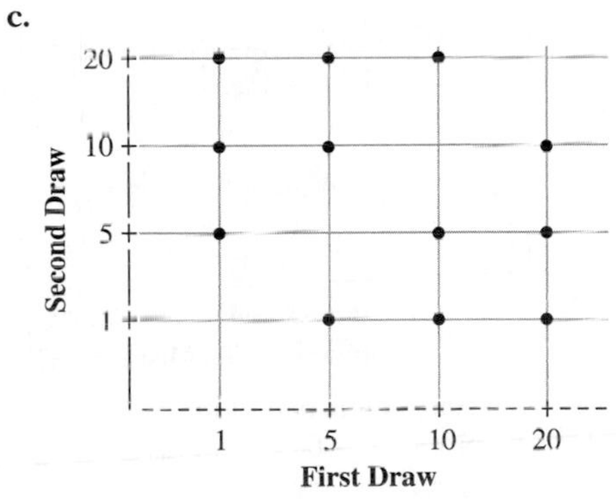

4.21 {JH, JC, JD, JS, QH, QC, QD, QS, KH, KC, KD, KS}
4.23 {yyy, yyn, yny, ynn, nyy, nyn, nny, nnn}
4.25 **a.** {HH, HT, TH, TT} and equally likely
b. {HH, HT, TH, TT} and not equally likely
4.27 {H1, H2, H3, H4, H5, H6, T1, T2, T3, T4, T5, T6}
4.29 Results will vary.
4.31 $P(5) = 4/36$; $P(6) = 5/36$; $P(7) = 6/36$; $P(8) = 5/36$; $P(9) = 4/36$; $P(10) = 3/36$; $P(11) = 2/36$; $P(12) = 1/36$
4.32 4/5
4.33 **a.** 1/13 **b.** 12:1
4.34 40/52
4.35 **a.** 0.0004267 **b.** 12:7 **c.** 37:1 (rounded)
4.37 $P(R) = 1/4$, $P(Y) = 1/4$, $P(G) = 2/4 = 1/2$
4.39 **a.** 1/6 **b.** 3/6 **c.** 4/6 **d.** 3/6
4.41 $P(A) = 1/7$, $P(B) = 2/7$, $P(C) = 4/7$
4.43 All are inappropriate.
4.45 **a.** 0.55 **b.** 0.40
4.47 **a.** 1 to 232, 232 to 1, 0.00429
b. 1 to 3699, 3699 to 1, 0.00027
c. 1 to 3999, 3999 to 1, 0.00025
d. 1 to 15, 15 to 1, 0.0625
e. 1 to 64, 64 to 1, 0.0154
f. 1 to 129, 129 to 1, 0.0077
4.49 0.90
4.51 **a.** yes **b.** no **c.** no **d.** yes **e.** no **f.** yes **g.** no
4.52 **a.** A & C and A & E are mutually exclusive.
b. 12/36, 11/36, 10/36
4.53 **a.** not mutually exclusive
b. not mutually exclusive
c. not mutually exclusive
d. mutually exclusive
4.55 There is no intersection.
4.57 **a.** 0.7 **b.** 0.6 **c.** 0.7 **d.** 0.0
4.59 no
4.61 **a.** 0.4412 **b.** 0.5000 **c.** 0.1667
d. 0.7745; 0.2255 **e.** Venn diagram:

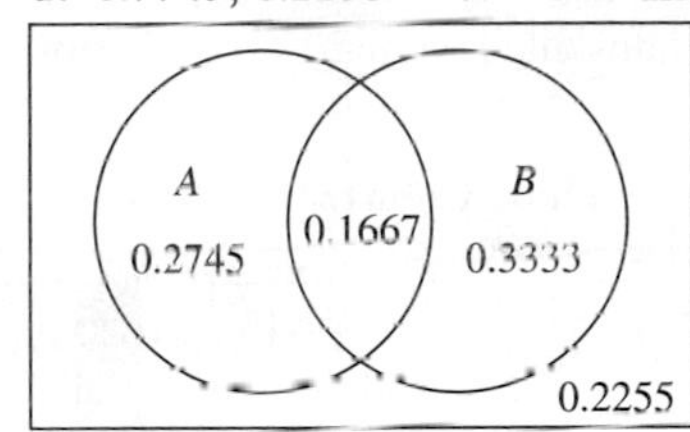

4.63 0.04
4.65 **a.** 0.45 **b.** 0.40 **c.** 0.55
d. no, $P(S) \neq P(S \mid F) \neq P(S \mid M)$
4.66 0.15
4.67 0.28
4.68 **a.** 0.40 **b.** yes
4.69 **a.** working separately **b.** 0.869
c. Working together is less reliable.
4.70 **a.** 4:44 or 1:11
b. $4/48 = 1/(1 + 11) = 1/12$ **c.** no
4.71 **a.** cannot occur at the same time or have no elements in common
b. occurrence of one has no effect on the probability of the other
c. Mutually exclusive has to do with whether or not the events share common elements; independence has to do with the effect one event has on the other event's probability.
4.73 **a.** independent **b.** not independent
c. independent **d.** independent
e. not independent **f.** not independent
4.75 **a.** 0.12 **b.** 0.4 **c.** 0.3
4.77 **a.** 0.5 **b.** 0.667 **c.** no
4.79 **a.** independent **b.** independent **c.** dependent
4.81 **a.** 0.51 **b.** 0.15 **c.** 0.1326
4.83 **a.** 0.0289 **b.** 0.6889 **c.** 0.0008
4.85 0.0741
4.87 **a.** 0.36 **b.** 0.16 **c.** 0.48
4.89 **a.** 3/5 **b.** 0.16; 0.48; 0.36
4.91 **a.** whether or not the events are independent
b. no **c.** 0.4074
4.93 $P(\text{owner} \mid \text{married})$, $P(\$200\text{+k income} \mid \text{married owner})$
4.95 0.7

4.97 **a.** 0.15 **b.** 0.65 **c.** 0.7 **d.** 0.5 **e.** 0.7
f. no, independent events

4.99 **a.** 20/56 **b.** 30/56 **c.** 6/56

4.101 **a.** P(A wins on 1st turn) = 1/2
P(B wins on 1st turn) = 1/4
P(C wins on 1st turn) = 1/8
b. P(A wins on 2nd turn) = 1/16
P(A wins on 1st try or 2nd try) = 9/16
P(B wins on 1st try or 2nd try) = 9/32
P(C wins on 1st try or 2nd try) = 9/64

4.103 8/30

4.105 **a.** 0.50 **b.** 0.50 **c.** 0.615 **d.** 0.385

4.107 **a.** 0.625 **b.** 0.25
c. P(satisfied | skilled woman) = 0.25;
P(satisfied | unskilled woman) = 0.667; not independent

4.108 **a.**

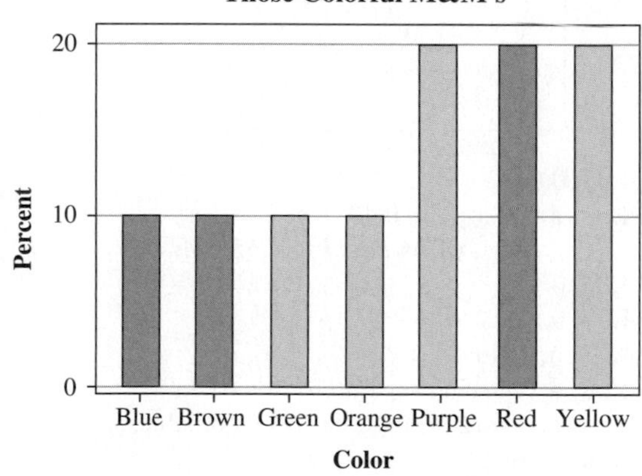

b. 40 times the percentage for each color
c. no, should expect the numbers to vary, but average, the numbers in (b)
d. Very surprised! To observe any predetermined set of numbers for each color would be a fairly rare event.

4.111 **a.** 0.0958 **b.** 0.1172 **c.** 0.1396 **d.** 0.4949
e. 0.1606 **f.** 0.2182 **g.** 0.0938

4.113 **a.** 0.30 **b.** 0.40 **c.** no
d. mutually exclusive events

4.115 **a.** false **b.** true **c.** false **d.** false

4.117 **a.** {GGG, GGR, GRG, GRR, RGG, RGR, RRG, RRR}
b. 3/8 **c.** 7/8

4.119 P(boy) ≈ 7/8

4.121 **a.** 0.3168 **b.** 0.4659 **c.** no **d.** no
e. "Candidate wants job" and "RJB wants candidate" could not both happen.

4.123 **a.** 0.429 **b.** 0.476 **c.** 0.905

4.125 0.300

4.127 **a.** 0.30 **b.** 0.60 **c.** 0.10 **d.** 0.60 **e.** 0.333
f. 0.25

4.129 **a.** 0.531 **b.** 0.262 **c.** 0.047

4.131 0.592

4.133 **a.** 0.60 **b.** 0.648 **c.** 0.710
d. (a) 0.70 (b) 0.784 (c) 0.874
e. (a) 0.90 (b) 0.972 (c) 0.997
f. The larger the number of games in the series, the greater the chance that the "best" team will win.
The greater the difference between the two teams' individual chances, the more likely the "best" team wins.

Chapter 5

5.1 **a.** 19%, 8% **b.** 5 nights
c. number of evening meals American adults cook at home in an average week
d.

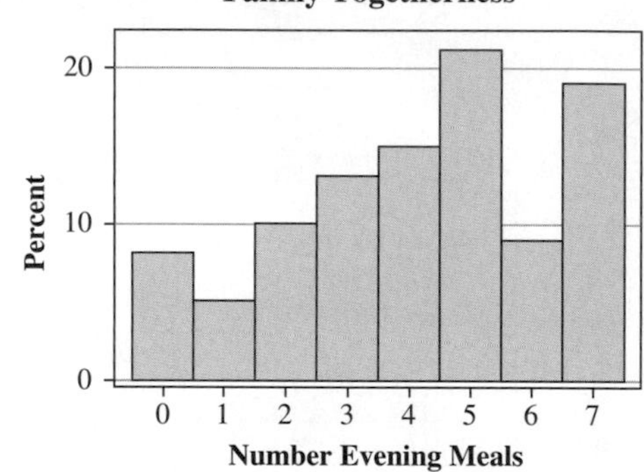

e. relative frequency distribution

5.2 number of siblings, $x = 0, 1, 2, 3, \ldots$
length of conversation, 0 to ? minutes

5.3 **a.** countable **b.** measurements

5.5 number of children; $x = 0, 1, 2, 3, \ldots, n$; discrete

5.7 distance; $x = 0$ to n, n = radius, continuous

5.9 **a.** number of home runs **b.** discrete, count

5.11

x	0	1
$P(x)$	1/2	1/2

5.12

x	1	2	3	4	5	6
$P(x)$	1/6	1/6	1/6	1/6	1/6	1/6

5.13 Each possible outcome is assigned a unique numerical value.

5.14 All possible outcomes are accounted for.

5.15 **a.** $P(x) = \frac{1}{6}$, for $x = 1, 2, 3, 4, 5, 6$

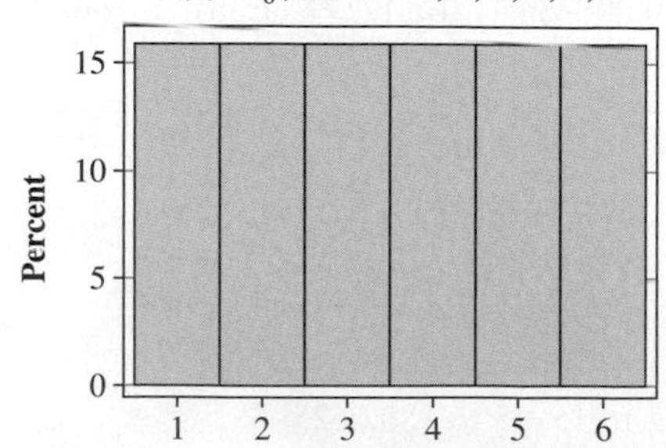

b. uniform or rectangular

5.16 a.

Number	Proportion
0	0.196
1	0.131
2	0.162
3	0.168
4	0.121
5	0.082
6	0.054
7 to 10	0.072
11 or more	0.014

b. Student now chooses between multiple acceptances.

5.17

x	0	1	2	3
$P(x)$	0.20	0.30	0.40	0.10

5.19 a.

x	$P(x)$
1	0.12
2	0.18
3	0.28
4	0.42
Σ	1.00

$P(x)$ is a probability function.

b. $P(x) = (x^2 + 5)/50$, **for** $x = 1, 2, 3, 4$

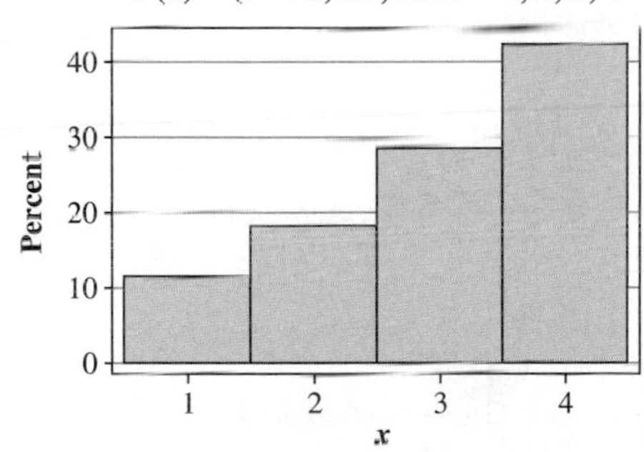

5.21 a. yes

b.

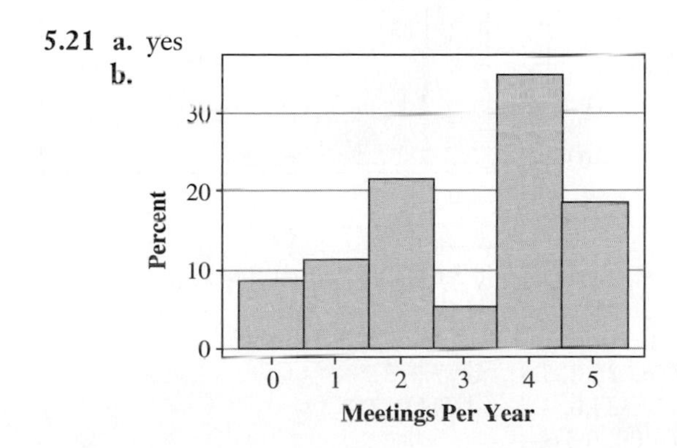

5.23 a. No. The information displays all properties of a probability distribution except one; the variable is attribute (not numerical).

b.

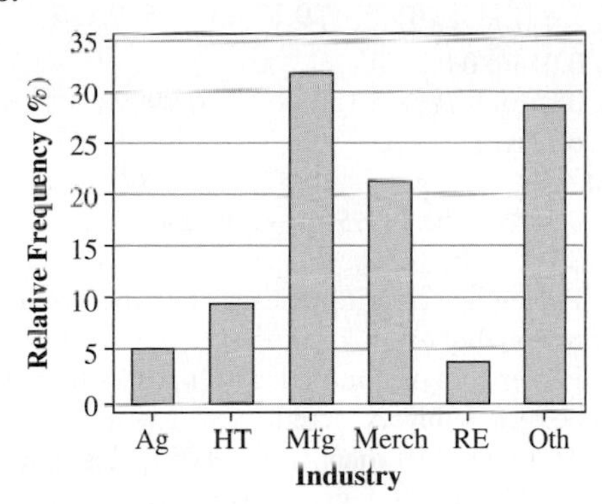

5.25 Answers will vary.

5.27 $\sigma^2 = \sum[(x - \mu)^2 \cdot P(x)]$
$= \sum [x^2 - 2x\mu + \mu^2) \cdot P(x)]$
$= \sum[x^2 \cdot P(x) - 2x\mu \cdot P(x) + \mu^2 \cdot P(x)]$
$= \sum[x^2 \cdot P(x)] - 2\mu \cdot \sum[x \cdot P(x)] + \mu^2 \cdot [\sum P(x)]$
$= \sum[x^2 \cdot P(x)] - 2\mu \cdot [\mu] + \mu^2 \cdot [1]$
$= \sum[x^2 \cdot P(x)] - 2\mu^2 + \mu^2$
$= \sum[x^2 \cdot P(x)] - \mu^2$ or $\sum[x^2 \cdot P(x)] - \{\sum[x \cdot P(x)]\}^2$

5.28 a.

x	$P(x)$	(b) $xP(x)$	$x^2P(x)$
1	1/6	1/6	1/6
2	2/6	4/6	8/6
3	3/6	9/6	27/6
Σ	6/6 = 1.0	(c) 14/6 = 2.33	36/6 = 6.0

d. $\mu = 2.33$ **e.** $\sigma^2 = 0.55556$ **f.** $\sigma = 0.745$

5.29 nothing of any meaning

5.30 a. vary, close to −\$0.20 **b.** −\$0.20 **c.** close, no, need mean = 0

5.31 2.0, 1.0

5.33 a.

Random Single Digits; $P(x) = 0.1$, for $x = 0, 1, \ldots, 9$

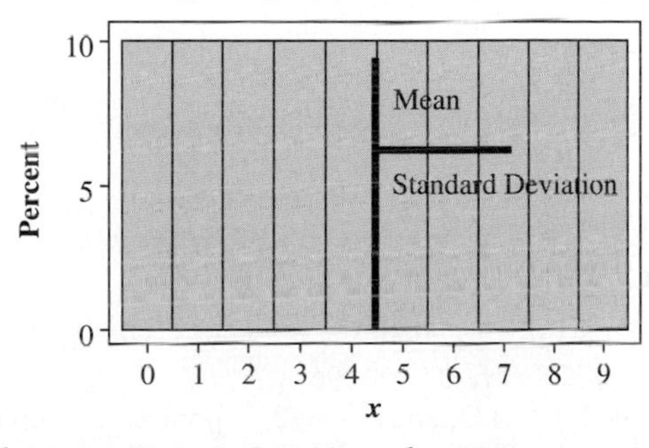

b. 4.5, 2.87 **c.** See (a). **d.** 100%

5.35 a. 2.0, 1.4 **b.** the numbers 1, 2, 3, and 4 **c.** 0.9

5.37 a. 1.623, 0.9 **b.** slightly smaller

5.39 Each question is a separate trial.

5.40 four different ways that one correct and three wrong answers can be obtained

5.41 1/3 is the probability of choosing the right answer; 4 is the number of questions; expect 1/3 correct

5.42 one trial is one coin, $n = 50$, independent; $p = P(\text{heads}) = \frac{1}{2}$, $q = P(\text{tails}) = \frac{1}{2}$; x = the number of heads, from 0 to 50

5.43 a. 24 **b.** 4

5.44 $P(x) = \binom{3}{x}(0.5)^x(0.5)^{3-x}$, 0.125, 0.375, 0.125

5.45 **a.** 0.0146, 0.00098
b. $0 \le$ each $P(x) \le 1$, $\sum P(x) = 0.99998 \approx 1$ (round-off error)

5.46 0.007

5.47 **a.** 0.3585 **b.** 0.0159 **c.** 0.9245

5.48 **a.** 0.240 **b.** 0.240

5.49 **a.** Percentage of minorities is "less than would be reasonably expected."
b. Percentage of minorities is "not less than would be reasonably expected."
c. 0.96, "not less than" **d.** 0.035, "less than"

5.51 **a.** 24 **b.** 5,040 **c.** 1 **d.** 360 **e.** 10 **f.** 15
g. 0.0081 **h.** 35 **i.** 10 **j.** 1 **k.** 0.4096
l. 0.16807

5.53 $n = 100$ trials (shirts)
two outcomes (first quality or irregular)
$p = P(\text{irregular})$
$x = n(\text{irregular})$; any integer value from 0 to 100

5.55 **a.** Trials are not independent
b. $n = 4$, the number of independent trials; two outcomes, success = ace and failure = not ace; $p = P(\text{ace}) = 4/52$ and $q = P(\text{not ace}) = 48/52$; $x = n(\text{aces drawn in 4 trials})$ and could be any number 0, 1, 2, 3 or 4

5.57

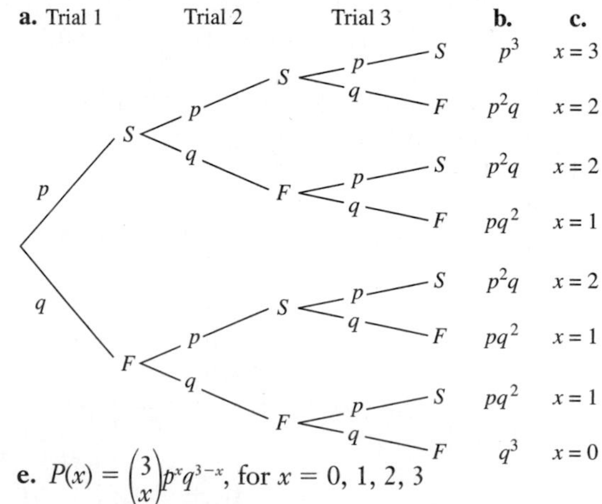

e. $P(x) = \binom{3}{x}p^x q^{3-x}$, for $x = 0, 1, 2, 3$

5.59 **a.** 0.4116 **b.** 0.384 **c.** 0.5625 **d.** 0.329218
e. 0.375 **f.** 0.0046296

5.61 $n = 5$, $p = 1/2$, and $q = 1/2$, exponents add up to 5, x, any integer value from zero to 5; $T(x)$ is a binomial probability distribution: 1/32, 5/32, 10/32, 10/32, 5/32, 1/32

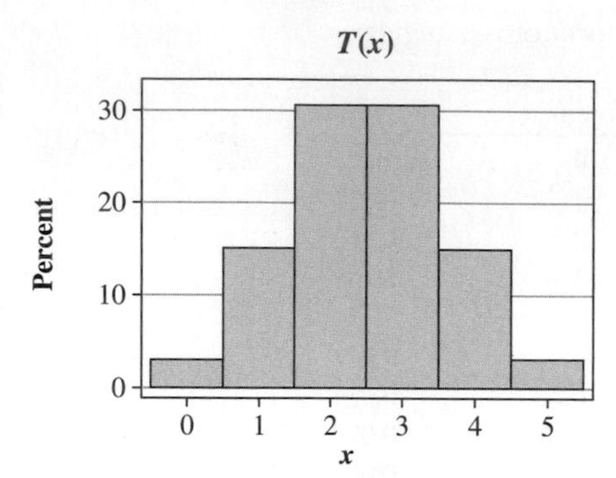

5.63 0.930

5.65 **a.** 0.444 **b.** 0.238 **c.** 0.153

5.67 **a.** 0.590 **b.** 0.918

5.69 0.0011

5.71 0.984

5.73 **a.** 0.3851 **b.** 0.0679 **c.** 0.0072

5.75 **a.** 0.2757 **b.** 0.1865 **c.** 0.5425 **d.** 0.6435

5.79 **a.** 0.2122 **b.** 0.6383

5.81 **a.** $P(2 \text{ of } 2) = 0.308$, $P(9 \text{ of } 15) = 0.274$
b. $P(2 \text{ of } 2) = 0.687$, $P(8 \text{ or more of } 10) = 0.763$

5.83 18, 2.7

5.84 **a.** 0.55, 0.72
b.

x	$P(x)$
0	0.569
1	0.329
2	0.087
3	0.014
4	0.001
5	0+

Binomial Probability Distribution, $n = 11$, $P = 0.05$

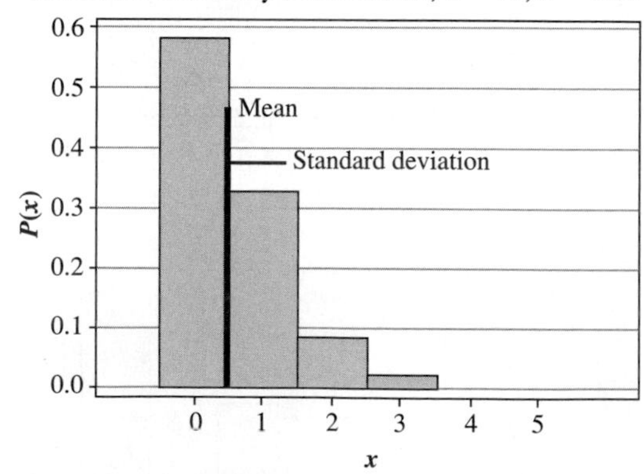

5.85 **a.** $\sum[xP(x)] = 0.549$, $\sum[x^2P(x)] = 0.819$, $\mu = 0.55$, $\sigma = 0.72$
b. the same

5.87 **a.** 25.0, 3.5 **b.** 7.7, 2.7
c. 24.0, 4.7 **d.** 44.0, 2.3

5.89 400, 0.5

5.91 0.021

5.93 **a.** 19% **b.** 8% **c.** 5 nights
d. number of evening meals American adults cook at home in an average week

e. 100% of the distribution; each sector of the circle is mutually exclusive of the others
f. See figure in 5.1d.
g. relative frequency distribution

5.99 a. 0.1 b. 0.4 c. 0.6
5.101 a. 0.930 b. 0.264
5.103 0.103
5.105 a. 0.012 b. 0.267 c. 0.549
5.107 0.063
5.109 not binomial, P(defective) changes from selection to selection
5.111 2600
5.113 a. 0.88 b. $B(50, 0.88)$
c.

x	Expected
39	1
40	2
41	3
42	4
43	6
44	7
45	7
46	5
47	3
48	2
49	0

5.115 tool shop: mean profit = 7,000.0
book store: mean profit = 6,000.0
5.117
$$\begin{aligned}\mu &= \sum[x \cdot P(x)]\\ &= (1)\cdot(1/n) + (2)\cdot(1/n) + \cdots + (n)\cdot(1/n)\\ &= (1/n)\cdot[1 + 2 + 3 + \cdots + n]\\ &= (1/n)\cdot[(n)(n+1)/2]\\ &= (n+1)/2\end{aligned}$$

Chapter 6

6.1 a. It's a quotient.
b. I.Q.: 100, 16; SAT: 500, 100; standard score: 0, 1
c. $z = (\text{I.Q.} - 100)/16$; $z = (\text{SAT} - 500)/100$
d. 2, 132, 700 e. same
6.2 0.4147
6.3 0.0212
6.4 0.9582
6.5 0.4177
6.6 0.0630
6.7 0.8571
6.8 0.2144
6.9 a. 0.84 b. −1.15 and +1.15
6.11 a. 0.4032 b. 0.3997 c. 0.4993 d. 0.4761
6.13 a. 0.4394 b. 0.0606 c. 0.9394 d. 0.8788
6.15 a. 0.7737 b. 0.8981 c. 0.8983 d. 0.3630
6.17 a. 0.5000 b. 0.1469 c. 0.9893 d. 0.9452 e. 0.0548
6.19 a. 0.4906 b. 0.9725 c. 0.4483 d. 0.9306
6.21 a. 1.14 b. 0.47 c. 1.66 d. 0.86 e. 1.74 f. 2.23
6.23 a. 1.65 b. 1.96 c. 2.33
6.25 −1.28 or +1.28
6.27 −0.67 and +0.67
6.29 1.28, 1.65, 2.33
6.31 2.88
6.32 a. 0.3944 b. 0.8944
6.33 a. no; amount varies from car to car
b. the mean and standard deviation of the distribution
6.34 a. 85.52, or 86 b. 89.28, or 89 c. 29,008
6.35 a. 0.7606 b. 0.0386 c. 0.2689
6.36 a. shifts to the right 3 units
b. short and wider
c. tall and narrow
6.37 Answers will vary.
6.38 Answers will vary.
6.39 Answers will vary.
6.40 Answers will vary.
6.41 0.2329 using computer or calculator; 0.2316 using Table 3
6.43 a. 0.5000 b. 0.3819 c. 0.0072 d. 0.2946 e. 0.9502 f. 0.0139
6.45 a. 0.0102 = 1.02% b. 0.1131 = 11.31%
6.47 a. 0.3240 or 32.4% b. 0.7519 or 75.19% c. 0.0885 or 8.9% d. 0.0582 or 5.8%
6.49 a. 0.4129 b. 0.3824 c. 0.0735 d. 25.32 points
6.51 a. 89.6 b. 79.2 c. 57.3
6.53 20.26
6.55 a.

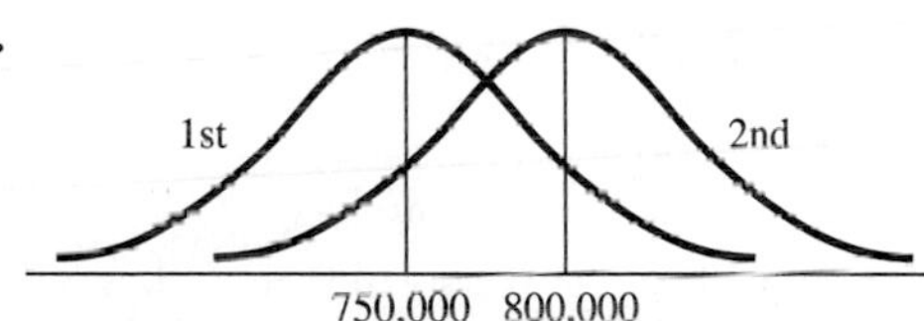

b. 0.7967 c. 0.7967
6.57 a. 0.056241 b. 0.505544 c. 0.438215
d. 0.0559, 0.5077, 0.4364 e. round-off errors
6.59 Answers will vary with a mean and standard deviation close to 100 and 16, respectively, and approximately normally distributed.
6.61 a.

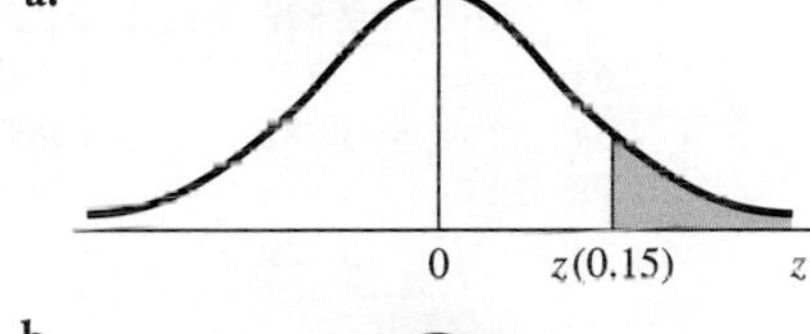

b.

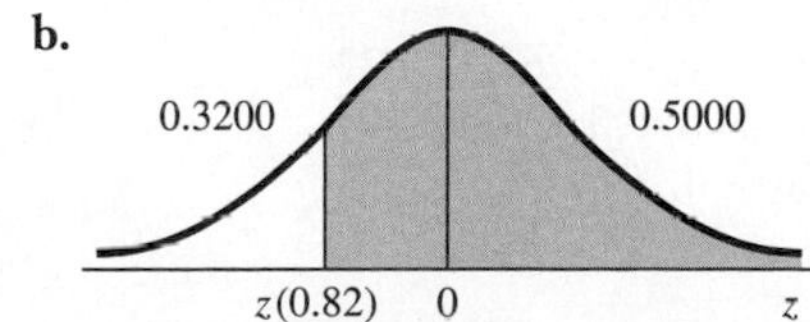

6.62 a. 1.04 b. −0.92
6.63 a. $z(0.03)$ b. $z(0.14)$ c. $z(0.75)$ d. $z(0.13)$ e. $z(0.90)$ f. $z(0.82)$
6.65 a. 1.96 b. 1.65 c. 2.33
6.67 a. 1.28, 1.65, 1.96, 2.05, 2.33, 2.58
b. −2.58, −2.33, −2.05, −1.96, −1.65, −1.28

6.69 **a.** area, 0.4602 **b.** z-score, 1.28
c. area, 0.5199 **d.** z-score, -1.65
6.71 no; $np = 2$, $nq = 98$
6.72 **a.** binomial probability function: 0.0149
b. normal approx.: 0.0202; differ by 0.0053
6.73 **a.** $np = 3$, not appropriate
b. $np = 0.5$, not appropriate
c. $np = 50$ and $nq = 450$, appropriate
d. $np = 10$ and $nq = 40$, appropriate
6.75 0.1822, 0.1770
6.77 0.9429, 0.9430
6.79 $x = n$(survive), 0.9999997
6.81 $\mu = 29.5$, $\sigma = 3.48$
a. 0.1251 **b.** 0.0107
6.83 $\mu = 27.63$, $s = 3.86$; 0.6844
6.85 **a.** $\mu = 395$, $\sigma = 18.237$; 0.9988 **b.** 0.9988
6.86 **a.** $z = (\text{I.Q.} - 100)/16$
b. -0.63, 0.63, 1.25
c. $z = (\text{SAT} - 500)/100$; -0.35, 0.75, 1.50
d. 2%, 98%
6.89 at least $^3/_4$, 0.9544
6.91 **a.** 1.26 **b.** 2.16 **c.** 1.13
6.93 **a.** 0.0930 **b.** 0.9684
6.95 **a.** 1.175 or 1.18 **b.** 0.58 **c.** -1.04 **d.** -2.33
6.97 **a.** 0.0158 **b.** 1.446 years
6.99 10.033
6.101 **a.** $n = 25$, $np = 7.5$, $nq = 17.5$
b. 7.5, 2.29
6.103 **b.** 0.77023 **c.** $\mu = 5$, $\sigma = 2.12$; 0.751779
6.105 **a.** $P(0) + P(1) + \cdots + P(75)$ **b.** 0.9856
c. $\mu = 60$, $\sigma = 6.93$; 0.9873
6.107 **a.** 5.534 **c.** Answers will vary. **d.** yes
6.109 **a.** 0.001 **b.** 1.000 **c.** 0.995
d. $\mu = 7.5$, $\sigma = 2.67$; 0.0668
6.111

Nation	$\mu = np$	$\sigma = \sqrt{npq}$	$P(x \geq 70)$
China	66	7.989	0.3300
Germany	10	3.154	0.0000+
India	130	11.024	0.9999+
Japan	8	2.823	0.0000+
Mexico	64	7.871	0.2420
Russia	34	5.781	0.0000+
S. Africa	98	9.654	0.9984
United States	14	3.729	0.0000+

6.113 **a.** $\mu = 37.5$, $\sigma = 3.062$; 0.7422
b. $\mu = 23.5$, $\sigma = 3.529$; 0.6103
c. 0.8320 **d.** 0.8469

Chapter 7

7.1 **a.**

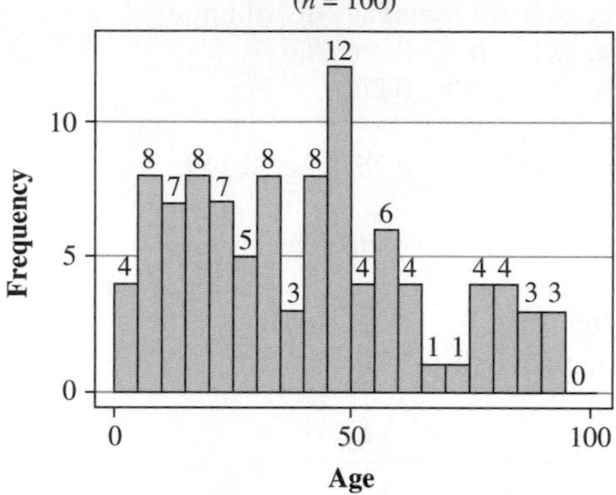

$\bar{x} = 39.31$, $s = 25.16$; mounded from 0 to 60, skewed to the right
b. looks like population **c.** skewed right
d. not exactly but fairly close
7.2 **a.** no **b.** variability
7.3 one of 25; 3 of 25
7.4 $\sum xP(x) = 3.0$, $\sum x^2P(x) = 11.0$; $\mu = 3.0$, $\sigma = \sqrt{2} = 1.41$
7.5 Classes: 1.8–4.6 using class width of 0.4; freq: 3, 5, 6, 6, 5, 4, 1
7.6 $\sum \bar{x}f = 89.4$, $\sum \bar{x}^2f = 278.20$; $\bar{\bar{x}} = 2.98$, $s_{\bar{x}} = 0.638$
7.7 **a.** Freq: 1, 1, 4, 2, 1, 3
b. Airlines have different size fleets.
7.9 **a.**

11	31	51	71	91
13	33	53	73	93
15	35	55	75	95
17	37	57	77	97
19	39	59	79	99

b.

$\bar{x}$	1	2	3	4	5	6	7	8	9
$P(\bar{x})$	0.04	0.08	0.12	0.16	0.20	0.16	0.12	0.08	0.04

c.

R	0	2	4	6	8
$P(R)$	0.20	0.32	0.24	0.16	0.08

7.11 Answers will vary.
c. normal, mounded, symmetric, centered around 4.5
d. same as part (c)
7.13 Answers will vary.
7.15 Answers will vary.
d. approximately normal, mounded, symmetrical, centered around 100
7.16 $25/\sqrt{16} = 6.25$; $25/\sqrt{36} = 4.167$; $25/\sqrt{100} = 2.50$
7.17 **a.** 1.0 **b.** As n increases, $\sigma/\sqrt{n}$ gets smaller.
7.18 **b.** very close to $\mu = 65.15$
d. approximately normal

7.19 b. very close to $\mu = 6.029$; $s_{\bar{x}}$ is approximately $10.79/\sqrt{4} = 5.395$; skewed right
c. Answers will vary.

	$\bar{\bar{x}}$	μ	$s_{\bar{x}}$	$\sigma/\sqrt{n}$	shape
25	5.974	6.029	2.087	$10.79/\sqrt{25} = 2.158$	skewed right
100	6.018	6.029	1.137	$10.79/\sqrt{100} = 1.079$	sl. lopsided
1000	6.03	6.029	0.335	$10.79/\sqrt{1000} = 0.341$	normal

d. See part (c).
7.21 a. approximately normal **b.** 2.4 televisions
c. $1.2/\sqrt{80} = 0.134$
7.23 a. 43.3 cents **b.** $7.5/\sqrt{150} = 0.61$ cents
c. approximately normal
7.25 Answers will vary.
d. $\bar{\bar{x}} \approx 20$; $s_{\bar{x}} \approx 4.5/\sqrt{6} = 1.84$; approximately normal
7.27 $P(0 < z < 2) = 0.4772$
7.28 0.9699
7.29 38.73
7.31 a. approximately normally, $\mu = 69$, $\sigma = 4$
b. 0.4013 **c.** approximately normally
d. $\mu_{\bar{x}} = 69$; $\sigma_{\bar{x}} = 4/\sqrt{16} = 1.0$ **e.** 0.1587
f. 0.0228
7.33 a. 0.3830 **b.** 0.9938 **c.** 0.3085 **d.** 0.0031
7.35 a. 0.2643 **b.** 0.0294
c. No, wind speed skewed to the right. Samples of size 9 are not large enough for the CLT.
d. most likely not as high
7.37 a. 0.9878 **b.** 0.0003 **c.** yes
7.39 a. 0.6826
7.41 a.

100 U.S. Citizens, 2002

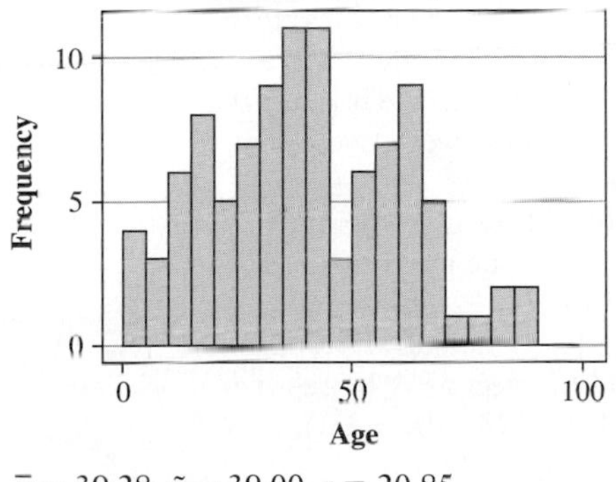

$\bar{x} = 39.28$, $\tilde{x} = 39.00$, $s = 20.85$
b. reasonably well
c. mounded 0 to 70, skewed to the right
d. The numerical statistics are very similar in value.
e. very similar
7.45 a. $e = 0.49$ **b.** $E = 0.049$
7.47 a. 0.1190 **b.** 0.0033
7.49 a. 0.1498 **b.** 0.0089
7.51 0.0228
7.53 0.0023
7.55 a. approximately 1.000 **b.** 0.9772
7.59 a. $\mu = 60$, $\sigma = 6.48$

Chapter 8

8.1 a. female health professionals
b. $\bar{x} = 64.7$, $s = 3.5$, mounded about center, approximately symmetrical, see graph in part (e)
c. The mean and standard deviation of the sample are expected to approximate the mean and standard deviation of the population; mean value that is approximated by the mean of the sample, has a standard error approximated by the standard deviation of the sample divided by square root of 50, and has an approximately normal distribution
d. 63.06 to 64.34 **e.** no

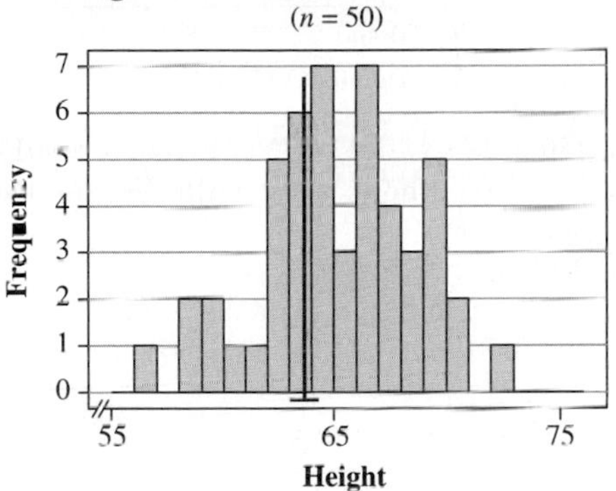

f. 0.0044; 44 in 10,000 samples drawn randomly from a population with mean 63.7 and standard deviation 2.75 will have a mean as large as 64.72.
g. seems highly unlikely
8.2 collector fatigue, cost of sampling, time
8.3 $\sigma_{\bar{x}} = \sigma/\sqrt{n} = 18/\sqrt{36} = 18/6 = 3$
8.5 a. between 10:14 A.M. and 10:34 A.M. **b.** yes
c. 90% occur within predicted interval
8.7 a. $24 = n$; $4'11'' = \bar{x}$ **b.** $16 = \sigma$
c. $190 = s^2$ **d.** $69 = \mu$
8.9 a. II has the lower variability.
b. II has a mean value equal to the parameter.
c. Neither is a good choice.
8.11 unbiased, $\mu_{\bar{x}} = \mu$; variability decreases as n increases, $\sigma/\sqrt{n}$
8.13 $3,133,948.80
8.15 Performance of an entire line tends to be normally distributed.
8.17 a. 75.92 **b.** 0.368 **c.** 75.552 to 76.288
8.18 Numbers are calculated for samples of snow.
8.19 a. 15.9, ≈68% **b.** 31.4; ≈95%;
c. 41.2; ≈99%
d. Higher level makes for a wider width.
8.21 49
8.23 Sampled population is normal or sample is sufficiently large for the CLT.
8.25 a. 25.76 to 31.64 **b.** yes, population is normal
8.27 a. 125.58 to 131.42 **b.** yes, CLT
8.29 a. speed **b.** 176.99 to 191.01
c. 175.67 to 192.33

8.31 **a.** 25.3 **b.** 24.29 to 26.31 **c.** 23.97 to 26.63
8.33 464.4 to 481.6
8.35 **b.**

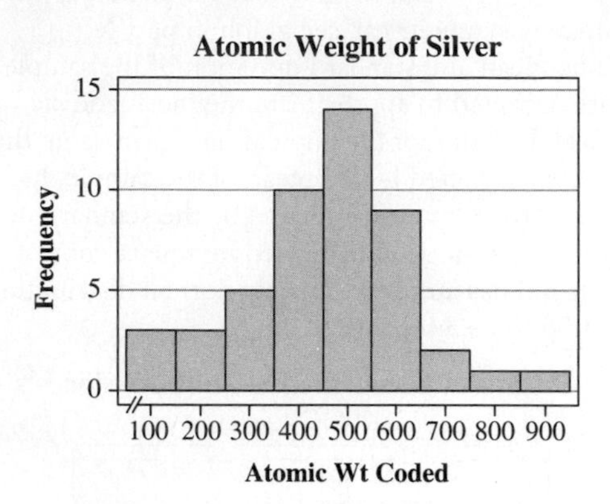

c. 450.6, 173.4 **d.** 107.86814506, 0.00001734
e. Histogram shows an approximately normal distribution.
f. Both apply. **g.** no
h. Use the sample standard deviation.
i. 107.86814015 to 107.86814997
8.37 27
8.39 28
8.41 Lobsters are or are not active at night.
8.42 H_o: The system is reliable
H_a: The system is not reliable
8.43 H_a: Teaching techniques have a significant effect on student's exam scores.
8.44 A: party will be a dud; Action: did not go
B: party will be a great time; Action: did go
I: party will be a dud; Action: did go
II: party will be a great time; Action: did not go
8.45 You missed a great time.
8.46 $\alpha = P$(rejecting a TRUE null hypothesis)
8.47 It is less likely to occur.
8.48 **a.** two groups of students, one is coached and the other is not
b. The coached group receives higher SAT scores.
8.49 **a.** H_o: Special delivery does not take too much time
H_a: Special delivery takes too much time
b. H_o: New design is not more comfortable
H_a: New design is more comfortable
c. H_o: Cigarette smoke has no effect
H_a: Cigarette smoke has an effect
d. H_o: Hair conditioner is not effective
H_a: Hair conditioner is effective
8.51 **a.** H_o: The victim is alive
H_a: The victim is not alive
b. A: alive, treated as alive
I: alive, treated as dead
II: dead, treated as alive
B: dead, treated as dead
c. Type I: very serious
Type II: not as serious
8.53 **a.** I: determine majority do not favor, when majority do favor
II: determine majority do favor, when they do not favor
b. I: determine low salt, when it is not
II: determine not low salt, when it is
c. I: determine building must be demolished, when it should not;
II: determine building must not be demolished, when it should
d. I: determine there is waste, when there is not waste
II: determine there is no waste, when there is waste
8.55 **a.** Type I **b.** Type II **c.** Type I **d.** Type II
8.57 **a.** very serious **b.** somewhat serious
c. not at all serious
8.59 **a.** α **b.** β
8.61 **a.** "See, I told you so."
b. "Okay, so this evidence was not significant, I'll try again tomorrow."
8.63 **a.** 0.1151 **b.** 0.2119
8.65 H_o: The mean shearing strength is at least 925 lb
H_a: The mean shearing strength is less than 925 lb
8.66 H_o: $\mu = 54.4$
H_a: $\mu \neq 54.4$
8.67 **a.** H_o: $\mu = 1.25$ ($\leq$)
H_a: $\mu > 1.25$
b. H_o: $\mu = 335$ ($\geq$)
H_a: $\mu < 335$
c. H_o: $\mu = 230{,}000$
H_a: $\mu \neq 230{,}000$
8.68 A: is at least, decide it is
I: is at least, decide it is less than
II: is less than, decide it is greater than or equal to
B: is less than, decide it is less than
II; buy and use weak rivets
8.69 −1.46
8.70 **a.** How likely are the sample results to have happened if the null hypothesis is true?
b. probability
8.71 **a.** 0.0107 **b.** 0.0359
8.72 **a.** fail to reject H_o **b.** reject H_o
8.73 **a.** 2.0, 0.0228 **b.** −1.33, 0.0918
c. −1.60, 0.1096
8.74 **b.** ≈0.0000 **c.** 0.0000 **d.** reject H_o
8.75 0.2714
8.76 **a.** H_a is two-sided. **b.** ≈ 0.0277
c. green areas in the tails **d.** reject H_o
8.81 **a.** H_o: $\mu = 210$ lb ($\leq$) vs. H_a: $\mu > 210$
b. H_o: $\mu = 570$ lb/unit vs. H_a: $\mu \neq 570$
c. H_o: $\mu = \$9.00$ ($\leq$) vs. H_a: $\mu > \$9.00$
8.83 **a.** decide mean is less, when mean is at least
b. decide mean is at least, when mean is less than
8.85 **a.** 1.26 **b.** 1.35 **c.** 2.33 **d.** −0.74

8.87 **a.** reject H_o; fail to reject H_o
b. Calculated p-value is smaller than or equal to α; calculated p-value is larger than α

8.89 **a.** fail to reject H_o **b.** reject H_o

8.91 **a.** 0.0694 **b.** 0.1977 **c.** 0.2420 **d.** 0.0174 **e.** 0.3524

8.93 **a.** 1.57 **b.** −2.13 **c.** −2.87 or 2.87

8.95 **a.** H_o: $\mu = 525$ vs. H_a: $\mu < 525$
b. fail to reject H_o **c.** $60.0/\sqrt{38} = 9.733$

8.97 **a.** mean test score
b. H_o: $\mu = 35.70$ (≥) vs. H_a: $\mu < 35.70$
c. −5.86, $P(z < -5.86) \approx +0.0000$ **d.** reject H_o

8.99 **a.** mean seconds in error per month
b. H_o: $\mu = 20$ (≤) vs. H_a: $\mu > 20$
c. normality is assumed, $n = 36$
d. $n = 36$, $\bar{x} = 22.7$
e. $z\bigstar = 1.78$; **P** = 0.0375 **f.** reject H_o

8.101 H_o: Mean is at least 925, H_a: mean is less than 925

8.102 H_o: $\mu = 9$ (≤), H_a: $\mu > 9$

8.103 **a.** H_o: $\mu = 1.25$ (≥), H_a: $\mu < 1.25$
b. H_o: $\mu = 335$, H_a: $\mu \neq 335$
c. H_o: $\mu = 230{,}000$ (≤), H_a: $\mu > 230{,}000$

8.104 A: is at least, decide it is at least
I: is at least, decide it is less than
II: is less than, decide it is greater than or equal to
B: is less than, decide it is less than
Type II error; buy and use weak rivets

8.105 −1.10

8.106 **a.** reject H_o **b.** fail to reject H_o

8.107 $z \le -2.33$

8.108 $z \ge 2.05$

8.109 2.2046 lb/kg, 119.9

8.111 **a.** H_o: $\mu = 16$ yr (≥) vs. H_a: $\mu < 16$
b. H_o: $\mu = 6$ ft 6 in (≤) vs. H_a: $\mu > 6$ ft 6 in
c. H_o: $\mu = 285$ ft (≥) vs. H_a: $\mu < 285$
d. H_o: $\mu = 0.375$ inch (≤) vs. H_a: $\mu > 0.375$
e. H_o: $\mu = 200$ units vs. H_a: $\mu \neq 200$

8.113 **a.** decide mean is greater, when it is not
b. decide mean is at most, when it is greater than

8.115 **a.** values that will cause reject H_o
b. boundary value of critical region

8.117 **a.** $z \le -1.65$, $z \ge 1.65$ **b.** $z \ge 2.33$
c. $z \le -1.65$ **d.** $z \le -2.58$, $z \ge 2.58$

8.119 247.1; 21,004.133

8.121 **a.** 3.0 standard errors **b.** yes

8.123 **a.** reject H_o or fail to reject H_o
b. calculated test statistic falls in the critical region; calculated test statistic falls in the non-critical region

8.125 **a.** H_o: $\mu = 15.0$ vs. H_a: $\mu \neq 15.0$
b. reject H_o **c.** $0.5/\sqrt{30} = 0.0913$

8.127 H_0: $\mu = 79.68$, H_a: $\mu > 79.68$, $z\bigstar = 1.07$, $z \ge +1.65$, fail to reject H_o

8.129 H_0: $\mu = 36.5$, H_a: $\mu < 36.5$, $z\bigstar = -2.47$, $z \le -2.33$, reject H_o

8.131 **a.** no **b.** See graph in part (e); $\bar{x} = 64.72$
c. Mean and standard deviation of the sample are expected to approximate the mean and standard deviation of the population; mean value that is approximated by the mean of the sample, has a standard error approximated by the standard deviation of the sample divided by square root of 50, and has an approximately normal distribution.
d. 62.94 to 64.46
e. (See histogram in 8.1e.) No; sample mean is not one of the 95% that are within the interval constructed in (d).

8.135 **a.** determines z, the number of standard errors
b. narrows

8.137 **a.** H_o: $\mu = 100$ **b.** H_a: $\mu \neq 100$ **c.** $\alpha = 0.01$
d. $\mu = 100$ **e.** $\bar{x} = 96$ **f.** $\sigma = 12$ **g.** 1.70
h. −2.35 **i.** 0.0188 **j.** fail to reject H_o
k. **P** = 0.0188

8.139 **a.** 39.6 to 41.6
b. H_a: $\mu \neq 40$, $z\bigstar = 1.20$, **P** = 0.2302, fail to reject H_o
c. H_a: $\mu \neq 40$, $z\bigstar = 1.20$, $\pm z(0.025) = \pm 1.96$, fail to reject H_o

8.141 **a.** 39.9 to 41.9
b. H_a: $\mu > 40$, $z\bigstar = 1.80$, **P** = 0.0359, reject H_o
c. H_a: $\mu > 40$, $z\bigstar = 1.80$, $z(0.05) = 1.65$, reject H_o

8.143 **a.** 9.75 to 9.99 **b.** 9.71 to 10.03
c. widened the interval

8.145 **a.** 69.89 to 75.31 **b.** yes

8.147 3.87 to 4.13

8.149 60

8.151 2.89, 0.0019

8.153 **a.** H_o: $\mu = 0.50$, H_a: $\mu \neq 0.50$
b. $z\bigstar = 1.25$, **P** = 0.2112 **c.** $z = \pm 2.33$

8.155 H_a: $\mu > 0.0113$, $z\bigstar = 2.47$, **P** = 0.0068 or $z \ge 2.33$, reject H_o

8.157 H_a: $\mu > 9$, $z\bigstar = 3.14$, **P** = 0.0008, $z \ge 2.05$, reject H_o

8.159 **a.** H_a: $\mu < 129.2$, $z\bigstar = -2.02$, **P** = 0.0217, fail to reject H_o
b. indicates the likelihood of being wrong

8.161 **a.** (2) Failure to reject H_o will result in the drug being marketed; burden of proof is on the old ineffective drug.
b. (1) Failure to reject H_o will result in the new drug not being marketed; burden of proof is on the new drug.

8.163 **a.** 125.10 to 132.95

8.165 **a.** H_a: $\mu \neq 18$; fail to reject H_o; population mean is not significantly different from 18

Chapter 9

9.1 **a.** 53.5 minutes, 25.68
b. approximately 53.5 minutes **c.** not really!

9.3 identical

9.4 **a.** 2.68 **b.** 2.07

9.5 **a.** -1.33 **b.** -2.82
9.6 ± 2.18
9.7 0.0241

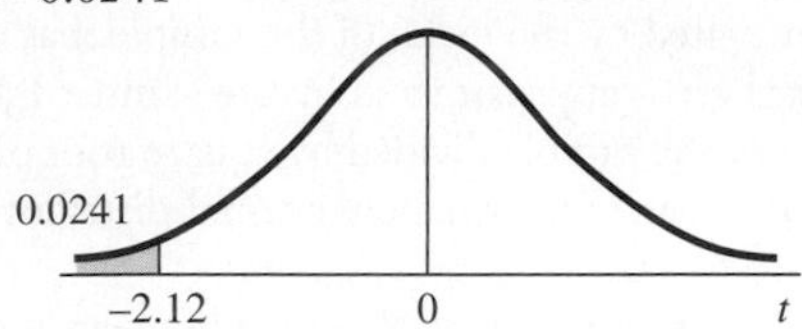

9.8 0.1402

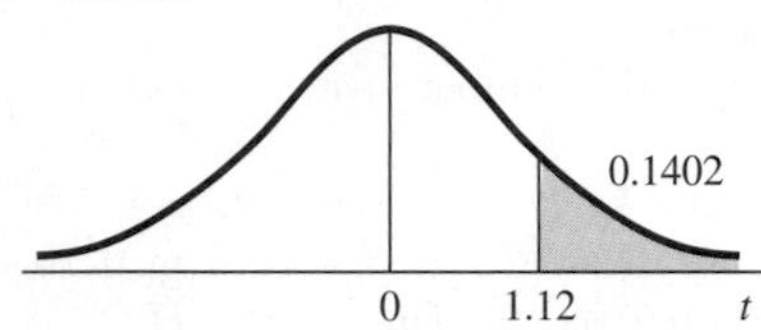

9.9 15.60 to 17.8
9.10 6.073 to 9.427
9.11 1.20
9.12 $\mathbf{P} = 0.124$, fail to reject H_o
9.13 $t \geq 1.75$, fail to reject H_o
9.14 -1.30
9.15 $\mathbf{P} = 0.220$, fail to reject H_o
9.16 $t \leq -2.20$, $t \geq 2.20$, fail to reject H_o
9.17 0.2202
9.18 $\mathbf{P} = 0.091$, fail to reject H_o
9.19 **a.** $P(t > 1.92 \mid df = 44) < 0.033$
b. $P(t > 3.41 \mid df = 44) = 0.001$
c. $0.064 < \mathbf{P} < 0.080$
9.21 **a.** -1.72 **b.** -2.06 **c.** -2.47 **d.** 2.01
9.23 7
9.25 **a.** -2.49 **b.** 1.71 **c.** -0.685
9.27 **a.** symmetric about mean: mean is 0
b. standard deviation, t has df.
9.29 51.84 to 56.16
9.31 **a.** \$47.86 **b.** \$14.32 **c.** 44.10 to 51.62
9.33 -2.60 to 10.50
9.37 **a.** H_o: $\mu = 11$ ($\geq$) vs. H_a: $\mu < 11$
b. H_o: $\mu = 54$ ($\leq$) vs. H_a: $\mu > 54$
c. H_o: $\mu = 75$ vs. H_a: $\mu \neq 75$
9.39 **a.** $t \leq -2.14$, $t \geq 2.14$ **b.** $t \geq 2.49$
c. $t \leq -1.74$ **d.** $t \geq 2.42$
9.41 H_o: $\mu = 25$ vs. H_a: $\mu < 25$, $t\bigstar = -3.25$, $\mathbf{P} < 0.005$ or $t \leq -2.46$, reject H_o
9.43 **a.** midrange is close to mean, approximately symmetrical
b. & c. H_o: $\mu = 38.4\%$ vs. H_a: $\mu \neq 38.4\%$, $t\bigstar = 1.08$, $0.20 < \mathbf{P} < 0.50$ or $t \leq -2.14$, $t \geq 2.14$, fail to reject H_o
9.45 $\bar{x} = 37.0$, $s = 4.817$
H_o: $\mu = 35$ vs. H_a: $\mu \neq 35$, $t\bigstar = 1.02$, $0.20 < \mathbf{P} < 0.50$ or $t \leq -2.57$, $t \geq 2.57$, fail to reject H_o
9.49 $\bar{x} = 18.8142$, $s = 0.0296$, se mean $= 0.0050$
H_o: $\mu = 18.810$ vs. H_a: $\mu \neq 18.810$, $t\bigstar = 0.84$, $\mathbf{P} = 0.41$; $t \leq -2.75$, $t \geq 2.75$, fail to reject H_o

9.51 **a.** yes **b.** Mean of p' is p.
9.53 **a.** 0.02104 **b.** 0.189 to 0.271
9.54 **a.** 0.031
b. moe is a rounded maximum error.
c. no bias present from the interviewing process
d. In section, the "success" outcome is clearly identifiable, whereas in polling the outcomes are more subjective and possible error may easily be introduced.
e. Measuring the error caused by nonresponse, interviewer error, and so on, is not possible.
9.55 2401
9.56 1277
9.57 1.78
9.58 $\mathbf{P} = 0.0375$, reject H_o
9.59 $z \geq 1.65$, reject H_o
9.60 $0.052 < p < 0.118$, 15% is not within interval
9.61 **a.** 0.0332 **b.** 0.0332 **c.** 0.0664
d. RC: "struck evenly or unbiased," H_o: $P(\text{H}) = 0.5$, BB: "not balanced," H_a: $P(H) \neq 0.5$
e. definition of the p-value
f. 0.498 to 0.622
g. 6% is a maximum error of estimate from a binomial experiment, not a survey
9.63 **a.** $p + q = 1$ **b.** They are equivalent.
c. 0.4 **d.** 0.727
9.65 **a.** $n = 1{,}027$, each person, saying "ketchup is their preferred burger condiment," $p = P(\text{say ketchup})$, x = number in sample who say ketchup
b. 0.47, statistic **c.** 0.031 **d.** same value
e. 0.439 to 0.501
9.67 0.206 to 0.528
9.69 0.250 to 0.306
9.71 could be biased
9.73 **a.** 0.028, 0.030, 0.026
b. variation caused by p **c.** yes
d. yields the maximum value **e.** $p = 0.5$
9.75 **a.–e.** The distributions do not look normal; they are skewed right.
f. The normal distribution should not be used.
9.77 **a.** 1048 **b.** 262 **c.** 2089
d. Increasing the maximum error decreases the required sample size.
e. Increasing the level of confidence increases the required sample size.
9.79 **a.** H_o: $p = 0.60$ ($\leq$) vs. H_a: $p > 0.60$
b. H_o: $p = 0.50$ ($\geq$) vs. H_a: $p < 0.50$
c. H_o: $p = 1/3$ ($\leq$) vs. H_a: $p > 1/3$
d. H_o: $p = 0.50$ ($\geq$) vs. H_a: $p < 0.50$
e. H_o: $p = 0.50$ ($\leq$) vs. H_a: $p > 0.50$
f. H_o: $p = 3/4$ ($\geq$) vs. H_a: $p < 3/4$
g. H_o: $p = 0.50$ vs. H_a: $p \neq 0.50$
h. H_o: $p = 0.50$ vs. H_a: $p \neq 0.50$
9.81 **a.** 0.1388 **b.** 0.0238 **c.** 0.1635 **d.** 0.0559
9.83 **a.** 0.017 **b.** 0.085 **c.** 0.101 **d.** 0.004
9.85 **a.** (1) **b.** 0.036 **c.** (4) **d.** 0.128

9.87 H_o: $p = 0.90$ $(\geq)$ vs. H_a: $p < 0.9$, $z\bigstar = -4.82$, $\mathbf{P} = 0.000003$ or $z \leq -1.65$, reject H_o

9.89 H_o: $p = 0.60$ vs. H_a: $p < 0.60$, $z\bigstar = -2.04$, $\mathbf{P} = 0.0207$ or $z \leq -1.65$, reject H_o

9.91 H_o: $p = 0.225$ $(\leq)$ vs. H_a: $p > 0.225$, $z\bigstar = 2.71$, $\mathbf{P} = 0.0034$ or $z \geq 1.65$, reject H_o

9.93 **a.** $z\bigstar = -1.72$
b. & c. H_o: $p = 0.35$ vs. H_a: $p < 0.35$, $\mathbf{P} = 0.0427$ or $z \leq -2.33$, fail to reject H_o

9.94 **a.** 23.2 **b.** 23.3

9.95 **a.** 3.94 **b.** 8.64

9.96 **a.** 0.8356 **b.** 0.1644

9.97 25.08

9.98 $0.05 < \mathbf{P} < 0.10$, fail to reject H_o

9.99 $\chi^2 \geq 33.4$, fail to reject H_o

9.100 60.15

9.101 $0.02 < \mathbf{P} < 0.05$, reject H_o

9.102 $\chi^2 \leq 24.4$, $\chi^2 \geq 59.3$, reject H_o

9.103 0.1208

9.105 **a.** 34.8 **b.** 28.8 **c.** 13.4 **d.** 48.3 **e.** 12.3 **f.** 3.25 **g.** 37.7 **h.** 10.9

9.107 **a.** 11.1 **b.** 11.1 **c.** 9.24

9.109 0.94

9.111 **a.** H_o: $\sigma = 24$ $(\leq)$ vs. H_a: $\sigma > 24$
b. H_o: $\sigma = 0.5$ $(\leq)$ vs. H_a: $\sigma > 0.5$
c. H_o: $\sigma = 10$ vs. H_a: $\sigma \neq 10$
d. H_o: $\sigma^2 = 18$ $(\geq)$ vs. H_a: $\sigma^2 < 18$
e. H_o: $\sigma^2 = 0.025$ vs. H_a: $\sigma^2 \neq 0.025$
f. H_o: $\sigma^2 = 34.5$ $(\leq)$ vs. H_a: $\sigma^2 > 34.5$

9.113 **a.** $0.02 < \mathbf{P} < 0.05$ **b.** 0.01 **c.** $0.05 < \mathbf{P} < 0.10$ **d.** $0.025 < \mathbf{P} < 0.05$

9.115 H_o: $\sigma = 8$ vs. H_a: $\sigma \neq 8$, $\chi^2\bigstar = 29.3$, $0.01 < \mathbf{P} < 0.02$ or $\chi^2 \leq 32.4$, $\chi^2 \geq 71.4$, reject H_o

9.117 **a.** allows use of chi-square **b.** examine sample
c. & d. H_o: $\sigma = 0.32$ $(\leq)$ vs. H_a: $\sigma > 0.32$, $\chi^2\bigstar = 66.04$, $\mathbf{P} < 0.005$ or $\chi^2 \geq 45.6$, reject H_o

9.119 $s^2 = 17.4595$; H_o: $\sigma = 3.50$ vs. H_a: $\sigma \neq 3.50$, $\chi^2\bigstar = 19.95$, $0.20 < \mathbf{P} < 0.50$ or $\chi^2 \leq 5.63$, $\chi^2 \geq 26.1$, fail to reject H_o

9.123 **a.** people in the health profession
b. $\bar{x} = 53.500$ (min), $s = 25.675$

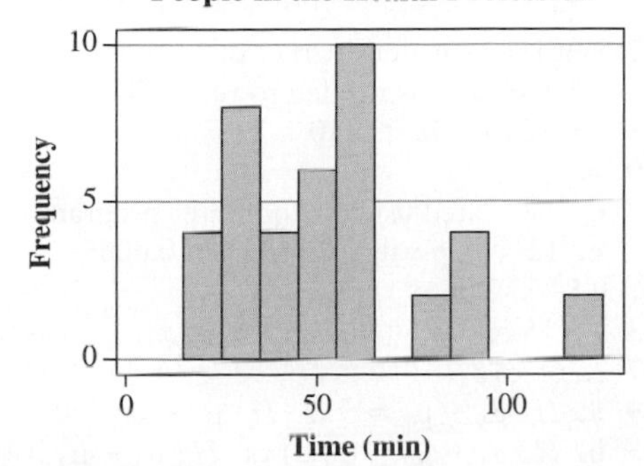

9.127 \$1,480 to \$1,620

9.129 **a.** $\bar{x} = 133.0$, $s = 47.3$

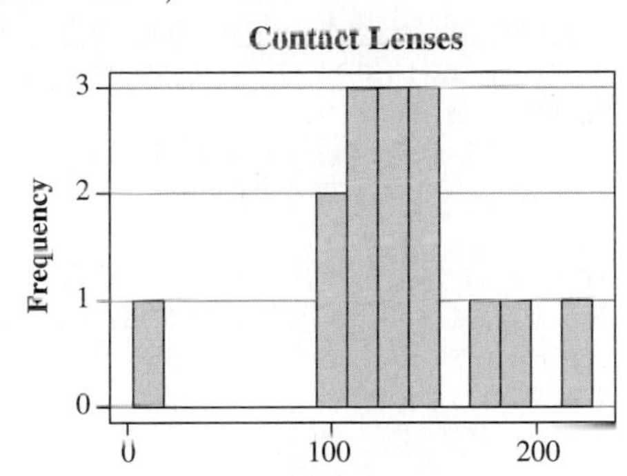

b.

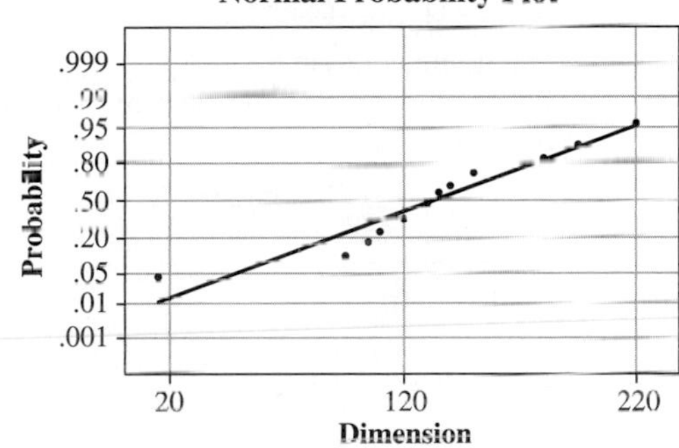

c. 106.89 to 159.11
d. With 95% confidence, the true mean dimension of the contact lens is between 106.89 and 159.11 units.

9.131 H_o: $\mu = 50$ $(\leq)$ vs. H_a: $\mu > 50$, $t\bigstar = 2.53$, $0.005 < \mathbf{P} < 0.01$ or $t \geq 2.42$, reject H_o

9.133 $\bar{x} = 3.44$, $s = 0.653$; H_o: $\mu = 3.8$ $(\geq)$ vs. H_a: $\mu < 3.8$, $t\bigstar = -1.91$, $0.025 < \mathbf{P} < 0.05$ or $t \leq -1.80$, reject H_o

9.135 **a.** appears to be approximately normally distributed

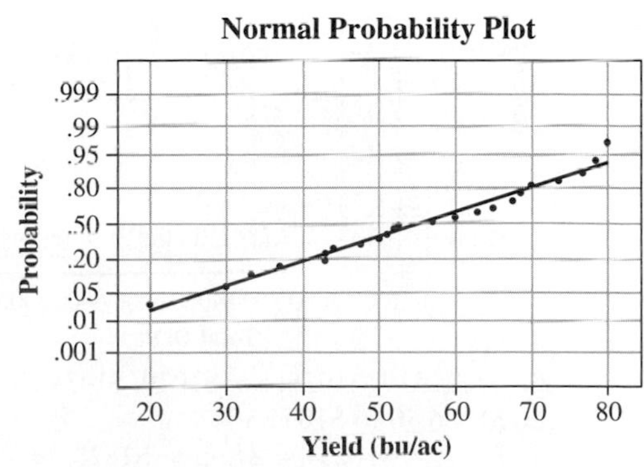

b. $\bar{x} = 55.44$, $s = 16.25$; H_o: $\mu = 60$ vs. H_a: $\mu < 60$, $t\bigstar = -1.37$, $0.05 < \mathbf{P} < 0.10$ or $t \leq -1.71$, fail to reject H_o

9.137 **a.** 0.50 **b.** 0.464 to 0.536 **c.** $50\% \pm 3.6\%$

9.139 0.088 to 0.232

9.141

$p = 0.1$	0.2	0.3	0.4	0.5	0.6	0.7	0.8	0.9
$pq = 0.09$	0.16	0.21	0.24	0.25	0.24	0.21	0.16	0.09

9.143 **a.** 995 **b.** 1068
c. yes, 75 smaller, cost less to obtain
9.145 0.0261
9.147 13
9.149 H_o: $\sigma = 12$ vs. H_a: $\sigma \neq 12$, $\chi^2\bigstar = 42.0$, $\mathbf{P} = 0.6844$ or $\chi^2 \leq 23.7$, $\chi^2 \geq 58.1$, fail to reject H_o
9.151 H_o: $\sigma = 0.5$ vs. H_a: $\sigma > 0.5$, $\chi^2\bigstar = 87.8$, $\mathbf{P} < 0.005$ or $\chi^2 \geq 42.6$, reject H_o

Chapter 10

10.1 **a.** Fr: 48.5%, So: 93.5%
b. mounded, but skewed with the right-hand tail being much longer
c. both mounded and skewed right; sophomores more dispersed

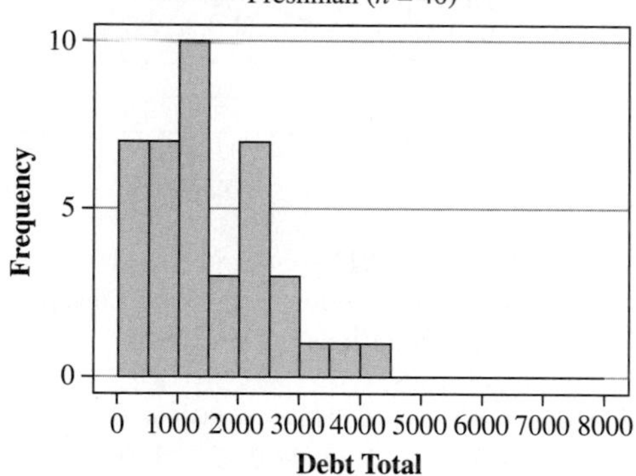

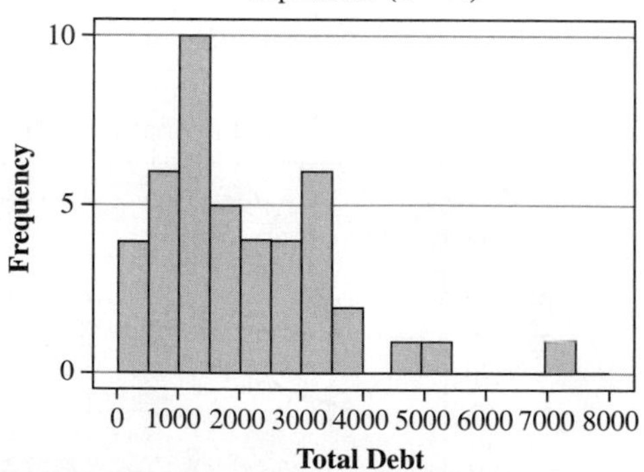

d. Fr: \$1519, \$1036; So: \$2079, \$1434
e. \$1186.50 to \$1851.50
f. H_o: $\mu = \$1825$ vs. H_a: $\mu > \$1825$, $t\bigstar = 1.17$, $\mathbf{P} = 0.123$ or $t \geq 1.68$; fail to reject H_o
10.2 Twins form pairs.
10.3 **a.** Divide the class into two groups, males and females; randomly select from each group.
b. Randomly select a set of students, obtaining the two heights from each.
10.5 dependent, each person providing one data for each sample
10.7 independent, separate samples
10.11 **a.**

Pairs	1	2	3	4	5
$d = A - B$	1	1	0	2	-1

b. 0.6 **c.** 1.14
10.12 2.13, number of standard errors
10.13 **a.** 4.24 to 8.36 **b.** narrower
10.14 d = Before − After: −0.51 to 6.51
10.15 H_o: $\mu_d = 0$ vs. H_a: $\mu_d > 0$, $t\bigstar = 2.45$, $0.033 < \mathbf{P} < 0.037$ or $t \geq 2.13$; reject H_o
10.16 $\mu_d = 0.00$ vs. $\mu_d < 0.00$, $t\bigstar = -1.30$, $\mathbf{P} = 0.13$; fail to reject H_o
10.17 H_o: $\mu_d = 0$ vs. H_a: $\mu_d \neq 0$, $t\bigstar = 1.35$, $0.234 < \mathbf{P} < 0.264$ or $t \leq -4.60$, $t \geq 4.60$; fail to reject H_o
10.18 **a.** The average difference is zero.
b. values used to make the decision
c. Test is two-tailed, t-distribution is symmetric, absence of negative numbers makes less confusing
d. "fail to reject the null hypothesis" in 12 of them
e. two methods are equivalent
f. revised Florida method for sampling accepted and implemented
10.19 **a.** 1.0 **b.** −1.53 to 3.53
10.21 $d = A - B$, $\bar{d} = 3.75$, $s_d = 5.726$; −1.03 to 8.53
10.23 $d = \text{I} - \text{II}$, $\bar{d} = 0.8$, $s_d = 1.32$; −0.143 to 1.743
10.25 **a.** H_o: $\mu_d = 0$ ($\leq$); H_a: $\mu_d > 0$; d = posttest − pretest
b. H_o: $\mu_d = 0$; H_a: $\mu_d \neq 0$; d = after − before
c. H_o: $\mu_d = 10$ ($\geq$); H_a: $\mu_d < 10$; d = after − before
d. H_o: $\mu_d = 12$ ($\geq$); H_a: $\mu_d < 12$; d = before − after
10.27 $\bar{d} = 5.5$, $s_d = 11.34$; H_o: $\mu_d = 0$ vs. H_a: $\mu_d > 0$, $t\bigstar = 3.067$, $\mathbf{P} \approx 0.002$ or $t \geq 2.44$; reject H_o
10.31 $\bar{d} = 1.35$, $s_d = 1.631$; H_o: $\mu_d = 0$ vs. H_a: $\mu_d > 0$ (less pain and discomfort), $t\bigstar = 3.70$, $\mathbf{P} = 0.001$ or $t \geq 2.54$; reject H_o
10.33 4.92
10.34 I: df will be between 17 and 40; II: df = 17
10.35 −6.3 to 16.3
10.36 1.24
10.37 With the smaller degrees of freedom, a higher calculated value is needed to reject H_o.
10.38 **a.** 0.210 **b.** ±2.10
10.39 **c.** 12 to 26
d. calculated using a computer program
e. 12 **f.** p-value is less than 0.0005
10.41 0.54 to 0.66
10.43 −0.75 to 8.75
10.47 2.265 to 3.781
10.49 **a.** H_o: $\mu_1 - \mu_2 = 0$ vs. H_a: $\mu_1 - \mu_2 \neq 0$
b. H_o: $\mu_1 - \mu_2 = 0$ ($\leq$) vs. H_a: $\mu_1 - \mu_2 > 0$
c. H_o: $\mu_1 - \mu_2 = 20$ ($\leq$) vs. H_a: $\mu_1 - \mu_2 > 20$

d. H_o: $\mu_A - \mu_B = 50$ ($\geq$) vs. H_a: $\mu_A - \mu_B < 50$
or equivalently
H_o: $\mu_B - \mu_A = 50$ ($\leq$) vs. H_a: $\mu_B - \mu_A > 50$

10.51 **a.** Table 6: $0.10 < \mathbf{P} < 0.25$, Table 7: $\mathbf{P} = 0.125$
b. Table 6: $0.01 < \mathbf{P} < 0.025$, Table 7: $\mathbf{P} = 0.012$
c. Table 6: $0.05 < \mathbf{P} < 0.10$, Table 7: $\mathbf{P} = 0.092$
d. Table 6: $0.05 < \mathbf{P} < 0.10$, Table 7: $\mathbf{P} = 0.084$

10.53 ≈ 0.18

10.55 H_o: $\mu_1 - \mu_2 = 0$ vs. H_a: $\mu_1 - \mu_2 > 0$, $t\bigstar = 4.02$, $\mathbf{P} < 0.005$ or $t \geq 2.44$; reject H_o

10.57 H_o: $\mu_2 - \mu_1 = 0$ vs. H_a: $\mu_2 - \mu_1 > 0$, $t\bigstar = 1.30$, $0.10 < \mathbf{P} < 0.25$ or $t \geq 2.47$; fail to reject H_o

10.59 **b.** $0.554 < \mathbf{P} < 0.624$ **c.** $0.560 < \mathbf{P} < 0.626$

10.61 $\bar{x}_A = 10.0$, $s_A^2 = 10.44$, $\bar{x}_B = 14.7$, $s_B^2 = 46.01$;
H_o: $\mu_B - \mu_A = 0$ vs. H_a: $\mu_B - \mu_A > 0$, $t\bigstar = 1.98$, $0.037 < \mathbf{P} < 0.047$ or $t \geq 1.83$; reject H_o

10.67 75, 250, 0.30, 0.70

10.68 **a.** $n_1p_1 = 36$, $n_1q_1 = 4$, $n_2p_2 = 45$, $n_2q_2 = 5$ **b.** no

10.69 -0.07 to 0.15

10.70 0.076, 0.924

10.71 $p'_p = 0.82$, $z\bigstar = 2.21$

10.72 0.0901

10.73 **a.** H_o: $p_n - p_h = 0$ vs. H_a: $p_n - p_h > 0$, $z\bigstar = 0.38$, $\mathbf{P} = 0.3520$ or $z \geq 1.65$; fail to reject H_o
b. size of each sample
c. Yes, one method is as good as the other.

10.75 -0.15 to 0.21

10.77 0.000 to 0.080

10.79 **a.** H_o: $p_m - p_w = 0$ vs. H_a: $p_m - p_w \neq 0$
b. H_o: $p_b - p_g = 0$ ($\leq$) vs. H_a: $p_b - p_g > 0$
c. H_o: $p_c - p_{nc} = 0$ ($\leq$) vs. H_a: $p_c - p_{nc} > 0$

10.81 H_o: $p_h - p_w = 0$ vs. H_a: $p_h - p_w > 0$, $z\bigstar = 3.21$, $\mathbf{P} = 0.0007$ or $z \geq 1.65$; reject H_o

10.83 H_o: $p_m - p_c = 0$ vs. H_a: $p_m - p_c \neq 0$, $z\bigstar = 1.42$, $\mathbf{P} = 0.1556$ or $z \leq -1.96$ and $z \geq 1.96$; fail to reject H_o

10.85 **a.** H_a: $p_m - p_w > 0$, $\mathbf{P} = 0.0018$ **b.** reject H_o

10.87 Divide inequality by σ_p^2.

10.88 3.03, 3.78

10.89 H_o: $\sigma_m/\sigma_p = 1$ vs. H_a: $\sigma_m/\sigma_p > 1$

10.90 1.52

10.91 3.37

10.92 0.495, smaller variance in numerator

10.93 Multiply by 2.

10.94 **a.** H_o: $\sigma_N^2 = \sigma_A^2$ ($\leq$) vs. H_a: $\sigma_N^2 > \sigma_A^2$
b. Sample results are very unlikely if H_o is true.
c. 0.0058

10.95 **a.** H_o: $\sigma_A^2 = \sigma_B^2$ vs. H_a: $\sigma_A^2 \neq \sigma_B^2$
b. H_o: $\sigma_I = \sigma_{II}$ vs. H_a: $\sigma_I > \sigma_{II}$
c. H_o: $\sigma_A^2/\sigma_B^2 = 1$ vs. H_a: $\sigma_A^2/\sigma_B^2 \neq 1$
d. H_o: $\sigma_C^2/\sigma_D^2 = 1$ vs. H_a: $\sigma_C^2/\sigma_D^2 < 1$
or equivalently,
H_o: $\sigma_D^2/\sigma_C^2 = 1$ vs. H_a: $\sigma_D^2/\sigma_C^2 > 1$

10.97 **a.** 2.51 **b.** 2.20 **c.** 2.91 **d.** 4.10 **e.** 2.67 **f.** 3.77 **g.** 1.79 **h.** 2.99

10.99 **a.** 0.10 **b.** 0.02

10.101 H_o: $\sigma_k^2 = \sigma_m^2$ vs. H_a: $\sigma_k^2 \neq \sigma_m^2$, $F\bigstar = 1.33$, $\mathbf{P} > 0.10$ or $F \geq 3.73$; fail to reject H_o

10.103 H_o: $\sigma_y^2 = \sigma_o^2$ vs. H_a: $\sigma_y^2 > \sigma_o^2$, $F\bigstar = 3.29$, $\mathbf{P} < 0.01$ or $F \geq 1.94$; reject H_o

10.105 1.60

10.109 **a.** Fr: 48.5%, So: 93.5% b. 45%
c. Fr: \$1519, So: \$2079
d. **Credit Card Debts**

Soph Debt

Fresh Debt

0 1000 2000 3000 4000 5000 6000 7000

e. \$560.

10.113 $\bar{d} = 3.583$, $s_d = 19.58$; -8.85 to 16.02

10.115 $\bar{d} = 5.2$, $s_d = 7.406$,
H_o: $\mu_d = 0$ vs. H_a: $\mu_d > 0$ (improvement), $t\bigstar = 2.22$, $0.022 < \mathbf{P} < 0.029$ or $t \geq 1.83$; reject H_o

10.117 -0.21 to 10.61

10.119 $\bar{x}_A = 57.0$, $s_A^2 = 209.111$, $\bar{x}_B = 62.2$, $s_B^2 = 193.289$; -9.14 to 19.54

10.121 $\bar{x}_A = 80.021$, $s_A^2 = 0.0005744$, $\bar{x}_B = 79.979$, $s_B^2 = 0.0009839$; 0.012 to 0.072

10.123 H_o: $\mu_2 - \mu_1 = 0$ vs. H_a: $\mu_2 - \mu_1 > 0$ (artif. turf yields a lower time), $t\bigstar = 0.988$, $0.164 < \mathbf{P} < 0.189$ or $t \geq 1.72$; fail to reject H_o

10.125 **a.** $\bar{x}_A = 15.53$, $s_A^2 = 1.98$, $s_A = 1.41$
b. $\bar{x}_B = 12.53$, $s_B^2 = 1.98$, $s_B = 1.41$
c. H_o: $\mu_A - \mu_B = 0$ vs. H_a: $\mu_A - \mu_B \neq 0$, $t\bigstar = 5.84$, $\mathbf{P} < 0.002$ or $t \leq -2.98$, $t \geq 2.98$; reject H_o

10.127 **a.** M: $\bar{x} = 74.69$, $s = 10.19$, F: $\bar{x} = 79.83$, $s = 8.80$

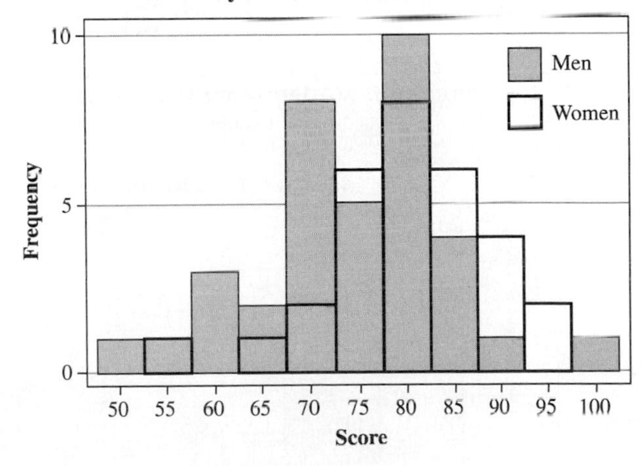

b. M: H_o: $\mu_M = 77$ vs. H_a: $\mu_M \neq 77$, $t\bigstar = -1.36$, $0.10 < \mathbf{P} < 0.20$ or $t \leq -2.03$, $t \geq 2.03$; fail to reject H_o
W: H_o: $\mu_F = 77$ vs. H_a: $\mu_F \neq 77$, $t\bigstar = 1.76$, $0.05 < \mathbf{P} < 0.10$ or $t \leq -2.05$, $t \geq 2.05$; fail to reject H_o
c. They're both not significantly different from 77.
d. H_o: $\mu_F - \mu_M = 0$ vs. H_a: $\mu_F - \mu_M \neq 0$, $t\bigstar = 2.19$, $0.02 < \mathbf{P} < 0.05$ or $t \leq -2.05$, $t \geq 2.05$; reject H_o
e. and **f.** no

10.129 **a.** -0.116 to 0.216
b. No; interval contains the value 0.

10.131 -0.044 to 0.164

10.133 **a.** no; $z\bigstar = 1.44$, $\mathbf{P} = 0.1498$
b. no; $z\bigstar = 1.77$, $\mathbf{P} = 0.0768$
c. yes; $z\bigstar = 2.04$, $\mathbf{P} = 0.0414$

10.135 H_o: $p_a - p_l = 0$ vs. H_a: $p_a - p_l \neq 0$, $p'_a = 0.5771$, $p'_l = 0.5091$, $z\bigstar = 1.26$, $\mathbf{P} = 0.2076$ or $z \leq -2.58$ and $z \geq 2.58$; fail to reject H_o

10.137 H_o: $\sigma^2_m = \sigma^2_f$ vs. H_a: $\sigma^2_m > \sigma^2_f$, $F\bigstar = 2.58$, $0.025 < \mathbf{P} < 0.05$ or $F \geq 2.53$; reject H_o

10.139 H_o: $\sigma^2_n = \sigma^2_s$ vs. H_a: $\sigma^2_n \neq \sigma^2_s$, $F\bigstar = 1.28$, $\mathbf{P} > 0.10$ or $F \geq 1.80$; fail to reject H_o

10.141 **a.** Cont: $\bar{x} = 0.005459$, $s = 0.000763$;
Test: $\bar{x} = 0.003507$, $s = 0.000683$

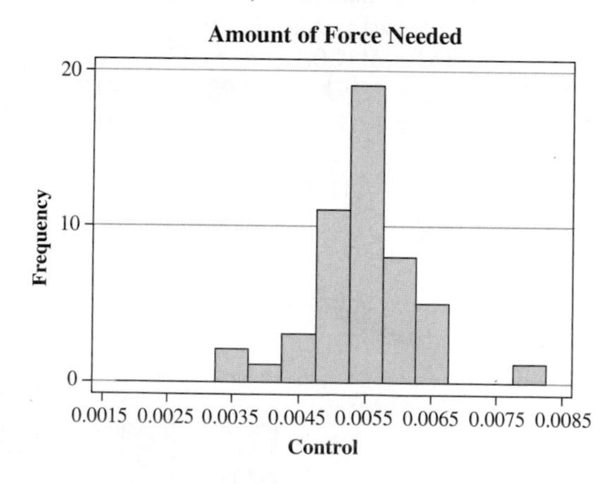

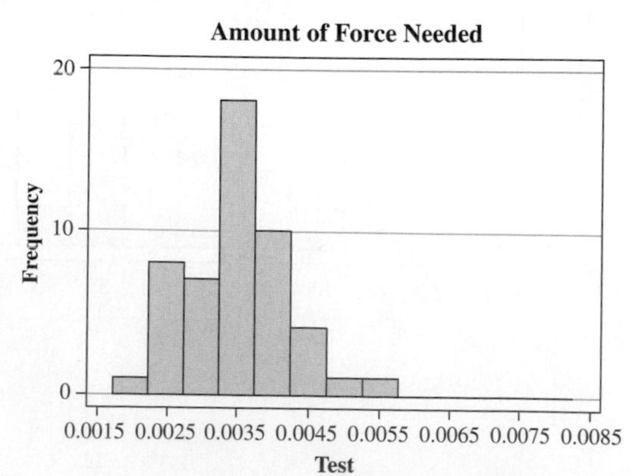

b. Both sets are approximately normal.
c. one-tailed—looking for a reduction
d. H_o: $\sigma^2_c = \sigma^2_t$ vs. H_a: $\sigma^2_c > \sigma^2_t$, $F\bigstar = 1.248$, $\mathbf{P} = 0.44$ or $F \geq 1.69$; fail to reject H_o
e. H_o: $\mu_{Cont} - \mu_{Test} = 0$ vs. H_a: $\mu_{Cont} - \mu_{Test} > 0$, $t\bigstar = 13.48$, $\mathbf{P} = 0.000+$ or $t \leq -1.68$, $t \geq 1.68$; reject H_o
f. The mean force has been reduced, but not the variability.

Chapter 11

11.1 **a.** and **b.**

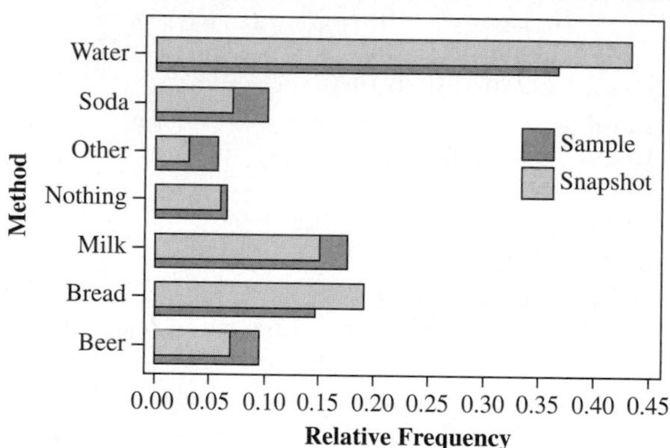

c. similar

11.2 **a.** 23.2 **b.** 23.3 **c.** 3.94 **d.** 8.64

11.3 **a.** 34.8 **b.** 28.8 **c.** 51.8
d. ≈69.95 **e.** 29.1 **f.** 14.6, 43.2

11.4 9 + 3 + 3 + 1 = 16 parts total

11.5 556(9/16), 315 − 312.75, $(2.25)^2/312.75$

11.6 **a.** each adult surveyed
b. where lesson learned
c. mistakes, school, unsure

11.7 Some use more than one method.

11.9 **a.** 0.0066 **b.** 0.0503

11.11 **a.** H_o: $P(A) = P(B) = P(C) = P(D) = P(E) = 0.2$
b. χ^2
c. H_a: preferences not all equal, $\chi^2\bigstar = 4.40$, $\mathbf{P} = 0.355$ or $\chi^2 \geq 7.78$; fail to reject H_o

11.13 H_a: Distributions are different, $\chi^2\bigstar = 7.586$, $\mathbf{P} = 0.18058$ or $\chi^2 \geq 11.1$; fail to reject H_o

11.15 H_a: There is a difference, $\chi^2\bigstar = 193.72$, $\mathbf{P} = 0+$ or $\chi^2 \geq 11.3$; reject H_o

11.17 H_a: The probabilities are not all equal, $\chi^2\bigstar = 6.00$, $\mathbf{P} = 0.740$ or $\chi^2 \geq 16.9$; fail to reject H_o

11.19 **a.** Calculated value of chi-square will be near zero.
b. "Little brother did not roll the die!"
He knew what to expect and reported it.
The calculated chi-square will be 'zero.'
You want your money back, he did not earn the $1.
The data and the theory are too much alike.

c. Hopefully the experimenter trusts himself to carry out the experiment randomly.

11.21 10

11.22 **a.** 113 **b.** 122 **c.** 300 **d.** 68.23

11.23 **a.**

	West	N. Cent.	South	N. East
Baked	55%	46%	47%	41%
Other	45%	54%	53%	59%

b. The information compares several distributions, a distribution for each region.
c. The distributions of "baked" and "other" are being compared for the 4 different regions.

11.25 Test of *independence* has one sample of data that is being cross-tabulated according to the categories of two separate variables; *test of homogeneity* has multiple samples being compared side-by-side.

11.27 H_a: not independent, $\chi^2\bigstar = 35.749$, $\mathbf{P} < 0.005$ or $\chi^2 \geq 13.3$; reject H_o

11.29 H_a: not independent, $\chi^2\bigstar = 3.390$, $\mathbf{P} = 0.335$ or $\chi^2 \geq 6.25$; fail to reject H_o

11.31 H_a: not independent, $\chi^2\bigstar = 8.548$, $\mathbf{P} = 0.074$ or $\chi^2 \geq 9.49$; fail to reject H_o

11.33 **a.** and **b.** H_a: proportions not the same, $\chi^2\bigstar = 24.84$, $\mathbf{P} < 0.005$ or $\chi^2 \geq 16.8$; reject H_o
c. Prof. #3

11.35 Guns: H_a: Distributions are different, $\chi^2\bigstar = 49.130$, $\mathbf{P} < 0.005$ or $\chi^2 \geq 6.63$; reject H_o.
Anger: H_a: Distributions are different, $\chi^2\bigstar = 5.042$, $\mathbf{P} = 0.025$ or $\chi^2 \geq 6.63$; fail to reject H_o

11.37 **a.** method used to cool mouth
b. adults professing to love hot spicy food, method of cooling
c. see graph in 11.1 **d.** similar

11.41 H_a: proportions are other than 1:3:4, $\chi^2\bigstar = 10.33$, $\mathbf{P} = 0.006$ or $\chi^2 \geq 5.99$; reject H_o

11.43 H_a: percentages are different, $\chi^2\bigstar = 6.693$, $\mathbf{P} = 0.153$ or $\chi^2 \geq 9.49$; fail to reject H_o

11.45 H_a: percentages are different, $\chi^2\bigstar = 17.92$, $\mathbf{P} = 0.003$ or $\chi^2 \geq 15.1$; reject H_o

11.47 **a.** H_a: Distributions are different, $\chi^{2*} = 6.195$, $\mathbf{P} > 0.25$ or $\chi^2 \geq 11.1$; fail to reject H_o
b. H_a: Distributions are different, $\chi^2\bigstar = 36.761$, $\mathbf{P} < 0.005$ or $\chi^2 \geq 11.1$; reject H_o
c. H_a: Distributions are different, $\chi^2\bigstar = 92.93$, $\mathbf{P} < 0.005$ or $\chi^2 \geq 11.1$; reject H_o
d. Sample sizes were too small.

11.49 H_a: not independent, $\chi^2\bigstar = 3.651$, $\mathbf{P} = 0.456$ or $\chi^2 \geq 9.49$; fail to reject H_o

11.51 H_a: not independent, $\chi^2\bigstar = 23.339$, $\mathbf{P} < 0.005$ or $\chi^2 \geq 13.3$; reject H_o

11.53 H_a: proportions not the same, $\chi^2\bigstar = 2.839$, $\mathbf{P} = 0.417$ or $\chi^2 \geq 7.82$; fail to reject H_o

11.55 H_a: Not independent, $\chi^2\bigstar = 2.494$, $\mathbf{P} = 0.114$ or $\chi^2 \geq 3.84$; fail to reject H_o

Chapter 12

12.1 **a.**

Reading the Newspaper

Age: 65–up, 50–64, 35–49, 30–34, 25–29, 18–24

5 15 25 35

Time Spent Daily Reading the Newspaper (minutes)

b. Mean amount of time spent reading the newspaper increases steadily.

12.2 **Units Produced per Hour at Each Temperature Level**

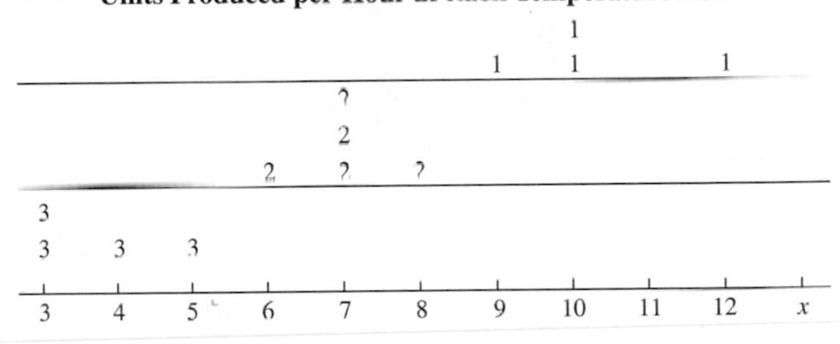

Yes, there appears to be a difference between the three sets.

12.3 H_o: $\mu_1 = \mu_2 = \mu_3$;
H_a: Means are not all the same

12.4 **a.** yes
b. Male and female categories encompass population.

12.5 "Amount of money donated" is categorized by age.

12.6 **a.** 0 **b.** 2 **c.** 4 **d.** 31 **e.** 393

12.7 **a.** H_o: $\mu_1 = \mu_2 = \mu_3 = \mu_4 = \mu_5$ vs. H_a: Means not all equal
b. H_o: $\mu_1 = \mu_2 = \mu_3 = \mu_4$ vs. H_a: Means not all equal
c. H_o: $\mu_1 = \mu_2 = \mu_3 = \mu_4$ vs. H_a: Means not all equal
d. H_o: $\mu_1 = \mu_2 = \mu_3$ vs. H_a: Means not all equal

12.9 **a.** $F \geq 3.34$ **b.** $F \geq 5.99$ **c.** $F \geq 3.44$

12.11 **a.** depends on whether it is larger or smaller than α
b. reject H_o **c.** fail to reject H_o

12.13 **a.** Test factor has no effect on the mean at the tested levels.
b. Test factor does have an effect on the mean at the tested levels.
c. $\mathbf{P} = P(F > F\bigstar) \leq \alpha$; $F\bigstar$ must fall in the critical region.
d. Tested factor has a significant effect on the variable.
e. $\mathbf{P} = P(F > F\bigstar) > \alpha$; $F\bigstar$ must fall in the noncritical region.
f. Tested factor does not have a significant effect on the variable.

12.15 **a.** 120 **b.** 3 **d.** yes; p-value is small
e. no; p-value is large

12.17 H_a: mean values are not all equal;

Source	df	SS	MS	$F\bigstar$
Work	2	17.73	8.87	4.22
Error	12	25.20	2.10	
Total	14	42.93		

P = 0.041 or $F \geq 3.89$; reject H_o

12.19 H_a: means are not all equal;

Source	df	SS	MS	$F\bigstar$
Rating cat.	2	48.11	24.055	4.58
Error	15	78.83	5.255	
Total	17	126.94		

P = 0.028 or $F \geq 3.68$; reject H_o

12.21 H_a: means are not all equal;

Source	df	SS	MS	$F\bigstar$
Division	2	284.46	142.23	2.732
Error	25	1301.93	52.07	
Total	27	1586.39		

P = 0.0845 or $F \geq 3.39$; fail to reject H_o

12.23 *Calories:* H_a: means are not all equal;

Source	df	SS	MS	$F\bigstar$
Category	3	6955.56	2318.52	6.088
Error	46	17519.02	380.85	
Total	49	24474.58		

P = 0.0014 or $F \geq 4.26$; reject H_o

Fat content: H_a: means are not all equal;

Source	df	SS	MS	$F\bigstar$
Category	3	22.46	7.49	1.745
Error	46	197.26	4.29	
Total	49	219.72		

P = 0.171 or $F \geq 4.26$; fail to reject H_o

12.25 a. H_a: At least one mean is different;

Source	df	SS	MS	$F\bigstar$
Factor	11	0.79	0.07	0.04
Error	117	230.58	1.97	
Total	128	231.37		

P = 1.000 or $F \geq 1.88$; fail to reject H_o

Boxplots of Jan.-Dec.
(means are indicated by solid circles)

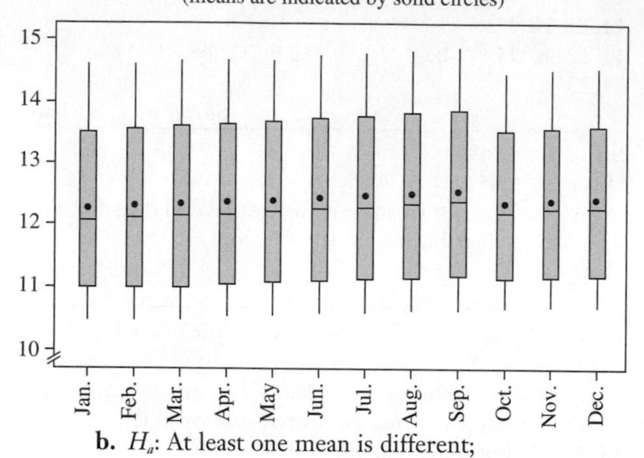

b. H_a: At least one mean is different;

Source	df	SS	MS	$F\bigstar$
Factor	10	229.4794	22.9479	1434.01
Error	118	1.8883	0.0160	
Total	128	231.3677		

P = 0.000 or $F \geq 1.92$; reject H_o

Boxplots of 1992-2002
(means are indicated by solid circles)

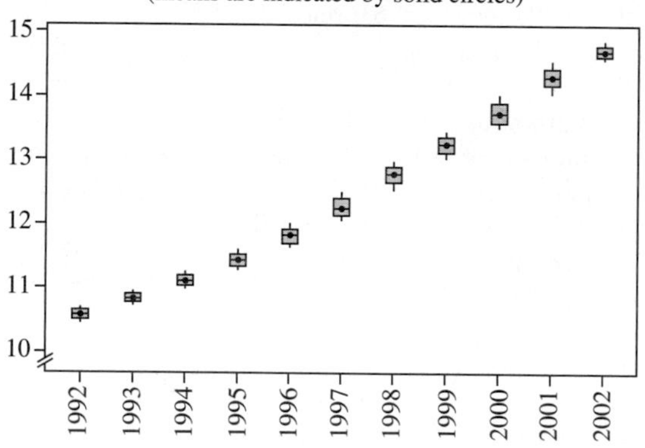

12.26 **a.**

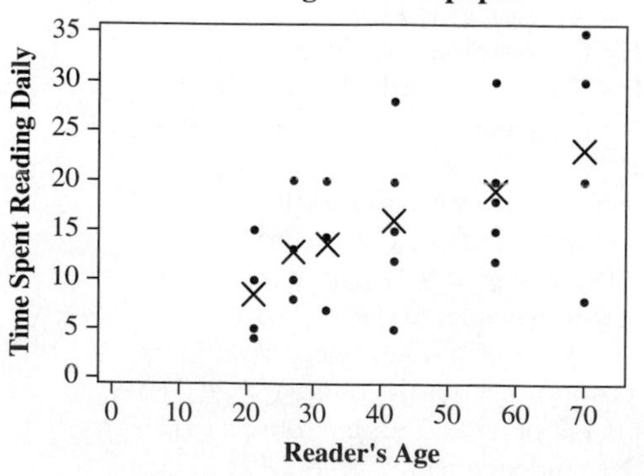

b. yes **c.** Yes, age appears to have an effect.

12.29 **a.** H_o: mean is the same;
H_a: mean is not the same
b. Samples were randomly selected; $F \geq 3.68$
c. fail to reject H_o, no significant difference
d. The sample data are quite likely to have occurred under the assumed conditions and a true null hypothesis.

12.31 H_a: mean is not same for all;

Source	df	SS	MS	$F\bigstar$
Drug	3	108.33	36.11	12.50
Error	12	34.67	2.89	
Total	15	143.00		

P = 0.001 or $F \geq 3.49$; reject H_o

12.33 H_a: mean amounts are not all equal;

Source	df	SS	MS	$F\bigstar$
Machine	4	20.998	5.2495	31.6
Error	13	2.158	0.166	
Total	17	23.156		

P = 0.000 or $F \geq 5.21$; reject H_o

12.35 H_a: mean points are not all equal;

Source	df	SS	MS	$F\bigstar$
Division	2	9,110	4555	1.39
Error	28	91,939	3284	
Total	30	101,049		

P = 0.266 or $F \geq 3.36$; fail to reject H_o.
H_a: mean points (opponents) are not all equal;

Source	df	SS	MS	$F\bigstar$
Division	2	7,353	3676	0.82
Error	28	125,808	4493	
Total	30	133,161		

P = 0.451 or $F \geq 3.36$; fail to reject H_o

12.37 H_a: mean is not the same;

Source	df	SS	MS	$F\bigstar$
Counties	2	1217.5	608.7	28.74
Error	24	508.4	21.2	
Total	26	1725.8		

P = 0.000+ or $F \geq 3.40$; reject H_o

12.39 H_a: brands do not test equally well;

Source	df	SS	MS	$F\bigstar$
Brand	5	75,047	15009.4	5.30
Error	36	101,899	2830.5	
Total	41	176,946		

P = 0.001 or $F \geq 2.48$; reject H_o

12.41 **a.** H_a: At least one mean is different;

Source	df	SS	MS	$F\bigstar$
Type	2	1671.56	835.78	118.06
Error	27	191.14	7.08	
Total	29	1862.70		

P = 0.000 or $F \geq 3.37$; reject H_o

b. H_a: At least one mean is different;

Source	df	SS	MS	$F\bigstar$
Type	2	197.1	98.6	7.78
Error	27	342.2	12.7	
Total	29	539.4		

P = 0.002 or $F \geq 3.37$; reject H_o

c. Type 0 has the shortest PW and the longest SW.
Type 1 has the longest PW and the middle SW.
Type 2 has the middle PW and the shortest SW.

12.43 29.3333
12.45 82.4752

Chapter 13

13.1 **a.** $r = 0.989$, $\hat{y} = -1.57 + 0.880x$

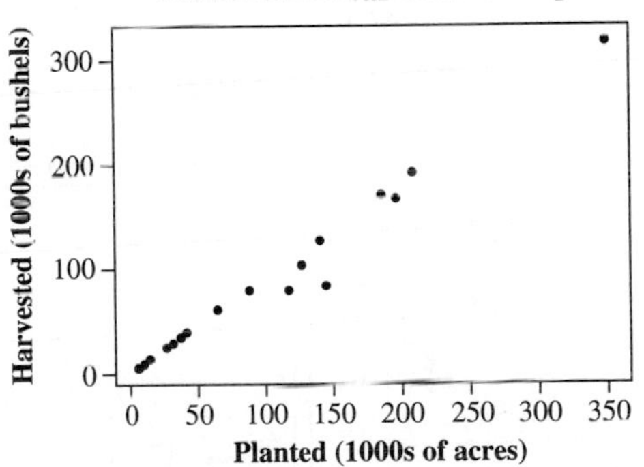

b.

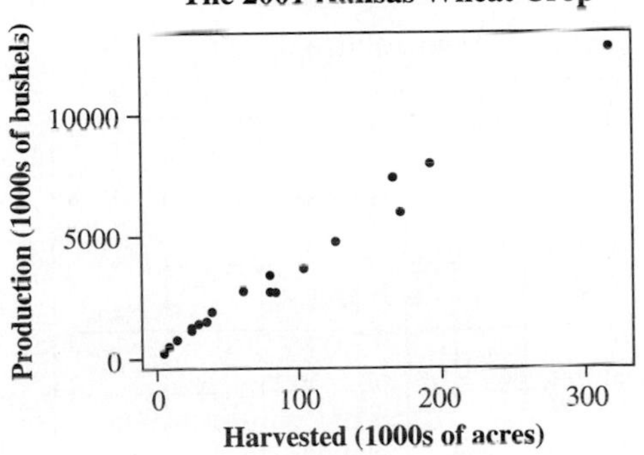

c.

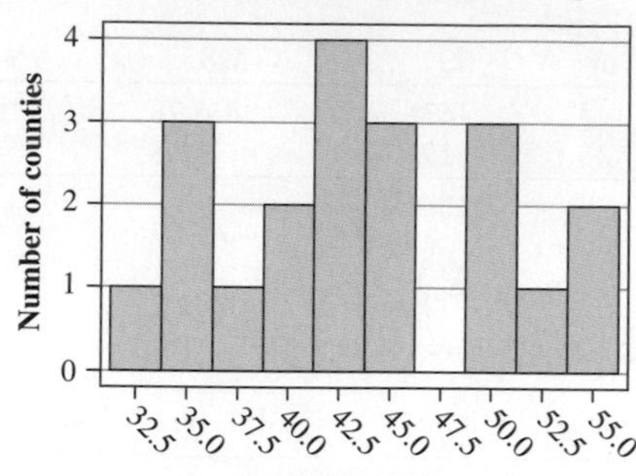

d. planted and harvested are highly linearly correlated, production and harvested are highly linearly correlated, yield rate is approximately normal

13.3 a.

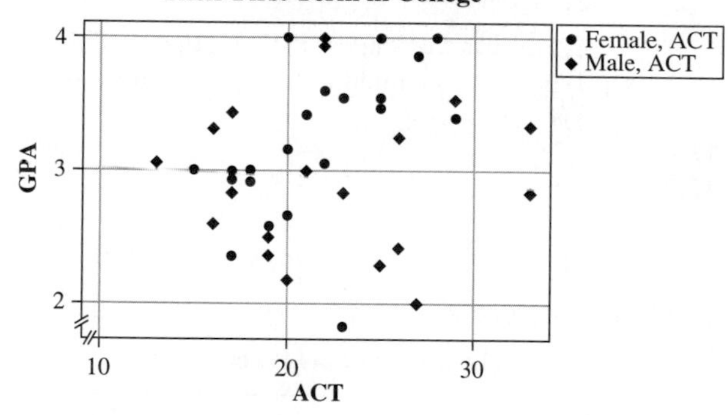

c.

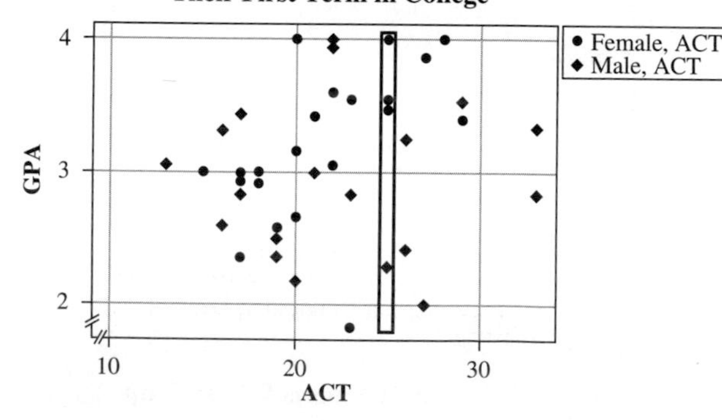

any value from 1.8 to 4.0

d. no

13.5 a.

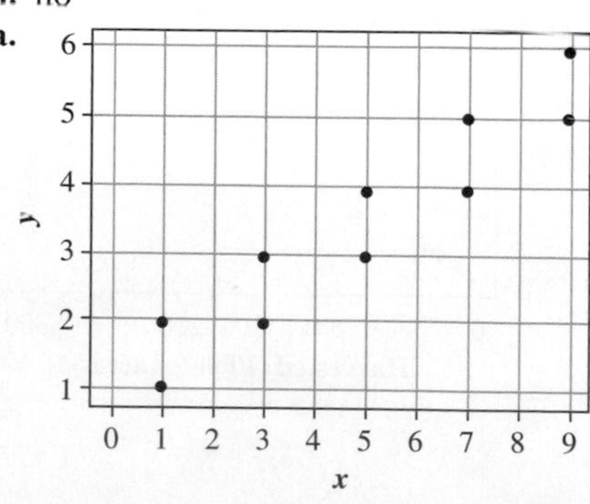

b. 4.44
$\sum x = 50$, $\sum y = 35$, $\sum x^2 = 330$, $\sum xy = 215$, $\sum y^2 = 145$;
c. 2.981, 1.581 d. 0.943 e. 0.943

13.9 a. −0.234
b. slight negative relationship
c.

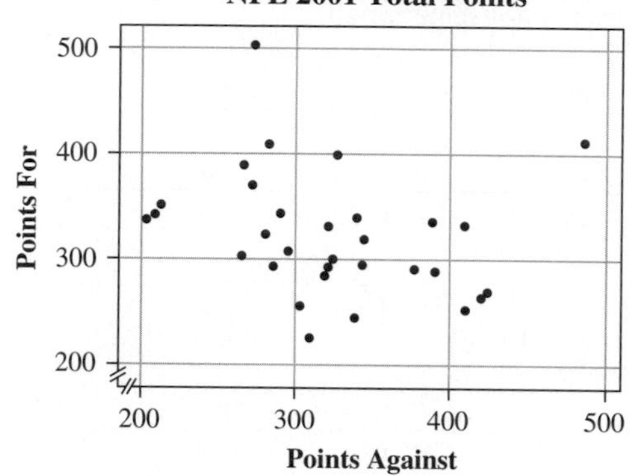

downward trend, points wide spread, weak correlation coefficient

13.11 a. 60 b. $s_x = 40.99$, $s_y = 20.98$
c. 0.07 d. The value for r is the same.

13.13 −0.02 to 0.63

13.14 a. $0.05 < \mathbf{P} < 0.10$ b. $0.025 < \mathbf{P} < 0.05$

13.15 a. ± 0.444
b. −0.378, if left tail; 0.378, if right tail

13.16 a. $r = 0.58$, is significant for $\alpha > 0.008$
b. 0.464 c. significant at $\alpha = 0.01$

13.19 $\sum x = 746$; $\sum y = 736$; $\sum x^2 = 57{,}496$; $\sum xy = 56{,}574$; $\sum y^2 = 55{,}826$; 0.955, 0.78 to 0.98

13.21 a. $\sum x = 1189$; $\sum y = 5482$; $\sum x^2 = 205{,}787$; $\sum xy = 928{,}950$; $\sum y^2 = 4{,}333{,}726$; −0.177
b. −0.78 to 0.62

13.23 a. H_o: $\rho = 0$ vs. H_a: $\rho > 0$
b. H_o: $\rho = 0$ vs. H_a: $\rho \neq 0$
c. H_o: $\rho = 0$ vs. H_a: $\rho < 0$
d. H_o: $\rho = 0$ vs. H_a: $\rho > 0$

13.25 H_o: $\rho = 0.0$, H_a: $\rho \neq 0.0$, $r\bigstar = 0.43$, $0.05 < \mathbf{P} < 0.10$ or $r \leq -0.378$ and $r \geq 0.378$; reject H_o

13.27 H_o: $\rho = 0.0$, H_a: $\rho \neq 0.0$, $r\bigstar = 0.16942$, $\mathbf{P} > 0.10$ or $r \leq -0.423$ and $r \geq 0.423$; fail to reject H_o

13.29 $\sum x = 26.2$, $\sum y = 82.5$, $\sum x^2 = 174.88$, $\sum xy = 256.41$, $\sum y^2 = 704.61$; H_o: $\rho = 0.0$, H_a: $\rho \neq 0.0$, $r\bigstar = 0.798$, $\mathbf{P} < 0.01$ or $r \leq -0.632$ and $r \geq 0.632$; reject H_o

13.31 a. $r = 0.613$
b. H_o: $\rho = 0.0$, H_a: $\rho \neq 0.0$, $r\bigstar = 0.613$, $\mathbf{P} > 0.10$ or $r \leq -0.754$ and $r \geq 0.754$; fail to reject H_o

13.35 a.

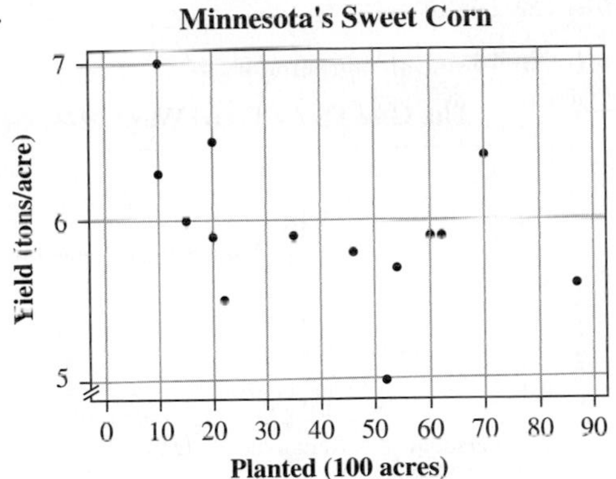

$\sum x = 563$, $\sum y = 83.42$, $\sum x^2 = 30583$, $\sum xy = 3289.15$, $\sum y^2 = 500.126$; -0.420, $\hat{y} = 6.29 - 0.00825x$, shows no relationship

13.37 a.

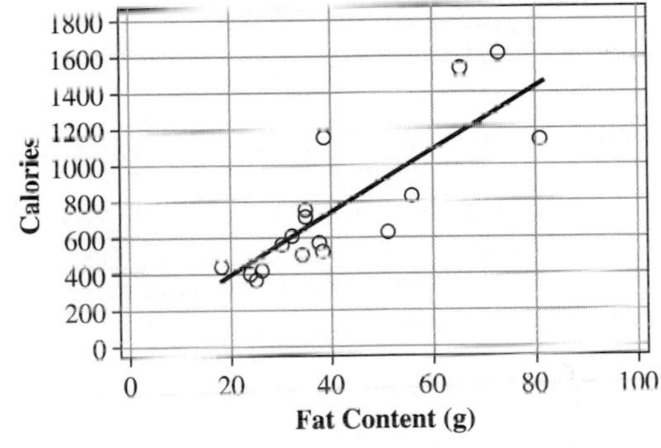

b. $\sum x = 696$; $\sum y = 12{,}655$; $\sum x^2 = 34{,}362$; $\sum xy = 617{,}415$; $\sum y^2 = 11{,}801{,}175$; $\hat{y} = 16.926x + 51.4$

c. 16.926 d. 46,524.8

13.39 a. $\sum x = 50$, $\sum y = 35$, $\sum x^2 = 330$, $\sum xy = 215$, $\sum y^2 = 145$; $\hat{y} = 1.0 + 0.5x$

b. 1.5, 2.5, 3.5, 4.5, 5.5

c. -0.5, 0.5, alternately

d. 0.3125 e. 0.3125

13.43 a. 0.0145 b. 0.0668 c. 0.0653

13.45 0.1894

13.47 a. $\hat{y} = -348 + 2.04x$ b. 1.60 to 2.48

13.49 a.

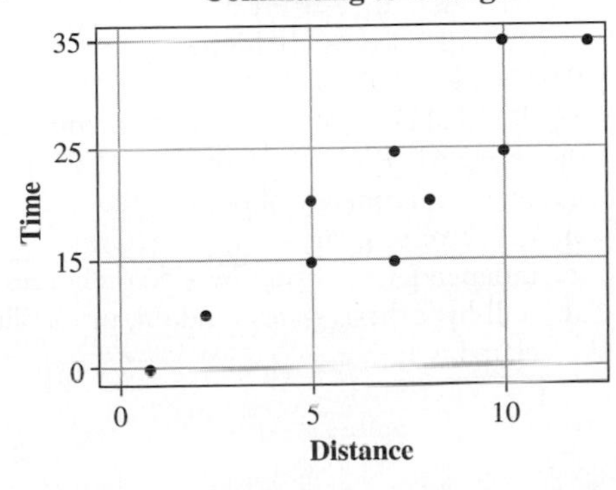

b. $\sum x = 68$, $\sum y = 205$, $\sum x^2 = 566$, $\sum xy = 1670$, $\sum y^2 = 5075$; $\hat{y} = 2.38 + 2.664x$

c. H_o: $\beta_1 = 0$, H_a: $\beta_1 > 0$, $t\bigstar = 6.55$, $\mathbf{P} < 0.005$ or $t \geq 1.86$; reject H_o

d. 1.48 to 3.84

13.51 a. $\sum x = 1105$; $\sum y = 93{,}327$; $\sum x^2 = 126{,}475$; $\sum xy = 10{,}447{,}010$; $\sum y^2 = 885{,}420{,}825$; $\hat{y} = 5937 + 30.73x$

b. 1135, 17.16

d. H_o: $\beta_1 = 0$, H_a: $\beta_1 > 0$, $t\bigstar = 1.79$, $0.05 < \mathbf{P} < 0.10$ or $t \geq 1.86$; fail to reject H_o

13.55 52.3 to 52.5

13.57 $\sum x = 16.25$, $\sum y = 152$, $\sum x^2 = 31.5625$, $\sum xy = 275$, $\sum y^2 = 2504$; $\hat{y} = 6.3758 + 5.4303x$

a. 15.4 to 19.1

b. 11.6 to 22.8

13.59 a. $\sum x = 256{,}593$; $\sum y = 454$; $\sum x^2 = 9{,}013{,}638{,}265$; $\sum xy = 15{,}173{,}202$; $\sum y^2 = 26{,}316$; $\hat{y} = 31.72 + 0.0007804x$

b. 59.04% to 66.84%

c. 53.5% to 72.38%

13.63 a. 0.974, $\hat{y} = 50 + 34.2x$

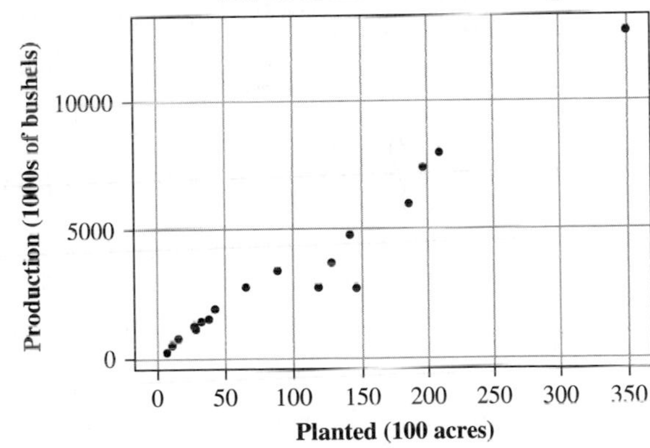

b. -0.604, $\hat{y} = 47.7 - 0.0439x$

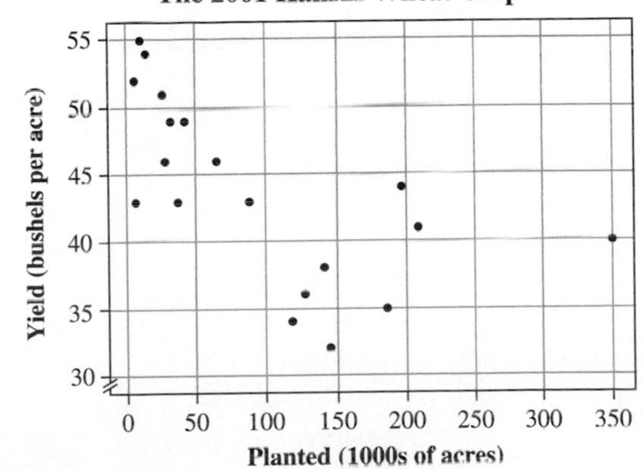

13.67 H_o: $\rho = 0.0$, H_a: $\rho > 0.0$, $r\bigstar = 0.69$, $\mathbf{P} < 0.005$ or $r \geq 0.29$; reject H_o

13.69 a. 0.019, 0.096 b. practically no correlation

13.71 **a.** H_o: $\rho = 0.0$, H_a: $\rho \neq 0.0$, $r\bigstar = 0.61$, $\mathbf{P} < 0.01$ or $r \leq -0.482$, $r \geq 0.482$; reject H_o
b. $\hat{y} = 1.8(50) + 28.7 = 118.7$
13.73 3.94 to 4.52
13.75 **a.**

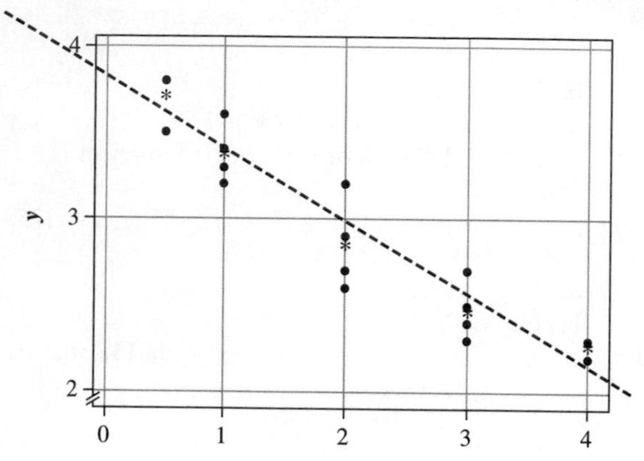

b. See dashed line on graph in (a).
c. See *'s on graph.
d. $\sum x = 37.5$, $\sum y = 52.7$, $\sum x^2 = 104.75$, $\sum xy = 98.75$, $\sum y^2 = 159.49$; $\hat{y} = 3.79 - 0.415x$
e. 0.21045
f. -0.502 to -0.328
g. At $x = 3.0$: 2.42 to 2.68; At $x = 3.5$: 2.18 to 2.50
h. At $x = 3.0$: 2.08 to 3.02; At $x = 3.5$: 1.87 to 2.81
13.77 $\sum x = 1177$, $\sum y = 567$, $\sum x^2 = 70033$, $\sum xy = 32548$, $\sum y^2 = 15861$;
a. H_o: $\rho = 0.0$, H_a: $\rho \neq 0.0$, $r\bigstar = 0.513$, $0.01 < \mathbf{P} < 0.02$ or $r \leq -0.433$, $r \geq 0.433$; reject H_o
b. $\hat{y} = 16.40 + 0.189x$
c. H_o: $\beta_1 = 0$, H_a: $\beta_1 > 0$, $t\bigstar = 2.61$, $0.005 < \mathbf{P} < 0.01$ or $t \geq 1.73$; reject H_o
d. 18.53 to 38.47
13.79 $\sum x = 16$, $\sum y = 38$, $\sum x^2 = 66$, $\sum xy = 145$, $\sum y^2 = 326$; $b_1 = 1.5811$, $r = 0.9973$

Chapter 14

14.1 **a.** "general" agreement
b.

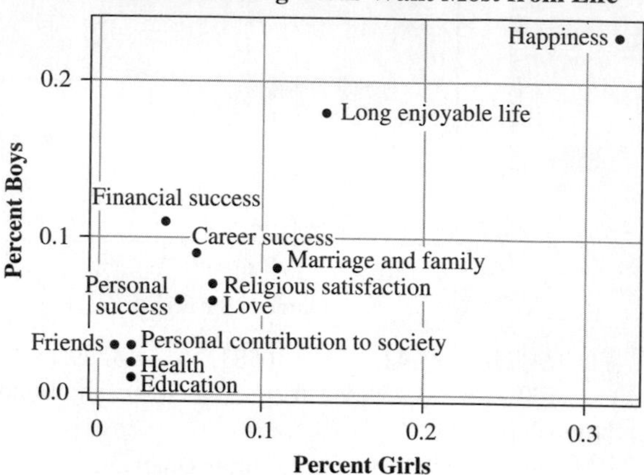

appear to correlate
14.2 75 to 94
14.3 39 to 47
14.5 -7 to 1
14.7 **a.** H_o: Median = 32 vs. H_a: Median < 32
b. H_o: P(prefer new recipe) = 0.50 vs. P(prefer) < 0.50
c. H_o: $P(+\text{gain}) = 0.5$ vs. H_a: $P(+) \neq 0.5$
14.9 H_a: Median ≠ 30 years, $x = n(+) = 40$; $0.05 < \mathbf{P} < 0.10$ or $x \leq 39$; fail to reject H_o
14.11 H_o: $P(+) = 0.5$ vs. H_a: $P(+) \neq 0.5$
If $\mathbf{P} \leq \alpha$, reject H_o; if $\mathbf{P} > \alpha$, fail to reject H_o
a. $x = n(+) = 20$, $\mathbf{P} < 0.01$
b. $x = n(+) = 27$, $0.01 < \mathbf{P} < 0.05$
c. $x = n(+) = 30$, $0.10 < \mathbf{P} < 0.25$
d. $x = n(+) = 33$, $\mathbf{P} > 0.25$
14.13 H_o: $M = 83{,}282$ vs. H_a: $M < 83{,}282$, $x = n(+) = 4$; $0.005 < \mathbf{P} < 0.025$ or $x \leq 5$; reject H_o
14.15 H_a: There is a preference for the new; $p > 0.5$; $z\bigstar = 1.74$, $\mathbf{P} = 0.0409$ or $z \geq 2.33$; fail to reject H_o
14.17 $x' = 712.04$
14.18 Sample B seems "slid" right about ten points.
14.19 Null hypothesis is "the same" or "no change"; when rejected there is evidence of change—when not rejected, no change can be claimed.
14.21 **a.** $U \leq 88$ **b.** $z \leq -1.65$
14.25 H_a: Group 1 scores are higher, $U\bigstar = 4.5$, $0.025 < \mathbf{P} < 0.05$ or $U \leq 6$; reject H_o
14.27 H_a: Rainfall amount is higher with cloud seeding, $z\bigstar = -2.61$, $\mathbf{P} = 0.0045$ or $z \leq -1.65$; reject H_o
14.30 **a.** relative frequency of occurrences
b. order, or sequence
c. independence of side-by-side outcomes
d. null hypothesis states random, probability associated is $1 - \alpha$

14.31 **a.** H_o: Occur in random order;
H_a: Do not occur in random order
b. H_o: Sequence is random;
H_a: Not random;
c. H_o: Order of entry was random;
H_a: Was not random

14.33 H_a: Hiring sequence is not random, $V\bigstar = 9$,
$\mathbf{P} > 0.05$ or $V \leq 4$, $V \geq 12$; fail to reject H_o

14.35 H_a: Lack of randomness, $V\bigstar = 8$,
$\mathbf{P} > 0.05$ or $V \leq 3$, $V \geq 10$; fail to reject H_o

14.37 **a.** 22.5, 8
b. H_a: The data did not occur randomly, $V\bigstar = 8$,
$\mathbf{P} > 0.05$ or $V \leq 3$, $V \geq 13$; fail to reject H_o

14.39 **b.** $z\bigstar = -3.76$, $\mathbf{P} = 0.0002$
c. yes, reject
d.

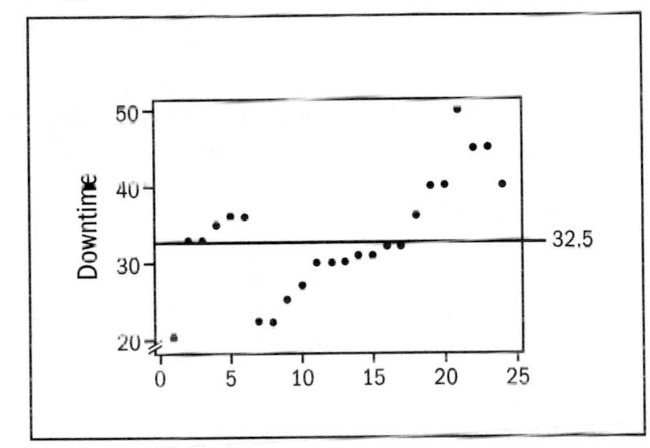

14.41 **a.** H_a: The data did not occur randomly, $V\bigstar = 9$,
$\mathbf{P} > 0.05$ or $V \leq 8$, $V \geq 20$; fail to reject H_o
b. H_a: Data did not occur randomly, $z\bigstar = -2.00$,
$\mathbf{P} = 0.0456$ or $z \leq -1.96$, $z \geq 1.96$; reject H_o

14.43 **a.** t-test for comparison of 2 dependent means
b. no linear correlation is rejected for all significance levels larger than 0.0001
c. no rank correlation is rejected for all significance levels larger than 0.0001
d. basically testing the same thing; Pearson's uses the VMI score, Spearman's uses ranks of the VMI scores

14.45 **a.** $r_s \leq -0.545$, $r_s \geq 0.545$ **b.** $r_s \geq 0.323$
c. $r_s < -0.564$

14.47 **a.** 0.133
b. H_o: $\rho_s = 0$ vs. H_a: $\rho_s > 0$, $r_s\bigstar = 0.133$,
$\mathbf{P} > 0.10$ or $r_s \geq 0.564$; fail to reject H_o

14.49 $n = 12$, $\sum d^2 = 70.5$, 0.753

14.51 H_o: $\rho_s = 0$ vs. H_a: $\rho_s > 0$, $r_s\bigstar = 0.736$,
$0.01 < \mathbf{P} < 0.025$ or $r_s \geq 0.564$; reject H_o

14.53 **a.**

Lee Rank	FL Rank	US Rank
1	1	1
6	3	2.5
3.5	6	6
3.5	3	4.5
8	8	7
2	3	4.5
5	5	2.5
9	9	8
7	7	*

b. H_o: $\rho_s = 0$ vs. H_a: $\rho_s \neq 0$, $r_s\bigstar = 0.8625$,
$\mathbf{P} < 0.01$ or $r_s \leq -0.700$, $r_s \geq 0.700$; reject H_o
c. H_o: $\rho_s = 0$ vs. H_a: $\rho_s \neq 0$, $r_s\bigstar = 0.619$,
$\mathbf{P} > 0.10$ or $r_s \leq -0.738$, $r_s \geq 0.738$; fail to reject H_o
d. H_o: $\rho_s = 0$ vs. H_a: $\rho_s \neq 0$, $r_s\bigstar = 0.869$,
$0.01 < \mathbf{P} < 0.02$ or $r_s \leq -0.738$, $r_s \geq 0.738$; reject H_o

14.55 **a.**

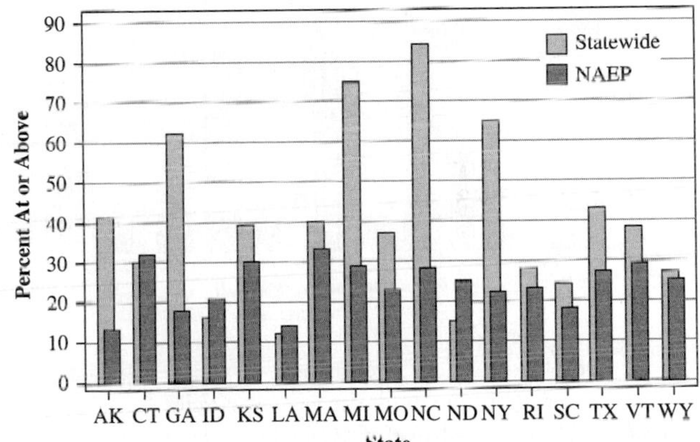

very little relationship

c.

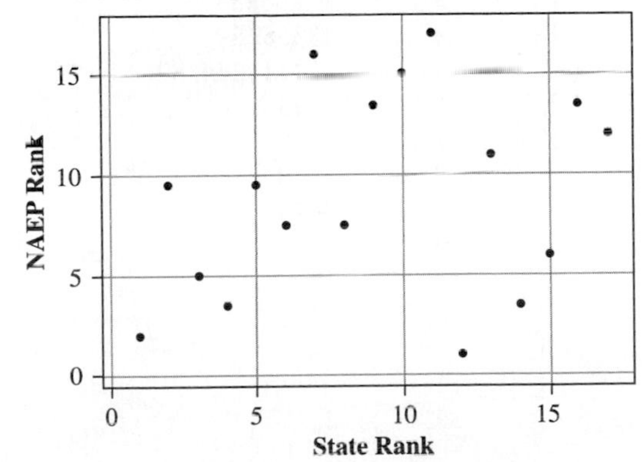

d. H_o: $\rho_s = 0$ vs. H_a: $\rho_s \neq 0$, $r_s\bigstar = 0.272$,
$\mathbf{P} > 0.10$ or $r_s \leq -0.490$, $r_s \geq 0.490$; fail to reject H_o

14.57 **a.**

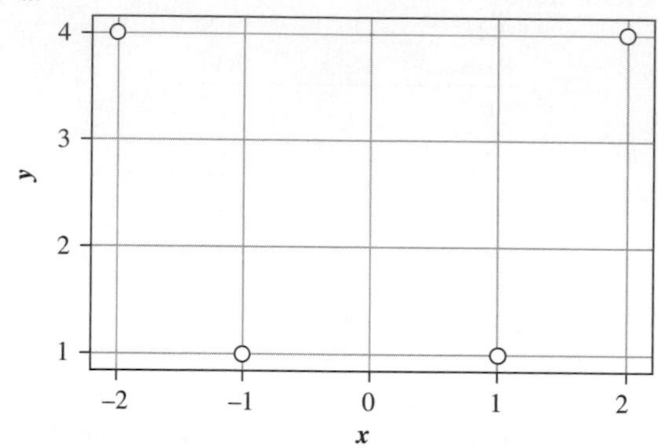

b. 0.10 **c.** 0.00

14.58 **a.** not exactly the same rate, but similar

b.

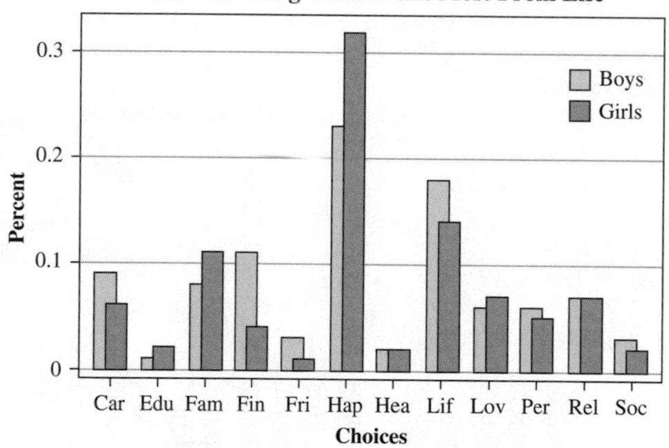

c. Information given is percentages, chi-square tests require the use of frequencies.

14.61 **a.** Stem-and-leaf of Hydrogen

N = 52

Leaf Unit = 0.10

3	2	078
10	3	1356688
18	4	13355689
(12)	5	001112444589
22	6	11137
17	7	2388
13	8	3378
9	9	4479
5	10	56
3	11	1
2	12	4
1	13	
1	14	
1	15	2

b. skewed right

c. 5.0 to 6.3

14.63 **a.** H_a: Median $\neq 50$, $x = n(+) = 10$, $\mathbf{P} \approx 0.10$ or $x \leq 9$; fail to reject H_o

b. H_a: Median < 50, $x = n(+) = 10$, $\mathbf{P} \approx 0.05$ or $x \leq 10$; reject H_o

14.65 H_a: Average time on B is less than on A, $x = n(-) = 2$, $\mathbf{P} > 0.10$ or $x \leq 1$; fail to reject H_o

14.67 reject for $U \leq 127$

14.69 H_a: There is a difference in line width, $U\bigstar = 2$, $\mathbf{P} \approx 0.05$ or $U \leq 2$; reject H_o.

14.71 **b.** *Batting Averages:*

H_a: Batting averages in AL are higher, $U\bigstar = 82.5$, $\mathbf{P} > 0.05$ or $U \leq 71$; fail to reject H_o

Earned Run Averages:

H_a: Earned run average for NL is lower, $U\bigstar = 77.5$, $\mathbf{P} > 0.05$ or $U \leq 71$; fail to reject H_o

14.73 H_a: Lack of randomness, $V\bigstar = 9$, $\mathbf{P} > 0.05$ or $V \leq 4$, $V \geq 10$; fail to reject H_o

14.75 **a.** Median = 22.5; Runs above: 6, Runs below: 6

b. H_a: Lack of randomness, $V\bigstar = 12$, $\mathbf{P} > 0.05$ or $V \leq 6$, $V \geq 16$; fail to reject H_o

c. not sufficient

14.77 H_a: $\rho_s > 0$, $r_s\bigstar = 0.880$, $\mathbf{P} < 0.01$ or $r_s \geq 0.399$; reject H_o

14.79 **a.** $r_{12} = 0.321$, $r_{13} = 0.298$, $r_{23} = 0.988$

b. H_o: $\rho_s = 0$ vs. H_a: $\rho_s \neq 0$

For (1) vs. (2): $r_s\bigstar = 0.321$, $\mathbf{P} > 0.10$ or $r_s \leq -0.400$, $r_s \geq 0.400$; fail to reject H_o.

For (1) vs. (3), $r_s\bigstar = 0.298$, $\mathbf{P} > 0.10$ or $r_s \leq -0.400$, $r_s \geq 0.400$; fail to reject H_o.

For (2) vs. (3), $r_s\bigstar = 0.988$, $\mathbf{P} < 0.01$ or $r_s \leq -0.400$, $r_s \geq 0.400$; reject H_o

ANSWERS TO CHAPTER PRACTICE TESTS

Part 1: Only the replacement for the word(s) in boldface type is given. (If the statement is true, no answer is shown. If the statement is false, a replacement is given.)

Chapter 1, Page 35

Part I

1.1 descriptive
1.2 inferential
1.4 sample
1.5 population
1.6 attribute or qualitative
1.7 quantitative
1.9 random

Part II

1.11 **a.** A **b.** B **c.** D **d.** C **e.** A
1.12 c, g, h, b, e, a, d, f

Part III

1.13 See definitions; examples will vary. Note: *population* is set of ALL possible, while *sample* is the actual set of subjects studied.
1.14 See definitions; examples will vary. Note: *variable* is the idea of interest, while *data* are the actual values obtained.
1.15 See definitions; examples will vary. Note: *data* is the value describing one source, the *statistic* is a value (usually calculated) describing all the data in the sample, the *parameter* is a value describing the entire population (usually unknown).
1.16 Every element of the population has an equal chance of being selected.

Chapter 2, Page 126

Part I

2.1 median
2.2 dispersion
2.3 never
2.5 zero
2.6 higher than

Part II

2.11 **a.** 30 **b.** 46 **c.** 91 **d.** 15 **e.** 1 **f.** 61 **g.** 75 **h.** 76 **i.** 91 **j.** 106 or 114
2.12 **a.** two items purchased
b. Nine people purchased 3 items each.
c. 40 **d.** 120 **e.** 5 **f.** 2 **g.** 3 **h.** 3 **i.** 3.0 **j.** 1.795 **k.** 1.34
2.13 **a.** 6.7 **b.** 7 **c.** 8 **d.** 6.5 **e.** 5 **f.** 6 **g.** 3.0 **h.** 1.7 **i.** 5
2.14 **a.** −1.5 **b.** 153

Part III

2.15 **a.** 98 **b.** 50 **c.** 121 **d.** 100
2.16 **a.** \$32,000, \$26,500, \$20,000, \$50,000
b.

20 30 40 50 60 70 80
mode median mean midrange
Salary (\$1000)

c. Mr. VanCott—midrange; business manager—mean; foreman—median; new worker—mode
d. The distribution is J-shaped.
2.17 There is more than one possible answer for these.
a. 12, 12, 12 **b.** 15, 20, 25
c. 12, 15, 15, 18 **d.** 12, 15, 16, 25 25
e. 12, 12, 15, 16, 17 **f.** 20, 25, 30, 32, 32, 80
2.18 A is right; B is wrong; standard deviation will not change.
2.19 B is correct. For example, if standard deviation is \$5, then the variance, (standard deviation)^2, is "25 dollars squared." Who knows what "dollars squared" are?

Chapter 3, Page 174

Part I

3.1 regression
3.2 strength of the
3.3 +1 or −1
3.5 positive
3.7 positive
3.8 −1 and +1
3.9 output or predicted value

Part II

3.11 **a.** B, D, A, C **b.** 12 **c.** 10 **d.** 175 **e.** N **f.** (125, 13) **g.** N **h.** P

3.12 Someone made a mistake in arithmetic, r must be between -1 and $+1$.

3.13 **a.** 12 **b.** 10 **c.** 8 **d.** 0.73 **e.** 0.67 **f.** 4.33 **g.** $\hat{y} = 4.33 + 0.67x$

Part III

3.14 Young children have small feet and probably tend to have less mathematics ability, while adults have larger feet and would tend to have more ability.

3.15 Student B is correct. -1.78 can occur only as a result of faulty arithmetic.

3.16 These answers will vary, but should somehow include the basic thought:
a. strong negative **b.** strong positive
c. no correlation **d.** no correlation
e. impossible value, bad arithmetic

3.17 There is more than one possible answer for these.
a. (1, 1), (2, 1), (3, 1) **b.** (1, 1), (3, 3), (5, 5)
c. (1, 5), (3, 3), (5, 1) **d.** (1, 1), (5, 1), (1, 5), (5,5)

Chapter 4, Page 230

Part I

4.1 any number value between 0 and 1, inclusive
4.4 simple
4.5 seldom
4.6 sum to 1.0
4.7 dependent
4.8 complementary
4.9 mutually exclusive or dependent
4.10 multiplication rule

Part II

4.11 **a.** $\frac{4}{8}$ **b.** $\frac{4}{8}$ **c.** $\frac{2}{8}$ **d.** $\frac{6}{8}$ **e.** $\frac{2}{8}$

f. $\frac{6}{8}$ **g.** 0 **h.** $\frac{6}{8}$ **i.** $\frac{1}{8}$ **j.** $\frac{5}{8}$

k. $\frac{2}{4}$ **l.** 0 **m.** $\frac{1}{2}$ **n.** no (e)

o. yes (g) **p.** no(i) **q.** yes (a, k) **r.** no (b, 1)
s. yes (a, m)

4.12 **a.** 0 **b.** 0.7 **c.** 0 **d.** no (c)

4.13 **a.** 0.14 **b.** 0.76 **c.** 0.2 **d.** no (a)

4.14 **a.** 0.7 **b.** 0.5 **c.** no, $P(E \text{ and } F) = 0.2$
d. yes, $P(E) = P(E \mid F)$

4.15 **a.** 0.4 **b.** 0.5 **c.** no, $P(G \text{ and } H) = 0.1$
d. no, $P(G)$ not equal to $P(G \mid H)$

4.16 0.51

Part III

4.17 Check the weather reports for a long period of time and determine the relative frequency with which each occurs.

4.18. Student B is right. *Mutually exclusive* means no intersection, while *independence* means one event does not affect the probability of the other.

4.19 These answers will vary, but should somehow include the basic thoughts:
a. no common occurrence
b. either event has no effect on the probability of the other
c. the relative frequency with which the event occurs.
d. probability that an event will occur even though the conditional event has previously occurred

4.20 **a.** No winners 15% of time; approx. 1 of every 7 drawings
b. 15% is not that rare
c. 0.0225 or 2.25%; approx. 1 of every 44 drawings
d. A fairly unlikely event, but will occur occasionally

Chapter 5, Page 269

Part I

5.1 continuous
5.3 one
5.5 exactly two
5.6 binomial
5.7 one success occurring on 1 trial
5.8 population
5.9 population parameters

Part II

5.11 **a.** Each $P(x)$ is between zero and 1, and the sum of all $P(x)$ is exactly one.
b. 0.2 **c.** 0 **d.** 0.8 **e.** 3.2 **f.** 1.25

5.12 **a.** 0.230 **b.** 0.085 **c.** 1.2 **d.** 1.04

Part III

5.13 n independent repeated trials of two outcomes; the two outcomes are "success" and "failure"; $p = P(\text{success})$ and $q = P(\text{failure})$ and $p + q = 1$; $x = n(\text{success}) = 0, 1, 2, \ldots, n$.

5.14 Student B is correct. The sample mean and standard deviation are statistics found using formulas studied in Chapter 2. The probability distributions studied in Chapter 5 are theoretical populations and their means and standard deviations are parameters.

5.15 Student B is correct. There are no restrictions on the values of the variable x.

Chapter 6, Page 309

Part I

6.1 its mean
6.4 one standard deviation
6.6 right
6.7 zero, 1
6.8 some (many)
6.9 mutually exclusive events
6.10 normal

Part II

6.11 **a.** 0.4922 **b.** 0.9162 **c.** 0.1020 **d.** 0.9082
6.12 **a.** 0.63 **b.** −0.95 **c.** 1.75
6.13 **a.** $z(0.8100)$ **b.** $z(0.2830)$
6.14 0.7910
6.15 28.03
6.16 **a.** 0.0569 **b.** 0.9890 **c.** 537 **d.** 417 **e.** 605

Part III

6.17 This answer will vary but should somehow include the basic properties: bell-shaped, mean of 0, standard deviation of 1.
6.18 This answer will vary but should somehow include the basic ideas: it is a z-score, α represents the area under the curve and to the right of z.
6.19 All normal distributions have the same shape and probabilities relative to the z-score.

Chapter 7, Page 336

Part I

7.1 is not
7.2 some (many)
7.3 population
7.4 divided by $\sqrt{n}$
7.5 decreases
7.6 approximately normal
7.7 sampling
7.8 means
7.9 random

Part II

7.11 **a.** 0.4364 **b.** 0.2643
7.12 **a.** 0.0918 **b.** 0.9525
7.13 0.6247

Part III

7.14 In this case each head produced one piece of data, the estimated length of the line. The CLT assures us that the mean value of a sample is far less variable than individual values of the variable x.
7.15 All samples must be of one fixed size.
7.16 Student A is correct. A population distribution is a distribution formed by all x values that make up the entire population.
7.17 Student A is correct. The standard error is found by dividing the standard deviation by the square root of the *sample size*.

Chapter 8, Page 408

Part I

8.1 alpha
8.2 alpha
8.3 sample distribution of the mean
8.7 type II error
8.8 beta
8.9 correct decision
8.10 critical (rejection) region

Part II

8.11 4.72 to 5.88
8.12 **a.** H_o: $\mu = 245$, H_a: $\mu > 245$
b. H_o: $\mu = 4.5$, H_a: $\mu < 4.5$
c. H_o: $\mu = 35$, H_a: $\mu \neq 35$
8.13 **a.** 0.05, z, $z \leq -1.65$
b. 0.05, z, $z \geq +1.65$
c. 0.05, z, $z \leq -1.96$ or $z \geq +1.96$
8.14 **a.** 1.65 **b.** 2.33 **c.** 1.18 **d.** −1.65
e. −2.05 **f.** −0.67
8.15 **a.** $z\star = 2.50$ **b.** 0.0062
8.16 H_o: $\mu = 1520$ vs. H_a: $\mu < 1520$, crit. reg. $z \leq -2.33$, $z\star = -1.61$, fail to reject H_o

Part III

8.17 **a.** No specific effect
b. Reduces it **c.** Narrows it **d.** No effect
e. Increases it **f.** Widens it
8.18 **a.** H_o – (a), H_a – (b) **b.** 3 **c.** 2
d. P(type I errror) is alpha, decreases: P(type II error) increases
8.19 The alternative hypothesis expresses the concern; the conclusion answers the concern.

Chapter 9, Page 467

Part I

9.2 Student's t
9.3 chi-square
9.4 to be rejected
9.6 t score
9.7 $n - 1$
9.9 $\sqrt{pq/n}$
9.10 z(normal)

Part II

9.11 **a.** 2.05 **b.** −1.73 **c.** 14.6
9.12 **a.** 28.6 **b.** 1.44 **c.** 27.16 to 30.04
9.13 0.528 to 0.752
9.14 **a.** H_o: $\mu = 255$, H_a: $\mu > 225$
b. H_o: $p = 0.40$, H_a: $p \neq 0.40$
c. H_o: $\sigma = 3.7$, H_a: $\sigma < 3.7$
9.15 **a.** 0.05, z, $z \leq -1.65$
b. 0.05, t, $t \leq -2.08$ or $t \geq +2.08$
c. 0.05, z, $z \geq +1.65$
d. 0.05, χ^2, $\chi^2 \leq 14.6$ or $\chi^2 \geq 43.2$
9.16 H_o: $\mu = 26$ vs. H_a: $\mu < 26$, crit. reg, $t \leq -1.71$, $t\star = -1.86$, reject H_o
9.17 H_o: $\sigma = 0.1$ vs. H_a: $\sigma > 0.1$, crit. reg. $\chi^2 \geq 21.1$, $\chi^2\star = 23.66$, reject H_o
9.18 H_o: $p = 0.50$ vs. H_a: $p > 0.50$, crit. reg. $z \geq 2.05$, $z\star = 1.29$, fail to reject H_o

Part III

9.19 If the distribution is normal, six standard deviations is approximately equal to the range.
9.20 B
9.21 They are both correct.
9.22 When the sample size, n, is large, the critical value of t is estimated by using the critical value from the standard normal distribution of z.
9.23 Student A
9.24 Student B is right. It is significant at the 0.01 level of significance.
9.25 Student A is correct.
9.26 It depends on what it means to improve the confidence interval. For most purposes, an increased sample size would be the best improvement.

Chapter 10, Page 532

Part I

10.1 two independent means
10.3 F distribution
10.4 Student's t-distribution
10.5 Student's t
10.7 nonsymmetric (or skewed)
10.9 decreases

Part II

10.11 **a.** H_o: $\mu_N - \mu_A = 0$, H_a: $\mu_N - \mu_A \neq 0$
b. H_o: $\sigma_o/\sigma_m = 1.0$, H_a: $\sigma_o/\sigma_m > 1.0$
c. H_o: $p_m - p_f = 0$, H_a: $p_m - p_f \neq 0$
d. H_o: $\mu_d = 0$, H_a: $\mu_d > 0$
10.12 **a.** z, $z \leq -1.96$ or $z \geq 1.96$
b. t, $t \leq -2.05$, $t \geq 2.05$
c. t, df $= 7$, $t \geq 1.89$
d. t, df $= 37$, $t \leq -1.69$
e. F, $F \geq 2.11$
10.13 **a.** 2.05 **b.** 2.13 **c.** 2.51 **d.** 2.18
e. 1.75 **f.** 1.69 **g.** −2.50 **h.** −1.28
10.14 H_o: $\mu_L - \mu_P = 0$ vs. H_a: $\mu_L - \mu_P > 0$, crit. reg. $t \geq +1.83$, $t\bigstar = 0.979$, fail to reject H_o
10.15 H_o: $\mu_d = 0$ vs. H_a: $\mu_d > 0$, crit. reg. $t \geq 1.89$, $t\bigstar = 1.88$, fail to reject H_o
10.16 0.072 to 0.188

Part III

10.17 independent
10.18 One possibility: Test all students before the course starts, then randomly select 20 of those who finish the course and test them afterwards. Use the before scores for these 20 as the before sample.
10.19 For starters, if the two independent samples are of different sizes, the techniques for dependent samples could not be completed. They are testing very different concepts, the "mean of the differences of paired data" and the "difference between two mean values."
10.20 It is only significant if the calculated t-score is in the critical region. The variation among the data and their relative size will play a role.
10.21 The 80 scores actually are two independent samples of size 40. A test to compare the mean scores of the two groups could be completed.
10.22 A fairly large sample of both Catholic and non-Catholic families would need to be taken, and the number of each whose children attended private schools would need to be obtained. The difference between two proportions could then be estimated.

Chapter 11, Page 569

Part I

11.1 one less than
11.3 expected
11.4 contingency table
11.6 test of homogeneity
11.8 approximated by chi-square

Part II

11.11 **a.** H_o: Digits generated occur with equal probability.
H_a: Digits do not occur with equal probability.
b. H_o: Votes were cast independently of party affiliation.
H_a: Votes were not cast independently of party affiliation.
c. H_o: The crimes distributions are the same for all four cities.
H_a: The crimes distributions are not all the same.
11.12 **a.** 4.40 **b.** 35.7
11.13 H_o: $P(1) = P(2) = P(3) = \frac{1}{3}$
H_a: preferences not all equal, $\chi^2\bigstar = 3.78$; $0.10 < \mathbf{P} < 0.25$ or crit. reg. $\chi^2 \geq 5.99$; fail to reject H_o
11.14 **a.** H_o: The distribution is the same for all types of soil.
H_a: The distributions are not all the same.
b. 25.622 **c.** 13.746
d. $0.005 < \mathbf{P} < 0.01$ **e.** $\chi^2 \geq 9.49$
f. Reject H_o: There is sufficient evidence to show that the growth distribution is different for at least one of the three soil types.

Part III

11.15 Similar in that there are n repeated independent trials. Different in that the binomial has two possible outcomes, while the multinomial has several. Each possible outcome has a probability and these probabilities sum to 1 for each different experiment, both for binomial and multinomial.
11.16 The test of homogeneity compares several distributions in a side-by-side comparison, while the test for independence tests the independence of the two factors that create the rows and columns of the contingency table.
11.17 Student A is right in that the calculations are completed in the same manner. Student B is correct in

that the test of independence starts with one large sample and homogeneity has several samples.

11.18 **a.** If a chi-square test is to be used, the results of the four questions would be pooled to estimate the expected probability.
b. Use a chi-square test for homogeneity.

Chapter 12, Page 599

Part I

12.2 mean square
12.3 SS(factor) or MS(factor)
12.5 reject H_o
12.7 the number of factor levels less one
12.8 mean
12.9 need to
12.10 does not indicate

Part II

12.11 **a.** T **b.** T **c.** F **d.** T **e.** T **f.** T **g.** F **h.** F **i.** F **j.** F **k.** T **l.** F **m.** F **n.** F **o.** T
12.12 **a.** 72 **b.** 72 **c.** 22 **d.** 4 **e.** 4.5

Part III

12.13 This answer will vary but should somehow include the basic ideas: It is the comparison of several mean values that result from testing some statistical population by measuring a variable repeatedly at each of the several levels for which the factor is being tested.

12.14 **a.** $x_{\bar{r},k} = \mu + F\text{ scrubber} + \epsilon_{k(r)}$
b. H_o: The mean amount of emissions is the same for all three scrubbers tested.
H_a: The mean amounts are not all equal.
c.

Source	df	SS	MS
Scrubber	2	12.80	6.40
Error	13	33.63	2.59
Total	15	46.44	

d. $F(2, 13, 0.05) = 3.81$, $F\bigstar = 2.47$, fail to reject H_o. The difference in the mean value for the scrubbers is not significant.
e. I

```
           .   .   .   :   .
--+---+---+---+---+---+---+--
  7       9      11      13
```

II

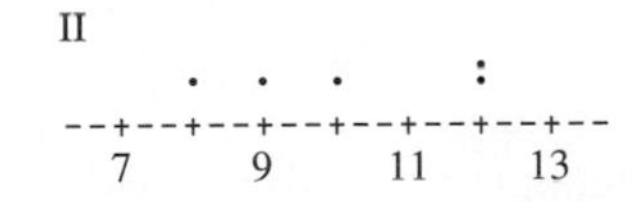

III

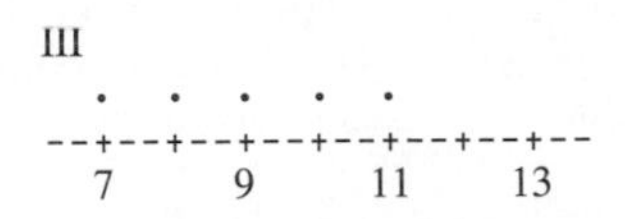

Chapter 13, Page 651

Part I

13.2 need not be
13.3 does not prove
13.4 need not be
13.6 the linear correlation coefficient
13.8 regression
13.9 $n - 2$

Part II

13.11

Amount of Wheat Harvest

Wheat

Fertilizer

13.12. $\sum x = 720$, $\sum y = 252$, $\sum x^2 = 49{,}200$, $\sum xy = 17{,}240$, $\sum y^2 = 6{,}228$
13.13 $SS(x) = 6000$, $SS(y) = 936$, $SS(xy) = 2120$
13.14 0.895
13.15 0.65 to 0.97
13.16 $\hat{y} = -0.20 + 0.353x$
13.17 See red line in figure in 13.11.
13.18 4.324
13.19 Yes; H_o: $\beta_1 = 0$ vs H_a: $\beta_1 > 0$, $t\bigstar = 6.33$, reject H_o
13.20 25.63 to 33.98
13.21 See blue vertical segment in 13.11.

Part III

13.22 Variable 1: The frequency of skiers having their bindings tested
Variable 2: The incidence of lower-leg injury. The statement implies that as the frequency with which the bindings are tested increases, the frequency of lower-leg injury decreases; thus the strong correlation must be negative for these variables.

13.23 A "moment" is the distance from the mean, and the product of both the horizontal moment and the vertical moment is summed in calculating the correlation coefficient.

13.24 A value close to zero, also. The formulas used to calculate both values have the same numerator, namely $SS(xy)$.

13.25 The vertical distance from a potential line of best fit to the data point is measured by $(y - \hat{y})$. The line of best fit is defined to be the line that results in the smallest possible total when the squared values of $(y - \hat{y})$ are totaled. Thus "the method of least squares."

13.26 The strength of the linear relationship could be measured with the correlation coefficient.
13.27 A random sample will be needed from the population of interest. The data collected need to be for the variables length of time on welfare and the measure of current level of self-esteem.

Chapter 14, Page 704

Part I

14.2 *t*-test
14.3 runs test
14.4 assigned equal ranks
14.7 Mann–Whitney *U* test
14.8 power
14.10 power

Part II

14.11 −2 to +7
14.12 H_o: No difference in weight gain.
H_a: There is a difference in weight gain, crit. val.: 23, $U\bigstar = 32.5$, fail to reject H_o
14.13 H_o: no correlation
H_a: correlated, crit. val.: ±0.683, $r_s\bigstar = -0.70$, reject H_o. Yes, there is significant correlation.
14.14 (+) = higher grade level than previous problem
(−) = lower grade level than previous problem
H_o: $P(+) = 0.5$
H_a: $P(+) = 0.5$, crit. val.: 7, $x = 11$, fail to reject H_o. Ths sample does not show a significant pattern.

Part III

14.15 The nonparametric statistics do not require assumptions about the distribution of the variable.
14.16 The sign test is a binomial experiment of *n* trials (the *n* data observations) with two oucomes for each data [(+) or (−)], and $p = P(+) = 0.5$. The variable *x* is the number of the least frequent sign.
14.17 The median is the middle value such that 50% of the distribution is larger in value and 50% is smaller in value.
14.18 The extreme value in a set of data can have a sizeable effect on the mean and standard deviation in the parametric methods. The nonparametric methods typically use rank numbers. The extreme value with ranks is either 1 or *n*, and neither changes if the value is more extreme.
14.19 d; $p = P(+) = P(\text{prefer seating arrangement A}) = 0.5$, no preference

INDEX

Formula Card for Johnson & Kuby, ELEMENTARY STATISTICS, Ninth Edition

Sample mean:
$\bar{x} = \frac{\sum x}{n}$ (2.1) or $\frac{\sum xf}{\sum f}$ (2.11)

Depth of sample median:
$d(\tilde{x}) = (n + 1)/2$ (2.2)

Range: $H - L$ (2.4)

Sample variance:
$s^2 = \frac{\sum(x - \bar{x})^2}{n - 1}$ (2.6)

or

$s^2 = \frac{\sum x^2 - \frac{(\sum x)^2}{n}}{n - 1}$ (2.10)

or

$s^2 = \frac{\sum x^2 f - \frac{(\sum xf)^2}{\sum f}}{\sum f - 1}$ (2.12)

Sample standard deviation:
$s = \sqrt{s^2}$ (2.7)

Chebyshev's Theorem: at least $1 - (1/k^2)$ (p. 109)

Sum of squares of x:
$SS(x) = \sum x^2 - ((\sum x)^2/n)$ (2.9)

Sum of squares of y:
$SS(y) = \sum y^2 - ((\sum y)^2/n)$ (3.3)

Sum of squares of xy:
$SS(xy) = \sum xy - ((\sum x \cdot \sum y)/n)$ (3.4)

Pearson's Correlation Coefficient:
$r = SS(xy)/\sqrt{SS(x) \cdot SS(y)}$ (3.2)

Equation for line of best fit: $\hat{y} = b_0 + b_1 x$ (p. 154)

Slope for line of best fit: $b_1 = SS(xy)/SS(x)$ (3.6)

y-intercept for line of best fit:
$b_0 = [\sum y - (b_1 \cdot \sum x)]/n$ (3.7)

Empirical (observed) probability:
$P'(A) = n(A)/n$ (4.1)

Theoretical probability for equally likely sample space:
$P(A) = n(A)/n(S)$ (4.2)

Complement Rule:
$P(\text{not } A) = P(\bar{A}) = 1 - P(A)$ (4.3)

General Addition Rule:
$P(A \text{ or } B) = P(A) + P(B) - P(A \text{ and } B)$ (4.4a)

Special Addition Rule for mutually exclusive events:
$P(A \text{ or } B \text{ or} \ldots \text{or } D) = P(A) + P(B) + \cdots + P(D)$ (4.4c)

General Multiplication Rule:
$P(A \text{ and } B) = P(A) \cdot P(B|A)$ (4.7a)

Special Multiplication Rule for independent events:
$P(A \text{ and } B \text{ and} \ldots \text{and } G) = P(A) \cdot P(B) \cdot \cdots \cdot P(G)$ (4.8b)

Conditional Probability:
$P(A|B) = P(A \text{ and } B)/P(B)$ (4.6b)

Mean of discrete random variable:
$\mu = \sum[xP(x)]$ (5.1)

Variance of discrete random variable:
$\sigma^2 = \sum[x^2 P(x)] - \{\sum[xP(x)]\}^2$ (5.3a)

Standard deviation of discrete random variable:
$\sigma = \sqrt{\sigma^2}$ (5.4)

Factorial: $n! = (n)(n - 1)(n - 2) \cdot \cdots \cdot 2 \cdot 1$ (p. 251)

Binomial coefficient:
$\binom{n}{x} = \frac{n!}{x! \cdot (n - x)!}$ (5.6)

Binomial probability function:
$P(x) = \binom{n}{x} \cdot p^x \cdot q^{n-x}, x = 0, \ldots, n$ (5.5)

Mean of binomial random variable: $\mu = np$ (5.7)

Standard deviation, binomial random variable:
$\sigma = \sqrt{npq}$ (5.8)

Standard score: $z = (x - \mu)/\sigma$ (6.3)

Standard score for $\bar{x}$: $z = \frac{\bar{x} - \mu}{\sigma/\sqrt{n}}$ (7.2)

Confidence interval estimate for mean, μ (σ known):
$\bar{x} \pm z(\alpha/2) \cdot (\sigma/\sqrt{n})$ (8.1)

Sample size for $1 - \alpha$ confidence estimate for μ:
$n = [z(\alpha/2) \cdot \sigma/E]^2$ (8.3)

Calculated test statistic for H_o: $\mu = \mu_0$ (σ known):
$z\bigstar = (\bar{x} - \mu_0)/(\sigma/\sqrt{n})$ (8.4)

Confidence interval estimate for mean, μ (σ unknown):
$\bar{x} \pm t(df, \alpha/2) \cdot (s/\sqrt{n})$ with $df = n - 1$ (9.1)

Calculated test statistic for H_o: $\mu = \mu_0$ (σ unknown):
$t\bigstar = \frac{\bar{x} - \mu_0}{s/\sqrt{n}}$ with $df = n - 1$ (9.2)

Confidence interval estimate for proportion, p:
$p' \pm z(\alpha/2) \cdot \sqrt{(p'q')/n}, \quad p' = x/n$ (9.6)

Calculated test statistic for H_o: $p = p_0$:
$z\bigstar = (p' - p_0)/\sqrt{(p_0 q_0/n)}, \quad p' = x/n$ (9.9)

Calculated test statistic for H_o: $\sigma^2 = \sigma_0^2$ or $\sigma = \sigma_0$:
$\chi^2\bigstar = (n - 1)s^2/\sigma_0^2, \quad df = n - 1$ (9.10)

Mean difference between two dependent samples:

Paired difference: $d = x_1 - x_2$ (10.1)

Sample mean of paired differences:
$\bar{d} = \sum d/n$ (10.2)

Sample standard deviation of paired differences:

$$s_d = \sqrt{\frac{\sum d^2 - \left[\frac{(\sum d)^2}{n}\right]}{n - 1}} \quad (10.3)$$

Confidence interval estimate for mean, μ_d:

$\bar{d} \pm t(\text{df}, \alpha/2) \cdot s_d/\sqrt{n}$ (10.4)

Calculated test statistic for H_o: $\mu_d = \mu_0$:

$t\bigstar = (\bar{d} - \mu_o)/(s_d/\sqrt{n})$, $\text{df} = n - 1$ (10.5)

Difference between means of two independent samples:

Degrees of freedom:

df = smaller of $(n_1 - 1)$ or $(n_2 - 1)$ **(p. 491)**

Confidence interval estimate for $\mu_1 - \mu_2$:

$(\bar{x}_1 - \bar{x}_2) \pm t(\text{df}, \alpha/2) \cdot \sqrt{(s_1^2/n_1) + (s_2^2/n_2)}$ (10.8)

Calculated test statistic for H_o: $\mu_1 - \mu_2 = (\mu_1 - \mu_2)_0$:

$t\bigstar = [(\bar{x}_1 - \bar{x}_2) - (\mu_1 - \mu_2)_0]/\sqrt{(s_1^2/n_1) + (s_2^2/n_2)}$ (10.9)

Difference between proportions of two independent samples:

Confidence interval estimate for $p_1 - p_2$:

$$(p_1' - p_2') \pm z(\alpha/2) \cdot \sqrt{\frac{p_1'q_1'}{n_1} + \frac{p_2'q_2'}{n_2}} \quad (10.11)$$

Pooled observed probability:

$p_p' = (x_1 + x_2)/(n_1 + n_2)$ (10.13)

$q_p' = 1 - p_p'$ (10.14)

Calculated test statistic for H_o: $p_1 - p_2 = 0$:

$$z\bigstar = \frac{p_1' - p_2'}{\sqrt{(p_p')(q_p')\left[\left(\frac{1}{n_1}\right) + \left(\frac{1}{n_2}\right)\right]}} \quad (10.15)$$

Ratio of variances between two independent samples:

Calculated test statistic for H_o: $\sigma_1^2/\sigma_2^2 = 1$:

$F\bigstar = s_1^2/s_2^2$ (10.16)

Calculated test statistic for enumerative data:

$\chi^2\bigstar = \sum[(O - E)^2/E]$ (11.1)

Multinomial experiment:

Degrees of freedom: $\text{df} = k - 1$ (11.2)

Expected frequency: $E = n \cdot p$ (11.3)

Test for independence or Test of homogeneity:

Degrees of freedom:

$\text{df} = (r - 1) \cdot (c - 1)$ (11.4)

Expected value: $E = (R \cdot C)/n$ (11.5)

Mathematical model:

$x_{c,k} = \mu + F_c + \epsilon_{k(c)}$ (12.13)

Total sum of squares:

$$\text{SS(total)} = \sum(x^2) - \frac{(\sum x)^2}{n} \quad (12.2)$$

Sum of squares due to factor:

SS(factor) =

$$\left[\left(\frac{C_1^2}{k_1}\right) + \left(\frac{C_2^2}{k_2}\right) + \left(\frac{C_3^2}{k_3}\right) + \cdots\right] - \left[\frac{(\sum x)^2}{n}\right] \quad (12.3)$$

Sum of squares due to error:

SS(error) =

$\sum(x^2) - [(C_1^2/k_1) + (C_2^2/k_2) + (C_3^2/k_3) + \cdots]$ (12.4)

Degrees of freedom for total:

$\text{df(total)} = n - 1$ (12.6)

Degrees of freedom for factor:

$\text{df(factor)} = c - 1$ (12.5)

Degrees of freedom for error:

$\text{df(error)} = n - c$ (12.7)

Mean square for factor:

MS(factor) = SS(factor)/df(factor) (12.10)

Mean square for error:

MS(error) = SS(error)/df(error) (12.11)

Calculated test statistic for H_o: Mean value is same at all levels:

$F\bigstar$ = MS(factor)/MS(error) (12.12)

Covariance of x and y:

$\text{covar}(x, y) = \sum[(x - \bar{x})(y - \bar{y})]/(n - 1)$ (13.1)

Pearson's Correlation Coefficient:

$r = \text{covar}(x, y)/(s_x \cdot s_y)$ (13.2)

or

$r = \text{SS}(xy)/\sqrt{\text{SS}(x) \cdot \text{SS}(y)}$ (3.2) or (13.3)

Experimental error: $e = y - \hat{y}$ (13.5)

Variance of error ϵ: $s_e^2 = \sum(y - \hat{y})^2/(n - 2)$ (13.6)

or

$$s_e^2 = \frac{(\sum y^2) - (b_0)(\sum y) - (b_1)(\sum xy)}{n - 2} \quad (13.8)$$

Standard deviation about the line of best fit:

$s_e = \sqrt{s_e^2}$ **(p. 623)**

Square of standard error of regression:

$$s_{b_1}^2 = \frac{s_e^2}{\text{SS}(x)} = \frac{s_e^2}{\sum x^2 - [(\sum x)^2/n]} \quad (13.11)$$

Confidence interval estimate for β_1:

$b_1 \pm t(\text{df}, \alpha/2) \cdot s_{b_1}$ (13.12)

Calculated test statistic for H_o: $\beta_1 = 0$:

$t\bigstar = (b_1 - \beta_1)/s_{b_1}$ with $\text{df} = n - 2$ (13.13)

Confidence interval estimate for mean value of y at x_0:

$$\hat{y} \pm t(n - 2, \alpha/2) \cdot s_e \cdot \sqrt{\frac{1}{n} + \frac{(x_0 - \bar{x})^2}{\text{SS}(x)}} \quad (13.15)$$

Prediction interval estimate for y at x_0:

$$\hat{y} \pm t(n - 2, \alpha/2) \cdot s_e \cdot \sqrt{1 + \frac{1}{n} + \frac{(x_0 - \bar{x})^2}{\text{SS}(x)}} \quad (13.16)$$

Mann–Whitney U test:

$U_a = n_a \cdot n_b + [(n_b) \cdot (n_b + 1)/2] - R_b$ (14.3)

$U_b = n_a \cdot n_b + [(n_a) \cdot (n_a + 1)/2] - R_a$ (14.4)

Spearman's rank correlation coefficient:

$$r_s = 1 - \left[\frac{6\sum d^2}{n(n^2 - 1)}\right] \quad (14.11)$$

Formula Card for Johnson & Kuby, ELEMENTARY STATISTICS, Ninth Edition

Sample mean:

$\bar{x} = \frac{\sum x}{n}$ (2.1) or $\frac{\sum xf}{\sum f}$ (2.11)

Depth of sample median:

$d(\tilde{x}) = (n + 1)/2$ (2.2)

Range: $H - L$ (2.4)

Sample variance:

$s^2 = \frac{\sum(x - \bar{x})^2}{n - 1}$ (2.6)

or

$s^2 = \frac{\sum x^2 - \frac{(\sum x)^2}{n}}{n - 1}$ (2.10)

or

$s^2 = \frac{\sum x^2 f - \frac{(\sum xf)^2}{\sum f}}{\sum f - 1}$ (2.12)

Sample standard deviation:

$s = \sqrt{s^2}$ (2.7)

Chebyshev's Theorem: at least $1 - (1/k^2)$ (p. 109)

Sum of squares of x:

$SS(x) = \sum x^2 - ((\sum x)^2/n)$ (2.9)

Sum of squares of y:

$SS(y) = \sum y^2 - ((\sum y)^2/n)$ (3.3)

Sum of squares of xy:

$SS(xy) = \sum xy - ((\sum x \cdot \sum y)/n)$ (3.4)

Pearson's Correlation Coefficient:

$r = SS(xy)/\sqrt{SS(x) \cdot SS(y)}$ (3.2)

Equation for line of best fit: $\hat{y} = b_0 + b_1 x$ (p. 154)

Slope for line of best fit: $b_1 = SS(xy)/SS(x)$ (3.6)

y-intercept for line of best fit:

$b_0 = [\sum y - (b_1 \cdot \sum x)]/n$ (3.7)

Empirical (observed) probability:

$P'(A) = n(A)/n$ (4.1)

Theoretical probability for equally likely sample space:

$P(A) = n(A)/n(S)$ (4.2)

Complement Rule:

$P(\text{not } A) = P(\bar{A}) = 1 - P(A)$ (4.3)

General Addition Rule:

$P(A \text{ or } B) = P(A) + P(B) - P(A \text{ and } B)$ (4.4a)

Special Addition Rule for mutually exclusive events:

$P(A \text{ or } B \text{ or} \ldots \text{or } D) = P(A) + P(B) + \cdots + P(D)$ (4.4c)

General Multiplication Rule:

$P(A \text{ and } B) = P(A) \cdot P(B|A)$ (4.7a)

Special Multiplication Rule for independent events:

$P(A \text{ and } B \text{ and} \ldots \text{and } G) = P(A) \cdot P(B) \cdot \cdots \cdot P(G)$ (4.8b)

Conditional Probability:

$P(A|B) = P(A \text{ and } B)/P(B)$ (4.6b)

Mean of discrete random variable:

$\mu = \sum[xP(x)]$ (5.1)

Variance of discrete random variable:

$\sigma^2 = \sum[x^2P(x)] - \{\sum[xP(x)]\}^2$ (5.3a)

Standard deviation of discrete random variable:

$\sigma = \sqrt{\sigma^2}$ (5.4)

Factorial: $n! = (n)(n - 1)(n - 2) \cdot \cdots \cdot 2 \cdot 1$ (p. 251)

Binomial coefficient:

$\binom{n}{x} = \frac{n!}{x! \cdot (n - x)!}$ (5.6)

Binomial probability function:

$P(x) = \binom{n}{x} \cdot p^x \cdot q^{n-x}, x = 0, \ldots, n$ (5.5)

Mean of binomial random variable: $\mu = np$ (5.7)

Standard deviation, binomial random variable:

$\sigma = \sqrt{npq}$ (5.8)

Standard score: $z = (x - \mu)/\sigma$ (6.3)

Standard score for $\bar{x}$: $z = \frac{\bar{x} - \mu}{\sigma/\sqrt{n}}$ (7.2)

Confidence interval estimate for mean, μ (σ known):

$\bar{x} \pm z(\alpha/2) \cdot (\sigma/\sqrt{n})$ (8.1)

Sample size for $1 - \alpha$ confidence estimate for μ:

$n = [z(\alpha/2) \cdot \sigma/E]^2$ (8.3)

Calculated test statistic for H_o: $\mu = \mu_0$ (σ known):

$z\bigstar = (\bar{x} - \mu_0)/(\sigma/\sqrt{n})$ (8.4)

Confidence interval estimate for mean, μ (σ unknown):

$\bar{x} \pm t(\text{df}, \alpha/2) \cdot (s/\sqrt{n})$ with $\text{df} = n - 1$ (9.1)

Calculated test statistic for H_o: $\mu = \mu_0$ (σ unknown):

$t\bigstar = \frac{\bar{x} - \mu_0}{s/\sqrt{n}}$ with $\text{df} = n - 1$ (9.2)

Confidence interval estimate for proportion, p:

$p' \pm z(\alpha/2) \cdot \sqrt{(p'q')/n}, \quad p' = x/n$ (9.6)

Calculated test statistic for H_o: $p = p_0$:

$z\bigstar = (p' - p_0)/\sqrt{(p_0q_0/n)}, \quad p' = x/n$ (9.9)

Calculated test statistic for H_o: $\sigma^2 = \sigma_0^2$ or $\sigma = \sigma_0$:

$\chi^2\bigstar = (n - 1)s^2/\sigma_0^2, \quad \text{df} = n - 1$ (9.10)

Mean difference between two dependent samples:

Paired difference: $d = x_1 - x_2$ (10.1)

Sample mean of paired differences:

$\bar{d} = \sum d/n$ (10.2)

Sample standard deviation of paired differences:

$$s_d = \sqrt{\frac{\sum d^2 - \left[\frac{(\sum d)^2}{n}\right]}{n - 1}} \quad (10.3)$$

Confidence interval estimate for mean, μ_d:

$\bar{d} \pm t(\text{df}, \alpha/2) \cdot s_d/\sqrt{n}$ (10.4)

Calculated test statistic for H_o: $\mu_d = \mu_0$:

$t\bigstar = (\bar{d} - \mu_o)/(s_d/\sqrt{n})$, df $= n - 1$ (10.5)

Difference between means of two independent samples:

Degrees of freedom:

df = smaller of $(n_1 - 1)$ or $(n_2 - 1)$ (p. 491)

Confidence interval estimate for $\mu_1 - \mu_2$:

$(\bar{x}_1 - \bar{x}_2) \pm t(\text{df}, \alpha/2) \cdot \sqrt{(s_1^2/n_1) + (s_2^2/n_2)}$ (10.8)

Calculated test statistic for H_o: $\mu_1 - \mu_2 = (\mu_1 - \mu_2)_0$:

$t\bigstar = [(\bar{x}_1 - \bar{x}_2) - (\mu_1 - \mu_2)_0]/\sqrt{(s_1^2/n_1) + (s_2^2/n_2)}$ (10.9)

Difference between proportions of two independent samples:

Confidence interval estimate for $p_1 - p_2$:

$$(p_1' - p_2') \pm z(\alpha/2) \cdot \sqrt{\frac{p_1'q_1'}{n_1} + \frac{p_2'q_2'}{n_2}} \quad (10.11)$$

Pooled observed probability:

$p_p' = (x_1 + x_2)/(n_1 + n_2)$ (10.13)

$q_p' = 1 - p_p'$ (10.14)

Calculated test statistic for H_o: $p_1 - p_2 = 0$:

$$z\bigstar = \frac{p_1' - p_2'}{\sqrt{(p_p')(q_p')\left[\left(\frac{1}{n_1}\right) + \left(\frac{1}{n_2}\right)\right]}} \quad (10.15)$$

Ratio of variances between two independent samples:

Calculated test statistic for H_o: $\sigma_1^2/\sigma_2^2 = 1$:

$F\bigstar = s_1^2/s_2^2$ (10.16)

Calculated test statistic for enumerative data:

$\chi^2\bigstar = \sum[(O - E)^2/E]$ (11.1)

Multinomial experiment:

Degrees of freedom: df $= k - 1$ (11.2)

Expected frequency: $E = n \cdot p$ (11.3)

Test for independence or Test of homogeneity:

Degrees of freedom:

df $= (r - 1) \cdot (c - 1)$ (11.4)

Expected value: $E = (R \cdot C)/n$ (11.5)

Mathematical model:

$x_{c,k} = \mu + F_c + \epsilon_{k(c)}$ (12.13)

Total sum of squares:

$$\text{SS(total)} = \sum(x^2) - \frac{(\sum x)^2}{n} \quad (12.2)$$

Sum of squares due to factor:

SS(factor) =

$$\left[\left(\frac{C_1^2}{k_1}\right) + \left(\frac{C_2^2}{k_2}\right) + \left(\frac{C_3^2}{k_3}\right) + \cdots\right] - \left[\frac{(\sum x)^2}{n}\right] \quad (12.3)$$

Sum of squares due to error:

SS(error) =

$\sum(x^2) - [(C_1^2/k_1) + (C_2^2/k_2) + (C_3^2/k_3) + \cdots]$ (12.4)

Degrees of freedom for total:

df(total) $= n - 1$ (12.6)

Degrees of freedom for factor:

df(factor) $= c - 1$ (12.5)

Degrees of freedom for error:

df(error) $= n - c$ (12.7)

Mean square for factor:

MS(factor) = SS(factor)/df(factor) (12.10)

Mean square for error:

MS(error) = SS(error)/df(error) (12.11)

Calculated test statistic for H_o: Mean value is same at all levels:

$F\bigstar$ = MS(factor)/MS(error) (12.12)

Covariance of x and y:

$\text{covar}(x, y) = \sum[(x - \bar{x})(y - \bar{y})]/(n - 1)$ (13.1)

Pearson's Correlation Coefficient:

$r = \text{covar}(x, y)/(s_x \cdot s_y)$ (13.2)

or

$r = \text{SS}(xy)/\sqrt{\text{SS}(x) \cdot \text{SS}(y)}$ (3.2) or (13.3)

Experimental error: $e = y - \hat{y}$ (13.5)

Variance of error ϵ: $s_e^2 = \sum(y - \hat{y})^2/(n - 2)$ (13.6)

or

$$s_e^2 = \frac{(\sum y^2) - (b_0)(\sum y) - (b_1)(\sum xy)}{n - 2} \quad (13.8)$$

Standard deviation about the line of best fit:

$s_e = \sqrt{s_e^2}$ (p. 623)

Square of standard error of regression:

$$s_{b_1}^2 = \frac{s_e^2}{\text{SS}(x)} = \frac{s_e^2}{\sum x^2 - [(\sum x)^2/n]} \quad (13.11)$$

Confidence interval estimate for β_1:

$b_1 \pm t(\text{df}, \alpha/2) \cdot s_{b_1}$ (13.12)

Calculated test statistic for H_o: $\beta_1 = 0$:

$t\bigstar = (b_1 - \beta_1)/s_{b_1}$ with df $= n - 2$ (13.13)

Confidence interval estimate for mean value of y at x_0:

$$\hat{y} \pm t(n - 2, \alpha/2) \cdot s_e \cdot \sqrt{\frac{1}{n} + \frac{(x_0 - \bar{x})^2}{\text{SS}(x)}} \quad (13.15)$$

Prediction interval estimate for y at x_0:

$$\hat{y} \pm t(n - 2, \alpha/2) \cdot s_e \cdot \sqrt{1 + \frac{1}{n} + \frac{(x_0 - \bar{x})^2}{\text{SS}(x)}} \quad (13.16)$$

Mann–Whitney U test:

$U_a = n_a \cdot n_b + [(n_b) \cdot (n_b + 1)/2] - R_b$ (14.3)

$U_b = n_a \cdot n_b + [(n_a) \cdot (n_a + 1)/2] - R_a$ (14.4)

Spearman's rank correlation coefficient:

$$r_s = 1 - \left[\frac{6\sum d^2}{n(n^2 - 1)}\right] \quad (14.11)$$

Glossary of Symbols

$\overline{A}$	Complement of set A
ANOVA	Analysis of variance
α (alpha)	Probability of type I error
β (beta)	Probability of type II error
$1 - \beta$	Power of a statistical test
β_0	y-intercept of the true linear relationship
β_1	Slope of the true linear relationship
b_0	y-intercept for the line of best fit for the sample data
b_1	Slope for the line of best fit for the sample data
${}_nC_r$	Number of combinations of n things r at a time
C_j	Column total
c	Column number or class width
d	Difference in value between two paired pieces of data or difference in the rankings
$\overline{d}$	Mean value of observed differences d
$d(\ )$	Depth of
df or df( )	Number of degrees of freedom
E	Expected frequency or maximum error of estimate
e	Error (observed)
ϵ (epsilon)	Experimental error
ϵ_{ij}	Amount of experimental error in the value of the jth piece of data in the ith row
F	F distribution statistic
$F(\text{df}_n, \text{df}_d, \alpha)$	Critical value for the F distribution
f	Frequency
H	Value of the largest-valued piece of data in a sample
H_a	Alternative hypothesis
H_o	Null hypothesis
i	Index number when used with $\sum$ notation
i	Position number for a particular data
k	Identifier for the kth percentile
k	Number of cells or variables
L	Value of the smallest-valued piece of data in a sample
m	Number of classes
MS( )	Mean square
μ (mu)	Population mean
μ_d	Mean value of the paired differences
$\mu_{\overline{x}}$	Mean of the distribution of all possible $\overline{x}$'s
$\mu_{y\mid x_0}$	Mean of all y values at the fixed value of x, x_0
μ_v	Mean number of runs for the sampling distribution of number of runs
M	Population median
n	(Sample size) number of pieces of data in one sample
$n(\ )$	Cardinal number of
$\binom{n}{r}$	Binomial coefficient or number of r successes in n trials
O	Observed frequency
P	Probability value or p-value
$P(A\mid B)$	Conditional probability, the probability of A given B
$P(a < x < b)$	Probability that x has a value between a and b
P_k	kth percentile
${}_nP_r$	Number of permutations of n things r at a time
p or $P(\ )$	Theoretical probability of an event or proportion of time that a particular event occurs
p' or $P'(\ )$	Empirical (experimental) probability or a probability estimate from observed data
p_p	Pooled estimate for the proportion
Q_1	First quartile
Q_3	Third quartile
q	$(q = 1 - p)$ probability that an event does not occur
q'	$(q' = 1 - p')$ observed proportion of time that an event does not occur
R	Range of the data
R_i	Row total
R^2	Coefficient of determination

(continued)

Glossary of Symbols (continued)

ρ (rho)	Population linear correlation coefficient
r	Linear correlation coefficient for the sample data or row number
r_s	Spearman's rank correlation coefficient
$\sum$ (capital sigma)	Summation notation
SS( )	Sum of squares
s^2	Sample variance
σ (lowercase sigma)	Population standard deviation
$\sigma_{\bar{x}}$	Standard error for means, the standard deviation of the distribution of all possible $\bar{x}$'s
$\sigma_{p'}$	Standard error for proportions
$\sigma^2_{\mu r}$	Variance among the means of the r rows (ANOVA)
σ_v	Standard error for the number of runs for the sampling distribution of number of runs
s	Sample standard deviation
s_d	Standard deviation of the observed differences d
$s^2_{b_1}$	Square of the standard error for repeated observed values of the slope for the line of best fit
T	Grand total
t	Student's t-distribution statistic
$t(\text{df}, \alpha/2)$	Critical value for Student's t-distribution
U	Mann–Whitney U statistic
V	Number of runs
χ^2	Chi-square statistic
$\chi^2(\text{df}, \alpha)$	Critical value of chi-square distribution
x	Value of a single piece of data or class mark
$\bar{x}$	Sample mean
$\tilde{x}$	Sample median
x_{ij}	Value of the jth piece of data in the ith row
x_0	A given value of the variable x
$\hat{y}$	Predicted value of y for a given x
y_{x_0}	Individual value of y at the x value of x_0
z	Standard score
$z(\alpha/2)$	Critical value of z
★ (star)	Identifies the calculated value of any test statistic
$\lvert\ \rvert$	Absolute value of a number
$=$	Equal to
$\neq$	Not equal to
$<$	Less than
$\leq$	Less than or equal to
$>$	Greater than
>	Followed by
$\geq$	Greater than or equal to
$\approx$	Approximately equal to
$\sqrt{\ }$	Square root